COMPREHENSIVE
MEDICINAL CHEMISTRY

IN 6 VOLUMES

COMPREHENSIVE
MEDICINAL CHEMISTRY

*The Rational Design, Mechanistic Study & Therapeutic
Application of Chemical Compounds*

Chairman of the Editorial Board
CORWIN HANSCH
Pomona College, Claremont, CA, USA

Joint Executive Editors
PETER G. SAMMES
Brunel University of West London, Uxbridge, UK

JOHN B. TAYLOR
Rhône-Poulenc Ltd, Dagenham, UK

Volume 5
BIOPHARMACEUTICS

Volume Editor
JOHN B. TAYLOR
Rhône-Poulenc Ltd, Dagenham, UK

PERGAMON PRESS

Member of Maxwell Macmillan Pergamon Publishing Corporation
OXFORD • NEW YORK • BEIJING • FRANKFURT
SÃO PAULO • SYDNEY • TOKYO • TORONTO

U.K.	Pergamon Press plc, Headington Hill Hall, Oxford OX3 0BW, England
U.S.A.	Pergamon Press, Inc., Maxwell House, Fairview Park, Elmsford, New York 10523, U.S.A.
PEOPLE'S REPUBLIC OF CHINA	Pergamon Press, Room 4037, Qianmen Hotel, Beijing, People's Republic of China
FEDERAL REPUBLIC OF GERMANY	Pergamon Press GmbH, Hammerweg 6, D-6242 Kronberg, Federal Republic of Germany
BRAZIL	Pergamon Editora Ltda, Rua Eça de Queiros, 346, CEP 04011, Paraiso, São Paulo, Brazil
AUSTRALIA	Pergamon Press Australia Pty Ltd., P.O. Box 544, Potts Point, N.S.W. 2011, Australia
JAPAN	Pergamon Press, 5th Floor, Matsuoka Central Building, 1-7-1 Nishishinjuku, Shinjuku-ku, Tokyo 160, Japan
CANADA	Pergamon Press Canada Ltd., Suite No. 241, 253 College Street, Toronto, Ontario, Canada M5T 1R5

First edition 1990

Library of Congress Cataloging in Publication Data
Comprehensive medicinal chemistry: the rational design, mechanistic study & therapeutic application of chemical compounds/ chairman of the editorial board, Corwin Hansch; joint executive editors, Peter G. Sammes, John B. Taylor. — 1st ed.
p. cm.
Includes index.
1. Pharmaceutical chemistry. I. Hansch, Corwin. II. Sammes, P. G. (Peter George) III. Taylor, J. B. (John Bodenham), 1939– .
[DNLM: 1. Chemistry, Pharmaceutical. QV 744 C737]
RS402.C65
615'.19—dc20
DNLM/DLC   89–16329

British Library Cataloguing in Publication Data
Hansch, Corwin
Comprehensive medicinal chemistry
1. Pharmaceutics
I. Title
615'.19

ISBN 0–08–037061–6 (Vol. 5)
ISBN 0–08–032530–0 (set)

Printed and bound in Great Britain by
BPCC Hazell Books Ltd, Aylesbury, Bucks, England

Contents

Preface

Medicinal chemistry is a subject which has seen enormous growth in the past decade. Traditionally accepted as a branch of organic chemistry, and the near exclusive province of the organic chemist, the subject has reached an enormous level of complexity today. The science now employs the most sophisticated developments in technology and instrumentation, including powerful molecular graphics systems with 'drug design' software, all aspects of high resolution spectroscopy, and the use of robots. Moreover, the medicinal chemist (very much a new breed of organic chemist) works in very close collaboration and mutual understanding with a number of other specialists, notably the molecular biologist, the genetic engineer, and the biopharmacist, as well as traditional partners in biology.

Current books on medicinal chemistry inevitably reflect traditional attitudes and approaches to the field and cover unevenly, if at all, much of modern thinking in the field. In addition, such works are largely based on a classical organic structure and therapeutic grouping of biologically active molecules. The aim of *Comprehensive Medicinal Chemistry* is to present the subject, the modern role of which is the understanding of structure–activity relationships and drug design from the mechanistic viewpoint, as a field in its own right, integrating with its central chemistry all the necessary ancillary disciplines.

To ensure that a broad coverage is obtained at an authoritative level, more than 250 authors and editors from 15 countries have been enlisted. The contributions have been organized into five major themes. Thus Volume 1 covers general principles, Volume 2 deals with enzymes and other molecular targets, Volume 3 describes membranes and receptors, Volume 4 covers quantitative drug design, and Volume 5 discusses biopharmaceutics. As well as a cumulative subject index, Volume 6 contains a unique drug compendium containing information on over 5500 compounds currently on the market. All six volumes are being published simultaneously, to provide a work that covers all major topics of interest.

Because of the mechanistic approach adopted, Volumes 1–5 do not discuss those drugs whose modes of action are unknown, although they will be included in the compendium in Volume 6. The mechanisms of action of such agents remain a future challenge for the medicinal chemist.

We should like to acknowledge the way in which the staff at the publisher, particularly Dr Colin Drayton (who initially proposed the project), Dr Helen McPherson and their editorial team, have supported the editors and authors in their endeavour to produce a work of reference that is both complete and up-to-date.

Comprehensive Medicinal Chemistry is a milestone in the literature of the subject in terms of coverage, clarity and a sustained high level of presentation. We are confident it will appeal to academic and industrial researchers in chemistry, biology, medicine and pharmacy, as well as teachers of the subject at all levels.

CORWIN HANSCH
Claremont, USA

PETER G. SAMMES
Uxbridge, UK

JOHN B. TAYLOR
Dagenham, UK

Contributors to Volume 5

Dr A. R. Boobis
Department of Clinical Pharmacology, Royal Postgraduate Medical School, Hammersmith
Hospital, Du Cane Road, London W12 0HS, UK

Professor A. M Breckenridge
Department of Pharmacology and Therapeutics, University of Liverpool, PO Box 147, Liverpool
L69 3BX, UK

Dr B. Bruguerolle
Laboratoire de Pharmacologie Médicale et Clinique, Faculté de Médecine, Université de
Marseille, 27 Boulevard Jean Moulin, F-13385 Marseille Cedex 4, France

Dr R. Bruno
Rhône-Poulenc Santé, Institut de Biopharmacie, 20 Avenue Raymond Aron, F-92165 Antony
Cedex, France

Dr J. Caldwell
Department of Pharmacology and Toxicology, St Mary's Hospital Medical School, London W2
1PG, UK

Dr J. Chamberlain
102 Chancellors Road, Stevenage, Herts SG1 4TZ, UK

Dr L. F. Chasseaud
Department of Metabolism and Pharmacokinetics, Huntingdon Research Centre Ltd, Huntingdon,
Cambs PE18 6ES, UK

Professor D. J. A. Crommelin
Rijksuniversiteit te Utrecht, Subfaculteit Farmacie, Croesestraat 79, 3522 AD Utrecht, The
Netherlands

Dr S. H. Curry
Division of Clinical Pharmacokinetics, Department of Pharmaceutics, J Hillis Miller Health
Centre, University of Florida (Box J-494), Gainesville, FL 32610-0494, USA

Dr H. Decousus
Service de Médecine Interne et Therapeutique, Hôpital de Bellevue, Boulevard Pasteur, F-42023
Saint-Etienne Cedex, France

Dr M. J. Dennis
Rhône-Poulenc Ltd, Rainham Road South, Dagenham, Essex RM10 7XS, UK

Professor J. B. Dressman
College of Pharmacy, University of Michigan, Ann Arbor, MI 48109, USA
Professor A. T. Florence
The School of Pharmacy, University of London, 29/39 Brunswick Square, London WC1N 1AX,
UK

Mr J. Gaillot
Rhône-Poulence Santé, Institut de Biopharmacie, 20 Avenue Raymond Aron, F-92165 Antony
Cedex, France

Dr N. J. Gooderham
Department of Clinical Pharmacology, Royal Postgraduate Medical School, Hammersmith
Hospital, Du Cane Road, London W12 0HS, UK

Dr J. Grevel
Department of Pharmacology and Division of Immunology and Organ Transplantation, The
University of Texas Medical School at Houston, 6431 Fannin, MSMB 6.252, Houston, TX
77030, USA

Professor R. H. Guy
Departments of Pharmacy and Pharmaceutical Chemistry, University of California, San Francisco, CA 94143-0446, USA

Dr G. W. Halbert
Department of Pharmaceutics, School of Pharmacy and Pharmacology, University of Strathclyde, Royal College Building, 204 George Street, Glasgow G1 1XW, UK

Dr D. R. Hawkins
Department of Metabolism and Pharmacokinetics, Huntingdon Research Centre Ltd, Huntingdon, Cambs PE18 6ES, UK

Dr C. Ionnides
Department of Biochemistry, University of Surrey, Guildford, Surrey GU2 5XH, UK

Dr D. B. Jack
Fidia Research Laboratories, Via Ponte della Fabbrica 3/A, I-35031 Abano Terme (Padova), Italy

Dr M. C. R. Johnson
Rhône-Poulenc Ltd, Rainham Road South, Dagenham, Essex RM10 7XS, UK

Professor M. R. Juchau
Department of Pharmacology, School of Medicine, University of Washington, Seattle, WA 98195, USA

Dr A. W. Kelman
Department of Clinical Physics and Bioengineering, West of Scotland Health Boards, 11 West Graham Street, Glasgow G4 9LF, UK

Dr G. Labrecque
Université Laval, Ecole de Pharmacie, Cité Universitaire, Bureau du Directeur, Quebec G1K 7P4, Canada

Dr M. S. Lennard
University Department of Pharmacology and Therapeutics, Royal Hallamshire Hospital, Glossop Road, Sheffield S10 2JF, UK

Dr F. Levi
Chronobiologie-Chronopharmacologie, Fondation Adolphe de Rothschild, 29 Rue Manin, F-75940 Paris Cedex 19, France

Dr S. J. Lewis
Rhône-Poulenc Ltd, Rainham Road South, Dagenham, Essex RM10 7XS, UK

Dr C. M. Macdonald
Hoechst UK Ltd, Walton Manor, Walton, Milton Keynes MK7 7AJ, UK

Dr S. C. Mitchell
Department of Pharmacology and Toxiology, St Mary's Hospital Medical School, London W2 1PG, UK

Dr G. Montay
Rhône-Poulenc Santé, Institut de Biopharmacie, 20 Avenue Raymond Aron, F-92165 Antony Cedex, France

Dr M. Ollagnier
Laboratoire Central de Pharmacologie et Toxicologie, Hôpital de Bellvue, Boulevard Pasteur, F-42023 Saint-Etienne Cedex, France

Dr L. K. Paalzow
Department of Biopharmaceutics and Pharmacokinetics, University of Uppsala, BMC Box 580, S-751 23 Uppsala, Sweden

Dr B. K. Park
Department of Pharmacology and Therapeutics, University of Liverpool, PO Box 147, Liverpool L69 3BX, UK

Professor D. V. Parke
Department of Biochemistry, University of Surrey, Guildford, Surrey GU2 5XH, UK

Dr A. Reinberg
Chronobiologie-Chronpharmacologie, Fondation Adolphe de Rothschild, 29 Rue Manin, F-75940 Paris Cedex 19, France

Dr G. Ridout
Syntex Research Centre, Research Park, Heriot-Watt University, Riccarton, Edinburgh EH14 4AP, UK

Dr D. Sesardic
Department of Clinical Pharmacology, Royal Postgraduate Medical School, Hammersmith Hospital, Du Cane Road, London W12 0HS, UK

Dr M. H. Smolensky
Room W 402, The University of Texas Health Science Center at Houston, School of Public Health, PO Box 20186, Houston, TX 77225, USA

Dr G. Storm
Rijksuniversiteit te Utrecht, Subfaculteit Farmacie, Croesestraat 79, 3522 AD Utrecht, The Netherlands

Dr M. Strolin-Benedetti
Farmitalia Carlo Erba, Via Carlo Imbonati 24, I-20159 Milan, Italy

Dr K. M. Thakker
Ciba-Geigy Corporation, Clinical Pharmacokinetics and Disposition Department, 444 Saw Mill River Road, Ardsley, NY 10502, USA

Professor J. Thomas
Australian Pharmaceutical Publishing Co. Ltd, Suite 5, Ground Floor, 174-180 Pacific Highway, North Sydney, NSW 2060, Australia

Professor T. N. Tozer
School of Pharmacy, University of California, San Francisco, CA 94143-0446, USA

Dr G. T. Tucker
University Department of Pharmacology and Therapeutics, Royal Hallamshire Hospital, Glossop Road, Sheffield S10 2JF, UK

Dr R. G. Turcan
Hoechst UK Ltd, Walton Manor, Walton, Milton Keynes MK7 7AJ, UK

Dr D. P. Vaughan
School of Pharmaceutical and Chemical Sciences, Faculty of Science, Sunderland Polytechnic, Galen Building, Green Terrace, Sunderland SR1 3SD, UK

Professor E. S. Vesell
Department of Pharmacology, The Milton S Hershey Medical Center, The Pennsylvania State University College of Medicine, PO Box 850, Hershey, PA 17033, USA

Professor B. Whiting
Department of Materia Medica, University of Glasgow, Stobhill General Hospital, Glasgow G21 3UW, UK

Professor H. F. Woods
University Department of Pharmacology and Therapeutics, Royal Hallamshire Hospital, Glossop Road, Sheffield S10 2JF, UK

Contents of All Volumes

Volume 2 Enzymes and Other Molecular Targets

Volume 4 Quantitative Drug Design

Volume 5 Biopharmaceutics

23.1

Absorption Processes

MICHAEL J. DENNIS
Rhône-Poulenc Ltd., Dagenham, UK

23.1.1 INTRODUCTION

Pharmacokinetics has been defined as the study of the kinetics of absorption, distribution, metabolism and excretion of drugs and their pharmacological, therapeutic or toxic response in animals and man.[1] Apart from those drugs which are administered directly into the systemic circulation, all other drugs and formulations are absorbed through membrane surfaces into the blood stream. As most patients find parenteral drug administration unpleasant and both physicians and patients find it inconvenient, there are considerable advantages in using a route which requires drug absorption. Thus for the vast majority of drugs, absorption will be an important aspect of the pharmacokinetic behaviour of the drug or its formulation.

The normally accepted definition of drug absorption encompasses those processes through which a drug may go after administration, up to the point at which the drug molecules reach the systemic circulation.[2] Thus loss of drug due to metabolism by the microflora of the gut, by the enzymes of the gut wall or by the first passage through the liver are all processes which will reduce absorption.

Bioavailability has been defined by numerous authors, although two definitions are now generally accepted. Absolute bioavailability is defined as the percentage of a drug contained in a drug product that enters the systemic circulation in an unchanged form after administration of the product.[3] Relative bioavailability is defined as the percentage of a test drug product absorbed relative to a standard preparation.[4]

Bioequivalence of two drug products is attained if the products do not differ significantly in the bioavailable dose or in the rate of supply.[1] Normally bioequivalence must be demonstrated by a second manufacturer wishing to market a drug product which is already available from an originating manufacturer.

The scientific study of drug absorption only began in earnest some 30 years ago[5] and was later stimulated by two major therapeutic problems. In the first case, phenytoin tablets manufactured in Australia were reformulated and the inactive tablet filler was changed from calcium sulfate to lactose. Phenytoin had been poorly available from the early formulation due to phenytoin binding to the filler and the dose had been adjusted to obtain a therapeutic effect. On changing the formulation the bioavailability improved, resulting in overdosing and causing toxicity in treated patients.[6]

In the second case, digoxin tablets were reformulated and in the new formulation the particle size of digoxin was reduced. This caused a more rapid dissolution of the digoxin crystals and thus resulted in an improved bioavailability. Subsequent dosing with the new formulation resulted in higher plasma drug levels and caused cardiac toxicity.[7]

The commonest route of drug administration is the oral route where the majority of the dose is absorbed from the small intestine into the portal vein, through the liver and into the systemic circulation. Smaller quantities of drug are absorbed through the stomach walls and through the large intestine walls. Solutions of drug are normally the most rapidly absorbed dosage form; however, tablets and capsules are more convenient and drugs are often more stable in these formulations. Recently plasma drug concentrations have been correlated with pharmacological effects for a number of drugs and in these cases the high plasma concentrations achieved during rapid absorption of a drug may be undesirable. For this reason, and to reduce the frequency of dosing for drugs with short half-lives, formulations which release their drug dose more slowly into the gastrointestinal tract have been designed. Such slow release or controlled release products[8] have found increasing acceptance by physicians and patients.

Drug absorption from the gastrointestinal (GI) tract is not a simple process and absorption will be influenced by many factors. The physiology of the GI tract itself is of primary importance as are the chemical characteristics of the gut contents. The structure of the drug molecule, its degree of ionization, solubility and lipophilicity will all influence the rate of absorption of a particular molecule. Other important factors in determining the rate at which the drug dissolves will be its crystal structure, polymorphic form and particle size. The formulation itself may also prevent or hinder drug absorption by not releasing the drug sufficiently quickly. Many other physiological factors such as food intake, stress or lying down can reduce absorption. Equally some factors such as warm liquids and movement may promote absorption. The coadministration of other drugs may either slow down or speed up absorption.

Absorption will not only be prevented by the drug molecule not being in the right form at the right time to allow absorption, but may also be prevented by the drug molecule being altered prior to absorption. The acidic medium of the stomach or the alkaline medium of the intestine can chemically alter some drug molecules. Alternatively the drug molecule may be metabolized by the intestinal flora prior to absorption, or it may be metabolized by the enzyme systems of the intestinal wall on passage through it. The main disadvantage of the oral route is that the majority of the absorbed dose passes directly through the liver before entering the systemic circulation. As the liver is the major metabolizing organ of the body, this can result in the total deactivation of easily metabolized drugs. This effect is normally known as the 'first pass effect'. In general, however, the oral route is the most convenient for the patient and for the majority of drugs for which systemic drug concentrations are required it is the route of choice.

The rectal route of drug administration is widely accepted as an alternative to the oral route on the continent of Europe but is little used in the United Kingdom. It can be a very efficient route of drug administration as quite large doses may be given and absorption can be quite rapid. An advantage of this route is that a proportion of the dose absorbed is not delivered directly to the liver and thus more parent drug may reach the systemic circulation for a drug which is easily metabolized. However absorption of drug from rectal formulations is often slow which may be a result of poor formulations, many of which originated prior to bioavailability comparisons.

Disadvantages of this route are that drug may be exposed to microbiological flora in the rectum and deactivated or possibly transformed into undesirable molecules. The timing of drug administration may be restricted to shortly after defaecation, especially for drugs which are slowly released or absorbed. In general, the rectal route is still commonly used for administration of drugs prior to and after surgery when the oral route may be less efficient.

Interest in the transdermal route has increased dramatically over the past 10 years, partly due to the increasing pharmacological activity of drug molecules and therefore the reduced size of dose administered, and partly due to an increased understanding of processes affecting absorption across the skin and the ability to control the rate of drug absorption. There are pharmacological advantages in being able to achieve a fairly constant plasma drug concentration over the dosing interval and constant rate drug administration usually allows the achievement of this objective. The release of drug from newer transdermal formulations is the rate-limiting step in the process and thus a constant rate of drug administration may be achieved over many hours. There are a number of advantages of the transdermal route over the oral route. Drug is not exposed to microorganisms or extremes of pH as in the GI tract and, in addition, drug is absorbed directly into the blood stream and any metabolizing capacity of the skin is usually less than that of the mucosal cells of the GI tract.

The inhaled route for administering drugs has been extensively used for the administration of gaseous anaesthetics. The lungs are, obviously, extremely efficient at transferring gaseous drugs to the blood stream and control of drug absorption can be extremely precise. However, there is now an increasing interest in administering other classes of drugs by this route. In general, the objective of drug administration to the lungs, with the exception of anaesthetics, has been to treat lung disorders. However, the lungs are very efficient at absorbing drugs into the blood stream. The advantage of this route is that very rapid absorption can be obtained, especially with aerosol formulations. The disadvantage is the difficulty of obtaining exact dosages and the relatively large proportion of drug which ends up in the stomach.

The other routes of drug administration such as vaginal, intranasal, buccal or sublingual are used much less frequently, but may have an advantage over other routes for a particular molecule or therapeutic indication. The buccal or sublingual route may be an alternative route when rapid absorption of a relatively small dose is required, but the drug is deactivated in the GI tract. The intranasal and vaginal routes are often used for local administration of drugs but a significant proportion of a dose may be absorbed into the systemic circulation as these areas are highly perfused with blood.

Methods of studying drug absorption usually fall into two classes: direct methods and indirect methods. Direct methods are methods in which the drug concentration remaining in the stomach or in the blood stream of the portal vein is measured. Many models of drug absorption are available using synthetic membranes or sections of animal intestine. However direct methods are usually more difficult to perform in man and therefore indirect methods are often used. Indirect methods usually involve comparing the blood concentration profile of a dosage form for which the rate of absorption is known, for example a bolus injection or constant rate infusion, with a dosage form for which the absorption profile is unknown. Mathematical techniques allow the rate and extent of absorption to be determined.

Bioavailability is the normal term used to express the extent of absorption of a formulation either absolute to an intravenous preparation or relative to a previously defined standard formulation. Studies of bioavailability are now normally required prior to marketing a new drug formulation for extravascular administration in order to establish the percentage of the formulation which is available to the circulation. In this way correct dosages for different routes of administration can be calculated.

Bioequivalence is a term which is used to express the similarity of the rate and extent of absorption of two formulations, usually of the same drug, which may be from the same or different manufacturers. Bioequivalence encompasses equal relative bioavailability with equality in the rate of absorption. Statistical approaches are normally used to demonstrate that two formulations are not significantly different in their absorption characteristics. It is, however, important to recognize the difference between two formulations which are equally bioavailable and two formulations which are bioequivalent. A slow release formulation and a rapid release formulation may deliver equal quantities of drug and be equally bioavailable but they cannot be bioequivalent if they deliver drug at different rates.

23.1.2 GASTROINTESTINAL ABSORPTION

23.1.2.1 Physiology of the GI Tract

The GI tract, which passes from mouth to anus, consists of four physiologically different layers (Figure 1). From the viewpoint of drug absorption, the inner two layers are important. The inner layer, known as the mucous layer, consists of an epithelial lining and an inner stroma containing glands and smooth muscle cells. The next layer, known as the submucous coat, consists of loose connective tissue, blood vessels and the lymphatic system. To be absorbed, drug must diffuse through the mucous coat and into the submucous coat capillaries. After swallowing a drug dosage form, it is propelled into the stomach by peristalsis in the oesophagus. Swallowing liquid with the dosage form ensures it does not become lodged in the oesophagus.

23.1.2.1.1 *The stomach*

The normal function of the stomach is to store food until it has been sufficiently broken down to allow it to proceed further down the GI tract. The anatomy of the stomach is shown in Figure 2, which also gives the terms used to describe the various regions of the stomach. Digestive juices such as hydrochloric acid and pepsinogens are secreted from cells in the body and fundus of the stomach. These are secreted either in response to central stimuli triggered by the sight or smell of food, or directly by the hormone gastrin, which is released by the presence of food in the stomach. The gastric mucosa are pitted by glands in which are situated the chief cells, which secrete pepsinogens, and the parietal cells, which secrete hydrochloric acid. Pepsinogens are converted by hydrochloric acid to the active pepsin enzymes. The pH of the lumen of the stomach under fasting conditions is normally within the range pH 1 to pH 3,[8] although directly after a meal the pH rises to between pH 3 and pH 5 for a period.[9] Thus the environment of the stomach is often quite hostile to drug substances and may cause chemical changes to the structure of the drug molecule. Thus benzylpenicillin and erythromycin undergo acid-catalyzed decomposition in the stomach.

The gastric mucosa consists of columnar cells which secrete mucous and alkaline bicarbonate.[10,11] Thus the lining of the stomach is covered with a thin layer of mucous, slightly alkaline on the mucosal side and acidic on the lumen side. This steep pH gradient has been referred to as the pH

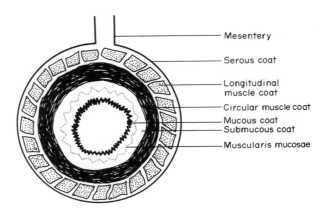

Figure 1 Diagrammatic representation of a cross-section through the gastrointestinal tract

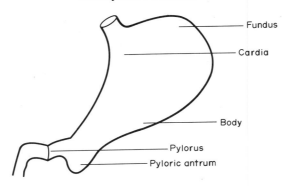

Figure 2 Gross anatomy of the human stomach

mucous barrier[12] and contributes to the protection of the mucosal cells from the acidic climate of the lumen. Bicarbonate production may be stimulated by secretion of acid and by prostaglandins, which may explain why some antiinflammatory substances which inhibit the synthesis of prostaglandins have gastric irritation and ulcers as side effects.

23.1.2.1.2 *The small intestine*

The small intestine is usually divided into three regions: the duodenum, the jejunum and the ileum. The duodenum consists of a section, approximately 20 cm in length, from the pyloric sphincter to the ligament of Treitz. Within this section the bile and pancreatic juices empty into the small intestine at the ampulla of Vater. The remainder of the small intestine, approximately 260 cm in length, is divided into the jejunum and ileum in the proportion 2:3. The mucosal surface of the small intestine is extremely suited to the absorption of small molecules. The mucosal surface consists of villi, which are finger-like projections about 1 mm in length and are at a density of 20–40 villi mm² of intestinal surface. Each villus is completely supplied with capillary blood vessels, lymph vessels and nerves (Figure 3). The surface of each villus consists of a single layer of columnar epithelial cells, which themselves have an indented surface about 1 μm high consisting of microvilli, which form a brush border to the cells.[13] Thus the surface area of the small intestine is increased some 600 times over its smooth surface area,[14] to give a total surface area of about 200 m². This fact is extremely important in designing drugs or formulations for oral absorption. The pH of the small intestine is more variable than that of the stomach, but, in general, the pH rises from the duodenum to the ileum. In the duodenum a mean pH in the range pH 5 to 6 is found which may be about one pH unit lower 1 to 2 h after food intake. The intestinal contents or chyme reaching the jejunum is close to neutral pH. The small intestine secretes about 3 L of fluid per day with a pH of 7.6, although a considerable proportion of this is reabsorbed before reaching the colon.

23.1.2.1.3 *The colon*

The ileum terminates in the ileo-caecal valve, the entrance to the colon which terminates at the rectum (Figure 4). After ingestion of food the valve opens regularly to allow intestinal contents to pass into the colon. There are no villi on the mucosal surface of the colon and thus the surface area is not as great as the small intestine. However, absorption of some drug molecules continues to take place in the colon. The mucosal surface consists of columnar epithelial cells and goblet cells and secretes mucous which neutralizes acids formed by bacteria in the colon.[15] About 1 L of chyme enters the colon each day but the majority of the water is reabsorbed in the caecum and ascending colon.

23.1.2.2 Gastrointestinal Transit

Gastrointestinal transit is of considerable importance when considering drug absorption. The time that slowly absorbed or poorly soluble drugs are in the small intestine, with its very large

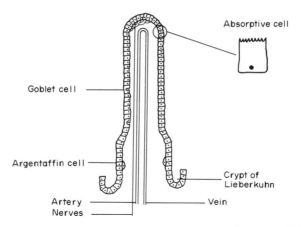

Figure 3 Diagrammatic representation of a villus from the human small intestine

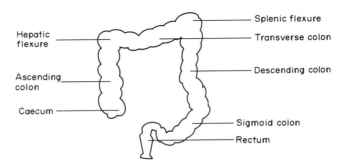

Figure 4 Gross anatomy of the human colon

surface area, may determine the amount of drug absorbed. In the fasting state gastrointestinal transit is influenced by cyclic motor activity, which shows four distinct phases.[16] Phase I has little contractile activity and lasts for about 1 h, phase II has intermittent and irregular contractions and occurs for about 30 min, phase III contractions occur for about 5–15 min at their maximal rate and phase IV contractions are brief and intermittent or may be absent altogether. Phase III activity has also been termed the 'housekeeper wave' due to the gastric cleansing action of the wave. Cyclic motor activity is observed in the lower oesophageal sphincter, stomach, small intestine and colon. It is also referred to as the interdigestive migrating motor complex (IMMC) as it migrates towards the colon. The migrating motor complex is disturbed by a meal of 350–400 kC (1 kC = 4.18 kJ) for between 1.5 and 4 h and is replaced by a different pattern of contractions.[17]

In the stomach, after feeding, peristaltic contractions of circular muscle from midcorpus to the pylorus forces gastric contents towards the antrum. The contractile wave consists of two components separated by about 3 s. The first contraction forces liquid material through the pylorus and shuts the narrow opening and the second stronger contraction squeezes the contents of the antrum and terminal antrum back into the corpus of the stomach.[18] The pylorus remains shut for 1–2 s after the antrum relaxes and begins to refill. Large particles will be drawn to the centre of the nonoccluded antrum and be forced further back into the corpus than small particles, which will be closer to the walls. These will then be squeezed into the small intestine when the pylorus opens again. During normal digestion, particles passing through the pylorus are about 1 mm or less in size. Larger particles of indigestible material are retained in the stomach after feeding and are emptied by the migrating motor complex of the fasting state. The gastric emptying of indigestible markers was delayed from about 2 h in the fasting state to about 6 h after a test meal. Two additional meals at 2 h intervals increased the time taken to empty the markers even further.[19] The half emptying time of a labelled soluble marker after a liquid fatty meal was about 1 h.[20]

Once emptied from the stomach both large and small particles as well as liquids appear to be transported at the same rate through the small intestine.[21] Intestinal transit time appears to be similar either in the fed state or in a fasting period and is usually between 3 and 5 h.[8]

Table 1 Mean Transit Time of Radio-opaque Markers through the Human Colon[22]

| | Colonic Transit (h) | |
	Men	Women
Caecum and ascending colon	6.0	7.2
Hepatic flexure	2.1	3.7
Transverse colon	3.9	5.5
Splenic flexure	1.7	5.0
Descending colon	3.6	5.7
Sigmoid colon and rectum	13.4	12.0

Colonic transit may be important for some poorly soluble drugs and slow release forms. The average transit time of indigestible markers was about 30 h in men and 39 h in women.[22] The mean time that the markers spent in each region of the colon is given in Table 1. After instillation of a liquid marker into the colon, 70% was excreted within 48 h.[23] In this case residence in the caecum and ascending colon was only about 1.5 h.

23.1.2.3 Theories of Drug Absorption

23.1.2.3.1 *The pH–partition hypothesis*

Two major concepts of drug absorption have developed during the last 50 years which have increased our understanding of drug absorption and stimulated interest in further research.

The first concept, developed from earlier work on the permeability of plant and animal cells to organic molecules,[24,25] suggested that the cell wall is composed of a lipid layer with aqueous pores. Nonpolar organic molecules would penetrate the cell by diffusion through the lipid membrane, whereas polar molecules would only penetrate the cell if they were small enough to pass through the pores.

Studies of this concept as applied to the absorption of drug molecules showed a positive correlation between the degree of lipid solubility of the molecule and the rate of transfer across the GI tract. This was demonstrated both for a homologous series of compounds[24] and for compounds of dissimilar structure.[25,26] The partition coefficient between chloroform/water or between heptane/water was used as a measure of the lipid solubility of the drug molecules. The use of the retention time of a compound on a C_{18} reverse phase HPLC column has also been used as a measure of lipophilicity.[27] Since many lipid soluble drugs are absorbed more readily than some small polar molecules, *e.g.* urea, it was suggested that the main route of absorption is through the lipid membrane and not through the aqueous pores.[25]

The second concept, which arose from the studies of drug absorption during the late 1950s, was that only the unionized form of the drug is able to cross the gastrointestinal membrane and the ionized form penetrates slowly, if at all. The evidence supporting this concept resulted from experiments in dogs, which showed that intravenously administered bases appeared in gastric juice in higher concentrations than in the plasma, whereas intravenously administered acidic drugs did not cross substantially into the gastric juice.[28] The theory is that the unionized form reaches equilibrium between plasma and gastric juice but a weak acid is mainly ionized in plasma in contrast to a weak base which is mainly in the nonionized form. The weak base will diffuse into the gastric juice where it becomes ionized and cannot diffuse back, but the weak acid will be ionized in the plasma and will not diffuse into the gastric juice. The equilibrium for a weak acid and a weak base is shown in Figure 5. The theory also predicts that acidic drugs should be more easily absorbed from the stomach than basic drugs. Acidic drugs will be mainly in the unionized form in the stomach and should diffuse more rapidly than the mainly ionized basic drugs. This was confirmed by studies of the absorption of a number of acid and basic drugs from the rat stomach. Acidic drugs, except highly ionized sulfonic acids, were readily absorbed, whereas basic drugs were poorly absorbed.[29] In addition, the absorption of basic drugs could be improved by raising the pH of the stomach contents, whereas at higher pH the absorption of acidic drugs was reduced.

Further experimental work, in which rat small intestine was perfused with both weak and strong acidic and basic molecules, demonstrated more rapid absorption for the weakly ionized molecules and poor absorption for the strongly ionized molecules.[25] Completely ionized quaternary ammonium compounds and sulfonic acids were very poorly absorbed.

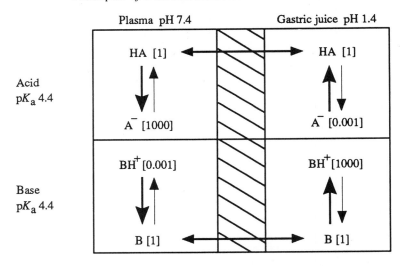

Figure 5 Equilibrium diagram for a weak acid and a weak base

23.1.2.3.2 *Limitations of the pH–partition hypothesis*

The unification of the two theories of drug absorption became known as the 'pH–partition hypothesis'. It has made an enormous contribution to our understanding of drug absorption as well as in the design of new drug molecules which will be well absorbed by nonparenteral routes. However, even during the work resulting in this hypothesis, a number of inconsistencies were observed. The lowest pK_a of an acidic drug which was rapidly absorbed was about 3, whereas the highest pK_a of a similarly absorbed basic drug was about 8. Thus at an intestinal pH of about 7 the ratio of ionized to unionized drug consistent with rapid absorption is very different for acids and bases.[26]

In order to explain this in terms of passive drug diffusion it was proposed that the pH close to the epithelial membrane was lower, perhaps about pH 5.3, than that in the lumen. The existence of an unstirred layer, probably formed by mucous secretion, has been demonstrated and both the pH and permeability of this layer are likely to be different from that of the lumen.[30]

The theory that only the unionized form of the drug can be absorbed came under scrutiny when it was shown that even in *in vitro* loop preparations significant amounts of quaternary ammonium compounds and even phenol red can be absorbed;[31] it had already been shown that absorption of quaternary ammonium compounds occurs *in vivo*.[32] The main consideration, which had not been taken into account in the original theory, was the extremely large surface area of the small intestine and the time over which absorption may take place. Thus even polar ionized molecules with very small nonionized fractions may be slowly absorbed, possibly through aqueous pore channels during a period of some hours.[33] Theoretically the pH–partition hypothesis indicates that acidic compounds should be absorbed more readily from the stomach than from the small intestine. That this does not occur was shown using acidic compounds such as aspirin, warfarin, barbitone and paracetamol.[30] The reason for this is, once again, that the surface area of the small intestine is much larger than that of the stomach.[34]

Another factor which has to be taken into account when considering the absorption of solid dosage forms is the solubility of the drug substance in the gastrointestinal fluids. Acidic compounds will tend to dissolve more slowly in the acidic contents of the stomach and basic compounds will dissolve readily. Some acidic compounds of low aqueous solubility may even be precipitated from solution on entry into the stomach.

Thus although the pH–partition hypothesis represents a first approximation for drug absorption, it has since been realized that other factors must also be taken into account. Firstly, at high octanol/water partition coefficients ($\log P > 3$) the absorption rate reaches a plateau level.[35,36] It has been suggested that this is due to a change in the rate-limiting factor from permeability control to aqueous diffusion control across the unstirred layer.[36] At low octanol/water partition coefficients ($\log P < -1$) the absorption rate may be slow and not increase with increasing partition coefficient. This may be due to partition coefficient independent absorption of these molecules through aqueous pores, which occur between the epithelial cells of the mucosal layer. The pore size of these junctions is about 6 Å in the jejunum and 3 Å in the ileum, with a population density about 10 times greater in the ileum.[37]

However, even with these explanations of absorption at either end of the partition coefficient scale, molecules with similar $\log P$ values may be absorbed at quite different rates.[38] Within homologous series of compounds much more predictable absorption profiles have been obtained[24] and a large amount of the variability can be accounted for by changes in partition coefficient, but for heterogeneous compounds often less than 50% of the variability can be accounted for by differences in partition coefficient.[31] Even in series of very similar compounds exceptions are found as is the case for acebutolol in the β-blocker series. Acebutolol is a moderately nonpolar compound with a $\log P$ value of 1.87. However, absorption from the rat intestine gave absorption rate constants similar to those of more hydrophilic β-blockers like atenolol ($\log P = 0.23$) and sotalol ($\log P = -0.79$), and much slower than β-blockers with similar $\log P$ values such as pindolol ($\log P = 1.75$) and timolol ($\log P = 2.10$).[35] Another example is that of *p*-aminobenzoic acid, which is polar and is almost completely ionized at the pH of the small intestine. However, considerable absorption of *p*-aminobenzoic acid takes place from the rat small intestine.[39] In this case it seems likely that some form of active transport is occurring, involving binding to the brush border membrane. Other molecules which do not obey the pH–partition hypothesis are some aminopenicillins such as ampicillin ($\log P = 0.23$) and amoxicillin ($\log P = 0.87$). These molecules exist as zwitterions or as anions at the pH of the small intestine, yet they are absorbed.[39] This has been attributed to the presence of active or carrier-mediated transport in the case of amoxicillin, although partitioning of the anion into the organic phase has also been demonstrated.

In spite of the above theoretical and experimental work, the majority (about 90%) of drugs or their salts with $\log P$ values between -2 and $+4$ in 'Clarkes Isolation and Identification of Drugs' are recorded as well absorbed. Those drugs recorded as poorly or variably absorbed usually have $\log P < 0$ or > 3 and are of low aqueous solubility (Table 2). Obviously, drugs with available water soluble salts will often be well absorbed, even with $\log P$ values below or above the range given above.

23.1.2.4 Rate-limiting Factors in Drug Absorption

A number of authors have studied theoretical models of intestinal drug absorption and rate-limiting factors in these models.[40,41] A differential equation proposed to describe drug flow and

Table 2 Partition Coefficients and Oral Absorption of some Marketed Drugs

Drug	Partition coefficient	Absorption in man
Aspirin	−1.1	Well absorbed
Amoxycillin	0.87	Well absorbed
Atenolol	0.23	About 50% absorbed
Atropine	1.8	Well absorbed
Caffeine	0.0	Well absorbed
Chlorpromazine	3.4	Slowly but complete
Chlorthiazide	−1.9	20–30% absorbed
Chloramphenicol	1.1	Well absorbed
Diazepam	2.7	Well absorbed
Ergometrine	−0.9	Fairly well absorbed
Haloperidol	4.3	Well absorbed
Ketoprofen	0.0	Well absorbed
Indomethacin	−1.0	Well absorbed
Metronidazole	−0.1	Well absorbed
Morphine	−0.1	About 60% absorbed
Nitrazepam	2.1	Well absorbed
Oxytetracycline	−1.6	Incomplete
Oxyphenbutazone	3.3	Well absorbed
Oxazepam	2.2	Well absorbed
Pethidine	1.6	Well absorbed
Practolol	−1.3	Well absorbed
Prazepam	3.7	Slow and variable
Prochlorperazine	2.4	Well absorbed
Sulfaguanidine	−1.2	Variable
Tetracycline	−1.4	Irregular and incomplete
Theophylline	0.0	Well absorbed

absorption from the small intestine is given in equation (1),

$$\frac{\partial C(x,t)}{\partial t} = \alpha\frac{\partial^2 C(x,t)}{\partial x^2} - \beta\frac{\partial C(x,t)}{\partial x} - \gamma C(x,t) \tag{1}$$

$$x \geqslant 0, \quad t \geqslant 0$$

where $C(x,t)$ is the drug concentration at any distance x in the intestine and at time t, α is the longitudinal spreading coefficient in $cm^2 s^{-1}$, β is the linear flow in $cm\, s^{-1}$ and $\beta = \Phi/\pi r^2$, where Φ is the bulk flow rate in $cm^3 s^{-1}$ and r is the radius of the intestinal lumen in cm, and where γ (the absorption constant in s^{-1}) $= P_e/r$, where P_e is the permeability coefficient in $cm\, s^{-1}$.

The three terms on the left hand side of equation (1) represent contributions to the drug concentration at any time and point in the intestine. The longitudinal spreading coefficient, within the expected range, appears to have relatively little effect on drug concentration for a drug in solution, whereas the bulk flow rate will have significant effects, especially for compounds with a low permeability coefficient. However the longitudinal spreading coefficient may be more important for drugs which remain in suspension on entry into the intestine.

The separation of the permeability coefficient of the aqueous boundary layer and the permeability coefficient of the membrane has proved useful in the rat through and through perfusion model.[42] The mathematical description of this model is given in equation (2), where $C(l)/C(0)$ is the fraction absorbed in the intestine, P_{aq} is the permeability coefficient of the aqueous boundary layer, P_m is the permeability coefficient of the membrane, l is the length of the intestine, and r and Φ are as previously defined. The two rate-limiting situations are when P_m is much greater than P_{aq}, that is aqueous boundary layer diffusion-controlled absorption, and when P_{aq} is much greater than P_m, that is membrane diffusion-controlled absorption.

$$\frac{C(l)}{C(0)} = \exp\left[\frac{-2\pi r l P_{aq}}{\Phi(1 + P_{aq}/P_m)}\right] \tag{2}$$

The rate-limiting equations are given in equation (3) for aqueous boundary layer rate-limited absorption and equation (4) for membrane rate-limited absorption.

$$\frac{C(l)}{C(0)} = \exp\left(\frac{-2\pi r l P_{aq}}{\Phi}\right) \tag{3}$$

$$\frac{C(l)}{C(0)} = \exp\left(\frac{-2\pi r l P_m}{\Phi}\right) \tag{4}$$

In Figure 6 is shown the relationship between the permeability coefficient P_e and the partition coefficient in octanol/water for a series of steroids. At high partition coefficient ($\log P > 3$), the aqueous boundary layer becomes rate limiting and the curve reaches a plateau. At low partition coefficient ($\log P = 0$ to 3), membrane diffusion becomes the rate-controlling step and flow rate in the intestine is less important. For an aqueous boundary layer controlled drug the flow rate in the intestine has been shown to be directly related to the permeability coefficient.[36] The thickness of the aqueous boundary layer will be dependent upon the flow rate in the intestine. In the rat jejunum the thickness has been estimated, at a flow rate of 5 mL min^{-1}, to be between 500 and 840 μm[42,36] while in humans, at a flow rate of 2 mL min^{-1}, the thickness has been estimated at between 632 and 950 μm.[43,36]

The concept of an anatomical reserve length of the intestine for drug absorption was defined as the length of intestine remaining after the theoretical length for absorption has been subtracted (equation 5; RL is the reserve length, L^* is the effective length of the small intestine and Φ, r and P_e are as previously defined).[44] From this relationship it was calculated that even for drugs with near optimal absorption characteristics a considerable length of the small intestine would be required for total absorption.

$$RL = L^* - (0.48\Phi/rP_e) \tag{5}$$

The physical parameters described above have been used together with the aqueous solubility of the drug and the fraction unionized at pH 6.5 to develop a dimensionless measure called absorption potential.[45] It was assumed that a relationship exists of the form given in equation (6), where F_{ab} is the fraction absorbed, S_0 is the solubility of the unionized species at 37 °C, X_0 is the dose, V_L is the

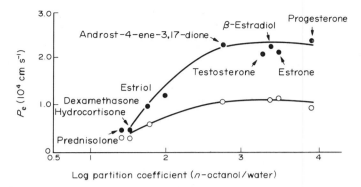

Figure 6 Comparison of *in situ* apparent permeability coefficients of various steroids for the rat jejunum with different stirring rates in the lumen and the n-octanol/water partition coefficients. Results obtained using modified Doluisio method. ○, Static, 150 s sampling intervals (slow stirring); ●, oscillation, 0.075 mL s^{-1} (rapid stirring) (reproduced from ref. 36, by permission of Adis Press)

volume of the lumen, F_{non} is the fraction unionized at pH 6.5, and P_m and P_{aq} are as previously defined.

$$F_{ab} \propto (P_m/P_{aq})^a (F_{non})^b [(S_0 V_L)/X_0]^c \tag{6}$$

By substituting absorption potential (*AP*) for fraction absorbed and making certain simplifying assumptions, the relationship given in equation (7) can be developed. This parameter was shown to be related to fraction absorbed for a number of dissimilar drugs.

$$AP = \log[(PF_{non}S_0 V_L)/X_0] \tag{7}$$

23.1.2.4.1 *Effect of alteration in gastric emptying*

It is now generally accepted that, due to the large surface area of the small intestine, the rate of absorption, even of acidic drugs, will be much greater from the small intestine than from the stomach. Thus, the rate of gastric emptying will have a pronounced effect on the speed of drug absorption.

One of the most important factors affecting gastric emptying is the presence or absence of food or liquids in the stomach. Gastric emptying of liquids is dependent upon the volume, pH and the constitution of the liquid. Drugs dissolved in larger volumes of a liquid (*ca.* 200 mL) are often emptied from the stomach more quickly than when dissolved in a smaller volume (*ca.* 50 mL). A 154 mM saline meal at pH 7 emptied rapidly (2 min), an acidic meal at pH 2 emptied less rapidly (11 min) and a fatty liquid meal emptied most slowly (18 min).[46] Solid meals empty, in general, more slowly than liquid meals and may range from 30 min for a light breakfast[47] to between 4 and 6 h for heavy meals containing high concentrations of fat or fibre.[48]

Currently, it is thought that the control of gastric emptying is from the duodenum and is inversely proportional to the 'load' which the duodenum receives. Larger indigestible particles will on average be emptied later than liquids or smaller particles, by the 'housekeeper wave'.

A number of review articles have detailed those drugs whose absorption is delayed or reduced by ingestion with or after food.[49,50] Many drugs exhibit delayed absorption and thus lower maximum plasma concentrations at later times when taken after food compared to the fasting state. Evidence that this is related to the rate of gastric emptying has been provided for paracetamol[51] and for aspirin.[52] A number of classes of drugs also exhibit reduced absorption when administered with food and this can often be attributed to longer exposure to stomach acids, which degrade the drug. As a result of the reduction in absorption of some drugs when administered with food, studies of bioavailability are normally carried out after an overnight fast and food is normally not consumed until 3 or 4 h after drug administration. However, relatively little is known about gastric emptying and gastrointestinal transit time in the fasting state. It is known, however, that the migrating motor complex, operative in the fasting state, may propel the contents of an almost empty stomach quite rapidly through the small intestine. In addition, drug administered after a migrating motor complex may be retained in the stomach for a period until the next phase II or phase III activity.[53] This motor activity in the fasting state may increase the variability of bioavailability studies and even lead to lower absorption for poorly absorbed drugs than in the fed state. As an example Figure 7 shows

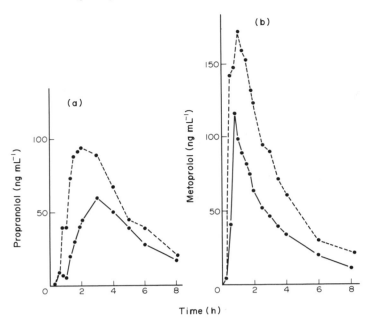

Figure 7 Concentrations of (a) propranolol (serum) and (b) metoprolol (plasma) in two healthy volunteers following ingestion of single doses (80 and 100 mg respectively) of the drugs on an empty stomach (●—●) and together with a standardized breakfast (●---●) (reproduced from ref. 54 by permission of the C. V. Mosby Company)

that metoprolol and propranolol absorption were both increased in subjects who took these drugs after a standard breakfast.[54] It is therefore suggested that variability may be reduced and absorption maximized if drugs are taken 1 h after a light nonfatty breakfast when studying bioavailability. Certainly, further study of this subject is required.

The effect of food on the gastric emptying and absorption of slow release products is of particular importance. Slow release products often contain large doses of drug, designed to be released over many hours. The rapid release of drug from such a formulation, known as 'dose dumping', may be therapeutically harmful. This problem was shown to occur with a theophylline slow release product which, when administered with food, was presumably retained in the stomach and release of drug from the formulation occurred.[57] Generally the study of the effect of food on the pharmacokinetics of slow release products is now a requirement of most regulatory agencies.

Gastric emptying is also influenced by posture and has been delayed by subjects in the supine position compared to ambulant patients.[55] In general posture should be standardized in bio-availability studies and generally an upright position, *e.g.* sitting, is preferred to supine.

In the clinical situation a number of other pathological and external factors may affect gastric emptying. Pain and trauma are thought to delay gastric emptying,[56] and in addition the coadministration of certain drugs will delay or increase the rate of gastric emptying. Drugs such as propantheline or atropine and narcotic analgesics delay gastric emptying, whereas metoclopramide increases the rate of gastric emptying.[56] Delays in gastric emptying may increase the amount of drug absorbed for poorly soluble drugs due to an increase in the time available for absorption or reduce the amount absorbed for drugs which are degraded in the stomach.

23.1.2.4.2 *Effect of microflora*

The content of the large intestine of most mammals is an ideal medium for the growth of many microorganisms. The particular distribution and number of types of aerobes and anaerobes vary from individual to individual. However, the principle organisms appear to be anaerobic bacteria and bifidobacteria, which are found mainly in the colon. Smaller numbers are found in the terminal ileum of man but the stomach and jejunum are normally sterile. Very small numbers of organisms are found in saliva. Those organisms commonly found are given in Table 3.

The bacteria of the large intestine are able to change many drug substances chemically and thus to reduce their absorption but the compounds formed are mostly different from cellular metabolic

Table 3 Organisms Commonly Found in the Gut Lumen or Faeces of Man and/or Animals[59]

Organism	Description	Distribution and density in the gut of man
Bacteroides, *i.e. B. fragilis,* *B. melaninogenicus*	Strictly anaerobic; Gram-negative; nonsporing; often encapsulated slender rods	Mainly present in colon/faeces (about 10^{10}–10^{11} organisms g^{-1}), smaller numbers throughout gut (up to 10^4 g^{-1})
Bifidobacteria, *i.e. Lactobacillus bifidus*	Anaerobic lactobacilli; Gram-positive; long slender rods	Mainly present in colon/faeces (about 10^{10}–10^{11} g^{-1}), smaller numbers elsewhere (10^4 g^{-1})
Clostridia, *i.e. C. putrefaciens,* *C. paraputrificum, C. perfringens,* *C. bifermentas*	Anaerobic; Gram-positive; large, broad rods	Present in colon/faeces (about 10^4 g^{-1}; very variable)
Enterobacteria, *i.e. Aerobacter aerogenes,* *Klebsiella, Proteus vulgaris*	Aerobic or facultative anaerobic; Gram-negative; nonsporing rods	Largely in lower small intestine and colon/rectum (10^6 g^{-1})
Enterococci, *i.e. Streptococcus salivarius,* *Streptococcus faecalis*	Aerobic or facultative anaerobic; Gram-positive; small cocci	Present throughout gut at low levels (10^3–10^4 g^{-1})
Lactobacilli, *i.e. L. acidophilus, L. brevis*	Microaerophilic; Gram-positive; slender rods	Present throughout gut at low levels (up to 10^4 g^{-1})

products. Metabolic reactions are normally oxidative or involve conjugation, whereas microbial action usually involves reduction or hydrolysis. In addition, most conjugates can be hydrolyzed by microorganisms, releasing the parent drug, which may be reabsorbed. This forms the basis of a cycle known as enterohepatic recycling in which conjugates are formed in the liver, eliminated in the bile, deconjugated by microorganisms and then reabsorbed.[58] Other functional groups which undergo hydrolysis include esters and amides. Reduction reactions occur for a number of functional groups including alcohols, aldehydes, ketones, alkenes, nitro compounds and *N*-oxides. Microorganisms are also able to replace functional groups, normally with hydrogen, and reactions such as dealkylation, deamination, decarboxylation, dehalogenation and dehydroxylation have been shown to occur. References to these varied chemical changes which microorganisms are able to perform have been tabulated.[59] Thus chemically the role of microorganisms may be thought of as the reverse of metabolic reactions, and the compounds formed may have pharmacological activity or toxicity.

The amount of an orally administered drug which is subjected to microbial degradation will be dependent upon a number of factors. Firstly, many drugs are quite rapidly absorbed from the duodenum and jejunum and hence will not reach the microorganisms of the large intestine. Drugs which are very polar or very nonpolar may be slowly absorbed and will pass into the ileum and colon. Some drugs are excreted as parent drug or metabolites in bile and so may reach the lower gut and other drugs may be secreted into the small intestine or colon from the blood stream.

The study of the effect of microorganisms on drugs is normally undertaken in experimental animals. Examination of the metabolic profile of the drug after oral and intravenous administration may lead one to suspect the involvement of gut flora. However confirmation can only be obtained in animals free of microflora or in which the microflora are suppressed. Suppression is normally achieved using a mixture of antibiotics. Neomycin is effective against aerobic organisms[60] and tetracycline, bacitracin or metronidazole may be used to suppress anaerobes. However the results of experiments in rats treated with antibiotics need to be treated with caution as the antibiotic treatment may affect absorption of the parent compound or may affect metabolism by the intestinal mucosa.[61]

The technique of comparing the metabolism of drugs in germ free animals and in normal animals is scientifically better but difficult to carry out. Germ free rats are maintained in metabolism cages protected by flexible film isolators.[62] Drugs are administered orally to both groups of rats and the absence of the metabolite in germ free rats which is present in normal rats implicates the microflora. However even in germ free rats physiological factors may confound the results. The germ free rat develops a larger caecum and smaller heart with reduced blood flow to small intestine and liver, thus possibly affecting rates of absorption and metabolism.[63] Confirmation of the role of microorganisms in the transformation of a drug is normally accomplished by incubation with caecal contents or a pure culture of one organism. These incubations are normally carried out in an atmosphere of hydrogen, carbon dioxide and nitrogen from which all traces of oxygen have been removed.[64] In general most meaningful results are obtained with relatively short incubations of fresh caecal contents where the distribution of organisms remains unaltered.

The study of the effects of microorganisms on drug absorption in humans is difficult. Patients who are receiving antibiotic therapy may be studied. Thus the inactivation of digoxin by reduction of the lactone ring by anaerobic bacteria was studied in children who were either concurrently taking antibiotic therapy or who were not. It was shown that bacterial inactivation rarely takes place before 15 months of age whether or not antibiotic therapy had been used, thus development of the normal adult microflora is a slow process.[65] Alternatively the metabolic profile of drugs in ileostomy patients may be studied. It was shown that in ileostomy patients, the sulfide metabolite of sulfinpyrazone, which reaches concentrations of $1.6\,\mu g\,mL^{-1}$ in normal subjects, was almost undetectable, suggesting that this metabolite is formed entirely by microbial action. However the sulfide metabolite of sunlindac peaked at a similar time and concentration in both groups, suggesting gut wall or liver metabolism. At later times, the half-life of the sulfide metabolite was much longer in normal patients than in ileostomy patients, again suggesting a significant activity of microorganisms, probably on drug excreted in bile.[66] Drugs whose activity or toxicity may depend upon the action of microorganisms to form an 'active metabolite' include sulfasalazine, metronidazole and cyclamate.[59]

23.1.2.4.3 *Effect of gut wall metabolism*

Gut wall metabolism may have a considerable effect on reducing the amount of certain drugs during absorption. It is normally classed together with metabolism by the liver as the 'first pass effect'. However for the purpose of this review it is briefly discussed, whereas liver metabolism is discussed in a subsequent chapter.

The lining of the gut wall consists of mucosal cells and it has been shown that enzymes which metabolize drugs occur in these cells with greatest concentration at the tips of the villi.[67] Many of the drug-metabolizing enzymes associated with hepatic tissue have been found in gut wall[68] and greatest activity appears to be in the duodenum and jejunum, although activity for some reactions occurs throughout the gut. In general the levels of enzymes in gut wall are lower than in hepatic tissue but the ratio for different enzymes is wide. For instance, the ratio of gut wall to hepatic activity for NADPH-cytochrome *P*-450 reductase and UDP-glucuronyl transferase is near unity but for epoxide hydrolase it is much smaller.[69]

It has been shown that metabolic enzymes in the intestine can be induced using standard compounds used for inducing hepatic enzymes, such as phenobarbitone and 3-methylcholanthrene. However the 3-methylcholanthrene induction of cytochrome *P*-448 appears to be much greater than the induction of cytochrome *P*-450 by phenobarbitone.[69] This may be due to differing capacities of these enzymes to be induced or to the state of induction of the enzymes prior to drug administration. It is well known that constituents of foods can induce intestinal enzymes in animals and man and thus it is difficult to obtain true control values. The effect of certain foods on intestinal enzyme levels in the rat is shown in Figure 8. Obviously constituents of different foods affect different enzymes. Hydrocarbon particles ingested during smoking also induce intestinal enzymes.

Very few studies have demonstrated inhibition of intestinal enzymes. Inhibition of the intestinal metabolism of tyramine to *p*-hydroxyphenylacetic acid has been demonstrated by administering selective and nonselective monoamine oxidase inhibitors.[70] In addition, a number of studies have demonstrated the saturable nature of specific enzyme activity within the intestinal mucosa. Thus salicylamide[71] and ascorbic acid[72] competitively inhibit the intestinal mucosal metabolism of isoprenaline. Isoprenaline is normally administered by aerosol, although a large proportion of the dose is swallowed. Drug reaching the intestinal tract would normally be inactivated by metabolism to the sulfate conjugate, but coadministration of a drug competing for intestinal enzymes would result in a potentiation of pharmacological activity.[73] A review of metabolism of a number of drugs including analgesics, steroids and cardiovascular drugs by intestinal mucosa has been presented.[74]

23.1.2.4.4 *Oral absorption of peptides*

The recent interest in the pharmacological activity of synthetic peptides results from the discovery of a number of endogenous peptides which have biochemical functions, such as enkephalins, peptide hormones and interferons. There is no doubt that synthetic peptide molecules can have significant pharmacological properties, but there are considerable difficulties in preparing orally absorbable forms. A detailed review of the absorption, metabolism and excretion of peptides and related drugs has been published recently.[75]

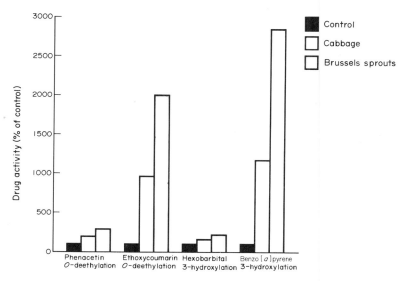

Figure 8 The effect of dietary constituents on intestinal metabolism in the rat (data drawn from E. J. Pantuck *et al.*, *J. Pharmacol. Exp. Ther.*, 1976, **198**, 278)

Proteins and larger peptides entering the GI tract will normally be broken down by enzymes secreted by cells in the stomach and small intestine. Thus pepsin, which is secreted by gastric cells, and trypsin, which cleaves at lysine and arginine residues, chymotrypsin, which cleaves at aromatic amino acid residues, and elastase, which cleaves at aliphatic acid terminals, are all secreted by pancreatic cells.[76] Any small peptides remaining may be degraded to amino acids by carboxy-peptidases in the microvilli and by di- and tri-peptidases in the cytosol of the absorptive cells.

The absorption of peptides by the small intestine is mostly limited to di- and tri-peptides. This uptake is not by simple diffusion but *via* facilitated diffusion by a specific saturable carrier system. The carrier system prefers bulky side chains and L stereoisomer amino acid residues at both C and N terminals. Methylation of the terminal amino group or conversion of the terminal acid to amide decreases absorption.[77]

The aminopenicillins and aminocephalosporins are effective antibiotics by the oral route. This is true in spite of a low partition coefficient and high degree of ionization at the pH of the small intestine. Saturable absorption, which can be inhibited by modification of —SH groups on the brush border, as well as competition between similar antibiotics and other dipeptides has been demonstrated in rats, which suggests a carrier-mediated process for some antibiotics.[78]

Chemical modification of peptide structures has proved moderately successful in improving absorption of some therapeutic classes. Thus L-alanyl L-1-aminoethylphosphonate, alafosfalin, is an antibacterial agent which acts as a prodrug for 1-aminoethylphosphonate. 1-Aminoethylphosphonate is very poorly absorbed but significant absorption of alafosfalin occurs in rats and man.[78] Improvement in absorption of the peptide inhibitors of angiotensin converting enzyme (ACE) was observed when the methyl ester of enalaprilat was prepared; thus enalapril was five times more bioavailable than enalaprilat in the dog.[79] Cyclization of some compounds has proved effective in increasing absorption, thus cyclosporin is well absorbed in rats when administered in olive oil. It has a cyclic oligopeptide structure and the peptide nitrogens are alkylated.[80] A cyclic dipeptide, cyclo-(L-Leu-Gly), was also enzymatically stable and well absorbed.[81]

A different approach, which has proved successful in some cases, is the use of formulation factors to promote absorption. Strategies which have been investigated include the use of oil/water emulsions, encapsulation in liposomes, entrapment in polymer particles in the nanometre range and coadministration with protease inhibitors or permeability enhancers.[75] An interesting formulation idea described recently[82] was to coat the drug with an azopolymer which would only be broken down by the microflora of the terminal ileum or colon, thus releasing drug into a less hostile environment.

The transport of larger peptides is unlikely to occur by carrier-mediated transport and in this case absorption probably occurs by other mechanisms. These may be: (i) diffusion through aqueous pores; (ii) passive diffusion through lipid membranes; or (iii) uptake into epithelial cells by endocytosis or pinocytosis. Many larger peptides demonstrate very poor absorption from the GI tract but recently a nonapeptide, Pro-His-Pro-Phe-His-Leu-Phe-Val-Phe, has been shown to be well absorbed across rabbit jejunum.[83]

23.1.2.4.5   Interactions with food, drink or other drugs

Food, drinks and other drugs have effects on gastric emptying and these effects have been described earlier. There are, in addition, a number of effects which are not related to gastric emptying. The absorption of tetracycline is reduced when administered with food, especially with milk or milk products. This is due to complex formation between tetracycline, polyvalent metal ions, especially calcium and magnesium, and protein molecules.[84] It was also shown that coadministration of EDTA with milk products improved the absorption of tetracyclines due to complexation of metal ions with EDTA.[85] A number of other antibiotics, including penicillin V and G and cephalexin, are less well absorbed when administered with milk products.[86] It has also been reported that the β-blocker solatol shows reduced absorption when administered with calcium gluconate solution, although it is not known if this is a specific interaction, or only the result of delayed gastric emptying.[87] This is in contrast to two other β-blockers, propranolol and metaprolol, whose absorption was increased in the presence of food.[54]

The absorption of a number of drugs of low aqueous solubility is improved if administered together with a fatty meal. It is suggested that delayed gastric emptying together with the presence of lipids increases the amount of drug which dissolves and thus absorption is increased. Increased absorption has been demonstrated for nitrofurantoin, diftalone[49] and griseofulvin.[88] The absorption of other drugs, such as hydrochlorthiazide and riboflavin, appears to be improved in the presence of food, possibly due to saturable absorption mechanisms in the fasting state which are not saturated during slower gastrointestinal transit in the presence of food.[89]

A number of studies have shown that fluid volume can influence the absorption of drugs. In most cases drugs are more completely absorbed from large volumes of fluid (*ca.* 200 mL) than from small volumes (*ca.* 20 mL). This appears to be due partly to a faster gastric emptying rate and partly to an increase in the mucosal to serosal fluid flux.[90]

The absorption of a number of drugs is affected by coadministration of other drugs or therapeutic agents. The coadministration of kaolin to reduce diarrhoea may reduce the absorption of antibiotics or other drugs due to adsorption of the drug on to the surface of the kaolin.[90] The same effect occurs upon administration of antacids with some antibiotics,[91] antiinflammatories[92] and psychotropics.[93] Cholestyramine, which is used to treat hyperlipidaemia, also binds to a number of drugs including warfarin and thiazide diuretics,[90] and this results in reduced absorption. Charcoal has very strong adsorbing properties and has been used to reduce absorption in overdose cases of phenothiazines[94] and other drugs.

23.1.2.4.6   Effect of blood and lymph flow on absorption

The combined vascular beds of the liver, spleen and gut are termed the splanchnic circulation. This consists of the celiac artery and the superior and inferior mesenteric arteries. At rest this system receives about one-third of the cardiac output (about 1.5 L min^{-1}). All the splanchnic circulation, except that which perfuses the lower rectum, passes through the liver either through the hepatic artery or the portal vein.[95] When parts of the GI tract become active, *e.g.* after a meal, blood flow may be increased by 50% over the resting state. It has been estimated that the smooth muscle layer may receive between 10 and 40 mL (100 g)$^{-1}$ min^{-1} and the mucosal cells between 50 and 400 mL (100 g)$^{-1}$ min^{-1} depending upon the state of activity of the small intestine.

Alterations in blood flow to the GI tract may affect drug absorption by various mechanisms. Blood flow acts as a drain for the mucosal absorptive cells and supplies oxygen to the cells which may be needed for active processes. Also blood supplies energy to the cells required to maintain an effective membrane.[96] For some drugs which are sufficiently water soluble and thus penetrate the unstirred layer but can also penetrate the mucosal membrane rapidly, blood flow may be rate limiting in determining the speed of absorption. This was demonstrated for tritiated water, antipyrine and even ethanol (Figure 9), whereas ribitol and sorbose were not affected by increasing the blood flow.[97]

Transport of drugs by the lymphatic system is usually a minor route for absorption due to the slow flow rate of lymph (1–5 mL min^{-1}). However the advantage of this route is that drugs avoid the 'first pass' effect through the liver and for highly metabolized drugs this may result in active drug

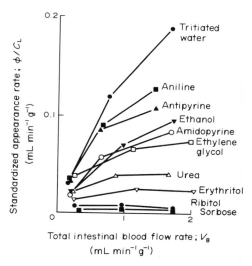

Figure 9 Dependence of appearance rate in rat jejunal blood on total intestinal blood flow rate (reproduced from ref. 97 by permission of Plenum Publishing Corporation)

reaching the circulation. In general, the more lipophilic a drug the more likely it is to be absorbed by the lymphatic system. Thus cholesterol, triglycerides, lipid soluble vitamins, DDT and other drugs may be absorbed by the lymphatic system.[107] Other drugs which are largely deactivated by the liver may be designed to be absorbed into the lymph. Thus testosterone undecanoate, when administered in a lipid rich formulation, is significantly absorbed by the lymph and is more active than orally administered testosterone.[98] However, long chain fatty acid 5-fluorouracil diglycerides were not selectively absorbed by the lymphatic system.[99]

A number of macromolecules may be exclusively transported by the lymph such as proteins, dextrans, dextran sulfate and enzymes. Complexes of these macromolecules with specific drugs have resulted in lymphatic directed absorption. Thus an ion-pair complex of dextran sulfate with bleomycin formulated in a mixed micelle of monoolein and taurocholate produced significant lymphatic absorption.[100] Similarly the lymphatic absorption of 1-hexylcarbamoyl1-5-fluorouracil (HCFU) was improved by a formulation with a β-cyclodextrin polymer ($M = 10\,000$) which contains cavities and thus forms an inclusion complex.[101]

23.1.2.5 Physical Formulation Factors

Problems with inadequate release of drug from its formulation or of poor solubility due to the crystal structure or particle size have resulted in a number of therapeutic failures and led to the introduction of widespread dissolution testing for oral dosage forms. With an increased understanding of these problems and suitable control of the formulation and its ingredients, the requirement for dissolution testing in routine manufacture may become obsolete. However, it will always be important in the development of any new formulation or in any alterations to existing formulations.

Interaction of the drug substance with inactive components of the formulation used as fillers, lubricants, disintegrants, *etc.* may seriously impair absorption. Thus calcium sulfate, used as a filler in phenytoin capsules, reduced the absorption of phenytoin. Magnesium stearate is often used as a lubricant in tablet formulations. It is, however, not very water soluble and if present in sufficient quantities may coat the drug particles and impair its solubility and hence reduce drug absorption.[8]

The particle size of the drug substance may be important in determining the rate of absorption as smaller particles will normally dissolve more rapidly than larger ones due to the larger relative surface area. This may be particularly important for drugs which have low aqueous solubility. These drugs are now often marketed in microcrystalline form or the crystals are micronized to give a smaller particle size. This factor was important in changes in the absorption of digoxin, which originally had a particle size of 20–30 μm. This material was slowly and incompletely absorbed. Reduction in the particle size to about 4 μm resulted in an increase in the rate and extent of absorption of digoxin.[102] Surfactants such as polysorbate 80 and sodium dodecyl sulfate may reduce or enhance drug absorption. Usually improved absorption occurs if the surfactant acts primarily as a wetting agent.[103]

Enteric coating of tablets to protect the drug against the environment of the stomach may lead to enhanced absorption over conventional tablets but if careful control of the formulation is not observed, reduced absorption may result. Enteric coated tablets are normally protected with a polymer layer which is insoluble at pH 1–3 but dissolves readily at pH 5–7; most are polymeric acids with ionizable carboxyl groups. The apparent pK_a of these polymers is important and should be between 4 and 7 for the coating to function correctly. [104] The thickness of the coating will be critical, as will the quantity of talc added to prevent 'sticking'. Talc is especially good at reducing the disintegration of enteric coated tablets.[105] The use of natural polymers such as shellac to prepare coatings, *e.g.* sugar coating, or slow release products should be avoided as these products tend to change their chemical structure and their physical properties during aging.[106] In general, sugar-coated tablets, which often show slow and variable disintegration times, have been replaced by film-coated tablets, which disintegrate rapidly.

The absorption of poorly absorbed drugs may be enhanced by the use of lipid adjuvants.[107] In general the use of lipids in formulations may be classified into three types: (i) formation of emulsions, particle size range 200–10 000 nm; (ii) formation of liposomes, particle size range 25–1000 nm; and (iii) formation of micelles, particle size range 3–100 nm.

In general oil/water emulsions are used to obtain a higher concentration of a lipophilic drug in solution than would be otherwise possible. Transport of the drug probably occurs *via* the aqueous phase or *via* micelles which may coexist. Certain lipids such as medium chain triglycerides and oleic acid may be absorbed and drug absorption may be related to the absorption of the lipid.

Liposomes have attracted interest as enhancers of drug absorption, but only a few reports have demonstrated increased absorption. In addition, liposomes are easily degraded by bile salts and there is no evidence that intact liposomes are absorbed. However the possibility of an interaction with the membrane wall or some similar mechanism may exist.

Micelles, the smallest of the lipoidal structures, have proved the most successful at enhancing drug absorption, particularly of poorly absorbed water soluble macromolecules. Micelles prepared from sodium taurocholate or glycocholate and monoolein improved the absorption of heparin markedly; normally heparin is not absorbed orally.[108] Other lower molecular weight drugs which are poorly absorbed, such as streptomycin, gentamycin and cefazolin, also showed improved absorption when administered in micelles and this absorption was more markedly increased in the large intestine than in the small intestine (Figure 10).

In general, unsaturated fatty acids and their monoglycerides enhanced absorption more than saturated fatty acids. It has been suggested that lipids with a polar head and low melting point chain are able to cause transient disorder in the membrane of the epithelial cells and thus the permeability of the membrane is increased. However, upon removal of the lipid adjuvant the membrane permeability reverts to normal.[107]

23.1.2.5.1 *Oral slow release dosage forms*

Over the past 10 years oral dosage forms which release drugs more slowly than conventional tablets and capsules have become popular. These dosage forms may have advantages over conventional formulations. Drugs with short half-lives may be administered less frequently, say once or twice a day rather than three or four times a day, which is usually more convenient for the patient.

Rapid release formulations may result in higher drug concentrations than required for therapeutic effects and this may lead to unnecessary side effects. However in the past many slow release preparations have been marketed without any demonstrable clinical advantage and this has led some authors to question the relevance of such preparations.[109] In addition in the past some slow release preparations were of dubious bioavailability and the effects of meals and other variables on their performance were unknown.

However there are circumstances where a slow drug release rate can improve therapeutic efficacy markedly. Thus nifedipine when administered rapidly causes an increase in heart rate and little fall in blood pressure. When it is administered slowly the desired fall in blood pressure is achieved without an increase in heart rate.[110]

There are many types of slow release formulation available. Polymer matrix tablets contain the drug dispersed in the matrix and drug diffuses out during passage through the GI tract. Such formulations usually exhibit marked variability in drug absorption as the formulation remains intact and may become lodged in the stomach in one individual or pass quickly through the small intestine

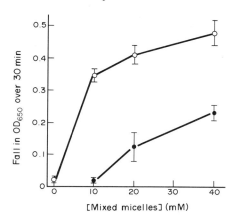

Figure 10 Effect of mixed micelles concentration on plasma clearing factor activity after administration of heparin. ●, Small intestine; ○, large intestine (reproduced from ref. 108 by permission of Elsevier Science Publishers, B.V. Biomedical Division)

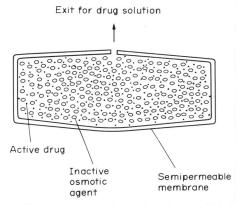

Figure 11 Diagrammatic representation of a single unit controlled drug delivery device

in another. Capsules containing coated granules or matrix granules may to some extent reduce interindividual variability as the granules are distributed statistically through the GI tract and absorption should be more predictable.

More advanced drug delivery systems include devices in which the rate of drug release is approximately constant. Devices such as the 'OROS' delivery system are surrounded by a semipermeable membrane. The drug plus an osmotic agent is contained in the device (Figure 11). Upon administration, water enters the device through the semipermeable membrane and the resulting pressure causes the drug solution to exit through one or more holes.[111]

All the above systems suffer from the limitation that unless the drug is absorbed from the large intestine within a period of about 8 h, they will be transported past the absorptive areas of the GI tract. Thus it is likely that some drugs with short half-lives will not be able to be administered once daily.[112]

One possibility for extending the period over which drug can be released to the small intestine is to retain the formulation in the stomach and release drug continuously. This can be achieved by a formulation which expands in the stomach and thus is difficult to empty or by a formulation which floats on the GI contents or by a combination of these attributes. This type of formulation has been used to administer riboflavin[113,114] and has shown that drug release can occur over 8 to 14 h. It is not known if such a formulation can deliver drug for 20 h or more, or if it is acceptable to patients.

23.1.3 RECTAL ABSORPTION

Although the rectal route is an accepted route of drug administration in some European and South American countries, it is not widely used in the United States or Japan. However, there are advantages for this route when a drug cannot be administered orally, for example if the patient is

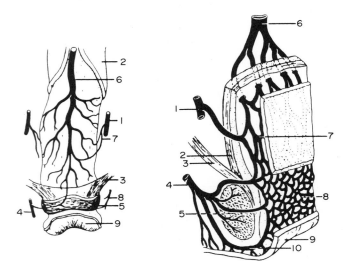

Figure 12 The venous drainage of the human rectum: 1. middle rectal vein; 2. tunica muscularis: stratum longitudinale; 3. levator ani; 4. inferior rectal vein; 5. external sphincter; 6. superior rectal vein; 7. and 8. submucous venous plexus; 9. skin; 10. marginal vein and subcutaneous plexus (reproduced from Angewandte und Topographische Anatomie, 5th edn., 1981, ed. G. Tondury, with permission from George Thieme Verlag)

vomiting of if a drug is deactivated on oral administration. It is obviously preferable to administer a drug rectally rather than parenterally if these are the only alternative routes, and rectal dosage forms are often easier to administer to young children than oral forms.

The physiology of the rectum is much less suitable for drug absorption than the small intestine and it is therefore surprising that some drugs are well absorbed by this route. The length of the ampulla recti is about 15–20 cm and the membrane surface is similar to that in the remainder of the large intestine, that is without villi and microvilli. Thus the absorbing surface of the rectum is probably less than 0.05 m^2 compared with 70 m^2 in the small intestine.[115] The rectum is normally empty unless the subject is constipated and there is little movement within it. Blood supply to the lower rectum drains into the inferior rectal and middle rectal veins, which empty into the inferior vena cava, thereby bypassing the liver, whereas the superior rectal vein drains into the portal system and blood passes through the liver before entering the general circulation (Figure 12). However, anastomoses occur between the rectal veins so that blood draining the lower rectum may enter the superior rectal veins and *vice versa*. The amount of liquid in the rectum is small, probably less than 5 mL, and certainly too small to dissolve sparingly soluble drugs. In most cases, therefore, drug will be best administered dissolved in a formulation. The pH of this liquid is close to pH 8, although the pH of the membrane surface is unknown. There is evidence to indicate that rectal secretions have only limited buffering capacity and can only slowly restore their normal pH after infusion of buffered solutions.

A proposed advantage of the rectal route is that absorbed drug may pass directly into the systemic circulation and not be subjected to metabolism in the liver. This should be particularly advantageous for drugs with high oral first pass effects. However evidence to support this advantage in humans is scarce.[116] Lignocaine was shown to be about 70% bioavailable by the rectal route as a solution but only about 30% bioavailable when administered as capsules by the oral route, thus supporting the hypothesis. Propranolol, on the other hand, demonstrated very variable bioavailability both by the oral and rectal route. Those subjects with low oral bioavailability seemed to show better rectal bioavailability and *vice versa*. Possibly, the absorption of propranolol from the rectum is incomplete and so offsets the advantage of avoiding the first pass effect. The bioavailability of salicylamide (1.5 g), which is metabolized by conjugation, was much lower by the rectal route than after oral administration, although urinary excretion of similar amounts of conjugates occurred, indicating similar amounts absorbed. As salicylamide is probably metabolized by gut wall and possibly by microflora, these effects may reduce markedly the advantage of partially avoiding first pass hepatic metabolism. Thus, for some highly metabolized drugs the rectal route may be a useful alternative but in each case the advantage must be proved.

The absorption of drugs from the rectum seems to follow, qualitatively at least, the pH–partition hypothesis.[26] Good agreement with this theory has been shown for sulfonamides and although quantitative deviations were shown to occur for salicylic acid in rats, these deviations were smaller than in the small intestine and colon.[117] Limited absorption of ionized species has been proposed to account for these differences, although for rapid drug absorption a measurable proportion of the unionized species in the pH range 7–8 seems desirable.

Solubility of the drug in aqueous media is probably more important for rectal dosage forms than other factors due to the limited amount of fluid present in the rectum. It has been demonstrated that the release of drug from suppositories is directly proportional to its water solubility.[118] In addition, the use of water soluble salts of acids or bases is preferable to that of the less soluble parent even if the pH of the solution moves to a less favourable region.[115]

In general, the vehicle used for the drug should encourage the drug to move out of the vehicle phase. Thus, water soluble substances are suspended in fatty vehicles and less water soluble substances are formulated in a more polar vehicle. Very nonpolar drugs will be slowly and poorly absorbed after rectal administration since they will have low solubility in the aqueous fluid.

The most widely used rectal dosage form is the suppository. This consists of a suspension of the drug in cocoa butter or a semisynthetic derivative prepared by the hydrogenation of coconut oil or esterification of glycerine with saturated fatty acids. These different fatty bases have differing physical properties and thus the essential requirement for the suppository to remain solid at room temperature but to melt quickly at 37 °C can be achieved by correct formulation.

Many studies have examined physical properties of suppositories and their spreading behaviour. The incorporation of additives may significantly alter the behaviour of the suppository and thus have a marked effect upon the drug release as is the case for surfactants, such as polysorbate, which are used to aid drug solubilization.[119] Thus, formulation changes in suppositories need to be controlled by bioavailability studies in the same way oral formulations are controlled. Other rectal dosage forms are microenemas and gelatin capsules, although these are only used for special cases.

23.1.3.1 Absorption-promoting Adjuvants

Interest has developed recently in the rectal route, as an alternative to parenteral administration, for drugs which are poorly absorbed orally. Unfortunately, most of these molecules are poorly absorbed rectally from traditional formulations. However, interest in increasing absorption in the GI tract has led to greater awareness of compounds which will promote absorption of drugs, and these compounds have proved particularly useful in promoting rectal absorption. Those drugs which have been studied include water soluble antibiotics, such as penicillin G and cefoxitin and peptide drugs such as insulin and pentagastrin. Compounds of quite different chemical classes have been found to be absorption enhancers in the rectum. Some of these are surface active agents and other have different properties (Table 4).

Surface active agents, such as sodium lauryl sulfate, have been used to increase rectal drug absorption but concern has been expressed about their nonreversible effects and possible damage which may be caused to membranes. However, mixed micellar solutions of monoolein with sodium taurocholate significantly enhanced the absorption of heparin from the large intestine in rats,[120] and at lower concentrations than required to enhance absorption from the small intestine. It was shown that lipids with polar head groups which are absorbed from the intestine are required to enhance absorption and those with unsaturated fatty chains such as oleic acid, linoleic acid and linolenic acid are particularly effective. In addition, these mixed micelles showed a transient effect upon the membrane, the increased absorption corresponding to disappearance of the fatty acid.[121]

Chelating agents such as EDTA and trisodium citrate increase intestinal permeability of a number of poorly absorbed drugs. Recently enamine derivatives have been shown to increase the rectal absorption of ampicillin, cefmetazole and cefalotin. It appears that this increased permeability is dependant upon calcium ion chelation as the enhancing activity of diethylethoxymethylene malonate can be inhibited by excess calcium ions.[122]

Antiinflammatory agents (NSAID) such as sodium salicylate, sodium 5-methoxysalicylate, indomethacin and phenylbutazone are extensively used as rectal permeation enhancers. Increased absorption has been observed for antibiotics, peptides and other drugs. The levels of salicylates or other NSAID required are below those causing any permanent changes to the rectal membrane and the effects appear to be transitory.[123] Addition of only 0.15 g sodium salicylate to a 1 g cefoxitin dose increased the rectal bioavailability from 3% to 12% in humans and addition of 0.5 g sodium salicylate gave 20% bioavailability.[124]

Table 4   Classes of Absorption-enhancing Agents[127]

Class	Examples	Model drugs
Chelating agents	EDTA	Heparin
	Trisodium citrate	Sulfanilic acid
	Enamine derivatives	Ampicillin
NSAIDS	Indomethacin	Ampicillin
	Diclofenac	Cefalotin
	Phenylbutazone	Cefmetazole
	Epirizole	Cefoxitin
	Salicylates	L-DOPA
		Lidocaine
		Insulin
Surfactants	Sodium lauryl sulfate	Cefoxitin
	Brij 35	Lincomycin
	Brij 58	Insulin
	Texafor B1	
Phenothiazines	Perphenazine	Cefoxitin
	Chlorpromazine	Gentamicin
Acylcarnitines	Palmitoylcarnitine chloride	Cefoxitin
	Palmitoylcholine iodide	Gentamicin
		Cytarabine
		Methyldopa
Miscellaneous		
fatty acids	Linolic acid	Insulin
acylamino acids	N-Caproyl-L-phenylalanine	Ampicillin
dicarboxylic acids	Nonamethylene dicarboxylic acid	Ampicillin

Acylcarnitines have been shown to be effective in increasing absorption of rectally administered drugs in rats and dogs. These compounds, which are involved in mitochondrial membrane transfer of fatty acids, also have reversible effects with no apparent tissue damage.[125]

Phenothiazines such as chlorpromazine and perphenazine have also been shown to enhance rectal absorption in the rat. The increase is concentration dependent, using doses from 1 mg kg^{-1}.[126] However, the pharmacological effects of these compounds may limit their use as absorption enhancers.

Many of the above and other absorption enhancers have only been studied in animal models and their effectiveness in rectal dosage formulations for humans is unknown. Limitations and requirements for absorption enhancers have been discussed[127] and much further work is required before absorption enhancers can be used in commercial formulations.

23.1.3.2   Lymphatic Absorption from Rectal Formulations

It has been shown that absorption into plasma becomes limiting with markedly increasing molecular weight. However this does not seem to be true for lymph and thus the lymph/plasma ratio increases with increasing molecular weight.[117] Thus, the use of absorption promoters such as mixed micelles of monoolein–taurocholate administered with a high molecular weight complex of bleomycin (a cationic glycopeptide) together with dextran sulfate (a high molecular weight anion) deliver a high concentration of bleomycin into lymph with a much lower concentration in plasma.[128] This compares with average concentrations in both plasma and lymph when free bleomycin was administered with the mixed micelles.

Interferon has also been shown to be absorbed by the lymphatic system when administered with mixed micelles of linoleic acid and a polyoxyethylene derivative of hydrogenated caster oil, whereas when administered alone it was not detected in serum or lymph.[129] There appears to be, therefore, considerable potential for exploiting this route for high molecular weight peptide drugs, which would otherwise require parenteral administration.

23.1.4   TRANSDERMAL ABSORPTION

23.1.4.1   Physiology of the Skin

The structure of the skin is the most significant factor in limiting the absorption of drugs by the transdermal route. A vertical section of skin (Figure 13) can be divided into four distinct layers.

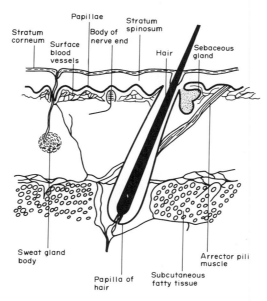

Figure 13 Diagrammatic representation of a vertical section through human skin

From the outside, these are the stratum corneum, the epidermis, the dermis or corium and the subcutis.

The stratum corneum is the surface layer consisting of dead cells which have become keratinized. This layer varies in thickness between 14 and 27 cells, each cell being about 0.5 μm thick and 30–40 μm long, thus giving a total thickness between 6 and 15 μm except for the palm of the hand and the sole of the foot which are much thicker.[130] Near the surface the cells, which are arranged in columns, become detached and are constantly shed. The cell membrane consists mainly of lipids and inside the dead cells α-keratin filaments are embedded in a protein matrix. The stratum corneum is covered by a thin waxy film consisting of aqueous and lipid components. The extracellular spaces are filled with a material consisting mainly of lipids which are excreted from cells in the epidermis. These lipids consist of high melting point mono- and di-glycerides, free fatty acids and phospholipids, which give rise to the firmness of the intercellular structure. The stratum corneum is normally the most difficult layer for drug molecules to penetrate and is, therefore, often the limiting factor in transdermal absorption.

The epidermis, which has a thickness of between 0.3 and 1 mm, consists of layers of cells which can be differentiated. From the outside, there is the stratum lucidum, a thin amorphous layer in which cell nuclei and cell boundaries are absent. Next, is the stratum granulosum, another thin layer, just a few cells thick. The cells in this layer are flattened and characterized by granules in the cells.[131] Below this layer is the stratum spinosum, a living layer of polyhedral-shaped cells, which become more flattened as the stratum granulosum is approached. The cells in this layer have a spiny or prickly appearance, caused by adhesions between cell membranes of adjacent cells. The lowest layer, the stratum basale, consists of a single layer of columnar cells, which form a boundary with the corium. The boundary layer is indented by the numerous folds of the corium (Figure 14). The stratum spinosum and the stratum basale together are sometimes known as the stratum germinativum, as the cells are continually dividing to produce new cells, which then migrate to the layers closer to the surface. The transit time of a cell until it is shed from the skin surface is about 28 days, which means that a layer of cells is lost from the stratum corneum every day.

The upper part of the dermis is thrown into numerous folds served by a network of blood vessels and nerves. This structure, known as the papillae of corium, allows rapid exchange of materials between the living epidermis and the vascular network. The dermis consists of collagen and elastin fibres embedded in a mucopolysaccharide matrix. The sweat glands and the follicles of fine hairs are found in the dermis. The subcutis consists of looser fibres together with stored fat cells. The follicles of thick hairs originate here.

In order, therefore, to obtain systemic concentrations of drugs following dermal administration, it is necessary for drug molecules to penetrate the stratum corneum and the epidermis to reach the blood vessels of the papillae of corium.

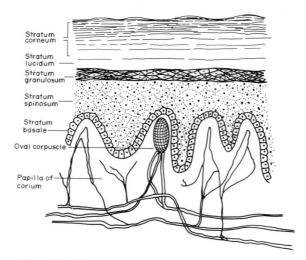

Figure 14 Diagrammatic representation of the epidermis

23.1.4.2 Mathematical Models

Drug transport through skin is, in the main, a process of diffusion. Molecules follow the path of least resistance and move down a concentration gradient. The path of least resistance is determined by the physicochemical nature of the membrane and by the relative affinity of the drug for the formulation and for the membrane. Most researchers who have developed models of transport have used Fick's first law of diffusion as a basis for developing equations (equation 8; dQ/dt is the rate of skin penetration, K_p is the partition coefficient of the drug between skin and vehicle, c is the concentration of drug in vehicle, h is the skin thickness (epidermis and stratum corneum) and D is the diffusion coefficient of drug through the stratum corneum) to describe the rate of skin penetration. This equation holds for steady state conditions and assumes sink conditions apply on the inside of the membrane.

$$dQ/dt = (DK_pc)/h \tag{8}$$

However, *in vivo* steady state conditions may not be achieved and the concentration gradient across the membrane may not be uniform, due to reservoir effects within the layers of the skin. In addition K_p is not easy to measure so that DK_p is often referred to as the permeability coefficient. In fact both D and K_p may be formulation dependent, thus alterations of dQ/dt for different drugs in different formulations may not necessarily be drug dependent.

In order to model drug transport under nonsteady state conditions Fick's second law must be applied (equation 9; D is the diffusion coefficient, x is the membrane thickness and c is the drug concentration).

$$dc/dt = D(dc^2/dx^2) \tag{9}$$

Solutions to this equation are complicated and depend upon the boundary conditions imposed.[132] However a plot of amount of drug which penetrates the skin membrane against time usually results in a curvilinear plot as in Figure 15. Extrapolation of the linear portion results in a lag time, L, and the diffusion coefficient, D, is given by equation (10).

$$D = h^2/6L \tag{10}$$

The above equations seem more suited to measurements in experimental membrane diffusion studies, where the concentration of drug in vehicle can be maintained constant, sink conditions can be maintained and both donor and receptor phases can be stirred to maintain homogeneity.

Another approach which has been used is the compartmental model (Figure 16). Drug diffusion through the stratum corneum may be represented by the rate constant k_1, and through the viable epidermis by k_2. The rate constant k_3 is the rate of drug transport back into the stratum corneum. Elimination from the body, assuming a one-compartment body model, is represented by k_4, which must be determined from intravenous data.[133] In many cases the stratum corneum will be the rate-limiting membrane and will act as a reservoir for the drug. A similar model has been used to predict

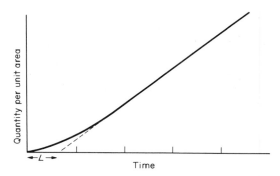

Figure 15 Theoretical graph of amount of drug penetrating skin with time; time L is the lag time

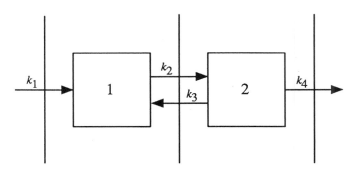

Figure 16 Compartmental model approach to transdermal absorption

the disposition of drug in the skin and plasma[134] and to interpret data obtained by multiple dosing to the same site.[135]

Due to the complicated structure of the skin, mathematical models are often limited in their ability to describe percutaneous absorption over a wide range of compounds. Models often assume the skin is homogeneous without pores and hair follicles, that the drug behaves ideally in solution and that the vehicle does not penetrate the skin or change in concentration by evaporation. The diffusion coefficient is assumed to be constant and unaffected by vehicle penetration or by the composition of the skin layer. The drug is assumed to remain constant in the formulation and not be degraded or metabolized and to be removed completely by the blood supply from the inner dermis.

Much of our present knowledge, therefore, has been obtained by the study of individual, and often simplified, processes and the application of this knowledge to the whole process of cutaneous absorption.

23.1.4.3 Mechanisms of Skin Permeation

23.1.4.3.1 Partition coefficient

The stratum corneum has for many years been thought of as a lipoidal barrier to skin permeability. The permeability of a range of homologous compounds was shown to be related to their oil/water partition ratio.[136] This concept was further developed with the introduction of thermodynamic activity coefficients to predict the maximum rate of permeation of the pure drug.[137] Thus, the solubility of a molecule in the stratum corneum could be determined from the entropy of fusion, S_f, and the melting point, T_m. The mole fraction solubility X_s is given in equation (11), where R is the gas constant and T is the surface temperature.

$$\ln X_s = [S_f(T - T_m)]/RT \tag{11}$$

This approach could be replaced by measurements of the solubility in stratum corneum or a suitable lipid solvent.[138] Thus the partition coefficient between octanol/water for a series of phenols was correlated with their permeability coefficient. Resorcinol had the lowest log partition coefficient

(0.8) and the lowest permeability coefficient (0.04×10^{-4} cm min^{-1}). The highest log partition coefficient was that of 2,4,6-trichlorophenol (3.69) and its permeability coefficient was next to highest (9.9×10^{-4} cm min^{-1}). A plot of partition coefficient against permeability coefficient gave a good correlation with a tendency to plateau at high partition coefficients.[139]

Other studies have shown good correlation between partition coefficient for stratum corneum/water or other solvent/aqueous systems and permeability coefficient for a series of steroids.[140] However even in these closely related series exceptions are relatively frequent, as is the case with estrogen steroids in the above series. In series of more divergent compounds, partition coefficients give a more limited indication of absorption potential (Figure 17).[141]

Studies of the effect of partition coefficient between membranes and water in series of homologues demonstrated a relationship between chain length and partition coefficient as given in equation (12), where K is the partition coefficient and n is the chain length. A plot of log K_n against n gave a straight line with slope π, which was a characteristic of the membrane. This value for biological membranes was typically between 0.3 and 0.5.[142] The above relationship only holds over a limited range and at high values of n a plateau is reached. At these high values it is assumed that the aqueous boundary layer to the membrane becomes rate limiting.

$$\log K_n = \log K_0 + \pi n \tag{12}$$

However, many more polar molecules do not fit this theory of penetration related to partition coefficient. Molecules such as methanol and ethanol had essentially the same permeability as water. Further studies of steroids demonstrated that the permeability of some polar compounds was far higher than could be expected from their partition coefficients. Thus, the theory began to evolve that an alternative pathway exists for more polar compounds, often referred to in this early work as aqueous pores.[143] This theory was substantiated by other workers, who showed that many polar compounds are absorbed at a similar rate, independent of partition coefficient. As drugs become more nonpolar the permeability rises and is proportional to partition coefficient. For very nonpolar molecules a plateau is reached when the transport through aqueous barriers, *e.g.* viable epidermis, becomes rate limiting.

It has been postulated that two parallel routes of absorption exist through the stratum corneum.[144] The lipid route would be between the columns of dead cells and therefore through the intercellular lipid material. The polar route would be through the cellular material and thus through the hydrated protein mass of the keratinocytes. Permeability coefficients for polar molecules are in the region of 10^{-5} to 10^{-6} cm h^{-1}, which is some 1000-fold lower than for reasonably lipid soluble molecules.

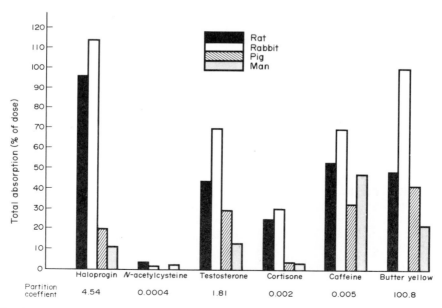

Figure 17   Total absorption of each test compound in rats, rabbits, swine and man (reproduced from ref. 141 by permission of Churchill Livingstone)

23.1.4.3.2 Effect of pH and pK$_a$

The ionization of a molecule generally decreases absorption through skin. Small ions such as sodium, potassium and bromine penetrate the skin with permeability constants of 10^{-6} cm h^{-1}.[145] The permeability of salicylic acid through the skin of pig's ear was found to be constant above pH 4 and to increase rapidly below pH 4, indicating more rapid absorption of the unionized form.[146] This was confirmed by a study using guinea pig skin in which the pH absorption profiles of salicylate and a weak base, carbinoxamine, were studied.[147] Both appeared very poorly absorbed in their ionized forms. A study of substituted phenols also supports the fact that the ionized molecule is absorbed much more slowly than the unionized form as the permeability coefficient decreased by a factor of 10 or more over a range of four pH units centred on the pK$_a$ of the molecule.[144]

23.1.4.3.3 Diffusion coefficient

The diffusion coefficient (D) of a drug is a measure of the resistance of the environment to its movement through it. The diffusional constant is influenced by the molecular volume of the drug and the viscosity of the surrounding medium. For ideal molecules, roughly similar in size and spherical in shape, the Stokes–Einstein equation can be applied (equation 13; T is the absolute temperature, r is the hydrodynamic radius of the drug molecule, and η is the viscosity of the environment).

$$D = kT/6\pi r\eta \tag{13}$$

Thus, increases in both molecular size of penetrant molecule and viscosity of the membrane will reduce diffusion.[148] This equation cannot be applied rigorously to skin however, due to the many other competing processes. Vehicle interactions with the barrier almost certainly affect the viscosity, while drug interactions with proteins may delay diffusional processes.

The diffusional resistance of the stratum corneum to many low molecular weight organic molecules was found to be relatively constant at about 10^{-9} cm^2 s^{-1}.[149] However the diffusion coefficient decreases with higher molecular weight and especially with the introduction of more polar groups, *e.g.* hydroxyl groups. Thus the diffusion coefficient for estriol is about 50 times smaller, and that of estradiol is 10 times smaller, than that of estrone.

23.1.4.4 Concentration of Drug in Vehicle

The amount of drug permeating the skin will be dependent upon the drug concentration in the vehicle if Fick's laws of diffusion hold. For many drugs, over certain concentration ranges, this can be shown to occur and even if the permeability rate is not directly proportional to concentration, very often an increase in the total dose absorbed will be effected by an increase in the drug concentration in the vehicle.[150]

When comparing drugs in different vehicles, the effective concentration is that determined by the thermodynamic activity. Thus, a certain drug concentration in differing vehicles may not have the same activity. This will depend upon the drug's affinity for the vehicle and, thus, may result in different effective concentrations.[148] Glycols which form hydrogen bonds with water will have lower effective concentrations, whereas only partially soluble molecules, *e.g.* methyl ethyl ketone, will have higher effective concentrations.

23.1.4.5 Vehicle Effects and Penetration Enhancers

The effect of the vehicle, in which the drug substance is applied to the surface of the skin, can result in considerable differences in the rate of permeation of the drug. The vehicle which has been most widely studied is water, either as a formulation or more often applied to hydrate the skin. The stratum corneum under normal conditions contains about 45% water and upon contact with water it can absorb up to five times its dry weight. Dry stratum corneum is about 10 times less permeable to polar molecules than normal stratum corneum and hydration increases the permeability some two- to three-fold.[151]

The importance of hydration of the skin was recognized during early work which showed that occlusive dressing of the skin increased the rate of absorption of corticosteroids.[152] In this way water loss from the surface of the skin was prevented and the stratum corneum became hydrated. Many topical formulations increase hydration of the stratum corneum, either by reducing water loss by covering the stratum corneum with a water impermeable layer such as a paraffin or wax base, or by the presence of a high degree of water in the formulation, as in the case of aqueous-based creams. The stratum corneum in contact with the viable epidermis is hydrated and that in contact with the air is drier, thus water continually diffuses down this concentration gradient and is lost from the skin surface. Increasing the relative humidity of the surface of the skin from 30% to 80% causes a doubling of the water content of the stratum corneum at the surface but has little effect on the water flux through the skin;[153] however, such a change in the water content of stratum corneum may be very significant for more polar molecules.

Propylene glycol has been shown to increase the penetration of hydrocortisone;[154] however, this was complicated by the increased solubility of hydrocortisone in the formulation and it is not known if diffusion was enhanced. The inclusion of a surface active agent, such as oleic acid with propylene glycol, was shown to increase markedly the penetration of salicylic acid, much more than the penetration in either vehicle alone.[155] It was suggested that this vehicle alters the viscosity of the lipid pathway through the stratum corneum. Another nonpolar surfactant, polysorbate, in combination with propylene glycol also enhanced hydrocortisone absorption.[156] Surfactants also appear to be effective in increasing penetration through the polar route. Thus decyl methyl sulfoxide increased the absorption of salicylic acid at pH 9.9 much more than at pH 2.7. Similarly sodium dodecyl sulfate increased urea penetration through the polar pathway but had little effect on pentanol, which is absorbed by the lipid route.[157]

Dimethyl sulfoxide (DMSO) has been studied extensively since it was shown that it can significantly increase the penetration of a number of drugs.[158] However, its irritant effects limit its use in commercial formulations. High concentrations of DMSO are usually used and these concentrations cause changes in the integrity of the skin which, together with high concentrations of the solvent in the stratum corneum, probably reduce its barrier properties. Dimethylacetamide and dimethylformamide have also been shown to be effective penetration enhancers, although their mechanisms of action are less certain. Recently, more interesting molecules have been found to enhance penetration, which produce minimal toxicity and are nonirritant. These include *N,N*-dimethyl-*m*-toluamide, *N*-methyl-2-pyrrolidone and 1-dodecylazacycloheptan-2-one (Azone). Azone is of particular interest as even small quantities (<2%) were able to greatly enhance penetration of 5-fluorouracil.[159] Interestingly, many of these molecules which enhance topical absorption are amides and, although they have widely differing molecular weights and partition coefficients, their dipole moments are within a narrow range close to 4 Debyes.[160]

Many other solvents commonly used in topical products have been reported to enhance percutaneous absorption,[19] including low molecular weight alcohols, acetone, cyclohexane and ethers but their effects are usually less marked than those molecules mentioned earlier.

23.1.5   INTRANASAL DRUG ABSORPTION

The intranasal administration of drugs has only relatively recently received the attention of the pharmaceutical industry. However this route has been used for the administration of nicotine (in snuff) and for drugs of abuse such as cocaine and hallucinogens for many hundreds of years. More recently, the recognition of the permeability of cutaneous membranes to drug substances and the problems of inactivation or metabolism of some drugs, especially steroids and peptides, during oral administration has led to an increased interest in the intranasal route. Intranasal administration of drugs is normally less acceptable to the patient than oral administration. This route will only be used for drugs which are poorly absorbed orally, inactivated in the GI tract or for which the oral route is precluded by some other factor such as vomiting. However the intranasal route may be preferable for some patients when compared with parenteral drug administration.

An understanding of the physiology of the nose is important if formulations are to be developed which maximize absorption potential yet will not affect the normal functioning of the nose. The main physiological functions of the nose are to adjust the temperature of the inspired air, to modify the water vapour content, to trap particulate matter and by the ciliary action deposit it in the oesophagus.[161] Because of these functions the anatomy of the nose ensures intimate contact of inspired air with the mucosal surface (Figure 18). The resistance of the nasal airway accounts for as much resistance to flow as the rest of the respiratory tract; this is due to the folds of the turbinates. Within the mucosa of the nasal passage are many goblet cells which secrete mucous and fluid on to the mucosal surface. Much of this fluid is transported to the nasopharynx, where it is swallowed.

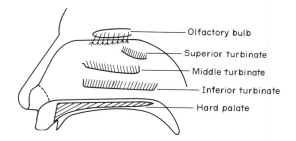

Figure 18 Diagrammatic representation of the human nasal passage

Factors which can affect absorption from the nose are related to the normal functioning of the nose. The presence of a cold or of hayfever may increase the flow of mucous and may therefore impair absorption through the membranes. The temperature or humidity of air which is drawn into the nose may affect blood flow through the capillaries and thus reduce or increase absorption by this route.[162] However, in one study the concentration of interferon in nasal wash samples before and after virus challenge was not different, suggesting that drug may not be removed by higher mucous flow.[163] As with other routes of administration physicochemical parameters of the drug molecule will influence the rate of absorption. The effect of pH and the pK_a of the drug has been examined for benzoic acid absorption.[164] Absorption of 44% benzoic acid in 1 h was shown to occur at pH 2.5; however, even at pH 7.2, 13% benzoic acid was absorbed in 1 h. At this pH, benzoic acid will be 99.9% ionized and thus it was calculated that the ionized form of benzoic acid is absorbed, although at a rate about four times slower than the unionized form.

The lipophilicity of drugs has been shown to be important in absorption through biological membranes and this has also been studied in nasal absorption using a series of barbiturates at pH 6.0. An increase in the chloroform/water partition coefficient of some fifty-fold only resulted in a four-fold increase in the percentage absorbed.[164] Similarly, other drugs with low partition coefficients such as L-tyrosine and quaternary ammonium compounds are well absorbed from the nasal cavity. The mechanism of absorption through nasal membranes thus may differ from that of other membranes. The volume of the formulation applied will also affect the rate of absorption and for effective absorption a small volume is better than a large volume. For most cases administration of volumes larger than about 1 mL to the nose will, in any case, be impractical.

The effect of structural modification of an amino acid, L-tyrosine, on the rate of nasal absorption has been examined. The nasal absorption of L-tyrosine was shown to be concentration dependent over the range $0.28-2.2 \times 10^{-3}$ M and to be a carrier-mediated process with Michaelis–Menten kinetics. It was suggested that esterification of the functional groups of tyrosine in a stepwise fashion would allow the contribution of each functional group to the absorption of L-tyrosine to be determined. Esterification of the phenolic group had little effect upon the absorption of L-tyrosine, despite the fact that the log P value increased. In these esters, the ionic nature of the amino acid was similar to the parent compound. Esterification of the acid group or amino group would affect the ionic nature of the molecule. Esterification of the amino group, however, had little effect upon the absorption rate. Esterification of the acidic group on the other hand caused a rapid increase in the rate of nasal absorption, which could not be attributed solely to the increased partition coefficient (Table 5). It was suggested that the increased nasal absorption was due to masking of the negatively charged carboxylate ion.[165]

The effect of molecular size on the intranasal absorption of some drugs has been studied in the rat. The absorption of 1-benzopyran-2-carboxylic acid, *p*-aminohippuric acid, inulin and dextran decreased as molecular weight increased; however, absorption of some high molecular weight molecules was observed. The development of peptides and small protein molecules which are poorly absorbed from the GI tract has led to an interest in the intranasal absorption of these molecules. Commercial formulations for nasal administration of peptide molecules used in the treatment of diabetes insipidus have been prepared. Lypressin (Diapid; Sandoz) and desmopressin (DDAVP; Ferring) were both effective and safe after administration by the intranasal route.[167] Other peptides which have been studied include insulin, luteinizing hormone releasing hormone (LHRH) and oxytocin.

Initial studies of the intranasal absorption of insulin showed that absorption was incomplete and variable;[168] however, recent studies have shown that absorption can be increased by formulating the insulin with bile salts or using an acidic medium.[169,170] Further studies in humans demonstrated

Table 5 Apparent Partition Coefficients and Nasal Absorption Rate Constants
for L-Tyrosine and Derivatives[165]

Compound	Apparent partition coefficient	Absorption rate constant (min^{-1})
L-Tyrosine	0.0256	0.0023
O-Acetyl-L-tyrosine	0.0468	0.0026
O-Valeryl-L-tyrosine	1.17	0.0029
L-Tyrosine methyl ester	1.97	0.0116
L-Tyrosine ethyl ester	5.20	0.0254
L-Tyrosine n-propyl ester	20.79	0.0230
L-Tyrosine t-butyl ester	62.50	0.0105
N-Acetyl-L-tyrosine	0.0256	0.0024

that insulin combined with sodium glycocholate or with deoxycholate was about 10–20% absorbed when the effects were compared with subcutaneously administered drug. Both normal volunteers and diabetics tolerated the intranasal formulation well.[171,172]

A considerable number of studies have been conducted on the absorption of natural and synthetic LHRH by the nasal route. This hormone is a decapeptide released by the hypothalamus to stimulate secretion of luteinizing hormone (LH) and follicle stimulating hormone (FSH). It is administered to treat a number of endocrine-related disorders. Intranasal administration in doses between 100 µg and 5 mg was shown to increase plasma LH concentrations, although comparison with intravenous administration indicated that bioavailability by the intranasal route was only 1–2% of that by the intravenous route.[173,174] However the convenience of this route of administration over parenteral administration may still make this an adequate bioavailability. Intranasal administration of a number of synthetic analogues of LHRH has also been studied. The desired pharmacological effects were obtained with buserelin, HOE 471 and nafarelin administered by the intranasal route,[175] although, in general, the bioavailability of these peptides was low.

Initial studies on the intranasal administration of an oxytocin extract began in the 1920s.[176] Since that time many studies have shown the clinical usefulness of intranasal oxytocin to induce or augment labour and to stimulate milk ejection. Comparison of the efficacy of a synthetic oxytocin product, syntocinon, indicated that a dose of 1000 mU was equivalent to about 50 mU by intramuscular injection and 10 mU by intravenous infusion over 5 min. Thus nasal bioavailability appeared to be about 1% of intravenous and 5% of intramuscular availability.[177]

Many other drugs have been administered by the intranasal route, including steroids which are inactivated in the GI tract,[178] β-blockers which show high first pass effects on oral dosing[179] and drugs such as metaclopramide used to control nausea and vomiting.[180]

Administration of drugs by the nasal route introduces new problems in formulation and in dosing design. Formulations have to be designed to be nonirritant to the nasal mucosa and not to reduce the ciliary beat frequency significantly. Administration of accurate doses is best carried out by aerosol formulations or by a metered-dose nebulizer. The aerosol tends to produce droplets of very small particle size which may be carried into the lungs, whereas the nebulizer produces large particles which will be trapped in the nose. Those factors which are important in the design of a nasal nebulizer have been presented.[181] The product should be able to deliver, reproducibly, an accurate dose for almost all the contents of the nebulizer. The device should be protected against contamination from outside and be made of materials which will not absorb the active ingredients. In addition the product should be so engineered that the dose cannot be altered by the way in which the nebulizer is operated. Also the nebulizer must not deteriorate on storage after filling even at elevated temperatures for periods of six months or more.

23.1.6 BUCCAL AND SUBLINGUAL ABSORPTION

Considerable interest has been shown in the membranes of the mouth for drug absorption, both for the administration of drugs and for the theoretical study of drug absorption. It is unlikely that there is any major difference between buccal and sublingual absorption and so the terms will be considered interchangeable.

In general, drugs would be administered by the buccal route only if there is an advantage over the oral route. Drugs which are deactivated by enzymes or the pH of the GI tract may be administered in this way, and will enter the systemic circulation directly without passing through the liver. This may be an advantage for highly metabolized drugs subject to a high first pass effect on oral administration. However the dose by this route is usually restricted to about 10 mg because if larger doses are employed, a larger proportion of the dose will probably be swallowed.

The buccal route has also been used as a model membrane for the study of differences in drug absorption between drugs of different pK_a and partition coefficient. The absorption of a series of basic drugs was examined and the fraction unionized was shown to correlate with the absorption across the buccal membrane.[182] The results from these absorption studies were also used to explain differences in the urinary excretion of the drugs based upon reabsorption across the kidney tubule membrane.

A series of acidic compounds was also studied in the buccal absorption test and the results demonstrated that increasing the partition coefficient increased buccal absorption up to a limiting state ($\log P \approx 3$). The unionized species was more rapidly absorbed until the lipid solubility of the ionized species was increased sufficiently as, for example, with dodecanoic acid.[183] Substitution of the straight chain with hydrocarbons reduced absorption, whereas substitution with halogens increased absorption in order of the molecular weight of the halogen.[184]

The buccal absorption of two β-blockers was studied and because of the different physical characteristics of the two drugs, the results are of considerable interest. Propranolol is a basic drug (pK_a 9.45) of relatively low water solubility (70 $\mu g\,mL^{-1}$) but with a relatively large partition coefficient ($\log P$ between 2 and 3 for many organic solvents). Atenolol is also basic (pK_a 9.6) but is relatively water soluble (9.5 $mg\,mL^{-1}$) and has a higher affinity for the aqueous phase ($\log P$ between 0 and -2 for a variety of organic solvents). At pH 9 a considerable proportion of the β-blocker is unionized and the absorption of propranolol is much more extensive than that of atenolol. However at pH 5 when the ionized species prevails, the absorption of both β-blockers is poor (Figure 19). The absorption of propranolol is very pH dependent, suggesting that the unionized species is absorbed through the lipophilic pathway, whereas the absorption of atenolol is pH independent indicating that both ionized and unionized species are more slowly absorbed by the aqueous pathway.[185]

Another interesting result of this work was that the intrinsic absorption rate constants for propranolol and for an acidic drug, *p*-hexylphenylacetic acid, were shown to be dependent upon the pH of the solution. It was concluded, therefore, that the absorption was influenced by an aqueous layer buffered to a pH just below 7. Further evidence for the existence of this aqueous layer was obtained from experiments with the hamster cheek pouch.[186] Stirring the mucosal solution increased the absorption of benzoic acid, salicylic acid and lauryl alcohol, presumably by partially destroying the resistance of the unstirred aqueous layer.

Earlier studies had also demonstrated that drug could be recovered by rinsing the mouth with a buffer favouring the ionized species, for example up to 50% of the absorbed dose of amphetamines was recovered by a pH 4 rinse[187] and up to 30% absorbed propranolol was recovered by a pH 5.2 rinse.[188] Thus drug retained either in the unstirred layer or in the lipophilic membrane could be recovered by favouring the concentration gradient in the reverse direction.

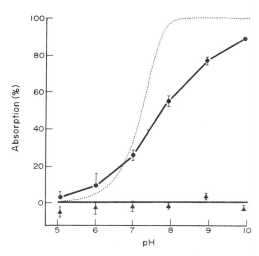

Figure 19 Buccal absorption of propranolol and atenolol at differing pH: ●, propranolol; ▲, atenolol (reproduced from ref. 185 by permission of The Pharmaceutical Society of Great Britain)

Studies of the buccal absorption of imipramine and its metabolites also demonstrated the important relationship between partition coefficient, pK_a and the rate of absorption through the membrane. Compounds with a high *n*-hexane/water partition coefficient ($\log P > 0$) demonstrated better absorption than those with lower partition coefficients. Compounds with pK_a of 8–9.5 showed pH dependent absorption over the range pH 5–9 and compounds with pK_a of less than 5 showed pH independent absorption.[189]

In all of the above studies, the drug had been presented in an aqueous solution and it was not necessary to take into account drug solubility. However, in a pharmaceutical dosage form such as a tablet, drug would be presented in a solid form. In this case a drug with a relatively high partition coefficient will probably have a low aqueous solubility. A prime consideration with such a formulation may be the preparation of a suitable salt or the use of other formulation methods to improve the aqueous solubility. In addition it appears that a greater degree of lipid solubility is required for extensive buccal absorption than for absorption across the mucosa of the small intestine and thus the balance between aqueous solubility and partition coefficient may be more difficult to achieve.

Little work has been carried out on compounds which might improve oral absorption, although recently it has been shown that inclusion complexes of steroids such as testosterone, progesterone and estradiol with 2-hydroxypropyl-β-cyclodextrin and poly-β-cyclodextrin markedly improved absorption from the oral mucosa. The β-cyclodextrin complexes are very water soluble molecules and form inclusion complexes with the steroids. These complexes can penetrate the aqueous unstirred layer and possibly release the drug at the membrane wall.[190]

A number of drugs which can be administered in small doses have been used clinically by buccal or sublingual administration. Thus 10 mg buccal tablets of morphine were found to give better bioavailability than 10 mg administered intramuscularly. Buprenorphine, also a strong analgesic, is normally administered sublingually, as the bioavailability is poor by the oral route. Another analgesic, nalbuphine, with low oral bioavailability has good bioavailability after buccal administration to rats.

Cardiac drugs have also been administered by the buccal route, especially for rapid effect or to avoid first pass metabolism. Nitroglycerine and isosorbide dinitrate are both well absorbed from the mucosa of the mouth, as is nicotine when chewed as tobacco sachets or chewing gum. The buccal absorption of propranolol had been demonstrated earlier, and recently, in a comparison of routes, a 10 mg sublingual tablet gave 63% bioavailability relative to the intranasal route, whereas the oral route gave only 25% bioavailability. The sublingual route also showed a sustained release effect, probably due to the reservoir effect of the membrane.[191]

23.1.7 ABSORPTION FROM THE VAGINA

The vaginal route has been used for many years for the administration of drugs, although this has been mainly for the treatment of local infections or conditions only possible in the female. Thus a number of antibiotics and antifungal agents have been administered by this route and more recently steroids and prostaglandins have also been administered in this way. However, it is unlikely that this will prove a popular route for drugs which act systemically, although evidence of systemic action of drugs administered by this route is plentiful. The vagina is anatomically and physiologically at least as suitable for absorption as the rectum, as it contains a dense network of blood vessels. The epithelial layer consists of lamina propria and a surface epithelium, which is thin before puberty and after the menopause, but thickens with the onset of puberty. The surface of the vaginal cell is composed of numerous microridges, which makes the cells well suited to drug absorption.[192]

For reasons similar to those mentioned for drugs administered rectally, drugs for vaginal administration should be fairly water soluble. Drugs administered by this route must traverse both an aqueous barrier, possibly composed of vaginal secretions, and a lipoidal or aqueous pathway through the membrane.[193] In general, therefore, solubility will be more important than the pH of the formulation although the unionized species will be more rapidly absorbed than the ionized species. The pH of the vaginal secretions is close to neutral up to puberty after which it rises to between pH 4 and pH 5.[194]

There is some evidence that intravaginal administration may be used to avoid first pass metabolism in the liver. Thus the application of 0.5 mg estriol intravaginally decreased luteinizing hormone levels to the same extent as 8 mg administered orally and conjugate levels were up to 24-fold higher after oral administration.[195] Testosterone is also highly metabolized after oral administration by glucuronidation and has been shown to be about 60% absorbed after intravaginal administration in the monkey. Prostaglandins administered by the intravaginal route are also absorbed systemically with about 25% bioavailability compared to intravenous doses.

Antibiotics are normally administered to the vagina to treat local conditions; however, a considerable number of studies have calculated the amount absorbed. In general, absorption of most of these drugs appears to be less than 10% of the administered dose and is often variable. One exception to this is metronidazole, which has a vaginal bioavailability of about 20% from pessaries or a cream, and thus plasma drug concentrations are achieved which will be effective against some anaerobes.

The absorption of a peptide analogue of LHRH, leuprolide, from the vagina of rats has been studied. The absorption of this peptide was enhanced by organic acids such as citric, succinic and tartaric acids and the bioavailability increased to 20%. The organic acids affected the permeability of the blood/vaginal epithelium barrier in a reversible manner.[197] A number of other peptides have been shown to be absorbed by this route including insulin in dogs and cats, thyroid stimulating hormone and TAP 144, a gonadotropin releasing hormone analogue, in rats.

Nonoxynol-9 is a nonionic detergent used in many spermicidal preparations, resulting in a dose of 50–130 mg per application. It has been shown to be well absorbed in rabbits and rats by the intravaginal route. In humans a similar decrease in serum cholesterol to that observed in treated rats was observed following use of the spermicidal preparations.[198]

23.1.8 ABSORPTION FROM THE LUNGS

The lungs are extremely permeable to drugs and other foreign molecules due to their extensive blood supply. The whole of the cardiac output passes through the alveolar capillary unit, which has an extensive capillary endothelium. In addition, the alveolar epithelial and endothelial layers are very thin to facilitate gas exchange. These are ideal conditions for the absorption of drugs.

Much of the work on the absorption of drugs with different physicochemical properties has concentrated on experimental animals. Hydrophilic molecules were shown to be absorbed by a nonsaturable diffusion process at rates which decrease in order of increasing molecular weight.[199] The hydrophilic compounds traverse the membrane through the aqueous pathway or through aqueous pores. In neonatal rats certain hydrophilic compounds with molecular weight between 122 and 1355 are absorbed more rapidly than in adult rats. However the administration of cortisone or thyroxine accelerates the development of the pulmonary epithelium to the adult status.[200]

Lipophilic compounds cross the alveolar epithelium more rapidly than hydrophilic compounds, a characteristic of all the viable membranes. Absorption is again by nonsaturable diffusion and is related to the partition coefficient of the molecule. The greater the lipid solubility, the more rapidly the compound is absorbed.[201]

Although many drugs are well absorbed by the lungs, often more rapidly than from other routes, this route is rarely used for systemic drug administration. The main reason for this is the difficulty of obtaining an accurate dose. Aerosol administration is the only practical method of dosing into the lungs, but even this method is very variable. A large proportion of the dose will remain in the mouth and throat and will eventually be swallowed. The size of the aerosol droplets must be about 0.6 μm to penetrate into the lungs, but even a proportion of this dose is exhaled in subsequent breaths.

This route is most commonly used to treat conditions of the respiratory tract when drug is administered to the site of action. Thus drugs to treat asthma such as the bronchodilator salbutamol and steroids such as betamethasone valerate are administered by aerosol.[202] In this way relatively small doses are required which may be an advantage if systemic concentrations cause other side effects. Pentamidine, which is used to treat pneumonia, an opportunist infection in AIDS patients, accumulates in tissues, especially liver and kidney, when administered systemically. If administered as an aerosol, however, high concentrations can be achieved in lung tissue without high systemic concentrations.[203]

Sodium cromoglycate, which is used to treat asthma, is poorly absorbed and ineffective after oral administration. However, absorption is much better from the lungs[204] and if the drug is administered to its site of action by aerosol or spin-haler formulations, treatment is very effective.

23.1.9 METHODS OF DETERMINING DRUG ABSORPTION

23.1.9.1 Direct Methods

Direct methods of determining drug absorption are methods in which the amount of drug absorbed or the amount remaining to be absorbed are measured directly. Such methods include models of drug absorption using artificial membranes, using *in vitro* animal preparations or using *in situ* animal preparations. Different methods will be applicable depending upon the route of drug administration.

23.1.9.1.1 Oral absorption

A number of methods for determining drug absorption or comparing drug absorption between series of drugs have used artificial lipophilic membranes. Although the pH–partition hypothesis indicates that such a membrane should give a reasonable indication of drug absorption, these membranes do not take into account the aqueous pathway, the unstirred layer or the virtual pH at the surface of the membrane. Thus in most cases these membranes have been fairly unsuccessful in accurately representing drug absorption, particularly when comparing chemically unrelated molecules.

The simplest models are three-phase systems comprising two aqueous phases, representing lumenal fluid and plasma, separated by a lipid phase, representing the membrane. The stability of this model may be improved by incorporating the lipid phase into a porous support. Membranes of bimolecular thickness have been prepared by painting a film across a small hole in a PTFE membrane.[205] Liposomes made from egg lecithin or from lipid extracts of intestinal mucosa have also been used to model drug absorption. Liposomes consist of lipid bilayer envelopes which enclose a small amount of drug solution. The rate of release of drug from the liposome was correlated with drug absorption rates.[206]

More successful correlations have been obtained by *in vitro* methods using intestinal tissue. Drug accumulation in intestinal rings was shown to be less variable than uptake into brush border membrane vesicles or isolated mucosal cells.[207]

The most commonly used *in vitro* technique is the everted intestinal sac. Small lengths of rat intestine are turned inside out, filled with fluid and ligated at both ends. The sac is then suspended in oxygenated fluid containing the drug of interest.[208] A sample from the sac is taken after sufficient time to allow drug transport (between 15–60 min). Modifications to allow further samples and comparison of everted and noneverted intestine have been reviewed in a previous article.[209] The viability of this preparation has been examined and it has been shown that progressive damage to the epithelial cells occurs from 15 min after the start of incubation.[210] In addition the absence of an intact blood supply results in drug accumulation on the serosal side and this may increase flow in the reverse direction.

The everted intestinal ring shows no advantages over the intestinal sac and viability may be 15 min or less. Thus the everted intestinal sac is probably the best *in vitro* method for studying drug absorption, although there are still few correlations with *in vivo* absorption either in the rat or man. This method will probably prove to be a useful high capacity screen for absorption of new drug candidates prior to carrying out *in vivo* experiments, and to differentiate between poor absorption and rapid metabolism in early studies.

Studies of drug absorption in anaesthetized animals are inherently better representations of the normal situation. In these *in situ* experiments, the blood supply to the intestine remains intact and flow through the intestine may be modelled by perfusion of drug solution. However the use of anaesthetized animals causes physiological changes in blood flow and intestinal motility and the method of perfusion may induce changes in the lumen.

In perfusion experiments, the rat is anaesthetized and the small intestine cannulated at the duodenal and ileal junctions. The stomach and large intestine are ligated and the intestine perfused with drug solution at 37 °C. The perfusate may be either collected after a single pass, recirculated or returned in the opposite direction. The single pass perfusion appears to maintain the unstirred aqueous layer relatively undisturbed and is viable for at least 30 min.[211] It is therefore likely to be closest to the *in vivo* situation and provide more accurate estimates of absorption (Figure 20). In most cases drug absorption is measured by disappearance from the perfusion fluid. Measurement of drug in blood requires an infusion of equal volumes of fresh blood.

An alternative *in situ* approach is the closed loop method. In this model the rat is anaesthetized, and the small intestine ligated to form a loop about 10 cm long. Drug is injected into the loop and the incisions closed. The animal may then recover, especially if a gaseous anaesthetic is used. At some later time the animal is sacrificed and the loop removed and a sample of lumenal fluid and blood may be analyzed for drug content.[212]

The above closed loop method has also been modified to allow multiple samples to be taken from the lumen with the animal under constant anaesthesia. This method, known as the Doluisio

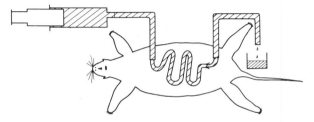

Figure 20 Diagrammatic presentation of the single pass intestinal perfusion experiment in the rat

method,[213] has been widely used and results have been shown to correlate well with the single pass perfusion method. However a major criticism of *in situ* absorption experiments is the limited comparative data with the *in vivo* situation, particularly when comparing molecules of differing physicochemical properties. In spite of the lack of information the *in situ* experiment appears to be the best model in which absorption can be measured, as far as possible, separate from other effects such as metabolism in liver and breakdown in stomach contents.

Intact animals have also been used as models for man to determine the amount of drug reaching the general circulation after oral administration. The rat has again been extensively used for these studies, especially when only small quantities of drug are available as is often the case in drug discovery. The major difficulties in using the rat apart from physiological differences from man, are the difficulty of taking repeated samples and using the animal in cross-over studies. A method for the cannulation of the jugular has been described and small samples (0.2 mL) may be taken repeatedly[214] or the tail vein may be used. If a larger sample is required, the animal must be sacrificed or a sample taken from the orbital venous plexus. The rabbit has also been used as a model of drug absorption but in normal circumstances it is a poor model. This is due to the long transit time of food in the rabbit intestine, partly due to the rabbits eating faeces direct from the anus (coprophagy). However it has been shown that rabbits conditioned on a special soft diet and muzzled to prevent coprophagy were better models for man.[215] The rabbit is probably the easiest animal to use for absorption studies due to the ease of blood sampling (1–2 mL) from the ear vein.

The dog has also been suggested as a model for oral absorption in humans. However the dose used is larger and thus the dog can often only be used at the later stages of drug discovery. The physiology of the GI tract of the dog is fairly similar to that in man, apart from a higher gastric pH and higher intestinal pH in the dog.[216] Stomach emptying is slower in the dog but this may relate to frequency of feeding and quantity of food eaten at each meal. No comparison with man of dogs fed thrice daily has been carried out. Small intestine transit time is faster in the dog and this is likely to be important for poorly absorbed drugs. It is unlikely that bioavailability studies in the dog would substitute for studies in healthy volunteers; although the dog is probably the best animal model for studies of drug absorption, such studies in the dog are not common.

23.1.9.1.2 *Other routes*

The rat has been used as a model for the absorption of suppository formulations. The usual suppository weight is about 50 mg and the percentage drug should be similar to the human dosage. The anus is closed by clips immediately after insertion. The production of faeces by the rat can be reduced by feeding a fibre free diet for a few days before dosing.[217] Studies of the vaginal absorption of drugs have been made in rats, in rabbits[115] and in monkeys[192] but few conclusions as to their applicability to man can be drawn.

Transdermal models of drug absorption are best carried out with animals with skins containing few hair follicles. Absorption from hairy skins is usually much more rapid than from skin with few hairs. Absorption from the skin of the mini-pig and the hairless mouse is usually found to resemble man as does that from the squirrel and the rhesus monkey.[151] Direct methods of measuring skin absorption can often be performed in man by swabbing the skin with an appropriate solvent to recover unabsorbed drug.

Buccal or sublingual absorption experiments are easily carried out in man by measuring the unabsorbed drug after rinsing the mouth with a solution of drug for a predefined period.[185] The hamster cheek pouch has also been used[186] to examine the effect of the unstirred aqueous layer and this may prove a useful model for absorption from the oral cavity. The rat has been used as a model for buccal absorption, although correlation with man has not been established.[218]

The rat is also the species of choice for absorption modelling of intranasal absorption. A number of authors have described the experimental conditions[164,219] required to study intranasal absorption. Two procedures are used, the *in situ* perfusion preparation (Figure 21) and the *in vivo* preparation. Both models may be used to screen the intranasal absorption of potential drug candidates. The *in situ* procedure has the advantage of sampling from the perfusion solution and may require a less sophisticated assay method, whereas the *in vivo* method is closer to the real situation, but samples of blood or plasma must be assayed.

The rat has also been used as a model for absorption from the lung, although the isolated perfused rat lung preparation is complicated.[219] *In vivo* studies have used the rat and the guinea pig,[199] although again there are few correlations with man since, as in man, it is very difficult to estimate the true dose reaching the lungs.

23.1.9.2 Indirect Methods

Indirect methods of measuring drug absorption are most often used in man due to the difficulty of measuring the residual amount of drug. However such methods will also be applicable to animal studies conducted under conditions with minimal anaesthetic or surgical interference. These indirect methods rely on an indifferent behaviour of the drug molecule to the route of administration once it reaches the general circulation, and, in addition, over the drug plasma concentration range of interest, the drug's behaviour must conform to linear pharmacokinetics. For practical purposes this means that repeated doubling of an intravenous dose over the plasma concentration range of interest will result in a doubling of the plasma concentration–time profile with no change in the half-life. Linear kinetics can be demonstrated for many drugs over limited concentration ranges. Another requirement which is sometimes ignored is that the pharmacokinetics of the intravenous dose must not vary with time. Thus the plasma profile should be the same if the drug is administered in the morning or the evening or at one week intervals. This requirement is necessary as the estimation of absorption is usually compared on different occasions with an intravenous dose or another standard preparation such as an oral solution dose.[220]

The plasma concentration–time profile after an extravascular dose is composed of two functions, the characteristic function and the absorption or input function. The characteristic function is the function describing distribution and elimination within the body and may be obtained by administering a bolus intravenous dose of unit size, *e.g.* 1 mg, or may be calculated from other known intravenous doses, *e.g.* a 10 mg bolus dose or a 1 mg min^{-1} intravenous infusion over 10 min. The plasma concentration–time profile, $R(t)$, resulting from any known absorption function may be

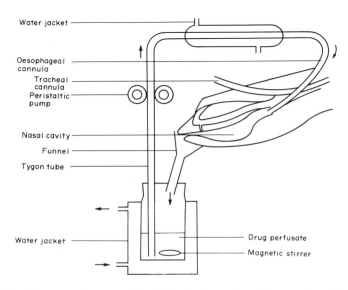

Figure 21 Diagrammatic presentation of the *in situ* nasal perfusion experiment in the rat

calculated by convolving the absorption function on to the characteristic function (equation 14; $G(\tau)$ is the characteristic function and $F(t)$ is the absorption function) using a standard method such as the Laplace transformation.

$$R(t) = \int_0^t F(t)G(t - \tau)d\tau \tag{14}$$

In order to study drug absorption in practice the reverse process is necessary. One measures both the characteristic function, *e.g.* plasma concentration–time profile after a bolus intravenous dose, and the response function or concentration–time profile after the drug dose of interest. The process of obtaining the absorption function from this data is known as deconvolution. A great many methods of carrying out this deconvolution have been proposed. Most involve finding a functional form of the characteristic function and in a number of methods a functional form for the response function must be defined.

In early work in pharmacokinetics simple approximations were made to obtain the absorption function. It was necessary to restrict the characteristic function to a mono- or bi-exponential function and the form of the absorption function was restricted to an exponentially declining function.[220] Another simple approach was to define the input function as a series of steps of varying height which would give exactly the response function or a least squares fit of the response function when convolved on to characteristic function.[221] A further approach was to define the response function as a series of points connected by straight lines and to obtain the input which would generate this function.[222] These approaches gave much more information about the shape of the absorption profile but suffered from problems arising from the approximations made and from a sensitivity to error in the response data points, which occasionally resulted in illogical or even negative absorption profiles.

Methods in which the characteristic function and the response function are fitted to general functions such as polynomials or to adaptive least squares cubic spline functions appear to be less sensitive to random error. In addition, if the characteristic function can be fitted to a polyexponential function, as is often the case, more accurate absorption functions may be obtained.[223,224] In other disciplines Fourier transforms have been widely used for deconvolution,[225] but this approach has found little favour in pharmacokinetics. A comparison of the currently accepted methods with Fourier transform methods would be of great interest, as would further comparison of the above methods with different data sets.

The measurement of the absorption profile of extravascular dose formulations, particularly of sustained release formulations, should be an integral requirement of its registration. Many formulations are manufactured based on dissolution data with little knowledge of the relationship between the *in vitro* dissolution data and the *in vivo* absorption profile. Generic formulations are marketed with a similar amount of bioavailability data but only a limited knowledge of the rate of absorption, characterized by the time and height of the maximum plasma concentration, which may give misleading information.

23.1.10 BIOAVAILABILITY AND BIOEQUIVALENCE

Definitions of bioavailability and bioequivalence have already been given in the introduction to this chapter. The first definition of bioavailability, then termed physiological .availability (equation 15), was given more than 40 years ago[226] and was later extended to the current definition for blood or plasma data (equation 16; C is the drug concentration in serum or plasma). The definition used by the FDA (US regulatory body) is even wider, *i.e.* 'the rate and extent to which the active drug or therapeutic moiety is absorbed from the drug product and becomes available at the site of action'. This definition allows, in theory at least, pharmacological or even clinical data to be used as a measure of bioavailability as well as blood or urine drug concentrations. A number of authors have suggested methods for determining bioavailability using pharmacological data.[227,228] In practice, however, bioavailability studies are almost always conducted using blood, plasma or urine drug concentrations. Urinary data may be less reliable if the pK_a of the drug is similar to the urinary pH, due to changes in the fraction reabsorbed. There are two main purposes for bioavailability testing. The first is to find out how much of the active drug reaches the systemic circulation when a drug is administered by different routes. The preceding sections of this chapter are relevant and indicate factors which will be important in determining the amount of drug reaching the circulation. However, even though one route may be advantageous in terms of bioavailability there may be medical reasons why another route is preferable. In this case many factors may influence the bioavailability of the drug, such as the ingestion of food and the age and disease state of the patients, and consequently it is necessary to document conditions which might significantly alter a drug's availability.

$$\text{Physiological availability} = \frac{\text{Amount of drug excreted in urine in a given time after test dose}}{\text{Amount of drug excreted in urine in the same time after giving drug in solution}} \times 100$$

$$(15)$$

$$\text{Bioavailability} = \left(D_{\text{ref}} \int_0^\infty C_{\text{test}} \, dt \right) \Big/ \left(D_{\text{test}} \int_0^\infty C_{\text{ref}} \, dt \right) \tag{16}$$

The second purpose for conducting bioavailability studies is to examine the amount of drug reaching the circulation from similar formulations, *e.g.* tablets from different manufacturers, a tablet and capsule from the same manufacturer or reformulation of an existing tablet. In these studies the objective is to determine whether formulation factors cause significant differences between the bioavailability of the different products. For these studies it is important to study the availability of the products under identical conditions. It is these studies which are normally conducted in healthy male volunteers with strict conditions regarding diet and consumption of fluids *etc.* Some authors critical of the way bioavailability studies are carried out fail to differentiate between these two purposes.[229]

Recommendations for when bioavailability studies are required have been given by the Australian Regulatory Authorities.[230] These include drugs with narrow therapeutic ranges or which are used in the treatment of life-threatening disease; drugs which show poor or variable absorption in animals or man; and drugs whose physicochemical properties indicate that formulation changes may affect bioavailability. In addition, bioavailability studies are necessary for special products, *e.g.* sustained release products, products containing drugs with a history of bioavailability problems and products containing a combination of drugs where one drug may affect the availability of other drugs.

Some examples of cases where bioavailability studies are not required are given in the EEC notes for guidance. These are if the drug is intended solely for intravenous use or if the product is for local application, *e.g.* a topical product. Bioavailability studies will also not be required when absorption of an oral product is not required, although studies of the passage of the drug into the circulation may be required. For a product of different dosage strengths where the ratio of active ingredients and excipients remains the same, the bioavailability of only one strength would be required.

An essential requirement for bioavailability studies is a suitable analytical method for the active drug which will allow measurement of drug concentrations over at least three, and preferably more, elimination half-lives. The use of equation (16) relies upon the linearity of the elimination pharmacokinetics and thus it is important to consider factors which may influence the linearity, such as enzyme induction or concomitant administration of other drugs. One method of overcoming changes in the pharmacokinetics between administration of different formulations is to use a labelled isotope in one formulation.[231] However this is often not practicable and in most studies formulations are administered in a latin square cross-over design with a washout period between doses.

The statistical treatment of data from bioavailability studies has been addressed by many authors. In general a sufficiently large population must be studied, based on preliminary knowledge of the variability in the parameters, to increase the likelihood of a certain difference, say 20%, between formulations being significant at the 95% confidence level. A likelihood of 80% is often used in bioavailability studies and this is denoted as the power of the test.[232] Alternative approaches use nonsymmetrical or symmetrical confidence limits.[233,234]

Recently there has been considerable interest in long-acting products and there are particular problems with bioavailability studies of these products. Often these products contain a larger dose than the product they are to replace and they are then administered less frequently. Bioavailability studies carried out with a single large dose of a conventional formulation may not give an accurate measure of the relative bioavailability of the sustained release product. A better approach is to administer the conventional formulation as divided doses, *e.g.* three times daily, and compare the total area under the curve with the sustained release product of the same total dose administered once daily. The use of the maximum and minimum plasma concentration of the conventional formulation as measures of the effectiveness of the controlled release formulations has also been presented.[235]

A considerable number of drugs have shown problems in their bioavailability, although fortunately very few have had serious clinical consequences. A selection of such 'problem' drugs is given in Table 6. Even newer sustained release preparations have caused problems due to the effect of

Table 6 Drugs for which Bioavailability or Bioequivalence Problems have been Reported

Dexamethasone	Phenytoin	Digoxin
Prednisone	Tolazamide	Chlorpromazine
Furosemide	Sulfonylureas	Levothyroxine
Warfarin	Haloperidol	Diazepam
Quinidine gluconate		

ingestion of food or as a result of other side effects. It is also unlikely that any *in vitro* test will show a sufficiently accurate correlation with *in vivo* bioavailability that it will replace human studies. However, in the future process control of the variables that influence dosage form manufacture may be sufficiently understood such that routine dissolution testing of each batch of tablets or capsules will be unnecessary and bioavailability testing will only be needed for new or reformulated products.

23.1.11 REFERENCES

1. S. Riegelman, L. Z. Benet and M. Rowland, *J. Pharmacokinet. Biopharm.*, 1973, **1**, 83.
2. M. Rowland, in 'Drug Absorption', ed. L. F. Prescott and W. S. Nimmo, Adis Press, Sydney, 1981, p. 285.
3. J. Koch-Weser, *N. Engl. J. Med.*, 1974, **291**, 233.
4. J. G. Wagner, in 'Biopharmaceutics and Relevant Pharmacokinetics', Drug Intelligence Publications, Hamilton, Illinois, 1971, p. 12.
5. P. A. Shoie, B. B. Brodie and C. A. M. Hogben, *J. Pharmacol. Exp. Ther.*, 1957, **119**, 361.
6. J. H. Tyrer, M. J. Eadie, J. M. Sutherland and W. D. Hooper, *Br. Med. J.*, 1970, **4**, 271.
7. B. F. Johnson, A. S. E. Fowle, S. Lader, J. Fox and A. D. Munro-Faure, *Br. Med. J.*, 1973, **4**, 323.
8. J. B. Dressman, *Pharm. Res.*, 1986, **3**, 123.
9. L. Ovesen, F. Bendtsen, U. Tage-Jensen, N. T. Pedersen, B. R. Gram and S. J. Rune, *Gastroenterology*, 1986, **90**, 958.
10. M. Feldman, *J. Clin. Invest.*, 1983, **72**, 295.
11. H. Forssell, B. Stenquist and L. Olbe, *Gastroenterology*, 1985, **89**, 581.
12. G. Flemstron and E. Kivilaakso, *Gastroenterology*, 1983, **84**, 787.
13. J. S. Trier and J. L. Madara, in 'Physiology of the Gastrointestinal Tract', ed. L. R. Johnson, Raven Press, New York, 1981, vol. 2, p. 925.
14. W. F. Ganong (ed.), in 'Review of Medical Physiology', Lange, Los Altos, 1981, p. 399.
15. C. A. Keele, E. Neil and N. Joels, in 'Samson Wright's Applied Physiology', Oxford University Press, Oxford, 1984, p. 437.
16. S. K. Sarna, *Gastroenterology*, 1985, **89**, 894.
17. J. E. Kellow, T. J. Borody, S. F. Phillips, R. L. Tucker and A. C. Haddad, *Gastroenterology*, 1986, **91**, 386.
18. J. H. Szurszewski, in 'Physiology of the Gastrointestinal Tract', ed. L. R. Johnson, Raven Press, New York, 1981, vol. 2, p. 1435.
19. H. J. Smith and M. Feldman, *Gastroenterology*, 1986, **91**, 1452.
20. P. Mojaverian, R. K. Ferguson, P. H. Vlasses, M. L. Rocci, Jr., A. Oren, J. A. Fix, L. J. Caldwell and C. Gardner, *Gastroenterology*, 1985, **89**, 392.
21. J. R. Malagelada, J. S. Robertson, M. L. Brown, M. Remington, J. A. Duenes, G. M. Thomforde and P. W. Carryer, *Gastroenterology*, 1984, **87**, 1255.
22. A. M. Metcalf, S. F. Phillips, A. R. Zinsmeister, R. L. MacCarty, R. W. Beart and B. G. Wolff, *Gastroenterology*, 1987, **92**, 40.
23. B. Krevsky, L. S. Malmud, F. D'Ercole, A. H. Maurer and R. S. Fisher, *Gastroenterology*, 1986, **91**, 1102.
24. L. S. Schanker, *J. Pharmacol. Exp. Ther.*, 1959, **126**, 283.
25. L. S. Schanker, D. J. Tocco, B. B. Brodie and C. A. M. Hogben, *J. Pharmacol. Exp. Ther.*, 1958, **123**, 81.
26. C. A. M. Hogben, D. J. Tocco, B. B. Brodie and L. S. Shanker, *J. Pharmacol. Exp. Ther.*, 1959, **125**, 275.
27. R. M. Carlson, R. E. Carlson and H. L. Kopperman, *J. Chromatogr.*, 1975, **107**, 219.
28. P. A. Shore, B. B. Brodie and C. A. M. Hogben, *J. Pharmacol. Exp. Ther.*, 1957, **119**, 361.
29. L. S. Shanker, P. A. Shore, B. B. Brodie and C. A. M. Hogben, *J. Pharmacol. Exp. Ther.*, 1957, **120**, 540.
30. M. Gibaldi, in 'Drug Absorption', ed. L. F. Prescott and W. S. Nimmo, Adis Press, Sydney, 1979, p. 1.
31. R. R. Levine and E. W. Pelikan, *Annu. Rev. Pharmacol.*, 1964, **4**, 69.
32. W. D. M. Paton and E. J. Zaimis, *Pharmacol. Rev.*, 1952, **4**, 219.
33. R. R. Levine, *Am. J. Dig. Dis.*, 1970, **15**, 171.
34. W. G. Crouthamel, G. H. Tan, L. W. Dittert and J. T. Doluisio, *J. Pharm. Sci.*, 1971, **60**, 1160.
35. D. C. Taylor, R. Pownall and W. Burke, *J. Pharm. Pharmacol.*, 1985, **37**, 280.
36. W. I. Higuchi, N. F. H. Ho, J. Y. Park and I. Komiya, in 'Drug Absorption', ed. L. F. Prescott and W. S. Nimmo, Adis Press, Sydney, 1979, p. 35.
37. D. C. Taylor, *Pharm. Int.*, 1986, **7**, 179.
38. Y. C. Martin, *J. Med. Chem.*, 1981, **24**, 229.
39. H. Sezaki, S. Muranishi, J. Nakamura, M. Yasuhara and T. Kimura, in 'Drug Absorption', ed. L. F. Prescott and W. S. Nimmo, Adis Press, Sydney, 1979, p. 21.
40. P. F. Ni, N. F. H. Ho, J. L. Fox, H. Leuenberger and W. I. Higuchi, *Int. J. Pharm.*, 1980, **5**, 33.

41. I. Komiya, J. Y. Park, A. Kamani, N. F. H. Ho and W. I. Higuchi, *Int. J. Pharm.*, 1980, **4**, 249.
42. D. Winne, *Experientia*, 1976, **32**, 1278.
43. N. W. Read, D. C. Barber, R. J. Levin and C. D. Holdsworth, *Gut*, 1977, **18**, 865.
44. N. F. H. Ho, J. Y. Park, G. E. Amidon, P. F. Ni and W. I. Higuchi, in 'Gastrointestinal Absorption of Drugs', ed. A. J. Aguiar, American Pharmaceutical Association, Washington, 1979.
45. J. B. Dressman, G. L. Amidon and D. Fleisher, *J. Pharm. Sci.*, 1985, **74**, 588.
46. C. P. Dooley, J. B. Reznick and J. E. Valenzuela, *Gastroenterology*, 1984, **87**, 1114.
47. R. C. Heading, P. Tothill, A. J. Laidlaw and D. J. C. Shearman, *Gut*, 1971, **12**, 611.
48. J. R. Malagelada, in 'Physiology of the Gastrointestinal tract', ed. L. R. Johnson, Raven Press, New York, 1981, p. 893.
49. R. D. Toothaker and P. G. Welling, *Annu. Rev. Pharmacol. Toxicol.*, 1980, **20**, 173.
50. A. Melander, *Clin. Pharmacokinet.*, 1978, **3**, 337.
51. J. A. Clements, R. C. Heading, W. S. Nimmo and L. F. Prescott, *Clin. Pharmacol. Ther.*, 1978, **24**, 420.
52. P. Mojaverian, M. L. Rocci, Jr., D. P. Conner, W. B. Abrams and P. H. Vlasses, *Clin. Pharmacol. Ther.*, 1987, **41**, 11.
53. P. Gruber, A. Rubinstein, V. H. K. Li, P. Bass and J. R. Robinson, *J. Pharm. Sci.*, 1986, **76**, 117.
54. A. Melander, K. Danielson, B. Schersten and E. Wahlin, *Clin. Pharmacol. Ther.*, 1977, **22**, 108.
55. W. S. Nimmo and L. F. Prescott, *Br. J. Clin. Pharmacol.*, 1978, **5**, 348.
56. W. S. Nimmo, in 'Drug Absorption', ed. L. F. Prescott and W. S. Nimmo, Adis Press, Sydney, 1981, p. 11.
57. L. Hendeles, P. Wubbena and M. Weinberger, *Lancet*, 1984, **2**, 1471.
58. K. C. Kwan, G. O. Breault, E. R. Umbenhauer, F. G. McMahon and D. E. Duggan, *J. Pharmacokinet. Biopharm.*, 1976, **4**, 255.
59. A. G. Renwick, in 'Presystemic Drug Elimination', ed. C. F. George, D. G. Shand and A. G. Renwick, Butterworth Scientific, London, 1982, p. 3.
60. J. C. Dacre, R. R. Sheline and R. T. Williams, *J. Pharm. Pharmacol.*, 1968, **20**, 619.
61. D. M. Gardner and A. G. Renwick, *Xenobiotica*, 1978, **8**, 679.
62. P. Goldman, in 'Drug Absorption', ed. L. F. Prescott and W. S. Nimmo, Adis Press, Sydney, 1981, p. 88.
63. H. A. Gordan and L. Pesti, *Bacteriol. Rev.*, 1971, **35**, 390.
64. L. V. Holdman, E. P. Cato and W. E. C. Moore, in 'Anaerobe Laboratory Manual', 4th edn., The Verginia Polytechnic Institute and State University Anaerobe Laboratory, Blacksburg, 1977.
65. L. Linday, J. F. Dobkin, T. C. Wang, V. P. Butler, Jr., J. R. Saha and J. Lindenbaum, *Pediatrics*, 1987, **79**, 544.
66. A. G. Renwick, H. A. Strong and C. F. George, *Biochem. Pharmacol.*, 1986, **35**, 64.
67. L. W. Wattenberg, *Toxicol. Appl. Pharmacol.*, 1972, **23**, 741.
68. K. Hartiala, *Physiol. Rev.*, 1973, **53**, 496.
69. J. Caldwell and M. V. Marsh, in 'Presystemic Drug Elimination', ed. C. F. George, D. G. Shand and A. G. Renwick, Butterworth Scientific, London, 1982, p. 29.
70. K. F. Ilett, C. F. George and D. S. Davies, *Biochem. Pharmacol.*, 1980, **29**, 2551.
71. P. N. Bennett, E. Blackwell and D. S. Davies, *Nature (London)*, 1975, **258**, 247.
72. J. B. Houston, H. J. Wilkins and G. Levy, *Res. Commun. Chem. Pathol. Pharmacol.*, 1976, **14**, 643.
73. J. B. Houston and S. G. Wood, in 'Progress in Drug Metabolism', ed. J. W. Bridges and L. F. Chasseaud, Wiley, New York, 1980, vol. 4, p. 116.
74. K. F. Ilett and D. S. Davies, in 'Presystemic Drug Elimination', ed. C. F. George, D. S. Shand and A. G. Renwick, Butterworth Scientific, London, 1982, p. 43.
75. M. J. Humphrey and P. S. Ringrose, *Drug Metab. Rev.*, 1986, **17**, 238.
76. A. K. Mitra, *Pharm. Int.*, 1986, **7**, 323.
77. D. M. Matthews, *Physiol. Rev.*, 1975, **55**, 537.
78. T. Kimura, *Pharm. Int.*, 1984, **5**, 75.
79. D. J. Tocco, A. deLuna, A. E. Duncan, T. C. Vassil and E. H. Ulm, *Drug Metab. Dispos.*, 1982, **10**, 15.
80. K. Takada, N. Shibata, H. Yoshimura, Y. Masuda, H. Yoshikawa, S. Muranishi and T. Oka, *J. Pharmacobio-Dyn.*, 1985, **8**, 320.
81. H. E. Gallo-Torres, E. P. Heimer, C. Witt, O. N. Miller, J. Meienhofer and C. I. Cheeseman, *Can. J. Physiol. Pharmacol.*, 1984, **62**, 319.
82. M. Saffran, G. S. Kumar, C. Savariar, J. C. Burnham, F. Williams and D. C. Neckers, *Science (Washington, D.C.)*, 1986, **233**, 1081.
83. K. Takaori, J. Burton and M. Donowitz, *Biochem. Biophys. Res. Commun.*, 1986, **137**, 682.
84. K. W. Kohn, *Nature (London)*, 1961, **191**, 1156.
85. H. Poiger and C. Schlatter, *Eur. J. Clin. Pharmacol.*, 1978, **14**, 129.
86. G. H. McCracken, Jr., C. M. Ginsburg, J. C. Clahsen and M. L. Thomas, *Pediatrics*, 1978, **62**, 738.
87. P. Kahela, M. Anttila, R. Tikkanen and H. Sundquist, *Acta Pharmacol. Toxicol.*, 1979, **44**, 7.
88. F. A. Ogunbona, I. F. Smith and O. S. Olawoye, *J. Pharm. Pharmacol.*, 1985, **37**, 283.
89. B. Beerman, in 'Drug Absorption', ed. L. F. Prescott and W. S. Nimmo, Adis Press, Sydney, 1981, p. 238.
90. P. G. Welling, in 'Progress in Drug Metabolism', ed. J. W. Bridges and L. F. Chasseaud, Wiley, New York, 1980, vol. 4, p. 144.
91. W. H. Barr, J. Adir and L. Garrettson, *Clin. Pharmacol. Ther.*, 1971, **12**, 779.
92. F. A. Ismail, N. Khalafallah and S. A. Khalil, *Int. J. Pharm.*, 1987, **34**, 189.
93. W. E. Fann, J. M. Davis, D. S. Janowsky, H. J. Sekerke and D. M. Schmidt, *Clin. Pharmacol. Ther.*, 1973, **13**, 388.
94. D. L. Sorby, *J. Pharm. Sci.*, 1965, **54**, 677.
95. C. A. Keele, E. Neil and N. Joels, in 'Samson Wright's Applied Physiology', 13th edn., Oxford University Press, Oxford, 1984, p. 144.
96. L. Ther and D. Winne, *Annu. Rev. Pharmacol.*, 1971, **11**, 57.
97. D. Winne, *J. Pharmacokinet. Biopharm.*, 1978, **6**, 55.
98. H. J. Horst, W. J. Höltje, M. Dennis, A. Coert, J. Geelen and K. D. Voigt, *Klin. Wochenschr.*, 1976, **54**, 875.
99. K. Takada, H. Yoshikawa and S. Muranishi, *Res. Commun. Chem. Pathol. Pharmacol.*, 1983, **40**, 99.
100. H. Yoshikawa, K. Takada and S. Muranishi, *J. Pharmacobio-Dyn.*, 1984, **7**, 1.
101. Y. Kaji, K. Uekama, H. Yoshikawa, K. Takada and S. Muranishi, *Int. J. Pharm.*, 1985, **24**, 79.

102. G. Levy and R. H. Gumtow, *J. Pharm. Sci.*, 1963, **52**, 1139.
103. G. L. Mattok, E. G. Lovering and I. J. McGilveray, in 'Progress in Drug Metabolism', ed. J. W. Bridges and L. F. Chásseaud, Wiley, New York, 1977, vol. 2, p. 267.
104. M. Gibaldi, in 'Biopharmaceutics and Clinical Pharmacokinetics', Lea & Febiger, Philadelphia, 1977, p. 50.
105. J. G. Wagner, in 'Biopharmaceutics and Relevant Pharmacokinetics', Drug Intelligence Publications, Hamilton, Illinois, 1971, p. 162.
106. C. E. Tarnowski, *Acta Tuberc. Scand.*, 1957, **34**, 76.
107. S. Muranishi, *Pharm. Res.*, 1985, **3**, 108.
108. K. Taniguchi, S. Muranishi and H. Sezaki, *Int. J. Pharm.*, 1980, **4**, 219.
109. J. Koch-Weser and P. J. Schechter, in 'Drug Absorption', ed. L. F. Prescott and W. S. Nimmo, Adis Press, Sydney, 1981, p. 217.
110. C. H. Kleinbloesem, P. Van Brummelen, M. Danhof and D. D. Breimer, *3rd World Conf. Clin. Pharmacol. Ther.*, 1986, Abs. 481.
111. F. Theeuwes, in 'Drug Absorption', ed. L. F. Prescott and W. S. Nimmo, Adis Press, Sydney, 1981, p. 157.
112. J. Hirtz, *Pharm. Int.*, 1986, **7**, 21.
113. G. S. Banker and V. E. Sharma, in 'Drug Absorption', ed. L. F. Prescott and W. S. Nimmo, Adis Press, Sydney, 1981, p. 194.
114. H. M. Ingani, J. Timmermans and A. J. Moes, *Int. J. Pharm.*, 1987, **35**, 157.
115. C. J. de Blaey and J. Polderman, in 'Drug Design', ed. E. J. Ariens, Academic Press, New York, 1980, vol. 9, p. 237.
116. A. G. de Boer and D. D. Breimer, in 'Drug Absorption', ed. L. F. Prescott and W. S. Nimmo, Adis Press, Sydney, 1981, p. 61.
117. S. Muranishi, *Methods Find. Exp. Clin. Pharmacol.*, 1984, **6**, 763.
118. R. Voigt and G. Falk, *Pharmazie*, 1968, **23**, 709.
119. K. Kakemi, T. Arita and S. Muranishi, *Chem. Pharm. Bull.*, 1965, **13**, 976.
120. K. Taniguchi, S. Muranishi and H. Sezaki, *Int. J. Pharm.*, 1980, **4**, 219.
121. N. Muranushi, M. Kinugawa, Y. Nakajimi, S. Muranishi and H. Sezaki, *Int. J. Pharm.*, 1980, **4**, 271.
122. T. Nishihata, M. Miyake and A. Kamada, *J. Pharmacobio-Dyn.*, 1984, **7**, 607.
123. L. Caldwell, T. Nishihata, J. Fix, S. Selk, R. Cargill, C. R. Gardner and T. Higuchi, *Methods Find. Exp. Clin. Pharmacol.*, 1984, **6**, 503.
124. S. S. Davis, W. R. Burnham, P. Wilson and J. O'Brien, *Antimicrob. Agents Chemother.*, 1985, **28**, 211.
125. J. A. Fix, K. Engle, P. A. Porter, P. S. Leppert, S. J. Selk, C. R. Gardner and J. A. Alexander, *Am. J. Physiol.*, 1986, **251**, 332.
126. J. A. Fix, P. S. Leppert, P. A. Porter and J. Alexander, *J. Pharm. Pharmacol.*, 1984, **36**, 286.
127. J. A. Fix and C. R. Gardner, *Pharm. Int.*, 1986, **7**, 272.
128. H. Yoshikawa, H. Sezaki and S. Muranishi, *Int. J. Pharm.*, 1983, **13**, 321.
129. H. Yoshikawa, K. Takada, S. Muranishi and Y. Satoh, *J. Pharmacobio-Dyn.*, 1983, **6**, S91.
130. H. Loth, *Acta Pharm. Technol.*, 1986, **32**, 109.
131. A. Durward, in 'Cunningham's Textbook of Anatomy', ed. G. J. Romanes, Oxford University Press, Oxford, 1964, p. 786.
132. R. H. Guy and J. Hadgraft, in 'Percutaneous Absorption', ed. R. L. Bronaugh and H. I. Maibach, Dekker, New York, 1985, p. 3.
133. R. H. Guy, J. Hadgraft and H. I. Maibach, *Int. J. Pharm.*, 1982, **11**, 119.
134. R. H. Guy and J. Hadgraft, *J. Pharm. Sci.*, 1984, **73**, 883.
135. R. H. Guy, J. Hadgraft and H. I. Maibach, *Int. J. Pharm.*, 1983, **17**, 23.
136. I. H. Blank, *Toxicol. Appl. Pharmacol.*, 1969, **Suppl. 3**, 23.
137. T. Higuchi, in 'Design of Biopharmaceutical Properties through Prodrugs and Analogs', ed. E. B. Roche, American Pharmaceutical Association, Washington, 1977, p. 409.
138. E. R. Cooper, *Pharm. Int.*, 1986, **7**, 308.
139. M. S. Roberts, R. A. Anderson and J. Swarbrick, *J. Pharm. Pharmacol.*, 1977, **29**, 677.
140. R. J. Scheuplein, I. H. Blank, G. J. Brauner and D. J. MacFarlane, *J. Invest. Dermatol.*, 1969, **52**, 63.
141. M. J. Bartek and J. A. LaBudde, in 'Animal Models in Dermatology', ed. H. I. Maibach, Churchill Livingstone, Edinburgh, 1975, p. 103.
142. S. H. Yalkowsky and G. L. Flynn, *J. Pharm. Sci.*, 1973, **62**, 210.
143. R. J. Scheuplein, *J. Invest. Dermatol.*, 1976, **67**, 31.
144. G. L. Flynn, in 'Percutaneous Absorption', ed. R. L. Bronaugh and H. I. Maibach, Dekker, New York, 1985, p. 17.
145. T. A. Loomis, in 'Current Concepts in Cutaneous Toxicity', ed. V. A. Drill and P. Lazar, Academic Press, New York, 1980, p. 153.
146. D. E. Loveday, *J. Soc. Cosmet. Chem.*, 1961, **12**, 224.
147. T. Arita, R. Hori, T. Anmo, M. Washitake, T. Yajima and M. Akatsu, *Chem. Pharm. Bull.*, 1970, **18**, 1045.
148. B. J. Poulsen, in 'Drug Design', ed. E. J. Ariens, Academic Press, New York, 1973, vol. 4, chap. 5, p. 161.
149. I. H. Blank and R. J. Scheuplein, *Br. J. Dermatol.*, 1969, **81**, Suppl. 4, 4.
150. R. J. Scheuplein and I. H. Blank, *Physiol. Rev.*, 1971, **51**, 702.
151. R. C. Wester and H. I. Maibach, in 'Percutaneous Absorption', ed. R. L. Bronaugh and H. I. Maibach, Dekker, New York, 1985, p. 231.
152. A. W. McKenzie and R. B. Stoughton, *Arch. Dermatol.*, 1962, **86**, 608.
153. I. H. Blank, in 'Percutaneous Absorption', ed. R. L. Bronaugh and H. I. Maibach, Dekker, New York, 1985, p. 97.
154. M. K. Polano and M. Ponec, *Arch. Dermatol.*, 1976, **112**, 675.
155. E. R. Cooper, *J. Pharm. Sci.*, 1984, **73**, 1153.
156. P. P. Sarpotdar and J. L. Zatz, *Drug Dev. Ind. Pharm.*, 1987, **13**, 15.
157. E. R. Cooper, in 'Solution Behaviour of Surfactants', ed. K. L. Mittal and E. J. Fendler, Plenum Press, New York, 1982, p. 1505.
158. H. I. Maibach and R. J. Feldmann, *Ann. N.Y. Acad. Sci.*, 1967, **141**, 423.
159. R. B. Stoughton, *Arch. Dermatol.*, 1982, **118**, 474.

160. E. J. Lien, in 'Progress in Drug Research', ed. E. Jucker, Birkhauser, Basel, 1985, vol. 29, p. 67.
161. D. F. Proctor, in 'Transnasal Systemic Medications', ed. Y. W. Chien, Elsevier, Amsterdam, 1985, p. 101.
162. D. S. Freestone and A. L. Weinberg, *Br. J. Clin. Pharmacol.*, 1976, **3**, 827.
163. R. J. Phillpotts, H. W. Davies, J. Willman, D. A. J. Tyrrell and P. G. Higgins, *Antiviral Res.*, 1984, **4**, 71.
164. C. H. Huang, R. Kimura, R. Bawarshi-Nassar and A. Hussain, *J. Pharm. Sci.*, 1985, **74**, 608.
165. A. Hussain, R. Bawarshi-Nassar and C. H. Huang, in 'Transnasal Systemic Medications', ed. Y. W. Chien, Elsevier, Amsterdam, 1985, p. 130.
166. A. N. Fisher, K. Brown, S. S. Davis, G. D. Parr and D. A. Smith, *J. Pharm. Pharmacol.*, 1987, **39**, 357.
167. K. S. E. Su, *Pharm. Int.*, 1986, **7**, 8.
168. Y. W. Chien and S. F. Chang, in 'Transnasal Systemic Medications', ed. Y. W. Chien, Elsevier, Amsterdam, 1985, p. 58.
169. S. Hirai, T. Ikenaga and T. Matsuzawa, *Diabetes*, 1978, **27**, 296.
170. S. Hirai, T. Yashiki and H. Mima, *Int. J. Pharm.*, 1981, **9**, 165.
171. A. E. Pontiroli, M. Alberetto, G. Dossi, I. Bosi, G. Pozza and A. Secchi, *Br. Med. J.*, 1982, **284**, 303.
172. A. C. Moses, G. S. Gordon, M. C. Carey and J. S. Flier, *Diabetes*, 1983, **32**, A70.
173. D. R. London, W. R. Butt, S. S. Lynch, J. C. Marshall, S. Owusu, W. R. Robinson and J. M. Stephens, *J. Clin. Endocrinol. Metab.*, 1973, **37**, 829.
174. G. Fink, G. Gennser, P. Liedholm, J. Thorell and J. Mulder, *J. Endocrinol.*, 1974, **63**, 351.
175. Y. W. Chien and S. F. Chang, in 'Transnasal Systemic Medications', ed. Y. W. Chien, Elsevier, Amsterdam, 1985, p. 23.
176. J. Hofbauer and J. K. Hoerner, *Am. J. Obstet. Gynecol.*, 1927, **14**, 137.
177. C. Hendricks and A. R. Gabel, *Am. J. Obstet. Gynecol.*, 1960, **79**, 780.
178. M. E. Dalton, D. R. Bromham, C. L. Ambrose, J. Osborne and K. D. Dalton, *Br. J. Obstet. Gynecol.*, 1987, **94**, 84.
179. A. Hussain, T. Foster, S. Hirai, T. Kashihara, R. Batenhorst and M. Jones, *J. Pharm. Sci.*, 1980, **69**, 1240.
180. M. L. Citron, J. R. Reynolds, J. Kalra, B. G. Kay, K. A. Nathan and N. D. Jaffe, *Cancer Treat. Rep.*, 1987, **71**, 317.
181. W. Petri, R. Schmiedel and J. Sandow, in 'Transnasal Systemic Medications', ed. Y. W. Chien, Elsevier, Amsterdam, 1985, p. 161.
182. A. H. Beckett and E. J. Triggs, *J. Pharm. Pharmacol.*, 1967, **19**, Suppl. 31S.
183. A. H. Beckett and A. C. Moffat, *J. Pharm. Pharmacol.*, 1968, **20**, Suppl. 239S.
184. A. H. Beckett and A. C. Moffat, *J. Pharm. Pharmacol.*, 1969, **21**, Suppl. 139S.
185. W. Schürmann and P. Turner, *J. Pharm. Pharmacol.*, 1978, **30**, 137.
186. Y. Kurosaki, S. Hisaichi, C. Hamada, T. Nakayama and T. Kimura, *J. Pharmacobio-Dyn.*, 1987, **10**, 180.
187. A. H. Beckett, R. N. Boyes and E. J. Triggs, *J. Pharm. Pharmacol.*, 1968, **20**, 92.
188. J. A. Henry, K. Ohashi, J. Wadsworth and P. Turner, *Br. J. Clin. Pharmacol.*, 1980, **10**, 61.
189. M. H. Bickel and H. J. Weder, *J. Pharm. Pharmacol.*, 1969, **21**, 160.
190. J. Pitha, S. M. Harman and M. E. Michel, *J. Pharm. Sci.*, 1986, **75**, 165.
191. G. S. Duchateau, J. Zuidema and F. W. Merkus, *Pharm. Weekbl., Sci. Ed.,* 1986, **8**, 98.
192. D. P. Benziyer and J. Edelson, *Drug Metab. Rev.*, 1983, **14**, 137.
193. C. J. de Blaey and J. Polderman, in 'Drug Design', ed. E. J. Ariens, Academic Press, New York, 1980, vol. 9, p. 262.
194. R. W. Kistner, in 'The Human Vagina', ed. E. S. E. Hafez and T. N. Evans, Elsevier, Amsterdam, 1978, p. 109.
195. I. Schiff, B. Wentworth, B. Koos, K. J. Ryan and D. Tulchinsky, *Fertil. Steril.*, 1978, **30**, 278.
196. M. M. Alper, B. N. Barwin, W. M. McLean and I. J. McGilveray, *Obstet. Gynecol.*, 1985, **65**, 781.
197. T. Kimura, *Pharm. Int.*, 1984, **5**, 75.
198. M. Chapil, C. D. Eskelson, W. Droegemueller, J. B. Ulreich, J. A. Owen, J. C. Ludwig and V. M. Stiffel, in 'Vaginal Contraception: New Developments', ed. G. I. Zatuchni *et al.*, Harper and Row, London, 1979, p. 165.
199. L. S. Shanker, *Biochem. Pharmacol.*, 1978, **27**, 381.
200. J. R. Bend, C. J. Serabjit-Singh and R. M. Philpot, *Annu. Rev. Pharmacol. Toxicol.*, 1985, **25**, 97.
201. S. J. Enna and L. S. Schanker, *Am. J. Physiol.*, 1972, **223**, 1227.
202. M. Gibaldi, in 'Biopharmaceutics and Clinical Pharmacokinetics', Lea & Febiger, Philadelphia, 1977, p. 73.
203. R. H. Waldman, D. E. Pearce and R. A. Martin, *Am. Rev. Respir. Dis.*, 1973, **108**, 1004.
204. G. F. Moss, K. M. Jones, J. T. Ritchie and J. S. G. Cox, *Toxicol. Appl. Pharmacol.*, 1971, **20**, 147.
205. K. Inui, K. Tabara, R. Hori, A. Kaneda, S. Muranishi and H. Sezaki, *J. Pharm. Pharmacol.*, 1977, **29**, 22.
206. T. Kimura, M. Yoshikawa, M. Yasuhara and H. Sezaki, *J. Pharm. Pharmacol.*, 1980, **32**, 394.
207. I. Osiecka, P. A. Porter, R. T. Borchardt, J. A. Fix, C. R. Gardner and L. Frost, *Pharm. Res.*, 1985, **6**, 284.
208. T. H. Wilson and G. Wiseman, *J. Physiol. (London)*, 1954, **123**, 116.
209. J. B. Houston and S. G. Wood, in 'Drug Metabolism', ed. J. W. Bridges and L. R. Chasseaud, Wiley, New York, 1980, vol. 4, p. 62.
210. R. R. Levine, W. F. McNary, P. J. Kornguth and R. Le Blanc, *Eur. J. Pharmacol.*, 1970, **9**, 211.
211. N. Schurgers, J. Bijdendijk, J. J. Tukker and D. J. A. Crommelin, *J. Pharm. Sci.*, 1986, **75**, 117.
212. R. R. Levine and E. W. Pelikan, *J. Pharmacol. Exp. Ther.*, 1961, **131**, 319.
213. J. T. Doluisio, N. F. Billups, L. W. Dittert, E. J. Sugita and J. V. Swintosky, *J. Pharm. Sci.*, 1969, **58**, 1196.
214. R. A. Upton, *J. Pharm. Sci.*, 1975, **64**, 112.
215. T. Maeda, H. Takenaka, Y. Yamahira and T. Noguchi, *J. Pharm. Sci.*, 1977, **66**, 69.
216. J. B. Dressman, *Pharm. Res.*, 1986, **3**, 123.
217. J. J. Rutten-Kingma, *Int. J. Pharm.*, 1979, **3**, 39, 179, 187.
218. M. A. Hussain, B. J. Aungst and E. Shefter, *J. Pharm. Sci.*, 1986, **75**, 218.
219. P. R. Byron, N. S. R. Roberts and R. Clark, *J. Pharm. Sci.*, 1986, **75**, 168.
220. D. Cutler, *J. Pharmacokinet. Biopharm.*, 1978, **6**, 227.
221. D. P. Vaughan and M. J. Dennis, *J. Pharm. Sci.*, 1978, **67**, 663.
222. D. P. Vaughan and M. J. Dennis, *J. Pharmacokinet. Biopharm.*, 1979, **7**, 511.
223. D. J. Cutler, *J. Pharmacokinet. Biopharm.*, 1978, **6**, 243.
224. P. V. Pedersen, *J. Pharm. Sci.*, 1980, **69**, 298.
225. C. M. Coulam, H. R. Warner, H. W. Marshall and J. B. Bassingthwaighte, *Comput. Biomed. Res.*, 1967, **1**, 124.
226. B. L. Oser, D. Melnick and M. Hochberg, *Ind. Eng. Chem. Anal. Ed.*, 1945, **17**, 405.
227. V. F. Smolen, *J. Pharmacokinet. Biopharm.*, 1976, **4**, 337.

228. N. H. G. Holford and L. B. Scheiner, *Clin. Pharmacokinet.*, 1981, **6**, 429.
229. W. A. Ritschel, *Methods Find. Exp. Clin. Pharmacol.*, 1984, **6**, 777.
230. J. P. Griffin, in 'Drug Absorption', ed. L. F. Prescott and W. S. Nimmo, Adis Press, Sydney, 1981, p. 334.
231. S. Riegelman, M. Rowland and L. Z. Benet, *J. Pharmacokinet. Biopharm.*, 1973, **1**, 83.
232. J. M. Lachin, *Cont. Clin. Trials*, 1981, **2**, 93.
233. B. E. Rodda and R. L. Davis, *Clin. Pharmacol. Ther.*, 1980, **28**, 247.
234. W. J. Westlake, *Biometrics*, 1976, **32**, 741.
235. J. J. Vallner, I. L. Honigberg, J. A. Kotzan and J. T. Stewart, *Int. J. Pharm.*, 1983, **16**, 47.

23.2

Pharmacokinetic Aspects of Drug Administration and Metabolism

THOMAS N. TOZER

University of California, San Francisco, CA, USA

23.2.1 PHARMACOKINETIC VIEW OF DRUG ADMINISTRATION

Within the last 25 years, a body of pharmacokinetic principles has evolved that has revolutionized the way in which one approaches the initiation and modification of the administration of drugs. The purpose of this chapter is to review and highlight these principles. To this end, references are given to textual sources and general reviews[1-9] as well as to specific papers.

This chapter begins with a single dose, progresses through constant-rate input and finishes with multiple-dose regimens, the most common mode of administration. For each area, the kinetics and the impact of the mode of administration on the response to a drug are presented. The purpose of controlled-release devices and formulations is also briefly explored.

23.2.1.1 A Single Dose

It is rare that drug therapy consists of only a single dose, but the information obtained therefrom can be used to design therapeutic regimens. The expectation after multiple single doses is, after all, simply the sum of the observations after each dose. This principle will be returned to in Section 23.2.1.3 on multiple-dose regimens. Out first goal is to summarize the observations after single intravascular and extravascular doses.

23.2.1.1.1 *Intravascular administration*

A drug, injected directly into a blood vessel, is rapidly distributed throughout the vascular system and, since its presence in the body is usually assessed from a blood sample, intravascular administration is easily distinguished from extravascular administration in that an input step is required for the latter. Thus, following an intravascular dose, one is only concerned with distribution, drug transfer to and from the various tissues, and drug elimination, irreversible loss from the body by excretion (loss of unchanged drug) and metabolism (conversion to another chemical species).

(i) Concept of a bolus

It is impossible to administer a drug instantaneously; but if the time of input is short relative to the time scale required for distribution and elimination to occur, the dose can be considered as a bolus or an impulse, especially if the input time is less than a few minutes. This statement must be modified for drugs that distribute and/or are eliminated in a time frame of minutes (examples are succinylcholine and nitroglycerin). Conversely, if the distribution and elimination processes are slow, an injection with an input time of an hour may still be considered as a bolus (examples are digoxin and phenobarbital).

(ii) Indices of exposure

The dose administered is a logical index of the exposure of the body to a drug after a bolus. However, it is an incomplete measure because it does not incorporate information on how long the drug remains in the body. Such information is only available by measuring the drug in the body, usually the concentration in blood or plasma.

(a) AUC *and clearance.* Clearance, CL, is a fundamental concept in pharmacokinetics. It is the proportionality constant that relates rate of elimination to the plasma (or blood) drug concentration, C, that is

$$\text{Rate of elimination} = CL \cdot C \tag{1}$$

On integrating this relationship from time zero to time infinity

$$Dose = CL \cdot AUC \tag{2}$$

where *Dose*, the amount given, is equal to the amount eliminated, $CL \cdot AUC$, and AUC is the total area under the plasma concentration–time curve.

AUC is a most useful and direct measure of exposure; its value depends on both the dose administered and how readily the drug is removed from the body (see equation 2). Figure 1 shows the effect of different (a) doses and (b) clearance values on the time course of the plasma drug concentration. Clearly, either an increase in dose or a decrease in clearance increases exposure to the drug.

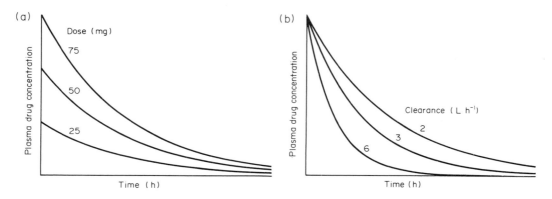

Figure 1 The area under the plasma drug concentration *versus* time curve (a) increases in proportion to dose administered and (b) decreases inversely with the clearance value (the concentration and time axes are arbitrary but are consistent within each graph)

Clearance has additional utility here. Note that when AUC of equation (2) is normalized to dose, $1/CL = AUC/Dose$, the value $1/CL$ is obtained. The reciprocal of clearance is then a normalized index of exposure. One need only to multiply by dose to obtain AUC, the more direct measure of drug exposure.

(b) Initial concentration. Another useful measure of exposure, particularly for reversibly acting drugs for which drug at the site of action is in rapid equilibrium with that in a plasma, is the initial plasma concentration. Following an intravascular bolus dose, the highest concentration occurs just after administration. The actual value depends on the initial apparent dilution of the drug as measured by the initial volume of distribution, V_1. The value of this parameter can be estimated from

$$V_1 = \frac{Dose}{C(0)} \tag{3}$$

where $C(0)$ is the initial concentration, obtained from extrapolation to time zero. This volume parameter, like clearance, is proportional to the dose administered. Thus, the initial concentration expected following a given dose can be estimated when V_1 is known. Figure 2 illustrates the effect of an increase in the volume of distribution on the time course of drug in the body. Note the decrease in the initial concentration and the increase in the time to eliminate the drug. Also note that the AUC values for all the curves are the same; clearance and dose were not changed.

(c) Mean residence time. A third practical measure of exposure is mean residence time, MRT, the average time a drug molecule resides within the body after intravascular administration. Like clearance, its value is independent of the dose administered. The product of dose and MRT is another useful measure of exposure as shown in Figure 3, but the value must be calculated and is not as directly measured as AUC can be.

The mean residence time is calculated[10] from the relationship

$$MRT = \frac{AUMC}{AUC} \tag{4}$$

where $AUMC$ is the area under the first moment of the plasma concentration–time curve, *i.e.* $\int_0^\infty t \cdot C \, dt$. It can also be calculated from urine data using the relationship

$$MRT = \frac{\int_0^\infty (Ae_\infty - Ae) \, dt}{Ae_\infty} \tag{5}$$

where Ae_∞ is the total amount excreted unchanged and Ae is the amount excreted unchanged up to a given point in time. This procedure can be reasonably accurate if renal excretion is the major route of elimination and/or renal clearance is constant with time (no pH or urine flow dependence and no saturable tubular secretion or reabsorption).

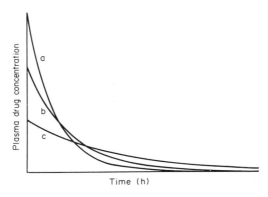

Figure 2 Clearance determines the area under the plasma concentration *versus* time curve after an intravenous bolus dose. Consequently, on examining three drugs with the same clearance, but increasing (a to c) volumes of distribution, the initial concentration decreases as does the rate of decline of the concentration with time but the area remains unchanged

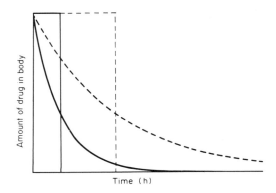

Figure 3 The exposure to a drug as measured by both the amount in the body and the time of its sojourn there can be assessed by the product of the mean residence time, denoted by the vertical lines, and the dose, the initial amount in the body. Note that the area of the corresponding rectangle formed by this product (solid curve) increases when the drug is eliminated more slowly (dashed curve). An increased exposure also occurs for an increase in dose (compare with Figure 1(a) by substituting amount with concentration on y axis)

(iii) Representation by exponentials

When the logarithm of the plasma drug concentration is plotted against time, one observes a decline as typified by the data for aspirin in Figure 4. On extrapolating the terminal straight line back to zero time and taking the difference between the observation and the value on the extrapolated line at the corresponding time, the *method of residuals*, it is apparent that the time course of the plasma concentration can be simulated by the sum of two exponential terms, that is

$$C = C_1 \cdot e^{-\lambda_1 t} + C_2 \cdot e^{-\lambda_2 t} \tag{6}$$

where C_1 and C_2 are the zero-time intercepts of the two lines and λ_1 and λ_2 are the slopes of the respective curves. More than two exponential curves may be needed to simulate the concentration–time curve, in which case the decline can be represented by the relationship

$$C = \sum_{i=1}^{n} C_i \cdot e^{-\lambda_i t} \tag{7}$$

where C_i and λ_i are the coefficients of the ith of n exponential terms. The decline of the plasma concentration may also be explained by the use of compartmental models as shown in Figure 5. However, such models have limited utility for two reasons. First, there are often multiple models to explain the observation. For example, elimination could occur from any one of the three compartments in Figure 5 or from any combination of two of the compartments or, alternatively, from all three compartments; there is usually no way to distinguish among them. Second, the body is a collection of many tissues, organs and organelles. Representing the body with only a few compartments does not make sense physiologically and one cannot give any meaningful interpretation to the rate constants obtained. The compartmental models are, however, useful to examine the general kinetic behavior expected as a result of distribution and elimination.

(iv) Intensity and duration of response

Among other factors, the intensity and duration of response after a single intravascular dose depend on the kinetics of a drug. This dependence is most readily established when the response is a fixed function of the plasma drug concentration. One such function is equation (8)[12] in which the intensity of the response, I, is related to the plasma drug concentration by

$$I = I_{max}\frac{C^{\gamma}}{C_{50}^{\gamma} + C^{\gamma}} \tag{8}$$

where I_{max} is the maximum response, C_{50} is the plasma concentration at which the response is one-half of the maximum and γ is a shape factor that determines the general steepness of the relationship between response and concentration.

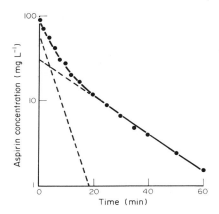

Figure 4 Following a 650 mg intravenous bolus dose, the decline in the plasma concentration of aspirin (solid line) can be characterized by the sum of two exponential terms (dotted lines) (reproduced from ref. 1 and the data in ref. 11)

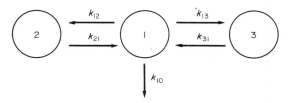

Figure 5 A compartmental model for the distribution of drug from compartment 1 to compartments 2 and 3. The k values represent the rate constants for the transfer of drug to and from these compartments. Elimination, with a rate constant equal to k_{10}, occurs from compartment 1

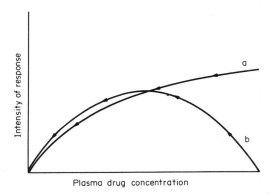

Figure 6 The intensity of the response to a drug is expected to increase with the plasma concentration: (a) after an intravenous bolus, the concentration decreases with time and, consequently, so does the intensity of the response; (b) a typical relationship between response and plasma concentration with time when there is a delay in the response with respect to the plasma drug concentration

If equation (8) holds at all times after an intravascular dose, one expects to see the response–concentration relationship shown by curve (a) in Figure 6. On the other hand, if there is a delay in the response, then a relationship like that of curve (b) is expected. There are many reasons, both pharmacokinetic and pharmacodynamic, for producing such a delay. Examples are given in Table 1.

When the response is a function of the concentration at all times (equation 8), the duration of response increases with the dose administered and with a decrease in clearance as shown in Figure 7 for a drug showing monoexponential kinetic behavior. The response reaches a defined lower limit when the plasma concentration drops to a corresponding value. Figure 8 shows duration of muscle paralysis as a function of the logarithm of the dose for succinylcholine (a), a drug that shows

Table 1   Mechanisms Producing a Delay in Drug Response after an Intravascular Bolus Dose

Mechanism	Example
Metabolite produces most of the response	Primidone → phenobarbital, antiepileptic therapy
Active site is in tissues	Digitalis glycosides. About 4 h is needed to develop maximal response even after an intravenous bolus dose of digoxin
Response measured is not the direct one	Oral anticoagulants. The direct effect is a decrease in the synthesis of clotting factors.
	A delay is observed in clotting, due to the time required for the clotting factors to be eliminated

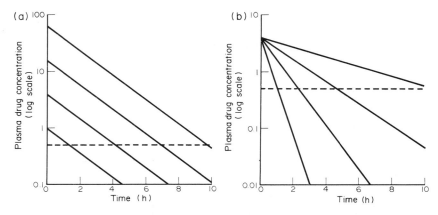

Figure 7   (a) Following a single intravenous bolus dose (one-compartment model), the time the concentration stays above a defined value (dotted line) increases with the dose administered. (b) With a fixed dose, the duration above the defined value also increases when the clearance (reflected by area and the time required for the decline to occur) is decreased

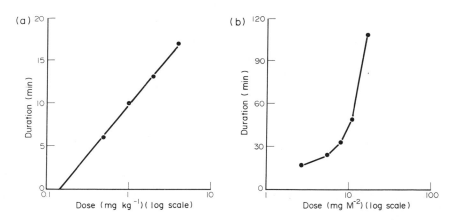

Figure 8   The duration of paralysis, as measured by the time to regain 50% of normal muscle twitch, increases linearly with the logarithm of the dose for succinylcholine (a), a one-compartment drug, but disproportionally with (+)-tubocurarine (b), a drug with multicompartmental characteristics (original data from ref. 15, with (a) and (b) being reproduced from refs. 13 and 14 respectively)

monoexponential decline in its plasma concentration, and for (+)-tubocurarine (b), a drug that shows multiexponential decay, after a single intravascular bolus dose. The disproportionate increase in the duration with the logarithm of the dose can be explained schematically by the time to decline to the minimum effective plasma concentration in the more rapid decline phase compared to that in the slower phase as shown in Figure 9.

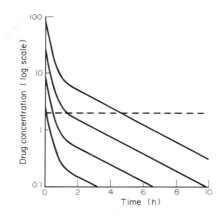

Figure 9 Schematic representation of a logarithmic plot of the plasma drug concentration of a drug with biexponential characteristics after different intravenous bolus doses. At a given concentration (dotted line), the effect wears off. At low doses the duration of response is seen to be primarily related to the decline within the rapid phase. For larger doses, the duration is more closely related to the decline in the terminal phase

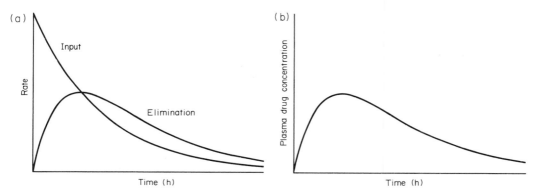

Figure 10 (a) After an extravascular dose, the rate of input declines with time when the process is first-order, because the amount remaining to be input decreases with time. The rate of elimination, however, increases as the plasma concentration increases. (b) When the rate of elimination equals the rate of input, the plasma concentration peaks. After this time, the rate of elimination is greater than the rate of input and so the plasma drug concentration declines

23.2.1.1.2 *Extravascular administration*

When a drug is administered by any route other than directly into the vasculature, input cannot be considered to be instantaneous. For extravascular routes of administration (for example, buccal, dermal, intramuscular, nasal, oral, rectal, subcutaneous and sublingual), time is required to transport drug to the blood which, in turn, delivers drug to the active site. This extra step adds another level of complexity to the concentration–time curve of drug in plasma. The more important kinetic considerations here are the rate and the extent (completeness) of this step.

(i) Absorption or input rate

Input is commonly modelled by first-order and zero-order processes, so called because the rates of the processes are proportional to the first and zero powers of the amount of drug at the site of input, respectively.

(a) First order. The kinetic behavior of a drug showing first-order input is shown in Figure 10. The rate of absorption declines as the amount at the absorption site decreases. The rate of elimination, paralleling the amount of drug in the body, increases until the rates of input and elimination are equal, at which time the plasma concentration peaks. Subsequently, the rate of elimination is greater than the rate of absorption and the plasma drug concentration declines.

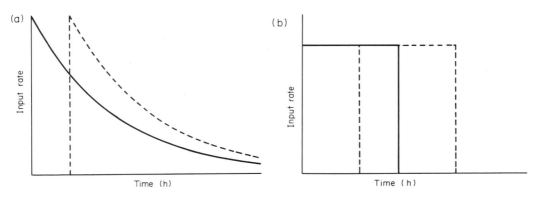

Figure 11 The input rate *versus* time profiles are quite different for (a) first-order and (b) zero-order processes. A *lag time* in the input produces rate–time profiles shown by the dashed lines

(b) Zero order. With zero-order input, the drug enters the systemic circulation at a constant rate until all of the drug that is going to be input has been. This model of drug input is simpler than that of first order and is often modeled for that reason.

(c) Lag time. After oral administration of a drug in a solid dosage formulation, particularly one with an acid-resistant coating to prevent gastric exposure, there is often a delay in starting drug input. Such a delay is shown in Figure 11 for both first- and zero-order input conditions.

(ii) Extent of input

The input of a drug from certain extravascular routes of administration, such as buccal, oral, rectal and sublingual, is often incomplete. Major reasons for this include: poor formulation of the drug product; competing reactions in the gastrointestinal tract (acid hydrolysis in stomach, degradation by bacterial flora or digestive enzymes); passage during input through the gut wall and liver; organs in which elimination occurs; and insufficient time at the absorption site (particularly true for the rectal route).

Bioavailability, F, is the fraction of a dose that reaches the systemic circulation. Its estimation is based on equation (9). Following extravascular administration the amount input is $F \cdot Dose_{ev}$ which equals the amount eliminated after a single dose ($CL \cdot AUC_{ev}$). Clearance can be obtained from an intravascular dose ($CL = Dose_{iv}/AUC_{iv}$). Thus

$$F = \frac{AUC_{ev}}{AUC_{iv}} \cdot \frac{Dose_{iv}}{Dose_{ev}} \tag{9}$$

Since the extravascular and intravascular doses are usually given on different occasions and the value of clearance varies from one person to another, and even within the same person, it is necessary to develop procedures by which these differences in clearance can be diminished or, on average, made the same. This is the goal of a bioavailability study protocol.

(iii) Mean input time

In certain cases, especially when the input process is slow compared to the elimination process, it can be useful to assess the mean input time, MIT. Similar to mean residence time, mean input time is the average time required to input a molecule into the body from the site of administration. The MIT cannot be measured directly, it is determined from the use of equation (4) after both extravascular administration and intravascular bolus injection. The time determined after extravascular administration is the sum of the mean times spent at the input site and in the body. The MIT is therefore obtained from

$$MIT = \frac{AUMC_{ev}}{AUC_{ev}} - \frac{AUMC_{iv}}{AUC_{iv}} \tag{10}$$

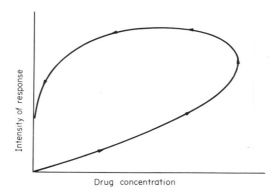

Figure 12 When there is a delay in response (see mechanisms in Table 1), the intensity of response is not related to concentration alone. Time, too, becomes important. At time zero there is not yet a plasma concentration and, of course, no response. With time (denoted by arrows) the concentration rises, peaks and falls. The peak effect occurs after the concentration peaks. The response continues even when there is little drug remaining in plasma. This lack of good correlation of response with concentration while the concentration is both increasing and decreasing is evidence of hysteresis in the relationship between response and concentration

(iv) Intensity and duration of response

As a result of prolonged input, the onset of response is always delayed after extravascular administration. The response is expected to peak at the same time as the plasma concentration and the response–concentration relationship should be the same both before and after the peak. The duration of response is difficult to quantify after extravascular administration because of the added input variable. If any one of the mechanisms that produces a delay in response is present (see the list in Table 1), the relationship between response and concentration then shows hysteresis. This kind of behavior is illustrated in Figure 12. The relationship between duration of response and dose is now even more difficult to predict.

23.2.1.2 Constant–rate Regimens

The therapeutic management of patients with drugs usually requires the maintenance of a drug concentration within a window, the *therapeutic window*, where the probability of therapeutic success is high. The upper bound is limited by toxicity and the lower bound by insufficient drug response.

The administration of drug at a constant rate should maintain a constant concentration once steady state, when rates in and out are equal, has been achieved. The questions now are: what determines the steady state and how can one adjust the input to maintain a plasma concentration within the therapeutic window?

23.2.1.2.1 Steady state

Letting $F \cdot R_0$ equal the constant rate of drug input and $CL \cdot C_{ss}$ equal the rate of drug elimination at steady state, it follows that

$$F \cdot R_0 = CL \cdot C_{ss} \tag{11}$$

Thus, the rate needed to maintain a given value of C_{ss} depends on the values of clearance and bioavailability. For an intravascular infusion, only clearance is pertinent since, by definition, $F = 1$ for this route.

23.2.1.2.2 Control of input

Intravascular infusions are painful, difficult to maintain for long periods, and inconvenient for both the patient and the person administering the infusion. To overcome these problems and yet to maintain constant concentrations chronically, a number of devices and delivery systems have been introduced and are under development. Table 2 lists examples of such controlled-rate systems.

Table 2 Representative Constant-rate Devices and Their Applications[1]

Type of therapeutic system	Drug	Rate specification	Application/comments
Transdermal	Nitroglycerin	1.5, 5 and 10, 15 mg over 24 h	Prophylaxis against attack of angina pectoris System aims to provide a constant plasma concentration of nitroglycerin Recommended application site is lateral chest wall
Transdermal	Clonidine	100, 200 and 300g day^{-1} for one week	Control of blood pressure System aims to provide constant plasma concentration of clonidine Recommended application site is upper part of body
Ocular	Pilocarpine	20 and 40 µg h^{-1} for one week	Control of elevated intraocular pressure System aims to provide a constant rate of input of pilocarpine into the eye
Oral	Phenylpropanolamine hydrochloride	25 mg immediate release and 3.4 mg h^{-1} for 16 h	Appetite suppressant System aims to provide a constant effective plasma concentration of phenylpropanolamine for 16 h
Uterine	Progesterone	65 µg day^{-1} for one year	Contraceptive System aims to provide a constant and effective uterine concentration of progesterone, an endogenous hormone

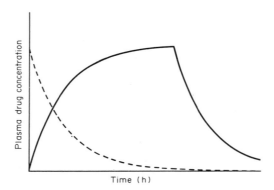

Figure 13 On administering a drug at a constant rate, the plasma concentration increases with time (solid line) until the rate of elimination matches the rate of input, achieving steady state. On discontinuing drug administration, the concentration falls. Note that if this fall-off curve is moved to the start of the infusion (dashed line) it becomes the complement (sum is steady-state concentration) of the infusion curve

23.2.1.2.3 *Time to steady state*

When a constant rate of drug input is started, the plasma concentration (or amount in the body) increases toward a steady-state value, as shown in Figure 13. The concentration falls when the input is discontinued. This curve is similar to that observed after a single intravascular bolus dose (equation 7), but has different coefficients as follows

$$C = \frac{F \cdot R_0}{CL} \sum_{i=1}^{n} a_i \cdot e^{-\lambda_i t} \tag{12}$$

where $\sum_{i=1}^{n} a_i = 1$. The concentration during the infusion is the complement of this curve; that is

$$C = \frac{F \cdot R_0}{CL} \left[1 - \sum_{i=1}^{n} a_i \cdot e^{-\lambda_i t} \right] \tag{13}$$

Thus, an index of the time to reach steady state (90% of steady-state value) is equal to the time for the concentration to drop to 10% of steady-state value after the end of an infusion to steady state.

23.2.1.3 Multiple-dose Regimens

Multiple discrete doses is, by far, the most common means of maintaining therapeutic plasma concentrations. The major questions here are: what determines the frequency of dosing (usually expressed by the dosing interval) and what is the size of the dose required to maintain therapy?

23.2.1.3.1 *Therapeutic regimens*

The two major determinants of therapeutic regimens are the width of the therapeutic window and the rapidity of drug elimination. If a single exponential term applies to the decline of drug after a bolus intravascular injection, the rapidity can be measured in units of half-life, the time for the concentration to drop in half. Table 3 shows selected examples of regimens required to maintain therapeutic concentrations. Multiple-dose regimens usually involve extravascular administration, but sometimes bolus or short-term intravenous infusions are used. The basic concepts of such regimens are most readily described in terms of multiple intravascular bolus doses.

(i) *Intravascular administration*

When doses are repetitively given, some drug remains at the time the next dose is given. This leads to accumulation of drug in the body as shown in Figure 14. When the dosing interval is large compared to the half-life, the fluctuation is large and little accumulation occurs, because little of the last dose remains at the end of a dosing interval. Conversely, when the dosing interval is small compared to the half-life, the accumulation is extensive. The amount in the body at steady state may be much greater than that of the single dose. An index of this accumulation (R_{ac}, maximum amount at steady state relative to the amount input from each dose) is

$$R_{ac} = \frac{1}{(1-e^{-k\tau})}. \tag{14}$$

where τ is the dosing interval and k (0.693/half-life) is the elimination rate constant of the one-compartment model.

When accumulation is extensive, serious consideration should be given to the use of a loading dose, D_L, if therapy is needed quickly. The loading dose required is the amount in the body achieved, at steady state, by repetitively administering a maintenance dose, D_M. Thus

$$D_L = \frac{D_M}{(1 - e^{-k\tau})} \tag{15}$$

(ii) *Extravascular administration*

The relationships above for intravascular bolus doses apply as well to multiple extravascular doses, with two additional considerations, the rate and extent of the input from each dose. These additional considerations are incorporated into Figure 15. First, extravascular administration decreases fluctuation because of prolonged input from each dose (curves b and c). Second, the average concentration is the same (curves a and b) for both routes of administration, but is decreased (curve c) if the bioavailability is less than one. The average concentration, C_{av}, during a dosing interval at steady state can be calculated from the following relationship

$$\frac{F \cdot Dose}{\tau} = CL \cdot C_{av} \tag{16}$$

This equation applies to both extravascular and intravascular routes of administration and to any combination of rate and extent of drug input as long as steady-state conditions prevail.

Table 3 Dosage Regimens for Continuous Maintenance of Therapeutic Concentrations[1]

Width of therapeutic window[a]	Half-life[b]	Ratio of initial dose to maintenance dose	Ratio of dosing interval to half-life	General comments	Drug examples
Medium to very wide	Very short (<20 min)	—	—	Candidate for constant-rate administration and/or short-term therapy	Nitroglycerin
	Short (20 min to 3 h)	1	3–6	To be given any less often than every 3 half-lives, drug must have a wide therapeutic window	Penicillin
	Intermediate (3 to 8 h)	1–2	1–3	Very common and desirable regimen	Tetracycline Sulfamethoxazole
	Long (8 to 24 h)	2	1	Once daily is practical. Occasionally given once weekly.	Chloroquine (suppression of malaria)
	Very long (>24 h)	>2	>1	Initial dose may need to be much greater than maintenance dose	
Narrow	Very short (<20 min)	—	—	Not a candidate except under very closely controlled infusion	Nitroprusside
	Short (20 min to 3 h)	—	—	Generally limited to infusion	Lidocaine
	Intermediate (3 to 8 h)	1–2	1	Requires 3–6 doses per day, but less frequently with a controlled-release formulation	Theophylline
	Long (8 to 24 h)	2–4	0.5–1		
	Very long (>24 h)	>2	>1	Requires careful control, since once toxicity is produced, drug level and toxicity decline slowly	Digitoxin

[a] Concentration range in which the probability of therapeutic success is high. [b] Descriptions of half-life are arbitrary.

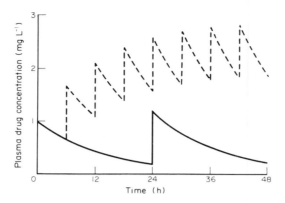

Figure 14 When a dose (producing an initial concentration of 1 mg L^{-1}) is repeatedly administered every 24 h (solid line), little accumulation occurs for a drug with a 10 h half-life and one-compartment distribution characteristics. However, on administering the same dose every 6 h (dashed line), the concentration continues to rise with subsequent doses because the fraction of the initial amount that is lost within a dosing is small. The accumulation continues until the amount lost within an interval is equal to the amount input, the dose here since bioavailability = 1

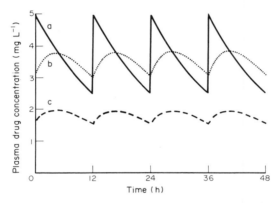

Figure 15 At steady state, the fluctuation in the plasma drug concentration for a given regimen is greater after intravascular administration (a) than after extravascular administration (b) because input continues with time during each dosing interval. When the bioavailability of the extravascular dose is less than one (c), the average concentration is reduced by this factor

23.2.1.3.2   Design of regimens

If one knows the values of bioavailability, clearance and volume of distribution, one can design a regimen to keep the plasma concentration within the therapeutic window. An example of this fundamental approach is shown in Table 4.

One can also design a regimen without knowledge of the values of a drug's pharmacokinetic parameters. If one determines the plasma concentration with time following a single dose, the concentrations after multiple doses can be predicted by simply adding up the curves that would result from repetitive dosing on any given regimen. Figure 16 shows such an approach. The predicted concentration–time course is the sum of the individual curves. Note that the concentrations from the first two doses contribute negligibly at steady state (approximated in last interval). Thus, the concentration remaining from these doses, and indeed from any doses further back in time had they been given, would not have to be added to determine the steady-state values.

23.2.1.3.3   Control of input

For drugs with narrow therapeutic windows and with half-lives less than about 6 h, a strong argument can be made for the use of controlled or prolonged release formulations and devices. Such controlled-input decreases the fluctuation and permits the use of regimens that are more convenient to the patient (or the nurse).

Table 4 Design of Multiple-dose Regimens[a]

Step	Description	Relationship
1	Establish maximum (C_{max}) and minimum (C_{min}) plasma concentrations desired	
2	Obtain population estimates of F, V and CL	
3	Calculate the maximum dosing interval τ_{max} to keep within the limits. Assume intravenous administration	$\tau_{max} = 1.44 \cdot t_{\frac{1}{2}} \cdot \ln(C_{max}/C_{min})$
4	Calculate the maximum maintenance dose, $D_{M,max}$ that can be given	$D_{M,max} = \dfrac{V}{F}(C_{max} - C_{min})$
5	Calculate the average dosing rate	$\text{Dosing rate} = \dfrac{D_{M,max}}{\tau_{max}}$
6	Adjust D_M to doses available and τ to a convenient frequency of administration, keeping approximately the same average dosing rate	$\dfrac{D_M}{\tau} = \dfrac{D_{M,max}}{\tau_{max}}$
7	If τ chosen is much less than one half-life and rapid initiation of therapy is desired, calculate a loading dose	$D_1 = \dfrac{V}{F} \cdot C_{ss}$
8	Appropriately adjust for other factors, such as slow absorption after extravascular administration, slow distribution to tissues, side effects on administering the calculated loading dose, and so on	

[a] To keep the plasma drug concentration within prescribed limits.

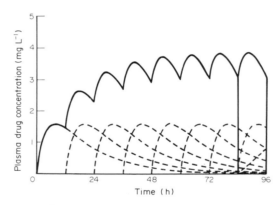

Figure 16 The plasma drug concentration expected (solid line) is the sum of the concentrations produced by each of the doses given (extravascular administration of a fixed dose at a fixed interval is shown). Note that the first two doses contribute negligibly to the expected concentration at steady state (last dosing interval), because so little of these doses remains at this time. Thus, one only needs to add the concentrations remaining from the last six doses

The value of such products is exemplified by the administration of theophylline in the usual rapid-release oral tablet to a child with severe asthma as shown under steady-state conditions in Figure 17. The child is in a hospital and is given the drug at 9 a.m., 1 p.m., 5 p.m. and 9 p.m., a common four-times-a-day regimen in an institutional setting. It is apparent that such a regimen in this child, whose kinetics are typical for his age, may result in potentially toxic levels at bedtime and subtherapeutic levels in the morning. The use of a controlled-release product that delivers the drug at a constant rate over 12 h can greatly reduce the fluctuation and, indeed, may permit dosing only twice a day as shown. Controlled release is a topic covered in Part 25 in this volume.

This section has only briefly reviewed the basic principles of drug administration. Further explanations and development of the concepts are given in refs. 1–15.

23.2.2   KINETIC ISSUES IN DRUG METABOLISM

This section focuses on kinetic issues in drug metabolism. It is divided into two principal sections, Metabolite Kinetics and Nonlinear Metabolism. For further information, the reader is referred to several excellent reviews in the area.[16–20]

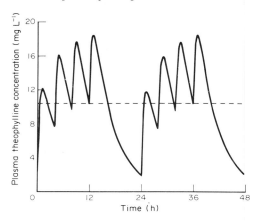

Figure 17 Even at steady state, the plasma theophylline concentration (solid line) fluctuates extensively in a 4-year-old child with typical pharmacokinetic parameter values when given 100 mg doses at '9–1–5–9', a regimen commonly used in both institutional and home-care settings. At 9 AM, time zero in the graph, the concentration is well below the therapeutic window, 10–20 mg L^{-1}, and minimal control of the child's asthma is expected. An increase in dose to overcome this problem would result in potentially toxic concentrations in the evening. One solution to the problem is to administer a controlled-release dosage form. Simulated here (dashed line) is the result of administering a 200 mg 12 h constant-rate controlled-release product every 12 h (reproduced from ref. 1 by permission of Lea and Febiger)

23.2.2.1 Metabolite Kinetics

Metabolites are of interest because many of them exhibit pharmacologic and toxic effects. Their kinetics are of interest because the onset, duration and time course of response to metabolites impacts on drug administration. This impact varies from negligible, when the metabolites are inactive, to being of primary concern, when most of the therapeutic response or toxicity resides with the metabolites.

Administration of an intravenous bolus dose of a metabolite is the best way to determine its kinetics. The principles for doing so are the same as those for drugs (see Chapter 23.15). Usually, however, administration of a metabolite has not been approved by the Food and Drug Administration (USA) or similar agency; kinetic information about an active metabolite must therefore often be obtained after drug administration. This context, also the therapeutic one, is emphasized throughout this section.

23.2.2.1.1 Prodrugs, drugs and active metabolites

Sometimes the compound administered produces neither a pharmacologic nor a toxic effect. It then must be converted to one, or more, active metabolite(s) to exhibit its effects. When the administered compound has virtually no effect, it is called a *prodrug*. This term has especially been applied when the prodrug was developed to improve the delivery, increase the stability or diminish the undesirable properties (*e.g.* bad taste) of the active principle.[20] Thus, two similar sets of terms, prodrug/drug and drug/active metabolite, are used. Subsequently in this section, *active metabolite* refers to any active chemical species derived from the compound administered. The administered compound may also produce a pharmacologic or toxic response.

23.2.2.1.2 Rate-limiting steps

Following a single intravenous dose of drug, the plasma concentration of a metabolite rises to a maximum value and then declines, as shown for two drugs in Figure 18. The steepness of the decline depends on the relative speed of the processes governing metabolite formation and elimination. To appreciate this dependence, consider Scheme 1, where D is the drug administered, M is the active metabolite and k_f, k_{other} and $k(m)$ are the first-order rate constants for the formation of the metabolite, the elimination of drug by routes other than formation of the active metabolite, and elimination of the metabolite, respectively.

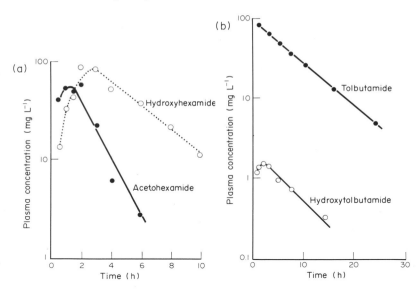

Figure 18 The time courses of (a) hydroxyhexamide and (b) hydroxytolbutamide, relative to their respective precursor drugs, acetohexamide[22] and tolbutamide,[23] are quite different. The decline of hydroxyhexamide is slower than that of aceto-hexamide after a 1 g oral dose of the drug, while the decline of hydroxytolbutamide is parallel to that of its precuesor after a 1 g intravenous dose of tolbutamide. For acetohexamide, the elimination of the metabolite is slower than that of the drug as measured by the terminal half-life or rate constant. The opposite apparently pertains to tolbutamide; because it is impossible to eliminate the metabolite before the drug, the two decline in parallel (the original data are from refs. 21 and 22, respectively, whilst the figures are reproduced from ref. 23 by permission of Lea and Febiger)

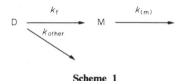

Scheme 1

When the metabolite elimination rate constant is much smaller than the formation rate constant, the drug disappears long before the metabolite does, as shown for acetohexamide in Figure 18(a). The disappearance of the metabolite, hydroxyhexamide, is then said to be rate limited by its elimination, or *elimination rate limited*. On the other hand, when the value of the metabolite elimination rate constant is much larger than the formation rate constant, k_f (and, more specifically, larger than the elimination rate constant of drug, $k_f + k_{other}$), the decline of the metabolite becomes *formation rate limited*. Under these circumstances, any metabolite formed is almost immediately eliminated, keeping the amount of the metabolite present low compared to that of the drug. The rates of formation ($k_f \cdot A$) and elimination [$k(m) \cdot A(m)$] are then virtually equal. Consequently

$$A(m) = \frac{k_f A}{k(m)} \qquad (17)$$

Because the value of k_f is much less than that of $k(m)$, the amount of metabolite [$A(m)$] in the body must be small compared to that of drug (A). Furthermore, $A(m)$ and A must decline in proportion, as shown for hydroxytolbutamide in Figure 18(b).

Thus, when drug and metabolite concentrations decline in parallel on a semilogarithmic plot after a single dose, the likely explanation is a rate limitation in metabolite formation. After a single oral dose it is, of course, possible that the absorption process of the drug itself is rate limiting. Persistence of a metabolite long after its precursor, the drug, has disappeared indicates a rate limitation in metabolite elimination and not in either drug absorption or metabolite formation. Such a limitation in the elimination of an active metabolite can explain the persistence of a response long after the drug itself has disappeared.

When formation is rate limiting, the metabolite concentration [$C(m)$] may actually be higher than that of drug (C) during the terminal decline phase after a single dose. This observation can occur when the volume of distribution of the metabolite [$V(m)$] is much smaller than that of the drug (V).

The principle of equation (17), that $A(m) \ll A$ even though $C(m) > C$, can be true if the ratio $V(m)/V$ is smaller that the ratio $A(m)/A$. This conclusion can be seen from the relationship

$$\frac{A(m)}{A} = \frac{V(m)}{V} \cdot \frac{C(m)}{C} \qquad (18)$$

23.2.2.1.3 *First-pass metabolism and its consequences*

When a drug is given orally, it must pass through the gut wall and liver to reach the general circulation. As these organs are often responsible for drug elimination, the amount reaching the general circulation may be less than the dose administered. This loss of drug during the input process is called the *first-pass effect*.[24] It applies to any route of administration in which elimination occurs between the site of administration and the site of measurement (usually a peripheral vein). Examples are percutaneous administration with metabolism in the skin, and rectal administration with metabolism in the rectum and the liver. The lungs are not usually included as first-pass organs because an intravenous dose is used as a reference. Furthermore, with the exception of intra-arterial administration into the organ from which intravenous samples are obtained, drug from all routes of administration must first pass through the lungs to reach the site of measurement.

First-pass metabolism has a number of therapeutic consequences that depend on whether most of the activity resides with the administered compound or an active metabolite. Examples of drugs showing extensive first-pass metabolism on oral administration are listed in Table 5.

(i) Drug active

When most of the activity resides with the drug, the maximum response after administration of a single oral dose is less than that after an intravenous dose, because the bioavailability (ratio of AUC values of the drug after single oral and intravenous doses with correction for differences in dose size) is decreased. The consequence here is the need for a larger oral dose to achieve the same intensity of response during chronic administration. Drugs with low oral bioavailabilities, because of first-pass metabolism, can show large route-dependent differences in the doses required to achieve a given level of response. For example, an oral dose of 250 mg of phenylephrine is required to achieve the same pressor response as that observed after a 0.8 mg intravenous dose.[25] Another consequence is that coadministration of an inhibitor of metabolism can, in this condition, greatly increase the input of active drug. An example is that of the 400 to 500% increase in the bioavailability of 6-mercaptopurine observed after pretreatment with allopurinol.[26] Conversely, coadministration of an enzyme inducer can decrease bioavailability as observed with labetolol when glutethimide is given.[27]

(ii) Metabolite active

When most of the activity resides with the metabolite, the therapeutic consequences can be quite different. First of all, the amount of metabolite formed is the same for both intravascular and extravascular routes of administration, if elimination occurs exclusively in the first-pass organ(s). The time courses of the metabolite following the two routes, however, may be very different. As shown in Figure 19, the concentration of the metabolite early in time after oral administration of a drug can greatly exceed that expected after an intravenous bolus of the same dose size. This may give rise, for example, to an initial effect greater after oral than after intravenous administration.

When the disposition of a metabolite is formation rate limited, another curious kinetic phenomenon may occur as shown for propranolol in Figure 20. The active metabolite, 4-hydroxy-

Table 5 Representative Drugs with High First-pass Hepatic Extraction

Alprenolol	Isoproterenol	Morphine
Arabinosylcytosine	Lidocaine	Nitroglycerin
Desipramine	Lorcainide	Pentazocine
Doxepin	Meperidine	Phenacetin
5-Fluorouracil	Methoxysalen	Propoxyphene
Hydralazine	Metoprolol	Propranolol

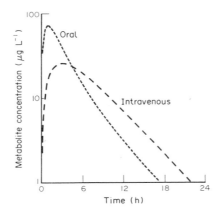

Figure 19 First-pass metabolism impacts on the onset and maximum intensity of response when a metabolite is responsible for the activity. The plasma concentration of an active metabolite can be much greater soon after oral administration compared to that after intravenous administration, because of the large amount of metabolite produced during the absorption of the drug. This effect is more pronounced when extraction in first-pass organ is high and absorption is fast relative to elimination. This occurs when the volumes of distribution of the drug and metabolite are large, a condition in which elimination is slowed. The areas under the curve (linear plot) are the same, because the same amounts of metabolite are formed from both routes of administration

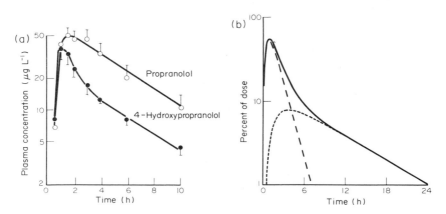

Figure 20 (a) The time course of one of the active metabolites of propranolol, 4-hydroxypropranolol, after an 80 mg oral dose of the drug shows unusual behavior. The metabolite concentration appears to decline more rapidly after reaching its peak than does the drug concentration. (b) The explanation of this behavior is simulated by a model in which a large fraction of the oral dose is converted to the metabolite during drug input. The metabolite so formed is rapidly eliminated (dashed line). Elimination of the metabolite formed from drug already in the body (dotted line) is formation rate limited. The sum of the two metabolite concentrations (solid line) is the overall observation (the data in (a) are from ref. 13, whilst both (a) and (b) are reproduced from ref. 23 by permission of Lea and Febiger)

propranolol, rapidly builds up and falls off in an unusual manner (Figure 20a). The kinetic behavior can be explained by simulating the simultaneous intravenous administration of both propranolol and its metabolite (Figure 20b). The metabolite concentration falls quickly as its elimination is associated with a short half-life. The decline of metabolite formed from the drug, however, is formation rate limited. This gives a biphasic appearance to the decline curve. Both propranolol and 4-hydroxypropranolol are pharmacologically active. This kinetic phenomenon explains why, at the same plasma propranolol concentration, β-blockade is observed at early times to be greater after oral rather than after intravenous administration.[29]

23.2.2.1.4 *Factors determining exposure to metabolites*

The exposure of the body to active metabolites can be considered in two major contexts: after a single dose and after chronic administration of multiple doses.

(i) Single dose of drug

Following a single dose, the area under the plasma metabolite concentration–time curve is a measure of the exposure of the body to the metabolite. The metabolite exposure, relative to that of the drug administered, can be assessed by examining the ratio of the areas of metabolite, $AUC(m)$, and drug, AUC. This ratio is equal to the ratio of the clearances of metabolite formation, CL_f, and elimination, $CL(m)$. This conclusion follows from analysis of Scheme 2.

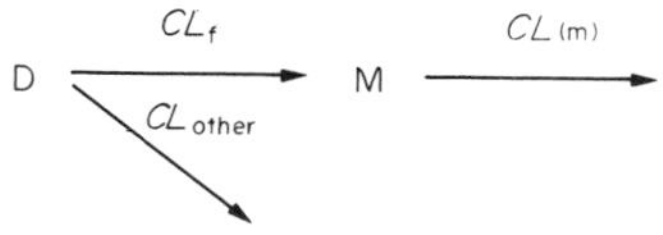

$$D \xrightarrow[CL_{other}]{CL_f} M \xrightarrow{CL_{(m)}}$$

Scheme 2

The rate of formation is $CL_f \cdot C$ and the rate of metabolite elimination is $CL(m) \cdot C(m)$. On integrating both rates from time zero to infinity and realizing that the amount of metabolite formed is equal to the amount eliminated, the following relationship is derived

$$\frac{AUC(m)}{AUC} = \frac{CL_f}{CL(m)} \tag{19}$$

Since the formation of metabolite accounts for only a fraction, fm, of total drug elimination, the value of CL_f is $fm \cdot CL$. Therefore

$$\frac{AUC(m)}{AUC} = \frac{fm \cdot CL}{CL(m)} \tag{20}$$

Thus, when the ratio of areas is determined to be greater than 1 after an intravenous dose, one can conclude that the value of CL is greater than $CL(m)$. This conclusion may also be valid when the ratio is much less than 1, if fm is smaller than the area ratio. The simple relationships of equations (19) and (20) are useful tools to analyze changes in the formation or elimination of active metabolites in a variety of conditions and disease states.

When drug is given orally or by any-route in which extensive first-pass metabolism occurs, the ratio of areas also depends on the formation of the metabolite while the drug is being absorbed. If the only cause of decreased bioavailability is first-pass metabolism and elimination only occurs in the first-pass organs, the ratio of areas after an oral dose can be shown to be

$$\frac{AUC(m)}{AUC} = \frac{CL}{F \cdot CL(m)} (fm \cdot F + 1 - F) \tag{21}$$

where F is the bioavailability of the drug.

If the metabolite extensively undergoes further metabolism during passage through the first-pass organ(s), the area ratio may also depend on the fraction of the metabolite that escapes this sequential metabolism. The ratio becomes a function of even more parameters, when elimination occurs in nonfirst-pass organs[16,30] and when nonlinearities exist.

(ii) Infusion

The exposure to an active metabolite during a short-term constant-rate infusion of drug depends on infusion rate, R, duration of infusion, t, fraction of drug converted to metabolite, fm, and clearance of metabolite, $CL(m)$.

$$AUC(m) = \frac{fm \cdot R \cdot \tau}{CL(m)} \tag{22}$$

The exposure relative to that of drug ($AUC = Dose/CL$) is

$$\frac{AUC(m)}{AUC} = \frac{fm \cdot CL}{CL(m)} \tag{23}$$

a relationship similar to that observed following an intravenous bolus dose of drug (equation 20).

Under chronic conditions, the steady-state concentration can be a useful index and correlate of the potential activity or toxicity of an active metabolite. The steady-state concentration of drug, C_{ss}, is equal to R/CL. Similarly, the metabolite concentration $C(m)_{ss}$ is equal to its rate of formation, $fm \cdot R$, divided by its clearance, $CL(m)$. Thus

$$\frac{C(m)_{ss}}{C_{ss}} = \frac{fm \cdot CL}{CL(m)} \tag{24}$$

The rate at which the metabolite concentration approaches steady state during an infusion of a drug depends on whether the formation or the elimination is rate limiting as shown in Figure 21. As with the infusion of drug, the metabolite concentration approaches steady state (curve b) at a rate that relates to the half-life of elimination of the metabolite when the drug is eliminated rapidly compared to the metabolite (see Figure 18a). Conversely, when the drug is slowly eliminated relative to the metabolite (see Figure 18b), the approach to steady state is virtually the same as that of the drug.

(iii) Multiple oral doses

The kinetic principles of metabolite behavior following intravenous drug infusion apply as well to multiple oral doses of drug with two exceptions. First, the first-pass contribution must be considered. Thus, the ratio of the average steady-state concentrations of metabolite, $C(m)_{av}$, and drug, C_{av}, are then, by analogy to equation (24) and the derivation of equation (21), as follows

$$\frac{C(m)_{av}}{C_{av}} = \frac{fm \cdot CL}{CL(m)} \tag{25}$$

Second, the drug and the metabolite concentrations may undergo extensive fluctuation within each dosing interval. The fluctuation of the drug is determined by the relative rates of drug absorption, distribution and metabolism, as well as the dosing interval. The fluctuation of the metabolite concentration tends to be less than that of its precursor. Figure 22 illustrates the metabolite concentration fluctuation expected for a given drug concentration–time curve after multiple oral doses. Note that the relative fluctuation within a dosing interval tends to be less when the rate-limiting step is metabolite elimination (b) rather than metabolite formation (a). The previously developed rule for the time required for metabolite concentration to reach steady state on drug infusion is, however, still the same; the half-life of the slower process (metabolite or drug elimination) governs the time.

(iv) Interconversion

The exposure to a metabolite is, of course, increased when its clearance, $CL(m)$, is decreased. Such decreases occur in the presence of hepatic, renal and cardiovascular diseases. Although this topic is

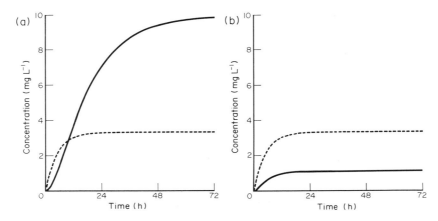

Figure 21 The time required for accumulation of a metabolite during infusion of a drug depends on whether the elimination or formation of a metabolite is rate limiting. When metabolite elimination is rate limiting (a), the drug concentration (dashed line) reaches its steady state long before that of the metabolite (solid line) does because the metabolite accumulation then primarily depends on the half-life of the metabolite elimination process. When formation is rate limiting (b), the metabolite and drug approach their respective steady-state concentrations with approximately the same half-life, namely, that of the drug

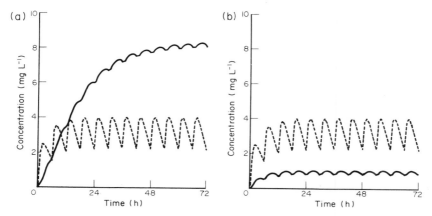

Figure 22 When administered twice each drug half-life, plasma concentrations of metabolite (solid line) and drug (dashed line) after successive oral doses show fluctuation but the general tendency is the same as that for constant-rate infusion. That is, if metabolite elimination is rate limiting (a; half-life of metabolite elimination is 3 times that of the drug), the time required for accumulation of the metabolite depends on the half-life of its elimination. Similarly, if the formation is rate limiting (b; half-life of metabolite is one-third that of the drug), the time to achieve steady state depends on the drug's half-life. Note that the relative fluctuation within each dosing interval is less when elimination rather than formation is rate limiting. The half-life of absorption of drug is 0.25 times the half-life of the drug

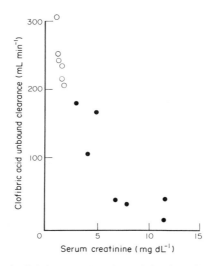

Figure 23 Although clofibric acid is only slightly excreted unchanged (fraction of intravenous dose recovered in urine is less than 0.1 in patients with normal renal function, ○), the clearance of the compound (based on unbound drug in plasma) decreases (●) as the renal function decreases (the abscissa, serum creatinine, is inversely related to renal function). This unexpected observation is explained by an interconversion between clofibric acid and its glucuronide conjugate. In renal disease there is more clofibric acid under steady-state conditions because there is a much higher concentration of the glucuronide. Thus, the less the renal function, the less the apparent clearance of the drug (data from ref. 31, whilst the figure is reproduced from ref. 23 by permission of Lea and Febiger)

beyond the scope of this chapter, the presence of interconversion between a drug and a metabolite warrants special consideration.

Usually the elimination of a drug that is not excreted unchanged in the kidney is not expected to be decreased in the presence of renal disease, but to be affected in hepatic disease, and the converse. Figure 23 illustrates one mechanism, interconversion, for which this principle is violated. In this example, the ethyl ester of clofibric acid (a prodrug) is converted to clofibric acid (drug) which, in turn, is conjugated with glucuronic acid (inactive metabolite). The conjugate is excreted in the urine and is partially hydrolyzed back to clofibric acid. Although clofibric acid is essentially not excreted unchanged, its disposition is affected dramatically by the presence of renal disease because of the interconversion. Indeed, when renal function approaches zero there is virtually no place for clofibric acid to go. Its clearance (based on drug unbound to plasma proteins in Figure 23) is then greatly reduced and its half-life lengthened (not shown).

23.2.2.2 Nonlinear Metabolism

Metabolism occurs by enzymatically mediated processes. As such, one expects metabolism to show capacity limitations, that is, for the rate of metabolism to approach a maximum value, V_m. The Michaelis–Menten enzyme kinetics model is often applied to data obtained *in vivo* when such nonlinearities (in which rate is not directly proportional to concentration) are observed. This model states that the rate of metabolism is related to the concentration at the metabolic site. The plasma drug concentration, assumed to be directly proportional to the concentration at the metabolic site, is used because measurement of the value at the site is not possible. The model is

$$\text{Rate of metabolism} = \frac{V_m \cdot C}{K_m + C} \tag{26}$$

where K_m is a constant equal to the concentration at which the rate is one-half the maximum value.

23.2.2.2.1 *Characteristic behavior*

Either the formation or the elimination of a metabolite may show nonlinear behavior. The consequences, shown subsequently for capacity-limited metabolism of the Michaelis–Menten type, can be quite different.

(i) Concentrations at steady state

To demonstrate the differences in capacity limitations in formation and elimination, consider Scheme 3 where the single arrow indicates a first-order (linear) process and the double arrow indicates a capacity-limited process. The first condition is a linear model in which the steady-state metabolite concentration is directly proportional to the rate of drug administration as shown in curve a of Figure 24. In the second condition, the formation of the metabolite is capacity limited, it therefore shows a limiting concentration as the rate of administration is increased (curve b). An example of this behavior is that of 4-hydroxyphenytoin when phenytoin is administered. The third condition is one in which the elimination of the metabolite is capacity limited. It could be represented by the administration of a prodrug of phenytoin. The steady-state phenytoin (metabolite) concentration would then increase disproportionately as the rate of prodrug administration is increased.

The ratio of the steady-state concentrations of metabolite and administered compound is shown as a function of the rate of administration in Figure 25. The nonlinearities in the system are readily identified from the observed change in this ratio. If capacity-limited metabolism is known to be the cause of the nonlinearity, the behavior permits one to identify whether formation or elimination of the metabolite is the likely cause of the observation.

(ii) Saturable first-pass metabolism

During the first pass through the liver, early plasma drug concentrations can be much greater than those occurring after intravenous administration of the same dose. This difference in concentration may produce nonlinear metabolism during the absorption phase, but not thereafter. As a consequence of such behavior, the bioavailability of a drug that is highly extracted in the liver increases with oral dose (or absorption rate). Drugs showing this behavior include: alprenolol,[39] 5-fluorouracil,[40] hydralazine,[41] lorcainide,[42] metoprolol,[43] phenacetin[44] and salicylamide.[45]

Even if the first-pass organs are the only sites of metabolism, the fraction of a specific metabolite produced during the first-pass may change. Consequently, the ratio of the areas under the concentration–time curves after a single dose or the ratio of steady-state concentrations of metabolite and drug may be influenced by both the change in bioavailability as well as a change in the fraction converted to a specific metabolite.

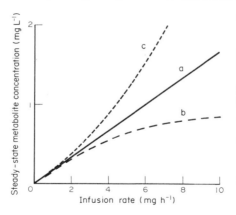

Figure 24 The steady-state concentration of a metabolite increases in direct proportion to the infusion rate of drug when drug and metabolite disposition are linear (first-order) as indicated in Scheme 3(a) and by line a above. When the formation of the metabolite is saturable (Michaelis–Menten in character) the metabolite concentration approaches a limiting value, determined by the maximum rate of its formation and its clearance as the rate of drug infusion increases (Scheme 3b, line b above). In contrast, when elimination of the metabolite is partially saturable (Scheme 3c, line c above), the steady-state concentration disproportionately increases with drug infusion rate

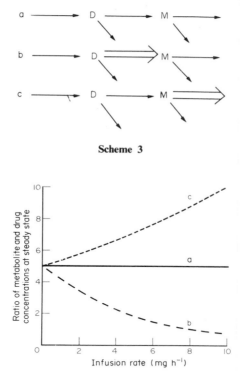

Scheme 3

Figure 25 The ratio of metabolite and drug concentrations at steady state is a convenient measure of metabolic nonlinearities. In linear kinetics (Scheme 3a, line a), the ratio is the same at all rates of drug infusion. When formation of the metabolite is saturable (Scheme 3b, line b), the ratio increases disproportionately. In contrast, when metabolite elimination exhibits saturable characteristics (Scheme 3c, line c), the ratio decreases at high drug infusion rates

23.2.2.2.2 *Other sources*

Table 6 lists selected sources, together with drug examples, of nonlinear kinetic behavior. For all these drugs, except acetaminophen, the consequences of the nonlinearities are apparent at therapeutic concentrations. Needless to say, nonlinear behavior becomes more common in drug overdose. In fact, here it may be the rule, rather than the exception that applies to the therapeutic range.

Table 6   Selected Causes of Nonlinear Metabolism

	Cause	Examples	Refs.
1	Capacity limitation	Phenytoin	32
2	Cofactor limitation	Salicylamide	33
3	Enzyme induction	Carbamazepine	34
4	Hepatotoxicity	Acetaminophen (in high doses)	35
5	Saturable plasma protein binding	Prednisolone	36
6	End-product inhibition	Lidocaine	37
7	Decreased hepatic blood flow	Propranolol	38

Many of these causes also exhibit *time-dependent* behavior, when the drugs are given repeatedly. For example, the cofactor limitation for salicylamide is exacerbated when the drug is repetitively administered. The increased synthesis of metabolizing enzyme after carbamazepine becomes apparent only after the enzyme has had time to accumulate. The end-product inhibition of lidocaine (metabolite inhibits its own formation) increases with time as the metabolite accumulates in the body.

In conclusion, considerable information can be obtained from measuring both drug and metabolite after drug administration. This information is useful for both analyzing the potential contribution of an active metabolite to the total activity and for predicting the therapeutic consequences of alterations in its formation or elimination.

23.2.3   REFERENCES

1. M. Rowland and T. N. Tozer, 'Clinical Pharmacokinetics: Concepts and Applications', 2nd edn., Lea and Febiger, Philadelphia, 1989.
2. L. Z. Benet, N. Massoud and J. G. Gambertoglio, 'Pharmacokinetic Basis for Drug Treatment', Raven Press, New York, 1984.
3. W. E. Evans, J. J. Schentag and W. J. Jusko, 'Applied Pharmacokinetics', 2nd edn., Applied Therapeutics, San Francisco, 1986.
4. M. Gibaldi and D. Perrier, 'Pharmacokinetics', 2nd edn., Dekker, New York, 1982.
5. M. Rowland and G. Tucker (eds.), 'Pharmacokinetics: Theory and Methodology', Pergamon Press, Oxford, 1986.
6. L. S. Goodman and A. Gilman, 'The Pharmacological Basis of Therapeutics', 7th edn., Macmillan, New York, 1985.
7. K. Heilmann, 'Therapeutic Systems. Rate-controlled Drug Delivery: Concepts and Development', 2nd revised edn., Thieme Verlag, Stuttgart, Thieme-Stratton, New York, 1983.
8. F. Theuwes, 'Drug Delivery Systems', *Pharmacol. Ther.*, 1981, **13**, 149.
9. M. E. Winter, 'Basic Clinical Pharmacokinetics', 2nd edn., Applied Therapeutics, Spokane, 1988.
10. S. Riegelman and P. Collier, *J. Pharmacokinet. Biopharm.*, 1980, **8**, 509.
11. M. Rowland and S. Riegelman, *J. Pharm. Sci.*, 1968, **57**, 1313.
12. N. H. G. Holford and L. B. Sheiner, *Clin. Pharmacokinet.*, 1981, **6**, 429.
13. G. Gevy, *J. Pharm. Sci.*, 1967, **56**, 1687.
14. M. Gibaldi, G. Levy and W. Hayton, *Anesthesiology*, 1972, **36**, 213.
15. L. F. Watts and J. B. Dillon, *Anesthesiology*, 1967, **28**, 371.
16. K. S. Pang, *J. Pharmacokinet. Biopharm.*, 1985, **13**, 633.
17. J. B. Houston, in 'Pharmacokinetics: Theory and Methodology', ed. M. Rowland and G. Tucker, Pergamon Press, Oxford, 1986, chap. 7.
18. S. Garattini, *Clin. Pharmacokinet.*, 1985, **10**, 216.
19. P. Jenner and B. Testa (eds.), 'Concepts in Drug Metabolism', Parts A and B, Dekker, New York, 1980.
20. T. Higuichi, in 'Bioreversible Carriers in Drug Design: Theory and Application', ed. E. B. Roche, Pergamon Press, New York, 1987.
21. J. A. Galloway, R. E. Mahon, H. W. Culp and E. C. Young, *Diabetes*, 1967, **16**, 118.
22. S. B. Matin and M. Rowland, *Anal. Letters*, 1973, **6**, 865.
23. M. Rowland and T. N. Tozer, 'Clinical Pharmacokinetics: Concepts and Applications', 2nd edn., Lea and Febiger, Philadelphia, 1989, chap. 21.
24. S. M. Pond and T. N. Tozer, *Clin. Pharmacokinet.*, 1984, **9**, 1.
25. L. S. Goodman and A. Gilman, 'The Pharmacological Basis of Therapeutics', 6th edn., Macmillan, New York, 1980, p. 165.
26. S. Zimm, J. M. Collins, J. D. O'Neill, B. A. Chabner and D. G. Poplack, *Clin. Pharmacol. Ther.*, 1983, **34**, 810.
27. T. K. Daneshmend and C. J. C. Roberts, *Br. J. Clin. Pharmacol.*, 1984, **18**, 393.
28. T. Walle, E. C. Conradi, K. Walle, T. C. Fagan and T. E. Gaffney, *Clin. Pharmacol. Ther.*, 1980, **27**, 22.
29. C. R. Cleaveland and D. G. Shand, *Clin. Pharmacol. Ther.*, 1972, **13**, 181.
30. P. J. M. Klippert and J. Noordhock, *Drug Metab. Dispos.*, 1985, **13**, 97.
31. P. Gugler, J. W. Kurten, C. J. Jensen, U. Klehr and J. Hartlapp, *Eur. J. Clin. Pharmacol.*, 1979, **15**, 341.

32. M. E. Winter and T. N. Tozer, 'Applied Pharmacokinetics: Principles of Therapeutic Drug Monitoring', 2nd edn., ed. W. E. Evans, J. J. Schentag and W. J. Jusko, Applied Therapeutics, Spokane, 1986, chap. 16.

33. J. A. Waschek, R. M. Fielding, S. M. Pond, G. M. Rubin, D. J. Effeney and T. N. Tozer, *J. Pharmacol. Exp. Ther.*, 1985, **234**, 431.

34. R. H. Levy, W. H. Pitlick, A.-S. Trouplin and J. R. Green, in 'The Effect of Disease States on Drug Pharmacokinetics', ed. L. Z. Benet, American Pharmaceutical Association, Washington, 1976, pp. 87–95.

35. L. F. Prescott, N. Wright, P. Roscoe and S. S. Brown, *Lancet*, 1971, **1**, 219.

36. T. N. Tozer, J. G. Gambertoglio, D. E. Furst, D. S. Avery and N. H. G. Holford, *J. Pharm. Sci.*, 1983, **72**, 1442.

37. J. A. Pieper and J. H. Rodman, in 'Applied Pharmacokinetics: Principles of Therapeutic Drug Monitorings', ed. W. E. Evans, J. J. Schentag and W. J. Jusko, Applied Therapeutics, Spokane, 1986, p. 662.

38. L. S. Olanoff, T. Walle, T. D. Cowart, U. K. Walle, M. J. Oexmann and E. C. Conradi, *Clin. Pharmacol. Ther.*, 1986, **40**, 404.

39. G. Alvan, M. Lind, B. Mellstrom and C. von Bahr, *J. Pharmacokinet. Biopharm.*, 1977, **5**, 205.

40. N. Christophidis, F. J. E. Vajda, I. Lucas, O. Drummer and W. J. Moon, *Clin. Pharmacokinet.*, 1978, **3**, 330.

41. T. Talseth, *Eur. J. Clin. Pharmacol.*, 1976, **10**, 395.

42. E. Jahnchen, H. Bechtold, W. Kasper, F. Kersting, H. Just, J. Heykants and T. Meinertz, *Clin. Pharmacol. Ther.*, 1979, **26**, 187.

43. C.-G. Regardh and G. Johnsson, *Clin. Pharmacokinet.*, 1980, **5**, 569.

44. J. Raafaub and U. C. Dubach, *Eur. J. Clin. Pharmacol.*, 1975, **8**, 261.

45. L. Fleckenstein, G. R. Mundy, R. A. Horovitz and J. M. Mazzullo, *Clin. Pharmacol. Ther.*, 1976, **19**, 451.

23.3

Distribution and Clearance Concepts

JEAN GAILLOT, RENE BRUNO and GUY MONTAY
Rhône-Poulenc Santé, Antony, France

23.3.1 GENERAL CONCEPTS OF DISTRIBUTION AND CLEARANCE

Four concepts define the biological disposition of xenobiotics: absorption, distribution, metabolism and excretion (ADME). This chapter discusses distribution and clearance (the latter concept describing the elimination processes, *i.e.* both metabolism and excretion). In contrast to the absorption processes, distribution and clearance are independent of the conditions of administration (dose, route, dosage form).

The concept of distribution is initially relevant in the first phases of the development of a new drug, when the testing of the drug candidate progresses from *in vitro* models to increasingly complex *in vivo* biological systems (isolated receptor preparations, cell systems or isolated perfused organs). However, the biological relevance of the concept of distribution acquires its full significance when the drug is administered to the whole living organism; in this case, the distribution governs the access of xenobiotics to the various physiological areas of the body (nonspecific affinity), and particularly to the effector biophase (specific affinity).

The concept of clearance includes all elimination processes which act to remove drug from the physiological areas and which are more efficient if affinity is weak. Clearance is achieved either by irreversible elimination of the xenobiotics or by their biotransformation, generally into derivatives with lower affinity characteristics (specific and nonspecific), thus increasing their elimination rate. Clearance is the factor which limits the duration of the active principle in the organism (and thus in the biophase), thereby often modulating the drug's duration of action.

Distribution and clearance can be optimized by modifying the molecular structure of xenobiotics. The criterion for optimization of distribution is the ratio of specific affinity to nonspecific affinity, whereas the criterion for optimization of clearance is the duration of the active substance in the biophase (as a function of the desired therapeutic schedule).

This chapter begins with an analysis of the concepts of distribution and clearance. This is followed by a review of critical factors influencing the characteristics of distribution and clearance of drugs, focusing mainly on physicochemical properties and molecular structure. The methods employed for the application of these concepts in the laboratory are dealt with in other chapters.

23.3.1.1 Distribution Concept

Distribution is a complex dynamic process whereby the active compound enters the body from the systemic circulation. This phenomenon occurs immediately after intravascular administration, or after absorption of the active molecule when administered orally. The dynamics of distribution are subsequently regulated by the rate at which the active molecule enters the general circulation and is therefore affected by conditions of administration and any presystemic phenomena.

Distribution results from processes which are governed initially by physiological factors such as the intensity of blood flow or the permeability of membranes and biological barriers, and by the drug's selective affinity for certain tissues.

The residence time of the active compound in a given tissue depends on the intensity of the systemic elimination processes (clearance) which determine (with the degree of tissue affinity) its return to the systemic circulation (change in concentration gradient). The dynamic interaction of these factors regulates the entry of the active compound into its effector biophase and determines the adequacy of *in vitro* and *in vivo* pharmacological responses. The design of a compound to ensure that its intrinsic properties result in optimal adequacy is one of the major challenges of medicinal chemistry.

23.3.1.1.1 *Distribution and affinity*

The distribution of a xenobiotic is largely affected by its affinity for various tissues. The compound's intrinsic properties, which govern its affinity, are both physicochemical and structural.

Physicochemical affinity directly depends on the relative solubility of the active principle in the various constituents of the biophase. All biological systems are characterized, to a certain extent, by a succession of aqueous (blood plasma, intra- and extra-cellular fluids) and lipid (membrane components) phases. Therefore the physicochemical affinity of drug molecules depends on the balance between their hydrophilic and lipophilic properties.

The physicochemical affinity is determined *in vitro* by measuring the partition coefficient P between a lipid phase and water, usually *n*-octanol/water (intrinsic lipophilicity), or *n*-octanol/pH 7.4

buffer for ionized compounds, taking into account the degree of ionization (effective lipophilicity).[1] More recently, reversed-phase high-performance liquid chromatography, which partitions drugs between mobile and stationary phases, has been used to determine the degree of lipophilicity (the less polar the substances, the longer they will be retained on the analytical column). In this case, lipophilicity is expressed by the retention index of the drug on the column under specified analytical laboratory conditions.[2]

Lipophilicity can be used as a measure of the transfer rate of xenobiotics through membranes. This transfer largely occurs by a process of passive diffusion, as described by Fick's first law (equation 1).

$$\frac{\mathrm{d}C}{\mathrm{d}t} = D\Delta C \tag{1}$$

This equation expresses the rate of diffusion ($\mathrm{d}C/\mathrm{d}t$) as a function of the concentration gradient (ΔC) and the diffusion coefficient (D), the value of which depends, on the one hand, on the geometry of the membrane (area and thickness) and, on the other hand, on the physicochemical properties of the compound (size, shape, polarity, degree of ionization, hydrophilicity and lipophilicity).

The major source of variation in the diffusion coefficient D is the partition coefficient P between the lipid membrane and the aqueous environment.[3] The rate of drug transfer through a biological membrane adjacent to an unstirred water layer is linearly related to the logarithm of the membrane–water partition coefficient until a maximum rate is reached. Only compounds with intermediate affinity are likely to cross both lipid and aqueous barriers and therefore are distributed extensively throughout the body, thus reaching their site of action.

These considerations form the basis of nonlinear models (parabolic and derived models) describing relationships between structure and activity (QSAR), as introduced by Hansch.[4] These relationships show that an optimal partition coefficient exists for a given activity (depending on the site of the effector biophase) when one examines series of homologues with widely ranging partition coefficients. More information is to be found in a review by Seydel and Schaper.[3]

Structure–pharmacokinetic relationships (QSPR) and, in particular, those concerning distribution are generally less well understood than structure–activity relationships (QSAR). In fact, QSPR data for series of homologues are less comprehensive than QSAR data; additionally, many factors other than physicochemical affinity affect distribution.

The affinity determined by molecular structure governs the nature of interactions (binding) between the xenobiotic on the one hand, and various endogenous protein or lipid macromolecules (biopolymers) on the other hand. To some extent these interactions account for variations in structural specificity. In extreme cases, the presence of asymmetric sites in the compound can cause stereospecificity of both drug disposition and biological response. Two types of interaction can be identified.

(i) Interactions usually highly specific with effector macromolecules, which are able to induce a biological response: receptors involved in desirable or undesirable pharmacological responses; enzymatic sites responsible for the biotransformation of the compound into its metabolites; and membrane proteins which act as transporters, thus permitting or facilitating transfer across membranes.

(ii) Interactions, usually not specific, with acceptor molecules which do not induce specific biological responses. These are nevertheless very important since they may significantly affect the compound's pharmacokinetics, particularly as regards distribution. This is especially true of plasma and tissue protein binding in which both ionic and lipophilic forces are involved, the free fraction being directly available to cross biological membranes and then to distribute itself.

Overall, the compound's affinity for a given tissue can be assessed by determination of the tissue to blood partition coefficient,[5] which is the resultant of the two types of affinity, the physicochemical and the structural.

23.3.1.1.2 *Physiological aspects*

The processes involved in blood/tissue disposition of xenobiotics are summarized in Figure 1.

Physiological factors can influence the distribution of drugs, in particular: (i) the intensity of blood flow (Q), which determines the rate of drug presentation (QC_A) and therefore the blood to tissue concentration gradient; (ii) the blood to tissue diffusion across the capillary endothelium, the cellular membrane and/or some specific membrane barrier; and (iii) the blood and tissue protein binding.

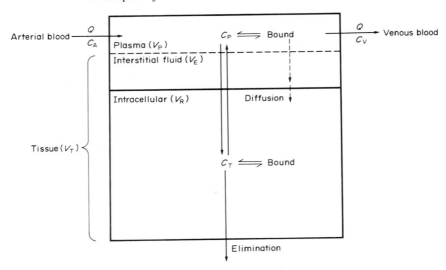

Figure 1   Schematic representation of tissue disposition of a xenobiotic

Depending on the absence or presence of blood to tissue diffusion limitation, the distribution of xenobiotics can be either perfusion-rate limited or diffusion-rate limited.

Assuming that perfusion limitation occurs, the mass balance equation describing the amount in the tissue A_T is given by equation (2). In this equation, C_A and C_V denote arterial and venous concentrations respectively.

$$\frac{dA_T}{dt} = QC_A - QC_V - \text{Elimination rate} \qquad (2)$$

For a tissue not involved in drug elimination and assuming a constant arterial concentration (*e.g.* during administration by constant rate infusion), the tissue concentration–time curve is then given by equation (3), in which K_p denotes the tissue to blood partition and k_T is a first-order rate constant.[6] K_p depends on both the physicochemical and structural affinities of the substance. Therefore k_T is proportional to blood flow and inversely proportional to the affinity (equation 4).

$$C_T = K_p C_A [1 - \exp(-k_T t)] \qquad (3)$$

$$k_T = \frac{Q}{K_p V_T} \qquad (4)$$

Therefore, given a constant rate of presentation of the drug, the amount taken up by the tissue increases until a steady state is achieved. The steady-state tissue concentration ($C_{T,ss}$) is directly related to the affinity of the drug for the tissue (equation 5). The time to reach the steady state is inversely related to the tissue equilibration rate constant k_T, the equilibrium therefore being achieved faster in highly perfused tissues with low affinity for the drug.

$$C_{T,ss} = K_p C_A \qquad (5)$$

It is interesting to note that the distribution in a given tissue, at steady state, depends only on the affinity of the substance for this tissue whereas the blood flow plays a major role in the early stage of distribution, particularly for those compounds readily diffusible through lipid membranes (*i.e.* those with a high lipid to water partition coefficient).

23.3.1.1.3   *Concept of distribution compartment*

Tissues and organs with similar distribution characteristics (affinity, equilibration rate) can be grouped together in a single distribution compartment. A distribution compartment is the space in which the drug distributes with the same dynamics to achieve a homogeneous steady state concentration. This space is characterized by an apparent volume (V) which relates the amount of drug (A) in the compartment to the concentration, $C = A/V$, at any time t. The distribution volume must vary initially unless distribution equilibrium is instantaneously achieved. After administration of an intravenous bolus dose, an initial distribution phase is generally observed which corresponds to the invasion of highly perfused tissues. The apparent volume of this initial distribution space generally comprises the plasma volume and the total body water volume, *i.e.* the so-called central compartment. Equilibration in this compartment is generally rapid because of the predominant influence of blood flow. Simultaneously the drug can be distributed into deeper compartments in which affinity governs the achievement of steady state. The total volume of distribution, generally called steady state distribution volume, is the sum of the volumes of each compartment at steady state.

23.3.1.2 Clearance Concept

23.3.1.2.1 *Physiological and kinetic aspects*

The clearance concept was first introduced in the 1930s to describe renal excretion of endogenous compounds (creatinine, urea). Clearance was defined as the ratio of renal excretion rate to the arterial concentration of the compound.

This concept, which was first introduced as a measure of kidney function, has been applied to other eliminating organs (such as the liver, the lung, *etc.*) and elimination pathways (biotransformation) for both endogenous and exogenous compounds, and has become central to the development of pharmacokinetics.[7]

From a physiological point of view, xenobiotics enter and leave the eliminating organ *via* the arterial and venous blood (Figure 1). The organ elimination rate of xenobiotics (the amount, A_E, eliminated per unit time, as shown in equation 6) is calculated from the general mass balance equation (equation 2) under steady-state conditions ($dA_T/dt = 0$). The steady-state arteriovenous gradient normalized to the arterial concentration defines the extraction ratio E (equation 7) which characterizes the intrinsic capability of the organ to extract a xenobiotic. From the definition of renal clearance and equations (6) and (7), the organ clearance, Cl, is the product of the blood flow and the organ extraction ratio (equation 8). Organ clearance can vary from zero ($E = 0$, non-eliminating organ) to the upper limit of organ blood flow ($E = 1$).

$$\frac{dA_E}{dt} = Q(C_A - C_V) \tag{6}$$

$$E = \frac{C_A - C_V}{C_A} \tag{7}$$

$$Cl = \frac{\text{Elimination rate}}{C_A} = QE \tag{8}$$

Clearance has dimension of volume per unit time and can be further defined as the volume of reference fluid (blood, plasma) cleared of a xenobiotic (total, unbound) per unit time. According to equation (8), the organ clearance is the proportionality constant between the elimination rate and the arterial concentration, *i.e.* the slope value of the elimination rate–arterial concentration relationship (Figure 2a). Such an elimination process obeys first-order kinetics.

In this situation the organ clearance is independent of the arterial concentration and consequently of the administered dose. However, as the concentration increases, the elimination process (or the diffusion from arterial blood) can become saturated (capacity limited); the elimination rate is no longer proportional to the arterial concentration and approaches a maximum value (V_{max}) (Figure 2b). The elimination rate–arterial concentration relationship obeys Michaelis–Menten kinetics (equation 9) in which K_m, the Michaelis–Menten constant, denotes the concentration at which the elimination rate equals half the maximum. Clearance (equation 10) is no longer constant but is inversely related to the concentration of the drug, and thus the administered dose.

$$\text{Elimination rate} = \frac{V_{max} C_A}{K_m + C_A} \tag{9}$$

$$Cl = \frac{V_{max}}{K_m + C_A} \tag{10}$$

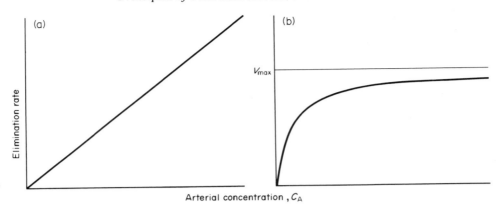

Figure 2   Kinetic aspects of organ elimination: (a) first-order elimination; and (b) capacity-limited elimination

However, when C_A is far below K_m, Michaelis–Menten kinetics collapse to first-order kinetics since the clearance can be approximated by the ratio of V_{max} to K_m, and the elimination rate is proportional to the concentration. Any elimination pathway therefore follows first-order kinetics (with constant Cl) within a given range of concentrations and doses.

Conversely, when C_A greatly exceeds K_m, then K_m can be disregarded in equations (9) and (10), and the elimination rate is quasi constant, approaching V_{max} (zero-order kinetics).

23.3.1.2.2   *Pharmacokinetic implications*

The total body clearance is the sum of all the partial clearances characteristic of each elimination pathway (assuming that all the elimination pathways are applied to the same reference fluid, generally the whole blood or the plasma). The various pathways contributing to the elimination of xenobiotics will be studied in subsequent parts of the present chapter. The two basic processes involved are: excretion (mainly renal, biliary) and metabolism (mainly in the liver).

If each elimination pathway obeys first-order kinetics, the overall elimination is first order and the xenobiotic exhibits linear pharmacokinetics. In this case, the plasma levels (area under the curve, mean steady-state plasma level following multiple dosing), are proportional to the administered dose (Figure 3a and equation 11).[8] In this case, the total body clearance is a pharmacokinetic parameter characteristic of the plasma level *versus* dose relationship for both the xenobiotic and the individual receiving it. It is independent of the dose, the route of administration and the dosage regimen.

$$AUC = \frac{1}{Cl}(Dose) \tag{11}$$

However, when an elimination pathway with concentration and dose-dependent clearance significantly contributes to the overall elimination, the total body clearance is concentration and dose-dependent and the plasma levels are no longer proportional to the dose[8] (nonlinear pharmacokinetics) (Figure 3b).

For each nonlinear pharmacokinetic system, there exists a range of doses in which the pharmacokinetic behaviour can be considered to be linear (*i.e.* the doses resulting in arterial concentrations compatible with first-order elimination). Most drugs exhibit linear pharmacokinetics in the therapeutic dose range. For drugs with dose-dependent kinetics at therapeutic levels, plasma level monitoring is crucial to clinical safety (*e.g.* antiepileptics such as phenytoin and valproic acid).

Total body clearance is therefore a parameter of primary importance in pharmacokinetics and nonlinearities in the pharmacokinetic behaviour must be recognized early during drug research and development[9] since they can affect the outcome of pharmacological and toxicological studies and the safety of clinical dose escalation studies (phase I tolerance studies).

23.3.2   CRITICAL FACTORS FOR DRUG DISTRIBUTION

In order to distribute from the circulatory system into tissues, a xenobiotic must cross the capillary membrane, diffuse through a tissue matrix and traverse cell walls. The rate and extent of drug distribution is therefore influenced by both physiological and compound-related factors.

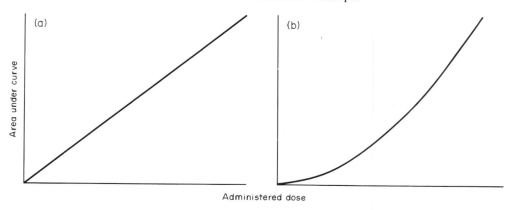

Figure 3 Plasma level *versus* dose relationships in (a) linear and (b) nonlinear pharmacokinetics

Biopharmaceutic factors nonspecifically related to the distribution process, such as the rate and extent of drug entry into the systemic circulation (administration route, absorption) and elimination (clearance), can also modulate the blood to tissue concentration gradient, thus indirectly the drug distribution.

We will, in this section, deal only with the following physiological factors which specifically determine the distribution of xenobiotics: organ blood flow; plasma and tissue protein binding; and transfer through biological membranes.

23.3.2.1 Blood Flow

As shown in the scheme of body vascularization (Figure 4), the blood flow is one of the primary determinants of drug distribution, influencing both the qualitative distribution of the drug in the body and the quantitative uptake by a given tissue.

The rate-limiting step of drug distribution can be either the blood flow rate, which determines the delivery of the drug to the tissues, or transmembrane transfer processes which are involved at both capillary and cellular levels.

In the absence of diffusion limitations, the rate of drug uptake by a tissue is limited by the perfusion rate of the tissue, which is expressed as the blood flow rate per unit volume of tissue (perfusion rate-limited distribution). When the membrane permeability for the drug tends to be less than the perfusion rate, tissue uptake is limited by transmembrane transfer (*i.e.* diffusion rate-limited distribution).

These considerations are basic to the development of physiological pharmacokinetic models of drug disposition.[10, 11]

Perfusion rate limitation often prevails in drug distribution since most xenobiotics are small lipid-soluble molecules. Moreover, capillary endothelium being loosely knit with large fenestrations in most tissues, xenobiotics have only to diffuse through interstitial fluid and cells membranes.

However, organs are physiologically different as regard capillary endothelium structure, capillary transit time and diffusion barriers. A given drug can therefore display perfusion rate limitation for some tissues and diffusion rate limitation for others. In this respect, the liver and the brain represent opposites since the structural arrangement of the former's capillary network (liver sinusoid) is optimized to favour hepatic uptake (see Section 23.3.4) whereas that of the latter constitutes a specific barrier (see Section 23.3.3.1). The influence of organ blood flow (Q) and perfusion rate (Q/V_T) on tissue uptake when perfusion rate limitation occurs, as expressed in equations (2)–(5) (Section 23.3.1.1.2), is particularly important both during the initial stage of tissue uptake and to determine tissue to blood equilibration time.

There exist marked inter-organ differences in blood flow[12] which may qualitatively influence the initial distribution of xenobiotics in the whole body. The lungs receive the total cardiac output and then the arterial outflow is distributed among the other organs as illustrated in Figure 5 and Table 1.

The main organs involved in drug elimination (liver, kidney) receive 52% of the cardiac output, which shows the primary importance of clearance processes in determining the disposition of xenobiotics. Muscle and brain are the other highly perfused organs (however, the brain uptake of most xenobiotics is diffusion rate limited). Inter-organ variability in perfusion rate and its influence on tissue equilibration half-life ($t_{1/2} = \log 2/k_T$) and equilibration time ($5 \times t_{1/2}$) are presented in

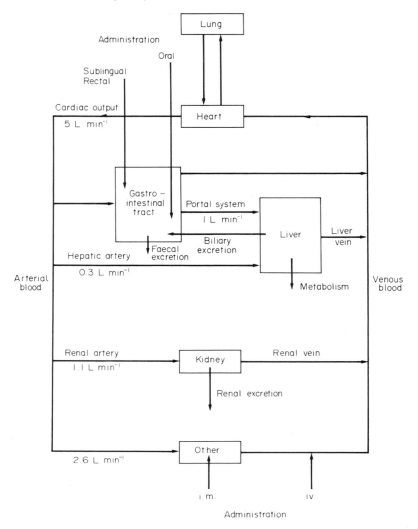

Figure 4   Schematic distribution of body vascularization in relation to administration and elimination routes

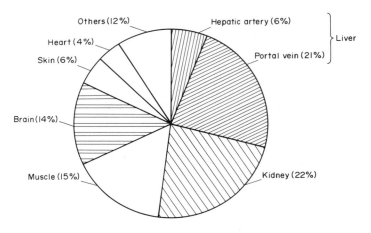

Figure 5   Schematic distribution of blood flow in man (% of cardiac output)

Table 1 Perfusion Rate and Equilibration Time of Various Organs[12]

Organ	Blood Flow (mL min^{-1})	Percent of cardiac output	Perfusion rate (mL min^{-1} mL^{-1} of tissue)	Equilibration half-life[a]	Equilibration time[a]
Lung	5000	100	10	4 s	20 s
Kidneys	1100	22	4	10 s	50 s
Liver	1350	27	0.8	0.9 min	4 min
Heart	200	4	0.6	1.2 min	6 min
Brain	700	14	0.5	1.4 min	7 min
Fat	200	4	0.03	23 min	2 h
				19 h[b]	4 days[b]
Muscle	750	15	0.025	28 min	2 h
Skin	300	6	0.024	29 min	2 h

[a] Assuming $K_p = 1$.   [b] Assuming $K_p = 50$.

Table 1. Assuming a blood to tissue partition coefficient of unity ($K_p = 1$), *i.e.* the tissue has no particular affinity for the xenobiotic, the equilibration time varies from some minutes in highly perfused organs to some hours in poorly perfused ones. The influence of the blood to tissue partition coefficient is also stressed; as an example, the equilibration time in fat of a drug with high affinity can be days rather than hours.

Therefore, these variations can markedly influence the development of pharmacological responses to the drug according to the topographical location of the biophase.

The very long equilibration time of tissues with high affinity for the drug can also limit tissue to blood return, and thus drug removal from the body, resulting in a long plasma half-life even for a drug which is eliminated efficiently. This is the case for amiodarone, a drug with a very high affinity for fat.[13] The diffusional clearance of amiodarone (165 mL min^{-1})[14] approaches the blood flow rate for the fat content of the body (200 mL min^{-1}), but the equilibration is very long because of the affinity for fat. This phenomenon is the primary determinant of the long plasma terminal half-life (40 days).

As well as inter-organ variability, blood flow shows inter-species variations. However, both blood flow and organ weight correlate with body weight, ensuring constancy of perfusion rate. For example, the liver perfusion rate is about 1 mL min^{-1} mg^{-1} of liver in all mammalian species.[15] Thus, flow-limited distribution models allow inter-animal scale-up of pharmacokinetics.[11, 15]

Finally, it should be mentioned that physiopathological status may alter blood flow. In particular, circulatory failure caused by a decrease in cardiac output may profoundly affect the distribution of flow-limited drugs. The available cardiac output is redistributed to preserve the perfusion rate to the brain and the heart, resulting in increased exposure of these organs.[16]

23.3.2.2 Protein Binding

The binding of a drug to macromolecules in the body is one of the main factors in its distribution. It depends on both the physicochemical properties of the xenobiotic and the conformation of the acceptor protein which jointly determine the drug's affinity for the acceptor. This affinity may or may not be specific. Of the non-specific affinities, serum protein binding is the most important.

The ability of a protein to bind to a xenobiotic is defined by n, the number of binding sites per mole of protein, and by K, the association constant which is characteristic of each site, according to equation (12) where C_b denotes the concentration of bound drug, C_u the concentration of free drug and R the protein concentration. The linearity or nonlinearity (saturation) of the xenobiotic binding to the protein is governed by the above parameters. At therapeutically effective concentrations, protein-binding capacity is rarely saturated.

$$C_b = \sum_{i=1}^{n} \frac{n_i K_i R C_u}{1 + K_i C_u} \qquad (12)$$

The effect of protein binding on tissue distribution, particularly in target areas, depends on the binding capacity (nK) of plasma vector proteins and on affinity for tissue proteins. In fact, on reaching the circulation, the xenobiotic distributes itself between the plasma proteins, plasma water and blood cells. Only the free form of the xenobiotic can penetrate, by passive diffusion, firstly into the interstitial fluid and the lymph, then into the cells. However, the circulating bound fraction constitutes a reservoir which is carried by the blood proteins, as a result of reversible binding and of the high association–dissociation rates of the drug–protein complex. Conversely, high affinity binding to plasma proteins or erythrocytes can be the limiting factor for the distribution of the xenobiotic.

We shall successively examine the proteins involved, the physicochemical properties of the xenobiotic responsible for the interaction and finally, how the distribution of a drug in the body is regulated by the nature and degree of this interaction.

The main protein in blood and tissues is albumin which is usually responsible for the nonspecific binding of most drugs. In addition, albumin carries endogenous compounds such as bilirubin and fatty acids. In blood, the distribution space of albumin is plasma. Outside the circulation, the distribution space of albumin consists of both tissues with a good blood supply (liver, kidneys, heart, lung, spleen, intestine) and tissues with a poorer blood supply (skin, muscle, adipose tissue). The amount of albumin in a tissue depends on the amount of extracellular fluid present in the tissue. About 59% of exchangeable albumin in the body is to be found outside the circulation.[17] The other main proteins binding xenobiotics in blood are α1-acid glycoproteins, globulins and lipoproteins.

Drug–protein interactions are divided into six groups depending on the physicochemical properties of the drugs (Table 2).[18] Some proteins possess very selective affinity since they specifically bind to a limited number of drugs which are structural analogues of endogenous compounds; these include transcortin, which is a carrier of corticosteroid hormones, and thyroid-binding globulin.

In tissues, other than albumin, proteins which can bind xenobiotics are: ligandin, which is present in the liver, kidney and intestine and may be involved in drug extraction and transfer; actin and myosin in muscular tissue; melanin in pigmented tissues; nuclear proteins and specific protein carriers, not wholly identified and involved in renal tubular secretion or transfer across the blood–brain barrier. It should be noted that tissue affinity is not completely explained by protein binding but may result from chelation (for example, tetracyclines by bone calcium), or from significant affinity for fat due to the lipophilicity of the compound.

The protein–xenobiotic complex results from binding in two main ways: (i) reversible binding (*i.e.* ionic, hydrogen, hydrophobic and Van der Waals); and (ii) covalent irreversible binding.

In some cases, the formation of irreversible drug–protein or reactive metabolite–protein complexes can be a factor in drug toxicity (cytotoxicity, genotoxicity, oculotoxicity, *etc.*). Such a formation of a drug–protein complex mainly depends on the charge of the compound and the hydrophobicity of the ionized form. Studies of the effect of structure modifications in a chemical series on protein binding repeatedly showed good correlation between the degree of binding to plasma proteins and the hydrophobic properties of xenobiotics.[3] The following few examples show the existence of such a relationship for both small basic or acidic molecules and for large molecules.

For β-blockers, which bind mainly to α-acid glycoprotein as well as albumin, it has been shown[19] that, in the case of pindolol and eight related basic compounds which are almost completely ionized at pH 7.4, there is a good correlation between the protein binding capacity, as expressed by the product nK, and the partition coefficient (Figure 6).

The study of a series of 11 organic acids derived from benzoic acid and phenylacetic acid has shown[20] that there is a relationship between fu, the free fraction of drug in the plasma, and P, the octanol/water partition coefficient. This relationship is characterized by equation (13) where a is constant for a given animal species (Figure 7).

$$fu = (1 + aP)^{-1} \qquad (13)$$

In the case of ardacins, which are glycopeptide antibiotics, an increase in serum protein binding was observed together with a drop in the isoelectric point (increasing negative charge) and an increase in lipophilicity.[21]

Depending on the binding capacity nK of circulating plasma proteins and of tissue proteins, the drug can be either restricted to the vascular space or mainly distributed in the tissues, or present in both the vascular space and some preferential tissues.

The dependence of the distribution volume (V) on plasma protein binding (fu, free fraction of drug in plasma) and on tissue binding (fu_R, free fraction of drug in intracellular space V_R) is clearly expressed in equation (14).

$$V = 7.2 + 7.8fu + 27\frac{fu}{fu_R} \qquad (14)$$

This equation is derived from the mass action law[22] and has been developed, taking into account the physiological values of the different components of the total body water in man (Figure 1):

Table 2 Classification of Drug–plasma Protein Interactions[18]

Types of drug	Acidic			Non-ionizable	Basic		Other
	I	II	III	IV	V	VI	
Reference drug	Warfarin Diazepam	Indomethacin	Phenytoin	Digitoxin	Erythromycin	Imipramine	Cyclosporin Polyene macrolides
Binding protein	HSA α1-AGP	HSA	HSA	HSA	HSA α1-AGP	HSA α1-AGP Lipoproteins	Lipoproteins
Possibility of plasma-binding saturation	+	−	−	−	+	−	

Abbreviations: HSA = human serum albumin; α1-AGP = α1-acid glycoprotein.

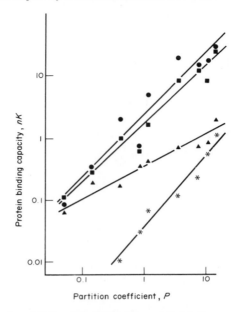

Figure 6 Correlation between the lipophilicity of nine β-blockers and the binding capacity obtained for serum (■), α1-AGP
(●), HSA (▲) and lipoproteins (∗) (reproduced from ref. 19)

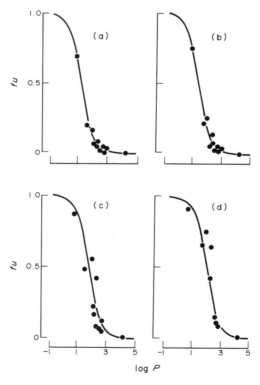

Figure 7 Relationship between plasma free drug fraction and lipophilicity of 11 organic acids obtained from man (a), rabbit
(b), rat (c) and mouse (d) (reproduced from ref. 20 by permission of the Pharmaceutical Society of Great Britain)

plasma volume ($V_p = 3$ L), extravascular extracellular fluid ($V_E = 12$ L) and intracellular water
($V_R = 27$ L). This equation serves to show the relationships between protein binding and distribution volume as follows: (i) drugs which do not enter intracellular space ($V_R = 0$) have a low volume of
distribution; this volume ranges from 7.2 L ($fu \neq 0$; compound highly bound to plasma proteins) to
a maximal value of about 15 L ($fu = 1$; compound not bound to plasma proteins); (ii) drugs which

enter intracellular space ($V_R \neq 0$) but are not bound to tissue proteins ($fu_R \# 1$) have an intermediate volume of distribution; this volume ranges from 7.2 L to 42 L (total body water volume) according to the plasma protein binding ($0 < fu \rightarrow 1$); and (iii) drugs which enter intracellular space and are bound to tissue proteins, have a high volume of distribution (> 50 L); the extent of the distribution depends on the relative affinity of the compound for plasma or tissue proteins, expressed by the ratio fu/fu_R; the higher the tissue binding, the greater the volume of distribution.

A high value for the distribution volume has no physiological significance when it is greater than that of the total body water (42 L), which is the case for most drugs. However, this volume provides a rough estimate of extravascular distribution in the body.

There is only a weak relationship between plasma binding and tissue binding, although the same physicochemical factors (*i.e.* lipophilicity, charge of the molecule, and hydrogen bonds) are involved in drug–protein interaction.[23] It is, however, interesting to note that drugs can be classified according to their distribution volume into three groups depending on the tissue to plasma-binding ratio (Table 3). These observations are in good agreement with the above theoretical considerations.

Within a series of homologous compounds, there is a relationship between plasma protein binding and distribution volume. Close relationships have been observed for cephalosporins[24] and tetracyclines.[18]

However, this relationship is apparently not evident when one considers drugs belonging to different therapeutic classes (Table 4). In fact, the distribution volume values predicted from equation (14) represent only the specific contribution of protein binding to the apparent total distribution volume. For a number of drugs, it has been possible to determine satisfactorily the relationship between the extent of distribution and lipophilicity by taking into account the distribution volume of free diffusible drug and solubility (expressed by the octanol/water partition coefficient P at pH 7.4). In this way, Ritschel and Hammer[25] predicted the distribution volume of 125 drugs widely ranging in polarity and therapeutic class (equation 15).

$$V = 0.1302P + 0.7799 \qquad (15)$$

$$n = 125, \quad r = 0.9375, \quad p < 0.001$$

Table 3 Classification of Drugs According to Tissue Binding (fu_R) to Plasma Binding (fu) Ratio[23]

	Type I $fu < fu_R$	Type II $fu \# fu_R$	Type III $fu > fu_R$
Drug	Sulfonamides Warfarin Salicylic acid Phenylbutazone	Barbiturates Phenytoin Diazepam	Tricyclic antidepressants Phenothiazines Propranolol Alprenolol
V (L kg^{-1})	0.1–0.3	0.6–1.1	>3
log P	<1	1–2	>1

Table 4 Distribution Volume and Plasma Protein Binding of Some Drugs in Man

Drug	Volume of distribution[a] (L)	Protein binding (%)
Warfarin	7.2	99
Aspirin	9.7	60
Gentamicin	16	5
Nalidixic acid	32	90
Diazepam	72	99
Trimethoprim	130	70
Propranolol	260	90
Flecainid	360	40
Imipramin	975	92
Haloperidol	2000	92
Chloroquine	15 000	60

[a] Volume is expressed for a 65 kg body weight.

Lastly, for a given drug, wide interindividual variations of distribution are often observed depending on physiopathological status (*e.g.* changes of protein concentration, in the number of binding sites, in lean body mass, in distribution of body fat and presence of additional distribution spaces.)

23.3.2.3 Membrane Transfer

As stated previously, the tissue distribution of a drug molecule will largely depend on its ability to cross biological membranes. Depending on both the physicochemical properties of the drug and the biochemical properties of the membrane, the three major processes of membrane transfer are: passive diffusion, carrier-mediated transport and pinocytosis.[26, 27]

23.3.2.3.1 Passive diffusion

A common feature of biological membranes is their lipid–protein structure, therefore the relative lipophilic/hydrophilic properties of the molecule, *i.e.* the affinity for the lipid constituents of the membrane will determine whether the molecule will readily cross membranes by passive diffusion.

In such a process, the driving force is the drug concentration gradient on both sides of the membrane. The diffusion rate is governed by Fick's law (equation 1), in which the diffusion rate constant D is directly proportional to the octanol/water partition coefficient P (see Section 23.3.1.1.1). Thus this process is unsaturable and follows first-order kinetics, with the pharmacokinetic consequences previously discussed for elimination processes (see Section 23.3.1.2). It is important to note that no structural specificity of the molecule is required for this type of transfer process. Only the unionized (lipid soluble) fraction of the molecule is transferable through the membrane (pH partition hypothesis). At the pharmacokinetic steady state, the concentration of unionized compound is therefore in equilibrium on both sides of the membrane. However, a difference in pH between the two compartments separated by the membrane could affect the distribution of partly ionized substances such as weak bases and weak acids.

The ratio (C_1/C_2) of total concentrations on both sides of the membrane at steady state is given by equations (16) (acidic compounds) and (17) (basic compounds) as a function of the pK_a of the molecule and of the pH of the biological media (pH_1 and pH_2). Some examples of these relationships will be presented subsequently.

$$\frac{C_1}{C_2} = \frac{1 + 10^{(pH_1 - pK_a)}}{1 + 10^{(pH_2 - pK_a)}} \tag{16}$$

$$\frac{C_1}{C_2} = \frac{1 + 10^{(pK_a - pH_1)}}{1 + 10^{(pK_a - pH_2)}} \tag{17}$$

Another diffusion mechanism, involving mainly filtration of small hydrosoluble molecules, seems to depend on the presence of aqueous channels through the membrane. In this case, diffusion is limited by both the molecular charge and the molecular volume of the solute relative to pore size. The existence of pores seems to be transient, their apparent diameter depending on cell metabolism and varying widely from membrane to membrane. The mean diameter is around 5–10 Å but considerably higher values are observed in capillary membranes (50–100 Å).

23.3.2.3.2 Carrier-mediated transport

The transfer of large lipid insoluble or ionized molecules can be facilitated by carrier-mediated systems in some membranes. A surface component of the membrane (carrier) is able to bind the solute molecule and the carrier–solute complex diffuses through the membrane. The solute is then released and the carrier returns to the opposite surface. The process can be either passive, if the driving force is still the concentration gradient of the solute (facilitated diffusion), or active, if the transport is associated with the production of energy, and when the solute can move against the concentration gradient (active transport).

Structural specificity can be high and these processes are subject to competitive inhibition by analogues or other molecules susceptible to binding by the carrier (including endogenous compounds). This feature can be responsible for drug–drug interactions. Such competition is also involved in the mechanism of action of some drugs such as Henle's loop diuretics.

The active transport processes require energy produced by cell metabolism and can be inhibited noncompetitively by substances which interfere with cell metabolism.

Both processes are capacity limited (saturable); they can, therefore, obey Michaelis–Menten kinetics and can be responsible for nonlinearities in overall pharmacokinetic behaviour as can the elimination processes (see Section 23.3.1.2).

23.3.2.3.3 *Pinocytosis*

Pinocytosis results from membrane invaginations, leading to intracytoplasmic vesicles, which entrap fluid and molecules from the medium surrounding the cells. Enzymatic degradation of the vesicles then liberates the compound into the cytoplasm of the cell. The biological significance of pinocytosis remains obscure; however, this process could be involved in the cellular uptake of peptides, proteins and other macromolecules. It is rather a slow process, nonselective, which requires energy and can therefore be noncompetitively inhibited by metabolic poisons.

23.3.3 DISTRIBUTIONS OF SPECIFIC INTEREST

23.3.3.1 Central Nervous System

The study of diffusion into the central nervous system (CNS) is of prime importance since the activity of many drugs is exercised in the CNS. CNS distribution enables drugs to reach either specific pharmacological receptors (psychotropics) or target cells within the CNS (antineoplastic agents, antibiotics). In other cases, CNS distribution is responsible for side effects of varying severity (antihistamines, β-blockers).

CNS distribution is subject to drugs crossing specific anatomical barriers such as the blood–brain barrier (BBB) and the blood–cerebrospinal fluid (CSF) barrier.

The main physiological properties of these two barriers and their implications for drug transfer mechanisms will be summarized below. Further information can be found in key papers by Oldendorf,[28,29] Rall,[30,31] Bonati *et al.*[32] and Pardridge.[33] The effects of the physicochemical properties of drugs on their ability to diffuse into the CNS will also be discussed.

23.3.3.1.1 *Physiological aspects*

The blood–brain barrier results from the structure of the cerebral capillary network which is characterized by: (i) a capillary endothelium consisting of a cell layer with a continuous tight intercellular junction (unlike other systemic capillaries, there are no pores), (ii) the presence of special cells, namely astrocytes. These astrocytes are elements of the supporting tissue, found at the base of the endothelial membrane, which form a solid envelope around the capillaries. They also secrete the trophic factors necessary for BBB biology.

In order to be distributed within the brain, compounds must cross a barrier consisting of several layers of cells by passive diffusion or active transport. Thus cerebral uptake is usually diffusion-rate limited rather than perfusion-rate limited.

The main purpose of the blood–brain barrier is to maintain the brain in a stable liquid environment, any modification of which may cause a neurological disturbance. The barrier exercises its protection against endogenous metabolism products such as neurotransmitters, whose cerebral concentration is regulated and must be protected against any changes in systemic concentration. Active transport systems by specific carrier proteins ensure both passage and control of polar metabolic substrates such as hexoses (glucose), amino acids, puric bases, nucleosides and water-soluble vitamins. In some cases, these carrier proteins can also act as receptors, as do those implicated in peptide transfer. Lastly, cerebral diffusion of steroid hormones through endothelial-induced conformational changes due to interaction with binding plasma proteins suggests that, in some cases, the bound fraction of a compound can also be transferred.

The blood–CSF barrier comprises mainly the epithelium of the choroid plexus responsible for the active secretion of CSF. Although the capillaries in the choroid plexuses have pores, plexus epithelial cells are joined by continuous tight junctions similar to those already mentioned. Most substances enter the CSF through choroid plexuses, the epithelium of which is similar in structure and function to renal epithelium. There are also, at that level, active transport systems.

The passage of compounds from the systemic circulation into the cerebral extracellular fluid (which accounts for 10–20% of brain tissue) and the CSF may be likened to passage into intracellular fluids. Protein binding in man is minimal because of the low protein concentration of approximately 0.15 g L^{-1}, *i.e.* about 200 times lower than that in plasma. This explains how, in most instances, there is a good relationship between the concentration of a drug in the CSF and its free concentration in plasma. Exchanges through the membranes (ependyma, pia glia) separating the cerebral and cerebrospinal fluids are achieved by free diffusion of molecules in both directions, there being no tight junctions here.

Although the mechanisms responsible for diffusion into the CNS and CSF are similar, the degree of drug uptake can vary significantly. In some cases, cerebral concentrations can be much lower than CSF concentrations; this applies to some antibiotics (cephalotin, trimethoprim, sulfamethoxazole).[34] Alternatively, cerebral concentrations can be markedly higher than CSF concentrations; this is true of β-blockers (Table 5).[35] The reasons for these findings could be: (i) a difference in exchange surfaces: the exchange surface of the BBB is more than 5000 times as great as that of the blood–CSF barrier; (ii) the absence of a steady state between the cerebral interstitial fluid and the CSF compartment in species with a large brain (dog, monkey, man); and (iii) the existence of carrier proteins specific for BBB.[36]

Drugs therefore enter the CNS as a result of passive or active transfer through special barriers, the permeability of which can be modified in a number of physiological or pathological situations, namely changes in CO_2 partial pressure, metabolic acidosis or alkalosis, hypertension, old age, tumours, convulsions, schizophrenia, stress and inflammatory processes, *e.g.* meningitis. Some peptides and some drugs (amphetamines, dexamethazone, xanthines, probenecid, some tranquillizers) can also cause changes in membrane permeability.

Lastly, mention should be made of specific cerebral sites for which the BBB does not exist, namely the trigger area and the median hypothalamic eminence which control emesis and prolactin secretion, respectively. Furthermore, direct diffusion of drugs into the CNS after intranasal administration is possible due to the continuity of the subarachnoid space of the olfactive lobe and the submucosal areas of the nose.

23.3.3.1.2 *Physicochemical determinants*

The physicochemical properties of a drug which facilitate its passive diffusion through the above-mentioned barriers are the same as those which generally allow cross-membrane transfer. Therefore, diffusion is usually greater for drugs with high liposolubility and low ionization at physiological pH. The significant part played by liposolubility has been demonstrated by Brodie *et al.*[37] who studied the diffusion of 14 drugs into the CSF after intravenous administration to the dog (Table 6). In the case of drugs which are highly ionized at pH 7.4, the degree of diffusion, as expressed by the permeability constant, is inversely proportional to the degree of ionization, whereas solubility in lipids becomes the limiting factor for diffusion of drugs which occur mainly in the nondissociated form in plasma. Lipid solubility is the most important determinant of substantial diffusion into the CNS, whereas the degree of dissociation is relatively unimportant because of the non-solubility of organic ions in lipids.

Study of the effect of the physicochemical properties of various anticonvulsants on their degree of penetration into dog CSF[38] showed that there was no correlation between the penetration rate and the fraction not bound to proteins. However, at steady state, the free fraction of the drug present in the CSF is approximately the same as that found in plasma (Table 7). A significant correlation was observed between the degree of penetration and the benzene–buffer partition coefficient. This

Table 5   Brain Uptake of some β-Blockers in the Rat[35]

	Propranolol	*Acebutolol*	*Sotalol*
IV dose (mg kg^{-1})	0.5–1.0	10	10
Brain/plasma	38	0.71	0.52
Brain/CSF	213	2.42	2.00

Concentration ratio determined 4 h after a single administration.

Table 6 Rate of Entry into CSF and Degree of Ionization of Drugs at pH 7.4[37]

Drug	pK_a	%Unionized at pH 7.4	Permeability constant $(min^{-1} \pm S.E.)$
Drugs mainly ionized at pH 7.4			
5-Sulfosalicylic acid	(strong)	0	0.0001
N'-Methylnicotinamide	(strong)	0	0.0005 ± 0.00006
5-Nitrosalicylic acid	2.3	0.001	0.001 ± 0.0001
Salicylic acid	3.0	0.004	0.006 ± 0.0004
Mecamylamine	11.2	0.016	0.021 ± 0.0016
Quinine	8.4	9.09	0.078 ± 0.0061
Drugs mainly unionized at pH 7.4			
Barbital	7.5	55.7	0.026 ± 0.0022
Thiopental	7.6	61.3	0.50 ± 0.051
Pentobarbital	8.1	83.4	0.17 ± 0.014
Aminopyrine	5.0	99.6	0.25 ± 0.020
Aniline	4.6	99.8	0.40 ± 0.042
Sulfaguanidine	>10.0	>99.8	0.003 ± 0.0002
Antipyrine	1.4	>99.9	0.12 ± 0.013
N-Acetyl-4-aminoantipyrine	0.5	>99.9	0.012 ± 0.0010

Table 7 Penetration of Antiepileptic Drugs into CSF of the Dog[38]

Drugs and active metabolites	% Non-ionized at pH 7.4	% Unbound to plasma proteins	CSF/Plasma ratio × 100	Benzene/pH 7.4 buffer distribution ratio
Phenytoin	89	22	23	6.02
Phenobarbital	51	55	57	0.33
Primidone	100	80	81	0.093
Phenylethylmalondiamine	100	92	86	0.13
Carbamazepine	100	28	36	48
Carbamazepine 10-11-epoxide	100	54	57	1.2
Valproic acid	0.16	22	21	0.064
Ethosuximide	99	95	99	0.48
Diazepam	>99	5	4	84
Desmethyldiazepam	>99	4	4.3	24
Oxazepam	>99	13	3.8	7.7
Chlorazepam	>99	18.5	21	38

demonstrates the importance of the lipophilic properties of these drugs, whose degree of ionization is nil or low at physiological pH. Valproic acid is an exception: it is practically totally ionized and its penetration is the result of active transport.

A study of the disposition of eight derivatives of l-arylpiperazine showed that the distribution volume and diffusion into the brain were both closely related to the lipophilicity of these compounds.[39] The degree of lipophilicity was calculated using the retention constant, log k, determined by reversed-phase high-performance liquid chromatography. This study is relevant in that it deals with the biotransformation of many centrally acting drugs (*e.g.* trazodone, buspirone) into derivatives of l-arylpiperazine which concentrate in the brain and may contribute to the pharmacological effects of these drugs by their affinity for serotoninergic receptors.

In other cases, cerebral diffusion of a series of homologous drugs has been negatively correlated to $\Delta \log P$, a parameter defined as the difference between $\log P$ values measured in the octanol/water and cyclohexane/water systems. This parameter takes into account the competition between protein binding in the peripheral blood, related to octanol $\log P$, and the partition process in the lipophilic areas of the brain, which is related to cyclohexane $\log P$.[40]

An important point concerning centrally acting drugs is to establish whether there is any relationship between their diffusion into the brain and their *in vitro* affinity for cerebral receptors. A study of a series of benzodiazepines[41] showed that diffusion of the free drug into rat brain is closely related to lipophilicity, as shown by the HPLC retention index (Figure 8). No correlation was observed between diffusion and affinity for benzodiazepine receptors. The role played by metabolites should also be investigated. Determination of the cerebral diffusion of metabolites and knowledge of their affinity for receptors are both essential for an accurate estimation of their contribution to pharmacological activity as compared to that of the unchanged product.

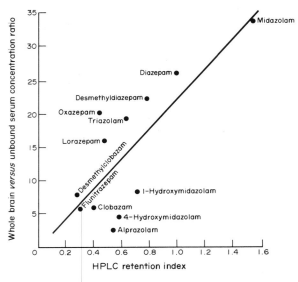

Figure 8   Relationship between diffusion of free drug in rat brain and lipophilicity of a series of benzodiazepines (reproduced from ref. 41 by permission of Springer-Verlag)

Whereas the degree of cerebral diffusion of benzodiazepines is related to their lipophilicity, their duration of action is related to their affinity for receptors. In fact, drug release by cerebral receptors is the limiting factor in clearance from the CNS. Thus, in a series of 1,4-benzodiazepines, a long elimination half-life in the CSF was associated with a marked affinity for receptors, both being correlated to the long duration of clinical and pharmacological effects.[42]

In some cases, it is possible to determine the diffusion of a drug into the CNS by assessment of its pharmacological or toxicological activity and to relate the findings to its lipophilic properties. Thus, the neurotoxic effects of some penicillins have been determined using semiquantitative EEGs in the rabbit.[43] These effects are assessed on the basis of the time of appearance of spike-wave paroxysms after administration of a range of doses and are related to the degree of lipophilicity of the penicillins expressed by the octanol/water partition coefficient P (equation 18). Penicillins with $P > 1$ are highly neurotoxic. Within a range of values of lipophilicity a linear relationship is observed between lipophilicity and cerebral diffusion, or between lipophilicity and the biological response for a series of homologous drugs.

$$\log(\text{neurotoxicity index}) \ = \ \log(11.38 \ + \ 14.2P^{2.2}) \tag{18}$$

$$n \ = \ 8, \quad r \ = \ 0.997$$

Beyond this range of values of lipophilicity there is a decrease in biological activity or diffusion. For example, a study of spiroperidol derivatives (dopaminergic antagonists) labelled with ^{18}fluorine in the rat showed that brain diffusion was significant in the case of compounds with a log P ranging from 3.18 to 4.0 and started to decrease for values beyond 4.2.[44] The lipophilicity range of most drugs with CNS activity was recently reviewed by Hansch *et al.*[45] who set the optimal value of log P for diffusion into the brain at approximately 2 (range: from 1.2 to 2.4).

Other factors which lessen the effect of lipophilicity are active transport and high molecular weight.

Active transport from the choroid plexuses to the CSF has been demonstrated for weak organic acids (penicillin, probenecid, salicylates) and, more recently, for weak bases (rimantadine, diphenhydramine). The affinity of ionized species of a drug for a carrier is significant in that the diffusion of a drug, which is highly or totally ionized at physiological pH, is facilitated.[46]

The effect of the molecular weight is illustrated by the following two examples. The first is a study of the diffusion of a range of substances into the brain of young and adult rats and rabbits[47] after intracarotid administration. Diffusion was measured using the brain uptake index (*BUI*), expressed as the ratio of the amount present in the brain to the amount administered, corrected for the amount present in cerebral blood vessels (measurement of the diffusion of indium-EDTA). The correlation between this index and the octanol/water partition coefficient P of 18 substances under study was good (equation 19), but it was better when account was taken of the molecular weight (equation 20 and Figure 9).

$$\log BUI = 0.36 \log P + 1.39 \tag{19}$$

$$n = 18, \quad r = 0.73$$

$$\log (BUI) \sqrt{MW} = 6.26 \log P + 17 \tag{20}$$

$$n = 18, \quad r = 0.86$$

After determination in the rat of the coefficient of cerebral capillary permeability (*PC*) of 27 compounds (natural products and drugs, mainly antineoplastic agents),[48] a good correlation was observed between this coefficient, the partition coefficient *P* and the molecular weight (*MW*) for 22 out of the 27 substances studied (equation 21). The 22 substances for which increased lipophilicity promotes diffusion had a molecular weight of less than 400, the other five being water and four compounds with a molecular weight higher than 400. In the latter cases, cerebral diffusion was almost nil (adriamycin, $MW = 543$; bleomycin, $MW = 1400$) or slight (epipodophylotoxin, $MW = 657$; vincristine, $MW = 825$). The limitation of diffusion caused by the high molecular weight explains why some drugs do not penetrate the CNS although they are highly lipophilic (*e.g.* cyclosporin, $\log P = 6.9$, $MW = 1202$).

$$\log PC = -4.60 + 0.411 \log[P \, (MW)^{-1/2}] \tag{21}$$

$$n = 22, \quad r = 0.91$$

As in the case of lipophilicity, the effect of molecular weight can nevertheless be counterbalanced by active transport, which has been demonstrated or suggested for compounds of high molecular weight such as natural peptides. Thus, insulin, insulin-like growth factors (IGF) and transferrin penetrate the CNS by transcytosis *via* carrier receptors found in the cerebral capillaries.[36]

23.3.3.2 Drug Transfer into Synovial Fluid

Transfer into synovial fluid is a prerequisite for a drug activity in joint diseases (osteoarthritis, rheumatoid arthritis, *etc.*) since in these disorders, the synovial fluid and/or the synovial tissue constitute the target biophases. This is mainly the case for antiinflammatory drugs and particularly for NSAIDs and antibiotics.

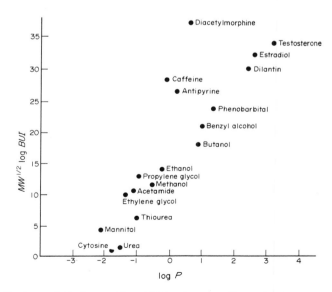

Figure 9 Correlation of brain uptake indices and lipophilicity of structurally unrelated compounds (reproduced from ref. 47 by permission of the American Physiological Society)

23.3.3.2.1 *Physiopathological aspects*

To enter the intrasynovial space, drugs in the perfusing blood must cross the synovium.[49, 50] The synovium is a double barrier comprising the capillary endothelium and the synovial lining layer composed of discontinuous cells lying over an interstitial matrix. The normal intrasynovial space contains a small volume of synovial fluid which represents a fluid phase extension of the extracellular interstitial matrix.

Since the fenestrated endothelial wall is not thought to limit the diffusion and filtration of drug molecules, the transfer of drugs through the synovium is limited by diffusion through the interstitial matrix (lipophilic and small hydrophilic molecules) and through the synovial cells (lipophilic molecules only).

The concentration gradient is governed by the free fraction. A facilitated transport system could enhance the transsynovial exchange of some molecules (glucose, anions, *etc.*). The transfer of proteins from plasma to synovial fluid occurs under normal conditions but the limiting barrier for these large molecules is the capillary endothelium rather than the synovial interstitium. Since proteins are removed from synovial fluid *via* the lymphatics at a higher rate, the protein level is lower in synovial fluid than in plasma.

The transfer of drugs into the normal synovial space is poorly understood. However, this question has little relevance to therapeutic use since drug treatment is usually administered to patients with synovial inflammation (synovitis) of various aetiologies (infectious synovitis, rheumatoid arthritis).

Synovitis significantly alters the properties of both the synovium (increased interstitial diffusion barrier, increased microvascular protein permeability) and the synovial fluid (marked increase in volume, decrease of pH). All these factors profoundly affect the transsynovial exchange of drug molecules depending on their physicochemical properties.

23.3.3.2.2 *Physicochemical determinants*

The physicochemical properties likely to influence drug transfer are the solubility of the drug and its pK_a which governs its ionization, its size and shape and the extent of plasma protein binding.[49, 50]

Nonsteroidal antiinflammatory drugs (NSAID) are acidic compounds with pK_a less than 5. Their acidic properties are likely to be of primary importance for *in vivo* antiinflammatory activity.

In a model of acute inflammation in the rat,[51] acidic compounds have been found to accumulate markedly in inflamed joints (concentration three to eight times higher than in control joints) whereas nonacidic ones did not. Moreover, doses of nonacidic congeners ten times higher than the doses of acidic drugs were required to achieve the same *in vivo* effect, the compounds having similar *in vitro* inhibitory activity against prostaglandin synthesis.

The acidic properties of inflammatory joint fluid would increase the lipophilicity of acidic compounds and favour their penetration into synovial tissue thereby improving their biological activity.

The diffusion of small hydrophilic molecules is limited primarily by the narrow diffusion path between the synovial lining cells. This barrier is further increased in inflamed synovium. Therefore, the size and molecular radius determine the rate of transsynovial exchange. Antiinflammatory agents and most antibiotics are small molecules with molecular weight less than 500 Da.

The high degree of plasma protein binding of NSAIDs could also favour their synovial penetration since, in pathological conditions, synovial inflammation augments the extravasation of plasma proteins (and bound drug) out of capillaries into the synovial fluid.

23.3.3.2.3 *Synovial fluid kinetics as compared to plasma kinetics*

The synovial fluid kinetic profile is dependent on the physiopathological status of the joint, the physicochemical properties of the drug and also its plasma pharmacokinetics.

Synovial fluid drug levels have been measured in many studies in patients with effusion of the knee which required aspiration. The reader is referred to reviews by Wallis and Simkins[50] and Furst.[52] Results of recent studies[53-60] are summarized in Table 8.

Briefly, the following common features can be described. Following oral or intramuscular administration, antiinflammatory drugs readily and rapidly enter the synovial fluid. However, the concentration peak in synovial fluid is delayed and lower than that in plasma. After an equilibration

Table 8 Synovial Fluid Kinetics of NSAID

Drugs	Dose	Dosage Duration (days)	Route	Time to peak SF/S (h)	Peak SF/S (mg L⁻¹)	Half-life SF/S (h)	Ref.
Flurbiprofen	100 mg b.i.d.	5	po	5.2/1.5	5 /13	5/ 3	53
Pirazolac	450 mg b.i.d.	7	po	3 /0.5	30 /59	66/31	54
Ketoprofen	100 mg	SD	IM	2 /0.5	1.3/ 6.6	5/ 4	55
Piroxicam	20 mg o.d.	7	po	3 /3	5.6/ 8	43/42	56
Naproxen	1100 mg	SD	po	7.5/ —	36 /—	31/15[a]	57
Etodolac	200 mg b.i.d.	7	po	3 /1	2.5/15.6	6/ 6	58
Tenoxicam	40 mg	SD	po	16 /6	1.4/ 4.3	—/40	59
Tiaprofenic acid	600 mg	7	po	8 /4	5.7/30.5	9/ 4	60

[a] Serum half-life from D. C. Rater, *J. Clin. Pharmacol.*, 1988, **28**, 518. Abbreviations: SF/S = synovial fluid/serum; po = oral; IM = intramuscular; SD = single dose.

time which depends on blood flow, synovial fluid volume and also on plasma half-life, the concentration in synovial fluid generally becomes higher than in plasma. The apparent elimination from synovial fluid either parallels, or is slower than, that from plasma. Following repeated administration, synovial fluid levels fluctuate much less than plasma levels.

Plasma kinetics of NSAIDs are therefore generally poorly predictive of the duration of action of these compounds. The slower dynamics of the drugs in synovial fluid could explain the duration of action, generally longer than predicted from the plasma half-life. Typical plasma and synovial concentration–time profiles are given for ketoprofen following a single intramuscular injection of 100 mg (Figure 10)[55].

The plasma to synovial fluid transfer is apparently nonstereoselective since equivalent concentrations of *S*- and *R*-enantiomers of tiaprofenic acid were measured both in plasma and synovial fluid.[61] However, in one study,[62] a pharmacokinetic model was developed to describe the plasma and synovial fluid kinetics of both *S*- and *R*-enantiomers of ibuprofen after dosing with the racemic mixture. The synovial levels of the active *S*-enantiomer exceeded those of the *R*-enantiomer, suggesting a stereoselective synovial uptake, but these observations could be the consequence of a stereoselective disposition of ibuprofen in plasma.

23.3.3.3 Bacterial Membrane

Antibacterial agents, whatever the extent of their distribution in the body, must cross an additional barrier in order to reach their target site, namely the bacterial wall, which is remarkable on account of its rigidity and permeability.

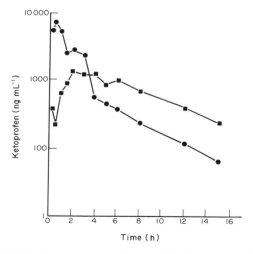

Figure 10 Serum (●) and synovial fluid (■) concentration–time profile of ketoprofen following administration of a single 100 mg intramuscular dose (reproduced from ref. 55 by permission of the CV Mosby Company)

In other words, to be effective, the ideal antibacterial agent, from the point of view of efficacy, should possess optimal ability to diffuse both into the invaded patient (vascular and/or extravascular sector) and into the infecting microorganisms.

Overall, bacteria are divided into two groups, Gram-positive and Gram-negative. The arrangement of the layers of the bacterial wall is summarized in Figure 11. Gram-negative bacteria possess an additional barrier: the outer membrane consisting of a bilipidic layer containing lipopolysaccharides and phospholipids with water-filled porin channels. On the other hand, Gram-positive bacteria have a thicker peptidoglycan layer.

Some bacteria have an additional external envelope or capsule, rich in polysaccharides or proteins. The exact composition of the membrane is characteristic of a given bacterial strain but modifications are seen in mutants.

The cytoplasmic membrane of Gram-negative and Gram-positive bacteria contains biological enzymatic systems which are involved mainly in the active transport of nutrients and in the synthesis of membrane constituents. Further information may be found in papers by Nikaido[63] and Chopra and Ball.[64]

One of the main features of the bacterial membrane is its ability to undergo evolution. Mutant strains become resistant mainly because of changes in membrane properties: changes in porins and active transport systems, increased synthesis of degradation enzymes (*e.g.* β-lactamases), and appearance of proteins which irreversibly bind to the antibiotic.

The progress of the antibacterial agent from outside the bacteria to the target enzymatic receptors depends on the site of the latter (membrane, cytoplasm). Penetration of the bacterial wall by antibacterial agents is effected by the same processes as passage through membranes. Approximately, diffusion (essentially passive, facilitated in some cases) takes place in the external membrane whereas active processes take place in the internal membrane or cytoplasmic membrane (Table 9).

The antibacterial agent crosses the external membrane of Gram-negative bacteria mainly by: (i) nonspecific passive diffusion through membrane channels (porin channel pathways); and (ii) diffusion across the bilipidic layer.

In the first instance, diffusion is conditioned by the following physicochemical properties of the drug: solubility in water, molecular weight (< 600), degree of ionization at neutral pH. The presence of many free carboxylic radicals in the external membrane of the capsule (when there is one) facilitates the passage of cationic molecules (*e.g.* aminosides). Differences in the diameter of the porin channels could account for differences in strains sensitivity.

Passage across the bilipidic layer appears to be more difficult but may be achieved: (i) by diffusion of hydrophobic molecules across areas linking the external and internal membranes (mechanism proposed for penicillins); and (ii) by means of a self-promoted pathway: some antibacterial agents (aminosides, polymyxin, quinolones) can, through various mechanisms (particularly chelation of divalent cations such as Mg^{2+}), disrupt bonds in lipopolysaccharide (LPS) chains, and thus

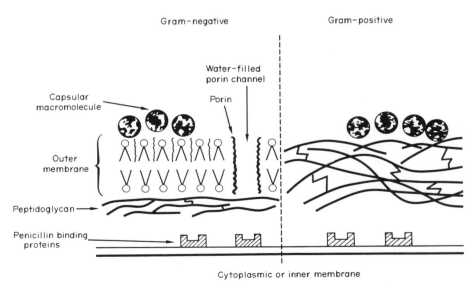

Figure 11   Schematic representation of the structure of bacterial membrane

Table 9 Mechanisms of Transfer of Antibacterials through Bacterial Membrane

Mechanism	Location	Antibacterials
Passive diffusion	Outer membrane	Aminosides
		Chloramphenicol
		Cycloserine
		β-Lactams
		Quinolones
		Tetracyclines
	Cytoplasmic or inner membrane	Chloramphenicol
Facilitated diffusion	Outer membrane	Fosfomycin
		Peptides
Active transport	Cytoplasmic or inner membrane	Aminosides
		Cycloserine
		Peptides
		Tetracyclines

penetrate the bilipidic layer. The affinity of aminosides for LPS is correlated to their ability to increase the permeability of the external membrane.[65]

Passage across the external membrane by facilitated diffusion of a transport-associated protein has been described for fosfomycin and for glycerol metabolism. Facilitated diffusion seems likely in the case of peptides.

Among the active transfer processes observed in internal membranes, the best known is that of aminosides, which is characterized by three phases,[66] of which only the last two are active. These are as follows. (i) *The ionic-binding phase*, which corresponds to the non-specific binding of drug to anionic groups of the cell surface. (ii) *EDP I (Energy Dependent Phase I)*: the transmembrane electric potential plays an important part in this phase which involves the electronic transfer chain (cytochromes). This phase corresponds to a slow rate of accumulation; its duration and intensity depend on the initial drug concentration in the membrane. (iii) *EDP II (Energy Dependent Phase II)*, which is characterized by a faster rate of accumulation than in the previous phase and depends on binding to the bacterial ribosome.

The other active processes described mainly concern peptides, cycloserine and tetracyclines. As regards macrolides, lincosamides and streptogramines, good correlation was observed between minimal inhibitory concentration and affinity, as shown by the competitive inhibition of [H^3] erythromycin binding to *Staphylococcus aureus*.[67]

As for other drugs, there are many studies of the structure–activity relationships of antibiotics. Activity is assessed mainly by determination of the drug's minimal inhibitory concentration (MIC) against the test strain. Studies dealing with the relationship between the transmembrane passage of the bacterial cell and the physicochemical properties of antibacterial agents are the most recent and are few in number.

Nikaido[63] reviewed the influence of these physicochemical properties for the β-lactam antibacterial agents. It is essential to distinguish the effect of the hydrophobic property of drugs from that of their electric charge. As regards the latter, the presence of hydration shells around charged groups causes hydrophilicity of the drug (case of β-lactam zwitterions), and facilitates permeation through porin channels. As already stated, the nature of porin channels facilitates cation passage in preference to anion passage.

Whereas hydrophobicity plays a limited role for zwitterionic compounds, the penetration rate of anionic cephalosporins is closely related to their hydrophobicity, expressed by the octanol/water partition coefficient. The mechanism by which the size of the anionic derivatives affect permeability becomes evident only with compounds of high molecular weight (slow rate of penetration of piperacillin and cefoperazone).

In all cases, it should be noted that considerable differences may occur depending on the type of bacterial strain being studied. Moreover, the balance in the periplasmic space between the 'input rate' and the hydrolysis rate appears to be most important in the case of compounds possessing significant resistance to β-lactamases. Thus, the poor activity of the compounds resistant to bacterial degradation is not due to low permeability.

To conclude, the relationship between the physicochemical properties of antibacterial agents and the efficacy of their diffusion is complex, mainly because of the part played by porin channels self-promoted pathways and the specificity of the bacterial wall (membrane constituents, degradation enzymes).

23.3.3.4 Feto–Placental Diffusion

Pregnancy is characterized by the existence of two entities, the mother and the fetus, separated by a barrier, the placenta, with a defined thickness, surface and blood flow. The fetus is fed and drugs pass through this barrier as a result of exchanges between maternal blood and fetal blood.

Gradual physiological changes during pregnancy affect mother, fetus and placenta. The nature of these changes, and the changes themselves cause modifications which affect the fate of xenobiotics in the mother, fetus and placenta.[68,69] It follows that the feto–placental diffusion of drugs is affected by their physicochemical properties, by the physiology of the placenta and, additionally, by changes in the mother, fetus and placenta as a whole.

In the mother, the two main factors affecting the distribution of xenobiotics are as follows. (i) An increase in cardiac output (from 30 to 40%) which causes an increase in renal blood flow, glomerular filtration rate (up to 50%) and pulmonary blood flow (30%). The clearance of drugs eliminated mainly by the kidneys and the lungs is thus increased. (ii) An increase in some body volumes: blood volume (up to 40–50%), total body water (17%) and fats (30%). This causes both an increase in distribution volume and a decrease in protein plasma concentration (*e.g.* a 30% decrease in albumin) which is responsible for the increase in the free concentration of the xenobiotic.

In the placenta, drugs are transferred from the mother mainly by simple diffusion according to Fick's principle, although the barrier is not permeable to compounds with a molecular weight exceeding 1000. The factors limiting transplacental passage are the same as those, previously described, which regulate the diffusion of drugs across membranes. Permeability varies during pregnancy as the thickness of the placenta gradually decreases (from 25 μm to 2 μm) whereas the exchange surface increases. There is little metabolism and active transport in the placenta.

In the fetus, the main factors affecting the distribution of xenobiotics are: (i) hemodynamic changes; (ii) changes in tissue affinity as the fetus grows (this affinity is the same as that of the mother, only at the end of pregnancy); (iii) the maturation of elimination functions (liver enzymes, glomerular filtration); and (iv) the presence of an exutory, the amniotic fluid, which contains the substances excreted by the kidneys and from which drugs and metabolites return to the placenta *via* the umbilical arteries but can be reabsorbed (there is, in the fetus, a direct amnion–gastrointestinal tract communication).

In practice, the physiological and pharmacokinetic changes occurring in the mother and in the fetus are the prime factors responsible for changes seen in the exposure and sensitivity of the fetus rather than changes in placental factor.

Recently, the importance of protein binding has been stressed.[70] The differences in drug binding between maternal and fetal proteins result from differences in the concentration of the two main proteins to which the drugs bind. In the fetus, albumin concentration is about 20% greater than in the mother whereas the concentration of α1 acid glycoprotein is only 30 to 40% that of the mother. The fetus/mother protein concentration ratio (f/m) can be used to estimate fetal protein binding according to equation (22), where B/F denotes the bound to free drug ratio. Thus, a good agreement between the observed and predicted fetal binding rates has been verified from maternal data for 23 different drugs. However, fetal albumin shows a significantly greater degree of binding when data on the binding of the same drugs in non-pregnant females are used.

$$B/F \text{ (fetal)} = [B/F \text{ (maternal)}] \, (f/m) \tag{22}$$

As a result of ethical and technical difficulties, data relating to drug transplacental transfer in humans are recent and fragmentary. However, relevant data have been published in general papers on anticonvulsants,[71] antiarrhythmics[72] and antibiotics[73] (Table 10).

23.3.3.5 Drug Transfer into Maternal Milk

The transfer of xenobiotics into milk may be considered to be not only a unique route of excretion but also an accessory absorption route in the infant.

Milk consists essentially of lactic secretions (compounds synthetized in the breast) and a few substances (water, some endogenous compounds) directly originating from the extracellular fluid, *i.e.* from the general circulation. Milk secretion (0.5–1 L daily) depends on a number of factors, both endogenous and exogenous (triggered by the infant, such as feed frequency). Authoritative data on mammary physiology and on the composition and secretion of milk are given by Rasmussen[74] and Wilson *et al.*[75]

Table 10 Fetal Diffusion of Some Drugs

Drugs	Plasma concentration ratio (*fetal/maternal*)
Antiarrhythmics	
Lignocaine	0.52–0.66
Quinidine	0.24–1.4
Disopyramide	0.39
Digoxin	0.38–1.0
Amiodarone	0.1 –0.15
β-Blockers	0.88–1.3
Antibiotics	
Dicloxacillin	0.1 –0.3
Cefoxitin	0.6
Moxalactam	0.4
Tetracycline	0.6
Sulfamethoxazole	0.7 –1.0
Aminosides	0.2 –0.5
Anticonvulsants	
Phenytoin	0.5 –1.0
Phenobarbitone	1.0
Carbamazepine	0.5 –0.8
Valproic acid	1.0 –2.0

The main factors involved in the transfer of drugs into milk are those governing their cross-membrane transfer. Three points deserve special attention: (i) milk is rich in lipids (3–4%, mainly triglycerides), which may explain differences in the distribution of drug between the fat and skimmed portion of milk; there are marked interindividual variations in the lipid content of milk; (ii) pH of milk is 7.0 (range: 6.4–7.6); and (iii) protein content is low: 8–11 g L^{-1}, *i.e.* seven to nine times less than that of plasma. The drug is therefore present mainly in the free form or slightly bound to milk.

Most drugs are transferred exclusively by simple passive diffusion of the nonionized form. Excellent quantitative correlations have been observed between physicochemical parameters and transfer into milk, as expressed by the M/P (concentration in milk/concentration in plasma) ratio. Meskin and Lien[76] have shown (Table 11) from human data that: (i) the excretion of basic drugs is inversely related to their partition coefficient (P) and degree of ionization (unionized to ionized drug ratio: U/D); and (ii) the excretion of acidic drugs is inversely related to their molecular weight (MW) and partition coefficient.

Strong binding to plasma proteins, which limits the amount of free drug able to diffuse from blood into mammary secretory cells and then into milk, could explain the negative correlation between excretion of acids and bases and their partition coefficient. In some cases, the influence of the partition coefficient is slight in comparison with the degree of ionization of the drug.[77]

The importance of the degree of ionization is in line with the concept of ion trapping. In many cases, distribution of the free drug between milk and plasma (M_u/P_u) may be calculated directly from equations (16) and (17) derived from Henderson–Hasselbach equations.

As the pH of milk is slightly more acid than that of blood, the transfer of bases into milk is facilitated at the expense of the transfer of acids (Figure 12).[74, 78] Thus the M/P ratio is lower or at best equal to 1.0 in the case of acidic drugs. In addition, the pH of milk accounts for the fact that the higher the dissociation of the drug in blood, the greater the transfer, particularly of weak bases.

Table 11 Correlation Between Transfer into Human Milk and Physicochemical Properties of Drugs[76]

	MW range	*log P range*	*pK$_a$ range*
Basic drugs $n = 15$	137/734 isoniazid/erythromycin	−1.47/7.30 tetracycline/imipramine	2.2/10.8 antipyrine/isoniazid
	$\log M/P = 0.265 - 0.153 \log P - 0.128 \log U/D$, $r = 0.801$		
Acidic drugs $n = 20$	128/454 thiouracil/methotrexate	−1.75/5.62 methotrexate/flufenamic acid	2.2/10.4 phenacetine/sulfanilamide
	$\log M/P = 2.068 - 0.162\sqrt{MW} - 0.185 \log P$, $r = 0.807$		

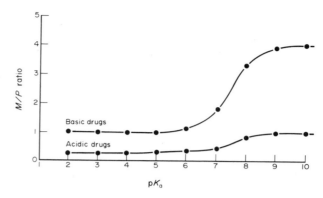

Figure 12   Milk to plasma concentration ratio as predicted from pK_a using equations (16) and (17) (milk pH = 6.8, plasma pH = 7.4) (reproduced from ref. 78 by permission of Blackwell Scientific Publication Ltd.)

When the observed M/P ratio does not correspond to the ratio calculated from the above equations, this may be mainly due to: (i) disregard of the degree of protein binding (the free drug concentration must be considered); and (ii) high molecular weight.

Moreover, the importance of the methodology and the reasons for the possible miscalculation of this ratio are discussed by Wilson *et al.*[79] The differences in milk composition and in the feeding patterns of young mammals highlight difficulties in interspecies extrapolation to secretion in milk and absorption of xenobiotics in the young. The secretion in milk of drugs for various therapeutic classes and their pharmacological or toxic effects on the infant have been the subject of some reviews.[79, 80]

23.3.3.6   Drug Transfer into Saliva

Saliva is secreted in the mouth by several paired salivary glands.[81, 82] It is produced by filtration of the plasma (secretory cells) and subsequent reabsorption (saliva ducts) at a flow rate of 0.6 mL min^{-1} under basal conditions. The composition of the secretion is variable between glands and is influenced by many factors including diet, age, sex and disease. Protein concentration in saliva is very low (no albumin); the pH is generally slightly acid (6.7), varying from 5 to 8, and increases following stimulation of saliva flow (by chewing).

Drugs are transferred between plasma and saliva primarily by passive diffusion through the epithelial membrane, and also by active transport in the case of some drugs such as lithium, penicillin, phenytoin and methotrexate. The transfer of most drugs is therefore largely related to their lipid solubility (diffusion), their molecular size (filtration through the aqueous pores of the membrane) and their acid–base properties (degree of ionization in plasma and saliva). In particular, basic drugs may be concentrated in saliva, as in milk, because of the slightly acidic pH (equation 17).

Drugs secreted in saliva are generally reabsorbed in the gastrointestinal tract. In fact, even for drugs concentrated in saliva, the process has little pharmacokinetic and therapeutic relevance.

Interest in drug–saliva transfer was motivated by therapeutic monitoring since, for some drugs, there is a correlation between saliva levels and protein-unbound concentration in plasma (*e.g.* anticonvulsant drugs). Therefore, saliva, which can be collected by noninvasive techniques, has been proposed as a substitute for plasma. However, this approach after initial interest has not been widely developed.

23.3.3.7   Experimental Evaluation of Drug Distribution

Two different approaches may be used for the experimental evaluation of drug distribution: on the one hand, indirect methods using descriptive pharmacokinetic models based on experimental data, generally from the systemic circulation; and on the other hand, specific methods for the qualitative and quantitative evaluation of tissue distribution, generally based on the measurement of tissue radioactivity after administration of labelled compounds.

Pharmacokinetic modelling is based upon the concept of multiple distribution compartments (see Section 23.3.1.1.3), and allows estimation of the following parameters: (i) the apparent volume of distribution in each identified compartment ($V_{app} = K_p \cdot V_T$); and (ii) the rate constants of inter-compartmental transfer (K_{ij}).

These methods are described elsewhere and the present paragraph will be limited to the basic description of direct methods.

The latter include the classical postmortem studies of tissue distribution in animals (auto-radiography) as well as the more recent imaging techniques allowing *in vivo* visualization of specific interactions between the drug and the biophase such as positron emission tomography (PET).

23.3.3.7.1 *Conventional techniques*

After treatment with a radiolabelled compound (^{14}C, ^{3}H), animals are sacrificed at different time intervals and the tissue distribution is estimated: (i) by counting radioactivity in each organ after dissection; and (ii) by autoradiography of the whole body (WBA) or of an individual organ.

In the WBA technique developed by Ullberg,[83] the sacrificed animal is immediately frozen and cut into thin slices (20–50 μm) which are dehydrated. The distribution of radioactivity is visualized across the whole section after several weeks of contact with a photographic film. This approach is limited to small animals (rodents, rabbits, small primates). It allows a qualitative and semi-quantitative evaluation of the tissue distribution with total compound-related radioactivity as well as the identification of target organs.

The dissection method allows a quantitative evaluation of the drug's global affinity for a number of organs selected *a priori* as well as the determination of the tissue to blood partition coefficient, K_p.

The latter method is limited by the constraints of dissection and, as a rule it only permits evaluation of the average concentration of radioactivity within a given organ.

The development of quantitation techniques for autoradiography by imaging analysis[84] allows the evaluation of radioactivity in fine structures that are difficult to dissect, such as individual brain structures (optic nuclei, hyppocampus, *etc.*) or the fetus.

23.3.3.7.2 *Biomedical imaging techniques*

The more recent biomedical imaging techniques allow the non-invasive *in vivo* study of the tissue distribution of drugs in animals and man.

The main approach, positron emission tomography (PET),[85] is based on the detection of radiation emitted by molecules labelled with short-lived isotopes such as ^{11}C ($t_{1/2} = 20$ min), ^{18}F ($t_{1/2} = 50$ min) or ^{76}Br ($t_{1/2} = 16$ min). PET is particularly useful in differentiating the tissue areas which display specific affinity for a given drug from those with non-specific affinity. PET allows visualization of the specific interaction of a drug with its receptor site(s) by observation of the displacement of a radiolabelled specific ligand after administration of the cold product.

The main limitation of the method is the relative lack of availability of specific ligands. Some adequate ligands exist for a number of receptors,[86] such as dopaminergic D_2 (haloperidol), benzodiazepine (flunitrazepam, suriclone), serotoninergic (serotonin), β-adrenergic (propranolol) and α-adrenergic receptors (prazosin).

To be suitable for PET studies, ligands must comply with some minimum criteria such as: feasibility of radiolabelling; high affinity for the receptor, in order to minimize the ratio specific/non-specific binding; saturable specific binding; and negligible *in situ* biotransformation.

As well as PET, some new approaches are currently being developed using for instance nuclear magnetic resonance (NMR); however, validation of these methods is still in progress.

The future development of PET and other imaging techniques appears very promising due to the high biological relevance of the data which allows evaluation of the specific and nonspecific aspects of drug distribution.

23.3.4 CRITICAL FACTORS FOR HEPATIC CLEARANCE

Hepatic clearance is a complex phenomenon depending on two distinct processes: firstly, metabolism which involves the chemical transformation of the xenobiotics, catalyzed mainly by the hepatic enzymatic systems; and secondly, the biliary excretion which involves elimination of the parent compound and/or its metabolites in the bile. The contribution of the liver to the overall biological drug behaviour is of prime importance because of the magnitude of hepatic blood flow and its remarkable metabolic capacity.

23.3.4.1 Physiological Aspects

23.3.4.1.1 First-pass effect

Among the physiological factors determining hepatic clearance, liver vascularization is particularly relevant to pharmacokinetics (Figure 4).

The liver is irrigated by both the hepatic artery and the portal system. The portal system is the necessary link between the gastrointestinal tract and the systemic circulation. Therefore, orally administered drugs are subject to hepatic elimination prior to reaching the systemic circulation. A first-pass presystemic elimination, generally referred as the hepatic first-pass effect,[87] may result.

The first-pass effect can have different origins. Xenobiotics administered orally can be degraded and/or metabolized in the gastrointestinal lumen, and also as they cross gastrointestinal membranes. However, the major presystemic source of loss of the unchanged drug is elimination by the liver. Knowledge of the first-pass effect process has been greatly facilitated by the development of *in vitro* experimental models, and particularly of perfused liver preparations.

Therefore, xenobiotics with high hepatic clearance undergo extensive first-pass elimination when administered orally, resulting in low absolute bioavailability associated with more rapid and extensive biotransformation (when compared to the i.v. route). Thus the metabolic ratio (metabolite to parent drug level) can be considerably higher following oral dosing; this can be responsible for differences in biological responses according to the route of administration. Prodrugs take advantage of this first-pass metabolism.

Extensive first-pass metabolism may result in pharmacokinetic problems with potential pharmacological and clinical implications related to marked variability of hepatic clearance. Hepatic clearance can vary both quantitatively and qualitatively (metabolic profile) from species to species.

Moreover, first-pass metabolism is dependent upon physiological factors (such as age, strenuous exercise and thermal stress) and pathological status (such as cardiac failure and hepatic disease), factors influencing the hepatic blood flow and/or the intrinsic activity of hepatic enzymes. Finally, genetic factors control some enzymatic reactions implicated in oxidative metabolism.

Compounds with high hepatic clearance can escape first-pass elimination when administered by alternative routes. This is the case when the venous vascularization at the sites of administration bypasses the portal system to directly access the systemic circulation. This situation occurs for the sublingual route and the percutaneous route, both routes being used to administer, for example, antianginal organic nitrates (nitroglycerin). The rectal route also offers, at least in part, the same advantage.

23.3.4.1.2 Hepatic extraction

The liver is irrigated by two different sources (Figure 4), the hepatic artery and the portal vein (hepatic inflow). The terminal hepatic arterioles and portal venules converge into a unique structure: the hepatic sinusoid. The blood perfusing liver tissue from these two sources is mixed before reaching the central vein (hepatic outflow).

The functional unit of the liver is the hepatocyte. Hepatocytes line the sinusoid and are in direct contact with plasma and therefore with the xenobiotics because of the absence of vascular tissue, as in other organs, and the existence of large fenestrations in sinusoidal epithelium. This unique feature is likely to contribute to the high extraction capability of the liver by favouring the hepatic uptake.[88]

Plasma protein binding should limit hepatocyte uptake since only the free fraction is available for membrane transfer, but hepatic uptake of substances tightly bound to plasma albumin is surprisingly efficient and there is evidence of direct uptake of bound drug. A direct interaction of the albumin–ligand complex with a binding site on the hepatocyte membrane could be involved in enhancing the dissociation of ligand from albumin.[7] Such an interaction is facilitated by the structural arrangement of the hepatic sinusoid.

The intrinsic ability of the liver to eliminate a drug is expressed by the intrinsic clearance (Cl_{int}) which is the maximal ability of the liver to irreversibly remove the drug by all routes (metabolic biotransformation as well as biliary excretion). However, the apparent hepatic clearance can be limited either by the intrinsic clearance or by factors that determine the availability of the drug to the hepatocyte, such as liver blood flow, diffusional barrier between blood and hepatocytes and plasma protein binding (Figure 1).

Diffusion between blood and liver tissue is therefore likely to be very rapid (not limiting) and hepatic clearance is generally considered to be perfusion-rate limited.[7]

Hepatic extraction is measured by the extraction ratio, E, and the hepatic clearance, Cl_H, is the product of the blood flow and the extraction ratio (equations 7 and 8). The physiological determinants of hepatic clearance are accounted for in physiological models of hepatic elimination.[7,87]

In the simplest case, the liver is considered as a well-stirred single compartment in which only the free concentration crosses the hepatocyte membrane and is available to metabolizing enzymes (well-stirred model). The liver can also be modelled as a large number of parallel tubes analogous to hepatic sinusoids. In this case, there is a drug concentration gradient in the perfusing blood across the tubes. The sinusoids can either be identical (undistributed sinusoidal perfusion model) or have functional heterogeneity (distributed sinusoidal perfusion model).

According to the well-stirred model, which is the most widely used, the hepatic extraction ratio E is expressed as a function of the intrinsic clearance of unbound drug (Cl_{int}^u), the fraction of unbound drug in the blood (fu) and the hepatic blood flow (Q_H), assuming that there is no diffusional barrier between the blood and the hepatocytes [the perfusion-limited case of equation (23)]. A fraction of dose equivalent to the extraction ratio is removed during initial transit through the liver of orally administered drugs (presystemic first-pass elimination). The fraction of a completely absorbed drug, escaping first-pass elimination, *i.e.* the apparent bioavailability (F) is given by equation (24). From these relationships, it becomes apparent that hepatic clearance, extraction ratio and apparent bioavailability will be differently influenced by physiological factors depending on the value of the total intrinsic clearance ($Cl_{int} = fu \cdot Cl_{int}^u$).

$$E = \frac{fu\,(Cl_{int}^u)}{Q_H + fu\,(Cl_{int}^u)} \tag{23}$$

$$F = 1 - E = \frac{Q_H}{Q_H + fu\,(Cl_{int}^u)} \tag{24}$$

For the highly extracted drugs ($E > 0.7$), *i.e.* drugs with a high intrinsic clearance relative to hepatic blood flow ($Cl_{int} \gg Q_H$), hepatic clearance is limited by the hepatic flow which represents the upper bound of hepatic clearance ($Cl_H \approx Q_H$ when $E \approx 1$). Lidocaine, propranolol and nitroglycerin are typical drugs with high hepatic extraction, extensive first-pass effect and flow-limited hepatic clearance. In their case, the intrinsic ability of the liver to eliminate these drugs is so high that plasma protein binding does not play a major role, *i.e.* bound as well as unbound drug are extracted (nonrestrictive elimination).

Conversely poorly extracted drugs ($E < 0.3$) have low intrinsic and hepatic clearances ($Cl_H \approx Cl_{int} \ll Q_H$). The hepatic clearance of these drugs is independent of the blood flow but is generally sensitive to plasma protein binding, particularly when the affinity for plasma proteins is high (restrictive elimination). The major determinant of both hepatic clearance and extraction ratio is the intrinsic clearance which is sensitive to induction or/and inhibition of enzymes involved in metabolism (enzyme-limited hepatic clearance). Procainamide, phenytoin and diazepam are typical drugs with low hepatic extraction ratio. The influence of physiological factors on hepatic clearance, extraction ratio and apparent bioavailability are summarized in Table 12.

Table 12 Influence of Physiological Factors on Hepatic Clearance, Extraction Ratio and Apparent Bioavailability.

	Highly extracted drugs $E > 0.7$, $Cl_{int} \gg Q$			Poorly extracted drugs $E < 0.3$, $Cl_{int} \ll Q$		
	Cl_H	E	F	Cl_H	E	F
Hepatic blood flow	+	0	+	0	+	0
Protein binding	0	0	±	+	+	0
Enzymatic activity (induction/inhibition)	0	0	+	+	+	0

+ = sensitive; 0 = unsensitive; ± = sensitive for highly bound compound.

23.3.4.2 Metabolic Clearance

Metabolism is a major route of drug clearance since only few drugs are excreted unchanged by the kidney. The most common pathways of drug metabolism, *i.e.* oxidation, reduction, hydrolysis (phase I) and conjugation (phase II) metabolic reactions, will be presented in detail in Chapter 23.5.

Metabolic processes generally lead to a decrease in lipophilicity of the parent compound[89] (in particular, oxidative metabolism and conjugation) favouring subsequent biliary and renal excretion of metabolites, which will be discussed further in Sections 23.3.4.3 and 23.3.5.

Metabolism is often responsible for biological inactivation of xenobiotics. However, the formation of pharmacologically and toxicologically active moieties is frequently observed.[90,91] The contribution of active metabolites to overall *in vivo* biological responses will depend on their intrinsic activity and on their own pharmacokinetic properties (distribution pattern in particular).

Since metabolism is mediated by enzymatic systems, it is a capacity-limited process. The enzymatic reaction rate is well described by the Michaelis–Menten equation (equation 9), characterized by a maximum reaction rate (V_{max}) proportional to both enzyme intrinsic activity and enzyme amount, and by the Michaelis constant (K_m) which is inversely related to the affinity of the drug for the enzyme system. The partial metabolic clearance ($Cl_{m,i}$) of the *i*th metabolic reaction is given by equation (25) (analogous to equation 10), whilst the overall intrinsic metabolic clearance is the sum of the partial clearances of all the metabolic pathways (equation 26).

$$Cl_{m,i} = \frac{V_{max,i}}{K_{m,i} + C} \tag{25}$$

$$Cl_{m,int} = \sum_{i=1}^{n} Cl_{m,i} \tag{26}$$

The pharmacokinetic implications of capacity-limited (Michaelis–Menten) elimination processes, already discussed, are particularly relevant for metabolic clearance since metabolism is the primary cause of nonlinear pharmacokinetics. Examples of drugs with nonlinear pharmacokinetics of metabolic origin include antiepileptics (phenytoin), anticancer drugs (5-fluorouracil) and β-blockers (propranolol). Drugs subject to both intensive first-pass effect and metabolic saturation can display dose- and time-dependent bioavailability (propranolol).[87]

Metabolism is one of the major sources of discrepancies between *in vitro* and *in vivo* biological activities of drug candidates. Therefore, the establishment of QSPR for metabolism could be important in drug design. However, data are scarce and concern mainly *in vitro* studies with isolated enzymes (see the review by Seydel and Schaper[3]). General trends indicate that lipophilicity favours the interaction of the xenobiotics with the metabolizing enzymes, probably by enhancing the microsomal localization and binding affinity of the drug for the enzyme.

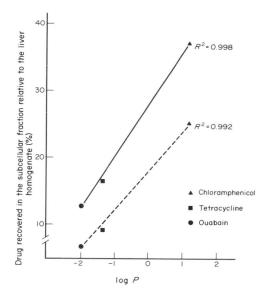

Figure 13   Correlation between the recovery of drug in subcellular fractions and lipophilicity: ——— 100 000 g pellet; - - - smooth microsomal fraction (reproduced from ref. 92 by permission of PJD publications Ltd.)

As an example, Ong *et al.*[92] found that lipophilicity did not influence hepatic uptake which was very rapid for nonpolar (chloramphenicol, $p = 13.8$) and polar or ionized substances (ouabain, $p = 0.01$ and tetracycline, $p = 0.03$). However, lipophilicity influences intrahepatocyte distribution since the concentration of drug in the microsomal fraction correlates with log P (Figure 13).

According to Seydel and Schaper,[3] the affinity for a microsomal NADPH-oxidase system ($1/K_m$) correlates closely with the lipophilicity of 14 structurally unrelated compounds. The acetylation of sulfonamides was also related to both the log P of the unionized form and the pK_a in *in vitro* as well as *in vivo* studies.

Strong correlation was observed[93] between the log P of a series of 13 β-adrenoceptor blocking drugs (β-blockers) and inhibition of lidocaine metabolism by rat liver microsomes. This is related to the high affinity of lipophilic β-blockers for cytochrome P-450.

Therefore, hydrophilic compounds are generally little metabolized and mainly eliminated by the renal and biliary routes whereas lipophilic ones are more extensively metabolized. The lipophilic/hydrophilic properties were found to determine the pharmacokinetic behaviour and particularly the clearance pattern of β-blockers administered intravenously[94] as well as orally[95] in humans. In particular, in both cases, the clearance of unchanged drug shifted from renal to nonrenal (essentially metabolic) as the lipophilicity of the compounds increased. As a consequence, following oral dosing, the intensity of the first-pass effect, negligible for log P values less than 1, increases markedly with the lipophilicity (Figure 14).

Similar relationships were observed for orally administered antiarrhythmic drugs in man.[95]

23.3.4.3 Biliary Clearance

Bile flow is the result of the secretory activity of the hepatocyte. The transfer of drugs across the biliary epithelium into bile is governed by active processes of secretion, distinct transport mechanisms being involved for organic anions, organic cations and neutral compounds.[88]

According to the above general definition of clearance, biliary clearance is expressed in equation (27). Bile flow is generally slow (less than 1 mL min^{-1} in man) but biliary clearance can be high as secreted compounds concentrate in bile and achieve bile to plasma concentration ratios higher than 100 (active secretion). Biliary clearance is therefore highly dependent on the efficacy of active and consequently capacity-limited processes. These processes may induce nonlinearities in the pharmacokinetics of a drug when biliary secretion is quantitatively prominent and rate limiting in overall elimination.

$$\text{Biliary clearance} = \frac{\text{Biliary excretion rate}}{C_p} = \frac{(\text{Bile flow})(\text{Bile concentration})}{C_p} \tag{27}$$

There are marked interspecies variations in bile flow and bile composition resulting in differences in excretion of xenobiotics.[88] Extrapolation of animal data to man is therefore difficult.

Drug (and metabolites) excreted in bile are stored in the gallbladder, then released into the intestine where they can be eliminated in faeces or reabsorbed through the intestinal membrane to

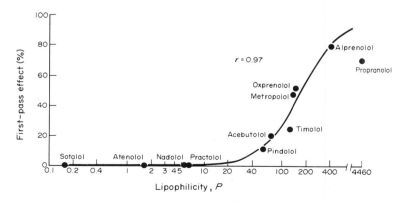

Figure 14 Relationship between first-pass effect and lipophilicity of β-blockers (reproduced from ref. 95 by permission of the New York Academy of Sciences)

complete an enterohepatic cycle (Figure 4). Xenobiotics which are excreted as conjugates can be reabsorbed following hydrolysis in the intestine (by enzymes of the intestinal flora, due to instability in alkaline intestinal pH). When quantitatively important, enterohepatic cycling is responsible for secondary peaks in the plasma profile with a reduction of total plasma clearance and a prolongation of apparent elimination half-life.[96]

The mechanisms of biliary excretion of xenobiotics remain unclear because of experimental difficulties; therefore structure–biliary excretion relationships are poorly understood.

Because of the low bile flow, xenobiotics must concentrate in bile in order to be excreted in significant amounts (bile to plasma ratio superior to 10). It is not the case for lipophilic molecules which are readily diffusible because they are subject to intensive reabsorption in the hepatic canaliculi. The presence of a strong polar group or a potentially ionizable moiety favours biliary excretion.[88]

Molecular weight is also a determinant of biliary excretion since molecules with a low molecular weight are reabsorbed when secreted into bile. The molecular weight should exceed a threshold (around 400–500) to allow biliary excretion[97] and xenobiotics with lower molecular weight are preferentially excreted in urine. In the rat,[98] for instance, excretion of compounds with high molecular weight ($MW > 500$) occurs mainly *via* the biliary route, and urinary excretion is low (5%). In contrast, high urinary excretion (90–100%) of unchanged compound as well as metabolites is observed for compounds with a molecular weight lower than 150.

Thus there is a cooperation between metabolism and biliary excretion since the polarity of metabolites is generally increased with respect to that of the parent drug. In particular, conjugation processes (with glucuronic acid, sulfate, *etc.*) increase both polarity and molecular weight favouring subsequent biliary excretion.

23.3.5 CRITICAL FACTORS FOR RENAL CLEARANCE

The kidney, like the liver, is a major route of excretion for most drugs whether they are excreted unchanged and/or as metabolites. The physiological processes involved in the renal excretion of drugs will be reviewed together with their main biological and physicochemical factors.

The renal clearance of a drug, in the context of clearance as defined above (Section 23.3.1.2.1), is the volume of plasma (blood) cleared of unchanged drug by the kidney per unit time. It is also the proportionality factor between the rate of renal excretion and the drug and the plasma concentration of that drug (equation 8).

Renal clearance, Cl_R, is the resultant of three basic physiological processes[99] involved in urine production: glomerular filtration, tubular secretion and reabsorption (equation 28). Comparison of this equation with overall renal clearance makes it possible to define the processes involved. Depending on active or passive nature of these processes, nonlinearities can occur with the pharmacokinetic consequences already described (clearance decreasing with increase in dose).

$$Cl_R = \frac{\text{filtration rate}}{C_p} + \frac{(\text{secretion rate} - \text{reabsorption rate})}{C_p} \tag{28}$$

The nephron is the functional unit which controls these processes. Each human kidney consists of about one million nephrons and each nephron is made up of a glomerulus linked to a tubule (Figure 15). The part played by glomeruli in drug filtration and that played by tubules in drug secretion and, later, reabsorption will be discussed.

23.3.5.1 Glomerular Filtration

Formation of urine first occurs in glomeruli by passive filtration of blood through a membrane wall made up of simple epithelium and having very wide pores (diameter from 75 to 100 μm). This means that only blood cells and proteins escape filtration, the exclusion limit corresponding to the molecular weight of haemoglobin, which is 68 000. Plasma is filtered at a flow rate of about 125 mL min^{-1}, typical of glomerular filtration rate (GFR). Only about 10% of renal blood flow (1100 mL min^{-1}) is therefore affected by filtration. The glomerular filter discriminates not only on the basis of molecular weight but also acts as a negatively charged selective barrier, promoting retention of anionic compounds.[100]

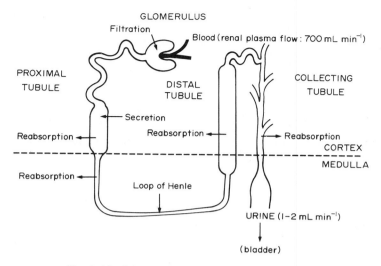

Figure 15 Schematic representation of the nephron

The only fraction of drug not bound to proteins, erythrocytes (or to other components of the blood), *i.e.* the free fraction present in plasma, is excreted by glomerular filtration.

The limiting value for clearance of a substance which is not protein bound, not secreted, not reabsorbed (or very slightly) and not metabolized, is therefore 125 mL min^{-1}. Such is the case for endogenous substances, *e.g.* creatinine, and certain rare exogenous substances, *e.g.* inulin, ^{51}Cr–EDTA and iothalamate, which serve as reference compounds for the quantitative assessment of glomerular filtration in man.

23.3.5.2 Tubular Secretion

Drug secretion occurs mainly in the proximal convoluted tubule; only the ionized fractions are secreted. It involves saturable active transport processes.

As a result of the active nature of these processes, drug binding to plasma proteins is not a factor which limits secretion. Substances transported by the same system compete with each other; this may account for drug interactions, some of which have pharmacological repercussions. Active secretion systems fall into two groups.[101,102,103] (i) For acids (organic anions) such as: *p*-amino-hippuric acid, probenecid and salicylates; loop diuretics and thiazide diuretics; most β-lactam antibiotics; sulfonamides and sulfonates; nalidixic acid and cinoxacin, and drug metabolites, *i.e.* glycine, glucuronic acid and sulfate conjugates. (ii) For bases (organic cations) such as: dopamine and *N*-methylnicotinamide (endogenous substances); quinine, quinidine, procainamide and *N*-acetylprocainamide; cimetidine and ranitidine; morphine and pethidine; quaternary ammonium compounds; and amiloride and triamterene.

Anion and cations are not secreted in the same segments of the proximal tubules. There may be some subsystems in addition to the above two systems.

Renal excretion of zwitterions is a more complex process since they may be secreted by both systems (this also applies to some cephalosporins). Mention should also be made of the special active system, within the distal tubules, which involves the sodium pump and is concerned with the elimination of cardiac glycosides (digoxin, ouabain).

23.3.5.3 Tubular Reabsorption

Reabsorption takes place all along the tubules and particularly in proximal tubules, Henle's loop, distal tubules and collecting tubules. The glomerular filtration rate is progressively reduced to 1–2 mL min^{-1} (urinary excretion rate) so that less than 1% of the filtrate is excreted. This marked reabsorption of water accounts for the concentration of drug in urine; thus, the concentration of a drug which is secreted but not reabsorbed is about 100 times as high in urine as it is in plasma.

Although there are active transport processes for some endogenous compounds or nutrients (electrolytes, glucose, amino acids, urate, vitamins) and for some drugs (oxopurinol), the reabsorption of most drugs is a passive process facilitated by the concentration gradient resulting from water reabsorption. The rate of reabsoption depends on the pH of urine, which governs the degree of ionization of the drug, and on urinary flow, which affects the concentration gradient of the drug.

23.3.5.3.1 Influence of urinary pH

The importance of urinary pH in xenobiotic excretion is explained by the fact that, unlike plasma pH which is highly stable, urinary pH ranges from 4.5 to 8.5. These pH variations can be normal (diurnal variations, effect of food), pathological (respiratory, metabolic acidosis or alkalosis) or secondary to drug treatment. Urinary pH, which affects the degree of ionization of the excreted drug, governs the passive reabsorption of the non-ionized fraction. In this way, urine acidification leads to reduced excretion of weak acids and promotes that of weak bases; alkalinization has the reverse effect. In fact, changes in urinary pH appear to be significant only for acids with a pK_a between 3.0 and 7.5 and for bases with a pK_a between 6 and 12. Outside these ranges, pH variations have no effect:[104] either (i) because the drug is almost completely ionized (pK_a of acids < 2; pK_a of bases > 12) and not reabsorbed; or (ii) because the drug is almost completely non-ionized (pK_a of acids > 8; pK_a of bases < 12) and actively reabsorbed whatever the pH.

The effect of pH on urinary excretion is illustrated by the well-known example of dexamphetamine,[105] urinary excretion of which ranges in man from 3 to 55% of the dose administered depending on the urinary pH (ranging from 5.0 to 8.0).

The same effect has also been fully reported in the case of sulfonamides. A study of the elimination of various sulfonamides in the rat (Table 13) showed that urine alkalinization promoted their excretion, the exception being sulfanilamide which, with a pK_a of 10.45, is always present mainly in the non-ionized form, whatever the urinary pH.[106]

The effect of pH on the renal excretion of a drug varies with the species. For instance, elimination of cinoxacin,[107, 108] a weak acid ($pK_a = 4.7$) excreted mainly in the unchanged form in urine, seems to be less sensitive to urinary pH variations in man than in the dog (Table 14). Renal clearance of cinoxacin in man is nonetheless highly correlated to urinary pH (equation 29).

$$Cl_R = 80.7 \text{ pH} - 354 \qquad (29)$$

$$n = 27, \quad r = 0.851$$

23.3.5.3.2 Effect of urinary flow

Drug reabsorption is inversely related to the intensity of the urinary flow; it also depends on the relative permeability, compared with that of water, of the drug across the tubular membrane.

This effect of diuresis on renal excretion has been demonstrated and used, particularly during diuretic treatment of barbiturate poisoning:[109] excellent correlations between urinary flow (ranging from 0.5 to 15 mL min^{-1}) and renal clearance of butobarbital, amylobarbital, cyclobarbital and phenobarbital (ranging from 0.5 to 40 mL min^{-1}) have been observed.

There are two types of reabsorbed compounds.[110] (i) Those with a capacity to diffuse through the tubular membrane equal to or greater than that of water (which is equivalent to reabsorption at equilibrium), *e.g.* phenobarbital. In this case, there is a linear relationship between the renal clearance of the drug and urinary excretion. (ii) Those, greater in number, which have a diffusion capacity less than that of water, *e.g.* theophylline. In this case, there is a convex curvilinear relationship between renal clearance and urinary excretion.

Table 13 Effect of Urinary pH on Elimination of Sulfonamides in the Rat[106]

| Drug | pK_a | Partition coefficient | | Elimination half-life (h) | |
		pH 5.4	pH 8.8	Acid urine[a]	Alkaline urine[b]
Sulfanilamide	10.45	0.0398	0.0358	1.67	1.57
Sulfathiazole	7.10	0.171	0.0056	2.31	1.17
Sulfamerazine	6.93	2.77	0.00515	6.90	3.84
Sulfasoxazole	4.62	0.806	0.0021	3.98	1.41

[a] pH = 5.4 $\pm$ 0.2 (oral administration of ammonium chloride). [b] pH = 8.6 $\pm$ 0.2 (oral administration of sodium bicarbonate).

Table 14 Influence of Urinary pH on the Elimination of Cinoxacin in the Dog and Man[107, 108]

	Urinary pH	$t_{1/2}$ (h)	Cinoxacin 24 h urinary excretion (% of dose)
Dog (10 mg kg^{-1} IV)	6.4	3.79	8.15
	5.5	15.8	7.8
	7.9	0.87	98.5
Man (0.5 g po)	6.0	1.1	65
	5.2	2.0	65
	7.7	0.6	80

The importance of the lipophilicity of the compound in reabsorption relates to its capacity to diffuse through membranes and will be discussed later.

23.3.5.4 Physicochemical Factors

The physicochemical properties affecting renal excretion of a drug are: (i) lipophilicity, which governs the degree of excretion in the unchanged form and possibly its tubular reabsorption rate; (ii) degree of ionization, which governs its tubular reabsorption rate; (iii) the molecular weight, which affects the renal excretion/biliary excretion balance and passage through the glomerular filter; and (iv) protein binding, which governs its glomerular filtration and tubular secretion rates (particularly when the latter is slight, and does not involve an active process).

The overlapping of physiological and physicochemical factors in renal excretion makes it necessary to control biological factors (pH, excretion rate) in order to study the effect of physicochemical parameters for a given chemical series.

The part played by the kidney in the overall elimination of the unchanged product or its metabolites depends on the molecular weight and on the hydrophilic or lipophilic properties of the compound. The following two examples illustrate the part played by lipophilicity in determining the rate of renal excretion. The first example is that of the renal excretion of 19 *N*-alkyl-substituted amphetamines in man:[111] urinary excretion of unchanged product decreases with the lipophilicity of the compounds (expressed by the apparent partition coefficient *n*-heptane/water at pH 7.4). The rate of urinary excretion is also inversely correlated to lipophilicity. Furthermore, metabolism by *N*-dealkylation is accelerated as lipophilicity increases. In this way, marked hydrophilia promotes rapid excretion and limited biotransformation.

The second example concerns the excretion of six barbiturates (5-alkyl-5-ethylbarbituric acids) studied in perfused isolated rat kidney.[112] There is a negative correlation between the renal clearance of these compounds and their lipophilicity (expressed by log *P* and calculated at pH 7.4). However, in this case, the decrease in renal clearance may be associated with an increase in its tubular resorption related to the lipophilicity of higher homologues.

The part played by the degree of ionization of a drug in its renal excretion has been demonstrated; it mainly affects its tubular reabsorption as a function of urinary pH. Its significance is illustrated in a study of the urinary excretion of various sulfonamides in the rat and rabbit.[106] It showed a poor relationship between the excretion rate constant and the partition coefficient of drugs at pH 7.4 whereas the correlations, which were negative, were excellent when the partition coefficients were determined at the urinary pH of the species under study (6.4 in the rat and 8.8 in the rabbit).

The degree of ionization is also important for metabolites. In many cases, in addition to modified intrinsic lipophilicity, metabolites have a degree of ionization which is different from that of the unchanged compound and this is likely to modify the intensity of the reabsorption processes. This is particularly true of tertiary *N*-alkyl amines, the metabolism of which by oxidation yields *N*-dealkylated or *N*-oxide metabolites with pK_a's different from those of the parent compound. In this way, metabolism of pethidine in man[113] results, in particular, in the formation of a more basic metabolite, norpethidine, and a more acidic one, pethidine *N*-oxide. The urinary excretion of both metabolites does not depend on pH value to the same extent (Table 15).

The effect of the rate of plasma protein binding on renal excretion is especially significant in the case of drugs excreted mainly by glomerular filtration. In fact, the glomerular filtration rate (*GFR*) (125 mL min^{-1}) is low in comparison with renal blood flow (1500 mL min^{-1}). Therefore, a drug eliminated solely by filtration has a low extraction coefficient and its renal clearance, Cl_R, depends

Table 15　pH Dependence of Urinary Excretion of Pethidine and Metabolites in Man[113]

		48 h Urinary excretion (% of the dose)	
	pK_a	Acid urine (pH 5.0)	Alkaline urine (pH 8.0)
Pethidine	8.5	27.6 ± 13.2	1.3 ± 0.4
Norpethidine	9.7	8.6 ± 2.4	1.7 ± 0.9
Pethidine N-oxide	4.9	1.3 ± 0.8	1.1 ± 0.7

mean $\pm$ s.d., $n = 6$.

directly on the free fraction *fu* (equation 30). This dependence has been demonstrated[17] in the case of antibiotics, particularly sulfonamides and tetracyclines (Table 16).

$$Cl_R = fu\,(GFR) \tag{30}$$

The effect of plasma protein binding has also been proposed to explain differences observed in the renal excretion of drug stereoisomers. This is true of moxalactam enantiomers and conjugates of mexiletine enantiomers which are mainly excreted by glomerular filtration. The case of enantiomers or diastereoisomers, such as disopyramide and quinidine/quinine, excreted by glomerular filtration and tubular secretion, appears to be more complex. If a difference in the rate of plasma proteins binding can account for a difference in filtration, it is also likely that a difference in affinity for the protein carrier is responsible for a difference in secretion.

23.3.6　GENERAL IMPLICATIONS OF DISTRIBUTION AND CLEARANCE IN DRUG RESEARCH AND DEVELOPMENT

The concepts of distribution and clearance allow one to express the affinity of a drug for the various sectors of the organism as well as the organism's capacity to eliminate that drug. The pragmatic exploitation of these concepts in the search and development of new molecules takes place both at explanatory and predictive levels.

During the early phases of drug research, the concepts of distribution and clearance (that are closely related to the physicochemical properties of molecules) may contribute to the selection of molecules within a chemical series, and eventually allow a feedback to drug design.

During the clinical development phase of a drug, these variables are also very important, as their behaviour in various physiopathological conditions allows one to categorize the populations of patients, to define risk populations and to establish the most appropriate (optimized) dosing schedule.

The study of these processes should therefore be started early in the research stage and be continued during drug development employing an integrated experimental approach involving both *in vitro* and *in vivo* techniques.[114].

Figure 16 shows the various levels of investigation of the processes involved in distribution and clearance, according to their potential to explain the *in vivo* dose/effect relationship. These are: (i) the peripheral or *systemic level*, which is the classical target of pharmacokinetic investigations (it is the most representative of the whole body, however, investigations at this site are more descriptive than

Table 16　Renal Clearance and Protein Binding[17]

Compound	fu (%)	Renal clearance (mL min^{-1})
Oxytetracycline	66	99
Tetracycline	45	74
Demethylchlortetracycline	32	37
Methacycline	22	31
Doxycycline	7	16

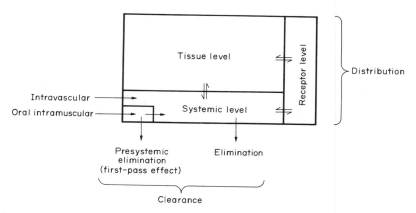

Figure 16 Various levels in experimental investigation of distribution and clearance of drugs

explanatory); (ii) the *tissue level* allows the observation of more specific distribution phenomena, with a pharmacological and toxicological content; and (iii) the *effector level* (biological target) where distribution is studied in the biophase and specific molecular interactions with receptor sites trigger both the pharmacological response (activity) and the biodynamic response (kinetics).

23.3.7 REFERENCES

1. C. Hansch and W. J. Dunn, III, *J. Pharm. Sci.*, 1972, **61**, 1.
2. R. M. Arendt, D. J. Greenblatt, R. H. de Jong, J. D. Bonin, D. R. Abernethy, B. L. Ehrenberg, H. G. Giles, E. M. Sellers and R. I. Schader, *J. Pharmacol. Exp. Ther.*, 1983, **227**, 98.
3. J. K. Seydel and K. J. Schaper, in 'Pharmacokinetics: Theory and Methodology', ed. M. Rowland and G. Tucker, Pergamon Press, Oxford, 1986, p. 311.
4. C. Hansch and J. M. Clayton, *J. Pharm. Sci.*, 1973, **62**, 1.
5. J. H. Lin, Y. Sugiyama, S. Awazu and H. Hanano, *J. Pharmacokinet. Biopharm.*, 1982, **10**, 637.
6. S. S. Kety, *Pharmacol. Rev.*, 1951, **3**, 1.
7. G. R. Wilkinson, *Pharmacol. Rev.*, 1987, **39**, 1.
8. J. G. Wagner, 'Fundamentals of Clinical Pharmacokinetics', Drug Intelligence Publications, Hamilton, IL, 1975.
9. B. Clark and D. A. Smith, *CRC Crit. Rev. Toxicol.*, 1984, **12**, 343.
10. K. J. Himmelstein and R. J. Lutz, *J. Pharmacokinet. Biopharm.*, 1979, **7**, 127.
11. M. Rowland in 'Pharmacokinetics: Theory and Methodology', ed. M. Rowland and G. Tucker, Pergamon Press, Oxford, 1986, p. 69.
12. M. Rowland and T. N. Tozer, 'Clinical Pharmacokinetics: Concepts and Applications', Lea & Febiger, Philadelphia, 1980, p. 34.
13. R. Latini, G. Tognoni and R. E. Kates, *Clin. Pharmacokinet.*, 1984, **9**, 136.
14. A. Iliadis, R. Bruno and J. P. Cano, *Comput. Biomed. Res.*, 1988, **21**, 203.
15. H. Boxenbaum, *J. Pharmacokinet. Biopharm.*, 1980, **8**, 165.
16. P. Pentel and N. Benowitz, *Clin. Pharmacokinet.*, 1984, **9**, 273.
17. W. J. Jusko and M. Gretch, *Drug Metab. Rev.*, 1976, **5**, 43.
18. J. P. Tillement, G. Houin, R. Zini, S. Urien, E. Albengres, J. Barre, M. Lecomte, P. d'Athis and B. Sebille, in 'Advances in Drug Research', ed. B. Testa, Academic Press, London, 1984, vol. 13, p. 60.
19. M. Lemaire and J. P. Tillement, *Biochem. Pharmacol.*, 1982, **31**, 359.
20. M. Laznicek, J. Kvetina, J. Mazak and V. Krch, *J. Pharm. Pharmacol.* 1987, **39**, 79.
21. R. W. Wittendorf, J. E. Swagzdis, R. Gifford and B. A. Mico, *J. Pharmacokinet. Biopharm.*, 1987, **15**, 5.
22. S. Oie and T. N. Tozer, *J. Pharm. Sci.*, 1979, **68**, 1203.
23. H. Kurz and B. Fitchl, *Drug Metab. Rev.*, 1983, **14**, 467.
24. W. A. Craig and P. G. Welling, *Clin. Pharmacokinet.*, 1977, **2**, 252.
25. W. A. Ritschel and G. V. Hammer, *Int. J. Clin. Pharmacol. Ther. Toxicol.*, 1980, **18**, 298.
26. L. S. Schanker, *Pharmacol. Rev.*, 1962, **14**, 501.
27. V. H. Cohn, in 'Fundamentals of Drug Metabolism and Drug Disposition', ed. B. N. La Du, H. G. Mandel and E. L. Way, Williams & Wilkins, Baltimore, 1971, p. 3.
28. W. H. Oldendorf, *Annu. Rev. Pharmacol.* 1974, **14**, 239.
29. W. H. Oldendorf and W. G. Dewhurst, in 'Principles of Psychopharmacology', 2nd edn., ed. W. G. Clark and J. del Giudice, Academic Press, New York, 1978, p. 183.
30. D. P. Rall, J. R. Stabenau and C. G. Zubrod, *J. Pharmacol. Exp. Ther.*, 1959, **125**, 185.
31. D. P. Rall, in 'Fundamentals of Drug Metabolism and Drug Disposition', ed. B. N. La Du, H. G. Mandel and E. L. Way, Williams & Wilkins, Baltimore, 1971, p. 76.

32. M. Bonati, J. Kanto and G. Tognoni, *Clin. Pharmacokinet.*, 1982, **7**, 312.
33. W. M. Pardridge, *Annu. Rev. Pharmacol. Toxicol.*, 1988, **28**, 25.
34. R. Norrby, *Scand. J. Infect. Dis.*, 1978, suppl. 14, 296.
35. R. M. Arendt, D. J. Greenblatt, R. H. de Jong, J. D. Bonin and D. R. Abernethy, *Cardiology*, 1984, **71**, 307.
36. W. M. Pardridge, *Endocr. Rev.*, 1986, **7**, 314.
37. B. B. Brodie, H. Kurz and L. S. Schanker, *J. Pharmacol. Exp. Ther.*, 1960, **130**, 20.
38. W. Löscher and H.-H. Frey, *Epilepsia (N.Y.)*, 1984, **25**, 346.
39. S. Caccia, I. Conti, A. Notarnicola and R. Urso, *Xenobiotica*, 1987, **17**, 605.
40. R. C. Young, R. C. Mitchell, T. H. Brown, C. R. Ganellin, R. Griffiths, M. Jones, K. K. Rana, D. Saunders, I. R. Smith and N. E. Sore, *J. Med. Chem.*, 1988, **31**, 656.
41. R. M. Arendt, D. J. Greenblatt, D. C. Liebisch, M. D. Luu and S. M. Paul, *Psychopharmacology (Berlin)*, 1987, **93**, 72.
42. W. A. Colburn and M. L. Jack, *Clin. Pharmacokinet.*, 1987, **13**, 179.
43. T. R. Weihrauch, H. Köhler, D. Höffler and J. Krieglestein, *Arch. Pharmacol.*, 1975, **289**, 55.
44. M. J. Welch, D. Yoon Chi, C. J. Mathias, M. R. Kilbourn, J. W. Brodack and J. A. Katzenellenbogen, *Nucl. Med. Biol.*, 1986, **13**, 523.
45. C. Hansch, J. P. Björkroth and A. Leo, *J. Pharm. Sci.*, 1987, **76**, 663.
46. R. Spector, *J. Pharmacol. Exp. Ther.*, 1988, **244**, 516.
47. E. M. Cornford, L. D. Braun, W. H. Oldendorf and M. A. Hill, *Am. J. Physiol.: Cell Physiol.*, 1982, **243**, (12), C161.
48. V. A. Levin, *J. Med. Chem.*, 1980, **23**, 682.
49. P. A. Simkin and K. L. Nilson, *Clin. Rheum. Dis.*, 1981, **7**, 99.
50. W. J. Wallis and P. A. Simkin, *Clin. Pharmacokinet.*, 1983, **8**, 496.
51. P. Graf, M. Glatt and K. Brune, *Experientia*, 1975, **31**, 951.
52. D. E. Furst, *Agents Actions*, 1985, suppl. **17**, 65.
53. L. Aarons, S. Salisbury, M. Alam-Siddiqui, L. Taylor and D. M. Grennan, *Br. J. Clin. Pharmacol.*, 1986, **21**, 155.
54. M. Kurowski, A. Dunky and M. Geddawi, *Eur. J. Clin. Pharmacol.*, 1986, **31**, 307.
55. P. Netter, B. Bannwarth, F. Lapicque, J. M. Harrewyn, A. Frydman, J. N. Tamisier, A. Gaucher and R. J. Royer, *Clin. Pharmacol. Ther.*, 1987, **42**, 555.
56. M. Kurowski and A. Dunky, *Eur. J. Clin. Pharmacol.*, 1988, **34**, 401.
57. R. Bruno, A. Iliadis, I. Jullien, M. Guego, H. Pinhas, S. Cunci and J. P. Cano, *Br. J. Clin. Pharmacol.*, 1988, **26**, 41.
58. M. Kraml, D. R. Hicks, M. McKean, J. Panagides and J. Furst, *Clin. Pharmacol. Ther.*, 1988, **43**, 571.
59. P. H. Hinderling, D. Hartmann, C. Crevoisier, U. Moser and P. Heizmann, *Ther. Drug Monit.*, 1988, **10**, 250.
60. F. E. Nichol, A. Samanta and C. M. Rose, *Drugs*, 1988, **35** (suppl. 1), 46.
61. N. N. Singh, F. Jamali, F. M. Pasutto, A. S. Russell, R. T. Coutts and K. S. Drader, *J. Pharm. Sci.*, 1986, **75**, 439.
62. R. O. Day, K. M. Williams, G. G. Graham, E. J. Lee, R. D. Knihinicki and G. D. Champion, *Clin. Pharmacol. Ther.*, 1988, **43**, 480.
63. H. Nikaido, *Pharmacol. Ther.*, 1985, **27**, 197.
64. I. Chopra and P. Ball, *Adv. Microb. Physiol.*, 1982, **23**, 183.
65. B. D. Davis, *Microbiol. Rev.*, 1987, **51**, 341.
66. H. W. Taber, J. P. Mueller, P. F. Miller and A. S. Arrow, *Microbiol. Rev.*, 1987, **51**, 439.
67. M. P. Fournet, R. Zini, L. Deforges, J. Duval and J. P. Tillement, *J. Pharm. Sci.*, 1987, **76**, 153.
68. G. W. Mihaly and D. J. Morgan, *Pharmacol. Ther.*, 1984, **23**, 253.
69. W. A. Parker, in 'Pharmacokinetic Basis For Drug Treatment', ed. L. Z. Benet, N. Massoud and J. G. Gambertoglio, Raven Press, New York, 1984, p. 249.
70. M. D. Hill and F. P. Abramson, *Clin. Pharmacokinet.*, 1988, **14**, 156.
71. H. Nau, W. Kuhnz, H. J. Egger, D. Rating and H. Helge, *Clin. Pharmacokinet.*, 1982, **7**, 508.
72. G. M. Mitani, I. Steinberg, E. J. Lien, E. C. Harrison and U. Elkayam, *Clin. Pharmacokinet.*, 1987, **12**, 253.
73. A. W. Chow and P. J. Jewesson, *Rev. Infect. Dis.*, 1985, **7**, 287.
74. F. Rasmussen, in 'Dietary Lipids and Postnatal Development', ed. C. Galli, G. Jacini and A. Pecile, Raven Press, New York, 1973, 231.
75. J. T. Wilson, R. D. Brown, D. R. Cherek, J. W. Dailey, B. Hilman, P. C. Jobe, B. R. Manno, J. E. Manno, H. M. Redetzki and J. J. Stewart, *Clin. Pharmacokinet.*, 1980, **5**, 1.
76. M. S. Meskin and E. J. Lien, *J. Clin. Hosp. Pharm.*, 1985, **10**, 269.
77. E. J. Lien, J. Kuwahara, R. T. Koda, *Drug Intell. Clin. Pharm.*, 1974, **8**, 470.
78. E. J. Lien, *J. Clin. Pharmacol.*, 1979, **4**, 133.
79. J. T. Wilson, R. D. Brown, J. L. Hinson and J. W. Dailey, *Annu. Rev. Pharmacol. Toxicol.*, 1985, **25**, 667.
80. H. C. Atkinson, E. J. Begg and B. A. Darlow, *Clin. Pharmacokinet.*, 1988, **14**, 217.
81. W. A. Ritschel and G. A. Tompson, *Methods Find. Exp. Pharmacol.*, 1983, **5**, 511.
82. M. Danhof and D. D. Breimer, in 'Handbook of Clinical Pharmacokinetics', ed. M. Gibaldi and L. Prescott, ADIS Health Science Press, Balgowlah, NSW, Australia, 1983, p. 207.
83. S. Ullberg, *Sci. Tools*, 1977, Special Issue, p. 2.
84. A. Schweitzer, A. Fahr and W. Niederberger, *Int. J. Rad. Appl. Instrum. (A)*, 1987, **38**, 329.
85. M. M. Ter-Pogossian, in 'Three-Dimensional Biomedical Imaging', ed. R. A. Robb, CRC Press, Boca Raton, Fl, 1985, vol. II, p. 41.
86. D. Comar, M. Berridge, M. Mazières and Ch. Crouzel, in 'Computed Emission Tomography', ed. P. J. Ell and B. L. Holman, Oxford University Press, New York, 1982, p. 42.
87. S. M. Pond and T. N. Tozer, *Clin. Pharmacokinet.*, 1984, **9**, 1.
88. C. D. Klaassen and J. B. Watkins, *Pharmacol. Rev.*, 1984, **36**, 1.
89. C. N. Manners, D. W. Payling and D. A. Smith, *Xenobiotica*, 1988, **18**, 331.
90. S. Garattini, *Clin. Pharmacokinet.*, 1985, **10**, 216.
91. D. V. Parke, *Arch. Toxicol.*, 1987, **60**, 5.
92. H. Ong, P. du Souich and C. Marchand, *Res. Commun. Chem. Pathol. Pharmacol.*, 1982, **35**, 237.
93. N. D. S. Bax, M. S. Lennard, S. Al-Asady, C. S. Deacon, G. T. Tucker and H. F. Woods, *Drugs*, 1983, **25** (suppl. 2), 121.
94. P. H. Hinderling, O. Schmidlin and J. K. Seydel, *J. Pharmacokinet. Biopharm.*, 1984, **12**, 263.

95. D. E. Drayer, *Ann. N.Y. Acad. Sci.*, 1984, **432**, 45.
96. W. A. Colburn, *J. Pharm. Sci.*, 1984, **73**, 313.
97. P. C. Hirom, P. Millburn, R. L. Smith and R. T. Williams, *Xenobiotica*, 1972, **2**, 205.
98. C. Fleck and H. Braünlich, *Pharmacol. Ther.*, 1984, **25**, 1.
99. E. R. Garrett, *Int. J. Clin. Pharmacol. Biopharm.*, 1978, **16**, 155.
100. B. M. Brenner, T. H. Hostetter and H. D. Humes, *Am. J. Physiol.*, 1978, **234**, F455.
101. J. V. Moller and M. I. Sheikh, *Pharmacol. Rev.*, 1982, **34**, 315.
102. B. R. Rennick, *Am. J. Physiol.*, 1981, **240**, F83.
103. A. Somogyi, *Trends Pharmacol. Sci.*, 1987, **8**, 354.
104. M. Rowland and T. N. Tozer, 'Clinical Pharmacokinetics: Concepts and Applications', Lea & Febiger, Philadelphia, 1980, p. 48.
105. A. H. Beckett, M. Rowland and P. Turner, *Lancet*, 1965, **1**, 303.
106. M. Yamazaki, M. Aoki and A. Kamada, *Chem. Pharm. Bull.*, 1968, **16**, 721.
107. J. F. Quay, R. F. Childers, D. W. Johnson, J. F. Nash and J. F. Stucky, *J. Pharm. Sci.*, 1979, **68**, 227.
108. R. H. Barbhaiya, A. U. Gerber, W. A. Craig and P. G. Welling, *Antimicrob. Agents Chemother.*, 1982, **21**, 472.
109. A. L. Linton, R. G. Luke and J. D. Briggs, *Lancet*, 1967, **2**, 377.
110. D. D. S. Tang-Liu, T. N. Tozer and S. Riegelman, *J. Pharm. Sci.*, 1983, **72**, 154.
111. B. Testa and B. Salvesen, *J. Pharm. Sci.*, 1980, **69**, 497.
112. J. M. Mayer, S. D. Hall and M. Rowland, *J. Pharm. Sci.*, 1988, **77**, 359.
113. K. Chan, *J. Pharm. Pharmacol.*, 1979, **31**, 672.
114. J. R. Gillette, *Food Chem. Toxicol.*, 1986, **24**, 711.

23.4

Sites of Drug Metabolism, Prodrugs and Bioactivation

CAMERON M. MACDONALD and ROBERT G. TURCAN

Hoechst Pharmaceutical Research Laboratories, Milton Keynes, UK

23.4.1 INTRODUCTION

Although the liver has long been considered to be the main site of drug-metabolizing activity in the body, there is now considerable evidence that many other tissues and organs also have the capacity to carry out a wide range of drug metabolism reactions. The extent to which these tissues and organs are involved can be influenced greatly by the route of administration. Thus, for example, the gastrointestinal tract may be involved in the metabolism of drugs given orally or rectally, whereas the lungs and skin might be involved when drugs are given by inhalation or by topical application. This presystemic metabolism of drugs is usually termed first pass metabolism and the extent of this in the exposed tissue will depend not only on the functional groups and structural elements of the drug but also on a number of physiological factors such as the intrinsic enzyme activity of the tissue, blood flow through the tissue, plasma protein binding, extraction ratio and, in the case of the gastrointestinal tract, gastrointestinal motility. Once a drug or its first pass metabolites reaches the systemic circulation it may be subjected to further metabolism, particularly by the liver. In some instances metabolism in other tissues such as blood, lung, kidney, brain and placenta may also be important, not only in relation to drug elimination but also in the control of pharmacological activity. Indeed, some specific reactions may only be carried out at one or two extrahepatic sites.

Although the levels of drug-metabolizing enzymes in other tissues are usually lower than the corresponding hepatic enzyme activities, the range of biotransformation reactions carried out appears to be similar. It is also recognized that the contribution of extrahepatic metabolism to total metabolism can be enhanced considerably by exposure to various inducing agents and environmental contaminants.

The first part of this chapter is therefore devoted to reviewing the different sites of metabolism in the body and takes into account the significance of routes of administration, the range of metabolic reactions carried out by the major tissues and organs (phase I and phase II reactions) and physiological factors which affect metabolic activity. In order to establish some sort of perspective for each tissue, an attempt has been made to compare hepatic and non-hepatic contributions to the overall metabolic process.

In the second part of the chapter, we move away from the traditional view of drug metabolism as an inactivating process by which drugs are rendered more soluble in order to facilitate elimination, to the role of metabolism in the bioactivation of drugs. The significance of metabolic activation and the relationship between chemical structure and toxicity is now better understood. This has led to the structural modification of drugs as a means of either blocking their metabolism (hard drugs) or enhancing metabolism along different routes, thereby reducing their toxic potential (soft drugs).

Bioactivation, however, can also have beneficial implications where the pharmacological effect is due wholly or in part to active metabolites. This has led to the design of prodrugs, this being the name given to compounds which must be transformed *in vivo* to exert their pharmacological effect.

An attempt has therefore been made to cover each of these aspects of bioactivation and examine how structural modification can influence toxicology and pharmacology.

23.4.2 SITES OF DRUG METABOLISM

23.4.2.1 Liver

No review covering sites of metabolism would be complete without some consideration being given to the role of the liver in the overall metabolic process. However, since other chapters in this work have devoted much of their content to hepatic processes, only the briefest sketch is required here to enable some perspective to be gained on the role of other tissues.

The liver is one of the largest organs in the body, typically weighing 1.2–1.8 kg in man and between 2 and 3 g $(100\ g)^{-1}$ body weight in the adult of most laboratory species. It is different from other tissues in that it has a dual blood supply, receiving approximately 25% and 75% of its total blood flow from the hepatic artery and portal vein, respectively. Since most of the gastrointestinal blood supply drains into the portal system, drugs taken orally must first pass through the liver before reaching the systemic circulation. The liver as the main organ of biotransformation may therefore have a considerable influence on the amounts of free drug or its metabolites reaching the extrahepatic tissues, and an assessment of hepatic drug disposition requires some examination of the influencing factors.

The extent to which a drug is metabolized by the liver or any other tissue is dependent not only on the metabolic activity of the tissue but also on several physiological and physicochemical factors.

Thus blood flow to the organ, the lipid solubility of the drug and the extent of protein binding, both in the plasma and within the cell, may influence the extent of metabolism. This is often best explained for the liver by the venous equilibrium model proposed by Rowland *et al.*[1-3] and Wilkinson and Shand,[4] which assumes that the concentration of the unbound drug leaving the organ is the same as that within it, and that the intrinsic capacity of the organ to metabolize and clear the drug is equal to the rate of elimination divided by the unbound concentration in the liver. When intrinsic clearance is low relative to blood flow, then the hepatic first pass effect is small. As intrinsic clearance increases relative to blood flow, then the first pass effect becomes more significant.

The uptake and metabolism of most drugs can also be related to the ease with which they can cross the hepatocellular membrane. Highly lipophilic drugs are often bound to intracellular membranes or organelles which may act as reservoirs from which the drug may be released either to undergo metabolism, biliary elimination or transfer to extrahepatic tissues. Apart from such passive mechanisms, an increasing number of foreign compounds have been found to be taken up by a saturable, sodium- and energy-dependent transfer system. Examples of these include ouabain, morphine and nalorphine.

As highlighted in Chapter 23.5, drug metabolism may involve a complex sequence of oxidative, reductive, hydrolytic and conjugative reactions, which may act sequentially or in competition with one another. The relative importance of each of these events will vary for different drugs even within a group of structural analogues. Within the hepatocyte, the enzyme systems primarily responsible for the biotransformation of drugs reside within the endoplasmic reticulum. These membrane-bound enzymes can easily be prepared for *in vitro* studies by differential centrifugation of liver homogenates. During the process the endoplasmic reticulum fragments and re-forms as small separate vesicles called microsomes.

The normal physiological role of the endoplasmic reticulum in the liver is associated with protein synthesis, especially serum albumin and various glycoproteins, synthesis and transformation of fatty acids (*e.g.* ω and $\omega - 1$ hydroxylation and desaturation) and biosynthesis of prostaglandins, cholesterol and the bile acids (*via* hydroxylation of cholesterol).

The drug-metabolizing enzymes of the endoplasmic reticulum are often divided up into phase I (oxidative/reductive) and phase II (conjugative) reactions. The phase I enzymes, or mixed function oxidases as they are also known, are capable of carrying out a wide range of drug-metabolizing reactions. Indeed, it is their non-specificity and versatility that makes it very difficult to make predictions about the metabolism of any particular compound.

There are at least two main electron transport chains associated with the mixed function oxidases. These are a NADPH-dependent system with cytochrome *P*-450 as the terminal oxidase, and a NADH-dependent system with cytochrome b_5 as the electron acceptor. A tremendous amount of work has gone into characterizing the properties and function of these enzyme systems and there are now available many excellent books and reviews on this topic.[5-7] In mammalian liver, cytochrome *P*-450 comprises around 20% of total microsomal protein and around 2% of total hepatic protein. Its earliest function in evolutionary terms may have been to protect the organism from the toxic effects of oxygen.

Of the phase II reactions carried out by the endoplasmic reticulum, glucuronide synthesis is quantitatively the most important in mammals. The enzyme responsible, UDPglucuronyltransferase, is present in the endoplasmic reticulum of most tissues, but the liver enzyme usually has the major role in the organism. A wide variety of drugs or their metabolites can be conjugated with glucuronic acid to form ether, ester, *N*- or *S*-glucuronides. Whilst it was originally thought that glucuronidation was a true detoxification reaction, it is now known that certain ester or *N*-hydroxyglucuronides are more reactive than the parent compound.

In addition to the endoplasmic-reticulum-catalyzed reactions, drugs may also be metabolized by a variety of other enzyme systems distributed throughout other parts of the hepatocyte. This may result in the exposure of the hepatocyte to substances which are activated at one site, but deactivated by further metabolism or conjugation at other sites. To highlight this point, Tables 1 and 2 summarize the localization of the phase I and phase II type reactions within the hepatocyte.

Chapter 23.5 highlights several examples of many of these biotransformations in greater detail.

23.4.2.2 Gastrointestinal Tract

The usual route of entry of drugs and other xenobiotics into the body is *via* the gastrointestinal tract. Since the metabolism of drugs has mainly been associated with the liver, the gastrointestinal tract has often been thought to be merely involved in the absorption process, and presystemic

Table 1 Intracellular Localization of Phase I Enzyme Systems[5, 8]

Subcellular Fraction		Major drug-metabolizing functions
Endoplasmic reticulum (microsomes)	Oxidation	Alcohols, carbonyl groups, aromatic carbon atoms, carbon–carbon double bonds, carboxylic acids, N, O and S dealkylation, deamination, nitrogen-containing groups
	Hydroxylation	Aliphatic and alicyclic carbon atoms
	Reduction	Alcohols, carbonyl groups, carbon–carbon double bonds, nitrogen-containing groups
	Hydrolysis	Esters, ethers, carbon–nitrogen bonds, non-aromatic heterocycles, dehalogenation
Mitochondria		Aldehyde dehydrogenase
		Monoamine oxidase
		Superoxide dismutase
		Esterases
Cytosol		Alcohol dehydrogenase
		Aldehyde dehydrogenase
		Azo and nitro group reductions
		Superoxide dismutase
		Xanthine oxidase
Lysosomes		Acid hydrolases
Peroxisomes		Catalase
Nuclei		Epoxide hydrase

metabolism by this tissue was thus considered to be of secondary importance. This neglect may in part be attributable to the nature of the gastrointestinal tract, comprising as it does 11 different anatomical sections and being between 400 and 450 cm long in adult man. Not only has this made it more difficult to work with, but it is also true that the preparation of suitable subcellular fractions has presented investigators with many problems. However, improvements in technology have now established that the gastrointestinal tract has important excretory and metabolic functions.

The mucosa of the small intestine possesses a wide range of phase I and phase II enzyme systems. Like the liver, these appear to be concentrated in the endoplasmic reticulum of epithelial cells. The monooxygenase system of the gastrointestinal tract is highly adaptive, in that activity may disappear in the absence of inducing agents but can be stimulated greatly, even to the level of the liver, when they are present (see Table 3). Since a normal diet, even those for laboratory animals, contains many inducing compounds, it is very difficult to obtain basal or control values for these enzyme systems. Moreover, cell turnover in the intestinal mucosa is very rapid, and the response to inducing agents can occur within 9–12 h of exposure.

The use of inducers has been particularly useful in examining enzyme activities throughout the intestine. Generally speaking, activities are highest at the proximal end of the small intestine, but activity has also been measured in the stomach, large intestine and caecum. It has also been shown that within the villi the highest activities are found in the villous tip cells and the lowest in the crypts.[10]

Whilst it is one thing to measure the intrinsic activity of these enzyme systems, it is quite another to estimate the overall contribution to the metabolic process. This has been attempted by Ullrich and Weber[11] who have calculated that the intestinal monooxygenase system may be responsible for up to 13% of the metabolism of a drug. However, interindividual variation may be so great as to cause considerable differences in therapeutic response.

Mucosal phase I biotransformation of numerous substances has now been established in animals and man. These include C hydroxylation of benzo[a]pyrene, biphenyl and aniline, N dealkylation of benzphetamine, perazine, ethylmorphine and aminoazobenzene, O dealkylation of phenacetin, deamination of 5-azacytidine, S oxidation of chlorpromazine, desulfuration of parathion and thiopentone, decarboxylation of L-DOPA and hydroxamate reduction of N-hydroxy-2-acetamido-fluorene.[9,12,13] In man there is also good evidence for mucosal metabolism of isoprenaline, isoethorine, rimiterol, salicylamide, tyramine, ethinylestradiol, talampicillin, pivampicillin and flurazepam.

Of the non-endoplasmic-reticulum-bound oxidative enzymes, alcohol dehydrogenase activity has been found in the stomach and small intestine and it is reported that the rates of alcohol metabolism

Table 2 Intracellular Localization of Phase II (Conjugating) Enzyme Systems[5, 8]

Conjugating system	Main enzyme systems	Intracellular localization	Typical substrates
Methylation	O-Methyl transferase	Cytosol	Catechols (*e.g.* dopamine)
	O-Methyl transferase	Endoplasmic reticulum	Catechols (*e.g.* dopamine)
	N-Methyl transferase	Cytosol	Phenolethanolamines (*e.g.* norepinephrine)
	N-Methyl transferase		Imidazoles (*e.g.* histamine)
	Transmethylation system	Endoplasmic reticulum	Sulfhydryl compounds (*e.g.* mercaptoethanol)
Acylation	N-Acyl transferases (transacetylation)	Cytosol	Aromatic amines, sulfonamides, hydrazines, aliphatic amines
	Amino acid transferases (*e.g.* conjugation with glycine)	Mitochondria	Carboxylic acids (*e.g.* substituted acetic acids)
Sulfation	Sulfotransferases	Cytosol	Phenols, alcohols, aromatic amino compounds
Glucuronidation	UDPglucuronyltransferases	Endoplasmic reticulum	Formation of ether, ester, N- or S-glucuronides (*e.g.* primary, secondary and tertiary alcohols, phenols, carboxylic acids, aromatic amines, sulfonamides, carbamates, heterocyclic compounds, sulfur-containing compounds)
Mercapturic acid formation	Glutathione S-transferases	Cytosol	Aryl compounds with labile halogen or nitro groups, epoxides, alkyl halides and nitroalkanes, aralkyl halides, β-unsaturated compounds

Table 3   Effect of Enzyme Inducers on Gastric Mucosal Enzyme Activities in Rat[9]

Enzyme system	Inducing agent	Increase in acitivity
Benzpyrene hydroxylase	3-Methylcholanthrene	14–20 ×
	Isosafrole	7 ×
	Piperonyl butoxide	11 ×
Biphenyl 4-hydroxylase	Isosafrole	2–3.5 ×
Biphenyl 2-hydroxylase	3-Methylcholanthrene	3 ×
Phenacetin metabolism	3,4-Benzopyrene	3–10 ×
7-Ethoxyresorufin deethylase	3-Methylcholanthrene	20 ×
7-Ethoxycoumarin deethylase	3-Methylcholanthrene	18 ×
	Phenobarbitone	2 ×
Ethylmorphine *N*-demethylase	Phenobarbitone	3.5 ×

Table 4   Phase II Reactions Occurring in the Gut[13]

Conjugation reaction	Substrate
Glucuronidation	*o*-Aminobenzoic acid
	Bilirubin
	Menthol
	4-Methylumbelliferone
	β-Nitrophenol
	Phenolphthalein
	Salicylamide
	Stilbestrol
Sulfation	Isoprenaline
	Salicylamide
	Paracetamol
	Estrogens
Glutathione conjugation	Sulfobromophthalein
	Ethacrynic acid
	Styrene oxide
	Octene oxide
	Benzo[*a*]pyrene 4,5-oxide
	1,2-Dichloro-4-nitrobenzene
Glycine conjugation	*p*-Aminobenzoic acid
	Salicylic acid
N acetylation	Sulfonamides
	p-Aminobenzoic acid
O methylation	Catecholamines

are similar to those of the liver.[15] Mitochondrial oxidative deamination is another important process. The most important biotransformations carried out by the gut wall, however, appear to be ester hydrolysis and conjugation reactions. Several hydrolytic enzyme systems have been characterized in the gut, the more important of which include non-specific esterases, deacetylases, β-glucuronidase, sulfatases and alkaline phosphatase. Epoxide hydrolase, which is an important detoxifying enzyme in liver, has a much lower activity in mucosal cells.

The activity of the conjugating enzymes of the gastrointestinal tract is such that when this activity is included, the overall metabolic capacity of the intestine approaches that of the liver. Table 4 comprises a summary of phase II reactions occurring in the gut as compiled by Caldwell and Marsh[13] for various species.

Sulfation of phenolic and steroid compounds is an important reaction in the gut wall. However, it is known that this reaction can easily be saturated and is subject to competitive inhibition. [16, 17] For example, the metabolism of the oral contraceptive steroid ethinylestradiol, which is metabolized to a greater extent by the gut wall than the liver, is reduced when paracetamol is coadministered.[18] Ascorbic acid is another compound which competitively inhibits the sulfation of drugs by the gut wall.

N Acetylation of sulfonamide drugs, particularly sulfadiazine, sulfathiazole and sulfamerazine, occurs readily in the intestinal wall.[15] It has also been shown that jejunal samples from man will acetylate *p*-aminobenzoic acid.[15]

Glutathione conjugation is an important protective mechanism against toxicity. In the absence of an active epoxide hydrolase, this enzyme system may have an important role to play in the intestine. It has been shown to be inducible by phenobarbital and polycyclic hydrocarbons and, at least in cattle, to have an activity of 60–100% of that in liver for chlorodinitrobenzene, ethacrynic acid and sulfobromophthalein.[12]

The enzyme UDPglucuronyltransferase, which is responsible for glucuronidation, is found throughout the gastrointestinal tract, with the highest activity in the duodenum. It is inducible by various agents including phenobarbital and 3-methylcholanthrene. Watkins and coworkers[15] have found that the rates of conjugation depend greatly on the aglycone. For sheep and cattle they reported enzyme activity of 20–50% of that for the liver using 1-naphthol, estrone, testosterone and diethylstilbestrol as substrates. In contrast to this, the ratio of glucuronidation between liver and intestine for rabbit has been estimated to be 1:4 for *o*-aminophenol and 1:10 for salicylic acid.[19]

O Methylation and conjugation with glycine are two further minor phase II pathways attributed to the gastrointestinal tract.

23.4.2.3 Gut Microflora

The microflora of the gut, representing as they do a diverse and highly adaptable range of organisms, have the potential to metabolize a wide range of drugs. In man more than 60 separate species have been characterized.

No part of the gastrointestinal tract is normally sterile. However, the acidic environment of the stomach is usually too hostile for most bacteria, although some such as *Lactobacilli* and *Streptococci* and some fungi which are associated with food may be present. Coprophagy in some species will also produce its own unusual gastric distribution. In man the microflora are usually confined to the lower gut, whereas the proximal small intestine in mice, rats and guinea pigs contains a greater number and variety of organisms. The number of organisms increases towards the lower gut, where there is less difference between the species. Facultative anaerobes such as *Coliforms*, *Staphylococci*, *Streptococci* and *Lactobacilli* and strict anaerobes such as *Bacteroides* and *Bifidobacteria* are found below the ileocaecal valve. These last two constitute the most abundant species within the gastrointestinal tract. The microfloral population of adult individuals is usually fairly stable, but can be influenced by diet, with those eating a vegetarian diet having a lower ratio of anaerobes to aerobes.

Much useful information on the role of the microflora in drug metabolism has been derived from the use of gnotobiotic or germ-free animals. Gnotobiotic animals are defined as animals in which all other organisms are known. Axenic animals have no microflora whereas mono-, di- or poly-associated gnotobiotics are associated with one, two or more known microfloral species. A simpler approach has been to reduce the intestinal microfloral populations with antibiotics, particularly neomycin and kanamycin.

The absence of microflora does, however, produce morphological changes in the intestine. In particular, the caecum of germ-free animals is greatly increased in size and has a much reduced motility. The intestinal wall is also much thinner and the faeces are changed in texture and quality.

Another more targeted method for establishing microfloral involvement in metabolism has centred on the use of specific inhibitors. For example, pretreatment of mice with lactose reduces the toxicity of the orally administered β-glycosidases by preventing hydrolysis of the glycoside linkages. Similarly, D-glucaro-1,4-lactone will inhibit the hydrolysis of drug glucuronides to the free aglycone.

The use of faecal or intestinal contents incubated under nitrogen, either directly with the drug in question or following cultivation of isolated strains or mixed cultures, has also been employed to define microfloral metabolism. Such experiments are really only valid after short periods of incubation, since prolonged anaerobic incubation can produce disproportionate changes in the bacterial population. Whilst much useful information has been derived in this way, many investigators have been unable to correlate their results with *in vivo* data.

It is usually assumed that presystemic metabolism by the microflora will be minimal for drugs which are quickly and completely absorbed, but will be more important for poorly absorbed drugs which remain in the gastrointestinal tract for a long time. This will be particularly true for man, since the drug will usually have to reach the hind gut before metabolism is likely. Drugs given *per rectum* as suppositories are inevitably more likely to be affected by microfloral presystemic metabolism than their oral counterparts.

Microfloral metabolism is mainly confined to three categories of reaction, namely hydrolytic reactions, reductions and reactions involving removal of certain groups. These are of considerable importance since they are virtually the opposite of the oxidative and conjugative reactions of the liver and most other tissues. The consequence of this to the drug is often greater lipid solubility, increase in pharmacological or toxicological activity and slower elimination from the body.

The realization of the importance of these microfloral conversions to the disposition and metabolism of drugs has prompted several excellent and detailed reviews on the subject.[20-23] It is therefore only necessary to summarize the main reactions at this point.

The major hydrolytic reaction carried out by gut microflora involves glucuronide hydrolysis. Almost all of this β-glucuronidase activity is associated with the anaerobic *Bifidobacteria*, *Bacteroides* and *Lactobacilli* of the large bowel. As a consequence of this, many drugs which are excreted in the bile as the inactive glucuronide may be liberated and reabsorbed as the pharmacologically active aglycone and may undergo further enterohepatic circulation. In germ-free rats or rats treated with antibiotics where the β-glucuronidase activity is very much reduced, there is far less reabsorption of drugs and a higher excretion of the glucuronide in the faeces. A few drugs such as chloramphenicol and prednisolone have specifically been administered as their glucuronides, which pass through to the large bowel unchanged where they are hydrolyzed to release the active antibiotic. An exception to this could be the ester glucuronides of carboxylic acids. Many of these have been shown to undergo transacylation involving migration of the aglycone from position 1 to positions 2, 3 and 4 of glucuronic acid at alkaline pH. These isomers are resistant to β-glucuronidase enzymes.

Other hydrolytic reactions carried out by the gut microflora include hydrolysis of glucosides, amides, sulfates and other esters. Typical examples of these and other hydrolytic reactions are shown in Table 5.

Glutathione conjugates may also be hydrolyzed by C—S lyases to form thiols, which in turn may be methylated to form methylthio, methylsulfinyl and methylsulfonyl metabolites.[24]

Of the wide range of reductive reactions carried out by the microflora, the most important involve nitro and azo group reduction, reductions of aldehydes, alcohols, *N*-oxides, sulfoxides, epoxides and double bonds (see Table 6).

Table 5 Hydrolytic Reactions Carried Out by the Gut Microflora[20-22]

Type of conjugate	Compound
Glucuronides	Bilirubin, chloramphenicol, diethylstilbestrol, digitoxin, diphenylacetic acid, phenolphthalein, estrone, metoclopramide, morphine, thyroxine, warfarin
Glucosides	Cardiac glycosides, cascara sagrada, amygadalin, cycasin
Sulfates	Sodium picosulfate, *N*-acetylaminosulfate, steroid sulfates
Sulfamates	Cyclamate, 3-methylcyclopentylsulfamate
Amides	Succinylsulfathiazole, 4-acetamidobenzoic acid
Esters	Carbenoxolone, pyrrolizidine alkaloids
Glutathione	Propachlor, 2,4,5-trichlorobiphenyl

Table 6 Reductive Reactions Carried Out by the Gut Microflora[20-23]

Group reduced	Compound
Nitro group	*p*-Nitrobenzene, *p*-nitrobenzene sulfonamide, 2-nitrotoluene, *p*-nitrobenzoic acid, metranidazole, nitrofurazone, misonidazole, chloramphenicol, dinitrotoluene
Azo group	Prontosil, amaranth, 4-dimethylaminoazobenzene, sulfasalazine, tartrazine
Aldehydes	Benzaldehyde, vanillin
Alcohols	4-Hydroxybenzyl alcohol
N-Oxides	Nicotine 1-*N*-oxide
Sulfoxides	5-Methylcysteine sulfoxide, sulfinpyrazone
Epoxides	Dieldrin
Ketones	Estrone, 4-hydroxyphenylpyruvic acid
Double bonds	Cinnamic acid, caffeic acid

Table 7 Other Reactions Carried Out by the Gut Microflora

Reaction	Compound
Dehydroxylation	L-DOPA, tyrosine, *p*-hydroxypropionic acid, protocatechuic acid, bile acids
Decarboxylation	L-DOPA, vanillin, amino acids, phenolic acids, hexobenidine
Deamination	L-DOPA, tyrosine, 5-fluorocytosine, urea
Dealkylation	3,4-Dimethylcinnamic acid, methamphetamine, eugenol
Dehalogenation	DDT, tetrachlorobenzene
Ring fission	Tartrazine, coumarins, flavenones
Aromatization	Quinic acid, steroids
Nitrosation	Dimethylamine
Methylation	Mercury

Nitro group reduction can be carried out by both mammalian cytochrome *P*-450 enzymes and by intestinal bacteria. However, comparative data from conventional and germ-free or antibiotic-treated rats have shown that most if not all the activity is associated with the microflora. For example, the conversion of *p*-nitrobenzenesulfonamide to *p*-aminobenzenesulfonamide is around 46% in conventional rats, but is a mere 6% in germ-free rats.[25]

The biological implications of nitro reduction can be considerable. It is believed to be the cause of nitrobenzene-mediated methemoglobulinaemia in conventional rats, the toxicity being absent in germ-free or antibiotic-treated rats.[26] In this instance the actual toxic agent is likely to be *p*-hydroxyaminobenzoic acid. Microfloral nitro reduction has also been attributed to causing the covalent binding of carcinogenic metabolites of dinitrotoluene in the liver,[27] the goitrogenic effects of chloramphenicol and the tumourogenic potential of the nitro heterocyclic compound, metronidazole.[29]

Similar comparative studies have shown that the microflora of the large intestine are also primarily responsible for the azo reduction of drugs. The conversion of sulfasalazine to sulfapyridine and 5-aminosalicylate is a typical example of this.[29] Azo reduction of prontosil has also been shown, though in this case it may be the biliary excreted *N*-glucuronide of prontosil which acts as substrate for the microflora.[30]

Amongst the other reactions now attributed to the microflora can be listed dehydroxylation, decarboxylation, dealkylation, dehalogenation, deamination, heterocyclic ring fission, aromatization, nitrosamine formation and methylation. All of these are covered in detail by Scheline.[20] Typical examples from this and other articles are listed in Table 7.

One further reaction which is worthy of mention is the cleavage of the N—S bond in the sweetening agent cyclamate to yield cyclohexamine. Whilst the gut microflora are known to be responsible for this, the mechanism remains unclear. Moreover, naive animals excrete only cyclamate. Only upon repeated administration do they acquire the ability to produce cyclohexamine.

23.4.2.4 Kidney

The kidney contains most of the drug-metabolizing enzymes found in the liver. Many of these, particularly those of the mixed function oxidase system, are located in the cortex on the smooth endoplasmic reticulum of the S3 cells of the proximal tubules. This localization is strategically important, since the proximal tubules are exposed to the greatest quantities of glomerular ultrafiltrate on the luminal side and blood-drug concentrations on the contraluminal side.

It is usually reported that drug-metabolizing enzyme activities are significantly lower in the kidney than in the liver (for examples see Table 8). However, many of these estimates have been based on *in vitro* assessment using isolated cells, homogenates or subcellular fractions such as microsomes. Since renal blood flow at 3 to 5 mL min^{-1} g^{-1} is considerably greater than that of the liver at 1 mL min^{-1} g^{-1}, and is almost entirely delivered to the cortex, others have suggested that kidney metabolism may have a greater significance *in vivo*.[31,32]

The kidney is thought to contain the same two microsomal cytochrome *P*-450 electron transport systems as the liver, though it is also reported that the mitochondria account for a significant proportion of this activity.[33] Little or no activity is associated with the medulla, though prostaglandin endoperoxide synthetase and fatty acid cyclooxygenases are found there in high concen-

Table 8 Comparative Drug-metabolizing Enzyme Activities in Liver and Kidney Preparations[5]

Species	Enzyme system		Ratio liver:kidney activities
Rat	Cytochrome *P*-450		6.5:1
	Aniline hydroxylation		20:1
	Hexobarbital hydroxylation		10:1
	Aminopyrene demethylation		14:1
	NADPH dehydrogenase		6.5:1
Rabbit	Cytochrome *P*-450		4.6:1
		C hydroxylation	7.3:1
	N-Methylaniline	*N* dealkylation	5.8:1
		N hydroxylation	6.3:1
	4-Aminobiphenyl	*N* hydroxylation	2.5:1
Man	Cytochrome *P*-450		9:1

trations.[34] Other enzymes found in the cortex are peptidases, *N*-acetyltransferase, demethylases, UDPglucuronyltransferase and sulfotransferase.[35]

Jones *et al.*[36] have listed the following substrates which have been shown to be metabolized by kidney microsomes: acetanilide, aminopyrene, aniline, benzo[*a*]pyrene, biphenyl, *p*-biphenylamine, decanoic acid, octanoic acid, chloromethylamine, coumarin, cyclohexane, ethylmorphine, hexobarbital, laurate, *N*-methylamine, 3-methylchloranthrene, myristic acid, nortriptyline, palmitic acid, phenitidine, stearic acid and testosterone. Apart from the cytochrome *P*-450-mediated reactions, the kidney is also capable of conjugation with glucuronic acid and sulfate (paracetamol, morphine), conjugation with amino acids (benzoic acid, salicylate), acetylation (cilastin, *p*-aminobenzoic acid), *N* acetylation (sulfisoxazole), deacetylation (cephapirin, cephalothin), methylation (isoproterenol), *N* oxidation (mepiridine), sulfoxide reduction (sulindac) and hydrolysis (imipenem).[31]

Glucuronide conjugation in the kidney usually only represents around 3–4% of that of the liver,[32] whereas the liver–kidney ratio for GSH *S*-transferase ranges from 1.5:1 to 8.8:1 for different substrates.[37] *N*-Acetylation is an important metabolic reaction in the kidneys, though the activity is usually somewhat lower than for the liver. However, citastitin is acetylated only within the liver.[38]

It is now also recognized that impairment of renal function not only affects the clearance of renally eliminated drugs but also alters the activities of the drug-metabolizing enzymes in the liver by causing reduction in cytochrome *P*-450 levels.[31] The mechanism is thought to involve a disturbance in heme biosynthesis, which can partly be overcome by administration of δ-aminolevulinic acid.[39]

23.4.2.5 Lung

It would appear that the lungs can metabolize as wide a range of xenobiotics as the liver, although the latter has a much larger capacity. This can be attributed to the lower concentration of cytochrome *P*-450-producing cells in the lung rather than to an intrinsically lower enzyme activity. The metabolic importance of the two organs is largely dependent on rate of delivery. Those compounds entering the body through the gastrointestinal tract will mostly be metabolized by the gut and the liver, whereas inhaled substances will be far more likely to be metabolized in the lungs.

In addition to the metabolism of xenobiotics, the lungs have an important role in the metabolism of endogenous compounds. These reach the lungs *via* the pulmonary circulation. Metabolism of such compounds seems to be highly selective. For example, the metabolism of biogenic amines such as noradrenaline (*O* demethylation and deamination) and *S*-hydroxytryptamine (deamination) by monoamine oxidases is important in regulating their effects by inactivating them. Similarly, PGE_2 is oxidatively inactivated by the enzyme PG-15-hydroxydehydrogenase. Other compounds may be activated by metabolism. A good example of this is the hydrolysis of the inactive decapeptide angiotensin I to its active octapeptide angiotensin II by angiotensin converting enzyme (ACE). Further comment on the biological significance of these and similar reactions can be found in various review articles.[40–42]

Whilst *in vitro* drug studies have been useful in characterizing the drug metabolism enzymes of the lung, such data may have very little meaning if drug and tissue are unlikely to come into contact *in vivo*.

The major oxidative reactions of drugs in the lung involve the cytochrome *P*-450 system. It has been estimated that the ratio of cytochrome *P*-450 in liver and lung microsomes may be 3:1 in rabbit,

Table 9 Compounds Metabolized by Drug-metabolizing Enzymes in the Lung [40–42]

Enzyme system	Typical substrates
Oxidative reactions	
N demethylation	Benzphetamine, aminopyrine, dimethylaniline, N-methylaniline, N-methyl-p-chloroaniline, ethylmorphine, imipramine
O demethylation	p-Nitroanisole
O deethylation	Ethoxycoumarin
Aromatic hydroxylation	Benzene, biphenyl, acetophenetidin, aniline, N-methylaniline, coumarin, benzo[a]pyrene, o-chloroaniline
Aliphatic hydroxylation	Xylene, decane
Oxidation of phosphothionates	Parathion
N oxidation	Aniline, N-ethylaniline N,N-dimethylaniline, perazine, imipramine
Reductive reactions	
Steroid dehydrogenases and hydrogenases	Cortisone
Nitro reduction	Chloramphenicol, p-nitrobenzoic acid
Azo reduction	Neoprontosil
Hydrolytic reactions	
Epoxide hydrolase	Epoxides
Ester hydrolysis	4-Nitrophenyl esters

10:1 in rodents and even higher in man. By comparison with the liver, the human lung is reported to possess less than 3% of the activities for microsomal oxidation of benzo[a]pyrene, phenacetin and 7-ethoxycoumarin, and 11.3% and 22.8% of the activities of epoxide hydratase and glutathione S-transferase respectively.[13] Amongst the reactions involved are demethylation, aromatic and aliphatic hydroxylations, N oxidation and oxidation of phosphothionates.[40] The lungs also carry out a range of reductive and hydrolytic reactions, including hydrolysis of epoxides which constitutes a major defence reaction against cellular toxicants. Examples of some of the more important reactions within these categories are listed in Table 9.

The lung is also involved in a wide range of conjugation reactions, including conjugation with glutathione, glucuronic acid, sulfate and N-acetylation. Particularly important within this group are the glutathione conjugates, where the enzyme glutathione S-transferase is responsible for protecting against electrophilic attack by reactive metabolites.

The local formation of toxic metabolites within the lungs usually appears to be the mechanism by which specific injury is caused in this tissue. Bleomycin is a glycoprotein anticancer agent which is frequently used because of its low incidence of other tissue reactions. Pulmonary toxicity has, however, been demonstrated both *in vivo* and *in vitro*. Recent findings now suggest that the toxicity may be related to the absence of bleomycin hydrolase in lung tissue.[44]

Nitrofurantoin also produces pulmonary toxicity in man, but in this case it may be metabolic activation which is responsible.[45] The toxic effects of paraquat have been ascribed to the peroxidation of cellular lipids in lung tissue through the involvement of 'active oxygen'. Both the poisoning and the lipid peroxidation can be partially inhibited by superoxide dismutase.[46,47] Volatile anaesthetics, on the other hand, which are administered by inhalation are not metabolized by the lungs and exert their toxicity mainly in the liver and kidney.[49]

23.4.2.6 Other Tissues

Drug-metabolizing enzymes have been characterized in most other tissues of the body. It is, however, unlikely that any contribute significantly to the overall systemic clearance of drugs. Nevertheless, specific reactions can have important biological implications if they are related to the pharmacological or toxicological effects.

The skin is the largest organ of the body and contains a wide range of cytochrome *P*-450-dependent monooxygenases, reductases, esterases, glutathione S-reductase, glucuronyltransferase and epoxide hydratase.[14] Much of this activity is probably located in the epidermis.[48,49] The nature of skin, presenting as it does a barrier to the external environment, discourages the percutaneous absorption of hydrophilic compounds. However, many compounds are administered topically and absorption can be enhanced by the matrix in which the drug is formulated. Cutaneous exposure of environmental chemicals may cause cytotoxic effects, and there is increasing evidence to indicate that topically applied polycyclic hydrocarbons may produce carcinogenic effects through their metabolic activation in the skin.[50]

The blood–brain barrier provides a protective envelope for the brain by limiting the uptake of many compounds, particularly if they are polar in nature. Notwithstanding this, various compounds are taken up by the brain and drug-metabolizing enzyme activities of both phase I and phase II types have been recorded. Inevitably, most work has been done on animal tissue, particularly from the rat. The greatest interest has focussed on centrally acting compounds and whether metabolism can influence their pharmacological actions.

Sulfation of biogenic amines appears to be one such reaction of importance. For example, 3-methoxy-4-hydroxyphenylglycol sulfate is the major metabolite of noradrenaline in the brain.[51] Sulfation of dopamine[52] and its metabolite 4-hydroxy-3-methoxyphenylethanol[53] has also been demonstrated. N acetylation has been demonstrated for 3,4-dimethoxyphenylethylamine, 5-methoxytryptamine, phenylethylamine and tryptamine,[54] and N methylation of normorphine and norcodeine to morphine and codeine.[55] In isolated brains from dogs and monkeys, drug metabolism was demonstrated by the demethylation and acetylation of aminopyrine and glucuronidation of oxazepam.[56] The rates were, however, considerably slower than for liver.

It has also been suggested that the metabolism of various endogenous amines which are structurally related to catecholamines may play a role in the actions of a number of centrally acting drugs. It may therefore be significant that phenylaniline can be converted to phenylethylamine and tyramine by rabbit brain, since these metabolites have been implicated in a number of cerebral mechanisms.[57]

The blood is an important source of esterase activity towards carboxylic acid esters.[58] Some of the enzymes involved, such as carboxylesterase, arylesterase, cholinesterase and 6-phosphogluconolactase, are now known. For some drugs such as procaine[59] and malathion,[60] hydrolysis in the blood is faster than for the liver, whereas for others such as methyl retinoate[61] and α-naphthyl acetate[62] it is much slower. Leinweber[58] in his review on carboxylic ester hydrolysis lists amongst others the following substrates for blood esterases: phorbol diesters, aspirin, tributyrin, phenyl butyrate, atracurium, cocaine, pilocarpic acid esters and morphine diesters including heroin.

N Acetylation of various drugs also occurs in the blood. Weber *et al.*[63] have covered much of this in their review and highlight species differences in addition to substrate specificity. Typical substrates include *p*-aminobutyric acid, procainamide, sulfanilamide and sulfamethazine. Oxidative deamination of biogenic amines probably also occurs in blood, since monoamine oxidase activity is present in platelets and plasma. The metabolism of 3,4-dihydroxyphenylacetaldehyde (DOPAL) and 5-hydroxyindole-3-acetaldehyde, the aldehydes derived from dopamine and 5-hydroxytryptamine, has also been investigated in human erythrocytes, leucocytes, platelets and plasma.[64] Reduction of the keto group of HWA 285, a xanthine derivative, has also been observed.[65]

Drug-metabolizing activities have been measured in most other tissues of the body. To avoid further repetition the interested reader is therefore referred to the reviews collated by Gram[42] which, in addition to providing invaluable material for this review, also contain articles on drug metabolism in the placenta, adrenals, lymphoreticular cells, gonads and ocular tissues.

23.4.3 PRODRUGS

23.4.3.1 Background

The goal of drug design is to correlate biological activity and physicochemical properties. However, drug design should not merely be aimed at increased pharmacological activity. Instead, it is more desirable to achieve a better ratio between the activity and toxicity of a drug; that is, to increase its selectivity. The majority of drugs in use today have been developed by a traditional empirical approach where several thousand substances are screened for biological activity in animal models. Of these only one may eventually become a new drug. Ideally, it should be possible to design a drug by a solely theoretical approach where a prediction is made of its pharmacodynamic, toxic and metabolic properties before synthesis. This type of design is not yet possible because of our lack of knowledge about structure–activity and structure–metabolism relationships. There are, however, three well-established qualitative design concepts which can be used to help optimize certain metabolic characteristics starting from a lead compound. These are the design of prodrugs, hard drugs and soft drugs.

In these approaches the design aims at predictable predetermined ways of drug transformation or excretion, which contrasts with conventional drug approaches in which there is generally a non-predictable biotransformation.

Prodrugs are intentionally formed derivatives of active substances. They can be used to improve drug stability, enhance absorption, allow site-specific delivery, mask side effects or extend the duration of action. Ideally, the prodrug itself is inactive and is converted to the active agent *in vivo* as soon as its target is achieved.

In the hard drug approach, a pharmacologically active molecule is produced which is resistant to biotransformation and is excreted unchanged.[66,67] Metabolic stabilization avoids the appearance of potentially toxic intermediates of biotransformation and implies a longer persistence in the body and less waste of active substance. Inter- and intra-subject variability of drug response and the risk of drug–drug interactions are reduced. Compared with conventional drugs, the hard drug approach may allow safer and simpler therapy. Metabolic stabilization can be achieved by protecting vulnerable positions of an active molecule or by replacing substituents by stable ones. Substitution of a hydrogen or a methyl group by a halogen is one such example. The half-life of tolbutamide has been prolonged in this manner. In man, the half-life of tolbutamide (**1**) is about 6 h.[68] Elimination is mainly by oxidation of the aromatic methyl group to a carboxyl function. If this group is replaced by a chlorine atom the product, chlorpropamide (**2**), has a longer half-life of 33 h.[69] Despite metabolic stabilization chlorpropamide is not a true hard drug, since the majority of the dose in man is metabolized by alternative transformations including hydroxylation in the alkyl side chain. This highlights the difficulty encountered in hard drug design, where it is often impossible to maintain optimal properties regarding absorption and excretion.

$$R \!-\!\! \langle\text{ring}\rangle \!-\! SO_2NHCONH(CH_2)_3Me$$

(**1**) R = Me
(**2**) R = Cl

A soft drug is pharmacologically active but undergoes a predictable and controlled conversion into non-toxic inactive metabolites after having exerted its therapeutic effect.[67] The ideal soft drug is designed to be inactivated by one metabolic step, preferably hydrolysis. Potentially toxic metabolites resulting from oxidative biotransformation can therefore be avoided and the therapeutic benefits to be expected are similar to those described for hard drugs. In practice, a metabolically sensitive bond or group is built into a non-critical position in the lead molecule. For instance, part of an *n*-alkyl side chain is replaced by an ester group to yield a soft analogue. The ester is readily cleaved *in vivo* by esterases. Atracurium (**3**), a neuromuscular blocking agent, possesses two types of sensitive group—quaternary nitrogen functions and ester groups—and can be considered a result of soft drug design. *In vivo*, the ester groups undergo hydrolysis catalyzed by plasma esterases, while a non-enzymatic Hofmann elimination takes place at the quaternary nitrogen functions.[70] Thus, atracurium is inactivated by two non-oxidative processes that are not dependent on liver and kidney function. This has contributed to its safe use in patients.

A review of the principles and methods used in the design of soft drugs can be found in Bodor.[67]

(**3**)

23.4.3.2 Prodrugs—Basic Concepts

The term prodrug or pro-agent was first used by Albert[71,72] to describe compounds which undergo biotransformation prior to eliciting a pharmacological response. Albert suggested that this approach could be used to alter the properties of certain drugs in a temporary manner to overcome chemical or biological barriers or to alter and/or decrease their toxicity. There are many obstacles that may prevent or limit a drug reaching its site of action. These include poor bioavailability, short duration of action, high first pass metabolism, instability and poor solubility. Chemical modification of a drug *via* the attachment of a promoiety generates the prodrug. The properties of the prodrug enable it to cross the limiting barrier and it is designed ideally to be cleaved efficiently by enzymatic or non-enzymatic processes.[73] This is followed by rapid elimination of the released promoiety. Other terms which have been applied to this approach include drug latentiation,[74,75] bioreversible derivatives and congeners. Prodrug is now the most commonly accepted term.[76-78]

Aspirin (**4**) and hexamine (**5**; methenamine) are examples of prodrugs developed as early as the late 19th century. Aspirin, a prodrug of salicylic acid, is less corrosive to the gastrointestinal tract. Following absorption, the active drug is released by the action of esterase enzymes. Formaldehyde could not be used as a urinary tract antibacterial until it was formulated as an enteric-coated tablet of hexamine. Following absorption, the prodrug is eliminated in the urine where formaldehyde is generated in the acidic environment.

(**4**)   (**5**)

23.4.3.3 Mechanisms for the Transformation of Prodrugs

The ideal prodrug should, after completion of its objective, quantitatively yield its drug.[79] For example, if a prodrug improves the palatability and taste of a drug, then rapid prodrug conversion should take place within the gastrointestinal tract and certainly no later than the arrival of the prodrug in the bloodstream. With the exception of target site enrichment or rate-limiting conversion for extended duration, circulating intact prodrug is contrary to the objectives of prodrug design. Indeed, the distribution of prodrug throughout the body has implications for clinical toxicology and pharmacology studies on the prodrug itself. The mechanisms employed to allow transformation of the prodrug to drug can be chemical or enzymatic.

23.4.3.3.1 Enzymatic reversal

The conversion of a prodrug to the corresponding drug depends on the presence of an enzyme capable of metabolizing the promoiety–drug linkage. Table 10 gives a list of enzymes that have been used in prodrug applications together with the prodrug linkage. Enzymes may be located in specific regions within the body, in which case their activity can be used in the site-specific delivery of drugs. In many cases, however, enzymes are ubiquitous, providing more limited prodrug applications.

Wide interpatient variability may be expected unless the enzyme is present in excess (*i.e.* esterases). In spite of this drawback, enzymatic reversal of prodrugs has found far greater application than the chemical reversal approach.

23.4.3.3.2 Chemical reversal

Prodrug conversion by non-enzymatic mechanisms shows less intersubject variability compared with enzymatic reversal. However, fewer chemical triggers exist for prodrug conversion to drug. The most common chemical reversal takes advantage of differential hydrolysis arising from differences in the pH in the body. For example, the pH of the stomach is 1–4, the intestine 5–8 and blood pH is 7.4. These differences have been utilized to design prodrugs with increased gastric stability followed by reversal in the intestine or blood.[80] Other examples include increased dry stability followed by chemical reversal in water for injection[81] and prolongation of drug action through controlled prodrug hydrolysis in the blood.[82]

In some cases prodrug reversal may represent a mixture of the chemical and enzymatic mechanisms, although it is common for one or the other to predominate.

Table 10 Some of the Enzymes Used in the Hydrolysis of Prodrugs[78]

Prodrug linkage	Hydrolyzing enzyme
Ester	
Short–medium chain	Cholinesterase
Aliphatic	Ester hydrolase
	Lipase
	Cholesterol esterase
	Acetylcholinesterase
	Aldehyde oxidase
	Carboxypeptidase
Long chain aliphatic carbonate	Pancreatic lipase
	Pancreatin
	Lipase
	Carboxypeptidase
	Cholinesterase
Phosphate, organic	Acid phosphatase
	Alkaline phosphatase
Sulfate, organic	Steroid sulfatase
Amide	Amidase
Amino acid	Proteolytic enzymes
	Chymotrypsins A and B
	Trypsin
	Carboxypeptidase A and B
Azo	Azo reductase
Carbamate	Carbamidase
Phosphamide	Phosphoramidases
β-Glucuronide	β-Glucuronidase
N-Acetylglucosaminide	α-*N*-Acetylglucosaminidase
β-Glucoside	β-Glucosidase

23.4.3.4 Prodrugs of Amines and Amides

A major problem in the general approach to the design of prodrugs is the limited possibilities available for making bioreversible derivatives of many drugs. When a drug possesses readily esterifiable groups such as hydroxyl or carboxyl functions, it is relatively simple to design prodrugs. Many drugs, however, have no apparent derivatizable group. New chemical approaches have therefore been developed to expand the application of the prodrug concept. One concept has focussed on the development of prodrug types for amines, amides, imides and various other NH-acidic compounds.[83] An example of this is the use of N-Mannich bases formed from the reaction of an NH-acidic compound with formaldehyde and a primary or secondary aliphatic or aromatic amine (equation 1). The process can be considered as *N* aminomethylation or *N* amidomethylation (in the case of the drug being an amide). The decomposition of the N-Mannich base derivative to yield the original NH-acidic compound follows first-order reaction kinetics in aqueous solutions and is not dependent on enzyme catalysis.[83] By appropriate selection of the amine component, it is possible to obtain prodrugs of a given drug with varying degrees of *in vivo* lability. In addition, other physicochemical properties such as aqueous solubility, dissolution rate and lipophilicity can also be modified for the parent compound.

$$RCONH_2 + CH_2O + R^1R^2NH \rightleftharpoons RCONHCH_2NR^1R^2 + H_2O \qquad (1)$$

The concept of N-Mannich base formation has been used to increase the water solubility of tetracycline through the formation of the prodrug rolitetracycline (**6**). This highly soluble N-Mannich base of tetracycline and pyrrolidine is decomposed quantitatively to tetracycline in neutral

(**6**)

Table 11   Review Articles on Prodrugs

Author	Article title	Ref.
Digenis and Swintosky (1975)	Drug latentation	140
Sinkula and Yalkowsky (1975)	Rationale for design of biologically reversible drug derivatives: prodrugs	78
Stella *et al.* (1980)	Prodrugs: the control of drug delivery *via* bioreversible chemical modification	118
Stella and Himmelstein (1980)	Prodrugs and site specific drug delivery	124
Notari (1981)	Prodrug design	79
Pitman (1981)	Prodrug of amides, imides and Amines	141
Bundgaard *et al.* (1982)	Book: 'Optimisation of Drug Delivery'	142
Gardner (1985)	Potential and limitations of drug targeting: an overview	126
Widder and Green (1985)	Book: 'Methods in Enzymology', vol. 112	143
Connors (1986)	Prodrugs in cancer chemotherapy	105
Azori (1987)	Polymeric prodrugs	144
Roche (1987)	Book: 'Bioreversible Carriers in Drug Design. Theory and Application'	145

aqueous solution with a half-life of 40 min at pH 7.4 and 37 °C.[84] In addition, the approach may be useful for improving the dissolution behaviour of poorly soluble drugs in an effort to improve oral bioavailability (*e.g.* phthalimide, phenytoin and allopurinol). Other derivatives of NH-acidic compounds have been investigated including *N*-hydroxymethyl, *N*-acyloxymethyl and *N*-acylated derivatives.

23.4.3.5   Obstacles to Drug Design

There are many obstacles to drug design. Some, such as the blood–brain barrier, are obvious when discussing the problems of drug delivery to the central nervous system. Some of the obstacles to the design of a commercially viable drug include: (a) aesthetic properties of the drug and/or its dosage form, *e.g.* odour, taste, pain upon injection, gastrointestinal irritability; (b) formulation problems, *e.g.* poor stability and/or solubility; (c) incomplete absorption, *e.g.* from the gastrointestinal tract or across the blood–brain barrier; (d) presystemic metabolism in the gastrointestinal lumen, mucosal cells and liver; (e) pharmacokinetic problems, *e.g.* slow rate of absorption or rapid plasma half-life; (f) poor site-specificity; and (g) toxicity associated with local irritancy or distribution into tissues other than the desired target organ. Examples will be provided of each of these types of problem. For further information the reader is referred to various articles on prodrugs shown in Table 11.

23.4.3.5.1   *Taste and odour*

Taste and odour present problems, particularly with paediatric patients. Chloramphenicol (**7**), for example, is extremely bitter. The palmitate ester of this drug (**8**), however, is practically tasteless because of its low aqueous solubility[85] and is efficiently hydrolyzed by the action of pancreatic lipase on solid chloramphenicol palmitate particles.[86]

$$O_2N - \langle \text{benzene ring} \rangle - CH(OH)CHCH_2R$$
$$\underset{NHCOCHCl_2}{|}$$

(**7**) R = OH
(**8**) R = $O_2C(CH_2)_{14}Me$

Molecules interact with taste receptors on the tongue to give bitter, sweet or other taste sensations when they dissolve in saliva.[87] Thus chloramphenicol palmitate is tasteless because of its low aqueous solubility. In lowering solubility to overcome taste problems, however, a balance between diminished solubility and reduced oral bioavailability of the drug must be sought. In addition, the prodrug must be rapidly converted to the parent compound by enzymatic or chemical means.

Another approach to overcoming taste problems is to alter the ability of the drug to interact with taste receptors. The bitter taste receptor is thought to consist of a hydrophobic portion together with an electrophilic and a nucleophilic centre.[87] The affinity of the prodrug for this receptor may be reduced compared with the parent drug if the electron density/charge distribution in the molecule is altered.

Odour, which is related to the vapour pressure of a substance, is also of concern in drug formulation. The tuberculostatic agent ethanethiol (9), used in the treatment of leprosy, has a boiling point of 35 °C and a strong disagreeable odour. The phthalate ester, diethyl dithioisophthalate (10), has a higher boiling point and is relatively odourless.[88,89] The ester has been used topically for the systemic delivery of ethanethiol, which is generated from the prodrug by the action of systemic thioesterases.

EtSH

(9)   (10)

23.4.3.5.2 Reduction of pain at injection site

Pain at an injection site, especially after intramuscular injection, can be caused by drug precipitation and transfer of the drug into the surrounding cells. Such pain is usually associated with haemorrhage, oedema, inflammation and tissue necrosis.[90] Some of these problems may relate to dose vehicle composition or pH.

The low aqueous solubility of the antibiotic clindamycin hydrochloride (11) is responsible for the pain experienced on intramuscular injection. The fiftyfold greater solubility of the 2-phosphate ester prodrug of clindamycin (12) results in little local pain or irritation on injection in animals and man.[91,92] The prodrug possesses little or no intrinsic antibacterial activity[93] but is rapidly converted to the parent drug by the action of phosphatases in the body. Only 1–2% of an intravenous dose is eliminated in the urine as unchanged prodrug.[94]

(11) R = H
(12) R = HPO_3^-

23.4.3.5.3 Reduction of gastrointestinal irritability

A wide variety of drugs can cause gastrointestinal disturbances.[95–98] Several mechanisms of action have been proposed to explain these, including injury to the gastric mucosa through direct contact of drug, damage due to stimulation of gastric acid secretion by circulating drug and interference with the protective gastrointestinal mucosal layer.[78] The mucosal layer is relatively resistant to proteolytic enzyme activity and protects the underlying epithelial tissue. This protective function is compromised by the localized gastric irritation produced by acetylsalicylic acid (aspirin). In an effort to reduce this gastric irritation a series of hydrophobic carbonate esters of acetylsalicylic acid was produced.[99] These prodrugs were absorbed over a greater area of the gastrointestinal tract, thus reducing localized irritation. Following absorption, they were rapidly hydrolyzed to parent drug. A number of other drugs which initiate gastric irritation have been dealt with in a similar manner, including nicotinic acid, 21-hydroxysteroids and oleandrin.[78]

23.4.3.5.4 *Alteration of drug solubility*

The solubility of a drug may be increased or decreased by the prodrug approach. Decreasing solubility to mask an unpleasant taste has already been discussed. The antibiotic chloramphenicol (**7**) provides an example where prodrug modifications have been used to increase solubility. While the palmitate ester of chloramphenicol has proved useful for oral formulations, a more water-soluble form, chloramphenicol sodium succinate (**13**), was developed for parenteral administration.[100] This prodrug has no antibacterial activity but is hydrolyzed in the body to free chloramphenicol by the action of esterases.

$$O_2N \underset{}{\overset{}{\bigcirc}} CH(OH)CHCH_2O_2CCH_2CH_2CO_2^- \ Na^+$$
$$NHCOCH_2Cl_2$$

(13)

A similar approach has been used with a number of steroids such as the glucocorticoids, betamethasone, prednisolone, hydrocortisone and dexamethasone which have been prepared as water-soluble disodium phosphate or sodium succinate prodrugs.[101] The sodium succinate salts, however, are only available as lyophilized powders for reconstitution because of their poor chemical stability compared with the phosphate esters. In the case of chloramphenicol sodium succinate and the succinate esters of the corticosteroids, significant amounts of prodrug are eliminated in the urine, thus complicating drug therapy.[102,103]

In some cases it is necessary to reduce the water solubility of a drug by the synthesis of a more fat-soluble prodrug. Anticancer agents such as 6-mercaptopurine (**14**) are also useful in the treatment of psoriasis but do not readily penetrate the skin. An alkylated derivative of the drug (**15**) can be used as a topical treatment of psoriasis since it is effectively transported through epidermal barriers followed by conversion to the active moiety thiopurine (**16**).[104] This is shown in equation (2).

$$\text{(2)}$$

(15)   **(14)**   **(16)**

23.4.3.5.5 *Increased chemical stability*

Chemical stability of a drug is an important requirement since a shelf life of at least two years is desirable for most products. If a drug is very unstable and the problem cannot be overcome by modifications in the formulation, it is sometimes possible to develop a prodrug with enhanced stability. Some anticancer agents are unstable in solution and this poses a problem when treatment requires infusion over a long period of time. Azacytidine (**17**), used in the treatment of acute myeloid leukaemia, cannot be administered as an intravenous bolus because of severe and dose-limiting gastrointestinal toxicity.[105] This toxicity could be eliminated if the drug was slowly infused over a five-day period. However, this is not practicable because of the instability of azacytidine in aqueous solution. Azacytidine is hydrolyzed reversibly to the ring-opened formyl derivative (**18**), which is then slowly and irreversibly converted to the guanylurea (**19**). Studies of the mechanism of decomposition showed that the addition of bisulfite across the 5,6-imine double bond produced a prodrug (**20**) stabilized at acidic pH but which rapidly reverted to azacytidine under physiological conditions (equation 3).[106] As well as being more stable and suitable for prolonged infusion, the prodrug is more soluble, thus requiring a small infusion volume.

$$\text{(3)}$$

(20)   **(17)**   **(18)**   **(19)**

23.4.3.5.6 Prodrugs that improve oral absorption

Most drugs are absorbed by passive diffusion and should therefore possess a degree of lipophilicity for efficient gastrointestinal absorption. For highly polar drugs the administration of a less polar, more lipophilic prodrug may help promote gastrointestinal absorption. This approach has been adopted with various penicillin antibiotics. Ampicillin (21), for example, is zwitterionic in the pH range of the gastrointestinal tract and as such is only about 20–60% absorbed following oral dosing.[107,108] Esterification of the carboxyl group of ampicillin to form the prodrugs pivampicillin (22),[109] bacampicillin (23)[110] or talampicillin (24)[111] alters the polarity of the molecule and successfully improves oral bioavailability. Following absorption, the prodrugs are rapidly cleaved by general esterase enzymes to give ampicillin. The products of the cleavage are acetaldehyde, carbon dioxide and ethanol in the case of bacampicillin and formaldehyde and pivalic acid for pivampicillin. The amount of formaldehyde produced is not considered sufficient to cause toxicity problems.[101] A similar approach to altering the polarity of zwitterionic molecules has been adopted with methyl-DOPA[112] and L-DOPA.[113]

(21) R = H or −
(22) R = $CH_2O_2CCMe_3$
(23) R = $CH(Me)O_2COEt$

(24) R =

23.4.3.5.7 Prevention of presystemic metabolism

Many drugs are efficiently absorbed from the gastrointestinal tract but undergo presystemic first pass metabolism or chemical inactivation before reaching the systemic circulation. Propranolol (25), for example, undergoes rapid first pass metabolism by phase I and phase II drug-metabolizing enzymes after oral administration. Microsomal metabolism requires a reasonably lipophilic drug. Propranolol, although significantly ionized at physiological pH, is very lipophilic, whereas the hemisuccinate prodrug (26) is zwitterionic at physiological pH and does not enter the microsomal environment, thus protecting the drug from first-pass metabolism.[114] Once in the general circulation, non-specific esterases cleave the prodrug to propranolol. Many steroids such as corticosteroids also undergo extensive first pass metabolism. This can be prevented by administration of acetonide, ester and ether prodrugs such as prednisolone and triamcinolone.[115–117]

(25) R = H
(26) R = $COCH_2CH_2CO_2^-$

23.4.3.5.8 Prolongation of drug action

The prodrug approach has been used to overcome the problem encountered with short half-life drugs which are rapidly cleared from the body. Frequent dosing of such drugs results in sharp peak–valley plasma-concentration–time profiles. While sustained-release preparations will sometimes overcome such problems, these are often administered in combination with a prodrug. For instance, lipophilic ester prodrugs of testosterone (27) such as the 17-cypionate ester (28) are administered in an oil vehicle by deep intramuscular injection. Following release from the intramuscular depot, the prodrugs are cleaved to testosterone.[118]

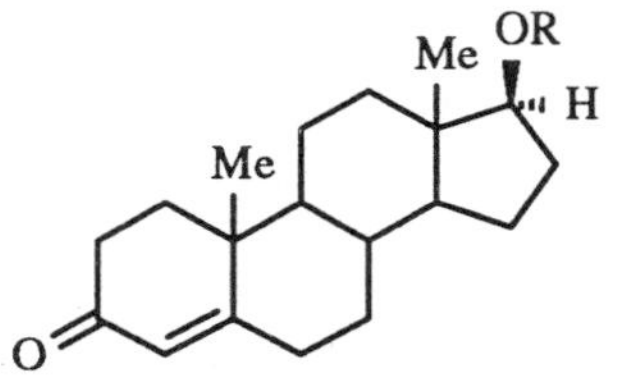

(27) R = H

(28) R = COCH$_2$CH$_2$ —⬠

This approach has also been used in the more effective control of schizophrenic patients where the intramuscular administration of the antipsychotic agent fluphenazine (29) as the heptanoate (30) or decanoate (31) ester in a sesame oil vehicle results in a longer duration of action compared with the parent drug.[119-123]

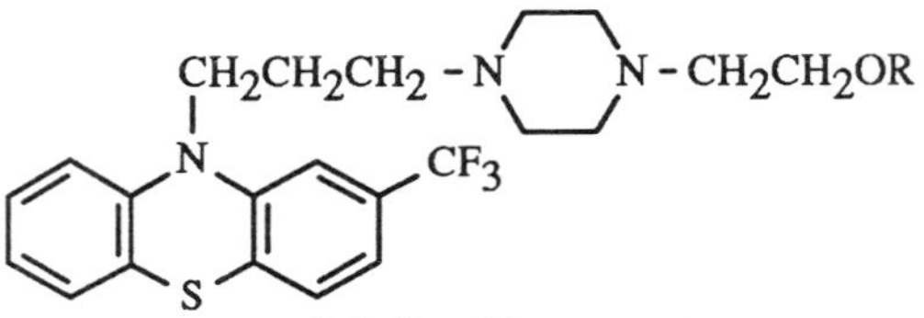

(29) R = H
(30) R = COC$_6$H$_{13}$
(31) R = COC$_9$H$_{19}$

This type of prodrug approach is not without its clinical dangers, since it is difficult to halt therapy if the patient experiences an adverse reaction to the drug.

23.4.3.5.9 *Site-specific drug delivery*

Site-specific drug delivery attempts to obtain very precise and direct effects at the 'site of action' without subjecting the rest of the body to significant levels of the active agent. Certain limitations to this approach have been highlighted.[124-126] These can be summarized as: (i) adequate accessibility of prodrug to target site; (ii) distribution, relative activity and specificity of reconverting enzymes; (iii) parent drug leakage from target site; (iv) access of parent drug to target site within target organ; (v) pharmacokinetics of prodrug and drug molecule; (vi) toxicity of prodrug moiety and its metabolites; and (vii) lack of information on selective location of potential reconverting enzymes. Whilst it is unlikely that any prodrug will provide total specificity of action, a number of examples exist where this approach has produced a more selective (and effective) drug action. The best known of these is the use of L-3,4-dihydroxyphenylalanine (32; L-DOPA) in the treatment of Parkinson's disease. Parkinsonism is a neurological disease resulting from the degeneration of dopaminergic neurons in the basal ganglia of the brain. Treatment with the neurotransmitter dopamine (33) is impracticable due to both a lack of absorption and metabolism in the intestine and at the blood–brain barrier. The natural precursor of dopamine, L-DOPA, is absorbed intact from the intestine, crosses the blood–brain barrier and is concentrated in the dopaminergic and adrenergic nerve terminals, where it is converted to dopamine, noradrenaline or adrenaline (equation 4).

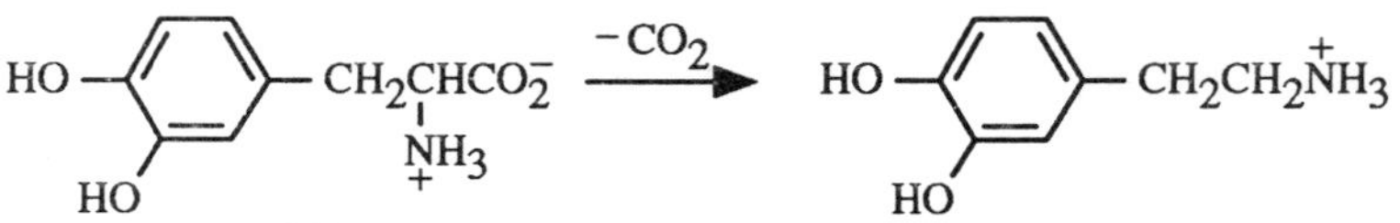

(32)   (33)   (4)

The conversion of dopamine in the corpus striatum of the brain produces the desired therapeutic effect. Peripheral transformation of L-DOPA leads to a number of side effects. These can be prevented by the additional administration of an inhibitor of aromatic amino acid decarboxylase such as carbiDOPA, which does not penetrate the central nervous system. Conversion of L-DOPA to dopamine is hence confined to the brain and peripheral side effects are avoided.[127]

An additional use for dopamine is as a renal vasodilatory agent. In this case administration of the prodrug L-γ-glutamyldopamine results in a preferential conversion to parent drug in the kidney due to localization of the enzyme γ-glutamyltranspeptidase in this organ. The use of the prodrug avoids affecting other dopamine-sensitive sites such as the cardiovascular or central nervous systems.[128]

A number of tumours have elevated levels of certain enzymes relative to those found in normal tissue. It is thus feasible to use such activity to activate prodrugs of antineoplastic agents selectively. This promising approach to site-specific drug delivery has been reviewed recently.[105] The principles employed in the design of prodrugs activated in tumour cells can be illustrated using the alkylating nitrogen mustards as examples. This class of antitumour agents acts by covalent binding to cellular macromolecules and causes cytotoxicity principally due to their ability to cross-link DNA. If these agents are not metabolized *in vivo*, a broad correlation can be seen between their chemical reactivity and cytotoxicity. Highly reactive nitrogen mustards are likely to be very toxic (both to tumour cells and normal cells) and poorly reactive ones much less toxic. Prodrugs can be designed to be non-toxic poor alkylating agents but which are enzymatically transformed to highly reactive cytotoxic agents. For the drugs to have selective action against tumours the prodrug must be a good substrate for an enzyme ideally localized uniquely in the tumour cell. Once formed from the prodrug, the drug should react quickly and not diffuse away from the tumour tissue. This can be achieved, for example, by ensuring the released drug is highly reactive or unstable or is charged so that it is trapped within the cell. Human hepatocellular carcinomas have been observed to have high azoreductase levels. As a consequence, certain azobenzene mustards, themselves poor alkylating agents and of relatively low toxicity, have been investigated as potential prodrugs in cancer chemotherapy.

N-Methylpyridinium-2-carbaldoxime chloride (**34**; 2-PAM) is used as an antidote to poisoning with cholinesterase inhibitors. However, because of its charge and hydrophilicity it does not penetrate the blood–brain barrier. Pro-2-PAM (**35**), the reduced dihydropyridine form of 2-PAM, readily enters the central nervous system. There it is oxidized to form 2-PAM, which remains trapped in the central nervous system since its charge reduces its rate of transfer from the brain back into the blood. Pro-2-PAM demonstrates the essential principles of a useful prodrug, namely greater accessibility to the target site where reconversion occurs followed by retention in the target area.[129] This concept has been extended to other drugs such as berberine[130] and to the use of dihydro-pyridines as carriers of drugs into the central nervous system.[131]

(34)   (35)

23.4.3.5.10 *Reduction in toxicity*

Therapeutic activity without toxicity is the goal of drug design. Some examples exist where a prodrug approach has resulted in a reduction in toxicity. Aspirin, for example, has already been mentioned as a less corrosive prodrug of acetylsalicylic acid. Similarly, the non-steroidal antiinflammatory sulindac (**36**) shows reduced local gastrointestinal irritation compared with the corresponding sulfide (**37**). The antiinflammatory activity and gastrointestinal intolerance of many antiinflammatory drugs seems to be related to their inhibition of prostaglandin synthetase.[132] As a sulfoxide, sulindac shows little or no such activity and this may account for the reduced local gastrointestinal irritation compared with the active sulfide. Sulindac shows little or no pharmacological activity but is rapidly reduced to the sulfide after absorption. In addition, the drug is irreversibly oxidized to the inactive sulfone, producing complex disposition kinetics.[133,134] Sulindac is a good example of a clinically important drug with decreased toxicity, improved absorption and a longer duration of action.

(**36**) R = SMe
(**37**) R = SMe

23.4.3.6 Alterations in Drug Metabolism by the Use of Prodrugs

Metabolism studies are a necessary part of the testing procedures that are applied when evaluating the safety and efficacy of drugs, pesticides and food additives, *etc.* In this section, the effect of prodrug structure on the pathways of metabolism which involve the formation of highly reactive and sometimes toxic products is considered.

Acetylation is one of the most common reactions used in prodrug design. In some instances this modification has resulted in a toxic product. Acetylation of isoniazid (**38**) provides an example of this, where the product acetylisoniazid (**39**) is hydrolyzed to give the hepatotoxin acetylhydrazine (**40**).[135] This is shown in equation (5).

$$+ \; MeCONHNH_2 \qquad (5)$$

The analgesic/antipyretic activity of phenacetin (**41**) is largely attributed to its oxidative deethylation product acetaminophen (**42**). The alteration in metabolism caused by this small structural modification is profound.[136,137] Acetaminophen is primarily conjugated at the phenolic oxygen, while a quantitatively less significant pathway of metabolism produces the reactive quinoneimine. In therapeutic doses this is removed by conjugation with glutathione. However, in overdose situations this metabolite causes hepatic necrosis. Phenacetin, in contrast, has never been reported to produce hepatic necrosis. Instead, it is predominantly hydrolyzed at the amide linkage to produce *p*-phenetidine, which is thought to yield further metabolites responsible for methemoglobulinaemia, haemolysis and kidney toxicities.

In addition, *N* hydroxylation of phenacetin produces additional reactive metabolites. Thus ethylation of the phenolic group of acetaminophen dramatically changes the metabolism and toxicity of the drug.

Terminal ethynyl groups have been employed as a precursor to the acetic acid moiety in biphenyl antiinflammatory agents.[138] Thus the prodrug 4'-ethynyl-2-fluorobiphenyl (**43**) is metabolized in the rat to 2-fluoro-4'-biphenylacetic acid (**44**). However, extensive studies have shown that terminal acetylenes can be oxidized by the cytochrome *P*-450 system to reactive intermediates that inhibit cytochrome *P*-450. It has been suggested that this is responsible for some of the adverse drug interactions observed with contraceptive steroids.[139]

Thus, subtle changes brought about by prodrug structure can significantly influence the metabolism and toxicity of the prodrug or its metabolites and consideration should always be given to this during prodrug design and development.

23.4.4 BIOACTIVATION

23.4.4.1 General Concepts

A significant advance in the understanding of how some relatively inert chemical compounds can cause harmful effects to various tissues in the body was made with the discovery that many such compounds undergo metabolism in the host to highly reactive products. The phenomenon was first reported by the Millers in 1966 [146, 147] to account for the activity of certain chemical carcinogens. This work has subsequently been extended through the efforts of numerous investigators to explain many different kinds of response in addition to carcinogenesis. These include, for example, cellular necrosis, mutagenesis, teratogenesis and blood dyscrasias. Relatively little is still known about the events between the formation of a reactive metabolite and the expression of the adverse biological effect. The reactive metabolite, however, may have several detrimental actions on the cell. These include: (a) covalent binding to structural, catalytic and/or informational cellular macromolecules, including proteins, nucleic acids and lipids; (b) covalent binding to low molecular weight cellular constituents; (c) generation of superoxide, H_2O_2 or other activated oxygen species; (d) stimulation of peroxidative decomposition of cellular lipids; and (e) cofactor depletion.

The covalent binding of reactive metabolites to nucleophilic tissue constituents such as macromolecules or low molecular weight substances may be an essential step to toxicity since these alkylation reactions may involve critical cellular components. As a consequence a wide variety of effects may occur, including disruption of energy production, changes in membrane permeability, reduced protein synthesis, *etc.* In the case of bromobenzene,[148] paracetamol[149] and carbon tetrachloride,[150] the extent of covalent binding to liver protein is related to the degree of hepatotoxicity induced by these agents. However, evidence to support the role of covalent binding in cell death is largely circumstantial and based on the correlation between the extent of binding and the severity of the accompanying lesion *in vivo*. The use of covalent binding as an indicator of drug toxicity has been reviewed by Gorrod.[151]

In instances where free radical metabolites are involved, the secondary production of reactive oxygen species and/or stimulation of peroxidative degradation of essential lipids may be crucial factors in the initiation of toxicity by reactive metabolites. Free radical metabolites may also be involved in redox recycling reactions, resulting in a depletion of cellular cofactors such as NADPH. This also occurs, for example, in the case of UDP-galactosamine, which is generated from galactosamine and results in the depletion of UTP levels within the cell.[152]

23.4.4.2 Formation of Reactive Metabolites

The biochemical mechanisms involved in the initiation of chemical toxicity are numerous, and toxicity may result from the direct action of the chemical (*e.g.* the inhibition of cytochrome oxidase by cyanide or acetylcholinesterase by paraoxon). Our concern in this section, however, is with metabolic activation of a chemical. The mechanisms of biological activation of toxic chemicals may be broadly classified into lethal synthesis, reductive activation, activation of tissue oxygen, oxidative activation and activation by conjugation.[153] Among the chemically unstable species that may be formed by cells are: (i) electrophiles, commonly formed from carbon-, nitrogen- or sulfur-containing compounds that are metabolized to species that are deficient or potentially deficient in an electron pair; (ii) free radicals, *i.e.* species that contain an odd number of electrons and may possess a positive, negative or neutral charge; and (iii) other neutral species, *e.g.* carbenes and neutrenes, in which two electrons are lost. There is as yet little information on the contribution of these species to cell injury.[154]

23.4.4.2.1 *Electrophiles*

In 1966, the Millers proposed that many carcinogens exert the effect after metabolic conversion to an electrophilic moiety that is able to bind covalently to DNA. Subsequent research has shown that cytochrome *P*-450 is the most important enzyme system responsible for generating electrophilic species, including epoxides, hydroxylamines, nitroso derivatives, nitrenium ions, azoxy derivatives and elemental sulfur. Other enzymes such as prostaglandin synthetase, nitroreductase and xanthine oxidase are also able to produce electrophilic metabolites, although their substrate specificity is narrower than that of cytochrome *P*-450. A review of the mechanism of formation of reactive electrophiles can be found in Guengerich and Liebler.[155]

23.4.4.2.2 *Free radicals*

Toxicity may occur as a result of chemicals being converted into free radicals or *via* the formation of superoxide as a by-product of their metabolism. A number of chemicals are known to produce radicals on metabolism, including quinones, nitroaromatics, carbon tetrachloride, bipyridyls and aromatic amines.[156,157] Free radicals are most commonly generated *via* NADPH cytochrome *P*-450 reductase or other flavin-containing reductases,[158] although cytochrome *P*-450 itself may be involved, as is the case in the reduction of carbon tetrachloride to form the radicals $\cdot CCl_3$ and $\cdot CCl_2O_2$. Many endogenous and exogenous chemicals, including adrenaline, DOPA, adriamycin and paraquat, that are reduced by intracellular enzymes are able to reduce oxygen by electron transfer to produce $O_2^- \cdot$ (superoxide) and H_2O_2 (hydrogen peroxide).

The superoxide radical is not itself thought to be particularly toxic because it is relatively stable chemically. However, it can be converted chemically and/or enzymatically to a number of more toxic forms of active oxygen, including $HO_2^\cdot$ (protonated form of superoxide), O^1 (singlet oxygen), $HO\cdot$ (hydroxyl radical) and H_2O_2. The hydroxyl radical is a highly reactive species and may cause direct damage to many endogenous molecules.[158,159] Besides the direct toxic effects of oxygen radicals, which can lead to membrane alterations by lipid peroxidation, destruction of intracellular and extracellular proteins including enzyme inactivation, or DNA damage, indirect toxic effects like mutagenicity and carcinogenicity including tumour promotion have been related to oxygen radical formation. In addition, inflammatory reactions might occur after drug-induced oxygen activation.[159]

Many free radicals can participate in recycling reactions, resulting in a sustained level of free radicals in the cell. Such redox recycling can occur with quinones, nitro compounds, nitroxides and azo compounds, as shown in Scheme 1, and can result in a depletion of reduced cofactors and hypoxia.[159] In addition, free radicals may initiate chain reactions leading to lipid peroxidation and membrane destruction. In view of the multiplier effect associated with free radicals, their potential toxicity is far greater than that of electrophiles.

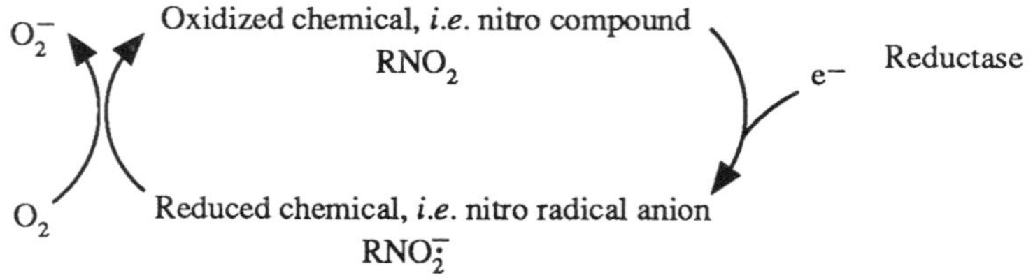

Scheme 1

23.4.4.3 The Site of Action of Reactive Metabolites

The half-life of the reactive metabolite determines where within the cell, tissue or body it will have its effect. Very reactive entities of ultrashort half-life will not escape their site of formation and will interact with the enzyme molecule that has produced them. In this case the reactive metabolites are likely to be suicide inhibitors of the activating enzyme. Progressively less reactive intermediates will travel further before interacting with a tissue component, such that some may reach other organs of the body to express their toxicity. In the case of the most stable intermediates, their toxicity may be expressed following excretion in the kidney, bladder or small intestine. For toxicity to occur in a particular tissue it must receive a sufficient dose of the parent compound or the reactive metabolite. Furthermore, if the tissue is exposed to the parent compound then the activating enzyme system must be present and the final expression of toxicity will depend on the balance between enzymatic activation and detoxication. Detoxication may arise as a consequence of either competing metabolic transformations or cellular defence mechanisms such as the protective role of glutathione and the glutathione-dependent enzymes glutathione S-transferase, glutathione peroxidase and glutathione reductase. In many instances it has now been established that cellular glutathione has to be depleted by reaction with the reactive electrophilic intermediates before the electrophile is able to interact with the crucial target for toxicity.[160] Glutathione, the most abundant cellular nucleophile, has several functions in the maintenance of cellular integrity including: (i) protection of SH groups, particularly those at the active sites of enzymes, against inactivation by heavy metals or oxidation by drugs such as organic mercurials; (ii) reactivation of SH enzymes that have been inactivated by oxidation; (iii) conjugation with toxic chemicals or metabolites; and (iv) detoxification of endogenous peroxides and reactive oxygen species. The protective role of glutathione in the cell has been the subject of a number of reviews.[161-163]

Detoxification of reactive oxygen species is an essential feature of aerobic life and the body has evolved many lines of defence acting to prevent, intercept and repair hazardous reactions initiated by oxygen metabolites. These defence mechanisms consist of non-enzymatic scavengers and quenchers together with enzymatic systems.[164] Non-enzymatic scavengers include α-tocopherol (vitamin E), ascorbate (vitamin C), β-carotene (vitamin A) and urate. The antioxidant enzymes include the superoxide dismutases and various hydroperoxidases such as glutathione peroxidase, catalase and other haemoprotein peroxidases. They are characterized, in general, by high specific cellular content, by specific organ and subcellular localizations and by a specific form of metal involvement in catalysis including copper, manganese, iron and selenium. These systems have a wide distribution in nature, highlighting their importance in dealing with the damaging effects of reactive oxygen metabolites in biological systems.

Reactive metabolites have been divided into six categories depending on their stability and their site of action. These are discussed below. Some examples of chemicals which undergo bioactivation are shown in Table 12.

23.4.4.3.1 Suicide inactivation

Two major classes of chemicals fall into the category of forming enzyme-bound intermediates which react only with the active site of the enzyme. The heme moiety of cytochrome P-450 is destroyed by both terminal alkenes (and alkynes) and cyclopropyl-substituted heteroatoms. The terminal alkenes can distort normal heme metabolism by forming N-alkylated porphyrins which inhibit 5-aminolevulinic acid synthetase and ferrochelatase, resulting in porphyrias.[165] In the case of the cyclopropyl-substituted heteroatoms, a one-electron oxidation of the heteroatom (nitrogen, oxygen, iodine or bromine) places a radical α to the cyclopropyl ring. Such systems rearrange to generate methylene radicals which are responsible for the destruction of the cytochrome P-450 heme.[166,167]

Suicide inactivation of enzymes other than cytochrome P-450 occurs due to interaction with critical amino acid moieties necessary for enzyme function or with other prosthetic groups.[168,169]

23.4.4.3.2 Interaction primarily with the enzyme generating the reactive metabolite

A number of examples are documented which involve cytochrome P-450. These are distinguished from suicide inactivation because products appear to be formed which then react with the enzyme rather than the reaction taking place in the enzyme-bound intermediate state. In the metabolism of parathion, for instance, sulfur from the substrate is released to bind irreversibly to cytochrome P-450.[170] In the case of chloramphenicol, clinical problems in humans arise when lysine residues in the cytochrome P-450 molecule are alkylated by acyl halide products of the drug.[171-173] Most reactive intermediates, however, are not this reactive, and binding amongst the microsomal proteins is relatively non-selective.[174,175]

Table 12   Activation of Chemicals to Toxic Metabolites

Activation reaction	Examples	Biological effect
Haloalkanes $\xrightarrow{P\text{-}450}$ acyl halides, radicals	Chloroform, carbon tetrachloride, halothane	Toxicity
Substituted alkanes $\xrightarrow{P\text{-}450}$ epoxides (?)	Ethyl carbamate, aflatoxin B_2	Carcinogenicity
Alkanes $\xrightarrow{P\text{-}450}$ epoxides	Styrene, acrylonitrile	Toxicity, carcinogenicity
Furans $\xrightarrow{P\text{-}450}$ diketones	4-Ipomeanol	Toxicity
Polycyclic aromatic hydrocarbons $\xrightarrow{P\text{-}450}$ epoxides, diol epoxides	Benzo[a]pyrene, 7,12-dimethylbenzanthracene, naphthalene	Carcinogenicity
Halobenzenes $\xrightarrow{P\text{-}450}$ epoxide radicals (?), quinones (?)	Bromobenzene, pentachlorophenol	Toxicity
Polyhalogenated biphenyls $\xrightarrow{P\text{-}450}$ epoxides (?)	Polychlorinated biphenyls, polybrominated biphenyls	Toxicity
Acetanilides $\xrightarrow{P\text{-}450,\ peroxidases}$ quinoneimines, semiquinone radicals (?)	Acetaminophen	Toxicity
Aromatic amines $\xrightarrow{P\text{-}450,\ flavin\ monooxygenase\ peroxidases}$ hydroxylamines and esters	2-Naphthylamine, benzidines	Carcinogenicity
Aminofluorenes $\xrightarrow{P\text{-}450,\ flavin\ monooxygenase\ peroxidases}$ hydroxylamines and esters, radicals, nitroso compounds	2-Aminofluorene	Carcinogenicity
Hydrazines $\xrightarrow{P\text{-}450,\ flavin\ monooxygenase}$ CH_3, CH_4^+, diazomethane (?)	Procarbazine, dimethylhydrazine	Toxicity, carcinogenicity
Thiocarbonyls $\xrightarrow{flavin\ monooxygenase}$ sulfenes, sulfines	Thioacetamide	Toxicity

23.4.4.3.3 *Products interacting within the cell in which they are produced*

Only a few examples exist of reactive metabolites which are stable enough to leave the enzyme where they are formed but are not able to leave the cell. The halogenated hydrocarbons, particularly carbon tetrachloride and halothane, yield the trichloromethyl[176] and 1,1,1-trifluoro-2-chloro-ethyl[177] radicals, respectively. These radicals, and the subsequently produced peroxy radicals from reaction with dioxygen, not only result in inactivation of cytochrome *P*-450 but also escape to initiate damage at other sites within the cell.[178] The escape of the trichloromethyl radical results in a variety of radical-initiated reactions, including covalent binding to membrane lipids and proteins, cofactor depletion and initiation of lipid peroxidation.[179] The process of lipid peroxidation leads to the generation of other reactive entities such as hydroxyaldehydes and malondialdehyde. Such a diversity of effect leads to the problem of distinguishing which event is responsible for the ultimate toxicity.

23.4.4.3.4 *Products interacting within the tissue of origin*

Many reactive metabolites are sufficiently stable to interact within the tissue of origin. Vinyl chloride, for example, causes tumours in extraparenchymal epithelial cells of the liver, presumably by DNA alkylation. Since these cells have little ability to activate vinyl chloride it is thought that activation to the epoxide occurs in hepatocytes. The epoxide then migrates to extraparenchymal cells.[180] While the epoxide is able to alkylate DNA in both cell types, the rate of repair is generally higher in hepatocytes.[181] Studies in which chemicals are administered to isolated hepatocytes have detected reactive intermediates outside the cells. In the case of benzo[*a*]pyrene,[182] dimethylnitrosamine,[183] trichloroethylene[184] and vinyl chloride,[185] more than 90% of the reactive intermediates which alkylate DNA can be trapped outside the hepatocytes.

23.4.4.3.5 *Reactive metabolites which leave the tissue of origin*

Evidence of reactive metabolites sufficiently stable to be included in this category is often circumstantial. However metabolites of both acrylonitrile (cyanoethylene oxide) and benzo[*a*]pyrene (benzo[*a*]pyrenediol epoxide) are both able to leave the liver and migrate to extrahepatic tissues.[155] Experimental evidence supports the suggestion that the pulmonary toxicity of the pyrrolizidine alkaloids such as monocrotaline is due to metabolic activation in the liver and then transport in the circulation to the lung.[186] This idea is supported by the observation that reactive metabolites may be formed by hepatic but not pulmonary enzymes. Studies in the rat have shown that the half-life of bromobenzene 3,4-epoxide in the blood following its formation in the liver is approximately 14 s. This is sufficient time to allow it to reach most organs of the body since the mean circulation time of blood in rats is less than 10 s.[187]

23.4.4.3.6 *Proximate toxic metabolite formed in one tissue and the ultimate toxic metabolite is generated in a second susceptible tissue*

Many toxic chemicals require multistep metabolic activation before generating their ultimate toxic metabolites. In some cases a stable proximate toxic metabolite is formed in one tissue and then transported to the target organ where it is metabolized to the ultimate toxic metabolite. Different enzymes may be involved at different stages of activation and this may be of importance in target organ toxicity. For example, a target tissue may not be able to metabolize the parent compound to its proximate toxic metabolite but would convert the proximate toxic metabolite to the ultimate toxic metabolite. One example of this is the case of hexachlorobutadiene, which is initially conjugated with glutathione in the liver, followed by transport to the kidney and activation to a nephrotoxic metabolite by the action of C—S lyase.[188] The bladder carcinogen 2-naphthylamine also proceeds in a similar manner. 2-Naphthylamine is metabolized in the liver to *N*-hydroxy-2-naphthylamine and its glucuronide conjugate. This may be reactivated in acidic urine to generate either reactive metabolites or compounds capable of being reactivated in bladder epithelial cells.[189]

23.4.5 REFERENCES

1. M. Rowland, *J. Pharm. Sci.*, 1972, **61**, 70.
2. M. Rowland, L. Z. Benet and G. G. Graham, *J. Pharmacokinet. Biopharm.*, 1973, **1**, 123.
3. K. S. Pang and M. Rowland, *J. Pharmacokinet. Biopharm.*, 1977, **5**, 625.
4. G. R. Wilkinson and D. G. Shand, *Clin. Pharmacol. Ther.*, 1975, **18**, 377.
5. B. Testa and P. Jenner, in 'Drug Metabolism. Chemical and Biochemical Aspects', ed. J. Swarbrick, Dekker, New York, 1976.
6. G. J. Mannering, in 'Concepts in Drug Metabolism. Part B', ed. P. Jenner and B. Testa, Dekker, New York, 1981, p. 53.
7. F. J. Peterson and J. L. Holzman, in 'Extrahepatic Metabolism of Drugs and other Foreign Compounds', ed. T. E. Gram, MTP Press, Lancaster, 1980, p.1.
8. B. G. Lake and S. D. Gangolli, in 'Concepts in Drug Metabolism. Part B', ed. P. Jenner and B. Testa, Dekker, New York, 1981, p. 167.
9. P. Wollenberg and V. Ullrich, in 'Extrahepatic Metabolism of Drugs and other Foreign Compounds', ed. T. E. Gram, MTP Press, Lancaster, 1980, p. 267.
10. H. Hoensch, C. H. Woo and R. Schmid, *Biochem. Biophys. Res. Commun.*, 1975, **65**, 399.
11. V. Ullrich and P. Weber, *Biochem. Pharmacol.*, 1974, **23**, 3309.
12. M. Laitinen and J. B. Watkins, III, in 'Gastrointestinal Toxicology', ed. K. Rozman and O. Hänninen, Elsevier, Amsterdam, 1986, p. 169.
13. J. Caldwell and M. V. Marsh, in 'Presystemic Drug Elimination', ed. C. F. George and D. G. Shand, Butterworths, London, 1982, p. 29.
14. M. D. Rawlins, in 'Drug Metabolism and Disposition. Considerations in Clinical Pharmacology', ed. G. R. Wilkinson and M. D. Rawlins, MTP Press, Lancaster, 1985, p. 21.
15. K. Hartiala, in 'Handbook of Physiology, Section 9: Reactions to Environmental Agents', ed. D. H. K. Lee, H. L. Falk, S. D. Murphy and S. R. Geiger, American Physiology Society, Washington, DC, 1977, p. 375.
16. G. Levy and T. Matsuzawa, *J. Pharmacol. Exp. Ther.*, 1967, **156**, 285.
17. J. B. Houston and G. Levy, *Nature (London)*, 1975, **255**, 78.
18. S. M. Rogers, D. J. Back and M. L. Orme, *Br. J. Clin. Pharmacol.*, 1987, **23**, 727.
19. J. Caldwell and R. L. Smith, in 'Formulation and Preparation of Dosage Forms', ed. J. Polderman, Elsevier, Amsterdam, 1977, p. 169.
20. R. R. Schelme, in 'Extrahepatic Metabolism of Drugs and other Foreign Compounds', ed. T. E. Gram, MTP Press, Lancaster, 1980, p. 551.
21. K. Pelkonen and O. Hänninen, in 'Gastrointestinal Toxicology', ed. K. Rozman and O. Hänninen, Elsevier, Amsterdam, 1986, p. 193.
22. A. G. Renwick, in 'Presystemic Drug Elimination', ed. C. F. George and D. G. Shand, Butterworths, London, 1982, p. 3.
23. Z. Gregus and C. D. Klaason, in 'Gastrointestinal Toxicology', ed. K. Rozman and O. Hänninen, Elsevier, Amsterdam, 1986, p. 57.
24. J. E. Bakke, A. L. Bergman and G. L. Larsen, *Science (Washington, D. C.)*, 1982, **217**, 645.
25. L. A. Wheeler, F. B. Soderberg and P. Goldman, *J. Pharmacol. Exp. Ther.*, 1975, **194**, 135.
26. B. S. Reddy, L. R. Pohl and G. Krishna, *Biochem. Pharmacol.*, 1976, **25**, 1119.
27. D. E. Rickert, R. M. Long, S. Krakowka and J. G. Dent, *Toxicol. Appl. Pharmacol.*, 1981, **59**, 574.
28. R. Q. Thompson, M. Sturtevant, O. D. Bird and A. J. Glazko, *Endocrinology (Baltimore)*, 1954, **55**, 665.
29. P. Goldman, in 'Drug Absorption', ed. L. F. Prescott and W. S. Nimmo, Adis Press, Balgowlay, Australia, 1979, p. 88.
30. R. Gingell and J. W. Bridges, *Xenobiotica*, 1973, **3**, 599.
31. T. P. Gibson, *Am. J. Kidney Dis.*, 1986, **8**, 7.
32. A. Aitio and J. Marniemi, in 'Extrahepatic Metabolism of Drugs and other Foreign Compounds', ed. T. E. Gram, MTP Press, Lancaster, 1980, p. 365.
33. J. G. Ghazarian, C. R. Jefcoate, J. C. Knutson, W. H. Orme-Johnson and H. F. DeLuca, *J. Biol. Chem.*, 1974, **249**, 3026.
34. B. B. Davis, M. B. Mattammal and T. V. Zenser, *Nephron*, 1987, **27**, 187.
35. M. W. Anders, *Kidney Int.*, 1980, **18**, 636.
36. D. P. Jones, S. Orrenius and S. W. Jacobson, in 'Extrahepatic Metabolism of Drugs and other Foreign Compounds', ed. T. E. Gram, MTP Press, Lancaster, 1980, p. 123.
37. L. F. Chasseaud, in 'Extrahepatic Metabolism of Drugs and other Foreign Compounds', ed. T. E. Gram, MTP Press, Lancaster, 1980, p. 427.
38. S. R. Norby, J. D. Rogers, F. Ferber, H. Jones, A. G. Zachie, L. L. Weidner, J. L. Demetriades, D. A. Gravallese and J. Y.-K. Hsieh, *Antimicrob. Agents Chemother.*, 1984, **25**, 747.
39. H. W. Leber, L. Gleumes and G. Shutterle, *Kidney Int.*, 1978, **13**, suppl. 8, S43.
40. G. E. R. Hook, in 'Presystemic Drug Elimination', ed. C. F. George and D. G. Shand, Butterworths, London, 1982, p. 117.
41. Y. S. Bakhle, in 'Presystemic Drug Elimination', ed. C. F. George and D. G. Shand, Butterworths, London, 1982, p. 147.
42. T. E. Gram, in 'Extrahepatic Metabolism of Drugs and other Foreign Compounds', ed. T. E. Gram, MTP Press, Lancaster, 1980, p. 159.
43. M. E. McManus, A. R. Boobis, G. M. Pacifici, R. Y. Frempong, M. J. Brodie, G. C. Kahn, C. Whyte and D. S. Davies, *Life Sci.*, 1980, **26**, 481.
44. J. S. Lazo and C. J. Humphreys, *Proc. Natl. Acad. Sci. U.S.A.*, 1983, **35**, 575.
45. R. F. Minchin, P. C. Ho and M. R. Boyd, *Biochem. Pharmacol.*, 1986, **35**, 575.
46. A. P. Autor, *Life Sci.*, 1974, **14**, 1309.
47. L. L. Smith, M. S. Rose and I. Wyatt, in 'Ciba Foundation Symposium 65', Excerpta Medica, Amsterdam, 1978, p. 321.
48. R. W. Vaughan, I. G. Sipes and B. R. Brown, Jr., *Life Sci.*, 1978, **23**, 2447.
49. D. R. Bickers and A. Kappas, in 'Extrahepatic Metabolism of Drugs and other Foreign Compounds', ed. T. E. Gram, MTP Press, Lancaster, 1980, p. 295.
50. D. R. Bickers, T. Dutta-Choudhury and H. Mukhtar, *Mol. Pharmacol.*, 1982, **21**, 239.
51. S. M. Schanberg, J. J. Schildkrant, G. R. Breese and I. J. Copin, *Biochem. Pharmacol.*, 1968, **17**, 247.

52. W. N. Jenner and P. A. Rose, *Biochem. J.*, 1973, **135**, 109.
53. D. Eccleston and I. M. Ritchie, *J. Neurochem.*, 1973, **21**, 635.
54. H. Y. Yang and N. H. Neff, *Mol. Pharmacol.*, 1976, **12**, 69.
55. D. H. Clouet, *Life Sci.*, 1962, **1**, 31.
56. G. Benzi, F. Berte, A. Crema and G. M. Frigo, *J. Pharm. Sci.*, 1967, **56**, 1349.
57. A. D. Mosnaim, R. Silkaitis and M. E. Wolf, *Life Sci.*, 1980, **27**, 557.
58. F.-J. Leinweber, *Drug Metab. Rev.*, 1987, **18**, 379.
59. B. N. La Du and H. Snady, in 'Handbook of Experimental Pharmacology', ed. B. B. Brodie and J. R. Gillette, Springer, New York, 1971, p. 477.
60. R. E. Talcott, *Toxicol. Appl. Pharmacol.*, 1979, **47**, 145.
61. C. C. Wang and D. L. Hill, *Biochem. Pharmacol.*, 1977, **26**, 947.
62. M. Salmona, C. Saronio, R. Bianchi, F. Marcucci and E. Mussini, *J. Pharm. Sci.*, 1974, **63**, 222.
63. W. W. Weber, H. E. Radtke and R. H. Tannen, in 'Extrahepatic Metabolism of Drugs and other Foreign Compounds', ed. T. E. Gram, MTP Press, Lancaster, 1980, p. 493.
64. A. Hellander and O. Tottmar, *Biochem. Pharmacol.*, 1987, **36**, 1077.
65. C. M. Macdonald, U. Gebert, K. Watson, T. A. Bryce and M. Omosu, *Jpn. Pharmacol. Ther.*, 1986, **14**, 2025.
66. E. J. Ariens and A. M. Simonis, in 'Strategy in Drug Research', ed. J. A. Keverling Buisman, Elsevier, Amsterdam, 1982, p. 165.
67. N. Bodor, *Med. Res. Rev.*, 1984, **4**, 449.
68. W. F. Beyer and E. H. Jensen, in 'Analytical Profiles of Drug Substances', ed. K. Florey, Academic Press, New York, 1974, vol. 3, p. 513.
69. J. A. Taylor, *Clin. Pharmacol. Ther.*, 1972, **13**, 710.
70. E. A. Neill, D. J. Chapple and C. W. Thompson, *Br. J. Anaesth.*, 1983, **55**, 23S.
71. A. Albert, *Nature (London)*, 1958, **182**, 421.
72. A. Albert, in 'Selective Toxicity', ed. A. Albert, Wiley, New York, 1973, p. 21.
73. S. M. Kupchan, A. F. Casy and J. V. Swintosky, *J. Pharm. Sci.*, 1965, **54**, 514.
74. N. J. Harper, *J. Med. Pharm. Chem.*, 1959, **1**, 467.
75. N. J. Harper, *Prog. Drug Res.*, 1962, **4**, 221.
76. T. Higuchi and V. Stella (eds.), 'Prodrugs as Novel Drug Delivery Systems', American Chemical Society, Washington, DC, 1975.
77. E. B. Roche (ed.), 'Design of Biopharmaceutical Properties through Prodrugs and Analogs', American Pharmaceutical Association, Washington, DC, 1977.
78. A. A. Sinkula and S. H. Yalkowsky, *J. Pharm. Sci.*, 1975, **64**, 181.
79. R. E. Notari, *Methods Enzymol.*, 1985, **112**, 309.
80. A. Tsuji, E. Miyamoto, Y. Terasaki and T. Yamana, *J. Pharm. Sci.*, 1979, **68**, 1259.
81. J. S. Wold, R. R. Joost, H. R. Black and R. S. Griffith, *J. Infect. Dis.*, 1978, **137**, suppl. S17.
82. L. E. Kirsch and R. E. Notari, *J. Pharm. Sci.*, 1984, **73**, 728.
83. H. Bundgaard, *Methods Enzymol.*, 1985, **112**, 347.
84. B. Vej-Hansen and H. Bundgaard, *Arch. Pharm. Chemi, Sci. Ed.*, 1979, **7**, 65.
85. A. J. Glazko, W. H. Edgerton, W. A. Dill and W. R. Lenz, *Antibiot. Chemother.*, 1952, **2**, 234.
86. H. Andersgaard, P. Finholt, R. Gjermundsen and T. Hoyland, *Acta Pharm. Suec.*, 1974, **11**, 239.
87. A. Sinkula, in 'Design of Biopharmaceutical Properties through Prodrugs and Analogs', ed. E. B. Roche, American Pharmaceutical Association, Washington, DC, 1977, p. 422.
88. G. E. Davies, G. W. Driver, E. Hoggarth, A. R. Martin, M. F. C. Paige, F. L. Rose and B. R. Wilson, *Br. J. Pharmacol.*, 1956, **11**, 351.
89. G. E. Davies and G. W. Driver, *Br. J. Pharmacol.*, 1957, **12**, 434.
90. J. E. Gray, *Arch. Pathol.*, 1967, **84**, 522.
91. J. E. Gray, R. N. Weaver, J. Moran and E. S. Feenstra, *Toxicol. Appl. Pharmacol.*, 1974, **27**, 308.
92. H. T. Edmondson, *Ann. Surg.*, 1973, **178**, 637.
93. T. F. Brodasky and C. Lewis, *J. Antibiot.*, 1972, **25**, 230.
94. R. M. DeHaan, C. M. Metzler, D. Schellenberg and W. D. Van Denbosch, *J. Clin. Pharmacol.*, 1973, **13**, 190.
95. J. T. Scott, I. H. Porter, S. M. Lewis and A. S. Dixon, *Q. J. Med.*, 1961, **30**, 167.
96. H. W. Davenport, *Gastroenterology*, 1964, **46**, 245.
97. H. W. Davenport, B. J. Cohen, M. Bree and V. D. Davenport, *Gastroenterology*, 1965, **49**, 189.
98. K. J. Ivery, *Gastroenterology*, 1971, **61**, 247.
99. L. W. Dittert, H. C. Caldwell, T. Ellison, G. M. Irwin, D. E. Rivard and J. V. Swintosky, *J. Pharm. Sci.*, 1968, **57**, 828.
100. A. J. Glazko, H. E. Carnes, A. Kazenko, L. M. Wolf and T. P. Reutner, *Antibiot. Annu.*, 1957, 792.
101. V. J. Stella, W. N. Charman and V. H. Naringrekar, *Drugs*, 1985, **29**, 455.
102. M. C. Nahata and D. A. Powell, *Clin. Pharmacol. Ther.*, 1981, **30**, 368.
103. J. C. Melby and M. St. Cyr, *Metabolism*, 1961, **10**, 75.
104. K. B. Sloan, M. Hashida, J. Alexander, N. Bodor and T. Higuchi, *J. Pharm. Sci.*, 1983, **72**, 372.
105. T. A. Connors, *Xenobiotica*, 1986, **16**, 975.
106. D. C. Chatterji and J. F. Gallelli, *J. Pharm. Sci.*, 1979, **68**, 822.
107. C. C. Loo, E. L. Foltz, H. Wallick and K. D. Kwan, *Clin. Pharmacol. Ther.*, 1974, **16**, 35.
108. P. Bolme, B. Dahlstrom, N.-A. Diding, O. Flink and L. Paalzow, *Eur. J. Clin. Pharmacol.*, 1976, **10**, 237.
109. W. V. Daehne, E. Frederiksen, E. Gundersen, F. Lund, P. Morch, H. J. Peterson, K. Roholt, L. Tybring and W. O. Godtfredsen, *J. Med. Chem.*, 1970, **13**, 607.
110. N. O. Bodin, B. Ekstrom, U. Forsgren, L.-P. Jalar, L. Magni, C.-H. Ramsey and B. Sjoberg, *Antimicrob. Agents Chemother.*, 1975, **8**, 518.
111. J. P. Clayton, M. Cole, S. W. Elson, K. D. Hardy, L. W. Mizen and R. Sutherland, *J. Med. Chem.*, 1975, **18**, 172.
112. W. S. Saari, M. B. Freedman, R. D. Hartman, S. W. King, A. W. Raab, W. C. Randall, E. L. Engelhardt, R. Hirschmann, A. Rosegay, C. T. Ludden and A. Scriabine, *J. Med. Chem.*, 1978, **21**, 746.
113. N. Bodor, K. B. Sloan, T. Higuchi and K. Sasahara, *J. Med. Chem.*, 1977, **20**, 1435.

114. Y. Garceau, I. Davis and J. Hasegawa, *J. Pharm. Sci.*, 1978, **67**, 1360.
115. H. P. Schedl and J. A. Clifton, *Gastroenterology*, 1961, **41**, 491.
116. J. Fried, A. Borman, W. B. Kessler, P. Garbowich and E. F. Sabo, *J. Am. Chem. Soc.*, 1958, **80**, 2338.
117. R. Gardi, R. Vitali and A. Ercoli, *J. Org. Chem.* 1962, **27**, 668.
118. V. J. Stella, T. J. Mikkelson and J. D. Pipkin, in 'Drug Delivery Systems, Characteristics and Biomedical Applications', ed. R. L. Juliano, Oxford University Press, New York, 1980.
119. A. G. Ebert and S. M. Hess, *J. Pharmacol. Exp. Ther.*, 1965, **168**, 412.
120. J. Dreyfuss, J. J. Ross, J. M. Shaw, I. Miller and E. C. Schreiber, *J. Pharm. Sci.*, 1976, **65**, 502.
121. J. Dreyfuss, J. M. Shaw and J. J. Ross, *J. Pharm. Sci.*, 1976, **65**, 1310.
122. H. L. Yale, *J. Med. Chem.*, 1977, **20**, 302.
123. H. L. Yale, *Drug Metab. Rev.*, 1978, **8**, 251.
124. V. J. Stella and K. J. Himmelstein, *J. Med. Chem.*, 1980, **23**, 1275.
125. V. J. Stella and K. J. Himmelstein, in 'Optimization of Drug Delivery—Alfred Benzon Symposium 17', ed. H. Bundgaard, A. B. Hansen and H. Kofod, Munksgaard, Copenhagen, 1982, p. 134.
126. C. R. Gardner, *Biomaterials*, 1985, **6**, 153.
127. B. Boshes, *Ann. Intern. Med.*, 1981, **94**, 364.
128. S. Welk, H. Mizoguchi and M. Orlowski, *J. Pharmacol. Exp. Ther.*, 1978, **206**, 227.
129. N. Bodor, E. Shek and T. Higuchi, *Science (Washington, D.C.)*, 1975, **190**, 155.
130. N. Bodor and M. E. Brewster, *Eur. J. Med. Chem.*, 1983, **18**, 235.
131. N. Bodor and J. W. Simpkins, *Science (Washington, D.C.)*, 1983, **221**, 65.
132. J. R. Vane, in 'Prostaglandin Synthetase Inhibitors: Their Effects on Physiological Functions and Pathological States', ed. H. J. Robinson and J. R. Vane, Raven Press, New York, 1974, p. 155.
133. D. E. Duggan, *Drug. Metab. Rev.*, 1981, **12**, 325.
134. M. R. Dobrinska, D. E. Furst, T. Spiegel, W. C. Vincek, R. Tompkins, D. E. Duggan, R. O. Davies and H. E. Paulus, *Biopharm. Drug Dispos.*, 1983, **4**, 347.
135. J. R. Mitchell, H. J. Zimmerman, K. G. Ishak, U. P. Thorgeirson, J. A. Timbrell, W. R. Snodgrass and S. D. Nelson, *Ann. Intern. Med.*, 1976, **84**, 181.
136. J. A. Hinson, *EHP, Environ. Health Perspect.*, 1983, **49**, 71.
137. S. D. Nelson, *J. Med. Chem.*, 1982, **25**, 753.
138. H. R. Sullivan, P. Roffey and R. E. McMahon, *Drug Metab. Dispos.*, 1979, **7**, 76.
139. K. Einarsson, J. L. E. Ericsson, J. A. Gustafsson, J. A. Sjovali, J. Zietz and E. Zietz, *Biochim. Biophys. Acta*, 1974, **369**, 278.
140. G. A. Digenis and J. V. Swintosky, in 'Handbook of Experimental Pharmacology', ed. J. R. Gillette and J. R. Mitchell, Springer, New York, 1975, part 3, p. 28, 86.
141. I. H. Pitman, *Med. Res. Rev.*, 1981, **1**, 189.
142. H. Bundgaard, A. B. Hansen and H. Kofod (eds.), 'Optimization of Drug Delivery', Munksgaard, Copenhagen, 1982.
143. K. J. Widder and R. Green (eds.), *Methods Enzymol.*, 1985, **112**.
144. M. Azori, *CRC Crit. Rev. Ther. Drug Carrier Syst.*, 1987, **4**, 39.
145. E. B. Roche (ed.), in 'Bioreversible Carriers in Drug Design. Theory and Application', Pergamon Press, Oxford, 1987.
146. E. C. Miller and J. A. Miller, *Pharmacol. Rev.*, 1966, **18**, 805.
147. E. C. Miller, *Cancer Res.*, 1978, **38**, 1479.
148. B. B. Brodie, W. D. Reid, A. K. Cho, G. Sipes, G. Krishna and J. R. Gillette, *Proc. Natl. Acad. Sci. U.S.A.*, 1971, **68**, 160.
149. W. Z. Potter, S. S. Thorgeirsson, D. J. Jollow and J. R. Mitchell, *Pharmacology*, 1974, **12**, 29.
150. C. P. Siegers, J. G. Filser and H. M. Bolt, *Toxicol. Appl. Pharmacol.*, 1978, **46**, 709.
151. J. W. Gorrod, in 'Testing for Toxicity', ed. J. W. Gorrod, Taylor and Francis, London, 1981, p.77.
152. J. L. Farber and S. K. El-Mofty, *Am. J. Pathol.*, 1975, **81**, 237.
153. D. V. Parke, *Arch. Toxicol.*, 1987, **60**, 5.
154. J. Berky (ed.), 'In Vitro Toxicity Testing', Franklin Institute Press, Philadelphia, 1978.
155. F. P. Guengerich and D. C. Liebler, *CRC Crit. Rev. Toxicol.*, 1985, **14**, 259.
156. M. J. Coon, A. H. Conney, R. W. Estabrook, H. V. Gelboin, J. R. Gillette and P. J. O'Brien (eds.), 'Microsomes and Drug Oxidations and Chemical Carcinogenesis 2', Academic Press, New York, 1980.
157. R. P. Mason and C. F. Chignell, *Pharmacol. Rev.*, 1981, **33**, 189.
158. H. Kappus and H. Sies, *Experientia*, 1981, **37**, 1233.
159. H. Kappus, *Arch. Toxicol.*, 1987, **60**, 144.
160. L. J. King, in 'Drug Metabolism—from Molecules to Man', ed. D. J. Benford, J. W. Bridges and G. G. Gibson, Taylor and Francis, London, 1987, p. 657.
161. W. B. Jackoby and W. H. Habig, in 'Enzymatic Basis of Detoxication', ed. W. B. Jackoby, Academic Press, New York, 1980, vol. 2, p. 63.
162. B. Ketterer, *Drug Metab. Rev.*, 1982, **13**, 161.
163. I. M. Arias and W. B. Jackoby (eds.), 'Glutathione: Metabolism and Function', Raven Press, New York, 1976.
164. H. Sies, in 'Oxidative Stress', ed. H. Sies, Academic Press, London, 1985, p. 1.
165. P. R. Ortiz de Montellano and M. A. Correia, *Annu. Rev. Pharmacol. Toxicol.*, 1983, **23**, 481.
166. F. P. Guengerich and T. L. Macdonald, *Acc. Chem. Res.*, 1984, **17**, 9.
167. F. P. Guengerich, R. J. Willard, J. P. Shea, L. E. Richards and T. L. Macdonald, *J. Am. Chem. Soc.*, 1984, **106**, 6446.
168. R. H. Abeles and A. L. Maycock, *Acc. Chem. Res.*, 1976, **9**, 313.
169. R. B. Silverman, *J. Biol. Chem.*, 1983, **258**, 14 766.
170. J. Halpert, D. Hammond and R. A. Neal, *J. Biol. Chem.*, 1980, **255**, 1080.
171. J. Halpert, *Biochem. Pharmacol.*, 1981, **30**, 875.
172. J. Halpert, *Mol. Pharmacol.*, 1982, **21**, 166.
173. J. Halpert, B. Näslund and I. Betner, *Mol. Pharmacol.*, 1983, **23**, 445.
174. F. P. Guengerich and P. G. Watanabe, *Biochem. Pharmacol.*, 1979, **28**, 589.
175. F. P. Guengerich, *Biochem. Pharmacol.*, 1977, **26**, 1909.
176. J. L. Poyer, P. B. McCay, E. K. Lai, E. G. Janzen and E. R. Davis, *Biochem. Biophys. Res. Commun.*, 1980, **94**, 1154.

177. J. L. Poyer, P. B. McCay, C. C. Weddle and P. E. Downs, *Biochem. Pharmacol.*, 1981, **30**, 1517.
178. T. F. Slater, in 'Free Radical Mechanisms in Tissue Injury', Pion, London, 1972.
179. R. O. Recknagel and E. A. Glende, in 'Handbook of Physiology, Reactions to Environmental Agents', ed. D. H. K. Lee, H. L. Falk, S. D. Murphy and S. R. Geiger, Williams and Wilkins, Baltimore, 1977, p. 591.
180. F. P. Guengerich, W. M. Crawford, Jr. and P. G. Watanabe, *Biochemistry*, 1979, **18**, 5177.
181. J. A. Swenberg, M. A. Bedell, K. C. Billings, D. R. Umbenhauer and A. E. Pegg, *Proc. Natl. Acad. Sci. U.S.A.*, 1982, **79**, 5499.
182. A. L. Shen, W. E. Fahl and C. R. Jefcoate, *Arch. Biochem. Biophys.*, 1980, **204**, 511.
183. D. R. Umbenhauer and A. E. Pegg, *Cancer Res.*, 1981, **41**, 3471.
184. R. E. Miller and F. P. Guengerich, *Cancer Res.*, 1983, **43**, 1145.
185. F. P. Guengerich, P. S. Mason, W. T. Stott, T. R. Fox and P. G. Watanabe, *Cancer Res.*, 1981, **41**, 4391.
186. M. R. Boyd, *CRC Crit. Rev. Toxicol.*, 1980, **7**, 103.
187. S. S. Lau, T. J. Monks, K. E. Greene and J. R. Gillette, *Xenobiotica*, 1984, **14**, 539.
188. J. A. Nash, L. J. King, E. A. Lock and T. Green, *Toxicol. Appl. Pharmacol.*, 1984, **73**, 124.
189. E. C. Miller and J. A. Miller, *Cancer (N.Y.)*, 1981, **47**, 2327.

23.5

Metabolic Pathways

JOHN CALDWELL and STEPHEN C. MITCHELL
St Mary's Hospital Medical School, London, UK

23.5.1 INTRODUCTION

The most cursory consideration of the literature on the metabolism of drugs and other foreign compounds serves to show the enormous number of metabolic transformations which they may undergo. A major insight, permitting simplification of the situation, was provided by Williams,[1] who classified these reactions into two distinct types, which generally occur in sequence. In this sequence,

143

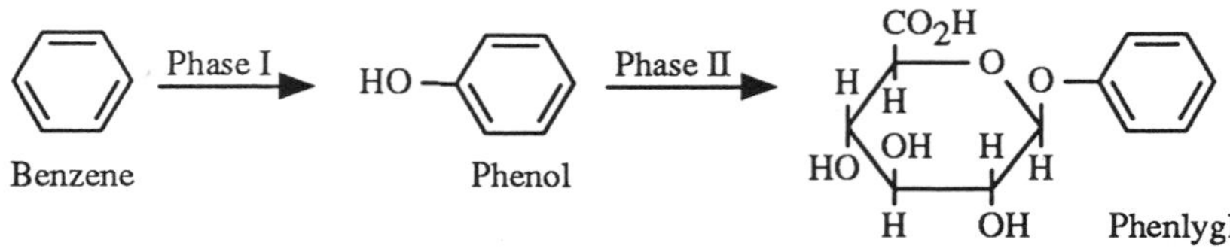

Figure 1 Biphasic sequence of drug metabolism

foreign compounds undergo first a phase I, or functionalization, reaction of oxidation, reduction or hydrolysis, giving a product which acts as substrate for a phase II, or conjugation, reaction. These latter are biosynthetic reactions in which the xenobiotic moiety (or exocon) is linked with an endogenous conjugating agent (or endocon) to give a product known as a conjugate. Exceptions to this biphasic sequence do occur: some phase I metabolites can be excreted as such, while other drugs may undergo conjugation directly. The ultimate excretion products of a xenobiotic may thus comprise the unchanged compound, phase I metabolites, phase II metabolites and the products of this biphasic sequence.[2] Quantitatively, conjugates are the most important metabolites of the majority of xenobiotics (Figure 1).[2]

In order that they may undergo metabolic conjugation, foreign compounds must possess within their structures appropriate functional groups for linkage to the particular endocon in question.[2,3] This is most commonly produced by phase I metabolism, by introduction of an oxygen atom, by reduction or by bond cleavage leading to the uncovering of a suitable functional group. This may occur by oxidation (*e.g.* dealkylation), reduction (*e.g.* azo bond cleavage) or hydrolysis (*e.g.* of esters and amides). The functional groups involved include both chemically stable groups (*e.g.* alcoholic, phenolic or carboxylic hydroxyls, amine, thiol or cyclic heteroatoms) and chemically reactive groups (arene oxide, carbonium or nitrenium ions).[3]

23.5.2 OXIDATIONS

23.5.2.1 Carbon

Carbon is the most abundant element in living systems and is also the major component of most of the drugs that are introduced to react with these systems. Carbon is unusual in that it tends to form multiple chemical bonds with other carbon atoms and the metabolic oxidation of a carbon centre can occur when it is present in a variety of different chemical combinations. Although these different chemical combinations may be legion, the criteria for metabolic differentiation, and simple classification, usually depend upon the saturation (single bond) or nonsaturation (usually double bonds, but also triple) of the adjacent carbon bonds.

23.5.2.1.1 Saturated bonds

Aliphatic hydrocarbons, open chain compounds resembling fatty acids, usually occur within drug molecules as alkyl side chains, which are attached to larger moieties containing ring structures. The oxidation of the side chains of thiopental and phenylbutazone to alcohols are examples of oxidative metabolism for this type of functionality. Further oxidation of the alcohols to aldehydes and carboxylic acids may then occur. Where several carbon atoms are present in an aliphatic chain, hydroxylation may occur at a variety of locations. When this occurs at the terminal carbon atom it is known as 'ω oxidation', and at the penultimate carbon atom it is termed '$\omega - 1$ oxidation'. Attack at the carbon atom which attaches the chain to the rest of the drug molecule, usually a ring structure, is known as 'α oxidation'. Oxidation at the next adjacent carbon is termed 'β oxidation', and so on along the chain. Oxidation of prochiral methylene groups creates new chiral centres. *n*-Propylbenzene is a compound in which the side chain undergoes metabolism at all possible positions, and serves as an example of this (Figure 2).[4] The hydroxylation of *n*-heptane by liver microsomes can also occur in all four possible positions.[5]

The mechanism by which aliphatic compounds are hydroxylated (by cytochrome *P*-450) is unknown. A stepwise process has been proposed involving the formation of (heme) iron peroxide,

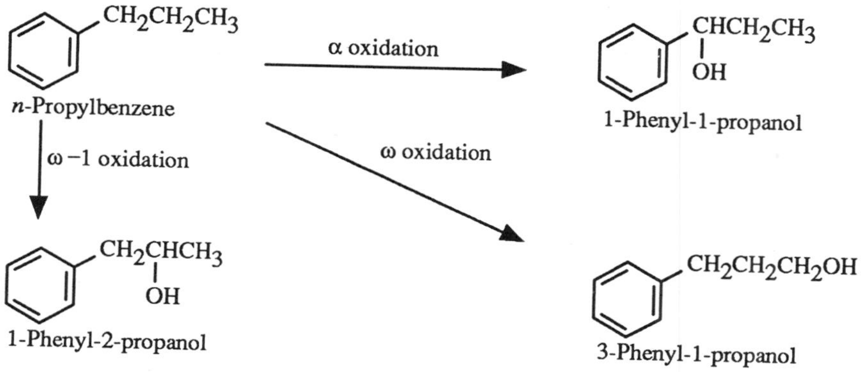

Figure 2 Side chain oxidations of *n*-propylbenzene

dehydrogenation to yield a free radical, followed by formation of a carbonium ion and the attack on the iron-bound carbonium ion or free radical by the substrate.[6] This radical abstraction method is similar to that observed in the hydroxylation of aliphatic compounds catalyzed by metal-containing peroxidases.[7]

An alicyclic compound is a molecule which contains a saturated ring system, possessing properties more similar to those of acyclic (aliphatic) than aromatic compounds. Cyclohexane is an example of a simple alicyclic compound. It is metabolized to cyclohexanol and then further to the 1,2-diol. The cyclohexene ring of hexobarbital is hydroxylated at a saturated position to produce an alcohol, which is further oxidized to produce a keto derivative.[4]

23.5.2.1.2 *Unsaturated bonds*

Aromatic compounds are cyclic molecules which are characterized by the presence of conjugated double bonds within the ring system. The hydroxylation of aromatic compounds produces phenols, that occurring with phenobarbitone being an example. Such oxidations generally proceed *via* the intermediate production of an epoxide, followed by intramolecular rearrangement to form a phenol. As an alternative, epoxides can be hydrated with the concomitant loss of a double bond, producing dihydrodiols. These are thought to be the proximate carcinogens for a number of polycyclic aromatic hydrocarbons found in nature, benzo[*a*]pyrene being an example of this.[8] Some epoxides are chemically stable, *e.g.* the metabolites of heptachlor and aldrin, the latter (dieldrin) being further metabolized to the corresponding dihydrodiol.[9]

When a compound contains both alicyclic and aromatic rings, the saturated ring is usually the more readily hydroxylated. An example of this is seen in the metabolism of the antihypertensive

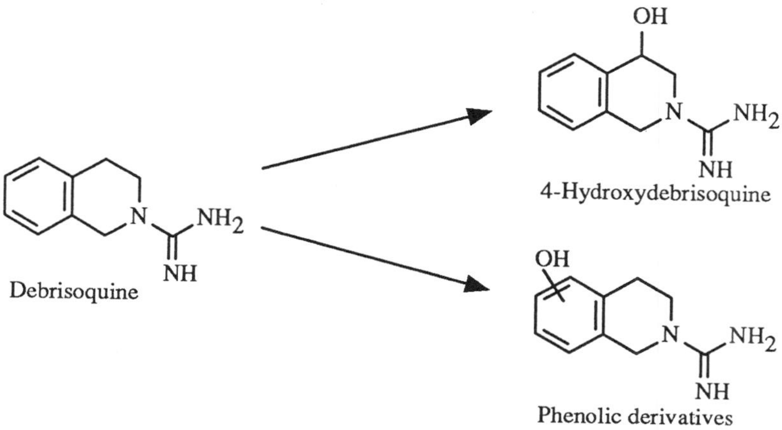

Figure 3 Aliphatic and aromatic oxidation of debrisoquine

drug debrisoquine, where the majority of individuals produce only small amounts of phenolic derivatives when compared to the major metabolite, 4-hydroxydebrisoquine, produced by aliphatic hydroxylation (Figure 3).[10] Unsaturated bonds do not always occur within ring systems. The side chain of anethole contains a double bond which undergoes epoxidation and hydration, yielding the corresponding diol derivatives.[11]

23.5.2.2 Nitrogen, Sulfur and other Heteroatoms

In comparison to oxidation at carbon centres, oxidation at nitrogen is quantitatively a relatively minor route of metabolism. The important oxidation reactions occurring at organic nitrogen centres can be subdivided into *N* hydroxylations and *N* oxidations. If a hydrogen atom is attached to the nitrogen centre (primary or secondary amine or amide), *N* hydroxylation can occur, *e.g.* aniline is converted into phenylhydroxylamine.[4] *N* Hydroxylation is believed to be an essential step in the metabolic activation of the carcinogen 2-acetylaminofluorene.[12] Tertiary amines are readily susceptible to *N* oxidation with the formation of their corresponding *N*-oxides. Trimethylamine, a compound which imparts the pungent characteristic odour to rotten fish, is metabolized to its innocuous *N*-oxide, and chlorpromazine is known to form a side chain *N*-oxide (Figure 4).[4]

Sulfur centres are similarly susceptible to oxidation. Compounds containing a sulfide moiety may undergo metabolic oxidation to the corresponding sulfoxide, which may in certain cases be further oxidized to the sulfone derivative. This can occur when the sulfide sulfur atom is present within a heterocyclic structure, as is the case with chlorpromazine (Figure 4), or when it is part of an aliphatic chain, such as dimethyl sulfide.

Several other heteroatoms (noncarbon) occur occasionally within drugs and other foreign compounds entering the body, and these too may be susceptible to oxidation. An example of this is the oxidative conversion of the arseno compound arsphenamine, the first synthetic chemotherapeutic agent, to the active oxoarsine, 3-amino-4-hydroxyphenyloxoarsine (Figure 5).[4]

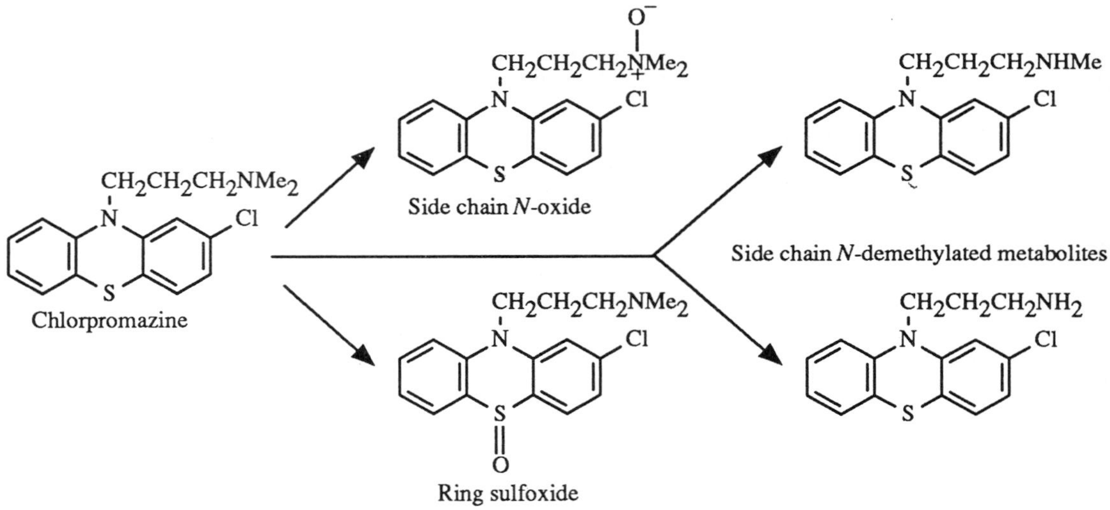

Figure 4 A few of the many metabolites of chlorpromazine, demonstrating the pathways of *N* oxidation, *S* oxidation and *N* demethylation

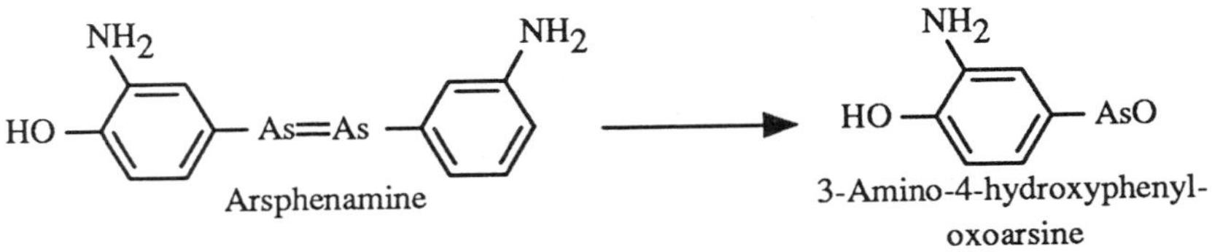

Figure 5 Oxidative cleavage of arsphenamine

23.5.2.3 Dealkylation and Deamination

A wide variety of drugs and foreign compounds undergo metabolic modification in the body by removal of *N*-, *O*-, or *S*-alkyl groups to yield the corresponding amines, phenols or thiols. This removal is accomplished by an oxidative process that converts the alkyl radicals into aldehydes. Only relatively small alkyl groups seem to be removed, with the majority of reports concerning demethylations and deethylations (Figure 4).[1]

The exact mechanism of *N* dealkylation, the most studied of these reactions, is uncertain. The major problem is whether the initial attack occurs at the nitrogen atom with the subsequent formation of an *N*-oxidized intermediate,[13] or whether the initial oxidation occurs at the carbon atom adjacent to the nitrogen to give a carbanolamine. Both of these unstable intermediates would then rearrange to release the amine and an aldehyde. This latter mechanism, involving α carbon oxidation, is assumed also to apply to *O* and *S* dealkylations.[14] Deamination, the removal of ammonia and its effective replacement by a keto group, is thought to proceed *via* a mechanism similar to that of dealkylation. The classification of a reaction as dealkylation or deamination depends very much upon the relative sizes of the substituents around the nitrogen atom, and whether the metabolite of interest is the product aldehyde or amine.

23.5.2.4 Dehalogenation

As with the dealkylation reactions, the detailed mechanisms involved in the removal of halogens (F, Cl, Br, I) from organic molecules have not been fully elucidated. The available evidence suggests that oxidative mechanisms are involved, but direct oxidation of the halogen atom appears unlikely. It is thought that there is a primary role for C—H bond oxidation in those dehalogenations, where an adjacent carbon atom undergoes the initial oxidation. The intermediates formed then rearrange with the release of a halide ion and an aldehyde, which may be further metabolized to an alcohol. This route, previously termed 'hydrolytic cleavage', is considered to be a major route of metabolism for halogenated compounds, but other processes involving epoxide formation, reductive reactions, interaction with glutathione and dehydrohalogenation are known to occur for certain compounds.[15]

23.5.3 ENZYMOLOGY OF XENOBIOTIC OXIDATION

23.5.3.1 Cytochromes *P*-450

The great majority of xenobiotic oxidations are effected by a hemoprotein, named cytochrome *P*-450 after the wavelength of the absorption peak obtained when it is complexed in the reduced form with carbon monoxide.

The cytochrome *P*-450 system is embedded within the phospholipid environment of the microsomes, artifacts created from the endoplasmic reticulum of living cells by homogenization and isolated by differential centrifugation. Lipid involvement is necessary to maintain the integrity and functional capacity of the enzyme system. Over the past decade the cytochrome *P*-450 system has been dissected into its component parts (NADPH–cytochrome *P*-450 reductase, numerous cytochrome *P*-450 isoenzymes, diacyl GPC), and reconstituted to give a functional system when supplemented with NADPH and oxygen.[16]

The activity of cytochrome *P*-450 can be increased (induction) or decreased (inhibition) by pretreatment with a variety of chemical agents. In some instances (*e.g.* phenobarbitone) this previous treatment with a drug can lead to enhanced metabolism with a concomitant decline in the therapeutic effectiveness of subsequent doses of the same or another drug. Complications in therapy will therefore ensue. Different substrate specificities of induced activities, dependent upon the inducing agent employed, led to the first suggestions that there may be more than one form of cytochrome *P*-450. It is now known that many multiple forms of cytochrome *P*-450 coexist, and that these isozymes have overlapping substrate specificities. The reader is advised to consult more specialist texts for further details.[17]

The reaction sequence of the oxidation of a drug by cytochrome *P*-450 involves first the reduction of cytochrome *P*-450 and the heme iron atom by the flavoprotein NADPH–cytochrome *P*-450 reductase. The iron then combines with molecular oxygen, which it activates. The iron(II)–oxygen complex rearranges, picking up two protons, and then decomposes, releasing the oxidized drug and water. The iron of the cytochrome is finally oxidized back to its iron(III) state. This sequence of events comprises a cycle, the cytochrome P-450 then being ready to oxidize another drug molecule.

The cytochromes P-450 are mixed function oxidases, which activate molecular oxygen and insert one atom into a variety of lipophilic substrates, the other being reduced to give water. The cytochromes P-450 have been implicated in virtually all biological oxidations of xenobiotics, including C, N and S oxidations, N, O and S dealkylations, deaminations and certain dehalogenations. In addition, they can catalyze reductive reactions under anaerobic conditions (nitro, N-oxide, azo and epoxide reductions and reductive dehalogenations), but it is not known to what extent these reactions may occur in living systems. It is possible that in certain situations, such as anaesthesia, the oxygen tension of the tissues may be lower than normal, permitting this type of reaction to take place.

Although the cytochrome P-450 system is the central catalyst in phase I drug metabolism, it is important to appreciate that other enzyme systems may play important roles.

23.5.3.2 Flavoprotein Oxidase (Ziegler's Enzyme)

Until about 20 years ago it was assumed that all the oxidative reactions which took place within the microsomes were effected by a 'mixed function oxidase', taken as being synonymous with cytochrome P-450. However, several anomalous observations were made and the microsomal fraction was subsequently shown to contain another enzyme system, a flavin-containing monooxygenase, free from cytochrome P-450. This flavoprotein oxidase is now known to catalyze the oxidation of diverse nucleophilic nitrogen and sulfur centres, including secondary and tertiary amines, hydrazines, sulfide-, disulfide- and thiol-containing molecules, as well as thioamides and thiocarbamides. Although the substrate specificity of this enzyme system has not been rigorously defined, a variety of factors are known to influence its activity and, as a general rule, compounds with a negative charge anywhere on the molecule are not substrates.

As with cytochrome P-450, oxidations catalyzed by this enzyme can also be represented as a cycle. The flavoprotein oxidase first binds NADPH, followed by oxygen and then the substrate. After rearrangements, the oxygenated product is released, followed by NADP, and the cycle begins again. The intimate details and rate-limiting steps within this cycle have not been fully elucidated as yet.

It can be appreciated that it is not only the specific chemical identity of an atom, but also factors such as its electromolecular environment, which will dictate which enzyme will undertake its oxidation. This can be exemplified by sulfur centres, which are now known to be substrates for several oxidative systems. Sulfur atoms within or adjacent to aromatic or heterocyclic ring systems, thereby losing partial control of the electronic shield through delocalization, are prone to oxidative attack by the cytochrome P-450 system, whereas those in aliphatic or alicyclic environments, which tend to retain their electronic cover and are thus more nucleophilic in character, are preferentially oxidized by the flavoprotein system.[18]

23.5.3.3 Molybdenum-containing Oxidases

Aldehyde oxidase and xanthine oxidase are metalloflavoproteins which coexist within the cytosol of the cell and are unusual in containing, in addition to iron as FeS clusters, the metal molybdenum. Substrate oxidation occurs at the molybdenum centre of these enzymes, which catalyze oxidations in which the oxygen linked to the substrate is derived from water rather than molecular oxygen. This mechanism differs from that of the previously mentioned 'mixed function oxidases', the cytochromes P-450 and flavoprotein oxidase, which employ molecular oxygen.[19,20]

Both of these molybdo-flavoprotein enzymes carry out the oxidation of a variety of nitrogen-containing heterocyclic compounds (purines, pyrimidines, pyridines, quaternary nitrogen compounds) as well as aliphatic and aromatic aldehydes. Their quantitative contribution to the oxidation of drugs and foreign compounds is yet to be fully established, but their wide substrate specificities make it likely that they play a significant role.[21]

23.5.3.4 Monoamine Oxidase

Monoamine oxidase is a flavoprotein which is widely distributed throughout most mammalian tissues, being tightly bound to the outer membrane of the mitochondria. It catalyzes the oxidative deamination of many amines, and its main function appears to be the breakdown of endogenous neurotransmitter amines, such as adrenaline and amines absorbed from the gut, present in the diet or produced by intestinal microbial metabolism.[22]

The enzyme is known to exist in many tissues in two forms (A and B), which possess different substrate specificities and inhibitor sensitivities.[23] It has been suggested that the A form metabolizes the amines which are generally believed to be neurotransmitters, whereas the B form is primarily concerned with the metabolism of intestinally absorbed amines. This rigid distinction has been criticized, and the true role of these two forms is still under debate.[24]

The deamination reaction involves abstraction of two hydrogen atoms, one from the nitrogen atom and the other from the α carbon, to form an imine intermediate which is subsequently hydrolyzed to release ammonia and an aldehyde.[25,26] The role of this enzyme in the metabolism of xenobiotics is presumably limited to that of oxidative deamination. A large number of substrates are known to be metabolized, including secondary and tertiary N-methylamines. The amine group itself must be attached to an unsubstituted methylene bridge, so that aniline and amphetamine are not substrates for this enzyme.[27-29]

23.5.3.5 Miscellaneous

Many other oxidative enzyme systems are known to exist, some of which undoubtedly play a role in the metabolism of particular drug molecules. Particular interest has focussed upon the fact that certain enzymes may employ drug substrates in cooxygenation reactions, where the sole function of the drug is to absorb the extra oxygen released during the reaction. For example, one oxygen atom from molecular oxygen would go to the true substrate of the enzyme system, the other surplus oxygen being 'discarded' to the drug cosubstrate. Thus the microsomal prostaglandin synthetase complex can use a variety of xenobiotics as cosubstrate reductants for the endoperoxidase precursors of prostaglandins and thromboxanes.[30]

23.5.4 REDUCTIONS

23.5.4.1 Carbon Centres

Conjugated (unsaturated) carbon–carbon bonds may undergo metabolic reduction to saturated bonds, as well as the oxidations previously mentioned. Pulegol is metabolized into menthol by

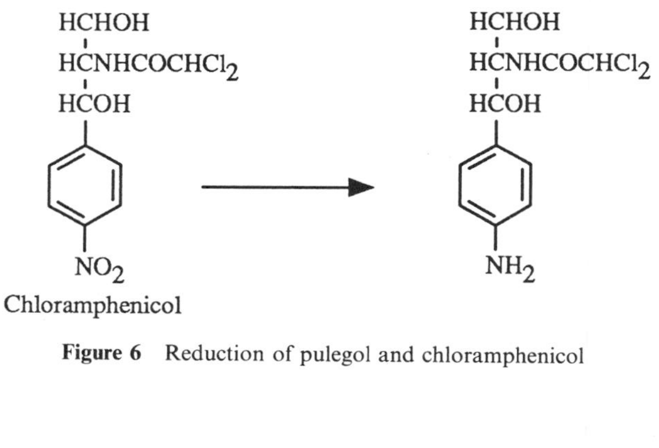

Figure 6 Reduction of pulegol and chloramphenicol

reduction of a side chain double bond (Figure 6) and this type of reaction is encountered metabolically within the ring system of a number of other unsaturated monocyclic terpenes.[4]

23.5.4.2 Nitrogen Centres

A variety of aromatic nitro compounds (*e.g.* nitrobenzene, chloramphenicol) can undergo reductive metabolism *via* nitroso and hydroxyamino intermediates to give the corresponding amines (Figure 6). It has been suggested that the formation of a nitro anion radical precedes the nitroso intermediate, and is the first step in nitro reduction.[31] When the nitro group is attached to a ring structure which is destabilized by the presence of a heteroatom, nitro reduction may result in ring opening as is seen with metronidazole, which yields an open chain derivative, *N*-(2-hydroxyethyl)oxamic acid, and acetamide.[32]

Compounds containing azo linkages are susceptible to reduction, first to hydrazo intermediates and then by reductive cleavage to yield two molecules of the appropriate amines. Again, it has been proposed that the first intermediate formed in some azo reductions is the azo anion radical, which is then reduced to the hydrazo compound. The inactive drug, prontosil, undergoes such reductive azo cleavage to form the active antibacterial compound sulfanilamide.[33]

A wide variety of aliphatic and aromatic tertiary amine *N*-oxides can be reduced to the corresponding tertiary amines. The nitrogen atom being reduced may be attached to or part of a ring system. Examples of such reductions are those of trimethylamine *N*-oxide and nicotine 1'-*N*-oxide.[34]

23.5.4.3 Sulfur Centres

Compounds containing a disulfide linkage may be reduced to their corresponding thiols in the animal body. An example is that of tetraethylthiuram disulfide (disulfiram), a drug used in the treatment of alcoholism, which is reduced to diethyldithiocarbamic acid, its biologically active metabolite.[4,35]

Sulfoxides, like *N*-oxides, may in part be reduced to sulfides, and dimethyl sulfoxide is known to undergo such a reaction.[36] In many cases, a redox equilibrium is set up in the body between sulfoxides and sulfides, *e.g.* sulindac[37] and ethionamide.[38] Sulfones may also be reduced to a marginal extent, but such a process remains to be evaluated and proved to be an important pathway.

23.5.5 ENZYMOLOGY OF REDUCTION

The enzymology of reduction is far less clear than that for oxidative processes.

Within mammalian tissues *N*-oxide reduction, nitro reduction and azo bond reduction activities have been shown to occur within the endoplasmic reticulum (microsomes) and cytosol of the cell, with some activity towards the last two reactions being observed within the mitochondria.[39-41] The enzyme carrying out these reactions in microsomes is thought to be the cytochrome *P*-450 system, and the associated enzyme NADPH–cytochrome *P*-450 reductase may also contribute. The cytosolic enzymes undertaking these reactions differ in not being inhibited by oxygen, and having a less strict requirement for NADPH as an electron donor. Several flavoproteins have been assigned these functions and aldehyde oxidase, xanthine oxidase and DT-diaphorase have all been implicated to varying degrees in these reactions.[42]

These metabolic conversions may also be carried out by anaerobic bacteria which reside within the gastrointestinal tract, although little is known about the enzymology. A wide variety of drugs containing oxidized centres are substrates for these microbial reductases, which can be responsible for their metabolic activation to toxic products, *e.g.* nitrobenzene[43] and metronidazole.[32]

Nonenzymatic reactions may also be involved. Bacterial NADH or NADH dependent flavoproteins can reduce flavins, which may then reduce azo compounds nonenzymatically. Other redox reagents, such as methyl viologen, crystal violet and menandione, can also act as electron shuttles. The bacterial ferredoxins, about 55 amino acids in length, and containing one to eight atoms of iron, are known to be widely distributed and to participate in several oxidoreduction reactions, including

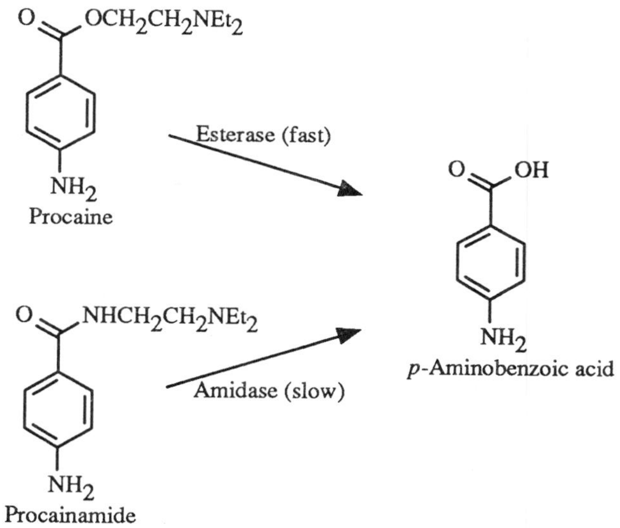

Figure 7 Hydrolysis of ester and amide structures

the hydrogenase system of many anaerobic microbes, in mitochondrial electron transport and in the reduction of nitrogen in nitrogen-fixing bacteria. The role of such nonspecific redox systems in the reduction of drug molecules has yet to be fully elucidated.[44]

The molecular mechanisms of the enzyme systems involved in the reduction of sulfoxides are not known. The existence of a thioredoxin dependent sulfoxide reductase system has been demonstrated both in mammalian tissues and microorganisms, and may have implications both in terms of initial metabolism and in enterohepatic cycling.[45,46]

23.5.6 HYDROLYSIS

Esters and amides are very similar in their chemical construction. The reaction of a carboxyl group with an alcohol will produce an ester, and with an amine it will give an amide. In both cases, a molecule of water is also formed. The mechanisms for hydrolysis are similar for both esters and amides. The C—O or C—N bonds adjacent to the carboxyl carbon atom are cleaved in a bimolecular reaction, catalyzed by acid or base.

Because of delocalization effects, esters and amides of aromatic carboxylic acids are generally more stable than the corresponding derivatives of aliphatic acids.[47] The great majority of simple esters undergo rapid and quantitative hydrolysis in the animal body, so that they serve as 'delivery systems' for the constituent acids and alcohols. In general the rate of hydrolysis of amides in the mammalian body is much slower than that for the corresponding esters. The ester local anaesthetic procaine undergoes rapid hydrolysis in the body, whereas procainamide is relatively stable (Figure 7). In many cases the hydrolysis of amide bonds is only a minor metabolic pathway with other functional groups around the molecule receiving more metabolic attention, *e.g.* local anaesthetics of the lignocaine class.[4,48]

23.5.7 ENZYMOLOGY OF HYDROLYSIS

Our knowledge of the biochemistry of enzymes carrying out these reactions is limited. In addition to specific enzymes, other biological catalysts, such as the basic side chains of plasma proteins, may be responsible for the hydrolysis of carboxylic esters and amides. Serum albumin is known to play a role in the cleavage of a wide variety of esters, although the rates are much lower than with reactions mediated by 'true' enzymes.[49] The stomach is the only area where acid-catalyzed hydrolysis may occur; the alkaline pH of the intestine presumably assists in base-catalyzed reactions.

Hydrolysis occurs in all mammalian tissues, with the highest capacities seen in the liver, blood and gastrointestinal tract. Large substrate dependent species differences are observed in hydrolytic capacity, human plasma being unable to hydrolyze atropine, whereas the plasma of some rabbit strains is highly active.[50] Differences between individuals may also occur, such as the rare patients with atypically low butyrylcholinesterase activity who are unable to hydrolyze succinylcholine rapidly, and therefore respond abnormally to this muscle relaxant with prolonged paralysis.

23.5.8 CONJUGATION REACTIONS

There occur in mammals six major conjugation reactions of foreign compounds, whose biochemistry and functional significance are (reasonably well) understood.[3] Each of these has in addition well-defined roles in the metabolism of endogenous substrates and/or biosynthesis.[3] In each case, the reaction is mediated by a transferase enzyme, which exhibits a very high specificity for the conjugating agent in question, but has a much broader specificity for the xenobiotic substrate.[51,52] Since each reaction involves the synthesis of a new bond, energy is required, and this may be obtained in one of two ways. In the majority of cases, the endogenous conjugating agent is used in the form of an activated intermediate (generally a nucleotide), but examples do occur where it is the xenobiotic which must first be activated before conjugation with the endogenous moiety can occur. Table 1 presents a listing of these major conjugation reactions, together with information about the activated intermediates involved and the endogenous role(s).

As well as these six major conjugation reactions, there are an enormous number of so-called 'novel' conjugations.[53,54] These all involve the linkage of a xenobiotic to an endogenous grouping, but differ from the major reactions listed in Table 1 in terms of our (relative) lack of knowledge. Thus, the novel conjugations are those which are, on the basis of present knowledge, restricted to particular combinations of compound and animal species. A listing of a number of novel conjugations is presented in Table 2. The distinction between these and the major reactions is an arbitrary one, and it is likely that, as more information is acquired about their substrate versatility, phylogenetic distribution and enzymic mechanisms, at least some will attain the status of major reactions. This is illustrated by the methylthio conjugation, first discovered in the mid 1970s, and noted then as a most unusual metabolic option,[55] which already seems to be established as a general route of metabolism.[56,57] The current literature is adding to the list of novel reactions at a considerable rate, which is a consequence both of the ever more rigorous examination of excreta and tissue residues, and the increasing sophistication of the analytical methodology applied.

23.5.9 GLUCURONIC ACID CONJUGATION

The glucuronic acid conjugation, although not the first conjugation reaction to be discovered, has been established for many years as the most versatile of the major reactions.[58] It involves the transfer

Table 1 Classification of the Major Conjugation Reactions[3,51]

Reaction	Conjugating agent	Functional groups involved	Endogenous roles
Reactions involving activated conjugating agents			
Glucuronidation	UDP glucuronic acid	OH, CO_2H, NH_2, NR_2, SH, CH	Biosynthesis, detoxication, *e.g.* bilirubin
Sulfation	PAPS	OH, NH_2, SH	Biosynthesis, *e.g.* chondroitin, steroid metabolism, detoxication, *e.g.* indoxyl
Methylation	*S*-Adenosyl-L-methionine	OH, NH_2	Biosynthesis, detoxication, *e.g.* catecholamines
Acetylation	Acetyl CoA	OH, NH_2	Biosynthesis, intermediary metabolism
Reactions involving activated foreign compounds			
Glutathione	Glutathione	Arene oxide, epoxide, alkyl and aryl halide	Maintenance of intracellular redox potential, leukotriene synthesis
Amino acid conjugation	Glycine, glutamine, ornithine, taurine	CO_2H	Detoxication of endogenous acids, especially in amino acidurias

Table 2 Some Novel Metabolic Conjugation Reactions[51,54,55]

Conjugation with:	Carbohydrates, *e.g.* xylose, ribose, galactose Acyl groups, *e.g.* formyl, succinyl, stearoyl Amino acids, *e.g.* histidine, alanine, glutamic acid, valine, serine
Formation of:	Methylthio conjugates Hybrid fatty acids Hybrid triglycerides

Table 3 Types of Compounds Giving Rise to Glucuronic Acid Conjugates[58,59]

Functional group	Example
Hydroxy	
Primary alcohol	Trichloroethanol
Secondary alcohol	Propranolol
Tertiary alcohol	*t*-Butanol
Alicyclic alcohol	Cyclohexanol
Terpenoid alcohol	Menthol
Phenol	Phenol
Terpenoid Phenol	Eugenol
Enol	4-Hydroxycoumarin
Aliphatic hydroxylamine	*N*-Hydroxychlorphentermine
Aromatic hydroxylamine	2-Naphthylhydroxylamine
Hydroxamic acid	*N*-Hydroxy-2-acetamidofluorene
Carboxylic acid	
Alkyl	2-Ethylhexanoic acid
Aromatic	Benzoic acid
Heterocyclic	Nicotinic acid
Arylacetic	Indole-3-acetic acid
Arylpropionic	Hydratropic acid
Aryloxybutyric	Clofibric acid
Carbamic acid	Tocainide carbamate
Amino functions	
Aromatic	Aniline
Azaheterocyclic	Sulfisoxazole
Carbamate	Meprobamate
Sulfonamide	Sulfadimethoxine
Hydroxylamine *N*-	*N*-Hydroxy-2-acetamidofluorene
Tertiary aliphatic	Cyproheptadine
Urea	Dulcin
Sulfur functions	
Thiol	2-Mercaptobenzothiazole
Dithiotic acid	*N,N*-Diethyldithiocarbamic acid
Carbon centres	
Pyrazolone ring	Phenylbutazone
Selenium centres	
Selenium	2-Selenobenzanilide (metabolite of ebselen)

of glucuronic acid from the nucleotide uridine diphosphate glucuronic acid (UDPGA) to a suitable functional group in the xenobiotic substrate, under the influence of the enzyme UDP glucuronyl transferase (UDPGT).[59] Table 3 gives a list of the functional groups to which glucuronic acid may be attached, and it is of interest that, although the occurrence of this mechanism has been known for more than a century,[59] new types of glucuronides are still being discovered.[58] Notable recent instances of this are the quaternary amine glucuronides, produced by the linkage of glucuronic acid to a tertiary amine through the nitrogen atom's lone pair of electrons, the C—C-linked glucuronides of pyrazolone drugs such as phenylbutazone, and the Se glucuronide formed in the metabolism of ebselen (Figure 8).[60]

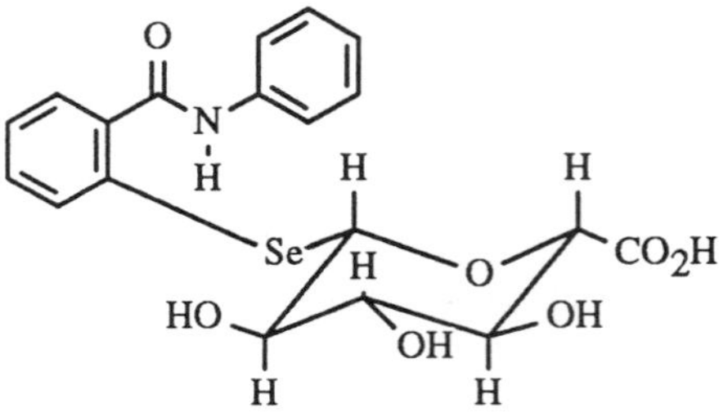

2-Glucuronylselenobenzanilide

Figure 8 Selenium glucuronide of ebselen

The formation of glucuronides involves the inversion of the anomeric centre of glucuronic acid, which is in the α form (protons on C-1 and C-2 in the *cis* configuration) in UDPGA, to the β configuration (these protons *trans* to each other). Biosynthetic glucuronides are thus always 1-*O*-substituted β-D-glucopyranosiduronates, although in the case of acyl (ester type) glucuronides there can occur a progressive base-catalyzed migration of the acyl function from C-1 to the hydroxyl groups of C-2, C-3 and C-4.[61,62]

UDPGT occurs in the endoplasmic reticulum of the liver and many other tissues. Its activity depends upon its lipid environment within the membrane, and is partly latent.[63] *In vivo*, UDPG-*N*-acetylglucosamine is a controlling influence, and a variety of chaotropic agents (natural and synthetic detergents, lipases, organic solvents, *etc.*) can influence its activity. It has long been recognized that UDPGT is functionally heterogeneous,[64] based on knowledge of species, strain and tissue differences, differential induction and inhibition and substrate selectivities. However, the molecular basis of this heterogeneity has proved harder to establish, but it is now clear that there are indeed a number of distinct UDPGTs differing in terms of substrate specificities, inducibility by xenobiotics and physical properties. At least eight separate isozymes have been identified in rat liver, five in human liver and smaller numbers thus far in other species.[65] There are considerable practical difficulties in purifying these membrane-bound enzymes in a catalytically active form[66] which are still hindering progress in this area, but more success has been obtained by the cloning of cDNAs for individual isozymes and transfecting these into mammalian cells, which provide the appropriate membrane environment for the activity of the encoded enzyme.[67] Mackenzie and Haque[67] have achieved this for four UDPGTs from rat liver, all active in the glucuronidation of steroids.

In general, conjugation with glucuronic acid results in the detoxication and facile elimination of the compound in question, since glucuronides are far more polar and water soluble than their aglycones,[59] with the consequence that glucuronidation capacity, which is determined by factors such as the size of the dose and drug–drug interactions and species, strain and genetic differences in enzyme activity, is an important determinant of toxicity.[3] This may be seen clearly when the acute toxicities of various glucuronidogenic compounds are compared in the rabbit, which is adept at this conjugation, and the cat, where this reaction is well known to be defective. Invariably, such compounds are markedly more toxic to the cat. In contrast to this detoxication role, in one case glucuronic acid conjugation has been shown to result in metabolic activation. Thus, morphine 6-glucuronide (but not the isomeric 3-glucuronide), a minor metabolite of morphine, contributes to the analgesic and other effects of morphine.[68] It is important to appreciate that this action is *not* a consequence of the hydrolysis of the conjugate.

In other cases, glucuronides of xenobiotics are responsible for target organ toxicity, by acting as delivery forms to particular sites in the body, where they are hydrolyzed and exert a localized effect, *e.g.* intestinal ulceration from indomethacin, which is excreted in the bile as its ester glucuronide.[69] The occurrence of bladder tumours following the administration of 2-naphthylamine is also noteworthy in this regard. This compound is *N* hydroxylated, and the hydroxylamine is carcinogenic to the liver subsequent to its *N,O*-sulfation (see below). However, it is also converted to an *N*-glucuronide, with a free OH group, and this is excreted in the urine, breaking down in the bladder to liberate the proximate carcinogen, the hydroxylamine.[70]

23.5.10 SULFATION

Sulfate conjugation involves the esterification of a hydroxyl group (or, rarely, an amino group), in a xenobiotic with the sulfate ion, giving a highly polar, highly ionized conjugate.[71] The source of the

Table 4 Xenobiotic Substrates for Sulfate Conjugation[3,71]

Functional group (hydroxy)	Example
Primary alcohol	Ethanol
Secondary alcohol	Butan-2-ol
Phenol	Phenol
Catechol	α-Methyl-DOPA
Alicyclic alcohol	Dehydroepiandrosterone
Heterocyclic alcohol	3-Hydroxycoumarin
Hydroxyamide	N-Hydroxy-2-acetamidofluorene
Aromatic hydroxylamine	2-Naphthylhydroxylamine
N-oxide	Minoxidil

inorganic sulfate incorporated in the conjugate is the high energy sulfate donor, 3'-phospho-adenosine-5'-phosphosulfate (PAPS),[72] and the transfer of sulfate to the xenobiotic involves a sulfotransferase.[73] The enzymes of sulfate activation and transfer are closely associated, and are found in the cytosol of liver and many other tissues.

The sulfotransferases are a family of isozymes, whose multiplicity has been known for many years. Early workers[74] were able to separate enzyme activities sulfating steroids and phenols by ammonium sulfate fractionation, and the application of modern techniques of protein chemistry has seen the resolution of each of these major groups of sulfotransferases into a number of distinct enzyme forms.[73] The phenol sulfotransferases comprise at least four distinct forms, conjugating phenols, N-hydroxyacetamides and estrone.

The xenobiotic substrates giving rise to sulfate conjugates generally contain hydroxyl groups, including alcohols, phenols, catechols and hydroxylamines, but certain aromatic amines also undergo sulfation, giving rise to sulfamic acids. Although hinted at in the early literature, there is no evidence that thiols can be sulfated. Table 4 classifies xenobiotics undergoing sulfate conjugation.

It will be seen from Table 4 that many substrates for sulfation may also undergo glucuronidation. Sulfation may be regarded as an alternative to glucuronidation,[3] and the relative extents to which a given compound will undergo these two options will be in part a consequence of its chemical structure. Conjugation with sulfate is particularly evident with hydrophilic, relatively small molecules, which are preferentially distributed to the cytosol, and in which the functional group involved is sterically unhindered.[75]

The second major determinant of the extent of sulfation is the size of dose administered. The recognition of the limited capacity of the sulfation mechanism was first made by Williams,[76] since when studies with many substrates have shown that the proportion of administered material undergoing sulfation falls with increasing dose. The sulfate conjugation is therefore particularly evident at small doses, but may become a minor reaction at high doses. The limited capacity of this conjugation is due to limitations on the supply of PAPS for conjugation.[77] In some cases, this may arise from a restricted availability of inorganic sulfate itself, in which case the capacity of the reaction may be increased by the administration of either inorganic sulfate or a sulfate precursor such as cystine or cysteine.[78] This is certainly effective in reversing the effects of competitive inhibitors (alternative substrates) but seems much less effective in increasing the sulfation capacity under normal circumstances. The tissues contain a variety of readily mobilized sulfate precursors, and it is likely that the saturability of sulfation arises, at least in part, from the kinetic features of the sulfotransferases.

The phylogenetic distribution of the sulfation mechanism is widespread, at least amongst mammals,[79] probably as a result of its great importance in the detoxication of the end-products of the metabolism of endogenous compounds. The pig is relatively deficient in the sulfation of certain phenols, but this is highly dependent upon the structure of the phenol in question.[79]

The formation of sulfate conjugates of xenobiotics gives products with high polarity and water solubility, which in the great majority of cases are inactive and are rapidly excreted. However, the sulfation of certain metabolically introduced hydroxyl groups can give rise to reactive intermediates which are responsible for the toxicity of the parent compound. This phenomenon was first recognized in the metabolism of 2-acetylaminofluorene and related carcinogenic aromatic amines and amides,[80] and has more recently been shown with safrole and related allylbenzenes.[81] These compounds undergo an initial hydroxylation (at the nitrogen atom in the cases of the amines and amides, and at C-1 of the allyl side chain in the cases of the safrole analogues) and are then sulfated at the newly introduced hydroxyl group. The *O*-sulfate moiety is a facile leaving group, readily giving rise to electrophilic intermediates, arylnitrenium ions from the *N,O*-sulfates and carbonium ions from the *C,O*-sulfates. The covalent interaction of these electrophilic reactive intermediates with cellular macromolecules, notably DNA, is thought to be critical in the aetiology of a variety of toxic manifestations.

23.5.11 METHYLATION

Methylation is a reaction of hydroxyl, catechol, thiol and various nitrogen centres in xenobiotics.[82] However, although methylation is common in the metabolism of endogenous compounds, this reaction is but rarely of quantitative importance in the disposition of xenobiotics. Functional groups in xenobiotics undergoing methylation include[3] primary amines (*e.g.* amphetamine), secondary amines (*e.g.* demethylimipramine), tertiary amines (*e.g.* dimethylaminoethanol), azaheterocycles (*e.g.* pyridine), phenols (*e.g.* 4-hydroxy-3,5-diiodobenzoic acid), catechols (*e.g.* isoprenaline), thiols (*e.g.* thiouracil), thiophenols (*e.g.* thiophenol) and dithioic acids (*e.g.* disulfiram). The significance of methylation in the fate of glutathione conjugates (the 'thiomethyl shunt') will be discussed below. Unlike the preceding reactions, methylation often serves to increase the lipid solubility and reduce the polarity of the substrates: an exception to this is offered by the metabolic *N* methylation (quaternization) of azaheterocycles.[83]

The methyl group transferred in this conjugation derives from the nucleotide *S*-adenosyl-L-methionine, and the reaction is effected by a methyltransferase: phenols undergoing methylation are generally either catechols or phenols with bulky *ortho* substituents, suggesting the involvement of catechol *O*-methyltransferase. A variety of *N*-methyltransferases are known,[84] while thiol methyltransferase has been intensively investigated in recent years.[85]

In comparison with the other principal conjugation mechanisms, there is little information regarding the zoological distribution of methylation. Catechol and azaheterocycle methylations, at least with respect to endogenous substrates, occur throughout the Mammalia.[3] A comparative study of the *N* methylation of pyridine in a range of mammalian species[83] showed a greater than 10-fold variation in the extent of this reaction; from rat, mouse and human, where it accounted for some 2–10% of the dose, up to the rabbit and cat, in which 50% of a dose was methylated. The *O* methylation of 4-hydroxy-3,5-diiodobenzoic acid is apparently restricted in its occurrence to primate species, being found at only trace levels in nonprimate laboratory species.[86]

Two instances of metabolic activation by methylation may be quoted: 3-*O*-methylisoprenaline, a metabolite of the β-adrenergic agonist isoprenaline, has β-adrenergic blocking activity and thus acts as an antagonist to the parent drug.[51] The simple azaheterocycle pyridine is methylated to the *N*-methylpyridinium ion, which has markedly greater acute toxicity than pyridine itself.[83]

23.5.12 ACETYLATION

Acetylation is a commonly occurring route of metabolism of the primary amino group (NH_2) in a wide variety of xenobiotics, including aromatic, alicyclic and aliphatic amines, amino acids, sulfonamides, hydrazines and hydrazides.[87] Acetylation is important in intermediary metabolism, in which OH groups (*e.g.* choline) and SH groups (*e.g.* coenzyme A) may be involved. There are apparently no reports of the acetylation of such centres in foreign compounds.[3]

The acetyl group transferred to a xenobiotic amine derives from the high energy compound acetyl CoA, which is ubiquitous in living cells. Indeed, the first recognition of the importance of this coenzyme arose from studies on the acetylation of foreign compounds.[88] The reaction is catalyzed by the enzyme *N*-acetyltransferase, and follows a ping-pong bi-bi mechanism, in which the enzyme is first acylated, with subsequent transfer to the acetyl group to the amine function and regeneration of the enzyme.[87]

Probably the most remarkable aspect of the acetylation of xenobiotics is the division of substrates into two classes, depending upon whether or not their conjugation exhibits a genetic polymorphism in humans.[89] Compounds undergoing acetylation are thus either monomorphic, typified by *p*-aminobenzoic acid, or polymorphic, such as isoniazid, sulfamethazine or dapsone. The distribution of acetylation capacity amongst the population for the first group is unimodal, but is bimodal with the second type. The polymorphic substrates are able to show the occurrence of two separate genetically determined forms of N-acetyltransferase, which catalyze the reaction at different rates. The inheritance of acetylation capacity follows simple Mendelian genetics, and is controlled by a pair of alleles, Hf for fast acetylation and Hs for slow, acting at a single locus. The population thus consists of three genotypes, homozygous fast acetylators (HfHf), homozygous slow acetylators (HsHs) and heterozygotes (HfHs). However, it is generally the case that only two phenotypes are seen, as the heterozygotes merge with the homozygous fast acetylators. Only by careful pharmacokinetic analysis is it possible to distinguish the heterozygotes.[90]

The enzymic basis of the genetic polymorphism of N acetylation has proved elusive. Various attempts have been made to show the existence of multiple forms of the transferase, but these have been fruitless.[87] Indeed, the best evidence at present indicates there to be only one N-acetyltransferase, which varies in terms of the induced fit of enzyme to substrate.[87]

The genetic polymorphism of N acetylation of various substrates is readily demonstrable in human and animal populations.[89] The distribution of the phenotypes exhibits a substantial ethnic variability, being 50% fast/50% slow in the USA and Europe, but in other regions of the world the fast acetylator phenotype is present in excess, up to 90% fast/10% slow in Eskimos. The occurrence of the polymorphism has also been established in outbred populations of rabbits and squirrel monkeys,[87] and animal models of the phenotypes are to be found in various inbred mouse and hamster strains.[87]

In addition to the genetic aspects of the control of N acetylation, there also exist two remarkable instances of species defective in particular N acetylations.[79] The dog and related canine species are unable to acetylate aromatic and other amines, hydrazines and hydrazides, but retain the ability to acetylate sulfonamides (at the sulfonamido nitrogen) and amino acids, notably the S-substituted cysteines, which are intermediates in the production of mercapturic acids from glutathione conjugates (see below). Guinea pigs are unable to N-acetylate such S-substituted cysteines.

Like methylation, acetylation reduces polarity and may reduce water solubility of the parent compound, and its function must thus be sought in terms other than facilitation of elimination. Indeed, the crystalluria and consequent nephrotoxicity of various sulfonamides is due to the precipitation of their N-acetylated metabolites in the lumen of the kidney tubule.[91]

The N-acetyl metabolites of a number of drugs have been shown to retain much of the activity of the parent compound, and in the cases of procainamide and sulfanilamide, the N-acetyl compounds are used as drugs in their own right.[51]

The metabolic activation of aromatic amines and amides by N hydroxylation and sulfation has already been mentioned. Many aromatic amines are N-acetylated prior to such activation, and in some cases, there can occur transfer of the acetyl group from N to O (the so-called N,O-acetyl group transfer), with the result that the O-acetyl moiety can act as a leaving group to yield the arylnitrenium ion, analogous with the process described for sulfation.[92]

In only a very few cases are conjugates further metabolized, but a good example with toxic consequences is seen with the antitubercular drug isoniazid (isonicotinic acid hydrazide). The free NH_2 group of the hydrazide moiety is acetylated, and the product, N-acetylisoniazid, is in part hydrolyzed, yielding isonicotinic acid and N-acetylhydrazine. The cytochrome P-450 dependent oxidation of N-acetylhydrazine yields an intermediate responsible for the hepatitis like toxicity of isoniazid.[93]

23.5.13 THE AMINO ACID CONJUGATIONS

The amino acid conjugations are a group of reactions in which the carboxyl group of a xenobiotic acid is linked with the amino group of one of a range of amino acids.[94] Only a small range of structures undergo amino acid conjugation, since only aromatic, heteroaromatic, cinnamic and arylacetic acids are substrates.[95] The extent of amino acid conjugation is heavily dependent upon the steric environment of the carboxyl group of the xenobiotic acid: the more hindered this grouping, the less extensive the amino acid conjugation.[95]

Major amino acid conjugations occur with glycine, glutamine, taurine and ornithine, while isolated cases of conjugation with alanine, serine, histidine, aspartic acid and glutamic acid and with various dipeptides (all glycyl–X) have been reported.[96] While the overall extent of amino acid conjugation is most significantly a function of the structure of the xenobiotic acid, the nature of the amino acid used depends upon the animal species in question.[95] Most versatile of all is the glycine mechanism, used by most species for aromatic, heterocyclic, cinnamic and arylacetic acids. Ornithine conjugation replaces this reaction in anseriform and galliform birds (ducks, chickens), while, for arylacetic and aryloxyacetic acids, glycine conjugation is replaced in primate species by glutamine. These latter acids are also substrates for taurine conjugation, which is particularly well developed in carnivorous species.

Although the formation of amino acid conjugates serves to decrease the toxicity and enhance the elimination of the parent acid, the changes to its properties are not dramatic.[98] It is likely that the key to the function of the amino acid conjugations lies in an appreciation of their enzymic mechanisms.[96] These reactions have three distinct steps, in which the acid is first activated by the formation of an acyl adenylate which is rapidly converted to an acyl CoA thioester. It is this latter high energy intermediate to which the amino acid is transferred, with the liberation of free CoA. Although the final product of this sequence (the amino acid conjugate) is a detoxication product, the intermediate acyl CoA is of considerable biological significance. Many xenobiotic acyl CoA species have inhibitory actions against a number of enzymes of lipid and carbohydrate metabolism, and the formation of these intermediates during the process of amino acid conjugation may assume considerable toxicological significance.[97] An excellent example of this is seen in the metabolism of hypoglycin A, the toxic principle of the unripe Ackee fruit, which is the agent responsible for Jamaican vomiting sickness, a frequently fatal syndrome involving hypothermia, hypoglycaemia and abnormal metabolism of amino acids.[98] The initial step in the fatty acid oxidation spiral is the formation of a thioester linkage between the fatty acid carboxyl group and the thiol group of CoA. The fatty acyl CoA is then dehydrogenated, hydrated and dehydrogenated again before being attacked by a second molecule of CoA to yield acetyl CoA. This sequence is repeated several times, depending upon the carbon chain length of the fatty acid. Hypoglycin A (methylenecyclopropylalanine) is converted to methylenecyclopropylacetyl CoA (MCPA CoA), by transamination and β oxidation, the latter involving the formation of the CoA ester. MCPA CoA is an irreversible inhibitor of a number of fatty acyl-CoA dehydrogenases, and this is directly or indirectly responsible for the various toxic effects of hypoglycin A.[98] This suggests that the toxicity of this agent is a consequence of the failure of MCPA CoA to undergo amino acid conjugation, perhaps due to limitations upon the availability of the endogenous cosubstrate glycine.

In other cases, the xenobiotic moiety of a xenobiotic acyl CoA can be transferred, in a fashion analogous to endogenous fatty acids, to a hydroxyl group of glycerol, leading to the formation of hybrid triglycerides and phospholipids.[97] The toxicological significance of these xenobiotic lipids is unknown, although it is clear that their behaviour can be markedly different from the normal triglycerides of the animal body.[97]

23.5.14 GLUTATHIONE CONJUGATION

The nucleophilic tripeptide glutathione is able to undergo conjugation with electrophilic centres in a wide variety of xenobiotics, which may be present *per se, e.g.* nitro- and halo-alkanes, or

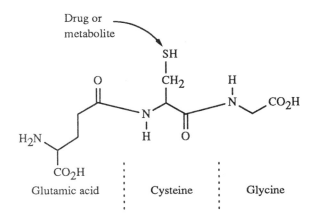

Figure 9 Structure of the tripeptide glutathione

introduced by metabolism, *e.g.* oxiranes (Figure 9).[99,100] In the great majority of cases, conjugation with glutathione serves to detoxify these electrophiles, preventing their interaction with nucleophilic centres in macromolecules such as proteins and nucleic acids, and in general leads to the elimination of a mercapturic acid (an *S*-substituted *N*-acetylcysteine) following metabolism of the *S*-substituted glutathione.[101] With some compounds, however, the reaction with glutathione does not produce conjugates but gives products resembling those of the phase I reactions, *e.g.* the glutathione dependent denitration of glyceryl trinitrate. Xenobiotics which undergo glutathione conjugation are electrophiles, a property which may be present *per se* or introduced by metabolism. Thus, various halo- and nitro-alkanes, benzenes and sulfonic acid esters may undergo displacement of the electron-withdrawing group by the sulfur atom of glutathione.[101] Glutathione may react in a Michael addition across substituted double bonds.[101]

Typical reactive electrophilic metabolites which undergo glutathione conjugation include oxiranes (arene oxides) of various types and *N* oxidation products.[101] Glutathione attacks electron-deprived centres, notably in the strained oxirane ring.

These various glutathione conjugations are catalyzed by the glutathione *S*-transferases, found in the cytosol of the liver and most other organs. These enzymes have long been known to be heterogeneous and are now known to belong to at least three multigene families.[102,103] They are dimeric proteins of $M \approx 50\,000$, and at least eight isozymes can be discerned in rat liver.[102,103] The distribution of these and other isozymes varies from tissue to tissue, being also influenced by development, exogenous chemicals and sex hormones.[104]

Glutathione conjugates are rarely, if ever, excreted in urine, since their molecular size and physicochemical properties favour extensive biliary excretion,[105] and they undergo very extensive metabolism by enzymes of the gut flora and the tissues.[106] *S*-Substituted glutathiones are hydrolyzed by γ-glutamyltranspeptidase and nonspecific hydrolases to yield the corresponding *S*-substituted cysteines.[107] These intermediates are then transformed by at least four reactions, transamination, *S* oxidation, C—S cleavage and *N* acetylation.[107] This last reaction produces mercapturic acids (*S*-substituted *N*-acetylcysteines) which are common urinary end products of the metabolism of glutathione conjugates,[101] but important interspecies differences do occur in the fate of *S*-substituted cysteines.[108] The C—S cleavage reaction produces free thiols, which are rapidly methylated giving methyl sulfides, which may in turn undergo *S* oxidation yielding sulfoxides and sulfones.[106] Although tissue enzymes with C—S lyase activity have been described,[109] it seems most likely that this reaction is, *in vivo*, a function of the enzymes of the gut flora.[107]

Glutathione conjugation most generally acts as a protective mechanism against the harmful effects of intracellular electrophiles.[101] However, there are now examples of the metabolic activation of xenobiotics occurring subsequent to glutathione conjugation, as is exemplified by 1,2-dibromo-ethane.[102] In this case, nucleophilic attack by glutathione on a C—Br bond results in the expulsion of Br and formation of a halo sulfide, analogous to a sulfur mustard, which dissociates to give a toxic carbonium ion. Other similar examples occur with a variety of dihaloalkanes. Other cases of metabolic activation as a result of glutathione conjugation involve the C—S cleavage pathway of catabolism of *S*-substituted cysteines, either yielding a toxic thiol, as is seen with the nephro-toxin dibromovinylcysteine,[110] or highly retained, harmful methyl sulfones of polyhalogenated aromatics.[107]

23.5.15 CLOSING REMARKS

Metabolic reactions are critical determinants of the activity and disposition of drugs in the animal body. In the majority of cases, the various metabolic pathways serve to reduce biological activity and facilitate elimination. The phase I or functionalization reactions bring about generally small changes to structure, which may result in chemical activation.[111] In contrast, the conjugation (phase II) reactions radically alter both structure and physicochemical properties, which greatly reduce activity (with a small number of exceptions), and facilitate elimination in urine and bile.[112] There exists a small number of examples of drugs retained in the body as a consequence of their metabolism, either by covalent binding of reactive intermediates to tissue macromolecules,[111] incorporation by biosynthetic pathways into macromolecules[54] or by conversion to products so lipophilic as to be retained in the tissues, *e.g.* the methyl sulfones produced from glutathione conjugates by C—S cleavage.[107]

The most important routes of elimination of drugs are the urine and hepatic excretion into the bile, ultimately resulting in faecal elimination. The physicochemical changes ensuing from metabolism,[105,113] increased polarity, water solubility and, in the case of bile, molecular weight, result in greater affinity for these excretory organs. Exhalation can be an important excretory mechanism for volatile compounds, while various other minor pathways may contribute to elimination, *e.g.* sweat, saliva, breast milk, *etc.*[2]

The pathways of metabolism which a drug will undergo determine both the chemical natures of the molecules actually present in the body after its administration and the time courses of their concentrations. The overall profile of the drug's activity is thus highly dependent upon its metabolism, with the consequence that factors influencing metabolism will influence drug action. Metabolic studies thus play a major role in drug development and in the therapeutic application of useful activities, and the delineation of metabolic pathways aids greatly in understanding both desirable and adverse drug action.

23.5.16 REFERENCES

1. R. T. Williams, 'Detoxication Mechanisms', 2nd edn., Chapman and Hall, London, 1959.
2. J. Caldwell, in 'Intermediary Xenobiotic Metabolism', ed. D. H. Hutson, J. Caldwell and G. D. Paulson, Taylor and Francis, London, 1989, p. 3.
3. J. Caldwell, in 'The Liver: Biology and Pathobiology', ed. I. Arias, H. Popper, D. Schachter and D. A. Shafritz, Raven Press, New York, p. 281.
4. D. V. Parke, in 'The Biochemistry of Foreign Compounds', Pergamon Press, Oxford, 1968, p. 34.
5. U. Frommer, V. Ullrich, H. Staudinger and S. Orrenius, *Biochim. Biophys. Acta*, 1972, **280**, 487.
6. J. R. Groves, G. A. McClusky, R. E. White and M. J. Coon, *Biochem. Biophys. Res. Commun.*, 1978, **81**, 154.
7. J. T. Groves and M. Van Der Puy, *J. Am. Chem. Soc.*, 1974, **96**, 5274.
8. P. Sims and P. L. Grover, *Adv. Cancer Res.*, 1974, **20**, 165.
9. F. Korte and H. Arent, *Life Sci.*, 1965, **4**, 2017.
10. J. R. Idle, A. Mahgoub, M. M. Angelo, L. G. Dring, R. Lancaster and R. L. Smith, *Br. J. Clin. Pharmacol.*, 1979, **7**, 257.
11. S. A. Sangster, J. Caldwell, R. L. Smith and P. B. Farmer, *Food Chem. Toxicol.*, 1984, **22**, 695.
12. J. W. Cramer, J. A. Miller and E. C. Miller, *J. Biol. Chem.*, 1960, **235**, 885.
13. F. H. Pettit and D. M. Ziegler, *Biochem. Biophys. Res. Commun.*, 1963, **13**, 193.
14. G. A. Hamilton, in 'Molecular Mechanisms of Oxygen Activation', ed. O. Hayaishi, Academic Press, New York, 1974, p. 405.
15. M. W. Anders, in 'Metabolic Basis of Detoxication', ed. W. B. Jakoby, J. R. Bend and J. Caldwell, Academic Press, London, 1982, p. 29.
16. M. J. Coon, J. L. Vermilion, K. P. Vatsis, J. S. French, W. L. Dean and D. A. Haugen, *ACS Symp. Ser.*, 1977, **44**, 46.
17. F. P. Guengerich (ed.), 'Mammalian Cytochromes *P*-450', CRC Press, Boca Raton, FL, vols. 1 and 2, 1987.
18. D. M. Ziegler, in 'Enzymatic Basis of Detoxication', ed. W. B. Jakoby, Academic Press, London, 1980, vol. 1, p. 201.
19. R. C. Bray, *Enzymes*, 1975, **12**, 299.
20. V. Massey, in 'Iron–Sulfur Proteins', ed. W. Lovenberg, Academic Press, New York, 1973, vol. 1, p. 301.
21. K. V. Rajagopalan, in 'Enzymatic Basis of Detoxication', ed. W. B. Jakoby, Academic Press, London, 1980, vol. 1, p. 295.
22. K. F. Tipton, in 'Handbook of Physiology', ed. H. Blaschko, A. D. Smith and G. Sayers, American Physiological Society, Washington, 1975, vol. 6, p. 677.
23. R. M. Denney, R. R. Fritz, N. T. Patel and C. W. Abell, *Science (Washington, D.C.)*, 1982, **215**, 1400.
24. C. J. Fowler, B. A. Callingham, T. J. Mantle and K. F. Tipton, *Biochem. Pharmacol.*, 1978, **27**, 97.
25. T. E. Smith, H. Weissbach and S. Udenfriend, *Biochemistry*, 1962, **1**, 137.
26. K. F. Tipton and T. J. Mantle, in 'Structure and Function of Monoamine Enzymes', ed. E. Usdin, N. Weinder and M. B. H. Youdim, Dekker, New York, 1977, p. 559.
27. H. Blaschko, *Pharmacol. Rev.*, 1952, **4**, 415.
28. M. D. Houslay and K. F. Tipton, *Biochem. J.*, 1974, **139**, 645.
29. K. F. Tipton, in 'Enzymatic Basis of Detoxication', ed. W. B. Jakoby, Academic Press, London, 1980, vol. 1, p. 355.
30. B. Samuelsson, M. Goldyne, E. Granstrom, N. Hamberg, S. Hammarstrom and C. Malmsten, *Annu. Rev. Biochem.*, 1978, **47**, 997.
31. R. P. Mason and J. L. Holtzman, *Biochemistry*, 1975, **14**, 1626.
32. R. L. Koch, E. J. T. Chrystal, B. B. Beaulieu and P. Goldman, *Biochem. Pharmacol.*, 1979, **28**, 3611.
33. R. P. Mason, F. J. Peterson and J. L. Holtzman, *Biochem. Biophys. Res. Commun.*, 1977, **75**, 532.
34. M. H. Bickel, *Pharmacol. Rev.*, 1969, **21**, 325.
35. T. M. Kitson, *J. Stud. Alcohol*, 1977, **38**, 96.
36. K. I. H. Williams, S. H. Burstein and D. S. Layne, *Arch. Biochem. Biophys.*, 1966, **117**, 84.
37. D. E. Duggan, K. F. Hooke, R. M. Noll, H. B. Hucker and C. G. Van Arman, *Biochem. Pharmacol.*, 1978, **27**, 2311.
38. J. P. Johnston, P. O. Kane and M. R. Kibby, *J. Pharm. Pharmacol.*, 1967, **19**, 1.
39. M. Sugiura, K. Iwasaki and R. Kato, *Mol. Pharmacol.*, 1976, **12**, 322.
40. P. R. Johnson and D. M. Ziegler, *J. Biochem. Toxicol.*, 1986, **1**, 15.
41. M. Sugiura and R. Kato, *J. Pharmacol. Exp. Ther.*, 1977, **200**, 25.
42. D. S. Hewick, in 'Metabolic Basis of Detoxication', ed. W. B. Jakoby, J. R. Bend and J. Caldwell, Academic Press, London, 1982, p. 151.
43. B. G. Reddy, L. R. Pohl and G. Krishna, *Biochem. Pharmacol.*, 1976, **24**, 1119.
44. K. T. Chung, G. E. Fulk and M. Egan, *Appl. Environ. Microbiol.*, 1978, **35**, 558.
45. M. W. Anders, J. H. Ratnayake, P. E. Hanna and J. A. Fuchs, *Biochem. Biophys. Res. Commun.*, 1980, **97**, 846.
46. M. W. Anders, J. H. Ratnayake, P. E. Hanna and J. A. Fuchs, *Drug Metab. Dispos.*, 1981, **9**, 307.
47. E. S. Gould, 'Mechanismus und Struktur in der Organischen Chemie', 2nd edn., Verlag Chemie, Weinheim, 1969, p. 372.

48. E. Heymann, in 'Enzymatic Basis of Detoxication', ed. W. B. Jakoby, Academic Press, London, 1980, vol. 2, p. 291.
49. Y. Kurono, T. Maki, T. Yotsuyanagi and K. Ikeda, *Chem. Pharm. Bull.*, 1979, **27**, 2781.
50. R. Ammon, *Arzneim.-Forsch.*, 1977, **27**, 944.
51. J. Caldwell, *Drug Metab. Rev.*, 1982, **13**, 745.
52. W. B. Jakoby (ed.), 'Enzymatic Basis of Detoxication', Academic Press, New York, 1980, vol. 2.
53. W. B. Jakoby, *Methods Enzymol.*, 1982, **77**.
54. G. B. Quistad and D. H. Hutson, *ACS Symp. Ser.*, 1986, **299**, 204.
55. G. B. Quistad, *ACS Symp. Ser.*, 1986, **299**, 221.
56. J. E. Bakke, *Chemosphere*, 1983, **12**, 793.
57. J. E. Bakke and J.-A. Gustafsson, *Trends Pharmacol. Sci.*, 1984, **5**, 517.
58. G. J. Dutton (ed.), 'Glucuronic Acid Free and Combined', Academic Press, New York, 1966.
59. G. J. Dutton, 'Glucuronidation of Drugs and Related Compounds', CRC Press, Boca Raton, FL, 1981.
60. H. Fischer, R. Terlinden, J. P. Lohr and A. Romer, *Xenobiotica*, 1988, **18**, 1347.
61. K. A. Sinclair and J. Caldwell, *Biochem. Pharmacol.*, 1982, **31**, 953.
62. E. M. Faed, *Drug Metab. Rev.*, 1984, **15**, 1213.
63. G. J. Dutton and B. Burchell, *Prog. Drug Metab.*, 1977, **2**, 1.
64. G. J. Wishart, *Biochem. J.*, 1978, **174**, 485.
65. T. R. Tephly, B. L. Coffman, C. N. Falany, M. D. Green, Y. Irshaid, J. F. Puig, S. A. Knapp and J. Baron, in 'Microsomes and Drug Oxidations', ed. J. O. Miners, D. J. Birkett, R. Drew and M. McManus, Taylor and Francis, London, 1988, p. 263.
66. B. Burchell, *Rev. Biochem. Toxicol.*, 1981, **3**, 1.
67. P. I. Mackenzie and S. J. Haque, in 'Microsomes and Drug Oxidations', ed. J. O. Miners, D. J. Birkett, R. Drew and M. McManus, Taylor and Francis, London, 1988, p. 271.
68. M. Mori, K. Oguri, H. Yoshimura, K. Shimomura, O. Kamata and S. Ueki, *Life Sci.*, 1972, **11**, 525.
69. H. B. Hucker, A. G. Zacchei, S. V. Cox, D. A. Brodie and N. H. R. Cantwell, *J. Pharmacol. Exp. Ther.*, 1966, **153**, 237.
70. F. F. Kadlubar, J. A. Miller and E. C. Miller, *Cancer Res.*, 1977, **37**, 805.
71. G. J. Mulder (ed.), 'Sulfation of Drugs and Related Compounds', CRC Press, Boca Raton, FL, 1981.
72. R. H. DeMeio, in 'Metabolic Pathways', ed. D. M. Greenberg, 3rd edn., Academic Press, New York, 1975, vol. 7, p. 287.
73. W. B. Jakoby, R. D. Sekura, E. S. Lyon, C. J. Marcus and J.-L. Wang, in 'Enzymatic Basis of Detoxication', ed. W. B. Jakoby, Academic Press, New York, 1980, vol. 2, p. 199.
74. Y. Nose and F. Lipmann, *J. Biol. Chem.*, 1958, **233**, 1348.
75. G. J. Mulder, in 'Metabolic Basis of Detoxication', ed. W. B. Jakoby, J. R. Bend and J. Caldwell, Academic Press, New York, 1981, p. 248.
76. R. T. Williams, *Biochem. J.*, 1938, **32**, 879.
77. G. Levy, in 'Conjugation Reactions in Drug Biotransformation', ed. A. Aitio, Elsevier, Amsterdam, 1978, p. 469.
78. K. R. Krijgsheld and G. J. Mulder, in 'Sulfate Metabolism and Sulfate Conjugation', ed. G. J. Mulder, J. Caldwell, G. M. J. Van Kempen and R. J. Vonk, Taylor and Francis, London, 1982, p. 59.
79. J. Caldwell, in 'Enzymatic Basis of Detoxication', ed. W. B. Jakoby, Academic Press, New York, 1980, vol. 1, p. 85.
80. J. R. DeBaun, E. C. Miller and J. A. Miller, *Cancer Res.*, 1970, **30**, 577.
81. E. W. Boberg, E. C. Miller, J. A. Miller, A. Poland and A. Liem, *Cancer Res.*, 1983, **43**, 5163.
82. R. T. Borchardt, in 'Enzymatic Basis of Detoxication', ed. W. B. Jakoby, Academic Press, New York, 1980, vol. 2, p. 43.
83. J. D'Souza, J. Caldwell and R. L. Smith, *Xenobiotica*, 1980, **10**, 151.
84. S. S. Ansher and W. B. Jakoby, *J. Biol. Chem.*, 1986, **261**, 3996.
85. R. A. Weisiger and W. B. Jakoby, *Arch. Biochem. Biophys.*, 1979, **196**, 631.
86. J. S. Wold, R. L. Smith and R. T. Williams, *Biochem. Pharmacol.*, 1973, **22**, 1865.
87. W. W. Weber and I. B. Glowinski, in 'Enzymatic Basis of Detoxication', ed. W. B. Jakoby, Academic Press, New York, 1980, vol. 2, p. 169.
88. F. Lipmann, *J. Biol. Chem.*, 1945, **160**, 173.
89. G. A. Ellard, *Clin. Pharmacol. Ther.*, 1976, **19**, 610.
90. D. J. Chapron, P. A. Kramer and S. A. Mercik, *Clin. Pharmacol. Ther.*, 1980, **27**, 104.
91. J. Caldwell, in 'Conjugation Reactions in Drug Biotransformation', ed. A. Aitio, Elsevier, Amsterdam, 1978, p. 111.
92. T. J. Flammang and F. F. Kadlubar, in 'Microsomes and Drug Oxidations', ed. A. R. Boobis, J. Caldwell, F. De Matteis and C. R. Elcombe, Taylor and Francis, London, 1985, p. 190.
93. J. R. Mitchell, H. J. Zimmerman, K. G. Ishak, U. P. Thorgeirsson, J. A. Timbrell, W. R. Snodgrass and S. D. Nelson, *Ann. Intern. Med.*, 1976, **84**, 181.
94. J. Caldwell, in 'Metabolic Basis of Detoxication', ed. W. B. Jakoby, J. R. Bend and J. Caldwell, Academic Press, New York, 1982, p. 271.
95. J. Caldwell, J. R. Idle and R. L. Smith, in 'Extrahepatic Metabolism of Drugs and Other Foreign Compounds', ed. T. E. Gram, SP Medical and Scientific Books, Jamaica, NY, 1980, p. 435.
96. P. G. Killenberg and L. T. Webster, Jr., in 'Enzymatic Basis of Detoxication', ed. W. B. Jakoby, Academic Press, New York, 1980, vol. 2, p. 141.
97. J. Caldwell, *Biochem. Soc. Trans.*, 1984, **13**, 852.
98. D. Billington, H. Osmundsen and H. S. D. Sherratt, *N. Engl. J. Med.*, 1976, **295**, 1482.
99. L. F. Chasseaud, in 'Glutathione: Metabolism and Function', ed. I. M. Arias and W. B. Jakoby, Raven Press, New York, 1976, p. 77.
100. W. B. Jakoby, J. Stevens, M. W. Duffel and R. A. Weisiger, *Rev. Biochem. Toxicol.*, 1984, **6**, 97.
101. L. F. Chasseaud, *Adv. Cancer Res.*, 1979, **29**, 175.
102. P. J. Van Bladeren, A. Van der Gen, D. D. Breimer and G. R. Mohn, *Biochem. Pharmacol.*, 1979, **28**, 2521.
103. C. B. Pickett, C. A. Telakowski-Hopkins, V. D.-H. Ding and R. G. King, in 'Microsomes and Drug Oxidations', ed. J. O. Miners, D. J. Birkett, R. Drew and M. McManus, Taylor and Francis, London, 1988, p. 313.
104. B. Ketterer, D. J. Meyer, B. Coles and J. B. Taylor, in 'Microsomes and Drug Oxidations', ed. J. O. Miners, D. J. Birkett, R. Drew and M. McManus, Taylor and Francis, London, 1988, p. 305.
105. R. L. Smith, 'The Excretory Function of Bile', Chapman and Hall, London, 1973.

106. W. B. Jakoby, in 'The Liver: Biology and Pathobiology', ed. I. M. Arias, W. B. Jakoby, H. Popper, D. Schachter and D. A. Shafritz, Raven Press, New York, 1988, p. 375.

107. J. E. Bakke and J.-A. Gustafsson, *Xenobiotica*, 1986, **16**, 1047.

108. J. Caldwell, Y. Tanaka and A. Weil, in 'Xenobiotic Metabolism and Disposition', ed. R. Kato, R. W. Estabrook and M. N. Cayen, Taylor and Francis, London, 1989, p. 217.

109. J. Stevens and W. B. Jakoby, *Mol. Pharmacol.*, 1983, **23**, 761.

110. A. J. Gandolfi, R. B. Nagle, J. J. Soltis and F. H. Plescia, *Res. Commun. Chem. Pathol. Pharmacol.*, 1981, **33**, 249.

111. M. W. Anders (ed.), 'Bioactivation of Foreign Compounds', Academic Press, New York, 1985.

112. J. Caldwell, in 'Concepts in Drug Metabolism, Part A', ed. P. Jenner and B. Testa, Dekker, New York, 1980, p. 211.

113. J. B. Pritchard and M. O. James, in 'Metabolic Basis of Detoxication', ed. W. B. Jakoby, J. R. Bend and J. Caldwell, Academic Press, New York, 1982, p. 339.

23.6

Drug Interactions

JACK THOMAS

University of Sydney, New South Wales, Australia

23.6.1   INTRODUCTION

A drug–drug interaction may be considered to occur when the effects of giving two or more drugs concomitantly are qualitatively or quantitatively different from the simple sum of the observed effects when the same doses of the drugs are given separately. These changes may be brought about either by one drug influencing the pharmacokinetics of the second drug, or by one drug modifying the pharmacodynamic actions of the other drug.

A number of points arise from this. First, the result of a drug–drug interaction may be for one drug to increase or decrease the pharmacological effects of another drug, *i.e.* quantitative changes. Second, the observed effects of giving two drugs concomitantly may bear no obvious relation to the pharmacological properties of the drugs when used individually, *i.e.* qualitative changes. The third point is the importance of both the sequence and the interval of time between taking the two drugs. This issue is a particularly complex one and it will be considered in some detail when individual drug interactions are described.

Apart from drug–drug interactions there are also many examples of drug–food interactions described in the literature. Again drug–food interactions may be associated with either quantitative or qualitative effects.

The term 'drug interaction' will be used to cover both drug–drug and drug–food interactions.

23.6.1.1   Significance of Drug Interactions

The significance of a drug interaction may be considered from two points of view: (i) clinical significance, or (ii) statistical significance. There is much debate in the medical literature as to the clinical significance of many statistically significant drug interactions. It is not the purpose of this chapter to enter into this and examples of interactions will be considered without undue regard to their established clinical significance, although comments will be made on the possible clinical relevance of some of the interactions discussed.

One of the many difficulties encountered when assessing the clinical significance of any particular drug interaction is that it might be clinically significant for an elderly person or for a person suffering from some condition but not be significant in a young healthy person. Also there is no clearly defined definition of what is meant by clinical significance. What might be considered as significant by one investigator, or for that matter one patient, may not be considered significant by another.

23.6.1.2   Mechanisms of Drug Interactions

Drug interactions are brought about by a number of different mechanisms, which may be classified into the following groups: (i) modification of drug absorption from its site of administration; (ii) modification of drug distribution about the body; (iii) modification of drug excretion; (iv) modification of drug metabolism; (v) modification of drug transport within cells; (vi) modification of the interaction between a drug and a receptor; (vii) modification of the effects of a drug by a second drug acting on the same physiological system; (viii) modifications in electrolyte or fluid balance; and (ix) miscellaneous.

Within these major classifications there are subgroups.

The drug–drug interactions discussed in this chapter will be limited to those which involve changes in: (i) absorption following oral administration, (ii) renal elimination, and (iii) protein binding. Drug–food interactions will also be considered.

23.6.2 INTERACTIONS RELATED TO ABSORPTION FOLLOWING ORAL ADMINISTRATION

23.6.2.1 Factors Involved in Absorption of Drugs from the Gastrointestinal Tract

Drugs are able to pass across biological membranes by a number of mechanisms, of which diffusion is the most common. The rate of diffusion is governed by the concentration gradient across the membrane, the fat solubility/water solubility balance of the molecule (usually expressed as the partition coefficient), and the area and the thickness of the membrane involved. Since movement results from the kinetic energy of the molecules, and no work is required by the system, this process is referred to as passive diffusion. It can be seen that characteristics of both the biological system and the drug molecules are involved, and both can therefore influence the rate of diffusion.

Other important mechanisms are active transport and facilitated diffusion. Both of these involve the drug molecule being transported through the membrane by a carrier system. In the case of active transport the drug molecules can be transported against a concentration gradient but with facilitated transport movement is down a concentration gradient. However, in both cases for a drug to be involved it must have structural features which allow it to interact with the carrier system. A result of this is that relatively few drugs are involved with either active transport or facilitated diffusion. On the other hand the vast majority of drugs can diffuse through membranes by passive diffusion.

Absorption of drugs can take place along the whole length of the gastrointestinal tract from the buccal cavity to the rectum but the main site of drug absorption, following oral administration, is the upper part of the small intestine.

When a particle of drug in solid form is placed in the small intestine two processes occur; (i) the drug particle passes into solution, and (ii) the drug, once in solution, diffuses through the epithelial cells of the intestinal mucosa into the circulation. The rate at which any particular drug is absorbed when placed in the small intestine is determined, therefore, by the rate at which it passes into solution (the dissolution rate) and its intrinsic ability to diffuse through the intestinal mucosal cells.

23.6.2.2 Gastric Emptying and Bioavailability

Since the absorption of drugs is primarily from the small intestine then the time spent in the stomach can have an effect on the rate of absorption and in some cases on the extent of absorption. Some drugs are degraded by the acidic conditions of the stomach or by enzymes present in gastric juice or gastric mucosa and consequently the longer the residence time in the stomach the greater will be the degree of decomposition and hence the bioavailability will be reduced. Examples of drugs which are sensitive to acidic conditions are some penicillins, erythromycin and its esters. Levodopa (L-DOPA) is an example of a drug which is degraded by the action of stomach enzymes.

23.6.2.3 Rate of Gastric Emptying

It is clear that the rate of gastric emptying can have a profound effect on the rate of drug absorption. This leads to a consideration of those factors which influence gastric emptying.

23.6.2.3.1 *Presence of food in the stomach*

The rate of gastric emptying is related to the volume and type of food in the stomach. The rate of gastric emptying is decreased if the viscosity of the meal is increased or if solid food is taken. Further proteins, starch and in particular fats and fatty acids reduce the rate of emptying. The temperature of food may have an effect; cold meals increase and hot meals decrease the rate of gastric emptying.

23.6.2.3.2 *State of the subject*

Patients suffering from duodenal ulcers or hyperacidity have a higher gastric emptying rate than normal healthy subjects. On the other hand patients suffering from achlorhydria have lower emptying rates. The body position of a subject also has an effect on gastric emptying. For example a patient lying on the left side empties his or her stomach more slowly than a patient lying on the right side. This is because the curvature of the gastric pouch is such that when lying on the left side there is an uphill path to the duodenum. Gastric motility can also be influenced by the emotional state of the patient either increased or decreased depending on whether the patient is excited or depressed.

23.6.2.3.3 *Effect of drugs*

A number of drugs can influence the rate of gastric emptying, either increasing it or decreasing it. For example, metoclopramide hydrochloride stimulates the motility of the upper part of the gastrointestinal tract. It can therefore increase the rate of absorption of other drugs given concomitantly. On the other hand propantheline bromide inhibits gastrointestinal motility. It can, therefore, reduce the rate of absorption of other drugs given concomitantly by reducing the rate of gastric emptying; it can also cause increased drug absorption in certain cases. This will be discussed later.

23.6.2.4 Dissolution Rate and Drug Absorption

Since a drug has to be in solution before it can be absorbed, it follows that the rate of dissolution may have a profound effect on absorption. Dissolution rate is controlled by: (i) the difference between the concentration of a saturated solution and the concentration of the drug in solution at the time, (ii) the surface area of the drug, and (iii) a constant which embraces many factors. From this it can be seen that dissolution rate is controlled to a large extent by the solubility of the drug in water, and its mean particle size.

23.6.3 DRUG–DRUG INTERACTIONS BASED ON CHANGES IN GUT MOTILITY

As discussed above the residence time in the stomach and transit time through the gastrointestinal tract can influence the rate and extent of drug absorption. Hence if one drug can change gut motility then it potentially can modify the absorption of a second drug taken concomitantly. Table 1 summarizes many drug–drug interactions related to changes in gastrointestinal motility.

23.6.3.1 Metoclopramide Hydrochloride

Metoclopramide hydrochloride has a positive effect on gastrointestinal motility. Gastric peristalsis is increased, leading to a decrease in gastric residence time. Duodenal peristalsis may be increased, which would accelerate the rate of transit through the duodenum. Metoclopramide may be administered orally or parenterally, and it is used clinically as an antiemetic in the treatment of some forms of nausea and vomiting and to increase gastrointestinal motility.

Metoclopramide increases the rate of absorption of aspirin,[1,2] ethanol,[5] diazepam,[6] levodopa,[10] lithium salts (carbonate),[10] mexiletine[11] and paracetamol,[12] while the absorption of chlorothiazide is decreased[4] and the effect on the absorption of digoxin varies with the nature of the digoxin preparation.[7,8]

The effect of metoclopramide on the absorption of aspirin in migraine is an interesting example of the therapeutic use of a drug interaction. The absorption of aspirin during a migraine attack is reduced and the resulting failure of oral therapy during such an attack may be due to this reduced absorption. The reduction in absorption is related to gastric stasis which accompanies migraine. It was found[1] that metoclopramide administered into the gluteus maximus during a migraine attack enhanced the absorption of aspirin significantly. Plasma salicylate levels obtained were about twice the levels attained in the same group of patients when aspirin was given alone. It was shown clinically, in a subsequent study,[2] that patients with migraine recover more quickly when given a combination of metoclopramide and aspirin than when given aspirin alone.

Metoclopramide (*i.v.*) increased the rate of absorption of oral diazepam profoundly in the initial stages of treatment.[6] While this effect is clinically useful when diazepam is used as preoperative medication, it is not necessarily of value when diazepam is used for long term medication. The data gave no information on the effect of metoclopramide on the amount of diazepam absorbed.

Table 1 Drug–Drug Interactions Related to Changes in Gastrointestinal Motility

Drug	Interacting drug	Effects	Ref.
Metoclopramide	Aspirin	Metoclopramide increased absorption of aspirin immediately prior to a migraine attack	1, 2
Metoclopramide	Cimetidine	Metoclopramide reduced area under curve of cimetidine but not peak plasma level. Clinical importance of interaction doubtful	3
Metoclopramide	Chlorothiazide	Absorption of chlorothiazide decreased	4
Metoclopramide	Ethanol	Increased rate of absorption	5
Metoclopramide	Diazepam	Metoclopramide increased plasma levels of diazepam (oral)	6
Metoclopramide	Digoxin	Serum levels of digoxin reduced when solid dose forms used	7
Metoclopramide	Digoxin	The absorption of digoxin was not modified when a product of high bioavailability was used	8
Metoclopramide	Droperidol	Three patients out of 14 developed akathisia within 90 min of receiving droperidol (20 mg) and metoclopramide (10 mg) orally as preoperative medication	9
Metoclopramide	Levodopa	Increased rate of absorption of levodopa	10
Metoclopramide	Lithium	Increased rate of absorption of lithium salts	10
Metoclopramide	Mexiletine	Metoclopramide accelerated absorption of mexiletine but did not affect extent of absorption. Interaction could be important when therapy started but not on long term treatment	11
Metoclopramide	Paracetamol	Absorption rate of paracetamol increased	12
Propantheline bromide	Chlorothiazide	Absorption of chlorothiazide increased	10
Propantheline bromide	Digoxin	The bioavailability of digoxin from tablets was significantly modified by propantheline bromide but not when digoxin was given as a liquid capsule formulation	13
Propantheline bromide	Digoxin	Serum levels of digoxin were increased when solid dose forms of digoxin were used but not when liquid formulations were administered	7
Propantheline bromide	Digoxin	The absorption of digoxin was not modified when a product of high bioavailability was used	8
Propantheline bromide	Ethanol	Rate of absorption of ethanol reduced	5
Propantheline bromide	Hydrochlorothiazide	The rate of absorption of hydrochlorothiazide was reduced but the extent of absorption increased	10
Propantheline bromide	Lithium	Absorption of lithium salts delayed	10
Propantheline bromide	Paracetamol	The rate of absorption of paracetamol was retarded	12
Atropine	Mexiletine	Atropine retarded the absorption of mexiletine but did not affect extent of absorption. Interaction could be important when therapy started but not on long term treatment	11
Benzhexol	Chlorpromazine	Reduced plasma concentrations of chlorpromazine	10
Diamorphine	Paracetamol	Intravenous diamorphine significantly reduced the rate of absorption of paracetamol but not the extent of absorption	14
Diphenhydramine	p-Aminosalicylate	Absorption of p-aminosalicylate delayed	10
Homatropine	Levodopa	Delayed stomach emptying and increased metabolism of levodopa in stomach led to reduced patient response	10
Pethidine	Paracetamol	Intravenous pethidine significantly reduced the rate of absorption of paracetamol but not the extent of absorption	14

The effect of metoclopramide on the absorption of digoxin provides an interesting example of the interplay between the influence of dissolution rate and residence time on absorption. Digoxin is practically insoluble in water and consequently the fraction of the dose absorbed from a particular pharmaceutical preparation is determined to a large degree by its dissolution rate. It has been reported[7] that the fraction of digoxin absorbed from standard commercial tablets was decreased when metoclopramide was administered. The reason for this is that metoclopramide decreases the total residence time in the gut and therefore reduces the time for dissolution, and hence absorption, to take place. In another investigation[8] the effect of metoclopramide on the absorption of digoxin from tablets prepared from standard particle size digoxin, micronized and large particle size digoxin was investigated. It was found, as expected, that micronized digoxin had a higher bioavailability than standard, which in turn was better than large particle digoxin. However, while metoclopramide reduced the absorption of large particle digoxin, it had no effect on the micronized preparation.

Mexiletine is an orally active antiarrhythmic agent. Its rate of absorption is increased by pretreatment with metoclopramide, but the degree of absorption is not affected.[11] The practical consequences of this are that for long term treatment with mexiletine the interaction would be of no concern, as steady state plasma levels would not be affected. However the interaction could be significant at the time of commencing oral mexiletine therapy, since it could contribute significantly to the marked antiarrhythmic effect of the initial dose.

23.6.3.2   Propantheline Bromide

Propantheline bromide diminishes gastric and intestinal motility. Consequently it tends to reduce the rate of absorption of other drugs, *e.g.* ethanol,[5] hydrochlorothiazide,[10] lithium salts (carbonate)[10] and paracetamol.[12]

The effect of propantheline bromide on the absorption of digoxin is an example of an interaction between propantheline and formulations of digoxin. The absorption of digoxin from tablets is increased by 33% to 50% when taken with propantheline bromide,[7] but no effect was seen when tablets made from micronized digoxin were used.[8] This effect is related to residence time in the gastrointestinal tract and dissolution rate.

The bioavailability of digoxin from liquid preparations is greater than from solid preparations. Up to about 70% of a dose of digoxin administered as standard tablets and 80% of a dose given as elixir is absorbed. The bioavailability of digoxin in solution in soft gelatin capsules can be as high as 90–100%. A study on the effect of propantheline bromide on the absorption of digoxin from tablets and soft gelatin capsules showed that it had no effect when capsules were used but that absorption was increased with propantheline bromide when tablets were used.

23.6.3.3   Miscellaneous Drugs

Other examples of drug interactions related to changes in gut motility have been recorded. Atropine reduces the rate but not extent of absorption of mexiletine,[11] and benzhexol reduced plasma concentrations of chlorpromazine.[10]

Both pethidine and diamorphine decrease the rate of stomach emptying.[14] Paracetamol given orally 30 min after placebo, pethidine or diamorphine injections had mean peak plasma levels at 22, 89.5 and 142 min respectively. It is considered that other narcotic analgesics will have similar effects.

Homatropine increased the gastric residence time of levodopa,[10] and this was accompanied by a reduction in the therapeutic effect. Levodopa is decarboxylated in the stomach with a consequence that its bioavailability is reduced the longer it is held in the stomach.

23.6.4   DRUG–DRUG INTERACTIONS BASED ON DRUG-INDUCED MALABSORPTION SYNDROMES

Some drugs have toxic effects on the intestine when they are taken orally over prolonged periods. This can result in a reduction in the efficacy of absorption of other drugs.

Oral neomycin may be associated with a malabsorption syndrome and steatorrhoea. The malabsorption syndrome is reversible and passes when neomycin is withdrawn. Neomycin has been reported to inhibit the absorption of vitamin A, cyanocobalamin, phenoxymethylpenicillin[10] and digoxin.[10,15]

Colchicine causes a malabsorption syndrome.[16] It reduces the absorption of cyanocobalamin;[10,17] reduced urinary excretion of cyanocobalamin was accompanied by increased faecal excretion.

Sodium aminosalicylate causes a malabsorption syndrome[16] and it has been reported to reduce the absorption of cyanocobalamin and folate.

23.6.5   DRUG–DRUG INTERACTIONS RELATED TO CHANGES IN GUT pH

23.6.5.1   Introduction

Changes in the pH of the gastrointestinal tract may change the rate of absorption of drugs by a number of different mechanisms.

Many drugs are either organic acids or organic bases. Since the rate of absorption is controlled, in part at least, by the proportion of the drug present as the neutral molecule and this in turn is influenced by the pH of the gastrointestinal fluids, then changes in the pH of these fluids can result in the rate of absorption of some drugs being altered. Raising the pH of the gastrointestinal tract will result in the proportion of basic drugs in the unionized form being increased, hence facilitating their absorption and the opposite effect will occur with acidic drugs. Table 2 provides a summary of many drug–drug interactions related to change in gastrointestinal pH.

Changes in the acidity of the gastrointestinal fluids may influence the dissolution rate of drugs and thus affect the rate of absorption. The pH of gastric fluids also influences the rate of gastric emptying. High acidity inhibits emptying, while slight alkalinity enhances it,[18] and it has already been discussed that this effect may influence drug absorption.

Antacids are a group of drugs which are given specifically to change the stomach pH and so it might be anticipated that this group would be involved in drug–drug interactions. Many of the antacids are also capable of adsorbing other drugs and thus reducing absorption. This will be considered in Section 23.6.6.2.

Antacids can increase the urinary pH[19,20] and this in turn may influence the rate of renal excretion of acidic and basic drugs. Drug interactions based on this mechanism are discussed in Section 23.6.10.1.

23.6.5.2 Examples

Aluminum hydroxide gel decreased bioavailability of propranolol;[21] animal model experiments on this interaction led to the suggestion that it was due to a decreased gastric emptying rate.[22] Aluminum magnesium hydroxide increased[23] the peak plasma level of metoprolol by 25% and bioavailability by 11%. With atenolol both peak level and bioavailability were reduced (37% and 33% respectively). The mechanism of the interaction with atenolol was considered to be a reduction in the dissolution rate due to elevation of gastric pH.

The bioavailability of tetracycline is reduced by many antacids. The mechanism is related to the formation of metal complexes with tetracycline (see Section 23.6.6.2). However it was reported that the absorption of tetracycline was reduced by approximately 50% when taken with sodium bicarbonate.[24] Since sodium does not form complexes with tetracycline, the reduced bioavailability cannot be related to this. The explanation proposed for the interaction was that the dissolution of

Table 2 Drug–Drug Interactions Related to Change in Gastrointestinal Tract pH

Drug	Interacting drug	Effect	Ref.
Aluminum hydroxide	Propranolol	Decreased bioavailability of propranolol	21
Aluminum hydroxide	Pseudoephedrine	Increased rate of absorption of pseudoephedrine related to raised pH and decreased proportion of ionized compound	33
Aluminum magnesium hydroxide	Atenolol	Reduction in bioavailability of atenolol	23
Aluminum magnesium hydroxide	Metoprolol	Increase in bioavailability of metoprolol	23
Aluminum magnesium hydroxide	Penicillamine	Absorption of penicillamine was significantly reduced by aluminum hydroxide. Recommend take two as far apart as possible	26
Aluminum magnesium hydroxide	Valproic acid	Possible increase in valproic acid bioavailability	22
Antacids	Aspirin	Rate of absorption of aspirin from enteric coated preparations increased	27
Antacids	Aspirin	Rate of aspirin absorption increased from antacid buffered preparations	28, 29
Antacids	Sulfonamides (sulphadiazine, sulphathiazole)	Increased gastric pH enhanced dissolution of acid form of sulfonamide and increased absorption rate	10, 22
Calcium carbonate	Iron(II) sulfate	Calcium carbonate significantly reduced plasma iron by 66%	22
Magnesium aluminum carbonates	Aspirin	Increased dissolution from solid dose forms	31
Magnesium aluminum hydroxide	Levodopa	Enhanced absorption in patients on levodopa	32
Sodium bicarbonate	Iron(II) sulfate	Sodium bicarbonate significantly reduced plasma iron by 50%	22
Sodium bicarbonate	Tetracycline	Absorption of tetracycline reduced	24, 25

tetracycline in the stomach was reduced when gastric pH was raised with sodium bicarbonate. This theory was challenged when it was observed that the release of tetracycline from the capsules was profoundly influenced by pH.[25] It was reported that under acidic conditions the capsules separated into strips and small fragments within 30 to 60 min and under these conditions the solubility of tetracycline was the limiting factor in overall dissolution. However, in neutral or alkaline solutions, the capsules did not dissolve but retained their original form even after 6 h, although a little leakage of tetracycline occurred through small holes in the capsules. These authors[25] suggested that the reason why tetracycline absorption from capsules is reduced by concomitant sodium bicarbonate is related to the inability of the capsules to release the tetracycline in the raised pH values of the stomach.

The above is another example of a drug–drug interaction which is more related to the formulation of a drug rather than the drug itself. This raises two issues: (i) that it is important to understand the nature of a formulation when investigating drug–drug interactions based on absorption, and (ii) that care should be taken in extrapolating a drug–drug interaction from one formulation to another.

The concomitant administration of antacids and penicillamine resulted in 66% reduction in plasma levels of penicillamine.[26] The mechanism of this interaction was considered to be decomposition of penicillamine in the raised pH of the stomach. Penicillamine is known to be less stable at high pH values.

Administration of antacids along with aspirin preparations may influence the absorption of aspirin in a number of ways. The rate of absorption from enteric coated aspirin tablets is increased significantly when taken with antacid preparations.[27] This is yet another example of an interaction related to formulation. Enteric coated products are designed to allow the tablet to pass through the stomach without disintegrating. This is achieved by coating the tablet with a material which is insoluble in gastric acid but which is soluble when the pH is raised. When an enteric coated product is taken along with an antacid, disintegration may commence in the stomach instead of in the small intestine, thus increasing the rate of overall absorption.

Aspirin can be formulated in effervescent antacid preparations. When a solution of such a product is taken, 94% of the acetylsalicylate salt is emptied from the stomach into the duodenum within 20 min of ingestion. A result of this is that the overall rate of absorption of aspirin is far more rapid when it is taken in a highly buffered formulation than when a normal aspirin tablet is taken. Since aspirin is an ester, it is rapidly hydrolyzed to salicylic acid and acetic acid, both enzymically and chemically, following ingestion.[28] The proportion of the dose present in blood as aspirin is significantly greater following use of an effervescent tablet than a normal tablet[29] and maximum plasma salicylate levels are attained much sooner.

While this interaction between aspirin and antacids is interesting, it really is of little consequence when aspirin is used as an analgesic except that response might be quicker because of more rapid absorption. There is also some suggestion that occult bleeding from the stomach is reduced by the buffered preparation. However a far more important effect of this interaction has been recently described. One of the pharmacological properties of aspirin is that it can inhibit the enzyme cyclooxygenase in platelets and this stops the production of thromboxane by platelets. Thromboxane has two main properties: (i) to cause vasoconstriction, and (ii) to increase the stickiness of platelets thus promoting aggregation. Since aspirin can inhibit aggregation, this has prompted the suggestion that it may be useful in preventing the complications of coronary artery disease. A number of studies have shown this to be the case.[30] The important point is that it is acetylsalicylate, not salicylate, which inhibits cyclooxygenase. Hence it is the plasma acetylsalicylate levels and not total salicylate levels which are of importance.

A full discussion of the effects of acetylsalicylate on blood clotting is beyond the scope of this review but the essential point from a formulation and drug interaction viewpoint is that it is the bioavailability of aspirin as such which is the cardinal factor.

The absorption of levodopa is increased by antacids. It is suggested that the mechanism of this interaction is the combined effect of decreased stomach residence time and increased gastric pH. Both these effects may decrease the degradation of levodopa.[32] It is claimed that there is no firm evidence of either clinical advantage or toxicity associated with this interaction.[22]

23.6.6 DRUG–DRUG INTERACTIONS BASED ON COMPLEX FORMATION IN THE GASTROINTESTINAL TRACT

23.6.6.1 Introduction

If two drugs react in the gastrointestinal tract in such a manner that the concentration of one of them in solution is significantly reduced, then there is the possibility that the rate and even the extent of absorption of this drug may be decreased. Drugs in solution may become adsorbed on to insoluble agents such as antacids, nonspecific adsorbents, *e.g.* kaolin and charcoal, or specific adsorbents, *e.g.* cholestyramine. Some drugs may react with metal ions to form insoluble salts or chelate complexes. Reactions of this type are the basis for many drug–drug interactions. Table 3 provides a summary of many drug–drug interactions related to complex formation in the gastrointestinal tract.

However, the effects of adsorbents on the absorption of drugs may be more complex than at first realized. Enterohepatic recycling occurs with some drugs; that is the drug is absorbed from the upper part of the small intestine in the usual manner; the drug, or a metabolite, can then be secreted into bile. When the bile passes into the intestine, it carries with it dissolved drug, which may then be reabsorbed. If a metabolite is present in the bile then this may be hydrolyzed in the gut to regenerate

Table 3 Drug–Drug Interactions Based on Complex Formation in Gastrointestinal Tract between Two Therapeutic Agents

Drug	Interacting drug	Effect	Ref.
Aspirin	Charcoal	Reduced absorption—adsorption	10
Carbamazepine	Charcoal	Reduced absorption—adsorption	10
Cephalexin	Cholestyramine	Reduced absorption—adsorption or steatorrhoea	10
Chloroquine	Kaolin, magnesium trisilicate	Reduced absorption—adsorption	40
Chlorothiazide	Colestipol	Reduced absorption—binding	45
Chlorpromazine	Aluminum hydroxide, magnesium hydroxide	Reduced absorption—adsorption	10, 47
Chlortetracycline	Antacids	Reduced absorption—adsorption	10
Clindamycin	Kaolin–pectin	Delayed absorption—adsorption	10
Diazepam	Antacids	Delayed absorption—adsorption	10
Dicoumarol	Magnesium hydroxide	Increased absorption—chelation	10
Diflunisal	Aluminum hydroxide	Reduced absorption—possibly adsorption	48
Digitoxin	Cholestyramine	Cholestyramine reduced half life of digitoxin—reduced reabsorption from intestine	41
Digoxin	Charcoal	Reduced absorption—adsorption	10
Digoxin	Cholestyramine	Reduced absorption with tablets but not with capsules	13
Doxycycline	Iron(II) sulfate	Reduced absorption—chelation influencing absorption	10
Hydrochlorothiazide	Cholestyramine	Reduced absorption	44
Hydrochlorothiazide	Colestipol	Reduced absorption	44
Iron(II) ion	Tetracycline	Reduced absorption—chelation	39
Griseofulvin	Phenobarbitone	Reduced plasma levels of griseofulvin when taken orally not when administered parenterally	51
Isoniazid	Antacids	Delayed and reduced absorption—adsorption and first pass metabolism	10
Methacycline	Iron(II) sulfate	Reduced absorption—chelation	10
Oxytetracycline	Iron(II) sulfate	Reduced absorption—chelation	10
Nitrofurantoin	Magnesium trisilicate	Reduced absorption—adsorption	50
Penicillamine	Iron(II) sulfate	Reduced absorption – chelation	10
Penicillamine	Aluminum hydroxide, magnesium hydroxide	Reduced absorption	10
Phenobarbitone	Charcoal	Reduced absorption and increased elimination rate, adsorption prevents reabsorption from intestine	46
Phenylbutazone	Charcoal	Reduced absorption and increased elimination rate, adsorption prevents reabsorption from intestine	10
Phenytoin	Charcoal	Reduced absorption—adsorption	10
Phenytoin	Magnesium trisilicate	Reduced absorption	22
Pseudoephedrine	Kaolin	Delayed absorption—adsorption	33
Promazine	Kaolin	Reduced absorption—adsorption	10
Rifampicin	*p*-Aminosalicylic acid	Delayed and reduced absorption—adsorption to bentonite in PAS granules	10
Salicylic acid	Aluminum hydroxide	Absorption rate reduced—adsorption?	49
Sulphamethoxazole	Cholestyramine	Reduced absorption—adsorption or steatorrhoea	10
Tetracycline	Iron(II) sulfate	Reduced absorption—chelation	10
Tetracycline	Magnesium aluminum hydroxide	Reduced absorption—chelation	37
Tetracycline	Iron(II) salts	Reduced absorption (variable depending on salt)—chelation	38
Tetracycline	Zinc sulfate	Reduced absorption—chelation	10
Thyroxine	Cholestyramine	Reduced absorption—adsorption	10
Trimethoprim	Cholestyramine	Reduced absorption—adsorption	10
Vitamin A	Aluminum hydroxide	Reduced absorption—adsorption oxidation	10
Warfarin	Cholestyramine	Reduced absorption—adsorption	10, 42

the original drug, which is reabsorbed. Glucuronide conjugates are the most frequently found metabolites in bile and when excreted into the intestine, they undergo hydrolysis by the β-glucuronidase enzymes present in resident flora. Adsorbents present in the intestine can interfere with this enterohepatic recycling process.

Drugs which are efficiently absorbed do not pass very far along the gastrointestinal tract and so the concentration of them in the lower part of the tract is virtually zero. In these circumstances the concentration of drug in the circulation is higher than that in the lower part of the gastrointestinal tract. Since drugs diffuse through membranes down a concentration gradient then drug can pass from the circulation into the lumen of the gut. Adsorbents, such as charcoal, can adsorb this drug and accelerate the rate of removal of it from the body by trapping the drug in the gut. This is particularly useful in the treatment of poisoning with drugs. That is, adsorbents not only reduce the rate and extent of absorption but also can assist in clearance.

23.6.6.2 Examples

It has been known for many years that the absorption of tetracycline is significantly decreased if metal ions are present in the intestine concomitantly. Calcium, iron, magnesium, zinc and aluminum ions have all been implicated in this interaction. The ions may be supplied by other drugs such as antacids, iron-containing medicines, or by milk and other dairy foods.

Tetracyclines are able to form chelate complexes with these heavy metal ions and it has been considered that chelation is the primary mechanism for this interaction. However it is not clear how chelation as such prevents absorption since it has been shown that the tetracycline–calcium chelate complex is more lipophilic than tetracycline alone.[34] It has been suggested that binding of tetracyclines to macromolecules in the gut is increased by chelate formation between tetracyclines and heavy metals, and that this is the main reason why absorption of tetracyclines·is reduced by calcium and other metal ions.[35] This concept is supported by the observation that at neutral pH the binding of tetracycline to isolated glycoproteins of the gastrointestinal mucus is increased in the presence of calcium and magnesium ions.[36]

The effect of antacids, other medications which contain metal ions, and food on the absorption of different tetracyclines varies. Changes in gastric pH have no effect on the bioavailability of tetracycline.[37]

A study of the interaction between iron salts and tetracyclines presents some interesting features. A number of iron salts are used for the treatment of iron deficiency anaemia and it has been found[38] that they do not all interact with tetracycline to the same degree. The reduction in areas under the tetracycline serum concentration curves compared with placebo was 80 to 90% with iron(II) sulfate, 70 to 80% with iron(II) fumarate, succinate and gluconate, 50% with iron(II) tartrate and 30% with iron(III) sodium edetate.

It should also be noted that not only does the concomitant use of iron salts and tetracyclines lead to a reduction in the bioavailability of the antibiotic, it also reduces the absorption of iron. A reduction of iron absorption of between 50 and 78% has been reported when tetracycline and iron(II) sulfate were administered together at normal therapeutic doses.[39]

Chloroquine AUC (area under curve) was decreased by 18.2% when taken with magnesium trisilicate and 28.6% when taken with kaolin.[40] It was considered that the mechanism of this interaction is possibly adsorption of the antimalarial by the antacid or adsorbent. *In vitro* studies support this.

Cholestyramine is a nonabsorbable anionic exchange resin. The effect of taking cholestyramine on the elimination of digitoxin provides an example of a drug interaction based on interference with enterohepatic cycling.[41] Digitoxin has a very long half-life of 96 to 144 h. Six subjects were given digitoxin daily for 30 days to attain steady state and then they took cholestyramine for a further 8 days along with digitoxin. It was found that cholestyramine reduced the plasma half-life from a mean of 141.6 to 84.4 h.

Cholestyramine also provides another example of an interaction which is related to the dose form. A study on the effect of cholestyramine on the absorption of digoxin from tablets and soft gelatin capsules showed that it had no effect when capsules were used but that absorption was decreased with cholestyramine when tablets were used.[13]

A case has been reported of the bioavailability of digoxin being reduced when taken with magnesium perhydrol.[42] However it appears that the mechanism of this interaction is unusual in that digoxin is decomposed by the hydrogen peroxide which is liberated from magnesium perhydrol by gastric juice.

Administration of cholestyramine orally concomitantly with intravenous warfarin decreased the plasma half-life of warfarin from a mean of 2 days to 1.3 days and increased the total clearance of warfarin.[43] A decrease in its anticoagulant effect was also observed. The mechanism of this interaction was considered to be interruption of the enterohepatic cycling of warfarin by cholestyramine.

Colestipol has properties similar to cholestyramine. Both cholestyramine and colestipol decreased the total urinary excretion of hydrochlorothiazide by 85% and 43% respectively.[44] Plasma levels of hydrochlorothiazide also confirmed a significant reduction in its bioavailability. Combined treatment with colestipol and chlorothiazide led to a 56% reduction in urinary recovery of chlorothiazide.[45]

Cholestyramine has been shown to reduce the bioavailability of cephalexin, sulfamethoxazole and thyroxine, and to reduce the rate of absorption but not the overall availability of trimethoprim.[10]

Activated charcoal significantly reduces the absorption of salicylates, dextropropoxyphene, nortriptyline, digoxin, phenytoin, phenobarbitone, carbamazepine, phenylbutazone and propantheline.[10] It can also reduce the blood levels of drugs after they have been absorbed, presumably by adsorption of drug which is secreted into the gut either *via* the bile or by back diffusion. Drugs which have been shown to be affected in this manner include carbamazepine, phenylbutazone[10] and phenobarbitone.[46] In each case the half-life of the drug was significantly reduced when charcoal was administered up to 10 h after ingestion of the drug.

Kaolin is an adsorbent and is often combined with pectin in preparations used for the symptomatic treatment of diarrhoea. Kaolin alone or kaolin–pectin mixtures are not such efficient adsorbents as charcoal but they can significantly reduce the bioavilability of a number of drugs including lincomycin, tetracycline and digoxin.[10] The rate of absorption of pseudoephedrine[33] and clindamycin[10] was reduced but their overall availability was not altered.

Insoluble antacids such as aluminum hydroxide, magnesium hydroxide and magnesium trisilicate can reduce the absorption of drugs by acting as nonspecific adsorbing agents. The bioavailability of chloroquine,[39] chlorpromazine,[10,47] clindamycin, diazepam,[10] diflunisal,[48] isoniazid and vitamin A[10] was reduced. The rate of absorption, but not the extent of absorption, of salicylic acid and diazepam[10] was reduced.

D'Arcy and McElnay[22] have critically reviewed drug–antacid interactions and assessed the clinical importance of the various interactions which have been reported.

Interactions of phenytoin with a variety of antacids, including aluminum hydroxide, magnesium hydroxide, calcium carbonate, magnesium trisilicate, calcium gluconate, magnesium oxide, and mixtures containing combinations of these agents, have been reported. The evidence is that phenytoin absorption can be reduced to a small degree when taken with antacids, presumably due to adsorption. A problem with phenytoin is that it has a narrow range of effective plasma concentrations, and so small changes in absorption can be potentially of importance. Sucralfate is not strictly an antacid but it is similar and it has been shown to reduce the availability of phenytoin, in a manner similar to antacids.

Quinidine is an antiarrhythmic drug which also has a narrow range of effective plasma concentrations. Studies on *in vitro* adsorption of quinidine on to antacids including aluminum hydroxide, magnesium hydroxide, kaolin–pectin suspension, magnesium carbonate, magnesium trisilicate and calcium carbonate, have shown that such adsorption can take place. *In vivo* studies on kaolin–pectin suspension, which was the most effective adsorbent *in vitro*, showed that even though the absorption rate was not affected there was a decrease in peak saliva quinidine concentrations and AUC of more than 50%.

An *in vitro* study of the adsorption of nitrofurantoin on to various antacids showed that there was a considerable variation in the ability of different antacids to adsorb it.[49] Magnesium trisilicate was outstanding in this regard, being as effective as activated charcoal. Under the conditions used, both substances adsorbed in excess of 99% of nitrofurantoin. *In vivo* magnesium trisilicate reduced both the rate and extent of nitrofurantoin absorption as indicated by excretion studies. The time during which the nitrofurantoin concentration in the urine was above the minimum effective concentration was also significantly reduced after oral administration of magnesium trisilicate. Nitrofurantoin is used for the treatment of urinary tract infections, and hence it is important that the concentration of it in the urine be maintained above the minimum antibacterial concentration.

It is clear that preparations containing magnesium trisilicate should be avoided when nitrofurantoin is taken but other antacids such as aluminum hydroxide, magnesium hydroxide or calcium carbonate may be used because they do not adsorb nitrofurantoin to anything like the extent of magnesium silicate.

This illustrates the point that even though one antacid may influence the absorption of a drug, this does not necessarily mean that all others will do so.

Penicillamine is a metal-chelating agent which aids in the elimination of certain metals. The absorption of penicillamine is significantly reduced by antacids (aluminum hydroxide and magnesium hydroxide), iron(II) sulfate and food.[10] The mechanisms of these interactions were not investigated but it was considered that iron(II) sulfate chelated with penicillamine, and antacid raised the pH of the gastric contents. Penicillamine is less stable at higher pH values.

Plasma levels of the antifungal agent griseofulvin are reduced when it is taken orally along with phenobarbitone but not when it is administered parenterally. Studies in animals led to the conclusion that the interaction is formulation dependent and is a result of diminished dissolution and consequently reduced absorption of griseofulvin,[50] and not induction of metabolism, which is the usual mechanism of drug–drug interactions involving phenobarbitone.

23.6.7 DRUG–FOOD INTERACTIONS

Food may increase, decrease, or have no effect on the absolute bioavailability of drugs. In general the effect of food is to retard the rate of absorption of drugs taken orally but not to reduce the fraction of the dose which is absorbed. Retardation of the rate of absorption occurs mainly because food slows gastric emptying. However, the nature of a meal influences its effects on gastric transit time. Hot meals, solutions of high viscosity, fat and to a lesser extent proteins and carbohydrates delay emptying. Solid meals almost double the emptying time as compared to liquid meals. Food also stimulates intestinal motility which may decrease the diffusion of drugs across the intestinal mucosa by accelerating their movement through the intestine. Increased absorption may occur as a consequence of increased residence time in the stomach, thereby allowing more time for the dissolution of solid particles of poorly soluble drugs. The increased motility of the intestine may also accelerate dissolution. Food, particularly fat, stimulates the flow of bile. Since bile contains surface-active agents, it enhances the dissolution of poorly soluble drugs, but it can also form insoluble salts with some drugs and thus reduce their absorption.

Food stimulates an increase in splanchnic blood flow and this may have a profound effect with drugs which have a high hepatic extraction. A number of explanations have been put forward to interpret why drugs such as lignocaine and propranolol have a higher bioavailability when taken with food than when taken fasting. One is that high extraction drugs pass rapidly through the liver when taken with food and hence escape hepatic removal thus leading to higher concentrations in the systemic circulation. Computer simulation of the effect of changes in hepatic blood flow induced by food indicated that the increase in hepatic blood flow was of too short a duration to explain the extent to which the bioavailability of propranolol rose. The influence of food on the presystemic clearance of drugs has been reviewed.[51]

Direct interactions between food and drugs may occur. For example tetracyclines may chelate with ions present in food, particularly calcium in dairy products.

Enteric coated preparations are a special case when considering food–drug interactions. Enteric coatings are resistant to the acidic conditions of the stomach and only release their contents when in the intestine at higher pH values. Consequently it may be anticipated that such preparations will be more susceptible to interactions with food as extended stomach transit time will delay drug release from the dosage form. Food may also influence plasma protein binding of drugs and their metabolism.

Patient factors such as age, sex, malnutrition, obesity and ethnic background may all contribute to changes in drug kinetics. Different responses may be observed in the same person at different times as dietary habits vary with the season and climate.

The complex interplay between food intake and drug disposition has been reviewed,[52] and so have food–drug interactions.[10]

The discussion which follows considers examples which illustrate many of the major factors involved in drug–food interactions.

23.6.7.1 Drug–Food Interactions in which the Rate, but not the Extent, of Absorption is Reduced

Coadministration of food with a single oral dose of fenoldopam significantly reduced plasma concentrations of fenoldopam and delayed its rate of absorption, but did not appear to reduce its bioavailability substantially.[53]

Following oral administration, midazolam is rapidly and completely absorbed and undergoes first pass metabolism, resulting in systemic availability of approximately 40% of the administered dose. Food did not influence the disposition of midazolam except when the drug was taken 1 h after a meal when the rate of absorption was reduced.[54] It was concluded that these changes may well be of some clinical significance as they may potentially delay the onset of sleep.

Authors of two studies on the clinical importance of the effect of food on nifedipine absorption came to different conclusions. Both studies showed that food delays the absorption of nifedipine, and reduces maximum plasma levels.[55, 56] Reitberg et al.[55] also demonstrated that food does not alter the amount of nifedipine absorbed and they concluded that the interaction was of no practical significance. Hirasawa et al.[56] found that the hemodynamic response to nifedipine significantly correlated with plasma concentration achieved. They stated 'The high C_{max} seen in the fasting state was associated with large decreases in blood pressure and tachycardia, which may be clinically relevant in producing side effects of hypotension, headache and flushing. The great transits in blood pressure and heart rate were absent after food, but the blood pressure reduction at 4 and 6 h was comparable. This may indicate adequate therapeutic effect after food. Therefore, ingestion of nifedipine after food may reduce vasodilator side-effects while retaining therapeutic efficacy.'

PUVA treatment for psoriasis is irradiation with long wave length UV light (UVA) 2 h after oxsoralen ingestion. For maximum effect it is considered that oxsoralen concentration in interstitial tissue of the skin should be at its height when the skin is irradiated with UVA. Food delayed the absorption of oxsoralen from the intestine.[57] Peak plasma levels and blister fluid levels were both delayed. Suction blister fluid was used as a model of the interstitial fluid of the epidermis. It was assumed that the delayed absorption was related to retarded stomach emptying.

Food significantly delayed the absorption of theophylline from a solution of choline theophyllinate.[58] The maximum plasma concentration was decreased but the AUC was not affected. Taking theophylline elixirs at night with food will increase its duration of action which is important because patients are often in the worst state in the early morning hours, and taking it on an empty stomach first thing in the morning will produce a rapid rise in the plasma concentrations. Children often take theophylline elixirs.

Food also reduces the rate of absorption and peak plasma levels of theophylline from a sustained release preparation.[59] This was associated with a reduction in adverse effects of theophylline in a group of children aged 6 to 14 years. Food however may have the opposite effect.[60] An 11-year-old girl had been treated with a delayed release theophylline (700 mg) preparation without problem. One day she ate cereal immediately following the dose and later the same day had headache and nausea. The following day she ate sweets within 15 min of the dose and later that day she had severe headache and projective vomiting. Serum levels of theophylline were 2.5 times higher than usual. It was considered that the interaction was due to theophylline being dumped rapidly in the small intestine, and the mechanism for the dose-dumping was likely to be the pH change in the duodenum which occurs in response to food. The coating on the beads rapidly dissolves at pH 7.4. Another similar case has been described.[61]

A high protein meal had no significant effect on the peak plasma level, plasma half-life or AUC of verapamil. Time to peak level was increased.[62]

23.6.7.2 Drug–Food Interactions in which the Extent of Absorption is Reduced With or Without an Effect on Rate

Food reduced both the absorption of endralazine and its antihypertensive effect but the blood pressure lowering effect was only weakly correlated with the food-related reduction in the plasma endralazine levels.[63]

Erythromycin stearate tablets, and erythromycin base enteric coated pellets when taken immediately before food, demonstrated equivalent bioavailability.[64] However, the bioavailability of erythromycin stearate when taken immediately after food was markedly reduced but the extent of absorption of the base was unaffected. The enteric coated base formulation offered protection against acid degradation, although its rate of absorption was reduced by the presence of food in the stomach.

The short intense diuresis after a frusemide tablet is a problem for many patients. It has been reported that food caused no change in the bioavailability of frusemide, although it did reduce the initial high plasma peak of it which was considered to be an advantage.[65] In a second study[66] it was found that the bioavailability and peak levels of frusemide were both decreased by approximately

30%. The reduced bioavailability caused a reduction in the diuretic effect. It was recommended that frusemide should not be given with food.

The question arises as to why such different results were obtained. It appears likely that it was related to the sensitivity of the analytical methods used for the determination of frusemide in blood and urine samples.

It was reported by Melander *et al.*,[67] using both coated and uncoated tablets, that food enhanced the bioavailability of hydralazine two- to three-fold; they recommended that hydralazine always be taken with food. Some years later Shepherd *et al.*[68] questioned the analytical methods used by Melander *et al.*[67] and reexamined the effect of food on hydralazine absorption in patients suffering from hypertension. It was found that taking hydralazine 45 min after a meal was associated with a reduction in peak blood levels of 46.2% and AUC of hydralazine fell by 45.7%. These changes in pharmacokinetic parameters were accompanied with reduced vasodepressor effects of 41.5%.

One of the major problems associated with the use of levodopa for the treatment of Parkinson's disease is the so-called 'on–off' phenomenon. That is patients rapidly alternate between mobility and immobility with no apparent correlation with timing of drug administration. A study of the interaction of levodopa and food has provided a possible explanation for the 'on–off' effect.[69] When levodopa and food were given together, it was found that food reduced the peak plasma levels by 29% on average and delayed absorption by a mean of 34 min, although there were considerable individual variations in these values. Since the plasma half-life of levodopa is very short (range 0.97 to 1.67 h) then the delay of 30 min or so is significant. When the problems of oral absorption were bypassed by giving levodopa as a constant intravenous infusion, the drug produced a stable clinical state for between 12 and 36 h. However, high protein meals or oral phenylalanine, leucine or isoleucine reversed the therapeutic effect of infused levodopa without reducing plasma levodopa levels. Glycine and lysine at the same dose had no effect. Since levodopa is transported from blood to the brain by an active transport process and the large neutral amino acids can compete with levodopa for the transport system, high plasma concentrations of these amino acids may block the transport of levodopa into the CNS.

A study on the effect of food on methotrexate levels[70] showed that peak serum levels were significantly reduced by a milky meal and both a milky meal and a citrus meal delayed drug absorption in most children. The AUC was also significantly reduced by the milky meal. The milky meal consisted of milk, cornflakes, sugar, bread and butter and the citrus meal consisted of orange juice, fresh oranges, bread and butter with jam.

The plasma levels of penicillamine were reduced to 52% when it was taken with food as compared to fasting.[26] It was considered that food increased the residence time in the stomach and thus increased the potential for decomposition there.

Food reduced the rate and extent of bioavailability of propantheline bromide.[71] Since propantheline bromide is a quaternary ammonium compound, its intrinsic ability to diffuse through membranes is low. It has an absolute oral bioavailability of 15%. It was suggested that the effect of food on the absorption of propantheline bromide may partly explain the variable and poor response to oral administration which has been observed in patients.

The effect of food on the clinical response to rifampicin has been reported.[72] Once daily rifampicin plus isoniazid was used for the treatment of tuberculosis, and the majority of patients showed clinical improvement after four weeks therapy. It was noted that those patients who had not improved were taking rifampicin immediately following breakfast. This was changed to 7 h after the last meal, and all these patients showed improvement within two weeks. Food reduced the bioavailability of rifampicin.

23.6.7.3 Drug–Food Interactions in which the Rate and/or Extent of Drug Absorption is Increased

Both plasma levels and AUC are significantly increased when chloroquine is taken with food, but the time to peak plasma level was unchanged.[73]

Etretinate undergoes first pass deesterification, forming etretin, which then isomerizes to form isoetretin. Both of these metabolites are active. One study has shown that the absorption of etretinate is increased when it is taken with milk.[74] A more detailed investigation[75] found that peak plasma concentrations and AUC of etretinate were significantly greater (three times higher) when it was administered with a high fat meal and whole milk than when it was taken with a high carbohydrate meal or during a complete fast. However, there was no increase in the plasma levels of the active metabolites following any of the meals. The mechanisms responsible for the increased etretinate plasma concentrations during a high fat meal or whole milk ingestion appear to be associated with the lipid nature of the diet. Since etretinate is a fat-soluble drug, it may become incorporated in the lipid portion of the bile acid micelles. These micelles are transported through the lymphatic system to the systemic blood, thus escaping first pass metabolism in the gut wall and/or liver.

Griseofulvin has a very low solubility in water and its rate of absorption is controlled by its dissolution rate. Absorption was enhanced significantly in children when it was taken with milk as compared to being taken without food.[76] It was recommended that griseofulvin be taken with milk or other fat-containing foods for the treatment of superficial fungal infections.

Food caused an increase in the mean systemic bioavailability of labetalol, but it did not alter the plasma half-life of it, irrespective of the route of administration.[77] It was suggested that food appears to increase labetalol bioavailability by decreasing the first pass metabolism, and that this change results from a transient alteration of hepatic blood flow.

Nitrofurantoin is a broad spectrum antibacterial agent, which is used for the treatment of urinary tract infections. It has a very low solubility at the normal pH of the gastrointestinal tract and so absorption is determined to a significant extent by its dissolution rate. A study has been carried out to investigate the effect of particle size and food on the absorption of nitrofurantoin.[78] The results showed that in the fasting state the maximum excretion rate for the macrocrystalline form was significantly lower than for the microcrystalline form. However, there was no difference in the excretion rate following administration of either form when they were taken with food. Food increased the bioavailability of the microcrystalline form on average by 30%, and by 80% for the macrocrystalline form. In another study[79] using five different nitrofurantoin products, including both the microcrystalline and macrocrystalline forms, it was shown that the mean duration of a therapeutic urinary concentration of nitrofurantoin was significantly increased when the medication was taken with food.

Food was found to enhance the rate of absorption of phenytoin with a consequence that the peak concentration of it was consistently and significantly increased when the drug was taken with a meal. The mean increase was about 40%. The absorption was two to four times greater in the first 4 h after ingestion when taken with food than when fasting.[80] Phenytoin was administered as the free acid in the micronized form.

There is considerable interpatient variation in plasma levels of both propranolol and metoprolol following oral administration. With propranolol up to a tenfold variation has been observed. Food enhances the bioavailability of both β-adrenergic blockers significantly.[81] However, even though these results suggest that propranolol and metoprolol should always be taken in a standard relation to meals, the more important practical question with their use is individualization of doses of them. The mechanism of this interaction has been investigated by giving propranolol orally and intravenously both with food and in the fasting state.[82] It was concluded that feeding reduces the first pass clearance of propranolol, which results in a higher fraction of the oral dose entering the general circulation. It was considered that this was associated with an increase in liver blood flow.

Propranolol is an optically active compound and it is the racemic form which is used clinically although the biologically active enantiomer is (−)-propranolol. The effect of food on the absorption of the laevo isomer and the racemic mixture is the same in humans. The increase in bioavailability was 39%.[83]

Following oral administration of spironolactone, only trace amounts of unchanged drug are found in the blood because it is rapidly metabolized to the active compound canrenone during the first pass through intestinal wall and liver. Food significantly increases the peak plasma concentrations and AUC of canrenone.[85] Since spironolactone has a low water solubility, it was suggested that delayed stomach emptying allowing greater dissolution and solubilization by bile acids were involved. It should be noted that the interaction did not influence the first pass effect.

Food has a dramatic effect on the absorption of tocopheryl nicotinate but has no effect on the rate of absorption.[86] Following a 600 mg dose the AUC and maximum plasma level were increased 29-fold and 32-fold respectively when the drug was taken with food as compared to fasting. Tocopherol, like the other fat soluble vitamins, is absorbed as part of the normal fat transport system, and hence its absorption was facilitated when it was taken with food.

23.6.7.4 Drug–Food Interactions in which the Pharmacokinetics of the Drug are Unaltered but the Therapeutic Effect is Modified

Iproniazid, isocarboxazid, phenelzine and tranylcypromine are monoamine oxidase inhibitor anti-depressants. The basis of interactions between monoamine oxidase inhibitors and food is sympathetic over-stimulation. The main results are occipital headache and hypertension, sometimes severe enough to produce subarachnoid haemorrhage and on rare occasions death. Tyramine, which is formed by decarboxylation of the amino acid tyrosine, is the main problem but histamine and dopamine may be involved. Normally tyramine, and other pressor amines, are not absorbed because they are metabolized by monoamine oxidase enzymes on their first pass during absorption. However when these enzymes are inhibited then they can exert their sympathomimetic effect, resulting in hypertension. Foods which have been implicated in this interaction include cheese, red wine, yeast extracts and pickled herrings. This is the famous 'cheese reaction' and was the drug–food interaction which stimulated investigations of the interplay between drug action and food. This interaction is unusual in that it is essentially the result of a drug altering the effects of a food instead of the other way round.

Even though both the monoamine oxidase inhibitors and isoniazid contain the hydrazine group, few reports have appeared which describe an interaction between isoniazid and foods rich in tyramine. A case has been reported of a monoamine oxidase type reaction occurring with isoniazid and red wine plus parmesan cheese.[86] The patient was being treated with isoniazid, rifampicin and pyridoxine. Half an hour after a meal which included parmesan cheese and red wine, she developed the following symptoms: palpitations, severe general flushing, conjunctival suffusion, headache, dyspnoea, tightness of the chest, moderate tachypnoea and sweating.

Tolbutamide is an orally active hypoglycaemic agent which reduces blood-sugar concentrations. The relationship between taking tolbutamide and taking food has been studied in relation to postprandial blood glucose levels.[87] Standard doses of tolbutamide were given to diabetic patients with food and 30 min before food. When tolbutamide was taken before meals then blood glucose levels were significantly lower than when it was taken with food. It was recommended that tolbutamide be taken 30 min before meals in order to maintain postprandial hyperglycaemia within a narrow and more physiological range. This interaction is related to the mechanism of action of tolbutamide rather than to food-induced changes in its absorption.

Warfarin sodium is an anticoagulant which depresses the hepatic vitamin K dependent synthesis of coagulation factors II (prothrombin), VII, IX and X. A number of reports have appeared in which resistance to warfarin has been associated with ingestion of liquid dietary supplements. In each case it was found that the supplement contained relatively high concentrations of vitamin K.[88] Two cases have been reported of the action of warfarin being inhibited by a diet containing large amounts of broccoli,[89] which contains vitamin K.

23.6.7.5 Drug–Food Interactions Based on Food-induced Changes in Renal Clearance

The effect of dietary protein on the pharmacokinetics of allopurinol and its active metabolite oxypurinol has been studied.[90] Three principal findings regarding the effect of dietary protein intake on the pharmacokinetics of allopurinol and oxypurinol have been reported. These were: (i) the renal clearance of oxypurinol was markedly diminished during protein restriction, and the change was disproportionate to the change in creatinine clearance (or glomerular filtration rate); (ii) the renal clearance of allopurinol decreased in proportion to the decrease in creatinine clearance (or glomerular filtration rate) during the low-protein diet; and (iii) the total body clearance of allopurinol decreased when the protein was restricted.

The potential clinical implications of these findings were considered to be: (i) malnourished patients, patients on low-protein diets, and those receiving prolonged intravenous dextrose infusions without protein or amino acid supplementation may have reduced oxypurinol clearance and increased plasma concentrations of oxypurinol during long-term therapy with allopurinol; and (ii) the decrease in urinary uric acid excretion during long term therapy with allopurinol may change the tubular reabsorption of oxypurinol, if uric acid and oxypurinol share the same transport mechanisms.

This shows the complex and subtle ways by which food can modify the action of drugs.

23.6.7.6 Drug–Food Interactions Based on Food-induced Changes in Plasma Protein Binding

Food decreases free fatty acids (FFA) in plasma and FFA can modify the binding of some drugs to plasma proteins. Hence FFA could be the modulators of meal-induced variations in binding of drugs to plasma proteins. At normal therapeutic concentrations diazepam is highly bound (approximately 98.7%) to plasma proteins. It has been reported that as the FFA levels fell following food, so the proportion of diazepam bound to plasma proteins increased.[91] Therefore it is likely that during normal use of diazepam when FFA decreases after a meal, the binding of diazepam increases causing diazepam in the tissues to enter the intravascular compartment and consequently total plasma diazepam concentration will rise.

Food decreases the rate of absorption and the peak plasma levels but not the extent of absorption of quinidine.[92] Food affects plasma protein binding of quinidine. The peak mean unbound quinidine serum levels were 34% lower and were reached later when quinidine was taken with food. The antiarrhythmic actions of quinidine are related to its plasma levels and it has a short half-life, which necessitates frequent dosage with normal tablets. Sustained release tablet formulations are often used, and it has been shown that food has little effect on the bioavailability from such a product.[93]

23.6.8 DRUG–DRUG INTERACTIONS RELATED TO DRUG-INDUCED DIARRHOEA

The actions of lithium are affected by body levels of both lithium and sodium. An increase in lithium levels or a decrease in sodium levels both tend to produce toxic effects characteristic of lithium. It has been reported that toxic effects of lithium occurred when tetracycline was given concurrently. One suggestion put forward to explain this interaction was that it was related to the nephrotoxic effect of tetracycline which led to a reduction in lithium renal excretion. A second suggestion was that tetracycline often produces diarrhoea and this caused sodium to be lost and hence its levels fell. In such a situation a patient would be susceptible to lithium toxicity.[94]

23.6.9 DRUG–DRUG INTERACTIONS WITH PREPARATIONS FOR SUBLINGUAL USE

Dry mouth is a common side effect of many drugs, including tricyclic antidepressants and anticholinergic drugs. In general this is accepted as a minor annoyance by most patients rather than an adverse effect of serious importance. It has been reported that a patient who was taking imipramine obtained delayed and limited relief of exertional angina when sublingual nitroglycerine tablets were used.[95] It was observed that a new nitroglycerine tablet placed under the tongue by the patient for 5 min remained virtually intact when it was removed. Withdrawal of imipramine was followed by resolution of dry mouth and more rapid relief of angina when sublingual nitroglycerine was used, associated with faster dissolution of the tablets.

23.6.10 DRUG–DRUG INTERACTIONS RELATED TO RENAL EXCRETION

23.6.10.1 Introduction

The basic unit of the kidney is the nephron. This consists of two structures, the glomerulus and the renal tubule which is divided into the proximal tubule, the loop of Henle, and the distal tubule. Blood flows through the glomerulus, some of the plasma is filtered and this then flows down the tubule. In health blood cells and high molecular solutes, *e.g.* proteins, are not filtered. Hence drugs in solution in plasma are present in the tubular fluid at the same concentration.

Many drugs are bound to some extent to plasma proteins so that the total plasma concentration of a drug consists of drug in free solution and bound drug, and so the degree of binding will have a profound effect on the glomerular filtration rate of drugs.

The efficiency of the renal tubules is such that approximately 99% of water filtered by the glomerulus is reabsorbed. Any drug or other substance present in the glomerular filtrate is therefore concentrated 100 times in the renal tubule. If the substance is water soluble, *e.g.* urea, it is excreted in urine. If the drug is lipid soluble then it will diffuse back across the epithelial cells of renal tubules into the circulation.

Many drugs are either organic bases or organic acids. Hence they ionize in aqueous solution so that there is present a neutral, usually lipid soluble, species and a charged conjugate acid or base. The rate of renal excretion of such a drug, whether acid or base, will be influenced by the degree of its ionization in the tubular fluid. This in turn is controlled by the pK_a of the substance and the pH of tubular fluid. Clearance of acidic drugs is favoured by high pH values of tubular fluid, and the converse is true for basic drugs. Since the pH of urine can range from about pH 5 to 8, the degree of ionization of drugs which have appropriate pK_a values may vary considerably with consequent effects on the rate of renal excretion.

Drug–drug interactions may occur as a result of one drug altering the pH of urine and thus either accelerating or retarding the rate of renal excretion of a second drug.

There are examples of drugs having an excretion rate greater than glomerular filtration rate. In such cases secretion of the drug from plasma to tubular fluid occurs. The renal tubules contain active transport processes which can either secrete substances from plasma to tubular fluid or reabsorb them. Secretion may be so extensive that virtually all the drug in the blood is removed whether or not it is bound to plasma proteins, and secretion is therefore far more important than filtration in such cases with regard to the rate of renal excretion. If one drug competes with a second drug for secretion then this can be a basis for a drug–drug interaction.

Other mechanisms, by which drug–drug interactions may occur related to renal excretion include an increased rate of renal excretion caused by diuretics, or alternatively a reduction of renal excretion.

23.6.10.2 Drug–Drug Interactions Based on Changes in Urinary pH

Hexamine is a urinary antiseptic, and it exerts its antibacterial action in the urinary tract by being hydrolyzed under the acidic conditions of urine to produce formaldehyde. If urine is rendered alkaline, the production of formaldehyde is retarded and so the activity of hexamine is reduced. Citrates, tartrates, bicarbonates and carbonic anhydrase inhibitor diuretics tend to make the urine alkaline, thus inhibiting the action of hexamine.

The effect of making urine alkaline has been shown to increase the proportion of doxycycline excreted, to increase the rate of excretion and to reduce the serum levels of doxycycline during both single and multiple dosing.[96]

Chlorpropamide is a weak acid (pK_a 4.8), and is cleared primarily by renal excretion. Sodium bicarbonate taken concomitantly raised the urinary pH (range 7.1 to 8.2) and reduced the plasma half-life from a mean of 50 h to a mean of 13 h. Ammonium chloride had the opposite effect, lowering the urinary pH (range 4.7 to 5.5) and prolonging the plasma half-life to a mean of 68.5 h.[97]

Diethylcarbamazine is a basic compound. The elimination half-life and AUC were significantly increased when urine was maintained alkaline (mean pH 7.5) with sodium bicarbonate as compared with the values of these parameters when the urine was maintained acidic (mean pH 5.5) with ammonium chloride.[98]

Since diet can influence urinary pH, *e.g.* a vegetarian diet tends to result in alkaline urine and an acid urine is usually found when the diet is predominantly rich in animal protein, then diet can influence the rate of renal excretion of drugs.

Methadone is a basic drug which is cleared both by renal excretion and by hepatic metabolism. The effect of raising the urinary pH from pH 5.2 (ammonium chloride) to pH 7.8 (sodium bicarbonate) was to double the plasma half-life and increase the volume of distribution. At pH 5.2 35% of the dose was excreted unchanged but less than 5% was excreted unchanged when the urine was alkaline (pH 7.8).[99] This illustrates the point that reduction in renal excretion may be compensated to some extent by increased metabolism.

23.6.10.3 Drug–Drug Interactions Based on Blocking of Tubular Active Transport Mechanisms

Penicillins and some cephalosporins are rapidly excreted by the kidney, primarily by renal tubule secretion. Probenecid inhibits this transport system and thus retards their rate of renal excretion. This interaction is used when prolonged high plasma levels of penicillins or cephalosporins are required. The plasma half-life of benzylpenicillin was increased from a mean of 40 min to 104 min when probenecid was taken concomitantly.[100]

Probenecid reduced the rate of excretion and increased the plasma concentration of indomethacin but the mechanism of this interaction was shown to be that probenecid inhibited nonrenal clearance of indomethacin, probably reducing biliary excretion.[101]

Cimetidine inhibited renal excretion of oral procainamide by 36%. The mechanism of this interaction was, in part at least, that cimetidine inhibited active secretion of procainamide by the kidney tubules and partly through inhibition of nonrenal clearance.[102] This illustrates the point that an interaction may occur through more than one mechanism.

It has been shown in single-dose studies that verapamil inhibits both renal and extrarenal elimination of digoxin, leading to elevated blood levels of digoxin.[103] The mechanism underlying the renal interaction appears to be suppression of active tubular secretion of digoxin. However it has been shown that, following prolonged administration of the two drugs, the suppression of active tubular secretion of digoxin by verapamil passes off with time.[104]

23.6.10.4 Drug–Drug Interactions Based on Increased Rate of Renal Excretion

Many diuretics increase the rate of potassium loss by the kidney and potassium supplements are often given to replace this. Metabolic alkalosis may be brought about in a number of ways, for example taking substances such as sodium bicarbonate or sodium citrate, or by hypokalaemia. One result of metabolic alkalosis is for the kidney to excrete potassium. Consequently the situation is that metabolic alkalosis induces potassium depletion and potassium depletion, in turn, induces and sustains metabolic alkalosis. Hence potassium replacement therapy will be ineffective if potassium is given along with any substance which can induce metabolic alkalosis, *e.g.* bicarbonate. Severe potassium deficiency has been reported in patients taking a diuretic along with potassium bicarbonate.[105]

Approximately 95% of a single dose of lithium is excreted in the urine. High sodium plasma concentrations produce a small enhancement of lithium excretion, but sodium depletion promotes a clinically important retention of lithium. Since lithium has a low therapeutic index, relatively small changes in the rate of renal excretion of lithium may be associated with toxic effects when a constant dosage is taken regularly. Drug interactions involving lithium may be caused by a change in sodium intake or by any drug which influences sodium levels. It has been pointed out that since many effervescent preparations contain large amounts of sodium bicarbonate, their use by patients taking lithium could lead to increased renal excretion and decreased plasma levels.[106]

23.6.10.5 Drug–Drug Interactions Based on Reduction of Renal Excretion

Amiloride, spironolactone and triamterene reduce the excretion of potassium by the kidney. Consequently if potassium supplements are taken while one of these potassium-sparing drugs is also being administered then blood potassium levels rise.

Renal prostaglandins are known to be important mediators of renal excretion of chloride, potassium and sodium and evidence is available that they may be important in the excretion of lithium as well. Diclofenac and indomethacin are prostaglandin synthetase inhibitors and it is considered that the effect they have on lithium excretion may be associated with their inhibition of prostaglandin synthesis. Concomitant administration of lithium and either diclofenac or indomethacin was associated with reduced renal excretion and increased serum lithium levels.[107]

Another study confirmed that indomethacin enhances the plasma levels of lithium significantly and decreased its renal excretion. On the other hand aspirin had no significant effect on either plasma levels or renal excretion of lithium.[108] However renal excretion of prostaglandin E_2 was suppressed to about the same degree by both indomethacin and aspirin, indicating that not all prostaglandin synthesis inhibitors have the same effect.

Toxic effects of lithium were observed following administration of spectinomycin to a patient taking lithium carbonate. Serum lithium levels rose significantly.[109]

Diuretics, *e.g.* thiazides, indirectly increase lithium reabsorption. This is because the reduction in body water caused by the diuretic forces the kidney actively to conserve sodium (and water), and as lithium ions are handled by the kidney in a manner similar to sodium, then lithium excretion is reduced. This combined with a smaller volume of body water for lithium to be distributed into tends to lead to higher plasma levels which may rise to toxic concentrations. Cases have been reported where patients have been stabilized on lithium for a considerable period, then serum lithium levels have risen unexpectedly. This has prompted the withdrawal of lithium which has resulted in the patients relapsing into mania. On investigation it was found that the rise in serum lithium levels was associated with diuretics being prescribed (bendrofluazide and bumetanide).[110] The effect of diuretics on lithium serum concentrations appears to be dose related.[111]

Urinary excretion of digoxin is reduced and plasma levels are elevated when diazepam is given concomitantly.[112] A detailed study of the interaction between quinidine and digoxin showed that quinidine reduced the systemic, renal, and nonrenal clearance of digoxin and approximately doubled its elimination half-life. Spironolactone had a similar effect on digoxin disposition.[113]

When indomethacin and triamterene were administered concomitantly, some subjects rapidly showed signs of kidney failure. It was suggested that the indomethacin may suppress the activity of prostaglandins which protect the kidney from nephrotoxic potentialities of triamterene.[114]

Frusemide is a potent diuretic which interestingly also reduces glomerular filtration rate and hence may influence the renal excretion of drugs cleared mainly by glomerular filtration. It was associated with a significant decrease in the plasma clearance and in higher plasma levels of both cephaloridine and gentamicin, the increase in antibiotic concentration being as much as 100% for cephaloridine and 70% for gentamicin 1 h after injection.[115]

23.6.10.6 Miscellaneous

Spironolactone has no effect on serum levels of warfarin when the two are given concomitantly. However, the hypoprothrombinaemic effect of warfarin was decreased when spironolactone was given.[116] A similar interaction was observed when chlorthalidone and warfarin were investigated.[117] In both cases it was observed that the hematocrit increased when a diuretic was taken, which indicates a decrease in the volume of plasma water. Chlorthalidone and spironolactone are diuretics and cause a loss of body water, in particular plasma water. This leads to an increase in the concentration of clotting factors in blood which offsets the anticlotting action of warfarin.

23.6.11 DRUG–DRUG INTERACTIONS RELATED TO PROTEIN BINDING

23.6.11.1 Introduction

Drugs can become reversibly adsorbed on to plasma proteins and tissue. Acidic drugs bind to albumin, and basic drugs to glycoprotein and lipoproteins. The rates of adsorption and desorption are usually very rapid so that bound and free drug may be assumed to be at equilibrium at all times. It is considered that only the free unbound drug is capable of diffusing into tissues and hence that it is this concentration which relates to the degree of pharmacological action exterted by the drug. The fraction of the drug free in plasma also influences glomerular filtration rate of the drug (see Section 23.6.10.1). It is possible for one drug to displace another drug from a common binding site, thus increasing the free concentration of the displaced drug. It is also possible for a drug to be displaced from its binding site by an allosteric change in the conformation of the protein caused by adsorption of the displacing drug at another site. Allosteric changes may result in increased affinity between drug and protein.

Even though displacement frequently occurs, there have to be particular conditions before it may be of significance as a basis for a drug–drug interaction. The first is that the drug must be extensively bound and hence the free concentration low. The second is that the displacing drug must have a high affinity for the binding sites and be present in sufficient concentration to occupy a significant proportion of binding sites.

When a drug is displaced from plasma proteins, the concentration of free unbound drug increases. The free drug then diffuses out of the vascular compartment into tissues to maintain the concentration of free drug the same in plasma and body water. So unless a drug is displaced from both plasma proteins and nonspecific binding sites in tissues, it is not likely to show enhanced pharmacological effects following the administration of a displacing drug.

An important practical result of drug–drug interactions involving plasma protein binding displacement is that the total plasma concentration of the displaced drug will fall and so the use of plasma concentrations to monitor dosage will be compromised.

Rowland and Tozer have discussed the pharmacokinetic basis of drug–drug interactions based on displacement from plasma proteins.[118]

23.6.11.2 Examples of Drug–Drug Interactions Based on Displacement

Both plasma and urine levels of digoxin are significantly increased when amiodarone is administered, but this is not associated with digoxin intoxication. It has been suggested that amiodarone displaces digoxin from tissue binding sites.[119]

Aspirin displaces phenytoin from plasma proteins but this is not accompanied by an increase in free concentration of phenytoin in plasma as it redistributes out of the vascular system into other body compartments. This results in a reduced total plasma concentration of phenytoin but no change in therapeutic effect. The relationship between total plasma concentration of phenytoin and activity will not hold when aspirin is being taken concomitantly.[120, 121]

Aspirin increased the steady-state free fraction of valproic acid from 12% to 43%. The plasma half-life of valproic acid rose as did plasma levels of both total and free valproic acid. It was suggested that salicylate displaces valproic acid from plasma proteins but also inhibits renal clearance and metabolism of it at the same time.[122]

It has been reported many times that digoxin serum levels are significantly increased when quinidine is given. This is caused by digoxin clearance being reduced,[113] and also by digoxin being displaced from tissue binding sites, and so during combined therapy serum digoxin levels rise while the cardiac tissue concentration and tissue to serum ratio of digoxin decline. Under these conditions the serum digoxin levels do not reflect the same relationship to cardiac performance as they normally do. Hence decreased cardiac effects despite elevated serum digoxin levels can be the result of this interaction.[123]

Enhancement of gentamicin-induced nephrotoxicity by frusemide has been demonstrated in animals and implicated in man. Frusemide administration was associated with a doubling of the amount of gentamicin excreted in urine. The most likely explanation for this effect of frusemide on gentamicin renal clearance, given that the diuretic had no effect on glomerular filtration rate or volume of distribution of gentamicin, is that frusemide displaces gentamicin from binding sites of plasma proteins to increase the amount of free gentamicin which is then filtered by the glomerulus.[124]

The interaction between phenytoin and valproic acid (sodium valproate) results in a decrease in the total plasma level of phenytoin and an increase in the fraction of free phenytoin. The general consensus is that this does not present a clinical problem in itself, but the major issue is that determination of the serum levels of phenytoin as a guide to dosage is misleading when valproic acid is also being taken.[125] The mechanisms involved are considered to be that valproic acid displaces phenytoin from protein-binding sites, and that it inhibits the metabolism of phenytoin.

23.6.12 REFERENCES

1. G. N. Volans, *Br. J. Clin. Pharmacol.*, 1975, **2**, 57.
2. G. Wainscott, T. Kaspi and G. N. Volans, *Br. J. Clin. Pharmacol.*, 1976, **3**, 1015.
3. R. Gugler, M. Brand and A. Somogyi, *Eur. J. Clin. Pharmacol.*, 1981, **20**, 225.
4. M. A. Osman and P. G. Welling, *Curr. Ther. Res.*, 1983, **34**, 404.
5. D. O. Gibbons and A. F. Lant, *Clin. Pharmacol. Ther.*, 1975, **17**, 578.
6. J. W. Dundee and J. A. S. Gamble, *Lancet*, 1975, **1**, 1032.
7. V. Manninen, M. A. Apajalahti, J. Melin and M. Karesoja, *Lancet*, 1973, **1**, 398.
8. B. F. Johnson, J. O'Grady and C. Bye, *Br. J. Clin. Pharmacol.*, 1978, **5**, 465.
9. T. R. E. Barnes, W. M. Braude and D. J. Hill, *Lancet*, 1982, **2**, 48.
10. P. G. Welling, *Clin. Pharmacokinet.*, 1984, **9**, 404.
11. L. M. H. Wing, P. J. Meffin, J. J. Grygiel, K. J. Smith and D. J. Birkett, *Br. J. Clin. Pharmacol.*, 1980, **9**, 505.
12. J. Nimmo, R. C. Heading, P. Tothill and L. F. Prescott, *Br. Med. J.*, 1973, **1**, 587.
13. D. D. Brown, J. Schmid, R. A. Long and J. H. Hull, *J. Clin. Pharmacol.*, 1985, **25**, 360.
14. W. S. Nimmo, R. C. Heading, J. Wilson, P. Tothill and L. F. Prescott, *Br. J. Clin. Pharmacol.*, 1975, **2**, 509.
15. J. Lindenbaum, R. M. Maulitz, J. R. Saha, N. Shea and V. P. Butler, *Clin. Res.*, 1972, **20**, 410.
16. G. F. Longstreth and A. D. Newcomer, *Mayo Clin. Proc.*, 1975, **50**, 284.
17. D. I. Webb, R. C. Chodos, C. Q. Mahor and W. W. Faloon, *N. Engl. J. Med.*, 1968, **279**, 845.
18. J. E. Thomas, *Physiol. Rev.*, 1957, **37**, 453.
19. M. Gibaldi, B. Grundhofer and G. Levy, *J. Pharm. Sci.*, 1975, **64**, 2003.
20. R. E. Gerhardt, R. F. Knouss, P. T. Thyrum, R. J. Luchi and J. J. Morris, *Ann. Intern. Med.*, 1969, **71**, 927.
21. J. H. Dobbs, V. A. Skoutakis, S. R. Acchiardo and B. R. Dobbs, *Curr. Ther. Res.*, 1977, **21**, 887.
22. P. F. D'Arcy and J. C. McElnay, *Drug Intell. Clin. Pharm.*, 1987, **21**, 607.
23. C. G. Regardh, P. Lundborg and B. A. Persson, *Biopharm. Drug Dispos.*, 1981, **2**, 79.
24. W. H. Barr, J. Adir and L. Garrettson, *Clin. Pharmacol. Ther.*, 1971, **12**, 779.
25. G. R. Elliott and M. F. Armstrong, *Clin. Pharmacol. Ther.*, 1972, **13**, 459.
26. M. A. Osman, R. B. Patel, A. Schuna, W. R. Sundstrom and P. G. Welling, *Clin. Pharmacol. Ther.*, 1983, **33**, 465.
27. S. Feldman and B. C. Carlstedt, *J. Am. Med. Assoc.*, 1974, **227**, 660.
28. A. R. Cooke and J. N. Hunt, *Am. J. Dig. Dis.*, 1970, **15**, 95.
29. J. R. Leonards, *Proc. Soc. Exp. Biol. Med.*, 1962, **110**, 304.
30. H. D. Lewis, J. W. Davis, D. G. Archibald, W. E. Steinke, T. C. Smitherman, J. E. Doherty, H. W. Schnaper, M. M. LeWinter, E. Linares, J. M. Pouget, S. C. Sabharwal, E. Chesler and H. DeMots, *N. Engl. J. Med.*, 1983, **309**, 396.
31. R. K. Nyak, R. D. Smyth, A. Polk, T. Herczeg, V. Carter, A. J. Visalli and N. H. Reavey-Cantwell, *J. Pharmacokinet. Biopharm.*, 1977, **5**, 597.
32. L. Rivera-Calimlim, C. A. Dujovne, J. P. Morgan, L. Lasagna and J. R. Bianchine, *Eur. J. Clin. Invest.*, 1971, **1**, 313.
33. R. L. Lucarotti, J. L. Colaizzi, H. Barry, III and R. I. Poust, *J. Pharm. Sci.*, 1972, **61**, 903.
34. K. Kakemi, H. Sezaki, M. Hayashi and T. Nadai, *Chem. Pharm. Bull.*, 1968, **16**, 2206.
35. H. Poiger and C. Schlatter, *Naunyn-Schmiedeberg's Arch. Pharmacol.*, 1979, **306**, 89.
36. P. Kearney and C. Marriott, *Int. J. Pharm.*, 1987, **35**, 211.
37. M. Garty and A. Hurwitz, *Clin. Pharmacol. Ther.*, 1980, **28**, 203.
38. P. J. Neuvonen and H. Turakka, *Eur. J. Clin. Pharmacol.*, 1974, **7**, 357.
39. P. J. Neuvonen, P. J. Pentikainen and G. Gothoni, *Br. J. Clin. Pharmacol.*, 1975, **2**, 94.
40. J. C. McElnay, H. A. Mukhtar, P. F. D'Arcy, D. J. Temple and P. S. Collier, *J. Trop. Med. Hyg.*, 1982, **85**, 159.
41. S. G. Carruthers and C. A. Dujovne, *Clin. Pharmacol. Ther.*, 1980, **27**, 184.
42. W. J. F. van der Vijgh, J. H. Fast and J. E. Lunde, *Drug Intell. Clin. Pharm.*, 1976, **10**, 680.
43. E. Jahnchen, T. Meinertz, H.-J. Gilfrich, F. Kersting and U. Groth, *Br. J. Clin. Pharmacol.*, 1978, **5**, 437.
44. D. B. Hunninghake, S. King and K. LaCroix, *Int. J. Clin. Pharmacol. Biopharm.*, 1982, **20**, 151.
45. R. E. Kauffman and D. L. Azarnoff, *Clin. Pharmacol. Ther.*, 1973, **14**, 886.
46. P. J. Neuvonen and E. Elonen, *Eur. J. Clin. Pharmacol.*, 1980, **17**, 51.
47. W. E. Fann, J. M. Davis, D. S. Janowsky, H. J. Sekerke and D. M. Schmidt, *J. Clin. Pharmacol. New Drugs*, 1973, **13**, 388.

48. R. Verbeeck, T. B. Tjandramaga, A. Mullie, R. Verbesselt and P. J. De Schepper, *Br. J. Clin. Pharmacol.*, 1979, **7**, 519.
49. V.F. Naggar and S. A. Khalil, *Clin. Pharmacol. Ther.*, 1979, **25**, 857.
50. F. Jamali and J. E. Axelson, *J. Pharm. Sci.*, 1978, **67**, 466.
51. A. Melander and A. McLean, *Clin. Pharmacokinet.*, 1983, **8**, 286.
52. E. S. Vesell, *Clin. Pharmacol. Ther.*, 1984, **36**, 285.
53. A. Clancy, J. Locke-Haydon, R. J. Cregeen, M. Ireson and J. Ziemniak, *Eur. J. Clin. Pharmacol.*, 1987, **32**, 103.
54. L. D. Bornemann, T. Crews, S. S. Chen, S. Twardak and I. H. Patel, *J. Clin. Pharmacol.*, 1986, **26**, 55.
55. K. Hirasawa, W. F. Shen, D. T. Kelly, G. Roubin, K. Tateda and J. Shibata, *Eur. J. Clin. Pharmacol.*, 1985, **28**, 105.
56. D. P. Reitberg, S. J. Love and M. Zinny, *Clin. Pharmacol. Ther.*, 1985, **37**, 223.
57. M. J. Herfst and F. A. De Wolff, *Eur. J. Clin. Pharmacol.*, 1982, **23**, 75.
58. J. H. G. Jonkman, W. J. V. Van der Boon, L. P. Balant and J. Y. Le Contonnec, *Eur. J. Clin. Pharmacol.*, 1985, **28**, 225.
59. S. Pedersen, *Br. J. Clin. Pharmacol.*, 1981, **12**, 904.
60. L. Hendeles, P. Wubbena and M. Weinberger, *Lancet*, 1984, **2**, 1471.
61. L. Vaughan, G. Milavetz, M. Hill, M. Weinberger and L. Hendeles, *Drug Intell. Clin. Pharm.*, 1984, **18**, 510.
62. B. G. Woodcock, N. Kraemer and N. Rietbrock, *Br. J. Clin. Pharmacol.*, 1986, **21**, 337.
63. J. Kindler, P. C. Ruegg, M. Neuray and W. Pacha, *Eur. J. Clin. Pharmacol.*, 1987, **32**, 367.
64. J. Rutland, N. Berend and G. E. Marlin, *Br. J. Clin. Pharmacol.*, 1979, **8**, 343.
65. M. R. Kelly, R. E. Cutler, A. W. Forrey and B. M. Kimpel, *Clin. Pharmacol. Ther.*, 1974, **15**, 178.
66. B. Beermann and C. Midskov, *Eur. J. Clin. Pharmacol.*, 1986, **29**, 725.
67. A. Melander, K. Danielson, A. Hanson, B. Rudell, B. Schersten, T. Thulin and E. Wahlin, *Clin. Pharmacol. Ther.*, 1977, **22**, 104.
68. A. M. M. Shepherd, N. A. Irvine and T. M. Ludden, *Clin. Pharmacol. Ther.*, 1984, **36**, 14.
69. J. G. Nutt, W. R. Woodward, J. P. Hammerstad, J. H. Carter and J. L. Anderson, *N. Engl. J. Med.*, 1984, **310**, 483.
70. C. R. Pinkerton, S. G. Welshman, J. F. T. Glasgow and J. M. Bridges, *Lancet*, 1980, **2**, 944.
71. D. K. Moses, B. G. Charles, P. J. Ravenscroft and I. M. Whyte, *Br. J. Clin. Pharmacol.*, 1983, **16**, 758.
72. G. V. Gill, *Lancet*, 1976, **2**, 1135.
73. A. Tulpule and K. Krishnaswamy, *Eur. J. Clin. Pharmacol.*, 1982, **23**, 271.
74. J. J. DiGiovanna, E. G. Gross, S. W. McClean, M. E. Ruddel, G. Gantt and G. L. Peck, *J. Invest. Dermatol.*, 1984, **82**, 636.
75. W. A. Colburn, D. M. Gibson, L. C. Rodriguez, C. J. L. Bugge and H. P. Blumenthal, *J. Clin. Pharmacol.*, 1985, **25**, 583.
76. C. M. Ginsburg, G. H. McCraken, M. Petruska and K. Olsen, *J. Pediatr. (St. Louis)*, 1983, **102**, 309.
77. T. K. Daneshmend and C. J. C. Roberts, *Br. J. Clin. Pharmacol.*, 1982, **14**, 73.
78. T. R. Bates, J. A. Sequeira and A. V. Tembo, *Clin. Pharmacol. Ther.*, 1974, **16**, 63.
79. H. A. Rosenberg and T. R. Bates, *Clin. Pharmacol. Ther.*, 1976, **20**, 227.
80. A. Melander, G. Brante, O. Johansson, T. Lindberg and E. Wahlin-Boll, *Eur. J. Clin. Pharmacol.*, 1979, **15**, 269.
81. E. Melander, K. Danielson, B. Schersten and E. Wahlin, *Clin. Pharmacol. Ther.*, 1977, **22**, 108.
82. A. J. McLean, C. Isbister, A. Bobik and F. J. Dudley, *Clin. Pharmacol. Ther.*, 1981, **30**, 34.
83. G. P. Jackman, A. J. McLean, G. L. Jennings and A. Bobik, *Clin. Pharmacol. Ther.*, 1981, **30**, 291.
84. A. Melander, K. Danielson, B. Schersten, T. Thulen and E. Wahlin, *Clin. Pharmacol. Ther.*, 1977, **22**, 100.
85. J. Hasegawa, Y. Tomono, T. Fujita, K. Sugiyama and K. Hamamura, *Int. J. Clin. Pharmacol. Biopharm.*, 1981, **19**, 216.
86. M. Toutoungi, R. L. A. Carroll, J.-F. Enrico and L. Perey, *Lancet*, 1985, **2**, 671.
87. A. Samanta, G. R. Jones, A. C. Burden and I. Shakir, *Br. J. Clin. Pharmacol.*, 1984, **18**, 647.
88. J. A. Zallman, D. P. Lee and P. L. Jeffrey, *Am. J. Hosp. Pharm.*, 1981, **38**, 1174.
89. S. J. Kempin, *N. Engl. J. Med.*, 1983, **308**, 1229.
90. W. G. Berlinger, G. D. Park and R. Spector, *N. Engl. J. Med.*, 1985, **313**, 771.
91. C. A. Naranjo, E. M. Sellers and V. Khouw, *Br. J. Clin. Pharmacol.*, 1980, **10**, 308.
92. E. Woo and D. J. Greenblatt, *Clin. Pharmacol. Ther.*, 1980, **27**, 188.
93. J. Spenard, G. Sirois and M. A. Gagnon, *Int. J. Clin. Pharmacol. Biopharm.*, 1983, **21**, 1.
94. U. Malt, *Br. Med. J.*, 1978, **2**, 502.
95. L. J. Robbins, *N. Engl. J. Med.*, 1983, **309**, 985.
96. J. M. Jaffe, R. I. Poust, S. L. Feld and J. L. Colaizzi, *J. Pharm. Sci.*, 1974, **63**, 1256.
97. P. J. Neuvonen and S. Karkkainen, *Clin. Pharmacol. Ther.*, 1983, **33**, 386.
98. G. Edwards, A. M. Breckenridge, K. K. Adjepon-Yamoah, M. L. 'E. Orme and S. A. Ward, *Br. J. Clin. Pharmacol.*, 1981, **12**, 807.
99. M.-I. Nilsson, E. Widerloy, U. Meresaar and E. Anggard, *Eur. J. Clin. Pharmacol.*, 1982, **22**, 337.
100. J. Kampmann, J. M. Hansen, K. Sierbaek-Nielsen and H. Laursen, *Clin. Pharmacol. Ther.*, 1972, **13**, 516.
101. N. Baber, L. Halliday, T. Littler, M. L. 'E. Orme and R. Sibeon, *Br. J. Clin. Pharmacol.*, 1978, **5**, 364P.
102. C. D. Christian, C. G. Meredith and K. V. Speeg, *Clin. Pharmacol. Ther.*, 1984, **36**, 221.
103. K. E. Pedersen, A. Dorph-Pedersen, S. Hvidt, N. A. Klitgaard and F. Nielsen-Kudsk, *Clin. Pharmacol. Ther.*, 1981, **30**, 311.
104. K. E. Pedersen, A. Dorph-Pedersen, S. Hvidt, N. A. Klitgaard and K. K. Pedersen, *Eur. J. Clin. Pharmacol.*, 1982, **22**, 123.
105. R. T. Taylor and M. J. T. Peaston, *Br. Med. J.*, 1973, **3**, 48.
106. J. J. Haggerty and D. A. Drossman, *Psychosomatics*, 1985, **26**, 277.
107. P. Hansten, *Drug Interactions Newsletter*, 1981, **1**, 47.
108. I. W. Reimann, U. Diener and J. C. Frolich, *Arch. Gen. Psychiatry*, 1983, **40**, 283.
109. F. J. Ayd, *Int. Drug Ther. Newsletter*, 1978, **13**, 15.
110. R. J. Kerry, J. M. Ludlow and G. Owen, *Br. Med. J.*, 1980, **281**, 371.
111. J. M. Himmelhock, J. F. Neil and A. G. Mallinger, *Drug Therapy*, 1978, **8**, 9.
112. J. R. Castillo-Ferrando, M. Garcia and J. Carmona, *Lancet*, 1980, **2**, 368.
113. P. E. Fenster, W. D. Hager and M. M. Goodman, *Clin. Pharmacol. Ther.*, 1984, **36**, 70.
114. L. Favre, P. Glasson and M. B. Vallotton, *Ann. Intern. Med.*, 1982, **96**, 317.
115. W. J. Tilstone, P. F. Semple, D. H. Lawson and J. A. Boyle, *Clin. Pharmacol. Ther.*, 1977, **22**, 389.

116. R. A. O'Reilly, *Clin. Pharmacol. Ther.*, 1980, **27**, 198.
117. R. A. O'Reilly, M. A. Sahud and P. M. Aggeler, *Ann. N.Y. Acad. Sci.*, 1971, **179**, 173.
118. M. Rowland and T. N. Tozer, in 'Clinical Pharmacokinetics', Lea and Febiger, Philadelphia, 1980, p. 249.
119. P. Douste-Blazy, J. L. Montastruc, B. Bonnet, P. Auriol, D. Conte and P. Bernadet, *Lancet*, 1984, **1**, 905.
120. D. G. Fraser, T. M. Ludden, R. P. Evens and E. W. Sutherland, *Clin. Pharmacol. Ther.*, 1980, **27**, 165.
121. J. W. Paxton, *Clin. Pharmacol. Ther.*, 1980, **27**, 170.
122. J. M. Orr, F. S. Abbott, K. F. Farrell, S. Ferguson, I. Sheppard and W. Godolphin, *Clin. Pharmacol. Ther.*, 1982, **31**, 642.
123. G. Das, C. E. Barr and J. Carlson, *Clin. Pharmacol. Ther.*, 1984, **35**, 317.
124. H. E. Barber, J. Petersen and P. H. Whiting, *Br. J. Clin. Pharmacol.*, 1981, **11**, 113P.
125. E. Perucca, S. Hebdige, G. M. Frigo, G. Gatti, S. Lecchini and A. Crema, *Clin. Pharmacol. Ther.*, 1980, **28**, 779.

23.7

Stereoselectivity in Pharmacokinetics and Drug Metabolism

MARTIN S. LENNARD, GEOFFREY T. TUCKER and H. FRANK WOODS
Royal Hallamshire Hospital, Sheffield, UK

23.7.1 INTRODUCTION

A high proportion of synthetic and naturally occurring chemicals have one or more chiral centres in their structure. Drugs are no exception and Ariens *et al.*[1] have estimated that nearly 60% of compounds in clinical use are optically active. Studies *in vitro* and *in vivo* have indicated that for most drugs with a single centre of asymmetry only one enantiomer has significant pharmacological activity.

Almost all naturally occurring or semisynthetic drugs are marketed as single isomers. However, the majority (70%) of drugs are synthetic[1] and, unless chiral reagents or chiral catalysts are employed, chemical synthesis of drugs with one chiral centre will produce a 50/50 mixture (racemate) of the two enantiomers. About 40% of synthetic drugs are chiral and of these almost 90% are used as racemates.[1]

There has been considerable debate[2] as to whether chiral drugs should be used only as the enantiomers. Thus, Ariens *et al.* have discussed the potential hazards of administering racemates containing 50% of one isomer which has no therapeutic benefit and which may be toxic or which may enhance the toxicity of the active isomer.[1] The example of thalidomide has been used to promote the case for pure enantiomer development and administration.[3] From studies showing that (+)- and (−)-thalidomide* were equipotent as sedatives in the mouse but only the (−) enantiomer was teratogenic in mice and rats,[4] it has been suggested that the thalidomide disaster might have been avoided if only the (+) enantiomer had been used therapeutically. However, this argument is weakened by the observation that (+)-, (−)- and (±)-thalidomide were all teratogenic in the New Zealand white rabbit.[5] This species is a better model than the rat for thalidomide teratogenicity in humans.[6]

Whatever the arguments for or against the development and therapeutic use of drug enantiomers, much useful information may be gained from monitoring the pharmacologically active drug species in whatever form it is administered. As a consequence of recent advances in analytical methodology, many data have been published on the pharmacokinetic and metabolic behaviour of drug enantiomers when given as racemates.

This chapter documents both the extent and complex nature of stereoselective drug disposition. It is not intended to be a comprehensive review of the literature but concentrates upon the phenomenon as it affects those drugs which are commonly used. The special case of polymorphic drug oxidation will also be considered.

23.7.2 STEREOSPECIFIC ANALYSIS OF ENANTIOMERS

A discussion of the analytical methodology available for drug enantiomer analysis is outwith the scope of this chapter. The reader is referred to the comprehensive text by Souter[7] and to other reviews.[8,9]

23.7.3 β-ADRENOCEPTOR ANTAGONISTS (see also Section 23.7.7)

All of the β-adrenoceptor antagonists contain at least one chiral centre located in the propanolamine side chain. With the exceptions of timolol and penbutolol, they are used clinically as racemates, although only the (−) forms possess significant β-blocking activity.[10]

23.7.3.1 Propranolol

23.7.3.1.1 Pharmacokinetics

The pharmacokinetics of the enantiomers of propranolol have been investigated by several groups following administration of a racemate or pseudoracemate (Figure 1).[12–15] All have found that the systemic and oral clearances of the (S) enantiomer were lower than those of the (R) enantiomer, resulting in higher plasma concentrations of the pharmacologically active form. However, the

---

* Where the absolute configuration of an enantiomeric pair has been published, the (RS) system of nomenclature[7] will be used throughout this chapter. Otherwise, enantiomers will be identified using the prefixes (+) or (−), which refer to the direction in which individual enantiomers rotate plane polarized light.

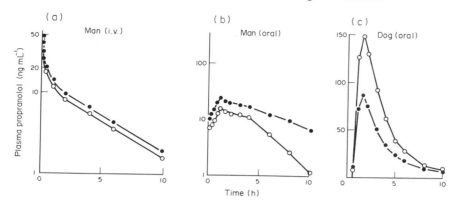

Figure 1 Plasma concentration–time profiles of (*S*)-propranolol (●) and of (*R*)-propranolol (○) after: (a) a single intravenous dose (0.1 mg kg^{-1}) of a deuterium-labelled pseudoracemate (mean data from five subjects); (b) oral administration of 160 mg of the racemate per day for 13 days (data from one subject); and (c) oral administration of 80 mg of a deuterium-labelled pseudoracemate every 6 h for one week (mean data from six dogs) (adapted from refs. 13, 14 and 27)

one- to two-fold variation in plasma propranolol enantiomer ratios should be put into context with the much higher interindividual variation (about 30-fold) in total (*R*)+(*S*) plasma drug concentrations.[15]

The inhibition of oral propranolol clearance by cimetidine is stereoselective for (*R*)-propranolol,[16] thereby minimizing the difference between the elimination of the enantiomers. Propranolol is itself an inhibitor of oxidative drug metabolism in the rat[41] and in humans.[42] However, the inhibition of lidocaine metabolism by (*R*)- and (*S*)-propranolol is not stereoselective.[43]

Propranolol is a 'high extraction' drug, the systemic clearance of which should be dependent mainly upon liver blood flow.[17] Since (*S*)-propranolol lowers cardiac output and liver blood flow, this isomer should decrease its own clearance and that of the inactive (*R*) enantiomer to the same extent when the racemate is administered. Accordingly, the systemic clearance of (*R*)-propranolol is lower when given as part of a racemate than when given alone.[18] However, the lower systemic clearance of (*S*)- compared to (*R*)-propranolol in humans following racemic drug administration[14] cannot be explained by differences in flow and must reflect differential enzyme activity.

The differences between the plasma enantiomer concentrations of propranolol were found to be much larger after oral than after intravenous administration of the racemate (Figure 1). As previously stated, the intravenous kinetics of propranolol are controlled more by liver blood flow than by hepatic enzyme activity, whereas the kinetics after oral dosing are enzyme controlled. Since differences in isomer clearance are due only to enzyme degradation, the stereoselectivity will be more apparent after oral dosing. This route dependent stereoselectivity together with the production of active metabolites on the first pass probably accounts for the two- to three-fold greater potency of propranolol when given orally.[19] In contrast, low extraction drugs, whose metabolism is enzyme controlled irrespective of route of administration, will show similar differences in enantiomer kinetics after both intravenous and oral dosing.

Stereoselectivity in the plasma protein binding of propranolol[20,21] partially explains an enantiomeric difference in volume of distribution.[14] The mean percentage unbound of (*R*)-propranolol is greater than that of (*S*)-propranolol (25% *vs.* 22%), which is compatible with the larger volume of distribution of the (*S*) isomer. α_1-Acid glycoprotein, the major binding protein for propranolol, also binds the (*R*) enantiomer preferentially, whereas the reverse stereoselectivity was observed for binding to human serum albumin. There is no difference in the uptake of the enantiomers into red blood cells. On the other hand, tissue binding may be stereoselective. There is evidence that the more active (*S*) enantiomers of propranolol and other β-adrenoceptor antagonists undergo preferential uptake into and release from adrenergic neurones in the heart.[69]

23.7.3.1.2 Metabolism

The metabolism of propranolol proceeds through three major pathways: direct glucuronidation, aromatic ring hydroxylation and side chain oxidation.[11] The oxidation products are further metabolized to their corresponding glucuronides and sulfates. Estimates of metabolic clearance *via* direct glucuronidation, a relatively minor pathway, indicate stereoselectivity in favour of (S)-propranolol.[13] Since the total clearance of (R)-propranolol is greater than that of (S)-propranolol, the opposite stereoselectivity would be expected for clearance by aromatic ring or side chain oxidation. An (S)/(R) ratio of 0.59 for the urinary recovery of total ring oxidation products[22] supports this view. This study also suggested that glucuronidation of 4′-hydroxypropranolol, a product of ring oxidation, is stereoselective. However, administration of 4′-hydroxypropranolol itself would be required to confirm this.

In vitro studies using human liver microsomes broadly support the *in vivo* findings. Von Bahr *et al.*[23] reported a more rapid formation of 4′-hydroxy- and N-deisopropyl-propranolol metabolites from the (R) than from the (S) enantiomer. In contrast, Nelson and Shetty[24] did not observe stereoselectivity in N dealkylation or 4′-hydroxylation but found a marked but variable preference for (R)-propranolol in the formation of 5′-hydroxypropranolol, another product of aromatic hydroxylation. The ratio of the 5′- to 4′-hydroxy metabolite was also highly variable between individuals. Nelson and Shetty[24] attributed the discrepancy between their data and those of von Bahr *et al.*[23] to possible coelution of the hydroxy metabolites during HPLC analysis.

23.7.3.1.3 *Comparison between species*

Humans and dogs show opposite stereoselectivity in their ability to clear propranolol. Thus, in the dog (S)-propranolol is eliminated more rapidly than (R)-propranolol after a single dose (Figure 1).[25,26] This stereoselectivity was less apparent on chronic dosing because of a preferential increase in the bioavailability of (S)-propranolol.[27]

The binding of propranolol enantiomers to plasma is similar in dogs and humans, with the former also having a higher free fraction of the (R) than of the (S) isomer.[26] If plasma binding was the only factor associated with the distribution of the drug, (R)-propranolol would be expected to have a higher total (free plus bound) volume of distribution. However, the opposite was observed in the dog, the volume of distribution being greater for (S)-propranolol. The presence of stereoselective extravascular binding suggested by this observation was confirmed by measurement of the volume of distribution of the free enantiomers. Thus, this value was higher for (S)- than for (R)-propranolol,[26] suggesting stereoselectivity in the tissue binding of propranolol.

Based upon measurements of urinary metabolite excretion, it has been suggested that the higher clearance of (S)-propranolol in the dog is due to stereoselectivity in glucuronidation and ring oxidation[31] and that the ring oxidation of (S)-propranolol is selectively impaired during chronic dosing.[27] Confirmation of these mechanisms awaits measurements of partial clearance down each pathway.

The *in vivo* data are supported by *in vitro* studies using dog liver microsomes showing a more rapid glucuronidation[28] and 4′-hydroxylation[23] of (S)-propranolol. In contrast, N deisopropylation was selective for the (R) enantiomer.[23]

In rat liver microsomes the metabolism of propranolol *via* 5′-hydroxylation also favoured the (S) enantiomer but stereoselectivity was not observed in 4′-hydroxylation or N deisopropylation.[29–31] The metabolism of propranolol enantiomers by reconstituted cytochrome P-448L from rat liver was similar to that observed in microsomes.[32] Complex substrate and product stereoselectivities were observed using rat liver microsomes for a new metabolic pathway of propranolol metabolism, in which the terminal carbon of the N-isopropyl group is hydroxylated to form a second chiral centre.[33] Christ and Walle[34] found marked species differences in the stereoselectivity of the sulfate conjugation of 4′-hydroxypropranolol. Using a 178 000 g liver supernatant they observed preferential conjugation of the (R) enantiomer in the hamster and the dog, whereas (S)-propranolol was sulfated more rapidly in the rat. The cat exhibited no stereoselectivity in this pathway.

23.7.3.2 Other β-Adrenoceptor Antagonists

Plasma concentrations of the (−) enantiomers of several other β blockers that are eliminated primarily by oxidative metabolism are higher than those of their (+) enantiomers. This has been demonstrated for penbutolol,[35] alprenolol,[36] metoprolol (see Section 23.7.7.1.1),[37] bufuralol (see Section 23.7.7.1.2),[38] moprolol[39] and xibenolol.[40] The pharmacokinetics of acebutolol, which is cleared mainly by hydrolysis and subsequent acetylation, are not stereoselective.[44]

Renal excretion of unchanged β blockers accounts for a variable proportion of their elimination.[11] Little information is available on the possible stereoselectivity of this process. However, differences in the renal clearances of the enantiomers of pindolol and metoprolol have been reported. Thus, Hsyu and Giacomini[45] found a 16% higher renal clearance of (−)-pindolol when compared to that of the (+) enantiomer. This increased to 30% when the net clearance by secretion was calculated. It was suggested that stereoselective renal transport may be responsible for these findings, as stereoselective renal metabolism was thought to be unlikely and there was no stereoselectivity in plasma drug binding. In contrast to the pindolol data, the renal clearance of (R)-metoprolol was found to be about 10% greater than that of the (S) enantiomer.[37] Assuming that the absolute configurations of pindolol and metoprolol are the same, this is difficult to reconcile with a common stereoselective renal transport process.

23.7.4 ANTIARRHYTHMIC DRUGS

23.7.4.1 Disopyramide

Stereoselectivity in the pharmacokinetics of disopyramide has been found after administration of both the racemate[46] (Figure 2) and a pseudoracemate.[47] The systemic and renal clearances of (R)-disopyramide, the enantiomer having negligible antiarrhythmic activity, were about 50% and 75% higher than those of (S)-disopyramide, respectively. The volume of distribution at steady state of the (R) isomer was also greater than that of the (S) isomer. These differences were explained by greater plasma protein binding of (S)-disopyramide. The binding of the enantiomers was also dose dependent.[48] Thus, at high plasma concentrations the free fraction of both enantiomers in the plasma would be expected to approach unity, resulting in a loss of stereoselectivity in the pharmacokinetics.

The elimination half-life of total (free plus bound) (R)-disopyramide was significantly longer than that of (S)-disopyramide, reflecting a greater stereoselective difference in distribution than in clearance. Theory predicts that stereoselective plasma protein binding alone would not cause a difference in the half-life of the enantiomers and it was suggested that stereoselective binding to tissues and/or a difference in the unbound clearance of the enantiomers may contribute. Support for the latter possibility comes from the observation of higher unbound systemic and renal clearances of (S)- compared to (R)-disopyramide after separate administration of the enantiomers.[49]

Giacomini et al.[47] did not detect stereoselectivity in elimination of either free or bound drug after separate dosing of the enantiomers. The difference between the results of this experiment and those of the pseudoracemate studies may be explained by a pharmacokinetic interaction between enantiomers involving competition for plasma binding sites.

The elimination of disopyramide isomers in the rat and dog[50] is similar to that in humans, with the (S) enantiomer being cleared more slowly and undergoing a more extensive first pass metabolism. The results of this study[50] were also consistent with stereoselectivity in N dealkylation in the dog but not in the rat and stereoselectivity of aromatic hydroxylation in the rat but not in the dog.

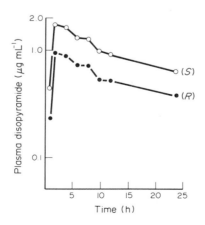

Figure 2   Plasma concentration–time profiles of (R)-disopyramide and of (S)-disopyramide after administration of a single oral dose (150 mg) of the racemate to one subject (reproduced from ref. 46 by permission of Elsevier)

23.7.4.2 Mexiletine

Following oral administration of the racemate of mexiletine, Grech-Belanger et al.[51] found that the plasma concentrations and half-life of (S)-mexiletine were higher than those of (R)-mexiletine. There was no significant difference between the renal clearance of the enantiomers. The conjugation of mexiletine, the products of which accounted for 28% of the dose recovered in urine, was virtually stereospecific for the (R) enantiomer. The renal clearances of the (R) conjugates were also higher than those of the (S) products. Mexiletine is a low extraction drug whose total clearance should be a function of both free fraction and intrinsic clearance. Since the total clearance of (R)-mexiletine is greater than that of the (S) form, yet plasma binding of (R)- exceeds that of (S)-mexiletine,[52] the intrinsic clearance of the (R) must be greater than that of the (S) enantiomer.

23.7.4.3 Tocainide

Stereoselectivity in the disposition of racemic tocainide has been described. Thus, Sedman et al.[53] reported consistently higher plasma (S)- than (R)-tocainide concentrations in seven patients during chronic oral therapy. The (S)/(R) ratio varied considerably both within and between individuals (range 1.3 to 4.0). No stereoselectivity was found in the volume of distribution at steady state[54] and only minor differences in the serum binding of the enantiomers were detected.[55] Therefore, it is unlikely that a difference in distribution contributed to the enantiomeric differences in plasma drug concentrations. A lower clearance of the (S) enantiomer was recorded in healthy volunteers after single oral doses[56] and in seriously ill patients receiving an intravenous infusion.[54] In these patients there was also a progressive increase in the plasma (S)/(R) concentration ratio with time. Up to 50% of a dose of tocainamide is cleared unchanged by the kidney with most of the remainder being metabolized to N-carboxytocainide and its glucuronide.[57] The renal clearances of the enantiomers were found to be similar[56] but urinary excretion studies have indicated a stereoselective formation of N-carboxytocainide from the (R) enantiomer.[58] Thus, it seems likely that stereoselective metabolism of tocainide explains the differences in plasma enantiomer kinetics.

The excretion of tocainide enantiomers into saliva has been studied by Pillai et al.[59] after intravenous infusion of the drug. A saliva to plasma ratio of greater than unity was found for both isomers and this ratio was pH dependent. In contrast to the plasma and urine data, salivary concentrations of (R)-tocainide were higher than those of (S)-tocainide for about 24 h after dosing. After this time the enantiomer ratio was reversed.

Species and sex differences in the urinary excretion of tocainide enantiomers and their metabolites have been observed.[60] The (R)/(S) ratio of renally excreted drug was higher in mice (> 1) than in rats (< 1). Pretreatment with phenobarbitone decreased the urinary recovery of (S)-tocainide in male and female mice and in male rats. The urine recovery of (R)-tocainide was decreased in male rats only. The excretion of tocainide enantiomers in female rats was unaffected by pretreatment with phenobarbitone. Reduction of 2-oxopropiono-2′,6′-xylidide, the achiral product of the oxidative deamination of tocainide, by rabbit and rat liver cytosolic fractions to the chiral alcohol, 2-hydroxypropiono-2′,6′-xylidide, was virtually specific for the (S) configuration.

23.7.4.4 Quinidine and Quinine

Quinidine and quinine are diastereoisomers with antiarrhythmic and antimalarial activities. Both drugs are metabolized extensively but a significant proportion of a dose is excreted unchanged in urine (quinidine 15–40%;[61] quinine 20%[62]). Notterman et al.[63] reported fourfold higher renal clearances of total and unbound quinidine compared to quinine when the drugs were given together or separately. Passive reabsorption was minimized by acidifying the urine. This difference in renal clearance was associated with higher total plasma concentrations of quinine. It was concluded that stereoselective renal clearance occurred through preferential net renal tubular secretion of quinidine.

In sheep the elimination of quinidine and quinine showed the opposite stereoselectivity to that in humans.[64] Plasma concentrations of quinidine were double those of quinine when the drugs were administered separately (10 mg kg^{-1}, intravenously). The effect of each isomer on the metabolism of the other has been examined in the isolated perfused rat liver after separate dosing.[65] Quinidine was cleared at a similar rate to that of quinine. However, after combined dosing the clearances of quinine and quinidine were decreased by 24% and 55%, respectively, indicating that the inhibition of quinidine metabolism by quinine was greater than the inhibition of quinine metabolism by quinidine. This observation is in marked contrast to the finding that quinidine is a more potent inhibitor than quinine of the metabolism of debrisoquine and metoprolol in humans (see Section 23.7.7.1.1).

23.7.4.5 Verapamil

After intravenous dosing with a pseudoracemate, plasma concentrations of (−)-verapamil, the more potent enantiomer *in vitro*, were found to be about half those of (+)-verapamil.[66] This was due to a twofold higher clearance of the (−) enantiomer. However, no difference in the half-lives of the enantiomers was observed. This may be explained by a compensatory effect of lower plasma protein binding and thus a higher volume of distribution of the (−) enantiomer. Since racemic verapamil is a 'high extraction' drug, the difference between the clearance of the enantiomers is magnified after oral administration compared to intravenous administration of the racemate. Thus, plasma concentrations of (−)-verapamil are five times less than those of (+)-verapamil after a single oral dose.[67] This accounts for the lower potency of oral compared to intravenous verapamil, since stereoselective first-pass metabolism favours the more active isomer.[68]

23.7.5 ANTICOAGULANT DRUGS

23.7.5.1 Warfarin

23.7.5.1.1 *Pharmacokinetics*

Marked stereoselective differences in warfarin pharmacokinetics and metabolism after single and chronic dosing have been observed when the enantiomers were given either separately or together.[70–73] The clearance of (S)-warfarin, the more active isomer, was found to be about 60% greater than that of the (R) form after oral dosing. The half-life of (S)-warfarin was correspondingly shorter than that of (R)-warfarin. On multiple dosing, stereoselectivity is enhanced as the half-life of (R)-warfarin increases.[70] No difference in the apparent volume of distribution of the enantiomers was observed, which is compatible with a lack of stereoselectivity in the binding of warfarin to human serum albumin.[74,75] However, Yacobi and Levy[76] noted that the free fraction of (R)-warfarin in human serum was higher than that of (S)-warfarin.

23.7.5.1.2 *Metabolism*

Urinary excretion studies show that the major pathways of warfarin metabolism in humans are hydroxylation of the aromatic ring and reduction to secondary alcohols.[77] 7-Hydroxylation is quantitatively more important for the (S) enantiomer, whereas alcohol formation accounts for most of the metabolism of (R)-warfarin in humans.[73] The latter reaction also introduces a second chiral centre into the molecule and seems to be specific for the (S) configuration.

Stereoselectivity in the cytochrome *P*-450-mediated hydroxylation of warfarin has been studied *in vitro* and the results complement those found *in vivo*. In human liver microsomes the formation of 6-, 8- and 10-hydroxywarfarin occurred more rapidly from (R)-warfarin, whereas 4′- and 7-hydroxylation favoured the (S) enantiomer;[78] 6- and 7-hydroxywarfarin were the major products. Considerable interindividual variability between livers in the ratio of enantiomers and also in the rates of total (R)+(S) metabolite formation was observed. In further studies, stereoselectivity and regioselectivity of warfarin metabolism have been used to define the role of purified forms of cytochrome *P*-450 in the overall metabolism of the drug.[79]

23.7.5.1.3 *Drug interactions*

Many drug interactions have been described involving warfarin, some of which have led to severe hypothrombinaemia.

The mechanism of the interaction with phenylbutazone is well understood.[73,77,80] The increased anticoagulant effect seen during coadministration of the drugs is associated with a marked decrease in the clearance of (S)-warfarin and an increase in the clearance of (R)-warfarin, resulting in little change in total (S)+(R) warfarin clearance. This finding suggested a simultaneous impairment of (S)-warfarin metabolism and induction of (R)-warfarin metabolism. However, displacement of

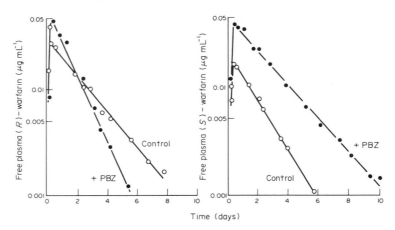

Figure 3 Plasma concentration–time profiles of unbound (R)-warfarin and of (S)-warfarin after oral administration of the racemate (1.5 mg kg^{-1}) to a subject before (control, ○) and at the end of treatment with phenylbutazone (100 mg three times per day for four days; PBZ, ●) (adapted from ref. 73)

warfarin from plasma albumin obscures the metabolic picture. Thus, correction of clearance values for plasma binding showed that there was a pronounced inhibition of the unbound clearance of the more active (S) isomer with a smaller inhibition of the unbound clearance of the (R) form (Figure 3).[73] These changes are accompanied by a decrease in the half-life of (R)- and an increase in that of (S)-warfarin.

The interactions between warfarin and sulphinpyrazone,[81] metronidazole,[82] cotrimoxazole[83] and ticrynafen,[84] which all result in increased anticoagulation, show similar characteristics. In all cases, decreased clearance leads to an elevation of the plasma concentrations of (S)-warfarin, whereas those of the (R) enantiomer were either unchanged or decreased. A combination of cimetidine and warfarin is also accompanied by prolongation of the prothrombin time.[104] However, in this case, inhibition of metabolism was virtually stereospecific for (R)-warfarin, the less active isomer.[86,87] Toon *et al.*[88] reported that enoxacin also preferentially lowers the clearance of (R)-warfarin as a result of inhibition of its 6-hydroxylation. However, no change in anticoagulation was detected. In contrast, amiodarone enhanced the anticoagulant effect of warfarin through a nonstereoselective mechanism.[89]

Interactions which lead to a lowered anticoagulant effect have also been reported. Thus, the decrease in hypothrombinaemia during warfarin and secobarbital[90] or rifampicin[91] coadministration was probably due to an increase in the clearance of both isomers of warfarin. In contrast, the augmentation of the anticoagulant effect of warfarin by clofibrate[92] and disulfiram[93] appears to have a pharmacodynamic rather than a pharmacokinetic basis.

23.7.5.1.4 *Comparison between species*

There is a striking species difference between humans and rats in their stereoselective disposition of warfarin. In the rat (S)-warfarin is cleared more slowly than (R)-warfarin,[94,95] whereas the opposite is observed in humans (see Section 23.7.5.1.1). Stereoselectivity in the metabolism of warfarin in the rat is in the same direction as, and amplifies the differences in, the intrinsic potencies of the enantiomers. By contrast, in humans the pharmacokinetic component of the difference in stereoselective action dampens the effect of the pharmacodynamic component.

There is no stereoselectivity in the apparent volume of distribution of warfarin in the rat[94] or in humans. However, an (R)/(S) ratio significantly greater than one has been reported for the free fraction of warfarin in rat and human plasma.[76] This could contribute to the higher clearance of (R)-warfarin in the rat but would not account for the higher clearance of the (S) isomer in humans (see Section 23.7.5.1.1).

The species difference in enantiomer clearance is associated with differences in the major pathways of warfarin metabolism. Although aromatic hydroxylation is the major pathway of metabolism of (S)-warfarin in both species, this reaction occurs mainly at the 4′-position in the rat.[96,97] In contrast to humans, 7-hydroxylation is the major route of (R)-warfarin metabolism in the rat.

The effects of chloramphenicol[98] and metronidazole[99] on warfarin enantiomer pharmacokinetics and pharmacodynamics have been studied in rats. Chloramphenicol potentiates anticoagulation by warfarin through a nonselective inhibition of the metabolism of both enantiomers. Metronidazole also increased the anticoagulant effect of warfarin as it did in humans. In this case, the mechanism underlying this interaction is probably more complex than that proposed for humans since both pharmacokinetic and pharmacodynamic changes are implicated.

23.7.5.2 Nicoumalone (Acenocoumarol)

Marked stereoselectivity in the disposition of nicoumalone has been observed following separate intravenous or oral administration of its enantiomers to healthy volunteers (Figure 4).[101,102] The total plasma clearance of (S)-nicoumalone is about 10 times that of the (R) isomer. At therapeutic doses of the racemate, plasma concentrations of the (S) enantiomer were barely detectable. Therefore, most of the anticoagulant activity of nicoumalone is contributed by the (R) isomer, which has the lower intrinsic potency.

The higher clearance of (S)-nicoumalone cannot be explained by a lower plasma binding of this isomer.[101] Racemic nicoumalone is extensively metabolized in humans by reduction of the aromatic nitro and keto groups as well as by 6- and 7-hydroxylation.[103] It is not known which of these reactions are stereoselective.

Following a single case report of enhanced anticoagulant activity of nicoumalone during cimetidine treatment,[104] controlled studies of this interaction were performed in healthy volunteers. Thijssen *et al.*[102] found cimetidine treatment (800 mg per day for four days) to have no effect on the elimination of single oral doses (5 mg) of the separate enantiomers of nicoumalone. On the other hand, Gill *et al.*,[106] using a stereospecific assay, demonstrated that cimetidine treatment (1000 mg per day for seven days) decreased the clearance of (R)-nicoumalone by 47% but did not influence that of (S)-nicoumalone following administration of a single oral dose (10 mg) of the racemate. Neither study showed any enhancement of the anticoagulant effect.

The (S) enantiomer of nicoumalone is preferentially eliminated in the rat,[107] although the degree of stereoselectivity was much less than in humans. Further work suggested stereoselectivity in the biliary excretion of the metabolites of nicoumalone[108] and in the acetylation of its amino metabolite.[107]

23.7.5.3 Phenprocoumon

The pharmacokinetics of the enantiomers of phenprocoumon display a stereoselectivity opposite to that of warfarin and nicoumalone. Thus, the plasma clearance of the (R) enantiomer is 20% higher than that of the (S) enantiomer.[109] The lower plasma binding of (R)- compared to (S)-phenprocoumon contributes to this difference. Based on pharmacokinetic and pharmacodyn-

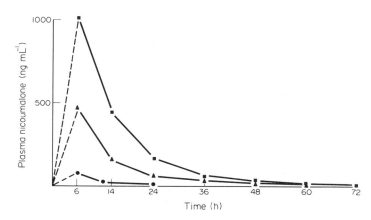

Figure 4 Plasma concentration–time profiles of nicoumalone measured by a nonstereospecific assay after the separate oral administration (0.25 mg kg^{-1}) of the racemate (▲), the (S) (●) and the (R) (■) enantiomers to a single subject (reproduced from ref. 101 by permission of Macmillan)

amic measurements, (S)-phenprocoumon was considered to be a more potent anticoagulant than (R)-phenprocoumon in humans. Stereoselectivity has also been observed in the urinary and faecal recoveries of unchanged drug and its hydroxylated metabolites.[110]

23.7.6 NONSTEROIDAL ANTIINFLAMMATORY DRUGS

Most nonsteroidal antiinflammatory drugs are 2-arylpropionic acids with a chiral centre. With the exception of naproxen, they are administered as racemates. A large discrepancy between the *in vitro* and *in vivo* potency of 2-arylpropionic acids has prompted studies into the metabolism of their enantiomers.[111] The discovery that the inactive (R) enantiomers of many of these compounds undergo chiral inversion has important implications for understanding the clinical and toxicological effects of these drugs.

23.7.6.1 Pharmacokinetics

Stereoselective metabolism is a major determinant of the pharmacokinetics of the 2-arylpropionic acids in animals and humans.[111] For example, the metabolic fate of ibuprofen has been studied in detail by Lee *et al.*[112] Separate oral administration of ibuprofen enantiomers resulted in a slower disappearance from the plasma of (S)-ibuprofen compared to (R)-ibuprofen (Figure 5). High plasma concentrations of (S)-ibuprofen were observed after dosing with (R)-ibuprofen, whereas those of the (R) enantiomer were negligible after administration of (S)-ibuprofen. When ibuprofen was given as a racemate, the mean AUC of the (S) isomer was 56% higher than that of the (R) form. The mean fraction of (R)-ibuprofen inverted to (S)-ibuprofen ranged from 0.57 to 0.71. This compares to no inversion of indoprofen[113] and to the almost complete inversion observed with fenoprofen.[114] Clearly, the pharmacokinetics of ibuprofen and other 2-arylpropionic acids have a profound damping effect on the differences in the intrinsic potencies of these drugs as inhibitors of prostaglandin synthesis.

The pharmacokinetics of ibuprofen are complicated by competition between the enantiomers for plasma binding sites. After correcting for the inversion process, Lee *et al.*[112] found that the AUCs of (R)- and (S)-ibuprofen after administration of the racemate were lower than those predicted from studies in which each isomer was given separately. This difference in AUC together with differences in the volumes of distribution of the enantiomers at steady state are compatible with the mutual competition of the enantiomers for plasma binding sites. This hypothesis is supported by *in vitro* studies showing that (R)-ibuprofen binds more avidly to human plasma than does (S)-ibuprofen.[115]

23.7.6.2 Metabolism and Mechanism of Inversion

The 2-arylpropionic acids are extensively metabolized by oxidation and conjugation.[111] Glucuronides of the carboxylic acid moiety of ibuprofen account for a significant proportion of (S)- but not

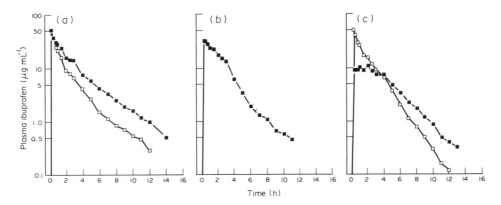

Figure 5 Plasma concentration–time profiles of (R)-ibuprofen (□) and of (S)-ibuprofen (■) after administration of: (a) 800 mg of the racemate; (b) 400 mg of the (S) enantiomer; and (c) 400 mg of the (R) enantiomer to one subject (reproduced from ref. 112 by permission of Macmillan)

of (R)-ibuprofen recovered in urine.[112] This finding may reflect a direct stereoselectivity in glucuronidation since, *in vitro*, (S)-ibuprofen is glucuronidated more rapidly than the (R) isomer.[116] Alternatively, there may be stereoselectivity in parallel oxidative pathways, in the renal clearance of the glucuronides or in the hydrolysis of glucuronide conjugates. It has been shown that ester glucuronides are susceptible to enzymatic hydrolysis[117] and recent studies in the rat[118] support the role of a 'futile cycle' involving glucuronidation and hydrolysis of 2-arylpropionic acids.

There is strong evidence that the inversion of (R)- to (S)-ibuprofen proceeds through the stereospecific formation of the coenzyme A thioester of (R)-ibuprofen.[111] It is thought that subsequent racemization of the thioester is mediated by methylmalonyl-coenzyme A racemase or a related enzyme.[121] The thioesters are then cleaved to yield (R)- and (S)-ibuprofen. Thus, since only the (R) enantiomer is a substrate for esterification, (R)- but not (S)-ibuprofen undergoes inversion. The site or sites of the inversion process have not been fully characterized but there are data to suggest that activity in the kidney is equal to or greater than that in the liver.[118]

23.7.6.3 Toxicological and Clinical Aspects

There is a growing awareness that the mechanisms involved in the stereoselective metabolism of nonsteroidal antiinflammatory drugs may be linked to the development of adverse effects of these drugs in patients. Recent studies have indicated that the coenzyme A esters of ibuprofen and other 2-arylpropionic acids can be incorporated into triacylglycerols by replacing natural fatty acids.[120] It has been suggested that these 'hybrid' triglycerides, which are eliminated very slowly and may accumulate leaving long-term residues, may interfere with normal lipid metabolism and membrane function. Furthermore, Williams *et al.*[121] have speculated that the CNS toxicity of some nonsteroidal antiinflammatory drugs, which is associated with abnormal lipids in the brain, may be mediated by such a mechanism. They have also demonstrated that (R)-ibuprofen is incorporated into rat adipose tissue to a much greater extent than (S)-ibuprofen.[122]

Meffin *et al.*[123] have discussed the implications of disease and drug interactions on the toxicity of nonsteroidal antiinflammatory agents. Decreased clearance of indoprofen, ketoprofen and benoxaprofen has been observed in patients with renal impairment and in elderly subjects with lowered renal function. Furthermore, probenecid impairs the clearance of ketoprofen and carprofen. The existence of a futile cycle, in which drug clearance is determined by the relative rates of formation, urinary excretion and hydrolysis of acyl glucuronide metabolites, has been proposed to explain these observations.[123] Thus, the renal clearance of acyl glucuronides will be decreased, leaving an increased fraction of the glucuronide of both enantiomers available for hydrolysis. This will lead to an apparent decrease in total drug clearance. Furthermore, in view of the increased amounts of the parent (R) enantiomer present, chiral inversion will cause an additional increase in the concentration of the (S) enantiomer. The mechanism of the toxicity of 2-arylpropionic acids in renal impairment is complex and its extent would be underestimated if predictions were based solely on total (R) + (S) concentrations. An animal model has been developed to study further the effect of renal impairment and other factors on the disposition of 2-arylpropionic acids.[123]

As a result of the toxicological implications of stereospecific coenzyme A thioester formation, there may be clinical advantages in administering the pure (S) enantiomers rather than the racemates of 2-arylpropionic acids. On the other hand, the use of the (R) enantiomer as a prodrug has been suggested to avoid the development of gastrointestinal ulceration due to prostaglandin synthetase inhibition by the more active (S) form. However, since the side effects of the nonsteroidal antiinflammatory drugs are thought to be produced systemically as well as locally, this approach may offer no advantage.

23.7.6.4 Comparison Between Species

Considerable species differences have been reported in the rate of chiral inversion of the 2-arylpropionic acids.[111] The most marked difference occurs with benoxaprofen for which the inversion half-life, as estimated from the time-course of the ratio of the enantiomers in the plasma, was found to be over 40 times longer in humans than in the rat.[124] It may be argued, therefore, that the toxicity testing of this compound in the rat was inadequate, since this species was exposed to much less of the (R) isomer than the human patient.

23.7.7 POLYMORPHIC DRUG OXIDATION

The existence of at least two genetic polymorphisms of drug oxidation has been established unequivocally. The first was discovered independently in the UK[125,126] and West Germany[127] and was shown to affect the oxidation of debrisoquine and sparteine. Poor metabolizers (PMs), characterized by an impaired ability to eliminate these drugs, constitute 5–9% of Caucasian populations. The remainder are designated as extensive metabolizers (EMs). The oxidation of debrisoquine and sparteine has been found to cosegregate with that of about 20 drugs and, in some cases, oxidation phenotype is an important determinant of their pharmacokinetics and pharmaco-dynamics.[128] A second polymorphism, independent of that of debrisoquine and sparteine, was described by Kupfer and colleagues[129] and is characterized by defective 4-hydroxylation of the anticonvulsant drug, mephenytoin, in 2–5% of Caucasians. The metabolism of relatively few other drugs appears to be under this type of genetic control.

Some of the drugs whose oxidation displays genetic polymorphism contain at least one chiral centre and are given as racemates. It has been of interest, therefore, to study the pharmacokinetics of their enantiomers and to determine whether stereoselective metabolism, if present, contributes to phenotypic differences in the elimination of total drug.

23.7.7.1 The Debrisoquine-type Polymorphism

23.7.7.1.1 *Metoprolol*

Stereoselectivity in metoprolol pharmacokinetics was first reported by Hermansson and von Bahr,[130] who found a large interindividual variability in the isomer ratio. Subsequently, the elimination of metoprolol enantiomers was investigated in EM and PM subjects following oral administration of the racemic drug.[37] A significantly higher clearance of the (*R*) enantiomer compared to that of the (*S*) form was observed in EMs, whereas in PMs a reversal of this stereoselectivity was apparent, the (*S*) enantiomer being cleared more rapidly than the (*R*) form (Figure 6). In a more limited study, these findings were confirmed in EMs but stereoselective differences were not detected in 3 h plasma enantiomer concentrations in PMs.[131]

Since metoprolol is a medium to high extraction drug in EMs, systemic clearance will tend to depend more on liver blood flow than on the intrinsic activity of the enzymes responsible for metabolism. This explains why the half-life of metoprolol is the same for both enantiomers in EM subjects (Figure 6). However, in PM subjects, where hepatic extraction of the drug is much lower, a difference in half-life of the enantiomers reflects stereoselectivity in enzyme activity.[37]

β-Blocking activity is thought to reside mainly in the (*S*) enantiomer of metoprolol. Thus, the observed shift to the left of the total plasma drug concentration–effect relationship in EMs compared to PMs[37] is consistent with a higher proportion of the active isomer in the plasma of EMs. However, since a greater pharmacological effect would be expected in EMs than in PMs, the

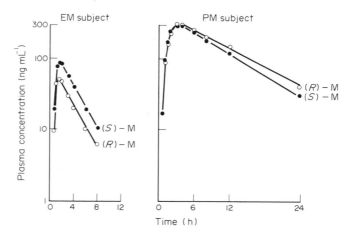

Figure 6 Plasma concentration–time profiles of (*S*)-metoprolol, (*S*)-M, and of (*R*)-metoprolol, (*R*)-M, after oral administration of the racemate (200 mg) to one extensive metabolizer and to one poor metabolizer of debrisoquine (reproduced from ref. 37 by permission of Mosby)

magnitude of the difference of the drug effect between phenotypes would tend to be less than that indicated by the difference in the pharmacokinetics of the sum of the enantiomers. Thus, the large absolute difference between EMs and PMs in β-blocking action as well as in total drug concentration cannot be explained by stereoselectivity in the pharmacokinetics of metoprolol. Nevertheless, although phenotypic differences in the disposition of metoprolol enantiomers are of limited clinical significance, a consideration of stereochemical factors is essential for the rigorous interpretation of relationships between pharmacological effect and drug concentration.

Although differences between EMs and PMs have been found in the urinary enantiomer ratio of metoprolol,[37] this index is not suitable for phenotyping. In a population study of 139 subjects, the (S)/(R)-metoprolol ratio was correlated with debrisoquine oxidation phenotype but the frequency distribution of this ratio did not display bimodality.[132] In the same study, metoprolol enantiomer ratios of this Caucasian population were compared to those from a Nigerian population. The median ratio was found to be significantly higher in Nigerians than in whites, although the magnitude of the difference was small.

The metabolic basis of stereoselectivity in the pharmacokinetics of metoprolol has been studied in human liver microsomes using the individual enantiomers as substrates. Consistent with the *in vivo* findings of a more rapid clearance of (R)-metoprolol in EMs,[37] Otton *et al.*[133] have observed that O demethylation, the major route of metabolism, is more rapid for the (R) than the (S) isomer. Furthermore, experiments using a liver incapable of oxidizing debrisoquine and sparteine but otherwise catalytically active, showed a lack of stereoselectivity for O demethylation and no detectable α hydroxylation. The latter route of metoprolol metabolism introduces a second chiral centre into the molecule and is highly selective for the (R) configuration in rat liver microsomes.[85]

Quinidine and its diastereoisomer, quinine, have been shown to inhibit stereoselectively the metabolism of metoprolol. Quinidine is a highly potent inhibitor of the 'debrisoquine' enzyme[135] and EMs are converted to apparent PMs by treatment with this antiarrhythmic drug. As well as producing a threefold increase in the total plasma concentration of metoprolol, quinidine abolished its stereoselective metabolism in EMs.[136] This effect appears to be particularly long-lasting.[136,137] Quinidine was found to be more than two orders of magnitude more potent an inhibitor than quinine in the oxidation of sparteine, debrisoquine and metoprolol in human liver microsomes.[133,135] Therefore, the use of this isomeric pair may prove to be a highly specific probe for identifying and characterizing those pathways catalyzed by the debrisoquine/sparteine cytochrome *P*-450.

23.7.7.1.2 *Bufuralol*

As with metoprolol, plasma concentrations of the pharmacologically active (−) isomer of bufuralol were shown to be higher in EMs than those of the (+) enantiomer (−/+ ratio between 1.6 and 1.8).[138] In PMs, however, the (−)/(+) plasma bufuralol ratio was higher than in EMs (range 2.55 to 2.65), which is opposite to the finding for metoprolol.[37] The results of urine metabolite excretion studies provide an explanation for this.[138] In EMs, the (−) enantiomer of bufuralol is preferentially hydroxylated on the aromatic ring, whereas the (+) enantiomer preferentially undergoes benzylic hydroxylation. In addition, unlike metoprolol, it is also glucuronidated. Since PMs are deficient in both bufuralol oxidation pathways and, in common with EMs, can glucuronidate the (+) but not the (−) enantiomer, a high plasma (−)/(+) bufuralol ratio occurs in PMs.

As a result of this increased ratio in PMs, a greater difference in β blockade between EMs and PMs than that predicted from differences in the pharmacokinetics of the sum of the enantiomers would be anticipated for bufuralol. This hypothesis has not been tested.

Mutual competitive inhibition studies using human liver microsomes have indicated that the major form of cytochrome *P*-450 catalyzing debrisoquine 4-hydroxylation is responsible for the 1′-hydroxylation of (+)- and (−)-bufuralol.[139] Stereoselectivity in the 1′-hydroxylation of bufuralol has been used as an *in vitro* probe for the debrisoquine polymorphism. 1′-Hydroxylation of bufuralol by liver microsomes from EMs is stereoselective for the (+) isomer. This stereoselectivity is decreased in PM liver microsomes.[140,143] In further work, two cytochrome *P*-450 enzymes were prepared, which were distinguished by this substrate stereoselectivity.[141] One, a low K_m enzyme termed *P*-450 bufI, had a marked preference for (+)-bufuralol metabolism (−/+ ratio = 0.15). The second was a high K_m form, bufII, which lacked stereoselectivity for 1′-hydroxylation (−/+ ratio = 1.03). These and other findings suggested that an enzyme with high stereoselectivity for (+)-bufuralol was almost completely absent in PMs. As shown for metoprolol oxidation, incubation of microsomes from EM livers with quinidine resulted in a complete loss of the stereoselectivity of

1'-hydroxybufuralol formation.[136] It has been shown that the inhibition of (+)-bufuralol 1'-hydroxylation by quinidine is biphasic in EM liver microsomes.[143] This finding provides additional evidence for the involvement of at least two isoenzymes in bufuralol 1'-hydroxylation.

23.7.7.1.3 *Debrisoquine*

4-Hydroxylation introduces a chiral centre into the debrisoquine molecule giving rise to two possible enantiomers. To determine whether 4-hydroxydebrisoquine formation is stereoselective and whether any differences in stereoselectivity exist between phenotypes, the urinary excretion of (R)- and (S)-4-hydroxydebrisoquine has been measured in panels of PMs and EMs.[144] In EMs, 4-hydroxylation was virtually stereospecific with more than 98% of the 4-hydroxydebrisoquine product having the (S) configuration. In PMs, however, metabolism was much less stereoselective and 10–36% of the 4-hydroxydebrisoquine product was in the form of the (R) enantiomer. A black Nigerian population was shown to be similar to Caucasian EMs in its ability to form (S)-4-hydroxydebrisoquine.[145] In both ethnic groups, a greater ability to 4-hydroxylate debrisoquine was associated with increased stereoselectivity.

23.7.7.1.4 *Other drugs*

The relationship between the debrisoquine oxidation phenotype and the elimination of the enantiomers of the antianginal agent perhexiline has been studied.[146] Following oral administration of the individual isomers, the (−) form was cleared more rapidly in PMs and in EMs. The influence of stereoselectivity in perhexiline pharmacokinetics on the known association between frequency of side effects and phenotype[147] has not been examined.

The antiarrhythmic drug N-propylajmaline is administered as a 55:45 mixture of its iso and neo diastereoisomers.[148] EMs clear the iso isomer about four times more rapidly than the neo isomer. This stereoselectivity was not present in PMs. However, the clearance of both diastereoisomers was much greater in EMs than in PMs, indicating that the differential stereoselective metabolism of N-propylajmaline in EMs and PMs explains only part of the large phenotypic difference in total plasma drug concentrations.

Relationships have also been observed between debrisoquine oxidation phenotype and the formation of the geometric isomeric hydroxylation products of nortriptyline. *In vivo* and *in vitro* studies show that nortriptyline is 10-hydroxylated preferentially to the (E) isomer.[149] Clearance to (E)- but not to (Z)-10-hydroxynortriptyline decreased in parallel with a diminishing ability to 4-hydroxylate debrisoquine. The lowered ability of PMs to (E)-10-hydroxylate nortriptyline probably contributes to the high plasma concentrations of unchanged drug seen in subjects of this phenotype.

23.7.7.2 The Mephenytoin-type Polymorphism

A marked substrate stereoselectivity has been observed in the pharmacokinetics and metabolism of mephenytoin in humans.[150,151] In most subjects, the (S) enantiomer undergoes rapid 4-hydroxylation, whereas the (R) enantiomer is slowly metabolized by N demethylation to nirvanol, a pharmacologically active product which accumulates during chronic dosing.[152] Thus, there is a 200-fold interphenotype difference in the ratio of the oral clearance of (S)- compared to that of (R)-mephenytoin. As a result of their inability to 4-hydroxylate (S)-mephenytoin, PMs can only eliminate this drug by demethylation. This results in an absence of stereoselectivity in PMs. In addition, the decrease in total (R) + (S) drug clearance taken together with a greater accumulation of nirvanol in PMs is associated with adverse effects in these subjects.[153]

The stereoselective metabolism of mephenytoin has been used in the development of an improved phenotyping test.[119] The urinary (S)/(R) enantiomer ratio, which is thought to reflect hepatic enzyme activity more closely than the conventional metabolic ratio,[153] has proved to be a good discriminator between EMs and PMs (Figure 7).

The influence of phenotype on the metabolism of nirvanol[100] and mephobarbital[105] is similar to that described for mephenytoin. Thus, stereoselective hydroxylation of the (S) enantiomers of both nirvanol and mephobarbital occurred in EMs but not in PMs. Further evidence that (S)-mephenytoin and (S)-mephobarbital hydroxylation are under similar genetic control comes from the

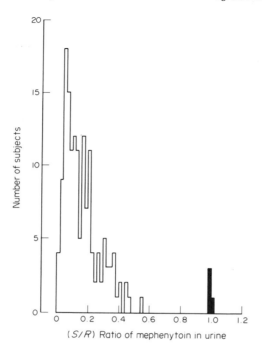

Figure 7 Frequency distribution of the $\log_{10}$ 0–8 h urinary $(S)/(R)$ ratio of mephenytoin after administration of a single oral dose (100 mg) of the racemate to 156 unrelated Caucasian subjects. The shaded area represents poor metabolizers of mephenytoin (adapted from ref. 19)

observation of higher AUC values of (S)-mephenytoin and an accompanying decrease in stereoselective metabolism of mephenytoin during concurrent mephobarbital treatment in EMs.[134]

Although the overall pharmacokinetics of phenytoin, a prochiral compound, are not influenced by mephenytoin phenotype,[154] a recent study has reported a relationship between the degree of stereoselectivity in the 4-hydroxylation of phenytoin and the ability to oxidize mephenytoin.[142] Phenytoin 4-hydroxylase metabolizes the (S) enantiomer preferentially in both phenotypes but (R)-hydroxyphenytoin is virtually absent from the urine of PMs. Bimodality in the distribution of the $\log(S)/(R)$-hydroxyphenytoin ratio in a population of epileptic patients and a complete concordance with mephenytoin oxidation phenotype suggest that this ratio is sufficiently discriminatory for phenotyping patients.

The results of *in vitro* studies using human liver microsomes from both phenotypes support the *in vivo* data showing a loss of stereoselectivity in mephenytoin 4-hydroxylation in PMs.[152] The findings suggest that the polymorphic oxidation of mephenytoin is a consequence of either the absence or the inactivity of a form of cytochrome *P*-450 which has high affinity for (S)-mephenytoin.

23.7.8 CONCLUSIONS

The evidence for stereoselectivity in drug disposition has been reviewed by selecting as examples those agents which are in widespread clinical usage.

Examples have been reviewed in which differences in enantiomer pharmacokinetics exert a considerable influence on the pharmacological response *in vivo*, either by damping or amplifying the pharmacodynamic component. In some cases, the contributions of pharmacokinetics and pharmacodynamics to *in vivo* drug potency are almost equal, but in other cases the kinetic contribution is much smaller.

The relationships between stereoselectivity in metabolism and factors such as the route of administration, polymorphic oxidation, chiral inversion, species differences and drug interactions have been investigated in detail. The results of many of these studies have led to significant advances in the interpretation and prediction of relationships between total $(R)+(S)$ drug concentration and pharmacological and toxicological effects. The results have also led to the development of probes for use in the study of molecular mechanisms of drug metabolism. A stereoselective approach is now essential for the rigorous study of the pharmacokinetics and metabolism of optically active drugs.

ACKNOWLEDGEMENT

At the time of writing M. S. Lennard was a Wellcome Trust Lecturer.

23.7.9 REFERENCES

1. E. J. Ariens, E. W. Wuis and E. J. Veringa, *Biochem. Pharmacol.*, 1988, **37**, 9.
2. Proceedings of Symposium on 'Stereochemistry in Drug Action', *Biochem. Pharmacol.*, 1988, **37**.
3. S. Mason, *New Sci.*, 1984, **101**, 10.
4. G. Blaschke, H. P. Kraft and F. Kohler, *Arzneim.-Forsch.*, 1979, **29** II, 1640.
5. S. Fabro, R. L. Smith and R. T. Williams, *Nature (London)*, 1967, **215**, 296.
6. H. Tuchmann-Duplessis, *Pharmacol. Ther.*, 1984, **26**, 273.
7. R. W. Souter, 'Chromatographic Separations of Stereoisomers', CRC Press, Boca Raton, FL, 1985.
8. R. Dennis, *Pharm. Int.*, 1986, **7**, 246.
9. D. W. Armstrong, *Anal. Chem.*, 1987, **59**, 84A.
10. T. H. Morris and A. J. Kaumann, *Naunyn-Schmiedeberg's Arch. Pharmacol.*, 1984, **327**, 176.
11. T. Walle, U. K. Walle and L. S. Olanoff, *Drug Metab. Dispos.*, 1985, **13**, 204.
12. G. P. Jackman, A. J. McLean, G. L. Jennings and A. Bobik, *Clin. Pharmacol. Ther.*, 1981, **30**, 291.
13. B. Silber, N. H. Holford and S. Riegelman, *J. Pharm. Sci.*, 1982, **71**, 699.
14. L. S. Olanoff, T. Walle, U. K. Walle, T. D. Cowart and T. E. Gaffney, *Clin. Pharmacol. Ther.*, 1984, **35**, 755.
15. C. von Bahr, J. Hermansson and K. Tawara, *Br. J. Clin. Pharmacol.*, 1982, **14**, 79.
16. K. H. Donn, J. R. Powell and I. W. Wainer, *Clin. Pharmacol. Ther.*, 1985, **37**, 191, abstract IIA.
17. G. R. Wilkinson and D. G. Shand, *Clin. Pharmacol. Ther.*, 1975, **18**, 377.
18. A. S. Nies, G. H. Evans and D. G. Shand, *J. Pharmacol. Exp. Ther.*, 1973, **184**, 716.
19. D. J. Coltart and D. G. Shand, *Br. Med. J.*, 1970, **3**, 731.
20. U. K. Walle, T. Walle, S. A. Bai and L. S. Olanoff, *Clin. Pharmacol. Ther.*, 1983, **34**, 718.
21. F. Albani, R. Riva, M. Contin and A. Baruzzi, *Br. J. Clin. Pharmacol.*, 1984, **18**, 244.
22. T. Walle, U. K. Walle, M. J. Wilson, T. C. Fagan and T. E. Gaffney, *Br. J. Clin. Pharmacol.*, 1984, **18**, 741.
23. C. von Bahr, J. Hermansson and M. Lind, *J. Pharmacol. Exp. Ther.*, 1982, **222**, 458.
24. W. L. Nelson and H. U. Shetty, *Drug Metab. Dispos.*, 1986, **14**, 506.
25. T. Walle and U. K. Walle, *Res. Commun. Chem. Pathol. Pharmacol.*, 1979, **23**, 453.
26. S. A. Bai, U. K. Walle, M. J. Wilson and T. Walle, *Drug Metab. Dispos.*, 1983, **11**, 394.
27. S. A. Bai, M. J. Wilson, U. K. Walle and T. Walle, *J. Pharmacol. Exp. Ther.*, 1983, **227**, 360.
28. B. K. Wilson and J. A. Thompson, *Drug Metab. Dispos.*, 1984, **12**, 161.
29. M. L. Powell, R. R. Wagoner, C.-H. Chen and W. L. Nelsen, *Res. Commun. Chem. Pathol. Pharmacol.*, 1980, **30**, 387.
30. T. Walle, J. E. Oatis, U. K. Walle and D. R. Knapp, *Drug Metab. Dispos.*, 1982, **10**, 122.
31. T. Walle, M. J. Wilson, U. K. Walle and S. A. Bai, *Drug Metab. Dispos.*, 1983, **11**, 544.
32. S. Fujita, S. Kariya, R. Ishida, R. Morimoto and T. Suzuki, Proceedings of the Xth International Conference of Pharmacology, Sydney, Australia, 1987, abstract no. O280.
33. H. U. Shetty and W. L. Nelson, *J. Med. Chem.*, 1986, **29**, 2004.
34. D. D. Christ and T. Walle, *Drug Metab. Dispos.*, 1985, **13**, 380.
35. H. R. Ochs, P. Hajdu and D. J. Greenblatt, *Klin. Wochenschr.*, 1986, **64**, 636.
36. J. Hermansson and C. von Bahr, *J. Chromatogr.*, 1982, **227**, 113.
37. M. S. Lennard, G. T. Tucker, J. H. Silas, S. Freestone, L. E. Ramsay and H. F. Woods, *Clin. Pharmacol. Ther.*, 1983, **34**, 732.
38. P. Dayer, T. Leemann, A. Kupfer, T. Kronbach and U. A. Meyer, *Eur. J. Clin. Pharmacol.*, 1986, **31**, 313.
39. C. Harvengt and J. P. Desager, *Int. J. Clin. Pharmacol., Ther. Toxicol.*, 1982, **20**, 57.
40. S. Honma, T. Ito and A. Kambegawa, *Chem. Pharm. Bull.*, 1985, **33**, 760.
41. C. S. Deacon, M. S. Lennard, N. D. Bax, H. F. Woods and G. T. Tucker, *Br. J. Clin. Pharmacol.*, 1981, **12**, 429.
42. N. D. Bax, M. S. Lennard and G. T. Tucker, *Br. J. Clin. Pharmacol.*, 1981, **12**, 779.
43. S. A. H. Al-Asadi, Ph.D. Thesis, University of Sheffield, 1984.
44. M. G. Sankey, A. Gulaid and C. M. Kaye, *J. Pharm. Pharmacol.*, 1984, **36**, 276.
45. P.-H. Hsyu and K. M. Giacomini, *J. Clin. Invest.*, 1985, **76**, 1720.
46. J. Hermansson, M. Eriksson and O. Nyquist, *J. Chromatogr.*, 1984, **336**, 321.
47. K. M. Giacomini, W. L. Nelson, R. A. Pershe, L. Valdivieso, K. Turner-Tamiyasu and T. F. Blaschke, *J. Pharmacokinet. Biopharm.*, 1986, **14**, 335.
48. P. J. Meffin, E. W. Robert, R. A. Winkle, S. Harapat, F. A. Peters and D. C. Harrison, *J. Pharmacokinet. Biopharm.*, 1979, **7**, 29.
49. J. J. Lima, H. Boudoulas and B. J. Shields, *Drug Metab. Dispos.*, 1985, **13**, 572.
50. C. S. Cook, A. Karim and P. Sollman, *Drug Metab. Dispos.*, 1982, **10**, 116.
51. O. Grech-Belanger, J. Turgeon and M. Gilbert, *Br. J. Clin. Pharmacol.*, 1986, **21**, 481.
52. K. M. McErlane, L. Igwemezie and C. R. Kerr, *Res. Commun. Chem. Pathol. Pharmacol.*, 1987, **56**, 141.
53. A. J. Sedman, J. Gal, W. Mastropaolo, P. Johnson, J. D. Maloney and T. P. Moyer, *Br. J. Clin. Pharmacol.*, 1984, **17**, 113.
54. A. H. Thomson, G. Murdoch, A. Pottage, A. W. Kelman and B. Whiting, *Br. J. Clin. Pharmacol.*, 1986, **21**, 149.
55. A. J. Sedman, D. C. Bloedow and J. Gal, *Res. Commun. Chem. Pathol. Pharmacol.*, 1982, **38**, 165.
56. B. Edgar, A. Heggelund, L. Johansson, G. Nyberg and C. G. Regardh, *Br. J. Clin. Pharmacol.*, 1984, **17**, 216P.

57. A. T. Elvin, J. B. Keenaghan, E. W. Byrnes, P. A. Tenthory, P. D. McMaster, B. H. Takman, D. Lalka, C. V. Manion, D. T. Baer, E. M. Wolshin, M. B. Meyer and R. A. Ronfield, *J. Pharm. Sci.*, 1980, **69**, 47.
58. K.-J. Hoffmann, L. Renberg and C. Baarnhielm, *Eur. J. Drug Metab. Pharmacokinet.*, 1984, **9**, 215.
59. G. K. Pillau, J. E. Axelson, C. R. Kerr and K. M. McErlane, *Res. Commun. Chem. Pathol. Pharmacol.*, 1984, **43**, 209.
60. J. Gal, T. A. French, T. Zysset and P. E. Haroldsen, *Drug Metab. Dispos.*, 1982, **10**, 399.
61. H. R. Ochs, D. J. Greenblatt and E. Woo, *Clin. Pharmacokinet.*, 1980, **5**, 150.
62. N. J. White, *Clin. Pharmacokinet.*, 1985, **10**, 187.
63. D. A. Notterman, D. E. Drayer, L. Metakis and M. M. Reidenberg, *Clin. Pharmacol. Ther.*, 1986, **40**, 511.
64. G. W. Mihaly, K. M. Hyman, R. A. Smallwood and K. J. Hardy, *J. Chromatogr.*, 1987, **415**, 177.
65. G. W. Mihaly, M. K. Hyman and R. A. Smallwood, Proceedings of the Xth International Conference of Pharmacology, Sydney, Australia, 1987, abstract no. P72.
66. M. Eichelbaum, G. Mikus and B. Vogelgesang, *Br. J. Clin. Pharmacol.*, 1984, **17**, 453.
67. B. Vogelsgang, H. Echizen, E. Schmidt and M. Eichelbaum, *Br. J. Clin. Pharmacol.*, 1984, **18**, 733.
68. M. Eichelbaum, P. Birkel, E. Grube, U. Gutgemann and A. Somogyi, *Klin. Wochenschr.*, 1980, **58**, 919.
69. T. Walle, J. G. Webb, E. E. Bagwell, U. K. Walle, H. B. Daniell and T. E. Gaffney, *Biochem. Pharmacol.*, 1988, **37**, 115.
70. A. Breckenridge, M. Orme, H. Wesseling, R. J. Lewis and R. Gibbons, *Clin. Pharmacol. Ther.*, 1974, **15**, 424.
71. D. S. Hewick and J. McEwen, *J. Pharm. Pharmacol.*, 1973, **25**, 458.
72. L. B. Wingard, R. A. O'Reilly and G. Levy, *Clin. Pharmacol. Ther.*, 1978, **23**, 212.
73. C. Banfield, R. O'Reilly, E. Chan and M. Rowland, *Br. J. Clin. Pharmacol.*, 1983, **16**, 669.
74. R. M. Sellers and J. Koch-Weser, *Pharmacol. Res. Commun.*, 1975, **7**, 331.
75. N. A. Brown, E. Jahnchen, W. E. Muller and U. Wollert, *Mol. Pharmacol.*, 1977, **13**, 70.
76. A. Yacobi and G. Levy, *J. Pharmacokinet. Biopharm.*, 1977, **5**, 123.
77. R. J. Lewis, W. F. Trager, K. K. Chan, A. Breckenridge, M. Orme, M. Rowland and W. Schary, *J. Clin. Invest.*, 1974, **53**, 1607.
78. L. S. Kaminsky, D. A. Dunbar, P. P. Wang, P. Beaune, D. Larrey, F. P. Guengerich, R. G. Schnellmann and I. G. Sipes, *Drug Metab. Dispos.*, 1984, **12**, 470.
79. P. P. Wang, P. Beaune, L. S. Kaminisky, G. A. Dannan, F. F. Kadlubar, D. Larrey and F. P. Guengerich, *Biochemistry*, 1983, **22**, 5375.
80. R. A. O'Reilly, W. F. Trager, C. H. Motley and W. Howald, *J. Clin. Invest.*, 1980, **65**, 746.
81. S. Toon, K. J. Hopkins, F. M. Garstang, L. Aarons, A. Sedman and M. Rowland, *Clin. Pharmacol. Ther.*, 1987, **42**, 33.
82. R. A. O'Reilly, *N. Engl. J. Med.*, 1976, **295**, 354.
83. R. A. O'Reilly, *N. Engl. J. Med.*, 1980, **302**, 33.
84. R. A. O'Reilly, *Clin. Pharmacol. Ther.*, 1982, **32**, 356.
85. H. U. Shetty and W. L. Nelson, *J. Med. Chem.*, 1988, **31**, 55.
86. I. A. Choonara, S. Cholerton, B. P. Haynes, A. M. Breckenridge and B. K. Park, *Br. J. Clin. Pharmacol.*, 1986, **21**, 271.
87. S. Toon, K. J. Hopkins, F. M. Garstang, B. Diquet, T. S. Gill and M. Rowland, *Br. J. Clin. Pharmacol.*, 1985, **20**, 245.
88. S. Toon, K. J. Hopkins, F. M. Garstang, L. Aarons, A. Sedman and M. Rowland, *Clin. Pharmacol. Ther.*, 1987, **42**, 33.
89. R. A. O' Reilly, W. F. Trager, A. E. Rettie and D. A. Goulart, *Clin. Pharmacol. Ther.*, 1987, **42**, 290.
90. R. A. O'Reilly, W. F. Trager, C. H. Motley and W. Howald, *Clin. Pharmacol. Ther.*, 1980, **28**, 187.
91. L. D. Heimark, M. Gibaldi, W. F. Trager, R. A. O'Reilly and D. A. Goulart, *Clin. Pharmacol. Ther.*, 1987, **42**, 388.
92. T. D. Bjornsson, P. J. Meffin and T. F. Blaschke, *J. Pharmacokinet. Biopharm.*, 1977, **5**, 495.
93. R. A. O'Reilly, *Clin. Pharmacol. Ther.*, 1981, **29**, 332.
94. A. Breckenridge and M. L. Orme, *Life Sci.*, 1972, **11**, 337.
95. A. Yacobi and G. Levy, *J. Pharmacokinet. Biopharm.*, 1974, **2**, 239.
96. L. R. Pohl, R. Bales and W. F. Trager, *Res. Commun. Chem. Pathol. Pharmacol.*, 1976, **15**, 233.
97. L. R. Pohl, W. R. Porter and W. F. Trager, *Biochem. Pharmacol.*, 1977, **26**, 109.
98. A. Yacobi, C.-M. Lai and G. Levy, *J. Pharmacol. Exp. Ther.*, 1984, **231**, 80.
99. A. Yacobi, C.-M. Lai and G. Levy, *J. Pharmacol. Exp. Ther.*, 1984, **231**, 72.
100. A. Kupfer, R. Patwardhan, S. Ward, S. Schenker, R. Preisig and R. A. Branch, *J. Pharmacol. Exp. Ther.*, 1984, **230**, 28.
101. J. Godbillon, J. Richard, A. Gerardin, T. Meinertz, W. Kasper and E. Jahnchen, *Br. J. Clin. Pharmacol.*, 1981, **12**, 621.
102. H. H. W. Thijssen, G. M. J. Janssen and L. G. M. Baars, *Eur. J. Clin. Pharmacol.*, 1986, **30**, 619.
103. W. Dieterle, J. W. Faigle, C. Montigel, M. Sulc and W. Theobald, *Eur. J. Clin. Pharmacol.*, 1977, **11**, 367.
104. M. J. Serlin, R. G. Sibeon, S. Mossman, A. M. Breckenridge, J. R. B. Williams, J. L. Atwood and J. M. T. Willoughby, *Lancet*, 1979, **2**, 317.
105. A. Kupfer and R. A. Branch, *Clin. Pharmacol. Ther.*, 1985, **38**, 414.
106. T. S. Gill, K. J. Hopkins and M. Rowland, *Br. J. Clin. Pharmacol.*, 1986, **21**, 564P.
107. H. H. W. Thijssen, L. G. M. Baars and M. J. Drittij-Reijnders, *Drug Metab. Dispos.*, 1985, **13**, 593.
108. H. H. W. Thijssen and L. G. M. Baars, *J. Pharm. Pharmacol.*, 1987, **39**, 655.
109. E. Jahnchen, T. Meinertz, H.-J. Gilfrich, U. Groth and A. Martini, *Clin. Pharmacol. Ther.*, 1976, **20**, 342.
110. S. Toon, L. D. Heimark, W. F. Trager and R. A. O'Reilly, *J. Pharm. Sci.*, 1985, **74**, 1037.
111. J. Caldwell, A. J. Hutt and S. Fournel-Gigleux, *Biochem. Pharmacol.*, 1988, **37**, 105.
112. E. J. D. Lee, K. Williams, R. Day, G. Graham and D. Champion, *Br. J. Clin. Pharmacol.*, 1985, **19**, 669.
113. V. Tamassiam, M. G. Jannuzzo, E. Moro, S. Stegnjaich, W. Groppi and F. B. Nicolis, *Int. J. Clin. Pharmacol. Res.*, 1984, **4**, 223.
114. A. Rubin, M. P. Knadler, P. P. K. Ho, L. D. Bechtol and R. L. Wolen, *J. Pharm. Sci.*, 1985, **74**, 82.
115. R. L. Nation, A. M. Evans and L. N. Sanson, Proceedings of the Xth International Congress of Pharmacology, Sydney, Australia, 1987, abstract no. P1284.
116. M. El Mouelhi, H. W. Ruelius, C. Fenselau and D. M. Dulik, *Drug Metab. Dispos.*, 1987, **15**, 767.
117. P. J. Meffin, D. M. Zilm and J. R. Veenendal, *J. Pharmacol. Exp. Ther.*, 1983, **227**, 732.
118. T. Yamaguchi and Y. Nakamura, *Drug Metab. Dispos.*, 1987, **15**, 529.
119. P. J. Wedlund, W. S. Aslanian, C. B. McAllister, G. R. Wilkinson and R. A. Branch, *Clin. Pharmacol. Ther.*, 1984, **36**, 773.
120. J. Caldwell and M. V. Marsh, *Biochem. Pharmacol.*, 1983, **32**, 1667.
121. K. M. Williams and R. O. Day, *Agents Actions (Suppl.)*, 1985, **17**, 119.

122. K. Williams, R. Day, R. Knihinicki and A. Duffield, *Biochem. Pharmacol.*, 1986, **35**, 3403.
123. P. J. Meffin, B. C. Sallustio, Y. J. Purdie and M. E. Jones, *J. Pharmacol. Exp. Ther.*, 1986, **238**, 280.
124. R. G. Simmonds, T. J. Woodage, S. M. Duff and J. N. Green, *Eur. J. Drug Metab. Pharmacokinet.*, 1980, **5**, 169.
125. A. Mahgoub, J. R. Idle, R. Lancaster and R. L. Smith, *Lancet*, 1977, **2**, 584.
126. G. T. Tucker, J. H. Silas, A. O. Iyun, M. S. Lennard and A. J. Smith, *Lancet*, 1977, **2**, 718.
127. M. Eichelbaum, N. Spannbrucker, B. Steincke and H. J. Dengler, *Eur. J. Clin. Pharmacol.*, 1979, **16**, 153.
128. W. Kalow, *Eur. J. Clin. Pharmacol.*, 1987, **31**, 633.
129. A. Kupfer, P. Desmond, R. Parwardhan, S. Schenker and R. A. Branch, *Clin. Pharmacol. Ther.*, 1984, **35**, 33.
130. J. Hermansson and C. von Bahr, *J. Chromatogr.*, 1982, **227**, 113.
131. P. Dayer, T. Leemann, A. Marmy and J. Rosenthaler, *Eur. J. Clin. Pharmacol.*, 1985, **28**, 149.
132. M. S. Lennard, G. T. Tucker, H. F. Woods, J. H. Silas and A. O. Iyun, *Br. J. Clin. Pharmacol.*, 1989, **27**, 613.
133. S. V. Otton, M. S. Lennard, G. T. Tucker and H. F. Woods, *Br. J. Clin. Pharmacol.*, 1989, **27**, 242.
134. E. Jacqz, S. D. Hall, R. A. Branch and G. R. Wilkinson, *Clin. Pharmacol. Ther.*, 1986, **39**, 646.
135. S. V. Otton, T. Inaba and W. Kalow, *Life Sci.*, 1984, **34**, 73.
136. T. Leemann, P. Dayer and U. A. Meyer, *Eur. J. Clin. Pharmacol.*, 1986, **29**, 739.
137. S. Kobayashi, C. J. Speirs, D. Watson, S. Murray, D. Sesardic and A. R. Boobis, *Br. J. Clin. Pharmacol.*, 1988, **25**, 637P.
138. P. Dayer, T. Leemann, A. Kupfer, T. Kronbach and U. A. Meyer, *Eur. J. Clin. Pharmacol.*, 1986, **31**, 313.
139. A. R. Boobis, S. Murray, C. A. Hampden and D. S. Davies, *Biochem. Pharmacol.*, 1985, **34**, 65.
140. E. R. Minder, P. J. Meier, H. K. Muller, C. Minder and U. A. Meyer, *J. Clin. Invest.*, 1984, **14**, 184.
141. J. Gut, T. Catin, P. Dayer, T. Kronbach, U. Zanger and U. A. Meyer, *J. Biol. Chem.*, 1986, **261**, 11 734.
142. S. Fritz, W. Linder, I. Roots, B. M. Frey and A. Kupfer, *J. Pharmacol. Exp. Ther.*, 1987, **241**, 615.
143. P. Dayer, T. Kronbach, M. Eichelbaum and U. A. Meyer, *Biochem. Pharmacol.*, 1987, **36**, 4145.
144. M. Eichelbaum, L. Bertilsson, A. Kupfer, E. Steiner and C. O. Meese, *Br. J. Clin. Pharmacol.*, 1988, **25**, 505.
145. M. S. Lennard, G. T. Tucker, H. F. Woods, A. O. Iyun and M. Eichelbaum, *Biochem. Pharmacol.*, 1988, **37**, 97.
146. N. S. Oates, R. R. Shah, J. R. Idle and R. L. Smith, *Br. J. Clin. Pharmacol.*, 1984, **18**, 307P.
147. R. R. Shah, N. S. Oates, J. R. Idle, R. L. Smith and J. D. F. Lockhart, *Br. Med. J.*, 1982, **284**, 295.
148. C. Zekorn, G. Achtert, H. J. Hausleiter, C. H. Moon and M. Eichelbaum, *Klin. Wochenschr.*, 1985, **63**, 1180.
149. B. Mellstrom, L. Bertilsson, J. Sawe, H.-U. Schulz and F. Sjoqvist, *Clin. Pharmacol. Ther.*, 1981, **30**, 189.
150. A. Kupfer, R. K. Roberts, S. Schenker and R. A. Branch, *J. Pharmacol. Exp. Ther.*, 1981, **218**, 193.
151. P. J. Wedlund, W. S. Aslanian, E. Jacqz, C. B. McAllister, R. A. Branch and G. R. Wilkinson, *J. Pharmacol. Exp. Ther.*, 1985, **234**, 662.
152. U. T. Meier, P. Dayer, P.-J. Male, T. Kronbach and U. A. Meyer, *Clin. Pharmacol. Ther.*, 1985, **38**, 488.
153. A. Kupfer and R. Preisig, *Eur. J. Clin. Pharmacol.*, 1984, **26**, 753.
154. E. Steiner, G. Alvan, M. Garle, J. H. Maguire, M. Lind, S.-O. Nilson, T. Tomson, J. S. McClanahan and F. Sjoqvist, *Clin. Pharmacol. Ther.*, 1987, **42**, 326.

23.8

Enzyme Induction and Inhibition

ALASDAIR M. BRECKENRIDGE and B. KEVIN PARK
University of Liverpool, UK

23.8.1 INTRODUCTION

Enzyme induction can be defined as an adaptive increase in the number of molecules of a specific enzyme, secondary either to an increase in its rate of synthesis or to a decrease in its rate of degradation. Enzyme activity may also be altered without change in the amount of enzyme; a variety of structurally unrelated biochemical agents are known to cause these effects by processes other than induction. Because of the diverse roles of induction in various biological processes, an understanding of this biochemical adaptation has considerable physiological and clinical importance. For example, enzyme-inductive effects appear to be an important part of normal development and differentiation, and enzyme induction and repression play an important role in the nutritional control of key metabolic pathways. The study of enzyme induction is equally important as a model for mechanisms whereby cells regulate the expression of their genetic inheritance. All cells in a multicellular differentiated organism possess the same set of genetic information, yet such organisms are characterized by diverse cell types which differ quantitatively and qualitatively in their complement.

Each cell thus expresses only a portion of its total genetic information and different cell types express different portions of their genome. A given cell may express different components of its genetic information during various stages of its differentiation and development; the fully differentiated cell may modify its pattern of gene expression in response to a variety of physiological and pharmacological stimuli. One of the most important aspects of enzyme induction is the way in which an organism reacts to a variety of foreign compounds, resulting in alteration of the handling of xenobiotics to which it is exposed.

The metabolism of exogenous lipophilic compounds by higher animals occurs predominantly in the liver and intestine but also in other tissues. This metabolism is normally a two-stage process. In the first phase (phase 1), a nucleophilic substituent group such as —OH, —NH$_2$, —CO$_2$H or —SH is inserted or revealed through the action of oxidation, reduction or hydrolytic enzymes. In the second phase (phase 2), this substituent group is conjugated, typically with a carbohydrate, amino acid or inorganic acid. The overall effect is to convert lipophilic compounds which tend to pass readily into cells and bind to cellular components into more hydrophilic products which can be excreted. The enzymes carrying out these two phases of metabolism are collectively referred to as the drug-metabolizing enzymes. In some respects this is a misnomer, since these enzymes metabolize a wide range of synthetic and naturally occurring xenobiotics. The true function of drug-metabolizing enzymes may be to facilitate the clearance from the body of exogenous lipophilic substances, or control the synthesis and degradation of endogenous bioactive compounds such as steroids, prostaglandins and similar substances.

The cytochrome *P*-450-containing microsomal mixed-function oxidases, described in other chapters of this volume, coupled with other metabolically linked enzymes provide an important pathway whereby the cell can metabolize and eliminate xenobiotics. These systems are responsible for detoxification as well as activation of drugs and other environmental agents to toxic, mutagenic and carcinogenic forms. The mixed-function oxidase drug-metabolizing enzyme systems are generally inducible, *i.e.* after exposure to certain chemicals the amount of enzyme activity increases. This regulatory process may thus provide an important mechanism for increasing the rate of metabolism of foreign compounds to detoxified forms or in some cases to harmful intermediates. An understanding of the factors and mechanisms involved in regulating drug-metabolizing enzyme activity may thus lead to a better understanding of the regulation, drug activation and detoxification of processes involved in chemical carcinogenesis, drug synergism, drug antagonism and toxicology.

23.8.2 HISTORICAL ASPECTS

The phenomenon of induction was described independently by two groups, that of Conney, Miller and Miller[1] and Remmer.[2] Conney and co-workers had been investigating the metabolism of methylated amino azo dyes and showed an increase in enzyme activity after the application of 3-methylcholanthrene. This was shown to be a powerful inducer of a specific form of cytochrome *P*-450, namely the variety that shows a spectral absorption of the reduced carbon monoxide complex at 448 nm. The authors showed that the increase in enzyme activity was due to an increase in biosynthesis of the enzyme protein, *i.e.* true induction. In 1959, on the other hand, Remmer showed that previous administration of phenobarbitone to animals resulted in a shorter sleeping time after dosing with hexobarbital and also to an increase in activity of the hepatic monooxygenase. Later it was demonstrated that the phenobarbitone-induced type of *P*-450 shows an absorption maximum at exactly 450 nm.

These two important observations initiated a series of studies demonstrating that enzyme induction is an important factor modifying the pharmacokinetics of many drugs and other foreign compounds. This has been investigated in animals and in man and both aspects will be described in this review, but more emphasis will be given to the latter aspect.

23.8.3 BIOCHEMICAL FEATURES OF ENZYME INDUCTION

Morphologically, the administration of enzyme inducers is accompanied by an increase in smooth endoplasmic reticulum, best demonstrated by electron microscopy. Within these membranes the monooxygenase system is arranged in such a way as to allow activation of molecular oxygen as well as reduction of cytochrome *P*-450. Various models for the arrangement of membrane-bound monooxygenase systems have been suggested; that of Greinert *et al.*[3] proposes that six cytochrome *P*-450 molecules surround one molecule of cytochrome *P*-450 reductase which rotates in such a manner as to distribute electrons to the heme moiety of the cytochromes.

Enzyme induction is only demonstrable *in vivo* and there is a delay before increased enzyme activity can be observed. This is to be distinguished from the phenomenon of activation of enzymes which can be observed *in vitro* and occurs virtually instantaneously. True enzyme induction is caused by an increase in cytochrome *P*-450 biosynthesis; as a corollary, inhibitors of protein biosynthesis prevent enzyme induction.

The induced form of cytochrome *P*-450 has a substrate affinity similar to the uninduced form; the maximal velocity of the reaction related to microsomal protein is, however, greatly enhanced.

As mentioned above, different inducers induce various forms of cytochrome *P*-450. The two most prominent variants produced by enzyme induction were previously known as the phenobarbitone-inducible and the methylcholanthrene-inducible forms of cytochrome *P*-450. Not only do they vary in the spectral properties of their carbon monoxide complexes, but also in their activity towards certain substrates. The methylcholanthrene form is relatively specific for benzopyrene as substrate, the phenobarbitone form has a wider range of activity including other barbiturates, aniline and ethylmorphine. Many different forms of cytochrome *P*-450 have now been identified and it is known that both the phenobarbitone- and methylcholanthrene-inducible forms are in fact mixtures of several of these subtypes.[4]

23.8.4 MECHANISM OF ENZYME INDUCTION

There are at least two mechanisms of induction within the liver cell, and probably within other tissues as well. These have been reviewed recently by Netter.[5] In the induction caused by phenobarbitone, increased concentrations of several isoenzymes of cytochrome *P*-450 and the corresponding mRNAs are found in the cell. Phenobarbitone induces a 2 kb mRNA which carries the code for translation of the phenobarbitone induction. Phenobarbitone increases this mRNA substantially and also increases the transcription of a respective gene. There is, however, no specific receptor responsible for the effect of phenobarbitone.

In contrast, the second type of induction, caused by polycyclic aromatic hydrocarbons, depends on the presence of a genomic locus which makes certain animal species responsive to this type of induction. 3-Methylcholanthrene and 2,3,7,8-tetrachlorodibenzodioxin (TCDD) are typical of this class of inducer. They act *via* a specific receptor which combines with the inducer, transfers it to the nucleus and initiates the induction process.

23.8.5 CLINICAL ASSESSMENT OF ENZYME INDUCTION AND INHIBITION

23.8.5.1 Introduction

It has already been mentioned that the enzymes concerned with both phases of drug metabolism exist in multiple forms which have relative but overlapping substrate specificities, and the reader is referred to the recent review by Gonzalez.[4] It is therefore important to classify individual enzymes with respect to representative model drug substrates so that predictions concerning enzyme induction and enzyme inhibition can be made. We are therefore faced with the problem of monitoring the activities of a large but, as yet, undefined number of enzymes within the constraints of what is ethical and feasible in man. Our current understanding of the hepatic mixed-function oxidase system is shown in simple form in Figure 1. According to this scheme, adapted from Breimer,[6] each circle represents an independent form of cytochrome *P*-450, and overlapping circles indicate overlapping substrate specificities.

23.8.5.2 Methods Used to Assess Drug-metabolizing Enzyme Activity

The three principal methods used in man are: (i) investigation of the pharmacokinetics of model compounds; (ii) investigation of drug metabolism *in vitro*; and (iii) measurement of changes in the disposition of endogenous compounds.

23.8.5.2.1 *Model drug substrates*

An ideal drug for assessment of drug-metabolizing enzyme activity should possess the following properties: rapid and complete absorption; simple pharmacokinetics; elimination solely by metabolism, and independently of blood flow and protein binding; all routes of metabolism should be

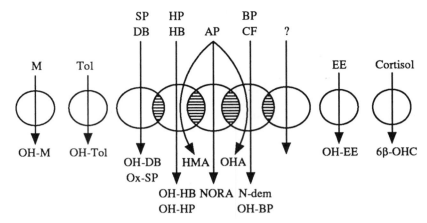

Figure 1 Graphic representation of the different forms of cytochrome *P*-450 (circles) in man with different but overlapping substrate specificates. The arrows indicate single metabolic pathways; M = mephenytoin, TOL = tolbutamide, DB = debrisoquine, SP = sparteine, HP = heptobarbitone, HB = hexobarbitone, AP = antipyrine, BP = benzo[*a*]pyrene, CF = caffeine, EE = ethinylestradiol (after ref. 6)

known; all metabolites should be excreted by the kidney; and the marker should be pharmacologically inert and nontoxic at the doses used. Not surprisingly, no single compound fulfils all these criteria, but several are widely used as probes for different forms of drug-metabolizing enzymes. The use of tests involving antipyrine, aminopyrine, caffeine and tolbutamide in studies of enzyme induction and inhibition has recently been reviewed by Park.[7] In addition, other model substrates such as debrisoquine, sparteine and mephenytoin are widely used for phenotyping individuals for their ability to metabolize drugs, but they have not yet been used in studies involving enzyme induction and inhibition.

Less information is available concerning the measurement of drug-conjugating (phase 2) enzyme activity in man than has been obtained for drug oxidation. Nevertheless, a picture of multiple forms of enzymes is emerging, particularly from studies on genetic polymorphisms in drug acetylation, sulfation and methylation. From animal studies, for example, there is evidence of at least five forms of glucuronyltransferase.

Two agents, oxazepam and paracetamol, are used as probes to investigate glucuronidation in man and their value has also been reviewed by Park.[7]

23.8.5.2.2 *Measurement of enzyme activity* in vitro

In theory, measurement of enzyme activity *in vitro* should be the ideal method for assessing enzyme induction and inhibition, especially when linked to parallel *in vivo* studies. However, its use is limited by both ethical and practical considerations, *e.g.* difficulties in obtaining human liver samples on more than one occasion from the same individual. The development of human liver banks is especially useful for studies of induction and inhibition for two reasons. First, such material can be used to determine precise enzyme kinetic constants for established and new drugs. Using this approach, the inhibitory potential of a new drug can be investigated prior to volunteer studies and *in vivo* animal studies can be dispensed with. Animal investigations are often irrelevant anyway because of species differences in drug handling. Second, *in vitro* experiments will allow the characterization and purification of individual forms of enzymes concerned and therefore help define the relationship between enzymes and model drugs used in the *in vivo* assessment of drug metabolism.

23.8.5.2.3 *Noninvasive methods*

Noninvasive methods for the assessment of drug-metabolizing enzyme activity are defined as those which do not require administration of a test compound. Such tests involve measurement of changes in the disposition of an endogenous compound or alterations in the activity of an accessible (blood) enzyme. Substances which are used for this purpose include γ-glutamyltransferase, urinary D-glucaric acid and urinary 6β-hydroxycortisol (6β-OHC). Of these, 6β-OHC, a minor metabolite of cortisol, seems the most appropriate, since it is formed primarily in the endoplasmic reticulum of

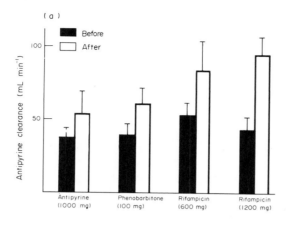

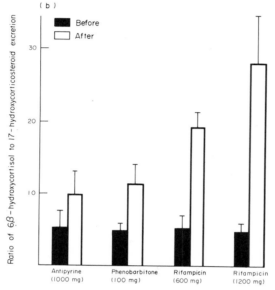

Figure 2 The effects of administration of antipyrine, phenobarbitone and rifampicin for 14 days on (a) antipyrine clearance and (b) 6β-hydroxycortisol excretion (reproduced from ref. 9 by permission of Springer-Verlag)

hepatocytes by a form of cytochrome P-450 and is excreted unconjugated. Urinary 6β-OHC is a sensitive marker of enzyme induction and can be used to monitor induction by nearly all known inducing agents. The effects of antipyrine, phenobarbitone and rifampicin are shown in Figure 2. The only inducing agents which do not affect 6β-OHC excretion are those agents which stimulate the induction of cytochrome P-448-type enzymes. Thus, cigarette smoking causes no change in 6β-OHC excretion. Of relevance is the finding in animal experiments that the aromatic hydrocarbon hydroxylase system may be induced independently of cortisol 6β-hydroxylase.

Two particular advantages of a noninvasive technique such as urinary 6β-OHC excretion are first that it can be used to investigate enzyme induction in sensitive groups (*e.g.* patients and children) and second it can be used to monitor the time course of induction; this is discussed later in this chapter.

23.8.6 ENZYME INDUCTION IN MAN—CLINICAL ASPECTS

As recently reviewed by Breckenridge,[8] enzyme-inducing agents of clinical importance can be considered to fall into two groups—substances that are used in therapeutics and substances found in the environment but to which man is exposed although not as therapeutic agents. Both of these broad types of inducing agent may influence the efficacy of co-administered drugs, and cause toxicity such as cancer, tissue necrosis or thrombotic lesions in their own right. These possibilities will be considered separately to demonstrate the scope of clinical enzyme induction.

23.8.6.1 Therapeutic Agents which Influence the Biological Action of Co-administered Drugs

One of the tenets of pharmacology is that the effect of many drugs is determined by their concentration at special sites of action or receptors. This, in turn, depends on factors such as their absorption, distribution and elimination. The rate of drug elimination is, in many instances, determined by rate of metabolism, and thus any factor modifying drug-metabolizing activity may result in modification of drug action.

As previously noted in this chapter and elsewhere in this volume, the main sites of drug-metabolizing activity in the body are the liver and the gut wall, although it should be mentioned at this juncture that white blood cells, placenta, skin and lung may play an important role in altering local drug efficacy and producing local toxicity. Such importance is magnified in the presence of the appropriate inducing agent.

This aspect of drug–drug interactions, namely enzyme induction produced by therapeutic substances, has received great attention over the past 20 years or so. In many respects, these studies have served only to show that, in overall biological terms, it represents one of the less important aspects of enzyme induction. There are at least three reasons for this statement.

(i) While it can be shown in experimental situations that many hundreds of chemically and pharmacologically unrelated chemicals have the ability to act as inducing agents in animal models, the number of drugs with important enzyme-inducing properties in man is small. Induction is in part a dose-related phenomenon, and since the doses used in clinical practice to produce a desired therapeutic effect are considerably less than those used in animal models to demonstrate an indicative property, most clinical effects will be of little importance. Further, there may also be species differences in drug handling between animal model and man and this may also be reflected in variations in inducing capacity in different species. Thus the only important inducing substances which are in widespread clinical use are anticonvulsants (*e.g.* phenobarbitone, phenytoin and carbamazepine) and the antibiotic rifampicin.

(ii) Enzyme-inducing agents will only cause major effects on the pharmacodynamics of a limited number of co-administered drugs, mainly those with a low therapeutic index (*e.g.* oral anticoagulants and antiarrhythmics) but also including oral contraceptives whose therapeutic index is reasonably high but whose dose, for safety reasons, has been progressively reduced over the years to a level where changes in plasma levels may result in therapeutic failure. From a clinical standpoint, the pharmacological effect of anticoagulant and antiarrhythmic agents must be monitored closely if enzyme-inducing agents are co-administered and, where appropriate, dose alterations made. In the case of contraceptive steroids, if an inducing agent is to be co-administered, either the dose of the steroid must be increased or alternative contraceptive methods used.

(iii) Unless the therapeutic potential of the drug is high, drug regulatory agencies tend to be wary of licensing agents which act as enzyme inducers. Reasons for this are discussed below, but the end result is that over the last 10 years or so few new drugs with potent inducing properties have been marketed.

Having said this, several aspects of clinical enzyme induction by therapeutic agents are of interest and importance, and fresh information has emerged over recent years.

23.8.6.1.1 Variation in the time course of induction

Methods of assessing the potency of various inducing agents have been discussed earlier in this chapter, and it was emphasized that one of the best ways of investigating the time course of induction is the use of the urine excretion of the endogenous substance 6β-hydroxycortisol. Ohnhaus and Park[9] used this method to investigate the time course of various inducing agents in man and showed (Figure 3) that augmentation of the steroid excretion was rapidly achieved with rifampicin, while the rate of increase using phenobarbitone or antipyrine as inducing agents was less rapid. These studies also demonstrated another important aspect of clinical enzyme induction, namely interindividual variability in response to various inducing agents. Reasons for this are not entirely clear, but it should depend principally on the rate of turnover of the enzyme being induced and, in different

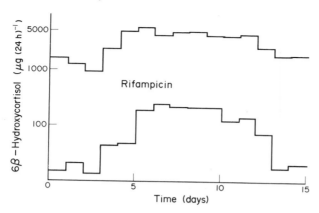

Figure 3 Urinary excretion of 6β-hydroxycortisol in two patients before, during and after treatment with rifampicin (reproduced from ref. 9 by permission of Springer-Verlag)

individuals, it seems probable that different forms of cytochrome *P*-450 are being induced at varying rates by various inducing agents.

23.8.6.1.2 Variation in the extent of induction

Breckenridge *et al.*[10] demonstrated many years ago that administration of the barbiturate quinalbarbitone to six patients on long-term warfarin therapy resulted in a fall in steady-state plasma warfarin concentration which ranged from 5 to 65% of the control value. Further, increasing the dose of quinalbarbitone resulted in a dose-dependent increase in the extent of induction. Ohnhaus and Park[9] have shown similar variability in induction and dose dependency in man using rifampicin as the inducing agent and measuring antipyrine clearance (Figure 2). This variability probably reflects factors similar to those described above.

23.8.6.1.3 Variation in metabolic pathways induced

It is now well established, and described elsewhere in this volume, that many different forms of cytochrome *P*-450 exist in human liver and other drug-metabolizing tissue. Many probes for these different forms have been proposed; one of the most widely used is antipyrine, three of whose urinary metabolites 4-hydroxyantipyrine, norantipyrine and 3-hydroxymethylantipyrine are formed by different forms of hepatic cytochrome *P*-450.[11] It has been suggested that, whereas the excretion of 4-hydroxyantipyrine is increased by administration of carbamazepine and phenytoin, excretion of norantipyrine is increased by rifampicin.

Reasons for these variations must include consideration of the drug and the individual. The physicochemical properties (*e.g.* lipid solubility) of the inducing agent, which will determine its accessibility to the liver, and the dose administered are both important considerations. With respect to the individual, Vesell and Page[12] showed that the inducing effect of phenobarbitone on antipyrine was proportional to the initial antipyrine half-life of the individual. A further important variable is the age of the subject. Twum-Barrima and co-workers[13] have clearly shown that, whereas rifampicin increased antipyrine clearance in a group of young subjects, it had no such effect in a group of elderly individuals.

23.8.6.2 Therapeutic Agents Causing Toxicity

Certain metabolites of drugs which are chemically more reactive than their parent compound may produce a range of adverse reactions by combining covalently with essential cellular components. In 1975, Mitchell and co-workers[14] explained the relationship between covalent binding of a drug or its metabolite and toxicity by involving four variables—the proportion of the drug converted to a reactive metabolite, the proportion of the reactive metabolite which becomes covalently bound, the proportion of the reactive metabolite that reacts with a critical site and finally the proportion of the critical site which cannot be repaired. The first of these terms depends on the activity of the enzymes needed to generate the reactive metabolite, although the ability of enzymes deactivating the chemically reactive metabolites will also be relevant.

Induction (and inhibition) of these enzymes by administration of other therapeutic agents will play an important role in determining drug toxicity caused in this way. Several important examples have now been presented. Bolt and co-workers[15] measured the covalent binding of ethinylestradiol in the liver microsomes of patients given rifampicin and of control subjects. Irreversible binding was some four times higher in the rifampicin-treated microsomes, suggesting that its administration had stimulated the production of a reactive metabolite. Similar experiments with guinea-pig liver microsomes have shown that rifampicin is a more potent inducer of the system, generating more of the toxic metabolite than other inducing agents such as phenobarbitone, β-naphthoflavone, antipyrine and spironolactone.[16] The enhanced vascular toxicity of contraceptive steroids in women who smoke cigarettes may have a similar explanation, since cigarettes contain potent inducing substances.

It is not only in liver cells that toxic metabolites can be produced. Mixed-function oxidase activity can be demonstrated in nonciliated bronchiolar (clara) cells in lung tissue, and Boyd[17] demonstrated that highly reactive metabolites of the pulmonary toxin 4-ipomeanol are both formed and bound in clara cells and that this process is inducible. The clinical and toxicological implications of these observations are not certain.

One group of patients who are subject to administration of inducing agents over long periods are epileptics. Phenobarbitone, phenytoin and carbamazepine are potent inducers of drug metabolic pathways. Phenobarbitone is known to cause proliferation and growth of hepatic cells in animal models, and, after initiating doses of carcinogens, induction promotes the growth of liver tumours. Whether a similar mechanism operates in man is a matter of considerable interest, but two studies[18,19] failed to show increased numbers of tumours in epileptics. Although the death rate in epileptics was some three times higher than predicted from an age-matched population, this was not due to the effects of enzyme induction but to apparently unrelated causes such as accidents and suicide. A further risk seen in epileptics on anticonvulsant-inducing agents is osteomalacia. Vitamin D in man is obtained largely by the effect of sunshine on the skin, dietary vitamin D being of minor importance. Vitamin D itself is inactive in calcium homeostasis and requires enzymic conversion. First it is converted in the liver to 25-hydroxyvitamin D by liver microsomal enzymes. This is then subject to two metabolic processes. It is converted in the kidney to 1:25-hydroxyvitamin D which has high activity in promoting calcium absorption from the intestine and calcium reabsorption from bone; this process is not inducible. The second possible metabolic process which is inducible is inactivation of 25-hydroxyvitamin D by glucuronidation within the liver and subsequent excretion in bile. Epileptics on anticonvulsants have low levels of 25-hydroxyvitamin D but normal levels of the 1:25-hydroxyvitamin. Dent and co-workers[20] proposed that the osteomalacia seen in treated epileptics was due to the increased biliary excretion of the metabolites of vitamin D. This mechanism has been contested by other workers but the original observation is not in doubt.

23.8.6.3 Environmental Agents Decreasing the Efficacy of Co-administered Drugs

The three agents which have been most widely investigated are cigarette smoking, diet and alcohol.

23.8.6.3.1 *Cigarette smoke*

Cigarette smoke contains many inducing agents, principally polycyclic aromatic hydrocarbons, and it was established some 25 years ago that the metabolism of many drugs was more rapid in smokers than nonsmokers. This topic was reviewed recently[21] and Table 1 shows those drugs whose metabolic elimination or pharmacological effects have been shown to be increased by cigarette smoking and those clinically important drugs not so affected. It must also be remembered that metabolism of xenobiotics in extrahepatic tissues may also be stimulated by cigarette smoking. Further, the work of Vestal and Wood[22] suggests that the effect of cigarette smoking on drug metabolism occurs predominantly in the young rather than in the elderly, in keeping with the observation[13] on the inducing effect of rifampicin referred to above. Thus, cigarette smoking has apparently a selective inducing effect and it is not possible at present to predict those agents whose

Table 1 Effect of Cigarette Smoking on the Metabolism and/or Action of Drugs[22]

Parameter	Drug
Decreased half-life and/or increased clearance	Antipyrine Caffeine Imipramine Warfarin Oxazepam Paracetamol
Increased urinary excretion	Nicotine
Decreased pharmacological action	Pentazocine Propoxyphene
Decreased adverse effects	Chlordiazepoxide Diazepam Theophylline
No effect	Nortriptyline Phenobarbitone Phenytoin

metabolism will be increased, although it is now known that reactivities of only certain forms of cytochrome *P*-450 are increased by smoking.[4]

23.8.6.3.2 Diet

Continued ingestion of such foods as charcoal broiled beef and brussel sprouts causes the rate of metabolism of various drugs to be increased in volunteers. Further, alteration in the balance of protein and carbohydrate in the diet determines the *in vivo* rate of metabolism of model drugs such as antipyrine and theophylline.[23] Whether the basis of these alterations is, in fact, enzyme induction as defined at the beginning of this chapter is not clear. The effect of malnutrition on drug metabolism has also been extensively studied in animals and to a lesser extent in man with somewhat conflicting results. This has recently been reviewed by Krishnaswamy.[24] Part of the confusion may be due to methods of expression of data. For some agents, *e.g.* contraceptive steroids, rates of elimination are apparently more rapid in malnourished women, but the basis of this effect is poorly understood.

23.8.6.3.3 Alcohol

The induction of xenobiotic metabolism by alcohol occurs predominantly in the liver rather than in extrahepatic tissues. *In vitro* and *in vivo*, the acute effect of alcohol on drug metabolism is inhibition; for an inductive effect chronic treatment is required.[21] Although the main effect of alcohol is on the monooxygenase system, it also affects some conjugation pathways. In animal models, certain forms of glucuronyltransferase in rabbit liver are induced by ethanol administration, but most attention has been paid to monooxygenase induction. It would now appear that ethanol induces its own type of cytochrome *P*-450, and this has now been purified and characterized in both rat and rabbit liver. The ethanol-inducible cytochrome *P*-450 is apparently important in the metabolism of nitrosamines which are well-known chemical carcinogens. How great a role the inductive effect of alcohol (a hepatotoxin in its own right) plays in the toxicity of carbon tetrachloride and paracetamol is not clear. Further, it is important to define whether studies have been carried out with ethanol or alcohol which may contain contaminants.

Alcohol (and ethanol) are both known to induce liver drug metabolism in man. Pelkonen and Sotaniemi[25] have investigated cytochrome *P*-450 levels in hepatic biopsies of alcoholics and, when compared to control subjects with normal liver histology, *P*-450 levels were higher than in equivalent controls. The same authors demonstrated that antipyrine clearance was higher even in alcoholics with normal liver histology, but investigation of this problem is difficult when the hepatotoxic effect of alcohol, producing fatty liver and cirrhosis, may intervene resulting in reduced rates of drug metabolism. It would appear that in man the extent of liver damage produced by alcohol is a more important determinant of xenobiotic metabolism by causing inhibition rather than any inductive effect.

23.8.6.4 Environmental Agents Producing Toxicity

This is potentially the most important clinical aspect of enzyme induction and, at the same time, the most difficult to study. It is suspected that environmental pollutants which are known to be inducing agents may cause disease in man by inducing microsomal enzymes. Even when populations have been exposed to carcinogenic agents such as polybrominated biphenyl (PBB), as in Michigan in the last decade, the associated morbidity from drug interactions leading to cancer is uncertain.[26] There are several reasons for this. First, many years may elapse before the disease expresses itself and clinical latency is extremely variable. The carcinogenic potential of more direct agents, such as vinyl chloride which produces angiosarcoma, β-naphthylamine causing bladder cancer, asbestos causing mesothelioma, leukaemia after exposure to benzene or ionizing radiation, all show marked clinical latency and intersubject variability. Contamination of a specific lot of dairy farm feed with PBB took place in the summer of 1973 in Michigan and the mistake was not evident for several months. By this time contaminated meat and dairy products had entered the food cycle. Detectable tissue levels of PBB were present in some 90% of those tested, with tissue to plasma ratios of 300:1. The second problem—again exemplified by the PBB case—is a lack of any apparent dose–response relationship between the toxin and its effect after induction. In pharmacology, responses, be they therapeutic or toxic, are usually expressed relative to dose and dose–response relationships are central to pharmacology. In experimental toxicology of the type described here, such a relationship is difficult to define. A third problem is the extrapolation from animal studies to man. This is exemplified by the work of Orrenius and co-workers[27] who showed that the effect of various types of inducing agents might have quite differing effects on toxicity. The toxicity of paracetamol, for example, is increased in mice by both phenobarbitone and methylcholanthrene, but while phenobarbitone will increase the toxicity due to carbon tetrachloride and bromobenzene, methylcholanthrene will decrease it. Thus, working with animals and making predictions for man is extremely difficult. The effects of enzyme induction depend not only on the inducing agent and the substrate but also on variations in the patient.

23.8.7 INHIBITION OF DRUG METABOLISM

23.8.7.1 Mechanisms of Inhibition

As information on pathways of drug metabolism was acquired in the 1950s, it became clear that a variety of chemical substances, including many used as therapeutic agents, could inhibit the enzymes involved. In 1954, two seminal papers appeared; Cook and co-workers from the Smith Kline and French laboratories in Philadelphia[28] and Cooper, Axelrod and Brodie from the National Institutes of Health[29] showed that an experimental compound, β-diethylaminoethyldiphenylpropyl acetate hydrochloride (better known as SKF 525A, originally designed as an antispasmodic drug but never developed as such), was a potent inhibitor of several drug metabolic pathways, *in vitro* and *in vivo*. SKF 525A has remained an important tool in our understanding of inhibition of drug metabolism, and, moreover, has led in many instances to the identification of drug hydroxylation as an essential reaction in the manifestation of biological effects *in vivo*. Thus, administration of SKF 525A and other inhibitors of drug metabolism has served to clarify the question of whether a compound or its metabolites are important in producing biological effects such as insecticidal action, mutagenesis or anaesthesia. Further, use of inhibitors of drug oxidation has increased our understanding of the mechanisms of drug oxidation, especially how a hydroxylated product is the end result of action of the cytochrome *P*-450 cycle.

As Netter has reviewed elsewhere,[30] inhibitors can act in several ways; these are summarized in Figure 4, which is taken from a recent review by that author.[5] This figure shows the outline of drug hydroxylation by hepatic monooxygenases and the sites of action of various types of inhibitor. First, a decrease of reducing equivalents necessary for drug hydroxylation can be achieved by redox agents that deplete intracellular NADPH or NADH. This, like many other inhibitory reactions, can be demonstrated both *in vivo* and *in vitro*. Second, destruction of important cofactors results in inhibition of drug hydroxylation. Third, competition with other electron acceptors such as cytochrome *C* will interrupt the reaction. Fourth, as the figure clearly shows, alteration of cytochrome *P*-450 itself by various agents will essentially prevent the transfer of activated oxygen to the substrate. Carbon monoxide, metyrapone and lack of oxygen inhibit drug oxidation in this manner. Fifth, competition for active sites on the enzyme by alternative substrates will inhibit the reaction; this is the most frequent basis of inhibition found in man and experimental animals. Some inhibitors (such as SKF 525A) bind to the same site as the substrate and are metabolized themselves. All the

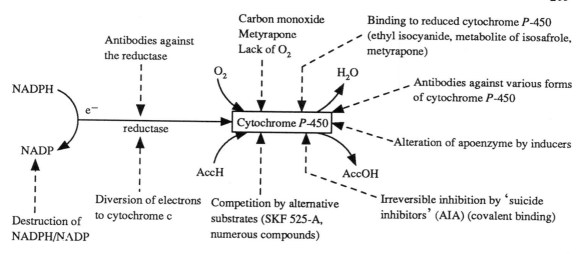

Figure 4 Schematic representation of the various modes of inhibitor interactions with cytochrome *P*-450

types of inhibition described so far are reversible. A sixth type of inhibitor, however, results in irreversible inhibition of the enzyme system concerned. Some of these inhibitors are themselves activated by cytochrome *P*-450, undergo subsequent covalent binding and thus cause an irreversible destruction ('suicide inhibition') of the enzyme. An example is 2-allyl-2-isopropylacetamide (AIA) which alkylates the porphyrin structure of cytochrome *P*-450 thus inhibiting its function.[5] Finally, antibodies against various components of the monooxygenase system have been raised and, provided these block appropriate parts of the enzyme, will act as inhibitors in experimental studies. Antibodies against various forms of cytochrome *P*-450 and against NADPH cytochrome *P*-450 reductase can also be used to investigate the importance of different parts of the enzyme system in metabolic reactions as well as acting as inhibitors of it.

23.8.7.2 Clinical Implications

The therapeutic problems associated with enzyme inhibition have undeservedly received much less attention than those associated with induction. In the context of clinical interactions due to inhibition, competition for the same substrate binding site seems the most likely mechanism and, as explained elsewhere in this chapter, *in vitro* studies can provide information relevant to the potential inhibitory nature of a drug *in vivo*. Efficacy as an inhibitor will depend on several factors, including the magnitude of the inhibitory constant (K_I) relative to the Michaelis constant (K_M) of the drug whose metabolism is inhibited. The relative maximal velocity of the metabolism of the inhibitor (V_{max}) and the drug are also important; other things being equal, inhibitory capacity increases to a maximum as V_{max} for metabolism of the inhibitor tends towards zero. Most drugs are metabolized by more than one route and Rowland[31] has developed a kinetic expression for the effect of inhibition on the ratio (R) of the new half-life of a drug in the presence of the inhibitor to the normal half-life. To determine R one needs to ascertain the fraction of the dose eliminated by a particular pathway of interest in absence of inhibitor (*fm*), the amount of inhibitor (I) and the inhibitor constant (K_I), as shown in equation (1).

$$R = t_{1/2}(\text{inhibited})/t_{1/2}(\text{normal})$$

$$= \frac{1}{fm/\{[1 + (I/K_I)] + (1 - fm)\}} \tag{1}$$

Using this model, Rowland was able to show that when all the drug is eliminated by the inhibited route (*fm* = 1) the ratio R changes dramatically with increasing inhibitor concentration. Below *fm* = 0.5, the maximum increase in the ratio is twofold and is inconsequential from a clinical standpoint; in such an instance, unless the therapeutic index of the drug is small, alteration in dosing is unnecessary. Interactions with oral hypoglycaemic agents, anticoagulant drugs and antihypertensive agents represent the most important and interesting clinical examples of inhibition of drug metabolism, partly because of the steep dose–response relationship of these agents. Imidazole

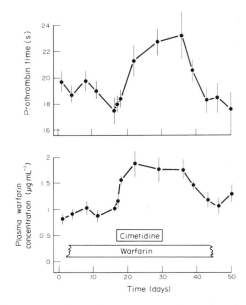

Figure 5 Cimetidine–warfarin interaction (reproduced from ref. 33 by permission of Lancet)

derivatives are specially important as inhibitors of drug oxidation, and Back and co-workers[32] have shown a clear structure–activity relationship with respect to position of substitution on the imidazole ring and the ability to inhibit drug oxidation. Cimetidine, a substituted imidazole and an H_2 antagonist, is an important inhibitor of drug oxidation (Figure 5).[33] Imidazole derivatives are thought to be potent inhibitors of cytochrome *P*-450 enzymes because they can bind to both the oxygen and substrate binding sites of these enzymes.

Drugs can alter the pharmacological action of other drugs by inhibiting nonmicrosomal pathways of metabolism and three examples of this are taken from a recent review by Park and Breckenridge.[34] Patients treated with monoamine oxidase inhibitors such as tranylcypromine are unusually sensitive to subsequent treatment with sympathomimetic agents which are metabolized by this mitochondrial enzyme. Severe hypertensive reactions can occur when individuals who have been treated with monoamine oxidase inhibitors eat cheese or other foods with a high tyramine content. The drug disulphuram used to treat alcoholism acts by inhibiting aldehyde dehydrogenase involved in the conversion of ethanol into acetic acid. When the alcoholic takes alcohol after ingestion of disulphuram there is acetaldehyde accumulation which produces nausea and flushing. A third example is allopurinol used in the treatment of gout by virtue of its ability to inhibit xanthine oxidase which catalyzes the conversion of hypoxanthine to xanthine and of xanthine to uric acid. This enzyme is also involved in the metabolism of synthetic xanthine analogues such as 6-mercaptopurine; allopurinol has been shown to increase both the toxicity and therapeutic actions of 6-mercaptopurine.

23.8.8 REFERENCES

1. A. H. Conney, E. C. Miller and J. A. Miller, *Cancer Res.*, 1956, **16**, 450.
2. H. Remmer, *Naunyn-Schmiedeberg's Arch. Pharmakol. Exp. Pathol. Pharmakol.*, 1959, **235**, 279.
3. R. Greinert, S. A. E. Finch and A. Stier, *Xenobiotica*, 1982, **12**, 717.
4. F. J. Gonzalez, *Pharmacol. Rev.*, 1989, **40**, 243.
5. K. J. Netter, *Pharmacol. Ther.*, 1987, **33**, 1.
6. D. D. Breimer, *Clin. Pharmacokinet.*, 1983, **8**, 371.
7. B. K. Park, *Pharm. Int.*, 1986, **7**, 123.
8. A. Breckenridge, *Pharmacol. Ther.*, 1987, **33**, 95.
9. E. E. Ohnhaus and B. K. Park, *Eur. J. Clin. Pharmacol.*, 1979, **15**, 139.
10. A. Breckenridge, M. L. Orme, L. Davies, S. S. Thorgeirsson and D. S. Davies, *Clin. Pharmacol. Ther.*, 1973, **14**, 514.
11. M. Danhof and D. D. Breimer, *Br. J. Clin. Pharmacol.*, 1979, **8**, 529.
12. E. S. Vesell and J. G. Page, *J. Clin. Invest.*, 1969, **48**, 2202.
13. Y. Twum-Barima, T. Finnigan, A. I. Habash, R. D. T. Cape and S. G. Carruthers, *Br. J. Clin. Pharmacol.*, 1984, **17**, 595.

14. J. R. Mitchell, W. Potter, J. A. Hinson, R. Snodgrass and J. R. Gillette, in 'Concepts of Biochemical Pharmacology', ed. J. R. Gillette, Springer, New York, 1975, p. 383.
15. H. M. Bolt, H. Kappus and M. Bolt, *Eur. J. Clin. Pharmacol.*, 1975, **8**, 301.
16. B. K. Park, A. D. Wittaker and M. R. Challiner, *Life Sci.*, 1978, **23**, 2463.
17. M. R. Boyd, *Nature (London)*, 1977, **269**, 713.
18. J. Clemmenson, V. Fuglsangsen and C. M. Plum, *Lancet*, 1974, **1**, 705.
19. S. J. White, A. E. M. McLean and C. Howland, *Lancet*, 1979, **2**, 458.
20. C. E. Dent, A. Richens, D. J. F. Rowe and T. C. B. Stamp, *Br. Med. J.*, 1970, **4**, 69.
21. K. Vahakangas, M. Pasanen, E. A. Sotaniemi and O. Pelkonen, in 'Enzyme Induction in Man', ed. E. A. Sotaniemi and R. O. Pelkonen, Taylor and Francis, London, 1987, p. 189.
22. R. E. Vestal and A. J. J. Wood, *Clin. Pharmacokinet.*, 1980, **5**, 309.
23. A. Kappas, A. P. Alvares, K. E. Anderson, E. J. Pantuck, R. Chang and A. H. Conney, *Clin. Pharmacol. Ther.*, 1978, **23**, 445.
24. K. Krishnaswamy, in 'Nutritional Toxicology', ed. J. N. Hathcock, Academic Press, New York, 1986, vol. 2, p. 105.
25. O. Pelkonen and E. Sotaniemi, *Pharmacol. Ther.*, 1982, **16**, 261.
26. I. J. Selikoff, *Ciba Found. Symp.*, 1980, **76**, 331.
27. S. Orrenius, H. Thor and B. Jernstrom, *Ciba Found. Symp.*, 1980, **76**, 25.
28. L. Cook, J. J. Toner and E. J. Fellows, *J. Pharmacol. Exp. Ther.*, 1954, **111**, 131.
29. J. R. Cooper, J. Axelrod and B. B. Brodie, *J. Pharmacol. Exp. Ther.*, 1954, **112**, 55.
30. K. J. Netter, *Pharmacol. Ther.*, 1982, **10**, 515.
31. M. Rowland, in 'Pharmacology and Pharmacokinetics', ed. T. Tevrell, R. L. Derrick and P. G. Condliffe, Plenum Press, London, 1975, p. 321.
32. D. J. Back and J. F. Tjia, *Br. J. Pharmacol.*, 1985, **85**, 121.
33. M. J. Serlin, R. G. Sibeon, S. Mossman, A. M. Breckenridge, J. R. Williams, J. L. Atwood and J. M. T. Willoughby, *Lancet*, 1979, **2**, 317.
34. B. K. Park and A. M. Breckenridge, *Clin. Pharmacokinet.*, 1981, **6**, 1.

23.9

Species Differences in Drug Metabolism

MONT R. JUCHAU
University of Washington, Seattle, WA, USA

23.9.1   INTRODUCTION: GENERAL CONSIDERATIONS OF SPECIES DIFFERENCES

Profound differences among species in rates and routes of biotransformation of small foreign organic molecules (xenobiotics) have been recognized for several decades. It is noteworthy that, in spite of apparent potential value to government regulatory agencies, to veterinary medicine and to human health, rigorous systematic investigations of species comparisons pertaining to xenobiotic biotransformation have been very sparse and progress toward understanding the *basis* for species differences has been remarkably slow. With an increasing deemphasis on experiments with whole animals, prospects for more rapid progress in the near future do not appear particularly encouraging. A number of other factors also bode for continued slow progress in this area of research. These include the sheer magnitude of the investigative area (innumerable species), a continuing lack of prospects that any experimental animal (or even combinations of animals) will be utilizable as a meaningful model for biotransformation in humans, and a prevalent perception that comparative investigations represent studies of lesser scientific merit. While species differences in drug biotransformation provide graphic examples of inherited (genetic) differences in drug-biotransforming capacities, a much more useful and rational approach toward the elucidation of the basis for genetic differences lies in the use of inbred *strains* from within a specific species rather than in comparisons of species differences *per se*. Research to date on species differences in drug biotransformation permits the following generalizations.

(i) Species differences *can* be very profound, with differences in measured rates by as much or more than three orders of magnitude when comparing the same organ (usually liver) of different species. In some species, specific enzymes appear to be entirely absent.

(ii) Measured *in vitro* differences are frequently reflected in measurements of the plasma half-lives of the parent xenobiotics, particularly when measuring rates of oxidation reactions. However, exceptions are sufficiently numerous to warn against reliance on this generalization.

(iii) Species differences are not currently predictable on the basis of phylogenetic considerations.

(iv) Species differences are not predictable on the basis of any other known criterion except that, in *very* general terms, rates appear inversely correlated with mean body weights among various species (Walker's rule). For example, the laboratory mouse often metabolizes drugs at rates that are considerably (one to two orders of magnitude) higher than rates of the same reactions in humans. Exceptions to this generalization, however, are very numerous.

(v) Strain differences can be as marked as species differences in many cases and in some instances are even more profound. This greatly complicates species comparisons because results from a comparison of species will depend upon the strains selected for study.

(vi) The underlying generalized basis for species specific differences in drug biotransformation lies primarily in differences among the nature, number and activities of *enzymes* that catalyze the individual reactions. Thus the question of a fundamental basis for difference is shifted to that of an acquisition of understanding of the species-specific complement of enzymes that possess the capacity to catalyze drug-biotransforming reactions. Desired, of course, is not merely a description of the complement/activities of drug-biotransforming enzymes in each of the myriad species but an understanding of the basic patterns and the phylogenetic development of each.

(vii) Various species differ markedly not only in basal levels and activities of biotransforming enzymes in normal adults but also in the *regulation* of such. This includes responses to inducing and repressing agents, sex differences, developmental/age differences, response to inhibitors and activators, response to hormonal factors and response to various nutritional and pathologic factors. Again, an underlying rationale for such regulatory differences among species is not currently available.

(viii)   The laboratory rat, which has been studied more intensively than any other species in terms of xenobiotic biotransformation, exhibits numerous and profound differences from the human and thus represents a particularly poor model for extrapolation of drug biotransformation data from experimental animals to humans.

(ix) A lack of *systematic* investigations of species differences renders rigorous quantitative statements difficult and also impedes the development of a theoretical underpinning for differences observed randomly in connection with experiments designed for other purposes.

At the outset it should be made clear that this review of the subject is in no way comprehensive. Entire books and book chapters[1-4] have been published on this topic and space limitations alone preclude such a formidable undertaking. The approach that has been elected is to point out some of the more recent examples and progress in this area in order to provide the reader with a concept of the enormous problems associated with attempts to generalize in this research field. A goal is to provide an introduction to the literature in the area to assist researchers, regulators and teachers in their attempts to gain varying degrees of mastery of this enormous subject. The topic is discussed with relation to the major groups of biotransforming enzymes (Table 1).

23.9.2   SPECIES DIFFERENCES IN OXIDATION OF XENOBIOTICS

23.9.2.1   *P*-450 Enzymes

By far the greatest amount of research relating to differences in drug biotransformation among species has dealt with cytochrome *P*-450 dependent monooxygenation, the associated isozymes and regulation thereof.[4,5] As of this writing, sufficient data on hepatic microsomal *P*-450 isozymes in

Table 1 Common Biotransformation Pathways[a] and Enzymes Involved in Biotransformation

Pathway	Enzymes
I. Oxidation	*I. Oxidoreductases*
(a) Monooxygenation	(a1) Cytochrome *P*-450 dependent systems
	(a2) Flavin dependent monooxygenases
(b) Dioxygenation	(b) Dioxygenases
(c) Dehydrogenation	(c) Dehydrogenases
(d) Peroxidation	(d) Peroxidases
(e) Monoamine oxidation	(e) Monoamine oxidases
II. Reduction	*II. Oxidoreductases*
(a) Nitro groups	(a) Nitro reductases
(b) Azo linkages	(b) Azo reductases
(c) Epoxides	(c) Epoxide reductases
(d) Carbonyl groups	(d) Dehydrogenases
(e) Unsaturated carbon–carbon bonds	(e) Dehydrogenases
III. Hydrolysis	*III. Hydrolases*
(a) Carboxylic acid esters	(a) Esterases
(b) Amides/peptides	(b) Amidases
(c) Sulfate esters	(c) Sulfatases
(d) Glucuronic acid esters	(d) Glucuronidases
(e) Epoxides	(e) Epoxide hydrolases
IV. Conjugation	*IV. Transferases*
(a) Glycosylation	(a) Glycosyl transferases
1. Glucuronidation	1. Glucuronosyl transferases
2. Riboside formation	2. Ribosyl transferases
3. Glucoside formation	3. Glucosyl transferases
(b) Sulfation	(b) Sulfokinases
(c) Acylation	(c) Acyl transferases
1. Acetylation	1. Acetyl transferases
2. Glycine/glutamine conjugation	2. Amino acid transferases
(d) Glutathione conjugation	(d) Glutathione transferases
(e) Methylation	(e) Methyl transferases

[a] Only the more common pathways are listed. For a more complete enumeration see ref. 2, volumes I and II.

rats, rabbits, mice and humans have accumulated to permit preliminary glimpses into the evolutionary acquisition of isozymic forms by various species. Recently, a group of researchers in the area proposed a standardized nomenclature for the increasingly large number of *P*-450 isozymes.[5] Rats, mice, rabbits and humans were the species for which enough information on hepatic microsomal *P*-450s was available for classificatory inclusion (Table 2). Other species for which similar information should become available in the relatively near future include certain marine species (*e.g.* trout, cod, scup, skate), guinea pigs, chickens, frogs, hamsters and perhaps dogs and monkeys. Interest in marine species has been relatively high[6,7] and it seems probable that definitive information on aquatic species will become available more rapidly than for certain well-known mammalian species such as swine, cats, sheep, cattle or horses. Interest in *P*-450 dependent biotransformation in nonvertebrate species has centered primarily in bacteria, yeast, fungi, houseflies, drosophilia and earthworms.

One of the most intensively investigated *P*-450 isoenzymes (P450CIA1, previously referred to as *P*-450$_{cam}$) has been isolated and purified from camphor-induced *Pseudomonas putida*. Detailed studies of this nonmembrane-bound isozyme at one phylogenetic extreme and comparisons with microsomal mammalian *P*-450 isoenzymes at the opposite end have led to the concept that all *P*-450 isozymes share a single basic monooxygenation mechanism, but vary greatly in physiologic/pharmacologic functions because of differences in substrate specificity and capacity for substrate turnover.[8] It is of considerable phylogenetic interest that the bacterial cytosolic *P*-450 dependent monooxygenation system shares with mammalian mitochondrial *P*-450 dependent systems the requirement for a nonheme iron protein. The eukaryotic *P*-450 isozymes associated with the endoplasmic reticulum, nuclear membranes, golgi membrane or cellular membranes do not utilize nonheme iron proteins in their respective hydroxylation reactions. All vertebrate *P*-450 systems studied to date prefer NADPH as initial electron donor, whereas cytosolic bacterial systems prefer NADH. Knowledge of the relationship between dependency upon specific carbon energy

Table 2 Standardized Nomenclature for the More Thoroughly Investigated *P*-450 Isozymes[a]

Standardized designation	Common designations, synonyms, orthologous forms	Major tissue source
P450IA1	Rat c, mouse P_1, rabbit 6, human P_1	Liver microsomes
P450IA2	Rat d, mouse P_3, rabbit 4, human P_3	Liver microsomes
P450IIA1	Rat a, UTF, 3	Liver microsomes
P450IIA2	Human *P*-450 (1)	Liver microsomes
P450IIB1	Rat b, rabbit 2	Liver microsomes
P450IIB2	Rat e	Liver microsomes
P450IIC1	Rabbit PBc1	Liver microsomes
P450IIC2	Rabbit PBc2, k	Liver microsomes
P450IIC3	Rabbit PBc3, 3b	Liver microsomes
P450IIC4	Rabbit PBc4, 1–8	Liver microsomes
P450IIC5	Rabbit 1	Liver microsomes
P450IIC6	Rat PB1	Liver microsomes
P450IIC7	Rat f	Liver microsomes
P450IIC8	Human 1	Liver microsomes
P450IIC9	Human mp	Liver microsomes
P450IIC10	Chicken PB15	Liver microsomes
P450IID1	Rat db1, human db1	Liver microsomes
P450IID2	Rat db2	Liver microsomes
P450IIE1	Rat j, rabbit 3a, human HLj	Liver microsomes
P450IIIA1	Rat p(pcn 1), rabbit 3c	Liver microsomes
P450IIIA2	Rat p(pcn 2), rabbit 3c	Liver microsomes
P450IIIA3	Human HLp, nf, 5	Liver microsomes
P450IVA1	Rat CLO, 452, LAω	Liver microsomes
P450XVIIA1	Bovine 17α, human 17α	Adrenal microsomes
P450XXIA1	Bovine C21A, mouse C21A, human C21A	Adrenal microsomes
P450XXIA2	Bovine C21B, mouse C21B, human C21B	Adrenal microsomes
P450XIA1	Bovine 11β, human 11β	Adrenal mitochondria
P450XXIIA1	Bovine scc, human scc	Adrenal mitochondria
P450LIA1	Yeast lan	Yeast
P450CIA1	*Pseudomonas putida*$_{cam}$	*Pseudomonas putida* cytosol

[a] Listing is adapted from ref. 5. For other common names, characteristics and synonyms, see Table 2, Chapter 23.10. Note that sex specific rat microsomal enzymes designated as *g*, *h*, *i* are not listed in this table.

sources for growth of microorganisms and the specific *P*-450 isozymes detected in the same organisms coupled with the known influence of diet/nutrition on differential regulation of vertebrate *P*-450s may lead to some general concepts of the basis for species difference in *P*-450 isozymes as well as other xenobiotic-biotransforming enzymes. Clearly, differences in *P*-450 isozyme complement among vertebrate species may also be strongly related to differences in endocrine (particularly steroidal) systems. Also, the degree to which various organisms are exposed to potentially lethal organic chemicals may be an important determinant. Whether the major consideration is nutritional, hormonal or toxicological, however, the *historical* exposure of the organism to specific small liposoluble organic chemicals (endogenous or exogenous) now appears to be the major determinant as to whether a specific *P*-450 isozyme or family of *P*-450 isozymes will appear in substantial quantities in a given species. Optimum survival would demand that such systems be in place. The best known examples for this principle are found in the bacterial systems in which specific *P*-450 isoenzymes (*e.g.* P450CIA1) do not appear unless the organism is grown upon the substrate as the sole carbon/energy source. Examples in mammalian species are also now well established, however, inasmuch as it is now known for rats that *P*-450s IA1, IIB1 and IIIA1 are not measurably present in hepatic tissues until after exposure to various respective inducing agents. Mammalian *P*-450s commonly regarded as constitutive are continuously exposed to an endogenous substrate (*e.g.* P450XXIIA1/cholesterol). Although these ideas seem somewhat simplistic, they may assist in assemblage of a foundation for the basis of species differences on *P*-450 dependent oxidation.

With respect to the use of various species as predictive 'models' for extrapolation of *P*-450 dependent drug biotransformation data from experimental animals to humans, the adult male rat has been the most intensively investigated. The utility of microorganisms as models has been discussed by Clark *et al.*[9] in a recent review. Rabbits, mice and marine vertebrates are also relatively commonly utilized. In terms of predicting the chemical *nature* of the various metabolites that could be generated from any specific xenobiotic, such models appear to have considerable utility. Problems arise in extrapolation of factors governing *rates* of generation of individual metabolites. These problems are not trivial because it is now clearly evident that the relative *rates* of various biotransformation reactions are crucial determinants of pharmacologic/toxicologic manifestations. This is possibly best illustrated with the utilization of the rat as a potential model. It is now known that sex specific *P*-450 isozymes account for the very notable sex differences observed in rats but in no other mammalian species studied thus far. Homologous sex specific, xenobiotic-biotransforming isozymes are not known to exist in humans, rabbits or mice. One of these male specific isozymes in rats is commonly referred to as *P*-450*h* (no standardized name assigned as yet) and is decreased markedly after exposure of rats to most of the common *P*-450 inducers. It is also decreased in response to a variety of physiological/pathological changes, *e.g.* starvation, diabetes, old age, *etc.* Thus many of the factors affecting rates of xenobiotic biotransformation in male rats do not apply to humans or even other experimental animals. It now seems clear that the male rat is unique in terms of the manner in which impinging factors alter rates of xenobiotic biotransformation and that this animal is not at all useful as a model for studies of the effects of such regulatory factors. Increased knowledge of species specific *P*-450 isozymes has already assisted (and will continue to assist) in understanding why this is the case.

An additional example of the unsuitability of the rat as a model for human xenobiotic biotransformation is provided by comparisons of responses of rats *vs.* humans to rifampicin. This macrolide antibiotic is a powerful inducer of certain *P*-450 isoenzymes in humans, pigs and rabbits—with concomitant increases in rates of several biotransforming reactions readily measurable.[10] Thus far, no one has succeeded in demonstrating that this antibiotic has any inducing activity in rats and guinea pigs.

From another standpoint, however, the rat may be regarded as a useful 'model' for understanding the various ways by which *P*-450 systems *can* be regulated. This information, of course, is of considerable basic importance, particularly from a more global point of view. An understanding of the many possible modes and mechanisms whereby *P*-450 isozymes *can* be regulated among various species should eventually provide the basis for a complete understanding of regulation in humans. Recent advances in molecular biology promise to provide rapid advances in the acquisition of the requisite data.

Major gaps to be filled are in the areas of species specific ontogenic and extrahepatic biotransformation. Formulation of a theory providing the general underlying basis for species differences in xenobiotic metabolism does not seem feasible in the absence of such data. Judging from past experience, however, the time required to accumulate sufficient meaningful pertinent data to permit the promulgation of general principles concerning species specific differences in ontogeny and extrahepatic biotransformation will be very long indeed.

An interesting example of the difficulty in extrapolating data on *P*-450 induction from species to species is provided by considerations of *Bacillus megaterium* (bacterium), *Salmo gairdneri* (trout) and mammals. The livers of trout and various other marine species do not respond well to phenobarbital as an inducing agent,[7] whereas all mammalian species thus far investigated respond quite profoundly (although variably). This would tend to lead to the prediction that species lower than trout on the phylogenetic scale would not respond to phenobarbital. However, there now exist numerous reports that a prokaryote (*Bacillus megaterium*) responds very well to the *P*-450-inducing effects of phenobarbital and other barbiturates (ref. 11 and refs. therein) and that the induced *P*-450 is a cytosolic protein with a molecular weight of 119 kDa. However, there have been no reports of additional barbiturate inducible *P*-450 dependent monooxygenase systems in prokaryotic species. The principle function of the NADPH dependent inducible bacterial protein (it does not appear constitutively) appears to be in the ω-2-hydroxylation of long chain fatty acids, amides and alcohols and the oxidation and/or hydroxylation of unsaturated fatty acids. The extent to which the observations concerning this *P*-450 and its inducibility can be fitted into a meaningful phylogenetic schema is not at all clear at the present time. The difficulty is also evident from the discussions of Wiseman's recent review of *P*-450s in microorganisms.[12] Other microbial species (in addition to *B. megaterium* and *P. putida*) in which *P*-450s have been investigated include *Corynebacterium*, *Streptomyces setonii*,[13] *Nocardia*, *Candida tropicalis*, *Cunninghamella*, *Eregane baineri*, *Rhizopus nigricans*, *Claviceps purpurea*, *Saccharomyces cerevesiae*, *Moroxella* spp., *Streptomyces erythraeus*, *Fusarium oxysporum* and *Aspergillus achracens*. *C. purpurea* also exhibits a phenobarbital inducible *P*-450 that appears to function in the biosynthesis of certain alkaloids. This suggests that responsiveness to phenobarbital induction can occur throughout the phylogenic spectrum but that certain species do not exhibit responsiveness for reasons still obscure. It is also of interest that prenatal liver in mammalian species responds either poorly or not at all to phenobarbital induction. By contrast, responsiveness to 'MC-type' inducers such as polycyclic aromatic hydrocarbons, various flavones and polyhalogenated aromatic hydrocarbons (TCDD as prototype) seems to occur (with certain notable exceptions) in virtually all cells and may well represent a very primitive and relatively undifferentiated but highly important system. Nevertheless, certain strains of mice (*e.g.* DBA/2N) exhibit minimal responsiveness to MC induction with no apparent survival disadvantage.

An example of a recent investigation of species differences in extrahepatic *P*-450 dependent biotransformation[14] provides excellent insights into the profound and unpredictable differences that can be observed in terms of both the constitutive presence and inducibility of various isozymes among rats, rabbits, mice, hamsters, monkeys and guinea pigs. The results dramatically illustrate the impossibility of drawing appropriate conclusions concerning extrahepatic species differences based on measurements of hepatic differences. Rabbit isozyme 5, which has not yet been named with the standardized nomenclature,[5] is detectable in the livers of only rabbits and hamsters (immunoblots), but is present in the lungs of all six species. The isozyme is also not phenobarbital inducible in the lungs of any of the six species. It is markedly induced in livers of rabbits but not of the other five species. The species differences in immunochemical measurements of isozyme 5 were paralleled by closely correlated differences in rates of *N* hydroxylation of aminofluorene.

An interesting example of an exception to the generalization ('Walker's rule') that rates of biotransformation in hepatic tissues are inversely related to body size[15] is provided by a recent comparative study on rates of biotransformation in rats *vs.* beagle dogs.[16] Control dog liver microsomes metabolized this substrate approximately 15 times more rapidly than rat liver microsomes. The study indicated that the marked species difference observed was due to the presence in dog liver microsomes of a *P*-450 isozyme with highly efficient metabolizing activity for the investigated substrate.

Other investigations that demonstrate the existence of species differences in *P*-450 dependent drug oxidation are far too numerous to mention in this brief review. Needed are rigorous comparisons in which the *basis* for observed differences is sought with techniques providing the highest available degree of sensitivity, reproducibility and insight. Investigations such as those of Schwen and Mannering,[17] which provide a working hypothesis for future studies, are to be highly encouraged. This comparative study of rat, trout, frog and snake suggests the hypothesis that phenobarbital inducibility is not functional in nonmammalian vertebrates. Many other studies would be needed to verify and extend this idea but a feasible testable working hypothesis is provided. If the hypothesis were to prove correct, the more interesting question is 'What survival advantage to nonmammalian vertebrates does absence of phenobarbital inducibility present?' For the molecular biologist, 'What evolutionary developments have resulted in the lack of capacity of phenobarbital to induce *P*-450s in certain bacteria and nonmammalian vertebrates yet presence of the same capacity in other bacteria, yeasts and mammalian vertebrates?' Ultimately, 'How can an understanding of the *basis* of such species differences assist in predicting the effects of chemicals on living organisms?' Clearly these are challenging yet formidable questions that will probably occupy scientists for decades to come.

23.9.2.2 Flavin Monooxygenases

In vertebrates, a flavin-containing monooxygenase (FMO, EC 1.14.14.8) is a microsomal enzyme that contains neither heme nor metal and catalyzes the oxygenation of a wide variety of nucleophilic nitrogen- and sulfur-containing xenobiotics to their corresponding *N*- or *S*-oxides. Important substrates include phenothiazines, secondary and tertiary amines and thioamides. FMOs exhibit some overlap in substrate specificity with the *P*-450 family but have been studied less intensively, due primarily to the relative lability of hepatic forms. The only known endogenous substrate for FMO is cysteamine, which is oxidized to cystamine, but a very large number of xenobiotic substrates have been identified, including several drugs.

Studies regarding species differences with respect to FMO have been relatively scarce. The existing earlier (prior to 1980) data concerning species differences for FMOs have been summarized by Ziegler.[18] These earlier investigations were performed by studying rates of oxidation of various substrates, particularly the *N* oxidation of dimethylaniline. Widely varying (as much as 50-fold) values were attained from different laboratories, rendering conclusions tentative at best. Of interest was the reported sex differences observed in mice with activities in females approximately five times that of males. When comparisons of activities in liver microsomal fractions measured in the same laboratory were considered, the differences among hogs, rats, rabbits, hamsters and guinea pigs were less than twofold at the extremes.

The great majority of the work concerning the characterization of FMO has been performed by Ziegler and his associates with purified porcine hepatic enzyme (ref. 19 and refs. therein). More recent investigations of the pulmonary enzyme[20,21] support the concept that FMO can exist in multiple forms. The latter study is of particular pertinence to this review because enzymes from mouse, rabbit and rat were studied in hepatic as well as extrahepatic tissues. These authors pointed to discrepancies in earlier studies concerning measurements of species differences and identified many of the problems associated with attempts to make rigorous species comparisons. They reported that pulmonary FMOs of the mouse, rabbit and rat appeared to have some catalytic properties quite different from one another in terms of stability to inactivation by alkaline conditions, detergents and heat. The mouse (but not rat or rabbit) pulmonary enzyme(s) was markedly increased in activity by the presence of N-octylamine but the same activator had no effect on the corresponding hepatic enzymes. Activities in renal microsomes resembled those in hepatic microsomes in these three species—at least in a qualitative sense. The authors also concluded that the substrate-metabolizing potential of specific FMOs among different species and tissues is phylogenetically conserved to a considerable extent. Sabourin and Hodgson[22] reported differences in pH optima, kinetic constants and immunoactivity when comparing purified FMOs from the livers of pigs and mice. To this author's knowledge, investigations in submammalian phyla have not been reported, although numerous flavin monooxygenase dependent reactions are known to occur in lower animals. It is clear that definitive investigations concerning the quantitative aspects of FMO-related species differences are only now beginning to emerge. These studies have dealt virtually exclusively with mammals and only a very few mammalian species have been studied, primarily pigs, mice, rats and rabbits, with some studies in humans, guinea pigs, hamsters, *etc.* A study by Dannan and Guengerich[23] showed a lack of immunochemical identity among FMOs from hogs, mice, rats, rabbits, dogs and humans, which greatly hindered efforts to compare enzyme content in tissues of these species. Antibodies were raised against the porcine hepatic enzyme and immunoquantitations were conducted with these same antibodies, rendering any conclusions strictly tentative. A more recent study by Tynes and Philpot is subject to similar problems.[24] Such studies have, however, provided further evidence for the wide tissue distribution and multiplicity of the FMOs and demonstrated their variable presence in tissues of several species, including humans. It seems apparent that a relatively large number of tissue specific and species specific isozymes may exist and that these can vary considerably in their substrate specificities as well as in other properties. Marked species differences now seem apparent, especially among pulmonary FMOs in comparisons of the mouse and rabbit. A huge amount of future research will be needed before definitive quantitative statements can be made concerning the phylogenetic aspects of this important group of xenobiotic-biotransforming enzymes.

23.9.2.3 Dehydrogenases

These enzymes are ubiquitous, being present in virtually all cells throughout all phyla. Isozymic forms of specific dehydrogenases are extremely numerous and the genetics of various mammalian dehydrogenases (particularly alcohol and aldehyde dehydrogenases) have been the subject of numerous investigations over the past several decades. The alcohol and aldehyde dehydrogenases have also been those of greatest interest in pharmacology/toxicology and are those which come to mind in discussions of xenobiotic biotransformations involving dehydrogenases, although, clearly, other dehydrogenases are also involved.

A consideration of the heavily investigated alcohol dehydrogenases is illustrative of the problems in this area. A problem still not resolved in questions of species differences in rates of alcohol oxidation is that of the importance of levels and activities of alcohol and aldehyde dehydrogen-ases.[25] One view is that the rate at which NAD is regenerated limits the rate of ethanol oxidation. An opposing view is that the level and/or activity of catalyzing of alcohol dehydrogenase(s) determines the rate. Until this question can be resolved satisfactorily, questions pertaining to the basis for species differences remain moot. It is of interest that scientists have regarded the two views as opposing and mutually exclusive. It is also of interest that the pharmacology/toxicology literature (textbook or journal) has given little or no recent attention to the species difference question, suggesting that species differences are of a magnitude that do not merit attention or that the questions surrounding such differences defy satisfactory resolution (or perhaps would be more easily resolvable by first approaching strain differences) at present. That the former is not tenable is indicated by the myriad studies of genetic differences in various animal *strains* and among various races of humans. Additionally, it is known that the rate of oxidative ethanol metabolism in humans proceeds two to three times faster than that in horses (approximately $0.3 \text{ g kg}^{-1} \text{ h}^{-1}$) despite a fivefold higher activity of the horse dehydrogenase(s).[26] Price Evans[27] and Atlas and Nebert[28] have discussed the problems encountered in relating alcohol dehydrogenase activities (among various polymorphic forms) to rates of ethanol elimination in humans. It seems clear that activities as assessed *in vitro* do not bear a simple relationship to rates of metabolism measured *in vivo*. It also seems almost incredible that there still exists virtually no basis of understanding for differences among various species and strains in terms of rates of metabolic clearance in spite of the practical uses to which such knowledge could be applied and the very long time that the problem has been recognized. A very recent review[29] suggests that although some progress has been made in finding solutions to these problems, they are still far from being satisfactorily resolved. The review suggests that scientists may finally be agreeing that *either/both* the availability of NAD *and/or* the levels/activities of alcohol dehydrogenases can limit rates of ethanol oxidation, depending upon specific conditions. It may be that, at lower ethanol concentrations, the dehydrogenase levels/activities are the more critical determinants, whereas at higher levels, availability of NAD becomes rate limiting. (At *very* high concentrations the *P*-450 dependent microsomal ethanol-oxidizing system (MEOS) may be of some importance.) It seems hopeful that substantial progress on this small but highly important area of species differences in dehydrogenation reactions will be made in the near future. Studies of species differences in drug oxidation reactions catalyzed by other dehydrogenases are too sparse and scattered to enable any meaningful discussion at present.

23.9.2.4 Other Oxidative Enzymes

In addition to the *P*-450s, FMOs and dehydrogenases, numerous other enzymes and enzyme systems are known to catalyze the oxidation of xenobiotic chemicals. These include peroxidases, dioxygenases, prostaglandin synthase, other *P*-450 independent monooxygenases, monoamine oxidase and other oxidases such as xanthine oxidase, aldehyde oxidase, *etc*. However, insofar as species differences in oxidation of *foreign* compounds are concerned, the data in the literature concerning these enzymes are sufficiently scattered and nondefinitive to render attempts to discuss the topic of their species distribution relatively meaningless at present. In addition, many examples of the corresponding reactions studied (*e.g.* peroxidative oxidation) have been demonstrated *in vitro* but not *in vivo*, rendering the significance of such discussions somewhat questionable. Very little attention has been given to pharmacologic/toxicologic implications of species differences in these enzymes. This is not to indicate that an attempt should not be made to summarize the existing data but rather to suggest that it may not be appropriate for this brief review.

23.9.3 SPECIES DIFFERENCES IN REDUCTION OF XENOBIOTICS

23.9.3.1 Nitro Group Reduction

Significant conversion of aromatic nitro groups to the corresponding nitroso, hydroxylamino and amino metabolites normally requires conditions of low oxygen tension because, in the presence of O_2, the initially generated nitro anion radical donates an electron to O_2 to convert it to superoxide anion and itself is reoxidized to the original nitro group. This process is referred to as redox cycling and is discussed more fully in Section 23.9.3.4. Because of the requirement for relatively anaerobic conditions, the importance of the gastrointestinal contents (which contain several obligate anaerobes as well as the requisite hypoxic conditions and a low redox potential) is now clearly evident among vertebrate forms. It was reported in the early 1970s that elimination of the gastrointestinal bacteria effectively blocked the conversion of *p*-nitrobenzoic acid to its corresponding primary amine in the rat.[30] This observation has been confirmed in numerous subsequent studies and has recently been reviewed by Rowland.[31] Importantly, it was also shown that specific enzymes are *not* required to effect a complete six-electron reduction to the primary amine.[32] The initial four electrons can be transferred by any flavin-containing protein as well as by free flavin (*e.g.* riboflavin). With NADPH or NADH as initial electron donors any heme-containing protein—as well as free heme—can catalyze the reductive conversion of the hydroxylamine to the primary amine. Clearly some enzymes are more effective than others in catalyzing these processes and, interestingly, certain bacterial enzymes will effect such conversions even in the presence of relatively high O_2 tensions.[33] Critically needed at present is a much better understanding of the phylogenetic distribution of this group of O_2 insensitive nitroreductase enzymes, their substrate specificities and normal housekeeping functions.

For mammals, species differences in the complement of bacteria present in various segments of the gastrointestinal tract would appear at present to represent the most important contributing factor to species differences in aromatic nitro group reduction. This issue has been discussed by Gillette,[34] who pointed out that rats, guinea pigs and mice have relatively high concentrations of bacteria in the upper portions of the gastrointestinal tract (stomach, duodenum, jejunum), whereas rabbits and humans have much higher proportions in the lower portion (ileum, cecum, colon). In addition, the gut flora of mice and rats differs in *composition* from that of rabbits, guinea pigs and humans. As an example, the gut flora of rats and mice contains more lactobacilli and streptococci. Thus, aromatic nitro compounds entering the gastrointestinal tracts of different species would encounter both qualitatively and quantitatively different populations of bacteria that could affect the degree to which conversion to reduced metabolites would occur prior to (and after) absorption. This, in turn, may be expected to influence differentially the biological effects of orally administered nitro compounds among various species. For this aspect of species differences in drug biotransformation, the critical issue appears to be bacterial populations rather than species specific biotransforming enzymes *per se*.

A recent review of many animals and humans has been given by Rowland and Walker[35] and a brief summary in the more accessible literature was presented by the same authors,[31] who concluded that rats, mice, hamsters, guinea pigs and marmosets were not good models for humans. It should be noted that very marked interindividual variations in gut flora (depending on age, diet, *etc.*) are evident in humans. Theoretically, however, the overall tendency would be for bioactivated nitro compounds to exert a greater toxicity on rodents than humans.

Miller[30-36] has discussed nitro group reduction in various microorganisms where the importance of a low oxygen tension is clearly evident. Measurement of bacterial toxicity indicated that inhibitory concentrations of nitro compounds for facultative anaerobes and aerobes were at least two orders of magnitude higher than for obligate anaerobes. The antimicrobial selectivity of various nitro compounds is clearly related to the capacity of the microorganism to bioactivate such chemicals reductively.

23.9.3.2 Azo Linkage Reduction

Although not as sensitive to O_2 inhibition as nitro group reduction, the reduction of azo linkages to their corresponding amines also is ordinarily much more efficient under conditions of low oxygen tension. It is now commonly believed that a very high percentage of azo linkage reduction occurring in vertebrates is effected by anaerobic bacteria present in the gastrointestinal tract. Thus, most of the statements made in Section 23.9.3.1 for nitro group reduction should likewise apply to the reduction of azo linkages and the factors that determine species differences should apply equally to each of these reactions. However, Rowland[31] compared the feces and cecal contents of rats, mice, hamsters, guinea pigs, marmosets and humans in terms of both nitroreductase and azoreductase activities and found the mouse preparation was roughly 10–15 times more active than the marmoset preparation in terms of nitro group reduction but approximately equal activities were observed for azo linkage reduction. Reasons for this difference were not provided. Thus, it appears that a better understanding of the factors governing reduction of nitro and azo groups is still needed for elucidation of mechanisms involved in species differences.

23.9.3.3 Carbonyl Group Reduction

Research results from the laboratory of Sawada and coworkers have been the principle recent source of information concerning enzymes that catalyze the reduction of carbonyl groups (aldehydes, ketones) to the corresponding alcohols.[37,38] Enzymes catalyzing these reactions are ubiquitous, occurring in the cellular cytosol of invertebrates as well as vertebrates, in virtually all tissue and organs and in microorganisms as well. A review of the species and organ distribution has been presented by Felsted and Bachur[39] and covers the literature up to 1979. They suggested that such enzymes may be universal components of living cells. These authors also discussed the *very* large number of enzymes (mostly dehydrogenases including alcohol dehydrogenase, discussed in Section 23.9.2.3) that *can* function as carbonyl reductases. This, of course, makes ubiquitous distribution understandable. It also renders discussions of the biochemical basis for species differences of lesser value in the absence of a discussion of each of the more important participatory enzymes. This would represent a heroic undertaking and is far beyond the scope of this review. It is clear, however, that marked species differences do exist with respect to the capacity to reduce the carbonyl groups of specific xenobiotic chemicals such as aflatoxin (very high in rabbit and trout, intermediate in rats, mice and primates, and absent in guinea pigs). Rabbits appear to possess relatively high overall capacity for the reduction of aldehydes and ketones but no striking examples of differences between rabbits and other species in terms of the pharmacologic/toxicologic effects of such agents (due to differences in carbonyl reduction) have been reported to my knowledge. Again, an enormous amount remains to be accomplished in this relatively unappreciated area of research.

23.9.3.4 Redox Cycling

The phenomenon referred to as redox cycling has received considerable attention in recent years due to increased recognition for the potential of the process to generate reactive toxic intermediates. An overview of the enzyme systems involved has been presented recently by Kappus,[40] who also discussed the mechanism whereby the process occurs, the classes of chemicals most heavily involved and the kinds of toxicity most likely to occur as a result of redox cycling. The process appears to occur in virtually all cells and tissues throughout all phyla—or at least no exceptions have been noted thus far. Quinones and aromatic nitro compounds (and chemicals that can be converted to such) represent by far one of the most common examples of redox cycling agents. However, nitroso and azo derivatives, *N*- and *S*-oxides and polyhalogenated aliphatics are also included. Since the process involves single-electron transfers (as opposed to two-electron transfers for carbonyl reduction), the enzymes that catalyze transfer of single electrons from NADPH, NADH or other electron donors to the xenobiotic in question are those upon which focus should be placed for understanding of species differences in redox cycling. These are most commonly flavoproteins, of which a large number may participate. NADPH-cytochrome *P*-450 reductase, xanthine oxidase and ferridoxin reductase are frequently mentioned but it is now clear that other flavoproteins are also involved. In bacterial systems, ferredoxins and associated reductases are prominent. Single-electron transfers from the reduced xenobiotic to O_2, GSH or other endogenous acceptors are normally regarded as uncatalyzed reactions and enzymes would therefore not participate. Therefore, species differences in the levels and activities of the flavoproteins, particularly in levels of those with relatively high catalytic activity for single-electron transfer to xenobiotics, should be the principal determinants of species differences in redox cycling. Other determinants would include differences in O_2 tension (of obvious importance in comparisons of aerobic *vs.* anaerobic bacteria), presence of endogenous antioxidants or radical-scavenging agents, levels of GSH and other reducing equivalents, pH and levels of enzymes that would catalyze two-electron reduction of the same substrate (*e.g.* DT diaphorase and other carbonyl reductases), thus competing for single-electron reduction leading to redox cycling.[41,42] It might be expected that species differences in NAPDH-cytochrome *P*-450 reductase would represent a major factor in species differences in redox cycling since it is commonly believed that this is the major enzyme catalyzing redox cycling of drugs.[40] Masters has provided comparisons of species differences in this important flavoprotein in a 1980 review of the subject.[43] Although the enzyme appears to be similar from species to species in terms of its catalytic activity and immunologic properties (will function interchangeably in reconstituted systems), rigorous species to species comparisons of tissue levels of the enzymes have not been conducted to my knowledge. It is to be expected that such species differences will exist and it will be of high interest to determine the extent to which such differences correlate with redox cycling and associated toxicity.

23.9.4 SPECIES DIFFERENCES IN HYDROLYSIS OF XENOBIOTICS

23.9.4.1 Esterases

Like carbonyl reductases, esterases are ubiquitous and also extremely numerous. Thus it is not surprising that relatively little interest has been given to the pharmacologic/toxicologic implications of species differences in xenobiotic-hydrolyzing esterase enzymes. Heyman[44,45] has provided brief reviews of the tissue and subcellular distribution of the 'B-type' esterases, which are known to be heavily involved in xenobiotic hydrolyses. He stressed high activities in the liver, gastrointestinal tract and blood and qualitative similarity of hydrolyses in tissues of vertebrates. He also provided references to several studies utilizing different species, but provided little data concerning direct quantitative species comparisons. The oft-cited absence of atropine esterase from human plasma was mentioned. Hattori *et al.*[46] compared the hydrolysis of several steroid hemisuccinate and acetate esters in the hepatic microsomes of seven species and found that the hamster exhibited very high activities relative to the other six species. Mice, cows and pigs were also generally high, guinea pigs and rabbits were relatively low and rats exhibited extremely low activities. For rats, much higher activities were obtained in tissues of the gastrointestinal tract than in the liver. Additional studies of this nature are needed to develop more quantitative statements concerning species differences in the hydrolysis of esters and amides. Ashour *et al.*[47] have provided a very recent example of species differences (rats *vs.* mice) in the response of esterases to inducers. Responses of hepatic, renal and testicular carboxylesterases to clofibrate induction were relatively weak and were fairly similar in the two species, although somewhat higher in mice.

Cholinesterase (EC 3.1.1.8), also referred to as pseudocholinesterase, butyrylcholinesterase, serum cholinesterase, nonspecific cholinesterase and acylcholine acyl hydrolase, is a hydrolase with unquestioned importance in drug biotransformation and exhibits highly interesting inherited differences in human populations. Its clinical importance was recognized when certain patients developed respiratory paralysis after receiving normal doses of succinylcholine. These patients hydrolyzed the drug very slowly because they had inherited an 'atypical' cholinesterase, which has a much lower affinity for succinylcholine than does the 'usual' cholinesterase. A relatively large number of genetic variants of this enzyme have been characterized and in some humans, *absence* of the active enzyme may be inherited as a genetic trait ('silent gene'). In view of these considerations, it would seem probable that large *species* differences in this esterase should also exist. Kutty[48] has stated that 'there are considerable species differences in the levels of cholinesterase activity' but no details of such differences were provided. Only very recently[49] has this enzyme (from human serum) been sequenced in spite of the recognition of its existence since 1932. The human enzyme exhibited considerable homology with torpedo acetylcholinesterase. Future studies should provide a much clearer picture of the taxonomic aspects of this important enzyme.

23.9.4.2 Epoxide Hydrolases

It is now clear that at least three epoxide hydrolases exist: a microsomal hydrolase (EC 3.3.2.3) that catalyzes the hydration of numerous xenobiotic epoxides; a cytosolic hydrolase that appears to function in a totally different capacity, with specificity for terpenoids, steroids and fatty ester epoxides; and a hepatic microsomal epoxide hydrolase that appears to be specific for cholesterol 5,6-oxides as well as other steroidal 5,6-oxides. This discussion will concern only the first of the three. Oesch,[50] who has extensively researched the microsomal epoxide hydrolases, has stated that 'between different organs of the same animal species and for a given organ of a different animal species, epoxide hydrolase activity varies by as much as 1000-fold'. Early studies have shown that epoxide hydrolase activity is measurable in a very wide variety of living species,[51] including insects, fungi, fruits and vegetables. In vertebrates, activity has been detected in the hepatic microsomes of all species investigated to date. In their review of the subject, Seigard and DePierre[51] report that the general trend of microsomal epoxide hydrolase activity in increasing order is fish < amphibia < birds < rodents < larger mammals. The rat has been by far the most extensively investigated species. It is of interest that the organ distribution of mice differs quantitatively (but not qualitatively) from the organ distribution in rats. Activity in gonadal tissues is generally high but in NMRI mice, testicular activity was higher than in the liver. Oesch *et al.*[52] have shown that 20-fold variations in hepatic microsomal epoxide hydrolase activities (and immunologically detectable quantities) were present among rats of 22 strains investigated. Mean values ranged from 4.3 (F344/Ola) to 12.7 (DA/Han) pmol mg^{-1} min^{-1}. They reported a 63-fold interindividual difference in humans. This, of course, illustrates the problems inherent in comparisons of species in terms of the particular species *strains* selected for comparison.

Comparisons[53] of the NH_2 terminal amino acid sequences and peptide maps of rat *vs.* human hydrolases showed that the enzymes exhibit marked similarities and provide further support for the idea that the major reason for observed species differences is quantitative amounts of enzyme rather than qualitative differences in structure of the protein. It now seems apparent that species will differ in terms of regulation of the microsomal hydrolase. Graichen and Dent, for example, showed marked differences among F-344 rats, mice, cotton rats and guinea pigs in terms of their responsivity to 2-acetylaminofluorene as an inducing agent.[54] Further investigation of regulatory differences among species will be of interest.

23.9.4.3 Other Hydrolases

It is very well known that a very large number of proteins can function as hydrolases. For example, serum albumin can function as carboxylesterase and, in spite of its relatively low activity as compared to more specific carboxylesterases, may be of high importance as a drug-metabolizing hydrolase because of the large quantities present in vertebrate species. Space does not permit a consideration of the species differences in all of the proteins with xenobiotic-hydrolyzing capacity. It should be mentioned, however, that the gastrointestinal flora possesses highly significant hydrolyzing capacity, some of which is of recognized established importance. For example, β-glucuronidase is known to play a very important role in enterohepatic cycling of drugs. For such hydrolases, the considerations of species differences discussed in terms of floral reductases in Section 23.9.3 would apply. Cleavage of glucuronides and sulfate esters generated *via* normal glucuronidation and sulfation of the pertinent xenobiotic acceptors by nonbacterial systems represents another potentially important facet of this area for which little is known concerning species differences and their implications for drug toxicity.

23.9.5 SPECIES DIFFERENCES IN CONJUGATION OF XENOBIOTICS

23.9.5.1 Glycosylation

Dutton[55] has extensively reviewed the literature up to 1979 on species variations in glycosylation (principally glucuronidation) of foreign aglycone acceptors. The most important species difference among mammals occurs in members of the cat family (Feloidae) and includes all members of this family investigated thus far, including the civet and genet. It is now apparent that cats are deficient only in specific glucuronosyl transferase isozymes and that several glucuronidation reactions occur as rapidly in these animals as in other species. An exact definition of the spectrum of the deficient isozymes has not yet been given but it is known that aglycones which are of low molecular weight and relatively planar (substrates for 'Group I' (GT_1) transferases in rats) are poorly converted to glucuronides in cats. A similar deficiency appears to exist in certain nonmammalian marine vertebrates but may be related to the thermolability of the enzyme in these species and/or different temperature optima. Chickens, on the other hand, exhibit a deficiency for glucuronidation of larger relatively nonplanar acceptors such as 4-methylumbelliferone and chloramphenicol. The deficient enzyme(s) in the chicken are similar to those categorized as Group II (GT_2) in rats. Bilirubin glucuronidation is also low in avian species.

Only very rarely has glucuronidation capacity (for exogenous aglycones) been exhibited by invertebrate species, which appear to substitute glucosidation for glucuronidation. It is of interest to note that treatment of slugs with phenobarbital results in increased glucosidation activity but no change in the undetectable glucuronidation. An attempted analysis of species differences in glucuronidation is especially fraught with problems because of a variety of factors, including difficulties in preparing active purified enzyme, enzyme latency, enzyme heterogeneity and multiplicity, variable susceptibility to activity losses and competition with other conjugation reactions. Thus, it is not extremely surprising that little quantitative information on species differences in glucuronidation is available at this time. A very recent review of the subject[56] has shed very little new light on the topic—quite reflective of the dearth of studies in species differences. Bock (refs. 56, 57 and refs. therein) has investigated UDP-glucuronosyltransferases in various species, including humans, but direct comparisons are still needed. Qualitative similarities in response to various inducing agents among rats, mice and humans were noted by this investigator and his coworkers.

One of the few recent studies in which direct comparisons among several mammalian species were made was published by Boutin *et al.*[58] The authors compared 16 aglycone acceptors in eight species (humans, monkeys, rats, mice, pigs, rabbits, guinea pigs and dogs). The studies were performed with Triton X-100 activated hepatic microsomes. Under the conditions of their assays, the activities observed in microsomes from humans were 'comparable' to those from the other species investigated. A general conclusion reached was that rates of rapidly glucuronidated aglycones (low molecular weight, planar phenols) would be five to six times lower in humans than in guinea pigs. It would now be of great interest to determine the extent to which these same species (and strains) would be comparable when generation of glucuronides of the small aglycones were measured *in vivo*. Hopefully, later investigations will provide such data.

Recent research on glucuronidation has provided a large body of increasing evidence for a relatively high number of glucuronosyl transferase isozymes, some of which have been characterized in their purified forms. A clearer picture of the various forms of such isozymes and their properties should enable the procurement of much more definitive information concerning species differences. However, it does not appear likely that rigorous quantitative comparative data will be available in the near future.

23.9.5.2 Sulfation

A comprehensive review of observed species differences in xenobiotic sulfation was provided by Mulder in 1981.[59] It is well recognized that genetic and species differences in the capacity to sulfate various acceptor molecules need not be a function of differences in the relevant sulfotransferase enzymes. This is because quantities of 3'-phosphoadenosyl 5'-phosphosulfate (PAPS) are frequently rate limiting and represent the principal control point for sulfation reactions *in vivo*. In turn, inorganic sulfate normally limits the tissue levels of PAPS and can be influenced by supply from a variety of sources, including diet.

Sulfation of xenobiotics occurs in all vertebrate species that have been examined to date and is also known to occur in arthropods, insects and molluscs. More recently, a bacterial arylsulfotransferase was found in Eubacterium A-44, one of the prominent bacteria in the human intestine, and was purified and characterized (ref. 60 and refs. therein). The investigators found that administration of antibiotics to rats markedly decreased the arylsulfotransferase activity in the feces of rats and humans. Arylsulfatase activities were also decreased to undetectable levels and the decreases were attributed to the lowered quantities of arylsulfotransferase enzymes (which also possess arylsulfatase activity). Major changes in quantities of sulfated metabolites were observed. The data suggested that the bacterial flora of the gastrointestinal tract can play an important role in the metabolism of phenolic compounds *via* sulfoconjugation.

A large number of the studies involving comparisons of species have approached the problem by giving oral doses of phenolic compounds and measuring the ratios of sulfated to glucuronidated metabolites excreted. The results are expressed as 'preference' for glucuronidation or sulfation, often without reference to mechanisms responsible for the observed ratios. It is now known that the cat 'prefers' sulfation, but not because of highly active sulfating systems, rather because of a relatively low complement of glucuronidating enzymes, particularly those which catalyze the glucuronidation of low molecular weight phenols as discussed in Section 23.9.5.1. If phenol is administered, pigs excrete nearly all as the glucuronide but none as the sulfate, whereas the opposite is seen in cats, lions, civets, genets and hyenas. Squirrel monkeys, fruitbats, capuchins and hamsters also exhibit low capacity to excrete sulfates when compared with glucuronides.[59] Pigs, in fact, excreted no detectable phenyl sulfate. Rabbits, guinea pigs and ferrets also tended to prefer glucuronidation over sulfation, whereas humans, rhesus monkeys, elephants, hedgehogs, chickens and, of course, cats favored sulfation. These observations are very interesting, but it is now time that the biochemical basis for such species differences be better understood.

It is now known that systems for sulfation of xenobiotics are saturated rather readily, due usually to a lack of tissue inorganic sulfate. Thus at low doses, sulfation/glucuronidation ratios tend to be much higher than at high doses, where the sulfation systems are saturated.

An example of a recent study which helps to shed some light on the question of observed species differences has been published by Waschek *et al.*[61] These investigators found that, in contrast to that observed in humans, the dose-dependent sulfoconjugation of salicylamide in dogs was not overcome by preventing the decrease in plasma inorganic sulfate. The differences between dogs and humans was attributed to normal lower plasma inorganic sulfate concentration in humans (approximately 0.3 mM) than dogs (approximately 0.9 mM). Although several alternative explanations for the observations could be invoked, the data were compatible with the idea that in the dog, much larger doses of the drug would be required to render sulfate concentrations sufficiently low to limit sulfoconjugation. Additional studies pursuing these questions will be needed to provide explanations for observed species differences. It is of interest that earlier studies reported[59] that humans excreted a higher ratio of sulfated/glucuronidated phenol than did dogs. Clearly, much remains to be learned concerning the sulfate conjugation/deconjugation of xenobiotics.

23.9.5.3 Glutathione Conjugation

The critical role of glutathione and its associated transferases (GSTs) in protecting cells against the effects of reactive electrophiles and free radical intermediates is increasingly appreciated. Glutathione *per se* is a soft nucleophile and thus particularly adept to interact with soft electrophiles in a similar manner to other soft cellular nucleophiles such as protein sulfhydryls. Thus, GSH itself is effective in protecting macromolecular soft nucleophiles against attack by soft electrophilic metabolites. The GSTs enable GSH to interact with hard electrophiles and thus protect hard macromolecular nucleophiles such as the nucleic acids from attack by such hard electrophilic species. The transferases that catalyze these reactions are very widespread phylogenetically, occurring in both animals and plants.[62] They appear to have evolved somewhat later than GSH itself, but are found in certain bacteria (including *Salmonella* tester strains) but were undetectable in *E. coli* and bakers yeast.[63] Higher organisms are particularly well endowed with regard to the intracellular concentrations of both glutathione and GSTs. Glutathione concentrations in most tissues are in the range of 3–10 mM. Among mammals, the livers of rodents appear to have the highest GST activities but gonads also frequently have remarkably high activity. Transferase activity is detectable in virtually all tissues of mammals and is even present in the blood, particularly after hepatocellular damage, due to release into the bloodstream. It is now evident that the GSTs are a family of multifunctional isozymes and exist in multiple forms in most mammalian tissues.

Warholm *et al.*[64] compared three mouse hepatic GSTs and reported similarities with rat and human enzymes. Awasthi *et al.*[65] reported that various GST forms isolated from cat and human lungs exhibited differential catalytic activities toward benzo[*a*]pyrene epoxides but did not provide data that would indicate that marked species differences exist. Reddy *et al.*[66] reported similarities between sheep and human liver GSTs. It is becoming clear that rigorous species comparisons will depend upon identification of a very large number of GST isozymes and comparative characterization of each isozyme. However, surprising progress in this area has been made in recent years and it seems likely that rigorous comparative data will rapidly become available. Comparative aspects of the *regulation* of each of these isozymes, however, should keep scientists occupied for many years to come, even though studies in this area have witnessed considerable progress as evidenced from studies such as those reported by Rothkopf *et al.*[67] An interesting study of the *microsomal* transferase activity[68] indicated that mammals possess the microsomal GSTs (humans, rhesus monkeys, chimpanzees, bulls, cows, pigs, rabbits, rats, mice, hamsters and guinea pigs) but that microsomal transferase activity was not detectable in nonmammalian species (roosters, hens, toads, pike and *Rhodospirillium rubrum*).

We are now beginning to understand the molecular basis of GST subunit multiplicity and this knowledge in turn will greatly assist in understanding the basis for differences among species. However, an understanding of the factors governing regulation of each subunit remains a formidable task.

23.9.5.4 Acetylation

Acetylation is of primary importance in the conjugation of primary amines but also (rarely) functions in the conjugation of hydroxyl and sulfhydryl groups. Arylamino, aliphatic amino, α-amino, hydrazino and sulfamino groups are commonly acetylated in biological systems. Insofar as foreign organic chemicals are concerned, arylamines and hydrazines are by far the most common substrates for acetylation. Although acetylation is very widespread phylogenetically, examples of marked species differences have been documented. One of the best known of these is the high capacity of rabbits to acetylate sulfanilamide on the N^4 position.[64] In rabbits, the *N*-acetyltransferase that catalyzes this reaction appears to be associated with reticuloendothelial cells and not hepatic cells, whereas in rat liver it is associated with both of these cell types.

As with virtually all other classes of drug-metabolizing enzymes, it has become clear that a large number of various isozymic forms of acetyltransferase enzymes exist and that the key to elucidation of the quantitative aspects and basic patterns of species differences lies in the identification/isolation/purification/characterization of the various multiple forms. Species specific *regulation* of individual isozymes at the molecular level is, of course, of primary importance and is a more formidable problem. The rabbit appears to be one of the few animals that exhibits the now classical isoniazid-acetylator polymorphism—even baboons are not suitable models for this well-established human polymorphism. Deermice and squirrel monkeys exhibit similar polymorphisms.

A relatively recent publication by Hein *et al.*[70] has provided a summary and shed some light on species differences among potential animal models for human polymorphic acetylation. In humans, it has been suggested that the specific *N*-acetylation capacity is inherited as a simple autosomal Mendelian trait with the rapid allele dominant to the slow. Genetic studies in animal models, however, appear consistent with a trimodal rather than bimodal distribution pattern. In rabbits, a bimodal pattern is exhibited *in vivo* but a trimodal pattern for the hepatic cytosolic transferase was observed *in vitro*. Trimodal patterns were also observed in mice. Thus, even these species may not be entirely suitable as models. Hein and coworkers found a unique genetic pattern in a strain of inbred hamsters which exhibited polymorphisms for the acetylation of *p*-aminobenzoic acid and *p*-aminosalicylic acid but not for isoniazid and phenelzine. (Humans and rabbits exhibit the opposite pattern.)

Studies on species differences and tissue distributions have shown that *O* acetylation activity is the highest in the hamster liver cytosol among several animals.[71] Other notable species differences in *N* acetylation capacity involve the dog and guinea pig.[72] The dog and related species such as the fox (Canidae) appear unable to acetylate most of the classes of substrates for this reaction, with the exceptions of *S*-substituted cysteines and the (N^1) nitrogen of sulfonamides. The guinea pig appears unable to *N* acetylate *S*-substituted cysteines, the breakdown products of glutathione conjugates, and thus does not excrete mercapturic acids. It seems clear that we still know very little concerning the phylogenetic distribution of acetyltransferases and their species-specific regulation. Very few scientists are currently actively engaged in the research that could provide answers to these questions in spite of its obvious importance in pharmacology, toxicology and environmental sciences.

23.9.5.5 Amino Acid Acylation

Rather spectacular differences among various species have been observed in terms of the principal amino acids utilized for acylation of carboxylic acids of foreign organic chemicals. Of the amino acid acylation reactions, glycine conjugation of benzoates appears to exhibit the most widespread phylogenetic distribution and occurs in virtually all vertebrate species studied to date with the notable exception of the fruitbat. An excellent concise review of the literature dealing with amino acid conjugations has been written by Caldwell,[73] who also provided a brief summary of the phylogenetic variations observed through 1979. The enzymes involved are normally mitochondrial and often localized heavily in renal tissues, although hepatic tissues also sometimes have respectable levels of the respective amino acid acyltransferases. Other tissues appear to have minimal or no activity. It has been speculated[74] that animals probably evolved these reactions to facilitate the removal of a variety of plant acids which they ingested but could not degrade or utilize completely. The major competing reaction is glucuronidation among higher animal forms and glucosidation among lower forms. These acylation reactions are often readily saturated and thus assume greatest importance at low substrate levels. This situation for amino acid acylation *vs.* glucuronidation of carboxylic acids is analogous to that for sulfation *vs.* glucuronidation of phenols. As expected, cats excrete much more as the amino acid conjugate than as the glucuronide in most cases. Also the larger more complex carboxylic acids tend to be excreted as glycosides, whereas the small simple acids (*e.g.* benzoic, arylacetic, salicylic) tend more toward excretion as the amino acid conjugates.

Glycine conjugation among mammals tends to follow the order: herbivores > omnivores > carnivores. Taurine conjugation is reportedly more prevalent in carnivores and during early stages of development of many mammals (due apparently to the high taurine content of maternal milk). Primates, but not prosimians, exhibit considerable conjugation with glutamine when phenylacetate is utilized as a substrate. Other species are not known to excrete significant quantities of glutamine conjugates. Avian species and reptiles exhibit conjugation with ornithine with metabolites excreted as the N^2,N^5-diacylornithine conjugates. An interesting comparison of amino acid conjugation is among cats (carnivores), rats (herbivores) and dogs (carnivores). Cats excrete relatively large proportions of α-methylphenylacetic and diphenylacetic acid as glucuronides (41 and 76% respectively) but undetectable amounts of the glucuronides of phenylacetic, 4-chlorophenylacetic, 4-nitrophenylacetic, 3-indolylacetic or 1-naphthylacetic acids are excreted. Metabolites of the latter five are excreted almost entirely as glycine or taurine conjugates.[75] Rats excrete the latter five almost exclusively as the glycine conjugates and the former two largely as the glucuronides. Rats excrete low or negligible quantities of taurine conjugates. Dogs are capable of excreting glucuronides as well as glycine and taurine conjugates. None of the three species excrete glutamine conjugates except for very small amounts in cats. This, however, may depend upon choice of substrates, as emphasized by Caldwell.[73]

23.9.5.6 Methylation

Methylation involving *N*-, *O*- and *S*-methyltransferases is increasingly recognized as a process which is of greater importance in the biotransformation of endogenous compounds than of xenobiotics. It is thus not surprising that these enzymes are very widespread phylogenetically and are present in virtually all cells of vertebrate and other species. The methylation of DNA as an important regulatory mechanism for gene activity has been recognized for several years (particularly among eukaryotes).[76] In drug biotransformation, the *O* methylation of endogenous catechols such as catecholamines and catecholestrogens is the most commonly mentioned reaction, although many others are also known to occur. The role of methylation in xenobiotic biotransformation has been reviewed by Borchardt,[77] Weisiger and Jakoby[78] and Caldwell.[72,73] Very little rigorous comparative work has been reported in terms of differences among species in capacity to methylate xenobiotic substrates. Rats, mice and humans exhibited less capacity to *N* methylate pyridine than cats, guinea pigs, gerbils, hamsters and rabbits. With *O*-methylmercaptoethanol as substrate, *S*-methylation was in the order guinea pig > rat = mouse = rabbit = sheep > cow ≫ chicken. Other examples are scattered in the literature. Possibly less attention to species differences has been given to methylation than for any of the other major drug biotransformation reactions.

23.9.6 SUMMARY AND CONCLUSIONS

A review of the major literature sources reveals that our current knowledge regarding differences among species with respect to xenobiotic-biotransforming processes is very primitive. Numerous examples of species differences have been described in the literature but in many cases it is not known whether the observed differences were a function of the specific species' strains utilized for comparisons. Far too often are investigators unappreciative of this complication, presuming, for example, that an observed difference between a given strain of mouse (*e.g.* Swiss-Webster) and a given strain of rat (*e.g.* Sprague-Dawley) represents a true species differences without further investigation. Enough data are available to make the statement that some very profound species differences do exist but the time is fully ripe that much more rigorous, much more quantitative and much more definitive investigations should be undertaken. It is not sufficient to be able merely to state that species differences exist. Needed now are data that permit us to make more quantitative statements, to generalize in terms of phylogenetic/evolutionary aspects and, particularly, to uncover the underlying *basis* for the genetic diversification manifest as species differences.

23.9.7 REFERENCES

1. D. V. Parke and R. L. Smith (eds.), 'Drug Metabolism: From Microbe to Man', Taylor and Francis, London, 1977.
2. J. Caldwell, in 'Enzymatic Basis of Detoxification', ed. W. B. Jakoby, Academic Press, New York, 1980, vol. I, p. 85.
3. J. Stegeman, in 'Foreign Compound Metabolism', ed. J. Caldwell and G. Paulson, Taylor and Francis, London, 1984, p. 149.
4. S. Black and M. Coon, in 'Cytochrome *P*-450. Structure, Mechanism, and Biochemistry', ed. P. R. Ortiz DeMontellano, Plenum Press, New York, 1986, p. 161.
5. D. W. Nebert, M. Adesnik, M. J. Coon, R. W. Estabrook, F. J. Gonzalez, F. P. Guengerich, I. C. Gunsalus, E. F. Johnson, B. Kemper, W. Levin, I. R. Phillips, R. Sato and M. R. Waterman, *DNA*, 1987, **6**, 1.
6. K. M. Kleinow, M. J. Melancon and J. J. Lech, *Environ. Health Perspect.*, 1987, **71**, 105.
7. J. J. Stegeman and P. J. Kloepper-Sams, *Environ. Health Perspect.*, 1987, **71**, 87.
8. D. W. Nebert and F. J. Gonzalez, *Annu. Rev. Biochem.*, 1987, **56**, 945.
9. A. M. Clark, J. D. McChesney and C. D. Hufford, *Med. Res. Rev.*, 1985, **5**, 231.
10. J. M. Van den Brock, *Pharm. Weekbl.*, 1983, **5**, 189.
11. L. Wen and A. J. Fulco, *J. Biol. Chem.*, 1987, **262**, 6676.
12. A. Wiseman, *TIBS*, 1980, **5**, 102.
13. J. B. Sutherland, *Appl. Environ. Microbiol.*, 1986, **52**, 98.
14. R. R. Vanderslice, B. A. Domin, G. T. Carver and R. M. Philpot, *Mol. Pharmacol.*, 1987, **31**, 320.
15. C. H. Walker, *Drug Metab. Rev.*, 1978, **7**, 295.
16. D. B. Duignan, I. G. Sipes, T. B. Leonard and J. R. Halpert, *Arch. Biochem. Biophys.*, 1987, **255**, 290.
17. R. J. Schwen and G. J. Mannering, *Comp. Biochem. Physiol. B*, 1982, **71**, 445.
18. D. M. Ziegler, in 'Enzymatic Basis of Detoxification', ed. W. B. Jakoby, Academic Press, New York, 1980, vol. I, p. 222.
19. K. L. Taylor and D. M. Ziegler, *Biochem. Pharmacol.*, 1987, **36**, 141.
20. D. E. Williams, S. E. Hale, A. S. Muerhoff and B. S. S. Masters, *Mol. Pharmacol.*, 1985, **28**, 381.
21. R. E. Tynes and E. Hodgson, *Arch. Biochem. Biophys.*, 1985, **240**, 77.
22. P. J. Sabourin and E. Hodgson, *Chem.-Biol. Interact.*, 1984, **51**, 125.
23. G. A. Dannan and F. P. Guengerich, *Mol. Pharmacol.*, 1982, **22**, 787.
24. R. E. Tynes and R. M. Philpot, *Mol. Pharmacol.*, 1987, **31**, 569.
25. A. G. Dawson, *TIBS*, 1983, **8**, 195.
26. M. S. Moss, in 'Drug Metabolism: From Microbe to Man', ed. D. V. Parke and R. L. Smith, Taylor and Francis, London, 1977, p. 276.
27. D. A. P. Evans, in 'Drug Metabolism: From Microbe to Man', ed. D. V. Parke and R. L. Smith, Taylor and Francis, London, 1977, p. 372.
28. S. A. Atlas and D. W. Nebert, in 'Drug Metabolism: From Microbe to Man', ed. D. V. Parke and R. L. Smith, Taylor and Francis, London, 1977, p. 400.
29. W. F. Bosron and T. Li, *Hepatology*, 1986, **6**, 502.
30. P. K. Zachariah and M. R. Juchau, *Drug Metab. Dispos.*, 1974, **2**, 74.
31. I. R. Rowland, *Biochem. Pharmacol.*, 1986, **35**, 27.

32. K. G. Symms and M. R. Juchau, *Drug Metab. Dispos.*, 1974, **2**, 194.
33. D. W. Bryant, D. R. McCalla, M. Leeksma and P. Laneuville, *Can. J. Microbiol.*, 1981, **27**, 81.
34. J. R. Gillette, in 'Drug Metabolism: From Microbe to Man', ed. D. V. Parke and R. L. Smith, Taylor and Francis, London, 1977, p. 149.
35. I. R. Rowland and R. Walker, in 'Toxic Hazards in Food', ed. D. M. Conning and A. B. G. Lansdown, Croom Helm, Beckenham, 1983, p. 183.
36. M. Mueller, *Biochem. Pharmacol.*, 1986, **35**, 27.
37. S. Usui, A. Hara, T. Nakayama and H. Sawada, *Biochem. J.*, 1984, **223**, 697.
38. T. Nakayama, K. Yashiro, T. Inoue, K. Matsuura, H. Ichikawe, A. Hara and H. Sawada, *Biochim Biophys. Acta*, 1986, **882**, 220.
39. R. L. Felsted and N. R. Bachur, *Drug Metab. Rev.*, 1980, **11**, 1.
40. H. Kappus, *Biochem. Pharmacol.*, 1986, **35**, 1.
41. H. Wefers, T. Komai, P. Talalay and H. Sies, *FEBS Lett.*, 1984, **169**, 63.
42. H. J. Prochaska, P. Talalay and H. Sies, *J. Biol. Chem.*, 1987, **262**, 1931.
43. B. S. S. Masters, in 'Enzymatic Basis of Detoxification', ed. W. B. Jakoby, Academic Press, New York, 1980, vol. I, p. 189.
44. E. Heymann, in 'Enzymatic Basis of Detoxification', ed. W. B. Jakoby, Academic Press, New York, 1980, vol. II, p. 298.
45. E. Heymann, in 'Metabolic Basis of Detoxification', ed. W. B. Jakoby, J. R. Bend and J. Caldwell, Academic Press, New York, 1982, p. 229.
46. K. Hattori, M. Kamio, E. Nakajima, T. Oshima, T. Satoh and H. Kitagawa, *Biochem. Pharmacol.*, 1981, **30**, 2051.
47. M. A. Ashour, D. E. Moody and B. D. Hammock, *Toxicol. Appl. Pharmacol.*, 1987, 361.
48. K. M. Kutty, *Clin. Biochem.*, 1980, **13**, 239.
49. O. Lockridge, C. F. Bartels, T. A. Vaughan, C. K. Wong, S. E. Norton and L. L. Johnson, *J. Biol. Chem.*, 1987, **262**, 549.
50. F. Oesch, in 'Enzymatic Basis of Detoxification', ed. W. B. Jakoby, Academic Press, New York, 1980, vol. II, p. 280.
51. J. Seidegard and J. W. Pierre, *Biochim. Biophys. Acta*, 1983, **695**, 251.
52. F. Oesch, A. Zimmer and H. R. Glatt, *Biochem. Pharmacol.*, 1983, **32**, 1753.
53. G. C. DuBois, E. Appella, D. E. Ryan, D. M. Jerina and W. Levin, *J. Biol. Chem.*, 1982, **257**, 2708.
54. M. E. Graicher and J. C. Dent, *Carcinogenesis*, 1984, **5**, 23.
55. G. J. Dutton, in 'Glucuronidation of Drugs and Other Compounds', ed. G. J. Dutton, CRC Press, Boca Raton, FL, 1980, p. 123.
56. K. W. Bock and G. Schirmer, *Arch. Toxicol. Suppl.*, 1987, **10**, 125.
57. K. W. Bock, W. Liliendum and C. von Bahr, *Drug Metab. Dispos.*, 1984, **12**, 93.
58. J. A. Boutin, B. Antoine, A. Batt and G. Siest, *Chem.-Biol. Interact.*, 1984, **52**, 173.
59. G. J. Mulder, in 'Sulfation of Drugs and Related Compounds', ed. G. J. Mulder, CRC Press, Boca Raton, FL, p. 154.
60. D. Kim and K. Kobashi, *Biochem. Pharmacol.*, 1986, **35**, 3507.
61. J. A. Wascheck, R. M. Fielding, S. M. Pond, G. M. Rubin, D. J. Effeney and T. N. Tozer, *J. Pharmacol. Exp. Ther.*, 1985, **234**, 431.
62. P. L. Grover and P. Sims, *Biochem. J.*, 1964, **90**, 603.
63. W. B. Jakoby and W. H. Habig, in 'Enzymatic Basis of Detoxification', ed. W. B. Jakoby, Academic Press, New York, 1980, vol. II, p. 83.
64. M. Warholm, H. Jensson, M. K. Tahir and B. Mannervik, *Biochemistry*, 1986, **25**, 4119.
65. Y. C. Awasthi, S. V. Singh, M. Das and H. Mukhtar, *Biochem. Biophys. Res. Commun.*, 1985, **133**, 863.
66. C. C. Reddy, J. R. Burgess, Z. Gong, E. J. Massaro and C. D. Tu, *Arch. Biochem. Biophys.*, 1983, **224**, 87.
67. G. S. Rothkopf, C. A. Telakowski-Hopkins, R. L. Stotish and C. B. Pickett, *Biochemistry*, 1986, **25**, 993.
68. R. Morgenstern, G. Lundquist, G. Andersson, L. Balk and J. W. DePierre, *Biochem. Pharmacol.*, 1984, **33**, 3609.
69. W. W. Weber and I. B. Glowinski, in 'Enzymatic Basis of Detoxification', ed. W. B. Jakoby, Academic Press, New York, 1980, vol. II, p. 169.
70. D. W. Hein, W. G. Kirlin, R. J. Ferguson and W. W. Weber, *J. Pharmacol. Exp. Ther.*, 1985, **233**, 584.
71. R. Kato, K. Saito, A. Shinohara and T. Kamataki, *Proc. Am. Assoc. Cancer. Res.*, 1984, **25**, 475.
72. J. Caldwell, in 'Metabolic Basis of Detoxification', ed. W. B. Jakoby, J. R. Bend and J. Caldwell, Academic Press, New York, 1982, p. 292.
73. J. Caldwell, in 'Metabolic Basis of Detoxification', ed. W. B. Jakoby, J. R. Bend and J. Caldwell, Academic Press, New York, 1982, p. 275.
74. P. G. Killenberg and L. T. Webster, Jr., in 'Enzymatic Basis of Detoxification', ed. W. B. Jakoby, Academic Press, New York, 1980, p. 141.
75. P. C. Hirom, J. R. Idle and P. Millburn, in 'Drug Metabolism: From Microbe to Man', ed. D. V. Parke and R. L. Smith, Taylor and Francis, London, 1977, p. 299.
76. W. Doerfler, *Annu. Rev. Biochem.*, 1983, **52**, 93.
77. R. T. Borchardt, in 'Enzymatic Basis of Detoxification', ed. W. B. Jakoby, Academic Press, New York, 1980, vol. II, p. 43.
78. R. A. Weisiger and W. B. Jakoby, in 'Enzymatic Basis of Detoxification', ed. W. B. Jakoby, Academic Press, New York, 1980, vol. II, p. 131.

23.10

Developmental Drug Metabolism

MONT R. JUCHAU

University of Washington, Seattle, WA, USA

23.10.1 INTRODUCTION: AN ANALYSIS OF DEVELOPMENTAL STAGES OF PARTICULAR RELEVANCE TO DRUG BIOTRANSFORMATION

It should be made clear from the outset that any attempt to divide the development of organisms into discrete stages must be regarded with reservations because of the nondiscrete nature (with certain exceptions) of the developmental processes. Nevertheless, arbitrary divisions will be utilized in the discussions of this chapter because, without them, the dialog becomes even more confusing. A recent discussion of prenatal aspects of developmental staging has been presented by Scialli and Fabro[1] and comparative aspects are detailed in Shepard's 'Catalog of Teratogenic Agents'.[2]

Various religious notions notwithstanding, it is clear that traceable beginnings of humans and other mammalian organisms (upon which this chapter will focus) extend prior to conception, that is, to the germ cells from which the conceptus is derived. That drugs and other small foreign organic chemicals (xenobiotics) can profoundly affect gametes to alter the subsequent development of the organism has been made abundantly evident from studies of mutagenesis. The role of biotransformation in chemical mutagenesis has received considerable attention in studies performed during the past decade or more and its overriding importance as a mutagenic determinant is now unquestioned. From the viewpoint of chemical mutagenesis alone, it seems imperative that a clear and detailed understanding of the xenobiotic-biotransforming capabilities of germ cells, their precursors and cells immediately adjacent should be sought. Impairment of germ cells in terms of normal processes of fertilization represents an additional aspect of concern.

After conception, the fertilized ovum (zygote), blastomere, morula and the preimplantation blastocyst are the subjects of focus. The extent to which these are capable of inactivating or bioactivating chemicals with which they come in contact is largely unexplored at present except for a few scattered reports concerning the preimplantation blastocyst. Implications for fertility and sterility are particularly pertinent for this particular stage of development but may possibly also be relevant in terms of the effects of chemicals on future development. Recent findings have provided the beginnings for investigations of the xenobiotic-biotransforming capabilities of the preimplantation blastocyst and have helped to explode previous notions that early embryos would not possess such capabilities. Because of the demonstration of apparent cytochrome *P*-450 dependent biotransformation of xenobiotics in preimplantation blastocysts, opportunity is provided for examination of developmental patterns of monooxygenation and regulation thereof during the very early stages of embryonic development with respect to their relationships to drug-induced embryotoxicity.

From the perspective of the traditional teratologist, the developmental period exhibiting the highest susceptibility to permanent or semipermanent morphologic damage by exogenous agents is the period of organogenesis. In the rat (see Table 1) this period extends arbitrarily from day eight to day 16 of gestation and provides a period in which sufficient tissue is available to perform a variety of meaningful experiments. Recent research has demonstrated that, at least under certain conditions, embryonic enzymes can catalyze bioactivation reactions to a sufficient extent that the generated reactive intermediates produce grossly observable morphological abnormalities in the selfsame embryos. Such observations ought to have considerable impact on studies of chemically elicited teratogenesis and should focus much attention on the presence of xenobiotic-biotransforming enzymes in embryonic targets and on the capacity of embryonic enzymes to catalyze the respective reactions.

Relatively speaking, a considerable body of information has now accumulated concerning the capacity of various fetal (postembryonic, prenatal) tissues to biotransform xenobiotics. Because of the quantity of information available, much of this review will be directed toward summarizing the findings pertaining to fetal drug biotransformation. This particular information has also been summarized in a number of other reviews. Recognition of the potential for drugs and chemicals to produce permanent or semipermanent functional deficits (*e.g.* abnormalities of behavior, mentation, reproductive function, motor function, immunologic function, *etc.*) in individuals exposed during relatively late prenatal development should provide a strong impetus for acquisition of a thorough understanding of xenobiotic biotransformation during these later stages of prenatal development.

The perinatal period (extending in humans from approximately one month prior to birth to one month subsequent to birth) is of great interest for several reasons: (i) the most rapid changes in levels and activities of drug-biotransforming enzymes occur during this period, in particular, immediately following birth; (ii) at birth, the newborn organism is suddenly solely dependent upon its own

Table 1   Arbitrarily Designated Beginnings of the Developmental Periods
Covered in This Review[a]

Period	*Approximate age, relative to birth* (d)			
	Human	Rat	Mouse	Guinea pig
Prefertilization[b]	−15 000	−1000	−700	−2000
Preimplantation	−267	−22	−20	−68
Early postimplantation	260	−16	−13	−62
Organogenesis	−249	−14	−10	−52
Fetal	−212	−8	−7	−38
Perinatal	30	−2	−2	−4
Early postnatal	0	0	0	0
Weaning	150	21	18	28
Puberty	4000	30	24	45
Pregnancy	6000	40	36	60
Senescence	22 000	1000	700	2000

[a] Periods begin at approximately the times indicated in the table and end at the beginning of the next designated developmental period. For several (*e.g.* prefertilization, puberty, pregnancy and senescence), the values are only very rough estimates and certain periods (*e.g.* fetal and neonatal) also overlap. [b] Ova are present in the offsprings' ovaries at the time of birth and thus can develop at least 40 years prior to fertilization.

xenobiotic elimination systems; and (iii) during labor and delivery, the tendency is for the maternal as well as the preborn organism to receive a large variety of medicinal agents that possess potential fetotoxic effects.

Postnatal development of drug biotransformation exhibits varied and unique patterns that are frequently unpredictable. A major challenge at present is to obtain sufficient knowledge concerning postnatal development such that patterns of developmental drug biotransformation could be predicted for specific drugs and chemicals. It would appear that progress toward this goal is very slow but reasonably steady.

It is well established that dietary factors can alter rates of drug biotransformation reactions, sometimes quite profoundly. At weaning, changes in the nature of the diet occur relatively rapidly and are often quite dramatic. Such changes might be expected to alter drug metabolic patterns during this phase of development and such changes will be analyzed in this review.

Likewise, hormonal factors are capable of exerting profound effects on drug metabolic reactions. This is especially notable in rats and certain fish but in most other species it is often difficult to demonstrate significant effects of endocrine influences on drug metabolism. Sexual maturation is accompanied by extensive hormonal changes that are known to profoundly alter rates and patterns of drug biotransformation in certain experimental animals. These will be reviewed and their relevance to the human situation discussed.

Pregnancy also results in highly significant changes in the hormonal milieu. In addition, pregnant humans are heavily exposed to prescribed and nonprescribed drugs. These two factors render an understanding of biotransformational capabilities during pregnancy of high priority. While it may be argued that pregnancy should not be regarded as a 'stage of development', it does in fact represent a stage(s) through which a highly significant fraction of the population passes and thus appears to deserve treatment in this context.

Only relatively recently has significant concern developed pertaining to drug biotransformation and disposition during aging. As the percentage of older people in the human population continues to increase, interest in such matters may also be expected to increase. Again, the goal of studies of changes in drug biotransformation as a function of aging should be to determine the extent to which recognizable patterns of developmental changes can be directly associated with the aging process. Ideally, this would permit predictions to be made for specific drugs or drug regimens. Considerable additional research will be required to achieve this ideal.

23.10.2 PREFERTILIZATION: DRUG BIOTRANSFORMATION IN THE GAMETES

Recognized effects of drugs and foreign chemicals on germ cells include primarily mutagenesis and cytotoxicity measurable as cell death. The former can give rise to inherited birth defects and the latter plays a role in fertility and sterility. Mutagenesis is of particular importance because 20–25% of all birth defects appear to be inherited and, presumably, thus originally resulted from a mutation in the germ line at some point in time. Additionally, a heritable drug-induced mutation in the germ line represents an effect that can persist for several generations, representing the effect of longest known duration of action producible by a drug.

Research during the 1970s demonstrated graphically that a very large number of drugs and chemicals, although incapable of acting as mutagens *per se*, could be metabolically converted to intermediates which were effective mutagens. From this knowledge arose the term 'promutagen' (or 'premutagen') and it is now recognized that the majority of chemicals capable of eliciting mutations in vertebrate species are in reality promutagens. Subsequent research has demonstrated that drugs can be metabolically converted to mutagenic intermediates by virtually all xenobiotic-biotransforming reactions heretofore described, including glucuronidation and glutathione conjugation—long regarded as solely detoxifying reactions. Of greatest importance for such bioactivating reactions, however, are oxidation and reduction reactions of which the superfamily of *P*-450 cytochromes and the corresponding reductase appear to play a highly prominent role. The most commonly utilized bioactivating systems in mutagenesis screening assays employ cytochrome *P*-450 dependent systems from hepatic tissues, illustrating the perceived high relative importance of the family of *P*-450 isozymes.

In tandem with the evolution of the above described concepts regarding mutagenesis, research has provided a clearer idea as to the nature and biologic disposition of reactive intermediates. It has become increasingly accepted that bioactivation (as well as inactivation) within the target (or immediately adjacent) cells *per se* is usually far more important than the occurrence of these processes at a remote site (such as the classically investigated liver) because of the instability of the majority of toxic intermediary metabolites and their rapid breakdown prior to transport to the target sites of interest.[4] As a result, studies of xenobiotic biotransformation/bioactivation in the lung (a target for various mutagenic carcinogens) and other extrahepatic tissues have become increasingly popular and are currently vigorously pursued.

From the above discussion, it would seem that the stage should have been set for a thorough evaluation of the capacity of germ cells to biotransform foreign organic chemicals, particularly those chemicals which could be regarded as bioactivatable promutagens or potential promutagens. However, at this writing, precious little can be documented concerning the xenobiotic-bio-transforming capabilities of germ cells. A number of studies have demonstrated that the tissues which contain the germ cells, testes[5-8] and ovaries,[9-11] each exhibit significant and interesting xenobiotic-biotransforming capabilities. Of particular interest are the relatively high levels of epoxide hydrolase activities observed in the testes of various species by Oesch and his co-workers[12,13] and other groups. This suggests that the male germ cells are protected from the potentially cytotoxic or mutagenic effects of chemicals that are convertible metabolically to reactive epoxides, usually *via* P-450 dependent monooxygenation. In a series of publications, Mattison and coworkers (see refs.10 and 11) have described the capacity of ovarian tissues to bioactivate various polynuclear aromatic hydrocarbons and thus cause damage to ovarian cells. In neither the testes nor the ovary, however, has it been definitively established that the respective germ cells *per se* participate in the biotransformation processes. Lee *et al.*[6] reported that aryl hydrocarbon hydroxy-lase activity and P-450 content were twofold greater in microsomes from interstitial cells than in microsomes prepared from the germ cell compartment, whereas epoxide hydroxylase and GSH-transferase activities in the germ cell compartment were about twice those in the interstitial cells. These observations are in harmony with the concept that expression of xenobiotic-biotransforming enzymes in germ cells would favor the inactivation of potential mutagens. It would appear that a major challenge to developmental toxicologists at present is to develop sensitive and accurate methodology to investigate systematically the capacity of both male and female germ cells to biotransform chemicals that are potentially mutagenic and/or cytotoxic to these cells.

23.10.3 DRUG BIOTRANSFORMATION IN THE ZYGOTE AND PREIMPLANTATION EMBRYO

Until very recently it has been a common belief that rodent embryos are capable of little or no significant xenobiotic-biotransforming activity. Such beliefs are reflected in relatively recent reviews[14-16] which also suggest that lack of such systems may constitute a protective mechanism, since many of the biotransforming enzymes (particularly the P-450 isozymes) catalyze bioactivation reactions. However, with specific reference to the preimplantation blastocyst, several recent stud-ies[17-19] have now demonstrated that this organism contains P-450 dependent systems that catalyze the monooxygenation of benzo[a]pyrene and, by inference, other xenobiotics at very early devel-opmental stages. Unfortunately, the limited studies performed thus far have not permitted a detailed characterization of the enzyme(s) responsible for the biotransformation detected. The activity was virtually nondetectable in the absence of preexposure to potent P-450-inducing agents such as benzo[a]pyrene or 3-methylcholanthrene (MC). Thus, it is presumed that the catalyzing enzyme(s) detected was probably not constitutive in nature and that such reactions would proceed measurably only after appropriate exposure to MC-type inducing agents. It seems probable that the induced enzyme(s) is either cytochrome $P-450_c$ (see Table 2) or a closely related isozyme(s), although further experimentation would be required for verification of this idea.

Since the early demonstrations by Lutwak-Mann, Keberle, Fabro and their coworkers (see refs. 20–22 and additional refs. therein) that several foreign organic chemicals readily pass from the maternal circulation into the preimplantation blastocyst *via* the uterine fluids, it has been evident that the potential for such chemicals to damage the early embryo is very real. Likewise, it seems apparent that biotransformational capacities of the blastocyst could play a critical role in determin-ing both the nature and the extent of such damage. At this writing, however, it is possible only to state that P-450 dependent biotransformation has been tentatively observed in the preimplantation blastocyst and that the catalyzing isozyme(s) is inducible by polycyclic aromatic hydrocarbons. Presumably, β-naphthoflavone, TCDD and *para/meta*-substituted chlorinated biphenyls and re-lated chemicals would also induce the same isozyme(s) which, in turn, would expectedly also increase rates of biotransformation of other xenobiotic substrates besides benzo[a]pyrene. As yet, other substrates have not been investigated to our knowledge. Likewise, no knowledge of the xenobiotic biotransformational capacity of the zygote, blastomere or morula appears to be available at this

Table 2 Commonly Utilized Terminology and Some Characteristics of Various Rat Hepatic Microsomal *P*-450 Isozymes

Isozyme	Common synonyms	Orthologous isozymes			Regulatory features[a]	Substrate specificity
		Rabbit	Mouse	Human		
a	UTF, 3	—	—	—	PB-, MC-induced	Testosterone (7α)
b	PBB, PB-4	2	116α	—	PB-induced	Testosterone (16β), pentoxyresorufin
c	BNF-B	6	P_1-450	8	MC-, ISF-induced	(*R*)-Warfarin (6,8), ethoxyresorufin
d	ISF-G	4	P_3-450	8	ISF-, MC-induced	Acetanilide, methoxyresorufin
e	PBD, PB-5	—	—	—	PB-induced	Same as *P*-450$_b$
f	—	—	—	—	Constitutive	Testosterone (16α)
g	RLM-3	—	—	—	Constitutive male	?
h	UTA, 2c, RLM-5	—	—	—	Constitutive male; PB-, MC-repressed	Testosterone (2α), (*S*)-warfarin (4')
i	UTI, 2d	—	—	—	Constitutive female	Steroid disulfate (15β)
j	—	3a	ALC	HLj	Ethanol-induced	Aniline
k	PBC, PB-1	—	—	—	PB-induced	Warfarin (7)
p	PCN-E, 2a	3c	—	5	PCN-, PB-induced	Warfarin (10)
—	PB-2	—	—	—	PB-induced	Testosterone (2α)
—	CLO, *P*-452	—	—	—	Clofibrate-induced	Laurate (ω)
—	UTH	—	—	DB	Polymorphic in humans	Debrisoquine

[a] PB = Phenobarbital, MC = 3-methylcholanthrene, ISF = isosafrole, PCN = pregnenolone-16a-carbonitrile. Isozymes *g*, *h* and *p* are male specific; isozyme *i* is female specific.

point in time. With the technology currently available, this information may begin to emerge in the near future.

23.10.4 DRUG BIOTRANSFORMATION IN THE EARLY POSTIMPLANTATION EMBRYO

Implantation provides a convenient demarcation point during early development and usually occurs between the third and sixth days of gestation in several mammalian species, including humans.[1,2] Organogenesis, for convenience, could be considered arbitrarily to extend from the appearance of the neural plate (within a few days subsequent to implantation) to the closure of the palate.[2] For humans, the period of organogenesis would thus extend from days 18–20 to days 55–60 of gestation (Table 1) and for rats from day 9.5 to day 16.

The time period between implantation and appearance of the neural plate, here arbitrarily designated as 'early postimplantation', is thus quite short. To this reviewer's knowledge, no attempts have been made to measure the drug biotransformational capacity of rat embryos during this particular interval of development. Judging from the previously mentioned data pertaining to preimplantation embryos and from the accumulated data on biotransformation during later stages of gestation (*vide infra*), we should not be surprised to discover *P*-450 dependent activities during early postimplantation as well.

23.10.5 DRUG BIOTRANSFORMATION DURING ORGANOGENESIS: IMPLICATIONS FOR TERATOGENESIS

Research performed in the past few years has provided a greatly renewed interest in the role of xenobiotic biotransformation in chemical teratogenesis. Much of the impetus has derived from the development of a system that permits the culturing of whole embryos during the period of organogenesis.[23] This developmental period is commonly accepted as the time in which the conceptus is most sensitive to a teratogenic insult.[1,2] The development of a culture system in which the growth and differentiation of the embryo *in vitro* is approximately equivalent to that observed *in utero* provides investigators with a tool for investigating mechanisms of dysmorphogenesis in the absence of potentially confounding maternal influences. This is particularly useful for investigating the role of biotransformation/bioactivation in embryotoxicity because biotransformation in the maternal organism would expectedly play a major determining role.

The first major hurdle in this research was to determine whether exogenous xenobiotic-biotransforming systems could be added to the culture medium without undue toxic effects to the cultured embryos. In 1979, with cyclophosphamide as a test substrate and potential proembryo-toxin, we found that such a system was both workable and practical.[24,25] This was rapidly confirmed in a number of other laboratories[26-28] and initiated a flurry of investigations which demonstrated that several chemicals could be metabolically converted to intermediates capable of altering developmental parameters measured in cultured embryos.[29] In initial investigations the bioactivating enzymes were exogenous and usually consisted of a microsomal preparation from inducer-pretreated adult male rats. However, we discovered[30-32] that day 10 rat embryos themselves contained enzymes that would catalyze the N and ring hydroxylation of 2-acetylaminofluorene (AAF), a chemical previously demonstrated[33-35] to act as a proembryotoxin in the culture system in the presence of an exogenous bioactivating system. The later investigations showed beyond a reasonable doubt that embryos preinduced *in utero* and subsequently explanted in the whole embryo culture system contained ample cytochrome *P*-450 to catalyze the bioconversion of AAF to reactive intermediates in quantities sufficient to produce grossly observable malfunctions in the selfsame embryos. The implications of these observations are enormous because they demonstrate that only very low levels of observed activity are sufficient to elicit major malformations and that the ratio of *P*-450 dependent bioactivation to inactivation and/or repair is probably a crucial determinant for many teratogens. This further implies that future studies should be directed toward the development of highly sensitive methods for assessing the xenobiotic-biotransforming activities of various embryonic tissues and the regulation thereof. Some methodology is currently available and has enabled investigators to explore certain aspects of embryonic drug metabolism. A summary of some of their findings is presented in a recent review (see ref. 36 and citations therein). Thus far, most of the information that has been published is of a relatively descriptive nature but should provide the foundation for more quantitative and mechanistic studies.

In studies with various nitroheterocycles, it was demonstrated that these agents could elicit a very striking asymmetrical malformation in which the right side of the embryo displayed profound hypoplasia and considerable associated necrotic degeneration.[37,38] The research suggested strongly that agents not bearing an aromatic nitro group (with an appropriately high single-electron redox potential) were not capable of causing this unusual defect.[39] This further suggested that reduction of the nitro group was in some manner responsible for the defect and that the affected embryos had effected reduction of the nitro group. In subsequent studies,[40] we have succeeded in demonstrating that both cultured embryos and embryo homogenates effectively reduced niridazole (the prototypic nitroheterocycle in these studies) and that appearance of one of the reduction products (1-thiocarbamoyl-2-imidazolidinone) was closely correlated with the incidence of asymmetrical abnormalities. Taken together, the evidence strongly indicated that day 10–11 rat embryos contain nitroreductases in sufficient quantities to bioactivate nitroheterocycles to the extent that gross malformations can be produced in the selfsame embryos. It will remain for future investigations to determine the exact nature of the embryonic enzymes involved in the nitroreduction reaction(s), the extent to which such phenomena can be generalized and the potential importance *in vivo*. In separate studies in which a somewhat different experimental protocol was utilized, Brown and Coakley have reported[41] that misonidazole, a nitroheterocycle not included in the aforementioned studies, produced an asymmetric defect involving the limb buds. The possibility that these agents may effect dysmorphogenesis *via* redox cycling has been discussed in a recent review.[42]

Recently we have demonstrated that the dysmorphogenic effects of not only xenobiotics but also endogenous chemicals can be markedly influenced by bioactivation processes.[43] Cytochrome *P*-450 dependent conversion of estradiol 17-β and estrone to the corresponding steroidal catechols resulted in marked increases in incidence and extent of estrogen-elicited prosencephalic hypoplasia with necrosis in the affected anatomy. Phenobarbital inducible *P*-450 isozymes were clearly very effective in catalyzing the bioactivation reaction whereas MC inducible and constitutive female isozymes were not significantly active. Thus far, it appears that embryonic *P*-450s are ineffective, although only preliminary investigations of this aspect have thus far been attempted. Other aspects of biotransformation of potential teratogens during organogenesis have been the subject of recent reviews.[43,44,45]

23.10.6 FETAL DRUG BIOTRANSFORMATION

Investigations of drug biotransformation during the fetal period of development have provided data sufficient to permit a number of generalizations, although some are quite tentative and may not apply to all species or all times during the fetal period.

Prior to the 1970s the prevailing concept of prenatal drug biotransformation reactions provided that these processes occurred at such minimal rates as to be virtually inconsequential, except possibly immediately prior to birth. Several investigations of drug biotransformation in human fetal tissues, however, demonstrated that several fetal organs contained enzymes capable of catalyzing a variety of xenobiotic-biotransforming reactions at unexpectedly rapid rates. For the most part, it was apparent that the prenatal tissue content of such enzymes was between one and three orders of magnitude lower than that of the adult counterparts. However, exceptions to this generality, the realization that relative rates were of greater potential toxicologic importance than absolute rates, and the knowledge that chemical insults during the fetal period can elicit permanent and semi-permanent developmental deficits each contributed as stimuli for further exploration of prenatal xenobiotic metabolism. The subject has been reviewed on a large number of occasions [14,15,16,46,47] and research performed in very recent years has served more to confirm the concepts resulting from studies performed in the 1970s than to provide new, additional concepts. Currently accepted concepts may be summarized as follows.

(i) Primates, particularly humans, appear to have a relatively high complement of xenobiotic-transforming *P*-450 isozymes during the early to mid-fetal period, whereas common laboratory species such as mice, hamsters, guinea pigs, rats and rabbits have a much lower complement as judged from measurements of various monooxygenase activities *in vitro*. Other species have been studied much less extensively.

(ii) The human fetal liver has a wide variety of *P*-450 isozymes as judged from several enzymological studies but the fetal adrenal gland displays surprisingly high monooxygenase activities for certain xenobiotic and steroid substrates and contains extremely high concentrations of *P*-450 cytochromes. Recent studies[48] have provided additional evidence for a major capacity of the human fetal adrenal gland to catalyze xenobiotic-biotransforming reactions. Fetal adrenal tissues exhibited enzymic activities for the conversion of valproic acid to 3-, 4- and 5-hydroxylated metabolites with rates comparable to those measured in livers of adult rats and macaques. Measured rates $(mol\,min^{-1}\,(mg\,protein)^{-1})$ were four times those observed in livers from the same fetuses and approximately 10 times the rates of the same reactions assayed in fetal lung and brain. Tissues from 12 fetuses taken at gestational days 50–77 were examined, again illustrating that human fetal tissues from that late embryonic and early fetal stages often display surprisingly high rates of biotransformation.

(iii) Fetal hepatic cells appear relatively resistant to the effects of transplacental induction. This is particularly true for phenobarbital and, presumably, other 'phenobarbital-type' inducing agents. Studies with experimental animals have demonstrated that fetal hepatic tissues of mammalian species thus far examined do not respond to the inducing effects of phenobarbital until the last days of gestation. This corresponds with observations of lack of response to 'phenobarbital-type' inducers of other rapidly proliferating populations of hepatocytes such as those of regenerating liver, maximal deviation hepatomas or hepatocytes proliferating in culture. Isozymes responsive to phenobarbital induction postnatally do not respond prenatally (except near term) for reasons as yet unknown. A simple inverse relationship between inducer responsivity and cellular proliferation rates, however, cannot be espoused because responsivity is more profound during the early postnatal period (when rates of hepatocyte proliferation are considerable) than in the fully mature animals when hepatocyte proliferation is at a near standstill.

(iv) Responsivity of fetal tissues *in utero* to MC and 'MC-type' inducing agents contrasts markedly to responsivity to 'phenobarbital-type' inducers and can be quite profound, even at very early stages of gestation. Response to 'MC-type' inducers is also observed in other populations of rapidly proliferating cells such as the hepatocytes of regenerating liver, maximal deviation hepatomas and proliferating hepatomas in culture. In addition, it now appears that virtually all cell types (with the possible exception of neurons) will respond to the inducing effects of 'MC-type' inducing agents. This is of particular importance for considerations of prenatal drug biotransformation because hepatocytes make their first appearance at relatively late stages of gestation in many species. For example, rat conceptuses do not display significant hepatocyte populations until approximately days 13–14 of gestation. Nevertheless, as discussed previously, both preimplantation and post-implantation conceptuses display marked increases in xenobiotic biotransformation following exposure to MC. Developmental changes in response to other classes of inducers have not been adequately investigated. A major future challenge for investigators is the acquisition of an understanding of the mechanisms controlling the developmental changes in response to inducing agents. At present, we do not even have a clear picture of the temporal aspects of changes in responsivity during development, much less an understanding of the mechanisms which govern such changes.

Only a very few inducers have been studied and it is only now starting to become clear as to which *P*-450 isozymes (and other drug-biotransforming enzymes) respond to specific inducing agents during postnatal life in a very limited number of animal species.

(v) It cannot be presumed that drug metabolic activities increase as a simple function of developmental age. Far too many exceptions to this outmoded notion have been found for any credibility to be longer warranted. The possibility remains that some rather dramatic changes may occur at developmental ages not yet investigated for specific isozymes. This represents yet another area of research for which an enormous amount of knowledge remains to be revealed in spite of the fact that we now have the wherewithal to readily obtain such knowledge. An examination of the dramatic 'on–off' switching of hemoglobin species during the early developmental stages of fish should alert us to the possibility (or even probability) that similar switching may occur with respect to various drug-biotransforming enzymes.

Neither should it be expected that responsivity to inducers increases progressively as a function of developmental age. Although much more definitive and quantitative data are needed (particularly for the stages of earlier prenatal development), it now seems apparent that response to inducers can also decrease with increasing age, even at relatively early stages of development.

A common error in studies of developmental drug metabolism is to perform measurements on 'fetuses' at relatively late stages of gestation (or at one particular stage) and then discuss the results as though they applied to all prenatal stages. It is becoming increasingly apparent that such generalizations are not acceptable. Particularly unacceptable are extrapolations from data gathered at or near term to all previous prenatal development.

(vi) Drug-biotransforming enzymes exhibiting inducibility during prenatal life tend to be those that are more closely associated with bioactivation, whereas the enzymes commonly deemed as those that catalyze highly important detoxifying reactions (such as glucuronidation, GSH conjugation, *etc.*) have not responded well to transplacentally administered inducers. This would suggest that transplacental exposure of the prenatal organism to xenobiotic-inducing agents would result in an increased probability of chemical damage due to an increased ratio of bioactivation/bioinactivation. Indeed, it has been shown that induced enzymes/enzyme systems present in human fetal and placental tissues can catalyze the generation of reactive metabolites capable of covalently binding to biomacromolecules and inducing bacterial mutations.[47]

(vii) There is currently some evidence to indicate that certain xenobiotic-biotransforming isozymes such as epoxide hydrolases, glutathione *S*-transferases and sulfokinases may be present in relatively high concentrations in prenatal hepatic tissues of some species (*e.g.* primates) but these have not been investigated in a systematic or extensive fashion. Independent confirmation of reports of such enzymatic activities have not been forthcoming in many cases and a review of the literature dealing with developmental drug biotransformation strongly suggests exercise of caution in the absence of such confirmation.[46]

Research reported during the past three to four years has begun to focus more specifically on the utilization of immunologic techniques to detect, identify and quantitate various isozymic forms that may be either constitutively present or inducible in fetal tissues. Cresteil and coworkers[49,50] have been particularly active in this area, although most of the work accomplished applies more appropriately to the stages of development referred to in Table 1 as neonatal and postnatal and are therefore discussed below. It is becoming increasingly clear that a multiple probe approach will be needed to explore drug-biotransforming isozymes during early prenatal life, particularly in experimental animals under investigation as models. This is because of the impracticality of isolating, purifying and characterizing the relevant purified isozymic forms that exist in very small tissue masses. In addition to the potential utility of immunologic probes, hybridization probes, substrate probes and specific inhibitors (including antibodies), activators, inducers and repressors are potentially of great value in elucidating the xenobiotic-biotransforming capacities of prenatal tissues, but will undoubtedly be of greatest value when used in combination in the multiple probe approach. For human fetal tissues and for those of large experimental animals, isolation and purification is feasible and research is currently in progress to isolate and purify various human fetal *P*-450 isozymes (see refs. 51 and 52 and cited refs.).

23.10.7 DRUG BIOTRANSFORMATION BY THE PERINATAL ORGANISM

During prenatal life, the conceptus enjoys the protection of the maternal biotransforming enzymes and also the placenta, although the latter is now known not to provide a significant biochemical barrier to xenobiotics except perhaps under unusual circumstances.[53,54] From a teleological standpoint, it is thus not surprising that neonatal mammals frequently exhibit rapid and marked increases in levels of drug-biotransforming enzymes with increases very shortly before or after birth. More difficult to understand are the varied patterns of changes that occur during subsequent postnatal development.[16] Past experiments have provided us largely with a relatively unpredictable complex and confusing pattern of enzymatic activities measured as a function of age in various species and with various substrates. Only recently has the need been recognized to examine developmental patterns for individual specific isozymic forms such that a meaningful picture of age dependent drug biotransformation can eventually be deduced. Recognition that the various individual metabolic activities may each be catalyzed by a relatively large number of isozymic species has served to set the stage for significant advances in this area of research. Major interest thus far has been in the quantitation of various isozymic forms of cytochrome *P*-450 as a function of postnatal development.[49,50,56,57] Thus far, few new concepts have evolved as a result of the research performed but it has become even clearer that different isozymic species within a given family (*e.g.* within various *P*-450 families) will exhibit independent developmental patterns and that different mechanisms can regulate the expression of the various genes that code for specific drug-biotransforming isozymes within a given family. Cresteil *et al.*[49] concluded that *P*-450s present in late fetal and early neonatal rats were replaced in older animals by immunologically different isoenzymes, some of them resulting from neonatal imprinting. Of interest are the various chemicals that can elicit precocious development of xenobiotic-biotransforming enzymes. The most prominent of these chemicals are various synthetic glucocorticoids, thyroxine and xenobiotic inducing agents. Leakey and coworkers have studied this phenomenon in some detail.[16,58-60] The molecular biology of precocious development of these enzymes (particularly glucuronosyl transferases and monooxygenases) is not yet understood but will undoubtedly come under careful investigation in the near future.

Other investigators have initiated studies of the pre- and post-natal development of mRNAs which translate specific drug-biotransforming isozymes.[61-63] The ultimate goal of these studies is to elucidate the mechanisms whereby the genes which code for various biotransforming enzymes are developmentally regulated. Research accomplished thus far has shown that changes in mRNA levels are usually in close parallel with changes in levels of the corresponding isozymes as measured immunologically or by assays of their enzymatic activities. It seems clear that exceptions to this rule can exist and that future reports of such exceptions can be expected. The molecular biology approach can ultimately be expected to divulge the various signals which govern developmentally regulated gene expression for drug-biotransforming enzymes and could thereby provide the means for additional and/or more effective modes of artificial regulation of the corresponding naturally occurring developmental processes. As pointed out by several previous reviewers, however, numerous other developmentally associated factors can effect changes in drug biotransformation in addition to simple changes in enzyme levels. These include changes in quantity and nature of the membranes of the endoplasmic reticulum, changes in the hormonal milieu and changes in associated cofactor-generating/degrading systems (*e.g.* NADPH, UDPGA, PAPS, *etc.*). Certainly, the elucidation of mechanisms of regulation of developmental drug biotransformation will provide a challenge for many years to come.

23.10.8 POSTNATAL PATTERNS OF DEVELOPMENT OF DRUG BIOTRANSFORMATIONAL PROCESSES

In *very* general terms quantities of drug-biotransforming enzymes tend to increase after the birth of a mammalian organism. However, the many and varied patterns of postnatal development of individual drug-biotransforming activities and exceptions to the general rule present a highly complex and puzzling picture. Such patterns are dependent upon the genetic background (species and strain of animal), specific substrate under investigation, specific reaction under study, pre- or peri-natal exposure to various chemicals and pathologic/nutritional alterations, endocrine factors, and probably several other parameters as yet unknown. Ideally, one would envision the desirability of ascertaining the exact postnatal developmental pattern of xenobiotic biotransformation for any given individual. Clearly, the complexity of the situation and the dearth of knowledge in this area of research are such that this ideal appears highly remote at present. One can only hope that future research will permit some generalizations to emerge and that a rationale for the puzzling complexity may be eventually forthcoming.

23.10.9 EFFECTS OF WEANING ON DRUG BIOTRANSFORMATION

The period of weaning represents a time in which the diet is changed from ingestion of maternal milk to the foodstuffs normally ingested by adults of the species in question. These foodstuffs (particularly certain plant foodstuffs) would be expected to contain a number of chemicals (*e.g.* indoles, flavones) that are capable of increasing rates of certain drug biotransformation reactions *via* induction. Such chemicals would normally be present in maternal milk at relatively low concentrations. These considerations cause one to expect that overall increased rates of drug biotransformation should accompany the weaning process. In addition, it is well established that nutritional factors can exert significant effects on levels and activities of various drug-biotransforming enzymes.[64,65] (It is now well recognized, however, that in mammalian species, such effects are prominent primarily in rats.) In spite of these facts, relatively little is known concerning the effects of weaning on drug biotransformation. Although several changes are known to be associated with weaning (*e.g.* increases in *P*-450 reductase activity during late suckling), rigorous control experiments to determine whether weaning is causal or merely associated by chance are usually lacking.

23.10.10 SEXUAL MATURATION AND DRUG BIOTRANSFORMATION

Puberty is associated with rather dramatic changes in the hormonal milieu, especially with regard to certain steroid hormones. Sex-associated steroids such as testosterone, androstenedione, estradiol-17β and estrone (which increase dramatically at puberty) frequently serve as substrates for the drug-biotransforming enzymes. It might be expected, therefore, that relatively marked changes in drug biotransformation would accompany puberty. That this is true for all strains of *rats* thus far investigated has been amply documented. However, other mammalian species, including humans, usually display minimal or no maturational changes specifically associated with puberty and sexual development. Some strains of mice display relatively minor sex differences after attainment of sexual maturity which, interestingly, are the reverse of those seen in rats, *i.e.* overall rates of xenobiotic oxidation in females are more rapid than in males. The only other species widely recognized to exhibit profound sex differences are nonmammalian vertebrates (*e.g.* trout, frogs, *etc.*).

The biochemical basis for the sex differences observed in rats is now becoming clear. Certain *P*-450 isozymes (see Table 2) commonly referred to as *g*, *h* and *p* are male specific and their regulation appears to be at least partially under the control of the hypothalamic/pituitary/gonadal axis, with estrogens, androgens, somatotrophic hormone and somatostatin playing major regulatory roles. Each of these *P*-450 isozymes can be regarded as constitutive (although certain forms of *p* are also highly inducible by synthetic glucocorticoids and pregnenolone-16α-carbonitrile) but are capable of attacking a rather wide variety of xenobiotic substrates as well. A female specific form, referred to as *P*450$_i$, has also been characterized. This form exhibits a rather high substrate specificity for certain disulfated steroids and does not appear to catalyze efficiently the oxidation of many xenobiotic substrates. Characterization of these sex specific isozymic forms has helped to explain the long-recognized, profound sex differences known to exist in rats. High interest in such isozymes derives from their unique modes of regulation as well as from the fact that the rat has been the most extensively utilized model for studies of drug biotransformation. The phenomenon has been studied extensively by Gustafsson and his coworkers,[66,67] by Waxman and his coworkers[68,69] as well as by several other groups. Because it seems logical to presume that the 'natural' substrates for these sex specific *P*-450s are steroids, further studies in this area could greatly enhance our understanding of modes and regulation of both steroid metabolism and drug biotransformation, even though studies in rats represent investigations of a highly unique mammalian species. It may be speculated that several sex specific isozymes exist in other mammalian species but that the *relative* drug-biotransforming activities of male-specific *versus* female-specific isozymes are counterbalanced to the extent that no remarkable overall sex differences in drug metabolism are normally encountered. As an example, in mice a *P*-450 isozyme referred to as *P*-450-15α is much higher in females than in males, but in this case the female specific form actively catalyzes the monooxygenation of several xenobiotic substrates, including benzphetamine, aniline, 7-ethoxycoumarin, benzo[*a*]pyrene and acetanilide.[70]

Recent research has also focused on analyses of the structures of various sex dependent *P*-450 isozymes in rats (*e.g.* see ref. 71 and cited refs.) and on the developmental profiles and regulation of such isozymes (*e.g.* see ref. 72 and cited refs.). This very active area of research promises to provide a plethora of information in the very near future.

23.10.11 EFFECTS OF PREGNANCY ON DRUG BIOTRANSFORMATION

Like puberty, pregnancy is also associated with dramatic hormonal changes. Particularly impressive are increases in estrogens and progesterone, steroids that are substrates for several drug-metabolizing enzymes. Again, changes in rates and/or routes of drug-metabolizing reactions might be expected to accompany the gravid state. In humans, it is abundantly clear that pregnancy can be accompanied by changes in drug disposition as assessed with classical pharmacokinetic measurements.[73-75] Because of multiple pregnancy-associated physiologic changes that affect or potentially affect not only biotransformation, but also absorption, distribution and other modes of elimination of drugs and chemicals, pharmacokinetic measurements are frequently unsatisfactory in terms of elucidating the degree to which biotransformational processes *per se* have undergone changes.[76,77] A review of the literature (see refs. 73–77 and those cited therein) suggests that rates of reactions tend to decrease with pregnancy. However, most of the well-controlled studies pertaining to this topic have been performed in rats. The utilization of the rat as an experimental *model* for studies of the effects of pregnancy on drug biotransformation may well be regarded as inappropriate since this is the only mammalian species known to exhibit marked sex differences in the activities of drug-metabolizing enzymes. It seems probable that the effects of pregnancy-associated hormonal changes would tend to be exaggerated in this species and would thus be more difficult to extrapolate to human pregnancy.

Pharmacokinetic studies in human patients have resulted in conflicting conclusions regarding the role of biotransformation on drug disposition in human pregnancy.[73-77] It is usually believed that, in general, rates may be decreased, but only in a very minor fashion and that other pregnancy dependent changes in drug disposition may tend to counterbalance the effects of slightly decreased biotransformation on measurements of pharmacokinetic parameters. Unless definitively shown otherwise, this appears to represent a reasonable first assumption as judged from long clinical experience of treating pregnant women with drugs. There is, in the vast majority of cases, no compelling evidence to suggest that drug dosage regimens for pregnant women should differ from those for nonpregnant women. However, lack of a substantial or well-controlled data base warns that caution should be advised, particularly in view of recent discoveries of pregnancy-specific changes in specific *P*-450 isozymes in rabbits[78] and mice.[79] At present, it seems necessary to regard each chemical individually and avoid broad generalizations that could arise as a result of studies on other chemicals. Obviously, interspecies extrapolations are unwarranted at present.

23.10.12 DRUG BIOTRANSFORMATION DURING AGING AND SENESCENCE

To date, studies of the effects of aging on xenobiotic biotransformation have received lesser attention, although there is very recent evidence that more attention is forthcoming. A number of factors appear to have limited past research progress. These include (i) the very gradual nature of the aging process (as opposed to some relatively rapid changes that occur during earlier developmental stages) making it difficult to focus on specific aging events that could regulate biotransformational processes; (ii) the widely promulgated generalization that levels of xenobiotic-biotransforming enzymes tend to diminish slowly as a function of age; and (iii) lack of widely publicized clinical catastrophes associated with administration of drugs to the elderly. Experiences with thalidomide, diethylstilbestrol, chloramphenicol, warfarin, folic acid antagonists and other drugs greatly stimulated research on the prenatal and neonatal periods but analogous experiences pertaining to older populations have not received notoriety. However, as the average age of the population continues to increase, pressure for additional research has been evident and will undoubtedly continue to increase.

Rikans has provided an excellent recent review of age-related changes in xenobiotic biotransformation and disposition during the aging process.[80] There appears to be no question that relatively marked changes in metabolic capacity can be demonstrated during this period, but the associated changes are (i) unpredictable from species to species, pathway to pathway, substrate to substrate and even from individual to individual; (ii) probably heavily dependent upon changes in dietary states, chronic illness, endocrine status, and long-term drug usage; and (iii) sometimes more heavily influenced by changes in hepatic blood flow and/or liver volume than by changes in enzyme levels. A number of investigators have succeeded in showing in experimental animals that individual isozymic *P*-450s can decrease as a function of aging[81-83] but the biochemical basis for the decreases remains a question. Are such decreases due to dietary or pathologic changes or are genes programmed to downregulate as a function of age? Importantly, Schmucker and Wang[84,85] have presented data to indicate that reported decreases in *P*-450 dependent metabolic activity in old age may be applicable to rodents (male rats in particular) but not necessarily to primates and perhaps other species. Their and other research has been unable to show age-related declines in *P*-450 dependent monooxygenase activities in nonhuman primates, and in fact reported sharp increases in NADPH–cytochrome *P*-450 reductase activities. These studies make it clear that much is yet to be learned concerning aging and drug metabolism. They also further demonstrate that the frequently utilized male rat is not an appropriate model for the study of effects of aging on drug biotransformation.

23.10.13 SUMMARY AND CONCLUSIONS

It is clear that many large gaps in our understanding of developmental xenobiotic biotransformation still remain. This review has attempted to highlight those gaps and point to areas that are in need of further research. The greatest needs appear to be for research during the period of organogenesis, when the developing organism is most sensitive to xenobiotic-elicited dysmorphogenesis. It has recently been shown that even very low reaction rates catalyzed by embryonic enzymes can generate sufficient reactive intermediates to cause profound morphologic alterations in the same embryos. These observations point to a critical need for a complete understanding of embryonic xenobiotic-biotransforming capabilities. However, the gaps for other developmental periods are enormous as well and should not be underemphasized. Research on gametes has been virtually nonexistent and should be vigorously pursued because of implications for inherited birth defects. We still know very little concerning the age dependent *changes* in biotransformation that occur during the fetal period. Rats have been utilized extensively for studies of age-related biotransformational changes during later periods of development and it is now clear that, because of the unique sex differences exhibited by this species beginning at puberty, the rat does not represent a highly useful model for such studies. The major conclusion to be drawn from this review is that our knowledge of developmental aspects of drug biotransformation is sufficiently meager that it may appropriately be regarded as shameful. It is hoped that this article may stimulate research designed to close some of the enormous gaps that remain.

23.10.14 REFERENCES

1. A. R. Scialli and S. Fabro, in 'Drug and Chemical Action in Pregnancy', ed. S. Fabro and A. R. Scialli, Dekker, New York, 1986, p. 191.
2. T. H. Shepard, 'Catalog of Teratogenic Agents', 6th edn., Johns Hopkins University Press, Baltimore, MD, 1989.
3. E. C. Miller and J. A. Miller, in 'Bioactivation of Foreign Compounds', ed. M. W. Anders, Academic Press, New York, 1985, p. 3.
4. J. R. Gillette, in 'Bioactivation of Foreign Compounds', ed. M. W. Anders, Academic Press, New York, 1985, p. 30.
5. S. L. Heinrichs and M. R. Juchau, in 'Extrahepatic Metabolism of Drugs and Other Foreign Compounds', ed. T. E. Gram, Spectrum Publications, Holliswood, NY, 1980, p. 319.
6. I. P. Lee, H. Mukhtar, K. Suzuki and J. R. Bend, *Biochem. Pharmacol.*, 1983, **32**, 159.
7. K. Suhara, Y. Fujimura, M. Shiroo and M. Katagiri, *J. Biol. Chem.*, 1984, **259**, 8729.
8. J. A. Goldstein and P. Linko, *Mol. Pharmacol.*, 1984, **25**, 185.
9. M. Bengtsson, J. Montelius, L. Mankowitz and J. Rydstrom, *Biochem. Pharmacol.*, 1983, **32**, 129.
10. D. R. Mattison and M. R. Nightingale, *Toxicol. Appl. Pharmacol.*, 1980, **56**, 399.
11. K. Shinomizu and D. R. Mattison, *Teratogen. Carcinog. Mutagen.*, 1985, **5**, 463.
12. F. Oesch, D. M. Jerina and J. Daly, *Biochim. Biophys. Acta*, 1971, **227**, 685.
13. F. Oesch, H. R. Glatt and H. Schmassmann, *Biochem. Pharmacol.*, 1977, **26**, 603.
14. D. Neubert and S. Tapken, in 'Role of Pharmacokinetics in Prenatal and Perinatal Pharmacology', ed. D. Neubert, H.-J. Merker, H. Nau, J. Langman, A. Bedürftig, R. Kreft, B. Stein and E. Gottschalls. Thieme, Stuttgart, 1978, p. 69.
15. O. Pelkonen, in 'Developmental Toxicology', ed. K. Snell, Croom-Helm, London, 1982, p. 167.
16. J. E. A. Leakey, in 'Biological Basis of Detoxication', ed. J. Caldwell and W. B. Jakoby, Academic Press, New York, 1983, p. 77.
17. R. Filler and K. J. Lew, *Proc. Natl. Acad. Sci. USA*, 1981, **78**, 6991.
18. S. M. Galloway, P. E. Perry, J. Meneses, D. Nebert and R. Pederson, *Proc. Natl. Acad. Sci. USA*, 1980, **77**, 3524.
19. R. A. Pedersen, J. Meneses, A. Spindle, K. Wu and S. M. Galloway, *Proc. Natl. Acad. Sci. USA*, 1985, **82**, 3311.
20. C. Lutwak-Mann, M. F. Hay and D. A. T. New, *J. Reprod. Fertil.*, 1969, **18**, 235.
21. H. Keberle, K. Schmid, J. W. Faigle, H. Fritz and P. Loustalot, *Bull. Schweiz. Akad. Med. Wiss.*, 1966, **22**, 134.
22. S. Fabro, in 'Fetal Pharmacology', ed. L. Boreus, Raven Press, New York, 1973, p. 443.
23. D. A. T. New, *Biol. Rev.*, 1978, **53**, 81.
24. A. G. Fantel, J. C. Greenaway, T. H. Shepard and M. R. Juchau, *Fed. Proc., Fed. Am. Soc. Exp. Biol.*, 1979, **38**, 437.
25. A. G. Fantel, J. C. Greenaway, M. R. Juchau and T. H. Shepard, *Life Sci.*, 1979, **25**, 67.
26. K. T. Kitchin, B. P. Schmid and M. K. Sanyal, *Biochem. Pharmacol.*, 1981, **30**, 59.
27. M. K. Sanyal, K. T. Kitchin and R. L. Dixon, *Toxicol. Appl. Pharmacol.*, 1981, **57**, 14.

28. J. C. Greenaway, A. G. Fantel, T. H. Shepard and M. R. Juchau, *Teratology*, 1982, **25**, 353.
29. B. P. Schmid, *Concepts Toxicol.*, 1985, **3**, 46.
30. M. R. Juchau, C. M. Giachelli, A. G. Fantel, J. C. Greenaway, T. H. Shepard and E. M. Faustman-Watts, *Toxicol. Appl. Pharmacol.*, 1985, **80**, 137.
31. M. R. Juchau, D. H. Bark, L. M. Shewey and J. C. Greenaway, *Toxicol. Appl. Pharmacol.*, 1985, **81**, 533.
32. E. M. Faustman-Watts, C. M. Giachelli and M. R. Juchau, *Toxicol. Appl. Pharmacol.*, 1986, **83**, 590.
33. E. M. Faustman-Watts, J. C. Greenaway, M. J. Namkung, A. G. Fantel and M. R. Juchau, *Teratology*, 1983, **27**, 19.
34. E. M. Faustman-Watts, H. Y. L. Yang, M. J. Namkung, J. C. Greenaway, A. G. Fantel and M. R. Juchau, *Teratogen., Carcinog. Mutagen.*, 1984, **4**, 273.
35. E. M. Faustman-Watts, J. C. Greenaway, M. J. Namkung, A. G. Fantel and M. R. Juchau, *Toxicol. Appl. Pharmacol.*, 1984, **76**, 161.
36. L. P. Brown, O. P. Flint, T. C. Orton and G. G. Gibson, *Drug Metab. Rev.*, 1986, **17**, 221.
37. J. C. Greenaway, P. E. Mirkes, E. A. Walker, M. R. Juchau, T. H. Shepard and A. G. Fantel, *Teratology*, 1985, **32**, 287.
38. A. G. Fantel, J. C. Greenaway, E. Walker and M. R. Juchau, *Teratology*, 1986, **33**, 105.
39. J. C. Greenaway, A. G. Fantel and M. R. Juchau, *Toxicol. Appl. Pharmacol.*, 1986, **82**, 307.
40. A. G. Fantel, R. E. Person and M. R. Juchau, *Teratology*, 1987, **35**, 72A.
41. N. A. Brown and M. E. Coakley, *Teratology*, 1987, **35**, 73A.
42. M. R. Juchau, A. G. Fantel, C. Harris and B. K. Beyer, *Environ. Health Perspect.*, 1986, **70**, 131.
43. B. K. Beyer and M. R. Juchau, *Biochem. Biophys. Res. Commun.*, 1987, **145**, 402.
44. M. R. Juchau, in 'Pharmacokinetics in Teratogenesis', ed. H. Nau and W. J. Scott, CRC Press, Boca Raton, FL, 1987, vol. II, p. 121.
45. M. R. Juchau, C. Harris, B. K. Beyer and A. G. Fantel, in 'Approaches to Elucidate Mechanisms in Teratogenesis', ed. F. Welsch, Hemisphere Publishers, New York, 1989, in press.
46. O. Pelkonen, *Dev. Pharmacol. Ther.*, 1984, **7**, Suppl. 1, 11.
47. M. R. Juchau, in 'Prenatal Drug Exposure: Kinetics and Dynamics', ed. C. N. Chiang and C. C. Lee, NIDA Research Monograph 60, US Department of Health and Human Services, Rockville, MD, 1985, p. 17.
48. A. E. Rettie, A. W. Rettenmeier, B. K. Beyer, T. A. Baillie and M. R. Juchau, *Clin. Pharmacol. Ther.*, 1986, **40**, 172.
49. T. Cresteil, P. Beaune, C. Celier, J. P. Leroux and F. P. Guengerich, *J. Pharmacol. Exp. Ther.*, 1986, **236**, 269.
50. T. Cresteil, E. LeProvost, J. P. Flinois and J. P. Leroux, *Biochem. Biophys. Res. Commun.*, 1982, **106**, 823.
51. R. Voutilainen and W. L. Miller, *J. Clin. Endocrinol. Metab.*, 1986, **63**, 1145.
52. M. Kitada, T. Kamataki, K. Itahashi, T. Rikihisa and Y. Kanakubo, *Biochem. Pharmacol.*, 1987, **36**, 453.
53. M. R. Juchau, *Pharmacol. Ther.*, 1980, **8**, 501.
54. M. R. Juchau, M. J. Namkung and A. E. Rettie, *Trophoblast Res.*, 1987, **2**, 235.
55. L. F. Gulyaeva, V. M. Mishin and V. V. Lyakhovich, *Int. J. Biochem.*, 1986, **18**, 829.
56. T. Cresteil, P. Beaune, P. Kremers, C. Celier, F. Guengerich and J. Leroux, *Eur. J. Biochem.*, 1985, **151**, 345.
57. K. T. Shiverick, A. G. Kvello-Stenstrom, W. H. Donnelly, A. S. Salhab, J. A. Goldstein and M. O. James, *Biochem. Biophys. Res. Commun.*, 1987, **141**, 299.
58. J. E. A. Leakey, Z. R. Althaus, J. R. Bailey and W. Slikker, Jr., *Biochem. Pharmacol.*, 1986, **35**, 1389.
59. Z. R. Althaus, J. R. Bailey, J. E. A. Leakey and W. Slikker, Jr., *Dev. Pharmacol. Ther.*, 1986, **9**, 332.
60. J. E. A. Leakey, Z. R. Althaus, J. R. Bailey and W. Slikker, *Biochem. J.*, 1985, **225**, 183.
61. M. R. Jackson, S. M. E. Kennedy, G. Lown and B. Burchell, *Biochem. Pharmacol.*, 1986, **35**, 1191.
62. C. M. Giachelli and C. J. Omiecinski, *J. Biol. Chem.*, 1986, **261**, 1359.
63. C. M. Giachelli and C. J. Omiecinski, *Mol. Pharmacol.*, 1987, **31**, 477.
64. J. N. Boyd and T. C. Campbell, in 'Biological Basis of Detoxication', ed. J. Caldwell and W. B. Jakoby, Academic Press, New York, 1983, p. 287.
65. W. R. Bidlack, R. C. Brown and C. Mohan, *Fed. Proc., Fed. Am. Soc. Exp. Biol.*, 1986, **45**, 142.
66. C. MacGeoch, E. T. Morgan, B. Cordell and J. Gustafsson, *Biochem. Biophys. Res. Commun.*, 1987, **143**, 782.
67. J. Gustafsson, A. Mode, G. Norstedt and P. Skett, *Annu. Rev. Physiol.*, 1983, **45**, 51.
68. D. J. Waxman, *J. Biol. Chem.*, 1984, **259**, 15481.
69. D. J. Waxman, G. A. Dannan and F. P. Guengerich, *Biochemistry*, 1985, **24**, 4409.
70. N. Harada and M. Negishi, *J. Biol. Chem.*, 1984, **259**, 1265.
71. H. Yoshioka, K. Morohashi, T. Sogawa, T. Miyata, K. Kawajiri, T. Hirose, Y. Fujii-Kuriyama and T. Omura, *J. Biol. Chem.*, 1987, **262**, 1706.
72. F. J. Gonzalez, S. Kimura, B. Song, J. Pastewka, H. V. Gelboin and J. P. Hardwick, *J. Biol. Chem.*, 1986, **261**, 10667.
73. W. A. Parker, in 'Pharmacokinetic Basis for Drug Treatment', ed. I. Z. Benet, N. Massoud and J. G. Gambertoglio, Raven Press, New York, 1984, p. 249.
74. H. Nau, W. Loock, M. Schmidt-Gollwitzer and W. Kuhnz, in 'Drugs and Pregnancy', ed. B. Krauer, F. Krauer, F. E. Hytten and E. del Pozo, Academic Press, New York, 1984, p. 45.
75. D. R. Mattison, in 'Drug and Chemical Action in Pregnancy', ed. S. Fabro and A. R. Scialli, Dekker, New York, 1986, p. 37.
76. M. R. Juchau and E. Faustman-Watts, *Clin. Obstet. Gynecol.*, 1983, **26**, 379.
77. M. R. Juchau, in 'Hazard Assessment of Chemicals', ed. J. Saxena, Academic Press, New York, 1983, vol. 2, p. 96.
78. D. E. Williams, S. E. Hale, R. T. Okita and B. S. S. Masters, *J. Biol. Chem.*, 1984, **259**, 14600.
79. G. H. Lambert, H. W. Lietz and A. N. Kotake, *Biochem. Pharmacol.*, 1987, **36**, 1965.
80. L. E. Rikans, *Life Sci.*, 1986, **39**, 1027.
81. S. Fujita, H. Kitagawa, M. Chiba, T. Suzuki, M. Ohta and K. Kitani, *Biochem. Pharmacol.*, 1985, **34**, 1861.
82. T. Kamataki, K. Maeda, M. Shimada, K. Kitani, T. Nagai and R. Kato, *J. Pharmacol. Exp. Ther.*, 1985, **233**, 222.
83. D. N. McMartin, J. A. O'Connor, Jr., M. J. Fasco and L. S. Kaminsky, *Toxicol. Appl. Pharmacol.*, 1980, **54**, 411.
84. D. L. Schmucker, *Pharmacol. Rev.*, 1985, **37**, 133.
85. D. L. Schmucker and R. K. Wang, *Drug Metab. Dispos.*, 1987, **15**, 225.

23.11

Pharmacogenetics

ELLIOT S. VESELL

The Pennsylvania State University College of Medicine, Hershey, PA, USA

23.11.1 OVERVIEW

Pharmacogenetics may be defined as the study of genetically controlled variations in xenobiotic response. Such variations can be either pharmacokinetic or pharmacodynamic in nature. Pharmacogenetic conditions are of great practical interest in understanding the causes of large differences among subjects in response to drugs and environmental chemicals. That explains why research in pharmacogenetics has become both diverse and complex. This chapter stresses fundamental principles, which are illustrated by selection for discussion of a few prototypic pharmacogenetic conditions. Furthermore, space limitations permit description of only representative entities among the approximately 60 pharmacogenetic conditions now recognized as causing large variations among human subjects in response to drugs and other environmental chemicals (xenobiotics).

23.11.2 HISTORICAL BACKGROUND

In 1962 and 1963 two books on pharmacogenetics appeared containing numerous studies in laboratory animals.[1,2] Because of many more recent investigations in this area, pharmacogenetic studies in laboratory animals require a fresh description. One reason for the development of experimental pharmacogenetics is recognition that pharmacological and toxicological effects often vary markedly not only with the strain or species,[3] but also even in a given strain or species as a result of only a single amino acid substitution in an enzyme controlling xenobiotic disposition. Therefore, different responses to an exogenous chemical by various species or strains can serve as sensitive probes to explore basic mechanisms of xenobiotic action. Because of pharmacogenetic differences, it is advisable to perform metabolic, toxicologic and carcinogenic studies of new drugs in at least two species and within a representative number of subjects from an outbred strain or species. Frequently encountered strain or species differences in xenobiotic disposition and toxicity diminish the hope of finding an animal model that resembles man in all these respects, while correspondingly increasing the problem of adequately testing drugs for deleterious effects in animals prior to human use. Several symposia and volumes illustrate the broad interest in pharmacogenetics and diverse approaches being pursued.[4-14]

23.11.3 TOXICOLOGICAL SIGNIFICANCE OF PHARMACOGENETICS

Pharmacogenetics is of toxicological importance because it reveals how certain subjects, due to their genetic constitution, develop toxicity on exposure to xenobiotics at doses well tolerated by subjects with different genetic constitutions. Technical advances permitting rapid, sensitive and accurate measurement of metabolites of drugs and environmental chemicals in human biological fluids are responsible for much recent pharmacogenetic progress.

23.11.4 RESTRICTION OF DEFINITION

This review restricts the term pharmacogenetic to conditions involving a protein or enzyme that participates *directly* in the metabolism or receptor binding of a drug or environmental chemical. Limitation of the definition in this way is necessary in order to exclude the hundreds of inborn errors of metabolism that are *indirectly* associated with some aberrant response to drugs and environmental chemicals. If the term pharmacogenetic were to encompass all inborn errors of metabolism, the word would lose specificity and hence utility.

23.11.5 RECENT PROGRESS IN PHARMACOGENETICS

Much recent progress in pharmacogenetics has ensued from the exciting discovery of several new genetic entities that produce large interindividual variations in drug oxidation. Different rates of drug elimination measured *in vivo* arise from different hepatic capacities to biotransform drugs; the relationship between rates of drug metabolism measured *in vivo* and *in vitro* has been most intensively investigated for the debrisoquine and mephenytoin polymorphisms. Some of the genetically discrete cytochrome *P*-450 isozymes that cause these polymorphisms have been isolated from human liver and characterized. The six physicochemically distinguishable cytochrome *P*-450 isozymes thus far identified in human liver account for only a small portion of the total hepatic content of cytochrome *P*-450, indicating not only more extensive heterogeneity of this protein than currently recognized but also the need for further biochemical work of this kind. Cloning of the cDNA for some cytochrome *P*-450 isozymes raises the possibility that ultimately in patients the genetically controlled capacity to eliminate drugs might be determined *in vitro* through restriction fragment length polymorphisms, thereby avoiding the more hazardous approach currently used of establishing phenotypes by oral administration of test drugs possessing pharmacological activity.

23.11.6 EXTRAGENETIC SOURCES OF PHARMACOKINETIC AND PHARMACODYNAMIC VARIATIONS

Genetic constitution is only one of many factors that can produce large differences among humans in xenobiotic response (Figure 1). Yet in some human subjects under certain conditions, genetic factors comprise the major or even sole cause of interindividual variations in xenobiotic response. For example, when age, sex, diet or exposure to environmental chemicals that induce or inhibit the hepatic drug-metabolizing enzyme system remain constant among human subjects, large interindividual variations in the disposition of and response to xenobiotics still occur and have been shown to arise from genetic factors. The magnitude of these interindividual variations in rates of elimination of a xenobiotic generally range from fourfold to as much as fortyfold, depending both on the particular xenobiotic and the population studied. In practice, these large differences in rates of xenobiotic elimination that exist among normal subjects, regardless of age or sex, mean that the same dose of a drug given by the same route can produce toxicity in one subject, a therapeutic effect in another and failure to obtain any pharmacologic response in a third. Such large interindividual differences in drug disposition present a formidable challenge to the practicing physician, who must often individualize therapy, especially for drugs with low therapeutic indices. Individualization of therapy means that the appropriate drug as well as the appropriate dose needs to be selected according to the particular needs of each patient to maximize therapeutic results from the drug, as well as to avoid the extremes of toxicity and ineffectiveness. In patients, particularly during the course of severe diseases, genetic and environmental factors dynamically interact, causing a subject's dosage requirements to change with time and clinical status.

Epidemiologic studies reveal that about one patient in five enters a teaching hospital in the United States for treatment of an adverse drug reaction. Further, 15–30% of all patients admitted to teaching hospitals have at least one such reaction.[15] Although wide disparity in patient's responses to drugs is only one of many causes of adverse reactions, including drug toxicity, it nevertheless constitutes a significant contribution to this major medical problem and demands individualization of the dosage of drugs with low therapeutic indices. Multiple factors have now been systematically investigated and identified as contributing to the large interindividual variations that in human subjects characterize disposition and response to xenobiotics. These factors are age, sex, time of day or season of drug administration, disease, hormonal and nutritional status, the status of heart, liver, kidney, endocrine organs, stress and exposure to inducers or inhibitors of the hepatic microsomal drug-metabolizing enzymes, including chronic administration of any one of several hundred drugs (Figure 1).[16–18] Also in the last 20 years, multiple genetic factors that directly affect xenobiotic disposition and response in humans have been discovered.[5,6,8,9,12–14,19]

The many therapeutic ramifications of large interindividual variations in disposition of numerous drugs make it necessary to identify mechanisms responsible for these differences among subjects. Such investigations are beset with many difficulties rooted in the extreme genetic and environmental heterogeneity of human beings with respect to factors affecting xenobiotic disposition. However, in laboratory animals such as rats or mice, these factors can be rigidly controlled. In such a laboratory setting, each variable can be manipulated independently of others. By this technique, the quantitative contribution of each factor to interindividual variations in drug disposition can be studied and

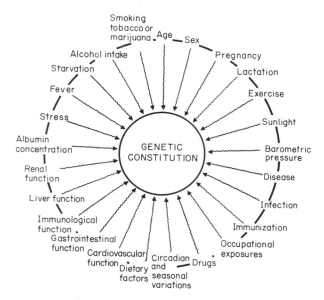

Figure 1 Schematic depiction of established or suspected environmental factors that can alter genetically controlled rates of drug elimination. Lines from each environmental factor to a central circle are wavy to suggest that modification of genetically controlled rates can occur at multiple levels. Such environmental effects need not occur directly at the genetic level. In the outer circle a line joins environmental factors to suggest that several are associated and interdependent, rather than independent (reproduced by permission of Mosby from ref. 18)

dose–response curves constructed for each factor. Over the past decade such studies have disclosed many factors that can affect drug disposition. Figure 1 schematically illustrates the multiple opportunities that exist in human subjects for interactions among such factors. For laboratory animals Table 1 presents only a partial list covering one of these factors, genetic constitution.[20] In human subjects, pharmacogenetic conditions can be categorized into those that affect how the body acts on drugs (pharmacokinetic pharmacogenetic conditions) and those that affect how drugs act on the body (pharmacodynamic pharmacogenetic conditions).

23.11.7 MONOGENICALLY INHERITED PHARMACOGENETIC CONDITIONS INVOLVING ENZYMES AND PROTEINS THAT DIRECTLY AFFECT DRUG RESPONSE

Pharmacogenetics deals with clinically significant hereditary variations in response to drugs and environmental chemicals. The term ecogenetics has been coined to designate genetically controlled variations in response to environmental chemicals.[21] Since the principles of ecogenetics, as well as most of the conditions illustrating ecogenetics, either closely resemble or duplicate those in pharmacogenetics, there seems to be no advantage to the word ecogenetics as opposed to pharmacogenetics. Furthermore, coining a new term, however fashionable it may seem or sound, creates unnecessary confusion in the literature and in the minds of students who are accustomed to apply the designation pharmacogenetics to these conditions and concepts.

23.11.8 MONOGENICALLY TRANSMITTED PHARMACOGENETIC CONDITIONS CHARACTERIZED BY ABNORMAL METABOLISM OF DRUGS BY THE BODY

23.11.8.1 Acatalasemia

In 1946, the Japanese otorhinolaryngologist Takahara discovered acatalasemia in an 11-year-old Japanese girl. In a series of classic studies he demonstrated that the defect was transmitted as an autosomal recessive trait.[22-24] His original patient lacked catalase activity in her oral mucosa and erythrocytes, as did three of her five siblings. The patient's parents were second cousins: con-

Table 1 Variations in Response to Xenobiotics Caused by Genetic Factors[20]

Mice
Chloroform susceptibility
Histamine sensitivity
High insulin tolerance
Phenothiazine protection against audiogenic seizures
Iproniazid lethality and hepatic injury
Ethanol preference
(+)-Amphetamine lethality
Chlorpromazine sensitivity
Serotonin toxicity
Tumorigenesis
Aging and spontaneous cancers
Teratogenesis
Aryl hydrocarbon hydroxylase induction by polycyclic
  hydrocarbons
δ-Aminolevulinic acid synthetase induction by
  3,5-dicarbethoxy-1,4-dihydro-2,4,6-trimethylpyridine
  (DDC)
2,3,7,8-Tetrachlorodibenzo-*p*-dioxin-produced
  teratogenesis
Skin ulcers caused by 7,12-dimethylbenz[*a*]anthracene
Cholesterol biosynthesis
NADPH-generating capacity
Induction of cytochrome P_1-450 by polycyclic
  hydrocarbons
Cadmium-induced testicular necrosis
3-Methylcholanthrene-initiated tumorigenesis
Induction of 7-ethoxycoumarin *O*-deethylase, *p*-nitro-
  anisole *O*-demethylase and 3-methyl-4-methylamino-
  azobenzene *N*-demethylase activities by polycyclic
  aromatic compounds
Susceptibility to ozone toxicity
Coumarin hydroxylase induced by phenobarbital
Zoxazolamine paralysis time
Phenacetin *O*-deethylase induction by polycyclic
  hydrocarbons
Acetylarylamine (2-acetylaminofluorene) *N*-hydroxylase
  induction by polycyclic hydrocarbons
Lung and aryl hydrocarbon hydroxylase induction by
  cigarette smoke

Acetaminophen-caused hepatic necrosis
Shortened survival time in mice exposed to polycyclic
  hydrocarbons, lindane and halogenated hydrocarbons
Induction of biphenyl-2-hydroxylase, biphenyl-4-hydro-
  xylase and acetanilide hydroxylase activities
  and naphthalene-1,2-dihydrodiol formation by polycyclic
  aromatic compounds
Reduced NAD(P):menadione oxidoreductase induction
  by polycyclic aromatic compounds
UDP glucuronyl transferase induction by polycyclic
  aromatic compounds

Rats
Thiourea toxicity
Trichlorobutanol stimulation of ascorbic acid and
  glucuronic acid excretion
Ethanol preference
Antipyrine metabolism
UDPG transferase deficiency (Gunn rat)
Serotonin toxicity
Morphine addiction
Sensitivity to warfarin
Aniline hydroxylation and ethylmorphine *N*-demethylation
Aldehyde dehydrogenase induction by phenobarbital
Cholesterol 7α-hydroxylase induction by phenobarbital
Serum esterases 1, 2 and 3
Serum esterases 4 and 5
Liver microsomal esterase

Rabbits
Bishydroxycoumarin inactivation
Cocaine esterase
Lactogenic effect of reserpine
Atropine esterase
Metabolism of nine drugs before and after phenobarbital
  treatment
Slow and rapid isoniazid inactivation (*N*-acetyltransferase)

sanguinity is a hallmark of autosomal recessive inheritance, being more frequent the rarer the abnormal gene.

Mild, moderate and severe expressions of acatalasemia have been described.[25] The mild form is characterized by ulcers of the dental alveoli; in the moderate type alveolar gangrene and atrophy occur; and in the severe form recession of alveolar bone develops with exposure to the necks—and eventual loss—of teeth. The enzyme is deficient in tissues such as mucous membrane, skin, liver, muscle and bone marrow. Trace levels of catalase activity occur in some patients, and the term 'severe hypocatalasemia' seems more appropriate than does acatalasemia.[26] Heterozygotes who usually have values of catalase activity between those of affected and normal persons would be classified as having 'intermediate hypocatalasemia'. In certain Japanese, some heterozygotes do not exhibit intermediate levels of catalase activity, but rather have values that overlap the normal range, suggesting heterogeneity.[27]

In 1959, a Korean patient was reported with acatalasemia, the first non-Japanese subject to be described.[28] Two years later Aebi and associates found three affected individuals by screening 73 661 blood samples from Swiss Army recruits.[29] All three were healthy and showed none of the dental defects typical of the Japanese cases. The Swiss 'acatalasemics', unlike the Japanese, exhibited residual catalase activity, possibly protecting them against the hydrogen peroxide formed by certain microorganisms thought to be responsible for the oral lesions. The catalase from Swiss subjects also differed from that of normal persons in electrophoretic mobility, pH and heat stabilities and sensitivity to certain inhibitors. These facts suggest that in Swiss families acatalasemia is attributable to a structural gene mutation.[30] Other variants most likely have a similar derivation, although more complex regulatory mutations cannot be excluded. Additional cases of hypocatalasemia observed in different countries together with physicochemical characteristics of the enzyme in affected subjects have been described by Aebi and Wyss.[31]

23.11.8.2 Polymorphism of *N*-Acetylation

Although isoniazid (INH) was synthesized in 1921, its bacteriostatic effect was not discovered until 1952. Soon thereafter marked differences among subjects were reported in rates of INH acetylation, but each patient maintained an unchanged pattern of INH excretion during long-term therapy.[32,33] Slow acetylators of INH show reduced activity of *N*-acetyltransferase, the soluble hepatic enzyme responsible for the metabolism of INH and of sulfamethazine,[34] as well as other monosubstituted hydrazines, such as phenelzine and hydralazine.[35] Toxic effects of these drugs occur chiefly in slow acetylators. Acetylation of procainamide is polymorphic, and thus the effect of this antiarrhythmic agent varies appreciably according to the genotype of the patient to whom it is administered.[36]

Toxicity that ensues from chronic INH administration takes the form of polyneuritis due to vitamin B_6 deficiency; this potential deficiency of pyridoxine, to which slow acetylators of INH are particularly susceptible, can be easily overcome by giving vitamin B_6 with INH whenever INH is administered for long periods. Another form of toxicity to which slow acetylators may be more liable than rapid acetylators is both the drug-induced and spontaneous forms of disseminated lupus erythematosis.[37] Thus, susceptibility to certain diseases may be caused by genetic differences among subjects in their capacity to biotransform drugs and other environmental chemicals. For example, 22 of the 23 subjects from the UK who developed malignant bladder tumors after chronic industrial exposure to benzidine were predominantly slow acetylators.[38,39] Therefore, the slow acetylator phenotype is at a considerably increased risk, and the initiating toxic event in such tumor bladder appears to involve the parent compound rather than its acetylated metabolite.

Acetylation of some drugs, such as *p*-aminosalicylic acid and sulfanilamide, is accomplished by a different *N*-acetyltransferase for which no genetic differences among subjects have been reported. Neither the slow nor the rapid phenotype for INH acetylation appears to be more liable to resistance to tubercle bacilli or reversion. The half-life of INH ranges from 45 to 80 min in the plasma of rapid acetylators, whereas the half-life extends from 140 to 200 min in slow acetylators.[1] Although slow acetylators excrete in urine approximately 70% of a dose of INH as metabolites, rapid acetylators excrete 97% of a dose as metabolites,[40] thereby being exposed to higher concentrations of potentially toxic reactive metabolites, but for a shorter time. Such toxic intermediates have been implicated as the cause of the severe hepatic necrosis and hepatitis that occur in approximately 1% of patients receiving INH chronically for prophylactic purposes.[41,42] These interpretations are now controversial, since more recent studies have shown no predominance of fast acetylator phenotype among patients who develop hepatitis after INH.[43] Transient elevation of serum transaminases due to liver damage in approximately 10–20% of patients receiving isoniazid or isoniazid plus rifampin chronically has been stated to occur predominantly in slow acetylators.[44,45]

Slow acetylation of INH is inherited as an autosomal recessive trait.[46] Although as yet incomplete, evidence suggests that in humans the different phenotypes may result from a structural, rather than a regulatory, gene mutation.[47] Diverse geographic and racial distributions of the gene are documented. Most uncommon in Eskimos (5%), slow acetylation is only slightly more frequent in Far Eastern populations, where it occurs in approximately 10% of Japanese subjects.[48] Slow acetylation is common in black races and in European populations, in nearly 80% of whom the recessive gene is present either in the homozygous or heterozygous state.[48–51] Approximately half the US population are rapid acetylators.

23.11.8.3 Succinylcholine Sensitivity due to Atypical Plasma Cholinesterase

Shortly after the muscle relaxant succinylcholine was introduced as a preanesthetic agent in 1952 and its use became widespread, patients occasionally were found to be extraordinarily sensitive to it; indeed, several deaths associated with its use were reported.[52] Normally, the duration of action of the drug is short (2–3 min). This brevity in normal subjects is due to the exceedingly rapid hydrolysis of succinylcholine by plasma pseudocholinesterase, which catalyzes the sequential removal of choline radicals. Serum pseudocholinesterase activity was reduced in the initially published reports of prolonged apnea. The difficulty can be reversed by transfusion of either normal plasma or a highly purified preparation of the human enzyme. The abnormality is the result of a structurally altered enzyme with kinetic properties markedly different from those of the usual enzyme.[53] The abnormal enzyme exerts no measurable effect on succinylcholine at concentrations of the drug usually present during anesthesia, whereas the normal enzyme shows marked hydrolytic activity[54].

The atypical enzyme is more resistant than the normal one to many pseudocholinesterase inhibitors, *i.e.* both fluoride and organophosphorus compounds inhibited the normal and atypical enzyme differentially.[55,56] Dibucaine, also a differential inhibitor of normal and atypical pseudocholinesterases, can distinguish three phenotypes: homozygous normals, heterozygotes and affected individuals who could not be satisfactorily separated simply by measuring the pseudocholinesterase activity of their plasma.[53] The percentage inhibition of pseudocholinesterase activity produced by 10^{-5} M dibucaine was designated the 'dibucaine number', or 'DN'. Whereas atypical pseudocholinesterase is inhibited only 20%, the normal enzyme is inhibited about 80% and heterozygotes exhibit 50–70% inhibition. Tetracaine, unlike other previously studied compounds, is hydrolyzed faster by atypical than by normal pseudocholinesterase, and an even larger separation of phenotypes apparently can be achieved with the procaine–tetracaine ratio than with the DN. The discovery of additional genetic variants resulted from using sodium fluoride as an inhibitor.[57]

In some families, the DNs do not follow the typical pattern of inheritance. These persons are thought to be heterozygous for a rare, so-called silent, gene. Heterozygotes for this gene exhibit two-thirds of the normal serum cholinesterase activity; they widely overlap normal values. A few rare individuals are presumably homozygous for the silent allele, reflecting complete absence of serum and liver pseudocholinesterase activity.[58] Otherwise apparently normal, these persons lack all four of the usual isozymes of serum pseudocholinesterase. Absence of antigenically cross-reacting material was revealed by immunodiffusion and immunoelectrophoretic studies.

This silent mutation may affect the controlling element of the gene, thereby completely disrupting protein production. Alternatively, a single structural mutation may affect both the active site and the antigenic determinants. Another 'silent' allele has been described in which there is some (about 2%) residual enzymatic activity, indicating further heterogeneity.[59]

Family studies suggest that inheritance of various types of atypical pseudocholinesterase occurs through allelic codominant or recessive genes at a single locus.[60,61] Symptoms may occur after treatment with succinylcholine in persons homozygous for any of the variant alleles and in some mixed heterozygotes.[26] At least four alleles have been definitely identified with the 10 resultant genotypes: $E_1^u E_1^u$, $E_1^u E_1^a$, $E_1^a E_1^a$, $E_1^s E_1^u$, $E_1^s E_1^s$, $E_1^s E_1^a$, $E_1^f E_1^u$, $E_1^f E_1^f$, $E_1^f E_1^a$ and $E_1^f E_1^s$, where E_1 signifies the pseudocholinesterase genetic locus and u, a, s and f indicate the 'usual', 'atypical', 'silent' and 'fluoride'-sensitive alleles, respectively. Another allele (E_1^j) has been described which apparently causes reduction of the usual (E_1^u) molecules by about 60%.[62]

The incidence of atypical pseudocholinesterase remains comparatively constant in different geographic areas. Persons who are homozygous recessive for the atypical allele number approximately 1 in 2000.[5] An exceptionally high incidence of the silent mutation was discovered in a population of southern Eskimos.[63] Prior to this survey in Alaska, only 10 individuals homozygous for the silent gene had been described. The gene frequency of 0.12 in this locality, extending from Hooper Bay to Unalakleet and centered on the lower Yukon River, suggested that 1.5% of this Alaskan population was sensitive to succinylcholine. The isolation and consequent inbreeding of these natives may have resulted in the high frequency of the rare silent gene in this region of Alaska, although only two of the 11 Eskimo families are known to be related.

Similarity of gene frequencies of atypical pseudocholinesterase in most populations suggests that either little selective advantage is conferred by the various genotypes or the contributing environmental factors are common to widely differing countries. In several abnormalities, such as thyrotoxicosis, schizophrenia, hypertension, acute emotional disorders and after concussion, plasma pseudocholinesterase activities may be elevated. Increases are also observed as a genetically transmitted condition without apparent clinical consequences, but associated with an electrophoretically slower migrating C_4 isozyme (the Cynthiana variant).[64] The person with this variant had plasma pseudocholinesterase activity more than three times higher than normal. Further investigation of his family revealed a sister and daughter, also with high values. The exceptionally high pseudocholinesterase activity was associated with resistance to the pharmacologic effects of succinylcholine.[65]

The Cynthiana variant may result from either a defect of a regulator gene controlling pseudocholinesterase activity or a duplication of a structural gene. Slightly higher than normal pseudocholinesterase activity associated with a retarded electrophoretic mobility of the main isozyme was found in roughly 10% of a random sample of the British population.[66] This slower-moving band was designated C_5. The greatly elevated total plasma pseudocholinesterase activity of the US variants distinguished them from the variants described in the UK.

23.11.8.4 Deficient Parahydroxylation of Phenytoin

Since its introduction, phenytoin has become one of the most popular anticonvulsants. However, it can cause multiple toxic reactions, including nystagmus, ataxia, dysarthria and drowsiness, reactions that are clearly dose related. The drug is metabolized in humans mainly by parahydroxylation of one of the phenyl groups to yield 5-phenyl-5′-*p*-hydroxyphenylhydantoin (PPHP), which is conjugated with glucuronic acid and then eliminated in the urine.[67] Many lipid soluble drugs, such as phenytoin, are rendered more water soluble, and hence more excretable, through metabolism by oxidative enzyme systems in liver microsomes. The earliest published example of a genetic defect of mixed function oxidases in human beings is deficient hydroxylation of phenytoin,[68] although only two affected families have as yet been described (Figure 2).

A study of two generations of this family (Figure 2a) revealed an affected mother and two affected siblings, suggesting that low activity of phenytoin hydroxylase may exhibit dominant transmission. Toxic symptoms developed in the propositus on a commonly used dosage of 4.0 mg kg^{-1}, but not on a dose of 1.4 mg kg^{-1}. Abnormally low urine levels of the metabolite PPHP occurred in combination with prolonged high blood levels of unchanged phenytoin. Apparently phenylalanine and drugs like phenobarbital are parahydroxylated by enzymes different from those that hydroxylate phenytoin, since the proband's capacity to parahydroxylate the former compounds was normal.

Another more recently described family contained four slow parahydroxylators of phenytoin (Figure 2b).[69] Intravenous administration of 5 mg kg^{-1} phenytoin revealed extremely low excretion of the hydroxylated metabolite PPHP in three family members representing three generations of the kinship in addition to the propositus, a 32-year-old woman. She received phenytoin in a usual dose of 300 mg d^{-1} to control her seizures, but developed the classical signs of phenytoin toxicity after only 10 d on this moderate dose. On these usual doses, her plasma phenytoin concentration of 52 μg mL^{-1} was in the highly toxic range. On much lower than normal doses of phenytoin the patient's plasma phenytoin concentration declined to the therapeutic, nontoxic range of 10–20 μg mL^{-1}. The members of this family exhibited a bimodal distribution of phenytoin metabolism and transmission of phenytoin sensitivity according to an autosomal dominant pattern (Figure 2b).

Slow inactivation of INH is a more important cause of phenytoin intoxication than heritable deficiency of parahydroxylase activity.[70] In 29 individuals receiving phenytoin and INH, all five

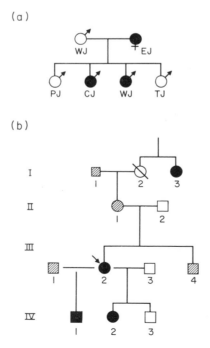

Figure 2 Two pedigrees illustrating deficient parahydroxylation of phenytoin: pedigree (a) shows propositus W. J. and his affected brother, C. J., and mother, E. J. (reproduced by permission of Lancet Publications from ref. 68). Pedigree (b) shows propositus (III 2) with affected grandmother (I 3) and two children (IV 1 and 2) (reproduced by permission of Mosby from ref. 69)

patients who developed symptoms of phenytoin toxicity were slow INH inactivators. Both INH and *p*-aminosalicylic acid interfered with parahydroxylation of phenytoin in rat liver microsomes. This example shows how genetic constitution can influence the clinical severity of a drug interaction.

23.11.8.5 Bishydroxycoumarin Sensitivity

Bishydroxycoumarin sensitivity occurred in a patient receiving the drug for an acute myocardial infarction.[71] On a dose of 150 mg the patient's plasma bishydroxycoumarin half-life was 82 h, compared to normal values of 27 ± 5 h. Although familial studies were not performed because of lack of cooperation, the patient's mother suffered a spinal cord hematoma and developed permanent paraplegia on a weekly dose of 2.5 to 5.0 mg warfarin initiated to treat a myocardial infarction. This unfortunate event in the treatment of the patient's mother suggests hereditary transmission of bishydroxycoumarin sensitivity.

Warfarin and bishydroxycoumarin are extensively hydroxylated in the rat.[72] Genetic factors influence responsiveness to anticoagulants in rabbits, as they do in rats. In rats resistance to warfarin as a rodenticide is transmitted as an autosomal dominant trait.[73] The metabolites in humans are not fully characterized, but the patient with bishydroxycoumarin sensitivity just described, and his mother, may have a metabolic defect involving deficiency of an hepatic microsomal cytochrome *P*-450 isozyme that hydroxylates bishydroxycoumarin.

Increased sensitivity to coumarin anticoagulants also can result from acquired conditions, including vitamin K deficiency, increased turnover of plasma proteins and numerous forms of liver disease that impair the subject's capacity to produce vitamin K dependent clotting factors. Various drugs can increase the prothrombinopenic response to coumarin anticoagulants. Cinchophen may damage liver cells; phenothiazines may produce cholestasis, thereby diminishing absorption of vitamin K; phenylbutazone increases sensitivity by displacing warfarin from plasma albumin; and phenyramidol inhibits the hepatic microsomal enzymes responsible for metabolism of coumarin drugs.

23.11.8.6 Acetophenetidin-induced Methemoglobinemia

Severe methemoglobinemia and hemolysis were reported in a 17-year-old girl after she had taken acetophenetidin.[74] Multiple studies excluded heritable erythrocytic disorders, including hemoglobinopathies, and extracorpuscular compounds seemed to be causing hemolysis. As much as one-half the patient's hemoglobin was occasionally in the form of methemoglobin. After administration of phenacetin, the patient excreted in her urine large amounts of 2-hydroxyphenetidin and 2-hydroxyphenacetin derivatives. In normal persons more than 70% of a dose of 2 g phenacetin appears in the urine as *N*-acetyl-*p*-aminophenol with only minute amounts of the hydroxylated products so prevalent in the patient's urine. One sister, a brother and both parents of the patient had a normal response to phenacetin, but another sister responded abnormally.

These observations suggest an autosomal recessive inheritance of a defect in which the patient's hepatic microsomal mixed function oxidases were deficient in deethylating capacity. Instead of being deethylated as in normal persons, phenacetin, in the patient and her 38-year-old sister, was hydroxylated.

The toxicity observed after phenacetin administration was probably produced by these hydroxylated products, since induction by phenobarbital of the hepatic microsomal phenacetin hydroxylating enzymes prior to administration of phenacetin exacerbated the conditions, *i.e.* severe neurologic symptoms, including bilateral positive Babinski responses, and profound methemoglobinemia developed. After the same phenobarbital pretreatment, a normal volunteer developed neither methemoglobinemia nor neurologic changes on receiving phenacetin.

23.11.8.7 Deficient *N*-Glucosidation of Amobarbital

Pursuing their initial observation based on a twin study that showed large interindividual variations in elimination rates of amobarbital to be predominantly under genetic control,[75] Kalow and associates investigated the family of one set of twins with a deficiency in *N*-hydroxylation, but not *C*-hydroxylation, of amobarbital.[76] Figure 3 shows this pedigree. Investigation of the family of these twins disclosed that this deficiency was most likely produced through autosomal recessive transmission,[76] although only urinary ratios of these metabolites were measured. Later this group reported that it had mistakenly identified the urinary metabolite as an N-hydroxyamobarbital and that instead the actual metabolite was N-β-D-glucopyranosylamobarbital.[77] This metabolite showed great variability in the urine of 129 unrelated volunteers studied after a single oral dose of amobarbital: one volunteer completely lacked the metabolite, whereas 14 subjects had it as the primary form.[78] Four of these 14 subjects were of Chinese origin, suggesting possible racial differences in the pattern and pathway of metabolite formation.[78]

These studies on amobarbital revealed several fundamental principles in pharmacogenetics: (i) the utility of searching for a monogenic origin of pharmacogenetic conditions; (ii) the necessity of performing genetic analyses in families; and (iii) the need, whenever more than a single major oxidative metabolite is produced from the parent drug, to measure rates of formation of each principal metabolite, rather than only disappearance of parent drug, since this permitted genetic analysis to approach more closely to the primary gene product, an isozyme of cytochrome P-450.

Subsequent studies in Oriental and Caucasian subjects revealed significant differences in the ratios of the two major metabolites of amobarbital in urine.[79] Orientals excreted the two metabolites in almost equal amounts, whereas Caucasians eliminated about two-thirds as the C-hydroxylated metabolite and one-third as the N-β-D-glucopyranosyl metabolite.

23.11.8.8 Polymorphic Hydroxylation of Debrisoquine

A landmark discovery was made in 1977 based on measurement of the main oxidative metabolite of debrisoquine in 94 unrelated subjects and in four families.[80] These measurements showed that approximately 7–9% of the population in the UK, Canada and the USA are deficient metabolizers of debrisoquine. Approximately 6% of the population of Ghana and 3% of the population of Sweden are slow metabolizers. In Ghana, heterozygotes were encountered in a higher frequency (36%) and homozygotes in lower frequency (58%) than elsewhere.[81,82]

After a 10 mg oral dose of debrisoquine, the phenotypes of extensive and poor metabolizers were distinguished by the ratio in an 8 h urine specimen between the parent drug and its principal metabolite, 4-hydroxydebrisoquine. Deficient (poor) debrisoquine metabolism is a phenotype controlled by a double dose of an autosomal gene. An affected subject receives one gene from each parent. The same genotype evidently controls a wide variety of different phenotypes characterized by deficient metabolism of numerous drugs. These include sparteine,[83–85] nortriptyline,[86,87] phenacetin,[88] phenformin,[89] guanoxan, dextromethorphan, certain β-adrenoceptor-blocking drugs such as metoprolol, alprenolol, timolol[90] and propranolol,[91] perhexiline[92] and encainide.[93]

Since subjects who are genetically deficient metabolizers of debrisoquine are also at increased risk of toxicity from decreased biotransformation of these other drugs, it is likely that a clinically useful predictive test based on drug-oxidizing capacity will soon become available. A subject with a phenotype of deficient debrisoquine metabolism has a similar retardation in capacity to biotransform approximately 21 other drugs presently recognized to be degraded mainly by the same hepatic drug-oxidizing pathway through which debrisoquine passes. This knowledge should help to prevent the dose dependent adverse reactions that occur mainly in the deficient phenotype when

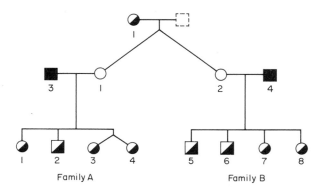

Figure 3 Deficiency of amobarbital N-glucosidation in a family. The propositi are the female twins in generation II. The marking of each subject's genotype is based on the assumption that the capacity for N-glucosidation is determined by two autosomal allelic genes, black field indicating the wild-type ('normal', high-activity) allele, a blank field the deficiency allele (reproduced by permission of Mosby from ref. 76)

usual doses of any of these drugs are given. In such subjects doses can be appropriately reduced or alternative drugs selected that are metabolized by other pathways.

Studies on liver biopsies from patients with the deficient phenotype showed absence of a single cytochrome *P*-450 isozyme that in extensive metabolizers could be clearly visualized on starch gel electrophoresis, even though it constituted only a very small proportion of the total liver content of cytochrome *P*-450.[94,95] Most drugs biotransformed by the debrisoquine cytochrome *P*-450 are also metabolized by several other pathways, each pathway being catalyzed by a different combination or family of cytochrome *P*-450 isozymes. Therefore, in the deficient phenotype, genetically controlled reduction of a specific cytochrome *P*-450 isozyme could have slightly different consequences for different drugs. For a specific drug the outcome depends not only on the contribution of the deficient pathway to the overall pattern of hepatic drug elimination, but also on whether this particular aberrant pathway involves additional cytochrome *P*-450 isozymes present in the patient in normal kind and amount. Even within each of the two clearly separable phenotypes of deficient and fast debrisoquine 4-hydroxylation, a broad range of values occurs.

23.11.8.9 Polymorphism Metabolism of Mephenytoin

Introduced in 1945 to treat epilepsy, mephenytoin has an antiseizure spectrum similar to that of phenytoin but is associated with significantly more toxicity which limits its use. Mephenytoin exhibits stereoselective hepatic metabolism. (*R*)-Mephenytoin is slowly demethylated, whereas the (*S*) enantiomer is rapidly demethylated. The demethylated product is pharmacologically active, but the (*S*) enantiomer is rapidly 4-hydroxylated by hepatic mixed function oxidases to an inactive product. However, in approximately 2–3% of Caucasian subjects an autosomal recessive condition exists causing slow or deficient 4-hydroxylation of the (*S*) enantiomer of mephenytoin.[96] In deficient hydroxylators, mephenytoin accumulates, even when given in low doses, to cause sedation.[97] This polymorphism of mephenytoin metabolism differs genetically from that affecting debrisoquine since subjects who are deficient metabolizers of one drug biotransform the other at normal rates.[97,98] Therefore, defective 4-hydroxylation of mephenytoin probably arises from an abnormality of a different cytochrome *P*-450 isozyme from that responsible for debrisoquine 4-hydroxylation. This cytochrome *P*-450 isozyme has been isolated and investigated, but in deficient metabolizers of mephenytoin no drug other than mephenytoin has been shown to exhibit retarded elimination. Like the debrisoquine polymorphism, the mephenytoin polymorphism is characterized by marked geographical differences in gene frequencies. But for the mephenytoin polymorphism, Chinese subjects exhibit a higher frequency than Caucasians for the gene that confers deficient metabolism. This appears to be the reverse of the situation for the debrisoquine polymorphism.

23.11.8.10 Polymorphic Methylation

Weinshilboum and Sladek[99] reported that erythrocyte thiopurine methyltransferase (TPMT) exhibited a trimodal distribution of activity in 298 randomly selected subjects. Furthermore, this distribution conformed to a Hardy–Weinberg equilibrium in which variation in TPMT activity is controlled by two alleles at a single locus with gene frequencies of 0.06 and 0.94. Family studies suggested transmission of the genetic control of TPMT activity as an autosomal codominant trait. Further studies are required to determine whether these *in vitro* results on TPMT activity exert effects *in vivo* on the metabolism of thiopurines administered as drugs, such as 6-mercaptopurine. In addition to TPMT, thiomethyltransferase and catechol *O*-methyltransferase in human erythrocytes show genetically controlled variations probably also attributable to two alleles at a single locus.[12] Thus, methylation phenotype may be one factor responsible for variations in the metabolism of drugs and environmental chemicals in humans.

23.11.8.11 Polymorphic Distribution of Residual Serum Paraoxonase Activity

The insecticide parathion, while inactive, has as its principal metabolite the potent insecticide paraoxon. Toxicity from paraoxon arises deliberately in insects and accidently in humans due to irreversible inhibition of acetylcholinesterase. This inhibition causes very high concentrations of acetylcholine at receptors near the neuromuscular junction with resultant paralysis. Like the previously discussed polymorphism of *S*-methylation, genetic differences among subjects with respect to residual paraoxonase activity have been demonstrated only *in vitro*. Genetic differences in this arylesterase, termed paraoxonase or cholinesterase, that have been established *in vitro* probably also possess toxicological significance *in vivo*. Studies performed *in vitro* on residual paraoxonase activity raise the possibility that certain subjects may be more or less susceptible to toxicity on exposure to paraoxon.

Geldmacher-v. Mallinckrodt, Hommel and Dumbach[100] incubated sera of 799 unrelated German subjects with paraoxon (3.2×10^{-8} M) at 37 °C for 60 min. The degree to which paraoxon inhibited serum cholinesterase activity under these conditions varied from 0% to 67% according to the particular subject. Residual serum cholinesterase activity exhibited a trimodal distribution that conformed to a Hardy–Weinberg equilibrium in which two alleles at a single genetic locus controlled the observed variation. The frequencies of the two alleles p and q at this locus were 0.24 and 0.76, respectively. Approximately 57.9% (q^2) of 799 unrelated German subjects had low residual serum cholinesterase activity, 36.1% (2 pq) had medium activity and 6% (p^2) had high activity. Low serum paraoxonase activity appeared to be transmitted as an autosomal recessive trait. Although it is unknown whether paraoxonase activity can be induced *in vivo* through chronic exposure to paraoxon and related chemicals, there appears to be no relationship between initial serum paraoxonase activity, age or sex and residual serum paraoxonase activity determined under the conditions described above.

23.11.9 MONOGENICALLY TRANSMITTED PHARMACOGENETIC CONDITIONS IN WHICH DRUGS OR METABOLITES ACT ABNORMALLY ON THE BODY DUE TO GENETICALLY ALTERED RECEPTOR SITES

23.11.9.1 Warfarin Resistance

Genetically controlled resistance to warfarin was found in a patient, age 71 years, receiving anticoagulants for a myocardial infarction.[101] Physical and laboratory examination showed no abnormalities other than a reproducible reduction in his one-stage prothrombin concentration to about 60% of normal. Anticoagulants were initially withheld because of the patient's low prothrombin time. They were administered after one month, at which time he proved to be resistant, rather than sensitive, to warfarin. A daily dose of 145 mg was required to reduce the prothrombin concentration to therapeutic levels, *i.e.* nearly 50 standard deviations above the mean.

Detailed studies showed that the drug was absorbed normally from the gastrointestinal tract; kinetic and binding studies were also normal. Even after administration of very high doses, warfarin was not excreted unchanged in the urine or stools, and amounts of a metabolite of warfarin similar to those recovered from the urine of normal subjects who were given equivalent amounts of the drug were recovered from the patient's urine. The patient also showed resistance to bishydroxycoumarin and the indanedione anticoagulant phenindione, but not to heparin.

An enzyme or receptor site with altered affinity for vitamin K or for anticoagulant drugs was postulated by O'Reilly *et al.* as the mechanism responsible for resistance to warfarin in this patient.[102] Five other members of both sexes of the patient's family over three generations were also resistant to warfarin, suggesting autosomal dominant transmission of the trait (Figure 4). A second large kindred of 18 patients with warfarin resistance in two generations has been reported (Figure 5).[103]

Various environmental conditions lead to resistance to coumarin anticoagulants as phenocopies of the genetic defect. Most commonly the resistance is related to the simultaneous administration of inducing agents that reduce the blood concentration of anticoagulant drugs by stimulating their metabolism, *e.g.* barbiturates, glutethimide, chloral hydrate and griseofulvin.

23.11.9.2 Glucose-6-Phosphate Dehydrogenase (G-6-PD) Deficiency

Formerly called primiquine sensitivity, or favism, G-6-PD deficiency is the commonest hereditary enzymatic abnormality in humans and is transmitted as an X-linked recessive disorder. More than 80 physicochemically discrete molecular variants have been described, each being associated with slightly different clinical features.[104] Ordinarily only the hemizygous male shows significant drug-related hemolysis. Female subjects may be affected mildly, as would be predicted from the Lyon hypothesis, or more severely, as in populations wherein the gene frequency is high enough that homozygosity is appreciable. A mild self-limited anemia is associated with the common variant of

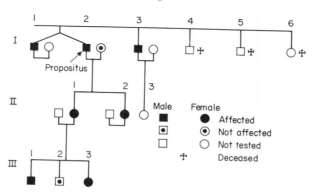

Figure 4 Transmission of warfarin resistance through three generations of a family (reproduced by permission of the Massachusetts Medical Society from ref. 101)

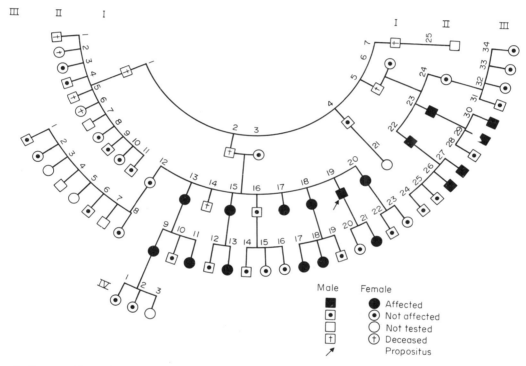

Figure 5 Larger pedigree of warfarin resistance in another family (reproduced by permission of the Massachusetts Medical Society from ref. 103)

G-6-PD found in black races, since only the susceptible, older erythrocytes are removed from the circulation by hemolysis; they are rapidly replaced by resistant younger cells.

In various Mediterranean G-6-PD variants, hemolysis affects a larger proportion of the total erythrocytic population and occurs more rapidly after administration of smaller doses of drugs. Several properties in addition to symptomatic severity can characterize the variants, including the total erythrocytic G-6-PD activity, enzymatic electrophoretic mobility and various kinetic measurements. The specific amino acid substitution in G-6-PD A^+ and in G-6-PD Hektoen has been elucidated by microfingerprinting techniques.[105,106]

The exact biochemical mechanisms by which a given drug or its metabolites cause hemolysis remain unknown. The metabolism of the erythrocyte is unusual since it must function without benefit of a nucleus. It still needs sources of energy to maintain concentration gradients of sodium and potassium and for continual reduction of methemoglobin. This energy source is supplied by the glycolytic and oxidative pathways of glucose metabolism. Certain enzymes, including G-6-PD, lose activity as the normal cell ages, and G-6-PD activity declines with cell age faster than normal in

G-6-PD deficient cells. Therefore, older cells of persons with mutations of their G-6-PD are more susceptible to hemolysis than are younger cells.

Reactions catalyzed by G-6-PD lead to the production of NADPH, a coenzyme necessary for maintenance of sulfhydryl substances, such as glutathione, in the reduced (GSH) state. Sufficient quantities of reduced glutathione are essential for erythrocytic membrane integrity. The sequence of events suggested in drug-induced hemolysis related to G-6-PD deficiency[107,108] includes first, the metabolism of the drug to a product more amenable to further oxidation. The erythrocyte converts this metabolite to an oxidant intermediate. The latter then damages the erythrocyte membrane (particularly in old cells), perhaps by oxidation of reduced sulfhydryl groups. The younger cells with their higher G-6-PD activities resist the osmotic and oxidant effects of various drugs and their metabolites, whereas the older, more sensitive cells with their greater relative deficiency are eliminated.

Determining *in vitro* the hemolytic potential of new drugs continues to be of prime importance, especially in geographic areas where the incidence of the disorder is high. Numerous tests have been devised, but none is suitable, in availability and cost, for routine screening of large populations.

Hemolysis apparently may occur spontaneously or during infection in certain G-6-PD variants. Obviously, enough stress can be placed on the metabolism of G-6-PD deficient erythrocytes to cause hemolysis by several environmental alterations in addition to those produced by drug administration. From the point of view of this discussion, an environmental alteration that can produce hemolysis in genetically susceptible G-6-PD deficient subjects is exposure to any one of many industrial chemicals. Surprisingly little evidence has been adduced to establish which environmental chemicals specifically cause hemolysis in G-6-PD deficient subjects.[21]

23.11.9.3 Drug Sensitive Hemoglobins

A life-threatening hemolytic anemia developed in a 2-year-old girl and her father after they received sulfa drugs.[109] Both subjects registered an abnormal hemoglobin content, electrophoretic mobility being between that of hemoglobins A and S.[110] Further studies showed an abnormality in the β chain, with arginine taking the place of the usual histidine residue at the 63rd position, where the heme group is attached.[111] Fifteen of the 65 relatives examined showed the abnormal hemoglobin, designated hemoglobin Zurich. This defect was transmitted as an autosomal dominant trait. In another family, discovered in Maryland, with the same substitution the severity of the hemolytic episodes was less than in the Swiss cases.[112]

Another drug sensitive hemoglobin, hemoglobin H, is a special form of α-thalassemia. Composed of four β chains, hemoglobin H is sensitive to the oxidant drugs described under G-6-PD deficiency. In certain regions, such as Thailand, the frequency of homozygous hemoglobin H is high, *i.e.* one in 300 individuals.

23.11.9.4 Phenylthiocarbamide Tasting Ability

The ability to taste phenylthiocarbamide (PTC) is transmitted as an autosomal dominant trait, and tasters may be either heterozygous or homozygous.[113] This polymorphism was discovered in 1932 when Fox, who synthesized the compound, noted that he could not detect a bitter taste from dust of the compound arising as it was poured into a container, whereas a colleague in the same room complained of the bitter taste.[114]

Although this polymorphism seems to be benign, some workers have related the ability to taste PTC to thyroid disease. Administration of PTC can produce goiter in the rat. Compounds related to PTC by possessing the N—C=S group, such as the antithyroid drugs methylthiouracil and propylthiouracil, also produce the same bimodality in taste perception observed with PTC. Forty-one percent of 134 patients with nodular goiter were nontasters,[115] an observation confirmed in 447 patients who underwent thyroidectomy for various reasons.[116] In male patients with multiple thyroid adenomas, a marked increase in nontasting frequency was also noted. Nontasters seem to be more susceptible to athyreotic cretinism and also to adenomatous goiter. These data suggest that nontasters may be more susceptible than tasters to environmental goitrogens. The physicochemical basis for the difference in taste perception in affected individuals is unknown.

The frequency of tasting capacity shows geographic variation in that 31.5% of Europeans, 10.6% of Chinese and 2.7% of Africans are nontasters.[117,118] As with the physicochemical findings, the explanation for these variations in the gene for PTC tasting are obscure.

23.11.9.5 Malignant Hyperthermia and Muscular Rigidity

In 1962 hyperthermia was reported to be the cause of death in 10 of 38 family members who had received anesthesia for various surgical procedures.[119] This was the first indication that the rare, hitherto seemingly sporadic, malignant hyperthermia afflicted persons exposed to various anesthetic agents might be genetically transmitted. More than 200 cases of malignant hyperthermia have been identified and shown to have a hereditary basis.[120] The condition is associated with muscular rigidity and appears to be transmitted as an autosomal dominant trait. It develops during anesthesia with nitrous oxide, methoxyflurane, halothane, ether, cyclopropane or combinations thereof. Malignant hyperthermia is commoner in association with the use of succinylcholine as a preanesthetic agent. In susceptible subjects the condition can also be triggered by stress. During anesthesia body temperature rises rapidly, occasionally reaching 112 °F (44 °C).

The incidence of malignant hyperthermia is in the range of 1 in 20 000 cases of general anesthesia. Malignant hyperthermia exhibits no sex preference, but occurs more frequently in younger than in older anesthetized patients. Approximately two-thirds of the patients die, usually from cardiac arrest. The degree of rigidity is variable, differing from patient to patient and sometimes being absent. This variability may indicate that the term malignant hyperthermia refers to several discrete diseases.

Occasionally, rigidity is so marked that the body literally becomes as stiff as a board, progressing without interruption into rigor mortis. Intravenous administration of procaine or procainamide was initially reported to alleviate the rigidity and fever in certain cases. Curare is ineffective. Interestingly, a limb under tourniquet does not become rigid, suggesting a peripheral rather than a central lesion. Animal models have been produced in dogs treated with halothane and dinitrophenol and in Landrace pigs.

Dantrolene sodium, a derivative of hydantoin and a skeletal muscle relaxant, has been shown to be effective in the prevention and treatment of malignant hyperthermia in susceptible swine[121] and humans.[122] Since malignant hyperthermia is probably caused by the unrestricted flow of ionic calcium from the sarcoplasmic reticulum of skeletal muscle into the cytoplasm,[123] dantrolene sodium may act by preventing the initial release of calcium. The symptoms of malignant hyperthermia arise from an excessively high concentration of mycoplasmic calcium, which in turn can generate heat by several mechanisms. Heat production is increased from increased catabolism of muscle glycogen and hepatic lactate, from accelerated hydrolysis of ATP by myosin ATPase, from uncoupling of oxidative phosporylation by calcium, and finally from a decreased heat loss secondary to peripheral vasoconstriction.

23.11.10 ASSOCIATIONS WITHOUT KNOWN MECHANISMS BETWEEN GENETIC FACTORS AND RESPONSE TO XENOBIOTICS

23.11.10.1 Ethanol Metabolism

Atypical alcohol dehydrogenase (ADH), a variant of the enzyme metabolizing ethanol, has been described in humans.[124] The enzyme occurs in sufficiently high frequencies in Swiss and British populations to be designated a polymorphism. Exceptionally active, the variant occurred in 20% of 59 liver specimens from a Swiss population and in 4% of 50 livers from a British population.

After intravenous infusion of ethanol, attempts were made to correlate rates of degradation of the drug with liver ADH types from biopsies obtained during surgical procedures.[125] Of 23 subjects, two had atypical ADH. Capacity to metabolize alcohol was not different in the male subject with atypical ADH than in male subjects with the typical enzyme, whereas capacity to degrade ethanol was greater in the female subject with atypical ADH than in a small group of female subjects with typical ADH. Liver ADH specific activity and isozyme pattern in biopsy specimens for seven American Indians and six whites revealed no racial differences.

The question of whether individuals with atypical ADH have increased capacity to degrade ethanol, and possibly to resist alcoholic cirrhosis of the liver, remains unresolved. The subject of ethanol toxicity is difficult to resolve experimentally, because of the considerable variation in ability of ethnic groups to metabolize ethanol. Natives of Far Eastern countries have less ethanol-metabolizing capacity than have those of Western countries. Possibly, too, major genetic differences are obscured by elevated rates of ethanol metabolism since induction of ethanol metabolism occurs with chronic ethanol consumption. There also appear to be racial differences in ethanol sensitivity since Japanese, Taiwanese and Koreans exhibit marked facial flushing and mild-to-moderate symptoms of intoxication after drinking amounts of ethanol that produce no detectable effect on whites.[126] These differences in ethanol responsiveness, present since birth, have been attributed to variations in autonomic reactivity.

Racial differences in alcohol sensitivity have been attributed to a polymorphism of liver aldehyde dehydrogenase (ALDH).[127] ALDH exhibited two major bands: an isozyme with faster electrophoretic mobility possessing a low K_m for acetaldehyde and an isozyme with slower electrophoretic mobility with a high K_m for acetaldehyde. Analysis of autopsy livers from Japanese subjects indicated a 52% frequency of the aldehyde dehydrogenase isozyme with slower electrophoretic mobility ('unusual' phenotype) and a 48% frequency for the faster-migrating isozyme ('usual' phenotype). Isozyme patterns from postmortem German liver samples showed only the usual phenotype. These results raise the possibility that sensitivity to alcohol common in subjects of Mongoloid origin might be caused by delayed oxidation of acetaldehyde rather than its higher than normal production by atypical alcohol dehydrogenase.

23.11.10.2 Blood Groups

To determine possible correlations between adverse reactions to drugs and genetic factors, a survey was made of many hospitalized patients.[128,129] In young women developing venous thromboembolism while taking oral contraceptives, a significant deficit was discovered of blood group O individuals relative to those possessing groups A and AB combined.

A correlation was found between ABO blood groups and the development of arrhythmias after administration of digoxin, with decreased risk in blood group O patients relative to non-O patients.

23.11.10.3 Diphenhydramine in Orientals and Caucasians

Spector *et al.*[130] showed that the kinetics and psychomotor effects of diphenhydramine differed in Orientals and Caucasians. After either oral or intravenous administration of diphenhydramine, Caucasians had not only plasma drug concentrations twice as high as Orientals at each time point tested, but also more sedation and deterioration in psychomotor performance than did Orientals. Plasma clearance and apparent volume of distribution, but not plasma half-life, of diphenhydramine were higher in Orientals than in Caucasians. Whether these kinetic and dynamic differences in response to diphenhydramine had a genetic cause or were related to such environmental differences as diet, smoking, *etc.* or a combination of both could not be resolved in this study.

23.11.10.4 Depression and Antidepressants

It has been suggested that symptoms of depression are produced by at least two genetically distinct entities and that 'endogenous' depressions more frequently benefit from treatment with imipramine, whereas 'reactive' depressions improve after administration of monoamine oxidase (MAO) inhibitors.[131] Data supporting this hypothesis are based on similarities in drug response between probands and relatives who also suffered from depression and who also received imipramine, MAO inhibitors or lithium carbonate. The concordance in drug response among depressed relatives and depressed probands was reported to be statistically greater than expected by chance alone. Studies in different patients tended to confirm these initial impressions.[132]

Striking differences may exist in the genetic background of lithium responders compared to nonresponders. Genetic determination of large interindividual variations in lithium ion distribution was observed in a study of monozygotic and dizygotic twins.[133]

23.11.10.5 Association between HLA Haplotype and Drug Response

The HLA genetic determinants on a single chromosome constitute a haplotype. Every person has two haplotypes, one inherited from each parent. At least four closely linked loci, only a few centimorgans apart on the short arm of human chromosome 6, control the synthesis of antigens on the surface of leukocytes. Because of their close linkage, the specific genes at these four loci only rarely separate through crossover between homologous chromosomes during meiosis. Of interest is the extreme genetic heterogeneity of these four loci, more than 100 alleles having been identified, theoretically enabling more than 1 000 000 different phenotypes. HLA loci and loci closely linked to them function in immune surveillance, protecting the body from several types of environmental assault. For example, these genes serve to combat viral and other pathogens and in the rejection response to foreign tissues.[134,135] Pertinent to this discussion, certain HLA haplotypes markedly increase susceptibility to diseases associated with immunological impairment, such as chronic active hepatitis, glomerulonephritis, ankylosing spondylitis, Reiter's syndrome, rheumatoid arthritis and disseminated lupus erythematosis.[134,135] In certain HLA haplotypes, risk of these disorders characterized by derangement of normal immunological mechanisms can increase by up to fortyfold above that of the population at large.[134,135]

For reasons as yet unclear, some uncommon alleles at the HLA loci may be much more strongly associated with certain drug responses than alternative alleles. For example, in schizophrenic patients HLA-A1 has been reported to be highly associated with a favorable response to chlopromazine.[136,137] By contrast, HLA-A2 appeared to be associated with an unfavorable response. In patients with affective disorders treated with lithium, HLA-A3 was associated with a high rate of relapse.[138]

With respect to adverse reactions to drugs, in patients with rheumatoid arthritis treated with sodium aurothiomalate, adverse reactions, particularly proteinuria, occurred 32 times more frequently in the HLA-DR3 haplotype.[139] Also, in patients with rheumatoid arthritis, agranulocytosis after levamisole was encountered much more frequently in subjects with the rare haplotype HLA-B27.[140] Hydralazine-induced systemic lupus erythematosis occurred twice to three times more frequently in hypertensive patients with HLA-DR4 haplotype who received the drug.[141] Another genetically determined phenotype, slow acetylation of hydralazine, is also associated with increased susceptibility to lupus.[142] Although the acetylation polymorphism itself does not appear to be linked to any HLA haplotype,[141] potential linkages between other HLA haplotypes and different pharmacogenetic polymorphisms remain to be established and merit study.

23.11.10.6 Oral Contraceptives and Jaundice

Many females still in their early teens take oral contraceptives. Oral contraceptives produce cholestatic jaundice only very rarely. When it does occur, this jaundice characteristically appears during the first few cycles of drug administration. Most oral contraceptives are combinations of a synthetic estrogen and progesterone, both of which are generally testosterone derivatives, thereby possessing cholestatic properties. Gallagher, Mueller and Kappas[143] reviewed the large number of steroids capable of producing liver injury. The extremely low incidence of jaundice associated with their use is attributable to some extent to the small doses of the offending hormones. Also there may exist a genetic predisposition on the part of those developing cholestasis. Firm evidence for a genetic predisposition is lacking, but both Sherlock[144] and Kalman[145] draw attention to this possibility and to the fact that those developing cholestasis on oral contraceptives are particularly liable to jaundice of pregnancy. The influence of estrogens on bile flow is known; in late pregnancy, the time when jaundice is observed, estriol production is increased 1000-fold.[146] Susceptibility to jaundice with oral contraceptives may be increased 2000- to 8000-fold in women who developed intrahepatic cholestasis during pregnancy.[147] Finally the geographical distribution of jaundice complicating oral contraceptive therapy suggests the possible operation of genetic factors, since the only large series of cases reported are from Scandinavia and Chile.[148]

23.11.11 GENETIC FACTORS AFFECTING THE DISPOSITION OF COMMONLY USED DRUGS

Because most of the conditions described above are rare and produce toxicity after only a few drugs, they probably contribute only to a relatively small extent to the current major medical problem of adverse drug reactions. However, another development in pharmacogenetics suggests that genetic differences directly affecting xenobiotic disposition play a prominent role in commonly encountered forms of drug toxicity. This idea was suggested by a series of experiments that examined rates at which healthy adult monozygotic and dizygotic twins eliminate commonly used drugs. These twins were in a basal state with respect to most of the factors depicted in Figure 1. This means that they were nonmedicated, living in different households, but not markedly exposed at work or at home to agents that either induce or inhibit the hepatic drug-metabolizing enzymes.

Twins are of two types: monozygotic twins have identical heredities since they arise from a single fertilized egg, whereas dizygotic twins arise from two fertilized eggs. They are no more alike genetically than siblings except that they are born at the same time. After phenylbutazone, bishydroxycoumarin, antipyrine, halothane, ethanol, phenytoin, nortriptyline or salicylate were administered, large interindividual variations that existed among unrelated people vanished within sets of monozygotic twins but were preserved in most, but not all, sets of dizygotic twins (Figure 6). The magnitude of interindividual variations in rates of drug elimination among unrelated people was thirtyfold for nortriptyline, tenfold for bishydroxycoumarin, sixfold for phenylbutazone and antipyrine, threefold for halthane and twofold for ethanol. Family studies on three of these drugs substantiated observations made in twins. In these family studies on bishydroxycoumarin,[48] phenylbutazone[149] and nortriptyline,[150] predominantly polygenic mechanisms appeared to control large interindividual variations in drug disposition. Precise modes of transmission could not be firmly established and attention focused on parent drugs rather than on rates of formation of individual metabolites.

Approaches based on formation rates of individual metabolites should permit a clearer understanding of the mode of genetic control over interindividual variations in the disposition of commonly used drugs that are biotransformed by multiple pathways. Two of the older type twin studies that examined mainly disappearance of parent drug have been published for tolbutamide[151] and theophylline.[152] Both studies raise the possibility of monogenetic, rather than polygenic, control of interindividual variations in the elimination rates of the parent drug. This possibility was supported recently for both antipyrine[153] and theophylline[154] by family studies that measured rates of production of individual metabolites. Rate constants for formation of individual metabolites are as close to primary gene products as presently possible to achieve safely *in vivo* in normal human subjects.

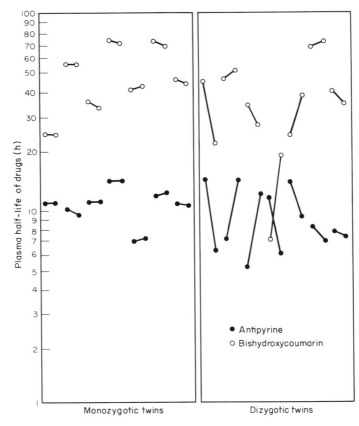

Figure 6 Plasma half-lives of bishydroxycoumarin and antipyrine were measured separately at an interval of more than six months in healthy monozygotic (identical) and dizygotic (fraternal) twins. A solid line joins the values for each set of twins for each drug. Note that intratwin differences in the plasma half-life of both bishydroxycoumarin and antipyrine are smaller in monozygotic than in most dizygotic twins

Monogenic control of the six- to eight-fold interindividual variations that occur in the rate constants of formation of the three main metabolites of antipyrine and theophylline provides additional evidence for the heterogeneity of the hepatic cytochrome *P*-450 isozymes that oxidatively biotransform these and many other xenobiotics. Regulation of such large interindividual variations in oxidation capacity by monogenic mechanisms has several toxicological implications since some of these metabolites are highly reactive and capable of forming covalent bonds to numerous cellular macromolecules. Recently, genetically controlled variations in aryl hydrocarbon hydroxylase (AHH) activity of cultured lymphocytes has been reconfirmed and differences in the inducibility of this enzyme have been associated with differences in susceptibility to certain cancers.[155,156] Specifically, high AHH activity has been demonstrated in cultured lymphocytes from patients with carcinoma of the lung.[155,156]

Precise determination of the mode of inheritance is usually relatively easy with single gene traits, especially if they are rare. Family studies are invaluable; alleles segregate during gamete formation and rejoin in offspring to form gene combinations that can be detected by their phenotypic expression.

Polygenic inheritance produces a different picture. Instead of alleles at only a single gene location, alleles at several sites on a chromosome or chromosomes contribute to the formation of the phenotype. The effect of each gene is not as profound as it is with single gene inheritance; instead, the final expression is a combination of genetic plus environmental effects. For drugs biotransformed by multiple oxidative pathways, such as antipyrine and theophylline, although the individual pathways show variation regulated by monogenic mechanisms, kinetics of the parent drug suggest that variation is under polygenic control.[153,154] That result follows the fact that the kinetics of the parent drug only distantly and insensitively indicate the kinetics of several separate monogenically controlled participating pathways. Thus, for pharmacogenetic purposes kinetic investigations performed only on parent drugs inadequately reveal the genetic mechanisms that regulate each separate metabolic pathway. Further complexity emerges from the fact that for many drugs biotransformed by multiple pathways, several genetically distinct isozymes of cytochrome *P*-450 probably participate in the formation of a single metabolite.

The possibility remains that the long sustained influence of certain environmental factors could be responsible in some measure for the particular rate of drug disposition of an individual who might otherwise appear to be in a basal state. This possibility is explored by careful inquiry into the individual's exposure at home and at work to these factors. The existence and operation of numerous environmental factors (Figure 1), each with different capabilities of altering the basal, genetically controlled rate of drug disposition, make it exceedingly difficult to attribute quantitatively different portions of the total interindividual variation to specific single environmental factors.

The task of partitioning the total interindividual variation in drug elimination of large heterogeneous populations into component parts is further complicated because seemingly pure 'environmental' factors such as smoking and diet are closely associated with other environmental factors, as well as with genetic factors. Extrapolation to a large population of the precise contribution of a particular trait, such as vegetarianism, to interindividual variations in rates of drug elimination observed in a selected, small study population can be hazardous. Such extrapolations should be accompanied by a demonstration that the frequency of the particular trait in the study group is similar to its frequency in the larger population. Without such demonstration, the quantitative contribution of the trait cannot be legitimately extrapolated from the small study group to the large population. The simultaneous contributions of multiple genetic and environmental factors to the particular drug-metabolizing activity of a given individual, as well as the change in relative importance of these different factors with time and condition, such as during aging, fever, disease, dietary change, drug administration, acute or chronic exposure to environmental chemicals, *etc.*, make it exceedingly difficult to quantitate the relative influence of numerous factors involved at any given time, other than in a transient, relatively basal state. Also, for these reasons, carefully controlled studies are required.

The hereditary control of large variations in xenobiotic disposition among healthy, nonmedicated volunteers in a basal state with respect to rates of drug elimination has several potentially useful implications. In the first place, determination of drug clearances in plasma or saliva might be ascertained before chronic drug administration as a guide to adjusting dosage according to individual requirements, thereby helping to reduce frequent occurrences of toxicity or undertreatment encountered when the same dose of a drug is given to all subjects. Secondly, if within a subject correlations existed in capacity to metabolize various drugs, drugs could be grouped into categories. Once the rate of metabolism of one drug in a category had been ascertained, the rate of handling all the others in that category could be readily calculated, and judgements as to dosage could be made. Several investigations suggest that correlations in drug-metabolizing capacity exist for a few, but not the majority of, chemically unrelated drugs. One study established correlations between the plasma half-lives of bishydroxycoumarin and phenylbutazone. Another investigation reported correlations between steady-state blood concentrations of demethylimipramine or nor-triptyline and the plasma half-life of oxyphenylbutazone. Therefore, for certain drugs, a common rate-limiting step may exist in the several reactions required for drug metabolism within the liver. Some enzyme(s) catalyzing a rate-limiting step common in the oxidation of several drugs could be responsible for these surprising correlations in rates of drug elimination. Attempts to develop further drug categories constitute a major effort in pharmacogenetics. For the reasons described above, these studies should attempt to correlate rates of formation of each principal metabolite of different drugs rather than just disappearance of the parent drugs themselves.[157]

Thirdly, physicians should be alert to the introduction of environmental factors that might change their patients' genetically determined drug-biotransforming capacity. For example, diseases affecting cardiovascular, thyroid, renal or hepatic status, environmental exposure to such agents as insecticides, nicotine, polycyclic hydrocarbons or caffeine, and finally chronic administration of many drugs either alone or in combination can influence the duration and intensity of drug action (Figure 1) . Not only in cases of polygenic or multifactorial inheritance are the influences of the environment of consequence. Although it is recognized that intelligence, which probably has some polygenic influence, can be altered by environmental insults such as malnutrition, viruses and toxic chemicals, it is also important to recognize environmental influences on single gene conditions. For example, acute intermittent porphyria is an autosomal dominant condition characterized by sudden attacks of burgundy-colored urine. Although bouts may occur without apparent environmental insult, administration of barbiturates is followed by explosive, clinical attacks. Similarly, acute exacerbations of diabetes mellitus and gout can be precipitated by environmental changes involving diet, drugs or both.

Of particular interest is the relationship, alluded to above, between genetic differences among subjects in rates of drug metabolism and the production of potentially toxic biotransformation products of a parent drug. In the past, the hepatic drug-metabolizing enzyme system has been regarded as a 'detoxification' system because this system converts lipid soluble parent compounds that could otherwise remain indefinitely in the body to more polar metabolites readily excreted in urine. More recently, however, it has been recognized that this enzyme system can produce potentially toxic, highly reactive metabolites that combine with tissue macromolecules, including DNA, to produce necrosis, immunological reactivity and mutations. Qualitative differences among subjects in pathways of drug metabolism, as well as quantitative differences in the activities of the enzymes that catalyze these reactions and pathways, could be involved in the regulation and control of such tissue damage. Thus, genetic differences among subjects can render certain individuals more and others less sensitive to toxicity from these different reactive metabolites.

23.11.12 FAMILIAL HYPERBILIRUBINEMIAS CAUSING DEFECTIVE GLUCURONIDE CONJUGATION OF VARIOUS DRUGS AND ENVIRONMENTAL CHEMICALS

Although bilirubin is not itself a drug, many drugs and environmental chemicals are conjugated like bilirubin with glucuronic acid by a hepatic glucuronyl transferase. Therefore, individuals possessing certain familial unconjugated hyperbilirubinemias, such as the Crigler–Najjar syndrome, exhibit decreased capacity to conjugate menthol,[158] chloral hydrate,[159] trichlorethanol,[159] salicylate,[159] N-acetyl-p-aminophenol[160] and some steroids.[161] The Crigler–Najjar syndrome, an almost uniformly fatal disorder, is inherited in an autosomal recessive fashion. A form of deficient bilirubin conjugation less severe than the Crigler–Najjar syndrome has been described.[162] In these reported eight patients, from 14 to 52 years of age, serum bilirubin was all of an indirect reacting or unconjugated type and ranged from 6.4 to 10.9 mg per 100 mL.[162] In addition to defective bilirubin conjugation, there was also substantial reduction in the glucuronidation of menthol, 4-methyl-umbelliferone and o-aminophenol.[162]

In Gilbert's disease, the mildest of the heritable hyperbilirubinemias, jaundice is intermittent. Frequently bilirubin levels are in the normal range and in 58 patients averaged 2.7 mg per 100 mL.[163] The bilirubin is of the indirect reacting type; *in vivo* impairment of glucuronide formation has not been demonstrated.[162-164] Moreover, glucuronide formation *in vitro*, determined by incubating liver tissue with such substrates as bilirubin, o-aminophenol and 4-methylumbelliferone, was normal.[162] Perhaps methods available at that time were insufficiently sensitive to detect very small decreases in glucuronide conjugating capacity. To be meaningful, tests of glucuronyl transferases require stepwise elevations in the dose of the various glucuronide receptors until excretion of the appropriate glucuronide is not increased.[165]

Inability to demonstrate a difference between jaundiced and normal subjects does not necessarily indicate a normal glucuronide conjugating system, since the limiting factor in the metabolism of the test substance may involve its rate of absorption, distribution or excretion, in addition to alternate pathways of metabolism.[165] Reductions in glucuronidation could arise either from slight decreases in the quantity of a normal transferase or from a qualitatively aberrant enzyme with kinetic properties different from those of the normal transferase. This distinction is important in genetics and concerns the location of the mutation that produces the human disease. The defect lies either at the site of a regulator gene in the former case or of a structural gene, in the latter situation.[166] The pathogenesis of many cases of mild unconjugated hyperbilirubinemia, formerly classified as Gilbert's disease, is heterogeneous and probably includes a postviral hepatitis syndrome, an autosomal dominant disorder, flavispidic acid toxicity causing impaired cytoplasmic binding of unconjugated bilirubin, mild hemolytic states of varied etiology and increased production of bile pigment from sources other than mature, circulating erythrocytes.[165]

A more biochemically and genetically satisfying classification of jaundiced patients with defective glucuronide formation *in vivo* and deficient hepatic glucuronyl transferase activity *in vitro* has been proposed.[165] On the basis of an examination of 16 such patients and their families, Arias *et al.*[165] proposed two forms of chronic nonhemolytic unconjugated hyperbilirubinemia with hepatic glucuronyl transferase deficiency. Group I manifested a more severe hyperbilirubinemia, frequently with kernicterus, colorless bile with only traces of unconjugated bilirubin, unresponsiveness to phenobarbital administration and a conjugation defect transmitted as an autosomal recessive character.

Group II, on the other hand, consisted of milder hyperbilirubinemia without kernicterus, exhibited pigmented bile containing bilirubin glucuronide, responded to phenobarbital administration by clearing of the jaundice and revealed an autosomal dominant mode of inheritance. Most previously described patients classified as having Crigler–Najjar syndrome seemed to belong to group I, but because some probably should be placed in group II,[165] the eponym has less specificity than the two categories proposed by Arias. In both group I and group II there occurred reduction of menthol glucuronide formation *in vivo* and of hepatic glucuronyl transferase activity determined by incubating liver biopsies with bilirubin, *o*-aminophenol and 4-methylumbelliferone. Although parents of group I patients had normal serum bilirubin concentrations, they each had abnormal menthol tolerance tests, an important observation from the point of view of this chapter, since it indicates defective glucuronide conjugation of various drugs in ostensibly normal heterozygous carriers.[165]

23.11.13 CHARACTERISTICS OF THE PATIENT THAT CAN AFFECT RATES OF DRUG ELIMINATION

In addition to the effects of genetic constitution on drug response, discussed above, many other factors of an environmental, developmental, nutritional or endocrinological nature can influence and alter a subject's rate of drug elimination. Such factors shown in the outer circle of Figure 1 are connected because they are frequently associated with each other in a given subject, rather than being independent. In fact, these factors often dynamically interact to change a subject's characteristic basal rate of drug absorption, distribution, metabolism, excretion or receptor interaction. Accordingly, effects of each of these factors on drug response can be complex and can change with time even in the same subject.[167, 168]

23.11.13.1 Effect of Age

The influence of age on drug disposition will be considered now because age illustrates the complexity of many factors in Figure 1. While individuals of different age tend to eliminate drugs at different rates, no more specific statements on the subject can currently be made with certainty and even this generalization requires qualification. While geriatric patients may exhibit retarded hepatic metabolism of some drugs, large differences in the magnitude of this effect have been demonstrated, age effects being marked for certain drugs and negligible for others, such as warfarin and ethanol. Furthermore, even for drugs whose hepatic metabolism is markedly impaired with age, interindividual variations caused by other factors outweigh these age effects so that some normal geriatric subjects display more active hepatic drug metabolism in all studies thus far published than a few of the 20- or 30-year-old subjects used. Thus, the correlation coefficient (r) between age and rate of hepatic metabolism of any drug thus far studied has been low. Use of age by itself to predict rate of drug elimination in any given patient depends on the value r^2 (not r), and hence the numerical value of this prediction (r^2) is much lower than that of the correlation coefficient (r). It can readily be appreciated that for clinical purposes, such predictions (accurate at best in approximately one patient out of four) are too often wrong to be useful. Nevertheless, the astute physician, alert to the possibility of greatly reduced capacity of certain elderly patients to metabolize and eliminate some drugs, takes this age factor into account. This can be done either by slightly reducing the dosage of potent drugs with low therapeutic indices given to elderly, as contrasted to younger, patients or by watching elderly patients who receive drugs especially carefully to assure therapeutic efficacy of prescribed medications and to detect early any undesirable drug-related side effects. Both methods should be employed for some drugs in certain geriatric patients.

In the aged patient, not only may drug metabolism be somewhat impaired, but there may also be concomitant changes in drug absorption, distribution, excretion and interaction with receptors. All of these may contribute in different ways to an altered response and increased sensitivity of the elderly to some drugs. Rather than being an isolated factor, age is associated with many other related factors that change with time such as diet, exercise, exposure to inducing or inhibiting environmental chemicals and function of critical organs.

Not only geriatric patients, but individuals in the fetal and neonatal, as well as pediatric, age groups show altered rates of drug elimination, compared to normal adults. Again, this can vary greatly with the particular drug and the particular subject. For example, antipyrine and phenylbutazone elimination in children age one to eight years old appeared approximately twice as fast as in normal adults.[169] While eight children without clinical symptoms of acute or chronic lead poisoning but with biochemical manifestations of plumbism had the same rates of antipyrine and phenylbutazone elimination as normal children, two children with both clinical and biochemical signs of acute plumbism exhibited prolonged plasma antipyrine half-lives. These retarded values returned to the normal range for children after institution of chelation therapy.[169] Similar results have also been obtained in other studies that disclosed more rapid elimination of diazoxide,[170] phenobarbital[171] and clindamycin[172] in children than in adults.

Compared to either children or adults, human fetuses and neonates have much reduced capacity to eliminate drugs, although human fetuses and neonates, in contrast to some rodents, exhibit measurable hepatic microsomal drug-metabolizing activity.[173,174] In premature infants the elimination rates of certain drugs appear to be even more retarded than in full-term healthy newborns. For example, in two premature infants plasma indomethacin half-lives were 21 and 24 h,[175] whereas in full-term healthy newborns the mean value was 14.7 h[176] and in healthy adults 2 to 3 h[177,178] or 7.2 h.[179] Obviously these marked differences in elimination rates of drugs among premature infants, full-term newborns, children and adults have practical significance in the calculation of appropriate dosages of drugs. They also indicate the need to investigate pharmacokinetic and pharmacodynamic processes for many more drugs in these age groups.

23.11.13.2 Effect of Dietary Change on Drug Disposition

Like other environmental factors that have been shown to affect drug disposition, the relationship between diet and rates of hepatic drug metabolism appears complex. Investigations in laboratory animals disclosed that starvation affects hepatic metabolism of drugs in different ways, some drugs being accelerated, others retarded in their elimination rate. Furthermore, a sex dependency exists for this starvation effect in rats.[180-183] Starvation of rats for 72 h diminished or abolished the sex differences in rates of metabolism that occur in normally fed rats for many drugs,[182] thereby suggesting that starvation effects on drug metabolism in rats may be mediated by interference with the stimulatory actions of the androgenic steroids.

In obese humans a starvation diet consisting of not more than 15 g carbohydrate each day for two weeks produced both weight loss and ketoacidosis but failed to alter elimination rates of either antipyrine or tolbutamide.[184] Nevertheless, more prolonged nutritional deprivation was associated with accelerated antipyrine metabolism in malnourished subjects without edema. Very slight retardation in antipyrine metabolism occurred when malnutrition was accompanied by edema.[185] On the other hand, malnutrition in patients with anorexia nervosa failed to alter plasma antipyrine half-lives or metabolic clearance rates.[186]

In carefully controlled experiments, effects on drug elimination of the relative proportion of carbohydrate to protein in an isocaloric diet were investigated.[187] Each nonmedicated, normal male volunteer, age 22 to 27 years, served as his own control. Compared to normal home diets, diets in which 44% of the total calories were in the form of protein and 35% of the total calories in the form of carbohydrate accelerated by approximately 40% rates of antipyrine and theophylline elimination.[187] However, when the high protein diet was reduced from 44% to 10% and carbohydrate raised from 35% to 70% of the total caloric intake, the reverse occurred: half-lives of these two test drugs were prolonged by about 40% from their values during the high protein period.[187] Each diet was maintained in this study for two weeks before effects on drug disposition were measured. These dramatic results are illustrated in Figures 7 and 8. Other interesting, carefully controlled dietary investigations performed in normal volunteers by this group revealed that charcoal broiling of beef[188] and also high consumption of cruciferous vegetables such as cabbage and brussels sprouts[189] enhanced rates of hepatic metabolism of antipyrine and theophylline.

23.11.13.3 Effect of Fever on Drug Disposition

Fever frequently accompanies disease. Yet, until recently, the effects of fever on drug distribution were unexamined. In 1975, it was shown that etiocholanolone-induced fever in normal volunteers was associated with a prolongation of plasma antipyrine half-life in 11 of 14 normal volunteers who developed fever.[190] These results have been confirmed by Trenholme and co-workers,[191] who demonstrated retardation of quinine elimination both in etiocholanolone-induced fever and in fever associated with experimentally induced malaria. Recently antipyrine clearance has been reported to be prolonged in patients with pneumonia.[192] The effects of fever on drug distribution need to be studied both with more drugs and in other febrile states before generalizations concerning the role of fever in drug disposition and appropriate dosage compensation can be made.

23.11.13.4 Effect of Thyroid Disease on Drug Disposition

An extensive literature based on experimental animals describes various effects produced by altered thyroid function on drug disposition. In 1973, patients with hyperthyroidism were shown to exhibit accelerated antipyrine metabolism, whereas patients with hypothyroidism displayed prolongation of plasma antipyrine half-life.[193] These observations were confirmed in 1974[194] and were

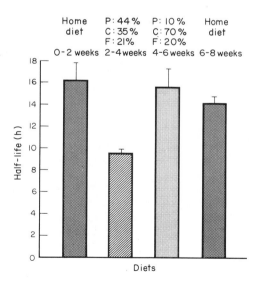

Figure 7 Antipyrine half-lives in six normal subjects maintained on their usual home diets and on two test diet periods. Each diet contained approximately 2500 calories per day. Each bar represents mean $\pm$ SE for the six subjects. The abbreviations are: P, protein; C, carbohydrate; and F, fat. After two weeks on their usual home diets (diet 1), subjects were maintained on the low CHO–high PRO diets (diet 2) for two weeks, followed by two weeks on the high CHO–low PRO diets (diet 3), followed by two weeks on their usual home diets (diet 4). The values for diets 1, 3 and 4 are not significantly different from each other. The value for diet 2 is significantly different from that of diet 1 ($p < 0.005$) and diet 3 ($p < 0.01$) (reproduced by permission of Mosby from ref. 187)

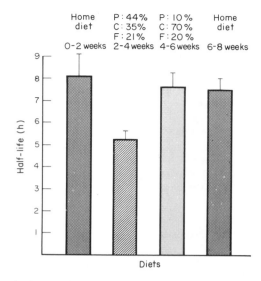

Figure 8 Theophylline half-lives in six normal subjects maintained on their usual home diets and two test diet periods. Each diet contained approximately 2500 calories per day. Each bar represents mean ± SE for the six subjects. The abbreviations are: P, protein; C, carbohydrate; and F, fat. The protocol for the study is described in the legend for Figure 7. The values for diets 1, 3 and 4 are not significantly different from each other. The value for diet 2 is significantly different from that of diet 1 ($p < 0.05$) and diet 3 ($p < 0.01$) (reproduced by permission of Mosby from ref. 187)

extended in 1975[195] to other drugs. Thus, for drugs with low therapeutic indices that are metabolized by hepatic mixed-function oxidases, the appropriate dose may need to be altered according to the functional status of the thyroid. In hypothyroid and hyperthyroid subjects effects of certain environmental chemicals metabolized by these enzymes may also differ from their effects in thyroid subjects.

23.11.13.5 Studies on Drug Disposition in Patients with Liver Disease

For many years uncertainty prevailed concerning the relationship between liver disease and drug disposition. In 1968, this ambiguity was shown to arise mainly from failure to consider how concomitantly administered drugs affected disposition of the drug under study.[196] In 1971, the metabolic pathway for hepatic biotransformation of prednisone was demonstrated to be affected by liver disease, the extent of this effect depending on the nature and severity of the disease.[197] In 1972, patients with severe bilharziasis were shown to have impaired metabolism of the schistosomacide niridazole (nitrothiamidazol), as revealed by higher blood concentrations of the drug and a higher incidence of adverse effects than in normal subjects.[198]

In 1973, serum antipyrine half-life was demonstrated to be prolonged in patients with liver disease,[199] an observation confirmed in 1974.[200, 201] Also in 1974, aminopyrine metabolism was shown to be deranged in a large percentage of patients with liver disease.[202] An aminopyrine breath test (ABT) was proposed in which $^{14}CO_2$ was measured in breath after the administration of 4-dimethyl-^{14}C-aminoantipyrine (^{14}C-aminopyrine) as a sensitive, clinically convenient test of hepatic function and drug-metabolizing capacity. The virtue of this method was that it measured production of a metabolite, rather than disappearance of the parent drug. It is now clear from numerous studies that the rate of disappearance of cold aminopyrine or antipyrine in plasma of a patient correlates highly with the rate of production of their major metabolites.[203]

It is now firmly established that hepatocellular disease is associated with decreased capacity to metabolize some drugs. However, it has not yet been determined whether the extent of depression of hepatic drug-metabolizing function in an individual with liver disease is similar for all drugs biotransformed by the liver.

Observations that rates of aminopyrine elimination are reduced in almost all patients with parenchymal liver disease and that liver disease is also accompanied by reduction in rates of hepatic antipyrine metabolism[199–201] need to be harmonized with observations that warfarin disposition during acute viral hepatitis is unchanged[204] and that oxazepam disposition is normal during acute viral hepatitis and cirrhosis.[205] Between these extremes, a group of drugs of which clindamycin is an example apparently exhibit intermediate or moderate changes in disposition in liver disease.[206] A wide range, from no change whatever in disposition to significant retardation, has been reported in liver disease, depending on the drug studied. For drugs with high hepatic extraction ratios (greater than 0.8), such as propranolol and lidocaine, alterations in blood flow accompanying liver disease can produce large changes in hepatic clearance of the compound. For drugs with very low hepatic extraction ratios (less than 0.2), such as antipyrine and aminopyrine, large variations in the extent to which hepatocellular disease alters their rates of metabolism cannot be due to abnormal liver blood flow. They may be attributed in part to multiple molecular forms of hepatic cytochrome *P*-450 and to the differential effects that a particular hepatic disorder might exert on these forms. Sequential measurements of hepatic drug-metabolizing enzymes during the course of various liver diseases can serve not only as guides to help individualize drug doses but also as probes to explore the pathogenesis of the disorders themselves.

23.11.14 REFERENCES

1. W. Kalow, 'Pharmacogenetics: Heredity and the Response to Drugs', Saunders, Philadelphia, 1962.
2. H. Meier, 'Experimental Pharmacogenetics', Academic Press, New York, 1963.
3. G. P. Quinn, J. Axelrod and B. B. Brodie, *Biochem. Pharmacol.*, 1958, **1**, 152.
4. B. N. LaDu and W. Kalow, 'Pharmacogenetics', ed. T. L. Hall, The New York Academy of Sciences, New York, 1968.
5. B. N. LaDu, *Annu. Rev. Med.*, 1972, **23**, 453.
6. G. S. Omenn and A. G. Motulsky, in 'Genetic Issues in Public Health and Medicine, ed. B. H. Cohen, A. M. Lilienfeld and P. C. Huang, Thomas, Springfield, IL, 1978.
7. P. Propping, *Rev. Physiol. Biochem. Pharmacol.*, 1978, **83**, 123.
8. E. S. Vesell, *Ann. N.Y. Acad. Sci.*, 1971, **179**, 1.
9. E. S. Vesell, *Prog. Med. Genet.*, 1973, **9**, 291.
10. E. S. Vesell, *Am. J. Med.*, 1979, **66**, 183.
11. W. Kalow, *Trends Pharmacol. Sci.*, 1980, **1**, 403.
12. R. M. Weinshilboum, *Fed. Proc., Fed. Am. Soc. Exp. Biol.*, 1984, **43**, 2295.
13. E. S. Vesell, *Fed. Proc., Fed. Am. Soc. Exp. Biol.*, 1984, **43**, 2319.
13A. W. Kalow, H. W. Goedde and D. P. Agarwall, in 'Progress in Clinical and Biological Research', Liss, New York, 1986, vol. 14.
14. G. S. Omenn and H. V. Gelboin, 'Banbury Report', Cold Spring Harbor Laboratory, Cold Spring Harbor, NY, 1984, vol. 16.
15. L. E. Cluff, G. L. Thornton and J. Smith, *Trans. Assoc. Am. Physicians*, 1965, **78**, 255.
16. A. H. Conney, R. Welch, R. Kuntzman, R. Chang, M. Jacobson, A. D. Munro-Faure, A. W. Peck, A. Bye, A. Poland, P. J. Poppers, M. Finster and J. A. Wolff, *Ann. N.Y. Acad. Sci.*, 1971, **179**, 155.
17. J. R. Gillette, *Ann. N.Y. Acad. Sci.*, 1971, **179**, 43.
18. E. S. Vesell, *Clin. Pharmacol. Ther.*, 1982, **31**, 1.
19. E. S. Vesell, *Adv. Pharmacol. Chemother.*, 1969, **7**, 1.
20. D. W. Nebert and J. S. Felton, *Fed. Proc., Fed. Am. Soc. Exp Biol.*, 1976, **35**, 1133.
21. E. J. Calabrese, 'Ecogenetics', Wiley, New York, 1984.
22. S. Takahara, *Lancet*, 1952, **2**, 1101.
23. S. Takahara and K. Doi, *Acta Med. Okayama*, 1959, **13**, 1.
24. S. Takahara, H. Sato, M. Doi and S. Mihara, *Proc. Jpn. Acad.*, 1952, **28**, 585.
25. S. Takahara, H. B. Hamilton, J. V. Neel, T. Y. Kobara, Y. Ogura and E. T. Nishimura, *J. Clin. Invest.*, 1960, **39**, 610.
26. J. B. Wyngaarden and R. W. Howell, in 'The Metabolic Basis of Inherited Disease', ed. J. B. Stanbury, J. B. Wyngaarden and D. S. Frederickson, McGraw-Hill, New York, 1966.
27. H. B. Hamilton and J. V. Neel, *Am. J. Hum. Genet.*, 1963, **15**, 408.
28. H. Yata, Nihon Shika Hyoron, 1959, **204**, 7.
29. H. Aebi, J. P. Heiniger, R. Butler and A. Hassig, *Experientia*, 1961, **17**, 466.
30. H. Aebi and H. Suter, *Adv. Hum. Genet.*, 1971, **2**.
31. H. E. Aebi and S. R. Wyss, in 'The Metabolic Basis of Inherited Disease', ed. J. B. Stanbury, J. B. Wyngaarden and D. S. Frederickson, McGraw-Hill, New York, 1978.
32. H. B. Hughes, *J. Pharmacol. Exp. Ther.*, 1953, **109**, 444.
33. H. B. Hughes, J. P. Biehl, A. P. Jones and L. H. Schmidt, *Am. Rev. Tuberc.*, 1954, **70**, 226.
34. D. A. Price Evans and T. A. White, *J. Lab. Clin. Med.*, 1964, **63**, 394.
35. D. A. Price Evans, *Ann. N.Y. Acad. Sci.*, 1965, **123**, 178.
36. M. M. Reidenberg, D. E. Drayer, M. Levy and H. Warner, *Clin. Pharmacol. Ther.*, 1975, **17**, 722.
37. M. M. Reidenberg, M. Levy, D. E. Drayer, E. Zylber-Katz and W. C. Robbins, *Arthritis Rheum.*, 1980, **23**, 569.
38. R. A. Cartwright, R. W. Glashan, H. J. Rogers, R. A. Ahmad, D. Barham-Hall, E. Higgins and M. A. Kahn, *Lancet*, 1982, **2**, 842.
39. R. A. Cartwright, in 'Banbury Report', ed. G. S. Omenn, H. V. Gelboin, Cold Spring Harbor Laboratory, Cold Spring Harbor, NY, 1984, vol. 16, p. 359.
40. J. Peters, in 'Transactions of Conference on Chemotherapy on Tuberculosis, 18th Conference', Excerpta Medica, Amsterdam, 1959, 37.
41. J. R. Mitchell, U. P. Thorgeirsson, M. Black, J. A. Timbrell, W. R. Snodgrass, W. Z. Potter, D. J. Jollow and H. R. Keiser, *Clin. Pharmacol. Ther.*, 1975, **18**, 70.
42. J. R. Mitchell, H. J. Zimmerman, K. G. Ishak, U. P. Thorgeirsson, J. A. Timbrell, W. R. Snodgrass and S. D. Nelson, *Ann. Intern. Med.*, 1976, **84**, 181.

43. A. Litwin, *Fed. Proc., Fed. Am. Soc. Exp. Biol.*, 1983, **42**, 3091.
44. S. Lal, S. N. Singhal, D. M. Burley and G. Crossley, *Br. Med. J.*, 1972, **1**, 148.
45. J. Smith, W. F. Tyrrell, A. Gow, G. W. Allan and A. W. Lees, *Chest*, 1972, **61**, 587.
46. D. A. Price Evans, K. Manley and V. A. McKusick, *Br. Med. J.*, 1960, **2**, 485.
47. W. W. Weber and D. H. Hein, *Pharmacol. Rev.*, 1985, **37**, 25.
48. A. G. Motulsky, *Prog. Med. Genet.*, 1964, **3**, 49.
49. H. W. Harris, 'Proceedings of the 16th International Tuberculosis Conference', Excerpta Medica, Amsterdam, 1961, vol. 2, p. 503.
50. D. A. Price Evans, *Med. Hyg.*, 1962, **20**, 905.
51. S. Sunahara, 'Proceedings of the 16th International Tuberculosis Conference', Excerpta Medica, Amsterdam, 1961, vol. 2, p. 513.
52. F. T. Evans, P. W. S. Gray, H. Lehmann and E. Silk, *Lancet*, 1952, **1**, 1229.
53. W. Kalow and K. Genest, *Can. J. Biochem.*, 1957, **35**, 339.
54. R. O. Davies, A. V. Marton and W. Kalow, *Cancer J. Biochem.*, 1960, **38**, 545.
55. H. Harris and M. Whittaker, *Nature (London)*, 1961, **191**, 496.
56. W. Kalow and R. O. Davies, *Biochem. Pharmacol.*, 1959, **1**, 183.
57. H. Harris and M. Whittaker, *Ann. Hum. Genet.*, 1962, **26**, 59.
58. W. E. Hodgkin, E. R. Giblett, H. Levine, W. Bauer and A. G. Motulsky, *J. Clin. Invest.*, 1965, **44**, 486.
59. H. W. Goedde and K. Altland, *Ann. N.Y. Acad. Sci.*, 1968, **151**, 540.
60. H. Harris, M. Whittaker, H. Lehmann and E. Silk, *Acta Genet.*, 1960, **10**, 1.
61. H. Lehmann and J. Liddell, *Prog. Med. Genet.*, 1964, **3**, 75.
62. P. J. Garry, A. A. Dietz, T. Lubrano, P. C. Ford, K. James and H. M. Rubinstein, *J. Med. Genet.*, 1976, **13**, 38.
63. B. B. Gutsche, E. M. Scott and R. C. Wright, *Nature (London)*, 1967, **215**, 322.
64. H. W. Neitlich, *J. Clin. Invest.*, 1966, **45**, 380.
65. A. Yoshida and A. G. Motulsky, *Am. J. Hum. Genet.*, 1969, **21**, 486.
66. H. Harris, D. A. Hopkinson, E. B. Robson and M. Whittaker, *Ann. Hum. Genet.*, 1963, **26**, 359.
67. E. W. Maynert, *J. Pharmacol. Exp. Ther.*, 1960, **130**, 275.
68. H. Kutt, M. Wolk, R. Scherman and F. McDowell, *Neurology*, 1964, **14**, 542.
69. M. R. Vasko, R. D. Bell, D. D. Daly and C. E. Pippenger, *Clin. Pharmacol. Ther.*, 1980, **27**, 96.
70. R. W. Brennan, H. Dehejia, H. Kutt and F. McDowell, *Neurology*, 1968, **18**, 283.
71. H. M. Solomon, *Ann. N.Y. Acad. Sci.*, 1968, **151**, 932.
72. M. Ikeda, B. Sezesny and M. Barnes, *Fed. Proc., Fed. Am. Soc. Exp. Biol.*, 1966, **25**, 417.
73. J. H. Greaves and P. Ayres, *Nature (London)* 1967, **215**, 877.
74. N. T. Shahidi, *Am. J. Dis. Child.*, 1967, **113**, 81.
75. L. Endrenyi, T. Inaba and W. Kalow, *Clin. Pharmacol. Ther.*, 1976, **20**, 701.
76. W. Kalow, D. Kadar, T. Inaba and B. K. Tang, *Clin. Pharmacol. Ther.*, 1977, **21**, 530.
77. B. K. Tang, W. Kalow and A. A. Grey, *Res. Commun. Chem. Pathol. Pharmacol.*, 1978, **21**, 45.
78. W. Kalow, B. K. Tang, D. Kadar and T. Inaba, *Clin. Pharmacol. Ther.*, 1978, **24**, 576.
79. W. Kalow, B. K. Tang, D. Kadar, L. Endrenyi and F.-Y. Chan, *Clin. Pharmacol. Ther.*, 1979, **26**, 766.
80. A. Mahgoub, L. G. Dring, J. R. Idle, R. Lancaster and R. L. Smith, *Lancet*, 1977, **1**, 584.
81. N. M. Woolhouse, B. Andoh, A. Mahgoub, T. P. Sloan, J. R. Idle and R. L. Smith, *Clin. Pharmacol. Ther.*, 1979, **26**, 586.
82. N. M. Woolhouse, M. Eichelbaum, N. S. Oates, J. R. Idle and R. L. Smith, *Clin. Pharmacol. Ther.*, 1985, **37**, 512.
83. M. Eichelbaum, N. Spannbrucker, B. Steincke and H. J. Dengler, *Eur. J. Clin. Pharmacol.*, 1979, **16**, 183.
84. M. Eichelbaum, L. Bertilsson, J. Sawe and C. Zekorn, *Clin. Pharmacol. Ther.*, 1982, **31**, 184.
85. T. Inaba, S. V. Otton and W. Kalow, *Clin. Pharmacol. Ther.*, 1980, **27**, 547.
86. L. Bertilsson, M. Eichelbaum, B. Mellstrom, J. Sawe, H.-U. Schulz and F. Sjoqvist, *Life Sci.*, 1980, **27**, 1673.
87. B. Mellstrom, L. Bertilsson, J. Sawe, H.-U. Schulz and F. Sjoqvist, *Clin. Pharmacol. Ther.*, 1981, **30**, 189.
88. T. P. Sloan, A. Mahgoub, R. Lancaster, J. R. Idle and R. L. Smith, *Br. Med. J.*, 1978, **2**, 655.
89. R. R. Shah, N. S. Oates, J. R. Idle and R. I. Smith, *Lancet*, 1980, **1**, 1147.
90. G. Alvan, C. von Bahr, P. Seideman and F. Sjoqvist, *Lancet*, 1982, **1**, 333.
91. R. R. Shah, N. S. Oates, J. R. Idle and R. L. Smith, *Lancet*, 1982, **1**, 508.
92. R. R. Shah, N. S. Oates, J. R. Idle, R. L. Smith and J. D. Lockhart, *Clin. Res.*, 1982, **284**, 295.
93. R. L. Woosley, D. M. Roden, H. J. Duff, E. L. Carey, A. J. J. Wood and G. R. Wilkinson, *Clin. Res.*, 1981, **29**, 501A.
94. D. S. Davies, G. C. Kahn, S. Murray, M. J. Brodie and A. R. Boobis, *Br. J. Clin. Pharmacol.*, 1981, **11**, 89.
95. P. J. Meier, H. K. Mueller, B. Dick and U. A. Meyer, *Gastroenterology*, 1983, **85**, 682.
96. A. Kupfer and R. Preisig, *Eur. J. Clin. Pharmacol.*, 1984, **26**, 753.
97. T. Inaba, M. Jurima, M. Nakano and W. Kalow, *Clin. Pharmacol. Ther.*, 1984, **36**, 670.
98. P. J. Wedlund, W. S. Aslanian, C. B. McAllister, G. R. Wilkinson and R. A. Branch, *Clin. Pharmacol. Ther.*, 1984, **36**, 773.
99. R. M. Weinshilboum and S. L. Sladek, *Am. J. Hum. Genet.*, 1980, **32**, 651.
100. M. Geldmacher-v. Mallinchkrodt, G. Hommel and J. Dumbach, *Hum. Genet.*, 1979, **50**, 313.
101. R. A. O'Reilly, P. M. Aggeler, M. S. Hoag, L. S. Leong and M. L. Kropatkin, *N. Engl. J. Med.*, 1964, **271**, 809.
102. R. A. O'Reilly, J. G. Pool and P. M. Aggeler, *Ann. N.Y. Acad. Sci.*, 1968, **151**, 913.
103. R. A. O'Reilly, *N. Engl. J. Med.*, 1970, **282**, 1448.
104. A. G. Motulsky, A. Yoshida and G. Stamatoyannopoulos, *Ann. N.Y. Acad. Sci.*, 1971, **179**, 636.
105. A. Yoshida, *Proc. Natl. Acad. Sci. U.S.A.*, 1967, **57**, 835.
106. A. Yoshida, *J. Mol. Biol.*, 1970, **52**, 483.
107. I. M. Fraser, B. E. Tilton and E. S. Vesell, *Ann. N.Y. Acad Sci.*, 1971, **179**, 644.
108. I. M. Fraser, B. E. Tilton and E. S. Vesell, *Pharmacology*, 1971, **5**, 173.
109. W. H. Hitzig, P. G. Frick, K. Betke and T. H. Huisman, *Helv. Paediatr. Acta*, 1960, **15**, 499.
110. P. G. Frick, W. H. Hitzig and K. Betke, *Blood*, 1962, **20**, 261.
111. C. J. Muller and S. Kingman, *Biochim. Biophys. Acta*, 1961, **50**, 595.
112. R. F. Reider, W. H. Zinkham and N. A. Holtzman, *Am. J. Med.*, 1965, **39**, 4.
113. A. F. Blakeslee, *Proc. Natl. Acad. Sci. U.S.A.*, 1932, **18**, 120.

114. A. L. Fox, *Proc. Natl. Acad. Sci. U.S.A.*, 1932, **18**, 115.
115. H. Harris, H. Kalmus and W. R. Trotter, *Lancet*, 1949, **2**, 1038.
116. F. D. Kitchin, W. Howel-Evans, C. A. Clarke, R. B. McConnell and P. M. Sheppard, *Br. J. Med.*, 1959, **1**, 1069.
117. N. A. Barnicot, *Ann. Hum. Genet.*, 1950, **15**, 248.
118. P. H. Saldanha and W. Becak, *Science (Washington, D.C.)*, 1959, **129**, 150.
119. M. A. Denborough, J. F. A. Forster, R. H. Lovell, P. A. Maplestone and J. D. Villiers, *Br. J. Anaesth.*, 1962, **34**, 395.
120. W. Kalow, *Ann. N.Y. Acad. Sci.*, 1971, **179**, 654.
121. G. G. Harrison, *Br. J. Anaesth.*, 1975, **47**, 62.
122. S. K. Pandit, S. P. Kothary and P. J. Cohen, *Anaesthesiology*, 1979, **50**, 156.
123. B. A. Britt, *Fed. Proc., Fed. Am. Soc. Exp. Biol.*, 1979, **38**, 44.
124. J. P. von Wartburg and P. M. Schurch, *Ann. N.Y. Acad. Sci.*, 1968, **151**, 936.
125. J. A. Edwards and D. A. Price Evans, *Clin. Pharmacol. Ther.*, 1967, **8**, 824.
126. P. H. Wolff, *Science (Washington, D.C.)*, 1972, **175**, 449.
127. H. W. Goedde, S. Harada and D. P. Agarwal, *Hum. Genet.*, 1979, **51**, 331.
128. H. Jick, D. Slone, B. Westerholm, W. H. M. Inman, M. P. Vessey, S. Shapiro, G. P. Lewis and J. Worcester, *Lancet*, 1969, **1**, 539.
129. G. P. Lewis, H. Jick, D. Slone and S. Shapiro, *Ann. N.Y. Acad. Sci.*, 1971, **179**, 729.
130. R. Spector, A. K. Choudhury, C.-K. Chiang, M. J. Goldberg and M. M. Ghoneim, *Clin. Pharmacol. Ther.*, 1980, **28**, 229.
131. C. M. B. Pare, *Humangenetik*, 1970, **9**, 199.
132. C. M. B. Pare and J. W. Mack, *J. Med. Genet.*, 1971, **8**, 306.
133. E. Dorus, G. N. Pandey, A. Frazer and J. Mendels, *Arch. Gen. Psychiatry*, 1974, **31**, 463.
134. A. Svejgaard, M. Houge, C. Jersild, P. Platz, L. P. Ryder, L. S. Nielsen and M. Thomsen, in 'Monographs in Human Genetics', ed. L. Beckman, U. M. Huage, Karger, Basel, 1975, vol. 7.
135. W. F. Bodmer, 'Inheritance of Susceptibility to Cancer in Man', Oxford University Press, Oxford, 1983.
136. E. Smeraldi, L. Bellodi, E. Sacchetti and C. L. Cazzullo, *Br. J. Psychiatry*, 1976, **129**, 486.
137. E. Smeraldi and R. Scorza-Smeraldi, *Nature (London)*, 1976, **260**, 532.
138. C. Perris, E. Strandman and L. Wahlby, *Neuropsychobiology*, 1979, **5**, 114.
139. P. H. Wooley, J. Griffin, G. S. Panayi, J. R. Batchelor, K. I. Welsh and T. J. Gibson, *N. Engl. J. Med.*, 1980, **303**, 300.
140. K. L. Schmidt and C. Mueller-Eckhardt, *Lancet*, 1977, **2**, 85.
141. J. R. Batchelor, K. I. Welsh, R. Mansilla Tinoco, C. T. Dollery, G. R. V. Hughes, R. Bernstein, P. Ryan, P. F. Naish, G. M. Aber, R. F. Bing and G. I. Russell, *Lancet*, 1980, **1**, 1107.
142. H. M. Perry, E. M. Tan, S. Carmody and A. Sakamoto, *J. Lab. Clin. Med.*, 1970, **76**, 114.
143. T. F. Gallagher, Jr., M. N. Mueller and A. Kappas, *Medicine*, 1966, **45**, 471.
144. S. Sherlock, *Br. J. Med.*, 1968, **1**, 227.
145. S. M. Kalman, *Annu. Rev. Pharmacol.*, 1969, **9**, 363.
146. A. Kappas, *Gastroenterology*, 1967, **52**, 113.
147. U. P. Haemmerli and H. I. Wyss, *Medicine*, 1967, **46**, 299.
148. J. M. Orellana-Alcalde and J. P. Dominguez, *Lancet*, 1966, **2**, 1278.
149. J. A. Whittaker and D. A. Price Evans, *Br. Med. J.*, 1970, **4**, 323.
150. M. Asberg, D. A. Price Evans and F. Sjoqvist, *J. Med. Genet.*, 1971, **8**, 129.
151. J. Scott and P. L. Poffenbarger, *Diabetes*, 1979, **28**, 41.
152. M. Miller, K. E. Opheim, V. A. Raisys and A. G. Motulsky, *Clin. Pharmacol. Ther.*, 1984, **35**, 170.
153. M. B. Penno and E. S. Vesell, *J. Clin. Invest.*, 1983, **71**, 1698.
154. C. A. Miller, L. B. Slusher and E. S. Vesell, *J. Clin. Invest.*, 1985, **75**, 1415.
155. R. E. Kouri, C. E. McKinney, D. J. Slomiany, D. R. Snodgrass, N. P. Wray and T. L. McLemore, *Cancer Res.*, 1982, **42**, 5030.
156. R. E. Kouri, A. S. Levine, B. K. Edwards, T. L. McLemore, E. S. Vesell and D. W. Nebert, in 'Banbury Report', ed. G. S. Omenn and H. V. Gelboin, Cold Spring Harbor Laboratory, Cold Spring Harbor, NY, 1984, vol. 16.
157. M. W. E. Teunissen, L. G. J. DeLeede, J. K. Boeijinga and D. D. Breimer, *J. Pharmacol. Exp. Ther.*, 1985, **233**, 770.
158. L. Szabo, Z. Kovacs and P. B. Ebrey, *Acta Paediatr. Hung.*, 1962, **3**, 49.
159. B. Childs, J. B. Sidbury and C. J. Migeon, *Pediatrics*, 1959, **23**, 903.
160. J. Axelrod, R. Schmid and L. Hammaker, *Nature (London)*, 1957, **180**, 1426.
161. R. E. Peterson and R. Schmid, *J. Clin. Endocrinol Metab.*, 1957, **17**, 1485.
162. I. M. Arias, *J. Clin. Invest.*, 1962, **41**, 2233.
163. W. T. Foulk, H. R. Butt, C. A. Owen, F. F. Whitcomb and H. L. Mason, *Medicine*, 1959, **38**, 25.
164. H. T. F. Barniville and R. Misk, *Br. J. Med.*, 1959, **1**, 337.
165. I. M. Arias, L. M. Gartner, M. Cohen, J. B. Ezzer and A. J. Levi, *Am. J. Med.*, 1969, **47**, 395.
166. W. C. Parker and A. G. Bearn, *Am. J. Med.*, 1963, **34**, 680.
167. E. S. Vesell, *Trends Pharmacol. Sci.*, 1980, **1**, 349.
168. E. S. Vesell, in 'Banbury Report', ed. V. R. Hunt, M. K. Smith and D. Worth, Cold Spring Harbor Laboratory, Cold Spring Harbor, NY, 1982, vol. 11, p. 107.
169. A. P. Alvares, S. Kapelner, S. Sassa and A. Kappas, *Clin. Pharmacol. Ther.*, 1975, **17**, 179.
170. A. W. Pruitt, P. G. Dayton and J. H. Patterson, *Clin. Pharmacol. Ther.*, 1973, **14**, 73.
171. L. K. Garrettson and P. G. Dayton, *Clin. Pharmacol. Ther.*, 1970, **11**, 674.
172. R. E. Kauffman, D. W. Shoeman, S. H. Wan and D. L. Azarnoff, *Clin. Pharmacol. Ther.*, 1972, **13**, 704.
173. O. Pelkonen, E. H. Kaltiala, T. K. I. Larmi and N. T. Karki, *Clin. Pharmacol. Ther.*, 1973, **14**, 840.
174. A. Rane and F. Sjoqvist, *Biochem. Pharmacol.*, 1972, **1**, 152.
175. Z. Friedman, V. Whitman, M. J. Maisels, W. Berman, Jr., K. H. Marks and E. S. Vesell, *J. Clin. Pharmacol.*, 1978, **18**, 272.
176. A. Traeger, H. Noschel and J. Zaumseil, *Zantralbl Gynaekol.*, 1973, **95**, 635.
177. D. E. Duggan, A. F. Hogans, K. C. Kwan and F. G. McMahon, *J. Pharmacol. Exp. Ther.*, 1972, **181**, 563.
178. H. B. Hucker, A. G. Zacchei, S. V. Cox, D. A. Brodie and N. H. R. Cantwell, *J. Pharmacol. Exp. Ther.*, 1966, **153**, 237.
179. L. Palmer, L. Bertilsson, G. Alvan, M. Orme, F. Sjoqvist and B. Holmstedt, in 'Prostaglandin Synthetase Inhibitors', ed. H. J. Robinson and J. R. Vane, Raven Press, New York, 1974, p. 91.

180. R. L. Dixon, R. W. Shultice and J. R. Fouts, *Proc. Soc. Exp. Biol. Med.*, 1960, **103**, 333.
181. R. L. Furner and D. D. Feller, *Proc. Soc. Exp. Biol. Med.*, 1971, **137**, 816.
182. R. Kato and J. R. Gillette, *J. Pharmacol. Exp. Ther.*, 1965, **150**, 279.
183. V. G. Zannoni, E. J. Flynn and M. Lynch, *Biochem. Pharmacol.*, 1972, **21**, 1377.
184. M. M. Reidenberg and E. S. Vesell, *Clin. Pharmacol. Ther.*, 1975, **17**, 650.
185. K. Krishnaswamy and A. N. Naidu, *Br. J. Med.*, 1977, **1**, 538.
186. O. M. Bakke, S. Aanderud, G. Syversen, H. H. Bassoe and O. Myking, *Br. J. Clin. Pharmacol.*, 1978, **5**, 341.
187. A. Kappas, K. E. Anderson, A. H. Conney and A. P. Alvares, *Clin. Pharmacol. Ther.*, 1976, **20**, 643.
188. A. Kappas, A. P. Alvares, K. E. Anderson, E. J. Pantuck, C. B. Pantuck, R. Chang and A. H. Conney, *Clin. Pharmacol. Ther.*, 1978, **23**, 445.
189. E. J. Pantuck, C. B. Pantuck, W. A. Garland, B. H. Min, L. W. Wattenberg, K. E. Anderson, A. Kappas and A. H. Conney, *Clin. Pharmacol. Ther.*, 1979, **25**, 88.
190. R. J. Elin, E. S. Vesell and S. M. Wolff, *Clin. Pharmacol. Ther.*, 1975, **17**, 447.
191. G. M. Trenholme, R. L. Williams, K. H. Rieckmann, H. Frischer and P. E. Carson, *Clin. Pharmacol. Ther.*, 1976, **19**, 459.
192. J. Sonne, M. Dossing, S. Loft and P. B. Andreasen, *Clin. Pharmacol. Ther.*, 1985, **37**, 701.
193. E. S. Vesell and G. T. Passananti, *Drug Metab. Dispos.*, 1973, **1**, 402.
194. M. Eichelbaum, G. Bodem, R. Gugler, C. Schneider-Deters and H. H. Dengler, *N. Engl. J. Med.*, 1974, **290**, 1040.
195. E. S. Vesell, J. R. Shapiro, G. T. Passananti, H. Jorgensen and C. A. Shively, *Clin. Pharmacol. Ther.*, 1975, **17**, 48.
196. A. J. Levi, S. Sherlock and D. Walker, *Lancet*, 1968, **1**, 1275.
197. W. G. E. Cooksley and L. W. Powell, *Drugs*, 1971, **2**, 177.
198. J. W. Faigle, *Acta Pharmacol. Toxicol.*, 1971, **29** (Suppl. 3), 233.
199. R. A. Branch, C. M. Herbert and A. E. Read, *Gut*, 1973, **14**, 569.
200. P. B. Andreasen, L. Ranek, B. E. Statland and N. Tygstrup, *Eur. J. Clin. Invest.*, 1974, **4**, 129.
201. P. B. Andreasen and E. S. Vesell, *Clin. Pharmacol. Ther.*, 1974, **16**, 1059.
202. G. W. Hepner and E. S. Vesell, *N. Engl. J. Med.*, 1974, **291**, 1384.
203. E. S. Vesell, G. T. Passananti, P. A. Glenwright and B. H. Dvorchik, *Clin. Pharmacol. Ther.*, 1975, **18**, 259.
204. R. L. Williams, W. L. Schary, T. F. Blaschke, P. J. Meffin, K. L. Melmon and M. Rowland, *Clin. Pharmacol. Ther.*, 1976, **20**, 90.
205. H. J. Shull, G. R. Wilkinson, R. Johnson and S. Schenker, *Ann. Intern. Med.*, 1976, **84**, 420.
206. G. R. Avant, S. Schenker and R. H. Alford, *Am. J. Dig. Dis.*, 1975, **20**, 223.

23.12

Chronokinetics

ALAIN REINBERG and FRANCIS LÉVI
Fondation Adolphe de Rothschild, Paris, France

MICHAEL H. SMOLENSKY
University of Texas, Houston, TX, USA

GASTON LABRECQUE
University of Laval, Quebec, Canada

MICHEL OLLAGNIER and HERVÉ DECOUSUS
Université de Saint-Etienne, France

and

BERNARD BRUGUEROLLE
Université de Marseille, France

23.12.1 INTRODUCTION

Chronopharmacology did not develop into a scientific domain of investigation until the early 1970s.[1] Chronopharmacology investigates the effects of drugs: (i) as a function of biological time, *e.g.* on the 24 h scale, and (ii) upon parameters characterizing endogenous bioperiodicities. A better understanding of periodic, and thus predictable, changes in drug effects can be attained through the consideration of three complementary concepts: (i) the *chronokinetics* of a drug (rhythmic, and thus predictable, changes in its pharmacokinetics); (ii) the *chronesthesy* (rhythmic changes in suscepti-bility of target biosystems to a drug); and (iii) the *chronergy* (rhythmic changes in the integrated overall effects of drugs).[1-3]

For conventionally trained pharmacologists, it may be surprising to read that predictable variations in both the disposition and effect of chemical agents (toxic substances, medications, hormones, *etc.*) are governed by endogenous biological rhythms, rather than by changes in external factors only. In the 24 h scale (as well as in the yearly scale) peaks and troughs of physiological functions and variables are not randomly distributed; their respective timings correspond to a highly ordered temporal organization controlled by a set of pacemakers (so-called biological clocks or oscillators), which are presumably interconnected and hierarchically integrated.[4] As a consequence, the metabolic fate of a pharmacological agent may differ from one dosing time to another, *e.g.* morning *versus* evening dosing in humans.

Despite the fact that this chapter is devoted to chronokinetics, we consider it relevant, for the benefit of the general audience, to overview briefly the definitions and properties of biological rhythms as well as to summarize especially pertinent findings related to chronesthesy. The latter concept represents the counterpart to the chronokinetics of a drug, and is discussed later (see Section 23.12.3).[3,4]

23.12.2 BIOLOGICAL RHYTHMS: DEFINITIONS, CHARACTERISTICS AND PROPERTIES

A biological rhythm can be characterized both by aspects of its curve pattern (*e.g.* raw data plotted over time) and the quantification of four of its defining parameters, as determined from a best-fitting mathematical function.[4-6]

The period, τ, is the duration of one complete cycle of a rhythmical variation. It is customarily expressed in units of time, *e.g.* minutes, hours or days.

The acrophase, ϕ, is the mathematically estimated span of time taken to reach the crest of the validated rhythm for the τ under consideration, with regard to a phase reference ϕ_0. When $\tau = 24$ h, ϕ can be expressed in hours and minutes (four figures) with midnight (0000), for example, as ϕ_0.

The amplitude, A, is the amount of variability, *i.e.* the crest to trough difference, for the τ under consideration.

The mesor, M, is the rhythm-adjusted mean for the τ under consideration. When the interval of time between data sampling (Δt) is constant, M equals the arithmetic mean ($\bar{X}$).

Halberg *et al.*[5] developed a computerized technique, the cosinor, to evaluate the statistical significance of periodic changes in a time series, and to provide estimates of ϕ, A and M when τ is known from complementary experimental and/or spectral analyses of the time series.[6] The least-squares method serves to determine the best-fitting function (usually a cosine) for approximating time series data.

Biological rhythms have similar properties in plants and animals, from the unicellular eukaryote to humans. Biological rhythms are genetically programmed. For example, specific genes which code for circadian (about 24 h) rhythms have been identified and located on chromosomes of *Drosophila*.[7] Moreover, studies on mono- *versus* di-zygotic twins further revealed the inheritability of rhythmical traits.[8]

Biological rhythms persist in constant environmental conditions without time cue or clue. Under these conditions, the period, τ, of circadian rhythms differs from exactly 24 h. In humans τ is around 25 h under these so-termed 'free-running' conditions.[4,9] To a great extent the circadian period, particularly in humans, is independent of environmental temperature.

For a given variable in a given species, for example cortisol secretion or body temperature in humans, several periods can be identified, indicating that the temporal organization of biological functions is highly ordered with regard to several spectral domains. This is made apparent by the identification of bioperiodicities of differing τ: ultradian rhythms with $\tau < 21$ h; circadian rhythms with $21 \leqslant \tau \leqslant 28$ h; circaseptan rhythms with $\tau \approx 7$ d; circamensual rhythms with $\tau \approx 30$ d and circannual rhythms with $\tau \approx 1$ year.

For a given period (*e.g.* $\tau = 24$ h) respective locations of crests (acrophases) and troughs of variables do not occur at random. On the other hand, the distribution in time of acrophases, for example over the scale of 1 d (circadian domain) and/or one year (circannual domain), represents a temporal organization, an 'anatomy in time', which complements the conventionally considered anatomy in space. Thus, the circadian temporal structure constitutes an important genetically determined characteristic of a species (human, rat, mouse, *etc.*).

Chronobiology can be defined as the study of the mechanisms and alterations of each organism's temporal structure under various conditions, including the administration of medications.

In arthropods as well as mammals, neural formations have been identified as locations of circadian oscillators. This is the case of the hypothalamic suprachiasmatic nucleus (SCN), which in mammals appears to be a biological clock controlling circadian changes in ACTH and prolactin secretions, among other variables.[10] In humans, biological clocks also may be located in the cortex of the brain.[11]

Organisms make use of periodic changes in environmental factors (light/dark or dusk/dawn, noise/silence, heat/cold alternations, imperatives of our social life, *etc.*) as signals to cue or reset their biological clocks and synchronize their circadian and perhaps other periodic rhythms. These periodic signals are called *Zeitgebers* (time givers) or synchronizers. Emphasis must be given to two facts: (i) *Zeitgebers* do not create rhythms; and (ii) the synchronization of subjects must be known to analyze correctly time dependent changes. From a practical point of view, this means that the timings of the onset and offset of lighting in animal experiments, and hours of activity and rest in human studies, must be known in any chronopharmacological study.

From an evolutionary perspective, chronopharmacological rhythms can be attributed to the temporal organization of biological activity and function, which today represent an aspect of an adaptive phenomenon of species to predictable environmental periodicities. Changes in the light/dark cycle and related phenomena are generated by rotation of the Earth around its axis as well as around the Sun.

Mayersbach[12] demonstrated over a decade ago that the circadian cytological reorganization of the rat liver is associated with functional readjustments. At the end of the activity span (onset of light) liver cells contain their highest amount of glycogen and lowest amount of protein. At the end of the resting span (onset of darkness) only a few glycogen granules can be visualized ultramicroscopically, while at the same time the liver albumin level is at its highest. These circadian changes are unrelated to food intake, since they persist, although with a reduced 24 h mean level, during fasting. Many other activities (*e.g.* DNA and RNA synthesis[13,14] and enzyme activities, including those responsible for drug metabolism[15,16]) exhibit large amplitude circadian changes in the hepatic cell. Moreover, the latter is one, among several, illustrative examples demonstrating coordinated ultrastructural, enzymatic and functional circadian rhythms at the subcellular level. From a basic point of view, it seems that many different cellular functions are programmed in time, presumably because the energy available is limited. Such a temporal organization favors a certain function at a certain time with a shift to another function at another time in coordination with other interrelated bioperiodic activities and requirements in this and other tissues of the body. As a result, the metabolism of a drug as well as its specific effect(s) are likely to depend on which stage of a programmed function is occurring when it or its metabolite(s) reach the cell. With regard to the chronobiology of target tissues, and in consideration of chronokinetic changes of an agent, metabolic pathways are neither open permanently, nor open to the same extent throughout the 24 h span. With regard to kinetic phenomena, processes such as absorption, transport, distribution, transformation and/or elimination of a drug may exhibit large amplitude circadian (and/or other periodic) changes. It is not appropriate to consider the chronokinetic attributes of medications alone; cellular, subcellular and molecular rhythms (circadian and others) of the target system must be taken into account too.

23.12.3 CHRONESTHESY

23.12.3.1 Definition

The term chronesthesy is used to designate rhythmical (predictable in time) differences in the susceptibility or the sensitivity of a target biosystem to an agent.[3,4] The target biosystem can be a human tissue: for example the circadian rhythm of human skin reactions in terms of the erythema and wheal responses to intradermal injections of histamine has a nocturnal crest;[17] and in the bronchial tree there is a circadian rhythm in the response to acetylcholine, among other tested agents, when inhaled in aerosol form.[2,4] However, the target biosystem may be located either at the molecular level of the receptors in different subcellular systems (membrane structure, enzyme systems, *etc.*) or elsewhere. Therefore the concept of chronesthesy not only represents changes dependent on the time of dosing in the pharmacokinetics (*i.e.* the effects of, and reactions to, drugs within the body), but also represents changes in other quantifiable phenomena found at the level of the target biosystem. With regard to chronesthesy, the circadian rhythm parameters of many specific variables can be altered at this level due to the response to a given drug at a certain time. Such alterations may be manifested by an acrophase shift or drift (so-called phase response curve) and/or a modification of the period length, mesor or amplitude.

23.12.3.2   Circadian Rhythms of Receptors

Circadian rhythms in various types of receptors and organs (brain and heart in rats, blood cells in humans) have been reported and recently reviewed.[18] In all cases, it is the B_{max} (number of binding sites) rather than the K_D (binding capacity of sites) which is disclosed as showing a biological rhythm. Statistically significant 24 h rhythms have been validated in all the receptors studied by Wirz-Justice and her group[18] in whole rat forebrain homegenates, namely in α_1-, α_2- and β-adrenergic, muscarinic, cholinergic, dopaminergic, 5-HT-1,5-HT-2, adenosine, opiate, benzodiazepine, GABA and imipramine receptors. In addition, the waveform, amplitude and acrophase of these circadian rhythms were found to vary with the time of year, even though the rats were kept synchronized with a defined and nonvarying light/dark cycle. These rhythms are endogenous since they persist in animals dwelling in constant environmental conditions without known *Zeitgebers*. However, they can be altered after lesioning of a putative circadian pacemaker in the suprachiasmatic nuclei. The pattern of a receptor rhythm may change from one brain region to another. As an example, the pattern can be different for the same ligand in different nuclei of the hypothalamus. In rats, receptor rhythms vary according to the strain and even within the same strain from different breeding lines. The binding of a chemical agent to a given ligand in a defined brain region varies with age, leading to changes in circadian rhythm parameters, such as amplitude, acrophase and/or 24 h mean.

It is of interest that in some systems thus far studied the circadian amplitude of certain receptors is not large enough to explain fully the impressive magnitude of changes in chronesthesy. For example, Lemmer *et.al.*[19] found that large amplitude circadian rhythms in the adenylatecyclase–phosphodiesterase system can coexist with the small amplitude 24 h rhythms (or even with the absence of such rhythms) shown by changes in B_{max} of rat brain receptors for [^{3}H]DHA. It is likely that the adenylatecyclase–phosphodiesterase system, which is activated after specific molecular binding, amplifies the circadian changes of the response to a greater extent than that predicted by the rhythm in B_{max}. The existence of a biological rhythm amplifier at the target level helps explain the observed circadian variations in certain drug effects, even if a steady state in their bioavailability can be achieved by technical means.

23.12.4   DEFINITION OF CHRONOKINETICS

Chronokinetics is defined as the predictable rhythmical changes, dependent on the time of dosing, in the parameters used to characterize the pharmacokinetics of a drug; such parameters are C_{max} (maximum concentration, *e.g.* in plasma), t_{max} (span of time to reach C_{max} with reference to dosing time), AUC (area under the concentration/time curve), $t_{1/2}$ (half life),[1,2,20–22] absorption and excretion rate constants and clearance volume of distribution indices.[23–25] An example from studies by Lemmer *et al.*[26] is given in Figure 1(a) and (b), regarding the chronokinetics of β-receptor-blocking drugs of differing lipophilicity. In accordance with much published data, the plasma half-lives of the more lipophilic drugs propranolol and metoprolol were shorter than those of the hydrophilic drugs sotalol and atenolol, following injection at both 0730 and 1930. A similar pattern was found for the half-lives in muscle. However, the most interesting finding was that the half-lives of the four β-blockers displayed a pronounced circadian dependency, in that the half-lives were longer when the drugs were injected at 0730 than when injected at 1930, *i.e.* in the activity period of the night active rats. The light/dark difference in $t_{1/2}$ was significant for propranolol in plasma, muscle, lung and brain, for metoprolol in heart, muscle, liver and kidney, for sotalol in all organs, and for atenolol in plasma and all organs except the brain. For metoprolol in plasma only an approximate half-life could be calculated, since elimination was very fast and plasma concentrations were below the detection limit of the analytical method 160 min after injection.

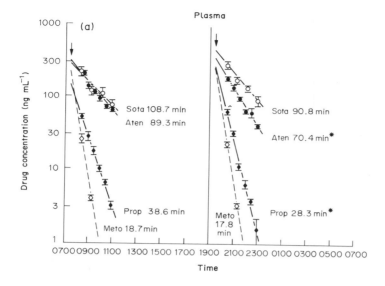

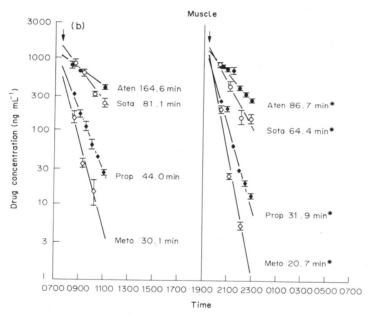

Figure 1 Chronokinetics of β-receptor-blocking drugs. Drug concentration and elimination half-life ($t_{1/2}$) for propranolol (prop), metoprolol (meto), sotalol (sota) and atenolol (aten) in (a) plasma and (b) rat muscle after intravenous injection of an equimolar dosage (6 mol kg⁻¹). Drugs were injected at either 0730 (during light) or at 1930 (during dark) as shown by arrows. Points are mean values ± SEM for five rats. * denotes a significant difference (p at least <0.05) between slopes of regression lines for light and dark[26,87]

By 1987, the circadian chronokinetics of at least 70 different molecules had been documented both in humans and in several animal models (reviewed by Bruguerolle[20]) and more than 40 articles had been devoted solely to the dependency on the time of dosing of the bioavailability of theophylline, both in children and adults (reviewed by Smolensky[21,22]). In most of the studies four to six fixed times of treatment with respect to the 24 h period were tested to validate statistically significant circadian changes in selected pharmacokinetic parameters for subjects who had been carefully synchronized. Ketoprofen serves as an example, as shown in Figure 2. In Figure 2 C_{max} and t_{max} are mean experimental data and the other parameters were calculated from the following equations

$$V = Ae^{-\alpha t} + Be^{-\beta t} + Ce^{-k_a t}$$

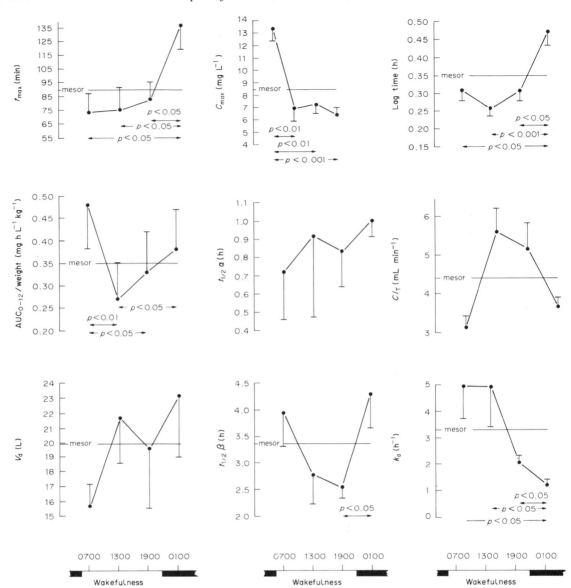

Figure 2   Circadian kinetics of oral ketoprofen. Chronograms of pharmacokinetic parameters (mean $\pm$ SEM) in absolute values, the horizontal arrows represent statistical comparison of values by Student's t-test, $n = 8$[85]

where A, B and C are exponential coefficients of distribution, elimination and absorption respectively, calculated by the method of residues. α, β and K_a are rate hybrid microconstants corresponding to distribution rate, elimination rate and absorption rate respectively. The area under the curve observed up to 12 h (AUC_{0-12}) was calculated by the trapezoidal method and was expressed as a value corrected for body weight in order to minimize interindividual variation. The apparent volume of distribution (V_d) was calculated from the equation

$$V_d = DF/\mathrm{AUC}_{0-12}$$

where D is the administered dose, and F the bioavailability coefficient. Lag time is the time for the appearance of ketoprofen in blood.[85]

Evidence suggests that chronokinetic phenomena are not restricted to the parameters of a single-compartment model but to the model itself (*i.e.* a mono- *versus* a multi-compartment model as a function of the drug dosing time). For instance, data reported on β-methyldigoxin, a cardiac glycoside,[27] and mequitazine, an anti-H₁ antihistamine,[28] revealed that the shape of the

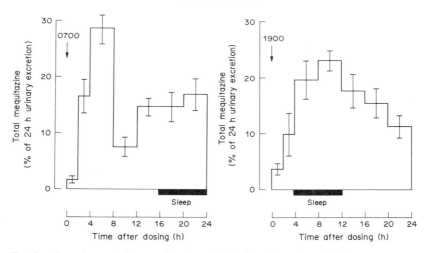

Figure 3 Circadian time dependency of pharmacokinetic models. A 5 mg dose of mequitazine was given orally to six healthy subjects, at 0700 and 1900 (as shown by arrows), one week apart. The subjects were synchronized with diurnal activity (from 0700 to 2300) alternating with nocturnal rest. Determinations were made by mass spectrometry in total urine collected at first every 2 h (twice) and thereafter every 4 h for 24 h. The results are given as the total mequitazine excreted per urinary collection period, expressed relative to the entire 24 h excretion[28]

concentration/time curve varied dramatically according to the clock hour of dosing (Figure 3). A statistical study confirmed the impression that the chronokinetics of mequitazine fits with a one-compartment model (kinetics characterized by one peak) when dosed in the evening at 1900 and a two-compartment model (kinetics characterized by two peaks) when dosed in the morning at 0700. At the present time, there is no explanation for the time dependent variation in the suitability of the two pharmacokinetic models. It could well be that circadian changes in the enterohepatic circulation play an important role, but this has yet to be demonstrated. Nonetheless, this aspect represents a new consideration in the updated definition of chronokinetics.

23.12.5 SPECIFIC METHODOLOGICAL REQUIREMENTS FOR CHRONOKINETIC STUDIES

23.12.5.1 Synchronization of the Subjects

The synchronization of the subjects must be controlled; this refers to the light/dark cycle in laboratory animal studies, and the activity/rest cycle in human studies. When reference is made to local time within this chapter, it was verified from the original publication that the clinical chronokinetics of a specified drug was evaluated in subjects synchronized to diurnal activity (usually from 0700 to 2300) and nocturnal rest.

23.12.5.2 Several Time Points are Necessary

Several (more than two) time points (dosing times) are necessary to define the chronokinetics of a drug. However, if only two dosing times can be studied, they must be selected for pertinent reasons (*e.g.* they must be close to both the expected crest and trough times of a marker rhythm) rather than by chance or for the convenience of the investigator. To determine which two times are the most pertinent, a small preliminary experiment dealing with four to six time points in the 24 h scale is recommended. For example, using such an approach, chronokinetic changes have been demonstrated in humans for exogenously administered cortisol[29] ($t_{1/2}$ was shorter when the dosing was at 1600 rather than at 0800) as well as for prednisone in rats[30] and humans.[31]

With only two dosing times, 0800 and 2000, selected by chance, it was not possible to discern a difference in prednisolone chronokinetics in humans.[32] This latter study is of no use from a chronopharmacological point of view. Therefore, reference to studies with two time points herein is restricted to investigations in which dosing times were selected for pertinent reasons based on experimental data.

Obviously all conventional, and thus expected, information about the daily dose and/or dosing time (or times), whether the administration is acute or chronic, and the route of administration must be given.[33] Also the time of year is relevant since seasonal modulations of chronopharmacological phenomena have been reported. [2,12,18,23,24,26] Timing of meals and changes in posture have to be taken into account from both a methodological and theoretical point of view; they are thus considered in subsequent sections. Since qualitative and quantitative aspects of chronokinetic phenomena can vary with the age, sex and health status of participating subjects, this information must be stated.[21,22]

The occurrence and magnitude of significant variations in drug kinetics, dependent on dosing time, can vary between formulations of a given chemotherapeutic agent (*e.g.* theophylline). Thus, it is necessary to stipulate 'new' properties of the agents studied for each given formulation and given dosing time.

With regard to the above points, caution must be exercised in interpreting findings so as not to overgeneralize from the results of a limited number of studies.

23.12.6 ROLE OF FOOD INTAKE AND ITS TIMING

Feeding and drinking behavior in rodents[34] and the squirrel monkey[35] appears to be driven by circadian oscillators. Therefore, spontaneous food intake and its circadian changes are part of the temporal structure of living beings. On the other hand, the manipulation of meal timings, for example by restricting access to food to a few hours during each day, is able to shift the phase of certain circadian rhythms in certain animal species.[36,37] For example, in humans, a large breakfast *versus* a large dinner may shift the crest time of plasma insulin and iron, but not of many other documented variables. To be specific, meal timing appears to have either a weak synchronizing effect or a masking effect[38] on certain circadianly organized variables. With regard to food intake and metabolic considerations both circadian[39] and circannual[36] changes in the blood sugar tolerance test have been reported in humans. This means that the concept of chronokinetics can be extended to nutrients in addition to xenobiotic agents. Let us emphasize that meal timing does not create circadian rhythms.

With regard to chronokinetics and the oral route of drug administration, it is critical to determine what role can be attributed to the presence (or absence) of nutrients in the digestive tract. In this regard, Goo *et al.*[40] studied circadian variations in the gastric emptying of meals in human subjects. Radionuclide-based studies of gastric emptying half-times were performed at 0800 and 2000 in healthy male volunteers. The duration of gastric emptying was greater for the evening meal than the morning one with regard to solids but not liquids. The increased gastric emptying time nocturnally may account for the delay in t_{max} following evening dosing of several drugs such as the nonsteroidal antiinflammatory drug (NSAID) indomethacin and the bronchodilator drug theophylline.

Changes in the pharmacokinetics of indomethacin dependent on time of dosing (acute administration, 100 mg by mouth, with a different dosing time evaluated each week over five weeks in a randomized order) have been substantiated in health young volunteers,[41] as shown in Figure 4. In Figure 4 nine healthy subjects (age range 19–29, seven males, two females) were synchronized with diurnal activity from 0700 to 0000, and, in a random order and at weekly intervals, they received a single oral dose (100 mg) of indomethacin at one of five fixed times (0700, 1100, 1500, 1900 and 2300). The subjects fasted for at least 4 h prior to each dosing time. Venous blood (sampled at 0, 0.33, 0.67, 1, 1.5, 2, 4, 6, 8 and 10 h after ingestion) was obtained for measuring plasma drug levels. Ingestion at 1900 and/or 2300 led to the smallest and latest peak, while ingestion at 0700 and/or 1100 led to the largest and quickest peak, greatest AUC and quickest disappearance rate. A circadian rhythm of both peak height and time-to-peak was detected.

Food intake neither creates nor explains these chronokinetic changes since the subjects were fasted for at least 4 h prior to each dosing time. Similar results were obtained for the same NSAID by Belanger *et al.*[42] in fasted rats, with a reduced plasma t_{max} and an increased C_{max} resulting from a dosing time corresponding to the beginning of the activity span as compared to the end of the activity span. That these findings are independent of meal timing and meal contents, even for the indomethacin sustained release (ISR) preparation, is shown by two complementary studies by Guissou *et al.*[43] and Bruguerolle *et al.*[44] In both ISR studies the chronokinetics was investigated at 0800, 1200 and 2000. In both t_{max} was again reduced, C_{max} increased and AUC reduced when dosing occurred in the morning (0800) rather than in the evening (2000). Similar concentration/time curves were produced, despite the fact that in one study[43] dinner was served at 1800 (2 h prior to ISR dosing at 2000), while in the other[44] dinner was served at 2000.

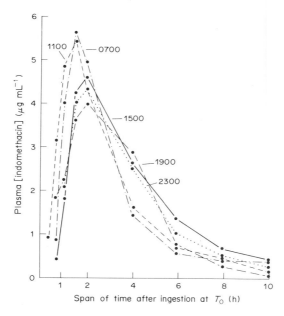

Figure 4 Clinical chronokinetics of indomethacin[41]

The relation between food intake and the chronokinetics of theophylline is more complex than that of ISR due to the fact that several different galenic forms of sustained release (SRT) preparations are involved. Let us state first that changes in plasma bioavailability of theophylline, dependent on dosing time, result mainly from differences in absorption *via* the digestive tract rather than from circadian rhythmic variations in the renal clearance of theophylline. It is clear from the review of Smolensky *et al.*[21] that large chronokinetic changes of SRT occur when the oral route is used, especially in children, while only minor chronokinetic changes are observed when theophylline is injected intravenously at constant rate.[21,46]

The release of theophylline from SRT can be influenced (either increased or decreased) by concomitant intake of food, depending upon the galenic preparation and the content and volume of the meal. Ultraslow release products are most vulnerable to food effects.[21,22,45,46] With some preparations, the composition of the meal, especially its fat content, determines the extent of the food effect.[21,45] However, all of the 12 consulted theophylline studies in which the effect of food intake has been assessed were performed with a morning dosing (reviewed by Smolensky *et al.*[21]). Even though a food effect has been observed for some SRT, dosing time related effects are still demonstrable, and this was especially marked for one particular galenic formulation.[46]

A circadian rhythm in the absorption of the anticonvulsant carbamazepine has been demonstrated by Valli *et al.*[47] in both fasted and fed rats. However, the circadian crest time (ϕ) of the absorption index differed by about 4 h between fed and fasted animals. In human subjects, Nakano *et al.*[48] documented large changes, dependent on dosing time, in plasma diazepam kinetics; and only minute differences in drug kinetics were observed between studies involving fasting and fed participants. Again in human subjects, Kabasakalian *et al.*[49] reported a large amplitude circadian rhythmicity in the kinetics of the antibiotic griseofulvin; a high-fat meal taken as either breakfast or dinner induced only minute changes in the kinetic behavior of the drug.

Based on the findings of all the studies thus far published, it can be concluded that meal timing and food intake: (i) neither create circadian rhythms in biological functions nor explain chronokinetic phenomena; (ii) may influence kinetic parameters of some drugs at certain dosing times (*e.g.* advance or delay t_{max} and/or increase or decrease C_{max}, perhaps reflecting day/night differences in gastric emptying and thus drug transit time); and (iii) have no or only minor effects on those changes in drug kinetics which are dependent on dosing time.

23.12.7 CIRCADIAN CHANGES OF DRUG ABSORPTION

In addition to periodic changes in gastric emptying[40] a circadian rhythm in intestinal absorption has been demonstrated for xylose[50] (as well as for its urinary excretion), a phenomenon which is presumably valid for many substances. However, drug solubility has to be taken into account as demonstrated by Bellanger *et al.*[51] In fasted rats, circadian variations in drug absorption occurred in substances with a low water solubility, such as indomethacin, furosemide and phenylbutazone, but not in those having a high water solubility, such as antipyrine, hydrochlorothiazide and paracetamol.

Kinetic parameters characterizing the absorption of the benzodiazepine lorazepam exhibit large differences between morning and evening dosing in humans.[52] The absorption $t_{1/2}$ was three times longer with evening rather than morning dosing in all study participants. This difference is related neither to food intake (a 7 h fast prior to dosing time was imposed) nor to posture (subjects were recumbent in bed). The results of Bruguerolle *et al.*[52] have been confirmed with the benzodiazepine triazolam for which the absorption $t_{1/2}$ in human subjects was found to be twice as long when dosing occurred at 2200 rather than at 0700.[53]

From our perspective, factors other than meal timing and content are responsible for the variations in kinetics which are dependent on the time of dosing. Endogenous circadian rhythms such as those in gastrointestinal pH, intestinal motility, digestive secretions and intestinal blood flow, among others, must be taken into account at least to explain the mechanisms of drug chronokinetics.

23.12.8   CIRCADIAN CHANGES IN DRUG DISTRIBUTION

Circadian rhythms in membrane permeability have been cited for many years as a candidate mechanism for the governing of biological clocks.[54-56] However, only a few studies have been devoted to circadian changes in membrane permeability to drugs. Taking the human red blood cell as a membrane model, Bruguerolle[57] performed around the clock *in vitro* permeability studies of lidocaine, a local anesthetic. The lidocaine content of erythrocytes sampled at 2200 was 74% of the plasma concentration, while it was only 48% when sampled at 1000.

The blood–brain barrier of rats exhibited rather large temporal variations in its permeability to a macromolecule, horse radish peroxidase.[58] Valpronate concentration in the cerebrospinal fluid of monkeys underwent diurnal changes parallel to those of the free fraction of the drug in the plasma. In human beings, circadian changes in the permeability of blood–brain barrier have not yet been documented.

23.12.9   CIRCADIAN CHANGES IN PLASMA PROTEIN BINDING OF DRUGS

In human beings most plasma proteins, such as albumin and acid α-1-glycoprotein, which are rather highly involved in drug transport, exhibit circadian rhythmicity.[59] Rhythms in these plasma constituents are independent of that of the plasma hematocrit and presumably reflect the 24 h pattern of protein production by the liver, rather than that of body fluids.[60] For example, the circadian rhythm of serum corticosteroid binding globulin is dramatically altered in patients suffering from alcohol-induced liver disease.[61] Although the difference between the diurnal crest and the nocturnal trough in protein concentration amounts to only 10% ($\pm 5\%$ of the 24 h mean) in young adults, it may be as great as 20% in elderly (~ 70 years) subjects.[60] In this group drug consumption is especially prevalent. The age-related increase of the circadian amplitude of this rhythm results from a large nocturnal dip in plasma proteins between 0000 to 0400, rather than increase in concentration at the diurnal crest.[60] It is anticipated, but as yet not demonstrated, that the free fraction of certain drugs at night may be larger in elder than in younger adult subjects.

In the rat, Bruguerolle *et al.*[20,24,52,57,62] have shown that the free (active) fraction of several medications is dependent on the time of dosing; the greatest protein binding for disopyramide (a heart antiarrhythmic),[62] lidocaine[57] and carbamazepine (an anticonvulsant)[19] occurs during the active span. The unbound fraction of these drugs varied from between 10 to 28% as a function of the different dosing times over the 24 h scale due to the circadian rhythm of plasma-binding proteins.

In humans, Angeli *et al.*[61,63] have reported that the binding of plasma transcortin to corticosteroids, as well as to natural adrenocortical hormones, exhibits circadian rhythmicity. Lowest binding capacity occurs during the night around 0400 when the cortisol secretion is physiologically nil or quite small; the highest capacity occurs around 1600. This means that higher free, and thus effective, plasma corticosteroid (both natural or synthesized) levels are achieved during the morning, the time of the peak cortisol secretion, and the early afternoon.

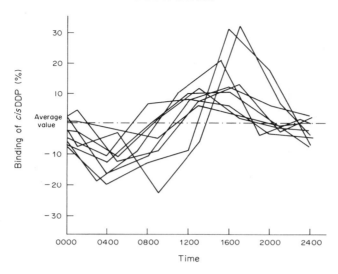

Figure 5 Circadian changes of human plasma protein binding for the anticancer drug *cis*-diamminedichloroplatinum (*cis*DDP) demonstrated *in vitro*[88]

Again in humans, circadian changes in plasma protein binding have been demonstrated for three other drugs: diazepam,[64] valproic acid[65] and carbamazepine.[66] For all these drugs the free fractions reach their respective crests in the early morning. In addition, Nakano *et al.*[48] have shown that the evening increase in the free fraction of diazepam parallels that of its clearance, while Narajano *et al.*[64] reported that diurnal changes of diazepam protein binding may be related to meal-induced changes in serum free fatty acids.

Circadian changes in the binding of the anticancer drug *cis*-diamminedichloroplatinum (*cis*DDP) have been demonstrated *in vitro* by Hecquet *et al.*,[8] as shown in Figure 5. The rates of *in vitro* *cis*DDP binding to plasma proteins were calculated every 4 h for 24 h in plasma from nine cancer patients. Protein binding was greater in plasma collected near 1600 than in that collected near 0400 (*p* from cosinor analysis < 0.001). These results were confirmed in nine additional cancer patients and were not affected by prehydration (2 L in 6 h). As a result free *cis*DDP plasma concentration may be greatest after dosing in the early morning.

23.12.10 CIRCADIAN CHANGES IN LIVER DRUG METABOLISM

The liver is a key organ in regard to drug chronokinetics. In 1968 Radzialowski and Bousquet[15] documented circadian rhythmicity in hepatic drug-metabolizing enzymes. In rats and mice of either sex, the enzymatic activity of aminopyrine *N*-demethylase, 4-dimethylaminobenzene reductase, and *p*-nitroanisole *O*-demethylase was greatest during the second half of the animals' activity span and lowest during the second half of the rest span. Some of these circadian rhythms appear to be dependent upon adrenocortical activity, since they are altered in adrenalectomized animals and thereafter restored in the latter by the administration of corticosteroids. Alteration of these enzyme rhythms also has been observed in rats pretreated for 4 d with phenobarbital, although the circadian rhythm of plasma corticosterone was not affected. Jori *et al.*[67] have shown in rats that circadian changes in the hepatic microsomal activity which controls the biotransformation of aminopyrine, hexobarbital, imipramine and *p*-nitroanisole is synchronized by the 12 h light/12 h dark cycle. In all cases the crest time (ϕ) of these microsomal oxidative enzymatic activities occurred during the course of the activity span. A 12 h shift of the L/D schedule (which is thereafter D/L) was followed, after several days, by a 12 h phase shift of the circadian ϕ of the enzymatic activities.

Liver conjugation represents a common metabolic pathway for a large variety of xenobiotics; Belanger *et al.*[68] demonstrated time dependent changes in the transferases and hydrolases involved in hepatic glucuronide and sulfate conjugation. In free-fed rats, V_{max} and K_m of sulfotransferases were respectively two and four times greater during rest than during the activity span. However, glucuronoconjugation, which is catalyzed by UDP glucuronosyl transferase, takes place more rapidly at the beginning of the activity span than during rest. Fasting does not alter the latter rhythm with regard to V_{max}.

Findings of circadian rhythmicity in hepatic enzyme activities agree well with observed chrono-kinetic variations in certain drugs. Antipyrine is commonly used by clinical pharmacologists to estimate the capacity of hepatic biotransformations in an individual through microsomal oxida-tions. In the rat, antipyrine clearance is greatest at the beginning of the activity span and is in phase, over 24 h, with the peak in microsomal oxidase activity.[16] Acetaminophen is metabolized mainly though the hepatic sulfoconjugation pathway. In the rat, the $t_{1/2}$ for acetaminophen is greater at the beginning of the activity span, when sulfotransferase activity is lowest. Dosing during the rest span is associated with a shorter $t_{1/2}$ and higher enzyme activity. Pertinent also is the report by Nair[69] on hexobarbital. The circadian rhythm of hexobarbital-induced sleep duration differs by 12 h from the rhythm in hepatic hexobarbital oxidase activity. Thus the end of the activity span corresponds in time to both the crest in induced sleep duration (and presumably in drug bioavailability) and the trough in activity of the hepatic hexobarbital oxidase.

The rhythm in liver perfusion, which results from circadian variations in hepatic blood flow, is also an important factor to be taken into account. This fact has been demonstrated by Lemmer *et al.*[26] for the changes in propranolol clearance related to the time of dosing. The clearance of this β-blocker is greater when injected in the rat at the end of the activity span than at the end of the rest span. The clearance of this drug depends predominantly upon the liver blood flow, which has been shown[16] to be greater at the end of the activity span than at any of the other three times which were tested.

23.12.11   CIRCADIAN CHANGES IN KIDNEY DRUG EXCRETION

Circadian rhythmicity in major renal functions, *e.g.* glomerular filtration, renal blood flow, urinary pH, volume and solutes, and either proximal or distal tubular reabsorption has been documented by Cambar and coworkers.[70,71] One or more of these rhythmic variations have been implicated in the differences, dependent on the time of dosing, in the urinary excretion of certain drugs and in the resulting chronokinetics.

As a first example, the acrophase of the urinary pH in human subjects occurs in the morning around the time of awakening, while its trough occurs between the beginning and middle of the rest span. Acidic drugs have therefore a faster excretion following an evening than a morning dosing. This is the case for sodium salicylate[72] and sulfasymazine (a sulfamide), but not for sulfanilamide, since the urinary excretion of the latter is not pH dependent.[73]

As a result of circadian variations in blood pressure and cardiac blood flow, circadian changes in renal flow ($\sim$ 20% of the cardiac flow) are not unexpected. In humans glomerular filtration is at its highest during diurnal activity and lowest during nocturnal sleep. However, the circadian amplitude of these rhythms is rather small, amounting to no more than $\pm 15\%$ of mesor values.

In addition, circadian changes in renal function are influenced by several hormones, *e.g.* cortisol, aldosterone (as well as renin and angiotensin), catecholamines and antidiuretic hormone, which also exhibit large amplitude 24 h rhythms in secretion and bioavailability. Thus, it is not surprising that the effectiveness of several types of diuretics, such as hydrochlorothiazide and furosemide, and of β-blockers is dependent on the time of dosing.[70,74,75]

The anticancer drug *cis*DDP provides a good illustrative example of the relationship between the chronokinetics of a drug and the chronotolerance of biosystems to it.[76,77] Circadian rhythms in urinary excretion, nephrotoxicity and antitumor effectiveness of *cis*DDP have been demonstrated, as shown in Figure 6. Eleven patients (seven men, four women) were given 60 mg (m^{-2} body surface area) of *cis*DDP over 30 min. Each patient received a dose of *cis*DDP (preceded 12 h earlier by adriamycin) either at 0600 or at 1800. The subsequent treatment was given four weeks later at the alternate circadian stage (crossover design). The urinary pharmacokinetics of free *cis*DDP was studied on two to six occasions in each patient (an overall total of 40 courses). Statistically significant differences were validated as a function of *cis*DDP timing for both peak urinary concentration and area under concentration–time curve. The urinary concentration showed both its greatest peak (C_{max}) and largest AUC when *cis*DDP was given to diurnally active patients at 0600. Lowest nephrotoxicity, as gauged by creatinine clearance, occurred when *cis*DDP was given at 1800, when C_{max} was lowest and AUC smallest.

In some studies the circadian rhythm of a tubular brush border lysozymal enzyme *N*-acetyl-glucose aminidase has been used as a marker to monitor the dependence on the time of dosing of the nephrotoxicity of *cis*DDP. The best tolerance of patients to this agent occurs at 1800, which corresponds also to its highest effective antitumor activity.[70,77]

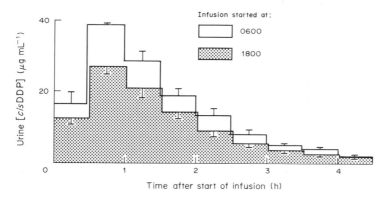

Figure 6 Drug timing and urinary concentration of *cis*DDP. Drug given as 30 min infusion[77]

23.12.12 NEW ASPECTS OF CHRONOKINETICS

Apart from basic definitions and concepts which have been derived from the study of rhythmic and thus predictable changes in the pharmacokinetics of a drug, dependent on the time of dosing, new aspects also have to be considered. They emerge from both practical and theoretical analyses of new experimental facts.

23.12.12.1 A Constant Drug Delivery Rate Does Not Result in a Constant Plasma Concentration

The delivery of some drugs at a constant rate does not result in a constant plasma level over the 24 h span. This has been shown for ketoprofen (Ket; a NSAID)[78] and two anticancer agents, 5-fluorouracil (5-FU)[79] and adriamycin (Adr).[80] Highly predictable large amplitude circadian changes in plasma concentrations result from the venous infusion of either Ket or 5-Fu at a constant rate.

In the case of Ket, five male subjects suffering from sciatica due to herniated disc were evaluated,[78] as shown in Figure 7. All subjects received Ket intravenously at the same dose (5 mg kg^{-1} d^{-1}) at a constant rate [0.2 mg h^{-1} (2 mL saline)$^{-1}$]. Venous blood samples were obtained every 2 h, and total urine voidings were collected every 4 h, over 24 h. The plasma concentration of Ket was

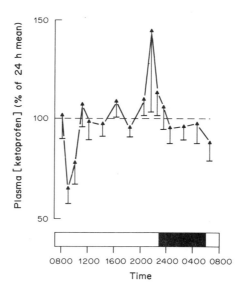

Figure 7 The continuous intravenous infusion of ketoprofen does not lead to a constant plasma concentration. Changes in concentration are expressed as an averaged percentage of each individual 24 h mean. Vertical bars represent SEM. ■ sleep, □ wakefulness[78]

not maintained constant but varied predictably with an evening peak at 2100 (3.95 ± 0.5, $\bar{X}$ ± SEM mg L^{-1}) and a morning trough at 0900 (1.99 ± 0.5 mg L^{-1} with $p < 0.01$).

Similarly, the mean urinary excretion of the drug followed a 24 h periodicity with a morning peak (17.5 ± 3.8 mg between 0800 and 1200) and two troughs, one in the afternoon (9.6 ± 1.5 mg between 1200 and 1600) and the other in the evening (10.4 ± 1.9 mg between 2000 and 2400). These differences were also statistically significant ($p < 0.01$).

The large amplitude circadian variation in the plasma concentration of Ket is presumably related, at least partially, to circadian changes in the rapid kidney clearance rate of this agent.[85] In fact, theophylline, which has a slow clearance rate, exhibits no circadian change in plasma levels when intravenously infused at a constant rate to children as well as adults.[46,81]

In a study performed by Petit *et al.*[79] on seven patients with stage C bladder carcinoma, a continuous venous infusion of 5-Fu (450–966 mg $m^{-2} d^{-1}$) was given *via* a volumetric pump for 6 d Blood samples were obtained every 3 h on days 2, 4 and 6. Regular and predictable changes in the plasma concentration of the drug were observed with a peak at 0100 (584 ± 160 ng mL^{-1}) and a trough at 1300 (254 ± 33 ng ml^{-1} with $F = 2.3$; $p < 0.03$). Other statistical analyses (ANOVA, cosinor, *etc.*) validated the existence of a large amplitude (± 25% of the 24 h mean) circadian rhythm in plasma 5-Fu. This drug also has a rapid kidney clearance rate.

23.12.12.2 Predictable-in-time *versus* Chaotic Changes in Plasma Concentration of Drugs

The constant rate delivery of Adr *via* the arterial route, chronically by an implanted programmable pump (Medtronic), to 13 patients suffering from advanced breast cancer[80,82] resulted in large unpredictable temporal variations in plasma concentration. In contrast, when for the same patients and the same 24 h dosage the infusion rate was varied in a sinusoidal fashion, with both peak and trough specifically programmed in the 24 h scale, the resulting temporal changes in plasma Adr concentration followed, as expected, the programmed pump delivery pattern. This finding is most important since, in order to ensure the best tolerance of this agent by cancer patients, the peak concentration of Adr must occur in the early morning around 0600.[80,86] Preliminary results from 13 evaluable patients indicate the feasibility of this approach and suggest that an improved tolerance of Adr by patients can be achieved by a circadian modulation of the infusion rate of Adr so that more drug is given around 0600 and much less at 1800.

The constant rate delivery of Adr results in unpredictable changes in its plasma concentrations. It could well be that such a dosing pattern induces circadian changes in its own metabolism leading to a 'chaotic' situation. From a physical point of view, a chaotic situation may result from the uncoupling of certain circadian organized oscillatory systems which are usually coupled.[83] The resulting chaos remains deterministic despite the fact that observed changes in C_{max} and C_{min} are likely to be randomly distributed with respect to their occurrence over 24 h. The understanding of chronokinetic changes obviously deserves to be analyzed using different models.[25] However, experimental evidence from chronokinetic studies of Ket, 5-Fu and Adr indicates that the constant rate infusion of these, and presumably other, drugs does not result in the expected constant plasma level.

Table 1 Available Programmable Drug Administration Systems for Cancer Chemotherapy (1986)

Manufacturer	Main device characteristics	Programmability	Present use for cancer chronotherapy
Medtronic Inc. (Minneapolis, USA)	Implantable; 20 mL reservoir; lithium battery (>1 year); monotherapy	Steps of various rates and time lengths; computer; telemetry	Adriamycin FUDR Methotrexate
Autosyringe Baxter-Travenol (Hooksett, USA)	External; ambulatory; 1–60 mL syringe; rechargeable batteries (>4 d); monotherapy	Sine curve (τ, M, A, ϕ); Keypad on pump	5-Fluorouracil
Intelligent Medicine (Denver, USA)	External; ambulatory; four 30 mL syringes independently programmable; polychemotherapy	Any delivery time curve; chip programmed by computer	5-Fluorouracil Theprubicin *cis*DDP

23.12.12.3 Programmable Delivery Systems

From our perspective, the optimization of treatment for certain diseases requires infusions to be timed specifically to certain stages of a biological rhythm in order to achieve the desired chrono-therapy. Based on the manner in which clinics and hospitals are organized, it was quickly appreciated by both chronopharmacologists and other health professionals that chronotherapeutics could be practical only when reliable and reasonably priced programmable drug delivery devices became available.[84] During the past few years, exciting new developments have occurred, enabling drug delivery by bedside pumps and ambulatory pumps. As of this date, we are aware of four different companies which are vigorously pursuing the design, manufacture and marketing of programmable infusion pump devices. Particulars of these drug delivery systems are summarized in Table 1. The choice of the particular pumps for discussion herein was made on the basis of information available to the authors when this chapter was written.

23.12.13 CHRONOTHERAPY: THERAPEUTIC IMPLICATIONS OF TIME DEPENDENCIES

Both the desired and undesired effects of therapeutic agents vary as a function of the time of day at which they are given. These chronopharmacological phenomena are related to time dependencies of both the chronokinetics of medicines and circadian changes in the susceptibility of target systems, including binding sites (chronesthesy). A set of drugs has been the subject of chronotherapeutic studies with two major aims in mind. The first aim is to provide experimental evidence that an agent delivered at a constant rate in the 24 h span may cause large variations in effects. This is the case for heparin which, when infused at a constant rate, leads to dramatic circadian changes in three common indices of blood coagulation. The second aim, as a consequence of the first, is to program the drug delivery in time in order to enhance the desired effects and/or reduce the undesired ones. Such a chronoptimization resulting from qualified timing of drug delivery has already been used for a rather large set of agents and diseases: (i) corticosteroids and (ii) theophyllines in patients suffering from allergic asthma; (iii) the NSAID indomethacin in patients suffering from osteoarthritis; (iv) mequitazine, an anti-H_1 agent, in patients with allergic rhinitis; (v) local anesthetics in stomatology; and (vi) anticancer agents (*cis*DDP, adriamycin) in patients with advanced cancers of either the ovary or bladder. Therefore, the rather old question 'when to treat? in the 24 h (and in the annual) scale begins to be given pertinent answers based on both experimental and clinical chronopharma-cologic data resulting fom studies designed according to a chronobiologic methodology.[89]

23.12.14 REFERENCES

1. A. Reinberg and F. Halberg, *Annu. Rev. Pharmacol.*, 1971, **11**, 455.
2. A. Reinberg and M. H. Smolensky, *Clin. Pharmacokinet.*, 1982, **7**, 401
3. A. Reinberg, M. H. Smolensky and G. Labrecque, *Annu. Rev. Chronopharmacol.*, 1986, **2**, 3.
4. A. Reinberg and M. H. Smolensky (eds.), 'Biological Rhythms and Medicine', Springer-Verlag, New York, 1983.
5. F. Halberg, F. Carandente, G. Cornelissen and G. S. Katinas, *Chronobiologia (Milan)*, 1977, **4** (suppl. 1), 1.
6. J. De Prins, G. Cornelissen and W. Malbecq, *Annu. Rev. Chronopharmacol.*, 1986, **2**, 27.
7. T. A. Bargiello and M. W. Young, *Proc. Natl. Acad. Sci. U.S.A.*, 1984, **81**, 2142.
8. A. Reinberg, Y. Touitou, A. Restoin, C. Migraine, F. Lévi and H. Montagner, *J. Endocrinol.*, 1985, **105**, 247.
9. R. Wever, 'The Circadian System of Man', Springer-Verlag, New York, 1979.
10. F. W. Turek, *BioScience*, 1983, **33**, 439.
11. A. Reinberg, Y. Motohashi, P. Bourdeleau, P. Andlauer, F. Lévi and A. Bicakova-Rocher, *Eur. J. Appl. Physiol. Occup. Physiol.*, 1988, **57**, 15.
12. H. von Mayersbach, in 'Biological Rhythms and Medicine', ed. A. Reinberg and M. H. Smolensky, Springer-Verlag, New York, 1983, p. 47.
13. C. P. Barnum, C. D. Jardetzky and F. Halberg, *Tex. Rep. Biol. Med.*, 1957, **15**, 134.
14. L. E. Scheving, J. E. Pauly, T. H. Tsai and A. Scheving, in 'Biological Rhythms and Medicine', ed. A. Reinberg and M. H. Smolensky, Springer-Verlag, New York, 1983, p. 79.
15. F. M. Radzialowski and W. F. Bousquet, *J. Pharmacol. Exp. Ther.*, 1968, **163**, 229.
16. G. Labrecque and P. Belanger, *Pathol. Biol.*, 1987, **35**, 917.
17. A. Reinberg, E. Sidi and J. Ghata, *J. Allergy Clin. Immunol.*, 1965, **36**, 273.
18. A.Wirz-Justice *Prog. Neurobiol. (Oxford)*, 1987, **29**, 219.
19. B. Lemmer, H. Bärmeier, S. Schmidt and P. H. Lang, *Chronobiol. Int.*, 1987, **4**, 469.
20. B. Bruguerolle, *Pathol. Biol.*, 1987, **35**, 925.
21. M. H. Smolensky, P. H. Scott, P. J. Barnes and J. H. G. Jonkman, *Annu. Rev. Chronopharmacol.*, 1986, **2**, 229.
22. M. H. Smolensky, G. E. D'Alonzo, G. Kunkel and P. J. Barnes, *Chronobiol. Int.*, 1987, **4**, 301.
23. B. Lemmer, in 'Topics in Pharmaceutical Sciences', ed. D. Breiner and P. Speiser, Elsevier, Amsterdam, 1981, p. 49.

24. B. Bruguerolle, in 'Chronopharmacologie', Ellipse Marketing, Paris, 1984.
25. B. Hecquet, *Pathol. Biol.*, 1987, **35**, 937.
26. B. Lemmer and K. Bathe, *J. Cardiovasc. Pharmacol.*, 1982, **4**, 635.
27. L. Carosella, D. Nardo, R. Barnabei, A. Cocchi and P. Carbonin, in 'Chronopharmacology', ed. A. Reinberg and F. Halberg, Pergamon Press, Oxford, 1979, p. 125.
28. A. Reinberg, F. Lévi, J. P. Fourtillan, C. Peiffer, A. Bicakova-Rocher and A. Nicolai, *Annu. Rev. Chronopharmacol.*, 1984, **1**, 57.
29. P. L. Morselli, V. Marc, S. Garattini and M. Zaccala, *Biochem. Pharmacol.*, 1970, **19**, 1643.
30. J. English and V. Marks, *IRCS Med. Sci.: Libr. Compend.*, 1981, **9**, 721.
31. J. English, L. I. Biol, M. Dunne and V. Marks, *Clin. Pharmacol. Ther.*, 1983, **33**, 381.
32. W. A. C. McAllister, D. M. Mitchell and J. V. Collins, *Br. J. Clin. Pharmacol.*, 1983, **33**, 381.
33. A. Reinberg, *Pathol. Biol.*, 1987, **35**, 909.
34. W. J. Rietveld, W. Hekkens and G. A. Groos, *Annu. Rev. Chronopharmacol.*, 1987, **3**, 29.
35. M. C. Moore-Ede, F. M. Sulzman and C. A. Fuller, in 'The Clocks that Time Us', Harvard University Press, Cambridge, 1982.
36. A. Reinberg, in 'Chronobiology and Nutrition in Biological Rhythms and Medicine', ed. A. Reinberg and M. H. Smolensky, Springer-Verlag, New York, 1983, p. 265.
37. A. Reinberg, F. Lévi and G. Debry, *Encycl. Med. Chir. Paris, Glandes Endocr. Nutrition*, 1984, **10 390 A 10**, 1.
38. D. S. Minors and J. M. Waterhouse, *Chronobiol. Int.*, 1984, **1**, 173.
39. R. J. Jarrett and H. Keen, *Br. Med. J.*, 1970, **4**, 334.
40. J. G. Moore, R. H. Goo, E. Greenberg and N. P. Alazaraki, *Chronobiologia*, 1987, **14**, 212.
41. J. Clench, A. Reinberg, Z. Dziewanoska, J. Ghata and M. H. Smolensky, *Eur. J. Clin. Pharmacol.*, 1981, **20**, 359.
42. P. M. Belanger, G. Labrecque and F. Doré, *Tribune Médicale*, 1982, **Suppl. 1**, 14.
43. P. Guissou, G. Cuisinaud, G. Llorca, E. Lejeune and J. Sassard, *Eur. J. Clin. Pharmacol.*, 1983, **24**, 667.
44. B. Bruguerolle, C. Desnuelle, D. Jadot, M. Valli and P. C. Acquaviva, *Rev. Int. Rhumatisme*, 1983, **13**, 263.
45. J. H. G. Jonkman, *Chronobiol. Int.*, 1987, **4**, 449.
46. M. H. Smolensky, P. H. Scott, R. B. Harrist, P. H. Hiatt, T. K. Wong, J. C. Baenziger, B. J. Klanck, A. Marbella and A. Meltzer, *Chronobiol. Int.*, 1987, **4**, 435.
47. M. Valli, B. Bruguerolle, L. Bouyard, G. Jadot and P. Bouyard, *J. Pharmacol.*, 1980, **11**, 201.
48. S. Nakano, H. Watanabe, K. Nagai and N. Ogawa, *Clin. Pharmacol. Ther.*, 1984, **36**, 271.
49. P. Kabasakalian, M. Katz, B. Rosenkrantz and E. Townley, *J. Pharm. Sci.*, 1970, **59**, 595.
50. A. Markiewicz and K. Semenowicz, *Chronobiologia (Milan)*, 1979, **6**, 129.
51. P. Belanger, G. Labrecque and F. Doré, *Int. J. Chronobiol.*, 1981, **7**, 208.
52. B. Bruguerolle, G. Bouvenot and R. Bartolin, *Annu. Rev. Chronopharmacol.*, 1984, **1**, 21.
53. R. B. Smith, P. D. Kroboth and J. P. Phillips, *J. Clin. Pharmacol.*, 1986, **26**, 120.
54. J. W. Hastings and H. G. Schweiger, in 'The Molecular Basis of Circadian Rhythms', Dahlem Konferezen, Berlin, 1976.
55. S. Njus, F. M. Sulzman and J. W. Hastings, *Nature (London)*, 1974, **248**, 116.
56. L. N. Edmunds, in 'Cell Cycle Clocks', Dekker, New York, 1984.
57. B. Bruguerolle and G. Jadot, *Chronobiologia (Milan)*, 1983, **10**, 295.
58. M. Mato, S. Ookawara, K. Tooyama and T. Ishizaki, *Experientia*, 1981, **37**, 1013.
59. B. Bruguerolle, F. Lévi, C. Arnaud, G. Bouvenot, M. Mechkouri, J. M. Vannetzel and Y. Touitou, *Annu. Rev. Chronopharmacol.*, 1986, **3**, 207.
60. Y. Touitou, C. Touitou, A. Bogdan, A. Reinberg, A. Auzeby, H. Beck and P. Guillet, *Clin. Chem. (Winston-Salem, N.C.)*, 1986, **32**, 801.
61. A. Angeli, F. Agrimonti, R. Frajria, P. L. Vioino, E. Barbadoro and F. Ceresa, *Int. J. Chronobiol.*, 1981, **7**, 199.
62. B. Bruguerolle, *Chronobiol. Int.*, 1984, **1**, 267.
63. A. Angeli, R. Frajria, R. De Paoli, D. Fonzo and F. Ceresa, *Clin. Pharmacol. Ther.*, 1978, **23**, 47.
64. C. A. Naranjo, E. M. Sellers, H. G. Giles and J. G. Abel, *Br. J. Clin. Pharmacol.*, 1980, **9**, 265.
65. I. H. Patel, R. Venkataramanan, R. H. Levy, C. T. Viswanathan and L. M. Ojemann, *Epilepsia (N.Y.)*, 1982, **23**, 282.
66. R. Riva, F. Albani, G. Ambrosetto, M. Contin, P. Cortelli, E. Perucca and A. Baruzzi, *Epilepsia (N.Y.)*, 1984, **25**, 476.
67. A. Jori, E. Di Salle and V. Santini, *Biochem. Pharmacol.*, 1971, **20**, 2965.
68. P. M. Belanger, M. Lalande, G. Labrecque and F. Doré, *Drug Metab. Dispos.*, 1985, **13**, 386.
69. V. Nair, in 'Chronobiology', ed. L. E. Scheving, F. Halberg and J. E. Pauly, Igaku Shoin, Tokyo, 1974, p. 182.
70. J. Cambar, C. Dorian and J. C. Cal, *Pathol. Biol.*, 1987, **35**, 977.
71. J. C. Cal, C. Dorian and J. Cambar, *Annu. Rev. Chronopharmacol.*, 1985, **2**, 143.
72. A. Reinberg, Z. Zagula-Mally, J. Ghata and F. Halberg, *Proc. Soc. Exp. Biol. Med.*, 1967, **124**, 826.
73. L. Dettli and P. Spring, *Helv. Med. Acta*, 1966, **33**, 291.
74. B. Lemmer, in 'Topics in Pharmaceutical Sciences', ed. D. D. Breimer and P. Speiser, Elsevier, Amsterdam, 1981, p. 49.
75. B. A. Gould, S. Mann, H. Kieso, V. B. Subramanian and E. B. Raftery, *Circulation*, 1982, **65**, 22.
76. W. Hrushesky, R. Borch and F. Lévi, *Clin. Pharmacol. Ther.*, 1982, **32**, 330.
77. F. Lévi, W. Hrushesky, E. Haus, F. Halberg and B. J. Kennedy, *Eur. J. Cancer*, 1982, **18**, 471.
78. H. Decousus, M. Ollagnier, Y. Cherrah, B. Perpoint, J. Hocqart and P. Queneau, *Annu. Rev. Chronopharmacol.*, 1986, **3**, 321.
79. E. Petit, G. Milano, F. Lévi, A. Thyss, F. Bailleul and M. Schneider, in 'Proceedings of the 4th European Conference on Clinical Oncology and Cancer Nursing, Madrid, 1987', no. 293. p. 78.
80. F. Lévi, F. Bailleul, G. Metzger, J. L. Misset, J. M. Vannetzel, C. Regensberg, A. Reinberg and G. Mathé, *Annu. Rev. Chronopharmacol.*, 1986, **3**, 237.
81. M. H. Smolensky, J. P. McGovern, P. H. Scott and A. Reinberg, *J. Asthma*, 1987, **24**, 91.
82. F. Bailleul, F. Lévi, G. Metzger, M. Benavides, A. Reinberg and G. Mathé, in 'Proceedings of the 4th European Conference on Clinical Oncology and Cancer Nursing, Madrid, 1987', no. 470, p. 124.
83. M. Markus and B. Hess, in 'Chronobiology and Chronomedicine', ed. G. Hildebrandt, R. Moog and F. Rasche, Lang, Frankfurt, 1987, p. 85.

84. A. Reinberg, M. H. Smolensky and G. Labrecque, *Annu. Rev. Chronopharmacol.*, 1986, **2**, 3.
85. M. Ollagnier, H. Decousus, Y. Cherrah, F. Lévi, M. Mechkouri, P. Queneau and A. Reinberg, *Clin. Pharmacokinet.*, 1987, **12**, 367.
86. W. J. Hrushesky, *Science (Washington, D.C.)*, 1985, **228**, 73.
87. B. Lemmer, H. Winkler, T. Ohm and M. Fink, *Naunyn-Schmeideberg's Arch. Pharmakol.*, 1985, **330**, 42.
88. B. Hecquet, J. Meynadier, J. Bonneterre, L. Adenis and A. Demaille, *Cancer Treat. Rep.*, 1985, **69**, 79.
89. A. Reinberg, M. Smolensky and F. Lévi, in 'Therapeutic Implications of Time Dependencies, Topics in Pharmaceutical Sciences', ed. D. D. Breimer and P. Speiser, Elsevier, Amsterdam, 1985, p. 191.

23.13

Population Pharmacokinetics

BRIAN WHITING

University of Glasgow, UK

ANDREW W. KELMAN

West of Scotland Health Boards, Glasgow, UK

and

JOACHIM GREVEL

University of Texas, Houston, TX, USA

23.13.1 INTRODUCTION

Pharmacokinetics traditionally describes the time course of drug absorption, distribution and elimination in terms of numbers of parameters, *e.g.* the rate and extent of absorption, the volume of distribution and the rate of elimination. Subject by subject, the values of these parameters differ to a greater or lesser extent due to biological variability and most, if not all, human pharmacokinetic studies try to evaluate this variability. Even relatively small-scale bioavailability studies designed to estimate the amount of the drug absorbed (F) intact into the systemic circulation will conclude with some form of statistic which may express either the average amount absorbed or a confidence interval expressing a range of F values. These results are then extrapolated to all other subjects (*i.e.* patients) receiving the drug.

From the point of view of controlling drug therapy—to maximize benefit and minimize harm—it is necessary for some drugs to achieve a relatively narrow range of target drug concentrations in the setting of a high degree of pharmacokinetic variability. In the absence of detailed pharmacokinetic studies performed in every patient, control can only be achieved if the pharmacokinetic variability associated with a particular drug can be anticipated. The crucial question is: in the absence of any unique pharmacokinetic observations (*e.g.* concentration measurements), what is the likely range of concentrations that will be achieved for a given dosage? In other words, what confidence can be placed on achieving a particular target concentration? The best way of answering this is to estimate the range of concentrations—say, the 68% confidence interval—which will be generated by combining the variabilities of all parameters of a particular pharmacokinetic equation, solved for the time period of interest. The concentration confidence interval can then be thought of as a joint probability distribution of the various parameters. A good example is provided by the broncho-dilator theophylline. When the distribution of the relevant parameters is taken into account, the 68% confidence interval of plasma concentrations at 12 h during steady-state therapy with a daily dose designed to achieve 15 mg L^{-1} at 12 h is approximately 8–24 mg L^{-1}. This wide range immediately justifies therapeutic drug monitoring (TDM), where a concentration measurement—by yielding more information about an individual's pharmacokinetics—increases the confidence with which target concentrations will be subsequently achieved.

In steady-state situations, TDM may be applied empirically; the dose rate may be increased or decreased proportionally (assuming a linear system) when drug concentrations are too low or too high. However, in many non-steady-state situations TDM works well in conjunction with a statistical strategy equivalent to 'learning from experience'; the statistical distribution of pharmacokinetic parameters obtained from a representative population of patients serves as *a priori* information for the parameters of an individual patient. The term 'population pharmacokinetics' refers to the assimilation, analysis and subsequent description of this kind of *a priori* information.

23.13.2 DIFFERENT PHARMACOKINETIC APPROACHES

23.13.2.1 The Traditional Approach

The focus of population pharmacokinetics is to quantitate variability. As indicated above in relation to bioavailability studies, this is often attempted even when the number of subjects is small, and this applies to all relatively small-scale pharmacokinetic studies. Each study often produces its own 'mean and standard deviation' of the parameters of interest. The usual approach is to analyze experimentally derived data from each subject and to obtain estimates of the individual parameters by model-dependent or model-independent means. The average parameter values (and their standard deviations) are then calculated and an attempt may be made to quantitate the relationships between these parameters and physiological or pathological variables, such as body weight, age, sex, renal function, *etc.*, in a separate stage. Because of the relatively small numbers involved and the restrictions normally imposed by experimental protocols, the amount of variability included in these studies will usually be limited, and will not be representative of the wider group of patients who are receiving, or who will eventually receive, the drug. The data analysis techniques employed may also present statistical problems as far as the quantitation of variability is concerned.[1]

Despite these problems, standard pharmacokinetic studies remain an integral part of drug development, establishing basic facts about the absorption and disposition of drugs, particularly during phase I and II studies. Population studies should be seen in the context of amplifying earlier findings by exploring further the effects of disease and alterations of organ function on drug disposition within the population of interest, most conveniently carried out during the course of phase III and possibly phase IV studies.

23.13.2.2 The Population Approach

Two sources of data have been successfully exploited over the past decade to demonstrate the utility of population pharmacokinetic data analysis—routine (non-experimental) clinical pharmacokinetic (TDM) data and data generated during the course of formal drug studies, not necessarily conforming to classical pharmacokinetic protocols. The TDM data has certainly lacked any experimental design, but, even so, useful information on the sources of variability have been obtained and it is still a matter of debate as to whether or not some form of experimental design should be imposed on studies which prospectively seek to obtain population pharmacokinetic information.

23.13.3 POPULATION PHARMACOKINETIC DATA

23.13.3.1 Content

Population pharmacokinetic data should always consist of (a) pharmacokinetic data *per se* (dosage regimens, drug concentrations, times, *etc.*) and (b) demographic data (sex, age, weight, biochemical indices, *etc.*). Although the requirement for individual patient concentration data is far less stringent than that required by traditional studies (where individual patient parameter estimation is often the focus of the analysis), each patient contributes a small (*e.g.* two or three concentration values) but unique fraction of the total data and precision and accuracy in reccrding sampling times and dosage regimens are of paramount importance.

The question 'How many patients should be included in a population pharmacokinetic study?' can only be answered in a pragmatic way at the moment. Data from as many patients as possible should be included. The minimal number depends very much on the degree of variability in the population studied. When it can be anticipated (*e.g.* as a result of earlier traditional studies) that a factor (*e.g.* weight) or a number of factors (*e.g.* weight, age and creatinine clearance) will explain pharmacokinetic variability, then common sense dictates that patients expressing as wide as possible a range in those factors should be included.

23.13.3.2 Statistical Structure

Traditional pharmacokinetic data analysis regards the *individual* as the unit of analysis. Parameters (*e.g.* clearance and volume of distribution) are estimated subject by subject from an appropriate amount of data, ideally consisting of four concentration measurements per parameter (an average of 8–24 points per subject depending on the model being considered). Population pharmacokinetics views the *population* as the unit of analysis. The dataset may then consist of several hundred concentration values obtained from a relatively large number of patients. Its statistical structure is such that, whereas concentration values from individuals within the population are mutually dependent (on that individual's pharmacokinetic parameters), concentrations from *different* individuals are independent of each other and are subject to systematic deviations generated by the distribution of pharmacokinetic parameters within the population. The focus of population pharmacokinetic analysis, therefore, is to estimate the distribution of these parameters— in terms of, say, central tendencies (average values) and variances—and to account for all residual variability in a separate error term. This by-passes any estimates of individual subject parameters and the analysis is therefore completed in a single stage. The single-stage concept can also be extended to the estimation of parameters linking basic pharmacokinetic parameters to other variables—*viz.* the demographic and pathophysiological information which is an essential component of the population pharmacokinetic dataset. In this way, models are developed which have the potential of explaining pharmacokinetic variability to a much greater extent. Body weight, for example, may explain a great deal of the variability in volume of distribution and/or clearance. The simple statement 'volume of distribution is proportional to body weight' may be included in the data analysis in the form of a simple equation

$$V = A \cdot \text{weight} \tag{1}$$

where the value of A and the distribution of V are estimated directly from the population data, given that V is included in the appropriate pharmacokinetic model, that appropriate concentrations are available and that information on each subject's weight is available. Much more complex equations of this kind may be developed and embedded into appropriate standard pharmacokinetic equations, the objective always being to increase the explanation of variability through the inclusion of identifiable demographic and/or pathophysiological variables. For example, a clearance relationship might be expressed as

$$\text{clearance} = A \cdot \text{weight} \times \text{age}^B \times \text{creatinine}^C \tag{2}$$

where A is a proportionality constant and B and C are power terms.

Summarizing, therefore, the variability within a population dataset is composed of (a) the relationship between concentration and time as described by a structural model consisting of a number of pharmacokinetic parameters, such as clearance, volume of distribution, rate of absorption, *etc.*; (b) the relationship between these parameters nd various identifiable features, as described by secondary models of the type shown in equations (1) and (2) above; (c) systematic deviations from mean parameter values (either in terms of (a) or (b) above) reflecting differences among the patients making up the population, described by a number of interindividual random-effect (variance) parameters; and (d) residual variability, composed principally of errors involved in making observations (assay error), spontaneous, random fluctuations in the kinetic parameters within individuals and errors between the true and hypothesized kinetic models (model misspecification errors).

23.13.3.3 Parameter Estimation

For reasons already developed, improved therapeutic control can be achieved if a sound knowledge of the expected variability in the pharmacokinetics of a drug is available. Quantitation of this variability and recognition of the width of a 'therapeutic range' will determine whether or not therapeutic drug monitoring is necessary or advisable. Relatively few groups have worked on the problem of parameter estimation in this setting, but major contributions have been made by the NONMEM project group at the University of California, San Francisco, and the Department of Biomathematics, INSERM U194, Paris. The San Francisco group has developed software (the NONMEM program[2-4]) which is exportable and relatively easy to implement and run on mainframe computers. The non-parametric maximum-likelihood (NPML) approach[5] being developed by the Paris group is still at the evaluation stage.

As most experience to date has been gained with the NONMEM program, it will now be discussed in a little more detail. It has been evaluated in a number of situations.[6-11] NONMEM yields direct estimates of all three types of population pharmacokinetic parameters.[12] Thus, from the heterogeneous kind of dataset envisaged above, the following are obtainable: (i) the average values of the structural model parameters (of type (a) or (b) above); (ii) the population standard deviations of each parameter, *i.e.* their variances; and (iii) an estimate of the residual error (which is usually approximately equal to the coefficient of variation of the drug assay involved).

Starting from the raw data which will consist of a large series of concentration–time measurements, c_{ij}, where the subscript '*ij*' refers to the *i*th concentration measurement in the *j*th individual, the expected concentration, $\hat{c}_{ij}$, will be

$$\hat{c}_{ij} = F(\text{dose}_j, \tau_j, t_{ij}, Cl_j, V_j, \ldots) \tag{3}$$

where dose_j and τ_j characterize the dosage history (τ_j being the dosage interval), t_{ij} is the time associated with c_{ij} and Cl_j and V_j are examples of structural model parameters (clearance and volume of distribution) for the *j*th individual. *F* signifies a structural model function that is in keeping with the type of kinetic data; the match between model and data must obviously be as exact as possible. Data collected during steady state situations, for example, are analyzed by the appropriate steady-state equation, specifying dose_j, τ_j, *etc.* Non-steady-state data must be analyzed by equations which use the precise dosage details associated with the concentration measurements.

If *Cl* and *V per se* are included in the model, the average values of *Cl* and *V* and their variances are obtained, but, as indicated above, these parameters can be related directly to one or more patient factors. As an example, drug clearance may be related to body weight and serum creatinine by the simple expression

$$Cl = A + B \cdot \text{creatinine} + C \cdot \text{weight} \tag{4}$$

The clearance of the *j*th individual could then be expressed as

$$Cl_j = A + B \cdot \text{creatinine} + C \cdot \text{weight} + \eta_{Cl_j} \tag{5}$$

where the assumption is that the η_{Cl_j} are unimodally and normally distributed with a mean of zero. The variance of η_{Cl_j} (ω^2_{Cl}) is then the variance of *Cl*. Similarly, *V* may be related to weight

$$V = D \cdot \text{weight} \tag{6}$$

and for the *j*th individual

$$V_j = D \cdot \text{weight} + \eta_{V_j} \tag{7}$$

where, again, the η_{V_j} have an average value of zero and the variance (ω^2_V) is the variance of *V*.

If it is assumed that there is a small random error associated with each measurement, and that this error is independent of the concentration, then each concentration measurement can be modelled as

$$c_{ij} = \hat{c}_{ij} + \varepsilon_{ij} \tag{8}$$

where the ε_{ij} will have a mean of zero and variance reflecting all sources of intrasubject variability.

NONMEM fits the model to the data for all subjects simultaneously and estimates A, B, C, etc., and therefore the kinetic parameters, their variances and the residual, intrasubject, variance. Powerful features of the program include the ability to test the statistical effect of including or excluding explanatory variables (age, sex, weight, creatinine concentration, etc.) and the opportunity of specifying different error (variance) models. Interindividual variability, for example, may be multiplicative (as distinct from additive as in equations 5 and 7), thus

$$V_j = V \exp \eta_{V_j} \tag{9}$$

Similarly, intrasubject variability may also be modelled in different ways.

23.13.3.4 Pragmatic Aspects of Data Analysis

NONMEM has been successfully used to derive population parameter estimates from several sets of 'routine' clinical pharmacokinetic data[1,6,8,13-20] and, more recently, from data generated during formal drug evaluation studies.[21-23] Experience has shown that, from a practical point of view, the overriding need is for extremely good organization at the data acquisition stage.[24] In order to ensure the right amount of heterogeneity, data collection during drug development should optimally take place during phase III studies, using an efficient computer-based filing system. The database may then contain three types of data—demographic, pharmacokinetic and pharmacodynamic—where the pharmacodynamic data may consist of both desirable and undesirable drug effects.

The choice of data depends on practical, economic and common-sense considerations. The temptation is to collect as many data as possible in order to increase the chance of serendipity, but the unexpected or chance finding has proved to be extremely elusive. From a recent review[25] it seems fairly clear that the relationships between patient factors and pharmacokinetic indices such as clearance and volume of distribution were understood or suspected at the outset of the population study. It seems advisable, therefore, to plan the data collection with sensible questions in mind and to use these questions as the basis of hypothesis testing during the analytical stage. Setting up these questions, of course, requires that prior knowledge of the pharmacokinetics of the drug in question is available from earlier work—from animal studies and from phase I and II human studies. However, these studies may not have revealed the true extent and nature of the relationships, or the subtle interactions between various explanatory factors. A sensible series of questions then suggests a logical approach to the choice of data, e.g. 'Is body weight related to clearance and/or volume of distribution?' and 'How is body weight related to these kinetic determinants?' If it is known that a drug is excreted primarily by the kidney, then the obvious questions focus on the relationship between drug clearance and serum creatinine or creatinine clearance. If a drug is wholly metabolized by the liver, then questions will focus on the relationship between drug clearance and indices of liver function such as the prothrombin time, serum concentrations of albumin, bilirubin, or the activity of alkaline phosphatase or other enzymes, and the presence or absence of other drugs acting as enzyme inducers or inhibitors.

It will be obvious by now that, notwithstanding the comment about the futility of collecting data indiscriminately, a population dataset will progressively accumulate a great deal of information and it is essential that this is well organized and reviewed regularly. This can be most easily achieved by using a spreadsheet program. The advantage of this is that graphical displays, e.g. histograms of weights, ages, creatinines, bilirubins, etc., can be readily generated and examined to assess the degree of heterogeneity in any particular factor included in the database. An example of such an approach was used in the population study of high-dose metoclopramide in cancer patients.[22]

23.13.4 CLINICAL APPLICATIONS OF POPULATION PHARMACOKINETICS

The analysis of population datasets compiled from routine clinical pharmacokinetic (TDM) data has been the subject of a comprehensive review[25] and only more recent applications will be discussed briefly here in order to illustrate the approach.

23.13.4.1 Gentamicin

A particularly interesting yet problematical group of patients to study is neonates, from whom it would be impossible to collect more than a very few blood samples. Gentamicin is routinely used in this 'small' population to prevent and treat serious Gram-negative infections. Effective treatment is associated with peak concentrations of 5 to 10 mg L^{-1}; toxicity is minimized if trough concentrations remain below 2 mg L^{-1}. To achieve these targets, dosage regimens should reflect the likely requirements of individual neonates as determined by measurable clinical characteristics. The relationships between these characteristics and the pharmacokinetic determinants clearance and volume of distribution were evaluated in 113 neonates[20] using NONMEM. Two hundred and seventy concentration measurements were available: 96 peaks, 154 troughs and 20 midway samples. Although 11 variables were tested for their influence on clearance (weight, height, postnatal age, gestational age, postconceptional age (PCA), serum creatinine concentration, Apgar score, sex, cardiac abnormality, infection and ventilation) only three were found to be of significance—weight, PCA age and Apgar score. Thus, clearance had an average value of 0.053 L h^{-1} kg^{-1} and was reduced by a factor of 0.83 if PCA was less than 34 weeks and by a factor of 0.82 if the 5 min Apgar score was less than 7. The clearance equation therefore had the form

$$Cl = 0.053 \, \text{L h}^{-1} \text{kg}^{-1} \times 0.83 \, (\text{PCA} < 34) \times 0.82 \, (\text{Apgar} < 7) \tag{10}$$

and the coefficient of variation of clearance (the intersubject variability) was 27%. Volume of distribution was related solely to weight

$$V = 0.47 \, \text{L kg}^{-1} \tag{11}$$

with a coefficient of variation of 16%.

As most of the babies used in this analysis were less than 10 days old, it would be inappropriate to extrapolate these results to older babies, but it is important to note that useful therapeutic guidelines could be developed for this particular group from concentration data already available. Dosage recommendations, based on these findings, were as shown in Table 1.

23.13.4.2 Metoclopramide

The prevention of nausea and vomiting is an essential component of cancer chemotherapy. There is every reason to administer an effective dose of an antiemetic agent and the value of relatively large, repeated intravenous doses of metoclopramide in this setting has been highlighted.[26] An infusion regimen utilizing pharmacokinetic principles to achieve a constant plasma concentration for 8 h after chemotherapy has subsequently been proposed[27] and the superiority of continuous infusion over intermittent infusions to combat the emesis induced by cisplatin has been demonstrated.[28]

Although a 'therapeutic range' for metoclopramide has not been defined, a lower limit of 800 μg L^{-1} seems to be required to prevent the nausea and vomiting caused by cisplatin.[29] There would therefore be considerable merit in designing infusion schemes for individual patients which would result in steady-state plasma concentration predictions with a lower 68% confidence limit of at least 800 μg L^{-1}. With this in mind, data from 47 patients who had received metoclopramide during 109 courses of chemotherapy in the Oncology Unit at the Newcastle General Infirmary, Newcastle-upon-Tyne, have been analyzed.[22] Between four and 10 blood samples were available from each patient, withdrawn during treatment, and for up to 24 h after the end of a maintenance infusion.

This study highlighted the use of a tailor-made database, the value of graphical overviews and the ability to construct a NONMEM input file on a personal computer and to subsequently transfer the data to a mainframe computer.

Table 1 Dosage Recommendations of Gentamicin in Neonates

Explanatory factors	Loading dose (mg kg^{-1})	Maintenance dose (mg kg^{-1})
>34 weeks PCA and 5 min Apgar>7	3.5	2.5 (12 hourly)
≤34 weeks PCA *or* 5 min Apgar>7	3.5	3.0 (18 hourly)
≤34 weeks PCA *and* 5 min Apgar>7	3.5	3.5 (24 hourly)

The analysis was somewhat complex, but it was concluded that, as far as metoclopromide clearance was concerned, body weight was an important explanatory variable, providing further support for the usual practice of administering metoclopramide on a $mg\,kg^{-1}$ basis. Alkaline phosphatase activity also had some explanatory power, but the influence of age was ambiguous. Typical pharmacokinetic parameters for an average individual (weighing 70 kg with an alkaline phosphatase of 100 i.u. L^{-1}) were: clearance, $20\,L\,h^{-1}$; volume of distribution at steady state, 190 L; elimination half-life, 8 h. Corresponding interindividual variabilities were 50%, 35% and 35%, respectively. Residual variability was 13%.

23.13.4.3 Lisinopril

Lisinopril is an angiotensin-converting enzyme (ACE) inhibitor at present undergoing evaluation as an antihypertensive drug. It is the lysine analogue of enalaprilat (the active metabolite of enalapril) but, unlike enalapril, is absorbed orally as the active moiety, not requiring hepatic activation. It is excreted by the kidneys and prolonged elimination half-lives have been observed in patients with renal impairment.[30] It has also been noted that age increases the area under the plasma concentration–time curve, even when this is corrected for differences in creatinine clearance.[31]

These 'indicators' of altered (reduced) clearance due to reduced renal function and increased age prompted exploration of these relationships in more detail on a population basis.[21] Steady-state concentration–time data were available from two multicentre trials: 40 elderly patients being treated for hypertension and 20 patients with renal impairment being treated for hypertension. Grouping all patients together, the age range was 21–85 years (mean 65 years), the weight range 51–115 kg (mean 72 kg). Twenty-one patients had serum creatinines above 120 mol L^{-1} and 13 patients had compensated cardiac failure. Estimated creatinine clearances ranged from 10 to 120 mL min^{-1}. This dataset, therefore, contained a representative amount of heterogeneity in terms of the questions being posed about the factors potentially influencing lisinopril clearance.

A total of 381 concentrations provided the basis for the NONMEM analysis. Not surprisingly, serum creatinine was the most important determinant of lisinopril clearance, but weight, age and cardiac failure were also significant. Clearance was best described by

$$Cl = 0.25 \cdot \text{weight (kg)} \times \left(\frac{\text{creatinine}}{70}\right)^{-0.89} \times \left(\frac{\text{age}}{65}\right)^{-0.45} \times 0.65\ (CF) \tag{12}$$

where CF indicates the presence of cardiac failure, albeit compensated, otherwise $CF = 1.0$. Despite the inclusion of these four factors, the intersubject variability remained relatively high at 52%. Residual intrasubject variability was 28%. This analysis, therefore, confirmed that linisopril clearance depends partly on renal function (as indicated by changes in creatinine concentration) and age, these factors modifying a basic weight relationship in a multiplicative way. The effect of 'cardiac failure' was surprising because this subset of patients did not have overt cardiac failure at the time of the study, but were classified as 'compensated cardiac failure', the result of previous, successful, treatment. This finding will certainly require confirmation and further exploration of the relationship between lisinopril clearance and varying degrees of heart failure and/or the drugs associated with its treatment.

23.13.5 CONCLUSIONS

The influence of disease and physiological variables on drug disposition can be clearly expressed in terms of a number of population pharmacokinetic parameters. Their identification and estimation should therefore be central to drug evaluation. Standard approaches, involving traditional small-scale experimental designs, often yield imprecise unrepresentative results. The alternative population approach should not suffer from these drawbacks if sufficient heterogeneity is included.

The examples cited above serve to illustrate three different kinds of data which are amenable to population analysis. Gentamicin concentrations—in neonates and older groups—are measured by default because of the risks of toxicity, but in neonates the number of measurements per baby are extremely limited. Yet with enough accumulated data, it is possible to explore the factors which significantly influence drug disposition and to develop rational dosage guidelines. This is an example of operational research where existing data is exploited to subsequently enhance the control of therapy.

The metoclopramide study again illustrates the desire to optimize therapy with an established drug by defining the factors which must be taken into account when choosing a dose to achieve a previously defined minimum effective concentration. In this case, advantage was taken of an existing experimental dataset.

The lisinopril analysis illustrates a population pharmacokinetic study undertaken during drug development to answer specific questions about clearance (and therefore about drug dosage) using data from two groups of patients. An unexpected finding emerged. Whether or not this will prove to be of clinical significance remains to be seen, but it does indicate the need for further investigation and this in itself is an extremely important part of safety evaluation during drug development.

The outcome of this kind of analysis is therefore best measured in terms of safety evaluation, the construction of nomograms and the estimation of initial parameter values used in pharmacokinetic forecasting procedures, such as Bayesian parameter estimation.[32-36]

A move towards the adoption of population pharmacokinetics as a routine procedure during drug development should now be encouraged. Existing routine data can be organized in such a way that valuable information on pharmacokinetic variability can be obtained and a similar effort should be made during drug development, with adequate prospective planning.

23.13.6 REFERENCES

1. L. B. Sheiner, B. Rosenberg and V. V. Marathe, *J. Pharmacokinet. Biopharm.*, 1977, **5**, 445.
2. S. L. Beal and L. B. Shiner, 'NONMEM Users Guides', NONMEM Project Group, University of California at San Francisco, 1989.
3. S. L. Beal, *Drug Metab. Rev.*, 1984, **15**, 173.
4. S. L. Beal and L. B. Sheiner, *Am. Stat.*, 1980, **34**, 118.
5. A. Mallet, Ph. D. Thesis, University of Paris, 1983.
6. L. B. Sheiner and S. L. Beal, *J. Pharmacokinet. Biopharm.*, 1980, **8**, 553.
7. L. B. Sheiner and S. L. Beal, *J. Pharmacokinet. Biopharm.*, 1981, **9**, 635.
8. L. B. Sheiner and S. L. Beal, *J. Pharmacokinet. Biopharm.*, 1983, **11**, 303.
9. H. Fluehler, H. Huber, E. Widmer and S. Brechbuehler, *Drug Metab. Rev.*, 1984, **15**, 317.
10. T. H. Grasela, E. J. Antal, R. J. Townsend and R. B. Smith, *Clin. Pharmacol. Ther.*, 1986, **39**, 605.
11. T. H. Grasela, E. J. Antal, L. Ereshefsky, B. G. Wells, R. L. Evans and R. B. Smith, *Clin. Pharmacol. Ther.*, 1987, **42**, 433.
12. S. L. Beal and L. B. Sheiner, *CRC Crit. Rev. Biomed. Eng.*, 1982, **8**, 195.
13. S. Vozeh, G. Katz, V. Steiner and F. Follath, *Eur. J. Clin. Pharmacol.*, 1982, **23**, 445.
14. S. Vozeh, M. Wenk and F. Follath, *Drug Metab. Rev.*, 1984, **15**, 305.
15. T. H. Grasela and L. B. Sheiner, *Clin. Pharmacokinet.*, 1984, **9**, 545.
16. A. W. Kelman, A. H. Thomson, B. Whiting, S. M. Bryson and D. A. Steedman, *Br. J. Clin. Pharmacol.*, 1984, **18**, 685.
17. L. B. Sheiner and T. H. Grasela, *Drug Metab. Rev.*, 1984, **15**, 293.
18. M. Rheeders, MSc Thesis, University of Potchefstroom, 1985.
19. D. R. Mungall, T. M. Ludden, J. Marshall, D. W. Hawkins and R. T. Talbert, *J. Pharmacokinet. Biopharm.*, 1985, **13**, 213.
20. A. H. Thomson, S. Way, S. M. Bryson, E. M. McGovern, A. W. Kelman and B. Whiting, *Dev. Pharmacol. Ther.*, 1988, **11**, 173.
21. A. H. Thomson and B. Whiting, *Gerontology (Basel)*, 1987, **33** (suppl. 1), 17.
22. J. Grevel, B. Whiting, A. W. Kelman, W. B. Taylor and D. N. Bateman, *Clin. Pharmacokinet.*, 1988, **14**, 52.
23. J. Grevel, P. Thomas and B. Whiting, *Clin. Pharmacokinet.*, 1989, **17**, 53.
24. J. Grevel, B. Whiting, A. W. Kelman and K. Kutz, *Pharm. Med.*, 1987, **2**, 127.
25. B. Whiting, A. W. Kelman and J. Grevel, *Clin. Pharmacokinet.*, 1986, **11**, 387.
26. R. J. Gralla, L. M. Itri, S. E. Pisko, A. E. Squillante and D. P. Kelsen, *N. Engl. J. Med.*, 1981, **305**, 905.
27. W. B. Taylor, S. J. Proctor and D. N. Bateman, *Br. J. Clin. Pharmacol.*, 1984, **18**, 679.
28. P. S. Warrington, S. G. Allan, M. A. Cornbleet, J. S. Macpherson and J. F. Smyth, *Br. Med. J.*, 1986, **293**, 1334.
29. B. R. Meyer, M. Lewin, D. E. Drayer, M. Pasmantier and L. Lonski, *Ann. Intern. Med.*, 1984, **100**, 393.
30. J. G. Kelly, G. Doyle, J. Donohoe, M. Laher, C. Long, D. R. Glover and W. D. Cooper, *Br. J. Clin. Pharmacol.*, 1987, **23**, 629P.
31. V. J. Cirillo, A. E. Till and H. J. Gomez, *Clin. Pharmacol. Ther.*, 1986, **39**, 187.
32. L. B. Sheiner, S. L. Beal, B. Rosenberg and V. V. Marathe, *Clin. Pharmacol. Ther.*, 1979, **26**, 294.
33. C. C. Peck, W. D. Brown, L. B. Sheiner, B. G. Schustser, in 'Proceedings of the Fourth Symposium on Computer Applications in Medical Care', ed. J. T. O'Neil, 1980, vol. 2, p. 989.
34. L. B. Sheiner and S. L. Beal, *J. Pharm. Sci.*, 1982, **71**, 1344.
35. S. Vozeh and J.-L. Steimer, *Clin. Pharmacokinet.*, 1985, **10**, 457.
36. A. W. Kelman, B. Whiting and S. M. Bryson, *Br. J. Clin. Pharmacol.*, 1982, **14**, 247.

23.14
Toxicokinetics

DENNIS V. PARKE and COSTAS IOANNIDES

University of Surrey, Guildford, UK

23.14.1   INTRODUCTION

Toxicokinetics is the study of the processes by which a chemical manifests toxicity and, conversely, the processes by which the same chemical is detoxicated and eliminated from the biological organism. These two opposing phenomena involve a number of simultaneously occurring processes including absorption, tissue distribution, metabolism, interaction with receptors eliciting toxicity, and excretion, all of which are characterized by specific rate constants. The overall rates of disposition of chemicals in a biological organism determine their toxicity to a considerable extent. These may range from rapid metabolism and/or elimination, high clearance values and short biological half-lives, characteristic of detoxication (*e.g.* the pesticide malathion in mammals), to very slow metabolism and elimination, low clearance values and long biological half-lives, characteristic of chronically toxic chemicals [*e.g.* the environmental pollutant 2,3,7,8-tetrachlorodibenzodioxin (TCDD)].

23.14.1.1   Causes of Toxicity

Chemical toxicity is the consequence of three fundamental factors: (i) the intrinsic molecular structure and physicochemical properties of the chemical; (ii) the genetic characteristics of the exposed biological organism which determine the natures and activities of its detoxicating and activating enzymes and receptors; and (iii) the environment of the biological organism, past and present, including its nutrition and exposure to other chemicals which can modulate the balance of detoxicating/activating enzymes. It is therefore oversimplistic to designate specific chemicals as 'toxic', 'mutagenic', 'carcinogenic' or 'teratogenic' without specifying the biological species involved, for a chemical which may be fatally toxic to one biological species may be innocuous to another (*e.g.* malathion is neurotoxic to insects but not to mammals). Indeed, these species differences in chemical toxicity form the very basis of 'selective toxicity' which has enabled the design of numerous pesticides, medicines and food additives.

23.14.1.2   Acute and Chronic Toxicity

For simplification, chemical toxicity may be rationalized into two distinct phenomena, although in practice both may occur simultaneously after exposure to various chemicals. The first type of chemical toxicity, usually associated with 'acute toxicity', involves the direct action of the parent chemical on the receptor mediating toxicity, for example the interaction of cyanide with cytochrome oxidase which inhibits oxidative phosphorylation; in this case metabolism is concerned primarily with detoxication (see Figure 1). The second phenomenon, usually associated with 'chronic toxicity', involves the metabolic activation of the parent chemical to 'reactive intermediates' which interact with nucleophilic receptors, such as sulfhydryl enzymes, proteins and DNA, mediating cytotoxicity, immunotoxicity, carcinogenicity, *etc.*, for example the mutagenic/carcinogenic activity of benzo[*a*]pyrene; detoxicating metabolism usually occurs simultaneously and, together with elimination, may prevent, limit or terminate the toxic action, depending on dosage (see Figure 1).

The toxicokinetic characteristics of a chemical in a specific biological species involve the summation of many different processes, including absorption by various routes, distribution to different tissues and non-covalent binding, detoxicating and activating metabolism, interaction and covalent binding with proteins, enzymes, DNA, *etc.*, elimination and excretion, enterohepatic circulation and accumulation, all of which are dealt with in detail in the following sections.

23.14.1.3   Toxicokinetic Measurements

The toxicokinetics of a given drug or chemical are defined in a particular biological species by the standard pharmacokinetic parameters of maximum plasma concentration (C_{max}), time of maximum

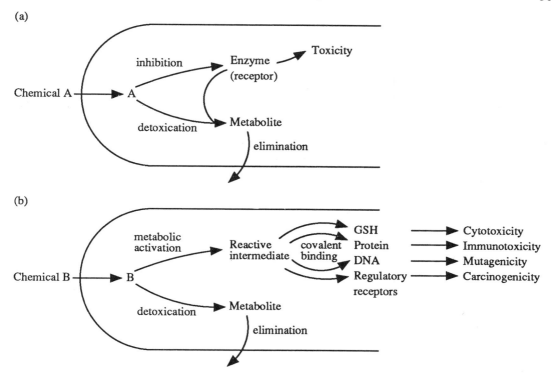

Figure 1 Typical mechanisms of acute toxicity and chronic toxicity. (a) In acute toxicity, chemical A exerts its toxic action by the direct inhibition of an enzyme, and is subsequently detoxicated and eliminated. (b) In chronic toxicity, chemical B is metabolically activated to a reactive intermediate which covalently binds to glutathione (GSH), proteins, DNA and possibly also to regulatory receptors

plasma concentration (t_{max}), plasma half-life ($t_{\frac{1}{2}}$), area under the curve (AUC$_0 \rightarrow \infty$), volume of distribution (V_d), clearance (Cl) and bioavailability (AUC$_{oral}$/AUC$_{i.v.}$) from determinations of the concentration of the unchanged drug and its metabolites in the blood plasma and urine at various time intervals after intravenous and oral dosing. Similar determinations for the drug and specific metabolites in urine, bile and specific tissues yield valuable information concerning the pathways and rates of metabolic detoxication and activation.

Such data, obtained in animals and man, are essential in the selection of appropriate animal species for chronic toxicity studies, and in the choice of suitable doses and form of dosing. For example, for drugs with a $t_{\frac{1}{2}}$ of less than 16.6 h, it is necessary to administer the drug more than once daily to obtain steady-state blood concentration; drugs with ultrashort half-lives, *e.g.* chloramphenicol with a $t_{\frac{1}{2}}$ of 25 min, may have to be administered continuously by infusion for toxicity testing.[1]

23.14.2 ABSORPTION

The routes, rates and extent of absorption of a chemical can markedly affect toxicity. In mammals, absorption of chemicals may occur *via* the gastrointestinal tract, by inhalation or through the skin, and drugs may be administered parenterally; the rate of absorption is usually determined by measuring the plasma concentration of the chemical and its metabolites as a function of time following exposure. Peak plasma concentration (C_{max}) and time of peak plasma concentration (t_{max}) are important in determining the extent and time of onset of toxicity where there is a relationship between toxicity and the plasma concentration of the chemical, and especially where a threshold plasma concentration occurs which has to be exceeded for toxicity to be manifested (see Figure 2).

23.14.2.1 Mechanisms of Absorption

The route of absorption may considerably affect chemical toxicity, depending on the nature of the chemical; for example cyanide is more toxic when inhaled than when administered orally. The

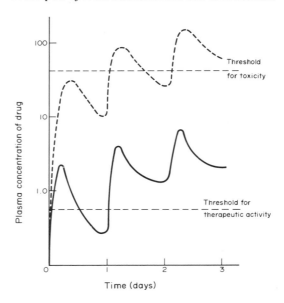

Figure 2 Hypothetical blood plasma concentration–time profiles of drugs or chemicals orally absorbed. The drugs are administered once daily. The drug indicated by a solid line is readily absorbed and fairly readily eliminated and achieves plasma concentrations above the threshold concentration for therapeutic activity but does not exceed the threshold concentration for toxicity. The homologue indicated by a broken line is readily absorbed but less readily eliminated and achieves plasma concentrations which quickly exceed the threshold concentration for toxicity

absorption of a chemical depends on its physicochemical properties, its molecular size and conformation, its lipid solubility and its ionization potential. Four major mechanisms of absorption are known: (i) passive diffusion, the principal mechanism, which depends on the concentration gradient of the chemical, its lipid solubility, degree of ionization and molecular size; (ii) filtration through aqueous membrane pores, which is governed mostly by molecular size and conformation; (iii) active transport, which involves energy-dependent saturable carrier mechanisms to transport large molecules, lipid-insoluble molecules and ions across the lipid membranes of the gastrointestinal and respiratory tract mucosa and the skin (chemical structure, size, conformation and ionic charge determine the affinity of a chemical for a particular carrier system which can be inhibited by poisons that interfere with cellular energy metabolism); and (iv) pinocytosis, by which macromolecules and particles are transferred across membranes by membrane mobility.

23.14.2.2 Gastrointestinal Absorption

The gastrointestinal tract is the most important route of absorption of most chemicals. Absorption from the buccal cavity is minimal but as chemicals are absorbed into the systemic circulation without being transported directly to the liver by the portal system they are not rapidly metabolized. Absorption from the stomach and intestines, because of pH differences, is dependent on the ionization potential and lipophilicity of the chemical; weak acids are readily absorbed from the stomach and weak acids and weak bases are absorbed from the intestines. Chemicals absorbed from the stomach and intestines are first transported by the portal system to the liver where they are metabolized, being detoxicated or activated by metabolism.

Inhibition of gastric secretion by H_2-receptor antagonist drugs such as cimetidine and neutralizing of gastric acid by antacids may decrease the absorption and bioavailability of certain drugs such as the tetracyclines.[2] The presence of food in the gastrointestinal tract may delay absorption of chemicals, and alterations in gastric/intestinal mobility can selectively change the rates of absorption of different chemcials, dependent on their ionization potential. Chemicals may be metabolized before absorption by the esterase activity of the digestive enzymes, by the mixed-function oxidases and conjugases of the intestinal mucosa, and especially by the enzymes of the intestinal microflora.

23.14.2.3 Pulmonary Absorption

Absorption from the lungs is usually very rapid, due to the high vascularity and large surface area of the pulmonary epithelium. Most rapidly absorbed are gases and aerosols of highly lipophilic chemicals of small particle size. Exposure to aerosols may also result in absorption from the gastrointestinal tract, due to adsorption of the chemical to bronchial mucus which is subsequently swallowed, as well as direct entry of the aerosol into the oesophagus. Metabolism of chemicals may also occur to a limited extent in the lung, before or during absorption, mostly in the Clara cells.

With volatile xenobiotics, two factors may limit their rates of metabolic clearance: (i) the metabolizing enzyme capacity, especially at high exposure concentrations; and (ii) the physiological parameters of respiration frequency and lung and liver perfusion, especially at low exposure. 1,3-Butadiene is markedly more carcinogenic to the mouse than to the rat, yet the metabolic elimination rate is greater in the mouse due to the greater frequency of respiration and the higher rate of metabolism to the reactive intermediate 1,2-epoxy-3-butene which is also exhaled. Although the greater carcinogenicity in the mouse may be attributed partly to the increased rate of metabolism, other factors such as further metabolism to the diepoxide, a bifunctional alkylating agent, may also play a role in the species difference.[3]

23.14.2.4 Percutaneous Absorption

Absorption through the skin is highly dependent on the lipophilicity of the chemical, as absorption requires penetration of the lipoprotein epidermis; the lower highly porous dermis is permeable to both lipid- and water-soluble compounds. Metabolism of chemicals may also occur in the skin during the process of absorption.

The skin has a dual role in toxicokinetics,[4] since it is a target organ as well as a site of absorption and metabolism, and drugs and chemicals may give rise to skin irritation, dermal toxicity, phototoxicity, sensitization and photoallergy, as seen with the drug benoxaprofen.

23.14.3 DISTRIBUTION

Distribution of chemicals from the blood plasma compartment into organs and tissues is dependent on transfer of the chemicals or their metabolites across membranous barriers, as in their absorption. The rate of tissue distribution is dependent on the physicochemical natures of the chemical and its metabolites which determine the rates at which membrane transfers occur, the rate of blood flow through the different organs and tissues, and the concentration gradient of the chemical and metabolites from plasma to tissues. Distribution is thus greatest to those tissues with high blood flow, such as the liver and kidney. Lipid-soluble chemicals tend to be distributed into adipose tissue where, unless they are readily metabolized, they tend to accumulate for long periods of time, *e.g.* the chlorinated pesticides DDT and dieldrin and the polychlorinated biphenyls, which are stored in body fat over periods of years after exposure.

2,2',4,4',5,5'-Hexabromobiphenyl (HBB), the major component of Firemaster BP-6, the fire-retardant chemical that was accidentally introduced into livestock feed in Michigan in 1973 and hence into the human population of North America, is almost completely resistant to metabolic degradation, is very slowly eliminated in the faeces and accumulates mostly in the adipose tissue and skin. A mathematical model of the tissue disposition kinetics of HBB, as the archetypal non-metabolizable highly lipophilic xenobiotic, has been developed for the rat, from which toxicokinetic parameters for humans have been predicted. The estimated body burden $t_{\frac{1}{2}}$ of a single dose in man is 6.5 years; as 30 to 70% of the body burden is contained in the adipose tissue, treatment with cholestyramine to increase faecal excretion is unlikely to be effective.[5]

The herbicide paraquat is selectively taken up from the circulating blood into the lungs, especially the alveolar epithelium and the Clara cells, by an energy-dependent diamine/polyamine transport process because of the structural similarity of paraquat to endogenous amines. The pulmonary accumulated paraquat undergoes an NADPH-dependent one-electron reduction to form the paraquat free-radical which then undergoes redox cycling in the oxygen-rich pulmonary atmosphere to form the toxic superoxide anion ($O_2^-\cdot$), which is responsible for the progressive damage to lung tissue characteristic of paraquat poisoning.[6] Hence, in the case of paraquat, the selective uptake of the chemical into a particular tissue is closely associated with the manifestation of toxicity.

Similarly, the orally administered pulmonary toxin 4-ipomeanol exhibits a good correlation between toxicity and the covalent binding of a reactive intermediate to lung tissue; induction of liver cytochrome *P*-448 by 3-methylcholanthrene shifts the site of toxicity from the lungs to the liver, where high levels of covalent binding are associated with hepatic necrosis.[7]

23.14.3.1 Blood–Brain and Placental Transfer

Passage of chemicals across other membranous barriers, such as the blood–brain barrier and the placental barrier is more difficult, because of the more complex nature of the membranes involved. Many chemicals fail to penetrate the blood–brain barrier which excludes most ionized and hydrophilic molecules, but the brain is much more permeable to highly lipophilic chemicals, *e.g.* thiopental.

The placental barrier similarly protects the foetus by preventing transfer of non-lipophilic chemicals from the maternal blood plasma into the placenta, but if the concentration gradient is high enough, by virtue of very high plasma concentrations, most chemicals may cross the placenta to some extent. Hence, high plasma concentrations (obtained by bolus intravenous injection) can result in foetotoxicity and teratogenicity with compounds that otherwise would not exhibit such toxic effects. Because of species differences in pharmacokinetics, it is desirable in selecting animal models for teratogenicity testing of drugs to determine whether peak plasma concentrations (as for valproic acid) or AUC values (as for cyclophosphamide) correlate with the teratogenic response.[8] Furthermore, it is now a requirement of most drug regulatory bodies to demonstrate that the pharmacokinetics of a drug in pregnant animals do not differ substantially from those in non-pregnant animals, and to determine the facility with which the drug is able to cross the placental barrier.[9]

Drugs and chemicals may also pass into the maternal milk. For the majority of chemicals the concentration found in the milk is lower than that in the blood, but some substances (*e.g.* propylthiouracil, meprobamate and erythromycin) are concentrated in the milk at levels higher than those found in the blood plasma, and these may be toxic to the neonate which lacks effective detoxication mechanisms. Many environmental contaminants, such as DDT, dieldrin, hexa-chlorobenzene, TCDD and polychlorinated biphenyls, are also found in milk, which constitutes a major pathway of elimination for these highly lipophilic, slowly metabolized chemicals.[10]

23.14.3.2 Macromolecular Binding

A major factor affecting the distribution of chemicals is their affinity to bind to specific proteins, phospholipids and other macromolecules. The binding of a chemical to a particular biological constituent effectively localizes the chemical to a certain tissue, thus lowering the plasma concentration, for example the herbicide paraquat has an affinity for the lung; a number of chemicals (electrophiles) that react with glutathione bind to ligandin (glutathione *S*-transferase) of the liver; the antiulcer drug carbenoxolone is confined almost entirely to the plasma compartment because it is almost completely bound to plasma albumin. The binding of a chemical to a tissue constituent may also effect the transfer of a chemical from one tissue to another, for example cadmium binds to metallothionein which effectively transfers the accumulation of cadmium from the liver to the kidney. This specific binding of chemicals to particular tissues is often responsible for tissue-specific toxicity, *e.g.* the pulmonary toxicity of paraquat and the nephrotoxicity of cadmium.

Many drugs and other chemicals are bound to plasma albumin. Both anionic and cationic compounds are bound because of several different binding sites, but competition for one particular binding site can profoundly alter the toxicokinetic profile of a chemical; for example the displacement of bilirubin from plasma albumin binding sites by sulfonamides and other drugs given during pregnancy may raise the maternal plasma free-bilirubin concentration, with the consequence that bilirubin enters the neonatal brain, resulting in kernicterus. Displacement of the anticoagulant drug warfarin from serum albumin by other highly protein-bound drugs, such as phenylbutazone, clofibrate and the metabolite of chloral hydrate, trichloroacetic acid, substantially increases the plasma concentration of unbound pharmacologically active warfarin, often with fatal haemorrhagic consequences.[11]

The teratogenicity of the synthetic estrogen diethylstilbestrol (DES) has been correlated with its tissue distribution in the foetus. In the foetal rat, concentrations of DES are relatively higher in the tissues than in the blood plasma, whereas the natural estrogen estradiol (E) is retained largely protein-bound in the foetal plasma. Foetal levels of E declined faster than those of DES, and both unchanged estrogens were retained longer in the target genital tissues than in other organs. The teratogenic urogenital malformations and malignancy appear to correlate with the concentration of free estrogen in the cell, and the more rapid metabolism and extensive protein binding of the natural estrogen E compared with DES is considered to account for the lower teratogenicity of the former.[12]

23.14.4 METABOLISM

It has been stated that chemicals are toxic because they are not readily detoxicated or quickly cleared from the body. An extension of this obvious truism is that chemicals are also toxic because they are activated by metabolism. This paradox by which chemicals (often the same chemical) can be either detoxicated or activated by metabolism, and often apparently by the same enzymes, holds the key to many of the problems of chemical toxicity and of chemical carcinogenesis.

23.14.4.1 Detoxication and Activation

Drugs and environmental chemicals may be metabolized in two phases, namely biotransformation or phase 1 metabolism (oxygenation, oxidation, reduction and hydrolysis) and then by conjugation or phase 2 metabolism (sulfation, glucuronidation, glutathione conjugation, acetylation, *etc.*), which increases the polarity of the molecule and accelerates its clearance from the body. Conjugation with glucuronic acid generally detoxicates phenols, dihydrodiols, alcohols and amines; sulfate conjugation similarly usually results in the detoxication of phenols and other nucleophiles, while acetyl conjugation generally detoxicates aromatic amines. Glutathione (GSH) and GSH-dependent enzymes provide a primary defence against toxic chemicals, not only by the conjugation of electrophilic reactive intermediates such as epoxides by the GSH-*S*-transferases, but also by detoxicating reactive oxygen such as hydroxyl radicals and superoxide anions (formed by redox cycling from quinones, nitroso compounds, *etc.*, or by futile cycling from cytochromes *P*-450) with GSH, GSH reductase and GSH peroxidase.

Chemicals which are detoxicated generally undergo both phases of metabolism (biotransformation and conjugation), whereas many chemicals which are activated appear to undergo only phase 1 metabolism (oxygenation), resulting in the formation of reactive intermediates. These reactive intermediates appear to be poor substrates for subsequent conjugation, and therefore react spontaneously with intracellular proteins, DNA, *etc.* to promote toxicity.[13] Recent studies have shown that in the great majority of cases the activation of chemicals to reactive intermediates is catalyzed by the cytochromes *P*-448 (cytochromes *P*450 I) or by free hydroxyl radicals, both of which can oxygenate chemicals in conformationally hindered positions.[14] The cytochrome *P*-448 family of enzymes have an active site which accepts only planar molecules, and the oxygenation of these chemicals in hindered positions makes the products (reactive intermediates) poor substrates for epoxide hydrolase and other phase 2 enzymes, and hence they are resistant to detoxication (see Figure 3).[13] The planarity of molecules with hindered oxygen functions also appears to be important in the promotion of malignancy, as these planar chemicals, such as benzo[*a*]pyrene and other polycyclic aromatic hydrocarbons and TCDD, interact with the cytosolic Ah receptor, and probably other receptors, to initiate transcription and replication of DNA, mitosis and tissue proliferation (see Figure 4).

Metabolic activation of a chemical is indicated if: (i) there is an increased effect following the induction of the metabolizing enzymes; and (ii) there is a delay between exposure or administration of the chemical and the onset of the toxic effects.

Toxicity of a drug may arise from a relatively minor pathway of metabolism. The antiepileptic drug valproic acid, which gives rise to occasional hepatotoxicity, is metabolized mainly by oxidation and glucuronidation to a variety of metabolites. A minor pathway of metabolism yields the Δ^4-unsaturated drug, 2-*n*-propyl-4-pentenoic acid, a hepatotoxic metabolite which is further metabolized to an electrophilic intermediate responsible for the ultimate hepatotoxicity of the drug.[15] It is therefore essential in the safety evaluation of a drug to determine the complete metabolic fate of the chemical and the toxicokinetics of all the metabolic pathways, and not to limit the study to only the major pathways of metabolism.

Although conjugation reactions generally result in drug detoxication, some conjugations are known to lead ultimately to increased toxicity, although this results strictly from deconjugation

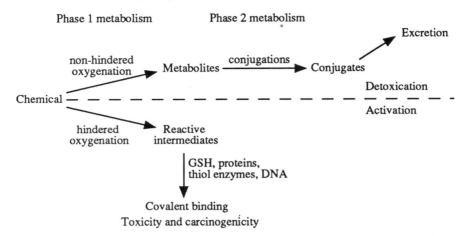

Figure 3 Metabolic detoxication and activation of drugs and chemicals. Non-hindered oxygenation of chemicals by the cytochromes *P*-450 yields metabolites which are readily conjugated and thereby detoxicated. Hindered oxygenation of chemicals which are planar by the cytochromes *P*-448 gives reactive intermediates which are not readily conjugated and eliminated and therefore interact non-enzymically with intracellular nucleophiles causing toxicity and carcinogenicity

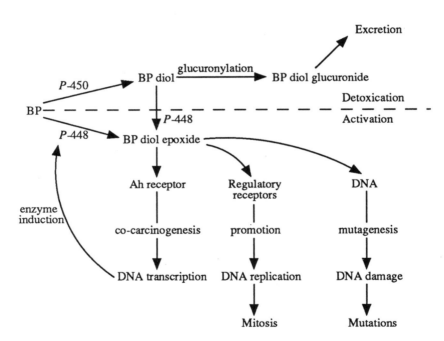

Figure 4 Mechanisms of metabolic activation of carcinogens. Benzo[*a*]pyrene (BP) is detoxicated by oxidative metabolism by cytochrome *P*-450 and epoxide hydrolase to BP diol, followed by conjugation to BP diol glucuronide. It is activated by oxidative metabolism by cytochrome *P*-448 to BP diol epoxide, which then interacts with DNA (mutagenesis), the Ah receptor (co-carcinogenesis) and possibly other regulatory receptors (promotion). BP itself also interacts with the Ah receptor

reactions.[16] The carcinogens 7-hydroxymethyl-12-methylbenz[*a*]anthracene and 4-aminoazobenzene and its *N*-methyl derivatives are activated by metabolism to their sulfate conjugates.[17,18] Similarly, the halogenated alkenes such as hexachlorobutadiene, tetrachloroethylene and trichloroethylene are nephrotoxic and carcinogenic because of their conjugation with GSH, subsequent deacetylation of the mercapturic acid and further cleavage to reactive and mutagenic intermediates by bacterial and mammalian β-lyase enzymes.[19,20] Likewise, the spontaneous decomposition of acylglucuronides or acyl sulfates of alcohol substrates results in the formation of reactive products and the acylation of nucleophiles (deconjugation), which have been postulated to account for the nephrotoxicity and hepatotoxicity of many drugs such as chlorambucil and clofibric acid.[16]

Many drugs or their metabolites which are carboxylic acids form CoA thioester derivatives which are then conjugated with amino acids such as glycine or taurine. Alternatively, the CoA derivatives may yield xenobiotic triglyceride or cholesteryl ester conjugates, which may form long-lasting tissue residues of the drug, inhibit lipid biosynthesis or enter membranes disturbing the microenvironment of membrane-bound enzymes. Several arylpropionic acid antiinflammatory drugs, *e.g.* ibuprofen, ketoprofen and fenoprofen, are known to form large amounts of these hybrid triglycerides *in vitro*, and consequently to inhibit lipid biosynthesis in rats.[21]

23.14.4.2 Sites of Metabolism

Metabolism of chemicals occurs mainly in the liver of mammalia, but can also occur in the gastrointestinal tract, lungs, skin, kidneys and most other tissues. Detoxicating metabolism occurs mostly in the liver and gastrointestinal tract, while activating metabolism occurs in both hepatic and extrahepatic tissues. In contrast to the liver, the mixed-function oxidase activities of the gastrointestinal mucosa are low, whereas the UDPglucuronyltransferase activities of this tissue can exceed those of the liver, and the GSH *S*-transferase activities of the gut are also high.[22] Metabolism of a drug by the enzymes of the gastrointestinal tract and the liver may be rapid and extensive (the first-pass effect) so that little unchanged drug may be found in the systemic circulation, *e.g.* glyceryl trinitrate and most β-adrenergic antagonists.

The mixed-function oxidases, the cytochromes *P*-450, which catalyze many aspects of phase I metabolism, are membrane-bound enzymes located in the endoplasmic reticulum of the cell, as also are the glucuronyltransferases, and some of the epoxide hydrolases. Oxidoreductases which also catalyze phase I biotransformations are found in the cell cytosol, together with the sulfotransferases and other conjugases.

Metabolism may also occur in the intestinal lumen, catalyzed by certain of the enzymes of digestion and, more importantly, by the xenobiotic-metabolizing enzymes of the gut microflora which can reduce azo and nitro compounds, dehydroxylate alicyclic alcohols, hydrolyze esters, *etc.*, these metabolic reactions being complementary to those oxygenations, oxidations and conjugations catalyzed by the enzymes of the liver and the intestinal mucosa. Hence, if biliary excretion occurs, a chemical and/or its metabolites may be subjected to a progressive cycle of metabolic events beginning, if orally administered, in the intestinal mucosa, continuing in the liver, extending to the intestinal microflora after biliary excretion, followed by reabsorption and further metabolism in the liver, eventually to be excreted as metabolites in the urine.

23.14.4.3 Diurnal Variation in Metabolism

Studies in rodents have shown that the rate of hepatic drug metabolism follows a diurnal rhythm, and diurnal variations in therapeutic efficacy and toxicity have been reported for a number of cytotoxic drugs, and for barbiturates, diazepam and alcohol. Drugs with short half-lives, such as the steroids, are especially affected by the time of administration and marked differences have been found in the protein binding, clearance, $t_{\frac{1}{2}}$, AUC and t_{max} for prednisolone administered orally to human volunteers at different times of the day, with maximal rates of absorption and metabolism occurring during the middle of the day.[23]

23.14.4.4 Resistance to Metabolism

Most highly polar compounds are resistant to metabolism since they are insufficiently lipophilic to reach the major site of metabolism, the endoplasmic reticulum, and they are excreted largely unchanged, *e.g.* the herbicide asulam.

Highly lipophilic compounds which are highly halogenated, such as DDT, TCDD and the polychlorinated and polybrominated biphenyls (PCBs and PBBs), are also resistant to metabolism and consequently have long biological half-lives, low clearance values and become stored in the tissues, particularly fat depots.

Some lipophilic compounds, such as phenobarbitone, which are resistant to metabolism may initiate 'futile cycling' by interacting with the cytochromes *P*-450, converting these to the high-spin state with subsequent activation of the oxygen bound to the *P*-450, release of activated oxygen, probably superoxy anion, and the subsequent induction of the cytochrome(s). The phthalate and adipate esters, which are resistant to metabolism, induce cytochrome *P*-452 (*P*450 IV) and are associated with the production of peroxide, may act similarly.

23.14.4.5   Non-linear Kinetics

Toxicokinetic processes are potentially saturable. Drugs and chemicals which exhibit saturable absorption, disposition, elimination or metabolism may therefore result in dose-dependent toxicokinetics.

Dose-dependent absorption and delayed uptake of drugs by the liver may occur if the drug affects gastric emptying, *e.g.* chloroquine, or affects the vascular supply to the gastrointestinal tract. The β_2-adrenergic drug rimiterol shows delayed uptake of the drug by the liver and gastrointestinal tract with increasing doses, due to α-adrenergic actions of the drug causing splanchnic shut-down.[24]

Plasma protein binding is also saturable, and at high doses the plasma concentrations of unbound drug may be disproportionally high, often resulting in toxicity with highly bound drugs, such as the hypertension, oedema and hypokalaemia seen in elderly subjects taking carbenoxolone.[25]

In renal excretion, the processes of active tubular secretion are saturable; the herbicide 2,4,5-trichlorophenoxyacetic acid is eliminated by active secretion, and at high exposure disproportionally higher plasma concentrations and longer half-lives are seen.[26]

Many aspects of xenobiotic metabolism can become saturated at high dosage of chemicals, so that steady-state plasma concentrations may be disproportionally increased with increasing dose, and alternative pathways of metabolism assume greater importance, often with resulting toxicity. When the dose of aspirin to man is increased from 300 mg to 10 g the plasma half-life increases from 3 h to 20 h, due to saturation of the glycine-conjugating and glucuronide-conjugating pathways. Many other drugs which are glucuronylated readily deplete the hepatic concentration of UDP-glucuronic acid (UDPGA), the cosubstrate for glucuronide formation, and salicylamide, valproic acid and chloramphenicol deplete UDPGA by more than 90% at doses of 1 to 5 mmol kg^{-1} administered to mice.[27] Sulfate conjugation may similarly be limiting, and ingestion of large amounts of ascorbic acid, which is eliminated as the sulfate conjugate, may impair the sulfate conjugation and detoxication of adrenaline, dopamine, isoprenaline and other catecholamine drugs.[28] Paracetamol, at therapeutic dosage, is detoxicated by glucuronide and sulfate conjugation, but at higher dosage the minor metabolic pathway of oxygenation by cytochromes *P*-450 to form the reactive quinoneimine comes into play, and redox cycling and conjugation with GSH occur with consequent oxygen radical generation, depletion of intracellular GSH and hepatotoxicity (see Figure 5).

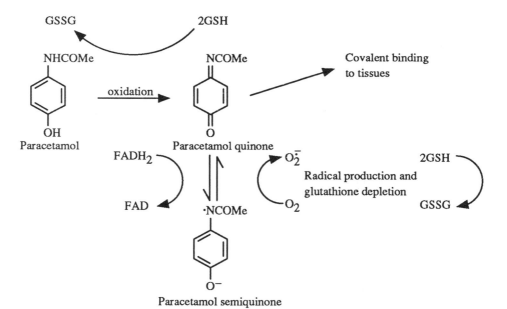

Figure 5   Redox cycling by paracetamol semiquinone. Paracetamol is oxidatively metabolized by cytochrome *P*-448 to the corresponding quinoneimine, which is then reduced by $FADH_2$ to the semiquinone radical. This free radical cyclically reduces oxygen to superoxide anion with concomitant reoxidation to the quinone

The natural flavouring material estragole is metabolized to the 1′-hydroxy analogue in a highly dose-dependent manner. The formation of the 1′-hydroxy metabolite, which is probably the proximate carcinogen, increases 70 000-fold for mice and 85 000-fold for rats when the dose of estragole is increased from 0.05 to 500 mg kg^{-1}, so that the carcinogenicity seen at very high dosage in rodents probably has little relevance to man for whom consumption is about 1 µg kg^{-1} day^{-1}.[29]

Similarly, certain drugs (*e.g.* the fungicide ketoconazole) and some metabolites may bind to the cytochromes *P*-450 inhibiting the enzymic activity, with the consequent inhibition of oxidative metabolism. In contrast, other drugs result in the induction of the cytochromes *P*-450 and other enzymes of drug metabolism, with consequent acceleration of metabolic detoxication or activation, as is seen in the increased deactivation of many barbiturates on repeat dosage (see also Sections 23.14.9.3 and 23.14.9.4).

These phenomena of enzyme saturation, inhibition and induction result in changes in the rates of xenobiotic metabolism and/or changes in the type of metabolism, so that the rates of detoxication and activation of chemicals may change considerably with increased dosage, resulting in non-linear kinetics (see Figure 6).

Non-linear kinetics may also result from capacity-limited clearance due to saturation of renal or biliary excretion, as well as to the saturation of metabolic pathways, and species differences in these effects may often give misleading indications of safety for man (*e.g.* the drug FPL 52757)[30] or misleading indications of toxicity.[31]

23.14.5 EXCRETION

Chemicals may be excreted unchanged, if they are sufficiently hydrophilic, or as metabolites or conjugates of the parent chemical or its metabolites. The major routes of excretion are the urine and bile, and, to a lesser extent, the expired air, sweat, milk, saliva and gastrointestinal secretions.

23.14.5.1 Renal Excretion

Three processes are involved: (i) glomerular filtration, which removes all xenobiotic chemicals and metabolites from the blood plasma, except those which are of high molecular weight or are highly bound to plasma proteins; (ii) active tubular transport, which comprises two saturable processes for organic anions and cations, respectively, which occur in the proximal tubule; the active secretion of organic anions is a major pathway of excretion for sulfates, glucuronides and other conjugates of xenobiotic chemicals; and (iii) passive tubular transport, by which lipid-soluble non-polar chemicals may be secreted or reabsorbed, so that highly lipophilic chemicals in the glomerular filtrate will be reabsorbed and stored in the body, whereas highly polar conjugates will be poorly reabsorbed and are consequently excreted.

Benoxaprofen, the antirheumatic drug which had to be withdrawn because of the many cases of skin photosensitization and fatal hepatotoxicity, is eliminated in the urine (60% dose) as the

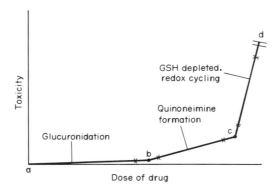

Figure 6 Non-linear toxicokinetics. The chemical (*e.g.* paracetamol) is detoxicated at low doses by conjugation with glucuronic acid and sulfate (ab). At high doses a new pathway of metabolic activation occurs (*e.g.* the formation of paracetamol quinoneimine, bc) with increasing toxicity. Later, further toxic events occur (redox cycling, redox depletion, cd) with ultimate cell death (d)

glucuronide conjugate, with the remainder being excreted unchanged in the faeces. The drug has a long half-life in man ($t_{\frac{1}{2}} = 25$–50 h) even in patients with normal renal function, and in subjects with impaired renal function, as judged by creatine clearance ($Cl_{CR} < 30$ mL min^{-1}) the half-life was markedly increased ($t_{\frac{1}{2}} > 70$ h). A significant correlation between renal function and benoxaprofen elimination kinetics has been established, so that to avoid accumulation of the drug in patients with renal insufficiency or impaired glucuronide conjugation, with all the consequent adverse effects, the dose of the drug should have been substantially reduced.[32]

23.14.5.2 Biliary Excretion

This is the major route of excretion of those chemicals and metabolites that are not readily removed from the blood plasma in the renal glomerular filtrate, *i.e.* high molecular weight chemicals (>300) and compounds that are highly bound to plasma proteins. Once secreted in the bile, compounds are not reabsorbed from the biliary tract, and are excreted into the gastrointestinal tract where they may be further metabolized by the gut microflora, reabsorbed from the intestines or excreted in the faeces.

23.14.5.3 Enterohepatic Circulation

In this phenomenon, biliary secretion is followed by reabsorption of the chemical from the gastrointestinal tract, its return to the liver in the portal system for further metabolism and finally biliary excretion again, with continuous recycling (see Figure 7). For compounds that are only slowly metabolized and are extensively biliary-excreted (*e.g.* carbenoxolone, which is $>99\%$ excreted *via* the bile), enterohepatic circulation may result in half-lives of several days, but for compounds which are highly resistant to metabolism and are also highly lipophilic and consequently biliary-secreted (*e.g.* TCDD, DDT, PCBs and PBBs), enterohepatic circulation results in half-lives of years that may even exceed the probable life-span of the animal organism under study.

Many pharmaceutical agents depend on repeated enterohepatic circulation for their therapeutic activity, for example oral contraceptive agents which are rapidly cleared *via* the bile have biological half-lives of 6 to 12 h because of enterohepatic circulation. Increased hepatic clearance by enzyme-inducing drugs such as rifampicin, phenobarbitone or phenytoin, or decreased activity of the intestinal microflora by chloramphenicol or other antibiotics, markedly decreases the effectiveness of oral contraceptives.

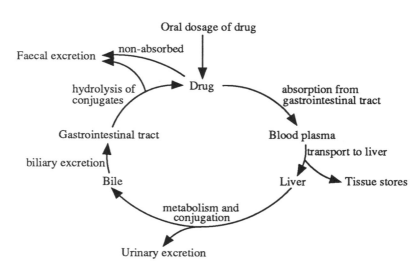

Figure 7   Enterohepatic circulation. The drug and/or its metabolites and conjugates is excreted in the bile, its conjugates are hydrolyzed by enzymes of the gastrointestinal microflora and the drug is reabsorbed to undergo further cycling

23.14.5.4 Pulmonary Excretion

Compounds which are lipophilic, non-polar and volatile, *e.g.* benzene and the halogenobenzenes, may be excreted unchanged *via* the lungs, the amount excreted by this route being a function of the steam volatility of the parent chemical.

23.14.5.5 Species Differences

The major route(s) of excretion of a chemical is dependent on the pathways of metabolism, for example carbenoxolone, an ester which in rodents is hydrolyzed by enzymes of the gut microflora and is excreted in both urine and bile in these species, in monkey, dog and man is mostly conjugated with glucuronic acid and is excreted completely in the bile.[33]

23.14.6 RECEPTOR INTERACTIONS

Chemical toxicity results ultimately from the consequence of interaction of the chemical or its reactive intermediate with 'biological receptors', which may be biological molecules vital to cellular integrity (*e.g.* glutathione), enzymes (*e.g.* the cholinesterases or the cytochromes *P*-450) or neurotransmitter or hormone receptors (*e.g.* acetycholine or steroid receptors). The consequent toxicity will depend on the tissue concentration of the receptor, the tissue concentration of the chemical or its reactive intermediate and the specific affinity of the receptor for the chemical, *i.e.* toxicity $\propto$ [receptor] [chemical] affinity. These three parameters will vary with different animal species; the tissue concentration of the specific receptors and their affinity for the particular chemical are genetically determined, and the tissue concentration of the chemical will be determined by the rates and routes of absorption, distribution, metabolism and excretion characteristic of the particular species and of the individual. Hence, in many instances the risk assessment of chemical toxicity has been decided ultimately by the comparative study of receptor affinities in different species of experimental animals and man.

Receptors may be detected, quantified and their affinities characterized by various binding assays, such as the scintillation proximity radioassay for membrane-bound receptors, which is based on the binding of ^{125}I ligands and has been used to study the interaction of α-bungarotoxin with the acetylcholine receptor.[34]

23.14.6.1 Glutathione

Intracellular reduced glutathione (GSH) is an essential redox buffer protecting the cell from the cytotoxic effects of reactive oxygen radicals and oxidants. Depletion of cellular GSH by diethyl maleate, phorone, buthionine sulfoximine, malnutrition or chemicals requiring GSH for conjugation greatly increases the hepatotoxicity of carbon tetrachloride, bromobenzene and other hepatotoxic chemicals, and substantially increases the spontaneous incidence of malignancy.[35,36]

23.14.6.2 Cholinesterases

Many pesticides exert their acute toxicity and some aspects of their chronic toxicity by inhibition of the various cholinesterase enzymes, including acetylcholinesterase and 'neurotoxic' esterase (NTE) which is responsible for 'delayed neurotoxicity'. In safety assessments of organophosphate and other pesticides, species differences in toxicity are predicted from the differing affinities of various cholinesterases for the pesticide, determined by inhibition kinetics of these enzymes in erythrocytes, brain tissue, *etc.* of different animal species.[37,38]

Delayed neurotoxicity produced by certain organophosphorus compounds, which involves selective degeneration of the central and peripheral nervous system, with a delayed neuropathy (hind-limb paralysis), is readily observed in chickens and cats but not in rodents. This species difference has been considered to be due to: (i) pharmacokinetic and metabolic differences, as the metabolic inhibitor piperonyl butoxide greatly increases rodent sensitivity to tri-*o*-cresyl phosphate (TOCP); or (ii) differences in the receptor (NTE) or its inhibition. Using TOCP in high dosage a rat model for delayed neurotoxicity has been validated, providing a valuable test system for the safety evaluation of new pesticides. NTE activity returns to normal values more rapidly in rats treated with TOCP than in chickens receiving an equipotent dose, indicating that, as rat and chicken NTE 'age' to the same extent, the rat enzyme is more rapidly resynthesized than the chicken enzyme.[39]

23.14.6.3 Steroid Receptors

Steroid hormones and antisteroids used as medicines (*e.g.* estradiol valerate in hormonal replacement therapy, diethylstilbestrol in the treatment of prostate carcinoma, or tamoxifen in the treatment of carcinoma of the breast) may interact with cytosolic hormone receptors as agonists or inhibitors, or may inhibit the specific cytochromes *P*-450 concerned with the biosynthesis of the steroid hormone. Similarly, many steroid hormones, both natural and synthetic, can modulate DNA replication and mitosis and may thus act as promoters of malignancy. Certain progestogens, unopposed by estrogen and at very high dosage, resulted in an increased incidence of carcinoma of the breast in the beagle dog; comparative receptor studies in dog and human showed that the dog was particularly susceptible because of very high levels of progestogen receptor in breast tissue and its very high affinity for the hormone, and that there was no significant risk for humans.

23.14.6.4 Ah Receptor

The polycyclic aromatic hydrocarbons and many other carcinogenic and co-carcinogenic chemicals (*e.g.* TCDD) which are substrates for the cytochromes *P*-448 have high avidity for the cytosolic Ah receptor, which regulates the induction of the cytochromes *P*-448 (*P*450 I) by increased DNA transcription. The inducer–receptor complex so formed translocates into the nucleus where it activates regulatory genes leading to increased transcription and synthesis of cytochrome *P*-448 and other proteins, and to DNA replication. Competitive inhibition of the binding of these chemicals and their reactive intermediates to the Ah receptor by epidermal growth factor, and the similarity of the mRNA produced by these chemicals and by various oncogenes, confirms that the Ah receptor may also regulate DNA replication, mitosis and tissue proliferation. Good correlations exist between the induction of cytochrome *P*-448 activity by various halogenated aromatic hydrocarbons, the avidity with which they bind to the Ah receptor and their molecular conformations (see Figure 8).

Strains of mice which are refractory to cytochrome *P*-448 induction have defective levels of the receptor and are less responsive to the carcinogenicity of chemicals whose activation requires

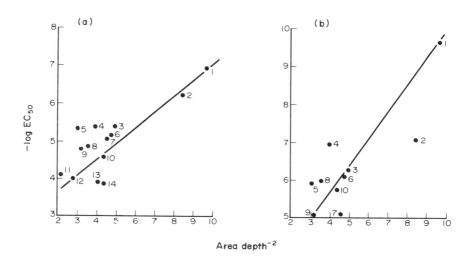

Figure 8 Correlation of (a) binding to the Ah receptor and (b) cytochrome *P*-448 induction with molecular conformation for a series of polychlorinated biphenyls. Data for (a) rat TCDD cytosolic receptor binding ($-\log \mathrm{EC}_{50}$) and for (b) cytochrome *P*-448 induction (EROD activity as $-\log \mathrm{EC}_{50}$) were taken from ref. 96. Molecular dimensions (area depth^{-2}) were determined by computer graphics as described in ref. 97. Identities of the PCBs are: (1) 3, 3',4,4',5-penta; (2) 3,3',4,4'-tetra; (3) 2,3,4,4',5-penta; (4) 2,3,3',4,4'-penta; (5) 2,3,3',4,4',5-hexa; (6) 2,3,3',4,4',5-hexa; (7) 2,3',4,4',5-penta; (8) 2',3,4,4',5-penta; (9) 2,3',4,4',5,5'-hexa; (10) 2,3,4,4'-tetra; (11) 2,2',4,4',5,5'-hexa; (12) 2,3',4,4',5',6-hexa; (13) 2,2',4,4'-tetra; (14) 2,3,4,5-tetra

metabolism by cytochrome *P*-448. Several laboratories are currently studying the toxic implications of the binding of chemicals to this receptor and their predictive value for chemical carcinogenicity.[40]

23.14.7 EFFECT OF CHEMICAL STRUCTURE

To a large extent, chemical toxicity is determined by the structure of the chemical. Organophosphate esters phosphorylate the active site of certain enzymes such as the cholinesterases resulting in their inhibition, oxidants such as benzoquinone may disrupt respiration and oxidative phosphorylation or lead to the production of active oxygen by redox cycling, and aromatic compounds may form reactive epoxides initiating covalent binding. Certain functional groups predispose to toxicity, for example nitro groups (*e.g.* chloramphenicol) are reduced *via* nitroso radicals to hydroxylamines, which are potent oxidants and often toxic and mutagenic; aromatic rings (*e.g.* benzene) can be hydroxylated to catechol and quinol derivatives which are oxidized to the highly toxic quinones; and halogen substituents may impair metabolic detoxication (*e.g.* fluorinated steroids) or undergo metabolic displacement to yield a highly toxic free-radical intermediate (*e.g.* carbon tetrachloride) or an unsaturated product which forms reactive epoxides (*e.g.* tetrachloroethane).

23.14.7.1 Stereochemical Aspects of Metabolism

The stereochemistry of a drug or chemical may also affect its rates of metabolism. The anticonvulsant drug mephenytoin (3-methyl-5-ethyl-5-phenylhydantoin) has a chiral centre at C-5 of the hydantoin ring; although both isomers are *N* demethylated, the (*S*) enantiomer is preferentially metabolized, resulting in some 200-fold difference in the clearances of the two enantiomers.[41] The 4-hydroxylation of (*S*)-mephenytoin exhibits polymorphism, independent of debrisoquine polymorphism and associated with a defective cytochrome *P*-450.[42] The (*R*) enantiomer of propranolol is more rapidly metabolized than the (*S*) enantiomer, and aromatic hydroxylation, but not *N* dealkylation, by human liver microsomes has been shown to be stereoselective.[43]

23.14.7.2 Activation by Cytochromes *P*-448

However, many aspects of chronic toxicity, mutagenicity and malignancy are the consequence of metabolic activation by the cytochromes *P*-448,[14,44] and it is significant that this is also dependent on chemical structure, for only rigid planar molecules are suitable substrates for these activating enzymes. Similarly, other specific groups of chemicals are preferentially metabolized by specific isozymes of cytochrome *P*-450; the nitrosamines are activated by cytochrome *P*450 II E, and the phthalate and adipate esters are hydroxylated by cytochrome *P*450 IV (see Table 1). Thus each of the *P*-450 isozymes appears to have an active site of characteristic spatial dimensions and electronic conformation, which defines the nature and structure of chemicals that are acceptable as substrates and hence may be metabolically detoxicated or activated (see Figure 9).

Table 1 Cytochromes *P*-450 and their Induction[89]

Family		Trivial name	Activity	Typical inducing agent
*P*450 I	A1	*P*-448 Rat *P*-450c	Activate mutagens and carcinogens	TCDD, 3-methylcholanthrene
	A2	*P*-448 Rat *P*-450d		Safrole
*P*450 II	A1	Rat *P*-450a		3-Methylcholanthrene, phenobarbital
	B1	Rat *P*-450b	Detoxication of drugs	Phenobarbital
	B2	Rat *P*-450e	and chemicals	Phenobarbital
	C1–10			
	D1 and D2			
	E	Rat *P*-450j	Activates nitrosamines	Ethanol, isoniazid, benzene
*P*450 III	A1 and A2	Rat *P*-450$_p$	Detoxicates drugs and chemicals	Pregnenolone-16α-carbonitrile, clotrimazole
*P*450 IV	A1	*P*-452	Peroxisomal proliferation	Clofibrate

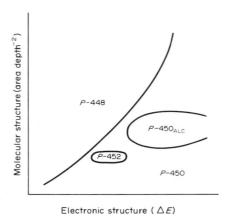

Figure 9 Plot of typical values for molecular structure (area depth^{-2}) against electronic structure (ΔE) for different families of cytochromes *P*-450. Cytochromes *P*-448 (*P*-450 I) have high values of area depth^{-2} (planar) and lower values for ΔE. Cytochromes *P*-450 (*P*450 II B) have lower values of area depth^{-2} and higher values of ΔE. Cytochromes *P*-450$_{ALC}$ (*P*-450j or *P*450 II E) and *P*-452 (*P*450 IV) have intermediate values (Data from refs. 46 and 97)

Substrates of the cytochromes *P*-448 have several important toxic attributes, which mostly interfere with cellular regulation: (i) they are metabolized to reactive intermediates which can covalently bind to proteins (neoantigens), enzymes (cytotoxins) and DNA (mutagens and initiators); (ii) they or their metabolites interact with the Ah cytosolic receptor and probably certain hormone receptors to bring about DNA replication, mitosis and tissue proliferation (non-genotoxic promoters); and (iii) they or their metabolites interact with the Ah receptor to effect the induction of cytochromes *P*-448 (co-carcinogens; see Tables 2 and 3). Human cancer causality at specific target organs is the result of initiation, co-carcinogenesis and promotion.[45] Metabolic activation by the cytochromes *P*-448 is therefore central to many toxic phenomena, including mutagenesis, carcinogenesis, immunotoxicity and teratogenesis. Furthermore, the natural causes of malignancy, namely ionizing and other radiations, oxygen radicals, viruses and steroid hormones, all seem to exert their tumourigenic effects by mechanisms similar to those manifested by the cytochromes *P*-448. Viral oncogenes initiate the transcription of mRNA similar to that induced by cytochrome *P*-448 metabolites; hydroxyl radicals produced by the action of radiation on water can, like cytochrome *P*-448, activate carcinogens and other toxic chemicals to reactive intermediates; hydroxyl radicals can also damage DNA and by their action on aromatic amino acids, *etc.* lead to the formation of heteropolycyclic compounds which induce the biosynthesis of cytochrome *P*-448 and possibly also deregulate the DNA to enhance mitosis and tissue proliferation.[13]

Substrates for the cytochromes *P*-448 are therefore likely to be mutagens, co-carcinogens, complete carcinogens or result in teratogenicity, immunotoxicity or tissue necrosis by virtue of their potential to form reactive intermediates and to interact with receptors that regulate the transcription and replication of chromatin. These substrates can be readily identified by computer-graphic determination of their molecular conformation and electronic structure (see Figure 9) and also by their specific induction of the cytochromes *P*-448, as measured by enhanced levels of 7-ethoxy-resorufin *O*-deethylase activity and other specific probes.[46] Substrates of other members of the cytochrome *P*-450 family of enzymes may similarly be identified, for example the substrates of cytochrome *P*-452 (*P*450 IV) such as clofibrate, ciprofibrate and the phthalate esters all appear to induce cytochrome *P*-452 which ω-hydroxylates lauric acid, and probably all are likely to be associated with the toxic phenomena of peroxisomal proliferation and peroxide formation.

Table 2   Mechanism of Formation of Reactive Intermediates and Oxygen Radicals

---

Oxygenation of chemicals in conformationally hindered positions by *P*-448 and oxygen radicals to form electrophiles

Oxidation of nucleophilic oxygenated metabolites (quinols and catechols) to quinones with redox cycling

Reductive dehalogenation of haloalkanes by *P*-450 and radical generation

Oxygen radical production by redox cycling, prostaglandin biosynthesis and *P*-450 futile cycling

Deconjugation of acylglucuronides and sulfates, and of GSH conjugates of haloalkanes

---

Table 3 Toxic Manifestation of Reactive Intermediates and Oxygen Radicals[a]

Depletion of intracellular GSH by reactive intermediates and oxygen radicals—loss of
    ultimate intracellular redox buffer and radical scavenger—necrosis
Covalent binding of reactive intermediates to proteins—neoantigens
Covalent binding of reactive intermediates to thio enzymes—necrosis
Covalent binding of reactive intermediates to DNA and action of oxygen radicals on
    DNA—mutations
Interaction of reactive intermediates with the Ah receptor—induction of cytochromes
    P-448 (co-carcinogens)
Interaction of reactive intermediates with Ah or other receptors—DNA replication and
    mitosis
Formation of reactive intermediates by oxygen radicals

[a] See refs. 13, 14, 35

The cytochromes *P*-448 (*P*450 I A1 and A2), corresponding to rat liver cytochromes *P*-450c and d, have been demonstrated in human liver by the use of specific polyclonal and monoclonal antibodies. Individual variations in the concentrations of these enzymes in human livers have been reported which may account for the different susceptibilities of human individuals to chemical carcinogenesis and spontaneous malignancy. Smokers and workers exposed to polychlorinated biphenyls were found to have higher levels of cytochrome *P*-448 in their liver, and pregnant women had higher levels in the placenta.

23.14.7.3 Oxygen Radical Production and Redox Cycling

A number of drugs and chemicals activate oxygen to oxygen radicals which may initiate lipid peroxidation, disrupt membranes, inactivate sulfhydryl enzymes and damage DNA. Of the different mechanisms of xenobiotic-mediated oxygen radical production, the two most important are (i) redox cycling and (ii) futile cycling of cytochrome *P*-450. In redox cycling, a chemical or its metabolite (quinone, nitro or azo compound) is reduced by a one-electron reduction to a radical intermediate (semiquinone, nitro radical anion or azo anion radical), which then transfers an electron to molecular oxygen forming superoxide anion radical ($O_2^- \cdot$) with the regeneration of the original chemical (see Figure 5).[47] Superoxide anion in the presence of traces of iron salts forms the more reactive hydroxyl radical ($\cdot OH$) which damages enzymes, DNA and other biological macromolecules, resulting in tissue necrosis, mutations and malignancy when the natural antioxidant and radical-scavenging defences (GSH, GSH peroxidase, superoxide dismutase, tocopherols, *etc.*) have been exhausted.[35,48,49] Adriamycin and other anthracycline antitumour drugs are associated with chronic glomerulonephritis and other tissue necrosis, which result from reductive activation of the quinonoid anticancer drug to the semiquinone free-radical intermediate by NADPH-cytochrome *P*-450 reductase and similar oxidoreductases, thereby initiating redox cycling, oxygen radical production and lipid peroxidation.[50] Some reactive intermediates such as epoxides may also exist in a free-radical form and hence initiate redox cycling.

In the futile cycling of cytochrome *P*-450, substrates which have an affinity for these enzymes and are able to raise the cytochrome to its high-spin state and allow the formation of the enzyme–substrate–activated-oxygen complex but do not readily undergo oxygenation, may result in the production of superoxide anion radicals or, alternatively, peroxide. Such substrates are usually inducing agents of cytochromes *P*-450 and of endoplasmic reticulum proliferation, as the oxygen radicals tend to destroy the cytochromes *P*-450 which are subsequently hyper-regenerated.

The production of oxygen radicals by redox cycling and futile cycling will be largely dependent on tissue O_2 uptake and is therefore likely to be more extensive the smaller the animal species, being more of a problem in rodents than in man.

23.14.8 EFFECTS OF GENETICS AND DEVELOPMENT

The toxicokinetics of a given chemical frequently differ in different animal species, which is attributable to genetic variations in the nature and activities of the drug-metabolizing enzymes, genetic differences in absorption and excretion, and especially to differences in the rates of basal metabolism and the rates of tissue oxygen uptake. Similarly, chemical toxicokinetics may differ markedly at different ages in the same animal species for essentially the same reasons as the species differences, namely differences in the activities of the drug-metabolizing enzymes, in the rates of absorption and excretion, and in basal metabolism.

23.14.8.1 Species Differences Due to Enzyme Variations

Species differences in toxicokinetics due to variations in the relative activities of the drug-metabolizing enzymes are numerous, hence the necessity to validate the choice of experimental animals used in toxicology studies for the evaluation of chemical safety. Some typical examples are: (i) the drug paracetamol is preferentially activated by cytochrome *P*-450 I (cyctochromes *P*-448) and the alcohol-inducible cytochrome *P*-450 II E, and alcohol-treated animals are more susceptible to paracetamol hepatotoxicity; in mice and hamsters which have high levels of this enzyme in the liver the drug is more hepatotoxic than in rats which generally have lower hepatic levels of this activating monooxygenase (see Table 4);[51] (ii) the pesticide malathion is highly neurotoxic to insects as it is metabolized primarily by cytochrome *P*-450-mediated desulfuration to the more toxic malaoxon, whereas malathion has very low toxicity in mammalia because it is largely metabolized by carboxylesterase to malathion diacid, a polar detoxication product;[38] (iii) the non-steroidal anti-inflammatory drug TAI-284 (6-chloro-5-cyclohexylindane-1-carboxylic acid) is readily hydroxylated in mice, rats and rabbits but not in guinea pigs; the $t_{\frac{1}{2}}$ of the parent drug in mice was 3.5 h and in guinea pigs was 4.5 days and species differences in the pathways and rates of metabolism correlate with species differences in pharmacological and toxicological activities (see Figure 10);[52] (iv) the hepatotoxicity of the antiasthmatic drug FPL 52757 in dog and man was shown to be due to slow oxidative metabolism and clearance compared with rodents and monkeys, leading to hepatic accumulation in dog and man; a comprehensive study of the toxicology and toxicokinetics in several different species revealed that the determination of drug clearance values was a better predictor of toxicity than were the extensive animal chronic toxicology studies;[30] (v) the species differences of the antiinflammatory drug feprazone illustrate the considerable diversity of metabolic pathways in different animal species and the difficulty in selecting an appropriate animal model for humans in toxicology studies; feprazone is oxidatively metabolized in rodents to 2'-hydroxy-, *p*-hydroxy-, and dihydroxy-feprazone, in dogs hydroxylation gives rise to different products, while in man hydroxylation is a very minor pathway, and the major pathway of metabolism is *C*-glucuronide formation (see Figure 11).[53]

23.14.8.2 Species Differences Due to Differences in Basal Metabolism

It has long been known that drugs such as barbiturates which are detoxicated by oxidative metabolism are deactivated more rapidly the smaller the animal species. Detailed studies of the *in vitro* and *in vivo* oxidative metabolism of a wide variety of chemicals in different animal species have shown that the rates of oxidative metabolism vary inversely with the body weight of the species, being highest in the mouse and lowest in man and larger animals such as the cow and horse.[54,55] This effect is illustrated with species differences in the rates of *in vivo* oxidative metabolism of $[^{14}C]$-aminopyrine in Figure 12.

Table 4　Species Differences in Paracetamol Hepatotoxicity

| Dose of paracetamol (mg kg^{-1}) | Incidence of fatal hepatic necrosis[a] (%) | | |
	Mouse	Hamster	Rat
200	—	20	0
500	76	100	0
750	99	100	0
1000	100	100	2
1500	100	100	6
Rate of covalent binding (nmol (mg protein)$^{-1}$ h^{-1})	1.6	2.1	0.8

[a] Data from refs. 99 and 100.

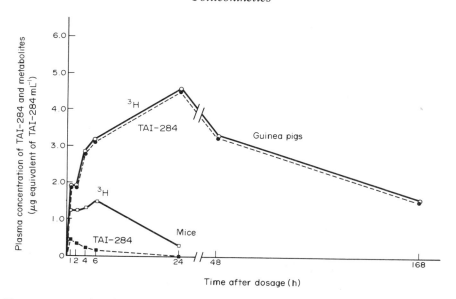

Figure 10 Plasma concentration–time profiles of [³H]TAI-284 in guinea pigs and mice. The drug TAI-284 is rapidly metabolized and eliminated by mice (unchanged drug, □ – – – – □; total ³H, ■——■), but is poorly metabolized by guinea pigs and accumulates unchanged (unchanged drug, ○ – – – – ○; total ³H, ●——●). The major ³H-labelled products in mouse plasma were hydroxylated metabolites, whereas in guinea pig more than 97% of the plasma total ³H was unchanged drug[52]

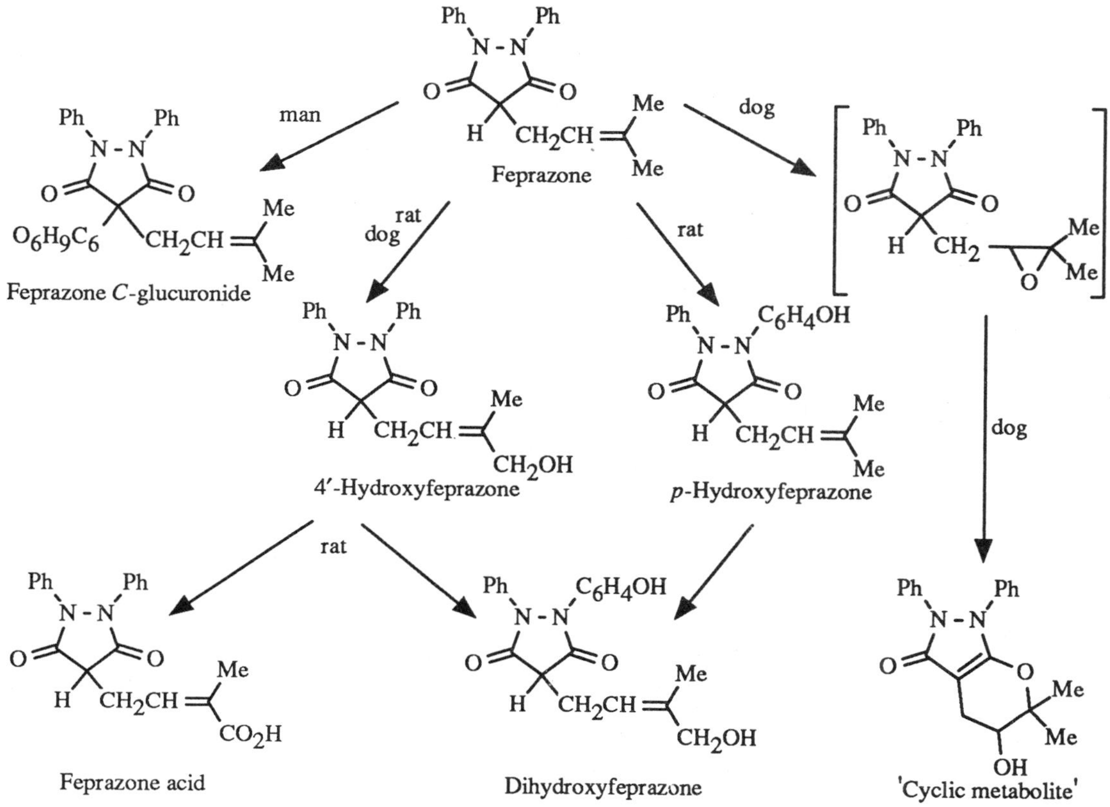

Figure 11 Species differences in the metabolism of feprazone. The non-steroidal antiinflammatory drug feprazone is metabolized by rat to *p*-hydroxyfeprazone, 4′-hydroxyfeprazone, dihydroxyfeprazone and feprazone acid, by dog to 4′-hydroxyfeprazone and the cyclic metabolite, and by man to the *C*-glucuronide (data from ref. 53)

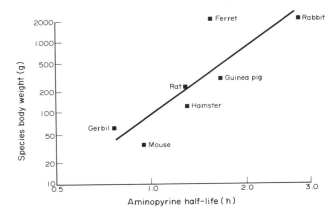

Figure 12 Species differences in the rates of oxidative metabolism of aminopyrine. The rates of oxidative metabolism (plasma $t_{\frac{1}{2}}$ values) in different species are shown to be related to body weight (data from ref. 98)

The reason for this phenomenon is that the oxidative metabolism of chemicals is partly dependent on the rate of tissue uptake of molecular oxygen, as well as on the tissue concentration of the chemical: rate of oxygenation = k [chemical] [O_2]. Because of the higher rate of basal metabolism in small animals and the dependence of basal metabolism on the oxidation of nutrient substrates, tissue uptake of O_2 varies inversely with body weight. Furthermore, the total cytochrome P-450 content of liver and other tissues varies inversely with body weight, possibly because the cytochromes P-450 are protective against the adverse effects of oxygen radicals which are formed more readily in the smaller animal; the higher rates of oxygenation of drugs in small mammals is thus seen *in vitro* as well as *in vivo*.

Hence, those chemicals which are detoxicated by oxidative metabolism are deactivated most rapidly in small rodents and consequently are less toxic than in man, whereas chemicals which are activated by oxidative metabolism are, conversely, generally more toxic in small rodents than they are in man. The generalization is, however, subject to the known variations in the abundance of the various cytochrome P-450 isozymes, some of which detoxicate while others activate chemicals.

23.14.8.3 Genetic Differences in Toxicokinetics

Individual susceptibility to drug and chemical toxicity has been shown to be due to genetic polymorphism in metabolism, such as differences in the activity of acetylation enzymes responsible for the fast and slow acetylators of isoniazid and isoniazid toxicity, and differences in esterase activity responsible for the slow deactivation and toxicity of ester drugs such as suxamethonium (see Table 5). More recently, human genetic differences in the rates of oxidative detoxication have been revealed with a number of drugs such as debrisoquine, mephenytoin, sparteine, demethylimipramine and perhexiline, and these have been shown to be due to genetic differences in the relative abundance of the cytochrome P-450 isozymes.[56] These human genetic differences in the relative

Table 5 Human Genetic Polymorphism Associated with Drug and Chemical Toxicity

Genetic polymorphism	Drug toxicity
Fast/slow acetylation	Isoniazid toxicity
Plasma cholinesterase deficiency	Suxamethonium toxicity
Glucose 6-phosphate dehydrogenase deficiency	Methaemoglobinaemia from antimalarial and sulfonamide drugs and nitro compounds
Warfarin resistance	Altered receptor affinity
Perhexiline hydroxylation	Perhexiline-induced peripheral neuropathy and hepatic necrosis
Debrisoquine hydroxylation	
Mephenytoin hydroxylation	
Cytochrome P450 I (P-448) polymorphism	Susceptibility to paracetamol hepatotoxicity and spontaneous malignancies ?

abundance of the various cytochrome *P*-450 isozymes may have much wider implications for genetic differences in toxicokinetics and chemical toxicity, as has been extensively demonstrated in the mouse. Genetic preponderance and inducibility of cytochrome *P*-448 could mean greater risks of malignancy, cardiovascular disease and other degenerative diseases associated with the formation of reactive intermediates of chemicals and with reactive oxygen radicals.

Genetic variations in the activity of glucose 6-phosphate dehydrogenase, an enzyme which produces the reduced coenzyme NADPH necessary for mixed-function oxidase activity, will also affect drug toxicokinetics. This major human genetic deficiency has been shown to diminish the oxidative metabolism of benzo[*a*]pyrene, including the formation of reactive intermediates.[57] Genetic abnormalities in the detoxication of phenytoin by epileptic patients, probably in the detoxication of phenytoin epoxide by epoxide hydrolase, has been associated with an increase in the incidence of major birth defects, but not with the incidence of minor defects.[58]

23.14.8.4 Developmental and Age Differences

The metabolic detoxication of chemicals is greatly impaired in the neonate, as was tragically discovered in the period 1945–1948 when the new antibiotic chloramphenicol was used in the treatment of the European post-war epidemic of infantile diarrhoea, with disastrous consequences. The high neonatal mortality was attributed to the very poor ability of infants to glucuronylate the drug, which is the major pathway of detoxication. The cause of this impairment of glucuronide conjugation, and of other detoxicating enzymes, is that these enzymes develop during the late foetal period and in the first few weeks after birth. Moreover, the enzymes develop at different times, resulting in markedly different toxicokinetics in the neonate from those observed in the adult animal.[59] Sulfate conjugation is present in the foetus, but glucuronide conjugation and other detoxicating conjugases develop postnatally in most mammalian species. Concerning the hepatic microsomal mixed-function oxidases, the cytochromes *P*-448 appear to be vestigial enzymes which are present at high levels in the late foetus and the young neonate, at least in the rat, whereas the other forms of cytochrome *P*-450 develop after birth and gradually replace the cytochromes *P*-448, achieving adult levels at about six weeks of age. Therefore, the neonate tends to activate toxic chemicals in preference to effecting their detoxication, and consequently is generally much more susceptible to chemical toxicity and carcinogenicity. This developmental deficiency in defence against toxic chemicals and carcinogens may be a contributing cause of childhood cancer.

A similar impairment in the detoxication of drugs and chemicals also occurs in old age. The aging process has been associated with the adverse effects of oxygen radicals which are known to impair the function of the endoplasmic reticulum and other membranes, resulting in decreased rates of clearance of toxic chemicals due to decreased rates of metabolism and decreased rates of excretion. This is of critical importance in the toxicity of drugs with low clearance values administered to elderly patients, especially where these drugs are enterohepatically circulated and tend to accumulate. As plasma protein binding is decreased in elderly subjects, due to lower plasma albumin concentrations, drugs which are highly bound to plasma proteins tend to exhibit increased toxicity in the elderly. Examples of these altered toxicokinetics in the elderly, resulting in drug toxicity, are: (i) the hypertensive hypokalaemic effects of the antiulcer drug carbenoxolone seen predominantly in the elderly ($t_{\frac{1}{2}}$ for > 65 years was 56 h, for < 40 years was 19 h; see Figure 13);[25] and (ii) the antirheumatoid drug benoxaprofen, which has a long half-life, tends to progressively accumulate in the elderly with hepatotoxicity and phototoxicity, often with fatal results.[32]

23.14.9 EFFECTS OF ENVIRONMENT

The rates of metabolism and clearance of drugs and chemicals may be profoundly affected by the environment of the individual, including the nutrition and previous and simultaneous exposure to other drugs and chemicals resulting in inhibition or induction of the drug-metabolizing enzymes. Adequate nutrition is essential to the detoxication of chemicals and to protection against the toxicity of reactive intermediates and of oxygen radicals, the production of which may be accelerated by drugs and other xenobiotics. Inhibition or induction of the drug-metabolizing enzymes, due to inadvertant exposure to environmental chemicals, may affect not only the rates of metabolic detoxication and activation but can also determine the predominant pathways of metabolism involved.[60]

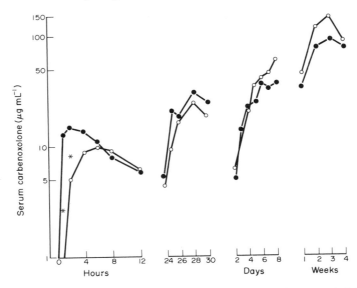

Figure 13 The effect of age on plasma concentration of carbenoxolone. Carbenoxolone was administered orally as a single dose, then three times daily. The plasma t_{max} is lower and later in the elderly (> 65 years, $\bigcirc$) than in the younger adult (< 40 years, $\bullet$). After three weeks of repeated dosing the plasma concentration in the elderly ($130 \mu g \, mL^{-1}$) is higher than that in the younger adult ($85 \mu g \, mL^{-1}$) and much higher than after a single dose ($10 \mu g \, mL^{-1}$) (data from ref. 25)

23.14.9.1 Nutrition

The rates of detoxication of drugs and chemicals, and hence their toxicity, are highly dependent on the nutrition of the individual. Adequate calories and dietary carbohydrate are necessary for the production of the reduced coenzyme NADPH required for the action of the cytochrome *P*-450 mixed-function oxidases, and for the formation of the coenzyme UDP-glucuronic acid (UDPGA) required for the formation of glucuronide conjugates. Polyunsaturated fatty acids and phospholipids are essential for the integrity of the endoplasmic reticulum and for the normal functioning of the microsomal drug-metabolizing enzymes, including the mixed-function oxidases and glucuronyltransferases, and deficiency may result in impairment of drug metabolism and uncoupling of the cytochrome *P*-450 from the reductase with electron leakage, increased activation of oxygen and production of oxygen radicals. Protein, especially sulfur amino acids, is needed for the synthesis of drug-metabolizing enzymes and enzymes protecting against oxygen toxicity (*e.g.* superoxide dismutase, GSH reductase), sulfate and peptide conjugations, and the synthesis of GSH required for GSH conjugation and mercapturic acid synthesis, and protection against the toxicity of quinones, and other oxidants and redox cycling agents. Vitamins are also essential for adequate drug metabolism and detoxication, with nicotinamide and ascorbic acid being required for mixed-function oxidation, riboflavin for the reduction of azo compounds (*e.g.* salazopyrine) by the intestinal microflora, folate is required for the synthesis of drug-metabolizing enzymes, and the tocopherols and ascorbic acid protect against oxygen radical toxicity. Malnutrition, with lowering of plasma albumin concentration, decreases the plasma protein binding of drugs such as phenylbutazone, increasing the unbound concentration of the drug, accelerating its metabolism and clearance and decreasing its plasma half-life. High levels of drug administration or high exposure to chemicals therefore require adequate nutrition or even nutritional supplementation, a fact generally ignored in toxicokinetics and therapeutics, often with toxic consequences.[61,62]

Important examples of the impact of nutrition on the toxicokinetics of drugs and chemicals include the following: bromobenzene, a model toxic chemical extensively studied in *in vitro* systems by toxicologists, is not particularly toxic to the well-nourished intact animal; it is readily detoxicated by intracellular GSH and the dose resulting in hepatotoxicity in the well-fed rat is a hundred fold greater than the minimum hepatotoxic dose in rats starved for 48 h (see Figure 14).[63] The drug toxicity, teratogenicity and haematopoietic toxicity of phenobarbitone and diphenylhydantoin in the treatment of epilepsy have been shown to be due to the enzyme-inductive effects of these drugs and the depletion of folate on a low folate diet, consequent upon the increased protein turnover associated with prolonged enzyme induction on chronic treatment.[64] Supplementation of hamsters

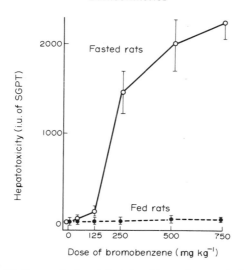

Figure 14 The effect of fasting on the hepatotoxicity of bromobenzene in the rat (data from ref. 63)

with vitamin A (2 μg g^{-1} retinyl acetate + 13-*cis*-retinoic acid) during and after dosing with benzo[*a*]pyrene reduced the incidence of benzo[*a*]pyrene-induced respiratory tract malignant tumours from 80% in control (0·3 μg g^{-1} retinyl acetate only) to 4%.[65] Lastly, the effect of dietary riboflavin may profoundly affect the metabolism and toxicity of azo compounds, such as the drug sulfasalazine; high levels of riboflavin in the diet accelerate the rate of reduction of azo compounds to the constituent aromatic amines by the intestinal microflora and, dependent on whether the parent azo compound or the aromatic amine metabolites are the more toxic/carcinogenic, decrease or increase the consequent toxicity.

23.14.9.2 Drug and Chemical Interactions

Interactions may occur if two drugs are ingested simultaneously or within a short period of each other. These may occur at sites of absorption, distribution, metabolism and/or excretion and normally are toxicologically important only when the therapeutic ratio is small.

Drug interactions at the site of absorption usually decrease the rate of absorption or the amount of drug absorbed, thus delaying or diminishing the pharmacological/toxicological effects. Changes in gastrointestinal pH may affect drug absorption by affecting the degree of ionization; the pharmacological effect of pentobarbital, a weak acid, is abolished by administration of antacids,[66] while that of pseudoephedrine, a weak base, is enhanced.[67] Drugs with anticholinergic activity slow gastric emptying and thereby delay absorption, while drugs that accelerate gastric emptying will hasten absorption; propantheline and metoclopramide delay and accelerate gastric emptying, respectively, and so modify the absorption of paracetamol.[68] Drugs may also interact chemically in the gastrointestinal tract to form poorly absorbed complexes, and the efficacy of tetracyclines is decreased when co-administered with iron because of complex formation.[69]

Drugs may also interact in their tissue distribution, and highly protein-bound drugs may displace each other from plasma proteins often with consequent toxicity; phenylbutazone may give rise to hypothrombinaemia and bleeding in patients receiving warfarin anticoagulants because of displacement of the latter from plasma albumin.[70]

Drugs and chemicals also interact by virtue of their metabolism. Inducers and inhibitors of the mixed-function oxidases and other drug-metabolizing enzymes accelerate or inhibit the metabolism of other chemicals leading to increased or decreased toxicity depending on the outcome of metabolism. Barbiturates, among the most potent inducing agents, have been implicated in many drug interactions some of which, *e.g.* with anticoagulants, expose the patient to a high risk of toxicity. When barbiturate–anticoagulant therapy is necessary, the dose of anticoagulant is increased to account for its enhanced deactivation, but when the barbiturate is withdrawn mixed-function oxidase activity returns to basal levels, and it is imperative that the dose of warfarin is reduced to avoid haemorrhage from excessive anticoagulant levels.[71] Mixed-function oxidase activity is also increased by exposure to tobacco smoke, and the bronchodilator theophylline, taken by smokers to treat obstructive lung disease, is more rapidly metabolized by smokers and large doses are required to produce a satisfactory response.[72] Chemicals may interact with the heme moiety of cytochrome *P*-450, thereby inhibiting the oxygenation of other substrates which consequently reach high concentrations. The antifungal agent ketoconazole is such a chemical, which gives rise to inhibition of several cytochrome *P*-450 dependent mixed-function oxidases with clinically significant interactions;[73] similarly, the H_2-antagonist cimetidine inhibits the metabolism of many concurrently administered drugs.[74]

Drugs may also compete in excretion, especially for the same carrier in renal tubular secretion; phenylbutazone potentiates the antidiabetic action of acetohexamide by inhibiting the renal excretion of its active metabolite hydroxyacetohexamide.[75] Similarly, drugs which undergo passive tubular resorption are affected by changes in urinary pH which change the extent of ionization; antacids raise urinary pH and accelerate the excretion of weak acids such as salicylic acid but impair the excretion of weak bases such as amphetamines.[76,77]

23.14.9.3 Enzyme Inhibition

The major aspects of enzyme inhibition by environmental chemicals affecting the toxicokinetics of other chemicals concern the cytochrome *P*-450 mixed-function oxidases and their inhibition by alcohol and carbon monoxide, and by specific imidazole inhibitors such as ketoconazole and cimetidine.

Carbon monoxide, a specific inhibitor of the various cytochrome *P*-450 isozymes, reaches significant levels in urban environments, due mostly to automotive emissions, and is also a major component of cigarette smoke, so that significant blood levels of carbon monoxide are achieved which consequently result in inhibition of hepatic cytochromes *P*-450. Similarly, the ingestion of alcoholic beverages results in inhibition of mixed-function oxidations, so that alcohol decreases the rates of metabolic detoxication of concomitantly administered barbiturates and many other drugs, resulting in prolonged and increased pharmacological activity, drug toxicity and sometimes fatal consequences. Alcohol is known to initiate the formation of oxygen radicals in the gastrointestinal tract and liver when orally ingested.[78] Lipid peroxidation and products of this process, the hydroxyalkenals and hydroxyl radicals, inhibit the mixed-function oxidases due to destruction of the cytochromes *P*-450. Therefore chemicals, like alcohol, and disease states, such as rheumatoid arthritis, which initiate lipid peroxidation, will inhibit cytochrome *P*-450 activities with consequent effects on the toxicokinetics of other drugs and chemicals.[79,80]

A number of drugs that are substituted imidazoles have been devised specifically to inhibit certain cytochromes *P*-450, for example ketoconazole, the systemic fungicide which inhibits the cytochrome *P*-450 of yeast responsible for the synthesis of fungal sterols. However, such drugs lack absolute specificity and also inhibit the cytochromes *P*-450 responsible for the biosynthesis of mammalian androgens, estrogens and corticosteroid hormones, which is the probable cause of the observed hepatotoxicity.[81] These and other imidazole drugs, such as cimetidine, metronidazole and etomidate, are therefore potential sources of drug interactions by their non-specific inhibition of the cytochromes *P*-450.

Cimetidine, the antipeptic-ulcer H_2-receptor antagonist, is an inhibitor of the oxidative metabolism of drugs by its inhibition of the cytochromes *P*-450. It increases the toxicity of hexamethylmelamine and other cytotoxic drugs,[82] decreases the clearance of warfarin and increases the plasma concentration and prothrombin time of warfarin and other oral anticoagulants, and inhibits the metabolic detoxication of aminopyrine, phenytoin, theophylline, propranolol, labetalol and benzodiazepines.[83] Ranitidine, a structurally related drug, is not only four- to five-fold more potent than cimetidine as an H_2 antagonist, but is almost without inhibitory action on cytochrome *P*-450-mediated oxidative drug metabolism, and is therefore the preferred antipeptic agent when other drugs are being administered simultaneously.[82]

Inhibition of enzymic activity may also arise from enzyme destruction, as occurs with cytochrome *P*-450 after exposure to carbon disulfide or administration of unsaturated suicide substrates. Carbon disulfide and various thiophosphonate pesticides such as parathion are metabolized by cytochrome *P*-450 which involves displacement of the sulfur by oxygen to yield CO_2 and paraoxon;[84] the displaced sulfene radical then results in the loss of cytochrome *P*-450 *via* denaturation to *P*-420.[85] Chemicals with terminal alkene or alkyne groups, such as the contraceptive steroid norethindrone, the antiinflammatory analgesic alclofenac, the anaesthetic fluoroxene or the allylbarbiturates, undergo metabolic activation by epoxidation of the unsaturated bond with destruction of the heme moiety of cytochrome *P*-450, the formation of green pigment in the liver and inhibition of ferrochelatase with accompanying porphyria.[86,87]

Inhibition of metabolism may also arise from the presence of impurities, arising either in the manufacture of the chemical or from its subsequent decomposition during storage. Impurities present in some productions of the pesticide phenthoate increase its mammalian toxicity more than fifty fold, due to inhibition of detoxication by the carboxylesterase enzymes,[88] and decomposition of malathion to isomalathion and other impurities on storage similarly increases the toxicity of malathion, due to inhibition of the detoxicating carboxylesterases by the impurities.[38]

23.14.9.4 Enzyme Induction

Both the phase 1 (mixed-function oxidases) and phase 2 (conjugases) enzymic activities may be enhanced by *in vivo* exposure to certain environmental chemicals. It was originally believed that there were two major classes of chemical inducing agents, namely planar molecules (*e.g.* 3-methylcholanthrene) which induce the cytochromes *P*-448 (*P*450 I) together with conjugases which accept planar substrates, and the globular inducers (*e.g.* phenobarbitone) of the other classes of the cytochromes *P*-450 and of many of the conjugases. However, with the resolution of the many different cytochrome *P*-450 families it is now realized that there are correspondingly many classes of inducing agents (see Table 1).[89]

Regarding the effects of enzyme induction on toxicokinetics, the induction of the cytochromes *P*-448 by cigarette smoke, pyrolysis food products, the polycyclic hydrocarbons of engine exhausts, TCDD and other halogenated polycyclic environmental contaminants is the most important as it results in a marked increase in the oxidative activation of chemicals to mutagens, carcinogens and other reactive intermediates. Moreover, apart from the induction of cytochromes *P*-448 (*P*450 I A1 and A2), an increase of the cytosolic Ah receptor also occurs, together with the promotion of DNA replication, mitosis and tissue proliferation, resulting in an overall increase in the processes of co-carcinogenesis and promotion, as well as mutagenesis.

In contrast, induction of other families of the cytochromes *P*-450 by a variety of drugs including phenobarbitone and diphenylhydantoin, pesticides such as DDT and dieldrin, and certain environmental contaminants such as the non-planar polychlorinated biphenyls generally shifts the toxicokinetics of a drug or chemical towards pathways of detoxication and increased clearance values.

The isozyme cytochrome *P*450 II E is specifically induced by alcohol, the drug isoniazid and by the chemicals trichloroethylene, pyrazole and acetone, and is the enzyme which specifically catalyzes the *N* demethylation of dimethylnitrosamine, a pathway which may determine the carcinogenicity of this chemical, together with the metabolic activation of paracetamol, activation of aromatic amines and the reductive dehalogenation of carbon tetrachloride and related hepatotoxic halo-alkanes. The isozyme *P*450 III is specifically induced by the 'catatoxic' steroid pregnenolone-16α-carbonitrile, the synthetic steroids spironolactone and dexamethasone, and by the antibiotics rifampicin and triacetyloleandomycin, and imidazoles such as clotrimazole, the induction being accompanied by an increased resistance to the toxicity of many drugs and chemicals, presumably due to the detoxicating activities of this particular cytochrome. Cytochrome *P*-452 (*P*450 IV) is specifically induced by a number of antihyperlipidaemic drugs, such as clofibrate and ciprofibrate, and by the industrial plasticizer di(2-ethylhexyl)phthalate and other related esters; this induction is accompanied by production of hydrogen peroxide and hepatic peroxisomal proliferation which, in rodents at least, results in hepatocellular malignancy. Consequently, the induction of the various isozymes of cytochrome *P*-450 by specific drugs and chemicals may exert a great diversity of effects on the toxicokinetics of a chemical, increasing the toxicity of some and decreasing the toxicity of others, depending on the molecular structure of the chemical concerned and the nature of the inducing chemical.[89]

Induction of the cytochrome *P*-450-dependent mixed-function oxidases may show marked species differences. Those most important to toxicokinetics concern the induction of cytochrome *P*-448 (*P*450 I), the enzyme responsible for the activation of most mutagens and carcinogens and many other toxic chemicals. Mice, hamsters and dogs have high basal levels of this enzyme, while rats and guinea pigs have low basal levels, but in the rat the enzyme is readily inducible.[90] Thus a single dose of the carcinogen 2-acetamidofluorene, which is activated by cytochrome *P*-448-mediated *N*-hydroxylation, is not highly toxic to the rat but is carcinogenic on repeated dosage; in contrast, the guinea pig, which does not exhibit the marked induction of *P*-448 seen in the rat, is resistant to the hepatocarcinogenicity of 2-acetamidofluorene.

Induction of the conjugating enzymes usually results in increased detoxication, self-regulating in the case of induction by drugs and fortuitous in the case of environmental chemicals. The UDPglucuronyltransferases, like the cytochromes *P*-450, appear to be a family of isozymes, selectively induced by specific chemicals, such as TCDD and 3-methylcholanthrene which induce the UDPglucuronyltransferase which glucuronylates planar phenols, phenobarbitone which induces UDPglucuronyltransferases conjugating globular phenols and terpenoid alcohols, and clofibrate which induces the glucuronylation of bilirubin.[91] Induction of the UDPglucuronyltransferases will modulate the toxicokinetics of a drug, generally towards increased detoxication, although there are known examples where glucuronide conjugation may lead to increased toxicity, as in the case of the acylglucuronides.[92] Induction and decreased activities of the microsomal and cytosolic epoxide hydrolases are known, which obviously will modulate the toxicokinetics of epoxides and chemicals metabolized by epoxidation. The microsomal enzyme(s) is induced by most classical inducers (phenobarbitone, Aroclor 1254) except 3-methylcholanthrene, and by alcohol, isoniazid and rifampicin, while butylated hydroxyanisole (BHA) produces a sixfold increase. The cytosolic epoxide hydrolase is increased particularly by clofibrate and other peroxisomal-proliferating chemicals, but is decreased by BHA.[91]

The antioxidant ethoxyquine fed to mice and rats increases liver GSH, GSH *S*-transferase activity and the cytochromes *P*-450. It protected mice against the hepatotoxicity of the pyrrolizidine alkaloid monocrotaline, probably by increased detoxication.[93] Butylated hydroxyanisole similarly protected rats against pyrrolizidine alkaloid toxicity.[94] Induction of detoxicating enzymes by ethoxyquine in rats also led to enhanced detoxication of the carcinogen aflatoxin B_1, decreased the covalent binding to DNA and decreased the incidence of neoplasia.[95]

23.14.10 REFERENCES

1. G. Zbinden, *Xenobiotica*, 1988, **18**, (suppl. 1), 9.
2. H. J. Rogers, F. R. House, P. J. Morrison and I. D. Bradbrook, *Lancet*, 1980, **2**, 694.
3. R. Kreiling, R. J. Laib, J. G. Filser and H. M. Bolt, *Arch. Toxicol.*, 1986, **58**, 235.
4. M. E. Andersen and W. C. Keller, in 'Cutaneous Toxicity', ed. V. A. Drill and P. Lazar, Raven Press, New York, 1984, p. 9.
5. D. B. Tuey and H. B. Matthews, *Toxicol. Appl. Pharmacol.*, 1980, **53**, 420.
6. L. L. Smith, *Rev. Biochem. Toxicol.*, 1987, **8**, 37.
7. J. W. Bridges and D. J. Benford, *Prog. Drug Metab.*, 1986, **9**, 53.
8. H. Nau, *EHP, Environ. Health Perspect.*, 1986, **70**, 113.
9. A. B. G. Lansdown, in 'The Future of Predictive Safety Evaluation', ed. A. Worden, D. Parke and J. Marks, MTP Press, Lancaster, 1987, vol. 2, p. 77.
10. D. Neubert, *Xenobiotica*, 1988, **18** (suppl. 1), 45.
11. J. P. Griffin and P. F. D'Arcy, 'A Manual of Adverse Drug Interactions', Wright, Bristol, 1984, p. 36.
12. E. C. Henry and R. K. Miller, *Biochem. Pharmacol.*, 1986, **35**, 1993.
13. D. V. Parke, *Arch. Toxicol.*, 1987, **60**, 5.
14. D. V. Parke, in 'The Future of Predictive Safety Evaluation', ed. A. Worden, D. Parke and J. Marks, MTP Press, Lancaster, 1987, p. 3.
15. A. W. Rettenmeier, W. P. Gordon, K. S. Prickett, R. H. Levy and T. A. Baillie, *Drug Metab. Dispos.*, 1986, **14**, 454.
16. F. C. Kauffman, *Fed. Proc., Fed. Am. Soc. Exp. Biol.*, 1987, **46**, 2434.
17. T. Watabe, T. Fujieda, A. Hiratsuka, T. Ishizuka, Y. Hakamata and K. Ogura, *Biochem. Pharmacol.*, 1985, **34**, 3002.
18. K. B. Delclos, E. C. Miller, J. A. Miller and A. Liem, *Carcinogenesis, (N.Y.)*, 1986, **7**, 277.
19. A. A. Elfarra, I. Jakobson and M. W. Anders, *Biochem. Pharmacol.*, 1986, **35**, 283.
20. S. Vamvakas, W. Dekant, K. Berthold, S. Schmidt, D. Wild and D. Henschler, *Biochem. Pharmacol.*, 1987, **36**, 2741.
21. R. Fears, *Prog. Lipid Res.*, 1985, **24**, 177.
22. O. Hänninen, P. Lindström-Seppä and K. Pelkonen, *Arch. Toxicol.*, 1987, **60**, 34.
23. J. English, M. Dunne and V. M. Marks, *Clin. Pharmacol. Ther.*, 1983, **33**, 381.
24. J. P. Griffin, J. R. B. Williams and E. Maughan, *Xenobiotica*, 1974, **4**, 755.
25. M. J. Hayes, M. Sprackling and M. J. S. Langman, *Gut*, 1977, **18**, 1054.
26. M. W. Sauerhoff, W. H. Braun, G. E. Blau and P. J. Gehring, *Toxicol. Appl. Pharmacol.*, 1976, **36**, 491.
27. S. R. Howell, G. A. Hazelton and C. D. Klaassen, *J. Pharmacol. Exp. Ther.*, 1986, **236**, 610.
28. J. W. Dunne, L. Davidson, R. Van Dongen, L. J. Beilin, A. M. Tunney and P. B. Rogers, *Br. J. Clin. Pharmacol.*, 1984, **17**, 356.
29. A. Zangouras, J. Caldwell, A. J. Hutt and R. L. Smith, *Biochem. Pharmacol.*, 1981, **30**, 1383.
30. A. J. Clarke, B. Clark, C. T. Eason and D. V. Parke, *Regul. Toxicol. Pharmacol.*, 1985, **5**, 109.
31. B. Clark and D. A. Smith, *CRC Crit. Rev. Toxicol.*, 1984, **12**, 343.
32. G. R. Aronoff, T. Ozawa, K. A. DeSante, J. F. Nash and A. S. Ridolfo, *Clin. Pharmacol. Ther.*, 1982, **32**, 190.
33. P. Iveson, W. E. Lindup, D. V. Parke and R. T. Williams, *Xenobiotica*, 1971, **1**, 79.
34. N. Nelson, *Anal. Biochem.*, 1987, **165**, 287.
35. N. E. Preece, P. F. Evans, L. J. King and D. V. Parke, *Xenobiotica*, 1989, in press.
36. R. Reiter and R. F. Burk, *Biochem. Pharmacol.*, 1988, **37**, 327.
37. J.-M. Chemnitius, K.-H. Haselmeyer and R. Zech, *Biochem. Pharmacol.*, 1983, **32**, 1693.
38. D. V. Parke and R. Truhaut, in 'Food Toxicology—Real or Imaginary', ed. G. G. Gibson and R. Walker, Taylor and Francis, London, 1985, p. 259
39. S. Padilla and B. Veronesi, *Toxicol. Appl. Pharmacol.*, 1985, **78**, 78.
40. D. W. Nebert and F. J. Gonzalez, *Annu. Rev. Biochem.*, 1987, **56**, 945.
41. P. J. Wedlund, W. S. Aslanian, E. Jacqz, C. B. McAllister, R. A. Branch and G. R. Wilkinson, *J. Pharmacol. Exp. Ther.*, 1985, **234**, 662.

42. S. A. Ward, F. Goto, K. Nakamura, E. Jacqz, G. R. Wilkinson and R. A. Branch, *Clin. Pharmacol. Ther.*, 1987, **42**, 96.
43. W. L. Nelson and H. U. Shetty, *Drug Metab. Dispos.*, 1986, **14**, 506.
44. C. Ioannides, P. Y. Lum and D. V. Parke, *Xenobiotica*, 1984, **14**, 119.
45. J. H. Weisburger and G. M. Williams, in 'Human Risk Assessment', ed. M. V. Roloff, Taylor and Francis, London, 1987, p. 163.
46. D. F. V. Lewis, C. Ioannides and D. V. Parke, *Chem.-Biol. Interact.*, 1987, **64**, 39.
47. H. Kappus, *Arch. Toxicol.*, 1987, **60**, 144.
48. L. J. Machlin and A. Bendich, *Fed. Am. Soc. Exp. Biol. J.*, 1987, **1**, 441.
49. S. Tong, H. A. Masson, C. Ioannides, W. D. Bechtel and D. V. Parke, *Xenobiotica*, 1986, **16**, 595.
50. E. G. Mimnaugh, M. A. Trush and T. E. Gram, *Biochem. Pharmacol.*, 1986, **35**, 4327.
51. R. Hyde, J. N. Smith and C. Ioannides, *Mutagenesis*, 1987, **2**, 477.
52. S. Tanayama, *Xenobiotica*, 1973, **3**, 671.
53. M. Gaetani, H. Yamaguchi, A. Vidi, Y. Hashimoto and A. Donetti, *Pharmacol. Res. Commun.*, 1979, **11**, 719.
54. C. H. Walker, *Drug Metab. Rev.*, 1978, **7**, 295.
55. J. Mordenti, *J. Pharm. Sci.*, 1986, **75**, 1028.
56. F. P. Guengerich, L. M. Distlerath, P. E. B. Reilly, T. Wolff, T. Shimada, D. R. Umbenhauer and M. V. Martin, *Xenobiotica*, 1986, **16**, 367.
57. L. Pirisi, R. Garcea, R. Pascale, M. E. Ruggiu and F. Feo, *Toxicol. Pathol.*, 1987, **15**, 115.
58. S. M. Strickler, L. V. Dansky, M. A. Miller, M. H. Seni, E. Andermann and S. P. Spielberg, *Lancet*, 1985, **2**, 746.
59. D. V. Parke, in 'Toxicology and the Newborn', ed. S. Kacew and M. J. Reasor, Elsevier, Amsterdam, 1984, p. 1.
60. D. V. Parke, *Regul. Toxicol. Pharmacol.*, 1987, 7, 222.
61. D. V. Parke and C. Ioannides, *Annu. Rev. Nutr.*, 1981, **1**, 207.
62. D. L. Frape, in 'The Future of Predictive Safety Evaluation', ed. A. Worden, D. Parke and J. Marks, MTP Press, Lancaster, 1987, vol. 2, p. 155.
63. D. Pessayre, A. Dolder and J. V. Artigou, *Gastroenterology*, 1979, **77**, 264.
64. D. Labadarios, J. W. T. Dickerson, E. G. Lucas, G. H. Obuwa and D. V. Parke, *Br. J. Clin. Pharmacol.*, 1978, **5**, 167.
65. P. M. Newberne and M. W. Conner, in 'Human Risk Assessment', ed. M. V. Roloff, Taylor and Francis, London, 1987, p. 65.
66. A. Hurwitz and M. B. Sheehan, *J. Pharmacol. Exp. Ther.*, 1971, **179**, 124.
67. R. L. Lucarotti, H. Barry and R. I. Poust, *J. Pharm. Sci.*, 1972, **61**, 903.
68. J. Nimmo, R. C. Heading, P. Tothill and L. F. Prescott, *Br. Med. J.*, 1973, **1**, 587.
69. P. J. Neuvonen and H. Turakka, *J. Clin. Pharmacol.*, 1974, **7**, 347.
70. H. M. Solomon, J. J. Schrogie and D. Williams, *Biochem. Pharmacol.*, 1968, **17**, 143.
71. M. G. McDonald, D. S. Robinson, D. Sylvester and J. J. Jaffe, *Clin. Pharmacol. Ther.*, 1969, **10**, 80.
72. J. Jenne, H. Nagasawa, R. McHugh, F. MacDonald and E. Wyse, *Life Sci.*, 1975, **17**, 195.
73. R. M. Ferguson, D. E. R. Sutherland, R. L. Simmons and J. S. Najarian, *Lancet*, 1982, **2**, 882.
74. M. J. Serlin, R. G. Sibeon, S. Mossman, A. M. Breckenridge, J. R. B. Williams, J. L. Atwood and J. M. T. Willoughby, *Lancet*, 1979, **2**, 317.
75. J. B. Field, M. Ohta, C. Boyle and A. Remer, *N. Engl. J. Med.*, 1967, **277**, 889.
76. G. Levy, T. Lampmann, B. L. Kamath and L. K. Garrettson, *N. Engl. J. Med.*, 1975, **293**, 323.
77. A. H. Beckett and M. Rowland, *J. Pharm. Pharmacol.*, 1965, **17**, 628.
78. E. Dicker and A. I. Cederbaum, *Alcohol.: Clin. Exp. Res.*, 1987, **11**, 309.
79. M. M. Iba, and G. J. Mannering, *Biochem. Pharmacol.*, 1987, **36**, 1447.
80. M. W. Lame and H. J. Segal, *Chem.-Biol. Interact.*, 1987, **62**, 59.
81. D. Feldman, *Endocr. Rev.*, 1986, **7**, 409.
82. K. Hande, G. Combs, R. Swingle, G. L. Combs and L. Anthony, *Cancer Treat. Rep.*, 1986, **70**, 1443.
83. A. Somogyi and R. Gugler, *Clin. Pharmacokinet.*, 1982, **7**, 23.
84. R. A. Neal, *Rev. Biochem. Toxicol.*, 1980, **2**, 131.
85. M. J. Obrebska, P. A. Kentish and D. V. Parke, *Biochem. J.*, 1980, **188**, 107.
86. L. M. Brown, and A. W. Ford-Hutchinson, *Biochem. Pharmacol.*, 1982, **31**, 195.
87. F. P. Guengerich and D. C. Liebler, *CRC Crit. Rev. Toxicol.*, 1985, **14**, 259.
88. S. G. K. Lee and T. R. Fukuto, *J. Toxicol. Environ. Health*, 1982, **10**, 717.
89. D. V. Parke, in 'Frontiers in Biotransformation', ed. K. Ruckpaul, Akademie-Verlag, Berlin, 1989, in press.
90. K. Iwasaki, P. Y. Lum, C. Ioannides and D. V. Parke, *Biochem. Pharmacol.*, 1986, **35**, 3879.
91. G. Siest, A. M. Batt, S. Fournel-Gigleux, M. M. Galteau, M. Wellman-Bednawska, A. Minn and A. Amar Costesec, *Xenobiotica*, 1988, **18**, suppl. 1, 21.
92. R. B. Van Breemen and C. C. Fenselau, *Drug Metab. Dispos.*, 1986, **14**, 197.
93. C. L. Miranda, H. M. Carpenter, P. R. Cheeke and D. R. Buhler, *Chem.-Biol. Interact.*, 1981, **37**, 95.
94. C. L. Miranda, D. R. Buhler, H. S. Ramsdell, P. R. Cheeke and J. A. Schmitz, *Toxicol. Lett.*, 1982, **10**, 177.
95. T. W. Kensler, P. A. Egner, N. E. Davidson, B. D. Roebuck, A. Pikul and J. D. Groopman, *Cancer Res.*, 1986, **46**, 3924.
96. S. Safe, S. Bandiera, T. Sawyer, B. Zmudzka, G. Mason, M. Romkes, M. A. Denomme, J. Sparling, A. B. Okey, and T. Fujita, *EHP, Environ. Health Perspect.*, 1985, **61**, 21.
97. D. V. Parke, C. Ioannides and D. F. V. Lewis, in, 'Toxicology in Europe in the Year 2000', ed. C. M. Hodel, Elsevier, Cambridge, 1986, p. 14.
98. P. J. Breacker, Ph.D. Thesis, University of Surrey, 1983.
99. D. S. Davies, *Xenobiotica*, 1988, **18** (suppl. 1), 3.
100. C. Ioannides, C. M. Steele and D. V. Parke, *Toxicol. Lett.*, 1983, **16**, 55.

23.15

Pharmacodynamics

LENNART K. PAALZOW

University of Uppsala, Sweden

23.15.1 INTRODUCTION

The advances in analytical chemistry laid the foundations for the rapid development of pharmacokinetics as a science focusing on the time course of the concentration of a drug and of its metabolite(s) in the body following the administration of single or repeated doses. Many of the achievements in clinical pharmacology are attributable to the increased understanding of the pharmacokinetic processes controlling drug concentrations in various tissues. The compartmental view of pharmacokinetics, which predominated in the first decades of its development, has changed in recent years to a more physiological approach, which has led to a better understanding of the physiological and pathophysiological factors of importance for the drug concentrations attained in the body. Of greater clinical relevance, however, are the relationships between these concentrations and the desired and adverse clinical effects. The relationship between drug concentration and effect is often referred to as the science of pharmacodynamics.

A common definition of pharmacodynamics is that it describes what the drug does to the body, while pharmacokinetics describes what the body does to the drug. In order to understand the action of a drug and to determine the optimal therapeutic use of drugs, it is obviously important to link

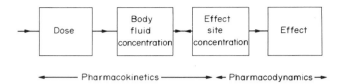

Figure 1 Schematic description of the relationship between pharmacokinetics and pharmacodynamics

these two branches of pharmacology together; the relationship between pharmacokinetics and pharmacodynamics may be regarded as the interconnection of a series of events, as depicted in Figure 1.

Pharmacokinetics describes the processes whereby a drug administered by a specific mode and in a specified dose is handled by the body, leading to specific drug concentrations in different tissues. Part of the drug will reach the site or sites of action (biophase) and exert its pharmacodynamic action there.

This chapter will review some of the proposed theories linking pharmacokinetic processes to pharmacodynamic effects and their application. Such a linkage may be looked upon as an integrated approach to the study of the overall relationship between dose and effect of drugs.

23.15.2 DEFINITIONS OF PHARMACODYNAMIC RESPONSES

The pharmacodynamic effects of most drugs result from an interaction with functional macromolecular components of the organism, which may lead to a series of biochemical and physiological changes that are characteristic of the action of the drug. If these changes can be measured through a continuously changing response to the drug (graded response), it may be possible to quantify the relationship between the dose (or concentration) of the drug and the effect produced. For instance, if the changes in blood pressure are measured over time after intravenous injection of different doses of norepinephrine in humans, a result as exemplified in Figure 2 may be obtained.

The magnitude of the increase in blood pressure is an expression of the *intensity* of the observed effect, which is often the measure used when the effect is plotted against the dose. Alternatively, the *duration* of a response may be used, since it is sometimes clinically more easy to judge when an effect appears and disappears than to estimate its degree. The duration, however, is dependent upon the definition of the endpoint for a specific effect. For instance, if we consider an increase in blood pressure of 10 mm Hg as a minimum increase (endpoint) above a placebo effect, we can see in Figure 2 that 5 μg of norepinephrine does not produce an increase which can be considered significant. The other two doses, on the other hand, have actions of a significant duration for the given endpoint, presented as a horizontal line in Figure 2, and 20 μg has a longer duration of action than 10 μg.

When a dose/response relationship is being evaluated, it is common to plot the maximum effect at each dose against the logarithm of the given dose. Usually maximum effects occur at the same point in time for all doses, but if the effects occur at different points in time, this is often an indication that the specific drug in question displays nonlinear pharmacokinetic behaviour, since, in linear kinetics, maximum drug levels in the body should appear at the same point in time independent of the dose

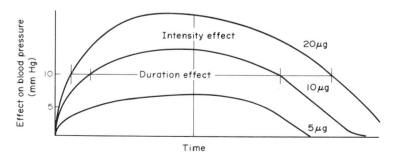

Figure 2 An example of the effects of an intravenous injection of different doses of norepinephrine in humans on the blood pressure

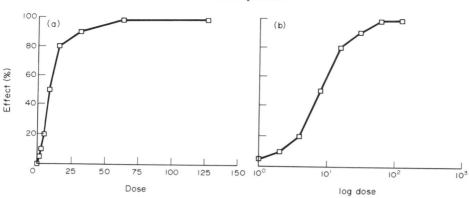

Figure 3 Plots of (a) dose–effect and (b) log dose–effect relationships

given. A typical dose–response relationship is shown in Figure 3, where the pharmacodynamic response is plotted against the dose (Figure 3a) and against the logarithm of the dose (Figure 3b).

In Figure 3(a) the dose is presented on an arithmetic scale and in Figure 3(b) the same data are presented on a semilogarithmic scale. The hyperbolic form of the curve in Figure 3(a) is transformed into a sigmoid form in Figure 3(b). In the latter figure, the nearly linear relationship between the intensity of the response and the log dose for effects between 20 and 80% of the maximum response can be observed. This logarithmic transformation makes it possible to evaluate the relationship between dose and effect more easily. The relevance of plotting the effect against the logarithm of the dose will be discussed later.

Sometimes the pharmacological response is either present or absent (an all or nothing event), for instance in the prevention or nonprevention of arrhythmia, convulsions or death. By determining the dose of a drug required to produce an effect of specified magnitude in a relatively large number of patients or animals, a *quantal dose–response* relationship can be obtained by plotting the cumulative frequency distribution of responders as a function of the log dose. The quantal dose–response curve is often characterized by defining the median effective dose (ED_{50}), which is the dose at which 50% of the individuals exhibit a specified quantal effect. If this effect is death of the animal, the median lethal dose (LD_{50}) is obtained. At present, instead of calculating from the dose, it is also possible, in many cases, to use the concentration in a specified tissue, such as plasma, which is related by some function to the concentration of the drug in the biophase, in order to determine the steady state plasma concentration for 50% of the maximal (100%) effective or lethal effect (EC_{50} or LC_{50}). The implications of using EC_{50} values will be discussed in detail later on.

23.15.3 DRUG–RECEPTOR INTERACTIONS: ANALYSIS OF THE GRADED DOSE–RESPONSE RELATIONSHIP

The oldest general theory of application of the law of mass action to the dose–response relationship was largely attributable to Clark (1885–1941).[1] The theory, as formulated by Clark, was extended and made more rigorous by Ariens.[2] The Clark–Ariens concept may appropriately be called the 'classical theory'. According to this, an observed biological effect is a reflection of the combination of drug molecules with receptors. The magnitude of the response (E) is assumed to be directly proportional to the degree of occupancy of receptors by drug molecules, the maximum effect (E_{max}) corresponding to the occupancy of all receptors.[3,4]

If a drug X combines with a receptor site R to yield a complex RX producing a biological response of magnitude E, this response will be proportional to the concentration (or amount) of complexes [RX], so that

$$X + R \xrightarrow{K_d} RX \tag{1}$$

$$E = k[RX] \tag{2}$$

where K_d is the dissociation constant of the complex and k is the proportionality constant between the effect and the concentration of occupied receptors.

According to the law of mass action

$$K_d = [R][X]/[RX] \tag{3}$$

If $[R_T]$ is the total concentration of receptors, which is equal to the sum of free $[R]$ and occupied receptors $[RX]$, *i.e.* $[R_T] = [R] + [RX]$, then equation (3) can be written as

$$K_d = ([R_T] - [RX])[X]/[RX] \tag{4}$$

which can be rearranged to

$$[RX]/[R_T] = [X]/(K_d + [X]) \tag{5}$$

If the maximum effect (E_{max}) is attained when all the receptors are occupied, equation (2) can be written as

$$E_{max} = k[R_T] \tag{6}$$

equation (5) then becomes

$$E/E_{max} = [X]/([X] + K_d) \tag{7}$$

K_d is an apparent dissociation constant, since the real concentration at the receptor is unknown. The reciprocal of K_d is a measure of the affinity of the drug for the receptor (affinity constant). The derivation of equation (7) is identical to that of the classical Michaelis–Menten equation, which describes the velocity of an enzymatic reaction as a function of the substrate concentration and the enzyme–substrate dissociation constant. The graph derived from equation (7) is a hyperbola and thus has the same shape as many dose–response curves; equation (7) has been accepted by many pharmacologists as generally applicable. If the concentration of drug required for a half-maximal effect is defined as X_{50}, and thus the left-hand side of equation (7) is equal to 0.5, then K_d is equal to X_{50}. When the dose is substituted for X, equation (7) acquires the more familiar form

$$E = E_{max} \times dose/(dose + ED_{50}) \tag{8}$$

This equation can be linearized by rearrangement so that

$$E/(E_{max} - E) = dose/ED_{50}$$

and by taking the logarithm of both sides, equation (8) becomes

$$\log[E/(E_{max} - E)] = \log dose - \log ED_{50} \tag{9}$$

A plot of the left-hand side of this equation against the log dose will yield a straight line with a slope $= 1$, and an intercept on the ordinate of $-\log ED_{50}$.

There are three assumptions underlying the derivation of equation (8).

(i) That the biological response is proportional to the receptor occupancy. This assumption has been called the occupancy theory. Several investigators have conducted experiments, however, in attempts to prove that a maximum response can be achieved without all the receptors being occupied, which led to the concept of 'spare receptors'.[4,5]

(ii) That one drug molecule combines with one receptor site. This is the simplest reaction mechanism, but if n molecules of the drug were to react with one molecule of receptor, the concentration term in equation (7) would have the power n, leading to another form of equation (7), which is usually called the Hill equation. This will be discussed later.

(iii) That a small fraction of the total drug concentration is combined with the receptors. In equation (7) the term $[X]$ is the concentration of uncombined (free) drug, and when this equation is plotted as the observed response E against the free drug level, equation (7) may fit the data better than if the total drug concentration is plotted. This is the basis of the commonly accepted hypothesis that the steady plasma concentration of unbound free drug is better correlated to the pharmacological effect than is the more frequently used total drug concentration in plasma.

(iv) According to equation (2), it is assumed that the biological response is directly proportional to the concentration of occupied receptors. This assumption may be too simple for certain drugs and alternative theories have been suggested, for example the 'rate theory' by Paton.[6] This author proposed that the magnitude of a drug effect depends on the rate of association of the drug and the receptor. Pros and cons of the 'rate theory' have been summarized in a review by Waud.[7] The proportionality constant k in equations (2) and (6) has been termed 'intrinsic activity' or 'efficacy' by Ariens and his colleagues.[2,8] They defined intrinsic activity as a proportionality constant α between the response as a fraction of maximal activity and the proportion of receptors occupied, as expressed by

$$E/E_{max} = \alpha[RX]/[R_T] \tag{10}$$

For an active agonist producing a maximal response by occupation of the receptors, $\alpha = 1$, for a partial agonist $1 > \alpha > 0$ and for an antagonist without agonist activity $\alpha = 0$.

As mentioned before, since the introduction by Clark of the theory of a simple proportionality between effect and concentration of the drug–receptor complex, many modifications have been put forward. One of the most well-known theories was developed by Stephenson.[5] This is sometimes called the 'stimulus theory' and comprises three postulates (for a review, see ref. 4): (i) a maximum response can be produced by an agonist without total receptor occupancy; (ii) the magnitude of a response is some unknown function of the number of receptors occupied; and (iii) the drug–receptor complex provides a stimulus to the tissue, and this stimulus is directly proportional to the number of receptors occupied. With these theories, Stephenson introduced the concept of spare receptors as mentioned above.

Besides the aforementioned theories of relationships between drug effects and the concentration of the drug–receptor complex, a large number of hypotheses have been introduced over the years and are being developed concurrently with the acquisition of knowledge about receptor mechanisms. These include the 'conformational pertubation theory' of Belleau;[9] the 'dynamic receptor hypothesis' of Bloom and Goldman;[10] the 'flux carrier hypothesis' of Mackay;[11] the 'allosteric two-state models' of Karlin[12] and Changeux;[13] and the 'receptor inactivation theory' of Gosselin.[14]

The discrepancy between the different theories of drug action currently in use arises from our lack of complete understanding of the links between the drug–receptor interaction and the response produced. The choice of a particular theory is often based on empirical considerations and on the way in which pharmacological principles are applied. In clinical practice, the occupational theory introduced by Clark,[1] with its modification, is still undoubtedly the most widely used. Nevertheless, the theories proposed by Clark,[1] Ariens,[2] Gaddum[15] and others[4] to model the interaction between a drug molecule and its receptor are based on the concept of a static receptor mechanism. This view has been modified over the years as experimental techniques have become more sophisticated; for example, it has been proposed that receptors undergo a conformational change induced by the drug molecule. There is also evidence that certain large ligands bind to and cap on cell membranes prior to internalization and become released in the cell, followed by a recycling of cell surface membrane components, including receptors. These mechanisms have important implications for the current understanding of the drug–receptor phenomenon and may shed some light not only on the end organ response but also on the regulation of the receptor itself (for a review, see ref. 16).

23.15.4 PHARMACODYNAMIC MODELS

A model is a simplified description of a true biological process. It enables selected factors that are considered important to be focused on. Models are usually mathematical forms with parameters and constants, which we use in attempts to determine such factors as the relationship between the concentration and effects of a drug.[17-19] Equation (8), which was discussed above, represents an empirical model for the relationship between the dose and effect of a drug, with E_{max} and ED_{50} as unknown parameters. In the following section, different pharmacodynamic models will be discussed, models which have been commonly used in relating drug concentrations to pharmacological effects.

Since in our pharmacological models we often use concentrations that are intended to reflect an unknown drug level in the biophase, we need to know how certain measurable reference concentrations, such as those in the plasma or blood, are related to the drug level at the site of effect. During the last decade, a considerable body of evidence has accumulated from experiments in animals and humans indicating that a pharmacodynamic response is better correlated with the plasma concentration or with the amount of drug in the body than with the dose administered. Pharmacokinetic models describe concentration as a function of both dose and time, whereas pharmacodynamic models are essentially independent of time and describe a time independent equilibrium relationship between effect and concentration. If plasma concentrations are used, the pharmacodynamic model is, in fact, a model at steady state plasma concentrations. In situations of time dependency, for example during development of tolerance to or sensitization against drug effects, this has to be taken into account when a suitable model is being developed to be fitted to experimental data. This will be discussed in detail later on.

23.15.4.1   The E_{max} Model

As mentioned previously, the E_{max} model was based on the classical theory put forward by Clark and modified by others. This model has been used in other areas such as enzyme kinetics (Michaelis–Menten) and in protein-binding studies (Langmuir adsorption isotherm). In studies of pharmacodynamic effects, it has been extensively applied and found useful both on empirical and on theoretical grounds.[8] By substituting concentration C for dose in equation (8), the E_{max} model is usually written as

$$E = E_{max} C/(EC_{50} + C) \tag{11}$$

where E is the observed pharmacodynamic response, E_{max} is the maximal effect obtained when C approaches infinity and EC_{50} is a constant expressing the steady state concentration producing 50% of the maximum response. This model has two important properties; it represents a hyperbolic relationship, which is commonly seen in pharmacological experiments (Figure 2) and, as such, it (i) predicts the maximal response a drug can produce and (ii) implies that there is no effect when no drug is present. By contrast, the most widely used relationship in pharmacology, the log concentration (dose)–effect relationship (the log-linear model) does not make these two important predictions.

Sometimes, it is of interest to estimate the effect when no drug is present and we therefore add a baseline effect (E_0) to equation (11)

$$E = E_0 + E_{max} C/(EC_{50} + C) \tag{12}$$

When the baseline effect, for instance the predrug blood pressure or pain threshold, is known, it can be subtracted from the effect observed in the presence of the drug, but sometimes it is not very well defined and has to be estimated. Whiting *et al.*[20] and Holford *et al.*[21] incorporated estimates of E_0 in their models of effects of disopyramide and quinidine, respectively, on the QT interval. They suspected that day-to-day variability in the placement of the electrodes for recording of the electrocardiograms produced systematic day-to-day variation in the baseline value, which was therefore estimated as a parameter.

Some drugs produce their effects by inhibition of a biological response, and, in such a case, equation (12) can be written as

$$E = E_0 - [E_{max} C/(IC_{50} + C)] \tag{13}$$

where IC_{50} is the concentration producing 50% inhibition of E_{max}. If the drug can reduce the baseline effect (E_0) to zero, then E_0 will be equal to E_{max}.

Another, not uncommon, situation is when two drugs are given simultaneously, where one (A) is an agonist and the other (B) is a pure antagonist. For such a case, equation (11) can be written as proposed by Holford and Sheiner[19]

$$E_{AB} = (E_{maxA} C_A)/[C_A + (EC_{50A}/EC_{50B})C_B] \tag{14}$$

where C_A and C_B are the concentrations of A and B, respectively.

As discussed above, the E_{max} model was developed on the basis of the receptor occupation theory, but it is mostly used on empirical grounds to describe the relationship between a pharmacodynamic effect and the dose (or concentration) of a drug. *In vitro*, it is used to estimate affinity for specific receptors, but *in vivo* it has also been applied to determine an EC_{50} value representing a true dissociation constant for a specific receptor; in the latter case one has to consider carefully all the factors which might influence its calculation. However, there are examples of correlation between, for instance, reduction of exercise-induced tachycardia in humans after administration of β-adrenoceptor blockers and binding of the same drug *in vitro* to membranes of rat reticulocytes and of rat parotid glands. Wellstein and Palm[22] investigated such relationships and found excellent accordance between EC_{50} values obtained *in vitro* and those calculated by the E_{max} model from *in vivo* data. For propranolol the plasma EC_{50} value was about $8–10 \, ng \, mL^{-1}$ (Figure 4).

The assumptions underlying their data treatment were (i) that plasma concentrations reflect the time course of the β blocker in the biophase; (ii) that no active metabolites are present; (iii) that there are no changes in the disposition of the drug during treatment or no changes in the concentration–effect relationship; and (iv) that the dissociation of the β blocker from the receptor site is a much faster process than its elimination from plasma. The example depicted in Figure 4 gives us an idea of the mechanistic background of the E_{max} model, but at the same time it reminds us of the caution we should observe when extrapolating *in vitro* data to the *in vivo* situation.

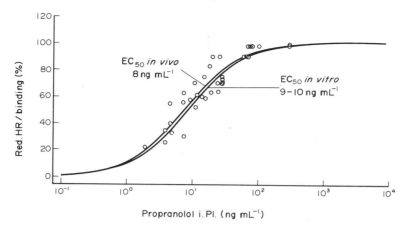

Figure 4 Plasma concentration–effect relationship of propranolol in healthy volunteers and occupancy of β adrenoceptors by the drug in receptor-binding studies *in vitro* (reproduced from ref. 22 by permission of Prous)

Some studies may be mentioned as illustrative examples of the clinical use of the E_{max} model. Mitenko and Ogilvie[23] investigated changes in normalized forced expiratory volume (FEV_1) in relation to serum concentrations of theophylline and obtained a predicted E_{max} value of 63% and an EC_{50} of 10 mg L^{-1}. A serum concentration of 20 mg L^{-1} (which is considered the upper limit of the therapeutic range) predicted a response of 67% of the maximal effect. The same approach was used by Oosterhuis *et al.*[24] to relate plasma concentrations of terbutaline to its bronchodilating effect.

The weakness of the E_{max} model is mainly of experimental origin, in that it is sometimes difficult to estimate the maximal effect of a drug *in vivo*. Due to the risk of side effects the dose cannot always be increased to a sufficiently high level to obtain the maximum (plateau) effects, which therefore have to be estimated by calculations as unknown parameters, a procedure which increases the uncertainty of the results. Sometimes, transformation of the E_{max} model to a form of a Lineweaver–Burk plot, as in equation (15), has been used to obtain a rough initial estimate of the E_{max} value for the subsequent computer fitting:

$$1/E = EC_{50}/E_{max}C + (1/E_{max}) \qquad (15)$$

A plot of reciprocal effects against reciprocal concentrations should yield a straight line with a slope of EC_{50}/E_{max} and a Y intercept of $1/E_{max}$. However, if one molecule combines with more than one receptor site, there will be a deviation from the linear relation of equation (15), and estimations of E_{max} values by extrapolation into unknown effect regions should be interpreted with great caution. This leads us to the next model, the sigmoid E_{max} model.

23.15.4.2 The Sigmoid E_{max} Model

The E_{max} model was originally developed on the theoretical assumption that one molecule of a drug forms a complex with one molecule of receptor. Hill[25] found empirically in studies on the binding of oxygen to hemoglobin that the experimental data were better described if a modified form of the E_{max} model was used according to the following equation (although Hill did write it somewhat differently)

$$E = E_{max}C^n/(C^n + EC_{50}^n) \qquad (16)$$

This equation has since been called the Hill equation and its usefulness was reemphasized by Wagner in 1968.[26] Since that time, it has been extensively used by Sheiner and associates and many others (for a review, see ref. 19). Holford and Sheiner[17] also named it 'the sigmoid E_{max} model'. The form of the equation is analogous to an equation describing a combination of a specific number of drug molecules (n) with one molecule of receptor. Theoretically, n should then be an integer, but very frequently nonintegral values are found. This does not mean that a nonintegral number of drug molecules combines with each receptor. *In vivo* this parameter, n, should merely be considered as a number influencing the slope of the concentration–effect curve and producing its sigmoid shape.

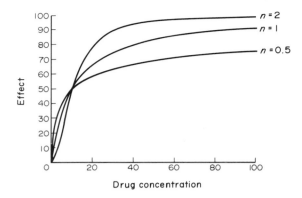

Figure 5 The influence of different values of n (equation 16) on the shape of the sigmoid E_{max} model (reproduced from ref. 18 by permission of Adis Press)

The influence of different values of n on the shape of the concentration–response curve is depicted in Figure 5. When $n = 1$, the curve has the hyperbolic form of the E_{max} model (Figure 2), but when $n > 2$, it becomes sigmoid in shape, with a steeper slope in its central portion. On the other hand, $n < 1$ produces a curve with a less steep central portion and the effect therefore approaches its maximum more slowly than in the simple hyperbolic curve ($n = 1$); however, the former curve has a steeper slope at low concentrations (Figure 5).

As with the E_{max} model, a term for baseline effect can be added to the Hill equation as shown in equations (12) and (13). The sigmoid E_{max} equation can be linearized by rearranging terms and inverting both sides of equation (16), which yields

$$E_{max}/E = (C^n + EC_{50}^n)/C^n = 1 + (EC_{50}^n/C^n)$$

By subtracting unity from both sides of this equation, collecting terms and again inverting both sides of the equation, we obtain

$$E/(E_{max} - E) = C^n/EC_{50}^n$$

Taking the logarithm of both sides, and changing n to s, since this is the notation used in many studies, yields a linear form of this equation

$$\log[E/(E_{max} - E)] = s \log C - s \log EC_{50} \tag{17}$$

A plot of $\log[E/(E_{max} - E)]$ *versus* $\log C$ will consequently give a straight line with a slope of s and an intercept on the ordinate of $-s \log EC_{50}$, as shown in Figure 6. By this linearization, the Hill equation is much simpler to analyze by ordinary linear regression and does not necessarily require access to computers when being fitted to experimental data. Despite this simple linear form of the sigmoid E_{max} model, it has not been used very extensively by pharmacologists, which is a pity as it was developed on principles much stronger than those of other commonly used relationships such as the log dose–response equation. The potential weakness of the Hill equation is the same as that of the E_{max} model, *i.e.* it is sometimes difficult to obtain values for maximal effects.

The sigmoid E_{max} model has been successfully shown to be appropriate for describing empirically the relationship between the concentration (dose) and effect of many drugs in animals and humans. A few representative human studies may be mentioned: Sheiner and associates[27] used the Hill equation to characterize the pharmacodynamics of (+)-tubocurarine. A similar approach has been used for other neuromuscular-blocking agents, for example pancuronium,[28] alcuronium,[29] atracurium[30] and vercuronium.[31] Scott *et al.*[32] successfully modelled the quantitative electro-encephalographic changes produced by the opioids alfentanil and fentanyl with an inhibitory sigmoid E_{max} model, including an estimation of baseline effects

$$E = E_0 - E_{max}C^s/(IC_{50}^s + C^s) \tag{18}$$

The same approach has been used for other drugs such as thiopental,[33] and equation (18) has also been applied by Holazo *et al.*[34] to study the antiarrhythmic effect of cibenzoline.

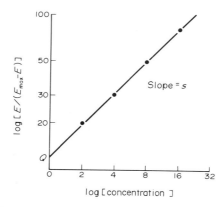

Figure 6 A plot of the linearized sigmoid E_{max} model

In animal studies, the Hill equation has been found to be useful for characterizing concentration (dose)–response relationships for effects such as the analgesic response of morphine,[35] phenoxybenzamine[36] and clonidine,[40] and the hyperalgesic response of drugs such as theophylline[37] and apomorphine[38] and for studying effects on blood pressure (clonidine[41] and apomorphine[39]).

Morino and associates[42] have also used the sigmoid E_{max} model to investigate the receptor-binding and brain concentration relationships for camazepam in the presence of active metabolites and to describe the pharmacological response of (+)-tubocurarine in the rat.[43]

As was shown in Figures 2 and 5, there is an almost linear relationship between concentration and effects at low concentrations. In this region, other concentration–response models have been applied, such as the linear effect model.

23.15.4.3 The Linear Model

If the concentration C in equation (12) is low and negligible in relation to the size of EC_{50}, this equation can be simplified to

$$E = (E_{max}/EC_{50})C + E_0 \qquad (19)$$

$$(E = \text{constant} \times C + E_0)$$

Thus, in the low concentration range a plot of the pharmacodynamic effect against concentration will yield a straight line with a slope which is equal to the ratio of E_{max} and EC_{50}.

This type of linear relationship has been applied in some clinical studies. Whiting *et al.*[20] used it to analyze the relation between QT prolongation on the ECG and changes in plasma concentration occurring after intravenous administration of disopyramide to volunteers. QT prolongation was also used by Holford *et al.*[21] to investigate the possibility of a linear relationship to plasma levels of quinidine. A difference between the slopes of the line after intravenous and oral doses was taken as an indication that active metabolites were formed during the absorption from the gut. Linear models were also used to investigate the pharmacokinetics and pharmacodynamics of trimazosin.[44]

23.15.4.4 The Log-linear Model

One of the most frequently applied dose–response relationships is the well-known empirical plot of response *versus* the logarithm of the dose or concentration. It was found at an early date that a logarithmic transformation of the abscissa yielded a linear part of the curve in the range of 20–80% of the maximal effect (Figure 3). The merit of such a plot was evident, since simple linear regression could be used by applying the following relationship

$$E = s \log C + i \qquad (20)$$

A plot of the intensity of an effect against, for instance, the steady state plasma concentration or the dose should yield a straight line with a slope of s and an intercept i of the ordinate (Figure 3). The obvious drawback of this equation is that it does not predict a maximum response (when C goes to infinity so does E) or a zero response, since the logarithm of zero is infinity. However, before digital computers and curve-fitting techniques became available its use was justified, since it could describe the slope of the line in a simple way and statistical tests could be applied to compare, for example, potencies of different drugs with parallel slopes or to investigate interactions between agonists and antagonists. It should not be forgotten that the log-linear model is valid only within a defined range of the dose–effect curve and that today we can easily apply more relevant relationships such as the E_{max} or sigmoid E_{max} models.

Levy[45,46] elegantly used the more simple log dose–response relationship to point out some important fundamentals of the time course of the pharmacological effect, which had previously been overlooked. Levy first assumed, for simplicity, that we administer a drug by the intravenous route and that only the unchanged drug, and no metabolite(s), produces a reversible pharmacodynamic response. Levy considered that if there is a relationship between the dose and effect, it is plausible that a relationship also exists between the amount of drug in the body (A_b) and the intensity of the produced effect (E) according to equation (20), which thus can be written as

$$E = s \log A_b + i \tag{21}$$

Levy based these calculations on the assumption that the amount of drug in the body (A_b) behaves kinetically in the most simple way, *i.e.* that it declines after the intravenous dose in a monoexponential manner

$$A_b = A_{b_0} e^{-kt} \tag{22}$$

where A_{b_0} is the amount of drug in the body immediately following its administration. Taking the logarithm of both sides of this function yields

$$\log A_b = \log A_{b_0} - (kt/2.303) \tag{23}$$

If we assume that a maximum response (E_0) is elicited by a maximum drug level in the body (A_{b_0}), we obtain

$$E_0 = s \log A_{b_0} + i \tag{24}$$

and inserting equations (21) and (24) into equation (23) yields

$$(E - i)/s = (E_0 - i)/(s - kt/2.303)$$

which can be simplified to

$$E = E_0 - kst/2.303 \tag{25}$$

This equation shows that, under the assumptions mentioned above, the intensity of the pharmacological response (E) declines over time (t) at a constant rate that is a function of the first-order rate constant of drug elimination (k) and the slope of the dose–response curve, s (equation 21). It is quite obvious that a pharmacological effect of a drug with rapid elimination (a large elimination rate constant, k) declines more rapidly, but it is not as immediately apparent that a steep dose–response curve also produces the same result. It should also be noted that according to equation (25) the pharmacological effect declines at a zero-order rate, while the amount of drug in the body declines at a first-order rate (equation 22). It is therefore incorrect to use the expression half-life for a pharmacological effect, since for a zero-order decline the 'half-life' changes continuously, while in a first-order process it has a constant value that is independent of dose or concentration. An illustrative example of the constant rate of decline of pharmacodynamic effects is given in Figure 7, which presents the finding by van Rossum and van Koppen[47] that the locomotor activity in rats declined at a zero-order rate during the postabsorptive phase of different doses of dexamphetamine sulfate. For further details on this subject the reader is referred to Gibaldi and Perrier.[48]

It should be recalled that outside the range of 20–80% of the maximal effect, the pharmacological effect does not decline at a constant rate, and to be able to model the whole range of the effect, other effect models, such as the E_{max} or sigmoid E_{max} model, have to be applied instead of the simple log-linear model (equation 21), as has been pointed out by Wagner.[26]

Levy[46] also drew attention to the common experience in pharmacology that the duration of a pharmacological effect (for definition see Figure 1) is related to the dose or amount of drug in the body. The duration of action (t_d) is assumed to be the length of time when the amount of drug in the

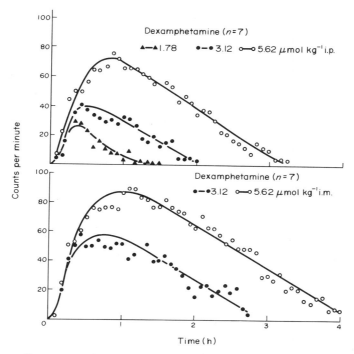

Figure 7 The time course of locomotor activity after intraperitoneal and intramuscular administration of dexamphetamine sulfate in rats (reproduced from ref. 48 by permission of Dekker)

body is above a certain minimum level (A_{min}). This leads to the following relation from equation (22)

$$A_{min} = A_0 e^{-kt_d}$$

After rearrangement and taking logarithms of this equation, the duration of a pharmacological response (t_d) could thus be written as

$$t_d = 2.303/k (\log A_0 - \log A_{min}) \tag{26}$$

According to this relationship a plot of the duration of a response against the logarithm of the dose (or amount in the body immediately following an i.v. dose) should yield a straight line with a slope of $2.303/k$ and an intercept of $-\log A_{min}$.

As an example of this, Levy[49] used the data presented by Walts and Dillon[50] to show that the time taken to recover from succinylcholine paralysis was proportional to the logarithm of the injected dose (Figure 8).

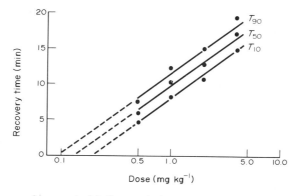

Figure 8 The duration of recovery from succinylcholine paralysis plotted against the logarithm of the dose injected. T_{90}, T_{50} and T_{10} indicate 90, 50 and 10 per cent recovery of muscle twitch (reproduced from ref. 58, by permission of Lea and Febiger)

23.15.4.5 Composed Pharmacodynamic Models

When studying the effects of the antihypertensive agent clonidine, Paalzow and Edlund[41] observed that low intravenous doses of this drug caused a decrease in arterial blood pressure in the rat, while immediately after injection of higher doses there was a phase of increased blood pressure followed by a decrease. Thus, clonidine is a drug which at low doses or low biophase concentrations can produce one type of pharmacodynamic effect and at higher biophase levels can have the opposite effect. On the basis of simultaneous studies of the kinetics of clonidine in plasma and brain tissue and the effects of this drug on blood pressure, Paalzow and Edlund[41,51] proposed a composite pharmacodynamic model to relate the steady state plasma concentration to the observed pharmacological effects. This was called a model for multiple receptor responses.[40]

In this model it was assumed that the observed intensity of a pharmacological effect (E) is a sum of two opposite effects induced by activity of receptors of two different opposing systems, one with the effect E_1 and the other with the effect E_2. The relationship between the net observed effect, E, and the steady state blood concentration, C, could thus be described by the sum of two Hill equations

$$E = \frac{E_{max1} C^{s_1}}{EC_{50(1)}^{s_1} + C^{s_1}} - \frac{E_{max2} C^{s_2}}{EC_{50(2)}^{s_2} + C^{s_2}} \tag{27}$$

$$(E = E_1 - E_2)$$

The first term describes the 'pure' effect of clonidine on receptors or the physiological system, which results in an increase in blood pressure, while the second term relates the concentration to the 'pure' decrease in pressure caused by clonidine. With this model the authors described the time course of the effects of this drug at different doses and later the description was also applied to human data.[52]

The same model was applied to the antinociceptive effects of clonidine in the rat, using, in equation (27), the given dose instead of the steady state blood concentration.[40] The results indicated that the antinociceptive effect of clonidine consisted in a sum of activities from at least two receptor sites and the fit of the composite Hill equation to the observed data is shown in Figure 9.

These results were discussed in terms of the possibility that separate adrenergic receptors mediate clonidine antinociception at different levels of pain transmission. Other authors have also pointed out the concentration or dose dependent effects of clonidine at separate receptor sites known as α_2- and α_1-adrenergic receptors.[53] At low drug concentrations the α_2-adrenergic receptor is usually activated and, with increasing biophase concentrations, the α_1 receptor additionally mediates the effect.[54] This theory also offers an explanation for the opposing effects of clonidine on the arterial blood pressure.

Other drugs which have been found to have composite dose–response curves for their effect on pain are apomorphine,[38] yohimbine[55] and promethazine.[56] These three substances produce hyperalgesia at low doses and analgesia at higher doses, and the composite effect could be resolved into its components by applying a model consisting of the sum of two Hill equations. The effect of apomorphine on the heart rate was found to be a composite one and was used to describe the way in which a drug can interfere with several physiological systems in producing its effect.[39] It is concluded that the body should be considered as a system of dynamic equilibria even in response calculations.

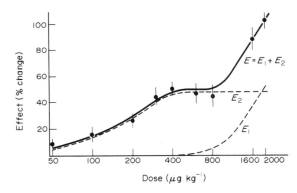

Figure 9 The dose–response relationship for clonidine·HCl on the vocalization threshold after subcutaneous administration of different doses. The observed effect (E) was expressed as the percentage change from the pretest threshold and was dissociated into two components, E_1 and E_2, representing two responses from different receptor interactions (reproduced from ref. 40 by permission of Am. Soc. Pharmacol. Exp. Ther.)

A different approach was used by Kelman and Whiting regarding composite pharmacodynamic models.[57] The responses of the antiarrhythmic agent disopyramide and the cardiac glycosides digoxin and β-methyldigoxin were described as a sum of multiple linear regressions, assuming a linear relationship between different concentrations in separate compartments and the effect according to the following equation

$$E = \Sigma_i [A_i X_i + B_i] \tag{28}$$

where X_i is the amount of drug in the *i*th compartment, A_i is the regression coefficient and B_i is the intercept. Assuming a two-compartment model for disopyramide and a three-compartment model for the glycosides, these authors were able to fit equation (28) successfully to their data. The results obtained following administration of disopyramide indicated that the effect was dependent on the drug concentrations in both central and peripheral compartments. The response to the glycosides, on the other hand, was dependent on the amounts of drug in the two peripheral compartments but not on that in the central compartment. This example of the use of a model to describe the kinetics of the drug levels in different theoretical compartments, instead of using different pharmacodynamic models, leads us to the next subject of discussion, *i.e.* the effect site (biophase) concentration and its relation to concentrations in accessible tissues such as plasma and blood.

23.15.5 EFFECT SITE CONCENTRATION

When a drug induces a pharmacological effect after interaction with a certain population of receptors, the time course of this effect can, in most cases, be measured. If the drug is given, for instance, by an intravenous bolus dose, the drug concentration in the blood will decline at a rate determined by the half-life of the drug. The half-life is a secondary parameter which is dependent on the primary pharmacokinetic parameters: steady state volume of distribution (V_d) and total blood clearance (Cl) by the following relationship[58]

$$t_{1/2} = 0.693 V_d / Cl \tag{29}$$

Despite an exponential decrease in the blood or plasma concentration after the intravenous dose according to the actual half-life, the resulting pharmacological effect may gradually increase with time, reaching a maximum and then declining. In many cases, this event can be explained pharmacokinetically—it takes some time for the drug to distribute to the site of action, which may be located in tissues outside the blood compartment. In many studies on the relationship between drug concentration data and pharmacological effects, unfortunately, only data points at which pseudoequilibrium (postdistribution) has been attained are considered. During that phase, the drug levels in most tissues of the body are considered to change in a parallel manner. Data collected in the postdistributive phase do not provide information about the kinetics of the drug at the site of action and the induced pharmacological effect during the distribution phase. As pointed out by Levy *et al.*,[59] a given tissue concentration, calculated or measured, should yield a pharmacological effect of essentially the same intensity during phases of rising and declining tissue levels. Recalling Figure 1, which shows the events from the point in time of administration of the dose until a pharmacodynamic effect is developed, the transfer of a drug from body fluids to its site of action is considered to be a pharmacokinetic process, while the potential lag time for the expression of the effect in the biophase is considered to belong to the area of pharmacodynamics (Figure 1). In the example given above, the lag time for the development of the maximal effect after an intravenous dose could thus be due either to pharmacokinetic or pharmacodynamic factors. Let us first consider the pharmacokinetic possibility.

23.15.5.1 The Central Compartment and Hysteresis Loops

A lag time attributable to pharmacokinetic factors could be revealed by performing a so-called hysteresis plot. If the measured plasma concentrations are plotted against the observed pharmacological effect after an intravenous dose and the points are joined in time sequence, an anticlockwise hysteresis plot may be obtained as shown in Figure 10.

It is evident from Figure 10 that as the plasma concentration decreases with time, the effect increases until a maximum is obtained and then both the effect and concentration decrease, since pseudoequilibrium is attained in the body. An anticlockwise hysteresis cannot, of course, prove that

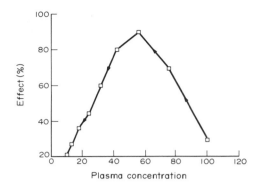

Figure 10 Anticlockwise hysteresis loop. The pharmacological effect is delayed by the time taken for equilibration between plasma and effect site. The arrows indicate the time sequence of observations

a delay in equilibration is causing a time lag in the development of the effect in the biophase unless more experiments are performed, for example with measurement of the potential effect site concentration. There are also other possible explanations for the anticlockwise hysteresis, such as the occurrence of opposing effects at high drug levels,[41] and sensitization of receptors, *i.e.* an increasing effect over time after a single dose or accumulation of pharmacologically active metabolite(s). However, as a starting point a hysteresis plot is useful and, by examining the way in which the effects change over time at a steady state plasma concentration of the drug, an equilibration delay between plasma and the biophase can be ruled out.[18] Techniques of relating effect data to plasma concentrations with hysteresis will be discussed later (Section 23.15.5.3).

Sometimes when hysteresis plots are performed after oral administration of a drug, a clockwise hysteresis, as illustrated in Figure 11, is seen.

As shown in Figure 11, during the absorption phase the plasma drug concentration will increase, and so will the effect, but during a decline in the drug concentration the effect will be weaker than during the absorption at an equal concentration. If we define tolerance as a reduced effect at a concentration which previously produced a greater effect, this plot indicates development of tolerance during a single oral dose. A clockwise hysteresis could, for instance, be explained by changes in receptor function, such as down-regulation of receptors, or by antagonism of the effect of the parent compound by accumulation of drug metabolites. Mathematical techniques for handling the development of tolerance in pharmacokinetic–pharmacodynamic modelling will be discussed in Section 23.15.8.

When relating the plasma concentration to an observed pharmacodynamic effect, using a pharmacodynamic model such as the E_{max} or sigmoid E_{max} model, it is assumed that in the simplest case the drug will reach the biophase almost instantaneously and the plasma level will thus reflect the time course of the drug at the site of action. In such a case, plasma concentrations can be used directly when fitting the pharmacodynamic model to the observed data. This situation is not uncommon when a drug is given by the oral route, as the absorption process is often slower than the

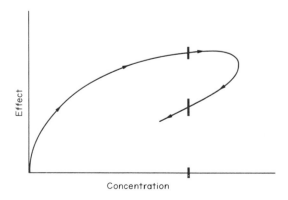

Figure 11 A clockwise hysteresis loop. The pharmacological effect is decreasing over time and, at a fixed concentration, the effect is weaker at a later than at an earlier point in time. The arrows indicate the time sequence of observations

distribution of the drug to the effect site as soon as the drug has been absorbed into the systemic circulation. If the drug is given intravenously, however, anticlockwise hysteresis is quite normal for the data points shortly after the injection. In such a case, one can either handle the hysteresis by applying a link model, as will be described later, or choose to ignore the data points during the distribution phase immediately following administration. The latter approach, although quite common, is strictly an erroneous procedure, as it gives the relationship between plasma drug level and effect only under pseudoequilibrium conditions, a fact which should not be forgotten, especially when applying the results in clinical practice.

Besides having pharmacokinetic reasons, a lag time in the development of a maximal effect can be due to pharmacodynamic factors; once the drug has reached the site of action, it will take time for the development of the cellular changes, which will lead to the final pharmacological response. This type of response is often called an indirect pharmacological response. Such a time lag has been suggested, for instance, as an explanation for the lack of relationship between the time course of analgesia and brain morphine concentrations.[60]

23.15.5.2 The Tissue Compartment

During the first decades of the development of pharmacokinetic science, lag times in pharmacological responses after intravenous administration were often treated by applying a compartmental approach. If the plasma concentrations decline in a biexponential manner, then an open two-compartment model with elimination from the central (plasma) compartment can be fitted to the experimental data. To the central compartment is connected a peripheral compartment, which is often called the 'tissue' compartment. By nonlinear least-squares regression, the function describing the plasma concentration profile is obtained and the time course of drug concentrations in the peripheral compartment can be calculated as shown in Figure 12.

With knowledge of the time course of drug concentrations in the tissue compartment (C_T), a sigmoid E_{max} model, for instance, may be fitted to the observed effect data points over time, with inclusion of the parameters (A, α and β) of the known tissue function as constants in some suitable digital computer fitting procedure. It is advisable to fit the pharmacokinetic data separately rather than to attempt a simultaneous fit of both pharmacokinetic and pharmacodynamic data,[27] and the following equations are thus used

$$C_T = Ae^{-\beta t} - Ae^{-\alpha t}$$

$$\beta < \alpha; \text{ coefficient and exponents are known}$$

$$E = E_{max} C_T^s / (C_T^s + EC_{T50}^s)$$

E is the dependent variable and t is the independent variable in the fitting procedure, yielding estimates of the unknown parameters s, EC_{T50} and eventually E_{max}.

The outcome of the described procedure may be successful, indicating that a theoretical tissue concentration is better related to the pharmacological effect than is the concentration–time profile in the central compartment. Although successful, the procedure is merely a descriptive way of

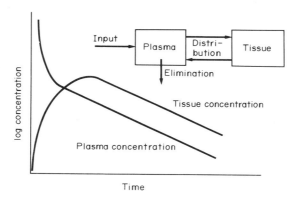

Figure 12 Description of the observed plasma concentration–time profile by a two-compartment open model following intravenous administration of a drug. The calculated tissue concentrations are also depicted

expressing a concentration–effect relationship, but it does not give any information about the actual type of tissues that are related to the 'tissue compartment'. For some drugs the above-mentioned approach is not applicable, since there is no *a priori* reason to assume that the time course of a drug concentration at the effect site must parallel the kinetics in other tissues, which mainly cause the multiexponential behaviour of the plasma concentration–time course. Nevertheless, the tissue compartment approach has been used successfully in some studies. Wagner *et al.*[61] related the time course of effects of lysergic acid diethylamide (LSD) to the mental performance of volunteers. A linear pharmacodynamic model was fitted to the data and a linear correlation was obtained between calculated 'tissue' concentrations and performance scores after an intravenous bolus dose. Kramer *et al.*[62] used a three-compartment open model to describe the pharmacokinetics of digoxin and using this model they obtained the best fit of the time course of the effect on electromechanical systole (QS2) of the heart, when relating it to the drug concentrations in the more shallow 'tissue' compartment. Paalzow,[63] on the other hand, found that the effect of theophylline hyperalgesia was better related to the time course of theophylline concentrations in the central (plasma) compartment rather than in the 'tissue' compartment. The peripheral or tissue compartment approach can of course only be applied to drugs with multicompartment kinetic characteristics. However, a time lag between drug levels and dynamic effects can also occur for drugs described by a one-compartmental model. Such characteristics have been reported for the effect of ergotamine on arterial tonus in migraine patients[64] and for the lymphocytopenic effect of prednisolone.[65] These problems were solved by applying a link model developed by Sheiner and associates.[27]

23.15.5.3 The Effect Compartment or the 'Link Model'

Segre[66] was the first to consider the possibility that the time course of a pharmacodynamic effect could be used by itself to describe the transfer rate of a drug to the biophase. In order to fit the blood concentration of norepinephrine–time data to the effect on the blood pressure in the cat, the lag time of effects was modelled by including two hypothetical tissue compartments between the plasma compartment and the pharmacodynamic response function. In a later study, Segre's approach was developed further by modelling the analgesic effect of morphine in the rat.[35] The pharmacokinetics of morphine in the brain and plasma was established and the brain concentrations were linked to the analgesic response over time. Despite the fact that the potential biophase drug levels were measured, maximal analgesia was obtained later than was predicted from the brain concentration, and the effect was also more sustained. To solve this problem, a separate effect compartment was linked to the brain compartment of the morphine pharmacokinetic model by a first-order rate constant. In order to avoid loss of drug substance to the hypothetical effect compartment, a negative rate constant was also added to the brain compartment, with the same magnitude as was the rate constant driving the effect compartment, and the parameters of the effect model (sigmoid E_{max} model) were successfully solved by the nonlinear regression program SAAM 25. The ideas of Segre were also applied by Forrester *et al.*[67] to describe the time course of effects of cardiac glycosides after intravenous bolus doses on the basis of data presented by Shapiro *et al.*,[68] who observed that the drug effects increased during the first hour following administration, while the plasma concentration continuously decreased. Forrester and coworkers used the time course of the onset of drug effects to estimate the half-time for attainment of the peak effect. Hull *et al.*[28] extended this idea further. In their studies of the effect of pancuronium on muscle twitch responses in humans, these authors assumed that the biophase volume was a negligibly small proportion of the plasma (central) compartment volume of the two-compartment open model describing the disposition of pancuronium. The biophase rate constants were determined and the biophase drug levels were obtained and related to the pharmacodynamic effects by a Langmuir-type equation.

To model the effect site as a separate compartment, a model linked to a traditional pharmacokinetic compartment model was subsequently elaborated by Sheiner *et al.*,[27] and this approach was later called 'the link model'.[69] This model postulates a hypothetical effect compartment which is linked to the plasma concentration by a first-order rate constant, k_{1e}, and with a rate constant for drug losses, k_{e0} (Figure 13).

The rate of change of the amount of drug in the hypothetical effect compartment can be expressed by the following differential equation

$$dA_e/dt = (k_{1e} A_1) - (k_{e0} A_e) \tag{30}$$

where A_1 is the amount of drug in the central compartment of a pharmacokinetic model and A_e is the amount of drug in the hypothetical effect compartment. It is assumed that the rate constant k_{1e} is

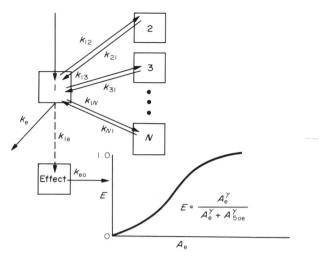

Figure 13 Diagrammatic presentation of the pharmacokinetic and dynamic model proposed for ($+$)-tubocurarine. A_e represents the amount of drug in the hypothetical effect compartment, and E is the fraction of the maximal response (reproduced from ref. 27 by permission of Mosby)

very small in relation to the other smallest rate constant of the pharmacokinetic model and the effect compartment will thus receive a negligible mass of drug. The introduction of this rate constant will consequently not affect the pharmacokinetic model describing the kinetics of the drug.

For instance, if the disposition of a drug in the plasma can be described by a one-compartment open model after a bolus intravenous input of a dose D, equation (30) will yield the following solution

$$A_e = [k_{1e}D/(k_{e0} - k)][\exp(-kt) - \exp(-k_{e0}t)] \tag{31}$$

where k is the elimination rate constant of drug losses from the plasma. The concentration of drug in the effect compartment (C_e) is obtained by dividing A_e by the effect compartment volume V_e and equation (31) can thus be rewritten as

$$C_e = [k_{1e}D/V_e(k_{e0} - k)][\exp(-kt) - \exp(-k_{e0}t)] \tag{32}$$

At equilibrium (steady state) no net transfer of drug into or out of any compartments takes place and thus the following equation holds for the effect compartment

$$k_{1e}A_1 = k_{e0}A_e$$

or

$$k_{1e}V_1C_1 = k_{e0}V_eC_e \tag{33}$$

where V_1 is the distribution volume of the plasma compartment with the steady state plasma concentration C_1. The steady state partition coefficient of the drug in the effect compartment, K_p, is defined as

$$K_p = C_e/C_1$$

and equation (33) can therefore be rewritten as

$$V_e = k_{1e}V_1/K_pk_{e0} \tag{34}$$

Substituting this expression for V_e in equation (32) yields

$$C_e = [k_{e0}DK_p/V_1(k_{e0} - k)][\exp(-kt) - \exp(-k_{e0}t)] \tag{35}$$

which is the function describing the drug concentration at the effect site over time. K_p cannot be estimated unless the effect site concentration can be determined, which is difficult since its tissue location is often unknown. However, at steady state the plasma concentration C_1 is equal to C_e/K_p according to the definition of K_p and, consequently, equation (35) can be rewritten to express the steady state plasma concentration (C_{ss} or C_1).

$$C_{ss} = [k_{e0}D/V_1(k_{e0} - k)][\exp(-kt) - \exp(-k_{e0}t)] \qquad (36)$$

This expression can then be incorporated into a pharmacodynamic model such as, for instance, the E_{max} model

$$E = E_{max}C_{ss}/(EC_{ss50} + C_{ss}) \qquad (37)$$

where C_{ss50} is the steady state plasma concentration producing 50% of the maximal effect.

Consequently, if we study the effects E of a drug over time t after a bolus dose D, and we have obtained from other studies the pharmacokinetic parameters of the drug (k, V_1), we can fit equation (37) simultaneously with equation (36) to the pharmacological data and obtain the unknown parameters (k_{e0}, EC_{ss50} and E_{max}) of the pharmacodynamic model. Sheiner and associates[27] recommend that the pharmacokinetic model be first fitted to the plasma concentration–time data to obtain the parameter estimates of that model and that these parameters should then be used as constants when fitting the pharmacodynamic model to the effect–time data.

The obtained rate constant k_{e0} characterizes the time dependent aspects of equilibration between plasma and the effect site concentrations. The rate of onset of the effect is thus controlled by the rate constant describing the loss of drug from the effect site (k_{e0}), in the same manner as the rate constant for drug elimination (k) determines the time taken for the plasma drug concentration to reach steady state during constant drug infusion. This approach as described by Sheiner and associates can thus be used to handle the anticlockwise hysteresis as discussed above (Figure 10). Using this link model, any pharmacokinetic model or drug input can be applied, and the different equations for the effect site concentration have been solved and described in detail.[18,70] The beauty of the approach of Sheiner and associates is that instead of relating a pharmacological response to the drug concentrations in some more or less well-defined tissues, they relate it to the steady state plasma drug level, which in clinical practice is of great importance.

The effect compartment approach of Sheiner and associates was applied to (+)-tubocurarine for the first time using the sigmoid E_{max} model, and these authors calculated that the equilibration half-time (ln $2/k_{e0}$) of this drug for muscle paralysis was 4 min.[27] Kelman and Whiting[57] obtained a half-time of 200 min for the effect of digoxin on the left ventricular ejection time. An extremely long equilibration half-time of about 10 h was found on calculation by Tfelt-Hansen and Paalzow[64] for the effect of ergotamine on cerbral arteries in migraine patients, and these authors considered this to reflect a slow dissociation of the drug from the biophase, rather than an equilibration delay in drug transfer from the blood to the effect site or the production of an active metabolite. In recent years, a growing number of studies have used the link model as proposed by Sheiner and associates to describe the relationship between steady state plasma concentrations of drugs and different pharmacodynamic effects.[20,21,29–31,33,34,39,43,44,65,71–76]

Besides permitting calculation of the temporal aspects of the effect site equilibration k_{e0}, the link model also gives the steady state plasma concentration for 50% of the maximum effect EC_{ss50}, which is the sensitivity component of the pharmacodynamic model (equation 37). This is a highly useful clinical parameter for estimating the interindividual sensitivity of patients to a specific drug. Thus, Sheiner et al.[27] found indications that renal failure may produce changes in EC_{ss50} and k_{e0} for (+)-tubocurarine. Vecuronium, another muscle relaxant, was found to give a lower EC_{ss50} value in infants than in children.[31] Age-related changes in the pharmacodynamic sensitivity to fentanyl and alfentanil have been observed by Scott and Stanski.[72] They found a decrease in brain sensitivity (as determined by EEG changes) with age but no age-related changes in the pharmacokinetic parameters. Considerable interindividual variability in EC_{ss50} values (40-fold) has been found for the effects of ergotamine on peripheral arteries, with the clinical implication that patients should be tested for their responses to ergotamine treatment.[64] A substantial interindividual variation in the relation between changes in plasma methadone concentration and degree of analgesia in patients with chronic pain receiving opioids has also been observed.[73] Fuseau and Sheiner[69] proposed a variation of the simultaneous modelling technique presented by Sheiner and associates, an approach which does not require any assumption of the pharmacodynamic model in order to estimate k_{e0}. These authors estimated this rate constant as the value which causes the hysteresis curve (C_e vs. effect points connected in order of time) to collapse to a single curve that represents the empirical C_e–effect relationship. In a more recent paper, Unadkat et al.[77] extended this further by introducing a nonparametric approach for both the pharmacokinetic and pharmacodynamic models. When the specification of the pharmacokinetic model is uncertain, this approach may be of advantage and an algorithm for obtaining a nonparametric estimate of k_{e0} was given.[77]

Meredith *et al.*[44] have also employed the effect compartment method of Sheiner and associates to incorporate effect compartments for both the parent drug and its active metabolite. Generally, a lag time in pharmacological effects, compared with the time course of the plasma concentration, can always be suspected to be caused by active metabolite(s). Such a possibility should always be taken into consideration before an investigation is extended to a more complex evaluation of pharmacodynamic–pharmacokinetic relationships.

23.15.5.4 Drug–Receptor Dissociation Time

In the model of Sheiner and associates the rate constant k_{e0} characterizes the time dependent aspects of equilibration between the plasma and the effect site drug concentration. For most drugs, it is assumed that this equilibration time is caused by the time needed for the transport of the drug between blood and the receptor environment. Other processes like the drug–receptor association and dissociation are thus considered to be more rapid than the disposition of the drug in the biophase. However, recent developments of *in vitro* receptor-binding techniques have shown that certain drugs dissociate slowly from the drug–receptor complex. For these slowly dissociating agents, the time of receptor occupation will be preponderant for the duration of action of the drug. For most other drugs having a rapid dissociation from the receptor sites, pharmacokinetic factors play a predominant role for the time course of drug response. A review of the dissociation times from receptor sites *in vitro* for a number of antagonists (serotonin-S2 antagonists, dopamine-D2 antagonists and histamine-H1 antagonists) indicated that the dissociation half-time was usually below 10 min. However, pimozide showed a half-time of about 200 min for the dissociation from the dopamine-D2 receptor preparation and ritanserin about 160 min from the serotonin-S2 receptor.[78] The same authors also reported an unusually long half-time from the latter receptor sites for the opiate lofentanil (260 min), compared to other opiates such as morphine, fentanyl and alfentanil, which all had half-times in the range of 1–3 min. However, before conclusions are drawn about the consequences of a slow dissociation from the receptor sites for the duration of the pharmacodynamic effects, the receptor kinetics must be interpreted in relation to the kinetics (persistence) of the drug in the tissues surrounding the biophase. For most drugs, it is commonly assumed that the rate of drug elimination from the biophase is a rate-limiting factor for the duration of effects, rather than the rate of dissociation from the receptor sites. For instance, a finding of a rather slow dissociation of lorazepam from specific *in vitro* binding sites of brain homogenates[79] was later found to be a misinterpretation of the data. The slow dissociation was merely the consequence of the overall persistence of this drug in the brain, which also controlled the time course of some of the pharmacodynamic effects of this drug.[80]

D'Hollander and Delcroix[112] have suggested a pharmacodynamic model for studying the *in vivo* effects of neuromuscular-blocking agents, a model which takes into account the apparent equilibration constant of the drug–receptor exchange. Contrary to the view of Sheiner and associates,[27] the equilibration between plasma and the receptor environment was considered to be a more rapid process than the receptor events. It is therefore suggested that incorporation of *in vitro* receptor-binding studies together with pharmacokinetic and pharmacodynamic studies *in vivo* should be performed in order to shed some additional light on the rate-limiting processes for the time course of drug action.

23.15.6 COMPLEX TIME COURSES OF EFFECTS

For situations in which discrepancies are obtained between the pharmacokinetics and pharmacodynamics, the approach proposed by Sheiner and associates for solving the problem is based upon the assumption that the time course of the mass of drug at the site of action is different from that derived from the pharmacokinetic model. However, during recent years Paalzow and associates[81,82] have pointed out that the temporal aspects of the pharmacological effect can, in addition, be characterized on pharmacological grounds, which these authors consider to be related to 'multiple receptor interactions'. When discussing a pharmacological effect, we thus have to be aware that one specific drug can bind to different receptors depending upon the concentration or the dose administered. As mentioned previously, when we register a response, this may be derived from a single type of receptor interaction or it may comprise a sum of responses from several types of interactions. As illustrated in Figure 14, apomorphine produces bradycardia in the rat at low plasma steady state concentrations, while at increasing apomorphine concentrations this effect is gradually

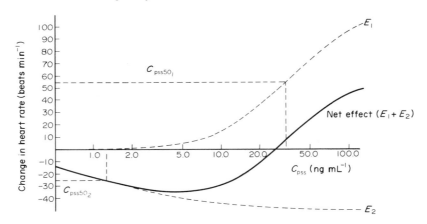

Figure 14 The relationship between the steady state plasma concentration of apomorphine (C_{pss}) and its effect on the heart rate. The relationship found between the observed net effect ($E_1 + E_2$) and C_{pss} was dissociated into two components, E_1 (tachycardia) and E_2 (bradycardia). The C_{pss50} values for the pure effects are given (reproduced from ref. 39 by permission of the Royal Pharmaceutical Society)

masked by the occurrence of tachycardia, which is the sole observed effect at a high steady state plasma concentration of the drug.[39] Thus, depending upon its concentration in the biophase, apomorphine is capable of interacting with different functional systems or receptors, producing different or opposing pharmacodynamic effects. It follows that any factor influencing the concentration at the site of action is critical not only for the degree of response, but it may also change the nature of the effect.

When, for instance, a drug is given by an extravascular route, the plasma concentrations rise during the absorption process and decline when the rate of elimination exceeds the rate of input to the body. If there is instantaneous equilibrium between the plasma and the biophase apomorphine concentrations, a large single dose given by the extravascular route will produce a temporal response that oscillates from bradycardia to tachycardia and back to bradycardia, since the concentrations given in Figure 14 move from left to right on the abscissa and then back again from right to left. The temporal fluctuations between opposing effects (Figure 15) observed experimentally after subcutaneous administration of a high dose of apomorphine can thus be explained by the composite concentration–response curve shown in Figure 14.

The existence of concentration–response curves of many different shapes has now been demonstrated in several clinical and experimental models, and problems of multiple responses during the time course of the effects of single doses have, in fact, been encountered in several studies, although no explanations for the findings have been given. Protais *et al.*[83] found that the yawning effect of

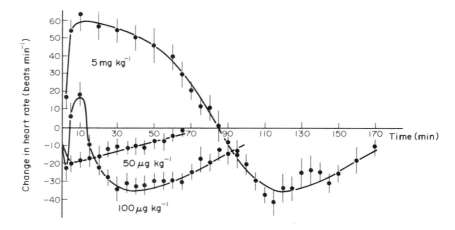

Figure 15 The observed effects of apomorphine on the heart rate (beats min^{-1}) after subcutaneous administration of different doses. The solid line represents the fit to a composite Hill equation (reproduced from ref. 39 by permission of the Royal Pharmaceutical Society)

apomorphine in rats had a bell-shaped dose–effect curve and, that during the time course of the effects of a subcutaneous dose, the yawns appeared, disappeared and reappeared. A bell-shaped dose–response curve has also been described for the effects of the mixed agonist/antagonist butorphanol on rat striatal DOPAC levels.[84] When plotting the improvement of acute schizophrenia against the steady state plasma levels of haloperidol a bell-shaped concentration–response curve was found, with a maximum effect in the range of 5–10 ng mL^{-1}.[85] Above 15 ng mL^{-1} the effects were negligible. The same type of curve was also found for nortriptyline effects in patients with endogenous depression.[86]

It has been suggested that the time dependent oscillations in motor function seen after oral administration of L-DOPA in patients with Parkinson's disease may be part of the dual effects of L-DOPA on motility; low doses decreasing and higher doses increasing motor activity.[87]

Dual effects of drugs are continuously being reported in the literature. For instance, low doses of morphine produce mainly excitatory effects on locomotor activity, and high doses inhibitory effects.[88-90] Accordingly, following the injection of a high dose of morphine, both suppression and stimulation of locomotor activity were recorded.[91] Kayser et al.[92] have also reported that morphine produces hyperalgesia in exceedingly low doses and analgesia at higher doses.

Thus, for many drugs the dose (concentration)–response curves are more complex than the classical sigmoid type of curve. Bell-shaped, U-shaped and multiphasic curves have been observed, indicating that a sum of activities from several receptor systems or physiological processes has an impact on the observed effect. In order to see this, the dose (concentration) range has to be extended and it is therefore important to study drugs over a dose range sufficiently large to permit the discovery of potential complex dose–effect relationships. As mentioned before, an anticlockwise hysteresis can also be observed for a drug with a bell-shaped or a U-shaped concentration–response curve upon intravenous administration, despite the lack of a time lag in the equilibration between the plasma and effect site concentrations.

A careful evaluation of the concentration–response relationship over a sufficiently wide dose range will prevent reports of confusing, paradoxical results, which are merely a consequence of the dose–response characteristics and the pharmacokinetics of the drug. Careful recordings of the time course of effects and drug levels will also give us an important insight into the mechanisms of drug actions and a means to improve drug therapy in the individual patient.

23.15.7 THE CONCENTRATION OF UNBOUND DRUG IN THE PLASMA

Over the years, a considerable body of evidence has accumulated indicating that the pharmacological response to a drug is better correlated with its plasma concentration or with the amount of the drug in the body than with the dose administered. The plasma concentration of unbound drug should be an even better correlate, since it more accurately reflects the drug concentration in the biophase. This becomes even more evident when individual plasma levels are related to pharmacological effects because of interindividual differences in protein binding, which can be further amplified by nonlinear binding. As illustrated in Figure 16, under steady state conditions, the concentration of unbound (free) drug in the plasma should be the same as the concentration of free drug in the biophase. Reidenberg et al.[93] observed, for instance, that phenytoin binding to plasma proteins was reduced in epileptic patients with renal failure. Lower doses were needed to control the seizures in these patients than in patients with normal renal function, but the resulting free plasma

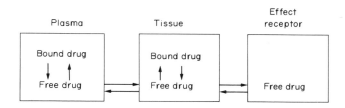

Total drug concentration = free drug concentration + bound drug concentration

fu = free drug concentration / total drug concentration

Figure 16 At equilibrium, the concentration of unbound (free) drug at the receptor site(s) will be equal to the concentration of free drug in the plasma, which is better related to the pharmacological effect than the total (free + bound) drug level in the plasma

concentrations were the same in both groups of patients. Disopyramide is an example of a drug exhibiting concentration dependent binding in plasma and, for this drug, the pharmacological effect was better related to the unbound plasma concentration than to the total plasma level.[94-96]

As pointed out previously, the commonly accepted hypothesis of today, that the steady state plasma concentration of unbound drug is better related to a pharmacological effect than the more commonly used total drug level in the plasma, also has a sound foundation in the theories of receptor occupancy.

23.15.8 TOLERANCE AND SENSITIZATION

If a drug produces its usual effect at unexpectedly low doses, the individual is said to be hyperreactive or hypersensitive.[97] Hyperreactivity to a drug is called supersensitivity if the increased sensitivity (sensitization) is a result of denervation or continued administration of a receptor antagonist. The opposite, hyporeactivity, is a manifestation of decreased sensitivity to a drug and is often described as tolerance when it is acquired as a result of prior exposure to a drug. Tolerance, which develops quickly after administration of only a few doses of a drug, is termed tachyphylaxis.[97]

Development of both tolerance and sensitization to the effects of certain drugs has been recognized for many drugs and the understanding of the basic cellular mechanisms underlying these changes of pharmacological effects has increased over the years. For instance, it has been possible to explain an altered response of an effect organ, resulting from repeated exposure to a drug, by estimating, *in vitro*, the changes in the numbers of receptors and in the receptor affinity for specific drugs. In studies of the relation between a pharmacological effect and the dose or concentration of the drug, all relationships described in this chapter assume a reversible pharmacodynamic response, which remains unchanged over the time course of drug treatment. However, as shown in Figure 11, clockwise hysteresis can be obtained when the effect is plotted against the concentration of the drug and the data points are joined in time sequence, indicating the development of tolerance during the time course of effects of a single dose. A hysteresis plot from *in vivo* data offers a possibility of describing acute tolerance during the time course of single doses. Such mathematical treatment of this phenomenon is rarely seen, however, and more research in this area is needed, since the data obtained today are, unfortunately, often only based on single-point measurements after long-term treatment. For instance, it is known that asthmatic patients receiving treatment with sympathomimetic agents progressively lose the effects of these drugs.[98] Another example is the loss of striatal dopamine receptors during long-term treatment with levodopa in Parkinson's disease.[99] Recently, however, there have been some attempts to describe the time dependent acute changes in the relationship between drug levels and the pharmacological response. Ellinwood *et al.*[100] reported how the intercept of the linear relationship between the concentration and effect of different benzodiazepines decreased exponentially with time. Hammarlund *et al.*[101] modelled the development of acute tolerance to the diuretic effects of furosemide. Using the sigmoid E_{max} model, it was seen that the sensitivity parameter (EC_{50} = furosemide excretion rate corresponding to 50% of maximal diuresis) increased exponentially as a function of the cumulatively excreted amounts of furosemide. In a subsequent study, the effect was re-established by compensating for fluid losses indicating that the observed tolerance to the diuretic effect of furosemide was of haemodynamic adjustment origin.[102]

23.15.9 BIOAVAILABILITY AND PHARMACODYNAMIC RESPONSE

Bioavailability is often defined as a term used to indicate the extent to and rate at which a drug reaches its site of action or the biological fluid from which the drug has access to its site of action. Sometimes, a narrower definition is used, such as that proposed by the American Pharmaceutical Association,[103] namely that it is the rate at and extent to which the active ingredient is absorbed from a drug product and reaches the systemic circulation. The latter definition overlooks the fact that measures of biophase drug concentrations can be quantified by monitoring pharmacological data. As discussed in this chapter, it is not uncommon that the time course of drug levels in the blood is quite different from that at the site of action. When considering the importance of bioavailability, it is essential to recognize the fact that the influence of this factor on the intensity and time course of the pharmacological effects is dependent on the characteristics of the dose (concentration)–response relationship. As pointed out by Levy,[104] and as is evident from Figure 17, a flat dose–effect curve and a low maximal efficacy will reduce the importance of a variable bioavailability within a class of drug products, whereas for drugs with steep dose–effect curves, even minor variations in bioavailability may imply considerable differences in responses. If, for instance, we have a drug with dose–response characteristics such as those in Figure 17, a 50% decrease in bioavailability can reduce the effect by only 10% at the top of the dose–response curve, but by as much as 80% at the bottom of this curve.

The rate of absorption can have both quantitative and qualitative effects on the drug responses. As illustrated in Figure 18, three products with the same active ingredient may have an identical extent of bioavailability (same areas under the plasma concentration–time curves), but different rates of absorption. Following oral administration, the product with the most rapid absorption will quickly exceed the minimum plasma level needed for the occurrence of effects and side effects, whereas the product with a slow absorption rate will not produce any effect at all.

If we have a complex concentration–response curve, as was discussed in Section 23.15.6, different degrees of bioavailability among products with the same active ingredient may result in qualitatively different effects. As is evident from Figure 14, if we take apomorphine as an example, a low bioavailability (low dose) may lead to bradycardia, while high bioavailability may cause tachycardia. The rate of absorption of a high dose of apomorphine will then, over time, determine the rate of a shift from bradycardia to tachycardia.[105] Among the problems associated with L-DOPA therapy in parkinsonism, a variable bioavailability in combination with a complex dose–response curve should be considered.[87] Again, the nature of the dose–response curve will determine how variation in bioavailability will influence the desired biological effect.

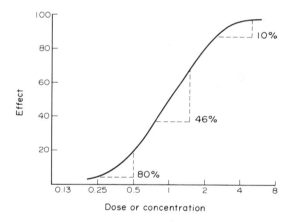

Figure 17 The influence of a 50% decrease in bioavailability on the pharmacological effect (reproduced from ref. 104 by permission of Raven Press)

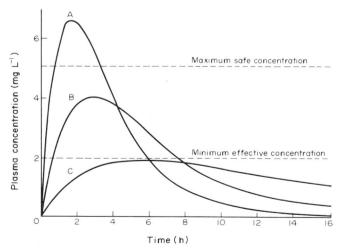

Figure 18 The influence of different absorption rate constants on the concentration–time profiles of three drugs with an identical extent of bioavailability (reproduced from ref. 111 by permission of the author)

23.15.10 PHARMACODYNAMICS IN DISEASE STATES

There is today a growing body of evidence indicating that the relationship between drug concentration and pharmacological effects can be altered by the disease state of the patient. It is well established, for instance, that some diseases of the liver or kidneys are associated with changes in the pharmacokinetics of many drugs, which may lead to an alteration of the pharmacodynamic response. However, in order to determine whether a changed pharmacological effect is due to altered drug–receptor events or to pharmacokinetic factors, the latter factors have to be evaluated separately from the pharmacodynamic ones. Such study designs are, unfortunately, rarely seen, and more investigations are urgently required, especially concerning the relationship between the concentration of free drug and the intensity of the pharmacological response. Information from such studies can also increase our knowledge of the inter- and intra-individual variability in pharmacodynamic responses. There are reasons to believe that this variability is of the same order of magnitude as the variability in pharmacokinetics.[82]

The need for investigation of both the pharmacodynamics and pharmacokinetics may be the reason for the relatively small number of reports that are available concerning the effects of disease states. Furthermore, the condition of a patient may improve or deteriorate during the therapy and, as a result, the response to a specific drug may also change.

The following examples of altered drug concentration–response relationships have been reported. Myerburg *et al.*[106] found that patients who had recently suffered an acute myocardial infarction needed lower steady state plasma concentrations of procainamide to suppress premature ventricular complexes than patients with chronic ischaemic heart disease. The effect of chlorpromazine on EEG activity in patients with cirrhosis of the liver indicated that this disease increased the sensitivity to the CNS depressant effect of the drug as compared with the findings in normal controls.[107] The pain threshold has been found to be considerably higher in hypertensive patients than in normotensive persons.[108] Galeazzi *et al.*[109] reported that the cardiac effects of pindolol in patients with uraemic renal failure were increased compared with those in normal volunteers. In a series of studies, Levy and associates have investigated the kinetics of drug actions in different animal models of disease states. They reported, for instance, that the concentrations of phenobarbital in the serum, brain and cerebrospinal fluid of rats with experimental renal dysfunction at the onset of loss of the righting reflex were considerably lower than the corresponding concentrations in normal animals.[110] The reason for the increased sensitivity of certain drugs during renal dysfunction, mentioned above, is unknown, but this finding opens up interesting perspectives regarding the question of how pathophysiological factors influence the intrinsic pharmacodynamic mechanisms underlying drug effects. Further studies in this area may provide a deeper insight into the molecular mechanisms of drug action.

23.15.11 REFERENCES

1. A. J. Clark, 'The Mode of Action of Drugs on Cells', Williams and Wilkins, Baltimore, 1933.
2. E. J. Ariens, 'Molecular Pharmacology', Academic Press, New York, 1964.
3. A. Goldstein, L. Aronow and S. M. Kalman, 'Principles of Drug Action', Wiley, New York, 1974, p. 82.
4. R. J. Tallarida and L. S. Jacob, 'The Dose-Response Relation in Pharmacology', Springer-Verlag, New York, 1979.
5. R. P. Stephenson, *Br. J. Pharmacol. Chemother.*, 1956, **11**, 379.
6. W. D. M. Paton, *Proc. R. Soc. London, Ser. B*, 1961, **154**, 21.
7. D. R. Waud, *Pharmacol. Rev.*, 1968, **20**, 49.
8. E. J. Ariens and A. M. Simonis, *J. Pharm. Pharmacol.*, 1964, **16**, 137.
9. B. Belleau, *J. Med. Chem.*, 1964, **7**, 776.
10. B. M. Bloom and I. M. Goldman, *Adv. Drug Res.*, 1966, **3**, 121.
11. D. Mackay, *Nature (London)*, 1963, **197**, 1171.
12. A. J. Karlin, *J. Theor. Biol.*, 1967, **16**, 306.
13. J. P. Changeux, J. C. Meunier, R. W. Olsen, M. Weber, J. P. Bourgeois, J. L. Popot, J. B. Cohen, G. L. Hazelbauer and H. A. Lester, in 'Drug Receptors', ed. H. P. Rang, Macmillan, London, 1973, p. 273.
14. R. E. Gosselin, *Br. J. Pharmacol.*, 1970, **39**, 215.
15. J. H. Gaddum, *Pharmacol. Rev.*, 1957, **9**, 211.
16. M. D. Hollenberg, *Trends Pharmacol. Sci.*, 1985, **6**, 242.
17. N. H. G. Holford and L. B. Sheiner, *CRC Crit. Rev. Bioeng.*, 1981, **5**, 273.
18. N. H. G. Holford and L. B. Sheiner, *Clin. Pharmacokinet.*, 1981, **6**, 429.
19. N. H. G. Holford and L. B. Sheiner, *Pharmacol. Ther.*, 1982, **16**, 143.
20. B. Whiting, N. H. G. Holford and L. B. Sheiner, *Br. J. Clin. Pharmacol.*, 1980, **9**, 67.
21. N. H. G. Holford, P. E. Coates, T. W. Guentert, S. Riegelman and L. B. Sheiner, *Br. J. Clin. Pharmacol.*, 1981, **11**, 187.
22. A. Wellstein and D. Palm, *Methods Find. Exp. Clin. Pharmacol.*, 1984, **6**, 641.
23. P. A. Mitenko and R. I. Ogilvie, *N. Engl. J. Med.*, 1973, **289**, 600.

24. B. Oosterhuis, M. C. P. Braat, C. M. Roos, J. Wemer and C. J. Van Boxtel, *Clin. Pharmacol. Ther.*, 1986, **40**, 469.
25. A. V. Hill, *J. Physiol. (London)*, 1910, **40**, iv.
26. J. G. Wagner, *Theor. Biol.*, 1968, **20**, 173.
27. L. B. Sheiner, D. R. Stanski, S. Vozeh, R. D. Miller and J. Ham, *Clin. Pharmacol. Ther.*, 1979, **25**, 358.
28. C. J. Hull, H. B. H. van Beem, K. McLeod, A. Sibbald and M. J. Watson, *Br. J. Anaesth.*, 1978, **50**, 1113.
29. J. S. Walker, C. Shanks and K. F. Brown, *Clin. Pharmacol. Ther.*, 1983, **33**, 510.
30. B. C. Weatherley, S. G. Williams and E. A. M. Neill, *Br. J. Anaesth.*, 1983, **55**, 39S.
31. D. M. Fisher, K. Castagnoli and R. D. Miller, *Clin. Pharmacol. Ther.*, 1985, **37**, 402.
32. J. C. Scott, K. V. Ponganis and D. R. Stanski, *Anesthesiology*, 1985, **62**, 234.
33. D. R. Stanski, R. J. Hudson, T. D. Homer, L. J. Saidman and E. Meathe, *J. Pharmacokin. Biopharm.*, 1984, **12**, 223.
34. A. A. Holazo, R. K. Brazzell and W. A. Colburn, *J. Clin. Pharmacol.*, 1986, **26**, 336.
35. B. Dahlström, L. K. Paalzow, G. Segre and A. Ågren, *J. Pharmacokin. Biopharm.* 1978, **6**, 41.
36. V. R. Spiehler and L. Paalzow, *Life Sci.*, 1979, **24**, 2125.
37. G. Paalzow and L. Paalzow, *Acta Pharmacol. Toxicol.*, 1973, **32**, 22.
38. G. H. M. Paalzow and L. K. Paalzow, *Eur. J. Pharmacol.*, 1983, **88**, 27.
39. L. K. Paalzow and G. H. M. Paalzow, *J. Pharm. Pharmacol.*, 1986, **38**, 28.
40. G. H. M. Paalzow and L. K. Paalzow, *J. Pharmacol. Exp. Ther.*, 1982, **223**, 795.
41. L. K. Paalzow and P. O. Edlund, *J. Pharmacokin. Biopharm.*, 1979, **7**, 495.
42. A. Morino, H. Sasaki, H. Mukai and M. Sugiyama, *J. Pharmacokin. Biopharm.*, 1986, **14**, 309.
43. A. Morino, N. Kitamura, K. Katayama, M. Kakemi and T. Koizumi, *J. Pharmacokin. Biopharm.*, 1983, **11**, 47.
44. P. A. Meredith, A. W. Kelman, H. L. Elliott and J. L. Reid, *J. Pharmacokin. Biopharm.*, 1983, **11**, 323.
45. G. Levy, *J. Pharm. Sci.*, 1964, **53**, 342.
46. G. Levy, *Clin. Pharmacol. Ther.*, 1966, **7**, 362.
47. J. M. Van Rossum and A. T. J. Van Koppen, *Eur. J. Pharmacol.*, 1968, **2**, 405.
48. M. Gibaldi and D. Perrier, 'Pharmacokinetics', 2nd edn., Dekker, New York, 1982, p. 227.
49. G. Levy, *J. Pharm. Sci.*, 1967, **56**, 1687.
50. L. F. Walts and J. B. Dillon, *Anesthesiology*, 1967, **28**, 372.
51. L. K. Paalzow and P. O. Edlund, *J. Pharmacokin. Biopharm.*, 1979, **7**, 481.
52. M. Frisk-Holmberg, L. Paalzow and L. Wibell, *Eur. J. Clin. Pharmacol.*, 1984, **26**, 309.
53. K. Starke, *Rev. Physiol., Biochem. Pharmacol.*, 1977, **77**, 2.
54. J. M. Cederbaum and G. K. Aghajanian, *Eur. J. Pharmacol.*, 1977, **44**, 375.
55. G. H. M. Paalzow and L. K. Paalzow, *Naunyn-Schmiedeberg's Arch. Pharmacol.*, 1983, **322**, 193.
56. G. H. M. Paalzow and L. K. Paalzow, *Psychopharmacology (Berlin)*, 1985, **85**, 31.
57. A. W. Kelman and B. Whiting, *J. Pharmacokin. Biopharm.*, 1980, **8**, 115.
58. M. Rowland and T. N. Tozer, 'Clinical Pharmacokinetics, Concepts and Applications', Lea and Febiger, Philadelphia, 1980, p. 138.
59. G. Levy, M. Gibaldi and W. J. Jusko, *J. Pharm. Sci.*, 1969, **58**, 422.
60. L. Paalzow and G. Paalzow, *Acta Pharm. Suec.*, 1971, **8**, 329.
61. J. G. Wagner, G. K. Aghajanian and O. H. L. Bing, *Clin. Pharmacol. Ther.*, 1968, **9**, 635.
62. W. G. Kramer, A. J. Kolibash, R. P. Lewis, M. S. Bathala, J. A. Visconti and R. H. Reuning, *J. Pharmacokin. Biopharm.*, 1979, **7**, 47.
63. L. K. Paalzow, *J. Pharmacokin. Biopharm.*, 1975, **3**, 25.
64. P. Tfelt-Hansen and L. Paalzow, *Clin. Pharmacol. Ther.*, 1985, **37**, 29.
65. B. Oosterhuis, R. J. M. ten Berge, H. P. Sauerwein, E. Endert, P. T. A. Schellekens and C. J. van Boxtel, *J. Pharmacol. Exp. Ther.*, 1984, **229**, 539.
66. G. Segre, *Farmaco*, 1968, **23**, 907.
67. W. Forester, R. P. Lewis, A. M. Weissler and T. A. Wilke, *Circulation*, 1974, **49**, 517.
68. W. Shapiro, K. Narahara and K. Taubert, *Circulation*, 1970, **42**, 1065.
69. E. Fuseau and L. B. Sheiner, *Clin. Pharmacol. Ther.*, 1984, **35**, 733.
70. W. A. Colburn, *J. Pharmacokin. Biopharm.*, 1981, **9**, 367.
71. J. Ham, D. R. Stanski, P. Newfield and R. D. Miller, *Anesthesiology*, 1981, **55**, 631.
72. J. C. Scott and D. R. Stanski, *J. Pharmacol. Exp. Ther.*, 1987, **240**, 159.
73. C. E. Inturrisi, W. A. Colburn, R. F. Kaiko, R. W. Houde and K. M. Foley, *Clin. Pharmacol. Ther.*, 1987, **41**, 392.
74. J. Dow, B. Laquais, J. Tisne-Versailles, B. Pourrias and M. S. Benedetti, *J. Pharmacokin. Biopharm.*, 1982, **10**, 283.
75. R. J. Hudson, D. R. Stanski, L. J. Saidman and E. Meathe, *Anesthesiology*, 1983, **59**, 301.
76. J. Grevel, J. Brownell, J. L. Steimer, R. C. Gaillard and J. Rosenthaler, *Br. J. Clin. Pharmacol.*, 1986, **22**, 1.
77. J. D. Unadkat, F. Bartha and L. B. Sheiner, *Clin. Pharmacol. Ther.*, 1986, **40**, 86.
78. J. E. Leysen and W. Gommeren, *Drug Dev. Res.*, 1986, **8**, 119.
79. M. L. Jack, W. A. Colburn, N. M. Spirt, G. Bautz, M. Zanko, W. D. Horst and R. A. O'Brien, *Prog. Neuro-Psychopharmacol. Biol. Psychiatry*, 1983, **7**, 629.
80. L. G. Miller, D. J. Greenblatt, S. M. Paul and R. I. Shader, *J. Pharmacol. Exp. Ther.*, 1987, **240**, 516.
81. L. K. Paalzow, *Drug Metab. Rev.*, 1984, **15**, 383.
82. L. K. Paalzow, G. H. M. Paalzow and P. Tfelt-Hansen, in 'Pharmacokinetics, A Modern View', ed. L. Z. Benet, G. Levy and B. L. Ferraiolo, Plenum Press, New York, 1984, p. 327.
83. P. Protais, I. Dubuc and J. Costentin, *Eur. J. Pharmacol.*, 1983, **94**, 271.
84. P. L. Wood, P. McQuade, J. W. Richard and M. Thakur, *Life Sci.*, 1983, suppl. 1, **33**, 759.
85. M. L. Mavroides, D. R. Kanter, J. Hirschowitz and D. L. Garver, *Psychopharmacology (Berlin)*, 1983, **81**, 354.
86. M. Åsberg, B. Cronholm, F. Sjöqvist and D. Tuck, *Br. Med. J.*, 1971, **3**, 331.
87. G. H. M. Paalzow and L. K. Paalzow, *Trends Pharmacol. Sci.*, 1986, **7**, 15.
88. M. Babbini and W. M. Davis, *Br. J. Pharmacol.*, 1972, **46**, 213.
89. E. F. Domino, M. R. Vasko and A. E. Wilson, *Life Sci.*, 1976, **18**, 361.
90. P. Schnur, F. Bravo, M. Trujillo and S. Rocha, *Pharmacol., Biochem. Behav.*, 1983, **18**, 357.
91. P. Schnur and V. R. Raigoza, *Life Sci.*, 1986, **38**, 1323.

92. V. Kayser, J. M. Besson and G. Guilbaud, *Brain Res.*, 1987, **414**, 155.
93. M. M. Reidenberg, I. Odar-Cederlöf, C. von Bahr, O. Borgå and F. Sjöqvist, *N. Engl. J. Med.*, 1971, **285**, 264.
94. J. J. Lima, H. Boudoulas and M. Blanford, *J. Pharmacol. Exp. Ther.*, 1981, **219**, 741.
95. J. D. Huang and S. Oie, *J. Pharmacol. Exp. Ther.*, 1982, **223**, 469.
96. M. Thibonnier, N. H. G. Holford, R. A. Upton, C. D. Blume and R. L. Williams, *J. Pharmacokin. Biopharm.*, 1984, **12**, 559.
97. E. M. Ross and A. G. Gilman, in 'Goodman and Gilman's The Pharmacological Basis of Therapeutics', ed. A. Goodman Gilman, L. S. Goodman, T. W. Rall and F. Murad, Macmillan, New York, 1985, p. 35.
98. H. G. Morris, *J. Allergy Clin. Immunol.*, 1980, **65**, 83.
99. P. Riederer, G. P. Reynolds and K. Jellinger, *Trends Pharmacol. Sci.*, 1984, **5**, 25.
100. E. H. Ellinwood, D. G. Heatherly, A. M. Nikaido, T. D. Bjornsson and C. Kilts, *Psychopharmacology (Berlin)*, 1985, **86**, 392.
101. M. M. Hammarlund, B. G. Odlind and L. K. Paalzow, *J. Pharmacol. Exp. Ther.*, 1985, **233**, 447.
102. P. A. Sjöström, B. G. Odlind, B. A. Beerman and M. M. Hammarlund-Udenaes, *Scand. J. Urol. Nephrol.*, 1988, **22**, 133.
103. Am. Pharm. Assoc., 'The Bioavailability of Drug Products', Washington DC, 1975.
104. G. Levy, *Pharmacology*, 1972, **8**, 33.
105. L. K. Paalzow, G. H. M. Paalzow and P. Tfelt-Hansen, in 'Variability in Drug Therapy: Description, Estimation and Control', ed. M. Rowland, L. B. Sheiner and J. L. Steimer, Raven Press, New York, 1985, p. 167.
106. R. J. Myerburg, K. M. Kessler and I. Kiem, *Circulation*, 1981, **64**, 280.
107. J. D. Maxwell, M. Carrella, J. D. Parkes, R. Williams, G. P. Mould and S. H. Curry, *Clin. Sci.*, 1972, **43**, 143.
108. N. Zamir and E. Shuber, *Brain Res.*, 1980, **201**, 471.
109. R. L. Galeazzi, M. Gugger and P. Weidmann, *Kidney Int.*, 1979, **15**, 661.
110. M. Danhof, M. Hisaoka and G. Levy, *J. Pharmacol. Exp. Ther.*, 1984, **230**, 627.
111. D. D. Breimer, Thesis, Nijmegen, Netherlands, 1974, p. 44.
112. A. A. D'Hollander and C. Delcroix, *J. Pharmacokinet. Biopharm.*, 1981, **9**, 27.

24.1

Use of Isotopes in Quantitative and Qualitative Analysis

LESLIE F. CHASSEAUD and DAVID R. HAWKINS
Huntingdon Research Centre Ltd, UK

24.1.1   INTRODUCTION

Elucidation of the nature of complex biochemical processes requires the application of sensitive and sophisticated methodology. Many of the discoveries made in biochemistry have followed or have accompanied improvements or new developments in methodology. Of the analytical techniques available, those involving the use of radioactive or stable isotopes are perhaps amongst the most versatile. Some of these isotope-based techniques were first utilized many years ago[1] but in time they have become more refined and their possible applications extended and expanded. However, even today, their full potential probably has still to be realized.

It is the purpose of this review to describe the uses of radioactive and stable isotope techniques in subject areas associated with biopharmaceutics. Even so, space does not permit exhaustive or even extensive coverage of this topic and the reader is consequently referred to the cited monographs for more detailed descriptions and for a source of other relevant literature. There are a number of earlier monographs on the production, measurement and use of radioisotopes[2-9] and more recent texts.[10-14]

24.1.2 RADIOACTIVE ISOTOPES

24.1.2.1 Commonly Used Radioisotopes

The radioactive isotopes (radioisotopes) used most frequently in biochemical research emit either beta (β) radiation (high velocity electrons) or gamma (γ) radiation (short wavelength electromagnetic radiation with similar properties to X-rays) or even a combination of both (Table 1). It is indeed fortuitous that two of the most important elements occurring in biological systems, carbon and hydrogen, are available as relatively long-lived low energy β-radiation-emitting isotopes (Table 1). Regrettably two others, nitrogen and oxygen, are not available in this experimentally convenient form. Rather short-lived radioisotopes of these latter two elements can be produced (Table 1) but their availability in this form is primarily confined to those few laboratories possessing the necessary equipment (cyclotron) for their preparation. Of the long-lived radioisotopes, ^{14}C is undoubtedly the most extensively employed for tracer applications, while tritium and radioactive iodine are often preferred for bioassays based on immunological or receptor-binding principles. The former isotope is particularly useful when high sensitivity and/or low energy are required, whereas the latter is notable for the relative ease with which it can be incorporated into certain types of compounds such as proteins.

A number of other experimentally useful radioisotopes are also available. These include ^{36}Cl, ^{32}P (and the less energetic ^{33}P) and ^{35}S which can be invaluable for studies of compounds that contain these elements. ^{51}Cr and ^{59}Fe have found application in the labelling of erythrocytes for assessing blood loss after administration of drugs such as non-steroidal antiinflammatory agents.[16] Studies of

Table 1 Radioactive Isotopes Used for Labelling[15]

Isotope	Principal radiation (MeV)	Half-life
^{45}Ca	0.25 β−	165 d
^{11}C	0.97 β+	20 min
^{14}C	0.16 β−	5760 y
^{36}Cl	0.71 β−	3×10^5 y
$^{51}Cr^a$	0.32 γ	28 d
^{18}F	0.65 β+	110 min
$^{125}I^a$	0.035 γ	60 d
$^{131}I^b$	0.36 γ	8 d
$^{59}Fe^b$	1.1 γ	45 d
$^{203}Hg^b$	0.28 γ	47 d
^{13}N	1.2 β+	10 min
^{15}O	1.7 β+	2 min
^{32}P	1.7 β−	14 d
^{33}P	0.25 β−	25 d
$^{75}Se^a$	0.27 γ	121 d
^{35}S	0.17 β−	87 d
$^{119m}Sn^c$	0.065 γ	245 d
^3H	0.018 β−	12 y

^{a} Electron capture, γ radiation energy shown. ^{b} Also emits β radiation. ^{c} Isomeric transition, γ radiation energy shown.

the metabolic fate of certain elements or organic compounds thereof are facilitated by the availability of suitable radioisotopes; these include ^{203}Hg, ^{75}Se and ^{119m}Sn. The absorption and biological fate of physiologically important elements can be followed using suitable radiolabelled forms such as ^{45}Ca and ^{59}Fe.

24.1.2.2 Advantages and Disadvantages of Radioisotopes

At least four advantages in using radioisotopes in biochemical research can be mentioned. The first and foremost is the relative ease with which radioactivity can be measured utilizing relatively inexpensive reliable easy-to-use commercially available equipment, the liquid scintillation counter for β radiation and the γ scintillation counter for γ radiation. The second is the high degree of accountability obtainable in studies of compounds radiolabelled in a metabolically stable position in the molecule, whereby the radiolabel remains associated with the compound and all its metabolites thereby enabling them to be tracked readily in biological systems; this makes the radioisotope ideal for tracer applications. Alternatively, where appropriate, the radiolabel can be sited at a metabolically labile part of the molecule so that a particular route of metabolism can be followed. Detection of metabolites using non-radiotracer methodology is often a matter of chance unless authentic reference compounds are available. The third advantage is the great sensitivity possible when radiolabelled compounds of high specific activity are used; certain relatively inexpensive radioimmunoassays can achieve limits of accurate measurement in the $pg\,mL^{-1}$ range. The fourth is that the 'pulse-dose' nature of radiolabelled compounds can be exploited whereby the behaviour of an individual dose of a compound can be followed despite an existing background of that compound, *e.g.* the absorption of calcium from the gastrointestinal tract can be measured using ^{45}Ca or the thyroid uptake of iodine determined using ^{131}I.

Three disadvantages associated with the use of radioisotopes can be mentioned. The first and most obvious is the risk of contamination that accompanies the use of radioisotopes. Careful laboratory work practices are essential so as to avoid contamination of the laboratory environment which could have deleterious consequences on all experimentation involving radioisotopes being conducted in that laboratory. The measurement of radioactivity should be regarded as non-specific unless this is demonstrated to the contrary, since, by its nature, it does not usually discriminate between the radioactivities associated with several different radiolabelled compounds that may be present, such as a drug and its metabolites, at least as far as the same radioisotope is concerned. Specificity can usually be assured, however, by introducing a chromatographic separation stage prior to measurement of radioactivity. A second disadvantage occurs with certain radioisotopes such as tritium which can exchange in biological systems resulting in partial loss of the radiolabel either by chemical or biochemical mechanisms, thereby causing an unknown alteration in specific activity. The third is the relative instability and short shelf-life of many radiolabelled compounds owing to radiolysis, more so at higher specific activities and with certain chemical classes. Thus the radiochemical purity of the radiolabelled compound needs to be checked regularly throughout its use. Several repurifications or even resynthesis may be necessary in a long-term research programme.

Radiation exposure of the laboratory worker must be considered another disadvantage, although soft β-emitters such as ^{14}C and tritium are not normally considered hazardous when used according to appropriate laboratory operating procedures. Indeed, under certain conditions, such radioisotopes are administered to human subjects.

The relatively high cost of radiolabelled compounds is viewed by some as another disadvantage but arguably this is not so, since the use of a radiolabelled compound can render many otherwise tedious and difficult experiments fairly straightforward. In such situations, the use of the radiolabelled compound is often the most cost-effective approach available.

As far as its biological fate is concerned, in general the radiolabelled compound can be expected to behave in an identical fashion to its non-radioactive ('normal') counterpart, hence its great usefulness in tracer applications (see Section 24.1.3.1 regarding isotope effects).

24.1.2.3 Synthesis of Radioisotope-labelled Compounds

It is important to appreciate that not all the molecules in a radioisotope-labelled (radiolabelled) compound contain the radiolabel. The greater the proportion that do the higher the specific activity, which is a measure of the amount of radioactivity present per millimole of compound. Common units of specific activity are millicuries per millimole ($mCi\,mmol^{-1}$) or microcuries per milligramme

(μCi mg^{-1}). The SI unit of radioactivity is the becquerel (Bq) which is one disintegration per second, and 1 μCi equals 3.7×10^4 Bq.

Radiolabelled compounds can be prepared by biosynthetic processes or chemical procedures. The former is limited in applications but has the advantage that for complex molecules, such as digitoxin, simple radiolabelled precursors can be used (*e.g.* $^{14}CO_2$) and the desired (naturally occurring) stereoisomer is obtained. The radiolabel is likely to be present in several different positions in the molecule, not all of which may be metabolically stable. The products of biosynthetic processes are usually of low specific activity, often insufficient for certain types of experiment. However, there may be no convenient alternative to biosynthetic processes for compounds that are normally produced in that way. The use of newer plant culture methods should improve the radiochemical yield and specific activity of biosynthetic processes. The increasing role of biotechnology in the search for new drugs will also result in the need to use such methods to prepare the corresponding radiolabelled compounds.

Chemical procedures are presently the methods of choice, permitting the production of relatively large amounts of the radiolabelled compound of the desired specific activity and position of labelling as well as easier product purification than is possible with biosynthetic processes. A variety of radiolabelled precursors are commercially available for use in chemical procedures and they can be introduced into the synthetic route at an initial stage (Scheme 1) or later on (Scheme 2). An ideal synthesis procedure involves a small number of high-yield reaction steps in which there are a minimum number of transfer and isolation stages.

The position of incorporation of the radiolabel merits careful consideration and its choice is governed by the proposed use of the radiolabelled compound together with synthesis considerations such as feasibility and cost. For metabolism studies, ^{14}C is normally the preferred radiolabel because usually it can be located in metabolically stable positions in the molecule such as aromatic or alicyclic ring systems, but if relatively high specific activity is needed then tritium labelling may be necessary. With the latter, regioselective incorporation of the tritium is more likely to lead to a radiolabelled compound that suffers little or no tritium exchange *in vivo* than that obtained using catalytic exchange with tritium gas or tritiated water which yield generally labelled compounds.

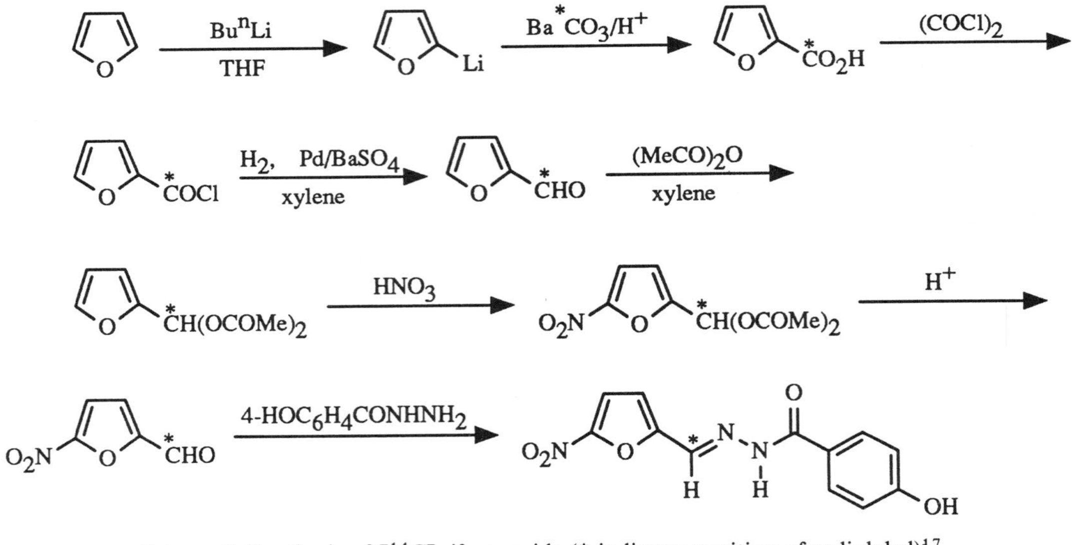

Scheme 1 Synthesis of [^{14}C]nifuroxazide (* indicates position of radiolabel)[17]

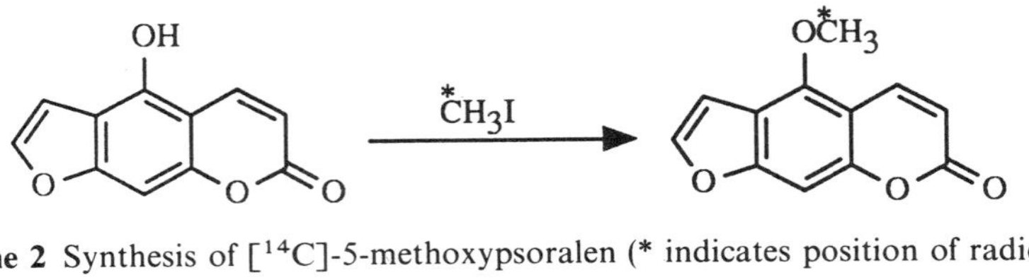

Scheme 2 Synthesis of [^{14}C]-5-methoxypsoralen (* indicates position of radiolabel)

Radioactive iodine is often used to radiolabel proteins, but it should be realized that the resultant product differs from the original protein in that it is an iodinated protein, and the two may not necessarily share the same biological fate. Alternatively, tritium and rarely ^{14}C may be used to radiolabel proteins. Studies of the metabolic fate of a sulfated glycopeptide utilized ^{35}S introduced by reaction of the precursor glycopeptide with chloro[^{35}S]sulfonic acid.[18]

On completion of its synthesis, both the radiochemical purity and chemical identity of the radiolabelled compound need to be verified, the former by TLC and/or HPLC with radioactivity detection, and the latter by spectral techniques including IR and NMR spectroscopy and MS. Certain types of compound, *e.g.* proteins, will require the use of other physicochemical techniques and perhaps even bioassays to confirm their identity. A radiochemical purity of at least 98% is desirable, whilst chemical purity should match that of the authentic standard.

More detailed accounts of radiosynthetic procedures are given in the references cited in Section 24.1.1 and elsewhere.[15,19-21] There is a specialist journal (*Journal of Labelled Compounds and Radiopharmaceuticals*) for this subject area.

24.1.2.4 Measurement of Radioisotopes

β Radioisotopes are readily measured using the technique of liquid scintillation counting which relies on ionizing radiation causing the emission of light (photons) of short duration when it strikes a scintillator. It is these photons that are detected in the liquid scintillation counter. The scintillator is dissolved in a solvent mixture in which the biological or other fluids to be measured, such as plasma or urine, are also miscible. Insoluble biological samples such as tissues and faeces or those such as blood that can cause severe light quenching and thus much reduced counting efficiency are first combusted (certain radioisotopes only) or solubilized before liquid scintillation counting. Combustion of such samples can be carried out in commercially available sample oxidizers; in the case of biological samples containing ^{14}C-labelled compounds, combustion yields $^{14}CO_2$ which is trapped in a base that is then mixed with the scintillator solution. Solubilization of biological samples is usually carried out in strong organic or inorganic bases. If necessary, the solubilized sample can be decolourized with hydrogen peroxide. These combustion or solubilization procedures permit the complete analytical recovery of radioactivity from all biological matrices, an objective seldom achieved when non-radiotracer methodology is used. Complete analytical accountability is a particularly advantageous property of radiolabelled compounds. There are several detailed accounts of liquid scintillation counting and related procedures.[22-29]

In metabolism studies of compounds labelled with β radioisotopes, metabolites are separated by TLC or HPLC, a practice often described as 'metabolite profiling'. Metabolites separated by TLC can be located and quantified directly on the thin-layer plate using linear analyzers that rely on gas flow proportional counting.[30] Radioactive metabolites on thin-layer plates can also be located more tediously by apposition autoradiography on X-ray film and quantified either after removing (or digesting) the relevant silica gel zones from the plate and adding them to a suitable scintillator solution, or after eluting radioactive metabolites from the silica gel zones and adding the eluates to scintillator solution. When HPLC is used, the system can include an on-line detector which can measure the radioactivity content of the effluent as it passes through either a solid scintillator flow cell or after mixing with a scintillator solution.[31] Alternatively the effluent can be collected and subjected to liquid scintillation counting.

Apposition autoradiography on X-ray film is still the method of choice.for the technique of whole-body autoradiography, where a thin section (*ca.* 20 μm) is taken from an animal (usually a rodent) killed after it has been dosed with a radiolabelled compound. The section is usually freeze-dried, applied to X-ray film and, after several days or weeks of exposure, apposition autoradiographs are obtained showing the location of radioactivity in the different organs/tissues of the body. This semiqualitative technique of whole-body autoradiography is now capable of quantitation employing computer-aided image analysis.[32] Whole-body autoradiography and other autoradiographic techniques have been reviewed.[33-35]

Measurement of γ radioisotopes is more straightforward than that of β radioisotopes. The sample, liquid or solid, is measured directly in the γ scintillation counter without the need for initial preparation.[26,27] Similarly, samples arising from chromatographic separation of γ-labelled compounds can be measured directly in the γ counter after they have been removed, excised or collected following chromatography. They can also be measured using linear analyzers.[30]

24.1.2.5   Use in Absorption, Distribution and Excretion Studies

Essential components of safety evaluation programmes for new chemical entities are absorption, distribution, metabolism and excretion (ADME) studies. Furthermore, the results of such studies should guide the medicinal chemist in designing new chemical entities possessing certain desirable features; such may include a longer half-life, specific tissue uptake, prolonged or improved absorption, *etc.* The use of the radiolabelled new chemical entity is essential to the success and timely completion of such studies. Use of non-radiotracer analytical techniques such as GLC or HPLC can seldom provide anywhere near the same volume of information as can be obtained when radioisotopes are utilized. However, the former are usefully employed in conjunction with the latter and are later usually sufficient once the basic ADME information has been generated. This basic information directs the development of appropriately sensitive non-radiotracer analytical methodology and enables the verification of its specificity; it may be necessary to measure not only the parent compound but also one or more of its metabolites first identified during the basic ADME studies. This subject area has been reviewed.[9,36-41]

The first consideration in an ADME study of a new chemical entity is the position of the radiolabel in the molecule. This is selected after review of the expected metabolism of the new chemical entity and the complexities involved in synthesizing it with a radiolabel in a desired position. It may be necessary to radiolabel more than one position or to synthesize more than one radiolabelled form. Even dual radiolabels such as tritium and ^{14}C have been incorporated into a single compound; despite its elegance, such an approach is not necessarily the best as the data obtained can be difficult to interpret. In structures (1) and (2) and those illustrated in Scheme 2 and Figure 3 (* = position of radiolabel), the radiolabel was placed: (i) in a metabolically stable position (Figure 3);[42] (ii) in a position that was easily radiolabelled and which turned out to be metabolically stable (Scheme 2) although this was not necessarily predictable at the outset;[43] (iii) in two positions in the molecule (1) owing to expected ester hydrolysis *in vivo* and production of two pharmacologically active moieties;[44] and (iv) in a position (2) dictated by the availability of a suitable precursor, in this case 17β-[6,7-^{3}H]estradiol.[45]

(1)   (2)

24.1.2.5.1   *Absorption*

Knowledge of their rate and extent of absorption is essential for the proper safety evaluation and development of new chemical entities.

Many drugs are administered by the oral route, and thus most studies of absorption are preoccupied with absorption from the gastrointestinal tract. For this reason this section emphasizes studies of orally administered compounds, but absorption through other portals of entry into the body such as the lung and skin is important for certain types of drug or during adventitious exposure to environmental chemicals. The use of radiolabelled compounds enables such studies to be more comprehensive.

The site of absorption of an orally administered compound is of some interest. A common method for its assessment in the rat involves ligating appropriate sections of the gastrointestinal tract, injecting the radiolabelled compound therein and determining the time course of the decline of radioactivity in the relevant section of the gut together with its concomitant appearance in blood (systemic and/or portal). Such studies usually demonstrate that absorption occurs mainly from the small intestine owing to its relatively greater surface area. Studies of this type have been reported for drugs such as camostat mesylate (3; FOY 305)[46] and 6-ethyl-3-(1*H*-tetrazol-5-yl)chromone.[47] Other techniques involve isolated gastrointestinal tissue ('everted sac preparations') and even model artificial membranes.[48,49]

$$\text{HN} = \overset{\text{H}_2\text{N}}{\underset{}{}} \text{N}-\text{C}_6\text{H}_4-\text{CO}_2-\text{C}_6\text{H}_4-\text{CH}_2\text{CO}_2\text{CH}_2\text{CONMe}_2 \cdot \text{MeSO}_3\text{H}$$

(3)

Information on the rate and extent of oral absorption of a radiolabelled compound is obtained from comparative studies involving intravenous and oral routes of administration (or any other extravascular route under study). Data obtained from the former serve as the reference for 100% absorption.

The rate of absorption is gauged from the time at which peak concentrations of radioactivity are achieved in plasma, but, for calculation of absorption kinetics, concentrations of the administered compound during the absorption phase need to be measured specifically; measurement of radioactivity alone is inadequate since the radioactivity may be associated mainly with metabolites and the data then merely attest that metabolism of the compound is rapid and extensive. It was shown that after oral doses of [^{14}C]bupranolol (4), peak concentrations of radioactivity in plasma were reached within 2 h in several species including man but that only the major metabolite, produced by oxidation (ring Me $\rightarrow$ CO$_2$H), could be detected.[50] This type of result is to be expected for a compound subjected to rapid and essentially complete first-pass metabolism in the gut and/or liver. Such compounds also include nitroglycerine,[51] benzarone[52] and, by design, prodrugs.

$$\text{Cl, Me-C}_6\text{H}_3-\text{OCH}_2^*\text{CH(OH)CH}_2\text{NHBu}^t$$

(4)

The ratio of areas under the plasma radioactivity concentration–time curves obtained after oral and intravenous doses is sometimes taken as an estimate of the extent of absorption of a radiolabelled compound from the gastrointestinal tract. However, such an approach is of doubtful validity unless rather similar proportions of the compound and its metabolites circulate in plasma after either route of administration when their combined clearances are likely to be the same. The calculation of area ratios is best applied to a discrete chemical entity such as the unchanged compound determined specifically, thereby providing a true estimate of its systemic availability. The area ratios calculated for a [^{14}C]drug in the rat and dog were about 1.0 and 0.7, respectively, when based on radioactivity concentrations but were 0.7 and 0.6, respectively, when based on parent drug concentrations.[53] The former thus failed to provide an accurate estimate of the systemic availability of the orally administered drug in the rat.

Similarity in the proportions of the radioactive dose excreted in the urine and faeces after administration of a radiolabelled compound by oral and intravenous routes is usually taken to indicate almost complete absorption of the oral dose from the gastrointestinal tract. This does not equate with complete systemic availability (of the unchanged compound) since the compound can be well absorbed from the gastrointestinal tract but it can also be subjected to extensive first-pass metabolism before reaching the systemic circulation, as in the case of [^{14}C]bupranolol (4) when more than 80% of the radioactivity was excreted in the urines of the species studied but the unchanged drug was not detected in plasma.[50] [^{14}C]Isosorbide dinitrate was apparently completely absorbed from the gastrointestinal tract of humans because more than 90% of the radioactivity was excreted in the urine.[54] However, measurement of unchanged isosorbide dinitrate in plasma after oral and intravenous administration showed that the systemic availability of oral isosorbide dinitrate was about 25%,[55] *i.e.* 75% was metabolized during the drug's first pass, presumably mainly in the liver. Even when similar proportions of radioactivity are excreted in the urine and faeces following oral and intravenous doses of a radiolabelled compound, if faecal excretion of radioactivity is extensive then the origin of this faecal radioactivity needs to be ascertained from studies in bile-duct cannulated animals. Radioactivity in faeces after oral doses could represent unabsorbed compound or absorbed compound and/or its metabolites excreted in the bile (or a combination of both), whereas after intravenous doses it would be due mainly to biliary-excreted compound and/or metabolites. Thus the sum of the proportions of radioactivity excreted in urine, bile and present in the body (excluding the gut) is the extent of absorption of the oral dose before first-pass metabolism, should this occur. Studies of the pentacyclic triterpene [³H]asiatic acid in the rat[56] illustrate this point—after an oral dose 31% of the radioactivity was excreted in 24 h bile, whereas after an intravenous dose 89% was excreted within 8 h. Urinary excretion of radioactivity was low. It was concluded that up to 50% of an oral dose of [³H]asiatic acid was absorbed, although much of this was probably subjected to first-pass metabolism and biliary excretion, thereby never reaching the systemic circulation as the unchanged drug.

24.1.2.5.2 Distribution

Radiolabelled compounds are invaluable in tissue distribution studies, where their major advantage is the completeness with which tissue radioactivity can be measured. Their use overcomes the problem of analytical recovery (see Section 24.1.2.4). Thus much of the information regarding the tissue distribution of compounds is based on total radioactivity measurements in a range of organs/tissues and, except in those tissues of pharmacological or toxicological interest, the chemical nature of the radioactive compounds present is seldom ever examined. If measurement of particular compounds (drug and/or metabolites) is desired, use of radiolabelled compounds enables their detection, separation, measurement and isolation. For this purpose, procedures such as tissue homogenization (perhaps also enzymic digestion), extraction and chromatography are used. Common findings in tissue distribution studies of radiolabelled compounds are: (i) that concentrations of radioactivity are usually greatest in organs/tissues associated with absorption and elimination, *i.e.* the gut, liver and kidneys, and in secretory organs such as the salivary and lacrimal glands; and (ii) that concentrations of radioactivity in many organs/tissues, *e.g.* pigmented regions of the eye,[57] decline more slowly than those in the corresponding plasma, presumably reflecting tissue binding of the radiolabelled compound and/or its radiolabelled metabolites.

Use of repeated doses of the radiolabelled compound is the ideal way to reveal accumulation of the test compound and/or its metabolites in organs/tissues. Thus a comparative study of L-[¹⁴C]tartaric acid or DL-[¹⁴C]tartaric acid administered orally to rats at equivalent dose levels for one week showed that the latter precipitated in the kidney (11 200 µg g⁻¹) as the calcium salt where its accumulation resulted in nephrotoxicity; no such accumulation in the kidney (1290 µg g⁻¹) was observed for the former and it was not nephrotoxic.[58] The calcium salt of the optically active form is some 10 times more soluble than that of the racemate.

In veterinary product residue studies, use of the radiolabelled compound is usually essential for revealing which tissues contain the greatest residues and in facilitating the chemical identification of the major residue therein so that this can be measured as an index of exposure. Studies of this type have been reported for [³H]ivermectin[59] and [¹⁴C]closantel.[60]

The availability of radiolabelled compounds has led to the development of the technique of whole-body autoradiography.[4,34,35,61] This technique depicts organ/tissue radioactivity data in the form of easily assimilated photographs which can show all the organs/tissues present (Figure 1), and relies on the detection of radioactivity by X-ray film applied to thin, often sagittal, sections (*ca.* 20 µm) of animals, usually rodents, dosed with the radiolabelled compound. Localization within

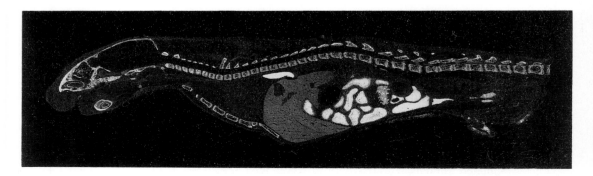

Figure 1 Whole-body autoradiograph of a rat obtained at 3 h after the last of seven daily oral doses of monosodium [¹⁴C]tartaric acid (2.73 g kg⁻¹ day⁻¹). Notable was the uptake of radioactivity (white areas) into bone probably owing to chelation of [¹⁴C]tartrate with bone calcium.[58] Also evident in this section was radioactivity in the liver and gastrointestinal tract. On this section, radioactivity is shown as white on a black background

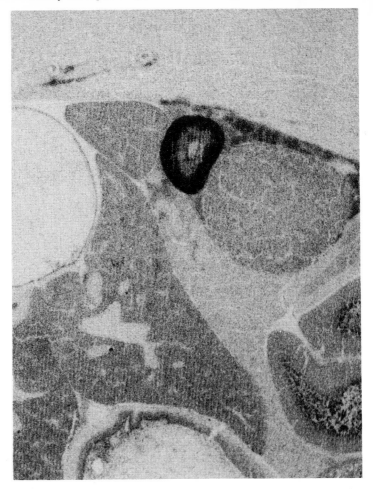

Figure 2 Enlargement of part of a whole-body autoradiograph obtained at 48 h after a single oral dose of [^{14}C]methoprene (25 mg kg^{-1}) to a rat. Notable was the much greater uptake of radioactivity into the adrenal cortex than in the adrenal medulla. Also evident in this section was radioactivity in the gastrointestinal mucosa, liver, kidney, perirenal fat and intestinal contents. On this section, radioactivity is shown as black on a grey background. Adrenal cortex radioactivity was presumed to originate from the incorporation of a metabolite(s) of [^{14}C]methoprene into cholesterol biosynthesis[62]

individual organs such as the brain or in smaller structures such as the foetus is readily delineated and the technique is most useful for studies of the placental transfer of compounds. The technique is particularly complementary to mechanistic or toxicological studies. Enlargements of the whole-body autoradiographs enables better visualization of localization in foetal tissues or in organs such as the eye, brain or adrenal (Figure 2). The information on autoradiographs can now be quantified using computer-aided image analysis.[32] For more detailed intraorgan/tissue data, such as uptake into skin,[63] the technique of microhistoautoradiography[33] can be advantageously utilized. Less energetic and, therefore, less penetrating isotopes such as tritium are ideal for this purpose on account of the better definition possible.

24.1.2.5.3 *Excretion*

The pivotal part of any metabolism investigation is the excretion balance study whose purpose is to account for the administered dose in terms of the proportions excreted or retained in the body, an objective which can be virtually impossible without the use of a radiolabelled compound. Excreted metabolites can then be quantified as a proportion of the administered dose (see Section 24.1.2.6). Information is also obtained on rates and routes of excretion.

Routes of excretion can exhibit marked species differences and can be dose-level dependent. A major sex difference in the biliary excretion and hepatic retention of a [^{14}C]sulfonylurea (**5**)

(5)

and/or its metabolites occurred in the rat which was not apparent in other species (Table 2).[64] The rat and dog, in which species molecular weight thresholds for biliary excretion are lower, tend to excrete a greater proportion of an absorbed dose in the faeces than does the monkey or man (Table 3). The extent of oral absorption of many compounds diminishes with increase in dose level, and often greater proportions of the dose are excreted in the faeces at higher dose levels (Table 3).

Use of the radiolabelled compound facilitates the study of enterohepatic circulation, whereby a compound excreted in bile is reabsorbed, usually after metabolism in the gut (Table 4).

Table 2 Hepatic Retention and Biliary Excretion of Radioactivity Following a Single Oral Dose ($0.2\ mg\,kg^{-1}$) of a [^{14}C]Sulfonylurea (5) to Various Animal Species[64]

Species	Biliary excretion (% dose)	Liver retention (% dose)	Time interval (h)
Rat ♂	71	4	0–48
Rat ♀	22	51	0–48
Dog ♂	45	7	0–24
Dog ♀	54	4	0–24
Rabbit ♂	3	5	0–48
Rabbit ♀	6	6	0–48

Table 3 Mean Excretion of Radioactivity (% Dose) Following Single Oral Doses of [^{14}C]Lormetazepam to Animals[65] and Humans[66]

Species (dose level mg kg^{-1})	Urine[a]	Faeces
Rat (0.25)	21	76
Dog (0.05)	65	30
Dog (250)	17	90
Monkey (0.07)	83	3
Monkey (360)	31	59
Humans (0.03)	86	—

[a] Includes cage wash in animals.

Table 4 Mean Excretion of Radioactivity (% Dose) in the Urine, Bile and Faeces During 48 h Following Single Intraduodenal Doses of [^{14}C]Benzarone ($2\ mg\,kg^{-1}$) to Male Rats, either Alone or in Tandem, whereby the First Rat was Dosed with [^{14}C]Benzarone and Bile Collected from the Second Rat[52]

Status of Rat	Urine	Bile	Faeces
Rats alone	4	76	16
Rats in tandem			
First rat	13	a	17
Second rat	4	46	17

[a] Bile passed directly into the duodenum of the second rat *via* a cannula.

24.1.2.6 Use in Metabolism Studies

Knowledge of the metabolism of a compound can provide information on its mode of action, facilitate interspecies comparisons and orientate the search for new drugs; a metabolite may be a suitable candidate for new drug development.

Investigations of the metabolism of a compound *in vivo* requires detection, separation, quantitation and isolation of the metabolites produced. These can be accomplished using the radiolabelled compound and associated techniques (see Section 24.1.2.4), provided that the radiolabel is located in a metabolically stable position in the molecule thereby remaining associated with the metabolites of the compound (*e.g.* Figure 3). Failure to place the radiolabel in a metabolically stable position in the compound can lead, after metabolism, to incorporation of a proportion of the radiolabel into endogenous compounds, and it may be exceedingly difficult and time-consuming to differentiate these from true metabolites of the compound under study. Occasionally, there may not be a metabolically stable position; for example in the terpenoid [^{14}C]plaunotol (**6**), although the radiolabel was placed in the middle of the molecule, far from sites of oxidative attack, at least 10% of the dose was metabolized to $^{14}CO_2$ in man[67] and presumably a proportion of the radiolabel was also incorporated into endogenous compounds which would not be regarded as true metabolites of (**6**).

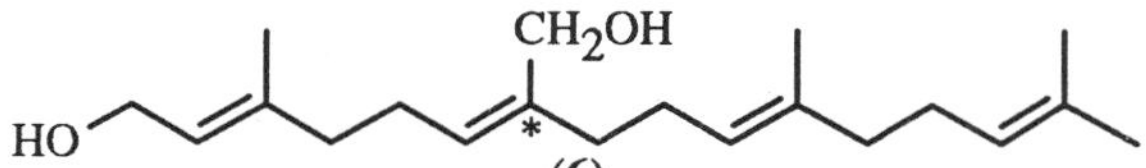

Comparison of metabolite profiles in biological samples (*e.g.* plasma, urine, bile and faeces) from different species is mostly conducted using TLC/radiochromatogram scanning (Table 5; Figure 4) and/or HPLC coupled to a radioactivity detector as exemplified during metabolism studies of [^{14}C]amlodipine in the rat and dog.[68] HPLC is advantageous where metabolite profiles may be

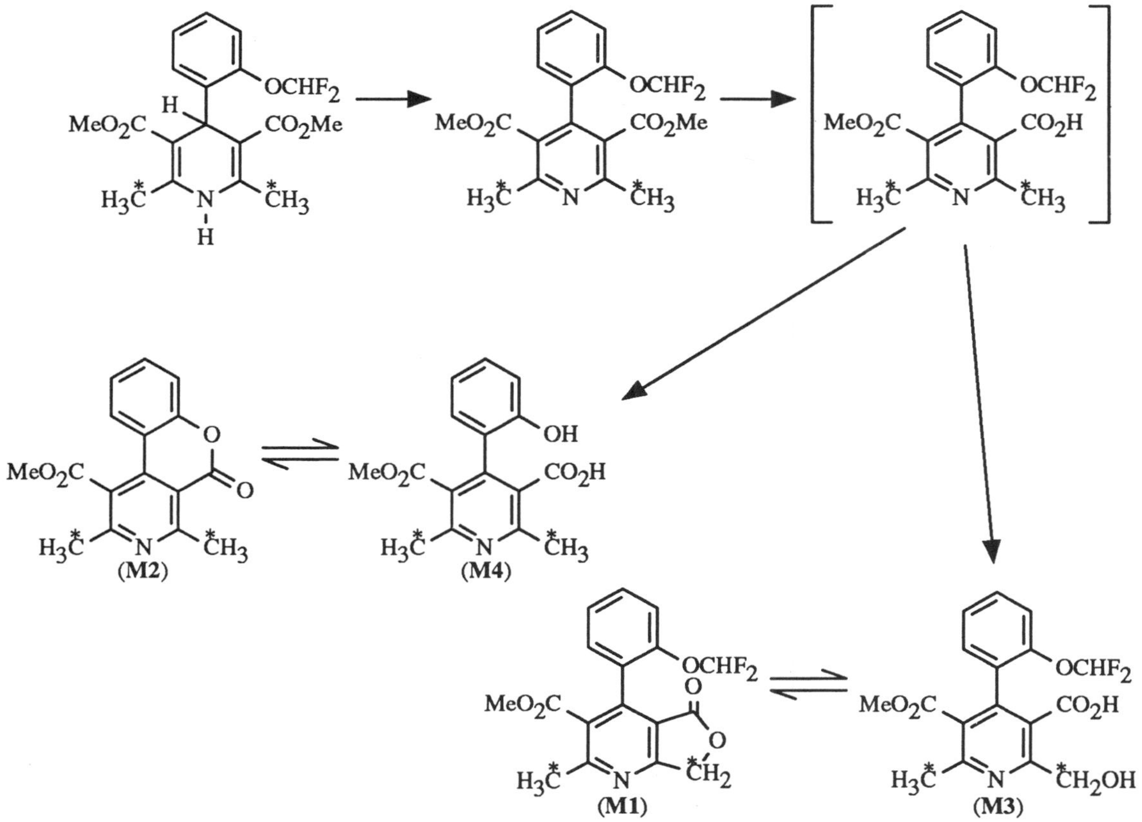

Figure 3 Metabolism of [^{14}C]ryosidine in man (* indicates position of radiolabel at carbons 2 and 6)[42]

Table 5 Urinary Metabolites (% Dose) of [^{14}C]Ryosidine Excreted During
0–24 h by Different Species[a,42]

Metabolite[c]	Untreated urine			Enzyme-treated urine[b]		
	Rat	Dog	Man	Rat	Dog	Man
M1	10	13	3	12	15	8
M2	6	15	12	5	15	13
M3	17	3	5	15	2	4
M4	3	7	4	3	9	4
Total	36	38	24	35	41	29
% Dose excreted in urine during 0–24 h	43	49	37	43	49	37

[a] The structures of the metabolites are shown in Figure 3. [b] After incubation with
β-glucuronidase/sulfatase. [c] Metabolite M1 (and/or M3) was excreted partly conjugated
(*cf.* untreated and enzyme-treated urine data).

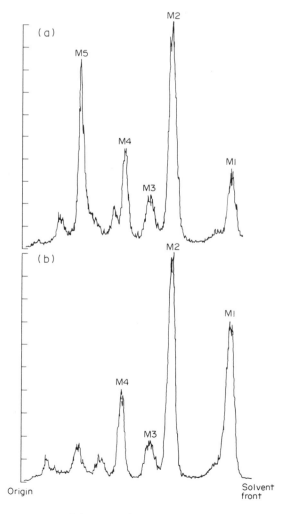

Figure 4 Thin-layer radiochromatogram of human urine (0–6h) (a) before and (b) after enzymic hydrolysis with
β-glucuronidase/sulfatase.[42] For structures of M1–M4, see Figure 3

complex, when better separation of metabolites can be achieved, using gradient elution and relatively long retention times, than would be possible with TLC. HPLC is also the technique of choice where volatile metabolites may be encountered, as during studies of [^{14}C]lofexidine (7).[69] Occasionally, other techniques such as electrophoresis may be used.

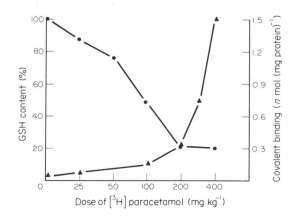

(7)

The metabolic interrelationships between species are judged on the basis of the qualitative and quantitative similarities or otherwise of the metabolite profiles obtained. However, it should be realized that the same chromatographic properties of individual radioactive zones in the metabolite profiles of different species do not necessarily equate with the same chemical identities; this needs to be established after isolation and characterization by appropriate physicochemical procedures. If a reference compound is available, isotope dilution[26,39,70] can be employed, whereby it is added in several-fold excess to the isolated radioactive metabolite and the mixture recrystallized or rechromatographed to constant specific activity. If the radioactivity remains mainly associated with the reference compound, this is evidence that it and the isolated radiolabelled metabolite have the same chemical identity. .

Another use of radioisotopes which is really only suitable for fundamental metabolism studies involves radiolabelling the body sulfur pool with a ^{35}S-labelled precursor such as [^{35}S]cysteine or [^{35}S]methionine. Any thioether metabolite formed from an administered compound will then be radiolabelled with ^{35}S and can be readily detected and measured. Studies of S-(N-benzylthiocarbamoyl)-L-cysteine[71] and 4-amino-3,5-dibromacetanilide[72] illustrate the use of this approach.

A great many metabolism studies are conducted *in vitro* using isolated tissues, cells or subcellular fractions. In all these, the radiolabelled compound, because of the ease with which it can be measured, can be invaluable in the determination of routes or mechanisms of metabolism or in the assessment of the importance or activity of a particular metabolic route in a selected tissue.

There is increasing use of hepatocytes with radiolabelled compounds in order to obtain rapid information on comparative metabolism in different species and to predict likely routes of metabolism *in vivo*.[73] Their use and that of subcellular fractions is often the only way to demonstrate the existence of active metabolites which by their nature are short-lived and not usually present in excreta. Such metabolites may covalently bind to biological macromolecules such as proteins, and this can be detected and measured using the radiolabelled compound (Figure 5);[74] covalent binding may initiate toxicity.

Radiolabelled substrates can be used in the determination of drug-metabolizing enzyme activities *in vitro*,[76] usually when the necessary assay sensitivity cannot be achieved using non-radiotracer methodology.

Figure 5 Depletion of hepatic glutathione (GSH) and covalent binding of radioactivity to liver proteins at 3 h after intraperitoneal administration of several doses of [^{3}H]paracetamol to hamsters[75]

24.1.2.7   Use in Bioanalysis

Radioisotopes are employed in bioanalysis in different ways; these include: (i) radiolabelled compounds which are studied directly in a specialized fashion; (ii) radioactive derivatizing agents which are used in the determination of other compounds; and (iii) radiolabelled substrates in more routine assay methods which rely on competition for binding sites whether these are antibodies (radioimmunoassay) or pharmacological receptors (radioreceptor assay; see Chapter 24.3).

24.1.2.7.1   Radiolabelled compounds

Most studies of the absorption, distribution, metabolism and excretion of new chemical entities involve the use of the radiolabelled form of the compound. In such studies, the parent compound can be quantified by a non-radiotracer method or by determination of its radioactivity content (after chromatographic separation). The latter approach is advantageous because it shows what concentrations of the parent compound are present, thereby dictating the type of non-radiotracer method that should be developed for subsequent routine use. It also demonstrates what proportions of radioactivity represent the parent drug and its respective metabolites from which their relative importance can be judged. For example, at 5 h after single oral doses of [^{14}C]isosorbide dinitrate in a sustained-release formulation to human subjects, peak plasma radioactivity concentrations were associated with about 2% of the parent drug and about 8, 44 and 46%, respectively, of metabolites,[77] two of which are pharmacologically active. Other studies have shown that the proportions of radioactivity representing the parent compound have ranged from the relatively large, *e.g.* [^{14}C]cinepazide,[78] [^{14}C]amitriptyline *N*-oxide[79] and [^{14}C]cetirizine,[80] to the quite small, *e.g.* [^{14}C]procetofene (fenofibrate),[81] [^{14}C]eterylate (**1**)[44] and [^{14}C]bupranolol (**4**).[50]

During pharmacokinetic studies of the vitamin pantothenate in dogs dosed with the ^{14}C compound, the unchanged vitamin was assayed by a thin-layer radiochromatographic procedure. The results obtained enabled derivation of basic pharmacokinetic parameters of pantothenate such as volume of distribution. Calculations showed that maximum proportions of the dose were reached in two peripheral compartments within 5 (50%) and 60 (30%) min, respectively, after dosing.[82]

Because of the great sensitivity that can be achieved and the ease of measurement of radioactivity, radiolabelled compounds are particularly useful in studies of drug protein-binding especially for extensively bound drugs when it may be difficult to measure the free (unbound) drug by non-radioisotope methods.

24.1.2.7.2   Radioactive derivatizing agents

One type of analytical technique involves employing a radiolabelled agent to derivatize a functional group(s) present in the compound to be measured. A refinement involves the use of two different radioistopes, one (*e.g.* tritium) for the derivatization process and the other (*e.g.* ^{14}C) is contained in the internal standard. Such methods incorporate extraction and chromatographic separation stages followed by measurement of radioactivity. Although employed for drug analysis in the past, *e.g.* maprotiline[83] and opipramol,[84] such methods are fraught with technical drawbacks and as a result they are now little used.

24.1.2.7.3   Radioimmunoassay

Of the different types of immunoassay now available, those involving radioimmunoassay were the first.[85] The components of the technique are the compound to be measured, the radiolabelled compound and an antibody capable of binding the compound. This antibody is produced by immunization of appropriate animal species with a form of the compound coupled to a protein, usually bovine serum albumin. The amount of the radiolabelled form binding to the antibody varies according to the ratio of the concentrations of the radiolabelled to the non-radiolabelled compound; thus the concentration of the (non-radiolabelled) compound can be determined with reference to a standard curve. The radioisotopes mostly employed in radioimmunoassays are tritium and ^{125}I or ^{131}I. Radioimmunoassay has been extensively reviewed (see also Chapter 24.3)[86-88] and earlier benchmark papers in this subject area collated.[89] Major advantages of radioimmunoassays are the great analytical sensitivities that can be achieved when high specific activity radiolabelled compounds are available, and the speed with which the assays can be conducted, particularly when automated. Radioimmunoassay may also be the only feasible way in which to analyze compounds that lack features detectable by other techniques. Disadvantages include, firstly, potential non-specificity as often some of the metabolites of the compound to be measured, or even other related compounds, also bind to the antibody thereby causing interference; secondly, the cost of the radiolabelled compound; and thirdly, the limited shelf-life of those compounds radiolabelled with short half-life isotopes such as those of iodine. The latter two disadvantages have encouraged the development of immunoassays not dependent on radioisotopes. The first can be overcome by introducing an extraction or chromatographic separation stage in the assay. This was necessary during radioimmunoassay of the antipsychotic agent bromperidol when extraction was required for human but not dog plasma before the assay was specific.[90]

Radioimmunoassay has been employed extensively for diagnostic applications (routine or otherwise) and for drug assays, often being selected when procedures based on GLC or HPLC failed to achieve the desired sensitivity; it is usually the method of choice for macromolecules such as proteins. Radioimmunoassay kits are commercially available for commonly used drugs such as digoxin, morphine and haloperidol, but often it has been specially developed on a 'one-off' basis for use in individual laboratories; drugs such as bromocriptine,[91] clonidine,[92] *cis*-(Z)-flupentixol[93] and loratadine[94] illustrate such development.

24.1.2.7.4 *Radioreceptor assay*

In principle, radioreceptor assays are basically similar to radioimmunoassays except that: (i) the receptors which bind the compound to be measured already exist, merely requiring isolation; (ii) binding to the receptor often leads to drug action; (iii) the assays are likely to be stereospecific; and (iv) the radiolabelled compound need not be the compound to be measured but one of the same class that binds to the same receptor. Radioreceptor assays share the same advantages and disadvantages as radioimmunoassays. The potential lack of specificity of the radioreceptor assay can be viewed as advantageous or disadvantageous depending on the use to which it is put. As, by their nature, radioreceptor assays measure pharmacologically active chemical species, *i.e.* drug and active metabolites, they provide an estimate of total pharmacological activity. This is useful for screening purposes or in efforts to correlate pharmacodynamic with pharmacokinetic properties.

Radioreceptor assays have fewer possible applications than radioimmunoassays because: (i) some of the utilizable receptors are as yet ill defined and consequently assays based on their use are suspect for many types of drug; (ii) the appropriate receptors have not been characterized and are thus not available; or (iii) the radioreceptor approach is not relevant, *e.g.* antibiotics.

Radioreceptor assays have been developed for drug classes such as the benzodiazepines, glucocorticoids, opiates, antihistamines, antipsychotics and β-adrenoceptor antagonists; this subject area has been reviewed,[95] and is discussed more fully in Chapter 24.3.

24.1.3 STABLE ISOTOPES

24.1.3.1 Commonly Used Stable Isotopes

Naturally occurring low abundance isotopes are generally stable and are known for all the major elements present in organic compounds, namely carbon, hydrogen, oxygen and nitrogen (Table 6). The natural abundance of these isotopes has allowed the separation and preparation of isotopically enriched compounds containing greater than 99% isotopic purities. With the exception of deuterium, the increase in atomic mass as a result of the introduction of a stable isotope in a molecule does not in general have any significant effect on its physicochemical properties. Thus a stable isotope-labelled drug containing ^{13}C, ^{18}O or ^{15}N atoms can be considered biologically equivalent to the non-labelled drug. The larger proportional increase in atomic mass when hydrogen is replaced by deuterium produces a corresponding change in the energy necessary to break the carbon–deuterium bond. Physicochemical interactions affected by changes in hydrogen bonding and molecular weight may also result in different properties for a deuterium-labelled compound. There are many examples where deuterium substitution results in changes in pharmacological activity and toxicity more particularly when carbon–deuterium bond cleavage is a rate-limiting process for biological activity.

Where a stable isotope label is being used to aid in the detection and identification of metabolites, ^{13}C is the most useful isotope and is entirely complementary to the use of the

Table 6 Stable Isotopes Used in Metabolism and Pharmacokinetic Studies

Element	Isotope	Natural abundance (%)
Hydrogen	^{1}H	99.985
	^{2}H (deuterium)	0.015
Carbon	^{12}C	98.893
	^{13}C	1.107
Nitrogen	^{14}N	99.634
	^{15}N	0.366
Oxygen	^{16}O	99.634
	^{17}O	0.037
	^{18}O	0.204

radioisotope ^{14}C. In many cases, use of mixtures of ^{13}C/^{14}C-labelled compounds is a valuable procedure whereby the complementary properties of the two isotopes maximize the amount of information that can be obtained from a single study. ^{18}O and ^{15}N are of less general use and where metabolic stability is important they may often not be a suitable choice. However, these isotopes may have value in mechanistic studies associated with biotransformation. Deuterium-labelled compounds represent an attractive proposition due to the range of options available for incorporation of deuterium into different positions. As with tritium labels, the metabolic stability is less certain and also isotope effects in metabolism may occur providing a distortion of the metabolic and pharmacokinetic profile compared to that of the unlabelled drug. Deuterium-labelled compounds may be particularly useful for the investigation of specific mechanistic aspects of metabolism.

24.1.3.2 Advantages and Disadvantages of Stable Isotopes

Invariably the use of stable isotopes is compared to radioisotopes but they should be regarded as a complementary tool used either in conjunction with radioisotopes or where their properties can be exploited to solve specific problems. Some of the advantages and disadvantages are summarized in Table 7. The biggest single disadvantage is the lack of routine procedures for quantitative measurement of stable isotope-labelled compounds, while the greatest perceived advantage over radioisotopes is the absence of radiological hazard. Detection and measurement rely primarily on two techniques, namely MS and NMR spectroscopy, the former being the more general technique since it relies on the intrinsically different property of stable isotope labels, namely mass. However, the nature of the detection procedures introduces a high level of specificity since the isotope(s) remains associated with the drug or metabolite molecules. The lack of any additional hazard associated with stable isotope-labelled compounds has been appealing for their use in man.

In general only limited studies using radiolabelled compounds are performed in man, providing there is no reason to suspect a significant radiological exposure to any body tissue. Such a situation might occur with a drug having a long half-life or one which is localized in a particular tissue. Use of stable isotopes does provide an alternative method of tracing a drug and its metabolites in the body.

Table 7 Advantages and Disadvantages of Stable Isotopes for Use in Metabolism and Pharmacokinetic Studies

Advantages	Disadvantages
(1) Availability of range of isotopes for labelling organic compounds	(1) Lack of routine procedures for detection and quantitation
(2) No specific hazards associated with handling	(2) Generally inferior detection sensitivity compared to radioisotopes
(3) Stability equivalent to non-labelled compounds	(3) Possible occurrence of isotope effects particularly with deuterium
(4) Lack of biological hazard facilitates use in man	
(5) Can greatly assist in obtaining structural identification of metabolites	

The techniques that could be used for detection and measurement of a drug and its metabolites in biological fluids are discussed below.

24.1.3.3 Synthesis of Stable Isotope-labelled Compounds

Procedures for the synthesis of stable isotope-labelled compounds with incorporation of deuterium and ^{13}C are similar to those using the corresponding radioactive isotopes tritium and ^{14}C.[2,96] The main difference is that the isotopic purity of the precursors and synthesized drugs are generally much higher (>90%). There are a wide range of commercially available labelled precursors containing ^{13}C and deuterium including some dual-labelled compounds. Some of the primary sources are carbonate and cyanide (^{13}C), ^2H_2O and deuterated metal hydrides, water and oxygen (^{18}O) and ammonia and nitrate (^{15}N). Criteria for selection of labelling position will depend upon the intended application but, for use in metabolism studies, ^{13}C would be the isotope of choice since, as for ^{14}C, it can be located in metabolically stable positions in the molecule. In general the cost of the labelled precursors is less likely to be a determining factor in selection of labelling positions than for ^{14}C.

Analytical characterization of the synthesized compounds would include the use of standard techniques to assess chemical purity. In addition, measurement of isotopic purity would be obtained by MS and quantitation of the abundance of ions containing the isotope. Confirmation of the location of an isotope in the molecule can be obtained by the use of ^{13}C NMR (direct) or ^1H NMR indirectly for deuterium by comparing the spectra obtained from unlabelled and labelled compounds.

24.1.3.4 Detection and Analysis of Stable Isotopes

The most powerful technique for the precise detection and measurement of isotopically labelled compounds is MS.[97,98] Essentially the molecular ion and any fragment ions of a compound containing an isotope label will exhibit a shift in the peaks due to the change in mass. All ions show an isotope peak ($M+1$) due to the natural abundance of isotopes such as ^{13}C. The isotopic composition can be calculated from a measurement of the peak heights and correction for contributions from natural abundance isotopes. MS can also provide information on the location of the label from an examination of those fragment ions containing the label. The qualitative and quantitative applications of MS in the use of stable isotopes will be described below.

NMR is another technique which is becoming increasingly important for ^{13}C compounds.[99] Unlike ^{12}C, ^{13}C has a non-zero magnetic moment and a nuclear spin of $\frac{1}{2}$ which means that magnetic resonances can be produced characteristic of the environment of each carbon atom. If spectra are recorded using proton decoupling, the ^{13}C atoms appear as sharp singlet signals. Quantitative estimates of different ^{13}C atoms in molecules are difficult due to the variable sensitivity of different types of carbon atoms. Perhaps the greatest value of ^{13}C NMR is for the structural elucidation of unknown compounds.

A detection system for deuterium is the microwave plasma detector which has been developed for use in conjunction with GLC. The eluate from a GLC column is subjected to a microwave-powered helium plasma resulting in an atomic emission discharge. The detector allows simultaneous selective detection and quantitation of certain elements including separate measurement of hydrogen and deuterium. This technique would therefore offer the possibility for detecting deuterated metabolites in a mixture, although only limited studies have appeared in the literature.[100]

24.1.3.5 Use in Bioanalysis and Pharmacokinetics Studies

24.1.3.5.1 Bioanalysis

The requirement for highly sensitive and specific methods to measure drugs and their metabolites in biological samples is of great importance in drug development. In many cases optimal sensitivity and specificity can be achieved by using the MS as a GLC detector operated in a selected scanning mode[98] and specialized systems such as the mass selective detectors have been developed commercially. The detection system is based on selective monitoring of the molecular ion or a fragment ion associated with the drug. Use of internal standards in bioanalysis is a routine procedure and a stable isotope-labelled compound represents an almost ideal choice owing to its close physico-chemical similarity to the unlabelled drug. Calibration curves for the analysis are constructed by measurement of the abundance ratio for two mass ions associated with the same entity (molecular or fragment ion) in the mass spectra. In most cases the drug and internal standard will have the same GLC retention time. The technique can allow measurement of compounds in plasma at the low to subnanogram mL^{-1} level, equivalent to that obtained with an electron capture detector.

24.1.3.5.2   *Bioavailability studies*

Due to the ability to analyze unlabelled and stable isotope-labelled drugs in the same sample by MS, the latter have been utilized as biological internal standards for drug bioavailability studies. This type of study traditionally involves administration of intravenous and oral formulations or different oral formulations of a drug each given in random order to the same group of subjects on separate occasions. The main assumption is that drug clearances are the same in each individual on the two separate dosing occasions. However, this assumption is not always valid, particularly for compounds subject to extensive variable first-pass metabolism, and large numbers of subjects may be required in order to satisfy certain statistical criteria with respect to differences in bioavailability. However, the use of a cross-over study can be avoided by simultaneous administration of the stable isotope-labelled compound as the reference formulation. There should be assurance that the combined dose of unlabelled and labelled drug does not result in non-linear pharmacokinetics. It is also, of course, vital that the labelled compound is biologically equivalent to the unlabelled drug and that there are no isotope effects which will modify its pharmacokinetics.

In relative bioavailability studies, the drug test formulation is administered together with a solution of the isotope-labelled compound. Comparison of the pharmacokinetic parameters obtained after measurement of plasma concentrations of each form allows calculation of the relative bioavailability of the test formulation. In several cases it has been shown that detection of 15–20% differences in bioavailability can be achieved with as few as four subjects.

Studies with 17α-methyltestosterone demonstrate the versatile use of stable isotope-labelled compounds in bioanalysis and bioavailability.[101] In order to establish bioequivalence of the unlabelled and 17-[$^2H_3$]methyl-labelled compound, plasma concentrations of each were compared after co-administration of a mixture containing equal amounts. The amounts of each in plasma were separately measured by GC-MS using $^2H_6$-labelled 17α-methyltestosterone as an internal standard. The ratio of concentrations were as expected and the half-lives and areas under the plasma concentration–time curves were identical. Bioavailability of the drug in a tablet formulation was assessed by comparison of plasma concentrations with those resulting from an equivalent dose of the deuterated compound co-administered in solution to human subjects (Figure 6). It was shown that the tablet formulation was bioequivalent to the oral solution and the rate of dissolution did not affect the extent of absorption. Evaluation of data from a conventional cross-over bioavailability

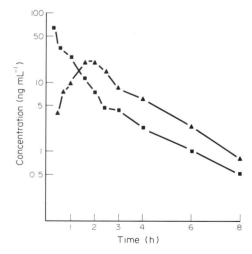

Figure 6   Serum concentrations of 17α-methyltestosterone (▲) and 17α-[$^2H_3$]methyltestosterone (■) after single 10 mg oral doses of each as a tablet and solution, respectively, to a human subject[101]

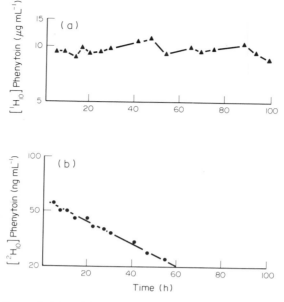

Figure 7 Plasma concentrations of (a) unlabelled phenytoin after multiple oral doses of 300 mg day^{-1} and (b) [^{2}H$_{10}$]phenytoin after a single intravenous dose of 4.77 mg to a human patient[103]

study showed that there were large interindividual differences in drug clearances on different dosing days. It was estimated that in order to detect a difference of 20% in relative bioavailability of 17α-methyltestosterone, 40 subjects would be required for a conventional cross-over study but only 12 subjects were necessary for the stable isotope co-administration approach.

Similar procedures can be adopted for assessment of absolute bioavailability of a drug by simultaneous administration of an oral dose of the formulated drug and an intravenous dose of a stable isotope-labelled form. A study with *N*-acetylprocainamide utilized a ^{13}C-labelled compound for intravenous administration, and plasma concentrations of labelled and unlabelled drug were measured by MS using the deuterated drug as an internal standard.[102] Bioavailability was assessed after deconvolution of the oral input function from the plasma concentration–time data obtained following intravenous and oral administration.

Administration of intravenous stable isotope-labelled drugs can also be used to obtain more accurate information on drug clearance under steady-state conditions during long-term treatment. Plasma concentrations of phenytoin have been measured after administration of a small intravenous dose of [^{2}H$_{10}$]phenytoin to patients maintained on daily oral doses of phenytoin for four weeks (Figure 7).[103] The AUC value for the intravenous dose was 3.48 µg h mL^{-1} and the calculated clearance was 1371 mL h^{-1}. The mean bioavailability value of the oral doses during the successive four days after the intravenous dose was 1.096 (coefficient of variation, 5.4%).

24.1.3.6 Use in Metabolism Studies

24.1.3.6.1 Detection and identification of metabolites

The role of stable isotopes as tracers in metabolism studies and in the detection and identification of metabolites has been established since the early 1970s and coincided with the rapid introduction and application of MS to biomedical research.[97] Due to the small amounts (nanogramme to microgramme range) of metabolites usually available for characterization and the difficulties of obtaining high chemical purities, MS has been the favoured physicochemical technique for investigation of metabolite structures. In most other techniques (such as NMR and IR), recorded spectra are representative of the whole sample while with MS some degree of separation can be achieved either by GLC, HPLC or by fractional distillation of samples volatilized directly into the ion source. Isolated metabolites are obtained in various states of purity and are often contaminated with endogenous biological components, and other impurities from environmental sources such as solvents and silica gel. When corresponding authentic reference compounds are not available, it can be difficult to recognize which recorded spectrum is derived from a metabolite. In some instances, this problem may be alleviated when a drug and its metabolites produce prominent characteristic fragment ions. For drugs containing atoms with relatively high natural abundance isotopes, $^{35}Cl/^{37}Cl$ (3:1) and $^{79}Br/^{81}Br$ (1:1), detection becomes much easier since mass spectra will contain characteristic doublets with a separation of two atomic mass units (amu) in their molecular ions and any fragment ions containing these isotopes. For compounds not containing these isotopes, artificial doublets can be produced by using mixtures of unlabelled and stable isotope-labelled drugs. It is important that the isotope is situated in position(s) which are stable both metabolically and in the mass spectrometer and also in a position such that the major fragments in the mass spectra also contain the label. These criteria are most readily achieved by the use of ^{13}C although multilabelled deuterium compounds are also qualitatively useful.

Identification of metabolites of roxatidine acetate (**8**) was facilitated by the use of the $[^2H_{10}]$piperidine-labelled compound in conjunction with the ^{14}C compound.[104] Thus metabolites containing the intact piperidine nucleus gave mass spectra showing doublet ions separated by 10 mass units. In addition two metabolites were identified resulting from C oxidation in the piperidine ring in which some of the deuterium atoms were lost. Other isotope labels have been used to aid in the detection and identification of metabolites of cambendazole (^{13}C)[105] and cinoxacin (^{15}N).[106]

(**8**)

Loss or retention of deuterium at specific positions in a molecule can provide valuable information on the sites of biotransformation such as hydroxylation. Complete loss of deuterium from an aliphatic carbon probably indicates hydroxylation at that carbon, although deuterium can also be lost at a carbon atom adjacent to a site of biotransformation due to tautomeric equilibration. The position of hydroxylation in an aromatic ring cannot usually be ascertained by MS analysis but use of specific deuterium-labelled compounds may provide positive evidence. However, aromatic hydroxylation involves intermediate arene oxides, and subsequent intramolecular rearrangement (NIH shift) can result in migration and retention of a deuterium to an adjacent position although this is not usually quantitative.

24.1.3.6.2 *Use of isotope-labelled derivatizing agents*

Reaction of unknown compounds with unlabelled and deuterium-labelled derivatizing reagents and comparison of the mass spectra of the products is a useful procedure to aid in the structural identification of metabolites. The shift in mass introduced by the isotope-labelled derivatization immediately provides the number of functional groups which have been derivatized. Assignment of the molecular structure of fragments and hence the structure of the metabolite is also facilitated by knowing which fragments contain derivatized functional groups. One of the most widely used derivatizations is silylation, where a tri(deuteromethyl)silyl reagent can react with hydroxy, amino or thiol groups giving an incremental increase in molecular weight compared to the unlabelled compound of nine mass units for each TMS group introduced.

Other derivatizations with deuterium-labelled reagents include acetylation, methylation and O-methyloximes of aldehydes and ketones. Since O and N demethylation and deacetylation are well-known metabolic processes, derivatization of metabolites with the deuterated methylating or acetylating reagents will provide unambiguous evidence for the presence or absence of a functional group in the metabolite, which would not be clear when the unlabelled derivatizing reagent was used. The above technique can be modified by using an equimolar mixture of labelled and unlabelled reagents to facilitate detection and structure elucidation of metabolites. In this case, doublet isotope clusters will be present in the mass spectra of derivatized metabolites.

24.1.3.6.3 *Measurement of oxidative metabolism*

Oxidative O, N or S demethylation involves C hydroxylation and liberation of formaldehyde which enters the one-carbon pool and appears ultimately as CO_2 in expired air. Thus measurement of CO_2 derived from the methyl group allows assessment of both the rate and extent of demethylation. Use of a standard compound enables a comparison of the extent of demethylase activity in different subjects. A procedure has been developed using radiolabelled material where $^{14}CO_2$ content is monitored in expired air.[107] However, the scope of this type of investigation in man is limited·due to the radiological exposure and methodology has been developed for the measurement of $^{13}CO_2$. Hepatic *N*-demethylase activity *in vivo* has been assessed by use of 4-(di[^{13}C]methylamino)antipyrine as a metabolic substrate and validated by comparison with the ^{14}C compound.[108]

24.1.3.6.4 Pseudoracemates

Many drugs are developed and used as a racemate of two enantiomers due to an asymmetric centre in the molecule. There is now an increased awareness that the pharmacological effects, pharmacokinetics and toxicity of enantiomers may be different. Differences in metabolism and pharmacokinetics may have pharmacological and toxicological implications. For a rigorous comparison of the disposition of enantiomers in a mixture, it is necessary to administer the racemic drug. Pharmacokinetics of drug enantiomers can be determined by use of a specific analytical method allowing separation of the enantiomers but this does not allow investigation of the metabolites derived from each enantiomer. These problems have been overcome by the use of pseudoracemates in which one of the enantiomers is labelled with a stable isotope, although there must be assurance of the absence of isotope effects. With this technique the pharmacokinetics of each enantiomer can be studied using mass spectroscopic methodology as described previously. Propranolol is a widely used cardiovascular drug consisting of two enantiomers, one of which contributes most of the pharmacological activity. Various studies have been carried out to investigate the stereoselectivity of its metabolism in laboratory animals and man using deuterium-labelled pseudoracemates.[109,110] The absence of isotope effects associated with the deuterium label in (+)-[$^2H_2$]-(ring labelled) (9) and (+)-[$^2H_6$]-(side-chain labelled) (10) was confirmed (* indicates the chiral carbon). Measurement of plasma and urine concentrations of each enantiomer after administration of the pseudoracemate showed differences in bioavailability for the enantiomers and between species. The bioavailability of (+)-propranolol in dogs was twice as high as that of its enantiomer, whereas in man the bioavailability of (−)-propranolol was 45% greater. Propranolol is metabolized by three major pathways, namely direct conjugation, ring oxidation and conjugation, and side-chain oxidation. Use of a pseudoracemate allows analysis of the enantiomeric composition of the metabolites and provides an insight into the stereoselectivity in metabolism and possible explanations for observed differences in bioavailability.

24.1.3.6.5 Mechanistic studies

There are many potential applications for the use of stable isotope labels in investigating mechanistic aspects of biotransformation; these are summarized below. Examples of these applications have been reviewed.[111]

(i) The loss or retention of deuterium labels located at key positions will provide information on the structure of intermediates in biotransformation. Information may be obtained to differentiate between alternative pathways to a metabolite.

(ii) Confirmation that an endogenous compound is also formed as a drug metabolite.

(iii) Elucidation of biotransformation pathways by confirmation that specific atoms are retained (^{13}C, ^{15}N).

(iv) Investigation of mechanisms of oxidation using ^{18}O precursors.

(v) Assessment of deuterium isotope effects to ascertain rate-limiting steps in biotransformation and effects on metabolite profile and modulation of pharmacological activity and toxicity.

24.1.3.7 Use of Nuclear Magnetic Resonance Spectroscopy

Proton NMR is an established valuable technique to aid in the structural characterization of organic compounds and its application to the identification of drug metabolites has been reviewed.[112,113] One particular value of this technique would be in the unambiguous assignment of the position of hydroxylation in aromatic or heterocyclic rings which cannot readily be achieved by MS. NMR spectra can be obtained for both tritium and the stable isotopes ^{13}C and deuterium.[10] For tritium the application to drug metabolism is restricted due to the limited sensitivity, although it may have value with *in vitro* studies and for confirmation of labelling positions in test compounds. Deuterium is also somewhat limited due to low sensitivity, broad peaks and restricted chemical shift range. ^{13}C has greater versatility owing to its natural abundance and the sensitivity of chemical shift to changes in the electronic environment which consequently vary over a much greater range than for protons. ^{13}C NMR spectra of the natural abundance isotopes are recorded using proton decoupling and, since the 1% abundance of ^{13}C makes it unlikely for two isotopes to be adjacent to each other, the resonances appear as single peaks. Thus ^{13}C NMR has been used extensively as a complementary tool to ^{1}H NMR for structural identification in organic chemistry but it has a limited application to drug metabolism due to low sensitivity. However, its usefulness can be considerably enhanced by the use of ^{13}C-enriched compounds synthesized with incorporation of ^{13}C at specified positions in the molecule. Since up to 99% atom incorporations can be achieved, the sensitivity is effectively enhanced by almost two orders of magnitude. This also enables spectra to be obtained on impure samples without interference from endogenous components. One application to metabolism studies is the detection and characterization of metabolites. If the ^{13}C labels are located at or are adjacent to positions of biotransformation, the chemical shifts of the ^{13}C resonances will be characteristic of the type of chemical modification occurring. This technique also allows information to be obtained on mixtures of metabolites in crude biological extracts[114] as well as on isolated purified metabolites.[115]

A further fruitful area of research is the application of ^{13}C NMR to follow the interactions and fate of a compound in active biological systems such as cell cultures, microsomal preparations, perfused organs and even *in vivo*.[116,117] However, there is no indication from the scientific literature that these techniques have been applied to drug metabolism. The potential has been demonstrated by experiments on the metabolism of exogenous ^{13}C-labelled precursors or substrates such as the gluconeogenesis of $[1,3-^{13}C]$glycerol in suspensions of rat hepatocytes and $[3-^{13}C]$alanine in hepatocytes and perfused liver.[118,119] By this procedure it was possible to follow the pathways for incorporation of the ^{13}C label into different endogenous components. The sequential formation of metabolites derived by oxidation of isobutenylmethyl groups in a chrysanthemate insecticide have been studied using the compound with ^{13}C-labelled methyl groups.[120] ^{13}C NMR spectra were obtained on metabolites formed in incubations with rat and mouse liver microsomes without isolation.

24.1.4 SPECIAL TECHNIQUES REQUIRING ISOTOPES

24.1.4.1 Positron Emission Tomography

Positron emission tomography is an imaging technique that can locate and measure positron-emitting radioisotopes in biological systems *in vivo*. The technique can be regarded as relatively nonspecific because, as it is total radioactivity derived from positron emission that is measured, this will include the parent compound and its metabolites. However, this possible limitation has not affected most of the recent applications of the technique. Positron emission tomography has been applied primarily to studies of physiological processes such as oxygen or glucose utilization in the brain, or ligand uptake in receptors such as those for dopamine, so as to better understand disease states and concomitant functional changes.[121,122] In studies of glucose utilization, $[^{18}F]$glucose has been used.

Positron emission tomography has obvious applications for studies of drug uptake at sites of drug action. Its major drawback is that it requires positron-emitting compounds that have to be rapidly synthesized and used owing to the rather short half-lives of positron-emitting isotopes (see Table 1), besides the specialized equipment needed to produce such isotopes. Its use will thus be likely to be restricted to fewer laboratories than most of the other isotope techniques mentioned in this review.

24.1.5 CONCLUSIONS

Radioactive and stable isotopes are now used routinely and to great effect in many of the subject areas associated with biopharmaceutics. This has been due to the versatility of isotope-based techniques which is leading also to newer applications. Without the availability of such isotopes and the means for their measurement, much of our present-day knowledge would not have been obtained. The use of radioactive and stable isotopes is now well established in many of the subject areas associated with biopharmaceutics, but with refinements in methodology, improvements in use will continue. However, it is in the understanding of drug action *in vivo* and in the earlier diagnosis of disease states where major advances in the use of radioactive and stable isotopes are likely to occur in the future.

24.1.6 REFERENCES

1. G. Hevesy, *Biochem. J.*, 1923, **17**, 439.
2. A. Murray and D. L. Williams, 'Organic Synthesis with Isotopes', Interscience, New York, 1958, vols. 1 and 2.
3. J. R. Catch, 'Carbon-14 Compounds', Butterworths, London, 1961.
4. L. J. Roth (ed.), 'Isotopes in Experimental Pharmacology', University of Chicago Press, Chicago, 1965.
5. E. A. Evans, 'Tritium and its Compounds', Butterworths, London, 1966.
6. S. Rothchild (ed.), *Adv. Tracer Methodol.*, 1968, **4**.
7. P. G. Waser and B. Glasson (eds.), 'Radioactive Isotopes in Pharmacology', Wiley, London, 1969.
8. Y. Cohen (ed.), 'Radionuclides in Pharmacology', Pergamon Press, Oxford, 1971, vols. 1 and 2.
9. E. A. Evans and M. Muramatsu (eds.), 'Radiotracer Techniques and Applications', Dekker, New York, 1977, vols. 1 and 2.
10. J. A. Elvidge and J. R. Jones (eds.), 'Isotopes: Essential Chemistry and Applications', Chemical Society, London, 1980.
11. W. P. Duncan and A. B. Susan (eds.), 'Synthesis and Applications of Isotopically Labeled Compounds, Proceedings of an International Symposium', Elsevier, Amsterdam, 1983.
12. R. R. Muccino (ed.), 'Synthesis and Applications of Isotopically Labeled Compounds, Proceedings, Second International Symposium', Elsevier, Amsterdam, 1986.
13. E. Buncel and J. R. Jones (eds.), 'Isotopes in the Physical and Biomedical Sciences: Labelled Compounds (Part A)', Elsevier, Amsterdam, 1987, vol. 1.
14. E. A. Evans and K. G. Oldham (eds.), 'Radiochemicals in Biomedical Research', Wiley, Chichester, 1988.
15. B. J. Wilson (ed.), 'The Radiochemical Manual', 2nd edn., Radiochemical Centre, Amersham, 1966.
16. J. Edelson and J. F. Douglas, *J. Pharmacol. Exp. Ther.*, 1973, **184**, 449.
17. L. F. Elsom and D. R. Hawkins, *J. Labelled Compd. Radiopharm.*, 1978, **14**, 799.
18. L. F. Chasseaud, B. J. Fry, V. H. Saggers, I. P. Sword and D. E. Hathway, *Biochem. Pharmacol.*, 1972, **21**, 3121.
19. I. P. Sword, in 'Foreign Compound Metabolism in Mammals', ed. D. E. Hathway, Chemical Society, London, 1972, vol. 2, p. 13.
20. E. A. Evans, in 'Isotopes: Essential Chemistry and Applications', ed. J. A. Elvidge and J. R. Jones, Chemical Society, London, 1980, p. 36.
21. E. A. Evans, in 'Isotopes: Essential Chemistry and Applications', ed. J. A. Elvidge and J. R. Jones, Chemical Society, London, 1980, p. 67.
22. M. A. Crook and P. Johnson (eds.), ' Liquid Scintillation Counting', Heyden, London, 1974, vol. 3 (and related volumes).
23. Y. Kobayashi and D. V. Maudsley, 'Biological Applications of Liquid Scintillation Counting', Academic Press, New York, 1974.
24. T. R. Roberts, 'Radiochromatography: The Chromatography and Electrophoresis of Radiolabelled Compounds', Elsevier, Amsterdam, 1978.
25. C. T. Peng, D. L. Horrocks and E. L. Alpen (eds.), 'Liquid Scintillation Counting: Recent Applications and Development', Academic Press, New York, 1980, vols. 1 and 2.
26. J. M. Chapman and G. Ayrey, ' The Use of Radioactive Isotopes in the Life Sciences', Allen and Unwin, London, 1981.
27. R. A. Faires and G. G. J. Boswell, 'Radioisotope Laboratory Techniques', 4th edn., Butterworths, London, 1981.
28. L. Botta, H. U. Gerber and K. Schmid, in 'Drug Fate and Metabolism', ed. E. R. Garrett and J. L. Hirtz, Dekker, New York, 1985, vol. 5, p. 99.
29. D. M. Wieland, M. C. Tobes and T. J. Mangner (eds.), 'Chromatographic Techniques in Radiopharmaceutical Chemistry,' Springer, New York, 1986.
30. H. Filthuth, in 'Analytical and Chromatographic Techniques in Radiopharmaceutical Chemistry', ed. D. M. Wieland, M. C. Tobes and T. J. Mangner, Springer, New York, 1986, p. 79.
31. M. J. Kessler, in 'Analytical and Chromatographic Techniques in Radiopharmaceutical Chemistry', ed. D. M. Wieland, M. C. Tobes and T. J. Mangner, Springer, New York, 1986, p. 149.
32. S. Nagatsuka, S. Hanawa, T. Honda and M. Hasegawa, *Xenobiotic Metab. Dispos.*, 1988, **3**, 121.
33. P. B. Gahan (ed.), 'Autoradiography for Biologists', Academic Press, London, 1972.
34. C. G. Curtis, S. A. M. Cross, R. J. McCulloch and G. M. Powell, 'Whole-body Autoradiography,' Academic Press, London, 1981.
35. S. Ullberg and B. Larsson, *Prog. Drug Metab.*, 1981, **6**, 249.
36. D. V. Parke, in 'Isotopes in Experimental Pharmacology', ed. L. J. Roth, University of Chicago Press, Chicago, 1965, p. 315.
37. L. F. Chasseaud, in 'Foreign Compound Metabolism in Mammals', ed. D. E. Hathway, Chemical Society, London, 1970, vol. 1, p. 34.
38. L. F. Chasseaud, in 'Foreign Compound Metabolism in Mammals', ed. D. E. Hathway, Chemical Society, London, 1972, vol. 2, p. 62.
39. D. R. Hawkins, in 'Isotopes: Essential Chemistry and Applications', ed. J. A. Elvidge and J. R. Jones, Chemical Society, London, 1980, p. 232.
40. D. R. Hawkins, in 'Radiochemicals in Biomedical Research', ed. E. A. Evans and K. G. Oldham, Wiley, Chichester, 1988, p. 14.

41. H. P. A. Illing (ed.), 'Xenobiotic Metabolism and Disposition: The Design of Studies on Novel Compounds', CRC Press, Boca Raton, FL, 1989.
42. I. Midgley, L. F. Chasseaud, I. W. Taylor, L. M. Walmsley, A. G. Fowkes, A. Darragh, R. F. Lambe and R. Bonn, *Xenobiotica*, 1985, **15**, 965.
43. B. A. John, S. G. Wood, L. F. Chasseaud and P. Forlot, in 'Psoralens in 1988: Past, Present and Future', ed. T. B. Fitzpatrick, P. Forlot, M. A. Pathak and F. Urbach, abstr., 1988, p. 56.
44. S. G. Wood, B. A. John, L. F. Chasseaud, I. Johnstone, S. R. Biggs, D. R. Hawkins, J. G. Priego, A. Darragh and R. F. Lambe, *Xenobiotica*, 1983, **13**, 731.
45. D. H. Moore, L. F. Chasseaud, A. Darragh, T. Taylor and D. G. Cresswell, *Steroids*, 1983, **41**, 15.
46. T. Nishihata, Y. Saitoh and K. Sakai, *Chem. Pharm. Bull.*, 1988, **36**, 2544.
47. Y. Kanai, Y. Nakai, N. Nakajima and S. Tanayama, *Xenobiotica*, 1979, **9**, 33.
48. P. K. Knoefel (ed.), 'Absorption, Distribution, Transformation and Excretion of Drugs', Thomas, Springfield, IL, 1972.
49. J. B. Houston and S. G. Wood, *Prog. Drug Metab.*, 1980, **4**, 57.
50. A. R. Waller, L. F. Chasseaud, R. Bonn, T. Taylor, A. Darragh, R. Girkin, W. H. Down and E. Doyle, *Drug Metab. Dispos.*, 1982, **10**, 51.
51. T. Taylor, I. W. Taylor, L. F. Chasseaud and R. Bonn, *Prog. Drug Metab.*, 1987, **10**, 207.
52. S. G. Wood, B. A. John, L. F. Chasseaud, R. Bonn, H. Grote, K. Sandrock, A. Darragh and R. F. Lambe, *Xenobiotica*, 1987, **17**, 881.
53. Huntingdon Research Centre, unpublished data.
54. L. F. Chasseaud, W. H. Down and R. K. Grundy, *Eur. J. Clin. Pharmacol.*, 1975, **8**, 157.
55. T. Taylor, L. F. Chasseaud, E. Doyle, R. Bonn, A. Darragh and R. F. Lambe, *Arzneim.-Forsch.*, 1982, **32**, 1329.
56. L. F. Chasseaud, B. J. Fry, D. R. Hawkins, J. D. Lewis, I. P. Sword, T. Taylor and D. E. Hathway, *Arzneim.-Forsch.*, 1971, **21**, 1379.
57. R. M. J. Ings, *Drug Metab. Rev.*, 1984, **15**, 1183.
58. W. H. Down, R. M. Sacharin, L. F. Chasseaud, D. Kirkpatrick and E. R. Franklin, *Toxicology*, 1977, **8**, 333.
59. S.-H. L. Chiu, R. Taub, E. Sestokas, A. Y. H. Lu and T. A. Jacob, *Drug Metab. Rev.*, 1987, **18**, 289.
60. M. Michiels, W. Meuldermans and J. Heykants, *Drug Metab. Rev.*, 1987, **18**, 235.
61. Autoradiography in Pharmacology and Toxicology, *Acta Pharmacol. Toxicol.*, 1977, **41** (suppl. 1).
62. D. R. Hawkins, K. T. Weston, L. F. Chasseaud and E. R. Franklin, *J. Agric. Food Chem.*, 1977, **25**, 398.
63. D. H. Moore, L. F. Chasseaud, D. Bucke and P. C. Risdall, *Food Cosmet. Toxicol.*, 1976, **14**, 189.
64. T. Taylor, C. Gotfredsen, L. F. Chasseaud, R. R. Brodie and E. Doyle, *Eur. J. Drug Metab. Pharmacokinet.*, 1977, **2**, 13.
65. R. Girkin, G. A. Baldock, L. F. Chasseaud, M. Humpel, D. R. Hawkins and B. C. Mayo, *Xenobiotica*, 1980, **10**, 401.
66. M. Humpel, V. Illi, N. Milius, H. Wendt and M. Kurowski, *Eur. J. Drug Metab. Pharmacokinet.*, 1979, **4**, 237.
67. L. F. Elsom, L. F. Chasseaud, D. R. Hawkins, A. Darragh and R. F. Lambe, *Prog. Med.*, 1985, **5** (suppl. 1), 863.
68. A. P. Beresford, P. V. Macrae and D. A. Stopher, *Xenobiotica*, 1988, **18**, 169.
69. I. Midgley, A. G. Fowkes, L. F. Chasseaud, D. R. Hawkins, R. Girkin and K. Kesselring, *Xenobiotica*, 1983, **13**, 87.
70. Y. Kobayashi and D. V. Maudsley, *Prog. Med. Chem.*, 1972, **9**, 133.
71. G. Brusewitz, B. D. Cameron, L. F. Chasseaud, K. Gorler, D. R. Hawkins, H. Koch and W. H. Mennicke, *Biochem. J.*, 1977, **162**, 99.
72. A. Prox, J. Schmid, J. Nickl and G. Engelhardt, *Z. Naturforsch. C: Biosci.*, 1987, **42C**, 465.
73. E. J. Rauckman and G. M. Padilla (eds.), 'The Isolated Hepatocyte: Use in Toxicology and Xenobiotic Biotransformations', Academic Press, Orlando, FL, 1987.
74. L. R. Pohl and R. V. Branchflower, *Methods Enzymol.*, 1981, **77**, 43.
75. D. J. Jollow, S. S. Thorgeirsson, W. Z. Potter, M. Hashimoto and J. R. Mitchell, *Pharmacology*, 1974, **12**, 251.
76. W. B. Jakoby (ed.), *Methods Enzymol.*, 1981, **77**.
77. L. F. Chasseaud, *Z. Kardiol.*, 1983, **72** (suppl. 3), 20.
78. B. D. Cameron, L. F. Chasseaud, D. R. Hawkins and T. Taylor, *Xenobiotica*, 1976, **6**, 441.
79. I. Midgley, D. R. Hawkins and L. F. Chasseaud, *Arzneim.-Forsch.*, 1978, **28**, 1911.
80. S. G. Wood, B. A. John, L. F. Chasseaud, J. Yeh and M. Chung, *Ann. Allergy*, 1987, **59**, 31.
81. R. R. Brodie, L. F. Chasseaud, F. F. Elsom, E. R. Franklin and T. Taylor, *Arzneim.-Forsch.*, 1976, **26**, 896.
82. T. Taylor, G. Bourne, L. F. Chasseaud and H. Partington, *Res. Vet. Sci.*, 1976, **20**, 151.
83. W. Riess, *Anal. Chim. Acta*, 1974, **68**, 363.
84. W. Riess, *Anal. Chim. Acta*, 1977, **88**, 109.
85. R. S. Yalow and S. A. Berson, *J. Clin. Invest.*, 1960, **39**, 1157.
86. V. Marks, G. P. Mould, M. J. O'Sullivan and J. D. Teale, *Prog. Drug Metab.*, 1980, **5**, 255.
87. D. Riad-Fahmy and G. F. Read, in 'Drug Fate and Metabolism', ed. E. R. Garrett and J. L. Hirtz, Dekker, New York, 1985, vol. 5, p. 51.
88. J. Landon and A. C. Moffat, *Analyst (London)*, 1976, **101**, 225.
89. R. S. Yalow (ed.), 'Radioimmunoassay', Hutchinson Ross, Stroudsburg, 1983.
90. E. Van den Eeckhout, F. M. Belpaire, M. G. Bogaert and P. de Moerloose, *Eur. J. Drug Metab. Pharmacokinet.*, 1980, **5**, 45.
91. J. Rosenthaler, H. Munzer and R. Voges, in 'Methodological Surveys in Biochemistry: Drug Metabolite Isolation and Determination', ed. E. Reid and J. P. Leppard, Plenum Press, New York, 1983, vol. 12, p. 215.
92. D. Arndts, H. Stahle and H. J. Forster, *J. Pharmacol. Methods*, 1981, **6**, 295.
93. A. Jorgensen, *Life Sci.*, 1978, **23**, 1533.
94. E. Radwanski, J. Hilbert, S. Symchowicz and N. Zampaglione, *J. Clin. Pharmacol.*, 1987, **27**, 530.
95. S. R. Nahorski and D. B. Barnett, in 'Cell Surface Receptors', ed. P. G. Strange, Horwood, Chichester, 1983, p. 270.
96. A. F. Thomas, 'Deuterium Labelling in Organic Chemistry', Appleton-Century-Crofts, New York, 1971.
97. S. Gaskell (ed.), 'Mass Spectrometry in Biomedical Research', Wiley, Chichester, 1986.
98. B. J. Millard, 'Quantitative Mass Spectrometry', Heyden, London, 1978.
99. D. M. Rackham, in 'Isotopes: Essential Chemistry and Applications', ed. J. A. Elvidge and J. R. Jones, Chemical Society, London, 1980, p. 97.
100. H. G. Hege, M. Hollmann, S. Kaumeier and H. Lietz, *Eur. J. Drug Metab. Pharmacokinet.*, 1984, **9**, 41.

101. Y. Shinohara, S. Baba, Y. Kasuya, G. Knapp, F. R. Pelsor, V. P. Shah and I. L. Honigberg, *J. Pharm. Sci.*, 1986, **75**, 161.
102. J. M. Strong, J. S. Dutcher, W.-K. Lee and A. J. Atkinson, *Clin. Pharmacol. Ther.*, 1975, **18**, 613.
103. Y. Kasuya, K. Mamada, S. Baba and M. Matsukura, *J. Pharm. Sci.*, 1985, **74**, 503.
104. S. Honma, S. Iwamura, R. Kobayashi, Y. Kawabe and K. Shibata, *Drug Metab. Dispos.*, 1987, **15**, 551.
105. W. J. A. Vanden Heuvel, in 'Stable Isotopes: Applications in Pharmacology, Toxicology and Clinical Research', ed. T. A. Baillie, Macmillan, London, 1978, p. 65.
106. R. L. Wolen, B. D. Obermeyer, E. A. Ziege, H. R. Black and C. M. Gruber, in 'Stable Isotopes: Applications in Pharmacology, Toxicology and Clinical Research', ed. T. A. Baillie, Macmillan, London, 1978, p. 113.
107. R. Platzer, R. L. Galeazzi, G. Karlaganis and J. Bircher, *Eur. J. Clin. Pharmacol.*, 1978, **14**, 293.
108. J. F. Schneider, D. A. Schoeller, B. Nemchausky, J. L. Boyer and P. Klein, *Clin. Chim. Acta*, 1978, **84**, 153.
109. T. Walle, M. J. Wilson, U. K. Walle and S. A. Bai, *Drug Metab. Dispos.*, 1983, **11**, 544.
110. T. Walle, *Drug Metab. Dispos.*, 1985, **13**, 279.
111. T. A. Baillie, *Pharmacol. Rev.*, 1981, **33**, 81.
112. D. E. Case, *Xenobiotica*, 1973, **3**, 451.
113. I. C. Calder, *Prog. Drug Metab.*, 1979, **3**, 303.
114. D. R. Hawkins and I. Midgley, *J. Pharm. Pharmacol.* 1978, **30**, 547.
115. L. I. Wiebe, J. R. Mercer and A. J. Ryan, *Drug Metab. Dispos.*, 1975, **6**, 296.
116. R. S. Norton, *Bull. Magn. Reson.*, 1980, **3**, 29.
117. P. W. Kuchel, *CRC, Crit. Rev. Anal. Chem.*, 1981, **12**, 155.
118. S. M. Cohen, R. G. Shulman and A. C. McLaughlin, *Proc. Natl. Acad. Sci. U.S.A.*, 1980, **76**, 407.
119. S. M. Cohen and R. G. Shulman, *Philos. Trans. R. Soc. London, Ser. B*, 1980, **289**, 407.
120. M. A. Brown, I. Holden, A. H. Glickman and J. E. Casida, *J. Agric. Food Chem.*, 1985, **33**, 8.
121. L. Battistin and F. Gerstenbrand (eds.), 'PET and NMR: New Perspectives in Neuroimaging and in Clinical Neurochemistry', Liss, New York, 1986.
122. H. N. Wagner, in 'Liver and Aging—1986, Liver and Brain', ed. K. Kitani, Elsevier, Amsterdam, 1986.

24.2

Chemical Analysis

DAVID B. JACK

Fidia Research Laboratories, Abano Terme, Italy

24.2.1 INTRODUCTION

24.2.1.1 Choosing a Method

The Shorter Oxford English Dictionary defines 'method' as a 'procedure for attaining an object . . . a way of doing anything especially to a regular plan'. There are two essential components of a method: to have a distinct aim and to reach it by systematic means. In chemical analysis a method is chosen with the aim of obtaining qualitative or, more usually, quantitative information on a given material. The choice of method will depend on the nature of the information desired and the complexity of the starting material; the analyst should begin by asking a series of simple questions.

(1) Is a quantitative answer necessary? Usually it is easier and cheaper to provide qualitative rather than quantitative information. It may be that in an industrial process it is necessary to limit the amount of compound A in a drug formulation. If a simple qualitative test can be found with an appropriate degree of sensitivity and selectivity, then a negative result for A may be all that is needed. Strictly speaking, this example is semiquantitative since the analyst will be able to say that A, if present, is there at a concentration less than x parts per million. In another context, a physician may have a patient who fails to respond to his prescribed medication. Is this because the patient has a resistant form of the disease, is it because he is unable to absorb the drug, is he receiving a sufficient dose, or is he simply not taking the drug? The last possibility can be investigated by obtaining a sample of urine from the patient at his next visit to the surgery, to test ostensibly for sugar, and applying a qualitative test to an aliquot to see if any unchanged drug or metabolites are present.[1,2] This will work only with drugs which are relatively rapidly eliminated and cannot be used with drugs like the phenothiazines that can be detected in the urine months after drug administration has ceased.

(2) How much is present in the material? In the case of a bulk sample of drug, B, we may be talking of 98, 99 or even 99.9% and it may be simpler to measure the other compounds present in trace amounts. If B is the active ingredient in a tablet formulation then it may account for less than 1% to more than 50% of the material in the tablet; measuring the amount of unchanged B in urine following an oral dose often means that a few p.p.m. have to be detected, while measuring the concentration B in blood, especially if it is extensively metabolized and given in low doses, means a search for a few parts in a thousand million. Obviously different methods are likely to be needed in each case.

(3) How many analyses are to be performed? If this will be a single analysis (like the one performed for the physician above) or only a few samples per week are expected, then it makes sense to choose a method that will be simple to set up and cheap to run (provided, of course, that the method has sufficient sensitivity). If it is expected that there will be a steady demand for such a test, say several hundred or more samples per week for a period of months, it will probably be necessary to set aside a piece of equipment dedicated to this analysis.

(4) How quickly are the results needed? If results are needed rapidly (within hours) on a one-off basis, then a relatively specialized form of analysis will be needed (*e.g.* ion selective electrodes) or a piece of equipment dedicated to this end (*e.g.* HPLC).

(5) What staff are available to carry out the analysis? If the method is complex and requires expensive non-automated equipment, then staff of high calibre will be needed. However, there are a number of robust techniques that will allow relatively junior staff to obtain satisfactory results with a minimum of training, even when relatively low concentrations are to be determined. Examples of such techniques are high performance thin layer chromatography (HPTLC), many high performance liquid chromatography (HPLC) techniques and some forms of gas–liquid chromatography (GLC).

It is no bad rule to adopt the simplest method available that will give the required sensitivity. However, it should be remembered that sensitivity is not all when dealing with complex mixtures. A method may be very sensitive when applied to the pure substance, but it may be useless when applied to a complex mixture if it lacks selectivity. Preliminary separation procedures, such as extraction, can be used to improve the selectivity of a method and these can be of great help but often a very efficient separation of the constituents of a mixture is needed before quantitation can be carried out. For this reason chromatographic methods are the most widely used in modern laboratories.

A number of publications describe different approaches to choosing an analytical method.[3–5]

24.2.2 SAMPLE PREPARATION

Some methods require only a minimum of sample preparation, especially where the compounds to be determined are present in high concentration. For example, to measure the amount of a drug substance in a tablet formulation it might well be sufficient to take 100 tablets at random from a batch, reduce them to a powder and submit a fraction of this to volumetric or gravimetric analysis, or UV or visible spectroscopy. Techniques requiring little sample preparation are discussed at the beginning of "Separation and Quantitation" (Section 24.2.3). In many cases, however, a relatively extensive pretreatment of the sample is necessary to remove potentially interfering compounds and also to concentrate the compound(s) to be determined. The 'classical' approach is to use liquid–liquid extraction where the material is usually presented dissolved in an aqueous liquid such as plasma or urine and the compound to be determined is extracted into an immiscible organic solvent. An alternative approach is liquid–solid extraction where the material is passed over a solid which retains the compounds in question while permitting the passage of much endogenous material. The compounds to be determined can be selectively eluted later using an appropriate solvent. This latter technique is generally more rapid and is more readily automated.

24.2.2.1 Liquid–Liquid Extraction

When a compound is shaken with two mutually immiscible solvents and the solvents allowed to separate, the compound will partition itself between the solvents depending upon its solubility in each. Usually one of the solvents is aqueous and the other organic, and the partition is usually described in terms of the relative concentrations in each phase

$$C_{organic}/C_{water} = K \tag{1}$$

When the drug is in its unionized form, this constant is called the partition coefficient, P. If the pH of the aqueous solution is such that the drug is ionized, then less will dissolve in the organic phase and the ratio will change but will still be a constant for this new set of conditions. When the drug is ionized the constant becomes, P', the distribution coefficient. With increasing ionization, P' will become progressively smaller. The relationship between the partition and distribution coefficients and the degree of ionization of any drug can be described as follows (where K_a is the dissociation constant of the drug)

For a weak base, $$P/P' = 1 + H^+/K_a \tag{2}$$

For a weak acid, $$P/P' = 1 + K_a/H^+ \tag{3}$$

Partition studies are usually carried out using a buffered aqueous phase and an organic phase which is usually octan-1-ol although chloroform, toluene and other solvents have been used. There is an extensive literature on the measurement of partition coefficients.[6]

The design of a liquid–liquid extraction scheme depends on the complexity of the material to be analyzed as well as on the concentration of the compounds to be determined. Obviously a fairly simple scheme will suffice to quantify an antibiotic in an aqueous solution for intravenous injection, while a more elaborate scheme will be needed to measure a drug in urine following its oral administration, especially if it demonstrates a high first-pass loss during passage across the small intestine and through the liver. The basic outline of the scheme is, however, the same and more selectivity can be introduced by a careful choice of pH and solvent polarity.

The most common organic solvents can be ranked in increasing order of polarity, as in Table 1.

A general scheme for the extraction of a drug from aqueous solution using liquid–liquid extraction is illustrated in Figure 1. This scheme can be simplified or elaborated depending on the complexity of the matrix and the concentration of the drug to be determined. Careful consideration of the physicochemical properties of the drug should be made during the planning stages. If a single known drug is to be quantified, then it is a relatively simple matter to choose an appropriate solvent. It is usually best to choose the least polar solvent that will efficiently extract the drug; this reduces the amount of endogenous material coextracted and reduces the chances of interference in the quantitation step. For example, ethyl acetate will easily extract many drugs from urine but it also removes many endogenous polar compounds. However, if a wide range of compounds are to be sought and

Table 1 Common Solvents Arranged in Increasing Polarity

Heptane	Dioxane
Hexane	Ethyl acetate
Carbon tetrachloride	Acetonitrile
Toluene	Pyridine
Benzene	Ethanol
Diethyl ether	Methanol
Chloroform	Acetic acid
Dichloromethane	Water
Acetone	

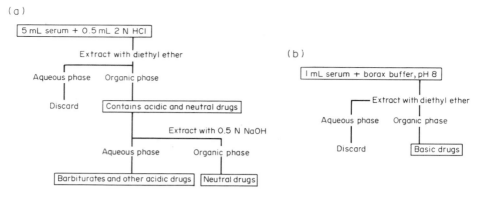

Figure 1 Extraction schemes: (a) acidic and neutral drugs; and (b) basic drugs

eventually quantitated, *e.g.* in a drug screen, then a relatively polar solvent must be used to be sure that as many drugs as possible are extracted. This means using solvents such as diethyl ether and the chlorinated hydrocarbons.

24.2.2.1.1 *Ion pair extraction*

Some drugs are very polar and not efficiently extracted by any of the common organic solvents. In this case the drug, in its ionized state, can be paired with a counterion of opposite charge to give an ion pair with an overall charge of zero. In this way polar quaternary ammonium compounds and others can be extracted into relatively non-polar solvents. By careful choice of counterion and extracting solvent, such extractions can be made very selective indeed yielding very clean extracts. The counterion can also be chosen because it has a strong UV absorption or fluorescence, something the drug itself may lack.

A series of typical examples of ion–pair extraction of drugs are given in Table 2 and detailed reviews are available.[7-9]

Care must be exercised in liquid–liquid extraction, however, since when drugs are extracted into organic solvents in their unionized form they often bind strongly to any glass object coming into contact with them, including Pasteur pipettes; this is particularly true of tertiary amines.[10]

24.2.2.1.2 *Glassware treatment in liquid–liquid extraction*

Great care should be taken to ensure that glassware is as clean as possible. Synthetic disposable test tubes can be used in some cases but preliminary tests should be carried out to see that no plasticizers are leached out by the organic solvent used for extraction. Thorough cleaning of all glassware (including Pasteur pipettes) in suitable detergent, followed by silanization and washing with alcohol is to be recommended. Direct fluorimetry following solvent extraction is probably about the most demanding technique as far as glassware is concerned. The most scrupulous care must be exercised when this technique is employed. As mentioned above, compounds such as tertiary amines bind to glass and silanization may not be effective. A number of authors recommend that glassware, previously cleaned thoroughly by standard methods, should be rinsed with 20% triethylamine in methanol followed by air drying.[11]

24.2.2.2 Liquid–Solid Extraction

This approach has been used for about fifty years with some success using a range of materials including carbon (charcoal), silica (celite), magnesium silicate (fluorosil), alumina and ion exchange resins.[12-15] There has been a reasonable degree of success particularly with the non-ionic ion exchange resin XAD-2 which has been used to remove both basic (alkaloid) and acidic (barbiturate) drugs from urine.[16,17] However, the technique was never widely adopted, although recent advances in HPLC technology in the last 10 years have meant that a range of chemically bonded silicas are now available in cartridge form, for solid phase extraction, from a number of commercial suppliers.

Table 2 Examples of Ion Pair Extraction of Drugs

Drug	Counterion	Solvent
Amitriptyline and nortriptyline	Chloride, bromide and perchlorate	$CHCl_3$, CH_2Cl_2, EtAc and methyl isobutyl ketone
Morphine, codeine, diAcmorphine and dihydrocodeinone	Bromothymol blue	CH_2Cl_2
2-Naphthyl sulfate	Tetrabutylammonium	$CHCl_3$
2-Naphthyl glucuronide	Tetrapentylammonium	$CHCl_3$
Piribenzil and metabolite	Bromothymol blue	CH_2Cl_2
Emepronium bromide	Perchlorate	CH_2Cl_2
Propantheline	Perchlorate	CH_2Cl_2
Acetylcholine	3,5-di-*t*-butyl-2-hydroxybenzene	CH_2Cl_2
Cholate and taurocholate	Tetrapentylammonium	$CHCl_3$
Deoxycholate and taurodeoxycholate	Tetrabutylammonium	$CHCl_3$

All the examples above have been taken from ref. 9

They are now used extensively for screening urine samples for drugs of addiction and are becoming popular in other areas because they allow highly reproducible extractions and are much more amenable to automation that liquid–liquid extraction.

Recent work has concentrated on the polar silica phases and the non-polar C_{18} bonded phases and their modifications. The range of phases available and the differences in approach are summarized in Table 3.

These solid phase extractants are commercially available in two types of cartridge; Sep-Pak from Waters Associates and Bond Elut from Analytichem International. The two types are similar in operation but their designs differ.

There are four stages in the use of these cartridges: priming, sample application, washing and elution. The exact technique varies depending on the nature of the cartridge and the polarity of the drug being extracted. A typical procedure for a C_{18} cartridge is described below.

The cartridges are individually packed in sealed foil to protect them from the air. They are opened immediately before use and primed by drawing or pumping methanol through them. This causes the hydrocarbon chains to swell with an increase in surface area. Sample application is carried out by drawing or pumping the aqueous sample through the cartridge. In practice this is often urine or diluted plasma and experience has shown that with the latter it is wiser to centrifuge the sample first to remove any precipitated fibrin (this happens when samples have been kept deep frozen for some time). If this is not done the large particles will clog the cartridge and, even if the flow is reestablished, recoveries are often poor.

Washing is carried out using water or a suitable buffer and elution is achieved using a polar solvent such as methanol, acetonitrile or a mixture of these. Obviously preliminary experiments need to be carried out to optimize conditions.

Following elution, the organic solvent can be removed by evaporation and the residue submitted to analysis. For example, it may be dissolved in mobile phase and injected directly on to an HPLC column or applied to an HPTLC plate.

A detailed review of the practice of solid phase extraction has been published.[18]

24.2.2.2.1 *Advantages and disadvantages of extraction techniques*

The advantages of liquid–liquid extraction are that it is selective, reasonably cheap and can be used effectively for batch extraction. It has a number of disadvantages, however, including emulsion formation which can be very difficult to remedy. It is impossible to quantitatively remove the organic phase and this leads to recoveries less than 100%, large amounts of waste organic solvents accumulate in the laboratory and their efficient disposal can be expensive. Liquid–liquid extraction is labour intensive and evaporation of organic solvents to concentrate the extracts poses its own problems as well as the risk of degradation of the drug or its interaction with the solvent.

Liquid–solid extraction is highly reproducible, efficient and rapid. It generates much less waste solvent and is less demanding on staff. It is easier to automate and several companies already offer equipment for automatic extraction. There are no major disadvantages and cartridge blockage is rare if samples are routinely cenrifuged before use. The cost per cartridge is relatively high but this is offset by the reduced labour and it may be possible, in certain cases, to reuse the cartridges.

Table 3 Use of Liquid–Solid Extraction Cartridges

Cartridge	Silica	C_{18}
Polarity	Polar	Non-polar
Sample applied in solvent of	Low polarity (*e.g.* hexane)	High polarity (*e.g.* water)
Wash solvent	Low polarity	High polarity
Eluting solvent	Increased polarity	Decreased polarity
	(*e.g.* aqueous buffer)	(*e.g.* methanol, acetonitrile)
Elution order	Non-polar–most polar	Most polar–non-polar

24.2.3 SEPARATION AND QUANTITATION

The choice of method depends both on the concentration of drug in the sample and on the complexity of the sample itself. When a drug is present in relatively high concentration in simple matrices such as a tablet, a suppository or an aqueous solution in an ampoule, then preparation of the sample for analysis will be relatively easy: where the drug is present in low concentration in a complex fluid like urine, extensive purification procedures will be needed. Techniques requiring little sample preparation are the classical techniques of volumetric and gravimetric analysis or the direct application of spectroscopic methods.

24.2.3.1 Non-spectroscopic Techniques

24.2.3.1.1 Gravimetric analysis

Every chemical laboratory has a sensitive chemical balance and this can be used to measure the amount of drug present in dosage forms following its removal from the formulation by solvent extraction or precipitation. Care must be taken to ensure that no other material is coextracted or coprecipitated with the drug in question, since this is obviously a very non-selective technique. However, by choosing the appropriate extraction procedure or chemical precipitant, the method can be successfully applied in a number of cases to determine, for example, many drugs present as their halide salts, cholesterol, histamine, gold preparations and others.

Details of the application of this technique can be found in many of the standard texts.[19]

24.2.3.1.2 Volumetric analysis

In this approach a reagent is chosen which will react chemically with the drug to be determined; it is dissolved in a suitable liquid and then titrated into the solution containing the drug and a suitable indicator. The volume of reagent consumed can then be used to calculate the amount of drug present. The basic method has been in use for over 100 years and requires very simple apparatus; its range and sensitivity can be extended using more modern equipment. Titrations can be carried out in non-aqueous media and a wide range of techniques are used to detect the end point, many of them electrochemical in nature such as coulometry, amperometry and conductivity.[20–22]

24.2.3.1.3 Polarography

This technique is applicable to metals and a range of organic compounds; it has the advantage that in some cases it can be carried out directly on plasma, urine or other body fluids. Derivatization of some drugs extends the use of this technique.

The apparatus is simple, and consists of a potentiometer linked through a microammeter to two electrodes. A commonly used configuration is to use a dropping mercury electrode (cathode) which is constantly renewing itself and a pool of the same metal as anode. The current is measured with changing applied potential and a characteristic half-wave potential is produced for the compound being determined. The concentration of the drug or metal in solution undergoing reduction can be determined from the Ilkovic equation, which links current and mercury flow, and where *i* is the average current, *n* is the number of electrons involved in the reduction, *D* is the diffusion coefficient,

C the concentration, m the weight of mercury flowing and t the time necessary for the formation of one drop.

$$i = 607 n D^{1/2} C m^{2/3} t^{1/6}$$

(4)

It is not possible here to give a full account of the applications of this technique but it has been used successfully to determine a wide range of both organic and inorganic drugs.[23,24]

24.2.3.1.4 *Ion selective electrodes*

These are now widely used to measure a range of cations such as Na^+, K^+ and Ca^{2+} directly in biological samples with no pretreatment. The basic design consists of a potentiometric sensor separated from the sample by a selective barrier of conducting material such as PVC. When a given drug is present in the sample, a potential difference develops across the selective membrane and the concentration of drug can be determined by means of the Nernst equation.

The design and operation of this type of electrode has been reviewed and the method is now being applied to a number of drugs, particularly antibiotics, in the range 10–40 μM.[25,26]

24.2.3.2 Spectroscopic Techniques

Much of the early work on pharmacokinetics was based on the measurement of drugs in body fluids using spectroscopic techniques; in many cases the methods lacked selectivity and both unchanged drug and metabolites were determined. These techniques are still widely used today but have been made much more selective by introducing a chromatographic separation before detection. This will be discussed in more detail in Section 24.2.3.3.

They can still be used without chromatography provided the problem is carefully chosen and precautions taken to ensure that interference is reduced to a minimum.

The electromagnetic spectrum extends from the high energy X-rays to the low energy radiofrequency waves (Figure 2).

When organic molecules are subjected to an energy source they undergo a series of changes. The relatively high energy in the UV and visible range raises the electrons, forming the chemical bonds in a given compound, to excited states. Absorption of lower energy IR radiation increases the vibrational energy of these bonds, while microwave radiation increases their rotational energy. All these changes can be measured by using the appropriate equipment.

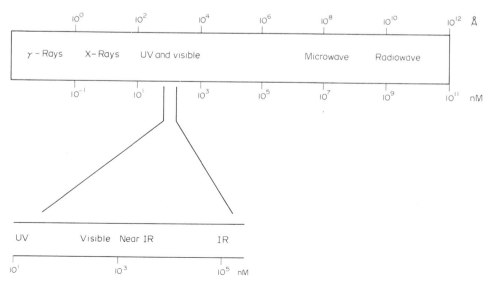

Figure 2 Electromagnetic spectrum

24.2.3.2.1   *Ultraviolet and visible spectroscopy*

UV and VIS spectroscopy are not widely used to obtain qualitiative information on drugs (except perhaps in drug screening) because many drugs and endogenous compounds possess the same chromophores. However, they are widely used for quantitative analysis of known compounds present in particular formulations.

The measurement of concentration depends upon the Beer–Lambert Law which relates the incident and transmitted monochromatic radiation (I_o and I) to the path length (b), concentration of the absorbing compound (C) and the absorbtivity of the system (K)

$$\log I_0/I = KCb \tag{5}$$

By preparing a series of standards containing known concentrations of drug and comparing their absorption with that of the sample, the drug concentration in the sample can be determined. Naturally, the range over which any drug obeys this law should be determined beforehand. Most modern instruments are of the double beam type and this eliminates any noise due to fluctuations in the radiation source. Some very sophisticated instruments are available with multichannel detection and computer control.[27] Multichannel detection will be discussed in more detail in Section 24.2.3.3.4(iv).

Difference spectroscopy is often employed, where absorbance is measured at a chosen wavelength of two samples containing different forms of the drug. This allows reliable results to be obtained even where interfering material is present, provided this material does not undergo the same change as the drug to be determined.

Derivative spectroscopy is also used and requires relatively sophisticated equipment to produce the first, second or higher derivatives of the absorbance. This is useful because very slight changes in the slope of the 'normal' (zero-order) absorbance become very pronounced changes or inversions on differentiation. This is used extensively to suppress interference and is particularly useful in biomedical analysis and a tenfold increase in the detection limit of some drugs has been reported.[28]

Chemically induced changes in the structure of the drug immediately before recording the absorbance spectrum is often useful and the change in absorbance of barbiturates and oestrogen with pH has been well documented.[29]

24.2.3.2.2   *Fluorimetry*

This is a more sensitive technique than either VIS or UV spectroscopy since the sample is irradiated at a selected wavelength (λ_{ex}) and emits at a higher one (λ_f). Relatively few drugs possess a natural fluorescence but this can be an advantage in certain cases since it reduces the possibility of interference from other components in the starting material. The applicability of the technique, however, can be extended by reacting drugs with derivatizing agents which are themselves fluorescent. This approach is widely used, particularly in HPLC and HPTLC, and will be discussed in detail there. It can be used for direct fluorimetry but means must be devised to remove unchanged derivatizing agent or any by-products that fluoresce.

Fluorescence is sensitive to a number of factors, such as pH, temperature and solvent, and for very sensitive work the sample cell should be thermostated. The glassware used must be scrupulously clean, and solvents and reagents must be carefully checked for interfering fluorophores.

In dilute solution the intensity of fluorescence, I_f, of a drug is related to its molar concentration, C, by equation (6) where K is a complex constant made up of factors describing the characteristics of the instrument used, the radiation intensity of the source, cell length, molar absorbtivity and quantum efficiency.

$$I_f = KC \quad \text{at} \quad \lambda_{ex}, \lambda_f \tag{6}$$

Although the linear range of fluorescent measurements is much narrower that for either UV or VIS spectroscopy, its greater sensitivity has ensured that it has been used to measure a number of drugs, including catecholamines, phenothiazines, β-adrenoceptor antagonists and some antibiotics.[30]

24.2.3.2.3   *Infrared spectroscopy*

When drugs absorb the relatively lower energies of the IR region they display changes in their vibrational and rotational bond energies. This means that IR spectra are very useful indeed as qualitiative tools to study the structure of molecules, as they allow the identification of the chemical groups present. Double and triple bonds, in particular, give rise to intense absorption bands and the use of this technique to identify structural features is described in many texts.[31-33]

IR can also be used as a quantitative tool for drugs possessing suitably strong absorption bands and this is, at least in principle, more versatile than either UV or VIS spectroscopy where most drugs display only one or two absorption peaks. The IR spectra of drugs display large numbers of peaks and it is often possible to choose one for a given drug that is absent from the other constituents present. Calibration curves of absorbance against concentration are prepared in the usual way.

The attraction of IR lies in its relatively simple requirements for sample preparation. The drug, or its extract from a formulation, can be dissolved in a suitable solvent that obviously must not absorb in the region to be studied; it may be suspended in liquid paraffin (as a mull) or it may be mixed with a halide such as KBr and compressed to give a transparent disk. The sample can later be recovered unchanged if required.

Older IR spectrometers were not very sensitive and mg quantities of drug were required; this was not always a disadvantage if the drug was present in high concentration in a tablet. For example, phenobarbitone is simply measured in tablets and the technique has also been applied to the simultaneous measurement of acetylsalicylic acid and phenacetin in a tablet formulation.[34] More sophisticated instruments such as Fourier Transform IR (FTIR) are much more sensitive and spectra can be recorded using μg quantities or less.[35]

24.2.3.2.4 *Atomic absorption*

This technique offers a very sensitive and selective means of measuring metallic elements in biological fluids with a minimum of sample preparation. It can be used to monitor plasma lithium concentration in patients being treated with lithium carbonate and it is also extensively used to determine trace metals such as zinc, copper and aluminum.[36]

24.2.3.3 Chromatographic Techniques

24.2.3.3.1 *Introduction*

Chromatography had its roots in the late 19th and early 20th century[37] but its most rapid development has been in the last fifty years when it has grown rapidly to become one of the most powerful tools available to the analytical chemist. The technique of chromatography allows mixtures of compounds to be separated by virtue of their different distribution between two phases, one stationary and the other moving. The technique can be subdivided into two different types: adsorption and partition chromatography. In the former, the stationary phase is a solid while in the latter it is a liquid; in both cases the moving, or mobile, phase may be either a gas or a liquid.

Adsorption chromatography is usually used to purify compounds on a preparative or semi-preparative scale, or to study the physical chemical properties of the solid materials used as the stationary phase. Partition chromatography is much more versatile and has given rise to a voluminous literature. The stationary phase can be packed into a column for gas–liquid chromatography (GLC) where the mobile phase is a gas, or high performance liquid chromatography (HPLC) where the mobile phase is a liquid or even spread as a thin coating on glass, plastic or other material to give thin layer chromatography (TLC). When the stationary phase is very carefully graded, according to particle size, this approach is referred to as high performance thin layer chromatography (HPTLC).

The strength of the chromatographic approach is two-fold: a wide range of stationary and mobile phases of differing polarities can be used in conjunction with a variety of different detectors.

24.2.3.3.2 *Theoretical aspects*

Any compound introduced into the mobile phase will partition between the stationary phase and the mobile phase depending on its solubility in both. For example, in GLC if a mixture contains both polar and non-polar compounds and the stationary phase used is polar then the non-polar compounds will be poorly soluble in the stationary phase and will spend most of their time in the mobile gas phase, eluting quickly from the column; the reverse will be true for polar compounds. It is important to remember that no separation takes place in the mobile phase so non-polar compounds, that have passed quickly through the column, will not be well separated from each other and will be detected as a single peak or a poorly resolved group of peaks. The time taken for a compound to pass through a column is called its retention time (t_R) and is constant for a particular set of operating conditions (temperature, composition and flow rate of mobile phase, column type and dimensions, *etc.*). If the compound is applied dissolved in a volatile solvent (as in GLC) the absolute time is not usually measured but the time relative to the elution of the solvent. In HPLC retention volumes (v_R) are often used and in TLC and HPTLC the distance travelled by the compound relative to that travelled by the mobile phase (the solvent front) is called its retardation factor (R_f).

A measure of the efficiency of a column is its number of theoretical plates, a concept coming from distillation theory. This is not an absolute number but depends on the drug being chromatographed. The calculation of the number of theoretical plates for a given compound is illustrated in Figure 3. The length of the column can then be divided by N to give the height equivalent of a theoretical plate (HETP). This again stems from the idea of the chromatographic column as a distillation column.

The factors that affect peak shape are the eddy and molecular diffusion and the resistance to mass transfer during passage through the column. These factors are described in the van Deemter equation which relates the HETP to the velocity of the mobile phase, μ (see Figure 4).

Other parameters that are important in chromatography are the resolution (R) of the column for two compounds and this is measured as shown in Figure 5. The solvent efficiency (α) is also important and its determination is illustrated in Figure 6.

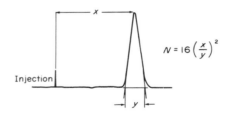

Figure 3　Calculation of theoretical plates

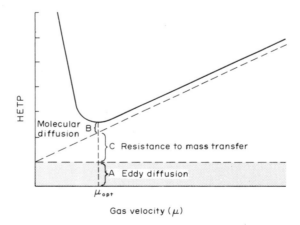

Figure 4　The change in HETP with gas velocity, μ, expressed mathematically in the van Deemter equation

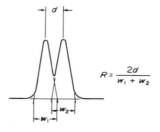

Figure 5　Calculation of the resolution of a column for two compounds

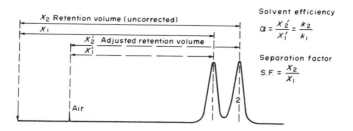

Figure 6 Calculation of the solvent efficiency

These parameters can then be used to calculate the number of plates needed for a given separation. The use of this equation and a more detailed treatment of the theory of chromatography can be found in a number of standard texts.[38,39]

The technique of GLC is better suited to relatively small, thermally stable, non-polar molecules but its use can be extended considerably by the use of derivatizing agents (see Section 24.2.3.3.3vi). HPLC complements GLC by being suitable for large, polar molecules which may also be thermally unstable. Of course, both techniques have been applied to a large number of drugs. A thin layer plate can be regarded as a two-dimensional column and HPTLC is closely related to HPLC.

The advantages and disadvantages of each will be treated in the appropriate sections.

24.2.3.3.3 *Gas–liquid chromatography (GLC)*

The column is the heart of any chromatograph be it GLC or HPLC and, no matter how sophisticated the instrument, if the wrong column is chosen the end result will be poor.

(i) *Columns*

In GLC two types of column are used, packed and capillary. The first, as the name suggests, are packed with a stationary phase, usually coated on an inert support. They are relatively short (1–4 m) and wide bore (2–4 mm). The empty columns can be made from stainless steel, nickel or glass and the latter is almost always used for drug analysis to minimize the possibility of decomposition of drug on the hot walls during chromatography. Glass columns have the additional advantage that they allow a visual inspection of the column packing to monitor any deterioration due to contamination or oxidation.

Capillary columns can be made of metal but for drug work are either glass or fused silica. Fused silica is much more robust and its relatively recent advent has been responsible for the increased use of capillary columns; for added strength the columns are coated on the outside with a polyimide heat resistant film.[40,41]

(a) Packed columns. Column packing consists of a stationary phase coated over an inert support which serves to provide a large surface area for the stationary phase so that a more efficient separation can be achieved.

The efficiency of packed columns is low, relative to capillary, being of the order of 500–2000 theoretical plates; this may not necessarily be a disadvantage, especially in much quantitative drug analysis where a fairly extensive 'clean-up' procedure may have taken place, and it might only be required to separate the drug from the internal standard. For drug screening, to identify unknown compounds, efficient separation is of paramount importance and capillary columns are now used routinely.

A wide range of column packings is available commercially and the stationary phase is usually applied to give a 1–3% coating on the support. Some chromatographers prefer to buy their supports and stationary phase separately and prepare their own packings.

Support. A number of supports are available commercially and the most widely used are derived from 'diatomaceous earths' which are composed almost entirely of pure silica with an excellent surface area to weight ratio. The manufacturer prepares them for GC use by heating them to a high temperature either alone or as a flux with sodium carbonate. Following this calcination, they are carefully graded to provide a range of closely controlled particle (mesh) sizes. Two types of diatomaceous earth support that are widely used are the Chromosorb and Gas Chrom series. Some relative properties are compared in Table 4.

Table 4 Properties of Some Chromosorb Supports

Chromosorb	P	W	G	Glass beads
Appearance	Pink	White	Pearl	Transparent
Activity	Partially inert	Inert	Inert	Inert
Packed density (g mL^{-1})	0.47	0.24	0.58[a]	1.4
Surface area (m^2 g^{-1})	4	1	0.5	0.2
Normal coating of stationary phase (%)	5–30	1–25	1–20	0.05–2
Theoretical plates/meter	800–2000	700–1500	1000–2000	200–500
Breakdown (%) after 5 min shaking	12	19	1	0

[a] Because of its higher density, a loading of 3% (w/w) on Chromosorb G is equivalent to a loading of 7% (w/w) on Chromsorb W.

These diatomaceous supports can be modified chemically before the stationary phases are applied and this treatment can take several forms. The support is often washed with acid to remove trace elements and then may be silanized to deactivate any hydroxyl groups present that could bind some drugs to give peak 'tailing'. The two most commonly used silanizing agents are dimethylchlorosilane (DMCS) and hexamethyldisilazane (HMDS). Suppliers, in their catalogues, will refer to supports in a way that describes this treatment, *e.g.* Chromosorb G AW–DMCS, signifies that this support has been both acid washed and silanized with dimethylchlorosilane.

Polymer supports are also used, the Porapak series being the most common in drug analysis and these are used for the separation of volatile polar compounds such as alcohols. The polymer beads themselves produce the separation and there is no need for a stationary phase. They are supplied in a range of polarity from *P*, the least, to *T*, the most polar.

A number of other supports have been used for specific purposes but the majority of drug applications require those discussed above.[42,43]

Stationary phase. A good stationary phase should be thermally stable, resistant to solvents and possess a low vapour pressure in order to prevent 'bleeding', which is the slow elution of some of the phase into the detector and is recognized by a steady upward drift of the recorder baseline.

There are over 200 stationary phases listed in some manufacturers' catalogues but only about 20 are widely used for drug analysis. Based on the principle that 'like dissolves like' non-polar compounds should, as a rule, be chromatographed on non-polar stationary phases. Some of the most popular stationary phases are based on the polysiloxane polymers which have the common skeleton of structure (**1**).

$$- \underset{|}{\overset{|}{Si}} - O - \underset{|}{\overset{|}{Si}} - O - \underset{|}{\overset{|}{Si}} - O - \underset{|}{\overset{|}{Si}} - O - \underset{|}{\overset{|}{Si}} - O - \underset{|}{\overset{|}{Si}} - O - \underset{|}{\overset{|}{Si}} - O - \underset{|}{\overset{|}{Si}} - O -$$

(**1**)

Different degrees of polarity can be introduced by varying the nature of the substituents. A typical series, the OV range from Supelco, used to be classified in increasing polarity, with increasing percentage of phenyl substitution, according to Table 5. However, the range has been extended and no longer can the number be used to gauge the polarity. For example OV 101 is similar in polarity to OV 1 but is a liquid rather than a gum. Stationary phase OV 210 contains the trifluoropropyl group while OV 225 and 275 contain cyanopropyl, phenyl and methyl substituents.

A knowledge of McReynolds constants is useful in order to compare and select stationary phases. The system was developed from the work of Kovats,[44] extended by Rohrschneider[45] and modified by McReynolds.[46] Each stationary phase can be characterized by studying how a series of test compounds behave when chromatographed on that particular phase and an idea of how the system works can be gained from an examination of Table 6. Squalane is the least polar stationary phase and all others are measured relative to it. The polarity of a phase is directly related to the magnitude of the McReynolds constants and an examination of the above table shows, for example, that there is little difference between OV 1 and OV 101; OV 7 is more polar, while XE 60 is more polar still, and so on.

The test compounds used to characterize a phase are benzene, butanol, 2-pentanone and the other compounds listed. If the drug to be chromatographed contains a basic nitrogen atom, attention should be paid to the pyridine value, if the drug contains a secondary alcoholic –OH group, then the value for 2-methyl-2-pentanol should be considered. The system is very useful if one wishes to set up

Table 5 Percentage Phenyl Substitution in OV phases

Phase	% Phenyl
OV 1	0
OV 101	0
OV 3	10
OV 7	20
OV 11	35
OV 17	50
OV 22	65
OV 25	75

a method that uses a stationary phase that is not available; by consulting the McReynolds constants of different phases it may be possible to select a suitable alternative.

The concept of retention index is useful since it can be used in the form of the McReynolds constants to characterize the stationary phase or to help identify drugs themselves. By determining when a drug elutes from a standard phase such as SE 30, relative to the nearest straight chain hydrocarbons, it can be assigned a retention index (I_x). The *n*-alkanes are assigned retention indices of $n \times 100$. Thus the drug oxprenolol elutes after C_{18} but before C_{19} and its retention index is calculated to be 1870. This is particularly useful in drug screening since the indices are relatively independent of temperature. The retention index, combined with some TLC and spectroscopic data, can often allow a drug to be quickly identified. Tables of retention indices are available.[47,48]

(b) Capillary columns. These are usually supplied in lengths of between 10 and 50 m with internal diameters of 0.2–0.7 mm. The stationary phase can be bonded directly to the (etched) inside walls of the column to give a wall coated open tubular column (WCOT) or a support, such as microcrystalline sodium chloride, can first be bonded to the walls and the stationary phase coated on this to give a support coated open tubular column (SCOT). Most small laboratories buy their capillary columns from commercial suppliers and do not attempt to prepare them for themselves.

The thickness of the film coating is important and ranges from 0.1–1.5 μm. The most efficient separations and shortest analysis times are obtained from thin films but higher sample capacity is obtained from the thicker. Capillary columns are also much more expensive than packed ones but this is offset by their much greater efficiency, often of the order of 10 000 theoretical plates or more. The use of wide bore capillary columns is becoming more widespread.[49,50]

(ii) Injection systems

In order to obtain the narrowest peaks and the best separation possible the sample, usually dissolved in a volatile organic solvent, should be introduced on column as a discrete 'plug'.

(a) Injection on to packed columns

Liquid injecion. With these columns the usual practice is to inject the sample in a volatile solvent using a glass syringe with a metal needle. This is an efficient and popular way of making an injection but it should be remembered that, since the injection port of the chromatograph is at a higher temperature (often a much higher temperature) there is always the possibility that sensitive compounds may decompose on the hot metal of the needle. To prevent this the sample is always drawn up into the glass body of the syringe and the injection made as rapidly as possible.[51]

Occasionally liquid injection can pose a problem if the drug itself is volatile. Under isothermal conditions the drug may elute quickly and appear as a peak on the tail of the solvent. Under such conditions quantitation of the drug can be difficult, especially at low concentrations. Temperature programming can overcome this; starting with the column at a low temperature until the solvent has appeared, then increasing the temperature. This may not be very successful if the drug elutes very close to the solvent. If the drug contains nitrogen or phosphorus atoms, then N/P detection can be used. If the drug contains neither of these or if an N/P detector is not available, solid injection can be employed.

Solid injection. The drug is dissolved in a suitable solvent and injected on to a fine needle (usually of glass) in a sealed device fitted outside the oven of the GC Carrier gas flows over the needle and rapidly evaporates the solvent leaving a thin film of drug on the needle tip. The needle is then lowered into the injection port of the GC where the drug is volatilized and swept on to the column. Usually only a very small solvent peak is observed and the technique has the added advantages that

Table 6   McReynolds Constants for Common Stationary Phases

	Benzene	Butanol	2-Pentanone	Nitropropane	Pyridine	2-Methyl-2-pentanol	1-Iodobutane	2-Octyne	1,4-Dioxane	cis-Hydrindane
Squalane	0	0	0	0	0	0	0	0	0	0
SE 30	15	53	44	64	41	31	3	22	44	−2
OV 1	16	55	44	65	42	32	4	23	45	−1
OV 101	17	57	45	67	43	33	4	23	46	−2
OV 7	69	113	111	171	128	77	68	66	120	35
OV 17	119	158	162	243	202	112	119	105	184	69
OV 25	178	204	208	305	280	144	169	147	251	113
XE 60	204	381	340	493	367	289	203	120	327	94

large volumes of solvent do not pass through the detector and any non-volatile material is retained on the needle to be removed by cleaning at suitable intervals. Solid injection can also be used with capillary columns.[52]

(b) Capillary columns

On column injection. This is the most efficient means of injection, the drug solution being introduced on to the capillary column itself. This is achieved by using a syringe with a very fine metal or fused silica needle which is narrower than the internal diameter of the column itself. The injection system is pneumatic since the fine needle is not strong enough to penetrate a silicone rubber septum. The column is at a temperature below the boiling point of the solvent and, after the injection has been made, temperature programming is started to elute the solvent rapidly with the drug eluting later as a very discrete band giving a very sharp peak. This technique allows several microliters of solution to be injected and, if a deactivated silica precolumn is used, larger volumes (up to 50 μL) can be injected.[53]

Split injection. Good results can also be obtained with this system. Here the drug, dissolved in a suitable solvent, is injected into a heated area through which there is a relatively large gas flow. Most of the volatilized material is vented to the atmosphere and only a small fraction enters the capillary column. The split ratio (the ratio of the amount reaching the column to the amount injected) is usually in the range 1:50–1:200. This system is used with capillary columns coated with a very thin layer of stationary phase. Non-volatile components do not enter the column but remain in the splitter which is cleaned periodically.[53]

Splitless injection. This approach is very useful for dilute solutions. The sample is dissolved in a solvent whose boiling point is much higher than the column temperature but lower than that of the solutes to be separated. The solution is then injected into a heated chamber flushed with carrier gas. The solvent vaporizes and is carried to the column where it condenses to behave as a thick film of stationary phase. The solutes are held in this film until they are released as a sharp band following evaporation of the solvent. Many capillary systems are designed to be operated in both split and splitless modes. Detailed treatements of their operation are available.[53]

Headspace analysis. This is used for the determination of volatile compounds in water, a good example being ethyl alcohol in blood. The specimen is placed in a vial which is sealed with a septum and placed in a thermostated bath and allowed to reach equilibrium. A sample of the gas phase (headspace) above the liquid is removed automatically and injected on to the column. This is a very rapid method since it involves no pretreatment of the sample, save perhaps diluting it with a solution of internal standard. More detailed information is available in the standard texts.[54]

(iii) Detectors

There are many types of detector in GLC, most commonly flame ionization, electron capture, thermal conductivity, photoionization and flame photometry. Of these flame ionization and electron capture are the most widely used for drug analysis. Mass spectrometry and other techniques can also be used and these are discussed in Section 24.2.3.3.6

(a) Flame ionization detector (FID). This detector consists of a hydrogen flame burning in air, the column effluent being mixed with the hydrogen as it leaves the column. When only carrier gas is emerging from the column there is a state of equilibrium, the flame burns steadily and there is a constant current between the flame jet and the collector electrode above it. The potential difference between the jet and the collector is about 150–200 V. When a combustible compound emerges from the column, it burns in the flame causing an increase in ionization and a rise in current flowing between jet and electrode. This is observed as a rise in the baseline of the chart recorder. When the compound has been completely consumed the current returns to the equilibrium value until another compound elutes from the column.

This detector is robust and is widely used for drug analysis. It is reasonably sensitive and submicrogram quantities can be detected on-column. Response is best for drugs with large numbers of carbon and hydrogen atoms and falls off as the number of nitrogen and oxygen atoms increases. An important advantage is its insensitivity to water and, provided the stationary phase is not adversely affected by water, direct aqueous injections can be performed. Another advantage is its large linear range, of the order of 10^6.[55,56]

(b) Nitrogen/phosphorus detector (N/PD). This is a modified version of the FID with a bead of a rubidium salt (usually the silicate) above the flame. When properly tuned, the detector shows a greatly enhanced response to compounds containing nitrogen or phosphorus at the expense of compounds containing only carbon, hydrogen and oxygen. The detector can be operated in either the nitrogen or phosphorus mode and its selectivity can be of the order of 10^4 compared with a

similar mass of a hydrocarbon. Since many drugs contain at least one nitrogen atom this is a very useful detector indeed and an appreciable response can be obtained for 1 ng of drug or less. A new benzodiazepine antagonist has been measured in plasma down to a concentration of 50 ng L^{-1} using this detector.[57] When combined with a capillary column this becomes a very powerful tool indeed for drug screening.[58]

(c) Electron capture detector (ECD). In this detector a radioactive source (usually ^{63}N) emits β-particles which excite the carrier gas to an unstable state. In reverting to a stable one, each carrier gas atom emits an electron which is picked up by the collector electrode. At equilibrium there is a steady current and a stable baseline is displayed by the recorder. When a compound emerges from the column that is capable of capturing electrons there is a fall in current that is detected as a sudden change in baseline; once the compound passes out of the detector the baseline returns to its original position. Although this is recorded as a 'peak' it is worth remembering that it is due to a fall in current, in contrast to the rise in current produced by the ionization detectors discussed above.

The linear range of an ECD is less than FID (about 10^4 in recent designs) but this is compensated by an increase in sensitivity. Halogenated compounds and those containing nitro groups and highly conjugated aromatic systems are strongly electron capturing but this is not limiting because there are a wide range of reagents that are able to convert many drugs to strongly electron-capturing derivatives (see Section 24.2.3.3.3iii).

Care must be taken, however, during sample preparation especially if halogenated hydrocarbons such as chloroform or dichloromethane are used for extracton. These must be completely removed from the extract to prevent contamination of the detector.

An ECD is relatively stable in operation but becomes contaminated with constant use; procedures are outlined in the manufacturers' manuals on how to monitor this. When switched off for a time, condensation of water vapour from the air can be a problem with some detectors and they may have to be cleaned before being used again.[59,60]

(d) Thermal conductivity detector (TCD). This detector used to be called a catharometer and was one of the earliest types of detectors used in GLC. It contains two heated filaments that comprise two arms of a Wheatstone bridge. A reference gas flow passes over one and the column effluent over the other. At equilibrium, with only carrier gas eluting from the column the bridge is balanced. Immediately a compound elutes from the column this equilibrium is disturbed and a signal produced. This is a 'universal' detector in that it will detect any substance that has a thermal conductivity different from that of the carrier gas. It is, however, relatively insensitive and depends on very precise temperature control. It is now rarely used for drug analysis.[61]

(e) Photoionization detector (PID). In this detector high energy UV radiation is used to produce ionized species from the compounds eluting from the column and the corresponding change in current is detected. Its sensitivity and linearity is similar to that of FID and it is frequently used for inorganic compounds although drugs can also be measured. It has not, however, become as popular as FID, N/P D or ECD.[62]

(f) Flame photometric detector (FPD). Here the column effluent burns in a reducing flame and a photomultiplier monitors the emission of ions. It is useful for phosphorus and sulfur containing compounds but is not widely used in drug analysis.[63]

(g) Multiple detectors. It is an easy matter to split the effluent from a column and pass it to more than one detector. It is usually easier to do this by having two or more detectors in parallel although some instruments allow two detectors to be mounted in series. Usually one of the detectors is an FID because of its good response to most types of organic compound and the other detector may be an ECD, N/P D or FPD. In this way valuable information can be gained about the structures of compounds, such as metabolites, unwanted impurities and by-products of reactions.[64] For quantitative work on a known drug a single detector is preferable.

More detailed discussions of detectors can be found in the standard texts.[38,39]

(iv) Pyrolysis gas chromatography

This is an indirect method since the drug is broken down at high temperature and one or more of the fragments detected. The sample is heated in the injection system to a temperature of 300–1000 °C in ms using one of several different methods; the Curie point pyrolyzer being one of the most frequently used. In order to yield reliable results the heating must be very reproducible and laser pyrolysis is now available.

All sulfonamides pyrolyze to give aniline and another aromatic amine whose structure depends on the original sulfonamide. The pattern of breakdown products can be used for identification and, sometimes, quantitation.

Reviews of this topic are available.[65,66]

(v) Supercritical fluid chromatography

This is a relatively new technique which has been claimed by some to combine the best characteristics of gas and liquid chromatography. A supercritical fluid is used as the mobile phase and pumped, using a modified HPLC pump, through the column of a capillary gas or high performance liquid chromatograph. Carbon dioxide has been used most frequently as a mobile phase because it can be relatively easily maintained above its critical point.

Ammonia has also been used, but it is very reactive and may damage stationary phases and seals.

The polarity of the supercritical fluid can be varied simply by altering its temperature and pressure. The main advantage of supercritical fluid chromatography is that it can be used with a range of GC and HPLC detectors including FID and UV, and its use with flame photometric and mass spectrometric detectors is also possible.[67]

(vi) Derivatization

A number of drugs such as tertiary amines and some benzodiazepines can be chromatographed directly but the majority are too polar and must be converted into less polar (and hence more volatile) compounds. This conversion is usually carried out by chemical means and is called derivatization. Often derivatization is carried out to allow detection at very low concentrations, *e.g.* by attaching groups that are very strongly electron capturing. Derivatization reactions can generally be classified as silylating, non-halogenating and halogenating. In the first case the drug is reacted with one or more powerful silylating agents to give a very volatile derivative. There are many different types of reagent available and an extensive literature exists. The non-halogenating reagents are generally used simply to make the drug more volatile and improve the peak shape by decreasing tailing. Flame ionization is generally used to detect both types of derivative. Halogenating reactions increase volatility and allow very low concentrations to be measured using electron capture detectors. A wide range of halogenated anhydrides, carbonyl chlorides and other reagents are available. A summary of the different types of reaction, applied to different drugs, is given in the Table 7.

The choice of reaction conditions is often critical for derivatization reactions: chloramphenicol reacts with the silylating agent, BSA, to give mono-, bis- or tris-trimethylsilyl derivatives depending on the solvent chosen. Temperature, time of reaction and catalysts are also important. For example, sterically hindered phenols may not be acetylated using the standard mixture of acetic anhydride in pyridine but yields of over 90% are obtained when 4-dimethylaminopyridine is substituted. All these variables should be optimized when any derivatization reaction is being studied. This topic has been reviewed in detail.[68-70]

24.2.3.3.4 High performance liquid chromatography (HPLC)

(i) Introduction

This technique is extremely versatile for the analysis of drugs, particularly in body fluids, and it can also be applied to drugs in pharmaceutical formulations with very little sample preparation.

Table 7 Derivatization Reactions for Gas Chromatography

Chemical type	Silylating	Non-halogenating	Halogenating
Acids			
Carboxylic	Fenoprofen[71]	Indomethacin[72]	Flurbiprofen[73]
Barbituric	Pentobarbitone[74]	Phenobarbitone[75]	Hexobarbitone[76]
Amino	Mixture of 20[77]	Mixture of 24[78]	Mixture of 20[79]
Bases			
Primary and secondary amines	Aminoglycoside antibiotics[80]	Mexilitine[81]	Amphetamines[82]
Alcohols and phenols	Morphine[83]	Clioquinol[84]	Catecholamines[85]
Steroids	Various[86]	Corticosteroids[87]	Oestrogens[88]
Other Reactions			
Tertiary amines	Thermal degradation of amitriptyline[89]	Dealkylation followed by derivatization[90]	
Quaternary ammonium and guanidine compounds	Reaction with Na benzenethiolate[91]	Conversion to primary amine[92]	
Ring closing of hydralazine[93]	Ring opening of benzodiazepines[94]	On column methylation of methimazole[95]	

Although it is difficult to make quantitative judgements, it appears that more drug methods are now being published using HPLC than GLC, although the latter is still widespread.

There are several different forms of chromatography used in HPLC: normal (including adsorption) and reverse phase partition, ion exchange and size exclusion.[96-98] The choice of column depends on the nature of the compounds to be separated and analyzed, and Table 8 indicates the factors to be considered.

(ii) Columns

(a) Normal phase. This approach is only 'normal' because it was developed before reverse phase. It was also called adsorption chromatography because it was originally believed that the separation of compounds in a mixture took place solely by differential adsorption on the silica or alumina stationary phase. However, it now seems as if partition plays at least as important a role, with the compounds interacting with the polar silanol groups on the silica or bound water molecules. Very good separations can be obtained using normal phase chromatography, particularly of drugs and their metabolites. It is true to say, however, that there are less parameters to be varied in this type of chromatography (although the introduction of bonded phases has been very useful) compared with reverse phase and for this reason there are more published methods using the latter. Normal phase columns can be prepared by hand without the need for expensive equipment and are very useful especially to laboratories with limited budgets.

(b) Reverse phase. While normal phase sees the passage of a relatively non-polar mobile phase over a polar stationary phase, reverse phase chromatography is carried out using a polar mobile phase (*e.g.* methanol, acetonitrile or water mixtures) over a non-polar stationary phase. A range of stationary phases are available and very selective separations can be obtained.[99] The pH of the mobile phase can be adjusted to suppress the ionization of the drug in question and thereby increase the retention on column. For some highly ionized drugs, this ion suppression is not successful and the technique of ion pairing can be used. Here the ionized form of the drug is paired with a counterion to form a species whose overall charge is zero. Often the counterion is added to the mobile phase. The use of ion pairing in reverse phase chromatography has been reviewed in detail.[8,9]

Ion exchange. This technique is now much less common that it used to be. It depends on the careful control of a number of factors including resin type, buffer and ionic strength and it has now been largely replaced by reverse phase.[100]

(c) Gel permeation chromatography. This approach depends on the different molecular sizes of the components of a mixture. Most drugs have molecular weights below 600 and gel permeation is restricted to the larger molecular weight drugs such as polypeptides.[101]

A detailed discussion of the different types of column is available.[102] Although preparative columns are available, they are relatively expensive and most applications of HPLC in the pharmaceutical sciences are directed at drug analysis where the trend is to more efficient small bore columns.[103]

(iii) Injection systems

These are relatively few in number. In the early days of HPLC injections were made through silicone rubber septa against a pressure gradient but today most injections are based on the loop

Table 8   Choosing an HPLC Column

Drug		Chromatography	Column	Mobile phase	Example
Water insoluble:	Non-polar	Partition, reverse phase	Bonded C_2–C_{18}	Polar	Antibiotics
	Weakly polar	Adsorption, solid phase	Silica, alumina	Weakly polar	Alkaloids
	Polar	Partition, normal phase	Bonded –CN, –NO_2, NH_2	Non-polar	Alkaloids
Water soluble:	Anionic	Ion exchange	Anion exchanger	Polar	Sulfonamides
	Both	Ion pairing	Bonded C_{18}	Polar	Wide range of drugs
	Cationic	Ion exchange	Cation exchanger	Polar	Cathecholamines
MW > 2000		Size exclusion	LiChrospher Spherosil Microgel	Polar	Proteins, biopolymers

system with some form of switching. The injection is made into a loop of fixed volume (usually 20–500 µL) in the 'load' position, the valve is then rotated to the 'inject' position and the sample swept on to the column. Switching can be performed manually or automatically. In fully automated systems, samples are placed in glass vials in a carousel and programs chosen to allow washing between samples and injection of suitable standards. It is also possible to switch columns during a run and this can be done automatically.[104]

The systems available have been reviewed.[105,106]

(iv) HPLC detectors

In HPLC the volume of the detector is kept as small as possible, often 10 µL or less, to reduce any peak broadening and any dead volume is kept to a minimum by careful design. The most popular are UV, fluorescence and electrochemical detectors. There is, unfortunately, no low cost 'universal' detector for HPLC except for the refractive index detector which is too insensitive for most drug work (see Section 24.2.3.3.6).

(a) Ultraviolet. This is still the most popular type of system for drug analysis because of its simplicity and stability. It is most suitable for drugs with strongly absorbing chromophores, although its use can be extended by derivatization (see Section 24.2.3.3.3vi). Detectors can either be fixed (*e.g.* 206, 226, 254, 280, 320 nm) or variable wavelength. Fixed wavelength detectors are generally more sensitive and stable with a low signal to noise ratio but lack versatility. Working at a low wavelength such as 206 nm allows the determination of a very wide range of drugs but interference from other compounds is a problem and great care must be taken to reduce this to a minimum.[107]

(b) Diode array. In this technique the UV absorbance can be measured at more than one wavelength simultaneously. It is particularly useful for detecting the presence of interfering compounds. It can also be used to obtain the complete UV spectrum of any compound as it passes through the detector and this makes it particularly useful in drug metabolism studies and for the identification of unknown compounds.[108]

(c) Fluorescence. A number of drugs possess natural fluorescence and these are well suited to this type of detection. Its use can be extended to other drugs by derivatization with suitable reagents. The method can be made selective by the careful choice of excitation and emission wavelengths. Fluorescence detection generally allows the determination of drugs at lower concentrations than UV but this will, of course, depend on the nature of the drug in question. Many drugs have been determined by this technique.[109,110]

(d) Electrochemical. This approach offers a very sensitive means of detection. The development of electrochemical detection methods has been rapid over the last fifteen years and reviews are published regularly. Most are electrolytic and depend on the detection of a current generated by the oxidation or reduction of the solute. Conductometric and capacitance detection is much rarer, at least for drug analysis. Because of the electrochemical nature of these detectors constraints are placed on the chromatography and solvent mixtures polar enough to dissolve suitable electrolytes must be used. In practice this means that normal phase chromatography is restricted and solvents with a polarity less than methylene chloride cannot be used. Any suitable electrolyte can be added postcolumn but in most cases reverse phase and ion pair chromatography are employed and, more rarely, ion exchange and size exclusion chromatography.

A mobile phase of relatively high ionic strength is generally used and careful control of pH is necessary: oxidation reactions take place more readily at high pH while reductions take place at lower pH.

Amperometric detectors are most frequently used and consist of three electrodes (working, auxilliary and reference) assembled in one of a number of different configurations. All EC detectors must be shielded from low frequency noise and this is done by housing them in a simple Faraday cage. The electrodes can be made from a range of materials that includes carbon, gold, platinum, mercury and amalgams. Glassy carbon is very popular, relatively inexpensive and is widely used for oxidations, amalgams are used for reductions and gold and platinum are used for detection in non-aqueous solvents.

Polarographic detectors are variants of amperometric systems where the working electrode takes the form of droplets of mercury (the dropping mercury electrode, DME). While this is excellent for operation in the reduction mode, prior deaeration of the solvent is necessary because any dissolved oxygen will be reduced at the DME.

Many amperometric detectors, especially the earlier designs, can be unstable and demonstrate sudden losses in sensitivity which can often be slow and difficult to rectify. Increased sensitivity can be obtained by preanodization.

Coulometric detectors are generally more sensitive than amperometric and, when operating under optimum conditions, the analyte (drug) is completely electrolyzed. The detector response is much less dependent on cell geometry, flow rate of the mobile phase and temperature.

Potentiometric detectors usually employ ion sensitive electrodes and react sluggishly. This is a major disadvantage for HPLC and most applications, so far, are directed at inorganic analysis. Some typical examples of the use of EC detection are summarized in Table 9 and a number of reviews are available.[111-113]

(v) HPLC derivatization

In gas chromatography derivatization is carried out to make the drug or metabolite more volatile and also to confer upon it some property that will allow its detection at lower concentrations (*e.g.* forming the trifluoroacetyl derivative for EC detection). In HPLC, derivatization is usually performed simply to increase the limit of detectability or selectivity for the compound in question. This usually means attaching some strongly UV-absorbing or fluorescent group. Details of the most common reagents and their use are given in Table 10.

As in all derivatization reactions, the experimental conditions must be carefully controlled and optimized for each drug. The choice of reaction solvent (or mixture), reaction time and temperature are very important.

Drugs containing a single derivatizable group, such as a primary or secondary amine, or a carboxylic acid, are relatively easy to derivatize. It is more difficult, however, to form single derivatives of drugs with several different functions such as a secondary amine, a phenolic and an alcoholic –OH group.

In most methods, derivatization is carried out before injection on to the column. This allows the derivative or derivatized extract to be purified by removing the excess derivatizing agent and by-products. It is also possible, however, to carry out postcolumn derivatization where the drugs or metabolites are converted to a strongly UV absorbing or fluorescent compound immediately before entering the detector. This reduces any chance of decomposition but offers a more limited range of derivatizing reactions. Reviews of derivatization for HPLC are available.[68,138,139]

(vi) Optimization of chromatographic conditions

Once a suitable column has been chosen, by considering the polarity of the components of the extract to be separated, optimization of the mobile phase must be carried out. This is achieved by conducting a series of experiments involving the properties of suitable solvents, the composition of the mobile phase, pH, ionic strength and the need to add surface active agents or other compounds to improve peak shape.

In the *single factor approach* all the above are investigated separately and optimized in turn. This is a very common method but assumes that each factor varies independently of all the others. This is

Table 9 Drugs Measured by LC–ECD

Alkaloids	*β-Agonists and Antagonists*
Morphine, nalorphine, diacetylmorphine[114]	Terbutaline[104]
Naloxone[115]	Salbutamol[126]
Nalbuphine[116]	Fenoterol[127]
	Nadolol[128]
Analgesics	Oxprenolol[129]
Paracetamol[117]	Mepindolol[130]
Paracetamol metabolites[118]	
	Tricyclic antidepressants
Antibiotics	Imipramine, desipramine[131]
Trimethoprim[119]	Mianserin metabolites[132]
Sulfonamides[120]	
Amoxicillin[121]	*Phenothiazines*
	Chlorpromazine[133]
Anticancer compounds	Fluphenazine, perphenazine[134]
Doxorubicin[122]	
Mitomycin C[123]	*Others*
Methotrexate[124]	Captopril[135]
Cisplatin[125]	Theophylline[136]
	Progabide[137]

Table 10 Derivatization for HPLC and HPTLC

UV Detection	*Examples*
1-Fluoro-2,4-dinitrobenzene (FDNB)	Primary and secondary amines
2,4-Dinitrobenzene sulfonic acid (DNBS)	Primary and secondary amines
2,4,6-Trinitrobenzenesulfonic acid (TNBS)	Primary amines
2,4-Dinitrophenylhydrazine (DNPH)	Aldehydes, ketones and quinones
3,5-Dinitrobenzoyl chloride	Alcohols and amines
N,N-Dimethylaminoazobenzenesulfonyl chloride (DABS-Cl)	Amines
N-(7-Dimethylamino-4-methylcoumarinyl)maleimide[168]	Thiols
Naphthyldiazomethane	Carboxylic acids

Fluorescence detection	*Examples*
Dansyl chloride (Dns-Cl) and related reagents such as Bns-Cl Mns-Cl, Dis-Cl and Nbd-Cl	Primary and secondary amines, alcohols and phenols
Fluorescamine	Primary amines
o-Phthalaldehyde (OPT)	Primary amines
Benoxaprofen chloride[174]	Primary and secondary amines, alcohols and phenols
9,10-Diaminophenanthrene	
4-Bromomethyl-7-methoxycoumarin	Carboxylic acids

Resolution of optical isomers	*Examples*
(+)-10-Camphorsulfonyl chloride	Primary and secondary amines
α-Methoxy-2-methyl-1-naphthaleneacetic acid	Primary and secondary amines
(−)-1,7-Dimethyl-7-norbornyl isothiocyanate	Amines
β-Naphthyl chloroformate	Amines
α-(4-N,N-Dimethylamino-1-naphthyl)ethylamine	Carboxylic acids

Detailed information on reaction conditions and properties of derivatives can be found in refs. 68, 138 and 139, and where indicated.

often not the case and the multifactor approach is more powerful although much more complicated to apply. However, software is available to enable this procedure to be carried out and data can be more easily interpreted by plotting in three dimensions (*e.g.* pH *versus* molarity *versus* retention time). Predictive models have also been developed and a number of reviews of this topic are available.[140]

Statistical approaches have also been developed in order to optimize mobile phase composition for isocratic operation using a technique called overlapping resolution mapping (ORM).[141] Where isocratic operation fails to achieve the desired separation, gradient elution can be employed. This is technically more complex and a programmer and second pump are needed. Detailed reviews of the theory and application of gradient elution are available.[142,143]

24.2.3.3.5 *Thin layer chromatography (TLC)*

(i) *Introduction.*

This technique has been used extensively for about the last thirty years because of its simplicity and cheapness and a wide variety of stationary phases are available on plastic, aluminum foil or glass. The two former are popular because they can easily be cut to the desired size and are economical. TLC can be regarded for most purposes as a two-dimensional form of HPLC, using very much smaller volumes of mobile phase. Indeed, TLC is often used to evaluate possible mobile phases for HPLC work since many different phases can be examined simultaneously on a range of plates. Usually only minor modifications have to be introduced after the transfer to HPLC is accomplished.

TLC plates are available with different types of silica gel (for 'normal' phase work) or with bonded hydrocarbons of different chain length (C_2, C_8 and C_{18}) for reverse phase operation using aqueous mobile phases. Different plate sizes are available, the most usual being 5×20, 10×20 and 20×20 cm. Plates can be supplied with inert spotting zones so that relatively large volumes of extract can be rapidly applied. When development is started the components of the extract are swept on to the stationary phase as a very narrow band and the separation proper takes place.

TLC plates can be supplied with or without fluorescent indicator and UV absorbing drugs easily detected; alternatively the plates can be sprayed with a number of reagents to visualize different classes of compound. An extensive literature on this topic is available.[144,145]

(ii) High performance thin layer chromatography (HPTLC)

Within the last ten years plates became generally available coated with a thin layer (200 μm) of silica of a very small and carefully graded particle size (about 5 μm). This offered a much better performance in terms of resolution and speed of development, often much less than 20 minutes compared with 60–90 minutes for conventional TLC. When used in conjunction with automatic sample application and scanning densitometry for detection, the technique is very powerful indeed. It is generally cheaper and quicker than HPLC and good results can be obtained without the need for highly skilled operators.

(a) Sample application. This is very important because a very small spot size (about 1 mm) is necessary if the high resolving power of the plate is to be obtained. This is impossible to achieve consistently by hand especially if large volumes (greater than 5 μL) are to be applied. Automatic devices are available that will allow the application of volumes ranging from the sub μL to as much as 50 μL. Up to 20 samples (extracts) can be applied simultaneously and, of course, they are all developed at the same time when the plate is placed in the development chamber. This offers a very rapid means of analysis, much more rapid than either GLC or HPLC although automation can compensate for this.[146]

(b) Development. Much of the HPTLC carried out is linear and the plates are often placed in simple glass tanks in the vertical position. The dimensions of these tanks vary and the volume of vapour phase can be drastically reduced by using specially designed tanks. Overpressure Layer Chromatography (OPC) is a relatively new technique in which linear development takes place with the total elimination of a vapour phase. The speed of development is claimed to be much faster than for HPTLC and the separated compounds can be collected efficiently.[147]

Circular development (CTLC) is also possible using the U-chamber technique in which the flow rate of the mobile phase can be carefully regulated and gradient elution can be performed. An analysis time of 3 min has been claimed for some separations.[148]

(c) Detection. Once a plate has been developed the separated drugs can be quantitated using scanning densitometry in the UV or fluorescence mode and nanogram quantities of many drugs can be determined. As in HPLC, derivatization techniques can be employed for drugs that do not fluoresce or absorb strongly in the UV region.

Reviews of HPTLC are available and a summary of typical methods is given in Table 11.[148,149]

24.2.3.3.6 Combination techniques

When the separating power of a chromatograph (GLC or HPLC) is coupled to a sensitive detector this is referred to as a combination technique. Examples such as GLC–FID and HPLC–UV are, strictly speaking, combination techniques but are really integrated in a single instrument. The term 'combination' or 'hyphenated' technique is usually reserved for one of the following couplings: GC–MS, GC–IR, GC–IR–MS, LC–MS, LC–FID.

Table 11   Application of TLC to Drug Analysis

Alkaloids
General[150]
Heroin and cocaine[151]

Antiasthmatics
Albuterol,[152] salbutamol[153]

Antibiotics
General[154,155]
Ampicillin[156]
Tetracyclines[157]

Anticonvulsants
Phenobarbitone, phenytoin, primidone,
 carbamazepine[158]
Carbamazepine, mephenytoin, phenobarbitone,
 phenytoin and primidone[159]
Phenytoin[160]

Antidepressants
Amitriptyline, nortriptyline, chlorpromazine[161]
Imipramine and desipramine[161,163]
Amitriptyline and nortriptyline[162]

Cardiovascular
Lidocaine, procaine, propranolol[164]
Oxprenolol[165]
Pindolol[166]
Diltiazem[167]
Captopril[168]
Hydrochlorothiazide[169]

Cytostatics
Adriamycin[170]
Nitrosoureas (alkylating activity)[171]
Sedatives and tranquillizers
Flurazepam[172]
Barbiturates[160]
Phenobarbitone and metabolite[173]

(i) Combinations with gas chromatography (GC)

(a) GC–MS. This is extensively used to obtain structural information on compounds such as drug metabolites or to provide a very specific means of quantitation. The gas chromatograph can be used with packed columns but usually capillary columns are chosen because of the improved resolution. In brief, the technique involves passing the effluent from the capillary column into the ionization chamber of the mass spectrometer where any compounds present are bombarded to produce a series of fragments.

In order to successfully interface GC with MS, some form of separation is needed because MS operates under a high vacuum (about 10^{-6} Torr). Different types of separation are available and they make use of the fact that the molecules of carrier gas (usually helium) are much lighter than the drug molecules and can be preferentially removed (jet or glass frit separation) or retarded by a silicone membrane (molecular separation).

Following this enrichment, the drug molecules must be ionized and two techniques are used: electron impact (EI) and chemical ionization (CI). In EI the molecules of drug are bombarded with high energy electrons to produce both positive and negative ions. Short-lived ions decompose and are not detected but the longer lived ions allow the molecular weight of the drug to be obtained from the molecular ion (although not all drugs display a molecular ion). CI is much 'softer' and this process allows a molecular ion to be generated while keeping further fragmentation to a minimum. In CI the sample is mixed with a low molecular weight hydrocarbon such as methane or isobutane before being bombarded by electrons. The methane or isobutane ionizes preferentially and a variety of chemically reactive species are produced which react with the drug molecules transferring a proton to give a molecular ion one unit greater than the drug itself.

CI is useful for obtaining the molecular weight of a drug or metabolite but the molecular ion is often too stable to fragment further and hence little structural information is obtained. EI is to be preferred when structural data is needed and a number of instruments are designed to operate in either mode so that the maximum amount of information can be obtained.

The ions generated are separated using a quadrupole filter or a magnetic sector, then detected and this signal amplified. This data is displayed on an oscilloscope and a permanent copy generated using the appropriate software.

Further refinements of GC–MS, such as ion trapping, are also available and reviews have been published.[175-177]

(b) GC–IR. Although this combination can be used for quantitative work, with detection at a single wavelength, its most useful application is to provide structural information on drugs and their metabolites. Both off- and on-line procedures have been developed the former being more time consuming and, on the whole, less attractive. The latter technique usually uses some form of heated glass light pipe connected to the GC by a heated transfer line. The pipe is about 20–100 cm in length with an internal volume of several mL. At both ends of the pipe are windows of KBr or Zn selenide; acceptable spectra can be obtained from quantities of between 10–1000 ng using Fourier Transform IR.[178]

(c) GC–IR–MS. The prospect of combining these three very powerful and complementary techniques has always been attractive to analytical and medicinal chemists and was suggested over 20 years ago.[179] Only recently, however, have the three been successfully integrated. The method of interfacing is outside the scope of this review but has been described in some detail using examples drawn from the analysis of peppermint oil (18 components) and by even more complex mixtures of up to 45 components.[180] With improvements in the limit of detection of IR, one can expect the technique to become more widely used particularly in difficult areas such as the unequivocal identification of novel substances (*e.g.* analogues of fentanyl and amphetamine) synthesized in attempts to circumvent existing drug laws.

(ii) Combinations with HPLC

(a) HPLC–MS. HPLC is frequently used to measure a wide range of drugs some of which can also be determined by GLC. Its strength lies in its ability to separate very polar, high molecular weight and thermally unstable compounds. The combination with MS is, therefore, an attractive one.

Although off-line techniques are used, operation on-line is more versatile. The major problem to be overcome is the removal of the large volume of mobile phase (often as much as 2 mL). The techniques used are the moving belt system or continuous sample preconcentration. Microbore columns have also been used to reduce the volume of mobile phase to be introduced into the MS.

Thermospray injection is frequently used especially for compounds of low volatility. The column effluent is first subjected to heat and a very low pressure then the vaporized material is passed through a device called a skimmer to the ion source.

A number of reviews of this combination are available.[181,182]

(b) HPLC–FID. Until recently the only universal detector for HPLC work was the refractive index detector which is very insensitive. FID, while not truly 'universal', provides a very versatile detector indeed since it responds to a large number of sample types and is linear over a very wide concentration range. It is particularly useful for drugs which possess no natural fluorescence or strong UV absorption and which cannot be readily derivatized (*e.g.* simple tertiary amines).

The main problem to be overcome with this combination is the obvious one of removing large volumes of mobile phase before the eluent can be passed to the detector. The approach most frequently adopted is to allow the column eluate to be applied to some form of moving belt, usually made of quartz fibre or metal wire. The mobile phase is removed from the belt by a current of heated air or by heating in an oven and the belt then carries the non-volatile components to the FID where they burn in a hydrogen flame to produce ions, detected as an increased current. The system can be operated using a very wide range of volatile solvents such as hydrocarbons, lower alcohols, halogenated hydrocarbons, diethyl ether, ethyl acetate and even aqueous acids and alkalis. The best results are obtained using non-polar solvents.

Memory effects have been reported using metal wire but better results seem to be obtained from quartz fibre belts.[183]

(c) HPLC–FTIR. Fourier transform infrared spectrometers are much more sensitive than the earlier prism or grating instruments and reasonable IR spectra can be obtained for compounds as they elute from the column. The technique is limited at present by the need to use mobile phases with large 'windows' transparent to IR, *e.g.* chloroform. Mobile phases containing water or other polar solvents cannot be used and this means that reverse phase columns are unsuitable. Work is being carried out in this area and improvements will, no doubt, be made.[184,185]

24.2.4 SPECIAL TOPICS

24.2.4.1 Separation of Optical Isomers

A large number of drug molecules contain at least one optically active centre and usually only one isomer is responsible for the pharmacological effect of the drug. From a manufacturing point of view, however, it is not usually cost effective to resolve the isomers and formulate the drug in an optically pure form. In some cases, unfortunately, it has been shown that the 'inactive' isomer contributes to, or is entirely responsible for, the major side effects of the drug. The classic case is that of thalidomide, where it has been shown that the $R(+)$ isomer was responsible for the desired sleep-inducing effect while the $S(-)$ isomer was teratogenic.[186]

For this reason a great deal of interest is being shown in analytical methods which will allow the optical isomers to be separated and measured in pharmacokinetic studies. Resolution of optical isomers can best be achieved on an analytical scale by chromatographic means and TLC, GLC and HPLC have all been used. The racemic drug mixture can be reacted with a suitable derivatizing agent to give diastereoisomers that can be separated and quantitated on conventional stationary phases. More information can be obtained in the Section 24.2.3.3.3vi and 24.2.3.3.4v.

Alternatively, chiral phases may be used to separate the isomers directly. This approach is receiving a great deal of attention because it is more efficient. Among the chiral phases used are the Pirkle (or Bakerbond) type, α-1-acid glycoprotein and bovine serum albumin.[187,188]

It is also possible to attach a polarimeter to a concentration detector to quantify the amount of each isomer without the need to achieve a separation.[189]

24.2.4.2 Automation and Data Processing

Most large laboratories have introduced at least some degree of automation to the chemical analysis of drugs and, increasingly, smaller laboratories are following their example. It is now relatively straightforward to control a separation process, such as GLC[190] or HPLC,[191] by computer. The output from the detector can also be processed by computer and displayed on a suitable monitor. This is particularly useful for combinations such as GC–MS. Some laboratories have dispensed with the conventional chart recorder and integrator and their entire output is stored and manipulated by computer.

Extraction of drugs and metabolites from body fluids is more difficult to automate but, even here, advances have been made particularly in the field of solid phase extraction.[18]

24.2.4.3 Quality Control

When any analytical method is set up, a number of steps should be taken to validate it and to monitor its performance, either continuously or at regular intervals.

The detector response should be examined using a standard solution of the compound to be determined. The limit of detection should also be established. Standards should be made up covering the range of concentrations expected and the concentration of internal standard (if one is to be used) chosen. These should be added in the appropriate concentrations to drug-free biological fluid (*e.g.* plasma) and extracted using the experimental conditions described by the method. As a general rule, five different concentrations, spanning the range expected, should be chosen. Additionally a number of replicates (*e.g.* 6) at a high and low concentration within the expected range should be determined. This will provide essential information about the coefficient of variation at these concentrations.

The least squares regression line of concentration *versus* detector response can be calculated. If the method being set up does not contain data on optimization of extraction, reaction conditions, *etc.*, then this should be obtained by experiment.

Standard samples should be made up using blank plasma, frozen and stored at $-20\,^{\circ}\text{C}$. These should be analyzed at regular intervals to monitor how long samples may be kept before analysis without significant deterioration. It is better to make up a series of such samples in order to avoid repeated thawing and freezing of a single sample.

Each time a series of samples is to be analyzed, standards and quality control specimens should be incorporated. The number of standards will depend on the linearity of the calibration curve but will usually be in the region of 4–6 with a high and low quality control. Naturally each laboratory will wish to adopt its own system. For drugs which are analyzed in large numbers in many different laboratories there may well be a regional, national or supranational quality control scheme. It is highly recommended that laboratories participate in such schemes. In the United Kingdom, for example, the Heath Control Scheme operates from Cardiff and offers schemes for anticonvulsants and a new scheme for opiates and sedatives is being introduced.

One of the strengths of interlaboratory control schemes is that unsatisfactory methods are rapidly identified and can be discarded in favour of more reliable ones. The identity of each laboratory in these schemes is protected from the others but each laboratory sees how it performs in relation to the others. This is usually demonstrated in histogram form with the mean value for all laboratories and the individual laboratory result being clearly indicated on every report issued for a particular analyte. This introduces a healthy element of competition.

There are a number of good introductions to the subject of quality control and its working in the clinical environment.[192–195]

24.2.5 REFERENCES

1. D. B. Jack, S. Dean and M. J. Kendall, *J. Chromatogr.*, 1980, **187**, 277.
2. D. B. Jack, S. Dean, M. J. Kendall and S. J. Laugher, *J. Chromatogr.*, 1980, **196**, 189.
3. J. A. F. de Silva, *J. Chromatogr.*, 1985, **340**, 3.
4. E. Reid, in 'Assay of Drugs and other Trace Compounds in Biological Fluids', ed. E. Reid, North Holland, Amsterdam, 1976, p. 55.
5. R. G. Cooper, *Methodol. Surv. Biochem.*, 1978, **7**, 1.
6. W. J. Dunn, III, J. H. Block and R. S. Pearlman (eds.), 'Partition Coefficient Determination and Estimation', Pergamon Press, Oxford, 1986.
7. G. Schill, *Ion Exch. Solvent Extr.*, 1974, **6**, 1.
8. G. Schill, R. Modin, K. O. Borg and B. -A. Persson, in 'Handbook of Derivatives for Chromatography', ed. K. Blau and G. King, Heyden, London, 1977, p. 500.
9. G. Schill, K. O. Borg, R. Modin and B. -A. Persson, *Prog. Drug Metab.*, 1977, **2**, 219.
10. I. D. Watson and M. J. Stewart, *J. Chromatogr.*, 1977, **134**, 182.
11. P. M. Edelbroek, E. J. M. de Haas and F. A. de Wolff, *Clin. Chem. (Winston-Salem, N.C.)*, 1982, **28**, 2143.
12. J. Meola and M. Vanko, *Clin. Chem. (Winston-Salem, N.C.)*, 1974, **20**, 184.
13. L. P. Hackett and L. J. Dusci, *J. Forensic Sci.*, 1977, **22**, 376.
14. V. P. Dole, W. K. Lim and I. Eglitis, *J. Am. Med. Assoc.*, 1966, 349.
15. W. J. Dekker, H. F. Combs and D. G. Corby, *Toxicol. Appl. Pharmacol.*, 1968, **13**, 454.
16. J. M. Fujimoto and R. I. H. Wang, *Toxicol. Appl. Pharmacol.*, 1970, **16**, 186.
17. L. B. Hetland, D. A. Knowlton and D. Couri, *Clin. Chim. Acta*, 1972, **36**, 473.
18. R. D. McDowall, J. C. Pearce and G. S. Harkitt, *J. Pharm. Biomed. Anal.*, 1986, **4**, 3.

19. A. H. Beckett and J. B. Stenlake, 'Practical Pharmaceutical Chemistry', 3rd edn., Part 1, Athlone Press, London, 1975, p. 272.
20. K. A. Connors, 'Textbook of Pharmaceutical Analysis', Wiley, Chicester, 1982, p. 3.
21. T. Higuchi and E. Brockmann-Hanssen, 'Pharmaceutical Analysis', Interscience, New York, 1961, p. 382.
22. A. H. Beckett and J. B. Stenlake, 'Practical Pharmaceutical Chemistry', 3rd edn., Part 1, Athlone Press, London, 1975, p. 126.
23. H. M. Abdou, in 'Remington's Pharmaceutical Sciences', ed. A. R. Gennaro, Mack Publishing Co., Easton, 1985, p. 619.
24. A. M. Boyd, 'Modern Polarographic Methods in Analytical Chemistry', Dekker, New York, 1980.
25. R. F. Burns and L. J. Russell, in 'Clinical Biochemistry Nearer the Patient', ed. V. Marks and K. G. M. M. Alberti, Churchill Livingstone, New York, 1985, p. 121.
26. Y. Shouzhuo and N. Lihua, *Anal. Proc.*, 1987, **24**, 338.
27. S. G. Schulman and B. S. Vogt, in 'Pharmaceutical Analysis Modern Methods', Part B, ed. J. W. Munson, Dekker, New York, 1981, p. 401.
28. A. F. Fell, D. R. Jarvie and M. J. Stewart, *Clin. Chem. (Winston-Salem, N.C.)*, 1981, **27**, 286.
29. A. F. Fell, in 'Clarke's Isolation and Identification of Drugs', ed. A. C. Moffat, 2nd edn., The Pharmaceutical Press, London, 1986, p. 221.
30. A. H. Beckett and J. B. Stenlake, 'Practical Pharmaceutical Chemistry' 3rd edn., Part 2, Athlone Press, London, 1975, p. 311.
31. J. C. P. Schwartz, in 'Physical Methods in Organic Chemistry', Oliver & Boyd, Edinburgh, 1964, p. 35.
32. D. I. Chapman, in 'Clarke's Isolation and Identification of Drugs', ed. A. C. Moffat, 2nd edn., The Pharmaceutical Press, London, 1986, p. 237.
33. L. J. Bellamy, 'Advances in Infrared Group Frequencies', 3rd edn., Chapman & Hall, London, 1975.
34. A. H. Beckett and J. B. Stenlake, 'Practical Pharmaceutical Chemistry', 3rd edn., Part 2, Athlone Press, London, 1975, p. 331.
35. J. R. During (ed.), 'Chemical, Biological and Industrial Applications of Infrared Spectroscopy', Wiley, Chicester, 1985.
36. S. G. Schulman and W. R. Vincent, in 'Pharmaceutical Analysis Modern Methods', ed. J. W. Munson, Dekker, New York, 1981, p. 359.
37. L. S. Ettre and A. Zlatkis (eds.), '75 Years of Chromatography', *J. Chromatogr. Libr.*, 1979, **17**.
38. C. F. Poole and S. A. Schuette, 'Contemporary Practice of Chromatography', Elsevier, Amsterdam, 1984, p. 1.
39. R. L. Grob, in 'Modern Practice of Gas Chromatography', Wiley, Chicester, 2nd edn., 1985, p. 49.
40. M. A. Kaiser, in 'Modern Practice of Gas Chromatography', ed. R. L. Grob, 2nd edn., Wiley, Chichester, 1985, p. 159.
41. W. G. Jennings, in 'Glass Capillary Chromatography in Clinical Medicine and Pharmacology', ed. H. Jaeger, Dekker, New York, 1985, p. 33.
42. D. B. Jack, 'Drug Analysis by Gas Chromatography', Academic Press, New York, 1984, p. 1.
43. W. R. Supina, in 'Modern Practice of Gas Chromatography', ed. R. L. Grob, 2nd edn., Wiley, Chichester, 1985, p. 117.
44. E. Kovts, *Helv. Chim. Acta*, 1958, **41**, 1915.
45. L. Rohrschneider, *J. Chromatogr.*, 1966, **22**, 6.
46. W. O. McReynolds, *J. Chromatogr. Sci.*, 1970, **8**, 685.
47. A. C. Moffat, A. H. Stead and K. W. Smalldon, *J. Chromatogr.*, 1974, **90**, 19.
48. R. E. Ardrey, 'Gas Chromatographic Retention Indices of Toxicologically Relevant Substances on SE-30 and OV-1', Report II of the DFG Commission for Clinical Toxicological Analysis, Special Issue of the TIAFT Bulletin, VCH Verlagsgesellschaft, New York, 1985.
49. W. Dunges, in 'Glass Capillary Chromatography in Clinical Medicine and Pharmacology', ed. H. Jaeger, Dekker, New York, 1985, p. 599.
50. A. G. de Boer, N. P. E. Vermeulen and D. D. Breimer, in 'Glass Capillary Chromatography in Clinical Medicine and Pharmacology', ed. H. Jaeger, Dekker, New York, 1985, p. 607.
51. W. G. Jennings and A. Rapp, in 'Sample Preparation for Gas Chromatographic Analysis', Huethig, Heidelberg, 1983, p. 5.
52. R. Schill and R. R. Freeman, in 'Modern Practice of Gas Chromatography', ed. R. L. Grob, 2nd edn., Wiley, Chichester, 1985, p. 313.
53. C. F. Poole and S. A. Schuette, 'Contemporary Practice of Chromatography', Elsevier, Amsterdam, 1984, p. 145.
54. B. Kolb (ed.), 'Applied Headspace Gas Chromatography', Heyden, London, 1980.
55. J. Sevcik, *J. Chromatogr. Libr.*, 1976, **4**.
56. M. Dressler, *J. Chromatogr. Libr.*, 1986, **36**.
57. M. Zell and U. Timm, *J. Chromatogr.*, 1986, **382**, 175.
58. M. J. O'Brien, in 'Modern Practice of Gas Chromatography', ed. R. L. Grob, 2nd edn., Wiley, Chichester, 1985, p. 211.
59. A. Karmen, *Adv. Chromatogr.*, 1966, **2**, 293.
60. A. Zlatkis and C. F. Poole (eds.), *J. Chromatogr. Libr.*, 1981, **20**.
61. M. J. O'Brien, in 'Modern Practice of Gas Chromatography', ed. R. L. Grob, Wiley, Chichester, 2nd edn., 1985, p. 225.
62. C. F. Poole and S. A. Schuette, 'Contemporary Practice of Chromatography', Elsevier, Amsterdam, 1985, p. 181.
63. D. H. Smith, *Adv. Chromatogr.* 1975, **12**, 177.
64. I. S. Krull, M. E. Swartz and J. N. Driscoll, *Adv. Chromatogr.*, 1984, **24**, 247.
65. C. E. R. Jones and C. A. Cramers (eds.), 'Analytical Pyrolysis', Elsevier, Amsterdam, 1977.
66. W. J. Irwin, 'Analytical Pyrolysis: A Comprehensive Guide', Dekker, New York, 1982.
67. R. M. Smith (ed.), 'Supercritical Fluid Chromatography', Royal Society of Chemistry, London, 1988.
68. K. Blau and G. S. King, 'Handbook of Derivatives for Chromatography', Heyden, London, 1977.
69. D. B. Jack, in 'Drug Analysis by Gas Chromatography', Academic Press, New York, 1984, p. 19.
70. J. Drozd, *J. Chromatogr. Libr.*, 1981, **19**.
71. J. F. Nash, R. J. Bopp and A. Rubin, *J. Pharm. Sci.*, 1971, **60**, 1062.
72. D. G. Ferry, D. M. Ferry, P. W. Moller and E. G. McQueen, *J. Chromatogr.*, 1974, **89**, 110.
73. D. G. Kaiser, S. R. Shaw and G. J. Vangiessen, *J. Pharm. Sci.*, 1974, **63**, 567.
74. H. V. Street, *Clin. Chim. Acta*, 1971, **34**, 357.
75. W. Dunges, *Anal. Chem.*, 1973, **45**, 963.

76. T. Walle, *J. Chromatogr.*, 1975, **114**, 345.
77. C. W. Gehrke and K. Leimer, *J. Chromatogr.*, 1971, **57**, 219.
78. C. W. Gehrke and H. Takeda, *J. Chromatogr.*, 1973, **76**, 63.
79. C. W. Gehrke, K. Kuo and R. W. Zumwalt, *J. Chromatogr.*, 1971, **57**, 209.
80. S. Omoto, S. Inouye and T. Niida, *J. Antibiot.*, 1971, **24**, 430.
81. J. G. Kelly, J. Nimmo, R. Rae, R. G. Shanks and L. F. Prescott, *J. Pharm. Pharmacol.*, 1973, **25**, 550.
82. R. B. Bruce and W. R. Maynard, Jr., *Anal. Chem.*, 1969, **41**, 977.
83. G. R. Wilkinson and E. Leong Way, *Biochem. Pharmacol.*, 1969, **18**, 1435.
84. D. B. Jack and W. Riess, *J. Pharm. Sci.*, 1973, **62**, 1929.
85. N. Sakauchi and E. C. Horning, *Anal. Lett.*, 1971, **4**, 41.
86. P. M. Simpson, *J. Chromatogr.*, 1973, **77**, 161.
87. S. J. Gaskell, C. G. Edmonds and C. J. W. Brooks, *Anal. Lett.*, 1976, **9**, 325.
88. D. Gupta, E. Breitmaier, G. Jung, G. von Lucadon, H. Pauschmann and W. Voeller, *Chromatographia*, 1971, **4**, 572.
89. H. V. Street, *J. Chromatogr.*, 1972, **73**, 73.
90. P. Hartvig and J. Vessman, *Acta Pharm. Suecica*, 1974, **11**, 115.
91. M. Shamma, N. C. Deno and J. F. Remar, *Tetrahedron Lett.*, 1966, **13**, 1375.
92. J. H. Hengstmann, F. C. Falkner, J. T. Watson and J. Oates, *Anal. Chem.*, 1974, **46**, 34.
93. D. B. Jack, S. Brechbuhler, P. H. Degen, P. Zbinden and W. Riess, *J. Chromatogr.*, 1975, **115**, 87.
94. J. A. F. De Silva, C. V. Puglisi and N. Munno, *J. Pharm. Sci.*, 1974, **63**, 520.
95. J. B. Stenlake, W. D. Williams and G. G. Skellern, *J. Chromatogr.*, 1970, **53**, 285.
96. L. R. Snyder and J. J. Kirkland, 'Introduction to Modern Liquid Chromatography', 2nd edn., Wiley, Chichester, 1979.
97. C. Horvath (ed.), 'High Performance Liquid Chromatography, Advances and Perspectives', Academic Press, New York, 1980.
98. H. Engelhardt, 'High Performance Liquid Chromatography', Springer Verlag, Berlin, 1979.
99. A. M. Krstulovic and P. R. Brown, 'Reversed-Phase High Performance Liquid Chromatography: Theory, Practice and Biomedical Applications', Chichester, Wiley, 1982.
100. J. S. Fritz, D. T. Gjerde and C. Pohlandt, 'Ion Chromatography', Huethig, Heidelberg, 1982.
101. T. Provdor (ed.), 'Size-exclusion Chromatography', American Chemical Society, Washington, 1980.
102. E. Grusshka (ed.), 'Bonded Stationary Phases in Chromatography', Ann Arbor Science, Ann Arbor, 1974.
103. R. P. W. Scott, *Adv. Chromatogr.*, 1983, **22**, 247.
104. L. -E. Edholm and B. -M. Kennedy, *Chromatographia*, 1982, **16**, 341.
105. C. F. Poole and S. A. Schuette, 'Contemporary Practice of Chromatography', Elsevier, Amsterdam, 1985, p. 353.
106. R. J. Kelsey and C. R. Loscombe, *Chromatographia*, 1979, **12**, 713.
107. R. P. W. Scott, *J. Chromatogr. Libr.*, 1986, **33**.
108. T. Amita, M. Ichise and T. Kojima, *J. Chromatogr.*, 1982, **234**, 89.
109. A. T. Rhys Williams, 'Fluorescence Detection in Liquid Chromatography', Perkin Elmer, Beaconsfield, 1980.
110. R. Weinberger, in 'Therapeutic Drug Monitoring and Toxicology by Liquid Chromatography', ed. S. H. Y. Wong, Dekker, New York, 1985, p. 151.
111. T. H. Ryan (ed.), 'Electrochemical Detectors, Fundamental Aspects and Analytical Applications', Plenum, London, 1984.
112. C. Lavrich and P. T. Kissinger, in 'Therapeutic Drug Monitoring and Toxicology by Liquid Chromatography', ed. S. H. Y. Wong, Dekker, New York, 1985, p. 191.
113. A. M. Krstulovic, H. Colin and G. A. Guiochon, *Adv. Chromatogr.*, 1984, **24**, 83.
114. B. Proksa and L. Molnar, *Anal. Chim. Acta*, 1978, **97**, 149.
115. C. L. Lake, C. A. DiFazio, E. N. Duckworth, J. C. Moscicki, J. S. Engle and C. G. Durbin, *J. Chromatogr.*, 1982, **233**, 410.
116. J. E. Wallace, S. C. Harris and M. W. Peck, *Anal. Chem.*, 1980, **52**, 1328.
117. D. J. Miner and P. T. Kissinger, *J. Pharm. Sci.*, 1979, **68**, 96.
118. J. M. Wilson, J. T. Slattery, A. J. Forte and S. D. Nelson, *J. Chromatogr.*, 1982, **227**, 453.
119. L. Nordholm and L. Dalgaard, *J. Chromatogr.*, 1982, **233**, 427.
120. M. A. Alawi and H. A. Ruessel, *Chromatographia*, 1981, **14**, 704.
121. M. A. Brooks, M. R. Hackman and D. J. Mazzo, *J. Chromatogr.*, 1981, **210**, 531.
122. J. A. Sinkule, C. Akpofure and W. E. Evans, *Curr. Separations*, 1982, **4**, 4.
123. U. R. Tjaden, J. P. Langenberg, K. Ensing, W. P. Van Bennekom, E. A. De Brujn and A. T. Van Oosterom, *J. Chromatogr.*, 1982, **232**, 355.
124. J. Lankelma and H. Poppe, *J. Chromatogr.*, 1978, **149**, 587.
125. I. S. Krull, X. D. Ding, S. Braverman, C. Selavka, F. Hochberg and L. A. Sternson, *J. Chromatogr. Sci.*, 1983, **21**, 166.
126. B. Oosterhuis and C. J. Van Boxtel, *J. Chromatogr.*, 1982, **232**, 327.
127. S. Bergquist and L. E. Edholm, *J. Liq. Chromatogr.*, 1983, **6**, 559.
128. W. Krause, *J. Chromatogr.*, 1980, **181**, 67.
129. M. Gregg, *Chromatographia*, 1986, **21**, 705.
130. S. Y. Chu, *J. Pharm. Sci.*, 1978, **67**, 1623.
131. R. F. Suckow and T. B. Cooper, *J. Pharm. Sci.*, 1981, **70**, 257.
132. R. F. Suckow, T. B. Cooper, F. M. Quitkin and J. W. Stewart, *J. Pharm. Sci.*, 1982, **71**, 889.
133. J. K. Cooper, G. McKay and K. K. Midha, *J. Pharm. Sci.*, 1983, **72**, 1259.
134. U. R. Tjaden, J. Lankelma, H. Poppe and R. G. Muusze, *J. Chromatogr.*, 1976, **125**, 275.
135. D. Perrett and P. J. Drury, *J. Liq. Chromatogr.*, 1982, **5**, 97.
136. M. S. Breenberg and W. J. Mayer, *J. Chromatogr.*, 1979, **169**, 321.
137. W. Yanekawa, H. J. Kupferberg and T. Lambert, *J. Chromatogr.*, 1983, **276**, 103.
138. J. F. Lawrence and R. W. Frei, *J. Chromatogr. Libr.*, 1976 **7**.
139. K. Imai, *Adv. Chromatogr.*, 1987, **27**, 215.
140. S. N. Deming, J. G. Bower and K. D. Bower, *Adv. Chromatogr.*, 1984, **24**, 35.
141. H. J. Issaq, *Adv. Chromatogr.*, 1984, **24**, 55.
142. C. Liteaunu and S. Gocan, 'Gradient Elution Chromatography', Wiley, Chichester, 1974.

143. P. Jandera and J. Churacek, 'Gradient Elution in Column Liquid Chromatography', Elsevier, Amsterdam, 1985.
144. H. M. Stevens, in 'Clarke's Isolation and Identification of Drugs', ed., A. C. Moffat, 2nd edn. The Pharmaceutical Press, London, 1986, p. 128.
145. E. Stahl (ed.), 'Thin-Layer Chromatography', 2nd edn., Springer Verlag, Berlin, 1969.
146. R. E. Kaiser, *J. Chromatogr. Libr.* 1977, **9**, 85.
147. E. Tyihak, E. Mincsovics and H. Kalasz, *J. Chromatogr.*, 1979, **174**, 75.
148. R. E. Kaiser, *J. Chromatogr. Libr.*, 1977, **9**, 73.
149. W. Bertsch, S. Hara, R. E. Kaiser and A. Zlatkis (eds.), 'Instrumental HPTLC', Huethig, Heidelberg, 1980.
150. A. Baerheim Svendsen and R. Verpoorte, *J. Chromatogr. Libr.*, 1983, **23A**.
151. E. Della Casa and G. Martone, *Forensic Sci. Int.*, 1986, **32**, 117.
152. F. Plavsic, *Clin. Chem. (Winston-Salem, N.C.)*, 1981, **27**, 771.
153. P. V. Colthup, F. A. A. Dallas, D. A. Saynor, P. F. Carey, L. F. Skidmore and L. E. Martin, *J. Chromatogr.*, 1985, **345**, 111.
154. G. H. Wagman and M. J. Weinstein, *J. Chromatogr. Libr.*, 1984, **26**.
155. A. Aszalos (ed.), 'Modern Analysis of Antibiotics', Dekker, New York, 1986.
156. A. Mrhar and F. Kozjek, *J. Chromatogr.*, 1983, **277**, 251.
157. H. Oka, K. Uno, K. -I. Harada and M. Suzuki, *J. Chromatogr.*, 1984, **284**, 227.
158. C. M. Davis and D. C. Fenimore, *J. Chromatogr.*, 1981, **222**, 265.
159. N. Wad, E. Weidkuhn and H. Rosenmund, *J. Chromatogr.*, 1980, **183**, 387.
160. U. Breyer and D. Villumsen, *J. Chromatogr.*, 1975, **115**, 493.
161. D. C. Fenimore, C. J. Meyer, C. M. Davis, F. Hsu and A. Zlatkis, *J. Chromatogr.*, 1977, **142**, 399.
162. P. Haefelfinger, *J. Chromatogr.*, 1978, **145**, 445.
163. N. Sistovaris, E. E. Dagrosa and A. Keller, *J. Chromatogr.*, 1983, **277**, 273.
164. K. Y. Lee, D. Nurok, A. Zlatkis and A. Karmen, *J. Chromatogr.*, 1978, **158**, 403.
165. M. Schaefer and E. Mutschler, *J. Chromatogr.*, 1979, **164**, 247.
166. H. Spahn, M. Prinoth and E. Mutschler, *J. Chromatogr.*, 1985, **342**, 458.
167. K. Kohno, Y. Takeuchi, A. Etoh and K. Noda, *Arzneim.-Forsch.*, 1977, **27**, 1424.
168. W. Cawello and R. Bonn, *Fresenius Z. Anal. Chem.*, 1987, **327**, 29.
169. M. Schaefer, H. E. Geissler and E. Mutschler, *J. Chromatogr.*, 1977, **143**, 615.
170. E. Watson and K. K. Chan, *Cancer Treat. Rep.*, 1976, **60**, 1611.
171. M. Asami, K. -I. Nakamura, K. Kawada and M. Tanaka, *J. Chromatogr.*, 1979, **174**, 216.
172. J. A. F. de Silva, I. Bekersky and C. V. Puglisi, *J. Pharm. Sci.*, 1974, **63**, 1837.
173. J. S. Levin, M. F. Schwartz, D. Y. Cooper and J. C. Touchstone, *J. Chromatogr.*, 1978, **154**, 349.
174. H. Spahn, H. Weber and E. Mutschler, *J. Chromatogr.*, 1984, **310**, 167.
175. J. Throck Watson, 'Introduction to Mass Spectrometry; Biomedical, Environmental and Forensic Applications', Raven Press, New York, 1976.
176. B. J. Gudzinowicz, M. J. Gudzinowicz and H. F. Martin, 'Fundamentals of Integrated Gas Chromatography–Mass Spectrometry', Dekker, New York, 1976, Part II, and 1977, Part III.
177. C. F. Poole and S. A. Schuette, 'Contemporary Practice of Chromatography', Elsevier, Amsterdam, 1984, p. 583.
178. M. D. Erickson, *Appl. Spectrosc. Rev.*, 1979, **15**, 261.
179. M. J. D. Low and J. K. Freeman, *J. Agric. Food Chem.*, 1968, **16**, 525.
180. C. L. Wilkins, *Anal. Chem.*, 1987, **59**, 571A.
181. C. Eckers and J. Henion, in 'Therapeutic Drug Monitoring and Toxicology by Liquid Chromatography', ed. S. H. Y. Wong, Dekker, New York, 1985, p. 115.
182. D. E. Grimes, *Adv. Chromatogr.*, 1983, **21**, 1.
183. V. L. McGuffin and M. Novotny, *J. Chromatogr.*, 1981, **218**, 179.
184. D. W. Vidrine, *J. Chromatogr. Sci.*, 1979, **17**, 477.
185. C. Combellas, H. Bayart, B. Jasse, M. Caude and R. Rosset, *J. Chromatogr.*, 1983, **259**, 211.
186. G. Blaschke, H. P. Kraft, K. Fickentscher and F. Kohler, *Arzneim.-Forsch.*, 1979, **29**, 1640.
187. R. W. Souter, 'Chromatographic Separation of Stereoisomers', CRC Press, Boca Raton, FL, 1985.
188. R. Dennis, *Pharm. Int.*, 1986, **7**, 246.
189. J. L. Di Cesare and L. S. Ettre, *J. Chromatogr.*, 1982, **251**, 1.
190. R. Schill and R. R. Freeman, in 'Modern Practice of Gas Chromatography', ed. R. L. Grob, Wiley, New York, 1985, p. 294.
191. K. F. Scott, in 'Therapeutic Drug Monitoring and Toxicology by Liquid Chromatography', ed. S. H. Y. Wong, Dekker, New York, 1985, p. 89.
192. T. P. Whitehead, 'Quality Control in Clinical Chemistry', Wiley, Chichester, 1977.
193. D. Burnett and J. Williams, in 'Clarke's Isolation and Identification of Drugs', ed. A. C. Moffat, The Pharmaceutical Press, London, 1986, p. 118.
194. T. F. Hartley, 'Computerized Quality Control', Ellis Horwood, Chichester, 1987.
195. A. Richens, in 'Clinical Pharmacology of Antiepileptic Drugs', ed. H. Schneider, Springer Verlag, Berlin, 1975, p. 293.

24.3

Biological Analysis

JOSEPH CHAMBERLAIN

Formerly of *Merck Sharp & Dohme, Hoddesdon, UK*

24.3.1   INTRODUCTION

The great biologist Sir Henry Dale once remarked that the ultimate goal of a biological assay was to do away with the need for biological assays. Thus, the biological or pharmacological activity of a crude extract of plant or animal tissue would be the first method of quantifying the active principle. As the preparation was purified in an attempt to isolate this principle, the biological assay would still be the only means of following the desired fraction, and would still be extensively used. Eventually, however, there would come a point when the pure chemical itself was isolated and identified and the better-defined physical or chemical properties of this compound would replace the biological assay as the method of identification and measurement.

Sir Henry was undoubtedly correct and there are very few pharmaceutical preparations that have their drug content specified in 'mouse units' or 'international units'. Bioassay, in the old sense of the word, is thus not commonplace in the analytical armoury of medicinal chemists. However, along with the advances in chemical determination of active drugs there has also been a parallel development and understanding of the action of drugs at the molecular level. A bioassay is no longer limited to a gross effect on the whole animal or isolated tissue, and the attendant uncertainties introduced by individual variation, multiple effects and a host of unknown factors, such as the behaviour of the drug as it interacts with precisely defined biological systems, such as enzymes or receptors, can be determined and utilized to develop analytical methods, which are accepted as bioassays, yet may be superior to chemical methods in sensitivity and specificity.

This chapter will deal with bioassays that find particular use in medicinal chemistry. For interactions at the molecular level, where the exact shape of the drug molecule as it binds to a specific biological macromolecule is crucial, these methods can be highly specific. In addition, the biological effect is the result of much amplification of the effect of a small number of molecules—nature's propensity for economy—and the methods can also be highly sensitive.

24.3.2   MICROBIOLOGICAL ASSAY

24.3.2.1   Background to Microbiological Methods

One does not have to be a bacteriologist, or even scientifically trained, to recognize that microorganisms, although invisible to the naked eye, become all too visible when allowed to grow and multiply in a favourable environment. The furry forgotten tomato, the mouldy wallpaper in a damp room attest to the presence of bacteria. Bacteriologists identify microorganisms in test samples by 'culturing' them with appropriate nutrients to the point where the 'colony' becomes visible. The characteristic native colours, or the colours appearing with particular staining agents, help in identification and classification.

Alexander Fleming was such a bacteriologist, studying and classifying the various pathological streptococci in his small laboratory at St. Mary's Hospital. In September 1928, so legend now has it, Fleming returned from a prolonged holiday to discover one of his culture plates had been spoiled by a green mould, presumably caused by a contaminating airborne spore. He was just about to throw away the useless plate when he noticed that around the edges of the green mould the streptococci had failed to grow. The mould was producing something that killed the streptococci. The 'something' turned out to be penicillin, the first and best known of the β-lactam antibiotics. This serendipitous discovery required much further work to establish it as one of the greatest medical discoveries of all time but, nonetheless, the basic observation was that the growth of a colony of bacteria was halted or retarded in the presence of a certain class of drug.

It was not long before it was established that the growth could be completely halted only when the drug was present in a sufficiently high concentration. For a given drug this required concentration—known as the minimum inhibitory concentration, or MIC—is different for different bacteria and, conversely, the same bacterium will exhibit different MICs for different drugs.

As the experimental conditions in determining MICs may be slightly different from laboratory to laboratory, it is not possible to quote absolute values for specific organisms and specific antibiotics. Table 1, used as an example, shows selected data from one laboratory only: more comprehensive data can be found in literature reviews.[1,2] The range of MICs for a particular antibiotic against a series of bacteria is often, unscientifically perhaps, referred to as its spectrum of activity. Hence, a drug may be a broad-spectrum antibiotic, effective against most organisms, or a narrow-spectrum antibiotic, particularly effective against individual organisms and useful in selective therapy where the infecting agent has been identified.

The ability of different drugs to inhibit bacterial growth and the determination of MICs against unknowns can also be used to identify and characterize test samples. From this quantitative use of microbiology, it is a logical step to use microbiological systems to assay the concentration of the antibiotic itself in test samples. Thus, the unknown, at different dilutions, is incubated with the appropriate organism to find the dilution threshold at which inhibition occurs; the original concentration can be determined by simple arithmetic. In general, microbiological assays can be made sensitive by choosing an organism which is the most sensitive to the particular antibiotic, and they can be made specific by choosing an organism that is uniquely sensitive to the antibiotic.

It should also be noted that the use of microbiological assays is not confined to antibiotics but can be used to assay any drug which inhibits the growth of the organism, such as antitumour agents.[3] Microbiological assays can also be used to assay vitamins, which are essential components of the nutrient medium.[4]

In general, two types of assay are popular for the microbiological assay of antibiotics, the solid phase or zone of inhibition method, and the serial dilution liquid phase method, which are discussed in more detail in the following sections.

Table 1 Minimum Inhibitory Concentrations (μg mL^{-1}) of Antibiotics Against Selected Organisms

Antibiotic	Organism			
	S. aureus	E. coli	P. aeruginosa	B. fragilis
Sulfamethoxazole	32	16	—	4
Trimethoprim	1	0.03	—	8
Benzylpenicillin	0.01	8	—	—
Ampicillin	0.03	1	—	8
Carbenicillin	0.25	1	32	16
Cefuroxime	0.5	0.12	1	8
Streptomycin	2	4	16	—
Neomycin	4	1	—	—
Kanamycin	0.5	2	32	—
Gentamycin	0.25	0.5	1	—
Tobramycin	0.25	0.25	0.25	—
Chloramphenicol	4	1	64	—
Tetracycline	0.12	0.25	4	0.25
Erythromycin	0.12	16	—	—
Fusidic acid	0.06	16	—	8
Rifampicin	0.007	8	16	0.06

24.3.2.2 Diffusion or Zone of Inhibition Method

This method bears a close resemblance to the methods used to detect the microbiological activity of candidate drugs. A solid phase agar gel containing the selected organism and nutrients to ensure healthy bacterial growth is prepared in Petri dishes. Samples of the antibiotic, standards or test samples, are placed onto the agar gel as discrete spots and the Petri dishes incubated for a specified time. Zones of inhibition, usually identified as rings about the sample where the gel remains clear, denote those samples where the antibiotic is present in excess of the MIC.

24.3.2.2.1 Preparation of agar plates

A typical agar-based nutrient medium[5] consists of peptone (6 g), pancreatic digest of casein (4 g), yeast extract (3 g), beef extract (1.5 g), dextrose (1 g), agar (15 g) and the chosen organism. The medium is made up to 1 L and maintained at 37 °C during dispensing into 10 cm diameter Petri dishes (20 mL per dish). The dishes are then placed in a refrigerator for long enough to cause solidification. The samples are placed onto or into the agar medium by a variety of methods, including cutting out cups in the medium and filling the cups with the test solution, the use of stainless steel cylinders placed on the surface and filled with solution, or by placing cellulose discs impregnated with sample onto the agar surface. In all cases, the plates are incubated at 30–35 °C for a specified period of time, typically 24 h, during which time the antibiotic diffuses into the surrounding medium.

24.3.2.2.2 Quantitation of antibiotic

Typically, in the diffusion method, each Petri dish has six spotted samples. These would consist of three standard solutions and three test solutions, the three standards chosen to be of similar concentration to those in the unknowns. This method requires a certain amount of guesswork or preliminary assays. A better method is to establish a standard curve from a range of standards and measure the diameters of the zones of inhibition. The unknown concentration can be determined by interpolation. The diameters can be read reasonably accurately using a projection magnifier, and for the standard curve it is usual to plot the diameter against the log of the concentration, although this appears to be an empirical relationship rather than a rigorously derived one. Zone area measurements are also sometimes used, requiring rather more sophisticated and expensive equipment; the increase in precision, usually as good as 5–10% for diameter measurements, would not seem to warrant the extra cost. By its very nature, the sensitivity of a microbiological assay is self-defining and self-limiting; very active compounds will be detected with high sensitivity but this has no inherent advantage over less active compounds as the former will be administered in correspondingly smaller doses.

24.3.2.3 Serial Dilution Method

In the serial dilution method for the assay of antibiotics, the antibiotic is incubated in a liquid nutrient containing a viable microorganism. As the organism multiplies the solution becomes cloudy and the concentration of antibiotic is varied to find the point at which the cloudiness is inhibited under standard assay conditions. The degree of cloudiness can be measured accurately to obtain a reasonably quantitative assay.

24.3.2.3.1 Preparation of tubes

As with all microbiological work, great care must be taken to avoid bacterial contamination. All glassware must be thoroughly cleaned, using nonionic detergent or chromic acid–sulfuric acid washing, followed by sterilization in an autoclave. All the following procedures should be carried out aseptically.

A standard stock solution of the antibiotic is prepared, the concentration depending on the potency of the antibiotic, and, from this, a range of secondary standards is prepared by successive dilutions in buffer or by appropriate dilution of the stock solution. At least three graded solutions of the test sample are similarly prepared to fall in the same range of concentration. Exactly 1 mL of each solution is added to rimless test-tubes (15 cm × 20 mm), usually in triplicate, and the assay medium containing the appropriate organism (9 mL) added. All the tubes are incubated in a water bath at 37 °C for 2–4 h.

24.3.2.3.2 *Quantitation of antibiotic*

The cloudiness, or turbidity, of the liquid nutrient medium on incubation is due to the increasing number of suspended microorganisms causing light scattering. Thus, nephelometry is required to quantitate the cloudiness in the standard samples and the test samples. In practice, measurements of optical density can be made with a visible spectrometer at an arbitrarily chosen wavelength, depending on the sensitivity required. As the absorbance will be greater at *lower* concentrations of antibiotic, a plot of the log of the concentration against the observed absorbance will have a negative slope (Figure 1).[6] Usually, such a plot will have a usable range of linearity but it is important to quantitate test samples by interpolation on the standard curve and never by extrapolation.

24.3.2.4 Organisms Used in Microbiological Assays

As mentioned above, microbiological assays for antibiotics were developed as a logical extension of the tests used to discover the antibiotic activity. Thus, it is apparent that microorganisms used for assay will be those used in the discovery process. A list of such commonly used organisms is given in Table 2.[7] In any one laboratory the test organism chosen can be any of those which is particularly susceptible to the antibiotic being assayed. However, where such parameters as sensitivity, specificity and linearity of response are being claimed, and the method is to be established in another laboratory, it is important to define the strain used. The organisms used in microbiological assays are best stored in the freeze-dried state. For current use they should be kept as cultures maintained at 4 °C and subcultured at weekly intervals.[8]

24.3.2.5 Advantages and Disadvantages of Microbiological Assay Methods

The analysis of antibiotics in biological samples was one of the earliest drug level assays of therapeutic value and microbiological assay was the method of choice. The prime reason for choosing a microbiological assay is that it measures the concentration of the biologically active species. It would be more correct to say it measures the biological effect as, due to protein binding and other effects, the standard curves are not identical for different matrices such as urine and serum.[9] The fact that metabolites that are also microbiologically active may also be included in the assay result may be considered both an advantage and a disadvantage. On the one hand, if the

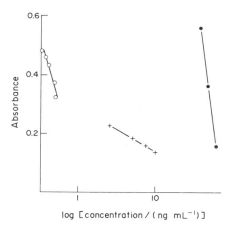

Figure 1 Standard curves for the quantitation of antibiotics by plotting absorbance against log concentration: ○, erythromycin; ●, gramicidin; and +, monensin

Table 2 Test Organisms Used in Microbiological Assays[7]

Drug	Test organism	Drug	Test organism
Aminoglycosides	*Klebsiella edwardsii*	Metronidazole	*Clostridium sporogenes*
Cephalosporins	*Staphylococcus aureus, Bacillus subtilis, Escherichia coli*	Nalidixic acid	*Escherichia coli*
Chloramphenicol	*Sarcina lutea*	Penicillins	*Bacillus subtilis, Sarcina lutea*
Clindamycin	*Sarcina lutea*	Polymyxins	*Bordetella bronchiseptica*
Erythromycin	*Sarcina lutea*	Rifamycins	*Sarcina lutea, Staphylococcus aureus*
Fusidic acid	*Corynebacterium xerosis, Staphylococcus aureus*		
		Sulfonamides	*Bacillus pumilus, Escherichia coli*
Lincomycin	*Sarcina lutea*	Tetracyclines	*Bacillus cereus*
Mecillinam	*Escherichia coli*	Trimethoprim	*Bacillus pumilus*

investigator is trying to demonstrate an effect, for example a long-lived antibiotic action of a particular dosed drug, then the total antimicrobial activity is an appropriate measurement. On the other hand, if the investigator wishes to demonstrate the basic pharmacokinetics of the new chemical entity, particularly in situations where the metabolite production and elimination may vary unpredictably, then an assay specific for the parent compound is essential. It is for this reason that, for research purposes, HPLC has almost superseded microbiological assays.[10,11] However, the microbiological assay still has the virtue of almost universal applicability within the field of antibiotics and can be readily adapted for any antimicrobial compound. Where increased specificity is required, a chromatographic step can be introduced, or a combination of methods using different organisms can be devised that will enable specific measurements of drug and metabolites,[12] or of more than one drug in the same sample. This latter method is particularly applicable in combination therapy where antibiotics with different profiles are used together in an attempt to broaden the spectrum or to take advantage of synergism.[13]

As with other assay methods which are inherently biological in nature, microbiological assays tend to produce large variations, need to be strictly controlled and are usually performed in triplicate. However, they do have the advantage that very little, if any, prior clean-up of the biological sample is required, unless increased specificity is required. The biological variation inherent in microbiological assays as generally practised may indeed be phased out as the basic biochemical findings related to the mechanism of action of the antimicrobials become clearer. Some investigators are now exploiting the interactions of antibiotics with microorganisms at the cellular and subcellular levels to devise novel assay methods, such as parameters depending on altered membrane permeability[14] and binding to microbial ribosomes,[15] rather than the gross lethal effect.

The diffusion method of microbiological assay generally takes 24 h to perform compared with minutes for many chromatographic methods. However, a large number of assays can be performed simultaneously using relatively simple equipment and there is much scope for automation. The shorter incubation time and the inherently more accurate endpoint of the serial dilution method would suggest that this method ought to be the more popular but, in practice, the diffusion method is considered more reliable.

Microbiological assays cannot be applied to all fluids and tissues without a thorough evaluation and construction of calibration curves. The effect of protein binding has already been mentioned. In particular, plasma and serum cannot be used interchangeably due to the use of anticoagulants in the preparation of plasma, which may cause unwanted changes in the agar gel matrix; microbiologists will always use serum rather than plasma for antibiotic assays.[16]

24.3.3 COMPETITIVE PROTEIN BINDING

Competitive protein binding is just one particular case of an analytical technique using the principles of saturation analysis. Other examples are discussed separately in other sections (immunoassay and receptor assay). However, as the name 'competitive protein binding' is so descriptive of the method, and as the technique was the first of the family to be used for small, drug-like molecules such as steroids, it is convenient to deal with this particular variation by way of introduction.

In fact, the principle of saturation analysis is considerably older than modern analytical chemistry, being used in an ingenious method for estimating the population in mediaeval Paris. The technique

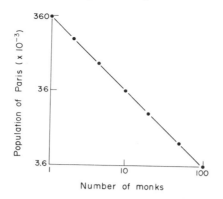

Figure 2 The 'monk method' for determining the population of Paris, to illustrate the principle of saturation analysis (total monk population = 360)

was to count out a fixed number of people passing a particular point in the city, at the same time noting how many of these were monks. Let us assume that 1000 people are counted, of which 18 are monks. Now the total number of monks in Paris would be known from church records and let us say this figure is 360. Then the population of Paris would be simply 360(1000/18) = 20 000. A calibration graph using the 'monk' method is shown in Figure 2.

Translated into the terms of saturation analysis, the saturable compartment, i.e. the compartment that will accommodate a limited number of molecules at any one time, is the fixed sample of 1000, and the label, i.e. the feature that enables the added molecules to be distinguished from those to be assayed, is the monk's habit. For a successful saturation analysis it is essential that the labelled molecules behave exactly the same way as the rest of the population and that they should mix rapidly with them, all molecules moving rapidly in and out of the saturable compartment. Thus, the method of saturation analysis essentially requires an equilibrium to be established; however, it is also permissible, where the movement out of the compartment is slow, to allow the labelled and unlabelled molecules to compete for the available sites.

The fundamental reactions of the competitive protein binding may be expressed by

$$B/F = K (P° - B)$$

where B and F are the concentrations of bound and unbound drug, K is the equilibrium constant in the direction of the bound complex and $P°$ is the molar concentration of the protein-binding sites. It can be seen that as the amount of free drug in the system increases as the fraction bound decreases, by separating the free drug from bound drug and determining the proportion of labelled drug in either of the two separated fractions, the total amount of the drug in the system can be calculated. The analytical procedure requires a method for the separation of bound and free fractions and a method for assaying the amount of label in one or the other.

One of the earliest assays for small drug-like molecules was that for thyroxine in 1960 using thyroid-binding globulin as the specific binding protein.[17] This was closely followed by a method for corticosteroid assay using the corticosteroid-binding globulin, transcortin.[18] The details of such assays can be illustrated by one of the several published methods for testosterone.[19]

24.3.3.1 Competitive Protein-binding Assay for Testosterone

In the published method of Liberti et al.,[19] the protein is obtained from late pregnancy plasma and is reacted with tritiated testosterone prior to reaction with the test sample when the testosterone in the sample displaces the bound tritiated testosterone. The free testosterone, labelled and unlabelled, is removed by absorption onto florisil and the tritium content of the bound fraction is determined by liquid scintillation counting. The detailed method is as follows.

24.3.3.1.1 Preparation of binding protein and displacement reaction

Tritiated-testosterone-bound protein is prepared by adding third-trimester human pregnancy plasma (1 mL) to distilled and deionized water (30 mL) followed by the addition of sufficient tritiated testosterone to give approximately 9000–12 000 c.p.m. per mL of plasma. The concentration of pregnancy plasma necessary to yield a satisfactory standard curve is determined experimentally for each new pregnancy sample. For the displacement reaction the prepared tritiated testosterone plasma (1 mL) is added to a dried extract of the test material and mixed for 5 s, placed in a water bath at 45 °C for 5 min, mixed again and placed on ice for 15 min.

24.3.3.1.2 *Separation of free and bound testosterone*

Florisil (80 mg) is added to the ice-cooled mixture, which is mixed for exactly 30 s and replaced in the ice bath. After precisely 30 min, a sample of supernatant (0.5 mL), which now contains no free testosterone, is withdrawn and added to a scintillation vial for counting.

24.3.3.1.3 *Quantitation of testosterone*

A standard curve is prepared by taking a range of known amounts of testosterone (0–6 mg) through the procedure and plotting the amount of bound testosterone against the amount of testosterone present. The amount of testosterone in the unknown sample is determined from this curve. In the complete method of Liberti *et al.*[19] an extensive prepurification procedure, including a TLC step, is used to ensure specificity, but even so a precision of 10% is claimed.

24.3.4 RADIORECEPTOR ASSAYS

In addition to competitive protein-binding assays for small hormone molecules such as thyroxine and the steroids,[18,19,20] assays have also been described for ACTH,[21] opioid peptides,[22] cAMP[23] and γ-aminobutyric acid.[24] The proteins used in these assays will often bind synthetic analogues used as drugs. It is not surprising that the proteins will bind closely related structures. What is surprising is a series of proteins, isolated from natural sources, which will bind synthetic drugs, such as neuroleptics,[25] β-blockers[26] and benzodiazepines,[27] which have no obvious structural analogues in nature. There is an apparent anomaly in nature having built these proteins waiting for humans to synthesize the relevant compounds. Some light is now being shed on this story where the use of molecular modelling is beginning to show previously unappreciated similarities between the synthesized drugs and natural ligands.

The development of radioreceptor assays as analytical tools is closely bound up with the biochemical pharmacology of drug discovery. Thus, the discovery that certain membrane receptors can be blocked by novel chemicals and, hence, stimulate or inhibit a biological event has led to intense research on the nature of the receptors and the search for compounds that show even stronger and more selective binding. For example, the specific benzodiazepine receptor was identified when it was demonstrated that radiolabelled diazepam binds to a specific fraction of mammalian brain tissues[28] and, to some extent at least, the ability of other benzodiazepines to displace the labelled diazepam from the receptor correlates with its potency.[29] Table 3 lists a number of receptor preparations developed in the course of pharmacological research that have proved useful in developing drug assay methods. Radioreceptor assays are typified by the following detailed description of the benzodiazepine receptor assay.[30]

Table 3 Biological Receptor Used in Drug Assays

Receptor	Drugs	Ref.
β-Endorphin receptor	Clobazam, loprazolam	30
	Diazepam	27
Dopamine receptor	Haloperidol, fluphenazine, trifluoperazine, chlorpromazine, thioridazine	25
β-Adrenoceptors	Propranolol	26

24.3.4.1 A Radioreceptor Assay for Benzodiazepines[30]

24.3.4.1.1 Preparation of membrane receptor

The membrane receptor is prepared from rat cerebral cortex according to the method of Mohler and Okada.[30] Samples of cortex are homogenized in 20 volumes of sucrose solution (0.32 M) and centrifuged at $1000 \times g$ for 10 min. The resulting supernatant is recentrifuged at $20\,000 \times g$ for a further 20 min and the resulting pellet resuspended in 20 volumes of Tris HCl (50 mM, pH = 7.4). After further centrifugation at $20\,000 \times g$ for 20 min the pellet is either resuspended in the same buffer for immediate use or stored at $30\,°C$. The final suspension contains approximately 1 mg protein in 2 mL.

24.3.4.1.2 Binding experiment and measurement of bound fraction

Tritiated diazepam (10^{-9} M) is incubated alone or in the presence of competitors at different concentrations with aliquots of the membrane suspension for 30 min at $0\,°C$. During this time the added benzodiazepine displaces tritiated diazepam from the binding sites to an extent depending on the relative binding affinities. The incubation is terminated by rapid filtration under vacuum through Whatman GF/C filters, which are washed with a further two portions (5 mL each) of ice cold incubation buffer. The filters are then left to stand overnight in scintillation fluid containing Triton X-100 to solubilize the protein and counted for radioactivity content. Nonspecific binding is estimated by incubation in the presence of a relatively large amount of unlabelled diazepam (10^{-6} M) and, where necessary, all other experiments are corrected for this binding.

24.3.4.1.3 Standard curves

The receptor preparation was developed to investigate the relative binding of different benzodiazepines, in an attempt to correlate the binding affinity with drug effect. Table 4 shows the wide variation in binding affinity expressed as the concentration of benzodiazepine necessary to displace 50% of tritiated diazepam in the standard experiment.[30] To use the preparation to assay a particular benzodiazepine, therefore, it is necessary to construct a standard curve for concentrations around that found to displace 50% of the tritiated diazepam. Figure 3 shows the different usable ranges of standard curves for a good displacer (diazepam, $IC_{50} = 7.1 \times 10^{-9}$ M) and a poor displacer (clobazam, $IC_{50} = 170 \times 10^{-9}$ M).

24.3.4.2 Advantages and Disadvantages of Radioreceptor Assays

As for microbiological assays, the major advantage of radioreceptor assays is stated to be that they measure the total activity of the sample, taking into account active metabolites. This view is only sustainable if it can be shown that the *in vivo* binding activity correlates with biological activity and this is not easy to prove because of complex factors involved. For example, metabolites present in the blood, which strongly bind to receptors, may not reach the receptors with the same ease as the parent drug.

Thus, the presence of metabolites of a drug, which also bind to receptors, is more probably a major disadvantage in using radioreceptor assays. Such cross-reactivity can be even more serious than the analogous contamination effect for chemical assays. Whereas for most chemical analyses the contribution of a closely related structure will be a function of its mass and may therefore be tolerable, a minute amount of an entity with a very strong binding affinity may give a very large error in the measurement.

Table 4 Binding Affinities of Benzodiazepines for a Rat Brain Membrane Preparation

Benzodiazepine	IC_{50} (10^{-9} M)	Benzodiazepine	IC_{50} (10^{-9} M)
Clobazam	170	Diazepam	7.1
Demethylclobazam	210	Nitrazepam	10
Demethyldiazepam	8.8	Oxazepam	20

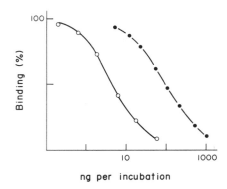

Figure 3 Displacement of tritiated diazepam from benzodiazepine-binding sites by a good displacer (diazepam, ○) and a poor displacer (clobazam, ●)

Receptor assays would be inappropriate for use in monitoring experiments where drugs of the same type, *e.g.* two benzodiazepines, are administered in combination.

On the other hand, receptor assays are generally very simple and convenient to carry out once the receptor preparation is available. The same assay is applicable for all drugs of a particular series, and, almost by definition, the sensitivity increases with decreasing therapeutic dose levels. In most cases, prior extraction and clean-up of biological samples is unnecessary, unless very high sensitivities are required.

Receptor assays of the classical type are only available for those classes of compounds which have been shown to bind to isolatable receptors, and therefore, on the face of it, the receptor assay is not applicable to all classes of drugs. However, as described in the next section, such 'designer receptors' can be biosynthesized in the laboratory.

24.3.5 IMMUNOASSAYS

Immunoassays, particularly radioimmunoassays, are the most widely used of those assays in the class of saturation analysis. In the previous section, it was pointed out that receptor assays were only applicable where nature has provided a convenient receptor which strongly binds drug molecules. The basis of an immunological reaction is that a particular protein, the antibody, will bind, and hence deactivate, a foreign body which has invaded the host; the protein itself is synthesized by the host in response to a previous challenge by the same or similar foreign body, the antigen. Thus, it can be seen that a 'receptor' can, in fact, be produced to any compound that will behave as an antigen. Antibodies can be readily raised to relatively large molecules, particularly proteins and the larger polypeptides. The first radioimmunoassay was described in 1960 for insulin.[32] However, smaller molecules, such as steroids and the more usual drug molecules, are not themselves antigenic and it was not until the idea of linking the small molecule to a carrier protein[33] and raising antibodies to the complex was perfected for estradiol that the technique—radioimmunoassay, or RIA—came into its own.

24.3.5.1 Radioimmunoassays of Drugs

24.3.5.1.1 Selection of conjugate

The first step in the total development of a RIA for drugs is to enlist the help of a synthetic, and sympathetic, chemist. The problem is to link the drug molecule through a reactive group to a large protein such as bovine serum albumin. Other proteins, such as thyroglobulin and chicken γ-globulin, have been used but bovine serum albumin is the most commonly used carrier protein. The availability of free amino and carboxylic acid groups on the protein suggests that a peptide bond is the most suitable type of conjugate and it is therefore desirable to have the corresponding function on the drug, or 'hapten' as it is termed in immunological nomenclature. Thus, conjugates can be directly prepared with drugs having an amino function, such as amphetamines, or with drugs having a carboxylic acid function, such as prostaglandins. Suitable methods for direct conjugation have been described.[34,35,36]

Reactive hydroxyl groups can also be utilized to form conjugates. This can be by direct reaction with phosgene to form the chlorocarbonate, which can then in turn form amides with lysine residues of the albumin, using a Schotten–Baumann reaction. Alternatively, the hydroxyl group can be reacted with succinic anhydride to form the ester with the free acid available for conjugation to lysine residues. Reactive carbonyl groups can also be utilized to form appropriate derivatives for subsequent conjugation to protein.

Variations of these general procedures are used for individual drug assays. For example, a reactive group can be produced in the terminal glucose residue of digoxin by periodate oxidation and subsequent direct conjugation to albumin, or the whole triose may be removed to reveal a reactive hydroxyl group.

The various methods of conjugation can be used to produce different bridge lengths and an optimal bridge length of four carbon atoms (*i.e.* succinic acid) has been suggested.[37]

All the above methods of conjugation utilize functional groups already present in the drug molecule, but this could well compromise the specificity of the immune reaction. This is because a characteristic feature of the drug molecule may be masked in the antigen, so that the corresponding antibody raised will not include this particular feature in its recognition kit. Thus, if possible the linkage should be through a position leaving the maximum number of antigenic determinants exposed. For example, conjugates of phenobarbitone have been prepared through a 4-aminobutyl analogue at position 5, leading to antibodies of very high specificity.[38] The general lack of such suitable analogues in large quantities is the reason why it is useful to consult with the originating chemist. In medicinal chemistry it is likely that a number of analogues will have been prepared as part of a series and it is possible that a suitable analogue can be chosen for producing specific antibodies.

Another chemical parameter that is important to the quality of antisera is the number of hapten residues that are linked to each protein molecule. This can be any number between two or three and nearly 100. There is some evidence to suggest that the smaller the number, the more antigenic is the conjugate, although it has also been suggested, for steroids at least, that this has little effect on either the affinity or specificity of the resulting antisera.[39] Other authors claim the optimum number is between 20 and 30 drug residues per molecule of bovine serum albumin.[37]

24.3.5.1.2 *Raising antibodies*

Once prepared, the conjugate is injected into animals to raise antibodies. The details of suitable procedures for raising antibodies are beyond the scope of this work. It is sufficient to note, however, that the immune response is by no means predictable and there is no precise immunization and harvesting procedure, nor is there a particular animal species that can be generally recommended. It is usual to inject animals first with a primary dose of conjugate, normally 100 μg, and then with periodic booster injections to prolong their exposure to the antigen. An immune response should appear after about two months. Animals are bled and the serum tested for the presence of antibodies to the hapten. When a response is found, it is necessary to dilute the serum to assess its quality; the measure of the strength of the antiserum is the degree of dilution necessary before a solution is obtained which binds 50% of the antigen, or the hapten. Although dilutions of several thousand are common for RIA's for large proteins, as far as drugs are concerned antisera with dilutions of more than 2000 are rare. It is for this reason that larger animals, such as sheep and goats, have found favour, as the bleeds will provide large amounts of reagent. This is important, not only for reasons of economy, but because every new bleed has to be reassessed for specificity and 'titre' (the dilution factor), due to the irreproducible nature of the immune response.

Although clear guidelines cannot be laid down for optimum antibody production, normal immunological practice suggests that it is desirable to use an animal of species and strain known to respond well, in good health, and not subject to other antigenic stimuli. Therefore, many factors are thought to contribute to the successful raising of antibodies, and yet the testing is so tedious that few detailed studies have been carried out to elucidate the clear rules.[40]

24.3.5.1.3 *The radiolabel*

The first RIA techniques for protein hormones invariably used ^{131}I and later ^{125}I as the radiolabels, as proteins can be iodinated without loss of immunoreactivity, and the iodine isotopes, being γ emitters, could be counted in γ counters without recourse to the more complex liquid scintillation techniques. However, iodination of drugs will drastically affect their chemical structure and, in general, iodine-labelled drugs would be unsuitable. Iodine labels can be used if they do not alter the structure too much and particularly if the label can be confined to the same point in the molecule which is attached to the protein when preparing the conjugate.[37] This will ensure that the drug and the labelled drug will be similarly recognized by the antibody. Alternatively, the protein of the conjugate can be iodinated and the labelled conjugate itself used as the tracer in the subsequent assay.

In drug development it is likely that a ^{14}C-labelled form of a candidate drug has already been prepared for metabolic studies, and this form is practically identical to the nonlabelled form. RIA procedures have been described for diazepam,[41] methadone[42] and phenytoin[43] using ^{14}C-labelled drugs as the tracer. Unfortunately the specific activity of ^{14}C-labelled drugs is a limiting factor; if sufficient label is added to the test system to obtain acceptable counting statistics, the mass added may be comparable to, or larger than, the amount being assayed, with consequent loss in performance of the assay. It is for this reason that tritium has found the most universal acceptance as the label in RIA of drugs. Tritiated drugs are relatively easy to prepare, and for RIA procedures metabolic exchange is not a problem. Tritium has a much longer half-life than either of the iodine isotopes (12 years compared with 8 days and 60 days for ^{131}I and ^{125}I) and, therefore, once the label has been prepared, shelf-life is not a problem. RIAs using tritium have been developed for almost all types of drug (Table 5). The main practical disadvantage of both tritium and ^{14}C is that, since they are β emitters, it is necessary to use liquid scintillation techniques for the measuring step.

24.3.5.1.4 *Optimization of the assay*

The crucial step in all RIA procedures is the incubation of the antiserum, the labelled tracer and the unknown substance. The method will be most useful if the three components are present in such amounts that a small change in the amount of the unknown produces a measurable change in the distribution of radioactivity between the free and bound fractions. Although various ways of plotting the binding curves have been proposed to help optimize the usable range of the RIA, the actual advantages are more apparent than real and sophisticated computer routines are essential for full optimization.[61] In general, it is agreed that the RIA is best operated under conditions where approximately half of the specifically bound radiolabel is displaced, and the usable range should be an order of magnitude above and below the concentration of the drug at which this happens. A typical example is shown in Figure 4. Concentrations outside the usable range should be reassessed using diluted sample or reagent.

24.3.5.1.5 *The separation step*

For RIA, it is not possible to measure the bound and unbound labels directly and the two fractions have to be physically separated prior to using standard counting techniques. Among the many separation techniques are differential migration methods, such as paper chromatoelectrophoresis and gel filtration, absorption methods using charcoal or silicates, fractional precipitation with salts or organic solvents, double antibody methods and solid phase methods.[62] However, for drug RIAs the most popular techniques are absorption methods, solid phase methods and the double antibody method. It is essential that the method chosen is rapid enough to effect the separation without altering the equilibration set up in the incubation step.

Charcoal was first used for separating free thyroxin from thyroxin bound to thyroid-binding globulin in the competitive protein-binding assay and was later applied to the RIA of insulin.[63] Charcoal absorbs small organic molecules from the solution and is readily separated from the remaining solubilized protein–label complex by centrifugation. The supernatant can then be counted for radioactivity content. The nature and amount of the charcoal used needs to be well controlled and reproducible. Activated charcoal with particle size of 60 μm is generally satisfactory. Raw activated charcoal is too severe for direct use, tending to bind the antibody itself or to strip the hapten from the antibody. To overcome this effect, the charcoal is coated with dextran to provide pores through which the free hapten will pass and be bound, but will exclude the larger complex.[64]

For solid phase methods, the antibody is covalently bound to the inside of the assay tube.[65] After incubation, the supernatant can be simply poured off, leaving the bound label attached to the antibody in the tube. The tube can then be counted directly, or after the addition of scintillation fluid. Alternatively, the antibody is linked to magnetizable particles, which can be subsequently separated with a magnet from the supernatant.[66]

Table 5 Radioimmunoassay of Drugs Using a Tritiated
Label

Drug	Sensitivity	Ref.
Amphetamines	2.5 ng	44
Antibiotics	20 ng mL^{-1}	45
Antihistamines	0.25 ng mL^{-1}	46
Antihypertensives	10 pg mL^{-1}	47
Antineoplastics	1 ng mL^{-1}	48
Barbiturates	0.2 pmol	49
β-Blockers	0.12 pmol	50
Benzodiazepines	10 ng mL^{-1}	51
Cannabinoids	2 ng mL^{-1}	52
Ergot alkaloids	0.5 ng mL^{-1}	53
Hyperglycaemics	—	54
Neuroleptics	1 ng mL^{-1}	55
Opiates	50 pg	56
Parasympatholytics	9 nM	57
Prostaglandins	—	58
Steroids	12.5 pg	59
Tricyclics	1 ng mL^{-1}	60

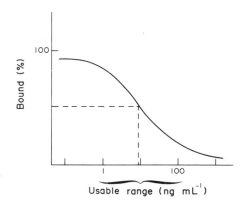

Figure 4 Usable range of radioimmunoassay

For the radioimmunoassay of large proteins, the antigen–antibody complex is often large enough to precipitate from solution and is easily removed by filtration or centrifugation. When the complex consists of drug–antibody only, then it may remain in solution and such simple separation methodology is not possible. However, if a second antibody, which has been raised against the globulins of the animal in which the drug–antibody has been raised, is added, then the resulting precipitate can be readily separated.

24.3.5.2 Nonisotopic Immunoassays

24.3.5.2.1 *Fluorescence label*

In this technique a fluorescent group is attached to the drug molecule to act as the labelled compound. Most popularly fluorescein and its derivitives have been attached to tobramycin, propranolol, morphine, digoxin, digitoxin, phenytoin, gentamicin and procainamide. The immunoassay is carried out in the usual way, separating free and bound fractions and measuring the fluorescence of either or both fractions.[67]

A variation of this technique is the fluorescence polarization immunoassay. If a molecule is excited using polarized light, then the emitted light will also be polarized in the same plane as the exciting light, provided that the molecule does not rotate in the time between excitation and emission. In solution, the free rotation of the molecule reduces the polarization of the emitted light and hence solutions of free drug do not normally exhibit fluorescent polarization. When bound to a macro-molecule, however, the rotation is slowed and the emitted light retains its polarization. When used in immunoassay, therefore, the free and bound molecules can be differentiated without a separation step. This method—described as a homogeneous rather than a heterogeneous immunoassay—is therefore extremely simple to carry out, though specialized equipment is needed.

24.3.5.2.2 *Enzyme label*

The principal of enzyme labelling is that when the enzyme is bound to a protein then its enzymic activity will be blocked; the enzyme attached only to the small drug molecule in solution retains its enzymic activity. Thus, the usual incubation experiment is followed by an assay to determine the activity. The technique was developed extensively by the Syva Corporation as the enzyme-multiplied immunoassay technique or EMIT.[68] In this procedure, the drug molecule is covalently linked to a stable enzyme such as glucose-6-phosphate dehydrogenase. Glucose-6-phosphate dehydrogenase catalyzes the conversion of glucose 6-phosphate to gluconolactone 6-phosphate, at the same time reducing NAD to NADH. The formation of NADH can be monitored spectro-photometrically by its absorption at 340 nm. Table 6 lists a number of drugs which have been successfully assayed by the EMIT technique.

The advantage of the EMIT technique over conventional RIA, apart from the obvious one of not needing a radioisotope, is that it is, like the fluorescence polarization immunoassay, a homogeneous assay method. On the other hand, the other major enzyme immunoassay method ELISA, or enzyme-linked immunosorbent assay, is a heterogeneous assay. In this method, the drug–enzyme conjugate retains its enzymic activity even after binding to the antibody[82] and therefore a separation step is necessary before measuring the enzyme activity. For example, in the assay for digoxin, using horseradish peroxidase conjugated to digoxin, the incubation step takes place in test-tubes coated with digoxin-specific antibodies. Free digoxin in the test sample and the enzyme-labelled digoxin compete for the available binding sites. When equilibrium is reached, the incubation mixture can be decanted off and either fraction is assayed for enzyme activity, determined by the colorimetric assay of hydrogen peroxide.

A variation on the enzyme-labelled assay is to label the substrate. In one such method the drug is covalently linked to umbelliferyl-β-D-galactoside.[83] When this complex is attacked by β-galac-tosidase, it produces the fluorescing entity, umbelliferone–drug complex (Figure 5).

In the assay procedure the umbelliferyl-β-D-galactoside–drug molecule competes with unlabelled drug for binding sites on the antibody so that the amount available for attack by β-galactosidase is proportional to the amount of drug in the sample. Substrate-labelled fluorescent immunoassay procedures have been described for gentamicin, tobramycin, phenytoin, theophylline, phenobarbi-tone and amikacin.[84]

24.3.5.2.3 *Spin label*

The spin label method, or free radical assay technique (FRAT), depends on the electron spin resonance (ESR) signal of free radicals such as the nitroxide radical.[85] In solution the free radical exhibits a simple three-line spectrum, even when linked to a drug molecule. If the radical is bound to a large protein molecule then there is slower rotation of the moiety and the spectrum becomes considerably flattened. Thus, in the absence of added drug the flattened signal only is seen. When

Table 6 Drugs Analyzed by EMIT Techniques

Drug	Ref.	Drug	Ref.
Benzodiazepines	69	Morphine	76
Cannabinoids	70	Phenobarbitone	77
Carbamezepine	71	Procainamide	78
Disopyramide	72	Quinidine	79
Ethosuximide	73	Theophylline	80
Gentamicin	74	Valproic acid	81
Methotrexate	75		

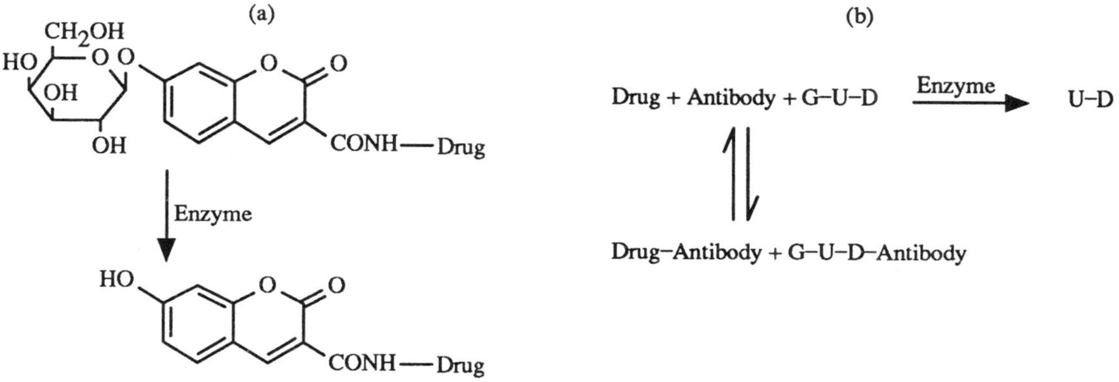

Figure 5 Reactions involved in substrate-labelled fluorescence immunoassays: (a) conversion of nonfluorescent galactosylumbelliferone–drug (G–U–D) reagent to fluorescent umbelliferone–drug (U–D) by β-galactosidase; (b) competitive reactions in an assay system

unlabelled drug is added it displaces part of the free-radical-labelled drug and the typical sharp spectrum becomes clear. For this technique there is the need for a very specialized piece of equipment, namely the ESR spectrometer, whereas other labels use standard spectrometers, fluorometers and scintillation counters.

24.3.5.3 A Radioimmunoassay for Glibenclamide

The various facets of the immunoassay procedures for the determination of drugs in biological fluids can be exemplified by the method described for the assay of the oral hyperglycaemic, glibenclamide.[86]

24.3.5.3.1 *Reagents*

The antiserum was raised in rabbits by the injection of glibenclamide coupled with bovine serum albumin. This antiserum needed a dilution of one in forty thousand, *i.e.* as only 20 μL diluted antiserum is required for a single assay, 1 mL of serum from the rabbit provides enough reagent for two million assays. The labelled glibenclamide was [^{3}H]glibenclamide with a specific activity of 24.7 Ci mmol^{-1} (1 Ci = 3.7×10^{10} Bq). A stock solution was prepared to give approximately 125 000 c.p.m. mL^{-1} for use in the assay. Standard solutions of glibenclamide were prepared over the range 1.5–200 ng mL^{-1} in methanol. A solution of 8-anilino-1-naphthalenesulfonic acid (ANS, 6 mg mL^{-1}) is used to block the nonspecific binding sites on the serum proteins. For the removal of unbound glibenclamide in the assay a suspension of dextran-coated charcoal (2.5% w/v Norit A + 0.25% w/v Dextran T-70) is prepared. All reagents were prepared and diluted in the assay buffer—0.2 M phosphate, pH = 7.4, containing 0.88% sodium chloride and 0.1% bovine serum albumin.

24.3.5.3.2 *Assay procedure*

The reagents are added to the assay tubes in the order indicated (Table 7). All analyses are performed in duplicate. After addition of the antiserum the contents of the tubes are mixed on a vortex mixer and incubated overnight at 4 °C. Dextran-coated charcoal suspension is added and the tubes centrifuged at 2500 r.p.m. for 10 min at 4 °C. The supernatant is then decanted into scintillation vials for determination of tritium content. A standard curve is constructed from the data provided by the standards and the concentration of glibenclamide in the unknowns determined by interpolation.

24.3.5.4 Special Problems with Immunoassays

In contrast with chromatographic, spectrophotometric and other methods depending on well-understood physical phenomena, radioimmunoassay and other related techniques need to be

Table 7 Protocol for the Radioimmunoassay of Glibenclamide All Volumes in μL.

Reagent	Total counts tube	Nonspecific binding tube	Standard tubes	Sample tubes
ANS	100	100	100	100
Buffer	200	200	—	100
Normal serum	20	20	20	—
Standards	—	—	100	—
Unknowns	—	—	—	20
Tracer	100	100	100	100
Antiserum	—	—	100	100
Vortex, incubate overnight at 4 °C				
Buffer	500	—	—	—
Dextran-coated charcoal	—	500	500	500
Centrifuge at 4 °C and 2500 r.p.m. for 10 min				
Decant supernatant for liquid scintillation counting				

treated with the caution needed for biological measuring systems. The unpredictability of the immune response is a real barrier to routine development of immunoassays of drugs, and the unpredictable properties of the antibodies additionally make it difficult to rely on specificity. For assays relying on naturally occurring binding proteins, such as competitive protein binding and receptor assays, at least the protein is reasonably well-defined and constant between laboratories; 'new' proteins raised in response to new chemical entities will not be guaranteed in their binding properties. It is for this reason that large animals are generally favoured for raising antibodies, so there is a large amount of consistent binding reagent available for many thousands of reproducible assays. The alternative of combining antisera from several animals to obtain a large pool is not favoured as the resultant 'blend' will be less specific.

The potential lack of specificity of antibodies may be partly explained by the same protein 'recognizing' closely related drug molecules. However, it is more likely that the lack of specificity is due to the presence of different proteins. The specificity of an antiserum can be improved by isolating the separate proteins and culturing the specific antibodies. Such monoclonal antibodies not only will have high specificity but will ensure a continuous supply as long as that specific antibody is required.

Given the appropriate dispensing and counting equipment, radioimmunoassay is a relatively simple technique and is suitable for development of kits, and the successful development of commercial kits for hormones, steroids, *etc.* has been partially repeated for some of the more important drugs. It is not wise, however, to rely on the manufacturer to package completely reproducible antisera and other reagents, and one of the drawbacks of such immunoassay methods is the need to always reassess and characterize kits from different suppliers by carrying out appropriate control assays.

The central question of all immunoassays is the specificity. For drugs this would mean testing the antisera against the known metabolites, but, of course, this is not possible for drugs in the early stages of development and the method usually has to be evaluated against a reference method of high specificity such as gas chromatography–mass spectrometry. In the absence of such a check, the specificity of a newly developed immunoassay will always be in doubt.

The problem of cross-reactivity is not confined to metabolites or endogenous material in the test sample. An impurity in the standard samples, with a high degree of cross-reactivity, can give a completely erroneous standard curve; normal -chemical quality control may be inadequate in detecting such impurities. Similar problems can be met with impure tracer, with the added hazard that a labelled impurity, active or inactive, may also give rise to an erroneous standard curve.

24.3.6 ENZYMATIC ASSAYS

The aim of modern medicinal chemistry is to design compounds that will have, at low concentrations, specific effects on biological systems. Hence, most drugs will inhibit or accentuate the activity of one enzyme or another and it is natural that drug assay systems should be developed that depend on the alteration of enzyme activity. The assay procedures, the enzymes and the detection methods may vary considerably but most popularly reaction rates are followed by spectroscopy or radiochemistry. For example, methotrexate is an inhibitor of dihydrofolate reductase, whose activity can be measured by the reduced UV absorption at 340 nm as NADPH is converted to $NADP^+$.[87]

An alternative enzymatic procedure is to monitor the production of a labelled form of a drug when the drug acts as a substrate for a particular enzyme. For example, chloramphenicol has been assayed by measuring [^{14}C]acetyl-CoA, the cofactor required in the enzymic acetylation of the drug.[91]

Table 8 summarizes for a number of drugs the enzymes that have been utilized in their assay.

An important advantage of enzymatic analysis is that it will often distinguish between enantiomeric forms of a drug. The major obstacle to their general use is that a very specific enzyme is required, or extensive purification of a resulting drug derivative is necessary.

Enzymatic methods of analysis are not widely used in medicinal chemistry.

24.3.7 QUALITY CONTROL

Quality control for the bioanalysis of drugs, whether chemical or biological, begins with the selection and evaluation of the analytical method, which is a necessary part of the establishment of facilities, procedures and laboratory practices, and is a continuing process to be applied every time an analysis is performed. It is a sound principle that no analytical result should leave the laboratory without its accompanying quality control having been carried out.

24.3.7.1 Description of the Analytical Method

The choice of an analytical method for the determination of drugs in biological fluids is usually made following a series of exploratory studies to determine optimal analytical conditions. Once a working method has been established, it is necessary to document the procedure in such detail that a second laboratory can reproduce the method, and to evaluate the procedure for its suitability for particular applications.

An analytical method is defined as a set of written instructions which describe the procedure, materials and equipment that are necessary for the analyst to obtain reproducible results. This description is also necessary for a full assessment of the method to be made by a second laboratory. The following lists the items which must be included in the description.

(i) Outline of the principle of the method with references to the appropriate literature.

(ii) Specifications of instruments and equipment, given in sufficient detail to allow the analyst to judge which other commercial instruments or models can or cannot be used. If necessary, procedures for testing the equipment should be described.

(iii) Specimen requirements, including storage and shipping conditions.

(iv) List of reagents with concentrations expressed in appropriate units, source, brand name and an indication of purity.

(v) A full description of all steps in the analytical procedure including clear indications of critical steps and the tolerances which may be allowed without affecting the results, *e.g.* temperature, exposure to light, volumes, times of critical steps, *etc.*

(vi) Any special safety precautions, which may be necessary regarding the handling of specimens, preparation of reagents, waste disposal and decontamination.

(vii) Calibration procedure and methods of calculating results, including full descriptions of standard, reference and control specimens.

Table 8 Enzymatic Methods in Drug Analyses

Drug	Enzyme used	Ref.
Acetazolamide	Carbonic anhydrase	88
Amikacin	Kanamycin acetyltransferase	89
Aminoglycosides	Aminoglycoside *N*-acetyltransferase	90
Chloramphenicol	Chloramphenicol *O*-acetyltransferase	91
Deoxycoformycin	Adenosine deaminase	92
Doxycycline	Luciferase	93
Enalapril	Angiotensin-converting enzyme	94
Ethanol	Alcohol dehydrogenase	95

(viii) Analytical range, that is the range of concentration in the specimen, over which the method is applicable without modification.

24.3.7.2 Evaluation of the Analytical Method

The developed method must be evaluated according to the criteria of precision, accuracy, specificity, detectability, linearity and practicability, these terms having the meanings defined in Table 9 and being more fully discussed below. In addition, the method must be shown to have been applied in at least one experimental project prior to being accepted as a recommended method.

24.3.7.2.1 Precision

Precision describes the closeness of agreement between replicate measurements. The term has no numerical value, and the term 'imprecision' has been recommended by the IFCC and other bodies for the numerical value. The 'imprecision' is the standard deviation or coefficient of variation of a set of replicate measurements. The experimental design for measuring imprecision must be clearly stated as the definition does not imply anything regarding the replicates other than that they are presumed to have the same 'true' value at the time of measurement. Thus, imprecision may be calculated from a series of replicates using the same method, the same analyst, the same instruments and in the same laboratory within a reasonably short time. This measure of precision is a measure of repeatability and must be included in the analytical method report. It should be determined as follows.

A number of specimens are prepared consisting of blank biological fluid, to which has been added known quantities of analyte. Six specimens are prepared to cover the expected analytical range of the method, each specimen divided into a number of aliquots suitable for analysis and stored frozen. A batch of samples (one of each concentration) is thawed and assayed on six separate occasions.

The standard deviation (SD) is calculated from the results at each concentration.

$$SD = [\Sigma(x - \bar{x})^2/(n - 1)]^{1/2}$$

The coefficient of variation (CV) is also calculated:

$$CV = 100(SD/\bar{x})$$

and the imprecision is reported for each concentration, *e.g.* CV = 30% at 0.01 µg mL^{-1} ($n = 6$); 8% at 0.1 µg mL^{-1} ($n = 6$); and 4.7% at 1.0 µg mL^{-1} ($n = 6$).

Table 9 Definitions Used in the Evaluation of Analytical Methods

Accuracy	Accuracy is the agreement between the estimate of a quantity and its true value. The term does not have a numerical value. See 'inaccuracy'
Analyte	The component to be measured
Batch	A set of concurrent or consecutive analyses performed without interruption, the results of which are calculated from the same set of calibration standards
Blank	Material identical to the test sample in all respects except that it is known not to contain analyte
Calibration	Procedure of relating the reading to the quantity required to be measured.
Control	Material used for quality control purposes
Detection limit	The smallest single result, which, with a stated probability, can be distinguished from a suitable blank
Detectability	The ability of an analytical method to detect small quantities of analyte. It has no numerical value. See detection limit
Error	Difference between a single estimate of a quantity and its true value
Imprecision	The standard deviation of a set of replicate measurements
Inaccuracy	The difference between the mean of a set of replicate values and the true value
Precision	The closeness of agreement between replicate measurements. It has no numerical value. See imprecision
Repeatability	The closeness of agreement between replicate measurements using the same method, the same analyst, the same instruments in the same laboratory and within a reasonably short time
Reproducibility	The closeness of agreement between replicate measurements using different methods, different analysts, different instruments or in a different laboratory or when the analyses take place over a relatively long time interval
Sample	The appropriately representative part of a specimen which is used in the analysis
Specimen	Material available for analysis

When imprecision is obtained from a set of replicate measurements where the factors mentioned above are not identical (method, analyst, instruments, laboratory and time), then the imprecision is a measure of the reproducibility. As these parameters are not usually investigated in great detail when the method is developed and evaluated, it is unrealistic to include the measure of reproducibility in the analytical method report. However, reproducibility is an important consideration when a method is in routine use and this is more fully discussed in Section 24.3.7.3.

24.3.7.2.2 Accuracy

Accuracy is the agreement between the estimate of a quantity and its true value. As for precision, the term does not have a numerical value and the term 'inaccuracy' is recommended by the IFCC and other bodies. Thus, the inaccuracy is the numerical difference between the mean of a set of replicate values and the true value. The inaccuracy can be calculated from the same sets of replicates utilized to calculate imprecision and must be within one standard deviation of the mean estimate, for acceptable accuracy. As for imprecision, the inaccuracy (bias) must be quoted for each concentration investigated.

24.3.7.2.3 Specificity

Specificity is the ability of an analytical method to determine solely the component(s) it purports to measure. It has no numerical value and is assessed on the evidence available on components which may contribute to the result. In principle, this assessment can never be complete as any food or drug must be suspected of contributing to the analytical result until the contrary has been demonstrated. In practice, it is usually sufficient, in the early phases of drug development, to demonstrate noninterference by a normal diet and by metabolites of the drug in question. This will be sufficient for tolerance studies and basic pharmacokinetic studies. As trials become more complex it becomes necessary to test for interference from pharmaceutical excipients, coadministered drugs (particularly for combination products) and even for endogenous disease-related components. The assessment of specificity needs continual review. Comparison of results obtained using the candidate method with results using a different method can give useful indications of specificity and should be included in the analytical method report whenever possible.

24.3.7.2.4 Detectability

Detectability is the ability of an analytical method to detect small quantities of the analyte. It has no numerical value. The measured value is referred to as the detection limit, which is the smallest single result (concentration or amount) which, with a stated probability (usually 95%), can be distinguished from a suitable blank. This value defines the point at which the analysis becomes feasible. For sample concentrations close to the detection limit, the imprecision approaches a limiting value and is not dependent on the measured value. For a probability of 95% and with $n = 6$, the detection limit is twice the standard deviation. The detection limit can thus be calculated from the sets of replicate analyses described in Section 24.3.7.2.1. It should be emphasized that the detection limit should be used as a warning sign and is not a justification for using a method continually at the lowest values it may be able to measure. Results below the detection limit must be reported as '< detection limit'.

24.3.7.2.5 Linearity

The linear range of an analytical method is that range in which, at the 95% confidence limits, the measured amount of analyte in a blank sample, to which has been added a known amount of analyte, is equal to the known amount. This definition does not imply that the response is directly proportional to the concentration of analyte as a mathematical manipulation may be required to convert response to concentration. This may be the case for some fluorimetric analytical methods and for displacement methods such as radioimmunoassay and radioreceptor assays. To test for linearity, as defined above, the observed values for a series of blanks, to which have been added known amounts of analyte, are plotted against these known amounts. The range over which all plots fall within the 95% confidence limits (*i.e.* ± 2 standard deviations of the mean observed value at each concentration) defines the linear range. It should be noted that this procedure is an alternative method of calculating the detection limit, as well as defining the upper limit where the method may need to be modified.

24.3.7.2.6 *Practicability*

The experienced analyst should be able to assess the practicability of an analytical method if all the aspects so far discussed have been fully presented. For a complete documentation of the method it is useful to state the following: (i) time to analyze a single specimen; (ii) number of analyses per hour, day or week; (iii) cost of analysis; (iv) technical skill required; (v) dependability of equipment, including note of breakdowns; and (vi) safety aspects.

24.3.7.2.7 *Application*

The crucial test of an analytical method is in its application. The analytical method report must include a description of its use in at least one project where the drug has been administered to a whole animal. This experiment will assist in demonstrating the specificity (with respect to metabolites) and will also indicate whether the method is applicable to samples from subjects receiving therapeutic doses.

24.3.7.3 Internal Quality Control

Internal quality control is the procedure followed by the analyzing laboratory to ensure the confidence of the analytical result. This is achieved by the simultaneous analysis of control samples in each batch of analytical samples in order to monitor precision and accuracy, leading to the recognition and minimization of errors. A control sample is a sample prepared from a specimen of material such that all control samples may be assumed to have the same true value. The control specimen may be one of two types; it may be obtained by adding known quantities of the analyte to biological fluid known not to contain the analyte (a blank specimen), or it may be obtained by pooling biological fluid from appropriate subjects, volunteers or patients. The latter material may be said to correspond more closely to the real test specimen.

24.3.7.3.1 *Preparation of control samples by addition of known amounts of analyte to blank specimen*

An accurately weighed amount of analyte is dissolved in the minimum amount of water to give the primary standard. In order to obtain a complete solution in a reasonably small amount of water, it may be necessary to modify the water by the addition of acid, alkali or organic solvents. An accurately apportioned amount of primary standard is added to a suitable vessel and blank biological fluid added to give the quality control specimen. A number of such specimens should be prepared for each type of analysis to cover the range of concentrations of analyte expected. The specimens are divided into control samples of appropriate size and stored frozen.

24.3.7.3.2 *Preparation of control samples by pooling biological fluid from subjects, volunteers or patients*

As in Section 24.3.7.3.1, it is desirable to prepare specimens at different concentration levels covering the range expected in real samples. Suitable material is usually obtained by combining the residues from trials where the drug concentration has been measured, by pooling appropriate samples from similar collection periods to obtain the appropriate levels; the actual samples to be combined will depend on the time profile of the analyte in the fluid. Such specimens do not give information on the accuracy of analysis as the true value is not known. Their function is to control the analysis by signalling changes in accuracy. As in Section 24.3.7.3.1, the control specimens are divided into control samples of appropriate size and stored frozen.

24.3.7.3.3 *Preliminary analysis of quality control samples*

The accuracy and precision of the analytical method should be established by the orginator of the method and are included in the analytical method report. However, the factors that make up the values for accuracy and precision include errors introduced by different analysts. Hence, before embarking on the analysis of samples in a particular project, the individual analyst must establish that their own accuracy and precision are acceptable. This is checked by the analysis of a batch of samples consisting of calibration samples and control only. Sufficient quality control samples (usually six at each level) are included to obtain a reasonable estimate of precision. The scientist in charge of the analysis will decide if the precision is acceptable before the analyst proceeds to analyze test samples. The precision values obtained are used to construct quality control charts.

24.3.7.3.4 *Analysis of quality control samples*

Quality control samples must be included with each batch of test samples. In the simplest application of quality control, only one concentration level is used.

If the value of a quality control sample is greater than three standard deviations from the mean value then the batch of analyses is rejected unconditionally.

If the value of a quality control sample is greater than two standard deviations from the mean value then the assay procedure must be examined to determine the cause and corrective measures taken.

If seven successive values fall above, or below, or on the mean value then the assay procedure must be examined to determine the cause and corrective measures taken.

If seven successive values show a persistent rise or fall then the assay procedure must be examined to determine the cause and corrective measures taken.

A more comprehensive control is achieved by including quality control samples of at least three concentrations with each batch of analyses.

The actual values obtained for the quality control samples are plotted directly on a chart and evaluated continuously, using all three levels for interpretation.

If the values for two quality control samples are greater than two standard deviations from the mean in the same direction then the batch of analyses is rejected unconditionally.

If three quality control samples are greater than one standard deviation from the mean in the same direction then the batch of analyses is rejected unconditionally.

If two successive plots from the same quality control specimen are greater than two standard deviations from the mean in the same direction then the assay procedure must be examined to determine the cause and corrective measures taken.

24.3.7.4 External Quality Control

It is recommended[96] that any laboratory engaged in the analysis of drugs, whether as a routine service or as part of a research project, should join an external assessment scheme in addition to establishing its own internal quality control scheme. For research projects with development drugs the opportunity to join an external assessment scheme is limited but it is a wise precaution for the originating laboratory to be aware of other laboratories which may become involved in such analyses and it should take the initiative in organizing such a scheme at the onset. This could save considerable work later on in disentangling discrepancies between laboratories acting completely independently. It is a sad fact that where the results of such 'ring trials' are published there are wide discrepancies amongst the values obtained by different laboratories analyzing the same samples. For antiepileptic drugs, the wide variation in reported results for the same samples prompted the establishment of several external assessment schemes for these drugs, and schemes have also been established for other classes of drugs where the 'therapeutic level' is crucial in controlled therapy.[97;98,99]

Unlike internal quality control schemes, external schemes cannot be used for deciding on acceptance or rejection of results as there is usually a substantial time interval before the results are known. In general, participants in such a scheme should be able to assess their performance in relation to others rather than use the scheme to control their own results. Thus, the reports to the participating laboratories should include the number of results used in the calculation, the mean, standard deviation, true value as originally prepared and the participants own result.

To some extent the participants will not need to be told if their position in the results chart is pleasing or gives cause for concern. However, a number of methods have been outlined for giving a 'performance index'. The Bartscontrol index for antiepileptic drugs has been described in detail.[96] In essence, the index is based on ranking the reported results in terms of the differences of individual results from the mean. However, this type of index compares the individual performance with the overall performance of the group and a poor overall performance will, by definition, give a high rating to at least some, basically poor, performances. This index and other methods of assessing performance have been discussed in detail.[96]

24.3.7.5   Good Laboratory Practice

In the early 1970s, investigations by the American Food and Drug Administration (FDA)[100] into the laboratory records of pharmaceutical and contract research organizations revealed enough bad housekeeping and even fraudulent reports for the FDA to issue legally binding rules on good laboratory practices (GLP). This set of rules gave in detail the procedures necessary for acceptable standards for carrying out any study designed to test or prove the safety of a new drug. Although the initial emphasis was on toxicological laboratories, the fact that the FDA stipulated studies involved in the safety of drugs has meant that the analytical laboratory comes, at least in part, under these regulations.

24.3.7.5.1   *Biopharmaceutical analytical testing compliance program*

Although the restyled guidelines of good laboratory practice regulations have been published,[100] the details are more appropriate to toxicology testing and do not transfer easily to an analytical laboratory. A more useful document to follow to enable the analytical laboratory to comply with the FDA's thinking is the manual to be used by inspectors involved in the Biopharmaceutical Analytical Testing Compliance Program. This manual (available through the Freedom of Information Act) is in the form of instructions or questions, which the inspecting officer would consider in a report. How a typical analytical laboratory copes with these detailed procedures has been described with respect to facilities, personnel, specimen handling and equipment.[100]

24.3.7.5.2   *Conclusions*

Analytical chemistry has long been recognized as a very skilled science and analytical chemists have justly taken great pride in their reputation. However, the advent of GLP has indicated that a high reputation is not sufficient either to ensure the correctness of the results or to provide acceptable records to the regulatory authorities. The correct application of quality control procedures is now recognized as an integral and necessary part of the analytical function. No self-respecting analyst would now issue results without outlining the procedures involved. The quality control aspects described here are procedures that have been developed and are applied to improve the quality of the assays undertaken. Good laboratory practice, on the other hand, is a system imposed from the outside, and, while it has brought attention to the possible shortcomings of laboratory work, has meant greater bureaucracy, which does little to add to the quality of the assays.

Nevertheless, the discipline of GLP as a bureaucratic necessity to further the cause of drug development is certainly here to stay. This is well recognized by pharmaceutical companies developing drugs and by contract houses.

The discussions in this chapter result in a suggested scheme probably now followed, in one way or another, by all those developing and applying methods for analysis of drugs in biological fluids.

The method is devised by a development scientist and evaluated for precision, accuracy, sensitivity and specificity. The method is written up in full detail and this is the official analytical method.

A standard operating procedure is written, which gives the full practical details of how to carry out the analysis.

A series of quality control samples are prepared, preferably by a skilled person outside the routine applications laboratory.

Any new analyst taking up the method is provided with the standard operating procedure and should carry out a number of assays using quality control specimens to establish their own ability to perform the assay.

On satisfactory completion of trial runs the analyst is provided with the appropriate study protocol for the samples to be analyzed. The analyses are then carried out exactly according to the standard operating procedure and the protocol, including appropriate quality control and calibration samples with each batch. All raw data and notebooks are checked and signed by the analyst and supervisor where appropriate. At the conclusion of the study all these data are archived and the report issued.

The full documentation of the analytical studies should be contained in only three officially reported documents—the method report, the standard operating procedure and the analytical results. These documents must contain sufficient references so all statements can be traced back to the properly archived raw data.

If these procedures are followed, then the analytical laboratory can be satisfied that it has provided an efficient and professional contribution to the particular study.

24.3.8 AUTOMATED ASSAYS

In this work, the emphasis has been on the inherent capabilities of the various analytical methods. However, of considerable importance to the modern analytical laboratory is the rate at which it can analyze samples and the cost of doing so. Thus, there have been many attempts to automate the various steps in the analytical procedure. Ideally, the analyst would like to feed in samples at one end of the laboratory and receive a typed report at a work-station, without human intervention. Considerable progress has been made on the automation of the measuring and computing aspects of analytical methods so that, for example, a gas chromatograph with automatic injector and computing integrator will outstrip the analyst's capacity to prepare samples in a suitable form. Thus, the development of automated systems to prepare samples, whether for subsequent chemical or biological assay, has been the focus of activity in more recent years.

Automated sample preparation for drug analysis can either attempt to imitate the action of the human operator—the robot chemist—or it can utilize procedures that are only possible with mechanical aids. An example of the latter is the Auto-Analyzer (Technicon),[101] in which the samples are set up as a train of tiny reaction vessels, separated by air bubbles. The samples flow through various mixing, separating and reaction devices in sequence, each device being designed to process the flowing segments presented to it in rapid succession rather than separate vessels that a robot chemist would handle. The Auto-Analyzer was designed and developed for clinical chemistry with a relatively straightforward and simple endpoint such as UV absorption or flame photometry. It has been little used for more complex assays, requiring component separation and solid phase manipulations.

The remainder of this section describes the principal successful automated procedures for sample preparation prior to final assay. Table 10 summarizes the overall automation possibilities for drug assays.

24.3.8.1 The PREP Automated Sample Processor

The PREP automated sample procesosr[112] was developed by Du Pont and is a centrifugally based, microprocessor-controlled instrument. The preparation principles depend on the popular

Table 10 Some Possibilities for Automation of Analytical Methods

Operation	Automation	Ref.
Liquid–liquid extraction	Multiple columns	102
	Autoanalyzer	103
	Flow injection	104
Liquid–solid extraction	Extraction columns	105, 106
Application to TLC	Autospotters	107
Injection on GC, HPLC	Autoinjectors; switching valves	
Sample handling	Centrifugal analyzers	108
Plating in microbiology	Rotary inoculator	109
Plate reading	Autoreaders	110
Immunoassay, competitive binding methods	Autodiluters, *etc.*	111
Counting of radioactivity	Scintillation counters	
Spectrometry readings	Sample changers	

solid phase extraction cartridge techique. For the PREP system the cartridge assembly is shown in Figure 6. In the extraction procedure sample aliquots are buffered and placed in the reservoir of the resin column. The cartridge is placed in the centrifuge and the entire extraction procedure is carried out in 10–30 min, depending on the chosen extraction program.

In typical applications of the device, the extraction variables are the adsorbent, the sample buffer, the wash solvent and the elution solvent, giving wide scope for the design of a specific extraction procedure for a given analyte. For example, in an assay for phenobarbital the sample is made alkaline with sodium hydroxide (1 mL, 0.1 M) and loaded into a strong anion exchanger cartridge. In the automated sequence the cartridge is washed with methanol and then methanol with boric acid ($pK = 9$). Phenobarbital is eluted with methanol/acetic acid ($pK = 4.75$). A recovery of 99% of added drug is claimed.

24.3.8.2   Advanced Automated Sample Processor (AASP)

The AASP system, developed by Varian, was designed to couple the cartridge extraction system with on-line HPLC.[106] The heart of the extraction is a cassette of 10 miniature columns packed with a suitable sorbent, generally silica-based, with bonded groups such as C_2 to C_{18} aliphatics, diol, amine, cyano, phenyl and ion exchange. The biological sample is applied to the top of the column and the cassette placed in a sealed chamber to allow positive pressure (air or nitrogen) to force the sample, with subsequent washes, through the sorbent bed. The analytes are retained on the columns and the cassettes then placed in the autoinjector assembly for HPLC, where the columns are incorporated into the analytical sequences *via* switching valves. Much of the system is readily automated and microprocessor controlled.

24.3.8.3   Centrifugal Analyzers

In the 1960s, Anderson introduced the concept of 'fast analyzers', where all operations take place in a very short time, preferably for a large number of samples processed in parallel.[108] This was made possible by the invention of methods for measuring, transferring, mixing and measuring the absorbance values of a large number of samples over very short intervals. By including standards in the parallel analyses, it is possible to make measurements before reactions have gone to completion. The objectives were realized by incorporating the analytical system into a centrifuge rotor, thereby enabling: (i) accurate measurement of volume using centrifugal force to fill, debubble and level menisci accurately in small vessels; (ii) quantitative transfer of fluids using centrifugal force; and (iii) rapid measurement of the absorbance of a large number of samples by rotating them rapidly past a beam of light. A number of ingenious devices have been described for analytical manipulations, all of

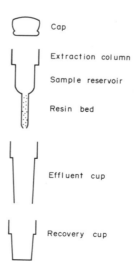

Cap

Extraction column

Sample reservoir

Resin bed

Effluent cup

Recovery cup

Figure 6   Components of the PREP cartridge

which can be carried out simultaneously on a large number of samples according to the number of places in the rotor. In a typical analysis, using the centrifugal analyzer, the absorbance values of all reception cuvettes are displayed simultaneously on a VDU, and a computer is used to perform appropriate calculations and report results. Centrifugal analyzers are mainly used in clinical chemistry laboratories analyzing large numbers of samples, where a short turn-round time is required. Centrifugal analyzers have been adapted for radioimmunoassays, including assays for digoxin and insulin, and there has been considerable effort in the development of centrifugal analyzers for enzyme immunoassays as all steps can be carried out without removal to the scintillation counter as would be necessary for radioimmunoassays.[113] In addition to the normal advantages of automated equipment, the use of microvolumes in such analyzers is extremely economical in the use of expensive reagents. Thus, an adaptation of the Centrifichem analyzer has been used to assay theophylline, phenobarbital, phenytoin, carbamazepine, primidone, ethosuximide and gentamicin, using sample volumes of 3 µL and total cuvette volumes of 210 µL. The imprecision is less than 1.5%, with drug recoveries of between 90 and 105% over the calibration range used. It is claimed that 600 tests can be carried out with a 100-test kit as normally provided.[114]

24.3.8.4 The Zymark Robot

Advances in precision engineering and computer controls have allowed the development of robotic arms which can carry out complex sequences of manipulations with great precision, repetitively. The use of such arms finds an obvious application as the 'robot chemist', where the actions of the human analyst are exactly reproduced. It is only when the systems designer attempts to do this, that it is realized how complex is even the simplest operation such as weighing a flask. Nevertheless, there are several designs of robot arms which are in use in analytical laboratories, of which the most highly developed is the Zymark Robot.[115] This system consists of a robot arm with a cylindrical working envelope, a number of stations placed within the envelope, and a computer controller (Figure 7).

The entire layout is arranged and programmed to mimic the operations carried out by a human analyst. Key components are shown in Figure 8. The arm of the robot can equip itself with a suitable

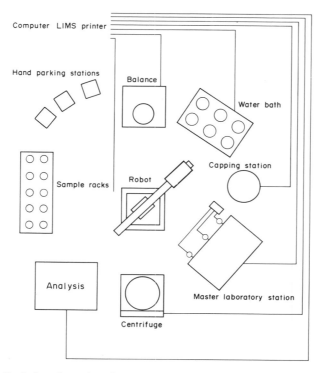

Figure 7 Typical configuration of a robot system for preparation of samples for analysis

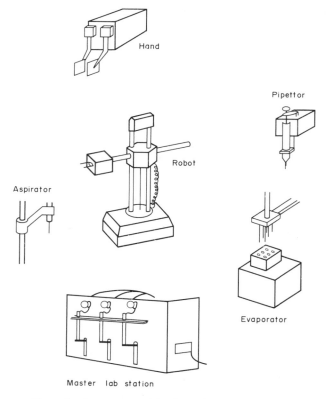

Figure 8 Components of the Zymark robotic system

hand from a hand-parking station, and can thence transfer samples and probes from station to station.

The first step in developing a program is to move the arm manually using the controller to the desired position and assign a unique descriptive name. The computer stores the appropriate coordinates under this name and a complete program (which itself can be named) can be defined as a list of the descriptive names.

Stations in the system can include sets of syringes for precise delivery of solvents or reagents, balances, centrifuges and test-tube racks.

The use of robotics will have an interesting effect on analytical laboratories. A complete robotic system needs access all round, and the robot itself needs to have all its stations within reach of its arm. Thus, laboratory benches will tend towards islands, even circular islands, rather than the traditional rectangular benches.

24.3.8.5 Chemometrics

The use of computers for rapid calculations which would previously have been done by hand has greatly aided the throughput of samples in analytical laboratories. Computers also play a large part in the control features of automatic analyzers of all types. However, these functions are basically, once again, the functions of a robot chemist; they imitate the traditional manipulations. Chemometrics has been defined as the development and application of mathematical and statistical methods to extract useful chemical information from chemical measurements.[116] Thus, chemometrics should enable us to make better use of modern equipment and techniques, by making the chemistry more effective and informative. Many of these refinements would be very difficult without computers, some would not even be possible without computers. For the techniques used in drug analysis the most useful application of chemometrics is in the enhancement of the signal to noise ratio. In general, noise is a high frequency component of the signal, whereas the required output is a low frequency component. An obvious example is in a 'noisy chromatogram' where the desired peak is obscured by a large number of 'spikes'. A Fourier transform of the data separates the high and low frequency components; a smoothing routine eliminates the high frequency component, and on retransformation the desired 'clean' signal is obtained. The availability of fast Fourier transformation chips means that such digital filtering can be done on line. In similar fashion, partially resolved chromatographic peaks can be resolved by mathematical analysis of the combined signal.

Chemometrics is also used to optimize experimental conditions, either by continually altering the conditions by a feedback loop, or by aiding in the setting up of the best conditions initially, such as in the simplex optimization approach to deciding on the optimum mobile phase for HPLC.[117]

24.3.8.6 Advantages of Automation

The initial outlay for any automated system can be significant and sometimes the evaluation and setting up can consume so much time that the analyst with urgent samples to process may question its worth. Nevertheless, successful implementation of automation can bring considerable economic and other benefits for the analytical laboratory.

The economic advantages of automation include the efficient use of expensive equipment, such as chromatographs, which can be operated unattended for between 16 and 24 h per day. Indeed, as a rule of thumb, merely automating the injection cycle for GC or HPLC may be said to double or even triple the productivity of a single technician without the purchase of an additional chromatograph. Thus, the payback period will be sufficiently short to justify automation.

To the analyst, the noneconomic advantages, however, are more important. The elimination of human error will itself improve precision, but the potential for including extra quality controls, calibration standards and duplicates in an automated system helps to identify the causes of errors and changes in experimental conditions.

Automation can generally standardize, if not decrease, the turn-round time for samples submitted to the laboratory. A typical work pattern would be to put all the samples received during the day onto the automated system in the afternoon with results available the next morning. Automated systems now usually include the calculation and documentation of results and the 'customer' can receive a report straight from the computer.

Operations that may be hazardous (radiochemicals and toxic derivatizing agents) can be safely carried out by an automated system by shielding or by enclosing the system in a fume-cupboard.

The reproducibility of an automated system does not lie only in the ability of the equipment to perform repetitive operations without tiring as the human analyst would, but in that the samples are all treated in exactly the same way and for the same time, *i.e.* even if derivatizations are incomplete or equilibria have not been attained then this will be true for all samples and the reproducibility will be retained.

Finally, an automated system is well suited to method development. This may be done by a high throughput of a series of planned analyses or even, with some types of automation, by programming the device to search for optimal conditions and propose the desired method.

24.3.9 REFERENCES

1. I. Phillips and J. D. Williams, in 'Laboratory Methods in Microbial Chemotherapy', ed. D. S. Reeves, I. Phillips, J. D. Williams and R. Wise, Churchill-Livingstone, Edinburgh, 1978, p. 3.
2. B. M. Barker and F. Prescott, 'Antimicrobial Agents in Medicine', Blackwell, Oxford, 1973, p. 214.
3. A. Brandberg, O. Almersjo, E. Falsen, B. Gustavsson, L. Hafstrom and G. B. Lindblom, *Acta Pathol. Microbiol. Scand., Sect. B: Microbiol.*, 1977, **85**, 227.
4. D. C. Garratt, 'The Quantitative Analysis of Drugs', 3rd edn., Chapman and Hall, London, 1964.
5. R. V. Smith and J. T. Stewart, 'Textbook of Pharmaceutic Analysis', Lea and Febiger, Philadelphia, 1981, p. 250.
6. R. A. Rippere, *J. Assoc. Off. Anal. Chem.*, 1979, **62**, 951.
7. D. S. Reeves, I. Phillips, J. D. Williams and R. Wise (eds.), 'Laboratory Methods in Antimicrobial Chemotherapy', Churchill-Livingstone, Edinburgh, 1978.
8. D. C. Garratt, 'The Quantitative Analysis of Drugs', 3rd edn., Chapman and Hall, London, 1964, p. 814.
9. M. Barber and P. M. Waterworth, *Br. Med. J.*, 1964 **ii**, 344.
10. I. Nilsson-Ehle, T. T. Yoshikawa, M. C. Schotz and L. B. Guze, *Antimicrob. Agents Chemother.*, 1976, **9**, 754.
11. R. P. Buhs, T. E. Maxim, N. Allen, T. A. Jacob and F. J. Wolf, *J. Chromatogr.*, 1974, **99**, 609.
12. L. O. White, H. A. Holt, D. S. Reeves, M. A. Bywater and R. P. Bax, in 'Current Chemotherapy and Infectious Disease, Proc. 11th. ICC and 19th. ICAAC', Am. Soc. Microbiol., Washington, D.C., 1980, p. 153.
13. P. M. Waterworth, in 'Laboratory Methods in Antimicrobial Chemotherapy', ed. D. S. Reeves, I. Phillips, J. D. Williams and R. Wise, Churchill-Livingstone, Edinburgh, 1978, p. 41.
14. R. F. Cosgrove, *J. Appl. Bacteriol.*, 1978, **44**, 199.
15. R. V. Smith, R. G. Harris, D. D. Maness and A. Martin, *Anal. Lett.*, 1977, **10**, 719.
16. R. V. Smith and J. T. Stewart, 'Textbook of Pharmaceutic Analysis', Lea and Febiger, Philadelphia, 1981, p. 254.

17. R. P. Ekins, *Clin. Chim. Acta.* 1960, **39**, 1157.
18. B. E. P. Murphy, W. Engelberger and C. J. Pattee, *J. Clin. Endocrinol. Metab.*, 1963, **23**, 293.
19. J. P. Liberti, C. H. Duvall, M. A. Mackler and G. R. Prout, Jr., *J. Lab. Clin. Med.*, 1970, **76**, 530.
20. E. V. Jensen and H. I. Jacobson, *Recent Prog. Horm. Res.*, 1962, **18**, 387.
21. R. J. Lefkowitz, J. Roth and I. Pastan, *Science (Washington, D.C.)*, 1970, **170**, 633.
22. R. Simantov, S. R. Childers and S. H. Snyder, *Brain Res.*, 1977, **135**, 358.
23. A. G. Gilman, *Proc. Natl. Acad. Sci. USA*, 1970, **67**, 305.
24. S. J. Enna and S. H. Snyder, *J. Neurochem.*, 1976, **26**, 221.
25. I. Creese and S. H. Snyder, *Nature (London)*, 1977, **270**, 180.
26. S. R. Nahorski, M. I. Batta and D. Barnett, *Eur. J. Pharmacol.*, 1978, **52**, 393.
27. F. Owen, R. Lofthouse and R. C. Bourne, *Clin. Chim. Acta*, 1979, **93**, 305.
28. R. F. Squires and C. Braestrup, *Nature (London)* 1977, **266**, 732.
29. T. Mennini and S. Garattini, *Life Sci.*, 1982, **31**, 2025.
30. P. Hunt, J.-M. Husson and J.-P. Raynaud, *J. Pharm. Pharmacol.*, 1979, **31**, 448.
31. H. Mohler and T. Okada, *Science (Washington, D.C.)*, 1977, **198**, 849.
32. R. S. Yalow and S. A. Berson, *J. Clin. Invest.*, 1960, **39**, 1157.
33. G. E. Abraham, *J. Clin. Endocrinol. Metab.*, 1969, **29**, 866.
34. B. F. Erlanger, F. Borek, S. M. Beiser and S. Lieberman, *J. Biol. Chem.*, 1957, **228**, 713.
35. M. J. Halloran and C. W. Parker, *J. Immunol.*, 1966, **96**, 373.
36. A. H. Korn, S. H. Feairheller and E. M. Filachione, *J. Mol. Biol.*, 1972, **65**, 525.
37. J. D. Robinson, B. A. Morris, E. M. Piall, G. W. Aherne and V. Marks, in 'Radioimmunoassay in Clinical Biochemistry', ed. C. A. Pasternak, Heyden, London, 1975, p. 81.
38. A. Castro, N. Monji, H. Ali, J. M. Yi, E. R. Bowman and H. McKennis, Jr., *Eur. J. Biochem.*, 1980, **104**, 331.
39. V. H. T. James and S. L. Jeffcoate, *Br. Med. Bull.*, 1974, **30**, 50.
40. J. H. L. Playfair, B. A. L. Hurn and D. Schulster, *Br. Med. Bull.*, 1974, **30**, 24.
41. B. Peskar and S. Spector, *J. Pharmacol. Exp. Ther.*, 1973, **186**, 167.
42. C. T. Liu and F. L. Adler, *J. Immunol.*, 1973, **111**, 472.
43. R. E. Tigelaar, R. L. Rapport, J. K. Inman and H. J. Kupferberg, *Clin. Chim. Acta*, 1973, **43**, 231.
44. J. W. A. Findlay, J. T. Warren, J. A. Hill and R. M. Welch, *J. Pharm. Sci.*, 1981, **70**, 624.
45. B. A. Faraj and F. M. Ali, *J. Pharmacol. Exp. Ther.*, 1981, **217**, 10.
46. C. E. Cook, D. L. Williams, M. Myers, C. R. Tallent, G. A. Leeson, R. A. Okerholm and G. J. Wright, *J. Pharm. Sci.*, 1980, **69**, 1419.
47. H. G. Eckert, S. Baudner, K. E. Weimer and H. Wissmann, *Arzneim.-Forsch.*, 1981, **31**, 419.
48. E. M. Piall, G. W. Aherne and V. M. Marks, *Br. J. Cancer*, 1979, **40**, 548.
49. A. Yamaoka and T. Takatori, *J. Immunol. Methods*, 1979, **28**, 51.
50. G. P. Mould, J. Clough, B. A. Morris, G. Stout and V. Marks, *Biopharm. Drug Dispos.*, 1981, **2**, 49.
51. R. Aderjan and G. Schmidt, *Z. Rechtsmed.*, 1979, **83**, 191.
52. E. P. Yeager, U. Goebelsmann, J. R. Soares, J. D. Grant and S. J. Gross, *J. Anal. Toxicol.*, 1981, **5**, 81.
53. S. Iwamura and A. Kambegawa, *J. Pharmacobio-Dyn.*, 1981, **4**, 275.
54. H. Suzuki, M. Miki, Y. Sekine, A. Kagemoto, T. Negoro, T. Maeda and M. Hashimoto, *J. Pharmacobio-Dyn.*, 1981, **4**, 217.
55. B. R. Clark, B. B. Tower and R. T. Rubin, *Life Sci.*, 1977, **20**, 319.
56. M. Michiels, R. Hendriks and J. Heykants, *J. Pharm. Pharmacol.*, 1983, **35**, 86.
57. R. Virtanen, J. Kanto and E. Iisalo, *Acta Pharmacol. Toxicol.*, 1980, **47**, 208.
58. M. A. Orchard, I. A. Blair, J. M. Ritter, L. Myatt, M. Jogee and P. J. Lewis, *Biochem. Soc. Trans.*, 1982, **10**, 241.
59. J. Morvay, K. Fotherby and I. Altorjay, *Clin. Chim. Acta*, 1980, **108**, 147.
60. K. K. Midha, J. C. K. Loo, M. Charette, M. L. Rowe, J. W. Hubbard and I. J. McGilveray, *J. Anal. Toxicol.*, 1978, **2**, 185.
61. R. P. Ekins, in 'Radioimmunoassay in Clinical Biochemistry', ed. C. A. Pasternak, Heyden, London, 1975, p. 3.
62. H. G. Eckert, *Angew. Chem.*, 1976, **15**, 530.
63. K. Dixon, in 'Radioimmunoassay in Clinical Biochemistry', ed. C. A. Pasternak, Heyden, London, 1975, p. 15.
64. V. Herbert, in 'Protein and Polypeptide Hormones', ed. M. Margoulies, Excerpta Med., Amsterdam, 1969, p. 55.
65. K. J. Catt, *Acta Endocrinol. (Copenhagen), Suppl.*, 1969, **142**, 222.
66. R. S. Kamel, J. Landon and D. S. Smith, *Clin. Chem. (Winston-Salem, N.C.)*, 1980, **26**, 1281.
67. J. Chamberlain, 'The Analysis of Drugs in Biological Fluids', CRC, Boca Raton, 1985, p. 124.
68. R. S. Schneider, *Clin. Chem. (Winston-Salem, N.C.)*, 1973, **19**, 821.
69. A. Poklis, *J. Anal. Toxicol.*, 1981, **5**, 174.
70. J. Fenton, M. Schaffer, N. W. Chen and E. W. Bermes, *J. Forensic Sci.*, 1980, **25**, 314.
71. F. Monaco and S. Piredda, *Epilepsia (N.Y.)*, 1980, **21**, 475.
72. A. Johnston and J. Hamer, *Clin. Chem. (Winston-Salem, N.C.)*, 1981, **27**, 353.
73. W. R. Kulpmann and M. Oellerich, *J. Clin. Chem. Clin. Biochem.*, 1981, **19**, 249.
74. P. R. Oeltgen, S. R. Hamann and R. A. Blouin, *Ther. Drug Monit.*, 1980, **2**, 423.
75. R. G. Buice, W. E. Evans, J. Karas, C. A. Nicholas, P. Sidhu, A. B. Straughn, M. C. Meyer and W. R. Crom, *Clin. Chem. (Winston-Salem. N.C.)*, 1980, **26**, 1902.
76. L. Von Meyer, G. Kauert and G. Drasch, *Beitr. Gerichtl. Med.*, 1981, **39**, 113.
77. A. K. N. Nandedkar, R. Williamson, K. Kutt and G. F. Fairclough, *Ther. Drug Monit.*, 1980, **2**, 427.
78. W. C. Griffiths, P. Dextraze, M. Hayes, J. Mitchell and I. Diamond, *Clin. Toxicol.*, 1980, **16**, 51.
79. H. R. Ha, G. Kewitz, M. Wenk and F. Follath, *Br. J. Clin. Pharmacol.*, 1981, **11**, 312.
80. G. Sivorinovsky, *Altex Chromatogram*, 1980, **3**, 1.
81. M. Leroux, D. Budnik, K. Hall, J. Irvine-Meek, N. Otten and S. Seshia, *Clin. Biochem. (Ottawa)*, 1981, **14**, 87.
82. W. Sadee and G. C. M. Beelen, 'Drug Level Monitoring', Wiley, New York, 1980, p. 90.
83. J. F. Burd, R. C. Wong, J. E. Feeney, R. J. Carrico and R. C. Boguslaski, *Clin. Chem. (Winston-Salem, N.C.)*, 1977, **23**, 1402.
84. J. Chamberlain, 'The Analysis of Drugs in Biological Fluids', CRC, Boca Raton, 1985, p. 124.

85. R. K. Leute, E. F. Ullman, A. Goldstein and L. A. Herzenberg, *Nature (London)*, 1971, **236**, 253.
86. J. D. Robinson, personal communication.
87. R. V. Smith and J. T. Stewart, 'Textbook of Pharmaceutic Analysis', Lea and Febiger, Philadelphia, 1981, p. 261.
88. G. J. Yakatan, C. A. Martin and R. V. Smith, *Anal. Chim. Acta*, 1976, **84**, 173.
89. E. Scarbrough, J. W. Williams and D. B. Northrup, *Antimicrob. Agents Chemother.*, 1979, **16**, 221.
90. P. Stevens and L. S. Young, in 'Microbiology-1975', ed. D. Schlessinger, Am. Soc. Microbiol., Washington, D.C., 1975, p. 64.
91. P. S. Lietman, T. J. White and W. V. Shaw, *Antimicrob. Agents Chemother.*, 1976, **10**, 347.
92. A. E. Staubus, A. B. Weinrib and L. Malspeis, *Biochem. Pharmacol.*, 1984, **33**, 1633.
93. H. Hojer and L. Nilsson, *J. Antimicrob. Chemother.*, 1978, **4**, 503.
94. D. J. Tocco, F. A. de Luna, A. E. W. Duncan, T. C. Vassil and E. H. Ulm, *Drug Metab. Dispos.*, 1982, **10**, 15.
95. M. D. Smith and C. L. Olson, *Anal. Chem.*, 1974, **46**, 1544.
96. D. Burnett and G. J. Ayers, in 'Therapeutic Drug Monitoring', ed. A. Richens and V. Marks, Churchill-Livingstone, Edinburgh, 1981, p. 83.
97. S. Boobis, N. Persaud and A. Richens, *Ther. Drug Monit.*, 1979, **1**, 257.
98. D. S. Reeves and H. J. Bywater, *J. Antimicrob. Chemother.*, 1975, **1**, 103.
99. W. S. Turner, P. Turano and S. Badzinski, in 'Pharmacokinetics of Psychoactive Drugs: Blood Levels and Clinical Response', Spectrum Publications, New York, 1976, p. 33.
100. J. Chamberlain, 'The Analysis of Drugs in Biological Fluids', CRC, Boca Raton, 1985, p. 61.
101. L. T. Skeggs, *Am. J. Clin. Pathol.*, 1957, **28**, 311.
102. G. B. Barlow, in 'Trace Organic Sample Handling', ed. E. Reid, Horwood, Chichester, 1981, p. 221.
103. P. B. Stockwell, in 'Trace Organic Sample Handling', ed. E. Reid, Horwood, Chichester, 1981, p. 227.
104. B. Karlberg and S. Thelander, *Anal. Chim. Acta*, 1979, **104**, 21.
105. W. Pacha and H. Eckert, in 'Trace Organic Sample Handling', ed. E. Reid, Horwood, Chichester, 1981, p. 240.
106. R. D. McDowall, J. C. Pearce, G. S. Murkitt and R. M. Lee, in 'Bioactive Analytes', ed. E. Reid, B. Scales and I. D. Wilson, Plenum Press, New York, 1986, p. 235.
107. J. P. Leppard, A. D. R. Harrison and J. D. Nicholas, *J. Chromatogr. Sci.*, 1976, **14**, 438.
108. N. G. Anderson, *Am. J. Clin. Pathol.*, 1970, **53**, 778.
109. C. H. Pearson and J. E. M. Whitehead, *J. Clin. Pathol.*, 1974, **27**, 430.
110. S. W. B. Newsom, in 'Laboratory Methods in Antimicrobial Chemotherapy', ed. D. S. Reeves, I. Phillips, J. D. Williams and R. Wise, Churchill-Livingstone, Edinburgh, 1978, p. 50.
111. M. E. Jolley, S. D. Stroupe, K. S. Schwenzer, C. J. Wang, M. Lu-Steffers, H. D. Hill, S. R. Popelka, J. T. Holen and D. M. Kelso, *Clin. Chem. (Winston-Salem, N.C.)*, 1981, **27**, 1575.
112. R. C. Williams, in 'Drug Determination in Therapeutic and Forensic Contexts', ed. E. Reid and I. D. Wilson, Plenum Press, New York, 1984, p. 53.
113. J. Chamberlain, 'The Analysis of Drugs in Biological Fluids', CRC, Boca Raton, 1985, p. 184.
114. D. Studts, G. T. Haven and E. J. Kiser, *Ther. Drug Monit.*, 1983, **5**, 335.
115. R. K. Brown, F. H. Zenie and G. Johnston, *Chem. Br.*, 1986, **22**, 640.
116. D. Betteridge, *Lab. Pract.*, 1983, **32**, 13.
117. J. L. Glajch, J. J. Kirkland, K. M. Squire and J. M. Minor, *J. Chromatogr.*, 1980, **199**, 57.

24.4

Methods in Drug Metabolism

ALAN R. BOOBIS, DOROTHEA SESARDIC and NIGEL J. GOODERHAM
Royal Postgraduate Medical School, London, UK

24.4.1 INTRODUCTION

The magnitude and duration of effect of many drugs and other chemicals is determined by their rate of metabolism. Although the products of metabolism are often less active and more water soluble than the parent compound, this is not always so, and in some instances they are biologically active and may even be toxic or carcinogenic. The activities of the enzymes of drug metabolism show considerable inter- and intra-individual variation, subject to influence by both environmental and genetic factors. As a consequence, it is important to define the routes and rates of metabolism of a new drug or other chemical early in its development and to try to obtain some indication of how such a compound will be eliminated in man.

Although much useful information can be obtained on the metabolism of foreign compounds by studies in whole animals, there are many situations where studies *in vitro* provide additional and supplementary information. One area where this is becoming increasingly important is in the use of human tissues since it is obviously not possible to administer a chemical to man prior to toxicity testing in animals.

Although obvious, it is important to distinguish between the substrate and the enzyme metabolizing it. For many researchers the metabolism of a particular compound is the subject of interest, whereas for others it is the enzyme involved and its characterization that is sought. In the development of drugs and chemicals, the primary consideration is of the compound itself and in this chapter general principles will be elaborated that can be applied to such studies. However, a knowledge of the enzymes may also be important, such as their specificity, regulation and similarity amongst species. For example, the knowledge that the isoenzyme of cytochrome *P*-450 catalyzing the 4-hydroxylation of debrisoquine is involved in the oxidation of a new drug would lead immediately to the suspicion that the drug might well show polymorphic oxidation in man.

One of the earliest studies performed on the metabolism of a new drug or other chemical is to determine the profile of metabolites produced on administration to animals. Here, there is a qualitative assessment, utilizing a variety of chemical techniques, of the metabolites eliminated in the urine and possibly in the faeces. The complexity of metabolic pathways often results in substantial degradation of the primary products of metabolism. As a consequence, it is not always possible to identify these from the elimination products. Studies *in vitro* often enable the primary products of metabolism to be identified and may give indications as to which further metabolites should be sought *in vivo*. Further, studies with human tissue may enable some indication of the likely metabolites in man to be obtained.

Having identified the metabolites produced, it is often necessary to determine their rates of production. *In vivo*, this would involve determining the kinetics of the parent compound and of the metabolite in question. Studies *in vitro* may well enable useful predictions to be made regarding the relative rates of formation of different metabolites, and in particular the concentration at which the enzymes involved are likely to become saturated. Such studies can also help in identifying the major sites of metabolism within the body, although it is important to realize that the demonstration that a given tissue can metabolize a compound *in vitro* does not necessarily mean that the tissue does contribute to the metabolism of the compound *in vivo*.

Studies *in vitro* on the metabolism of foreign compounds play an essential role in elucidating the reaction mechanisms involved in producing a given metabolite, particularly where such a metabolite is toxic. Often, these are electrophilic species and, as such, so short lived that they are not eliminated intact. Studies *in vitro* may enable the route of formation of such a metabolite, and its rate of production, to be determined.

Although, in such studies, most emphasis is placed on identifying the routes and rates of formation of the metabolites, there may be some concern to identify the particular enzymes involved. This would provide some indication as to whether the activity is likely to be induced, show polymorphism or sex differences in man, and also which other compounds are likely to inhibit activity. Obviously, much useful information on these questions can be obtained from studies *in vitro*.

As already indicated above, one area where studies *in vitro* on drug metabolism are of paramount importance is in comparison of species with man. Early in drug or chemical development it is possible to examine the metabolism of the compound by human-derived tissue and to compare the routes and rates therein with those obtained in animal studies. This should provide an early indication of the likely pharmacokinetics in man, and also whether a potentially toxic metabolite produced in animals is likely to be produced in the human population. This particular aspect of studies *in vitro* is still relatively young but will certainly develop in the coming years.

It is ironic that the longest-established technique for studies *in vitro*, that of the isolated perfused liver, is perhaps the least used today of the methods available for studying the metabolism of foreign compounds *ex vivo*. Attempts to perfuse the liver date back to the last decades of the 19th century, but modern methods are all based on the studies of Brauer *et al.*[1] and Miller *et al.*,[2] who, in 1951, simultaneously and independently published methods for perfusing rat liver, the basic method for which has not changed since that time. Important modifications have included the introduction of heterologous erythrocytes, replacement of the medium with a totally synthetic one, and perfusion at a constant flow rate rather than at constant pressure. Although methods to perfuse the liver had been available for some time, it was not until the mid 1960s that the metabolism of foreign compounds was studied using this technique, when groups such as Kalser *et al.*[3] and Stitzel *et al.*[4] investigated the elimination of atropine, hexobarbitone, procaine, pentobarbitone and many others.

Methods have been available since 1931 for the perfusion of the isolated lung, but it was not until the 1970s[5] that these found application in studies of drug metabolism. More recently, methods have been developed for the perfusion of the isolated kidney (1959)[6] and various sections of the intestine (1967),[7] but to date these have found only limited application in studies of drug metabolism.

Some of the early limitations of perfusing intact organs led to the exploration of alternative techniques and the first of these was probably that of tissue slices, pioneered by such as Warburg (1923).[8] This has found limited but continuing application in studies of drug metabolism. Although the technique has been largely supplanted by that of dispersed isolated cells, kidney slices still provide a relatively valuable means of studying active uptake processes in that organ.

Many methods to isolate hepatocytes have been attempted, early efforts being met with very limited success. Modern techniques date from studies of Berry and Friend in 1969[9] and by Seglen in the early 1970s.[10,11] The pioneering work of these groups has led to the development of a highly reproducible technique for the isolation of cells of excellent viability and well preserved metabolic activities. These have been used extensively, either as a dispersed suspension or in culture, for studies of drug metabolism. Culturing the cells provides the advantage of long term maintenance but against that the activities of many of the enzymes of drug metabolism decrease quite rapidly after commencing culture.[12] One important advantage of cultured cells is that protein synthesis can be studied so that induction of enzymes may be investigated.

As with many of the techniques in drug metabolism, the development of methods for *in vitro* investigation using subcellular fractions depended upon a considerable prior body of work. During the 1930s and 1940s the techniques of tissue fractionation were developed and the microsomal fraction, derived from the endoplasmic reticulum, was characterized. However, it was not until the mid-1950s that this fraction was applied to the study of the metabolism of foreign compounds.[13] Since that time, of course, the use of such subcellular preparations has achieved extremely widespread application and is undoubtedly the most widely used technique for studying drug metabolism *in vitro* today. Other subcellular fractions have also been investigated, particularly the soluble or cytosolic fraction, and to a lesser extent the mitochondrial and nuclear fractions.

Following the relatively reactionary tradition of drug metabolism towards technical innovation, developments in protein purification were only relatively belatedly applied to the investigation of the biotransformation of foreign compounds. Thus, it was not until the 1970s that methods were developed for the purification to apparent homogeneity of many of the enzymes of drug metabolism. The 1980s have seen the introduction and development of the techniques of monoclonal antibody production and molecular biology as applied to studies of drug metabolism.

This chapter will consider systematically the different preparations that might be used for studies *in vitro* of drug metabolism, including isolated organs, tissue slices, isolated cells, subcellular fractions, purified enzymes, and the more recent techniques in drug metabolism, including molecular biology probes.

24.4.2 ISOLATED ORGANS

24.4.2.1 Introduction

Isolated perfused organs, that is, isolated from the systemic circulation, provide a unique preparation by which to study the biotransformation of foreign compounds. The organ is morphologically intact, it retains the normal anatomical arrangement of the arterial and venous blood supply, it maintains the ability to excrete products of biotransformation, results are amenable to classical pharmacokinetic analysis, and it is possible to alter the levels of cofactors or enzymes by prior treatment of the donor animal. The technique is not without its disadvantages, of course. It requires a degree of surgical expertise, experiments are labour intensive, organs have to be prepared fresh at the time of use, and they are not very suitable for mechanistic studies on the enzymes involved in biotransformation reactions. However, as a means of obtaining information on the nature of the metabolites produced from a given compound and its relative distribution amongst several metabolites, the isolated perfused organ preparation has much to commend it.

The liver is the organ most involved in biotransformation reactions, and as such has been most extensively studied by this method. However, techniques have been developed for the perfusion of other organs and these have been applied to a limited extent in studies of drug metabolism. The perfused organ system is perhaps the most suitable technique for determining whether a given tissue makes a significant contribution to the overall biotransformation of a foreign compound in a way that isolated cells or subcellular fractions cannot. Thus, the isolated perfused intestinal loop preparation has found considerable application in determining the relative contribution of the liver and intestine to the presystemic metabolism of certain compounds. Similarly, the perfused kidney preparation enables determination of the contribution of that organ to the appearance of metabolites in the urine.

The basic principle of organ perfusion is extremely simple, it involves the isolation of the arterial and venous blood supply from that of the systemic circulation and cannulation of any excretory duct, if appropriate. However, the execution of this is somewhat more complex. The organ may be perfused *in situ* or removed from the donor and perfused in purpose-built apparatus. The method utilized will depend upon the donor species and the organ in question. Thus, the liver is frequently removed from the donor, except in very small and very large species. In contrast, the isolated intestinal loop preparation is most readily perfused whilst still within the donor animal.

The organ should be perfused with a medium as close to blood as possible. The essential characteristics are a balanced salt composition, buffered pH, appropriate osmotic species and oxygen-carrying capacity. The specific medium chosen will depend upon the approaches adopted to achieve these goals. Thus, it is possible to use a totally synthetic medium with a precisely defined chemical composition in which the oxygen-carrying capacity is provided either by the solution of oxygen in the aqueous phase[14] or by the addition of fluorocarbons.[15] However, it is also possible to use a semisynthetic medium based on a salt solution supplemented with red cells to provide a means of oxygen transport. The cells may be from homologous donors[2] or from heterologous donors such as ox[16] or man.[17]

The medium should be delivered to the organ under conditions as close to physiological as possible, with respect to perfusion pressure, flow rate and pulsatile flow. The flow rate will be determined to some extent by the oxygen-carrying capacity of the medium, as the oxygen supply to the tissue is more important than the overall flow rate. It is the delivery of the perfusion medium to the tissue that requires most in the way of apparatus. Obviously, one requires a pump, usually a peristaltic model, with appropriate pulse-dampening equipment, some means of oxygenating the medium, a filter for the medium, bubble traps to prevent emboli and some means of bringing the temperature of the medium to 37 °C. Where the tissue is being perfused in a purpose-built apparatus, there must be some means of maintaining the temperature of the organ and also of preventing the surface of the tissue from becoming dehydrated. Some consideration should also be given to the means by which the perfusion medium is to be returned to the reservoir and also to the collection of any excretory fluid.

Obviously, as in any study of drug metabolism, the choice of animal species and the control of its environment are critical factors in the study. In isolated perfused organ studies, the choice of species is further constrained by the practicalities of the technique, as is the age of the animal at which the study can be performed.

24.4.2.2 The Perfused Liver

There is insufficient space here to describe in detail all that is involved in perfusion of the liver. The interested reader is referred to several excellent comprehensive descriptions of the techniques.[18-21] Below is a brief description of the individual steps necessary and some of the more critical points involved.

The perfusion apparatus itself may either be purchased purpose built (*e.g.* from MRA, Clearwater, FL, USA) or assembled in the laboratory from a combination of purpose-designed equipment and general purpose items. The materials from which the apparatus is made are, to some extent, a matter of individual preference. It is possible to construct a perfusion system in which the perfusion medium comes into contact only with inert plastics but in other systems extensive use is made of glass apparatus. This should perhaps be avoided unless it is siliconized, a tedious procedure, to minimize haemolysis during long perfusions.

The perfusion medium is held in an appropriate reservoir which may be a polycarbonate or polyethylene bottle or a glass round-bottomed flask. The medium should be stirred if it contains erythrocytes, either by a stirring device in the vessel[2] or by rotating the vessel slowly at an angle.[22] At least one inlet and one outlet will be required to enable recirculating perfusion to occur. An additional outlet can be useful to help in the sampling of the medium.

Some means must be provided to oxygenate the medium, and this may be through combining the functions of the reservoir with that of an oxygenator[22] or by having a separate piece of apparatus such as a falling film oxygenator.[17] A particularly interesting design has been reported by Alexander *et al.*,[23] which allows a high gas transfer rate to be obtained. The gas should be humidified by passage through a flask of 0.9% saline prior to entering the oxygenator. This prevents evaporation of the perfusate. The gassing mixture itself is often 95% O_2:5% CO_2, but there have been suggestions that 95% air:5% CO_2 might be preferable, as higher oxygen tensions might cause degenerative changes due to increases in lysosomal activity.[24]

Perfusion medium is drawn from the reservoir into a bubble trap consisting of a plastic tube of low volume (1 or 2 mL for rat liver) from which the perfusate leaves at a lower level than its point of entry. In this way, any small bubbles in the medium will be trapped at the surface of the fluid column. A similar device should be placed in the circuit immediately prior to entry to the liver. There are two main systems for perfusion: constant pressure[2] and constant flow.[25] The constant flow method is perhaps preferable and can be achieved by having a closed circuit and a peristaltic pump, such as a Watson–Marlow HR constant flow peristaltic roller type (MHRE 3) or Watson–Marlow model 502S. If constant pressure perfusion is required, this can be achieved by providing a constant head device which is connected to the liver by a vertical tube. In such a system, it will be unnecessary to include a second bubble trap as the constant head device will serve this purpose. In this system, the flow rate can be determined by measuring the time it takes to fill a calibrated tube on the outlet side of the circuit. In a constant flow system the pressure is monitored by means of a side arm to the circuit, just before it enters the liver, which is then connected to a pressure transducer, such as a Gould Statham, by a saline-filled Tygon tube. The transducer is connected to a potentiometric recorder. Oscillations in the perfusion pressure due to the roller pump may be dampened by an air buffer comprising a column of air (10 mL for the rat liver) in a vertical plastic tube connected to the circuit *via* a T-piece.

The perfusion medium is filtered through a filter unit in the circuit, containing filtering material with an exclusion of 50–100 µm, such as nylon bolting cloth.[18] The perfusate must be heated to 37 °C prior to entry into the liver. This is accomplished by a heat exchanger, which may be of a condenser style or by jacketing the perfusion tubing with wider bore (10×14 mm) PVC tubing suitably sealed at both ends. The temperature of the water in the heat exchanger is maintained by a commercial device such as a Churchill pump.

The liver itself sits within a perfusion chamber made of Perspex or glass, on a support which prevents undue compression of the lobes. This may be a piece of nylon bolting cloth, or other perforated device. This allows the perfusate to leave the perfusion chamber and return to the reservoir, usually by gravity. However, it is possible to have a completely closed system in which the outflow of the perfusate is achieved *via* a cannula in the vena cava.[25] A tube from this then runs back to the reservoir. In this system it is not necessary for the liver to sit on a perforated surface and alternatives include a blood-warming bag made of inert plastic filled with water at 37 °C. The liver should be kept moist and warm throughout the perfusion, and this can be accomplished by placing the entire perfusion apparatus within a humidified chamber,[2] or by lightly covering the liver with moist gauze.[18] Another approach to this has been taken by Powis,[25] where the chamber is filled with warm liquid paraffin, which helps to keep the lobes of the liver from becoming compressed and prevents fluid evaporation from the surface of the organ. This system can be used only with a completely closed perfusion circuit.

The catheter in the bile duct is placed below the level of the liver and bile is collected into a suitable vessel (such as an LP3, 64×11 mm polycarbonate tube). It is also possible to record the rate of bile flow by using an electronic drop counter connected to a potentiometric recorder, or by timing flow into a capillary tube.

The materials for the perfusion circuit should be of inert plastic wherever possible and care should be taken to avoid using those types of tubing which have high rates of gas transfer such as silicone rubber.[26] Polyethylene or polypropylene tubing (such as supplied by Portex, Hythe, Kent, UK) is ideal for most of the circuit (PP270 or 800/100/500, 2×3 mm) and for the bile duct cannula (PP10 or 800/100/100, i.d. 0.28 mm; or PP25 or 800/100/140, i.d. 0.46 mm). Connections can be made *via* pushfit nylon or by short lengths of silicone rubber tubing (Silescol). For the tubing within the perfusion chamber, Tygon (2.5×4 mm) may be more suitable as this provides more flexibility than polyethylene or polypropylene.

24.4.2.2.1   *The perfusion medium*

Many different perfusion media have been described for use with the isolated perfused liver.[18] However, most comprise the same essential components, a relatively simple balanced salt solution, an energy source, a plasma expander and some means of enhancing oxygen-carrying capacity. Other possible additions include anticoagulants and antibiotics. For the vast majority of perfusions a balanced salt solution such as Krebs–Henseleit[27] or Krebs–Ringer[28] solution is sufficient. These are buffered by the bicarbonate present in the medium and CO_2 in the gassing mixture. Buffering capacity is sometimes supplemented by increasing the phosphate content. It is possible to use more complex media such as tissue culture medium 199 but this is unnecessary in most instances. The calcium concentration should be adjusted to take into account the addition of any albumin, which will bind some of the calcium (approximately 50% with 4% (w/v) albumin).[18] The commonest energy source is glucose, usually at a concentration of 5–10 mM. The plasma expander is usually serum albumin (fraction V, at a concentration of 2.5 to 5% w/v). Most commercial preparations of albumin are quite acidic and the pH should be adjusted following the addition of the albumin to the salt solution. Neutral fraction V powder is available from some companies, and this might be preferred.[17] Several other plasma expanders have been used but of these only Ficoll 70 appears to serve as a good substitute.

The oxygen-carrying capacity of buffer alone is approximately 2.8% v/v O_2, which is obviously far below that of blood. To maintain adequate oxygenation of the tissue, therefore, it is necessary to perfuse at high flow rates, often four to five times physiological rates. To overcome this, it is possible to include an additional oxygen-carrying component in the medium, such as erythrocytes. Although homologous cells can be used,[2] this involves sacrificing a number of animals immediately prior to perfusion. In addition, rat blood is extremely rapidly coagulating and releases vasoconstrictor substances on isolation. In contrast, heterologous cells can be used, and these appear to function entirely adequately. Donors can be ox,[16] sheep, goats, or even humans.[17] Aged human red cells from a blood bank have been used by many investigators and, if washed thoroughly prior to addition to the medium, serve the purpose well. The concentration of red cells is usually 25% v/v. The albumin is dissolved in a portion of buffer first, the pH adjusted, if necessary, and then the red cells are added and the volume made up to the appropriate total. If necessary, anticoagulants such as heparin or citrate can be added to the buffer prior to preparation of the complete medium, although it is possible to avoid this if the animal is anticoagulated during the operation. In perfusions that last for more than 2 or 3 h it is often advisable to add antibiotics to the perfusate to prevent bacterial growth. The antibiotic itself should have minimal toxicity and avoid interacting with the enzymes of drug metabolism. Compounds that appear to meet these requirements include neomycin (50 μg mL^{-1}) and gentamicin (220 μg mL^{-1}).[21]

24.4.2.2.2   *Isolation of the liver*

The surgical procedure involved in the isolation of the liver is not particularly complex, but it does require a degree of expertise before it is possible to remove the liver rapidly and without undue damage.

The commonest approach involves cannulation of the portal vein and the bile duct with perfusate outflow *via* the severed inferior vena cava.[2] As an additional refinement, it is possible to cannulate this vein to achieve a completely closed circulation.[25] Detailed instructions of the surgery involved in the isolation of the liver can be found in several excellent reviews on the topic.[2,18,21] Only a brief description of the procedure is provided here.

Rats are anaesthetized with a suitable agent, usually a barbiturate such as pentobarbitone,[17] ether[2] or halothane/nitrous oxide mixture.[29] If possible, barbiturates should be avoided in studies of drug metabolism due to the possibility of inhibition. Halothane/nitrous oxide provides good control of anaesthesia, but care must be taken when using halothane due to its potential hepatotoxicity in man, particularly on repeated exposure. Thus, a suitable ventilated hood should be used for the operative procedure.

If necessary, aseptic procedures can be used for removing the liver,[1] though this is unnecessary for perfusions lasting less than 2 h. The abdomen is opened and the portal vein and bile duct identified. The bile duct is cannulated with polyethylene or polypropylene tubing (Portex PP10 or 800/100/100, 0.28 × 0.61 mm). The portal vein is isolated by blunt dissection and side vessels tied off. The vein is then cannulated, using Portex polyethylene or polypropylene tubing, either PP60 (800/100/240, i.d. 0.7 mm), or PP270 (800/100/500, 2 × 3 mm) drawn out over a flame to a diameter of 2 mm and cut square at the tip. Points to watch in the cannulation of this vessel include avoiding pushing the tip of the cannula beyond the bifurcation of the vein, and tying the cannula into the vessel securely, inadvertent loss of the cannula being one of the commonest causes of failure in liver perfusion. Prior to insertion of this cannula, the rat may be heparinized by i.v. injection of 250 i.u. heparin. After carefully removing any bubbles in the venous cannula it is connected to the perfusion circuit and perfusion commenced slowly at a rate of 2–3 mL min^{-1}. The hepatic artery is then ligated.

From this point on the liver must be removed as quickly as possible from the animal. The thorax is opened and if necessary the inferior vena cava is cannulated by directing a bevel-cut cannula through the wall of the right atrium and slipping this along the length of the vein through its point of passage through the diaphragm.[17] If cannulation of this vessel is not required, the vena cava is simply severed with a pair of scissors and the remaining connecting tissue cleared from the liver. The diaphragm is dissected to leave a cuff of tissue around the vena cava and the liver is then removed from the animal and placed in the perfusion chamber. The perfusion rate is then increased, after ensuring that perfusate flow is not obstructed, to approximately 1 mL g^{-1} of liver min^{-1} (for media containing red cells). The back pressure should be approximately 8–12 cm H$_2$O. Bile is collected at a slight negative pressure, below the level of the perfusion chamber.

24.4.2.2.3 *Drug metabolism studies in the perfused liver*

Following isolation of the liver the organ is left to equilibrate for several minutes in the perfusion system. Some criterion of viability is necessary, so that the results of experiments obtained with a dying liver are not included in the final analysis. Several simple macroscopic indicators provide useful information on the state of the liver. For example, soon after starting perfusion the surface of the liver should change in appearance. If perfusion is with simple salt solution, the liver should become blanched. With a red-cell-based perfusate the liver should become red/pink in colour. There should be few, if any, areas of patchy perfusion, indicative of embolization of small vessels. There should be a perceivable inflow–outflow difference in oxygenation of the red cells, which should be obviously bright red in appearance on the inflow side and almost bluish on the outflow side. Bile flow is also indicative of functioning of the liver, although this is by no means definitive. However, a healthy liver should produce bile, particularly when perfused with red cells.[1] The normal flow rate is initially 0.5–0.75 mL h^{-1}, but this will decline with time unless bile salts are recirculated.[30]

Biochemical assessment of the liver can also be achieved, by removing aliquots of the perfusion medium at regular intervals. In this way one can monitor the lactate/pyruvate ratio,[16] reflecting the redox state of the liver, specific O$_2$ utilization and CO$_2$ production,[31] glucose utilization,[17] erythrocyte lysis, the appearance of cytosolic enzymes such as ALT, AST and sorbitol dehydrogenase,[32] and the synthesis of proteins.[33]

The study of drug metabolism in the perfused liver can be achieved using either a recirculating system[17] or a nonrecirculating system.[20] The kinetic considerations differ in the two methods. Normally, the substrate is added as a bolus to the perfusate reservoir and aliquots of the outflow medium sampled at regular intervals for assay of the substrate and possible metabolites. In addition, bile can be sampled regularly and analyzed for the presence of drug and its metabolites. It is also possible to infuse a solution of drug continuously into the portal venous cannula to achieve a measure of steady state kinetics.

24.4.2.2.4 *Kinetics of the recirculating system*

The recirculating system can be considered a two-compartment model. The rate of disappearance of the drug from the reservoir is described by the biexponential equation[34]

$$C = Ae^{-\alpha t} + Be^{-\beta t}$$

where C is the concentration of the drug in the perfusion fluid at time t, A and B are the zero-time intercepts and α and β are the elimination rate constants, respectively. This equation can be solved by normal curve-fitting procedures for biphasic elimination (*e.g.* NONLIN, MKMODEL) to yield values for the rate constants k_{12}, k_{23} and k_{21}. Similarly, values for the apparent volume of distribution and the clearance of the drug can be obtained.[35] If the amount of drug distributed to the liver is small or, as is more likely, distribution is rapid relative to elimination, the elimination pattern will approximate that of a monoexponential system.[36] Thus, the equation would reduce to

$$C = Ae^{-kt}$$

where k is the elimination rate constant. Under these circumstances

$$V = \text{dose}/C_0$$

where V is the apparent volume of distribution, C_0 is the extrapolated y-axis intercept of the log concentration–time plot, and

$$Cl = Vk$$

where Cl is the clearance. The elimination rate constant can be obtained from the slope of a plot of log concentration against time $(0.693/t_{1/2})$.

24.4.2.2.5 *Kinetics of the nonrecirculating system*

This system can be analyzed in a manner analogous to that used to determine the intrinsic clearance of drugs eliminated by presystemic metabolism.[35] Thus

$$Cl = QE$$

where Q is the perfusate flow rate and E is the extraction ratio of the drug (ratio of inflow–outflow concentration difference to inflow concentration).

24.4.2.2.6 *Other considerations*

Useful information can also be obtained by monitoring the appearance of a drug in bile. Further, it is possible to vary other parameters such as the method of administration of the compound (bolus or continuous infusion), amount of protein in the system to determine the effect of protein binding, and to alter the perfusion rate or the perfusion volume to determine the influence that these might have on the elimination of the compound.

The recirculating system provides a useful means of generating a full range of metabolites that might be produced in the liver. These include both phase I and phase II compounds, and further metabolites might be identifiable in the bile where these are eliminated directly from the hepatocytes. Both radiolabelled and stable labelled compounds can be used to obtain labelled metabolites, which can then be detected using appropriate methodology.

The metabolism of a wide variety of drugs and other foreign compounds has been investigated using the isolated perfused liver. Some examples are shown in Table 1. Although the rat is the most widely used donor for such studies, it is also possible to perfuse the liver of the hamster, guinea pig, rabbit, dog, mouse and indeed a lobe of human liver. In most cases similar techniques are applicable, but for mouse liver it may be more convenient to perfuse the organ *in situ*,[17] the animal itself having been killed by pneumothorax, when the inferior vena cava is cannulated. Human liver sections are more frequently perfused as a means of obtaining hepatocytes,[80] but it would be possible to use a similar preparation to study the metabolism of compounds by the perfused tissue.

24.4.2.3 The Isolated Perfused Kidney

The kidney, normally the right from the rat, can be perfused by techniques similar to those described for the liver,[6,81–84] preferably at constant flow. The medium is similar to that used for the perfused liver, most frequently containing washed erythrocytes for oxygen transport[81–84] and bovine serum albumin as plasma expander.[81,82,84] The function of the perfused kidney can be improved by the addition of supplements to the medium[85] and the removal of waste products.[86] Perfusate enters the organ *via* a cannula in the right renal artery, which is cannulated *via* the superior mesenteric artery. Outflow can be *via* the cannulated renal vein with closed recirculation or from the cut vein using an open recirculation system. The perfusion pressure in the kidney is higher than in the liver, typically 80–160 mmHg. To maintain adequate oxygenation, flow rates of 10–25 mL min^{-1} g^{-1} tissue with erythrocyte-free media and 3–5 mL min^{-1} g^{-1} with erythrocyte-containing media are required.[18] The ureter is cannulated with polyethylene or polypropylene tubing (*e.g.* PE10) and urine is collected into suitable vials. It is also possible to record urine flow rate using a drop counter or other flow measurement device.

Table 1 Examples of Compounds Used in Studies of Drug Metabolism with the Isolated Perfused Liver

Substrate	Ref.	Substrate	Ref.
2-Acetylaminofluorene	37	Mestranol	62
Allylpropylacetamide	38	Methadone	61
Aminopyrine	39	Methanol	63
Androstenedione	40	Methocarbamol	64
Aniline	41	Mitoxantrone	65
Antipyrine	42	Neostigmine	66
Asulam	43	Nitrazepam	46
Atropine	3	p-Nitroanisole	67
Benzothiazines	44	Nitropyrene	68
Bromsulfthalein	45	Nitrosobenzene	27
Carbamazepine	46	Noracetylmethadol	64
Chlorinated ethylenes	47	Norbenzphetamine	64
Chlormethiazole	48	Norethindrone	69
Chlorpromazine	49	Nortriptyline	42
Cyclohexanecarboxylate	50	Oxotremorine	42
Cyclophosphamide	51	Pentobarbitone	70
Deoxycytidine	52	Phenol	71
Deptrone methiodide	53	Phenylbutazone	42
Demethylimipramine	54	Phenylhydroxylamine	70
Diazepam	55	Procaine	72
Erythromycin estolate	56	Procarbazine	73
Ethanol	57	Propranolol	74
7-Ethoxycoumarin	58	p-Nitrophenol	75
Halothane	59	Shikimate	76
Harmol	60	Styrene oxide	77
Hexobarbitone	39	Sulfanilamide	72
Hydroxyamphetamine	61	Tetrahydrocannabinol	78
Imipramine	54	Warfarin	79

As in the liver, the perfusate may be sampled on either side of the organ to determine clearance across the kidney. In addition, samples of urine may be taken for analysis of drug and its metabolites. Glomerular filtration rate may be determined by the addition of [^{3}H]inulin or creatinine (1 mg mL^{-1}) to the perfusate. The clearance of these two compounds by glomerular filtration in the rat kidney is almost identical and can be used interchangeably.[87]

The isolated perfused kidney can be used to measure the contribution of this organ to the metabolism of a foreign compound in a manner similar to that described for the liver. Substrates that have been studied in this way include paracetamol,[88] isoprenaline[89] and 4-dimethylaminophenol.[90] In addition, the active secretion of foreign compounds can be investigated, utilizing p-aminohippuric acid and tetraethylammonium (TEA) as prototype organic acids and organic bases, respectively. Similarly, this preparation could be used to investigate the extent to which compounds are reabsorbed from the tubular filtrate. This process is usually passive for foreign compounds, but a number of endogenous materials, and their analogues, are actively reabsorbed.[87] These include glucose and a number of amino acids.

24.4.2.4 The Perfused Lung

Many of the considerations for the perfusion of the liver apply to the perfused lung, with the additional complication that a means of respiring the organ is required. The lungs of any small laboratory animal, such as rabbit, rat, or guinea pig may be perfused.[5,18,91-93] The animals are anaesthetized with pentobarbitone sodium by i.p. injection. The trachea is cannulated and the lungs ventilated with a small animal respirator at a suitable rate (55–60 breaths min^{-1} for the rat, 50 min^{-1} for the rabbit). In the rat, respiration is at 8 cm H$_2$O inspiratory pressure and 2.5 cm H$_2$O positive end expiratory pressure. The inspired gas comprises 95% air:5% CO$_2$ at 37 °C, humidified and heated by bubbling it through warm saline. The thorax is opened, heparin is injected into the right ventricle and the pulmonary artery is cannulated as is the pulmonary vein *via* an incision in the left ventricle. The heart and lungs are then carefully excised and removed together to the perfusion apparatus.

The perfusion apparatus comprises a humidified chamber in which the lungs are suspended, the temperature being maintained at 37 °C by a water jacket around the chamber. The perfusate may be homologous blood, or one of the media described for perfusing the liver. A common physiological salt solution in use comprises 119 mM NaCl, 4.7 mM KCl, 1.17 mM $MgSO_4$, 22.6 mM $NaHCO_3$, 1.18 mM KH_2PO_4, 1.6 mM $CaCl_2$, 5.5 mM glucose and 4% (w/v) bovine serum albumin or Ficoll 70 (70 000 molecular weight).[94] The perfusate is adjusted to pH 7.4 with NaOH prior to use and equilibrated with the respiratory gas mixture. The perfusate is delivered to the lungs at a constant flow rate of 0.03 mL to 0.1 mL g^{-1} body weight min^{-1}, although it is also possible to perfuse at constant pressure.[5] In a constant flow system, the pressure of the perfusate is monitored with an electronic transducer (Statham) connected to the arterial side of the circuit. Perfusion pressure should be 17–18 mmHg. The perfusate is normally recirculated and can be sampled from the venous side for the determination of drug and metabolites, as well as blood gases and pH.

It is also possible to utilize a nonrecirculating system to perfuse the lung. Here, drug may be administered either as a bolus to the reservoir or by continuous infusion into the arterial side of the circuit. In such studies, it is possible to measure efflux from the lung by terminating the infusion and then measuring the appearance of compound in the effluent.[95] An additional possibility with this system is the addition of compound *via* the inspired air, as is sampling of the expired air. Suitable modifications of the apparatus will be required to enable the collection of gas samples from a side arm in the gas delivery system.

The viability of the perfused lung preparation can be determined from measurements of blood gases, the weight of the lung at the end of the experiment, increased weight indicating oedema,[18] and by determining the area of perfusion by the addition of either blue dextran to the perfusate during the experiment or of India ink at the end of the experiment.[95] Obviously, the use of dyes should be avoided if this might interfere with any proposed assays of either the lung tissue itself or of the perfusate.

The disposition and metabolism of a number of compounds have been studied in the perfused lung, most frequently after their addition to the perfusate. Examples include benzo[a]pyrene,[96] imipramine,[95] aldrin and dieldrin,[97] amphetamine, methadone and a number of other xenobiotic amines.[5]

24.4.2.5 The Perfused Intestine

A variety of techniques have been described for the perfusion of various segments of the intestine.[18,98] In addition to perfusion *in situ* and *ex vivo*, it is possible to study absorption and metabolism by segments of the intestine supplied by the systemic circulation, using a once-through perfusion system. A particular concern in the use of such preparations is the effect that intestinal motility has on drug absorption. A number of drugs have been used in an effort to overcome this, including the antihistamine promethazine, the α-adrenoceptor antagonist phenoxybenzamine and the smooth muscle relaxant papaverine.[18]

Most studies on perfusion of the intestine have been performed in the rat. The perfusion medium is similar to that used to perfuse the liver, and may comprise diluted freshly drawn rat blood[99] or a semisynthetic medium containing heterologous erythrocytes and albumin.[22] In perfusion of the small bowel, the superior mesenteric artery is cannulated either directly[99] or through the aorta.[7] Outflow is from a cannula in the superior mesenteric vein[99] or the portal vein.[104] In some studies, the mesenteric lymph duct is also cannulated.[99] A suitable segment of the intestine is ligated and the sac thus formed is either cannulated at both ends for continuous irrigation[99] or is injected with test material, which is then left static in the sac.[100] The rat colon may be perfused by cannulating the left internal iliac artery, with outflow from the superior mesenteric vein cannulated through the portal vein.[22] The colon may be irrigated by perfusion through cannulae placed at each end of a sac isolated by ligation.

A novel procedure has been described for the perfusion *in situ* of an isolated loop of the dog small intestine.[101] Dogs are anaesthetized with pentobarbitone and respiration is maintained with a suitable respiratory pump. Following laparotomy, a loop of the terminal duodenum and proximal jejunum drained by a single branch of the mesenteric vein is identified. The loop is isolated between two balloon catheters and the contents of the loop washed out by irrigation with 0.9% w/v saline. The venous effluent from the loop is collected from a cannula placed in the branch of the mesenteric vein draining the loop, and circulation is from the normal arterial supply to the intestine. Thus, this preparation has a relatively limited lifespan, usually restricted to 30–45 min. Drugs are added in solution through one of the balloon catheters into the lumen of the gut loop.

The mammalian intestine is normally populated with a diverse range of microorganisms, which can contribute to the metabolism of foreign compounds, either prior to their absorption following oral administration, or after their excretion into the intestine *via* the bile. It is possible to study metabolism by these organisms by a variety of techniques, including closed cultures and continuous flow systems, containing either a mixture of species or a single species.

24.4.3 TISSUE SLICES

For a considerable time, tissue slices have provided a useful intermediate preparation between that of an intact organ and isolated dispersed cells.[8] Their use has been largely superseded by the use of isolated cell preparations because of the disadvantages inherent in the use of slices. These include the difficulty of maintaining adequate oxygen tension in the cells, the slice being several cells thick, and the presence of numerous damaged cells on the surface of the section. Nevertheless, tissue slices provide a readily available means of studying metabolism in intact cells without going to the lengths necessary to obtain isolated dispersed cells.

In theory, slices could be obtained from any solid organ from any species, although in practice for drug metabolism studies this usually means the liver or kidney. Human tissue samples are particularly amenable to this technique.[102,103] The tissues must be kept cool at all times, particularly during slicing, and the slices themselves should be kept at 4 °C until used. The basic technique for the preparation of tissue slices is as reported by Umbreit *et al.*[104] Although it is possible to obtain purpose-designed apparatus (*e.g.* a Stadie–Riggs stage) for preparing tissue slices, adequate slices can be produced using a ground glass plate, a frosted glass microscope slide and a single-edged razor blade to slice the tissue. The organ is sliced on the ground glass plate, which should be cooled on a bed of ice. Gentle, but firm pressure is applied, using the microscope slide parallel to the surface of the plate. By varying the pressure used to compress the organ it is possible to alter the thickness of the slice. Slices are obtained by cutting close to the under surface of the slide with a sharp, single-edged razor blade. The slices should be as thin as possible, without producing fragmentation (usually 0.3–0.5 mm in thickness). Some practice will be required to produce slices of uniform thickness, which is essential to obtain useful results with the technique. Slices are weighed and incubated in a balanced salt solution, frequently phosphate-buffered saline, for several hours at 37 °C in a shaking water bath. It is important to gas the flasks with an atmosphere of 100% O_2, as air alone provides insufficient oxygenation. The medium can be sampled at regular intervals and analyzed for remaining parent substrate or for the appearance of metabolites, both phase I and phase II. In addition, it is possible to recover the slices themselves and to identify any drug-derived material that might remain within the tissue. This might be particularly useful when using compounds that could give rise to reactive metabolites, where covalent binding could occur. Although liver slices cut by the traditional method are now rarely used in drug metabolism studies, there has been recent interest in the possible application of precision cut slices[105] for this purpose.

Although liver slices are rarely used, kidney slices are still widely studied. This is due to the greater difficulty in obtaining homogeneous populations of cells from this organ and the fact that cortical slices retain the ability to accumulate compounds by active secretory and absorptive uptake. Thus, kidney slices[106] have found particular application in studies of excretory transport. In such cases the results are usually expressed as the ratio of material present in the slice to that in the medium (slice/medium, or S/M, ratio).

24.4.4 ISOLATED CELLS

24.4.4.1 Introduction

The use of tissue slices has been largely superseded by the use of isolated cells. These offer many advantages over the former technique and are amenable to a wider range of studies. Such cell preparations provide the means of investigating the metabolism of compounds in an intact cellular system, with both phase I and phase II metabolism proceeding. It is possible to manipulate both enzyme and cofactor levels by pretreatment of the donor animal or by culturing the cells in the presence of appropriate modifying agents. The cells are free of interfering factors that might occur *in vivo* such as alterations in blood flow, hormonal effects and other physiological factors. As the cells are dispersed, oxygenation is not a problem, adequate gas transfer occurring by simple diffusion. Cells can be prepared from many tissues and from many species. Human-derived cells provide the most readily accessible means for studying the metabolism of compounds in man, in a relatively intact system, without the necessity for administering them to members of the population.

Isolated cell systems, however, are not without disadvantages. The nature of the preparation is such that organization of the tissue is lost, which might be important in intertissue interactions *in vivo*, such as the processing of metabolites by more than one organ. Further, it cannot be assumed that isolated cells maintain a full complement of cofactors and enzymes at the levels present *in vivo*. This should be verified for the particular system under investigation. Although culturing of cells provides a useful means of manipulating protein synthesis, it should be recognized that during culture there are profound alterations in both the morphology and biochemistry of the cells and this may have consequent implications for studies of drug metabolism. The most dramatic of these is perhaps the decline in cytochrome *P*-450 mediated oxidation that occurs following the first few hours of culture, particularly in rodent hepatocytes, in which the *P*-450 content can decline to approximately 20% of control values within 24–48 h.[12]

In addition to the use of isolated cells it is possible to study explant culture from human and other species.[107] This provides an intermediate step from whole animal studies to studies in the isolated cell. This technique has not been widely applied to studies of drug metabolism, although a number of groups have used it extensively to investigate the metabolic activation of potentially carcinogenic compounds.[107]

In studies of drug metabolism, the most widely used cells are obviously those from the liver, the hepatocytes. However, a variety of cell types from a number of tissues have been used including nonparenchymal cells, various cell types from the kidney, the lung, blood-derived cells, cells from the intestinal tract, particularly epithelial cells, and mucosal cells from the bladder. Cells have been isolated from numerous species, both mammalian, including man, and nonmammalian and the limitations are generally practical rather than theoretical. For example, isolation of hepatocytes from large species is extremely costly and would require such scale up as to render the exercise impractical. However, alternative techniques are available, although they often yield fewer cells; these include incubation of tissue slices in dispersing enzyme mixture[108] or scraping cells such as mucosal or epithelial cells from the appropriate tissue.

Cells in solid organs such as the liver and kidney are retained within a biomatrix comprising collagen and other components. In order to isolate such cells from these tissues it is first necessary to degrade this matrix. This is usually accomplished by perfusion of the organ with a collagenase-containing medium.[9–11] The biomatrix requires calcium for maintenance of its properties so that initial removal of extracellular calcium using the specific chelator ethylenedi-(oxyethylenenitrilo)tetraacetic acid (EGTA) considerably aids in the action of the collagenase. It should be noted, however, that the enzyme itself requires the presence of calcium ions which therefore have to be added to the perfusion medium. However, these do not replace those removed by prior perfusion with EGTA.[10,11] Although collagenase is now the enzyme of choice for hepatocyte isolation, the isolation of other cells often requires a different enzyme, either alone or in addition to collagenase; these include trypsin, pronase and hyaluronidase.

If cells are to be incubated for other than short intervals (less than 1 or 2 h), sterile techniques should be adopted to ensure that the cells are bacteria free. It is also possible to include appropriate antibiotics in the perfusion medium and in the subsequent incubation medium. The complexity of the medium in which the cells are suspended will vary, depending upon the particular application in question. In many circumstances a simple balanced salt solution, for example Earle's balanced salt solution or Krebs–Henseleit buffer,[109] will be sufficient when supplemented with a few essential nutrients, whereas full growth media are sometimes used, particularly when culturing the cells.[110] These latter include a full range of amino acids, vitamins and other additives. Examples include tissue culture medium 199, Williams E and Leibovitz L15. The pH of the medium is generally maintained by bicarbonate–CO_2, the latter being present in the gaseous phase, usually at 5%. It is also possible to modify the medium to include phosphate and/or 4-(2-hydroxyethyl)-1-piperazineethanesulfonic acid (HEPES) to provide an entirely aqueous buffering system. It is normal to use a mixture of buffers under such circumstances, the bicarbonate–CO_2 buffering system almost always being one component. Most commercial media incorporate a pH indicator such as phenol red, so that the pH of the medium can be determined by simple visual inspection. However, care must be taken that the dye does not affect the metabolism of the compound under study. It is frequently necessary to add serum, usually foetal calf or newborn horse, to the medium, particularly when culturing the cells. The serum provides a variety of growth factors, many of which are present in extremely low concentrations and some of which have yet to be identified. There are several defined, completely serum free media now available, and these are adequate for the maintenance of certain cell types.[110]

Metabolic activity is usually expressed relative to the number of viable cells present, although alternative methods include expression per mg of total protein or per mol of DNA. Viability can be determined by a number of techniques, based largely on either the permeability of the cell membrane or the intermediary metabolism of the cells. The most widely used indices of viability are those that are based on determination of the integrity of the plasma membrane. These can be divided into two subcategories, those in which exclusion of an exogenous compound is determined, and those in which the measurement of an endogenous component normally retained within the cell is measured in the supernatant.

The commonest test of viability is exclusion of trypan blue.[109] An aliquot of the cells is mixed with a solution of trypan blue in buffered salt solution and within a few minutes the cells are examined under a microscope using a haemocytometer chamber. Nonviable cells will be stained blue, particularly the nucleus. The total number of cells is determined and the number of viable, nonstained, cells expressed as a percentage of this. Other exclusion tests include the use of naphthalene black and NADH.[111]

Leakage assays widely used include determination of lactate dehydrogenase (LDH).[109] This has been shown to provide results comparable to those of the trypan blue test.[112] Aliquots of a cell suspension are taken and the cells in one portion are lyzed, either by repeated freezing and thawing, or by the addition of detergent such as Triton X-100. The cells in a second portion are sedimented by low speed centrifugation and the supernatant is removed for the analysis of LDH, using a simple spectrophotometric assay dependent upon the change in absorbance occurring on the addition of NADH.[109] Leakage of enzyme into the supernatant is expressed as a percentage of activity in the total cell fraction. The activity of many other enzymes can be used to determine the viability of the cells, and these include virtually any cytosolic enzyme, including sorbitol dehydrogenase, AST, ALT and isocitrate dehydrogenase. A variant on this technique is the chromium release assay:[113] Cr^{3+} is readily taken up by cells. However, once in the cell the chromium is reduced to Cr^{2+}, which is retained by viable cells. Thus, if cells are incubated with $^{51}Cr^{3+}$, subsequent leakage of $^{51}Cr^{2+}$ into the supernatant, determined by γ counting, will provide a measure of the loss of viability.

Biochemical tests of viability include determination of the cellular content of ATP,[114] the lactate/pyruvate ratio and, where cells are to be cultured, the plating efficiency provides a very good estimate of cell viability. Plating efficiency will decrease as the viability of the cells decreases.

Although all of these methods are suitable for cells in suspension, in culture some methods are less suitable than others. For example, dye exclusion methods depend upon determining the number of stained cells relative to the total number of cells present. However, counting cells in culture, particularly when they have been cultured for some time and have become extremely flattened on the dish, can be very difficult. Release methods, such as LDH release and ^{51}Cr release, are perhaps more suitable for this purpose.

To study the metabolism of drugs and other chemicals with isolated cells, the substrate is added in an appropriate vehicle to cells in suspension or to a flask of cells in culture and incubated for intervals of up to several hours, and perhaps even for 24 h in the case of cultured cells. Aliquots of cell suspension or, in the case of cultured cells, of medium, are removed at regular intervals. If necessary, the cells are removed by centrifugation, and the medium is then analyzed by appropriate methods for the presence of metabolites. It is also possible to wash the cell pellet and then lyze the cells to explore the possibility that some metabolites are retained within the hepatocytes, or to investigate covalent binding or sequestration by cellular macromolecules.

24.4.4.2 Hepatocytes

Hepatocytes from a number of species, including man,[80,115,116] rat,[9-11] mouse,[117] guinea pig,[118] hamster,[119] rabbit[120] and dog,[121] are used widely in studies of drug metabolism and toxicity. Their isolation is almost always accomplished by a modification of the two-stage collagenase perfusion technique of Berry and Friend.[9-11] Essentially, this involves perfusing the liver by a simplification of the technique described in Section 24.4.2, with EGTA to remove extracellular calcium, followed by collagenase to disperse the hepatocytes. For larger species, such as dog and man, it is possible to take a biopsy from the periphery of the lobe and perfuse this through one or more cut branches of the portal vein.[115] The remainder of the technique is essentially the same as that for the isolation of hepatocytes from smaller species. Mouse liver is usually perfused *in situ*, that of rat, guinea pig and hamster may be perfused either *in situ* or *ex vivo*, and the liver of large species, such as rabbit, dog and human is normally perfused *in situ*.[80]

Experimental animals are anaesthetized, rodents receiving pentobarbitone by i.p. injection. If the cells are to be cultured, aseptic technique should be used. Otherwise, if the cells are to be used in suspension for only 2 or 3 h, it is not necessary to adopt rigorous aseptic procedures. The portal vein is cannulated using a similar operative technique to that described for the perfused liver. It is not necessary to cannulate the bile duct, and recirculation of the perfusate is *via* the cut superior vena cava.

The liver is then either removed to the perfusion apparatus or perfused *in situ*. The apparatus is similar to that used for the perfusion of the isolated organ as described in Section 24.4.2. As the medium does not contain erythrocytes, it is possible to include a filter with a smaller pore size than is used for the isolated perfused liver preparation. As most perfusions take only 15–30 min, it is usually sufficient to cover the surface of the organ with a moist gauze to prevent dehydration.

The essential feature of the first perfusion medium is that it contains no calcium. Frequently a simple balanced salt solution, such as Ca^{2+},Mg^{2+}-free Earle's, which contains glucose and phenol red as a pH indicator[116] is used. The pH of this medium is maintained by the addition of 26.8 mM $NaHCO_3$ and gassing with 5% CO_2. It is also possible to use a salt solution buffered with HEPES.[109] While some workers rely upon the absence of calcium from the perfusion medium to remove extracellular calcium from the liver, it is often preferable to include 0.5 mM EGTA at this stage.[10,11] Removal of extracellular calcium is critical to the successful dissociation of liver. When the hepatocytes are to be used for longer term incubations or to be cultured, the medium can be supplemented with gentamicin. The liver is perfused with calcium-free medium using a non-recirculating system at flow rates of 4–10 mL min^{-1} g^{-1}. This perfusion is for 5–10 min.

Some workers proceed to the collagenase perfusion at this stage, but it might be preferable to perfuse for 5 min or so with the same medium without EGTA. This is because collagenase requires calcium, and any EGTA remaining in the liver would chelate calcium in the medium. This perfusion is also without recirculation.

Hepatocytes are dissociated by perfusion with medium containing collagenase and 4 mM $CaCl_2$. This medium should be filtered prior to use to remove any collagenase not in solution. As with the other media, this is gassed with 95% O_2:5% CO_2 before use, to establish the appropriate pH and to saturate it with oxygen. The source of collagenase is of some importance. Whilst type I collagenase from Sigma is often adequate, this varies from batch to batch, whereas preparations from Boehringer have been found to be more reliable between batches. The concentration of collagenase and the time of perfusion will depend upon the species. The concentration of collagenase should be kept as low as possible whilst still providing adequate dissociation within a reasonable time. For rat liver, concentrations of 0.025–0.035% give complete dissociation within 10–15 min of perfusion. Hamster hepatocytes can be isolated with slightly lower concentrations of collagenase and those from mouse require only 0.02%. Human liver is more difficult to dissociate and higher concentrations of collagenase, often 0.05%, are required, with longer perfusion times, up to 50 min, before evidence of dispersion of the biomatrix is apparent.[115,116]

Following collagenase perfusion, the collapsed organ is removed to a Petri dish for dispersion of the hepatocytes into an enriched medium such as Leibovitz L15 enriched medium containing 0.2% bovine serum albumin,[122] or EBSS supplemented with 10% foetal calf serum, 10 mM HEPES, 8 mM $NaHCO_3$, 12 mM sodium phosphate, 1.5 mM $CaCl_2$, 0.73 i.u. mL^{-1} insulin, 0.1 mM hydrocortisone 21-hemisuccinate and 0.1 mg mL^{-1} gentamicin sulfate.[123] Rodent hepatocytes are dissociated by disruption of the Glisson's capsule and gentle stroking of the surface of the liver with a spatula or glass rod. The cells should readily disperse into the medium.

Human hepatocytes are less easily dispersed and usually require removal of the capsule over the perfused area with scissors.[116] In some cases the cells will disperse into the medium as for animal liver, but otherwise it will be necessary to stroke the surfaces with a glass rod fairly vigorously. This produces small clumps of cells, which are then finally dispersed into single cells by rolling the suspension at 45 cycles min^{-1} for 2 min.

The crude cell suspension is then filtered through nylon bolting cloth (50–250 μm mesh). The cell suspension is then allowed to sediment for up to 20 min and washed three times by centrifugation at 50–100 g_{av} for 1 min to remove cell debris, nonparenchymal cells, erythrocytes and nonviable hepatocytes, which remain in the supernatant.[9] It has been found that preincubation of hepatocytes at this stage in the dispersion medium for 15–30 min at 37 °C in an oxygen atmosphere improves the quality of the final cell preparation.[123] This may be due in part to repair of damage incurred during isolation, and also dissociation of any small clumps of cells. Remaining damaged cells are lyzed during this incubation period.

It is possible to obtain populations of hepatocytes of even higher viability by subjecting the cell suspension to density gradient centrifugation on Percoll.[124] The cells are sedimented by centrifugation at $100g_{av}$ for 1 min and then resuspended in fresh dispersion medium at a density of 5×10^6 cells mL^{-1}. 10 mL of this suspension are layered on to 20 mL of Percoll ($d = 1.056$, comprising Ca^{2+},Mg^{2+}-free EBSS containing 10 mM HEPES, 10 mM $NaHCO_3$ and 370 mL L^{-1} Percoll ($d = 1.130$), pH 7.4). The cells are isolated by centrifugation for 1 min at $100g_{av}$. The intact viable cells will sediment through the Percoll to form a pellet, whereas damaged and dead cells will remain in suspension and are removed by aspiration. The cell pellet is then resuspended and kept at 4 °C until use. It is possible to keep such cells at 4 °C for up to 24 h without loss of monooxygenase activity.

If the hepatocytes are used in suspension, then the techniques are as described above in Section 24.4.4.1. To culture hepatocytes, there is a wide choice of techniques. The majority of these are a compromise between technical feasibility and deterioration of biochemical function. One of the major difficulties in using cultured hepatocytes in studies of drug metabolism is the loss of monooxygenase activity, related to the decline in cytochrome P-450 content, that occurs soon after attachment of the cells to the culture matrix.[12] The cells are diluted to a density of 0.5×10^6 mL^{-1} in culture medium (see below). Then $8–10 \times 10^4$ cells cm^{-2} are plated out into tissue culture vessels such as 24 cm^2 flasks, with a suitable substratum, for example coated with rat tail collagen,[125] or Falcon Primaria flasks. The hepatocytes are then incubated for an initial period of 3–4 h at 37 °C during which time healthy cells will attach to the substratum, plating efficiencies of 60–70% being common.[126] Nonattached, nonviable cells are then removed by careful aspiration of the medium, followed by replacement with fresh medium. Culture continues at 37 °C in a humidified atmosphere of 90% air:10% CO_2, most conveniently accomplished in a tissue culture incubator. The culture medium should be changed after overnight incubation and then regularly every 2 d thereafter. Cells can be maintained with a reasonable degree of differentiation for 5–7 d, but after that time they undergo progressively more extensive dedifferentiation.[125,127]

There have been many different suggestions as to how the activity of the enzymes of drug metabolism, particularly those dependent upon cytochrome P-450, can be maintained during culture of hepatocytes; these range from the addition of ligands,[128] alterations in the media supplements,[129] growth factors,[130] improved culture substrata[131] to, most recently, coculture with rat liver epithelial cells.[132] This last technique has shown considerable promise, although it is technically demanding to isolate the appropriate epithelial cells. Readers are referred to the relevant references[132] for further information, this topic being too complex to consider in detail here.

Hepatocytes divide only very slowly in culture, and as a consequence have a limited lifespan, usefully of 5–7 d.[127] This, of course, is even further reduced when the experiments necessitate use of fresh suspensions, where the usable lifespan of the cells, when maintained at 4 °C, is of the order of 24 h. Thus, to preserve hepatocytes, in particular those from human liver, and perhaps as a means of securing supplies of a standardized preparation of animal hepatocytes, cryopreservation of the cells has been investigated by a number of groups.[133–138] Whilst several methods are now available that enable the recovery of hepatocytes of relatively high viability and moderately good metabolic activity, these cells are still much more fragile than before preservation, and show relatively poor attachment in culture, although this can be improved by coculture with rat liver epithelial cells.[139] Thus, cryopreservation has yet to be developed to a situation where it is reliable for the long term storage of hepatocytes for routine use in studies of drug metabolism. The technique involves equilibrating cells in cryoprotectant, the most effective of which is dimethyl sulfoxide, above the temperature of ice nucleation, followed by slow freezing of the cells at 1 °C min^{-1} to -80 °C and then more rapid freezing and storage in liquid nitrogen. The cells are recovered by very rapid thawing and rapid dilution of the cryoprotectant. Activity of these fractions is then studied using the techniques described in Section 24.4.5.

Detailed comparisons have been performed of drug metabolism by hepatocytes from different species and there is increasing interest in the use of hepatocytes for metabolic profiling early in drug development. Hepatocytes are capable of producing a wide range of both phase I and phase II metabolites, including sequentially derived products. Thus, conjugates of oxidation metabolites are frequently identifiable. Hepatocytes are capable of synthesizing their own cofactors, provided that appropriate precursors are included in the incubation medium.[109] A wide variety of substrates has been investigated using hepatocytes.[140–144]

The content of the enzymes of drug metabolism can be manipulated either by pretreating the donor animal or, in the case of cultured cells, by incubating the cells in the presence of a suitable inducer.[131,145] In the latter case, care must be taken in interpreting the data as the cytochrome P-450 content will be declining as the cells are cultured and so it may not be possible to increase the specific content of cytochrome P-450 to beyond the initial value, despite evidence of protein synthesis and induction.[145]

Hepatocytes may be used to determine both the qualitative and quantitative patterns of metabolism of foreign compounds.[121,146] Where the liver is the principal organ of biotransformation *in vivo*, hepatocytes often provide an accurate indication of the profile of metabolites that will be observed in the intact animal.[121] Similarly, the kinetics of metabolism often agree closely with those *in vivo*. Rates of formation of metabolites are determined at varying substrate concentrations and classic Michaelis–Menten analysis is then performed.[147] However, care must be taken for some compounds where uptake is limiting, or the release of the metabolites is not rapid relative to their formation. In the latter case, the problem can be overcome by determining the rate of production relative to aliquots of the complete cell suspension, medium plus cells, so that any metabolite within the hepatocytes is measured along with that which is released.

24.4.4.3   Hepatic Nonparenchymal Cells

Although the majority of biotransformation reactions occur within the hepatocyte, there is interest in the possible role of other hepatic cell types in drug metabolism. In particular, the Kupffer cell has been implicated in some reactions. Methods have been developed for the isolation of these cells and these are outlined below.

Although, in theory, nonparenchymal cells could be isolated from the liver of any species from which hepatocytes may be isolated, in practice the rat has been most frequently used for this purpose to date. The technique is very similar to that for the isolation of hepatocytes to the point of obtaining the initial cell suspension. Some workers have found better yields of nonparenchymal cells by replacing collagenase with pronase.[148] The liver is perfused with pronase (0.2% w/v) for 2–5 min, following which the tissue is minced with scissors and the fragments resuspended in buffer containing 0.2% pronase. The liver is then incubated at room temperature in an atmosphere of 95% O_2:5% CO_2 with vigorous stirring. DNAase (0.5 mg) is added at 20 min and 40 min to prevent cell agglutination and to enhance dispersion. The pH is maintained at 7.3 throughout by the addition of NaOH as necessary. After 60 min the cell suspension is filtered through nylon bolting cloth (50–80 μm). The cells are then left to settle at $1g$ for 15–30 min. The denser parenchymal cells will sediment as a discrete layer. The nonparenchymal cells will remain in suspension above the hepatocyte layer. This suspension is carefully removed and centrifuged for 1 min at $50g_{av}$ to remove any contaminating hepatocytes. This procedure is repeated and the nonparenchymal cells are then sedimented by centrifugation at $500g_{av}$ for 3 min. This is then repeated and the pellet of nonparenchymal cells is finally resuspended in buffer.

It is possible to further purify the cells by gradient centrifugation on metrizamide or Nycodenz. The latter is perhaps preferable as it is slightly less toxic. The Nycodenz is prepared in 5 mM Tris buffer, pH 7.6, containing 3 mM KCl and 0.3 mM EDTA.[149] It is diluted in 7.45% (w/v) sucrose in a similar buffer. The cells in buffer are overlaid on to a two-step, discontinuous gradient comprising 4.6% (v/v) and 13.8% (v/v) Nycodenz solution and then centrifuged for 45 min at $150g_{av}$. The larger parenchymal cells, together with damaged cells and erythrocytes, sediment at the bottom of the tube. The nonparenchymal cells form a layer at the interface of the 4.6%:13.8% Nycodenz. The cells are carefully aspirated and then washed twice in fresh buffer. The majority of cells (approximately 50%) comprise endothelial cells, with Kupffer cells representing the other main cell type (15–40%), with variable numbers of leucocytes and occasional Pit cells and Ito cells. The cells may be further purified by centrifugal elutriation.[150]

Nonparenchymal cells, either as a semipurified preparation comprising a relatively heterogeneous mixture of cell types, including Kupffer cells, Ito cells and endothelial cells, or following purification by elutriation can be used as a suspension or cultured. The only important difference in culturing such cells from culturing hepatocytes is that the former are seeded at a slightly higher density.[151] The nonparenchymal cells can be used to study the metabolism of foreign compounds exactly as described for hepatocytes, although in most cases activity will be much lower.[148,149] In addition to metabolism, the phagocytic activity of the Kupffer cells can be investigated, either by chemical or microscopic techniques.[152]

24.4.4.4   Isolated Renal Fragments

Although isolated renal cells can be obtained by enzymatic digestion with collagenase,[153] due to the heterogeneity of the kidney, the suspensions comprise a mixture of different cell types. It is possible to produce a preparation enriched in proximal tubule cells by subjecting this cell population to centrifugation, but this is still far from homogeneous.[154] Renal cells isolated by enzymatic digestion are active in the metabolism of xenobiotics,[155] and also exhibit active secretory uptake.

As an alternative to enzymatic dispersion, fragments of kidney highly enriched in particular segments of the nephron, including the tubules and the glomerulus, can be obtained by mechanical disruption and selective sieving.[156] Kidneys are removed from anaesthetized rats and the medulla is dissected free of the cortex. The cortical tissue is then used for the isolation of glomerular and tubular fragments. The tissue is chopped with scissors and then forced through a number of stainless steel sieves ranging in size from 250 μm, which dissociates the tissue, through to 150 μm, which traps fragments of the proximal tubules but not the glomeruli, which are finally collected on a 75 μm sieve. The tubular fraction is resuspended in Tyrode's solution and washed several times, the final preparation containing >98% tubular fragments. The glomerular fraction is similarly washed and reisolated by mild centrifugation. This preparation should contain >95% pure glomeruli. The preparations can be used fresh in suspension and maintain many of the differentiated functions of the respective segments of the nephron.

Glomeruli may be cultured following aseptic isolation, using RPMI 1640 medium supplemented with 20% foetal calf serum. Unattached glomeruli are removed by changing the medium after 24 h.[157]

Cortical tubular epithelial cells may also be isolated by differential attachment during culture.[158] Cortical cells can be cultured in a variety of media, the essentials of which are the inclusion of D-valine, which prevents fibroblast overgrowth, and the omission of arginine.[158]

Renal medullary interstitial cells may be isolated by suspending finely minced medullary tissue in hypertonic Hank's solution containing collagenase at 0.3% (w/v) and incubating at 37 °C for approximately 40 min.[159] The dispersed cells are filtered through 100 μm nylon bolting cloth and isolated by centrifugation. The cells are finally resuspended in Eagle's minimal essential medium containing 20% foetal calf serum. The cells can be studied fresh or cultured for longer term investigation.

24.4.4.5 Isolated Lung Cells

Although the lung contains over 40 different cell types, only a few of these have been obtained in sufficient purity to enable their study independent of other cells. These include the alveolar macrophage, the Clara cell and the type II pneumocyte. The rabbit has been used most frequently for the isolation of these cell types, utilizing techniques developed by Devereux *et al.*,[160–164] who have described the procedure in detail, which is only summarized here.

The lungs are cleared of blood by perfusion *in situ* with warm (37 °C) Krebs–Ringer bicarbonate buffer containing 4.5% (w/v) bovine serum albumin and 5 mM glucose. The entire lungs are then removed and the macrophages isolated as follows. The lungs are lavaged with ice cold balanced salt solution containing HEPES. This is repeated five times and the cells in the lavage fluid isolated by centrifugation for 10 min at $800g_{av}$. The cells are resuspended and purified of blood cells and cell debris by subjecting the suspension to elutriation, using 10^8–10^9 cells. The macrophages are obtained at 1200 r.p.m. (flow rate 22 mL min^{-1}) and are >95% homogeneous. This fraction is not contaminated with Clara cells or type II cells.

The lung itself is then used to isolate the Clara cells and the type II cells. This is accomplished by instilling a solution of HEPES-buffered balanced salt solution, pH 7.4, containing 0.15% protease and 0.5 mM EGTA. The lungs are incubated at 37 °C for 15 min and then degassed for 30 s. The tissue is minced finely with scissors and suspended in HEPES-buffered balanced salt solution containing F12K growth medium 3:1 and 0.5% DNAase and 0.5% BSA. This suspension is stirred at 4 °C for 15 min. Debris is removed by filtering the suspension, which is then centrifuged at $800g_{av}$ for 8 min and resuspended in a small volume of buffer. The cell fractions are obtained by centrifugal elutriation into four initial fractions. This is achieved by a stepwise increase in the flow rate and/or a decrease in the centrifugation speed. The alveolar type II cells are obtained at a rotor speed of 2000 r.p.m. and a flow rate of 15 mL min^{-1}. The alveolar type II cells are further purified by subjecting this fraction to density gradient centrifugation on metrizamide or Percoll. The Clara cells are isolated in fraction 4, obtained at a rotor speed of 1200 r.p.m. and a flow rate of 13 mL min^{-1}. These cells are further purified by density gradient centrifugation on Percoll and a second centrifugation step. The type II pneumocyte preparation should be >80% pure and relatively free of Clara cells, macrophages and lymphocytes. The Clara cell fraction is usually less enriched, containing 40–60% Clara cells, with appreciable numbers of type II cells (5–10%), and macrophages (20–30%).

24.4.4.6 Foetal Cells

Foetal cells are incompletely differentiated and rapidly growing. As a consequence, when cultured they readily grow and divide. Thus, for example, it is possible to obtain a replicating culture of hepatocytes. Against this, many of the enzymes of drug metabolism are low or even absent in such cells. Thus, their use in studies of drug metabolism must be considered carefully if the results are to be extrapolated to the situation *in vivo* in man. One of the major advantages of using foetal cells is that foetal tissue may be obtained relatively easily from abortuses.[165] Such tissue has been utilized by a number of workers to prepare cultured cell preparations. For details, the reader is referred to the work of Pelkonen *et al.*[165,166] Foetal cells can be isolated from a number of tissues and from many species.

Foetal liver cells can be isolated from both rat and human liver using a similar technique. Although trypsin was the first enzyme utilized for this purpose,[167] collagenase is now widely used.[168] Although this has often been combined with trypsin or hyaluronidase, it has been shown to be adequate without further supplementation.[169-172] The foetal liver (for the rat, livers from a number of different foetuses are pooled) is chopped finely with scissors or a scalpel. It is possible to isolate hepatocytes from rat foetal liver without prior mincing. The tissue is washed several times with HEPES buffer, pH 7.6. Cells are isolated by incubation of the liver in a balanced salt solution containing 0.05% (w/v) collagenase, from Boehringer. The mixture is aerated by gassing with 95% O_2:5% CO_2. The mixture is incubated at 37 °C for 10 min with gentle stirring after which time the supernatant, which contains mainly blood cells, is removed and discarded. Further dissociation medium is added and the cells incubated for a further two 10 min periods. The remaining undissociated tissue may be separated by taking up several times into a glass syringe or pipette. The cell suspension is then filtered through nylon bolting cloth and the filtered cells collected by centrifugation at $50g_{av}$ for 2 min. They are then washed twice with balanced salt solution and finally resuspended. Although the cells may be used in suspension at this stage, it is more common to culture them by resuspending in Eagle's minimal essential medium, deficient in arginine, containing 10% (v/v) foetal calf serum and antibiotics (penicillin, streptomycin and fungizone). For liver from young foetuses (in the rat, < 16 d of gestation) the cells are isolated by repeated aspiration of the tissue into a wide bore Pasteur pipette or syringe. The dissociated tissue is then processed as above.

24.4.4.7 Intestinal Cells

Mucosal cells can be isolated from the small intestine.[173,174] A suitable segment of the intestine is removed, irrigated with ice cold saline or balanced salt solution to remove the intestinal contents and then everted. The segments are tied and filled with a balanced salt solution containing EDTA. The mucosal cells are obtained by vibration of the segments.

24.4.4.8 Blood Cells

Whole blood, blood fractions and isolated blood-derived cells can be employed in studies of drug metabolism. Although whole blood can be used to examine reactions such as hemoglobin oxidation by aromatic amines, the complexity of whole blood and the fact that it clots demands the inclusion of anticoagulants which may interfere with the metabolism of foreign compounds. Thus, for most applications it is desirable to employ isolated cells. In addition, plasma contains a number of esterases and amidases that can hydrolyze a variety of foreign compounds.

24.4.4.8.1 Erythrocytes

Erythrocytes may be prepared by centrifugation ($2500g_{av}$, 15 min) of whole blood containing anticoagulant (*e.g.* citrate, heparin, EDTA). The plasma supernatant is removed by aspiration and can, itself, be used for studies of drug metabolism. The erythrocytes are resuspended in buffered isotonic saline (0.9% w/v) and then sedimented by centrifugation ($2500g_{av}$, 15 min). The pellet is usually washed twice more with buffered saline (0.9%). Erythrocytes may be quantified with a haemocytometer. Oxyhemoglobin can act as an oxidizing agent, substrates including aromatic amines.[175] Red cells also contain high concentrations of glutathione, which can form adducts with xenobiotic electrophiles, formed within the erythrocyte or diffusing from a remote site of formation.

24.4.4.8.2 *Lymphocytes*

Lymphocytes, although deficient in their ability to oxidize chemicals unless stimulated with mitogens in culture, have been shown to possess a number of enzyme systems capable of metabolizing xenobiotics. Several of these enzymes are detoxifying (*e.g.* glutathione transferase, epoxide hydrolase). Thus, lymphocytes can be used as a model to study the detoxication of cytotoxic agents in intact human target cells.[176] Lymphocytes are readily prepared by centrifugation. Blood containing lithium heparin as an anticoagulant is diluted (1:1) with isotonic saline (0.9%) then layered on to a mixture of sodium metrizoate and Ficoll (9.6% and 5.6% w/v respectively, $d = 1.077\,\mathrm{g\,mL^{-1}}$) and centrifuged ($400g_{av}$ for 40 min at ambient temperature). The lymphocyte layer, located at the interface, is removed, suspended in isotonic saline (0.9%) then sedimented by centrifugation ($500g_{av}$, 15 min). Platelet contamination is eliminated by three washes with HEPES buffer and centrifugation. The final pellet of lymphocytes is resuspended in HEPES buffer.

24.4.4.8.3 *Neutrophils*

Neutrophils (polymorphonuclear leucocytes) possess enzyme systems with powerful oxygen activation mechanisms which are involved in the bactericidal functions of these cells. These enzyme systems are also capable of oxidizing xenobiotics, often to reactive species which can form adducts with nucleophiles.[177,178] Such oxidation is dependent on the neutrophil being stimulated to undergo what is known as respiratory burst, a cyanide resistant respiration with the production of activated oxygen species. Agents which trigger 'resting' neutrophils to undergo respiratory burst include bacterial cell walls, opsonized zymosan and phorbol esters. Neutrophils can be isolated by centrifugation with greater than 95% purity.[179] Blood (40 mL) containing citrate as anticoagulant is centrifuged ($300g_{av}$, 20 min) to separate platelet rich plasma, which is removed by aspiration, then recentrifuged ($2500g_{av}$, 15 min) to obtain platelet poor plasma which is removed but retained for use later. Dextran ($M = 500\,000$, 6 mL of 6%) is added to the remaining cells and the volume made up to 50 mL with isotonic saline (0.9%). The cell suspension is gently mixed, then left to stand for 30 min to allow erythrocyte sedimentation. Leucocyte-rich plasma so obtained is centrifuged ($275g_{av}$, 6 min) and the resultant pellet resuspended in platelet poor plasma. Plasma–Percoll gradients are prepared using Percoll (90% in saline) and platelet poor plasma. Leucocytes suspended in platelet poor plasma are underlayered with 42% plasma–Percoll, which in turn is underlayered with 51% plasma–Percoll, and the complete gradient centrifuged ($275g_{av}$, 10 min). Neutrophils, which form a band at the interface of the 42% and 51% Percoll layers, are recovered, suspended in platelet poor plasma and sedimented ($275g_{av}$, 10 min) before washing and resuspending in Hank's balanced salt solution. The neutrophil preparations obtained using this method are reported to be minimally influenced by the isolation procedure, retain original shape and are not 'primed', which can occur using alternative preparative procedures.

24.4.4.9 Established Cell Lines

Freshly isolated primary cells, particularly in suspension, although to some extent in culture, retain many of the differentiated functions of the tissue of origin. This is particularly important in studies of drug metabolism where there are multiple enzymes which show profound changes during differentiation. However, there are circumstances where this approach is extremely difficult or indeed impossible, such as in the study of human-derived tissue. Further, it may be desirable to obtain a homogeneous cell population for a large series of studies and this is not possible using primary cells, as these will vary from animal to animal. Many of these problems are overcome when an established, continuously dividing cell line is used. The major disadvantage is that this will often be poorly differentiated, and further, may have a substantially altered phenotype from the progenitor cell due to continuous subculture for many years. Nevertheless, in recent years a number of cell lines have found use in studies of drug metabolism and these have been derived from a variety of tissues and species, including man. Some of the more widely used examples of such cell lines are listed in Table 2.

The techniques involved in the use of such cells are relatively straightforward. Master stocks of the cells are stored in liquid nitrogen, preferably at several different locations, to insure against loss. The cells for study are established in culture using appropriate growth conditions by standard tissue culture techniques. The cells can then be maintained by subculturing, often indefinitely. It is

Table 2   Examples of Cell Lines Used in Drug Metabolism Studies

Cell line	Tissue of origin	Ref.
HeLa	Human portio carcinoma	180
C2Rev7	Hepatoma	181
H4IIEC3/G$^-$	Hepatoma	182
NCI-H322	Human lung tumour	183
NCI-H358	Human lung tumour	184
BL9L	Liver	185
LLC-PK$_1$	Proximal tubule	186
MDCK	Distal tubule	187
HepG2	Human hepatoblastoma	188
Hepa-1	Mouse hepatoma	189

recommended that the phenotype of the cells be checked regularly to ensure that there has not been important genetic drift during this process. For example, cells derived from the proximal tubule of the kidney can be tested for the ability to accumulate *p*-aminohippuric acid, as a measure of secretory transport.

Drug metabolism is studied in the cells by the addition of the substrate to the culture dish with removal of the supernatant at appropriate intervals for analysis of parent compound and metabolites. It is also possible to scrape portions of cells from the plate at intervals, lyze the cells and then analyze the cell contents for the presence of drug and its metabolites. Alternatively, the cells can be harvested from the culture dish by scraping into medium, if necessary following trypsinization. The cells can then be used either in suspension or lyzed and subjected to differential ultracentrifugation to isolate the appropriate subcellular fractions. The metabolic activity of these fractions is then studied using the techniques described in Section 24.4.5.

24.4.5   SUBCELLULAR FRACTIONS

24.4.5.1   Introduction

Although the techniques described above enable much useful information to be obtained on the routes and rates of metabolism of foreign compounds, it is extremely difficult to apply them to any investigation on the specific enzymes involved. In addition, they are all relatively time consuming and require a degree of technical expertise. These problems can be overcome by the use of specific subcellular fractions. It has been possible for some time to obtain pure preparations of most of the major organelles of the cell, such as the nuclei, mitochondria, microsomes and the cytosol itself. More recently, methods have been developed for the isolation of relatively pure fragments of plasma membrane and other subcellular structures. The two main techniques for the isolation of the major organelles are differential centrifugation and density gradient centrifugation. In both cases, the difference in size of the organelle is used as a discriminating factor in its isolation. The latter technique, that of density gradient centrifugation, enables better-defined fractions to be isolated but in many cases the simpler technique of differential centrifugation is sufficient to obtain a relatively pure fraction for studies of drug metabolism. The preparations most widely used in such studies are the cytosolic and microsomal fractions.

Prior to centrifugation, the tissue must be disrupted and individual cells lyzed to enable their organelles to be isolated. This is usually accomplished in one step by homogenization. Several different methods are available, including the use of a homogenization tube with close-fitting pestle, the tube usually being made of glass and the pestle of either glass, plain or ground, or of Teflon.[190] For tissues that are more difficult to homogenize, although it has general applicability and is now being widely used, an alternative is a high-speed tissue disrupter such as that made by Ultra-turrax or Kinematica.[191] The authors profess a particular preference for the latter type of homogenizer as it provides a rapid and flexible means of preparing subcellular fractions, particularly the microsomal and cytosolic fractions. Cell lysis can be further improved by subjecting the initial homogenate to ultrasonication, although this is not usually necessary for the tissues in most common use. However, when preparing subcellular fractions from isolated cells, ultrasonication will often be necessary, as the other techniques do not enable adequate cell disruption to be achieved.[190]

Homogenization is carried out in a relatively simple medium of salt solution, which must be buffered and often contains potassium chloride to help in removing hemoglobin. When cells are disrupted, they lose their endogenous protectants against oxidizing species and for that reason it is quite frequent to include an antioxidant in the homogenization buffer, such as EDTA and/or dithiothreitol, butylated hydroxytoluene, at concentrations of 0.1–1.0 mM.[192] For some tissues, such as the intestine and most foetal tissues, the homogenate will tend to form a gelatinous mass which is difficult to separate by centrifugation. This can be overcome, in many cases, by the addition of a few units of heparin and rehomogenizing for a few seconds.[193]

Once the cell is disrupted, many of the proteins are relatively labile and should be maintained at 4 °C at all times when not being incubated. Thus, immediately after isolating the organ from the donor, the material is kept on ice, homogenization being performed, if necessary, in an ice bath and the ultracentrifuge is maintained at 4 °C.

Following its isolation, the organelle of interest is resuspended in appropriate buffer, most usually with the aid of a Teflon–glass homogenizer operated by hand.[194] It is often necessary to store the suspension of organelles for some time so that not all studies have to be performed on the one occasion, particularly important when using human tissue. Conditions of storage must be selected to provide adequate maintenance of the activity of the enzymes of interest. For most preparations, this will require low temperature storage, at −75 °C or lower, and the inclusion of a suitable cryoprotectant such as glycerol in the resuspension medium.[192,194] Glycerol serves more as an antioxidant than as a cryoprotectant and hence more traditional cryoprotectants, such as dimethyl sulfoxide, are not suitable for this purpose. Further, the concentrations that would be required to provide adequate cryoprotection would inhibit enzyme activity and it could not be readily removed on thawing the sample. In contrast, glycerol can remain in the sample and is diluted during the incubation.

24.4.5.2 Whole Homogenates

It is possible to investigate the metabolism of drugs and other foreign compounds using whole homogenates of tissue. The homogenate is prepared by the methods described above. However, such preparations have very limited application as they provide few of the advantages of a specific subcellular fraction and the type of information obtained would be more readily achievable by using whole cells where the cellular structure is maintained. The most common application of such preparations is in the determination of the amount, rather than the activity, of specific proteins or cofactors. Circumstances where this might be necessary are where one wishes to avoid or to determine any losses that might occur during centrifugation, or where the amount of tissue available is extremely limited and subcellular fractionation would not be possible. This would include occasions where only very small biopsy samples are obtained[195] or with foetal tissue where the size of the organ is limiting.

Storage of whole homogenates for the determination of enzyme activities is not recommended.

24.4.5.3 Postmitochondrial Supernatant (S9)

The S9, or postmitochondrial supernatant, is widely used in studies of drug metabolism. It comprises the microsomes, together with the cytosolic fraction. Its major advantage is that it can be prepared rapidly without recourse to an ultracentrifuge. Further, both soluble and membrane-bound enzymes are present, although this may not always be an advantage. The tissue is homogenized as above in an appropriate buffer,[194,196] which should not contain high levels of glycerol as the homogenate has to be centrifuged. Since it is the supernatant that is required, it is not possible to incorporate glycerol at a later time unless dilution of the fraction is acceptable. However, the inclusion of EDTA or other antioxidant in the homogenization buffer enables the preparation to be stored for some time without loss of drug-metabolizing activities. However, the stability of the activity in question should be established before the use of such preparations.

The homogenate is centrifuged at $15\,000g_{av}$ for 10 min or other appropriate g–time combination. The supernatant is carefully decanted, care being taken not to contaminate it with the pellet, which will contain mainly cell debris and nuclei, together with other large organelles. If desired, the fraction can be centrifuged for a second time to remove any contaminating material. The fraction is then used fresh or divided into suitable aliquots, of a size sufficient for a single experiment, and frozen at either −80 °C or in liquid nitrogen. The S9 preparation is frequently used as an activating system in the

Ames test and an extensive literature is available for further details on the isolation and storage of this preparation (*e.g.* see Maron and Ames[197]).

Although the S9 preparation contains the soluble fraction of the cell, the concentrations of cofactors for the enzymes of drug metabolism are insufficient to support such activities during incubations *in vitro*. Thus, as with all such reactions, the preparation must be supplemented with the cofactors appropriate to the reaction under study. For oxidation reactions, this will be NADPH, or a generating system comprising NADP$^+$ and a substrate such as isocitrate or glucose 6-phosphate, the enzyme itself for the generating system being present at more than adequate concentrations in the S9 preparation. Other cofactors include active sulfate (phosphoadenosine phosphosulfate) or its precursors, UDP-glucuronic acid and CoASAc. Although many of the enzymes of drug metabolism coexist in this preparation, it is often not possible to study the activity of more than one enzyme at a time, because the conditions for one reaction may not be optimum for another.

Incubations are performed in an air atmosphere at 37 °C in a shaking water bath, the time of incubation and the amount of protein having been determined to be on the linear portion of the respective activity curves, in pilot studies. Normally, the amount of protein will be up to 1–2 mg and the time of incubation up to 30–60 min. The reaction is normally terminated by denaturing the protein with acid,[198] precipitating salts, such as barium hydroxide and zinc sulfate,[199] or with organic solvent, such as methanol. The products of the reaction are then assayed chemically, either following their extraction or directly in the deproteinized sample. Reactions which can be investigated in this way are the same as those that can be studied with the microsomal and cytosolic fractions, and some examples are listed in Table 3.

It is possible to obtain estimates of the kinetic constants for the various drug-metabolizing reactions catalyzed by the S9 preparation. This is accomplished in the normal fashion by using a range of concentrations bracketing the K_m concentration. However, due to the complexity of this system, it might be difficult to obtain reliable estimates of these parameters, which might be better obtained using a purer subcellular fraction.

24.4.5.4 Microsomal Fraction

The microsomal fraction is derived from the disruption of the endoplasmic reticulum.[252] It comprises closed vesicles which contain all of the constituents of the endoplasmic reticulum.[253] Normally, the vesicles form with the external membrane on the outer surface of the vesicle, although it is possible to obtain vesicles in which the outer reticulum membrane is on the internal surface. Microsomes are obtained from both the rough and the smooth endoplasmic reticulum.

The microsomal fraction is the site of many of the enzymes of drug metabolism, including the mixed function oxidase system,[13] the flavin-containing monooxygenase system,[254] epoxide hydrolase,[255] UDP-glucuronyltransferases,[256] deacetylase and other esterases.[257] The presence of the mixed function oxidase system in this subcellular fraction has made it the most extensively studied fraction from any tissue. There is a voluminous literature on various aspects of the use of this preparation and only a relatively brief outline can be provided here.

The microsomal preparation is obtained by differential centrifugation.[191,253] However, several other methods have been described for the isolation of this fraction and these include chromatography on Sepharose 2B[258] and low speed centrifugation to isolate a calcium aggregate.[259] Although these methods provide some advantages (*e.g.* there is no requirement for an ultracentrifuge), these do not outweigh their disadvantages and the majority of workers continue to use the differential centrifugation technique to prepare the microsomal fraction.

The source of the microsomal fraction can be virtually any tissue from any species. Laboratory animals can be pretreated as appropriate to modify the levels of specific enzymes within the endoplasmic reticulum prior to sacrifice. Further, many groups have been able to obtain human tissue in relatively fresh condition.[260] This is best suited to the isolation of subcellular fractions such as the microsomal fraction. When using laboratory animals, they are often fasted overnight to deplete hepatic glycogen. Whilst there is evidence that a high glycogen concentration in the microsomal preparation may affect some metabolic activities, it is now known that fasting can also alter the levels of specific forms of cytochrome *P*-450. Laboratory animals are killed humanely, using a suitable technique so as to cause the minimum effect on enzyme activities. Thus, the use of anaesthetic agents should be avoided if possible. The technique in this laboratory for animals with body weight below 1 kg is stunning followed by exsanguination. UK workers are reminded that the method of sacrifice must follow the regulations laid down in the new Animals (Scientific Procedures) Act 1986 for the use of animals in scientific investigation.

Table 3 Examples of Assays for Drug-metabolizing Activity of Subcellular Fractions

Reaction	Substrate	Enzyme[a]	Metabolite	Cofactor[b]	Assay method	Ref.
C-Oxidation, aliphatic	Coumarin	P-450	Hydroxylated coumarins	NADPH	Fluorimetric	200
	Lauric acid	P-450	Hydroxylated lauric acid	NADPH	HPLC/radiometric	201
	Testosterone	P-450	Hydroxylated testosterones	NADPH	HPLC	202
	Ethanol	ADH	Acetaldehyde	NAD+	Fluorimetric	203
C-Oxidation, aromatic	Aniline	P-450	4-Aminophenol	NADPH	Spectrophotometric	204
	Benzo[a]pyrene	P-450	Hydroxylated benzo[a]pyrenes	NADPH	Fluorimetric	205
	Warfarin	P-450	Hydroxylated warfarins	NADPH	HPLC	206
	Acetanilide	P-450	4-Hydroxyacetanilide	NADPH	Spectrophotometric	207
	Aldrin	P-450	Aldrin epoxide (dieldrin)	NADPH	GLC	208
	Xanthine	XO	Uric acid	NAD+	Spectrophotometric	209
	Quinoline	AO	Quinolone	NAD+	Spectrophotometric	210
N-Dealkylation	Ethylmorphine	P-450	Formaldehyde	NADPH	Spectrophotometric	211
	Aminopyrine	P-450	Formaldehyde	NADPH	Radiometric	212
	Benzphetamine	P-450	Formaldehyde	NADPH	Spectrophotometric	213
O-Dealkylation	Ethoxycoumarin	P-450	7-Hydroxycoumarin	NADPH	Fluorimetric	214
	Ethoxyresorufin	P-450	Resorufin	NADPH	Fluorimetric	215
	Nitrophenetole	P-450	4-Nitrophenol	NADPH	Spectrophotometric	216
	Phenacetin	P-450	Paracetamol	NADPH	GC–MS	217
S-Dealkylation	Methitural	P-450	Formaldehyde	NADPH	Spectrophotometric	218
Oxidative deamination	Amphetamine	P-450	Phenylacetone oxime	NADPH	GLC	219
	Benzylamine	MAO	Benzaldehyde	—	Spectrophotometric	220
N-Oxidation, amines primary	Amphetamine	P-450	N-Hydroxyamphetamine	NADPH	GLC	221
	Aminobiphenyl	P-450	N-Hydroxyaminobiphenyl	NADPH	GLC	222
secondary	Metamphetamine	P-450	N-Hydroxymetamphetamine	NADPH	GLC	223
	Benzphetamine	P-450	N-Hydroxybenzphetamine	NADPH	GLC	224
	N-Methyl-4-aminoazabenzene	P-450/FMO	N-Hydroxy-N-methyl-4-aminoazabenzene	NADPH	Spectrophotometric	225
	N-Benzylaniline	P-450/FMO	α,N-Diphenylnitrone	NADPH	HPLC	226
tertiary	Guanethidine	FMO	Guanethidine N-oxide	NADPH	HPLC	227
	N,N-Dimethylaniline	FMO	N,N-Dimethylaniline N-oxide	NADPH	Spectrophotometric	228
	N-Ethyl-N-methylaniline	FMO	N-Ethyl-N-methylaniline N-oxide	NADPH	GLC	229
N-Oxidation, amides	Acetylaminofluorene	P-450	N-Hydroxyacetylaminofluorene	NADPH	HPLC	230
N-Oxidation, imines	2,4,6-Trimethyliminoacetophenone	P-450	2,4,6-Trimethylacetophenone oxime	NADPH	GLC	231
N-Oxidation, heterocycles	Pyridine	P-450	Pyridine N-oxide	NADPH	GLC	232
N-Oxidation, hydrazines	Procarbazine	P-450/FMO	N-Isopropyl-α-(2-methylazo)-p-toluimide	NADPH	HPLC	233

Table 3 (*Contd.*)

Reaction	Substrate	Enzyme[a]	Metabolite	Cofactor[b]	Assay method	Ref.
S-Oxidation	Thiobenzamide	P-450/FMO	Thiobenzamide S-oxide	NADPH	Spectrophotometric	234
Desulfuration	Parathion	P-450	Paraoxon	NADPH	Radiometric	235
Nitro reduction	Nitrobenzoic acid	NO_2-red	4-Aminobenzoic acid	NADPH	Spectrophotometric	236
N-Oxide reduction	N-Ethyl-N-methylaniline N-oxide	P-450-red	N-Ethyl-N-methylaniline	NADPH	GLC	237
Azo reduction	Prontosil	P-450-red	Sulfanilamide	NADPH	Spectrophotometric	238
Hydrolysis, esters	Procaine	Esterase	4-Aminobenzoic acid	—	Spectrophotometric	239
Hydrolysis, amides	Procainamide	Amidase	4-Aminobenzoic acid	—	Spectrophotometric	240
Glucuronidation	Nitrophenol	UDPGT	4-Nitrophenol glucuronide	UDPGA	Spectrophotometric	241
	Morphine	UDPGT	Morphine glucuronide	UDPGA	Radiometric	242
	Methylumbelliferone	UDPGT	4-Methylumbelliferone glucuronide	UDPGA	Fluorimetric	243
N-Acetylation	4-Aminobenzoic acid	NAT	N-Acetylaminobenzoic acid	CoASAc	Spectrophotometric	244
	Isoniazid	NAT	N-Acetylisoniazid	CoASAc	Spectrophotometric	245
Glutathione	Various	GST	Glutathione conjugates	GSH	Spectrophotometric	246
Epoxide hydration	Styrene oxide	EH	Styrenediol	—	Radiometric	247
	Benzo[a]pyrene 4,5-oxide	EH	Benzo[a]pyrene-4,5-diol	—	Radiometric	248
Sulfation	Nitrophenol	PST	Nitrophenol sulfate	PAPS	Spectrophotometric	249
	2-Naphthol	PST	2-Naphthol sulfate	PAPS	Spectrophotometric	249
Methylation	Serotonin	MT	N-Methylserotonin	SAM	Radiometric	250
	Adrenaline	MT	O-Methyladrenaline	SAM	Radiometric	250
Peptide bond	Benzoic acid	AT	Hippuric acid	CoA	Spectrophotometric	251

[a] Enzyme abbreviations: P-450, cytochrome P-450; FMO, flavin-containing monooxygenase; ADH, alcohol dehydrogenase; XO, xanthine oxidase; AO, aldehyde oxidase; MAO, monoamine oxidase; NO_2-red, nitroreductase; P-450-red, NADPH-cytochrome P-450 reductase; UDPGT, UDP-glucuronyltransferase; NAT, N-acetyl transferase; GST, glutathione-S-transferase; EH, epoxide hydrolase; PST, phenyl sulfotransferase; MT, methyl transferase. [b] Cofactor abbreviations: UDPGA, uridine diphosphoglucuronic acid; PAPS, phosphoadenosine phosphosulfate; SAM, S-adenosylmethionine.

The organ of interest, usually the liver, is removed as quickly as possible and placed in ice cold buffer. If a hemoglobin free preparation is required, the organ should be perfused with saline or buffer just before or after removal from the donor. This can be accomplished *via* the main vessel supplying the tissue, the portal vein in the case of the liver. Many different buffers have been used in homogenization of tissues for the preparation of the microsomal fraction. In our laboratory, a high molarity phosphate buffer is used.[261] This comprises 0.25 M potassium phosphate buffer, pH 7.25, 0.15 M KCl and 1 mM EDTA. The potassium salts are used in preference to the sodium salts in the preparation of this buffer because of their greater solubility at low temperatures. High molarity sodium phosphate buffers are prone to precipitate during storage at 4 °C. The inclusion of potassium chloride aids in the removal of hemoglobin from the microsomal fraction during its isolation. EDTA is added as an antioxidant to prevent lipid peroxidation during the isolation procedure. One of the main differences in the other homogenization media that have been used is replacement of the KCl with isotonic sucrose, which has arisen from the classical medium used for subcellular fractionation. However, KCl is to be preferred for the isolation of microsomes due to its superior ability to remove hemoglobin. Further, there is some evidence that lipid peroxidation proceeds more readily in a sucrose-based medium than in a salt solution. There is also some variation in the choice of buffer and its molarity. Thus, in addition to the phosphate buffer described above, it is possible to use a Tris buffer or a HEPES buffer, with molarity 5–50 mM.[192,196]

The liver is dissected free of extraneous tissue and then minced coarsely with scissors. The pieces are swirled gently and the buffer carefully poured off, retaining the pieces of liver in the vessel with a glass rod. This washing procedure is repeated two or three times until the supernatant is relatively clear. This helps in obtaining a hemoglobin free preparation. A suitable volume of homogenization buffer is then added to the liver fragments, usually three or four volumes (*e.g.* 4 mL g^{-1} for a 20% homogenate). When preparing the microsomal fraction, the original dilution of homogenate is of relatively minor importance, as the final pellet can be resuspended in any desired volume. When the amount of tissue available is limited, such as from small organs, foetal or biopsy samples, particularly from human tissue, it is often desirable to add a large excess of buffer relative to the weight of tissue. For example, for a 50 mg needle biopsy sample of human liver, one might add 100–200 volumes of the buffer.[261] This does have the advantage that it provides a relatively clean microsomal pellet without the necessity of a washing step, impractical with such a small sample. The alternative is to homogenize in a lower volume of buffer and then to use an adaptor for the ultracentrifuge to enable small volumes to be fractionated.

The sample is homogenized using either a motor-driven pestle, either glass or, more frequently, Teflon, in a glass homogenizing tube. Most tissue samples can be homogenized adequately with five to ten return strokes of the pestle. For some samples, particularly human tissue, it is necessary to keep the homogenization vessel on ice throughout the procedure. In such circumstances the pestle should be precooled prior to homogenization.[261] An alternative, and perhaps preferable, technique for homogenization is disruption with a Polytron tissue disrupter.[191] This is set to an intermediate speed (number 4 or 5) and homogenization is performed for 15 s intervals, with a 1 min break between them to allow the sample to cool. For most tissue samples, adequate homogenization is achieved within 30–45 s. For tissues prone to aggregate, such as the intestine and foetal liver, the addition of 3 i.u. mL^{-1} of heparin to the homogenate at this stage aids in fractionation.[196] The sample is briefly rehomogenized after the addition of heparin.

The homogenate is decanted into ultracentrifuge tubes. We have found that the most suitable are made of polycarbonate; these are transparent and resist collapse at high *g*-force. The tubes should be capped and balanced. The postmitochondrial supernatant is then obtained by centrifuging the samples for 20 min at $12\,000g_{av}$ or for 15 min at $15\,500g_{av}$ at 4 °C. This sediments cell debris, nuclei and mitochondria. Isolation of the postmitochondrial supernatant may be performed using an intermediate speed centrifuge, such as that made by Sorvall. In some protocols, the homogenate is first subjected to low speed centrifugation, $800g_{av}$ for 10 min, to sediment the cell debris and nuclei. The supernatant is then used to prepare the postmitochondrial supernatant as described above. This protocol has largely been abandoned in favour of the simpler, two-step isolation procedure.

The postmitochondrial supernatant is carefully decanted into fresh tubes, which are capped and balanced. These are then subjected to ultracentrifugation on a fixed angle rotor with an appropriate centrifugal force/time combination. A variety of such combinations have been used, often chosen on the basis of the availability of rotor and machine. The most frequently used is $105\,000g_{av}$ for 1 h, which is generally suitable for the isolation of the microsomal fraction. The choice of rotor will be dictated by the centrifuge available and the volume of supernatant. In practice, rotors with a tube capacity of 10–50 mL are most widely used. Smaller tubes (10–12 mL) can be centrifuged to $140\,000g_{av}$ at which speed 45 min is sufficient to sediment the microsomes. It may only be possible to centrifuge higher capacity tubes (35–50 mL) to $75\,000g_{av}$, at which speed 90 min is required for sedimentation of the microsomes.[253]

Following sedimentation of the microsomes, any lipid on the surface of the supernatant is carefully removed with a Pasteur pipette. The supernatant is then aspirated. Any lipid adhering to the surface of the tube is removed with a tissue and any remaining cytosol is drained from the pellet. Although it is possible to resuspend the microsomes and use them at this stage, it may be desirable to wash them in fresh homogenization buffer. The microsomal pellet is resuspended using a hand-operated Teflon–glass homogenizer or a Polytron as above, although the time of homogenization will only be for a few seconds. The fractions are subjected to recentrifugation as above to obtain a clean microsomal pellet. Washing the microsomes removes any remaining hemoglobin and also any residual cytosol. The supernatant is then decanted, the pellets drained by blotting the tube on to filter paper and the microsomal pellet resuspended.

The choice of buffer for resuspension varies considerably amongst laboratories. It is possible to use the same buffer as for homogenization or any other suitable combination of buffer and salt or sucrose. However, a number of groups, including our own, have a preference for a glycerol-based resuspension medium, as this has been shown to have superior properties in maintaining the activity of microsomal enzymes during long term storage.[261] This medium comprises 0.25 M potassium phosphate buffer, pH 7.25, containing 30% v/v glycerol. An appropriate volume of buffer is added to the microsomal pellet in the centrifuge tube. For adult rodent liver, two volumes of buffer relative to the initial weight of tissue are used. For other tissues and for foetal liver, one volume or even 0.5 volumes of buffer are used relative to the initial weight of tissue. The aim is to obtain a final microsomal suspension with a protein content of $>10\ \mathrm{mg\,mL^{-1}}$. Microsomes with lower protein concentrations are less stable during storage at low temperatures. It also has the practical advantage of enabling the aliquots to be stored in a smaller space. The microsomal pellet is dislodged from the base of the centrifuge tube with the tip of a spatula or by gentle vortex mixing for a few seconds, particularly when there is a glycogen pellet. The contents of the tube are swirled and then poured into a glass homogenization vessel. The microsomes are resuspended with a few strokes of a Teflon pestle by hand or, when larger amounts of material are being handled, homogenization with a Polytron for a few seconds. The microsomal suspension is then used fresh or aliquoted for storage for later use.

Sufficient material is placed into vials suitable for low temperature freezing, such as J & J glass vials or cryovials of polypropylene. The samples are best stored at temperatures of below $-70\,°C$. Storage in liquid nitrogen is also possible, but this appears to provide no additional advantage for maintaining activity. However, storage at higher temperatures such as $-20\,°C$ is not as successful in the maintenance of all drug-metabolizing activities. It should be stressed, however, that the stability of the enzymes of drug metabolism shows considerable variability[262] and that the stability of the particular activity in question should be established prior to assuming that long term storage is possible. The amount of material in each aliquot should be sufficient for one experiment, as the suspension should not be refrozen after thawing. The vials should be labelled clearly to enable their identification at a later date, with freezer resistant marker. We have found that Scotch vinyl matt self-adhesive tape wrapped completely around the vial and labelled with indelible marker provides a suitable means of permanently labelling the vials. The labels should be applied to the vials before they are put on to ice. The samples can be flash frozen in a mixture of solid carbon dioxide and alcohol, and then transferred to the freezer for storage.

When the original sample is very small, such as from a biopsy or a foetal organ, the microsomal pellet is resuspended in a small volume of buffer and this requires a microhomogenizer. The other manipulations necessary are described in detail elsewhere.[261]

The microsomal fraction has been used to study the metabolism of a wide range of compounds by a number of different enzymes of drug metabolism. A number of factors have to be considered, whatever the reaction under investigation. These include the requirement for cofactors, the reaction kinetics, the optimum pH, the optimum protein concentration and duration of incubation. Activity can be quantified either by disappearance of substrate, where this is appropriate, or by appearance of a specific product. In the latter case the appearance of product may be monitored continuously[263] or after a fixed interval.[264] In the majority of cases the samples are incubated for a fixed time, the reaction is terminated and the product then quantified.

The rigour with which the conditions of the reaction are characterized will depend to some extent upon the overall aim of the study. For example, if the objective is a qualitative screen of possible metabolites, then linearity with protein concentration and time of incubation may not be as critical as when one wishes to determine the kinetics of the enzyme reaction. Most of the enzymes of drug metabolism require a particular cofactor, for example, the monooxygenase system and the microsomal amine oxidase require NADPH, the UDP-glucuronyltransferases require UDP-glucuronic acid, whereas others have no requirement for additional cofactors, for example epoxide hydrolase and esterases. Where NADPH is required, this may be added preformed or as its precursor (NADP$^+$), together with a suitable system for generating the reduced form of the cofactor. The most frequently used of such generating systems comprise glucose 6-phosphate plus glucose-6-phosphate dehydrogenase, or isocitrate plus isocitrate dehydrogenase. The latter appears to have marginal advantages, but is more costly. The pH optimum for the reaction should be determined. For example, the UDP-glucuronyltransferases often have a higher pH optimum than the mono-oxygenases (7.6 *cf.* 7.4).[265] However, the pH may vary not only with the enzyme under study, but also with the substrate. The linearity of the reaction with protein concentration and time of incubation should be determined. For some activities the turnover number is very rapid and short incubation times are necessary, from 1–5 min, whereas other reactions proceed at much lower rates and incubation times in excess of 1 h may be required. A standard incubation time would be 10 or 15 min. Long incubation times can lead to a decrease in the reaction rate due to peroxidative damage to the microsomal membrane, although this can be minimized by the inclusion of an antioxidant. When using preformed cofactor, this may become limiting with time. A similar problem arises when using high protein concentrations. Ideally, one should choose a concentration of microsomal protein where product formation is first order with respect to protein concentration.

Substrates should be added to the incubation medium in a solvent which has no effect on activity. Ideally, this would be buffer or water, but many substrates have only very limited solubility in aqueous medium. To overcome this, the compound is dissolved at high concentration in organic solvent and added in a very low volume, typically 1–10 μL mL^{-1} but in some instances up to 50 μL mL^{-1}. Studies should be performed to establish that the solvent used has minimal effect on the activity in question. Solvents that are used relatively widely include dimethyl sulfoxide, ethanol, methanol and acetone. For the monooxygenase system, methanol appears to have the least effect on most activities, but not all compounds will be soluble in this solvent.[266] When determining reaction rates, the incubation conditions should be such that the percentage loss of substrate throughout the incubation is low, usually < 10%. At higher rates, Michaelis–Menten kinetics cease to apply.

The reaction should be characterized with respect to its kinetics. This is achieved by adding a range of concentrations of substrate, over perhaps a 100-fold range, and determining the reaction velocity. The results are then subjected to classical Michaelis–Menten analysis, preferably using iterative nonlinear least squares model-fitting procedures, rather than the more error prone linear regression analysis of the Lineweaver–Burk plot. Some enzyme reactions show multiphasic kinetics, due to the contribution of different components with very different values of K_m. A number of methods have been described enabling the kinetics of such systems to be analyzed,[267,268] and commercial software packages such as NONLIN, MLAB, StatGraphics, GrafPad, for a variety of computers, both mainframe and micro, are available. In single concentration assays it is standard practice to use a substrate concentration several times above K_m, so that activity will depend upon the amount of enzyme present. At lower concentrations, activity will be affected by competing pathways. There may, of course, be circumstances where the study requires that this is so.

In a continuous monitoring assay, often used where the turnover number of the enzyme is very high, the incubation is performed in a cuvette placed in the measuring device, most frequently a spectrophotometer or fluorimeter. For some assays, the cuvette has to be thermostatted to 37 °C, but in others the reaction rate is so rapid that to obtain adequate velocities it is necessary to monitor activity at room temperature (25 °C), for example the *O*-deethylation of ethoxycoumarin[269] (although there are more reliable methods for determining this activity now available[214]). More often, it will be necessary to quantify the products after terminating the reaction. The reaction can be terminated by rapidly deproteinizing the sample or by rapidly cooling to below 4 °C. Deproteiniz-ation can be accomplished by the addition of acid, for example trichloroacetic acid (final concentra-tion 6–10%) or perchloric acid,[198] precipitating salts such as zinc sulfate to a final concentration of approximately 2% followed by an equal volume of saturated barium hydroxide[199] or water miscible organic solvent, acetone or methanol, to a final concentration of at least 50%. Reactions can also be stopped by the addition of several volumes of the same immiscible organic solvent (for example chloroform or diethyl ether) used for any extraction, or by rapidly heating the samples to high temperature (> 60 °C) to deproteinize them.

The procedure followed for incubation normally involves adding the appropriate reagents to test-tubes on ice, those reagents common to all of the samples being prepared in bulk as a 'pool', which is then distributed to all of the samples. One of the reactants is omitted from the samples and the reaction is initiated by the addition of that reactant after preincubation of the other components at 37 °C in a shaking water bath for 2 or 3 min to equilibrate the samples. The reaction is started by the addition of the cofactor, the microsomal protein, or the substrate, depending on the nature of the study. If, for example, the study involves the use of a range of substrate concentrations, it would be best to omit the protein or the cofactor, since this will be added in common to almost all the samples to start the incubation.

The vessels used for incubating these samples should be broad-based to enable adequate oxygen transfer at the surface. Suitable vessels include Erlenmeyer flasks and glass scintillation vials. Samples are usually incubated in an air atmosphere in a water bath, which is agitated by mechanical shaking at 30–50 cycles min^{-1}. After equilibration of the samples the initiating reactant is added at suitable intervals, every 10–15 s, and the incubation is terminated after incubation for the desired time by the addition of the denaturing reagent, again at 10–15 s intervals. In this way all of the samples will be incubated for exactly the same length of time. The denaturant is often added ice cold and the samples are then transferred to an ice bath to await further processing.

Incubation blanks should always be included. The two main types of blank are a substrate blank and an enzyme blank. The blanks contain all of the reactants save the one for which they are controlling. They are then processed exactly as for the test samples, and the missing reactant is added after terminating the reaction.

Where deproteinization has been performed, the aggregated protein is removed by low speed centrifugation ($1500g_{av}$, 10 min). The supernatant is removed and subjected to chemical analysis. A wide variety of techniques can be utilized and these include radiometry, spectrophotometry, either of the compound of interest itself or after derivatization to provide a suitable chromophore, spectrophotofluorimetry, HPLC, GC or GC–MS. In some cases a combination of techniques is utilized, for example, HPLC with radiometric detection. Examples of some of the assays that have been used with the microsomal fraction are given in Table 3.

During incubations *in vitro* the percentage conversion of substrate is often extremely low, <1%, and thus when measuring the metabolite this is in the presence of a considerable excess of the structurally similar parent compound. It may thus be necessary to remove this by differential extraction prior to analysis of the metabolite.[270] This may be necessary even when using a highly specific technique such as GC–MS.[271]

In the past, relatively simple methods were used for detection of products but more sophisticated analytical techniques are now being applied, with the same rigorous criteria for analysis as applied to other situations. This can only be a desirable development and, for example, the use of internal standards for quantification is to be encouraged wherever possible.

24.4.5.5　Cytosol

The cytosolic, or soluble, fraction of the cell comprises that fraction which is not sedimented when a tissue homogenate is centrifuged for 1 h at $105\,000g_{av}$. Thus, the cytosol is normally prepared utilizing the same techniques as described above for obtaining the microsomal fraction.[191,253] After sedimenting the microsomes, the supernatant comprises the cytosol. A number of additional considerations are necessary, however. These include the amount of buffer in which the tissue is homogenized, since the concentration of cytosol will be determined by the extent to which tissue is initially diluted with homogenization buffer. Dilution factors of two to four volumes relative to wet weight are normal. Following the initial sedimentation of the microsomes, any lipid at the surface of the cytosol is aspirated with a Pasteur pipette, and the clear supernatant is then aspirated from the tubes, care being taken not to disturb the microsomal pellet. The cytosol can then be used as it is or, preferably, it is then recentrifuged, using the same conditions as those used for its initial isolation. It is then used fresh or stored frozen in aliquots at $-80\,°C$ or in liquid nitrogen.

The cytosol contains the majority of the glutathione transferases,[272] sulfotransferases,[273] N-acetyltransferases[274] and a number of esterases and amidases.[275] Examples of substrates for these reactions are given in Table 3. The considerations described above for the determination of enzyme activity with the microsomal fraction apply to such studies with the cytosol. The cofactors necessary include reduced GSH for the glutathione transferases, active sulfate (phosphoadenosine phosphosulfate; PAPS) for the sulfotransferases and CoASAc for the N-acetyltransferases. The cytosol also contains a number of NADPH-dependent dehydrogenases, including isocitrate dehydrogenase and glucose-6-phosphate dehydrogenase, and can thus be used as one component of an NADPH-generating system. Although the cytosol does contain endogenous cofactors, such as NADPH, CoASAc and UDP-glucuronic acid, these will have been diluted to such an extent that they will not support the corresponding metabolic activities without additional exogenous supplementation. However, if necessary the cytosol can be dialyzed against buffer to remove residual endogenous cofactors and other soluble components which might otherwise interfere in some studies.

24.4.5.6 Other Organelles

A number of other cellular organelles may be isolated for use in studies of drug metabolism, although such use is likely to be extremely limited. These organelles include mitochondria,[190,276] lysosomes and peroxisomes,[277–279] nuclei[280,281] and fragments of the plasma membrane.[282–284]

24.4.6 PURIFIED ENZYMES

Most of the enzymes of drug metabolism exist in multiple forms, often with overlapping specificity. This has made the identification of the specific isoforms involved in a particular reaction extremely difficult. Considerable effort has been spent on purifying individual isoenzymes in an attempt to overcome this. Once purified, most of the enzymes are stable when stored under appropriate conditions. Activity can be reconstituted, with appropriate coenzymes where necessary. Thus, in theory at least, such preparations provide an ideal means of determining the specificity towards a given substrate. However, there are some difficulties with this approach. Many of the enzymes in question are membrane bound, and have proved extremely difficult to purify to homogeneity. This almost always involves the use of detergent, which cannot be removed completely from the final preparation. Further, the stoichiometry of the components of a multienzyme system has proved difficult to reproduce in reconstitution. In the case of the cytochrome *P*-450 system, there is still debate as to the role of the cytochrome b_5 system. If one isoenzyme is shown to be very active towards a substrate compared to other purified forms, it is still not possible to conclude that this is the only form, or indeed the major form, involved in catalyzing the reaction in the intact cell. The possibility always exists that another form, not yet purified, is responsible. This is true, even when estimates are based on the specific content of the isoenzyme and the kinetics of the reaction, as the kinetics of purified enzymes are sometimes different from those of the membrane-bound form. In addition, due to the hydrophobicity of many of the proteins in question and the fact that they exist as multiple isoenzymes, it has proven difficult to purify them to homogeneity.

Considerable advances have been made in the last few years in purifying the enzymes of drug metabolism, and many have now been resolved to homogeneity. Whilst this still does not overcome all the problems listed above, such preparations are ideally suited to the production of specific inhibitory antibodies, which can then be used with the intact subcellular fraction to determine the contribution of the immunizing isoenzyme to the reaction in question.

The strategy involved in the purification of one of the enzymes of drug metabolism will obviously depend upon whether the protein is soluble or membrane bound. In the case of the *P*-450 system, these are all membrane bound in eukaryotes. This would also be the case with the microsomal epoxide hydrolase, the glucuronyltransferases, the mixed function amine oxidase and a number of esterases and deacetylase. Soluble enzymes include many of the sulfotransferases and the glutathione transferases. There is a membrane bound glutathione transferase and a soluble epoxide hydrolase and these too have been purified (Table 4).

The methods utilized in the purification of the enzymes of drug metabolism are all based on traditional chromatographic resolution. Although newer methods have been developed, with some success, the majority of workers continue to use the more established techniques. These involve open column chromatography on ion exchange resins (both anion and cation), affinity chromatography, particularly cofactor dependent (of considerable use in the purification of the glucuronyltransferases and the glutathione transferases, as well as reductases for cytochrome *P*-450 and b_5), hydrophobic interaction chromatography, particularly on modified Sepharose matrices, such as *n*-octylamino-Sepharose for cytochrome *P*-450, mixed chromatography on hydroxyapatite (particularly useful for the removal of excess detergent) and chromatofocusing. Purification to homogeneity often requires the use of several of these techniques.

In the case of membrane-bound proteins, the subcellular fraction is subjected to detergent solubilization, a large number of detergents having been used for this purpose, including cholate and some of the Triton derivatives. The proteins are then partially concentrated by selective precipitation using ammonium sulfate or polyethylene glycol. Where hydrophobic interaction chromatography is used, it is common to elute with a nonionic detergent such as Renex 690, Emulgen 911 or Lubrol PX. Purification is monitored by a variety of techniques, including absorbance at specific wavelengths, such as 420 nm for heme-containing proteins, the presence of protein of the desired molecular weight, determined by SDS–polyacrylamide gel electrophoresis and specific enzyme activity. In this last case, it may be necessary to remove detergent and reconstitute with appropriate coenzymes. Should an antibody to the protein of interest be available, an

Table 4 Purification of Enzymes of Drug Metabolism

Enzyme	Ref.
Rat cytochromes *P*-450 (at least 20 isoenzymes)	285–287
Rabbit cytochromes *P*-450 (at least 13 isoenzymes)	288
Human cytochromes *P*-450 (at least six isoenzymes)	289
Cytochromes *P*-450 from other species (mouse, guinea pig, ox, dog, hamster and fish)	290–295
Flavin-containing monooxygenase	253
Microsomal epoxide hydrolase	296, 297
Cytosolic epoxide hydrolase	298
Rat UDP-glucuronyltransferases (at least six isoenzymes)	299, 300
Human UDP-glucuronyltransferases (at least two isoenzymes)	301
Rat cytosolic glutathione transferases (at least eight forms)	302, 303
Human cytosolic glutathione transferases (at least eight forms)	302
Microsomal glutathione transferase	304
Microsomal carboxylesterases (at least four isoenzymes)	305
Sulfotransferases (at least four isoenzymes)	306, 307
N-Acetyltransferases	308, 309
Cytochrome b_5	310
NADPH-cytochrome *P*-450 reductase	311
NADH-cytochrome b_5 reductase	312
Cysteine conjugate β-lyase	313

additional possibility for monitoring the progress of purification is Western blotting. Dot blotting or ELISA could also be used, but these techniques are less reliable under such circumstances. If it is possible to label the protein of interest in some way, such as with a suicide substrate, or by photoaffinity labelling, the presence of label could be used to monitor progress during purification. Attempts to use this approach for cytochrome *P*-450 have proved unsuccessful to date. However, for some other proteins this has proved a very satisfactory means of identifying the protein of interest.

Once the protein has been purified, its purity must be checked and claims of homogeneity must be vigorously confirmed by several techniques, including SDS–polyacrylamide gel electrophoresis, isoelectric focusing, two-dimensional gel electrophoresis, N-terminal sequencing and specific content. This last may be of little practical value due to the difficulties of determining protein content or specific activity in the presence of trace amounts of detergent.

When determining metabolic activity with the purified enzyme, it may be necessary to include essential coenzymes. This is the case with the cytochrome *P*-450 system, where it is necessary to add NADPH-cytochrome *P*-450 reductase together with lipid. It is also possible to include cytochrome b_5 and NADH-cytochrome b_5 reductase. These components are added in low volume and incubated for a few minutes prior to the addition of the other reactants. Monooxygenase activity can then be determined using the techniques described above for the microsomal fraction.

Enzymes of drug metabolism have also been purified using fast protein liquid chromatography (FPLC)[314] and by HPLC.[315] Whilst of considerable benefit in the purification of soluble proteins, these techniques have proved less successful in the purification of membrane-bound proteins, due to the difficulties of obtaining detergents suitable for these particular techniques. However, there is increasing use of HPLC-profiling of cytochrome *P*-450, used to obtain information on the number and amount of different isoenzymes present, rather than for purification.[316] From a practical point of view, the purification of large amounts of protein for studying metabolic activity is still probably best confined to the more traditional open column methods.

24.4.7 NEWER TECHNIQUES IN DRUG METABOLISM

As in the rest of biology there has been an explosion in the application of molecular biological techniques to the study of drug metabolism and the genes for many of the enzymes of drug metabolism have now been cloned.[317] This has been paralleled by advances in protein purification and in antibody production.[287] Thus, the purification of the enzyme of interest, followed by the production of an antibody (either polyclonal or monoclonal) against it, enables the identification of those clones expressing the protein in λGT11.[318] This enables cDNAs to be obtained. The nature of monoclonal antibodies has meant that preparations of protein that are not entirely homogeneous can still be used for immunization and yet a specific antibody to the protein of interest can be obtained. This is because a single hybridoma clone produces only one species of antibody and the antigen with which this reacts can be identified by Western blotting and immunoaffinity purification. A further test of specificity is N-terminal sequencing of the protein recognized by the antibody.[319]

With the acquisition of cDNA probes to some of the members of a family of drug-metabolizing enzymes, there is usually an exponential increase in the identification of further genes due to the homology that exists amongst members of any gene family.[320] Indeed, the difficulty now is sorting out which sequences are for proteins and which are silent, which are allelic variants and which are from primary loci. A further advantage of the availability of cDNAs is the ability to explore the sequences in a different species, including man. Thus, the use of probes against rat and rabbit forms of drug-metabolizing enzymes in human samples has enabled the identification of many of the homologous genes.[321]

With respect to determining metabolic activity, molecular biology is helping by providing the genes to clone into cells which express them functionally. Whilst this was first carried out using yeast cells,[322] a number of groups have now successfully inserted the genes into mammalian cells, such as V79 Chinese hamster fibroblasts,[323] and this should prove a powerful technique for exploring drug-metabolizing activity. However, one word of caution is perhaps in order. It is still difficult to identify the major enzyme involved in a particular reaction if that enzyme has not yet been purified or cloned, even if other enzymes have a high activity in the reaction of interest. The solution to this problem will depend upon obtaining reliable kinetic data from the cloned enzymes. If these can be expressed in a suitable environment in which the reaction kinetics resemble those of the enzyme within its native subcellular fraction, then this approach will provide the necessary answers. A further application of the availability of the cloned genes is the ability to introduce site specific mutations to explore the basis of the reaction mechanism and specificity of the enzyme.[324] In addition, once the primary sequence has been obtained, it is possible to predict the amino acid sequence and thereby synthesize a peptide of defined specificity. Antibodies against this peptide have the expected immunoreactivity profile.[325] As the structure–function correlates of the enzymes of drug metabolism are resolved, it will become possible to synthesize peptides, against which antibodies will be inhibitory and of highly defined specificity. These could then be used against the native enzyme still within its subcellular fraction to determine its specificity.

24.4.8 CONCLUSION

The range of techniques available for studies of drug metabolism *in vitro* has never been wider, ranging from the isolated perfused organ to cloned enzymes within a mammalian cell. It is important to choose the technique suitable for the problem at hand, recognizing any limitations inherent in the system. Many of the advances in the last few years have enabled studies to be performed with human enzymes, thus enabling us to determine the extent to which extrapolation from animals to man is valid. The increasing availability of human tissue fractions, immunological probes to the enzymes of drug metabolism, and of their cloned genes, should make it possible to study the metabolism of a new compound by human enzymes at an earlier stage of development of the compound than has previously been possible.

24.4.9 REFERENCES

1. R. W. Brauer, R. L. Pessotti and P. Pizzolato, *Proc. Soc. Exp. Biol. Med.*, 1951, **78**, 174.
2. L. L. Miller, C. G. Bly, M. L. Watson and W. F. Bale, *J. Exp. Med.*, 1951, **94**, 431.
3. C. S. Kalser, E. J. Kelvington, M. M. Randolph and D. M. Santomenna, *J. Pharmacol. Exp. Ther.*, 1965, **147**, 260.
4. R. E. Stitzel, M. W. Anders and G. J. Mannering, *Mol. Pharmacol.*, 1966, **2**, 335.
5. T. C. Orton, M. W. Anderson, R. W. Pickett, T. E. Eling and J. R. Fouts, *J. Pharmacol. Exp. Ther.*, 1973, **186**, 482.
6. A. Rosenfeld, A. Sellers and J. Katz, *Am. J. Physiol.*, 1959, **196**, 1155.
7. H. Kavin, N. W. Levin and M. M. Stanley, *J. Appl. Physiol.*, 1967, **22**, 604.
8. O. Warburg, *Biochem. Z.*, 1923, **142**, 317.
9. M. N. Berry and D. S. Friend, *J. Cell Biol.*, 1969, **43**, 506.
10. P. O. Seglen, *Exp. Cell Res.*, 1972, **74**, 450.
11. P. O. Seglen, *Exp. Cell Res.*, 1973, **76**, 25.
12. D. M. Bissell, L. E. Hammaker and U. A. Meyer, *J. Cell Biol.*, 1973, **59**, 722.
13. B. B. Brodie, J. Axelrod, J. R. Cooper, L. Gaudette, B. N. LaDu, C. Mitoma and S. Udenfriend, *Science (Washington, D.C.)*, 1955, **121**, 603.
14. R. Scholz, W. Hansen and R. G. Thurman, *Eur. J. Biochem.*, 1973, **38**, 64.
15. H. Brown and W. G. M. Hardison, *Surgery (St. Louis)*, 1972, **71**, 388.

16. H. Schimassek, *Life Sci.*, 1962, **1**, 69.
17. R. Hems, B. D. Ross, M. N. Berry and H. A. Krebs, *Biochem. J.*, 1966, **101**, 284.
18. B. D. Ross, 'Perfusion Techniques in Biochemistry: A Laboratory Manual in the Use of Isolated Perfused Organs in Biochemical Experimentation', Clarendon Press, Oxford, 1972.
19. L. L. Miller, in 'Isolated Liver Perfusion and its Applications', ed. I. Bartošek, A. Guaitani and L. L. Miller, Raven Press, New York, 1973, p. 11.
20. R. G. Thurman, L. A. Reinke and F. C. Kauffman, in 'Reviews in Biochemical Toxicology', ed. E. Hodgson, J. R. Bend and R. M. Philpot, Elsevier/North Holland, New York, 1979, p. 249.
21. R. S. Jones, in 'Biochemical Toxicology: A Practical Approach', ed. K. Snell and B. Mullock, IRL Press, Oxford, 1987, p. 23.
22. D. S. Parsons and G. Powis, *J. Physiol. (London)*, 1971, **217**, 641.
23. B. Alexander, T. von Arnem, M. Aslam, P. S. Kolhe and I. S. Benjamin, *Lab. Invest.*, 1984, **50**, 597.
24. R. Abraham, W. Dawson, P. Grasso and L. Goldberg, *Exp. Mol. Pathol.*, 1968, **8**, 370.
25. G. Powis, *Proc. R. Soc. London Ser. B*, 1970, **174**, 503.
26. P. M. Galletti, M. T. Snider and D. Silvert-Aiden, *Res. Eng.*, 1966, **5**, 20.
27. P. Eyer, H. Kampffmeyer, H. Maister and E. Rosch-Oëhme, *Xenobiotica*, 1980, **10**, 499.
28. A. Guaitani, P. Villa and I. Bartošek, *Xenobiotica*, 1983, **13**, 39.
29. A. R. Boobis and G. Powis, *Biochem. Pharmacol.*, 1974, **23**, 3377.
30. C. Enteman, R. J. Holloway, M. L. Albright and G. F. Leong, *Proc. Soc. Exp. Biol. Med.*, 1968, **127**, 1003.
31. H. Michelakakis and C. J. Danpure, *Biochem. Pharmacol.*, 1984, **33**, 2047.
32. B. C. Serrou, C. Solasol, L. Meiss, C. Gelis and C. Romieu, in 'Isolated Liver Perfusion and its Applications', ed. I. Bartošek, A. Guaitani and L. L. Miller, Raven Press, New York, 1973, p. 127.
33. B. M. Mullock, R. S. Jones, J. Peppard and R. H. Hinton, *FEBS Lett.*, 1980, **120**, 278.
34. T. G. Richards, V. R. Tindall and A. Young, *Clin. Sci.*, 1959, **18**, 499.
35. M. Gibaldi and D. Perrier, 'Pharmacokinetics', 2nd edn., Dekker, New York, 1982.
36. M. Rowland, *Eur. J. Pharmacol.*, 1972, **17**, 352.
37. K. Miyata, Y. Noguchi and M. Enomoto, *Jpn. J. Exp. Med.*, 1972, **42**, 483.
38. A. Smith, *Biochem. Pharmacol.*, 1976, **35**, 2429.
39. J. Kvetina, M. Simkova, M. Citta and F. Deml, in 'Isolated Liver Perfusion and its Applications', ed. I. Bartošek, A. Guaitani and L. L. Miller, Raven Press, New York, 1973, p. 235.
40. H. Eriksson, J.-A. Gustafsson and A. Pousette, *Eur. J. Biochem.*, 1972, **27**, 327.
41. A. R. Boobis and G. Powis, *Drug Metab. Dispos.*, 1975, **3**, 63.
42. C. von Bahr, B. Alexanderson, D. L. Azarnoff, F. Sjöqvist and S. Orrenius, *Eur. J. Pharmacol.*, 1970, **9**, 99.
43. W. M. H. Heijbroek, D. F. Muggleton and D. V. Parke, *Xenobiotica*, 1984, **14**, 235.
44. S. J. Lan, A. V. Dean, B. D. Walker and E. C. Schreiber, *Xenobiotica*, 1976, **6**, 171.
45. C. Bel, *C.R. Seances Soc. Biol. Ses Fil.*, 1969, **162**, 1949.
46. S. Garattini, A. Guaitani and I. Bartošek, in 'Isolated Liver Perfusion and its Applications', ed. I. Bartošek, A. Guaitani and L. L. Miller, Raven Press, New York, 1973, p. 225.
47. G. Bonse, Th. Urban, D. Reichert and D. Henschler, *Biochem. Pharmacol.*, 1975, **24**, 1829.
48. G. Herbertz, T. Metz, H. Reinauer and W. Staib, *Biochem. Pharmacol.*, 1973, **22**, 1541.
49. A. Cordelli, M. Ferrari and E. Savonitto, *Arch. Int. Pharmacodyn. Ther.*, 1969, **180**, 121.
50. D. Brewster, R. S. Jones and D. V. Parke, *Xenobiotica*, 1977, **7**, 601.
51. I. Bartošek, M. G. Donelli, A. Guaitani, T. Colombo, R. Russo and S. Garattini, *Biochem. Pharmacol.*, 1975, **24**, 289.
52. G. B. Gerber, B. Zicha, J. Deroo, J. P. De Cock and E. Geyer, *Biophysik (Berlin)*, 1972, **8**, 333.
53. U. I. Lavy, W. Hespe and D. K. F. Meijer, *Naunyn-Schmiedeberg's Arch. Pharmacol.*, 1972, **275**, 183.
54. M. H. Bickel and R. Minder, *Biochem Pharmacol.*, 1970, **19**, 2425.
55. J. Kvetina, F. Marcucci and R. Fanelli, *J. Pharm. Pharmacol.*, 1968, **20**, 807.
56. J. Kendler, S. Anuras, O. Laborda and H. J. Zimmerman, *Proc. Soc. Exp. Biol. Med.*, 1972, **139**, 1272.
57. D. R. Van Harken and G. J. Mannering, *Biochem. Pharmacol.*, 1969, **18**, 2759.
58. S. Ji, J. J. Lemasters and R. G. Thurman, *Mol. Pharmacol.*, 1981, **19**, 513.
59. R. A. Van Dyke and C. L. Wood, *Anesthesiology*, 1973, **38**, 328.
60. H. Koster, I. Halsema, E. Scholtens, K. S. Pang and G. J. Mulder, *Biochem. Pharmacol.*, 1982, **31**, 3023.
61. J. R. Gumbrecht and M. R. Franklin, *Xenobiotica*, 1979, **9**, 547.
62. H. M. Bolt and H. Remmer, *Horm. Metab. Res.*, 1973, **5**, 101.
63. D. R. Van Harken, T. R. Tephly and G. J. Mannering, *J. Pharmacol. Exp. Ther.*, 1965, **149**, 36.
64. R. M. Thompson, N. Gerber and R. A. Seibert, *Xenobiotica*, 1975, **5**, 145.
65. G. Ehninger, B. Proksch, F. Hartmann, H. V. Gärtner and K. Wilms, *Cancer Chemother. Pharmacol.*, 1984, **12**, 50.
66. S. M. Somani and J. H. Anderson, *Drug Metab. Dispos.*, 1975, **3**, 275.
67. R. G. Thurman, D. P. Marazzo, L. S. Jones and F. C. Kauffman, *J. Pharmacol. Exp. Ther.*, 1977, **201**, 498.
68. J. A. Bond, M. A. Medinsky and J. S. Dutcher, *Toxicol. Appl. Pharmacol.*, 1984, **75**, 531.
69. D. J. Back, C. M. Macnee, M. L'E. Orme, P. H. Rowe and E. Smith, *Biochem. Pharmacol.*, 1984, **33**, 1595.
70. S. C. Kalser, M. P. Kelly, E. B. Forbes and M. M. Randolph, *J. Pharmacol. Exp. Ther.*, 1969, **170**, 145.
71. G. M. Powell, J. J. Miller, A. H. Olavesen and C. G. Curtis, *Nature (London)*, 1974, **252**, 234.
72. S. C. Kalser, E. J. Kelvington and M. M. Randolph, *J. Pharmacol. Exp. Ther.*, 1968, **159**, 389.
73. M. Baggiolini, B. Dewald and H. Aebi, *Biochem. Pharmacol.*, 1969, **18**, 2187.
74. S. Keiding and E. Steiness, *J. Pharmacol. Exp. Ther.*, 1984, **230**, 474.
75. N. Hamada and T. Gessner, *Drug Metab. Dispos.*, 1975, **3**, 407.
76. D. Brewster, R. S. Jones and D. V. Parke, *Biochem. J.*, 1978, **170**, 257.
77. J. Van Anda, J. R. Bend and J. R. Fouts, *Pharmacologist*, 1977, **19**, 191.
78. M. M. Halldin, H. Isaac, M. Widman, E. Nilsson and A. Ryrfeldt, *Xenobiotica*, 1984, **14**, 277.
79. R. J. Losito and C. A. Owen, *Mayo Clin. Proc.*, 1972, **47**, 731.
80. C. Guguen-Guillouzo, J. P. Campion, P. Brissot, D. Glaise, B. Launois, M. Bourel and A. Guillouzo, *Cell Biol. Int. Rep.*, 1982, **6**, 625.
81. B. D. Ross, *Clin. Sci. Mol. Med.*, 1978, **55**, 513.

82. T. Maack, *Am. J. Physiol.*, 1980, **238**, F71.
83. J. F. Newton and J. B. Hook, *Methods Enzymol.*, 1981, **77**, 94.
84. P. H. Bach and E. A. Lock, in 'Nephrotoxicity: Assessment and Pathogenesis', ed. P. H. Bach, F. W. Bonner, J. W. Bridges and E. A. Lock, Wiley, Chichester, 1982, p. 128.
85. F. H. Epstein, J. T. Brosnan, J. D. Tange and B. D. Ross, *Am. J. Physiol.*, 1982, **243**, F284.
86. H. J. Schurek, E. Schlatter, M. Weier, R. Zick, G. Dorn, R. Hehrmann and E. Stolte, *Int. J. Biochem.*, 1980, **12**, 237.
87. R. H. Bowman and T. Maack, *Am. J. Physiol.*, 1972, **222**, 1499.
88. K. R. Emslie, M. C. Smail, I. C. Calder, S. J. Hart and J. D. Tange, *Xenobiotica*, 1981, **11**, 43.
89. S. J. Szefler and M. Acara, *J. Pharmacol. Exp. Ther.*, 1979, **210**, 295.
90. R. Elbers, H. J. Kampffmeyer and H. Rabes, *Xenobiotica*, 1980, **10**, 621.
91. W. Niemeier and E. Bingham, *Life Sci., Part 2*, 1972, **11**, 807.
92. A. F. Junod, *J. Pharmacol. Exp. Ther.*, 1972, **183**, 341.
93. I. F. McMurtry, A. B. Davidson, J. T. Reeves and R. T. Grover, *Circ. Res.*, 1976, **38**, 99.
94. J. Haynes, S. W. Chang, K. G. Morris and N. F. Voelkel, *J. Appl. Physiol.*, 1988, **65**, 1921.
95. T. E. Eling, R. D. Pickett, T. C. Orton and M. W. Anderson, *Drug Metab. Dispos.*, 1975, **3**, 389.
96. H. Vainio, P. Uotila, J. Hartiala and O. Pelkonen, *Res. Commun. Chem. Pathol. Pharmacol.*, 1976, **13**, 259.
97. H. M. Mehendale and E. A. L. El-Bassiouni, *Drug Metab. Dispos.*, 1975, **3**, 543.
98. J. R. Dawson and J. W. Bridges, *Biochem. Pharmacol.*, 1979, **28**, 3299.
99. H. G. Windmueller, A. E. Spaeth and C. E. Ganote, *Am. J. Physiol.*, 1970, **218**, 197.
100. G. B. Gerber and J. Remy-Defraigne, *Arch. Int. Physiol. Biochim.*, 1966, **74**, 785.
101. C. F. George, E. W. Blackwell and D. S. Davies, *J. Pharm. Pharmacol.*, 1974, **26**, 265.
102. K. S. Roth, P. Holtzapple, M. Genel and S. Segal, *Metab. Clin. Exp.*, 1979, **28**, 677.
103. G. Powis, D. J. Moore, T. J. Wilke and K. S. Santone, *Anal. Biochem.*, 1987, **167**, 191.
104. W. W. Umbreit, R. H. Burris and J. F. Stauffer, in 'Manometric Techniques', 4th edn., Burgess, Minneapolis, 1964, p. 114.
105. O. Dale, A. J. Gandolfi, K. Brendel and S. Schuman, *Br. J. Anaesth.*, 1988, **60**, 692.
106. S. Kacew and G. H. Hirsch, in 'Toxicology of the Kidney', ed. J. B. Hook, Raven Press, New York, 1981, p. 77.
107. H. Autrup and C. C. Harris, in 'Human Carcinogenesis', ed. C. C. Harris and H. Autrup, Academic Press, New York, 1983, p. 169.
108. K. Miyazaki, R. Takaki, F. Nakayama, S. Yamauchi, A. Koga and S. Todo, *Cell Tissue Res.*, 1981, **218**, 13.
109. P. Moldeus, J. Hogberg and S. Orrenius, *Methods Enzymol.*, 1978, **52**, 60.
110. J. A. McGowan, in 'Isolated and Cultured Hepatocytes', ed. A. Guillouzo and C. Guguen-Guillouzo, Libbey Eurotext, London, 1986, p. 13.
111. J. Hogberg and A. Kristoferson, *Eur. J. Biochem.*, 1977, **74**, 77.
112. H. O. Jauregui, N. T. Hayner, J. L. Driscoll, R. Williams-Holland, M. H. Lipsky and P. M. Galletti, *In Vitro*, 1981, **17**, 1100.
113. R. Zawydiwski and G. R. Duncan, *In Vitro*, 1978, **14**, 707.
114. P. E. Stanley and S. G. Williams, *Anal. Biochem.*, 1969, **29**, 381.
115. J. A. Reese and J. L. Byard, *In Vitro*, 1982, **17**, 935.
116. L. B. G. Tee, T. Seddon, A. R. Boobis and D. S. Davies, *Br. J. Clin. Pharmacol.*, 1985, **19**, 279.
117. K. W. Renton, L. B. Deloria and G. J. Mannering, *Mol. Pharmacol.*, 1978, **14**, 672.
118. I. J. Arinze and D. L. Rowley, *Biochem. J.*, 1975, **152**, 393.
119. C. J. Maslansky and G. M. Williams, *In Vitro*, 1982, **18**, 683.
120. G. L. Corona, G. Santagostino, R. M. Facino and D. Pirillo, *Biochem. Pharmacol.*, 1973, **22**, 849.
121. R. J. Chenery, A. Ayrton, H. G. Oldham, P. Standring, S. J. Norman, T. Seddon and R. Kirby, *Drug Metab. Dispos.*, 1987, **15**, 312.
122. C. Guguen-Guillouzo and A. Guillouzo, in 'Isolated and Cultured Hepatocytes', ed. A. Guillouzo and C. Guguen-Guillouzo, Libbey Eurotext, London, 1986, p. 1.
123. L. B. G. Tee, A. R. Boobis, A. C. Huggett and D. S. Davies, *Toxicol. Appl. Pharmacol.*, 1986, **83**, 294.
124. *Percoll: Methodology and Applications Bulletin*, Pharmacia LKB, Uppsala.
125. G. Michalopoulos and H. C. Pitot, *Exp. Cell Res.*, 1975, **94**, 70.
126. R. I. Freshney, 'Cultures of Animal Cells: A Manual of Basic Techniques', Liss, New York, 1984.
127. L. M. Reid, M. Narita, M. Fujita, Z. Murray, C. Liverpool and L. Rosenberg, in 'Isolated and Cultured Hepatocytes', ed. A. Guillouzo and C. Guguen-Guillouzo, Libbey Eurotext, London, 1986, p. 225.
128. B. G. Lake and A. J. Paine, *Biochem. Pharmacol.*, 1982, **31**, 2141.
129. A. J. Paine and L. J. Hockin, *Biochem. Pharmacol.*, 1980, **29**, 3215.
130. G. M. Decad, D. P. H. Hsieh and J. L. Byard, *Biochem. Biophys. Res. Commun.*, 1977, **78**, 279.
131. G. Michalopoulos, C. A. Sattler, G. L. Sattler and H. C. Pitot, *Science (Washington, D.C.)*, 1976, **193**, 907.
132. A. Guillouzo, P. Beaune, M. M. Gascoin, J. M. Bégué, J. P. Campion, F. P. Guengerich and C. Guguen-Guillouzo, *Biochem. Pharmacol.*, 1985, **34**, 2991.
133. B. J. Fuller, G. J. Morris, L. H. Nutt and B. D. Attenburrow, *Cryo-Lett.*, 1980, **1**, 139.
134. B. J. Fuller, B. W. Grout and R. J. Woods, *Cryobiology*, 1982, **19**, 493.
135. M. J. Gómez-Lechón, P. Lopez and J. B. Castell, *In Vitro*, 1984, **20**, 269.
136. B. A. Jackson, J. E. Davies and J. K. Chipman, *Biochem. Pharmacol.*, 1985, **34**, 3389.
137. C. Chesné and A. Guillouzo, *Cryobiology*, 1988, **25**, 323.
138. T. Seddon, A. R. Boobis and D. S. Davies, *Br. J. Clin. Pharmacol.*, 1985, **20**, 546P.
139. C. Chesné, P. Gripon and A. Guillouzo, in 'Liver Cells and Drugs', ed. A. Guillouzo, Libbey Eurotext, London, 1988, vol. 164, p. 343.
140. J. R. Fry, in 'Reviews on Drug Metabolism and Drug Interactions', ed. A. H. Beckett and J. W. Gorrod, Freund, Tel Aviv, 1982, vol. 4, p. 99.
141. P. Moldeus, B. Jernstrom and J. R. Dawson, in 'Reviews in Biochemical Toxicology', ed. E. Hodgson, J. R. Bend and R. M. Philpot, Elsevier/North Holland, New York, 1983, vol. 5, p. 239.
142. M. T. Smith and S. Orrenius, in 'Drug Metabolism and Drug Toxicity', ed. J. R. Mitchell and M. G. Horning, Raven Press, New York, 1984, p. 71.

143. A. Guillouzo, in 'Isolated and Cultured Hepatocytes', ed. A. Guillouzo and C. Guguen-Guillouzo, Libbey Eurotext, London, 1986, p. 313.
144. A. E. Sirica and H. C. Pitot, *Pharmacol. Rev.*, 1980, **31**, 205.
145. A. R. Steward, G. A. Dannan, P. S. Guzelian and F. P. Guengerich, *Mol. Pharmacol.*, 1985, **27**, 125.
146. A. Guillouzo, L. Grislain, D. Ratanasavanh, M. T. Mocquard, J.-M. Bégué, P. Du Vignaud, N. Bromet, P. Genissell and B. Beau, *Xenobiotica*, 1988, **18**, 757.
147. R. E. Billings, R. E. McMahon, J. Ashmore and S. R. Wagle, *Drug Metab. Dispos.*, 1977, **5**, 518.
148. D. P. Praaning-Van Dalen and D. L. Knook, *FEBS Lett.*, 1982, **141**, 229.
149. K. J. Rich, MSc Thesis, University of Kent, 1986.
150. A. Brouwer, R. J. Barelds, R. De Zanger and D. L. Knook, in 'Centrifugation, a Practical Approach', ed. D. Rickwood, 2nd edn., IRL Press, Oxford, 1984, p. 183.
151. K. Cain and D. N. Skilleter, *Biochem. J.*, 1983, **210**, 769.
152. D. Skilleter, K. Cain, D. Dinsdale and A. Paine, *Xenobiotica*, 1985, **15**, 687.
153. D. P. Jones, G.-B. Sundby, K. Ormstad and S. Orrenius, *Biochem. Pharmacol.*, 1979, **28**, 929.
154. J. I. Kreisberg, A. M. Pitts and T. J. Pretlow, *Am. J. Pathol.*, 1977, **86**, 591.
155. J. R. Fry and N. K. Perry, *Biochem. Pharmacol.*, 1981, **30**, 1197.
156. P. H. Bach, C. P. Ketley, I. Ahmed and M. Dixit, *Fd. Chem. Toxicol.*, 1986, **24**, 775.
157. J. I. Kreisberg and M. J. Karnovsky, *Kidney Int.*, 1983, **23**, 439.
158. M. A. Smith, D. Acosta and J. V. Bruckner, *Fd. Chem. Toxicol.*, 1986, **24**, 551.
159. S. E. Benns, M. Dixit, I. Ahmed, C. P. Ketley and P. H. Bach, in 'Alternative Methods in Toxicology', ed. A. Goldberg, Liebert, Baltimore, 1985, vol. 3, p. 221.
160. T. R. Devereux and J. R. Fouts, *Methods Enzymol.*, 1982, **77**, 147.
161. T. R. Devereux, J. J. Diliberto and J. R. Fouts, *Cell Biol. Toxicol.*, 1985, **1**, 57.
162. T. R. Devereux and J. R. Fouts, *In Vitro*, 1980, **16**, 958.
163. T. R. Devereux, C. J. Serabjit-Singh, S. R. Slaughter, C. R. Wolf, R. M. Philpot and J. R. Fouts, *Exp. Lung Res.*, 1981, **2**, 221.
164. B. A. Domin, T. R. Devereux and R. M. Philpot, *Mol. Pharmacol.*, 1986, **30**, 296.
165. O. Pelkonen, P. Korhonen, P. Jouppila and N. T. Kärki, *Life Sci.*, 1975, **16**, 1403.
166. O. Pelkonen and P. Korhonen, in 'Microsomes and Drug Oxidations', ed. V. Ullrich, Pergamon Press, Oxford, 1977, p. 411.
167. C. Plas, *C.R. Hebd. Seances Acad. Sci., Ser. D*, 1969, **268**, 143.
168. H. L. Leffert and D. Paul, *J. Cell Biol.*, 1972, **52**, 559.
169. J. van Cantfort, F. Goujon and J. E. Gielen, *Chem.-Biol. Interact.*, 1979, **28**, 147.
170. P. Kremers, in 'Isolated and Cultured Hepatocytes', ed. A. Guillouzo and C. Guguen-Guillouzo, Libbey Eurotext, London, 1986, p. 285.
171. C. Guguen-Guillouzo, J. Marie, D. Cottreau, N. Pasdeloup and A. Kahn, *Biochem. Biophys. Res. Commun.*, 1980, **93**, 528.
172. C. Guguen-Guillouzo, L. Tichonicky, M. F. Szajnert and J. Kruh, *In Vitro*, 1980, **16**, 1.
173. P. J. M. Klippert, P. J. A. Borm and J. Noordhoek, *Biochem. Pharmacol.*, 1982, **31**, 2545.
174. P. J. A. Borm, A. Frankhuijzen-Sierevogel and J. Noordhoek, *Biochem. Pharmacol.*, 1982, **31**, 3707.
175. M. Kiese, 'Methemoglobinemia: A Comprehensive Treatise', CRC Press, Cleveland, 1974.
176. S. P. Spielberg, *J. Pharmacol. Exp. Ther.*, 1980, **213**, 395.
177. M. A. Trush, J. L. Seed and T. W. Kensler, *Proc. Natl. Acad. Sci. U.S.A.*, 1985, **82**, 5194.
178. Y. Tsuruta, V. V. Subrahmanyam, W. Marshall and P. J. O'Brien, *Chem.-Biol. Interact.*, 1985, **53**, 25.
179. C. Haslett, L. A. Guthrie, M. M. Kopaniak, R. B. Johnston and P. M. Henson, *Am. J. Pathol.*, 1985, **119**, 101.
180. B. Ekwall, *Toxicology*, 1980. **17**, 273.
181. F. J. Wiebel, F. Kiefer and U. Murdia, *Chem.-Biol. Interact.*, 1984, **52**, 151.
182. F. J. Wiebel, S. S. Park, F. Kiefer and H. V. Gelboin, *Eur. J. Biochem.*, 1984, **145**, 455.
183. S. S. Lau, J. B. MacMahon, M. G. McMenamin, W. C. Hubbard, H. M. Schuller and M. R. Boyd, *Cancer Res.*, 1987, **47**, 3757.
184. F. J. Wiebel, F. Kiefer, G. Krupski and H. M. Schuller, *Biochem. Pharmacol.*, 1986, **35**, 1337.
185. D. N. Skilleter, A. R. Mattocks and G. E. Neal, *Xenobiotica*, 1988, **18**, 699.
186. R. N. Hull, W. R. Cherry and D. W. Weaver, *In Vitro*, 1976, **12**, 670.
187. A. Hassid, *J. Cell Physiol.*, 1983, **116**, 297.
188. T. Cresteil, A. K. Jaiswal and H. J. Eisen, *Arch. Biochem. Biophys.*, 1987, **253**, 233.
189. A. G. Miller, D. Israel and J. P. Whitlock, Jr., *J. Biol. Chem.*, 1983, **258**, 3523.
190. K. Cain and D. N. Skilleter, in 'Biochemical Toxicology: A Practical Approach', ed. K. Snell and B. Mullock, IRL Press, Oxford, 1987, p. 217.
191. A. R. Boobis, D. W. Nebert and J. S. Felton, *Mol. Pharmacol.*, 1977, **13**, 259.
192. F. P. Guengerich, in 'Principles and Methods of Toxicology', ed. A. W. Hayes, Raven Press, New York, 1982, p. 609.
193. S. J. Stohs, R. C. Graftström, M. D. Burke, P. W. Moldeus and S. G. Orrenius, *Arch. Biochem. Biophys.*, 1976, **177**, 105.
194. S. A. Atlas, A. R. Boobis, J. S. Felton, S. S. Thorgeirsson and D. W. Nebert, *J. Biol. Chem.*, 1977, **252**, 4712.
195. K. W. Bock, G. Brunner, H. Hoensch, E. Huber and G. Josting, *Eur. J. Clin. Pharmacol.*, 1978, **14**, 367.
196. B. G. Lake, in 'Biochemical Toxicology: A Practical Approach', ed. K. Snell and B. Mullock, IRL Press, Oxford, 1987, p. 183.
197. D. M. Maron and B. N. Ames, *Mutat. Res.*, 1983, **113**, 173.
198. J. B. Schenkman, H. Remmer and R. W. Estabrook, *Mol. Pharmacol.*, 1967, **3**, 113.
199. J. Cochin and J. Axelrod, *J. Pharmacol. Exp. Ther.*, 1959, **125**, 105.
200. M. Jacobson, W. Levin, P. J. Poppers, A. W. Wood and A. H. Conney, *Clin. Pharmacol. Ther.*, 1974, **16**, 701.
201. G. G. Gibson, T. C. Orton and P. P. Tamburini, *Biochem. J.*, 1982, **203**, 161.
202. J. M. Tredger. H. M. Smith and R. Williams, *J. Pharmacol. Exp. Ther.*, 1984, **229**, 292.
203. H. Theorell and R. Bonnichsen, *Acta Chem. Scand.*, 1951, **5**, 1105.
204. Y. Imai, A. Ito and R. Sato, *J. Biochem. (Tokyo)*, 1966, **60**, 417.
205. D. W. Nebert and J. E. Gielen, *Fed. Proc., Fed. Am. Soc. Exp. Biol.*, 1972, **31**, 1315.

206. L. S. Kaminsky, F. P. Guengerich, G. A. Dannan and S. D. Aust, *Arch. Biochem. Biophys.*, 1983, **225**, 398.
207. T. Shimaya, *Biochim. Biophys. Acta*, 1965, **105**, 377.
208. J. M. Tredger, H. M. Smith, M. Davis and R. Williams, *Biochem. Pharmacol.*, 1984, **33**, 1729.
209. W. R. Waud and K. V. Rajagopalan, *Arch. Biochem. Biophys.*, 1976, **172**, 354.
210. W. E. Knox, *J. Biol. Chem.*, 1946, **163**, 699.
211. N. E. Sladek and G. J. Mannering, *Mol. Pharmacol.*, 1969, **5**, 174.
212. A. Poland and D. W. Nebert, *J. Pharmacol. Exp. Ther.*, 1973, **184**, 269.
213. P. Kremers, P. Beaune, T. Cresteil, J. DeGraeve, S. Columelli, J. P. Leroux and J. E. Gielen, *Eur. J. Biochem.*, 1981, **118**, 599.
214. W. F. Greenlee and A. Poland, *J. Pharmacol. Exp. Ther.*, 1978, **205**, 596.
215. M. D. Burke and R. T. Mayer, *Drug Metab. Dispos.*, 1974, **2**, 583.
216. H. Shigematsu, S. Yamano and H. Yoshimura, *Arch. Biochem. Biophys.*, 1976, **173**, 178.
217. D. Sesardic, A. R. Boobis, R. J. Edwards and D. S. Davies, *Br. J. Clin. Pharmacol.*, 1988, **26**, 363.
218. P. Mazel, J. F. Henderson and J. Axelrod, *J. Pharmacol. Exp. Ther.*, 1964, **143**, 1.
219. H. B. Hucker, B. M. Michniewicz and R. E. Rhodes, *Biochem. Pharmacol.*, 1971, **20**, 2123.
220. C. W. Tabor, H. Tabor and S. M. Rosenthal, *J. Biol. Chem.*, 1954, **208**, 645.
221. A. H. Beckett and R. Áchari, *J. Chromatogr.*, 1977, **135**, 200.
222. J. L. Radomski and E. Brill, *Arch. Toxikol.*, 1971, **28**, 159.
223. H. Yamada, T. Baba, Y. Hirata, K. Oguri and H. Yoshimura, *Xenobiotica*, 1984, **14**, 861.
224. A. H. Beckett and G. G. Gibson, *Xenobiotica*, 1978, **8**, 73.
225. F. F. Kadlubar, J. A. Miller and E. C. Miller, *Cancer Res.*, 1976, **36**, 1196.
226. N. J. Gooderham and J. W. Gorrod, *J. Chromatogr.*, 1984, **309**, 339.
227. M. E. McManus, P. H. Grantham, J. L. Cone, P. P. Roller, P. J. Wirth and S. S. Thorgeirsson, *Biochem. Biophys. Res. Commun.*, 1983, **112**, 437.
228. D. M. Ziegler and F. H. Pettit, *Biochem. Biophys. Res. Commun.*, 1964, **15**, 188.
229. J. W. Gorrod, D. J. Temple and A. H. Beckett, *Xenobiotica*, 1975, **5**, 453.
230. A. Åström and J. W. DePierre, *Carcinogenesis*, 1982, **3**, 711.
231. C. J. Parli, N. C. Wang and R. E. McMahon, *J. Biol. Chem.*, 1971, **246**, 6953.
232. J. W. Gorrod and L. A. Damani, *Xenobiotica*, 1979, **9**, 209.
233. P. Wiebkin and R. A. Prough, *Cancer Res.*, 1980, **40**, 3524.
234. J. R. Cashman and R. P. Hanzlik, *Biochem. Biophys. Res. Commun.*, 1981, **98**, 147.
235. T. Kamataki, D. A. Belcher and R. A. Neal, *Mol. Pharmacol.*, 1976, **12**, 921.
236. B. N. La Du, H. G. Mandel and E. L. Way (ed.), 'Fundamentals of Drug Metabolism and Disposition', Williams and Wilkins, Baltimore, 1971.
237. L. H. Patterson and J. W. Gorrod, in 'Biological Oxidation of Nitrogen', ed. J. W. Gorrod, Elsevier/North Holland, New York, 1978, p. 471.
238. J. R. Fouts, J. J. Kamm and B. B. Brodie, *J. Pharmacol. Exp. Ther.*, 1957, **120**, 291.
239. W. Kalow, *J. Pharmacol. Exp. Ther.*, 1952, **104**, 122.
240. L. C. Mark, H. J. Kayden, J. M. Steel, J. R. Cooper, I. Berlin, E. A. Rovenstine and B. B. Brodie, *J. Pharmacol. Exp. Ther.*, 1951, **102**, 5.
241. K. J. Isselbacher, M. F. Chrabas and R. C. Quinn, *J. Biol. Chem.*, 1962, **237**, 3033.
242. E. Sanchez and T. R. Tephly, *Drug Metab. Dispos.*, 1974, **2**, 247.
243. I. M. Arias, *J. Clin. Invest.*, 1962, **41**, 2233.
244. W. W. Weber, *Biochim. Biophys. Acta*, 1968, **151**, 276.
245. J. W. Jenne and P. B. Boyer, *Biochim. Biophys. Acta*, 1962, **65**, 121.
246. W. H. Habig, M. J. Pabst and W. B. Jakoby, *J. Biol. Chem.*, 1974, **249**, 7130.
247. F. Oesch, D. M. Jerina and J. Daly, *Biochim. Biophys. Acta*, 1971, **227**, 685.
248. A. Y. H. Lu and W. Levin, *Methods Enzymol.*, 1978, **52**, 193.
249. J. D. Gregory and F. Lipmann, *J. Biol. Chem.*, 1957, **229**, 1081.
250. J. Axelrod, *J. Pharmacol. Exp. Ther.*, 1962, **138**, 28.
251. I. K. Brandt, *Biochem. Pharmacol.*, 1966, **15**, 994.
252. A. Claude, *Science (Washington, D.C.)*, 1939, **90**, 213.
253. J. W. DePierre and G. Dallner, *Biochim. Biophys. Acta*, 1975, **415**, 411.
254. D. M. Ziegler and L. L. Poulsen, *Methods Enzymol.*, 1978, **52**, 142.
255. A. Y. H. Lu and G. T. Miwa, *Annu. Rev. Pharmacol. Toxicol.*, 1980, **20**, 513.
256. G. J. Dutton, in 'Glucuronic Acid', ed. G. J. Dutton, Academic Press, New York, 1966, p. 185.
257. M. Jarvinen, R. S. S. Santti and V. Hopsu-Havu, *Biochem. Pharmacol.*, 1971, **20**, 2971.
258. K. Staron and Z. Kaniuga, *Biochim. Biophys. Acta*, 1971, **234**, 297.
259. S. A. Kamath and K. A. Narayan, *Anal. Biochem.*, 1972, **48**, 53.
260. A. R. Boobis and D. S. Davies, *Xenobiotica*, 1984, **14**, 151.
261. A. R. Boobis, M. J. Brodie, G. C. Kahn, D. R. Fletcher, J. H. Saunders and D. S. Davies, *Br. J. Clin. Pharmacol.*, 1980, **9**, 11.
262. J. M. Tredger and R. S. Chhabra, *Drug Metab. Dispos.*, 1976, **4**, 451.
263. M. D. Burke and R. T. Mayer, *Drug Metab. Dispos.*, 1974, **2**, 583.
264. B. G. Lake and A. J. Paine, *Xenobiotica*, 1983, **13**, 725.
265. D. Zakim, J. Goldenberg and D. A. Vessey, *Eur. J. Biochem.*, 1973, **38**, 59.
266. F. J. Wiebel, J. Leutz, L. Diamond and H. V. Gelboin, *Arch. Biochem. Biophys.*, 1971, **144**, 78.
267. A. R. Boobis, G. C. Kahn, C. Whyte, M. J. Brodie and D. S. Davies, *Biochem. Pharmacol.*, 1981, **30**, 2451.
268. M. E. McManus, R. F. Minchin, N. Sanderson, D. Schwartz, E. F. Johnson and S. S. Thorgeirsson, *Carcinogenesis*, 1984, **5**, 1717.
269. D. W. Nebert, N. Considine and I. S. Owens, *Arch. Biochem. Biophys.*, 1973, **157**, 148.
270. G. C. Kahn, A. R. Boobis, I. A. Blair, M. J. Brodie and D. S. Davies, *Anal. Biochem.*, 1981, **113**, 292.
271. G. C. Kahn, A. R. Boobis, S. Murray, M. J. Brodie and D. S. Davies, *Br. J. Clin. Pharmacol.*, 1982, **13**, 637.
272. W. P. Jakoby, *Adv. Enzymol.*, 1978, **46**, 383.

273. A. B. Roy, *Adv. Enzymol.*, 1960, **22**, 205.
274. D. J. Hearse and W. W. Weber, *Biochem. J.*, 1973, **132**, 519.
275. E. Heymann, in 'Enzymatic Basis of Detoxication', ed. W. B. Jakoby, Academic Press, New York, 1982, vol. 2, p. 229.
276. D. Johnson and H. Lardy, *Methods Enzymol.*, 1967, **10**, 95.
277. R. Wattiaux, S. Wattiux-de Coninck, M. F. Ronveaux-Dupal and F. Dubois, *J. Cell Biol.*, 1978, **78**, 349.
278. R. Wattiaux and S. Wattiaux-de Coninck, in 'Iodinated Density Gradient Media—A Practical Approach', ed. D. Rickwood, IRL Press, Oxford, 1983, p. 119.
279. M. Dobrota, in 'Biochemical Toxicology: A Practical Approach', ed. K. Snell and B. Mullock, IRL Press, Oxford, 1987, p. 255.
280. R. Berezney, L. K. Macaulay and F. L. Crane, *J. Biol. Chem.*, 1972, **247**, 5549.
281. S. Sakai, C. E. Reinhold, P. J. Wirth and S. S. Thorgeirsson, *Cancer Res.*, 1978, **38**, 2058.
282. M. A. Wisher and W. M. Evans, *Biochem. J.*, 1975, **146**, 375.
283. J. O. Tsokos-Kuhn, E. L. Todd, J. B. McMillin-Wood and J. R. Mitchell, *Mol. Pharmacol.*, 1985, **28**, 56.
284. M. I. Sheikh and J. V. Møller, in 'Biochemical Toxicology: A Practical Approach', ed. K. Snell and B. Mullock, IRL Press, Oxford, 1987, p. 153.
285. F. P. Guengerich, in 'Mammalian Cytochromes *P*-450', ed. F. P. Guengerich, CRC Press, Boca Raton, 1987, vol. 1, p. 1.
286. A. Åström and J. W. DePierre, *Biochim. Biophys. Acta*, 1986, **853**, 1.
287. P. E. Thomas, L. M. Reik, S. L. Maines, S. Bandiera, D. E. Ryan and W. Levin, in 'Biological Reactive Intermediates III', ed. J. J. Kocsis, D. J. Jollow, C. M. Witmer, J. O. Nelson and R. Snyder, Plenum Press, New York, 1986, p. 95.
288. G. E. Schwab and E. F. Johnson, in 'Mammalian Cytochromes *P*-450', ed. F. P. Guengerich, CRC Press, Boca Raton, 1987, vol. 1, p. 55.
289. L. M. Distlerath and F. P. Guengerich, in 'Mammalian Cytochromes *P*-450', ed. F. P. Guengerich, CRC Press, Boca Raton, 1987, vol. 1, p. 133.
290. M.-T. Huang, S. B. West and A. Y. H. Lu, *J. Biol. Chem.*, 1976, **251**, 4659.
291. M. Kitada, C. Yamazaki, K. Hirota and H. Kitagawa, *Biochem. Biophys. Res. Commun.*, 1980, **93**, 1020.
292. J. A. Bumpus and K. M. Dus, *J. Biol. Chem.*, 1982, **257**, 2696.
293. D. B. Duignan, I. G. Sipes, T. B. Leonard and J. R. Halpert, *Arch. Biochem. Biophys.*, 1987, **255**, 290.
294. M. Watanabe, H. Fujii, I. Sagami and M. Tanno, *Arch. Toxicol.*, 1987, **60**, 52.
295. A. E. Klotz, J. J. Stegeman, B. R. Woodin, E. A. Snowberger, P. E. Thomas and C. Walsh, *Arch. Biochem. Biophys.*, 1986, **249**, 326.
296. P. Bentley and F. Oesch, *FEBS Lett.*, 1975, **59**, 291.
297. A. Y. H. Lu, D. Ryan, D. M. Jerina, J. W. Daly and W. Levin, *J. Biol. Chem.*, 1975, **250**, 8283.
298. J. Haeggstrom, J. Meijer and O. Radmark, *J. Biol. Chem.*, 1986, **261**, 6332.
299. B. Burchell, M. R. Jackson, M. W. H. Coughtrie, D. Harding, S. Wilson and J. R. Bend, in 'Drug Metabolism—From Molecules to Man', ed. D. J. Benford, J. W. Bridges and G. G. Gibson, Taylor and Francis, London, 1987, p. 40.
300. T. R. Tephly, B. L. Coffman, C. N. Falany, M. D. Green, Y. Irshaid, J. F. Puig, S. A. Knapp and J. Barron, in 'Microsomes and Drug Oxidations', ed. J. Miners, D. J. Birkett, R. Drew and M. McManus, Taylor and Francis, London, 1988, p. 263.
301. Y. M. Irshaid and T. R. Tephly, *Mol. Pharmacol.*, 1987, **31**, 27.
302. T. Mantle, C. B. Pickett and J. D. Hayes (ed.), 'Glutathione *S*-Transferases and Carcinogenesis', Taylor and Francis, London, 1987.
303. T. B. Boyer and W. C. Kenney, in 'Biochemical Pharmacology and Toxicology', ed. D. Zakim and D. A. Vessey, Wiley, New York, 1985, vol. 1, p. 297.
304. R. Morgenstern and J. W. DePierre, *Eur. J. Biochem.*, 1983, **134**, 591.
305. D. Heymann and R. Mentlein, in 'Metabolism of Xenobiotics', ed. J. W. Gorrod, H. Oelschläger and J. Caldwell, Taylor and Francis, London, 1988, p. 189.
306. S. S. Singer, *Biochem. Soc. Trans.*, 1984, **12**, 35.
307. S. S. Singer, in 'Biochemical Pharmacology and Toxicology', ed. D. Zakim and D. A. Vessey, Wiley, New York, 1985, vol. 1, p. 95.
308. R. Kato, K. Saito, A. Shinohara and T. Kamataki, in 'Biological Reactive Intermediates III', ed. J. Kocsis, D. J. Jollow, C. M. Witmer, J. O. Nelson and R. Snyder, Plenum Press, New York, 1986, p. 551.
309. R. Kato and Y. Yamazoe, in 'Xenobiotic Metabolism and Disposition', ed. R. Kato, R. W. Estabrook and M. N. Cayen, Taylor and Francis, London, 1989, p. 383.
310. J. Ozols, *Biochemistry*, 1974, **13**, 426.
311. Y. Yasukochi and B. S. S. Masters, *J. Biol. Chem.*, 1976, **251**, 5337.
312. L. Spatz and P. Strittmatter, *J. Biol. Chem.*, 1973, **248**, 793.
313. J. Stevens and W. B. Jakoby, *Mol. Pharmacol.*, 1983, **23**, 761.
314. N. Muto and L. Tan, *Biochem. Cell. Biol.*, 1986, **64**, 184.
315. A. N. Kotake and Y. Funae, *Proc. Natl. Acad. Sci. U.S.A.*, 1980, **77**, 6473.
316. P. L. Iversen and M. R. Franklin, *Toxicol. Appl. Pharmacol.*, 1985, **78**, 1.
317. R. Kato, R. W. Estabrook and M. N. Cayen (ed.), 'Xenobiotic Metabolism and Disposition', Taylor and Francis, London, 1989.
318. M. R. Jackson and B. Burchell, *Nucleic Acid Res.*, 1986, **14**, 779.
319. F. K. Friedman, S. S. Park, B. J. Song, K. C. Cheng, T. Fujino and H. V. Gelboin, in 'Biological Reactive Intermediates III', ed. J. J. Kocsis, D. J. Jollow, C. M. Witmer, J. O. Nelson and R. Snyder, Plenum Press, New York, 1986, p. 145.
320. D. W. Nebert, M. Adesnik, M. J. Coon, R. W. Estabrook, F. J. Gonzalez, F. P. Guengerich, I. C. Gunsalus, E. F. Johnson, B. Kemper, W. Levin, I. R. Phillips, R. Sato and M. R. Waterman, *DNA*, 1987, **6**, 1.
321. R. H. Tukey, S. Okino, U. R. Pendurthi, E. F. Johnson and L. Quattrochi, in 'Microsomes and Drug Oxidations', ed. J. Miners, D. J. Birkett, R. Drew and M. McManus, Taylor and Francis, London, 1988, p. 55.
322. H. Murakami, Y. Yabusaki and H. Ohkawa, *DNA*, 1986, **5**, 1.
323. J. Doehmer, S. Dogra, T. Friedberg, S. Monier, M. Adesnik, H. Glatt and F. Oesch, *Proc. Natl. Acad. Sci. U.S.A.*, 1988, **85**, 5769.
324. T. Shimizu, K. Hirano, M. Takahashi, M. Hatano and Y. Fujii-Kuriyama, *Biochemistry*, 1988, **27**, 4138.
325. R. J. Edwards, A. M. Singleton, D. Sesardic, A. R. Boobis and D. S. Davies, *Biochem. Pharmacol.*, 1988, **37**, 3735.

24.5

Isolation and Identification of Metabolites

MARGHERITA STROLIN-BENEDETTI

Farmitalia Carlo Erba, Milan, Italy

and

JOHN CALDWELL

St Mary's Hospital Medical School, London, UK

24.5.1   THE ANALYTICAL CHALLENGE OF DRUG METABOLISM

The metabolism of drugs and other xenobiotics in the mammalian body is, in chemical terms, most generally a biphasic process,[1] with a first phase of a functionalization reaction of oxidation, reduction or hydrolysis, followed by a second phase of conjugation or biosynthesis (Figure 1). To a greater or lesser extent, there will occur excretion of the unchanged compound and of the unconjugated products of functionalization. In many cases, the compound itself may contain functional groups able to undergo conjugation directly, so that the final metabolites of a drug may include all of these various possibilities.[1,2]

The functionalization reactions serve to introduce or reveal functional groups within the drug molecule. These reactions are most typically oxidations, notably hydroxylations; reductions and hydrolyses also occur.[3] In general, these reactions do not bring about dramatic changes to the physicochemical properties of drugs, a typical change being the oxidative demethylation of tertiary amines to secondary amines. However, aromatic hydroxylation introduces a phenolic hydroxy group, which may be ionized, and this will have a particular effect in the case of amines converted to phenolic amines, *e.g.* the 4-hydroxylation of amphetamine, which produces a zwitterionic phenolic amine markedly different from the parent drug.[4] The hydrolysis of esters and amides may also cause significant changes to physicochemical properties.

In comparison with the functionalization reactions, the conjugations produce very marked alterations to the physicochemical properties of the drug.[5] The conjugations are biosynthetic reactions in which the drug, or one of its phase I metabolites, is linked with an endogenous conjugating agent.[2] These are generally large, very polar molecules, extensively ionized at physiological pH, and are typified by glucuronic acid.[5] Other examples of conjugations leading to increases in polarity and ionization include linkage with glutathione, glycine and other amino acids and sulfate ion, all of which serve to facilitate the elimination of the drug from the body.[5]

Contrasting with the above are two conjugation mechanisms which serve to 'mask' a particular functional group within the drug molecule.[5] These are: (i) methylation of phenols, largely exemplified by the action of catechol *O*-methyltransferase, which effectively terminates the biological actions of noradrenaline and related catecholamines; and (ii) acetylation of amino groups, which may serve to reduce the formation of potentially toxic metabolites by *N* oxidation.

The overall consequence of the biphasic sequence of drug metabolism is that any drug, which may be acidic, basic, neutral or zwitterionic to begin with, will be transformed in a variety of ways which cause profound modifications to its chemistry, giving the range of products mentioned above.[2] The analyst is thus presented with the very substantial challenge of characterizing metabolites in

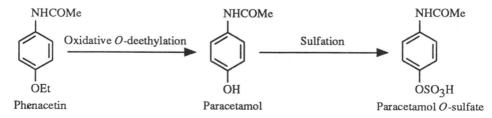

Figure 1   The biphasic sequence of drug metabolism, illustrated with reference to the analgesic drug phenacetin

complex mixtures, containing numerous potential endogenous interferences,[6] together with structurally closely related groups of metabolites whose physicochemical properties differ greatly from group to group. This may be exemplified by the case of amphetamine, mentioned earlier;[5,7] the challenge is to separate: (i) the phenolic amines produced by hydroxylation, (ii) their zwitterionic conjugates with glucuronic acid and/or sulfate, (iii) other metabolites produced from the parent amine by oxidation of the three-carbon side chain, and (iv) the basic parent drug. As well as defining the covalent structures of these various metabolites, it is necessary to address issues such as the positional isomerism amongst the phenolic amines and the enantiomeric composition of chiral drugs and their metabolites.

It is clear that no single analytical technique is able to approach these problems, so the drug metabolism scientist draws upon a range of methodologies. These may be divided into separation technologies,[8] including extraction, fractionation and chromatography, and chemical characterizations,[9] most often by means of spectroscopic instrumentation. In general, metabolic problems require resort to a wide range of techniques for their effective resolution.

Metabolic studies involve the separate and sequential use of the techniques mentioned, starting with fractionation, and leading up to mass spectrometry and NMR spectroscopy. However, this situation is being transformed by the increased use of the so-called hyphenated techniques, which involve the online combination of an informative spectroscopic technique with a separation method. The most widely used of these combined approaches is, of course, GC–MS, but others are gaining in applicability and availability (see Section 24.5.7).

24.5.2 GENERAL PRINCIPLES

The isolation and identification of metabolites were carried out in the past mainly on phase I metabolites as such, or as liberated by different types of hydrolysis, whereas nowadays there is an increasing interest in isolating and characterizing intact conjugates.[10,11] We will present here the general principles of a three-step procedure: fractionation, separation and characterization of metabolites, starting from biological material such as urine, blood, plasma or serum, and bile.

Fractionation is the step by which the whole metabolites present in a biological sample are isolated from the biological matrix, and separated into a number of families (fractions), each containing metabolites with some common characteristic(s). Typical is extraction at different pH values, allowing the separation of strongly acidic, weakly acidic, neutral and basic metabolites. Separation means here the procedure adopted to separate the metabolites present in a given fraction; separation methodology is almost always chromatographic. Finally characterization is used here to indicate the identification of the structures of metabolites, and is almost exclusively based on spectroscopic techniques.[11]

When it is desired to isolate intact conjugates, the same three-step procedure is adopted with suitable modification, or the biological sample is submitted to chemical or enzymatic hydrolysis, followed by the three-step procedure.[11]

When the metabolism of new drugs is being studied, it is almost essential to use radiolabelled drugs if the metabolites present in biological fluids are to be isolated and identified without an inordinate amount of effort. Stable-isotope-labelled drugs, although welcome in metabolic studies (see Sections 24.5.6.3 and 24.5.6.4), do not have the universal application of radiolabelled compounds.

24.5.3 HYDROLYSIS OF CONJUGATES

As mentioned above conjugation of drugs may occur with glucuronic acid, sulfate, methyl and acetyl groups, amino acids and glutathione.[12] Probably the most widespread conjugation reaction is that with glucuronic acid, giving rise either to ether or ester linkages between the aglycone and the sugar (Figure 2). Ester glucuronides are hydrolyzed by the enzyme β-glucuronidase, as well as by mild alkali treatment. If the degree of hydrolysis of a conjugate obtained following mild alkaline treatment of a suspected ester glucuronide is the same as that given by properly controlled treatment with β-glucuronidase, this gives excellent evidence for the presence of a glucuronic acid conjugate. However it is commonly the case that enzymatic treatment results in less hydrolysis than does alkali, and this is an indication that the ester glucuronide might have undergone intramolecular acyl migration, the products of which would not be substrates for the enzyme.[13-15] Of course acid hydrolysis can also be used, but it is not selective at all.

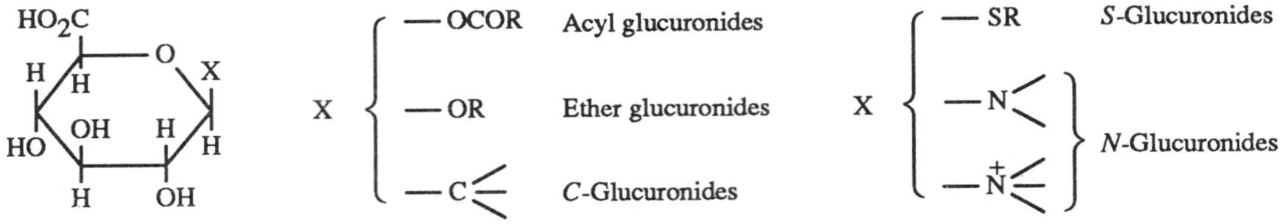

Figure 2 Structures of 1-*O*-β-D-glucuronides

24.5.3.1 Enol Glucuronides

The enol (pseudo-ester) glucuronides are formed from compounds, *e.g.* warfarin or steroids, in which a carbonyl group can enolize; they have similar stability to ester glucuronides,[15] but do not undergo intramolecular acyl migration.[13]

24.5.3.2 Ethereal Glucuronides

In the case of ethereal glucuronides, provided appropriate controls are included, hydrolysis by β-glucuronidase, which is specific for 1-*O*-substituted β-D-glucopyranosiduronates, and inhibition by saccharo-1,4-lactone are sufficient for characterization and no further confirmatory work is required. These glucuronides are stable to mild alkali, but are hydrolyzed by acid.

24.5.3.3 Hydroxylamine Glucuronides

An example is the conjugate of *N*-hydroxy-2-acetylaminofluorene (Figure 3), which is hydrolyzed by bacterial β-glucuronidase.[16]

24.5.3.4 *N*-Glucuronides

N-Glucuronides are formed by the reaction of glucuronic acid with compounds containing an aromatic or aliphatic amino group, a sulfonamide or a heterocyclic nitrogen atom. The stability of *N*-glucuronides in acidic solution depends upon the nature of the aglycone and the mode and site of attachment of the glucuronic acid moiety to the aglycone. Conjugates of aromatic amines are acid labile, whereas those from sulfonamides are stable in acid. Tertiary amines can be conjugated with glucuronic acid through the pair of electrons on the nitrogen forming a quaternary ammonium compound, as has been described for tripelennamine, cyproheptadine and amitriptyline (Figure 2). Treatment with β-glucuronidase hydrolyzes such conjugates to a lesser extent than hot alkaline treatment.[12,17]

24.5.3.5 *S*-Glucuronides

S-Glucuronides are formed by compounds containing thiol or dithioic acid groups.[18] Information on their susceptibility to different hydrolytic procedures is limited, and indicates considerable variability between compounds.[18]

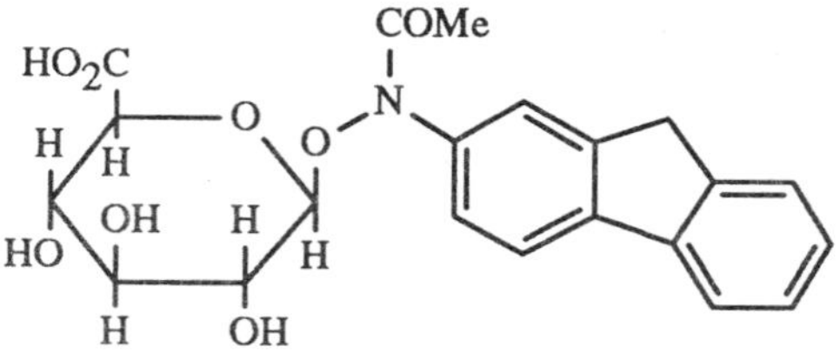

Figure 3 Structure of the glucuronide derivative of *N*-hydroxy-2-acetylaminofluorene

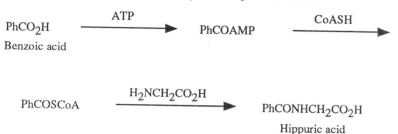

Figure 4 Mechanism of the conjugation of benzoic acid with glycine to form hippuric acid

24.5.3.6 *C*-Glucuronides

C-Glucuronides have been shown to be formed from tetrahydrocannabinol, phenylbutazone and sulfinpyrazone.[12] In this unexpected conjugation reaction, a carbon atom in the benzene or pyrazolidine ring is directly attached to glucuronic acid through a carbon–carbon bond (Figure 2), giving a conjugate which is resistant to β-glucuronidase.

24.5.3.7 Conjugates of Carboxylic Acids with Amino Acids

Xenobiotic carboxylic acids undergo conjugation with glucuronic acid or with amino acids (Figure 4).[5, 13] The hydrolysis of such ester glucuronides and their unusual chemical properties have been briefly mentioned above. Any conjugates remaining after treatment of a sample of interest with β-glucuronidase and mild alkali are most likely to be amino acid conjugates.[13] These chemically stable conjugates may be separated into classes on a predictive basis by TLC and HPLC, and characterized by MS, NMR and IR spectroscopy. General approaches to the differential analysis of glucuronic acid, glycine, glutamine and taurine conjugates of xenobiotic acids have been described by Caldwell and Hutt.[13]

24.5.3.8 Sulfate Conjugates

The conjugation of drugs with sulfate has been reviewed recently by Mulder.[19]

Most phenyl sulfates will be hydrolyzed by aryl sulfatases, although some are highly resistant.[19] More or less pure aryl sulfatase preparations can be obtained commercially. An important contaminant in most preparations is β-glucuronidase, which must be inhibited with saccharo-1,4-lactone before the diagnostic use of such sulfatases can be accepted.[20] Aryl sulfate esters are acid labile, and can be hydrolyzed in aqueous media under relatively mild, acidic conditions, while alkyl sulfate esters are somewhat more resistant.[19]

Sulfate esters of alcohols readily undergo solvolysis, although the same chemical mechanism is involved in both hydrolysis and solvolysis. In solvolysis, the reaction takes place in an organic solvent like acetone or ethanol; only a trace of water is necessary for the reaction. Solvolysis is used frequently with bile acid conjugates and steroid sulfates, yielding convenient derivatives, and could be considered as an alternative in the cases of conjugates where the aglycone may degrade during hydrolysis.[21]

Additional information on the resistance to hydrolysis of the different types of sulfate and glucuronic acid conjugates is provided in a recent review.[22]

24.5.4 FRACTIONATION METHODS

24.5.4.1 Solvent Extraction

The isolation of organic compounds from biological fluids or tissue homogenates most often involves extraction into suitable water immiscible organic solvents after the pH of the biological sample has been adjusted appropriately. Water miscible organic solvents can occasionally be used, the organic layer being separated by snap freezing. Fenofibric acid, the major metabolite of fenofibrate, is extracted from plasma into acetonitrile in acidic conditions, after having obtained two phases by leaving the mixture at $-20\,°C$ for 10 min (Guichard, personal communication).

24.5.4.1.1 *Concentration of sample*

When large volumes of urine or other biological fluids are investigated, it is sometimes necessary to concentrate the sample prior to solvent extraction. The use of heat is undesirable, since labile compounds may be destroyed and volatile compounds lost at elevated temperatures. Evaporation under reduced pressure and freeze-drying (lyophilization) are more suitable methods.[23]

Sometimes inorganic salts are added in high concentration before solvent extraction. The purpose, if not merely to minimize the problems of emulsification, is to obtain effective transfer by 'salting out' into a relatively low volume of organic phase and thereby achieve concentration without having necessarily to reduce the volume by evaporation. Hucker *et al.*[24] employed this technique, resulting in a 25-fold concentration of a large urine sample. The urine was saturated with anhydrous sodium sulfate, extracted with 2-propanol and the 2-propanol evaporated. Strolin-Benedetti *et al.*[25] also used this technique to extract acipimox derivatives such as 2-methoxymethyl-5-methylpyrazine 4-oxide. Horning *et al.*[26] have used salt–solvent pairing to obtain excellent recoveries of drugs and metabolites from urine, plasma and breast milk. Saturation of the water-diluted biological fluid with ammonium carbonate, followed by extraction with ethyl acetate, resulted in 84% or greater recoveries of various weakly acidic, neutral and basic drugs. With this method, the amphoteric drug morphine was recovered in 88–100% yield.

24.5.4.1.2 *Choice of solvent*

Commonly used solvents include chloroform, diethyl ether, ethyl acetate, dichloromethane, benzene, toluene and 2-propanol and mixtures of these. Isopropyl chloride is not often used in extraction procedures, but consideration of its properties (b.p. $36\,°C$; dielectric constant 9.82; water solubility 1.3%) suggests that it might be an excellent solvent which should be used more often.[27] Solvent toxicity is an important consideration in devising extraction procedures. Thus toluene should be used rather than benzene where feasible and chloroform or, better, ethylene dichloride, rather than carbon tetrachloride.

Solvents should be distilled immediately prior to use so as to ensure that they are free of trace quantities of solutes, which could interfere with subsequent analyses. Phthalates, for example, are ubiquitous, as is readily seen in the mass spectrometer, and some solvents, especially diethyl ether, may contain antioxidants such as butylated hydroxytoluene (BHT) that could be erroneously identified as metabolites. Traces of peroxides also form rapidly in diethyl ether and can convert drugs and metabolites to oxygenated products which may be erroneously identified as additional metabolites.[28] A simple purification procedure that should be a routine precaution is passage of ethers through alumina to remove peroxides.

24.5.4.1.3 *Extraction procedures*

Since drugs and metabolites may be neutral, acidic, basic or amphoteric compounds, it is necessary to adjust the pH of the solution which is being extracted to effect as complete an extraction as possible. Such a procedure is not without its disadvantages. Some drugs or metabolites are labile compounds, sensitive to changes in pH, and degradation products can easily be mistakenly identified as metabolites. If the substrate is a neutral compound or an acidic drug, chemical degradations occur less frequently than with basic substrates. The majority of metabolites formed from neutral or acidic drugs are *C*-oxygenated products, especially phenols, carboxylic acids, ketones and aldehydes, which are relatively easy to isolate from neutralized or acidified solutions with many organic solvents, so that good to excellent recoveries of drugs and metabolites are usually possible. If, however, the substrate is a basic compound, a variety of metabolites (basic, neutral, acidic and amphoteric) is possible. A common practice is to adjust the pH of the solution successively to 12.0, 8.5–9.5 and 1.0 and extract with a suitable solvent at each of these pH values. Such a procedure will extract, respectively, stable basic and neutral compounds, amphoteric products and acidic compounds. Sulfates and other water soluble drug conjugates are not extracted and the extraction of amphoteric compounds is rarely complete. In addition, many primary

N-oxygenated metabolites of basic drugs and chemicals are sensitive to changes in pH and undergo isomerization and/or chemical degradation during extraction.[28]

A major disadvantage of solvent extraction procedures is that, regardless of the pH to which the solution has been adjusted, some drugs and metabolites (*e.g.* sulfate conjugates) cannot be extracted with solvents from aqueous solutions. If metabolites are present in conjugated form, it is often the practice to treat an aliquot of the biological solution with an enzyme (β-glucuronidase and/or sulfatase) or with hot mineral acid prior to extraction. Both processes liberate metabolites from their conjugates, which may then be removed from aqueous solution by conventional solvent extraction techniques. In fact sulfates can be isolated from aqueous solutions by means of an anion exchange column.[29]

Xenobiotics and their metabolites are often present at concentrations of the order of $1 \mu g \, mL^{-1}$, or even much lower, therefore important losses due to phenomena such as adsorption can occur during workup. In the case of anthracyclines, adsorption phenomena were completely eliminated by the use of a secondary amine, desipramine, for quantitative reextraction of anthracyclines from the organic to the aqueous phase.[30-32]

Protein normally remains in the aqueous phase or, especially at acidic pH, is precipitated, during traditional solvent extraction. Indeed, methods often entail an initial protein precipitation step, *e.g.* with phosphotungstate, trichloroacetic acid or ethanol. For traces of protein in urine, ultrafiltration offers a means of removal which is convenient but prone to loss of the analyte on to the membrane.

24.5.4.1.4 *Ion pair extraction*

The techniques of ion pair extraction can be used to remove hydrophilic, ionizable compounds from aqueous solution, which is often difficult with conventional liquid–liquid extraction. For weak bases and acids such as amines, carboxylic acids and phenols, the ion pair technique is a valuable somewhat under-used alternative to the more traditional approach of extraction in ionized form. By careful selection of the counterion, a readily extractable ion pair will result, which may be extracted in good yield into a suitable solvent. Morphine and bromothymol blue, for example, form strong ion pairs which are extractable from aqueous solution into dichloromethane.[33] The method may be used to remove both hydrophilic drugs and their metabolites from biological fluids, although no single counterion will be suitable for pairing with all the metabolites present.

HPLC separations of several biogenic amines and some of their metabolites have been described following ion pair extraction.[34] Alternatively, the ion pair extract may be analyzed by gas chromatography (GC); the complexes usually dissociate in the injection port or on the column, particularly if the stationary phase is appropriately chosen. For example, a base–sulfonic acid ion pair dissociates, allowing GC analysis of the base on an alkaline column.[35]

Hydrophilic anions such as glucuronides and amino acids have been extracted from aqueous solution by forming ion pairs with large quaternary ammonium ions,[36,37] although the technique is currently little used in metabolism studies. Catecholamines have been extracted from aqueous solutions in excellent yield by the use of adduct-forming reagents which also serve as counterions. The extraction of synephrine with a 0.05 M chloroform solution of di(2-ethylhexyl)phosphoric acid is typical.[38]

Two publications describe in some detail the theory and uses of ion pair extraction in drug analysis and metabolism studies.[39,40]

In addition to the use of this principle for extraction, both TLC and HPLC may be used in the ion pair mode by incorporation of the ion-pairing reagent in the mobile phase.

24.5.4.2 Fractionation on Phases

24.5.4.2.1 *Anion exchangers*

Acidic metabolites can often be extracted from biological fluids by the solvent extraction method. An alternative method, now being used more extensively, is anion exchange. Using this method, excellent recoveries of organic acids from urine have been reported.[41] DEAE–Sephadex is the most widely used anion exchanger, but others are also suitable such as Dowex $3 \times 4A$ resin (chloride form) and AG-1-X-2 (hydroxy form).[29,42] The fluid of interest, adjusted to pH 7–8, is passed down the column, which is then washed with water. The acids retained on the column are eluted with aqueous pyridinium acetate[41] or hydrochloric acid,[29] the effluent is lyophilized, and the residue derivatized as appropriate for analysis. A comparison has been made of methods commonly used to isolate organic acids from urine, which showed that more organic acids and fewer interfering substances were isolated by anion exchange than by solvent extraction.[43]

24.5.4.2.2 Cation exchangers

Cation exchangers have been used to some extent to isolate basic metabolites from metabolism mixtures. Amphetamine, 4-hydroxyamphetamine and hydroxynorephedrine, for example, were removed from a tissue homogenate preparation by adjusting the pH to 6.5 and passing it down a Dowex 50 cation exchange column. Subsequent elution of the column with 4 M ammonium hydroxide eluted the three bases.[44]

24.5.4.2.3 Nonionic resins

A continued need for a rapid method of screening urine samples for drugs and metabolites has stimulated a search for improved extraction methods. Of the column materials in use, the nonionic resin Amberlite XAD-2 (a synthetic cross-linked polystyrene polymer with a high surface area) has been used to the greatest extent. Whilst the mechanism of uptake of metabolites by XAD-2 resin is not clearly defined, a wide range of molecules, containing lipophilic and hydrophobic groups, including some glucuronides, have been isolated using XAD-2 resin. Extraction methods for biological samples (including urine, blood, serum, bile, gastric contents and tissues) using XAD-2 resin have been reviewed.[45,46] Efficiency of extraction is dependent on the pH, volume and flow rate of the sample, and on the nature and flow rate of the eluant used to recover the adsorbed drugs from the column. Weakly acidic, neutral and basic drugs (but not acetylsalicylic acid, a stronger organic acid), were efficiently removed to the extent of at least 89% from urine (20 mL) buffered to pH 8.5 ± 0.5 when the flow rate through the XAD-2 column was maintained at 2.5 mL min^{-1} and the eluting solvent was acetone or methanol/chloroform.[47] As mentioned above, XAD-2 resin removes many conjugated and most nonconjugated metabolites from biological fluids. The metabolites may then be eluted from the resin with methanol. If the eluate is concentrated before being partitioned between larger volumes of chloroform and water, the former solvent will contain nonconjugated metabolites, whereas the conjugated metabolites will be found in the latter.[48,49]

24.5.4.2.4 Activated charcoal

This efficient adsorbent is employed to a limited extent in the removal of drugs of abuse from urine,[50] but is not used routinely in drug metabolism studies. Perhaps it could be used to more advantage in metabolism studies since it is claimed[50] that most drugs bind avidly to small amounts of charcoal [< 500 mg per 10 mL urine] and may easily be eluted by small amounts of solvent.

However, investigators who try charcoal without full awareness of the pioneer literature may encounter disappointing recoveries through overlooking a stratagem which minimizes the risk of tenacious binding, that is pretreatment of the charcoal with an agent such as stearic acid. Charcoal thus 'coated', as used for metadrenaline, deserves revived popularity.[51] However, uncoated charcoal has coped well with glucuronides,[52] *e.g.* in a multistep procedure where, as is amplified later, XAD-2 resin and ion exchange columns were also used,[53] presumably on the sound principle[54] that a variation of approaches is advantageous for clean up to be successful. The use of charcoal calls for alertness to possible trace metal contaminants, especially with autoxidizable analytes. In this connection[51] mercaptoethanol may be a useful protective agent, as in the hydrolysis of catecholamine conjugates.[55] In general, charcoal can be a useful agent, but should not be expected to show marked selectivity.

24.5.4.2.5 Solid phase extraction

Over the past decade, a substantial number of solid phase extraction methods have been developed for the concentration of organic solutes out of dilute aqueous media. This technology is based upon silica-based solid extractants, which include silica itself and silica modified by a range of different functionalities, packed into cartridges suitable for the application and elution of solutions.[56] This technique has developed along with developments in HPLC, using the same solid phases of larger (50 μm) particle size, and is now reaching the point where the application of chromatographic principles allows the development of highly selective and specific extractions. Early solid phase extractions used C_{18}-bonded silica, which removed hydrophobic molecules from solution. After washing with water, these in turn could be eluted from the solid phase with an organic solvent such as methanol.[57] Now, essentially all the bonded silicas used for HPLC are also available in cartridges for solid phase extraction, including C_2-, C_8- and C_{18}-alkyl, phenyl, cyclohexyl, amino, diol and CN, which, combined with suitable eluting solvents, allow great versatility in the use of this technique.[56] A number of manufacturers supply a full range of these cartridges, suitable for use in routine assays and with equipment permitting multiple elutions. The technique is ideally suited to laboratory automation and the application of robotics,[58] and represents a major advance in the extraction of compounds not readily extracted into organic solvents, or causing other problems, *e.g.* those which are polar, ionized, unstable at extremes of pH or during concentration.[57]

24.5.4.2.6 *Alumina*

Subtle separation may be achievable with alumina,[38] a conventional chromatographic medium which may show selectivity for analytes with vicinal hydroxy groups. Such adsorptive separation, based on chemical affinity, can in fact be a powerful tool in specific situations. Its key role in catecholamine methodology is discussed by Weil-Malherbe.[55]

Alumina has also been used to isolate salsolinol and derivatives from biological fluids.[59,60] Non acid-washed aluminum oxide generally results in poor and erratic recoveries of catecholamines and related compounds. Therefore acid-washed alumina (Al_2O_3) is recommended;[61] catechol compounds are retained on it when the biological fluids are at a basic pH (about 8.4), then they are eluted with acid.

24.5.4.2.7 *Immobilized phenylboronic acid*

This has been used for the selective extraction of catecholamines, catechol compounds and *cis* diols from biological fluids and foods.[62,63] The immobilized phenylboronic acid (on silica) is first equilibrated with an alkaline solution to obtain the reactive boronate form $RB(OH)_3$. The diol is applied and is bound with the concomitant release of water. Once the compound of interest is retained, contaminants may be washed from the bonded phase provided that an alkaline pH is maintained. Finally, the compound of interest is eluted by acidification of the boronate complex which releases the catechol-containing compound and renders the immobilized phenylboronic acid neutral [$RB(OH)_2$]. Phenylboronic acid gels are now available in cartridge form, and have been successfully applied to diol metabolites of polycyclic aromatic hydrocarbons as well as catechols. It should be noted that the reaction is specific for *cis* diols, and their *trans* isomers will not bind.[64]

24.5.4.3 Steam Distillation

This is an approach which is seldom worth serious consideration. Essentially the operation consists in volatilizing sparingly soluble metabolites by passing steam into the sample of interest. Providing the metabolites have an appreciable vapour pressure, they will distil with the steam. This technique has been used by Strolin-Benedetti and coworkers to extract a tertiary amine (*N*, *N*-dimethyl-4-(2-fluorophenyl)-2-naphthalenethanamine) and its monodemethylated derivative from rat brain homogenates.[65]

24.5.5 CHROMATOGRAPHIC SEPARATION

The vast majority of separations of mixtures of drugs and their metabolites are accomplished using gas chromatography (GC), thin layer chromatography (TLC) and high performance liquid chromatography (HPLC). Other techniques such as column chromatography (adsorption, partition and gel), paper chromatography, countercurrent distribution and electrophoresis are less frequently encountered.

24.5.5.1 Gas Chromatography (GC)

Numerous reviews on the subject are available.[66–69]

GC separations are accomplished on a glass, metal or Teflon column containing a nonvolatile liquid (the stationary phase), usually coated on to an inert solid support material with a large surface area. The components of the mixture are carried through the heated or occasionally cooled column by an inert carrier gas. They separate from one another according to their different partition coefficients between the carrier gas and the stationary phase.

When a component elutes from the column, it is detected and displayed as a peak on a recorder. The retention time of the peak (*i.e.* the time interval between the point of injection and the apex of the recorded peak) is characteristic of, but not unique to, the component giving rise to it, under the GC conditions employed. A peak may also be characterized by its relative retention time. The retention time of a reference compound is determined and that compound is assigned a relative retention time of 1.00. The relative retention times of all other compound are obtained by dividing the retention time of each by that of the reference compound. GC data from different sources are more readily compared if relative retention times are quoted.

24.5.5.1.1 GC columns

GC columns made of glass, stainless steel, copper, aluminum or Teflon are available commercially. In drug metabolism glass columns are usually employed since they are more inert than columns made of metal and can be used at much higher temperatures than Teflon columns. Glass-lined metal columns are also available. They are more robust than all-glass columns, but equally inert. Failure by researchers until 1974[28,70] to identify hydroxylamines as metabolites of primary and secondary amines, for example, was due, at least in part, to metal ion-catalyzed degradation of these metabolites on heated metal columns.

For conventional GC analysis, packed columns are employed. These are typically 1–2 m in length and 2–4 mm internal diameter, and are packed with an inert solid support, coated with stationary phase.

Capillary columns, however, are increasingly supplanting packed columns. They are made of glass or stainless steel, 15–200 m in length and 0.25–0.75 mm in internal diameter. With capillary (open tubular) columns the stationary phase takes the form of a thin coating on the etched or porous column wall. This ensures an adequate carrier gas flow rate.

Solid supports are not widely used, although micropacked capillary columns are available. The major attraction of the capillary column is that much better resolution of components of a mixture is achieved, compared to what is possible with conventional packed columns. For example, the identification of 38 of the many constituents of *Cannabis* samples was achieved using a capillary column in a temperature-programmed GC linked to a mass spectrometer.[71]

Chromatographic separations can be performed isothermally or by temperature programming in which the column temperature is increased at a preset rate during the analysis with, if desired, preprogrammed isothermal periods before and after the temperature increase. The latter procedure permits a greatly increased analysis speed when solutions containing a wide range of compounds are analyzed. Capillary columns are rarely used in the isothermal mode.

24.5.5.1.2 Stationary phases and solid support materials

There are literally hundreds of stationary phases used in GC,[72,73] and various inert solid supports are encountered.[72] The amount of stationary phase used to coat the solid support can vary from what approximates to a monomolecular layer to as much as 30% w/w. Thus the number of different possible prepared columns is virtually unlimited.

In practice, however, a small number of well-studied stationary phases are used in the majority of instances: those most commonly employed in drug metabolism studies or gaining in popularity have been identified by Moffat.[74] A few support materials are used commonly. In drug metabolism studies, those most often encountered are Chromosorb G, Chromosorb W, Gas Chrom Q, Haloport F and Chromosorb 750. The most popular gas chromatographic supports are prepared from diatomaceous earth (kieselguhr), which is very porous and has a high surface area. Supports differ in their densities and hence in the amount of stationary phase that can be applied. Chromosorb W, for example, is a low density support which can be loaded with up to 30% of stationary phase. Chromosorb G has a high density with a maximum liquid phase loading capacity of 5% and is recommended for very polar samples. Chromosorb 750 is a relatively new support material which has a density midway between Chromosorbs W and G; the particles are hard and generate virtually no undesirable fines during coating or packing. It is claimed that only minimal adsorption of polar compounds and minimal decomposition of sensitive compounds occur with this support.

Glass beads are also used as GC supports because of their inertness and their ability to be coated with low loadings of the stationary phase. Commercially available glass beads are of a uniform size and are acid-etched to roughen the surface. This permits the application of a uniform film of the stationary phase and eliminates peak tailing.

24.5.5.1.3 Detectors

Only five GC detectors are used to any extent in drug metabolism. The use of one of these, the mass spectrometer, as a sophisticated detector, is discussed later. The flame ionization (FI) detector and the thermal conductivity (TC) detector, are in widespread use. They are nonselective, that is, the signal produced by them is proportional only to the quantity of an organic material eluting from the GC column, but not its nature.

Conversely, the electron capture (EC) and nitrogen–phosphorus (NP) detectors are selective. The response generated by them is dependent upon both the chemical nature and the quantity of the material emerging from the column.

(i) Thermal conductivity detector

The TC detector is not often used in drug metabolism studies because of its poor sensitivity compared to the FI detector. The minimum detectable amount for a TC detector is around 1 μg, which is some 100 to 1000 times less sensitive than the FI detector. TC detectors give a linear response over a wide range of sample amounts (*ca.* 10^5); this range, however, is somewhat less than that of most flame ionization detectors. An original application of this detector in drug metabolism and excretion balance studies with ^{15}N- and ^{13}C-labelled compounds has been described.[75,76]

(ii) Flame ionization detector

Flame ionization detection is the method most widely used in drug metabolism studies. In principle, the effluent from the GC column, mixed with an equal volume of hydrogen, is passed through a metal jet and burned in an atmosphere of air. Combustion of the components which elute from the GC column produces positive and negative ions, and these species are collected on a polarized 'collector', situated immediately above the jet. The resulting current is amplified by an electrometer and displayed as a function of time on a recorder. All compounds which combust with ionization in the hydrogen/air flame will give a response in a FI detector. Notable compounds *not* detected include water, ammonia, carbon disulfide, carbon dioxide, other simple gases and inert gases.

The response of the FI detector to an organic compound is approximately proportional to the number of carbon atoms it contains and is linear over a slightly wider range (*ca.* 10^5 to 10^7) than TC detectors. Sample quantities down to about 1 ng can be detected, depending upon the design of the detector, notably its internal volume. FI detectors require three gas supplies, the optimal relative flow rates of which are quite critical to obtain maximum sensitivity and stable detector operation.

(iii) Electron capture detector

Due to their selective nature, EC detectors are used much less in drug metabolism studies than FI detectors. EC detectors respond only to halogenated compounds and those containing other functional groups, such as nitro and various heterocyclic systems. However, their selectivity, together with their greater sensitivity, make them ideal in appropriate circumstances, since they can detect as little as 1 pg of an organic compound under optimum conditions.

Electron capture detectors contain a ^{63}Ni or ^{3}H source which emits relatively high energy β particles that collide with carrier gas molecules (normally 95% argon/5% methane), to produce a large number of low energy secondary electrons. When a potential is established in the detector, a small current, called the 'standing current', is produced. Molecules with electron-capturing groups eluting from the GC column absorb some of these electrons and reduce the magnitude of the standing current, which returns to its original level when the sample has left the detector. This change in current is amplified and inverted to give a positive peak on the recorder.

The sensitivity of the detector varies enormously depending on the ability of the eluting compound to absorb electrons. For example, the response of chlorobenzene is 1200 relative to benzene ($=1$), bromobenzene is 7500, 1-iodobutane is 1.5×10^6, chloroform is 1×10^6, and 2,3-butanedione is 0.8×10^6. The suitability of the EC detector for the analysis of drug metabolites depends on whether they contain one or several halogens or other functions that might easily absorb electrons, or whether they can be derivatized with halogen-containing reagents such as trifluoroacetic anhydride or heptafluorobutyric anhydride. Derivatization with these reagents often gives volatile products which are particularly well suited for analysis by GC with EC detection.

(iv) Nitrogen–phosphorus detector

NP detectors are extremely sensitive towards most nitrogen- or phosphorus-containing organic compounds. Because of their instability, they have not until recently been widely used to detect drugs and metabolites.

Greatly improved NP detectors are now available, and their use for the detection of drugs and metabolites in urine and other biological fluids is rapidly increasing.

In some respects, the NP detector is similar in design to the FI detector. The effluent from the GC column is mixed with a much smaller volume (about one-tenth) of hydrogen, and the gas mixture is passed into the detector chamber and heated electrically in the presence of an alkali source (usually a rubidium salt) to form a low temperature plasma rather than a discrete flame. This treatment produces a minute electric current which is amplified and recorded. Low picogram quantities of N- and P-containing compounds are easily detected. The detector is 20 000–40 000 times more sensitive towards nitrogen and 40 000–80 000 times more sensitive towards phosphorus than to carbon.

The NP detector is particularly useful if mixtures of nitrogenous drugs and metabolites which retain nitrogen are being monitored. Heterocyclic drugs would be particularly suitable, since they are unlikely to lose the heteroatom during metabolism. In contrast, acyclic nitrogenous drugs are often deaminated, so that use of an NP detector to analyze their metabolites would be unwise.

24.5.5.1.4 Derivatization techniques

Most drug metabolites are more polar than the drug from which they were derived and have considerably reduced volatility. Polar metabolites often have long GC retention times and produce asymmetric, tailing peaks. They may undergo 'on-column' degradation, sometimes they fail to elute from the column. However, such metabolites may be analyzed by GC methods provided they are converted into stable, more volatile derivatives. The choice of the best derivatization reagent depends on the structure of the polar compound. Crippen and Smith[77] have constructed a detailed list of derivatives of organic compounds suitable for GC analysis. They identified numerous functional groups which require derivatization, and suggested appropriate derivatives in each case. In a more recent review Ahuja[67] has identified eight derivatization procedures: on-column reactions, reactions with dialkyl acetals, silylation, esterification, acylation, hydrazone formation, ion pair formation, and derivatization for EC detection (see above). Earlier, silylation, alkylation and acylation were identified as the most common procedures, but the value in many cases of special derivatives and of on-column pyrolysis was also recognized.

(i) Silylation

This is the most common method of derivatizing polar metabolites. Silylation is the substitution of a trialkylsilyl (usually trimethylsilyl; TMS; $SiMe_3$) group for the active hydrogen atom of compounds containing OH, SH and NH groups, including alcohols, phenols, acids, steroids and carbohydrates and their thio analogues, amines, amino acids, amides, imides, imines and related compounds. TMS derivatives are not difficult to prepare, provided anhydrous reaction conditions are employed. They are safe to handle and most of them have excellent chromatographic properties, although isomerization and other undesirable reactions occur occasionally.[67]

(ii) Alkylation

Acidic drugs and metabolites are generally polar compounds and must be derivatized prior to GC analysis. Diazomethane (CH_2N_2) has been used by many to convert organic acids to methyl esters and phenolic metabolites to methyl ethers, but there are disadvantages to its use, principally the hazards of this reagent and the fact that many reactive impurities are also methylated and thus interfere with GC analysis.[78] Esterification of organic acids is readily accomplished by treating the acid with a Lewis acid, such as boron trifluoride etherate, as catalyst and an alcohol.[79] The conditions required are very mild and the presence of other functional groups in the molecule does not interfere with the reaction.

Another versatile group of reagents which can be used to alkylate various metabolites are the dimethylformamide (DMF) dialkyl acetals,[80] which have the general structure $Me_2NCH(OR)_2$. The reagents which are commercially available have R = Me, Et, Pr^n, Bu^n and CD_3. DMF–dialkyl acetals readily react with acids, alcohols, phenols, primary amines, secondary amines and other compounds.

Quaternary hydroxides (such as phenyltrimethylammonium hydroxide[63]) in polar solvents form salts with organic acids.[81] These salts react with primary alkyl iodides to give esters of the organic acid. Numerous alkyl groups (such as pentafluorobenzyl) can be introduced by this method.

(iii) Perfluoroacylation

It is now possible to detect and quantify picogram or smaller quantities of amines and related products by converting them to suitable fluorinated derivatives (see Section 24.5.5.1.3.iii). Ahuja[67] lists the most commonly used derivatives as trifluoroacetyl, heptafluorobutyryl, pentafluoropropionyl and perfluorobenzyl. The electron capture sensitivity of N-(pentafluorobenzoyl)amphetamine is exceptionally high,[67] enabling picogram quantities of this base to be easily determined.

24.5.5.1.5 Application of GC to drug metabolism studies

Examples of the use of GC in drug metabolism studies may easily be found.[68]

One application of GC in drug metabolism studies which is worthy of emphasis is its use in determining which enantiomers are formed when the metabolic process is one which produces a centre of asymmetry in the metabolite,[82] such as the reduction of the carbonyl group in diethylpropion. The reduced metabolites were derivatized with N-trifluoroacetyl-L-prolyl chloride and the diastereoisomers obtained analyzed by GC.

24.5.5.2 High Performance Liquid Chromatography (HPLC)

The need for analytical methods of increased sensitivity and stability for drug metabolism studies has coincided with the introduction of HPLC as a routine method of separating and quantifying drugs and metabolites. Three particular advantages of HPLC are: (i) water soluble compounds can be analyzed without prior extraction; (ii) it is generally performed at room temperature, which permits the analysis of thermally labile metabolites; and (iii) it is nondestructive. A disadvantage of the use of HPLC in drug metabolism studies was that it did not directly allow the identification of metabolites of unknown structures but now also HPLC, as GC, can be linked with a mass spectrometer (see below).

As in TLC, HPLC separations of solute molecules depend on the distribution of these molecules between a stationary and a liquid mobile phase. The relative affinities of solutes for the two phases determine the separation characteristics. Affinity for the stationary and mobile phases may involve one or more of adsorption, partition, ion exchange or a solvation mechanism. The stationary phase is contained within a short, small bore column through which the liquid mobile phase is pumped at high pressure. An efficient pumping system is required to draw the mobile phase from storage reservoirs and to maintain pulseless constant flow rates against the back pressure of the column.

The instrumental and theoretical aspects of HPLC are well documented[83–85] and will not be further elaborated here.

24.5.5.2.1 HPLC columns

Since they have to withstand high pressures (up to 350 atm) HPLC columns are generally made of stainless steel. They are commonly 25 cm in length (or shorter) and up to 8 mm in internal diameter. Wider bore columns may be used for semipreparative work, *e.g.* isolation of metabolites for physicochemical characterization, while the so-called microbore columns are of value when sample size is a limitation.

Several different types of column packing are available, utilizing various separation principles.

Adsorbent packings are generally totally porous and consist of particles in the 5–10 μm range. Because of their small particle size and high surface area, they offer high capacity and high resolution. Silica is the most commonly used adsorbent. Pellicular packing materials are now only occasionally used for precolumn preparation; they are spheroidal in shape, 30–40 μm diameter and can be tightly packed in the column.

Polar bonded phase packings for partition HPLC are prepared by chemically bonding a suitable organic moiety, such as a propionitrile group, to a spheroidal solid core material through a Si—C bond or aminocyano groups Si—O—Si bonded to irregular silica gel. These column packings are useful for the separation of polar compounds. They have been used by Marrari *et al.* in the HPLC determination of acipimox and its metabolite in human plasma.[86]

Reversed phase packings are made in the same way by bonding hydrocarbon residues of various chain lengths (commonly C_2, C_8 and C_{18}) to the silica. These column-packing materials are most useful for the separation of relatively nonpolar, nonionic compounds including aliphatic and aromatic hydrocarbons, steroids, pesticides and halogenated compounds. These packings can also be used for the separation of polar compounds by using ion pair chromatography (see Section 24.5.5.2.3). Reversed phase HPLC is now greatly used for the detection of catecholamines, serotonin and their metabolites, as described in a recent review on HPLC of neurotransmitters and their metabolites.[87]

Anion exchange HPLC packing materials are produced by chemically bonding a suitable ion exchange group (*e.g.* quaternary ammonium) through a Si—C bond, on to a spheroidal solid support. Cation exchange packing materials are similarly prepared by bonding sulfonic acid residues on to the solid support. Ion exchange HPLC is of particular value in the separation of ionic and other polar materials in biological fluids. This application, with an emphasis on the separation of urinary components, has been reviewed by Scott[88] and more recently discussed by other authors.[87]

24.5.5.2.2 *Detectors*

After resolution on an appropriate column, the separated components of a mixture pass directly through a detection system. Fixed (254 nm) or variable wavelength UV detectors and refractive index detectors are commonly used, as are fluorescence and electrochemical detectors.[84,85] Although derivative formation is used much less commonly in HPLC than in GC, derivatives are frequently formed for fluorescence detectors. For example, *o*-phthaldialdehyde (OPA) has been used as a derivatization reagent for glycinamide, the major metabolite of milacemide, in order to detect it by HPLC with a fluorescence detector (Figure 5).[89] Another fluorescent reagent for primary and secondary amino acids as well as for amines, is 9-fluorenylmethyl chloroformate (Fmoc-Cl).[89]

24.5.5.2.3 *Ion suppression and paired ion HPLC*

Extraction of ionic compounds from aqueous biological media can be time-consuming and is generally incomplete. In addition, mixtures of ionic compounds are not usually separated satisfactorily using conventional GC or adsorption or partition HPLC. These problems promoted a search for improved HPLC methods of separating and analyzing the large numbers of ionic body fluid components. High performance ion exchange chromatography was the initial result; ion suppression and paired ion HPLC are alternative techniques.

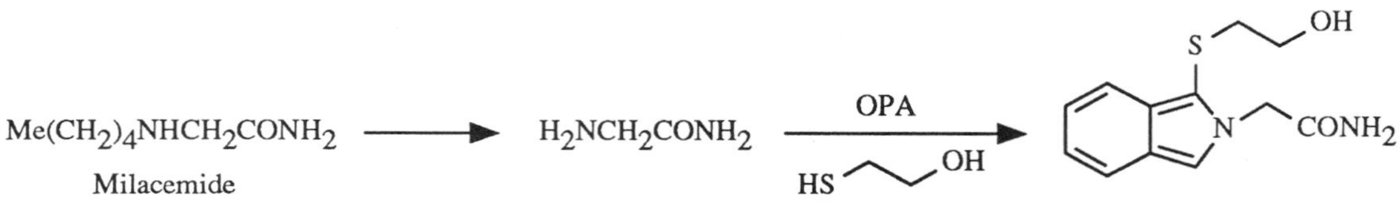

Figure 5 Detection of glycinamide as a metabolite of milacemide by derivatization with *o*-phthaldialdehyde (OPA) and 2-mercaptoethanol

Ion suppression HPLC is applicable to the separation of mixtures of organic acids ($pK_a > 2$) and mixtures of organic bases ($pK_a < 8$). It is commonly performed on reversed phase columns, which operate best with a mobile phase in the pH 2–8 range. A mobile phase buffered to about pH 3.5 ensures that weak organic acids are in their nonionic form (RCO_2H) and therefore lipophilic. Similarly, if the mobile phase is buffered to about pH 7.5, weak bases will be nonionized (lipophilic) and may be rapidly separated from a mixture containing other solutes. Strong acids ($pK_a < 2$) and strong bases ($pK_a > 8$), however, remain ionized in the pH 2–8 range and cannot be separated by ion suppression HPLC.

Paired ion HPLC permits the separation of strongly ionic compounds by reversed phase chromatography. A large organic counterion is added to the mobile phase and forms a reversible ion pair complex with the ionized sample. This complex behaves as a neutral, nonpolar (lipophilic) compound, and can be rapidly chromatographed as a sharp, symmetrical peak on a reversed phase HPLC column. The lipophilicity of the ion pair complex formed depends on the sample and on the counterion employed. The more lipophilic is the complex, the greater will be its attraction for the nonpolar stationary phase and the longer will be its retention time. Some examples are heptane-sulfonic acid and pentanesulfonic acid buffered to pH 3.5, which form ion pair complexes with organic bases, and tetrabutylammonium phosphate buffered to pH 7.5, which forms ion pair complexes with organic acids. A review of ion suppression and paired ion HPLC, containing a list of 40 drugs which can be chromatographed using paired ion chromatography, has been prepared by Waters Associates.[90]

24.5.5.2.4 *Applications of HPLC to drug metabolism studies*

Gas chromatography can be used for the separation of drugs and metabolites, provided the compounds are lipophilic, are capable of being extracted from biological preparations and solutions, and are thermally stable (after chemical derivatization, if necessary). When, however, metabolites are difficult or impossible to extract from aqueous solutions (*e.g.* amphoteric compounds, quaternary amines, glucuronides, sulfates and other conjugates), are sensitive to the pH changes used during extraction procedures, or are thermally labile, then HPLC can be used to advantage. In their reviews on the use of HPLC in drug metabolism, Skellern[91] and Tomăsić[52] suggest that the technique is particularly suitable for the quantitative analysis of conjugates of drugs and metabolites. Skellern describes the use of HPLC techniques to resolve and quantify metabolites of drugs such as acetaminophen, carbimazole, chlordiazepoxide and indomethacin. A review of HPLC by Tomlinson[92] includes its applicability to the quantitation of drugs, with brief comments on metabolites.

Clinical applications of HPLC have also been reviewed.[93,94] Examples are the separation of salicylic acid and its metabolites (salicyluric and gentisic acid) in a urine sample using paired ion HPLC, and the analysis of the amphoteric compound aminobenzoic acid by paired ion HPLC using two different counterions. The number of references in the literature to the use of HPLC in drug metabolism studies is increasing enormously. HPLC techniques have been used to resolve and quantitate the major human metabolites of the antitumour drugs doxorubicin, epirubicin and idarubicin.[31,95] HPLC has also been used by Strolin-Benedetti and coworkers to isolate the urinary metabolites of rifabutin in humans.[96] Finally one application of HPLC in drug metabolism studies is, as mentioned before for GC, its use in determining which enantiomers are formed when the metabolic process produces a centre of asymmetry in the metabolite, such as the reduction of the carbonyl group in the molecule of fenofibric acid.[97] The reduced metabolites were derivatized with (R)-1-(naphthen-1-yl)ethylamine and the diastereoisomeric naphthylethylamides obtained were analyzed on a reversed phase HPLC column. Alternatively the two enantiomers were directly injected on a chiral column Cyclobond I (cyclodextrin). Two excellent recent reviews on chiral stationary phases for the direct HPLC separation of enantiomers and on chemical derivatization for diastereomer formation are those of Pirkle[98] and Imai.[99]

24.5.5.3 Supercritical Fluid Chromatography (SFC)

There is currently great interest in the application of supercritical fluid technology to various bioanalytical problems.[100] Supercritical fluids are liquefied gases (typically CO_2), held under pressure, which may be used both for extraction (SFE) and chromatography (SFC). Extraction with supercritical fluids is rapid, avoids problems of thermal lability, uses nontoxic solvents which do not present disposal problems, and may be controlled in terms of selectivity by varying the pressure of the solvent. SFC uses a supercritical fluid as the mobile phase for chromatography, which may use capillaries of the GC type or packed columns resembling HPLC columns. These columns are held in GC style ovens, and compounds eluted by the supercritical fluid are detected by a variety of methods, typically by FID. To optimize separations, organic modifiers are commonly added to liquid CO_2 to form the mobile phase.

24.5.5.4 Thin Layer Chromatography (TLC)

TLC may be employed for various purposes in drug metabolism studies, most generally for the separation and purification of metabolites prior to their characterization by other analytical methods such as ultraviolet (UV), infrared (IR) and nuclear magnetic resonance (NMR) spectroscopy and mass spectrometry (MS). TLC may be used to help identify drugs or their metabolites, if authentic reference samples are available for direct comparison of R_F values obtained on two or three different TLC systems. In addition, TLC has been used for the quantitation of compounds which are not amenable to GC analysis, either because of their instability or poor GC properties. TLC can also be of assistance in determining structural features introduced into the drug molecule during the metabolism reaction. A judicious choice of spray reagents can often reveal the chemical nature of the introduced functional group or the type of conjugate formed. For example, naphthoresorcinol is used for glucuronides,[14] potassium dichromate–silver nitrate has been used for divalent sulfur and, together with ninhydrin, for the detection of glutathione conjugates.[101] The metabolite of heptaminol, 6-amino-2-methyl-1,2-heptanediol, was visualized on thin layers with the reagent which detects 1,2-diols (sodium metaperiodate benzidine) and with amine group reagents (ninhydrin, vanillin–potassium hydroxide);[102] finally iron(III) chloride and potassium ferricyanide are regularly used to detect phenolic compounds.[103]

24.5.5.4.1 *Sorbents*

Silica gel is the most commonly used sorbent in TLC; alumina, cellulose and other sorbents are used infrequently. As is true in GC and liquid column chromatography, a narrow and well-defined particle size distribution, a controlled pore size and pore volume and a specific and graded particle surface area are necessary to minimize spot tailing and maximize solute resolution and sensitivity. Improvements in these factors have led to the development of high performance thin layer chromatographic (HPTLC) plates, designed to be used for the separation of components in the low nanogram or even picogram range, the actual limit being dependent on the nature of the particular solute and the detection system used. TLC plates which allow reversed-phase chromatographic separation of mixtures of highly polar compounds are also available. The sorbent is manufactured by reacting an appropriate hydrocarbon-silane (*e.g.* octadecasilane) with the surface hydroxy groups of a special silica gel. Apart from enabling separations of polar compounds to be made with simple TLC systems, the method is suggested to have some value for working out analytical conditions preparatory to reversed phase HPLC.

24.5.5.4.2 *Solvent purity*

The purity of solvents used both for the TLC system itself and for eluting material from isolated TLC spots is of paramount importance. Since eluates are often concentrated up to 1000 times for GC or MS analyses, trace quantities of impurities in the solvent will also be concentrated by this factor and will interfere. All solvents should therefore be purified.

24.5.6 SPECTROSCOPIC CHARACTERIZATION OF DRUG METABOLITES

Five major spectroscopic techniques have been applied to the characterization of drug metabolites, namely UV, fluorescence, IR and NMR spectroscopy and mass spectrometry. Although all are different, with their own advantages and disadvantages, each of these is sufficiently sensitive to be applied to a range of metabolic problems and/or sufficiently informative to contribute to the structural characterization of metabolites. A detailed discussion of these various techniques is clearly outside the scope of the present coverage, and the paragraphs below simply serve to highlight the advantages and disadvantages of each of these methods in their application to metabolic studies.

24.5.6.1 Ultraviolet and Fluorescence Spectroscopy

These related techniques will be considered together. Both find widespread use, at least in the early stages of metabolic studies. Both are able to provide information on solutions, the ideal sample state for metabolic studies, and can be very sensitive, especially in the case of fluorescence. However, the amount of structural information which may be derived from a UV or fluorescence spectrum is not great, and it is only in very rare cases that a metabolite can be characterized solely by these means.[104] UV spectroscopy finds special application to compounds containing conjugated diene systems, notably steroids (the well-known Woodward and Fieser Rules).[105]

Both the UV and fluorescence spectra of compounds containing ionizable groups can be influenced by pH, and this is very well established for phenols.[106] The demonstration of spectral shifts with pH can be very helpful in the early stages of metabolic studies, since these indicate the strong likelihood that a phenol may be present.

24.5.6.2 Infrared Spectroscopy

IR spectra are highly diagnostic, giving a great deal of structural information which can often be sufficient for the characterization of a novel molecule.[107] Until recently, IR spectrophotometers were not sensitive enough for widespread application to metabolic studies, requiring very large samples, but the technique has been revolutionized by the application of Fourier transform mathematics to data processing.[108] FTIR is increasingly contributing to metabolic studies by virtue of its sensitivity and diagnostic capability.

The greatest limitation to the application of IR and FTIR techniques is that of sample presentation. These are frequently solid state methods, giving the best results from samples in KBr discs, and are thus not readily compatible with most purification and separation methods used in drug metabolism studies.[109] However, this limitation is now being overcome by the use of FTIR online detectors for GC and SFC (see below).

24.5.6.3 Nuclear Magnetic Resonance Spectroscopy

NMR spectra are extremely informative and NMR is a near to ideal technique for metabolic studies using samples in solution.[110] Its major limitation is that of sensitivity, although instrumental developments, like FT data processing and higher fields (instruments up to 600 MHz are now commercially available), are progressively overcoming this. The presentation of the sample is ideal for metabolic studies, solutions being introduced directly into the instrument. Indeed, there are numerous reports of the direct NMR examination of biological samples without any work up at all, blood or urine simply being placed in the NMR sample tube.[111] Other applications of NMR are outside the scope of the present coverage, but it should be noted that it is possible to examine metabolism in real time by NMR, in cells, whole organs, whole animals and even whole humans.[112] Obviously these techniques will never reach routine use due to the demands they make on instrument time, but they are available for the resolution of specialized problems.[113]

NMR spectra may be obtained of a variety of magnetic atoms (with a nuclear spin of 1/2) within a molecule. The most important of these are protons and the ^{13}C isotope of carbon. Of other magnetic atoms, ^{19}F spectra are often of value in metabolic studies of fluorinated drugs or derivatives,[113] while ^{15}N spectra have also been used. Proton NMR spectra can be very difficult to interpret, due to complex interactions between protons, but ^{13}C spectra are extremely informative. Although the sensitivity of the method falls off with increasing atomic number, instrumental improvements have minimized the impact of this on sample size requirements.

Although most NMR spectra are obtained at the natural abundance of an atom, generally an isotope of low abundance, the sensitivity of the technique can be greatly increased by the use of drugs appropriately labelled with stable isotopes, notably ^{13}C. If in turn ^{13}C is introduced at a site liable to metabolic modification, it may be possible to gain profound insight into these processes very simply. In a recent study, the fate of [*carboxyl*-^{13}C]phenylacetic acid was examined in the horse simply by recording the ^{13}C NMR spectrum of the urine.[114] The chemical shift of the carboxyl carbon is uniquely altered by each of the metabolic pathways it may undergo, *i.e.* esterification with glucuronic acid, amide formation with amino acids *etc.*, and quantitative information on each of these options was readily obtained. This approach is limited only by the natural abundance of the isotopes in endogenous compounds, which may interfere with the assay. This was the case in the studies on phenylacetic acid, since this acid and its conjugates are normally present in horse urine (and that of other species).

24.5.6.4 Mass Spectrometry

For the past 20 years, mass spectrometry (MS) has represented the standard technique for the characterization of drug metabolites and is well suited for this by virtue of its extreme sensitivity and the provision of highly diagnostic structural information.[115] Its applicability is only limited by the fact that it is a high vacuum technique and thus requires volatile samples. Despite the need for high vacuum it has been successfully interfaced with a number of chromatographic techniques, principally GC.[116] Additionally, an increasing number of HPLC–MS interfaces are now available. GC–MS, using both packed columns and capillaries, has been immensely useful in metabolic studies,[117] and its use is limited only by the types of compounds amenable to GC separations. However, even this problem may be overcome by appropriate derivatization: with the aid of such procedures even polar, hydrophilic conjugates such as glucuronides can be characterized by GC–MS as pertrimethylsilyl derivatives.[118] LC–MS techniques are discussed elsewhere.

The structural information presented in a mass spectrum comprises the molecular ion, which is found at the molecular weight (assuming single charges on each ion) of the compound plus or minus a proton, and various charged fragment ions. The fragmentation pattern is explicable in terms of the structure of the compound and the ease with which substituents are lost.[119] In the past, the application of MS was often limited by the problem of excess fragmentation in the electron impact mode, many compounds not giving a molecular ion at all. However, the development of progressively softer ionization techniques has largely overcome this restriction;[120] particular successes have been achieved with chemical ionization using a range of reagent gases, fast atom bombardment and thermospray. These methods are generally successful in enhancing the molecular ion but paradoxically this can reduce the amount of structural information to be derived from the mass spectrum. Judicious adjustment of mass spectral conditions may be required so as to give both molecular ions and suitable fragments.

MS techniques, notably hyphenated techniques such as GC–MS, may be used in three ways in drug metabolism studies,[121] for: (i) quantitative analysis using either stable isotope labelled or homologous internal standards (the former are preferable); (ii) characterization of metabolites; and (iii) for metabolic studies of stable isotopically labelled compounds using the mass spectrometer to determine the isotopic composition of each peak. This approach may be used simply to 'flag' those spectra which genuinely belong to the compound. However, in many cases suitably labelled compounds are now used to delineate metabolic pathways by targetting on selected metabolites and putative intermediates.[122] This approach may be extended to the cofactors required for metabolic pathways: examples here include the use of deuterated NADPH to investigate reduction and its tissue origin and the use of $^{18}O_2$ and $H_2^{18}O$ to evaluate directly the source of oxygen used in an oxidative pathway.

Information on metabolic mechanisms may also be obtained by such labelling. The placement of a stable isotope into a site of metabolism may alter the rate and/or route of metabolism.[123] For a given bond the ease of bond breaking depends upon the isotopic masses of the two atoms linked and this is especially marked for carbon–hydrogen bonds. Replacement of hydrogen by deuterium can slow metabolism substantially; in cases where C—H cleavage is the rate-limiting step in metabolism, the rate can be up to nine times slower when deuterium is substituted. The discernment of these 'kinetic isotope effects' is often extremely valuable in studies on metabolic pathways, and may be of such magnitude as to cause 'metabolic switching' away from the slowed pathway towards routes not subject to the kinetic isotope effect.

24.5.7 HYPHENATED TECHNIQUES

The term 'hyphenated techniques' is nowadays commonly used to describe the combination of two (or more) methods to accomplish analyses more easily and quickly. In general, the term applies to combinations of chromatographic methods, which serve to separate complex mixtures into their components, and spectroscopic techniques which provide online characterization of the individual components.[124] The success of such approaches will depend upon: (i) the compatibility of the output characteristics of the chromatographic method with the input requirements of the spectroscopic technique; and (ii) the rate at which the chosen spectroscopic information can be acquired.

The first, and still most widely used hyphenated technique, was the combination of gas chromatography and mass spectrometry.[116] The output of a GC column presents samples in a state readily amenable for MS analysis, and the MS is able to produce considerable information within the width of a typical GC peak. The combination of these techniques was first achieved in the mid-1960s with the development of the first concentrating interfaces able to separate the samples of interest from the high volumes of carrier gas required for the packed column GCs then in use. These interfaces were the Biemann–Watson jet separator and the fritted glass interface of Ryhage and Holmstedt.[125] GC–MS has been greatly facilitated by the development of capillary GC, which enhances chromatographic resolution and cuts by 10–20-fold the gas flow compared with a packed column. Substantial improvements to the pumping of MS sources mean that nowadays the output of a capillary GC column may be fed directly into an MS source without a specialized interface other than a heated transfer line.[126] Once in the MS source, the sample (which is, of course, in the vapour phase) may be subjected to EI or CI and tandem/triple MS, just as samples from any other source.

In comparison with GC–MS, the coupling of HPLC with MS has proven very troublesome and is only now achieving reliability.[127] Although HPLC is a most advantageous technique for drug metabolism studies, its utility is especially evident for polar, hydrophilic compounds which are not particularly amenable to MS. The mobile phases in which the samples are found are incompatible with the high vacuum, high temperature conditions within the MS source. Over the years a great deal of ingenuity has been expended upon the development of interfaces able to remove HPLC solvents and present samples to the MS source in a form suitable for ionization.[127] These have included various moving belt and moving wire designs with complex heaters, pumping and vapour locks, but none of these have proved reliable enough for widespread application. In addition, these interfaces permit only EI or CI MS of the HPLC eluate, methods of limited usefulness for polar involatile compounds.

A major step forward in the development of a universal HPLC–MS interface came with the chance discovery that when the eluate of a microbore HPLC column was sprayed into the source of an MS, ions were produced even when the EI source was switched off.[128] This so-called thermospray ionization remains poorly understood in physical terms, but provides an extremely useful means of obtaining mass spectra of polar, hydrophilic nonvolatile molecules which may be directly applied to drug metabolism studies. Thermospray interfaces may be applied to both quadrupole and magnetic sector mass spectrometers, and the ions may be examined with single or multiple sector analyzers.[128]

The widespread application of HPLC in bioanalysis has depended very much upon the use of UV detectors, this detection method having the required sensitivity, but UV detection at a single wavelength provides little or no structural information about the compounds eluting from the HPLC column. However this situation has changed recently with the availability of diode array spectrophotometers. These instruments, instead of using monochromators to scan spectral wavelengths, irradiate the samples with light of all wavelengths, generally between 190 and 700 nm, and use a diode array detector, with elements sensitive to individual wavelengths over that range. The output of the diode array detector is electronically assembled, with correction for energy differences in the output of the light source across the spectrum, into a UV absorption spectrum. Thus, the entire spectrum is recorded at the same time, and typically some 20 ms are required. Early photodiode array spectrophotometers were restricted by their low sensitivity, but they may now be applied to HPLC systems.[129] This enables the recording of a three-dimensional chromatogram, relating both absorption and wavelength to time: essentially the whole spectrum of the sample may be obtained at any time during the HPLC run. This technique requires computerized data acquisition to enable its benefits to be fully realized.[129] It is especially useful for the determination of peak purity in analysis, and may be of value for metabolite characterization, depending upon the amount of structural information which may be derived from the UV spectrum of the parent compound and the impact of the changes produced by metabolism upon this.

In comparison with UV, IR spectroscopy is extremely informative about the structure of molecules. However, scanning IR spectrophotometers generally require large amounts of sample (*ca.* 10 µg) and take 10–15 min to present satisfactory spectra. This situation has been dramatically altered by the application of interferometer technology and Fourier transform (FT) mathematics for the rapid accumulation of data, with the result that IR is now available as an online technique for the spectral characterization of chromatographic peaks. The nature of the mobile phases makes HPLC unsuitable for this application, but a number of GC–FTIR instruments have become available commercially.[130] These instruments, obviously coupled to a data acquisition system, are seemingly well suited to metabolic studies, but there are as yet no well-documented examples of their application.

Recent developments in NMR spectroscopy have made very important improvements to the sensitivity and speed of this technique, mainly by the application of superconducting magnets, allowing ever higher radio frequencies, and Fourier transform mathematics in the data processing. However, these marked improvements at present still do not fit NMR for application as an online technique.

24.5.8 FUTURE PERSPECTIVES

Even those scientists whose interest in analytical methodology is only peripheral will be all too well aware of the pace of current developments in equipment and applications. As well as refinements of existing methodologies and the linkage of these in new hyphenated techniques, genuinely new principles are being introduced. Thus, the major chromatographic techniques of GC and HPLC are constantly being improved in terms of equipment, notably contributing towards their automation and data processing with microcomputers, and in column technology. These two methods are being linked in new ways with physicochemical instruments to give new hyphenated techniques, which greatly facilitate metabolite characterization, e.g. GC–FTIR.

At the time of writing, there are two new separation techniques which are likely to have an impact on the ways in which drug metabolism studies are carried out, these being the application of supercritical fluid technology[100] and capillary electrophoresis.[131] This latter offers the opportunity to apply the great separative potential of electrophoresis to biomolecules, while reducing the impact of irreproducibility and poor quantitation inherent in most manually intensive electrophoretic separations. By performing electrophoresis in chromatographic capillaries, the difficulties of the preparation and staining of gels are overcome, electrophoretic separation can be combined with other principles in the capillary, and analytically acceptable detection methods, e.g. conductivity, UV and fluorescence may be used. Capillary electrophoresis has even been combined with MS to give a new hyphenated technique.[132] At the moment, capillary electrophoresis is limited in terms of the sample size it can accept, but this very limitation can almost be an advantage when working with body fluids and other biological samples, by reducing the sample size required and minimizing background contaminants.

Like many other new analytical techniques, SFC has great promise of facilitating work on drug metabolism problems. Like many other new techniques, SFC suffers at present from a lack of documented applications to biological problems. The method has potential applicability for hyphenated techniques, notably SFC–MS, where the mobile phase requires no concentrating interface, and SFC–FTIR, where the use of CO_2 gives great advantages due to its IR transparency.[133] At the present time, SFC seems to work best with hydrophobic molecules, compounds which do not present an excessive challenge to the drug metabolist. However, this position may be expected to change as more applications are reported.

Drug metabolism has presented, and continues to present, formidable challenges to the analyst, of compounds of vastly differing properties, often in very dilute solutions rich in contaminants. While the major interest in drug metabolism will always concern what has been found, it is important not to overlook the methodological and instrumental developments required to produce the data.[134] The addition of new methodologies to the armamentarium of the drug metabolism scientist will depend upon the satisfactory demonstration of their value with well-worked applications.

24.5.9 REFERENCES

1. R. T. Williams, 'Detoxication Mechanisms', 2nd edn., Chapman and Hall, London, 1959.
2. J. Caldwell, in 'Intermediary Xenobiotic Metabolism', ed. D. H. Hutson, J. Caldwell and G. D. Paulson, Taylor and Francis, London, 1989, p. 3.
3. W. B. Jakoby (ed.), 'Enzymatic Basis of Detoxication', Academic Press, New York, 1980, vols. 1 and 2.
4. L. Lemberger, E. D. Witt, J. M. Davis and I. J. Kopin, *J. Pharmacol. Exp. Ther.*, 1970, **174**, 428.
5. J. Caldwell, in 'Concepts in Drug Metabolism, Part A', ed. P. Jenner and B. Testa, Dekker, New York, 1980, p. 211.
6. R. P. Maickel, in 'Drug Determination in Therapeutic and Forensic Contexts', ed. E. Reid and I. D. Wilson, Plenum Press, New York, 1985, p. 3.
7. J. Caldwell, *Drug Metab. Rev.*, 1976, **5**, 219.
8. W. Muecke, *ACS Symp. Ser.*, 1986, **299**, 108.
9. E. Reid and J. P. Leppard (eds.), 'Drug Metabolite Isolation and Identification', Plenum Press, New York, 1983.

10. C. Fenselau and L. Yellet, *ACS Symp. Ser.*, 1986, **299**, 159.
11. V. J. Feil, *ACS Symp. Ser.*, 1986, **299**, 177.
12. M. Strolin-Benedetti, *Actual. Chim. Ther.*, 1980, **7**, 357.
13. J. Caldwell and A. J. Hutt, in 'Progress in Drug Metabolism', ed. J. W. Bridges and L. F. Chasseaud, Taylor and Francis, London, 1986, vol. 9, p. 11.
14. A Weil, J. Caldwell and M. Strolin-Benedetti, *Drug Metab. Dispos.*, 1988, **16**, 302.
15. J. A. R. Mead, J. N. Smith and R. T. Williams, *Biochem. J.*, 1958, **68**, 61.
16. C. C. Irving, in 'Biological Oxidation of Nitrogen in Organic Molecules', ed. J. W. Bridges, J. W. Gorrod and D. V. Parke, Taylor and Francis, London, 1972, p. 75.
17. M.-L. Dahl-Puustinen and L. Bertilsson, *Pharmacol. Toxicol.*, 1987, **61**, 342.
18. J. Caldwell and H. M. Given, in 'Sulphur Containing Drugs and Related Organic Compounds', ed. L. A. Damani, Ellis Horwood, Chichester, 1989, vol. 1b.
19. G. J. Mulder, *Prog. Drug Metab.*, 1984, **8**, 35.
20. A. B. Roy, in 'Sulfate Metabolism and Sulfate Conjugation', ed. G. J. Mulder, J. Caldwell, G. M. J. Van Kempen and R. J. Vonk, Taylor and Francis, London, 1982, p. 299.
21. L. E. Martin and E. Reid, *Prog. Drug Metab.*, 1981, **6**, 197.
22. F. M. Kaspersen and C. A. A. Van Boeckel, *Xenobiotica*, 1987, **17**, 1451.
23. M. Strolin-Benedetti, A. Donath, A. Frigerio, K. T. Morgan, C. Laville and A. Malnoe, *Ann. Pharm. Fr.*, 1978, **36**, 279.
24. H. B. Hucker, A. J. Bailetto, J. Demetriades, B. H. Arison and A. G. Zacchei, *Drug Metab. Dispos.*, 1977, **5**, 132.
25. M. Strolin-Benedetti, P. Marrari, E. Moro, V. Tamassia and R. Roncucci, *Int. Congr. Pharmacol., 10th*, 1987, 75.
26. M. G. Horning, P. Gregory, J. Nowlin, M. Stafford, K. Lertratanangkoon, C. Butler, W. G. Stillwell and R. M. Hill, *Clin. Chem. (Winston-Salem, N.C.)*, 1974, **20**, 282.
27. A. Zlatkis and K. Kim, *J. Chromatogr.*, 1976, **126**, 475.
28. R. T. Coutts and A. H. Beckett, *Drug Metab. Rev.*, 1977, **6**, 51.
29. C. Lin, Y. Li, J. McGlotten, J. B. Morton and S. Symchowich, *Drug Metab. Dispos.*, 1977, **5**, 234.
30. S. Eksborg, H. Ehrsson, B. Andersson and M. Beran, *J. Chromatogr.*, 1978, **153**, 211.
31. E. Moro, M. G. Jannuzzo, M. Ranghieri, S. Stegnjaich and G. Valzelli, *Chromatography*, 1982, **230**, 207.
32. E. Moro, V. Bellotti, M. G. Jannuzzo, S. Stegnjaich and G. Valzelli, *J. Chromatogr.*, 1983, **274**, 281.
33. G. Schill, *Acta Pharm. Suec.*, 1965, **2**, 13.
34. B. A. Persson and B. L. Karger, *J. Chromatogr. Sci.*, 1974, **12**, 521.
35. P. F. G. Boon and A. W. Mace, *J. Chromatogr.*, 1969, **41**, 105.
36. B. Fransson and G. Schill, *Acta Pharm. Suec.*, 1975, **12**, 107.
37. T. Norfgren and R. Modin, *Acta Pharm. Suec.*, 1975, **12**, 407.
38. R. Modin and M. Johansson, *Acta Pharm. Suec.*, 1971, **8**, 561.
39. G. Schill, K. O. Borg, R. Modin and B. A. Persson, *Prog. Drug Metab.*, 1977, **2**, 219.
40. G. Schill, R. Modin, K. O. Borg and B. A. Persson, *Prog. Drug Metab.*, 1977, **1**, 135.
41. A. M. Lawson, R. A. Chalmers and R. W. E. Watts, *Clin. Chem. (Winston-Salem, N.C.)*, 1979, **22**, 1283.
42. E. Nakamura, L. E. Rosenberg and K. Tanaka, *Clin. Chim. Acta*, 1976, **68**, 127.
43. J. A. Thompson and S. P. Markey, *Anal. Chem.*, 1975, **47**, 1313.
44. B. B. Brodie, A. K. Cho and G. L. Gessa, in 'Amphetamines and Related Compounds', ed. E. Costa and S. Garattini, Raven Press, New York, 1970, p. 217.
45. A. Stolman and P. A. F. Pranitis, *Clin. Toxicol.*, 1977, **10**, 49.
46. Technical Bulletin No. 10, Applied Science Laboratories, P.O. Box 440, State College, PA, 1980.
47. M. P. Kullberg and C. W. Gorodetzky, *Clin. Chem. (Winston-Salem, N.C.)*, 1974, **20**, 177.
48. A. Karim, J. Hribar, W. Aksamit, M. Doherty and L. J. Chinn, *Drug Metab. Dispos.*, 1975, **3**, 467.
49. Y. Kishimoto, S. Kraychy, R. E. Ranney and C. L. Grant, *Xenobiotica*, 1972, **2**, 237.
50. J. M. Meola and M. Vanko, *Clin. Chem. (Winston-Salem, N.C.)*, 1974, **20**, 184.
51. A. A. A. Aziz, J. P. Leppard and E. Reid, *Methodol. Dev. Biochem.*, 1976, **5**, 159.
52. J. Tomăsić, *Drug Fate Metab.*, 1978, **2**, 281.
53. T. T. L. Chang, C. F. Kuhlman, R. T. Schillings, S. F. Sisenwine, C. O. Tio and H. W. Ruelius, *Experientia*, 1973, **29**, 653.
54. B. L. Karger, L. R. Snyder and C. Horvath, 'An Introduction to Separation Science', Wiley, New York, 1973.
55. H. Weil-Malherbe, in 'Analysis of Biogenic Amines and their Related Enzymes', ed. D. Glick, Interscience, New York, 1971, p. 119.
56. 'Sorbent Extraction Technology', ed. K. C. Van Horne, Analytichem International, Harbor City, CA, 1985.
57. R. Whelpton and P. R. Hurst, in 'Bioanalysis of Drugs and Metabolites', ed. E. Reid, J. D. Robinson and I. D. Wilson, Plenum Press, New York, 1988, p. 289.
58. H. M. Hill, L. Delehean and B. A. Bailey, in 'Bioanalysis of Drugs and Metabolites', ed. E. Reid, J. D. Robinson and I. D. Wilson, Plenum Press, New York, 1988, p. 299.
59. M. W. Duncan, G. A. Smythe and M. V. Nicholson, *J. Chromatogr.*, 1984, **336**, 199.
60. G. Dordain, P. Dostert, M. Strolin-Benedetti and V. Rovei, in 'Monoamine Oxidase and Disease', ed. K. F. Tipton, P. Dostert and M. Strolin-Benedetti, Academic Press, New York, 1984, p. 417.
61. A. H. Anton and D. F. Sayre, in 'A Study of the Factors Affecting the Aluminum Oxide Trihydroxyindole Procedure for the Analysis of Catecholamines', ed. A. H. Anton and D. F. Sayre, 1962, pp. 138, 360.
62. M. Strolin-Benedetti, V. Bellotti, E. Pianezzola, E. Moro, P. Carminati and P. Dostert, *J. Neural Transm.*, 1989, in press.
63. A. H. B. Wu and T. G. Gornet, *Clin. Chem. (Winston-Salem, N.C.)*, 1985, **31**, 298.
64. D. R. Thakker, W. Levin, H. Yagi, M. Tada, D. E. Ryan, P. E. Thomas, A. H. Conney and D. M. Jerina, *J. Biol. Chem.*, 1985, **257**, 5103.
65. J. F. Rumigny, M. Strolin-Benedetti, H. Dupont and M. Dedieu, in 'Collegium Internationale Neuropsychopharmacologicum, 14th C.I.N.P. Congress', Fidia Research Biomedical Information, Florence, 1984, p. 417.
66. J. Drozd, *J. Chromatogr.*, 1975, **113**, 303.
67. S. Ahuja, *J. Pharm. Sci.*, 1976, **65**, 163.
68. W. J. A. VandenHeuvel and A. G. Zacchei, *Drug Fate Metab.*, 1978, **2**, 49.
69. J. Arthur, F. de Silva and C. V. Puglisi, *Drug Fate Metab.*, 1983, **4**, 245.

70. A. H. Beckett and S. Al-Sarraj, *J. Pharm. Pharmacol.*, 1973, **25**, 328.
71. M. Novotny, M. L. Lee, C.-E. Low and A. Raymond, *Anal. Chem.*, 1976, **48**, 24.
72. Catalog 18 (June 27, 1976), Analabs, Inc., 80 Republic Drive, North Haven, CT.
73. S. T. Preston, *J. Chromatogr. Sci.*, 1978, **8**, 18A.
74. A. C. Moffat, *Proc. Anal. Div. Chem. Soc.*, 1976, **13**, 355.
75. M. Strolin-Benedetti and B. Pataky, *Adv. Mass Spectrom. Biochem. Med.*, 1976, **1**, 73.
76. M. Strolin-Benedetti, S. Perczel, P. Strolin, A. Assandri, J. Quaglia and D. A. Larue, *Proc. Int. Conf. Stable Isot.*, *1975*, 1976, 511.
77. R. C. Crippen and C. E. Smith, *J. Gas Chromatogr.*, 1965, **3**, 37.
78. J. Horner, S. S. Que Hee and R. G. Sutherland, *Anal. Chem.*, 1974, **46**, 110.
79. P. K. Kadaba, *J. Pharm. Sci.*, 1974, **63**, 1333.
80. Pierce Handbook and General Catalog, Pierce Chemical Co., Box 117, Rockford, 1977–1978, 253.
81. R. H. Greeley, *J. Chromatogr.*, 1974, **88**, 229.
82. B. Testa and P. Jenner, *Drug Fate Metab.*, 1978, **2**, 143.
83. B. F. H. Drenth and R. A. de Zeeuw, *Drug Fate Metab.*, 1985, **5**, 185.
84. R. Weinberger, in 'Therapeutic Drug Monitoring and Toxicology by Liquid Chromatography: Chromatographic Science Series', ed. S. H. Y. Wong, 1985, vol. 32, p. 151.
85. C. Lavrich and P. T. Kissinger, in 'Therapeutic Drug Monitoring and Toxicology by Liquid Chromatography: Chromatographic Science Series', ed. S. H. Y. Wong, 1985, vol. 32, p. 191.
86. P. Marrari, V. Bellotti, E. Moro and G. Valzelli, in 'International Conference, Developments in Analytical Methods in Pharmaceutical, Biomedical and Forensic Sciences', Verona, 1986, Abstracts book, p. 75.
87. E. Gelpi, *Adv. Chromatogr. (New York)*, 1987, **26**, 321.
88. C. D. Scott, *Science (Washington, D.C.)*, 1974, **186**, 226.
89. M. Strolin-Benedetti, P. Marrari, E. Moro, P. Dostert and R. Roncucci, *Pharmacol. Res. Commun.*, 1988, **20**, suppl. 4, 135.
90. Waters Associates, 'Paired-Ion Chromatography—An Alternative to Ion-Exchange', Milford, MA, 1976.
91. G. G. Skellern, *Proc. Anal. Div. Chem. Soc.*, 1976, **13**, 357.
92. E. Tomlinson, *Pharm. J.*, 1976, **220**, 13.
93. I. M. House and D. J. Berry, in 'High Pressure Liquid Chromatography in Clinical Chemistry', ed. P. F. Dixon, C. H. Gray, C. K. Lim and M. S. Stoll, Academic Press, New York, 1976, p. 155.
94. I. D. Watson, *Adv. Chromatogr. (New York)*, 1987, **26**, 118.
95. S. Eksborg, H. Ehrsson and I. Andersson, *J. Chromatogr.*, 1979, **164**, 479.
96. G. Cocchiara, M. Strolin-Benedetti, G. P. Vicario, M. Ballabio, B. Gioia, S. Vioglio and A. Vigevani, *Xenobiotica*, 1989, in press.
97. A. Weil, J. Caldwell, M. Strolin-Benedetti and P. Dostert, in 'Drugs Affecting Lipid Metabolism', ed. R. Paoletti *et al.*, Springer, Berlin, 1987, p. 324.
98. W. H. Pirkle and T. C. Pochapsky, *Adv. Chromatogr. (New York)*, 1987, **27**, 73.
99. K. Imai, *Adv. Chromatogr. (New York)*, 1987, **27**, 215.
100. K. D. Bartle, M. P. Burke, A. A. Clifford, I. L. Davies, J. P. Kithinji, M. W. Raynor, G. F. Shilstone and A. Williams, *Eur. Chromatogr. News*, 1988, **2**, 12.
101. A. Malnoe, M. Strolin-Benedetti, R. L. Smith and A. Frigerio, in 'Biological Reactive Intermediates', ed. D. J. Jollow, J. J. Kocsis, R. Snyder and H. Vainio, Plenum Press, New York, 1977, p. 387.
102. F. Chanoine, M. Strolin-Benedetti, J.-F. Ancher and P. Dostert, *Arzneim.-Forsch.*, 1981, **31**, 1430.
103. A. Malnoe and M. Strolin-Benedetti, *Xenobiotica*, 1979, **9**, 281.
104. J. Renson and P. Kremers, in 'Fundamentals of Biochemical Pharmacology', ed. Z. M. Bacq, Pergamon Press, Oxford, 1971, p. 3.
105. A. I. Scott, 'Interpretation of the Ultraviolet Spectra of Natural Products', Pergamon Press, Oxford, 1964.
106. D. H. Williams and I. Fleming, 'Spectroscopic Methods in Organic Chemistry', McGraw-Hill, London, 1980.
107. F. S. Parker, 'Applications of Infrared Spectroscopy in Biochemistry, Biology and Medicine', Hilger, London, 1971.
108. J. R. Ferraro and L. J. Basile (eds.), 'Fourier Transform Infrared Spectroscopy', Academic Press, New York, 1978, vol. 1.
109. A. D. Cross and R. A. Jones, 'An Introduction to Practical Infrared Spectroscopy', 3rd edn., Butterworths, London, 1969.
110. J. K. Nicholson and I. D. Wilson, in 'Drug Metabolism from Molecules to Man', ed. D. J. Benford, J. W. Bridges and G. G. Gibson, Taylor and Francis, London, 1987, p. 189.
111. J. K. Nicholson, J. A. Timbrell, J. R. Bales and P. J. Sadler, *Mol. Pharmacol.*, 1985, **27**, 634.
112. B. M. Hitzig, J. W. Pritchard, H. L. Kantor, W. R. Ellington, J. S. Ingwall, C. T. Burt, S. I. Helman and J. Koutcher, *FASEB J.*, 1987, **1**, 22.
113. K. E. Wade, J. Troke, C. M. Macdonald, I. D. Wilson and J. K. Nicholson, in 'Bioanalysis of Drugs and Metabolites', ed. E. Reid, J. D. Robinson and I. D. Wilson, Plenum Press, New York, 1988, p. 383.
114. J. Caldwell, B. P. Nutley, A. J. Hutt and M. V. Marsh, *Br. J. Cancer Res.*, 1986, **25**, 457.
115. M. E. Rose and R. A. W. Johnson, 'Mass Spectrometry for Chemists and Biochemists', Cambridge University Press, Cambridge, 1982.
116. G. H. Draffan, J. D. Gilbert and M. T. Gilbert, in 'Drug Metabolism in Man', ed. J. W. Gorrod and A. H. Beckett, Taylor and Francis, London, 1978, p. 193.
117. W. McFadden, 'Techniques of Combined Gas Chromatography Mass Spectrometry', Wiley, New York, 1973.
118. C. Fenselau and L. P. Johnson, *Drug Metab. Dispos.*, 1980, **8**, 274.
119. J. H. Beynon, R. A. Saunders and A. E. Williams, 'The Mass Spectra of Organic Molecules', Elsevier, Amsterdam, 1968.
120. K. Biemann and S. A. Martin, *Mass Spectrom. Rev.*, 1987, **6**, 1.
121. G. H. Draffan, J. D. Gilbert and M. T. Gilbert, in 'Drug Metabolism in Man', ed. J. W. Gorrod and A. H. Beckett, Taylor and Francis, London, 1978, p. 207.
122. T. A. Baillie, *Pharmacol. Rev.*, 1981, **33**, 81.
123. M. I. Blake, H. L. Crespi and J. J. Katz, *J. Pharm. Sci.*, 1975, **64**, 367.
124. T. Hirschfeld, *Anal. Chem.*, 1980, **52**, 297A.

125. B. J. Millard, 'Quantitative Mass Spectrometry', Heyden, London, 1978.
126. R. E. Ardrey, in 'Clarke's Isolation and Identification of Drugs', ed. A. C. Moffat, J. V. Jackson, M. S. Moss and B. Widdop, Pharmaceutical Press, London, 1986, p. 251.
127. A. P. Bruins, in 'Bioanalysis of Drugs and Metabolites', ed. E. Reid, J. D. Robinson and I. D. Wilson, Plenum Press, New York, 1988, p. 339.
128. C. R. Blakley and M. L. Vestal, *Anal. Chem.*, 1983, **55**, 750.
129. A. F. Fell, in 'Clarke's Isolation and Identification of Drugs', ed. A. C. Moffat, J. V. Jackson, M. S. Moss and B. Widdop, Pharmaceutical Press, London, 1986, p. 221.
130. W. Herres, 'Bruker FT-IR Application Note 15', Bruker Instruments, 1988.
131. M. J. Gordon, X. H. Huang, S. L. Pentoney, Jr. and R. N. Zare, *Sciences (N.Y.)*, 1988, **242**, 224.
132. J. A. Olivares, N. T. Nguyen, C. R. Yonker and R. D. Smith, *Anal. Chem.*, 1987, **59**, 1230.
133. P. Morin, M. Caude, H. Richard and R. Rosset, *Chromatographia*, 1986, **21**, 523.
134. J. Lederberg, 'Beckman Symposium on Biomedical Instrumentation', Beckman Instruments, Fullerton, CA, 1986.

24.6
Systems Analysis in Pharmacokinetics

DONALD P. VAUGHAN

Sunderland Polytechnic, UK

24.6.1 INTRODUCTION

A drug is distributed around the body *via* the circulation of blood, and partitioning of the drug into tissues is a random process. Similarly, the eventual elimination of a drug is also a random process, dependent on the probability that a drug molecule will reach the elimination organ and the probability that it will interact with the biochemical process responsible for irreversible elimination. Consequently, on first consideration, the time-dependent phenomena of drug distribution and elimination must be considered as potentially complicated and intractable problems. Fortunately, it is mathematically feasible to translate a random process into a deterministic one. The resulting deterministic equations can then be manipulated using the wealth of analytical mathematics.

A similar logic step is invariably used when describing chemical reactions, nuclear reactions, biological growth, 'predator–prey relationships' and mathematical economics. All the latter phenomena are essentially random or stochastic in nature, but when the number of elements is large these stochastic processes are adequately defined by well-behaved functions. For example, the decay of radioactive nuclei, where each nucleus behaves independently and each nucleus has a probability of decay in some observational period, is represented by an exponential function. This analytical solution is dictated by the assumption that the rate of decay is proportional to the number of radioactive nuclei present at any particular time.

Historically, pharmacokinetics originates in the description of observed drug concentrations by single exponential time functions. The latter is synonymous with a mathematical model for the body which is composed of one homogeneous compartment. More detailed observations of plasma drug concentrations necessitated an upgrading of the latter model to a two- and then to a three-compartment system. However, the purpose of a mathematical model is not just to provide an overall description of observed events. Its purpose is to derive global statements consistent with feasible boundary conditions, to predict the functional behaviour of the system under a variety of forcing functions and to define input functions that will control the system's response within some predefined error interval.

In keeping with such a philosophy a natural progression is to generalize the mathematical model, so that global solutions are independent of the exact description or topological structure. Such a progression has occurred in pharmacokinetics, and compartmental systems are replaced by an abstract system or 'black-box', the properties of which depend only on the assumption of time invariance and linearity. More specific information is obtained by considering a linear time-invariant system that mimics the recirculation of a drug within the body. Such a representation harmonizes with detailed physiological pharmacokinetic models.

Any theory should provide insight into the fundamentals and should be directly applicable. The latter is unreservedly true for pharmacokinetic theory which encompasses all aspects of drug design, drug metabolism, drug absorption, drug administration and drug toxicity.

In the words of Gibaldi and Levy[1] 'Pharmacokinetics is concerned with the study and characterization of the time course of drug absorption, distribution, metabolism and excretion, and with the relationship of these processes to the intensity and time course of therapeutic and adverse effects of drugs. It involves the application of mathematical biochemical techniques in a physiologic and pharmacologic context.'

In the current text some general pharmacokinetic theory is given and, wherever possible, practical conclusions and applications are referred to. In particular, reference is made to therapeutic effects and the related phenomenon of drug toxicity, the ramifications of which in toxicity testing, formulation design and dosage adjustment, particularly in the elderly and frail, should as a result become transparent.

24.6.2 DRUG DISTRIBUTION AND COMPARTMENTAL MODELS

24.6.2.1 Single-compartment Models

For a few drugs the plasma drug concentrations, after a rapid intravenous injection of a single dose, *i.e.* a bolus or impulse input, appear to decay away as a single exponential function, as is shown from the linear regression of the log(concentration) data against time.

Similarly, after oral or intramuscular administration, a large number of drugs exhibit, after an initial absorption phase, exponential decay. Examples of exponential decay, *i.e.* linear log(concentration)–time functions, for prednisolone, metoprolol and norephedrine are illustrated in Figures 1, 2 and 3, respectively. For intravenous drug administration, the obvious description of

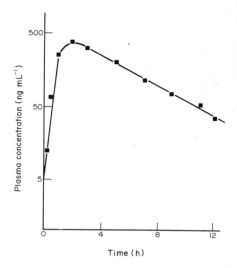

Figure 1 Plasma prednisolone concentration–time profile following oral administration of 20 mg prednisolone to a kidney transplant patient (reproduced from ref. 88 by permission of Plenum Press)

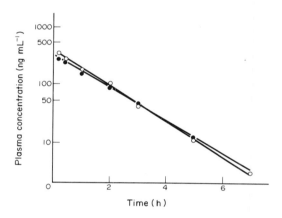

Figure 2 Plasma concentration of metoprolol after administration of metoprolol in two dogs (reproduced from ref. 89 by permission of Plenum Press)

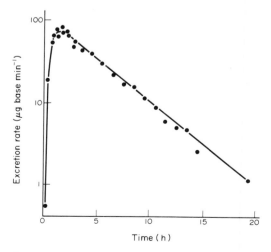

Figure 3 Urinary excretion rate of norephedrine after oral administration to a healthy adult subject (reproduced from ref. 90 by permission of Williams and Wilkins)

Figure 4   Representation of a single-compartment system

such data is

$$\ln C = -\alpha t + \ln C_0 \tag{1}$$

where C is the observed concentration of drug and C_0 the extrapolated concentration, or initial concentration, and α the gradient of the linear function. Taking inverse logarithms then

$$C = C_0 e^{-\alpha t} \tag{2}$$

and the first derivative of C with respect to time is

$$dC/dt = -\alpha C \tag{3}$$

A pictorial representation of the latter equation is a compartment, with a homogeneous concentration of drug, from which drug is eliminated (Figure 4), the elimination rate constant being α (units of time^{-1}). A compartmental representation must have an associated volume V with $D = VC_0$.

The compartment has no direct physiological meaning, save that in the present context the amount of drug in the body $X(t)$ is defined by $VC(t)$. The compartmental system is simply a mathematical model and the correspondence of a multiple-component system to a single-component system is due to the inability to distinguish between the multiplicity of the operational processes that occur in the body. The description is simply a representation of the average behaviour of the drug in the body, $X(t)$. Converting to drug mass then gives

$$X(t) = De^{-\alpha t} \tag{4}$$

24.6.2.1.1   *Half-life*

In a single-compartment system there is only one parameter, α, which relates to the elimination of the drug from the body. For ease of interpretation the decay half-life is usually considered (equation 5).

$$\text{Half-life} = t_{1/2} = \ln 2/\alpha = 0.693/\alpha \tag{5}$$

Some drugs, *e.g.* penicillin and aspirin, are rapidly eliminated from the body, whereas other compounds, *e.g.* demethyldiazepam, digoxin and phenobarbitone, dwell in the system for days.

The rapid elimination of penicillin is essentially due to active renal tubular excretion of the drug into urine, and the rapid decay of aspirin is due to enzymatic hydrolysis to salicylic acid. The latter may prompt questions concerning the role of salicylic acid in the chronic antiinflammatory action of aspirin, since salicylic acid does not inhibit cyclooxygenase but is in itself an effective antiinflammatory agent. A small collation of drug half-lives is presented in Table 1. It is of particular interest and importance that the half-life of a drug can vary considerably between individuals, and that chronological age as well as physiological state has a profound influence on the half-life value. Under the strict assumption that the body behaves as a single-compartment system, possible interpretations of the half-life data are that the variation in the elimination of phenobarbitone is determined by variations in the hepatic oxidation of the drug by P-450 enzymes, such variations being either controlled by genetic factors or by enzymatic induction resulting from environmental factors. Similarly, the inability of the newborn to eliminate drugs can be related to the immaturity of drug-metabolizing enzymes in the liver. Digoxin and amoxycillin are principally eliminated from the body by renal excretion. Since there is a gradual reduction in renal function (normal renal function, creatinine clearance 120 ml min^{-1}) with age, the increased half-lives of these drugs in the elderly can be attributed to reduced renal functions.

Considering the above explanations it is evident that α is a composite rate constant representing the summation of renal and hepatic drug elimination constants.

Table 1 Approximate Half-lives of Some Drugs in Plasma[a]

Drug	Half-life (h)	Drug	Half-life (h)
Demethyldiazepam (pregnant women)	188	Nortriptyline (mother)	17
Demethyldiazepam	65	Nortriptyline	33 (very variable)
Digitoxin	106–142	Nitrazepam (young subjects)	40
Barbitone	96–120	Nitrazepam (geriatrics)	29
Phenobarbitone (newborn infants)	111	Diphenylhydramine	8
Phenobarbitone (nursing mothers)	79	Oxazepam	8
Phenobarbitone	48–144	Heptobarbitone	8
Digoxin (elderly patients)	73	Dihydrocodeine	4
Digoxin (young patients)	51	Chloramphenicol	3–5
Digoxin	44	Amoxycillin (patients with reduced renal function, creatinine clearance $< 50\,\text{mL min}^{-1}$)	1.3
Diazepam	32	Amoxycillin	0.5
Nortriptyline (newborn infants)	56	Acetylsalicylic acid	0.1 (approx.)

[a] Human data.

In the context of a single compartment, non-intravenous drug administration requires a drug input function to drive the system.

24.6.2.1.2 *Drug metabolites*

Single-compartment systems can be extended to include drug metabolites. For example, consider an irreversible catenary system consisting of drug absorption, drug metabolism to a principal metabolite which is further metabolized to an active agent. In compartmental terms the catenary chain is represented as in Figure 5, where K_a is the absorption rate constant for drug absorption (assumed first order), K_1 is the rate constant for conversion of the drug, $X_1(t)$, to the first metabolite, $X_2(t)$, and K_2 is the rate constant for conversion of the first metabolite to the second metabolite, $X_3(t)$. The elimination of the ultimate metabolite from the body is represented by K_3.

The associated set of differential equations for the catenary system are

$$dD/dt = -K_a D \tag{6}$$

$$dX_1/dt = K_a D - K_1 X_1 \tag{7}$$

$$dX_2/dt = K_1 X_1 - K_2 X_2 \tag{8}$$

$$dX_3/dt = K_2 X_2 - K_3 X_3 \tag{9}$$

which have the solution set

$$D(t) = D_0 e^{-K_a t} \tag{10}$$

$$X_1(t) = D_0 K_a \left[\frac{e^{-K_a t}}{(K_1 - K_a)} + \frac{e^{-K_1 t}}{(K_a - K_1)} \right] \tag{11}$$

$$X_2(t) = D_0 K_a K_1 \left[\frac{e^{-K_a t}}{(K_1 - K_a)(K_2 - K_a)} + \frac{e^{-K_1 t}}{(K_a - K_1)(K_2 - K_1)} + \frac{e^{-K_2 t}}{(K_a - K_2)(K_1 - K_2)} \right] \tag{12}$$

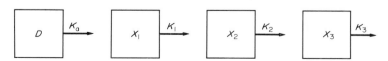

Figure 5 Catenary chain of single compartments

$$X_3(t) = D_0 K_a K_1 K_2 \left[\frac{e^{-K_a t}}{(K_1 - K_a)(K_2 - K_a)(K_3 - K_a)} + \frac{e^{-K_1 t}}{(K_a - K_1)(K_2 - K_1)(K_3 - K_1)} \right.$$

$$\left. + \frac{e^{-K_2 t}}{(K_a - K_2)(K_1 - K_2)(K_3 - K_2)} + \frac{e^{-K_3 t}}{(K_a - K_3)(K_1 - K_3)(K_2 - K_3)} \right] \tag{13}$$

where D_0 is the dose of drug administered, $D_0 \equiv D(0)$.

The absorption of diazepam and its conversion to the active metabolite demethyldiazepam, which is subsequently converted to the active metabolite oxazepam prior to elimination as inactive glucuronide, is an example of a strict precursor chain sequence. A simulation of the amounts of these drugs is given in Figure 6. The rate of drug absorption K_a is unity, $K_1 = \ln 2/32$, $K_2 = \ln 2/65$ and $K_3 = \ln 2/8$. Inspection of the equations illustrates that the functions $X_1(t)$, $X_2(t)$ and $X_3(t)$ are biexponential, triexponential and quartexponential functions respectively. In each case the limiting exponential function is determined by the component with the smallest exponential coefficient. In the case of $X_1(t)$ this is K_1 and in the case of $X_2(t)$ this is K_2. However, for $X_3(t)$ the smallest modular exponential coefficient is K_2 and not K_3. Thus in a precursor chain the limiting half-life of oxazepam is 65 h, *i.e.* the half-life of demethyldiazepam, and not the half-life of oxazepam (8 h) which is observed when oxazepam is administered. The parallel decline of any two log(concentration) functions indicates that the product of metabolism is formation limited, whereas divergence in exponential decline is indicative of elimination limitation for the metabolite.

Integrating the differential equations and solving for the total areas then gives

$$\int_0^\infty X_1(t)\,dt = D/K_1 \tag{14}$$

$$\int_0^\infty X_2(t)\,dt = D/K_2 \tag{15}$$

$$\int_0^\infty X_3(t)\,dt = D/K_3 \tag{16}$$

For repeated administration of diazepam at an interval of τ h steady-state concentrations $X_{i,\,ss}$ result. Thus

$$X_{1,\,ss} = \frac{D}{K_1 \tau} \equiv \frac{32D}{\tau \ln 2} \qquad \text{for diazepam} \tag{17}$$

$$X_{2,\,ss} = \frac{D}{K_2 \tau} \equiv \frac{65D}{\tau \ln 2} \qquad \text{for demethyldiazepam} \tag{18}$$

$$X_{3,\,ss} = \frac{D}{K_3 \tau} \equiv \frac{8D}{\tau \ln 2} \qquad \text{for oxazepam} \tag{19}$$

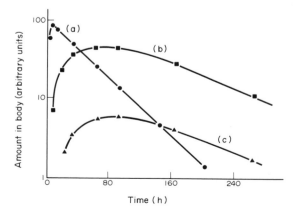

Figure 6 Simulation of the amounts of diazepam (a), demethyldiazepam (b) and oxazepam (c) in the body after oral administration

The latter illustrates how the therapeutic effect or indeed a toxicological effect can be determined by a metabolite. The degree of this effect is dictated by the elimination of the metabolite.

Cummings and co-workers[2,3] first recognized the importance of drug metabolites in therapeutic/toxicological responses. However, for many years drug metabolism was considered synonymous with the production of pharmacologically inactive and toxicologically inert compounds and was consequently ignored. This traditional view is no longer tenable.[4-6] Indeed, the seemingly innocuous ester glucuronides can acylate protein macromolecules and are potentially immunogenic.

24.6.2.1.3 Prodrug design

A simplistic chain of compartments is also of value in the design of prodrugs. A prodrug is a chemical derivative which is either chemically or enzymatically reversible within the body, *e.g.* aliphatic esters, phosphate esters and azo linkages.[7-10] Prodrugs are designed to overcome poor or negligible absorption of a drug from the gastrointestinal tract.

Assuming negligible absorption of the parent drug from the intestinal tract, the absorption and metabolism of the prodrug combine to generate an optimal chemical reversal rate. For first-order absorption from the gastrointestinal tract, say at rate K_a, an oral prodrug will undergo chemical degradation prior to absorption, say at rate K_1, and will be subjected to irreversible metabolism once absorbed, say at rate K_2, as well as being converted into the drug.

For such a scheme the compartmental system shown in Figure 7 is applicable. Solving for $\int_0^\infty X_1(t)\,dt$ then gives

$$\int_0^\infty X_1(t)\,dt = \frac{DK_a}{(K_1 + K_a)}\frac{K_1}{(K_1 + K_2)} \equiv PQ \tag{20}$$

where D is the prodrug dose.

The factor Q is a strictly increasing asymptotic function of K_1 with limit unity. However, the factor P is a strictly decreasing function of K_1 with limit zero. Consequently, the products of these two factors combine to generate a non-negative function with a single maximum. A readily reversible chemical linkage will fail to generate sufficient drug because of generation of non-absorbable drug in the gastrointestinal tract. Similarly, a strong chemical linkage in the prodrug will fail to generate sufficient drug as a result of metabolic or elimination processes not involved in drug generation.

24.6.2.1.4 Exponential decay and pharmacological response

The simple exponential decay of drug concentrations is responsible for a log-linear relationship between the administered drug dose and the duration of a pharmacological response. The latter follows by assuming that a minimum amount of drug in the body or plasma, X_{min}, is required to elicit any pharmacological response. For a single-compartment system the duration of drug action,

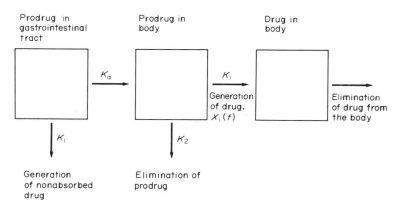

Figure 7 Compartmental representation of a prodrug

say $\hat{t}$, is the time required for the initial drug mass X_0 to decline to X_{min}. Thus

$$\ln X_{min} = \ln X_0 - \alpha \hat{t} \tag{21}$$

or

$$\hat{t} = \frac{\ln(\text{dose})}{\alpha} - \frac{\ln X_{min}}{\alpha} \tag{22}$$

where dose $\equiv X_0$.

General analysis of an *n*-dimensional compartmental system demonstrates that the linear relationship between the duration of drug action and the logarithm of the administered dose is not a property generated by the restrictive assumption of a single-compartment system. An example relating to the duration of anaesthesia to the logarithm of the intravenous dose is given in Figure 8.

The time dynamics of a pharmacological response is complicated in some instances by the existence of a finite drug receptor population and drug binding to these receptors. Usually receptor occupancy is tacitly assumed to relate to an elementary Langmuir binding isotherm, and because the receptor population is infinitesimal in comparison to the drug population, drug binding does not perturb the drug population.[11] Assuming a receptor binding capacity of $\hat{X}$, the dynamics of receptor occupancy $X_2(t)$ is

$$dX_2/dt = K_1(\hat{X} - X_2)X_1 - K_2X_2 \tag{23}$$

where X_2 is the receptor occupancy and $(\hat{X} - X_2)$ is the available binding capacity. The rate constants K_1 and K_2 are the first-order rate constants for the formation and dissociation of drug–receptor complexes respectively. The function $X_1(t)$ is the mass of unbound drug.

Under the reasonable assumption that the dynamics of equilibrium are rapid in relation to the time dependency of $X_1(t)$, equilibrium can be assumed, thus

$$K_1(\hat{X} - X_2)X_1 = K_2X_2 \tag{24}$$

and

$$X_2 = \frac{\hat{X}X_1}{\left(\dfrac{K_2}{K_1} + X_1\right)} \tag{25}$$

Considering the pharmacological response to be directly proportional to the occupancy X_2, then a hyperbolic relationship exists between the amount of drug in the body, X_1, and the intensity of pharmacological response. Since the maximal intensity of a pharmacological response, I_{max}, is defined by $\hat{X}$ and K_2/K_1 is defined by the value of X_1 that generates 50% of the maximal response,

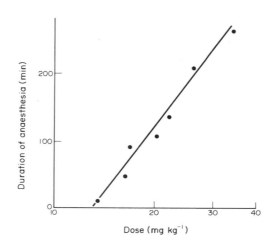

Figure 8 Relationship between the duration of anaesthesia in monkeys and the intravenous dose of pentobarbital. Minimum dose to elicit anaesthesia is 13 mg kg^{-1} (reproduced from ref. 91 by permission of C. V. Mosby and Co.)

X_{50}, the relationship can be cast in the form

$$I = \frac{I_{max}X_1}{X_{50} + X_1} \tag{26}$$

It is interesting to note that the functional form of I against the logarithm of X_1 is sigmoidal and essentially linear in the range between 20 and 80% of I_{max} (see Figure 9).

The expectancy of a log-linear decline in $X_1(t)$ with respect to time, for a single-compartment model coupled with an essentially log-linear relationship with pharmacological response suggests that the decline of a measured response will be linear with respect to time. The dynamics of pharmacological responses support the latter conclusion (see Figure 10). However, there are other seemingly contradictory observations in which the logarithm of a pharmacological response is linearly related to time (see Figure 11). The paradox vanishes when $\hat{X}$ is essentially large with respect to $X_1(t)$, which can easily occur when the uptake of drug into a tissue is limited. In such cases the intensity of the pharmacological response, I, is proportional to $X_1(t)$.[12]

As a global principle the screening of drugs based on their duration of action does not rank them in order of potency but simply reflects the elimination kinetics. It is erroneous to assume that a series of experimental compounds have identical elimination rates.

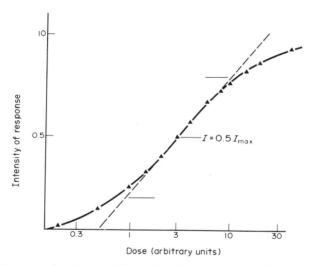

Figure 9 Typical log(dose)–response function according to the relationship $I/I_{max} = X_1/(X_{50} + X_1)$: $X_{50} = 3$ units. The plot is essentially linear in the range 20 to 80% of the maximal (reproduced from ref. 92 by permission of S. Karger AG)

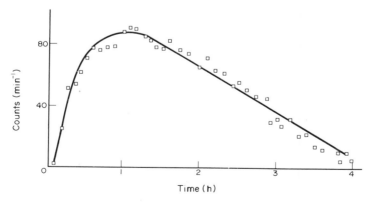

Figure 10 Time course of central nervous system response (locomotor activity measured in counts min^{-1}) after intramuscular administration of dexamphetamine sulfate to rats (5.62 mg kg^{-1}); the effect of the drug declines at a constant rate during the postabsorptive phase (reproduced from ref. 93 by permission of North Holland Publishing Co.)

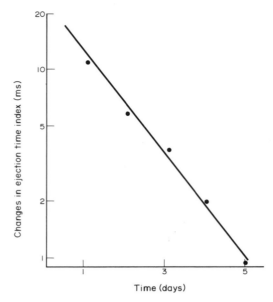

Figure 11 Time course of cardiac response (change in ejection time index) after intravenous administration of digoxin. Exponential decline of response is also observed with other cardiac glycosides, *e.g.* ouabain, deslanoside C and digitoxin; see also Figure 16 (reproduced from ref. 94 by permission of Dun-Donnelly Publishing Co.)

24.6.2.2 Two-compartment Models

In general, observed plasma drug concentrations do not exhibit single exponential decay. Observed functions are frequently biphasic or triphasic and eventually for sufficiently large values of time decay as single exponential functions, *i.e.* the limiting behaviour is monoexponential. Examples of such functions are illustrated in Figures 12 and 13.

The streptomycin data have been deliberately chosen to illustrate the fact that the acceptance of a single exponential decay was untenable even in the early days of pharmacokinetic theory (*ca.* 1946).

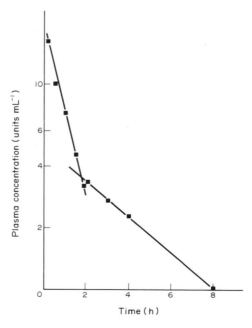

Figure 12 Plasma concentration of streptomycin after intravenous administration (100 000 units) to a patient with pulmonary tuberculosis (reproduced from ref. 95 by permission of the American Medical Association)

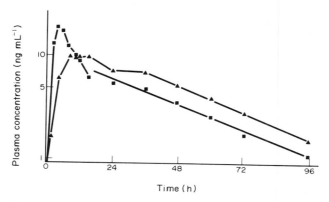

Figure 13   Mean amitriptyline (■) and nortriptyline (▲) plasma concentrations in four subjects after a single oral dose of 75 mg amitriptyline hydrochloride (reproduced from ref. 96 by permission of Hall Associates)

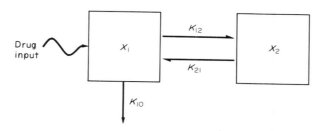

Figure 14   A two-compartment system

An interpretation of these data, given in 1960, is that they probably reflect two concurrent exponential processes. One of these is the diffusion of the drug into the tissue spaces, which causes a rapid fall in the blood level in the early stages. The other is the gradual elimination of the drug by inactivation and excretion.[13] However, it was as late as 1968 when the shortcomings of conceiving the body as exhibiting the properties of a single compartment were clearly expounded.[14]

In compartmental terms a biexponential plasma function can be represented by Figure 14, where $X_1(t)$ represents the mass of drug in the blood and all the tissues that are in rapid equilibrium with the blood, and $X_2(t)$ represents the mass of drug in poorly perfused tissues or tissues that slowly equilibriate with the drug in blood. The constants K_{12} and K_{21} represent the rate of drug exchange between the two distinct compartments and the ratio K_{12}/K_{21} can be regarded as a partition coefficient between the two compartments. Irreversible loss of drug from the system *via* metabolism and excretion is represented by the first-order rate constant K_{10}.

24.6.2.2.1   *Mathematical analysis*

Mathematical analysis of a two-compartment system is in itself instructive in that it illustrates a duality between all biexponential functions of the form

$$X_1(t) = Ae^{-\alpha t} + Be^{-\beta t} \qquad A, B, \alpha, \beta > 0 \qquad (27)$$

and all two-compartment models. All two-compartment models generate a function composed of the summation of two positive exponential functions. Any arbitrary positive biexponential function can be used to generate a compartmental system. Consequently, the ability to fit plasma concentration data using a two-compartment system cannot be regarded as evidence for the validity of the mathematical model.

The set of first-order differential equations for a two-compartment system is

$$dX_1/dt = -(K_{10} + K_{12})X_1 + K_{21}X_2 \qquad (28)$$

$$dX_2/dt = K_{12}X_1 - K_{21}X_2 \qquad (29)$$

With impulse input into compartment 1 at $t = 0$, the initial conditions are $X_1(0) = D$ and $X_2(0) = 0$, where D is the dose of drug.

Using standard Laplace transform methods to linearize the differential equations,[15,16] the set of equations become in Laplace space

$$sx_1 - D = -(K_{10} + K_{12})x_1 + K_{21}x_2 \tag{30}$$

$$sx_2 = K_{12}x_1 - K_{21}x_2 \tag{31}$$

or in matrix notation

$$\begin{pmatrix} s + K_{10} + K_{12} & -K_{21} \\ -K_{12} & s + K_{21} \end{pmatrix} \begin{pmatrix} x_1 \\ x_2 \end{pmatrix} = \begin{pmatrix} D \\ 0 \end{pmatrix} \tag{32}$$

where x_1 and x_2 are the Laplace transforms of $X_1(t)$ and $X_2(t)$ respectively and s is the Laplace variable. Solving the linear set of equations for x_1 and x_2 then gives

$$x_1 = \frac{D(s + K_{21})}{(s + K_{10} + K_{12})(s + K_{21}) - K_{12}K_{21}} \equiv \frac{P(s)}{Q(s)} = \frac{D(s + K_{21})}{(s + \alpha)(s + \beta)} \tag{33}$$

$$x_2 = \frac{DK_{12}}{Q(s)} = \frac{DK_{12}}{(s + \alpha)(s + \beta)} \tag{34}$$

where $-\alpha$ and $-\beta$ are the roots of $Q(s)$.

In general $Q(s)$ could have a single pair of complex conjugate roots, two distinct real roots or a single real root of multiplicity two. The two roots are always real and distinct, which follows from the construction of a root-location diagram.

For $s = 0$ the value of $Q(s)$ is $K_{10}K_{21}$ and for all $s > 0$ the functional value of $Q(s)$ is positive. For $s = -(K_{10} + K_{12})$ or $s = -K_{21}$ the functional value of $Q(s)$ is negative. However, for sufficiently large negative values of s the polynomial will become positive. Consequently the locus of $Q(s)$ must intersect the axis twice, *i.e.* there are two real roots. The first intersection is located between zero and the negative of the smallest modular value $|K_{21}|$ or $|K_{10} + K_{12}|$. The second intersection lies between the negative of the largest modular value $|K_{21}|$ or $|K_{10} + K_{12}|$ and $-(\infty)$ (see Figure 15).

Partial fraction expansion of the equation for x_1 gives

$$x_1 = \frac{D(s + K_{21})}{(s + \alpha)(s + \beta)} = \frac{DA}{(s + \alpha)} + \frac{DB}{(s + \beta)} \tag{35}$$

where

$$A = \frac{(K_{21} - \alpha)}{(\beta - \alpha)} \quad \text{and} \quad B = \frac{(K_{21} - \beta)}{(\alpha - \beta)} \tag{36}$$

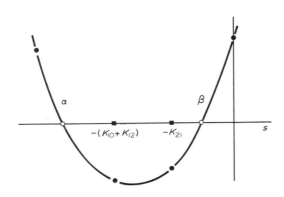

● Values of $Q(s)$

○ Roots of $Q(s)$

Figure 15   Root location diagram for $Q(s)$

Inspection of Figure 15 will verify that both A and B are strictly positive. Inverse transformation generates

$$X_1(t) = DAe^{-\alpha t} + DBe^{-\beta t} \tag{37}$$

Thus all two-compartment models generate a positive biexponential function. The inverse is equally valid. Consider an arbitrary positive biexponential function representing a plasma concentration function, say $C(t)$

$$C(t) = D\hat{A}e^{-\alpha t} + D\hat{B}e^{-\beta t} \qquad \hat{A}, \hat{B}, \alpha, \beta, D > 0 \tag{38}$$

Taking Laplace transforms

$$c(s) = \frac{D\hat{A}}{(s + \alpha)} + \frac{D\hat{B}}{(s + \beta)} = \frac{\left(s + \dfrac{\hat{A}\beta + \hat{B}\beta}{\hat{A} + \hat{B}}\right)(\hat{A} + \hat{B})D}{(s + \alpha)(s + \beta)} \tag{39}$$

Set $(\hat{A} + \hat{B})$ to the volume of compartment 1, V_1, and $(\hat{A}\beta + \hat{B}\alpha)/(\hat{A} + \hat{B})$ to K_{21}. Realizing that $X_1 = CV_1$ then the concentration function is transformed into a compartmental system. Consequently, a compartmental system is just another way of representing an observed biexponential function, the existence of which is determined by the initial abstract assumption of the biexponential nature of the observed concentration dataset.

It is immediately observable that the half-life of the slowest exponential function, $t_{1/2}\beta$, is not equivalent to the rate of loss, K_{10}, from the system. The coefficient β is a root of $Q(s)$ and its value is inextricably linked with the values of all the other rate constants in the model. The latter is diametrically opposed to the impression gained by conceiving the body as a single compartment. However, for the particular model considered

$$\int_0^\infty X_1(t)\,dt = \frac{D}{K_{10}} \tag{40}$$

and

$$\int_0^\infty X_2(t)\,dt = \frac{K_{12}}{K_{21}}\frac{D}{K_{10}} \tag{41}$$

which demonstrates that the total exposures to the drug in compartments 1 and 2 are inversely related to the elimination rate of the drug, K_{10}. The function $X_2(t)$ is a rise and decay biexponential given by

$$X_2(t) = \frac{DK_{21}}{(\alpha - \beta)}e^{-\alpha t} - \frac{DK_{21}}{(\alpha - \beta)}e^{-\beta t} \tag{42}$$

In more general terms, $X_2(t)$ is the convolution of $X_1(t)$ with a second exponential function. Since

$$x_2 = \frac{K_{12}}{(s + K_{21})}x_1 \tag{43}$$

and the product of two Laplace functions in the time domain is their convolution, then

$$X_2(t) = \int_0^t K_{12}e^{-K_{21}\tau}X_1(t - \tau)\,d\tau \equiv K_{12}e^{-K_{21}t} * X_1(t) \tag{44}$$

where the symbol $*$ represents the convolution operator. The trajectory of $X_1(t)$ *vs.* $X_2(t)$ is obviously non-linear, and if the pharmacological effect is associated with receptors not in rapid equilibrium with the plasma, attempted correlations will fail.

24.6.2.2.2 *Pharmacological response*

There is ample evidence to demonstrate that some pharmacological effects cannot be considered to be associated with receptors located in the compartment composed of tissues in rapid equilibration with the blood. For example, the reduction in ventricular rate with cardiac glycosides takes approximately 12 h to achieve maximal reduction (see Figure 16) despite the fact that the heart is perfused with the entire circulatory volume of blood. Similarly, the pain relief induced by morphine is not directly related to plasma concentrations (see Figure 17).

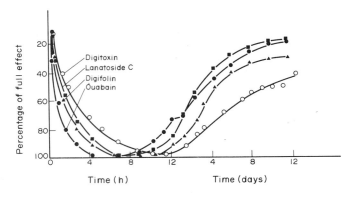

Figure 16   Onset and duration of action, reduction of ventricular rate percentage of full effect, of cardiac glycosides following intravenous injection (reproduced from ref. 96 by permission of Hall Associates)

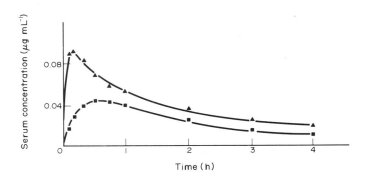

Figure 17   Relationship between serum morphine concentrations (▲) and degree of analgesia (■) in patients receiving intramuscular morphine, 0.14 mg kg^{-1} (reproduced from ref. 98 by permission of C. V. Mosby and Co.)

These time delays between pharmacological effects and plasma drug concentrations most probably reflect a slow equilibrium between the plasma and the effect tissue (*i.e.* heart and brain respectively).

24.6.2.3   Multiple-compartment Models

There is no guarantee that an observed biexponential plasma drug concentration function distinguishes all the processes involved in drug distribution. In general, when additional body tissues are sampled, the need to include further compartments in the mathematical model is revealed (see Figures 18 and 19).

For thiopentone, the mass of drug in the highly perfused liver organ reflects the plasma drug concentrations, presumably because of rapid uptake of thiopentone by the liver. In contrast, the mass of thiopentone in the muscle exhibits a typical time function associated with a second compartment. Concentrations of the drug in fat indicate the necessity for a three-compartment mathematical model despite the fact that a two-compartment model is adequate to define the biexponential plasma concentration function. On this basis a three-compartment system with elimination from the central compartment seems adequate. Central compartment elimination assumes that the liver and kidney are in rapid equilibrium with the blood and that these body organs are included within the central compartment. Even this assumption can be challenged.

For the polar drug phenol red, the plasma drug concentration function does not parallel the concentrations of the drug in the liver and kidney. The phenol red concentrations in the liver and plasma are typical of a two-compartment model with the liver as a component of the second compartment. However, the liver and kidney concentrations of phenol red are not parallel. In fact, the kidney concentration function is triexponential and not biexponential. Consequently a third compartment is required to accommodate the kidney.

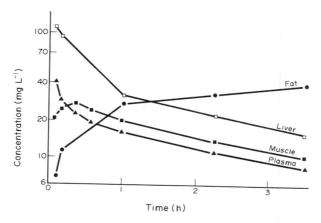

Figure 18 Thiopentone concentrations in various tissues of a dog after intravenous administration of 25 mg kg^{-1} (reproduced from ref. 99 by permission of the Federation of American Societies for Experimental Biology)

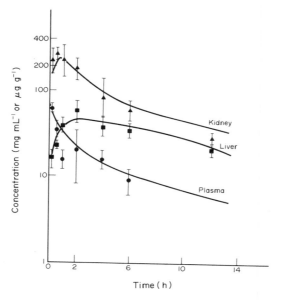

Figure 19 Tissue and plasma concentrations of phenol red in the dogfish shark after intravenous administration (reproduced from ref. 100 by permission of Plenum Press)

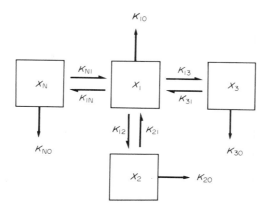

Figure 20 General mammillary model

In view of the latter observation a more general mathematical model is required. A general N-compartment mammillary model has, in most respects, the desired properties (see Figure 20).

24.6.2.3.1 *Number of observable exponential terms*

A general N-compartment mammillary model can define a positive function for $X_1(t)$ composed of the summation of N positive exponential functions

$$X_1(t) = \sum_{i=1}^{N} A_i e^{-\alpha_i t} \tag{45}$$

The observability of an N-exponential function for $X_1(t)$ is not a prerequisite for such a model. Indeed, the optimal number of exponential terms necessary to define the plasma concentrations of sodium amobarbital after the same intravenous dose to different volunteers varies from one to three (see Table 2).

The variations are directly attributable to the relative magnitudes of the exponential coefficients in the various subjects. The variation does not imply that different subjects exhibit different disposition kinetics. Rather, the collapse or loss of exponential terms reflects the fact that when two exponential coefficients are less than a factor of two different the resulting function cannot be resolved into its individual components.[17]

The problem is further compounded by multiple minima in the least-squares surface used in the optimization of an observed dataset. Although the mathematical solutions of classical pharmacokinetic models are given as exponential sums, it is necessary to optimize the parameters of these equations to fit some given dataset, f_i^{obs}, of concentrations which is subject to an unknown random error. Usually optimization is taken to mean the generation of a set of functional values from the defining equations, f_i^{cal}, such that the generated set is close to the observed set. Frequently closeness is defined by the least-square criterion

$$F(x) = \sum_{i=1} (f_i^{obs} - f_i^{cal})^2 \tag{46}$$

and the best set of f_i^{cal} values is that set which minimizes $F(x)$. The surface of $F(x)$ will invariably exhibit multiple minima and it may not be possible to locate the global minimum.

Vaughan and Walton[18] have described in detail the mathematical principles involved in optimization algorithms. An illustration of the inability to define a global minimum and hence the requisite number of exponential terms to define a given set of concentration data has also been provided by them.[18]

Least-square fits for pancuronium plasma concentrations using three different algorithms are given in Table 3. All three algorithms give different results for the biexponential function (Table 3, part a) indicating that none has located a global minimum. Increasing the number of variables to a triexponential function should improve the least-square values or result in a collapse to a biexponential function. However, two of the algorithms gave worst least-square values and no algorithm has located a global minimum (Table 3, part b). Consequently it is not possible to state with confidence that the function is either biexponential or triexponential. For another dataset (Table 3, part c) all three algorithms generate the same least-square value and some confidence can be placed on the assumption of a biexponential fit to the observed data.

Vanishing exponential terms in an N-compartment mammillary model were predicted by Shaney and co-workers,[19] Wagner,[20] and Ronfeld and Benet.[21] However, an N-compartment mammillary

Table 2 Computer Fitting of Amobarbital Concentration Data to Exponential Functions[16]

Optimum number of exponential terms	Number of subjects	Percentage of subjects
1	5	18
2	16	57
3	7	25
Total	28	100

Table 3 Least-squares Fits to Pancuronium Concentrations after Intravenous Injection[17]

Exponential Parameters[a]	Fletcher's VM	Algorithm[b] NONLIN	FUNFIT
(a) A_1	2.372	1.983	10.66
m_1	-0.200	-0.237	-0.400
A_2	0.568	0.659	0.620
m_2	-0.00619	-0.00745	-0.00686
$F_{\text{least squares}}$	3.193×10^{-2}	5.102×10^{-2}	2.199×10^{-2}
(b) A_1	2.372	1.452	4.924
m_1	-0.199	-0.287	-0.303
A_2	0.395	0.455	0.0408
m_2	-0.00687	-0.0302	-0.169
A_3	0.176	0.399	0.601
m_3	-0.00502	-0.00417	-0.00663
$F_{\text{least squares}}$	3.191×10^{-2}	7.131×10^{-2}	2.410×10^{-2}
(c) A_1	0.9464	0.9464	0.9464
m_1	-0.03716	-0.0376	-0.03761
A_2	0.2373	0.2373	0.2374
m_2	-0.00293	-0.002093	-0.002095
$F_{\text{least squares}}$	0.008194	0.008194	0.008194

[a] Parts a and b list bi- and tri-exponential fit to experimental data from subject I.A. respectively. Part c lists biexponential fit to experimental data from another subject.
[b] NONLIN and FUNFIT parameters are taken from ref. 112 and the experimental data from ref. 113.

model can analytically collapse to any number of exponential terms and can ultimately collapse to a biexponential function.[22] All that is required is for the summation of exit rate constants from two compartments to be equal. The latter readily follows from the general solution for the Laplace transforms of $X_1(t)$, *i.e.* $x_1(s)$, in an N-compartment mammillary system after a unit impulse input into compartment 1, *i.e.* the central compartment (equation 47).[15,16]

$$x_1(s) = \frac{\prod_{j=2}^{N}(s + E_j)}{\prod_{j=1}^{N}(s + E_j) - \sum_{\substack{j=2}}^{N} K_{1j}K_{j1}\prod_{\substack{i=2 \\ i \neq j}}^{N}(s + E_i)} \equiv \frac{P(s)}{Q(s)} = \frac{\prod_{j=2}^{m-1}(s + E_j)}{\prod_{i=1}^{m}(s + \alpha_i)} \qquad m \leq N \qquad (47)$$

where $E_j = (K_{j0} + K_{j1})$. When all the E_j values are distinct each value of $-E_j$ strictly separates two real negative roots of $Q(s)$, and $X_1(t)$ is therefore of the form

$$X_1(t) = \sum_{i=1}^{N} A_i e^{-\alpha_i t} \qquad A_i > 0 \qquad (48)$$

Whenever two $E_j, j > 1$, values are equal there is a pole-zero cancellation in $P(s)$ and $Q(s)$ and the order of the denominator polynomial, $Q(s)$, is decreased by a factor of one. This generates one less real root and hence one less exponential term for $X_1(t)$. For p equal $E_j, j > 1$, values there are $(p-1)$ common factors in $P(s)$ and $Q(s)$ and pole-zero cancellation then generates $N-(p-1)$ exponential terms. For all $E_j, j > 1$, values equal a biexponential function results.

24.6.2.3.2 First-pass effect

Regarding the body as an N-compartment system in which the eliminating organs, *e.g.* the liver, are considered as distinct compartments the effect of different routes of drug administration on the function $X_1(t)$ can be observed. For example, an intravenous dose may be considered to enter the central compartment, whereas an oral dose is delivered to the liver, say compartment 2, in which drug metabolism occurs. Assuming complete absorption of an oral drug dose then

$$\frac{\int_0^\infty X_1(t)\,dt_{\text{oral}}}{\int_0^\infty X_1(t)\,dt_{\text{iv}}} = \frac{\begin{array}{c}\text{Area of drug concentration function}\\\text{(blood) after oral dose}\\\hline \text{Area of drug concentration function}\\\text{(blood) after intravenous dose}\end{array}}{} = \frac{K_{21}}{(K_{21} + K_{20})} = \delta < 1 \qquad (49)$$

when the oral and intravenous doses are equivalent.

The ratio of the oral area to the intravenous area is termed the 'first-pass' effect.[15,16,23-26] Factors affecting the magnitude of the 'first pass' effect have been reviewed.[27,28]

The function δ is the fraction of an oral drug dose that traverses the liver intact and reaches the central compartment. Subsequently it will be demonstrated that the area relationship is globally valid and independent of the assumption of a mammillary model. Further, the result is also independent of the assumption that the liver can be represented as a single homogeneous pool or compartment.

24.6.2.3.3 *Pharmacological response*

When the pharmacological effect originates in a tissue that slowly equilibrates with the plasma the 'biophase' drug concentrations will be related to $X_1(t)$ by a convolution integral in a mammillary system. This relationship may be represented as

Plasma drug concentrations ↝ Biophase concentrations ↝ Pharmacological response

$$C_1(t) \qquad\qquad C_1(t) * k_{1i}e^{-k_{i1}t} \qquad\qquad\qquad\qquad (50)$$

The pharmacological response may be linearly related to the biophase concentration of drug or related by the Langmuir binding equation. Frequently the derivative of a Langmuir binding equation is too small to account for the rate of change of a pharmacological response with respect to the concentration and is replaced by the Hill equation (equation 51).

$$I = \frac{I_{\text{max}}C^N}{C_{50} + C^N} \qquad\qquad (51)$$

An interesting example illustrating these principles is the relationship between the plasma concentrations of (+)-tubocurarine and the neuromuscular blocking action of the drug (see Figure 21).

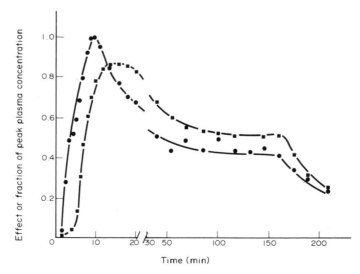

Figure 21 Best fits of plasma concentrations of (+)-tubocurarine (●) and neuromuscular blocking effect (■) after intravenous infusion of (+)-tubocurarine to a patient (reproduced from ref. 101 by permission of C. V. Mosby and Co.)

In this model the plasma concentrations are assumed to be generated by a two-compartment model with a stepped intravenous infusion. The effect compartment is a single isolated compartment driven by the plasma compartment and the biological response generated by binding to a finite receptor pool. A separate effect compartment reversibly connected to the plasma compartment is an unnecessary abstraction since a two-compartment plasma concentration model only represents the observability of a multiple-compartment system with each compartment reversibly connected to a central compartment.

24.6.3 GLOBAL PROPERTIES OF GENERAL COMPARTMENTAL SYSTEMS

24.6.3.1 Matrix Representation

Consider a system whose state at any time is completely defined by N functions $X_1(t), \ldots, X_N(t)$. Assuming that the rate of change with respect to time of the state variables only depend on the functional values of the state variables at time t, then a set of differential equations of the form

$$dX_i/dt = f(X_1, X_2, X_3 \ldots X_n) \qquad i \in 1, 2, \ldots, N \tag{52}$$

with an associated set of initial conditions

$$X_i(0) = C_i \qquad i \in 1, 2, \ldots, N \tag{53}$$

define the system's state.

By assumption, the relationship between the variables does not involve functions of time. Consequently the set of differential equations can be expressed in matrix form as

$$\dot{x} = A_1 x + A_2 x^2 + A_3 x^3 \ldots \tag{54}$$

where the matrices A are constant coefficient matrices and x the vector of state variables with first derivative vector $\dot{x}$. Ignoring the higher order terms the system's behaviour is determined by a linear set of differential equations with constant coefficients

$$\dot{x} = Ax \qquad x(0) = c \tag{55}$$

where $A \equiv A_1$. Reasonable assumptions concerning drug distribution and elimination within the body are (i) for any drug input into the body, all drug concentrations at any point within the body are non-negative and (ii) for a finite drug input into the body, the processes of drug metabolism and elimination eventually guarantee the condition that for sufficiently large time all concentrations tend towards zero.

Placing a non-negative boundary condition on the state variables, $X_i(t)$, for $t \geq 0$ whenever the initial conditions are non-negative generates the necessary and sufficient condition that all the off-diagonal elements of matrix A are non-negative. Further, a necessary and sufficient condition for all the solutions $X_i(t)$ to approach zero as $t \to \infty$ is that all the roots of the polynomial $|A - \lambda I|$ have negative real parts. Coupling these two constraints on the elements of matrix A, generates additional constraints on the diagonal elements of matrix A. These constraints are that the diagonal elements of matrix A are negative and that the modular value of the diagonal elements must be greater than or equal to the summation of the other column elements. Additionally since $X_i(t) = 0$ for $t \to \infty$ at least one column summation must be an inequality.[29,30] Consequently, matrix A is of the type

$$\begin{pmatrix} -a_{11} & a_{12} & a_{13} \ldots a_{1N} \\ a_{21} & -a_{22} & a_{23} \ldots a_{2N} \\ a_{31} & a_{32} & -a_{33} \ldots a_{3N} \\ \vdots & & \\ a_{N1} & a_{N2} & a_{N3} \ldots -a_{NN} \end{pmatrix} \tag{56}$$

where $a_{ij} \geq 0$ and

$$a_{ii} \geq \sum_{\substack{j=1 \\ j \neq 1}}^{N} a_{ji} \qquad i \in 1, 2, \ldots, N \tag{57}$$

with inequality for at least one i.

The latter matrix is precisely that of an N-compartment system when the elements a_{ij} are replaced by K_{ji}, the structure of which has been generated from quite general considerations that are unconnected with any hypothetical statements or concepts concerning drug distribution or elimination. In this respect a general compartmental system for drug distribution and elimination is a natural representation provided the assumptions of first-order relationships and time invariance are valid.

A general compartmental system has no restricting boundary conditions imposed such as the interconnection of compartments or the sites of drug elimination. Also, there is no suggestion of either rapidly equilibriating tissue spaces or of well-perfused tissues. For precisely the latter reasons global information concerning pharmacokinetic behaviour are best derived from the general system matrix A, thus avoiding the imposition of topological constraints that could potentially generate a specific result rather than a global result.

24.6.3.2 Solution Set for $\dot{x} = Ax$

Given a general compartmental matrix and the defining relationships

$$\dot{x} = Ax \qquad x(0) = c \quad \text{with} \quad c_i \geqslant 0 \tag{58}$$

the solution for a particular $X_i(t)$ is obtainable by taking Laplace transforms and applying Cramer's rule, whence

$$X_i(t) = \mathscr{L}^{-1}\left[\frac{|sI - B|}{|sI - A|} \equiv \frac{P(s)}{Q(s)}\right] \tag{59}$$

where s is the Laplace variable and I the identity matrix. The determinant $|sI - A|$ is a polynomial in s of degree N. Matrix $(sI - B)$ is obtained from matrix $(sI - A)$ by replacing column i with the initial condition vector. Its determinant $|sI - B|$ is a polynomial in s of degree $N - 1$.

Factorizing $Q(s)$ in terms of its roots gives

$$Q(s) = (s + \lambda_1)^{k_1}(s + \lambda_2)^{k_2}\ldots(s + \lambda_m)^{k_m} \tag{60}$$

where $N = k_1 + k_2 \ldots k_m$ and $-\lambda_i$ $(i = 1, 2, \ldots, m)$ are the roots of the polynomial $Q(s)$. Since all the roots of $Q(s)$ must have negative real parts, for $X_i(t) \geq 0$, complex roots must occur as complex conjugate pairs. Suppose $\lambda_1, \lambda_2 \ldots, \lambda_q$ are real and the remaining λ are complex, with real parts α and imaginary parts β. Resolving $P(s)/Q(s)$ into partial fractions and taking the inverse transformation the real roots will contribute the terms

$$\sum_{i=1}^{m}\left(\sum_{j=1}^{k_i} \frac{A_{ij}e^{-\lambda_i t}t^{(k_i - j)}}{(k_i - j)!}\right) \tag{61}$$

while the complex conjugate pairs of roots will contribute

$$\sum_{\substack{i=q+r \\ r=1,3,5\ldots}}^{m-1} \frac{F_i(\beta_i, t)e^{-\alpha_i t}t^{(k_i - j)}}{(k_i - j)!} \tag{62}$$

where $F_i(\beta_i, t)$ is a periodic sinusoidal time function.

The general solution is certainly not restricted to a summation of exponential functions and the possibility exists that the final or limiting decay function could be an oscillatory function dampened by a negative exponential, as opposed to a positive exponential decay.

24.6.3.3 Limiting Exponential Decay

A necessary condition for the existence of limiting exponential decay in a general N-compartment system is the existence of a real distinct root of $|sI - A|$ which has a modular value less than the modular values of all the remaining roots. Such a root of $Q(s)$ always exists and there can be no pole zero cancellation of this root in $P(s)/Q(s)$.

The theorems of Perron on a positive matrix which occupies a central position in mathematical economics are applicable to the compartmental matrix A. The fundamental result of Perron[31] is that if B is a positive matrix, there is a unique characteristic root of B, $\lambda(B)$, which has greatest absolute value.

For simplicity, assume all the off-diagonal elements of matrix A are strictly positive, in which case matrix B is a positive matrix, equation (63)

$$A + \begin{pmatrix} \varepsilon & 0 & 0 \ldots 0 \\ 0 & \varepsilon & 0 \ldots 0 \\ 0 & 0 & \varepsilon \ldots 0 \\ 0 & 0 & 0 \ldots \varepsilon \end{pmatrix} \equiv B \tag{63}$$

whenever $\varepsilon > \max|-a_{ii}|$. Since the roots of $|sI - B|$ are $\lambda_i + \varepsilon$ where λ_i is a root of $|sI - A|$ and matrix B must have a characteristic root with greatest absolute value, it follows that matrix A must have a characteristic root of smallest absolute value which is unique, *i.e.* a simple distinct real root. In the much more involved case where only the condition $a_{ij} \geqq 0$ is imposed, the Perron result is still valid.[32]

Recalling that $Q(s)$ has only roots with negative real parts and the existence of the Perron root, the slowest transitional function for any solution, $X_i(t)$, with any set of positive initial conditions is a single exponential function of the form $Ae^{-\lambda t}$, with A and $\lambda > 0$. Applying the latter to observable drug concentration functions in the body after a finite intravenous drug dose, all tissue concentrations of the drug will eventually decay away as single exponential functions, and all the limiting decay functions have exactly the same exponential coefficient, indicating a limiting parallel decline of all the corresponding log concentration/mass functions. The data for thiopentone (Figure 18) and phenol red (Figure 19) illustrate the latter phenomena. For thiopentone, the liver, muscle and plasma concentrations decline in a parallel manner. However, the concentrations of thiopentone in fat are still increasing at 3.5 h. This is simply due to an insufficiency in the duration of the experiment. Ultimately the concentrations in fat must decline and it is anticipated that this limiting decay will parallel the limiting decline observed in all other tissues.

The attainment of thiopentone concentrations in fat tissue that are greater that those in plasma, during the limiting decay period, results from the affinity of the drug for fat tissues. The latter can be established as described below.

In simplistic terms consider the fat compartment or compartments to be reversibly connected to the plasma phase. When the plasma phase is declining as a single exponential the mass of drug in fat, $X_j(t)$, is defined by

$$\dot{X}_j = K_{1j}Ae^{-\lambda t} - K_{j1}X_j \qquad X_j(0) = 0 \tag{64}$$

The necessary condition for a positive derivative $\dot{X}_j(t)$ is

$$Ae^{-\lambda t} > \frac{K_{j1}}{K_{1j}}X_j \tag{65}$$

which is achievable if $K_{1j} \gg K_{j1}$. Also

$$\int_0^\infty X_j(t)\,dt = \frac{K_{1j}}{K_{j1}}\int_0^\infty X_1(t)\,dt \tag{66}$$

where X_1 is the plasma space mass-time function. With $K_{1j} \gg K_{j1}$, and

$$\int_0^\infty X_j(t)\,dt > \int_0^\infty X_1(t)\,dt \tag{67}$$

the function $X_j(t)$ will intersect the function $X_1(t)$ at some time point when $X_j(t)$ is strictly increasing.

No special significance can be attributed to the limiting exponential function's half-life. It simply reflects the dynamics of the system which dictates that no part of a connected system of compartments can decay at a rate greater than the smallest modular value of the diagonal elements in matrix A. The origin of the Perron roots as a characteristic root of the polynomial $|sI - A|$ inextricably links its value in a complicated manner with all the other elements of matrix A.

One very useful property of the final exponential decay and its commonality to all tissues is that in approximately ten half-lives all concentration functions will be essentially zero. The $t_{1/2}$ metric then determines the approximate dwelling time in the body of a single finite impulse drug dose.

A crucial observation is the fallacy of assuming that once plasma drug concentrations are small or close to detectable limits then all other tissue concentrations are also essentially zero. Drugs can and do accumulate in tissues without accumulation in the plasma space.

24.6.3.4 Linearity and Time Invariance

The response $X_i(t)$ of a general compartmental system to a unit impulse input at some point q is, as before, given by the solution of

$$\dot{x} = Ax \qquad x(0) \geqq 0 \tag{68}$$

Assume, without loss of generality, that $q = 1$ in which case

$$x(0) = \begin{pmatrix} 1 \\ 0 \\ 0 \\ \vdots \\ 0 \end{pmatrix} \tag{69}$$

and the solution is given by

$$x_i(t) = \mathcal{L}^{-1} \frac{-1^{(1+i)}|B_{1i}(s)|}{|sI - A|} \equiv \mathcal{L}^{-1} G_i(s) \tag{70}$$

where $|B_{1i}(s)|$ is the determinant of the resulting matrix obtained by deleting the first row and column i from the matrix $(sI - A)$.

Denoting the individual responses $X_i(t)$ as the characteristic responses of the system to a unit impulse input at point q, say $G_i(t)$, the responses of the system to any drug input function $f_1(t)$ are the solution of

$$\dot{x} = Ax + f \tag{71}$$

with $x(0)$ as a null vector and

$$f = \begin{pmatrix} f_1(t) \\ 0 \\ 0 \\ \vdots \\ 0 \end{pmatrix} \tag{72}$$

The solution in Laplace space is

$$\hat{x}_i(s) = \frac{f_1(s)[-1^{(1+i)}|B_{1i}(s)|]}{|sI - A|} = f_1(s) G_i(s) \tag{73}$$

where $\hat{x}_i(s)$ is the Laplace transform of the response to the input function $f_1(t)$. These responses are the product of the Laplace transforms of the characteristic responses and the Laplace transform of the input function. Since the product of two functions in the Laplace domain corresponds to the convolution of the two functions in the time domain, the solution is given as

$$\hat{X}_i(t) = \int_0^t f_1(t - \tau) G_i(\tau) d\tau \equiv f_1(t) * G_i(t) \tag{74}$$

We may consider $f_1(t)$ to be a summation of functions say $f_1(t) = \hat{f}_1(t) + \hat{f}_2(t) \ldots$, in which case

$$\begin{aligned} \hat{X}_1(t) &= \hat{f}_1(t) * G_i(t) + \hat{f}_2(t) * G_i(t) \ldots \\ &\equiv [\hat{f}_1(t) + \hat{f}_2(t) \ldots] * G_i(t) \end{aligned} \tag{75}$$

With this representation the total response of system $\hat{X}_i(t)$ is the summation of the response produced by the individual input functions and there is no interaction between them save the summation. In general terminology the system is linear to the inputs, *i.e.* the response is a linear combination of the individual responses, and the characteristic responses are linear operators.

A further property of the general compartmental system is time-invariance. Regarding $f_i(t)$ as a time-translated function $\bar{H}(t-\rho)f_1(t-\rho)$ where $\rho > 0$ and $\bar{H}(t-\rho)$ is the unit step function which is zero for $t \le \rho$ and one for $t > \rho$ then this function has the Laplace transform $e^{-s\rho}f_1(s)$. In this case

$$\hat{X}_i(s) = e^{-s\rho}f_1(s)G_i(s) \tag{76}$$

and

$$\hat{X}_i(t) = \bar{H}(t-\rho)\int_0^{t-\rho} f(t-\tau)G_i(\tau)d\tau \tag{77}$$

The response is merely the time-translated response of the system to the unshifted input function. The characteristic responses are therefore static and the system parameters are time invariant, which is of course the same as stating matrix A is a constant coefficient matrix.

The general compartmental matrix therefore represents a linear and time invariant operator. As such it is a particular member of the general class of linear time invariant operators. Its global properties should be evident from the properties of the general operator, in which case the tedious evaluation of determinants is obviated.

24.6.4 GENERAL LINEAR TIME-INVARIANT NON-COMPARTMENTAL SYSTEMS

24.6.4.1 System Analysis

The response of a system, say $Q(t)$, to an input $In(t)$ can be represented by

$$Q(t) = \Phi In(t) \tag{78}$$

where the symbol Φ represents an operator which converts the input function to the response. If

$$\Phi[In_1(t) + In_2(t)\ldots] = \Phi In_1(t) + \Phi In_2(t)\ldots \tag{79}$$

the operator is linear. For a linear operator

$$\Phi[c\,In(t)] = c\Phi In(t) \tag{80}$$

where c is any positive scalar. If the linear operator is also time invariant then

$$\Phi\bar{H}(t-\rho)In(t-\rho) = \bar{H}(t-\rho)Q(t-\rho) \tag{81}$$

The response of a linear time invariant system to any input function can be constructed from the unit impulse response of the system $G(t)$, as described below.

Let the system's response to a unit impulse input at $t = 0$ be $G(t)$. Since the system is linear the response to an impulse of magnitude c is $c \cdot G(t)$ and from the time-invariance property the response to an impulse of magnitude $c_i(\tau)$ occurring at time $t = \tau_i$ is $\bar{H}(t-\tau_i)\cdot c_i(\tau_i)\cdot G(t-\tau_i)$. By virtue of the linearity the system's response to an impulse sequence is

$$\sum_{i-1} \bar{H}(t-\tau_i)c_i(\tau)G(t-\tau_i) \tag{82}$$

Setting $c_i(\tau_i) = In(\tau_i)\Delta\tau$ then the response to an impulse sequence constructed from an input function is

$$\sum_{i=1} \bar{H}(t-\tau_i)In(\tau_i)G(t-\tau_i)\Delta\tau \tag{83}$$

As we pass to the limit $\Delta\tau \to 0$ the approximation to the system's response to a continuous input becomes exact. In the limit this becomes

$$Q(t) = \int_0^t In(\tau)G(t-\tau)d\tau$$

$$\equiv \int_0^t In(t-\tau)G(\tau)d\tau$$

$$\equiv In(t)*G(t) = G(t)*In(t) \tag{84}$$

which is precisely the convolution of the input function and the characteristic response.

The properties of linearity and time invariance are fundamental to most pharmacokinetic principles and applications. Some applications are given below.

24.6.4.2 Bioavailability

In system analysis terms the response of a system to a non-negative input or forcing function is

$$Q(t) = In(t) * G(t) \tag{85}$$

and in Laplace space

$$Q(s) \equiv \int_0^\infty e^{-st} Q(t)\,dt = \int_0^\infty e^{-st} In(t)\,dt \int_0^\infty e^{-st} G(t)\,dt \equiv In(s)\,G(s) \tag{86}$$

Provided the integrals of $In(t)$ and $G(t)$ on $[0, \infty]$ exist, we can let $s \to 0$ which gives

$$\int_0^\infty Q(t)\,dt = \int_0^\infty In(t)\,dt \int_0^\infty G(t)\,dt \tag{87}$$

In terms of pharmacokinetics we may regard $In(t)$ as the rate of drug absorption into the body and $Q(t)$ as the blood or plasma concentration time function of drug that results from the drug input. The function $G(t)$ is then the concentration time function of drug observed in the blood or plasma after a unit impulse input of the drug.

Since $G(t)$ is a positive function which, because of drug metabolism and elimination, tracts to zero for large time, $G(\infty) = 0$, the integral on $[0, \infty]$ of $G(t)$ exists and is a finite number. The integral of $In(t)$ on $[0, t]$ is the amount of drug that has been absorbed at time t. Provided a finite drug dose has been administered the integral of $In(t)$ on $[0, \infty]$ also exists and it is the total amount of drug that is eventually absorbed into the body.

The total amount of drug that is absorbed is frequently substantially less than the administered drug dose. When a suspension of crystalline particles is administered orally the particle size, the polymorphic form and suspending agent determine the rate and extent of drug dissolution. Consequently it is inadvisable to screen a series of compounds for biological activity by administration of oral suspensions without particle size and dissolution control.

The relative bioavailability of two oral formulations can be defined from the linear integral property. Thus if two oral formulations generate the responses $Q_1(t)$ and $Q_2(t)$ respectively then

$$\text{Relative bioavailability} = \frac{\displaystyle\int_0^\infty Q_1(t)\,dt}{\displaystyle\int_0^\infty Q_2(t)\,dt} \cdot 100 \equiv \frac{\displaystyle\int_0^\infty In_1(t)}{\displaystyle\int_0^\infty In_2(t)} \tag{88}$$

When the two formulations have the same stated drug content, the relative bioavailability is the percentage of the drug dose that is absorbed from the first formulation relative to the drug dose absorbed from the second formulation. This is the basis of bioavailability testing.

Numerous production factors, *e.g.* degree of compaction, wetting agents, lubricants and diluents have a pronounced effect on the extent of drug absorption and there are many examples of so-called equivalent products that fail to generate equivalent concentrations of the drug in blood or plasma. Notable examples are products containing digoxin,[33] chloramphenicol[34] and oxytetracycline.[35]

In general only a fraction of an oral drug dose will dissolve and become available for absorption. During absorption some of the drug may be metabolized by enzymes in the gastrointestinal wall and additional metabolism may occur as the absorbed drug traverses the liver for the first time. Comparing the area under the plasma concentration time function after an oral drug dose, D_{oral}, to the area obtained after an intravenous dose defines the absolute availability of an oral drug dose, whence

$$\text{Absolute availability of oral drug dose } D_{\text{oral}} = 100 \frac{\displaystyle\int_0^\infty Q(t)_{\text{oral}}\,dt}{\displaystyle\int_0^\infty Q(t)_{\text{iv}}\,dt} = \frac{f_1 f_2 f_3\, D_{\text{oral}}}{D_{\text{iv}}} 100 \tag{89}$$

where the subscripts oral and iv refer to oral and intravenous administration respectively. The absolute availability is determined by the product of three fractional amounts, *viz.* f_1 the fraction of the administered dose that is absorbed, f_2 the fraction of the absorbed dose that traverses the intestinal wall intact and f_3 the fraction of the drug dose delivered to the liver that traverses this organ intact for the first time.

If the fraction f_1 is principally determined by chemical degradation in the acidic environment of the stomach, *e.g.* penicillins and erythromycin, the use of a poorly soluble salt, *e.g.* erythromycin stearate, or an enteric coated product may suffice to increase the availability. Whenever the fractions f_1 and f_2 control drug availability a possible solution to poor oral availability is to convert the compound chemically into a prodrug.

The use of linear time invariant systems analysis offers a general proof of Dost's law of corresponding areas[36] which was stated as 'the ratio of the area beneath the blood level–time curve, after oral administration to that following intravenous administration of the same dose, is a measure of the absorption of the drug administered'.[37] Without defining the boundary conditions of the system Dost's law is an empirical expectation, but in the context of linear time-invariant systems it becomes a valid assumption.

24.6.4.3 Determination of Absorption–Input Functions

24.6.4.3.1 *Defined inputs*

Provided the intravenous unit impulse function $G(t)$ is known the response to any defined input function is defined by the convolution integral. For example the response to first order input is

$$D K_a e^{-K_a t} * G(t) \tag{90}$$

where D is the dose that is absorbed and K_a the first-order absorption rate constant.

Similarly the response to a zero-order input function of rate K and duration τ is in Laplace space

$$\frac{K}{s}(1 - e^{-\tau s}) G(s) \tag{91}$$

which in the time domain is

$$K \int_0^t G(t)\,dt - \bar{H}(t - \tau) K \int_0^{t-\tau} G(t)\,dt \tag{92}$$

with $\bar{H}(t - \tau)$ as a unit step function which is zero for $t \leq \tau$ and unity for $t > \tau$.

Observed response functions are errant datasets and to obtain best values for the unknown input parameters the problem is cast as a least-squares problem with best values determined by optimization.[18] Crude estimates, or starting values, for optimization methods can be determined by analytical manipulation of the response function as follows: For a zero-order input function the observed response data $R(t_i)$ are defined by

$$R(t_i) = K_i \int_0^{t_i} G(t)\,dt \qquad t_i \leq \tau \tag{93}$$

which defines a set of estimated values, K_i values, for the unknown parameter K and the mean value defines a suitable initial value.

For first-order absorption the value of K_a can be derived from the limiting exponential decays of $G(t)$ and $R(t)$. For sufficiently large time the limiting decays are

$$G(t) \to A e^{-\alpha t} \qquad A, \alpha > 0 \tag{94}$$

and

$$R(t) \to \frac{D K_a}{(K_a - \alpha)} A e^{-\alpha t} \equiv \hat{A} e^{-\alpha t} \qquad K_a > \alpha \tag{95}$$

since

$$D = \frac{\displaystyle\int_0^\infty R(t)\,dt}{\displaystyle\int_0^\infty G(t)\,dt} \tag{96}$$

an estimate of D is obtainable by area analysis of the response function. Also an estimate of α is defined by the terminal half-life of the two functions, $\alpha = \ln 2/t_{1/2}$. Thus

$$\frac{1}{K_a} = \frac{\hat{A} - AD}{\hat{A}\alpha} \tag{97}$$

Whenever $\alpha > K_a$ the limiting decay of $R(t)$ is determined by K_a. In this case $\ln R(t)$ and $\ln G(t)$ are divergent linear functions for sufficiently large values of t and K_a is directly obtained from the limiting half-life value of $R(t)$.[38,39]

In general the use of rigidly defined input functions is unduly restrictive and inappropriate. Drug release from conventional tablets and suppositories is rarely first-order and sustained release products seldom release the drug by a strict zero-order process.

The preferred approach is to derive an approximate input function by numerical deconvolution of the response function, $R(t)$, and the characteristic response, $G(t)$.

24.6.4.3.2 Wagner–Nelson method

When $G(t)$ is a single exponential function numerical deconvolution is particularly simple and encompasses the Wagner–Nelson method for the determination of the cumulative amount of drug absorbed.[40] In this particular case

$$R(t) = \text{In}(t) * Ae^{-\alpha t} \tag{98}$$

and

$$R(s) = \text{In}(s)\frac{A}{s + \alpha} \tag{99}$$

which gives

$$\text{In}(s) = \frac{R(s)s}{A} + \frac{R(s)\alpha}{A} \tag{100}$$

Since the derivative of a function, $\dot{F}(t)$, in Laplace space is $sF(s) - F(+0)$ and $R(+0) = 0$ the input function is given on inversion as

$$\text{In}(t) = \frac{\dot{R}(t)}{A} + \frac{R(t)\alpha}{A} \tag{101}$$

Approximating the dataset $R(t_i)$ by a polynomial or a cubic spline will then generate an approximate input function. Further

$$\frac{\text{In}(s)}{s} = \frac{R(s)}{A} + \frac{R(s)\alpha}{sA} \tag{102}$$

which on inversion gives the cumulative drug input

$$\int_0^\infty \text{In}(t)\,dt = \frac{R(t)}{A} + \frac{\alpha}{A}\int_0^\infty R(t)\,dt \tag{103}$$

The latter is analogous to the Wagner–Nelson which was originally derived from mass-balance considerations within a single compartmental disposition function. Mass-balance requires

Total amount of drug absorbed at time t ≡ Amount eliminated at time t + Amount in the body at time t (104)

For a single compartmental system with volume V and elimination rate K_{10} the mass balance equation becomes

Total amount of drug absorbed at time t = $K_{10}V\int_0^t C(t)\,dt + VC(t)$ (105)

Noting that $\alpha \equiv K_{10}$ and $1/A \equiv V$ the Wagner–Nelson method is cast as a deconvolution method.

24.6.4.3.3 *Loo–Riegelman method*[41]

For a two-compartment disposition function with elimination from the central compartmental mass balance considerations result in the Loo–Riegelman expression for the amount of drug absorbed at time t_i. The resulting equations can be written as

$$A_{abs}(t_i) = V_1 C(t_i) + K_{10} V_1 \int_0^t C(t) dt + X_2(t_i) \tag{106}$$

where V_1 is the volume of the central compartment and $C(t_i)$ are the observed concentrations. To estimate the amount of drug in compartment two, $X_2(t_i)$, a linear approximation for $C(t)$ over each data interval is assumed, and $X_2(t_i)$ is given by

$$X_2(t_i) = X_2(t_{i-1}) e^{-K_{21} \Delta t} + V_1 \frac{K_{12}}{K_{21}} C(t_{i-1})[1 - e^{K_{21} \Delta t}] + V_1 \frac{K_{12}}{(K_{21})^2} \frac{[C(t_i) - C(t_{i-1})]}{\Delta t} (e^{-K_{21} \Delta t} + K_{21} \Delta t - 1)$$

$$\tag{107}$$

where $\Delta t = t_i - t_{i-1}$ and K_{12} and K_{21} are the first-order rate constants for drug transfer from compartment 1 to 2 and from compartment 2 to 1 respectively. $C(t_i)$ is the observed concentration at t_i. Each $X_2(t_i)$ is determined from the previous value $X_2(t_{i-1})$, with $X_2(0) = 0$.

Paradoxically the Loo–Riegelman method yields the correct $A_{abs}(t_i)$ functions even when the incorrect two compartmental model is used to generate the observed dataset.[42-44] However, the Loo–Riegelman method is a particular case of a general deconvolution algorithm. Under the assumptions that the characteristic response, $G(t)$, is a summation of positive exponential functions and the response function is approximated, by linear interpolation, as a trapezoidal function a general deconvolution algorithm can be specified.[45] This linear system representation depends on the functional form of the characteristic response but not on any compartmental representation. Realizing that all compartmental systems can be transformed to a compartmental system with elimination from the central compartment and with the same characteristic response the paradox of the original derivation vanishes.[45,46]

A notable application of the Loo–Riegelman method is the determination of the cumulative absorption of griseofulvin after oral administration (see Figures 22 and 23). The prolonged absorption of griseofulvin over a period of 24 h and the variable extent of total absorption (approximately 20–60% of the administered dose), suggest poor dissolution of the drug from its crystalline state despite the fact that a micronized powder was administered.

Formulation of griseofulvin as molecular dispersions in water-soluble materials has eliminated the variable absorption of the drug and has permitted a 50% dose reduction.

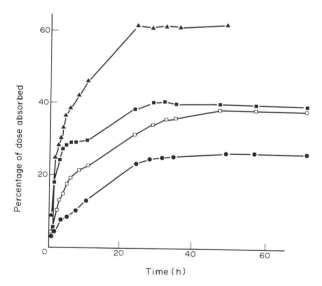

Figure 22 Cumulative percent of administered dose of griseofulvin absorbed following oral administration, 0.5 g micronized griseofulvin, to four subjects (reproduced from ref. 102 by permission of the American Pharmaceutical Association)

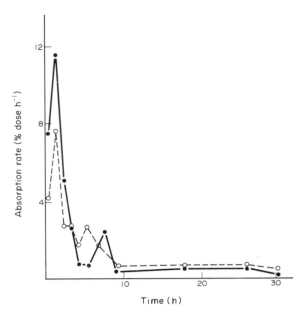

Figure 23 Rate of absorption, *i.e.* input function of griseofulvin after oral administration, 0.5 g micronized griseofulvin, to each of two males (reproduced from ref. 102 by permission of the American Pharmaceutical Association)

24.6.5 DECONVOLUTION ALGORITHMS INDEPENDENT OF THE FUNCTIONAL FORM OF THE CHARACTERISTIC RESPONSE

Both the Wagner–Nelson and Loo–Riegelman methods require the intravenous administration of a drug to define the characteristic response and this response must be composed of positive exponential functions. The drug input function determined by these two methods is essentially the convolution of three functions: the drug dissolution function, the drug absorption function and the transport function from the absorption site to the point of observation. Consequently, the dissolution of a drug from its formulation cannot be examined directly, by defining $G(t)$ as the observed response after an oral solution drug dose.

Two methods that are independent of the functional form of $G(t)$ are the 'point-area' deconvolution method and 'orthogonal polynomial' method.

24.6.5.1 'Point-area' Method

Assuming the input function is a staircase the response vector R is given by

$$r \equiv \begin{pmatrix} R(t_i) \\ R(t_2) \\ R(t_3) \\ \vdots \\ R(t_n) \end{pmatrix} = \bar{G} \begin{pmatrix} K_1 \\ K_2 \\ K_3 \\ \vdots \\ K_n \end{pmatrix} \equiv \bar{G}k \tag{108}$$

where matrix $\bar{G}$ is a lower triangular matrix with elements

$$g_{ji} = \int_{t_j - t_i}^{t_j - t_{i-1}} G(t)\,\mathrm{d}t \qquad j \leq i \tag{109}$$

and

$$g_{ji} = 0 \qquad i > j \tag{110}$$

and where the magnitude of an input step in the interval $t_i - t_{i-1} = K_i$. Solving for the vector k then

$$(\bar{G})^{-1}r = k \tag{111}$$

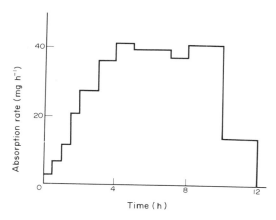

Figure 24 Staircase absorption function for metronidazole after rectal administration as a suppository (500 mg), point-area deconvolution method

Since the value K_j depends on all the previous values of $K_i (i < j)$ the error in estimating K_j depends on the error in previous estimates which in turn depends on the errors in the estimates of $R(t_i)$.

Assuming the exact response function to a staircase input function is subject to random error, *i.e.* $R(t_i) + \varepsilon_i = R_{obs}(t_i)$, the error structure is

$$\hat{k} - k = (\bar{G})^{-1} \begin{pmatrix} \varepsilon_1 \\ \varepsilon_2 \\ \vdots \\ \varepsilon_N \end{pmatrix} \tag{112}$$

with $\hat{k}$, the vector of estimate step heights, determined from the set of errant data. Obviously the error depends on the observational errors ε and the inverse integral matrix $(\bar{G})^{-1}$.

The point-area method of deconvolution[47] is a rationalization and generalization of a method elaborated by Rescigno and Segre.[48] In the latter method the integral functions g_{ji} are crudely approximated by $G(t_j - t_{i-1})(t_i - t_{i-1})$ which can lead to large errors.[47] Similarly the empirical deconvolution method of Chiou[49] is an approximation to the 'point-area' method in which the integrals g_{ji} are approximated by $G(t_i/2)(t_i - t_{i-1})$ where $G(t_i/2)$ is the value of $G(t)$ at the mid-point of the time interval $(t_i - t_{i-1})$.[50]

Whenever the intervals $(t_i - t_{i-1})$ are constant the methods of Rescigno and Segre[48] and Chiou[49] decompose into a representation of the input function as a train of impulses. Cutler[51] incorrectly casts the Chiou approximation of the 'point-area' method as an impulse sequence. Other methods of approximating the integral functions g_{ji} say by linear interpolation between a set of observed values $G(t_i)$ or by polynomial interpolation could obviously provide approximate algorithms to the 'point-area' method.

An example of an input function determined by the 'point-area' method is given in Figure 24. The practical information gained from this example is that the formulation of a drug as a suppository requires particular attention if the loss of a substantial amount of the drug dose by defaecation is to be avoided.

24.6.5.2 Orthogonal Polynomials

Representing an unknown input as a polynomial function would on first considerations seem an obvious deconvolution method. In principle the unknown coefficients of an *n*th order polynomial could be defined as a least-square problem. In this case the coefficients are optimized such that the convolution of the polynomial function with $G(t)$ generates a set of calculated values for $R(t_i)$ that are close to the observed values, closeness being defined by the minimal least square value. However, ill-conditioning of the resulting equations produces a problem and the degree of the input polynomial cannot be determined.

Cutler[52] has employed the general curve fitting method of orthogonal polynomials[53] to deconvolution. In this method the input function is approximated by a set of *n* orthogonal polynomials,

$\sum_{i=1}^{n} a_i f_i(t)$. Since the number of polynomials is unknown a suitable value of n is determined from the residual mean square

$$\delta_n^2 = \frac{S_n}{m - n} \tag{113}$$

where S_n is the least-square value when n terms are used and m is the number of data points. In ideal cases δ_n^2 decreases as n increases to a plateau value and the first n on this plateau is considered appropriate.[52] However there is no guarantee that a plateau value exists and in many cases n increases so that the calculated least-square response 'hunts' the observed data. In such a case the advantage of the least-squares method vanishes.

24.6.6 MULTIPLE DOSING AND TOXICITY

24.6.6.1 Multiple Dose Function in Man

Frequently drug doses are administered at a regular time interval. The time variance of the plasma or blood concentrations of the drug and its metabolites is crucial in differentiating therapeutic and toxicological effects. Defining a suitable human dose based on toxicological observations in animals is erroneous unless due consideration is given to the frequency of drug administration and to the time variance of the drug concentrations in both animals and man.

For a linear time-invariant system the multiple dose function can be derived as follows: Suppose that the response $G(t)$ of the body to a single drug dose is known, for example the blood or plasma concentrations of drug after an oral dose, and that this dose is repeatedly administered at a regular time interval of $j\tau$ ($j = 0, 1, 2, \ldots$). Since the system is time invariant each dose produces a response

$$\bar{H}(t - j\tau) G(t - j\tau) \tag{114}$$

where the step function $\bar{H}(t - j\tau) = 0$ for $t \le j\tau$ and $\bar{H}(t - j\tau) = 1$ for $t > j\tau$.

For a linear system the total response is the summation of the individual responses produced by each input. In which case the total response function, $R_n(t)$, is given as

$$R_n(t) = \sum_{j=0}^{n-1} \bar{H}(t - j\tau) G(t - j\tau) \tag{115}$$

where n is the total number of drug doses administered. If $G(t) > 0$ for $t > \tau$ the second drug dose will result in an increased response. We may suppose that each term in $R_n(t)$ is positive for $t > (n-1)\tau$ in which case $R_n(t)$ increases in value in the time interval $[(n-1)\tau, n\tau]$ as n increases. Quite naturally the latter generates the question of divergency or convergency of the function $R_n(t)$ in the time interval $[(n-1)\tau, n\tau]$ as $n \to \infty$. The function $R_n(t)$ is convergent in this interval, which is established as described below.

In order to avoid the problem of considering an infinite time interval as $n \to \infty$ a change of variable generates an equivalent problem. Setting $\rho = t - (n-1)\tau$ then

$$\begin{array}{ccc} R_n(t) & \equiv \hat{R}_n(\rho) & = \sum_{j=0}^{n-1} G(\rho + j\tau) \\ t = (n-1)\tau \text{ to } t = n\tau & \rho = 0 \text{ to } \rho = \tau & \end{array} \tag{116}$$

and with a further trivial change of variable, $\rho = t$, then

$$\begin{array}{cc} \hat{R}_n(t) & = \sum_{j=0}^{n-1} G(t + j\tau) \\ t = 0 \text{ to } \tau & \end{array} \tag{117}$$

and for $n \to \infty$

$$\begin{array}{cc} \hat{R}_\infty(t) & = \sum_{j=0}^{\infty} G(t + j\tau) \\ t = 0 \text{ to } \tau & \end{array} \tag{118}$$

Any positive function is bounded from above in some interval $[a, b]$ if its integral on $[a, b]$ is finite.

Since no analytical structure has been assigned to $G(t)$ the test for convergency of $\hat{R}_\infty(t)$ must rely on general considerations. Noting that any positive function is bounded from above in any interval [a, b] if its integral on [a, b] is finite, an appropriate test for convergency is an integral test. Thus if

$$\int_0^\tau \hat{R}_\infty(t)\,dt = \int_0^\tau \left[\sum_{j=0}^\infty G(t + j\tau) \right] dt \tag{119}$$

is finite the infinite series remains bounded from above and the series cannot diverge. The properties of infinite series are quite different from those finite series. Indeed it is unclear whether the series can be integrated term by term.

Since $G(t)$ is a positive function, bounded from above, such that $G(t) = 0$ for $t \to \infty$ a comparison test establishes that the terms of $\hat{R}_\infty(t)$ can be integrated term by term. From the properties of $G(t)$ we may state

$$Ae^{-\alpha t} > G(t) \tag{120}$$

for some A and $\alpha > 0$, whence

$$\int_0^\tau \hat{R}_\infty(t)\,dt < \int_0^\tau \left(\sum_{j=0}^\infty Ae^{-\alpha(t + j\tau)} \right) dt \equiv \frac{A}{\alpha} \tag{121}$$

which established that $\int_0^\tau \hat{R}_\infty(t)\,dt$ is finite and can be integrated term by term, thus

$$\int_0^\tau \hat{R}_\infty(t)\,dt = \sum_{j=0}^\infty \int_{j\tau}^{(j+1)\tau} G(t)\,dt \equiv \int_0^\infty G(t)\,dt \tag{122}$$

Consequently the multiple dose function is bounded in the interval $[(n-1)\tau, n\tau]$ and has a limiting periodic steady-state response R_{ss}, where $R(t) = R(t - \tau)$.

The convergence time t_i is governed by the time required for

$$\int_0^{t_i} G(t)\,dt \cong \int_0^\infty G(t)\,dt \tag{123}$$

In general the function $G(t)$ will exhibit a limiting exponential decay and the convergence time will be determined by the time this function takes to become essentially zero. To all intents an exponential function is zero after a time period of 10 half-lives. In this way the convergence time to steady-state is given by analysis of a single dose function. Additionally, the approximate magnitude of the steady-state response is determined from the single dose function by applying the first mean value theorem of integration. Thus

$$\text{First mean value at steady state} = \frac{\int_0^\tau R_\infty(t)\,dt}{\tau} = \frac{\int_0^\infty G(t)\,dt}{\tau} \equiv \text{Mean steady state concentration} \tag{124}$$

and the mean value in the ith dosage interval is

$$\text{Mean value in } i \text{ dosage interval } i = 1, 2 \ldots = \frac{\int_0^{i\tau} G(t)\,dt}{\tau} \tag{125}$$

Using the mean values an accumulation index can be defined as

$$\text{Accumulation index} = \frac{\int_0^\infty G(t)\,dt}{\int_0^\tau G(t)\,dt} \tag{126}$$

Whenever τ is such that $\int_0^\tau G(t)\,dt \cong \int_0^\infty G(t)\,dt$ no accumulation of the drug occurs upon multiple dosing, but in all other instances some accumulation will occur. The accumulation index is an integral function and it is erroneous to equate this with the terminal exponential of $G(t)$. Expanding $G(t)$ in terms of an exponential function

$$G(t) = G(t) - \bar{H}(t - t_i)G(t) + \bar{H}(t - t_i)G(t_i)e^{-\alpha t} \tag{127}$$

where $G(t) \cong G(t_i)e^{-\alpha t}$ for $t > t_i$, the contribution of the final exponential decay is apparent. Thus

$$\text{Accumulation index} = \frac{\int_0^{t_i} G(t)\,\mathrm{d}t \; + \; \left(\dfrac{G(t_i)}{\alpha}\right)}{\int_0^{\tau} G(t)\,\mathrm{d}t} \qquad (128)$$

There is no guarantee that whenever $\tau > t_i$ the accumulation index approximates to $(1 - e^{-\alpha\tau})^{-1}$, which is frequently stated.

An example of a multiple dose function is given in Figure 25. This example is atypical in the sense that steady-state conditions are rapidly achieved. Other examples are given in Figures 26, 27 and 28. In the latter examples the full plasma drug concentration profiles have not been determined, but the expected minimal concentrations, *i.e.* $R(j\tau)\, j\in 0, 1, 2, \ldots$, are presented.

In the case of nortriptyline and phenobarbitone a marked intersubject variation in the steady-state concentrations is evident. In fact one subject on phenobarbitone has not achieved steady-state concentrations within the period of drug administration.

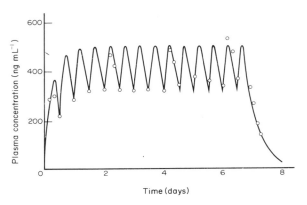

Figure 25 Plasma concentration of pseudoephedrine after oral administration of a slow-release product twice a day (reproduced from ref. 104 by permission of Springer-Verlag)

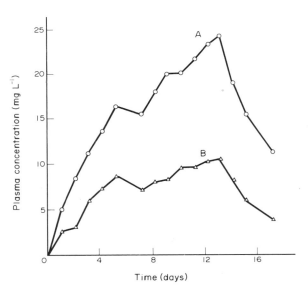

Figure 26 Plasma concentration of phenobarbital from administration of 12 oral doses given once a day. Subject A $4\,\text{mg}\,\text{kg}^{-1}$, Subject B $2\,\text{mg}\,\text{kg}^{-1}$. Note steady-state conditions are not achieved in subject A (reproduced from ref. 105 by permission of Williams and Wilkins)

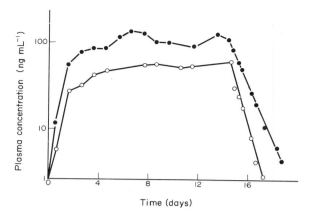

Figure 27 Plasma nortriptyline in two normal subjects who received 0.4 mg kg^{-1} three times a day for two weeks (reproduced from ref. 106 by permission of Springer-Verlag)

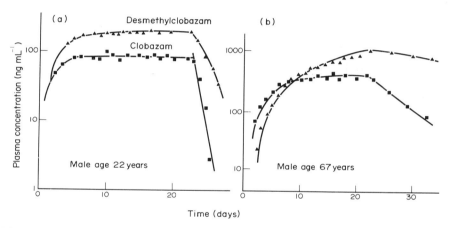

Figure 28 Plasma concentrations of clobazam and its major metabolite demethylclobazam during a multiple dose study in a young and an elderly male. Note the excessive accumulation in the elderly male and that steady-state concentrations of demethylclobazam are not achieved in the elderly male. Excessive accumulation does not occur in elderly females (reproduced from ref. 107 by permission of ADIS Press)

In clinical practice there is a tendency to dismiss reported side-effects in patients who have tolerated drug administration for some weeks previously. Such an attitude ignores the dynamics of drug accumulation and the very real probability of such a phenomenon.

Further, the large intersubject variations in steady-state drug concentrations indicate the desirability of correlating the frequency and intensity of adverse drug reactions with mean steady-state drug concentrations in clinical trials. A global frequency of side-effects fails to identify vulnerable groups.

Frequently the elderly exhibit excessive drug accumulation, which is illustrated by the accumulation of benzodiazepines and their active metabolites demethylbenzodiazepines (see Figure 28). It is naïve not to expect drug-induced sedation and confusion in elderly patients when they are administered the so-called regular dose of a benzodiazepine.

The reasons for the variations in steady-state drug concentrations in the elderly are discussed subsequently.

24.6.6.2 Multiple Dosing in Animals

Most compounds are more readily eliminated in small mammalian species, *e.g.* mice, rats, hamsters. As a general principle the plasma drug concentration function in these species is scaled

Table 4 Approximate Gestational Period when Typical Teratogenic Defects
Occur[110,111]

| Type of fetal defect | Days of gestation when teratogenic effects are produced | | | | |
	Humans	Hamsters	Sheep	Cattle	Pigs
Microphthalmia	16–17	6.5	12–13	16–18	10–11
Cyclopia	21–23	7.0	14	21	12
Exencephaly	26	7.5	16.5	26–27	15.5
Spina bifida	28–29	8–8.5	18	29–31	17
Shortened limbs	36	9.5	29–31	42	20
Total gestation period	267	16	147	283	115

both in amplitude and time. Thus if $G(t)$ is the concentration function in say man for a particular drug, the expected function in a smaller species is $bG(at)$ where $a > 1$ and $b < 1$.

Since

$$b \int_0^\infty G(at)\,\mathrm{d}t = \frac{b}{a} \int_0^\infty G(t)\,\mathrm{d}t \tag{129}$$

it is necessary to scale the doses in animals by $a/b > 1$ to obtain the same mean steady-state value in the test animal and in man; whenever the dosage intervals are the same in both species. Amplitude scaling is partially corrected for in toxicity studies by adjusting the doses in test animals on a $\mathrm{mg\,kg^{-1}}$ basis. However, this latter does not correct for the magnitude of the oscillation about the mean-steady-state value. The latter is controlled by the time scaling element, a. To achieve the same minimal steady-state in a test animal as that in man it is essential to decrease the frequency of drug administration in the smaller species. An initial estimate of an appropriate dosage interval in the test species is τ/a, where τ is the dosage interval in man. A full solution of the dose and frequency problem is readily obtained by casting the problem as a constrained least-square problem, with imposed constraints equal to the upper and lower limits of the steady-state concentration in man.

Ignoring any of these design concepts will lead to a serious problem in the interpretation of chronic animal toxicity studies.

The accumulation of drugs and their metabolites to steady-state also poses a serious problem in teratogenicity testing. Organogenesis occurs at distinct intervals (see Table 4). Failure to achieve similar drug and metabolite concentrations during these crucial periods in all species leads to the possible underestimation of the teratogenic potential in humans. The use of constant rate infusions of the drug and its metabolite avoids these pitfalls.[54]

The importance of multiple dose functions in therapeutics was originally advocated by Krüger–Thiemer.[55-59] These derivations of steady-state functions assume polyexponential functions but protein binding and free drug concentrations are also considered. Wagner has also stated the mean steady-state equations.[60,61] Accumulation of drugs that are non-linear in their pharmacokinetics has been considered by Chau[62] and by Lam and Chiou.[63]

24.6.7 RECIRCULATORY PHARMACOKINETIC MODEL AND CLEARANCE

Linear-systems analysis as such provides no information as to what parameters determine the steady-state concentration in particular individuals. A general time-invariant linear system which mimics the recirculation of blood *via* the heart is particularly useful in casting the drug concentration time function in terms of elimination parameters. The latter then provides a ready explanation for excessive drug accumulation in the elderly.[64-66]

24.6.7.1 Linear Model

Arterial blood perfuses the body tissues and the mixed venous blood is then recirculated by the heart–lung–heart circulation system. Drug elimination can occur in the lungs and in some of the body tissues. An intravenous input function is delivered to the entrance of the heart–lung–heart system. A block flow diagram for the body is given in Figure 29.

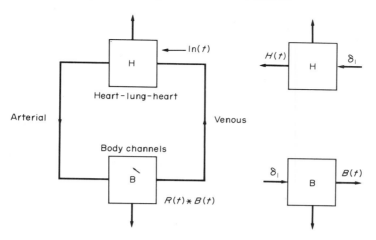

Figure 29 Blood flow diagram of the body

We may state the rate of drug exit (mass/time) from the combined body channels as $B(t)$ after a unit impulse input when there is no recirculation, *i.e.* $B(t)$ is the open loop output. The open loop output of the heart–lung–heart system is similarly defined at $H(t)$.

Since drug elimination occurs, only a fraction of a unit impulse input will eventually leave the body channels or the heart–lung–heart system. Thus we may define the total output of the open loop by

$$\int_0^\infty B(t)\,dt = \delta_B < 1 \tag{130}$$

and

$$\int_0^\infty H(t)\,dt = \delta_H < 1 \tag{131}$$

For the recirculatory system define $R(t)$ as the rate of drug appearance on the arterial side. The system equations are readily determined, provided the system is linear and time invariant, as follows. The input to block B is $R(t)$ and since its characteristic response is $B(t)$ its output is

$$R(t) * B(t) \tag{132}$$

Input into block H is the sum of a recirculatory input function, $R(t) * B(t)$, and exterior input function from the administered dose, $\text{In}(t)$. Consequently the output of block H, which has been defined as $R(t)$, is the convolution of the combined inputs with the function $H(t)$, thus

$$R(t) = H(t) * (R(t) * B(t)) + H(t) * \text{In}(t)$$
$$\equiv H(t) * R(t) * B(t) + H(t) * \text{In}(t) \tag{133}$$

Applying the limit theorem for the integral of convolution functions on the interval $[0, \infty)$,[36] then with appropriate substitution

$$\int_0^\infty R(t)\,dt = \delta_H \int_0^\infty R(t)\,dt\,\delta_B + \delta_H \int_0^\infty \text{In}(t)\,dt \tag{134}$$

The integral of the exterior input function, $\text{In}(t)$ is the dose of drug that is delivered to the system, D, whence

$$\int_0^\infty R(t)\,dt = \frac{D\delta_H}{1 - \delta_H\delta_B} \tag{135}$$

The function $1 - \delta_H\delta_B$ is the fraction of a unit impulse that is eliminated in a single cycle through the system and its reciprocal is the average number of cycles a drug molecule undergoes before its elimination from the system. Converting $R(t)$ to a concentration function, then

$$CO \int_0^\infty C_A(t)\,dt = \frac{D\delta_H}{(1 - \delta_H\delta_B)} \tag{136}$$

where C_A is the arterial drug concentration and CO is the cardiac output, *i.e.* cardiac blood flow. From physiological considerations, block B is essentially composed of a parallel set of body organs that are perfused with arterial blood, see Figure 30.

A fraction of cardiac output, f_i, perfuses the ith body channel. For a unit impulse of drug into the open loop system the total amount of drug delivery to each channel is therefore f_i, with $\sum_{i=1}^N f_i = 1$. The total output of drug from the open-loop body channel system is thus

$$\delta_B = \sum_{i=1}^N f_i\delta_i \tag{137}$$

where δ_i is the total output of drug from an open loop body channel after a unit impulse input.

Assuming that only the liver and kidney eliminate the drug then δ_4 and δ_3 are less than unity where δ_4 and δ_3 are the fractions of a unit impulse into open loop liver and kidney channels that traverse these channels intact respectively. All the other fractions δ_i are unity. In this case

$$(1 - \delta_B) = f_3 + f_4 - f_3\delta_3 - f_4\delta_4$$
$$\equiv f_3(1 - \delta_3) + f_4(1 - \delta_4) \tag{138}$$

and

$$\delta_B = 1 - f_3(1 - \delta_3) - f_4(1 - \delta_4) \tag{139}$$

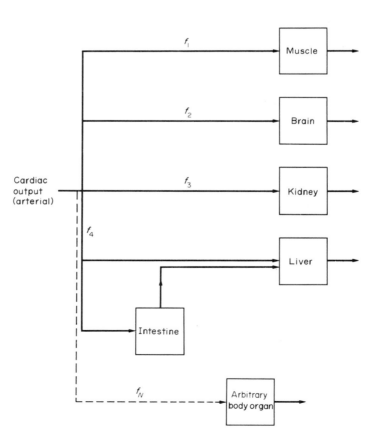

Figure 30   Blood flow diagram to body organs

which on substitution in the expression for $\int_0^\infty R(t)\,dt$ gives

$$\int_0^\infty R(t)\,dt = \frac{D}{\left(\dfrac{1 - \delta_H}{\delta_H}\right) + f_3(1 - \delta_3) + f_4(1 - \delta_4)} \tag{140}$$

or

$$\int_0^\infty C_A(t)\,dt = \frac{D}{CO\left(\dfrac{1 - \delta_H}{\delta_H}\right) + COf_3(1 - \delta_3) + COf_4(1 - \delta_4)} \tag{141}$$

Since the fractions of cardiac output, COf_i, are the blood flow to particular channels we may cast the expression in a more suggestive form as

$$\int_0^\infty C_A(t)\,dt = \frac{D}{CO\left(\dfrac{1 - \delta_H}{\delta_H}\right) + \text{Renal blood flow} \cdot (1 - \delta_R) + \text{Hepatic blood flow} \cdot (1 - \delta_L)} \tag{142}$$

where $\delta_R \equiv \delta_3$ and $\delta_L \equiv \delta_4$. It is important to realize that the terms $(1 - \delta_R)$ and $(1 - \delta_L)$ are the fractions of drug that are lost in a single passage of drug through the kidneys and liver respectively.

The representation of the integral of arterial drug concentration displays the dependence on elimination and the independence of distribution into tissues that are non-eliminating.[64] For a non-eliminating organ the blood flow in and out of the channel are equal, *e.g.* muscle tissues, in which case

$$\int_0^\infty C_A(t)\,dt = \int_0^\infty C_V(t)\,dt \tag{143}$$

where C_V is the venous concentration of the drug.

24.6.7.2 Relationship to Clearance Terminology

Historically renal physiologists used the term renal clearance to define the elimination of a compound *via* the kidney.[67] Renal clearance is defined as the 'effective volume' of arterial blood that has the drug completely removed from it per unit time, *e.g.* mL min^{-1}, during its passage through the kidney under the condition of a constant input and effluent concentrations (*i.e.* steady-state conditions).

According to the strict definition

$$\text{Renal clearance} = \frac{Q[C_A - C_V]}{C_A} = Q\left(1 - \frac{C_V Q}{C_A Q}\right) \equiv Q(1 - \delta_R) \tag{144}$$

where Q is the renal blood flow and C_A and C_V are the arterial and venous concentrations of the substance respectively. Arithmetical manipulation of the strict definition gives

$$\text{Renal clearance} \equiv \text{Renal blood flow } (1 - \delta_R) \tag{145}$$

Adopting the terminology of renal physiologists and extending it to include other organs then gives

$$\text{Hepatic clearance} = \text{Hepatic blood flow } (1 - \delta_L) \tag{146}$$

and

$$\int_0^\infty C_V(t)\,d \equiv \int_0^\infty C_A(t)\,d = \frac{D}{CO\dfrac{(1 - \delta_H)}{(\delta_H)} + \text{Renal clearance} + \text{Hepatic clearance}} \tag{147}$$

In pharmacokinetics the inverse function

$$D \bigg/ \int_0^\infty C_V(t) = \text{Total body clearance} \equiv \text{Clearance} \equiv Cl \tag{148}$$

is now almost universally used in deriving pharmacokinetic expressions. However this loose terminology should be deprecated since it is only exact when there is no elimination of the drug *via* the lungs. In addition, there is a further complication when there is intestinal drug elimination, since

$$f_4(1 - \delta_L) = \hat{f}_i(1 - \delta_L) + \hat{f}_j(1 - \delta_g\delta_L) \tag{149}$$

where $\hat{f}_i$ and $\hat{f}_j$ are the fractions of hepatic blood flow that directly perfuses the liver and intestines respectively and $\hat{\delta}_L$ and $\hat{\delta}_g$ are the fractions of a drug input that traverse the liver and intestines respectively. Thus

$$\text{Hepatic clearance} = \text{Direct liver clearance} + \text{Intestinal blood flow } (1 - \hat{\delta}_g\hat{\delta}_L) \tag{150}$$

The importance of these separate factors is illustrated by the elimination of phenol in the rat. From the carotid arterial concentrations of phenol after oral, intraportal, intra-arterial and intravenous administration[68] the various fractions can be calculated. For phenol $\hat{\delta}_g = 0.08$, $\hat{\delta}_L = 0.94$ and $\delta_H = 0.38$ which implies extensive metabolism in the intestinal tract and in the lungs but only minor metabolism in the liver.

Intuitively, a terminology based on effective volumes of blood that are cleared of drug during the passage of blood through organs should have an upper limiting value equivalent to the cardiac output. However, for intravenous drug administration a value of $\delta_H < 0.5$ guarantees a so-called clearance value greater than the cardiac blood flow.

24.6.7.3 Relationship to Compartmental Models

Under the assumption that drug elimination does not occur in the intestine and the lungs, the fraction of an oral drug dose that is absorbed, F, can be estimated from the renal clearance of the drug, RC, and the hepatic blood flow, Q, by

$$F = \frac{Q A_{\text{oral}}}{A_{\text{iv}}(Q + \text{RC}) - D} \tag{151}$$

where A_{oral} and A_{iv} are the total area under the concentration time function after oral and intravenous administration of dose D respectively. Originally the expression for F was derived by the application of a mammillary compartmental system.[26] However the equation is quite simply obtained from the recirculatory equations, since

$$\int_0^\infty C_{\text{oral}}(t)\,dt = \frac{F D \delta_L}{\text{RC} + Q(1 - \delta_L)} \equiv A_{\text{oral}} \tag{152}$$

and

$$\int_0^\infty C_{\text{iv}}(t)\,dt = \frac{D}{\text{RC} + Q(1 - \delta_L)} \equiv A_{\text{iv}} \tag{153}$$

then

$$A_{\text{oral}}/A_{\text{iv}} = F\delta_L \tag{154}$$

Isolating δ_L from the intravenous equation and substituting into the equation for the ratio of areas give on rearrangement of the required result.

Similarly by assuming $F = 1$ but admitting the terms representing elimination by the lungs in the equations for A_{iv} and A_{oral}, then

$$A_{\text{iv}} - A_{\text{oral}} = A_{\text{iv}}(1 - \delta_L) \tag{155}$$

and

$$Q(A_{\text{iv}} - A_{\text{oral}}) = Q A_{\text{iv}}(1 - \delta_L) \equiv \text{Amount of drug metabolized in the liver after an intravenous dose} \tag{156}$$

The latter identity follows from noting the total drug input into the liver is $Q A_{\text{iv}}$ and $(1 - \delta_L)$ is the fraction lost in traversing the liver. Again this simple equation was originally derived using a complicated compartmental system.[25] However, not all general equations derived using compartmental systems are valid.[64]

24.6.8 STEADY-STATE CONCENTRATIONS IN THE ELDERLY

The total area under the blood concentration–time function is inversely proportional to the sum of factors that determine drug elimination from the system. Any deterioration in the functional ability of the eliminating organs to dissipate the drug will result in an increased steady-state drug concentration upon repeated administration.

The renal clearance of both exogenous and endogenous compounds is composed of filtration through the basement membrane of the glomerulus and secretion or reabsorption of the compound in the renal tubule. The renal clearance of inulin, a macromolecule, is equivalent to the filtration rate, since its molecular weight precludes secretion or reabsorption (inulin clearance = 120 mL min^{-1} $\equiv$ glomerular filtration rate). In contrast, diodrast, an X-ray opaque material for kidney imaging, is almost completely extracted from the blood in a single passage through the kidneys. Consequently δ_R is close to zero and the renal clearance of diodrast approximates to the renal plasma flow of 700 mL min^{-1}.

With increasing age the number of glomeruli in the kidney decreases, such that by the seventh decade there is only 30–50% of the number found in young adults.[69] This gradual destruction, tissue death, of functional glomeruli is paralleled by a reduction in both the inulin and diodrast clearance with increasing age (see Figure 31). Urea clearance also declines with age (see Figure 32).

Since the kidney is essentially a filtration/secretion organ, linear correlations between the renal clearance of any drug and the renal clearance of a reference compound, *e.g.* inulin, creatinine or diodrast are anticipated. Replacing the renal clearance of a drug with a measure of renal function (*e.g.* creatinine clearance (120 ml min^{-1}), young adults), the relationship between steady-state drug concentrations is, in general form,

$$\text{Steady-state concentration} = \frac{\text{Dose}}{(a + b \cdot \text{renal function})\tau} \tag{157}$$

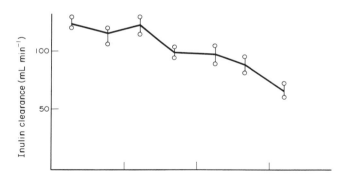

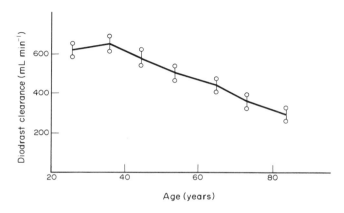

Figure 31 Average inulin and diodrast clearances in relation to age (reproduced from ref. 108 by permission of Rockefeller University Press)

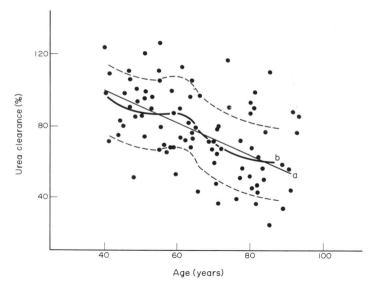

Figure 32 Urea clearance in relation to age; (a) is linear regression, 40–89 years; (b) is smoothed regression of five-year means, with limits (dashed lines) of standard error of estimate, 40–89 years (reproduced from ref. 109 by permission of the American Physiological Society)

For drugs that are partially excreted *via* the kidney, *i.e.* with an excretion fraction greater than 0.3, numerous correlations supporting the latter relationship have been observed in man.

Noteworthy articles on excessive drug accumulation and renal function which may be referenced are Dettli,[70,71] Jelliffe,[72] and Orme and Cutler.[73,74]

Since renal function declines with age and many elderly patients have creatinine clearances of less than 50 mL min^{-1} it is distinctly advisable to assume at least mild renal impairment when defining appropriate drug doses for the elderly. Failure to adjust the doses simply results in excessive drug accumulation and toxicity. For example ototoxicity with ethacrynic acid, myopathy with clofibrate, ototoxicity and nephrotoxicity with aminoglycosides and nephrotoxicity with cephalothin.

Additionally renal blood flow is reduced in hypertensive states due to increased resistance caused by progressive destruction of renal parenchyma. Blood flow may be as low as 10% of normal,[75] which further compounds the effects of a reduced glomeruli number.

On the subject of hepatic elimination of drugs, no information is available with respect to general deterioration with age. However, the excessive accumulation of benzodiazepines and their pharmacologically active metabolites (see Figure 28) should caution against assuming that elevated concentrations only occur with renally excreted drugs. An important noteworthy observation may be made that many antiinflammatory analgesics are not excreted unchanged in urine but accumulate excessively in the elderly. This result is most probably caused by a reduced renal clearance of the ester–glucuronide metabolite which then reverts to the parent drug.[76]

24.6.9 FACTORS AFFECTING DRUG ELIMINATION

The recirculating linear model does not provide any information about which physiological variables determine the fraction δ. Two quantitative models are currently used to define δ. The first is the 'well-stirred' model[77–80] which defines δ as

$$\delta = \frac{\text{Flow}}{\text{Flow} + \text{Intrinsic clearance}} \tag{158}$$

Under the arbitrary assumption that drug bound to plasma protein is not available for elimination, an assumption originally made by renal physiologists, an arbitrary correction factor for protein binding has been introduced; thus

$$\delta = \frac{\text{Flow}}{\text{Flow} + FU_b \cdot \text{Intrinsic clearance}} \tag{159}$$

where FU_b is the fraction of unbound drug in the plasma, which is assumed constant with respect to time.[78-80]

The other is the 'parallel tube' or 'sinusoidal perfusion' model.[84-86] The differences between the two models are only significant when $\delta \cong 0$. There are experimental data which both support and refute either model.[87]

24.6.10 REFERENCES

1. M. Gibaldi and G. Levy, *J. Am. Med. Assoc.*, 1976, **235**, 1864.
2. A. J. Cummings and B. K. Martin, *Nature (London)*, 1963, **200**, 1296.
3. A. J. Cummings, B. K. Martin and G. S. Park, *Br. J. Pharmacol. Chemother.*, 1967, **29**, 136.
4. D. E. Drayer, *Clin. Pharmacokinet.*, 1976, **1**, 426.
5. D. J. Jollow, J. J. Koscio, R. Snyder and A. Vainio, 'Biological Reactive Intermediates: Formation, Toxicity and Inactivation', Plenum Press, New York, 1977.
6. A. R. Hansen, K. A. Kennedy, J. J. Ambre and L. J. Fisher, *N. Engl. J. Med.*, 1975, **292**, 250.
7. A. A. Sinkula and S. J. Yalkowsky, *J. Pharm. Sci.*, 1975, **64**, 181.
8. A. Tsuji, Y. Intantani and T. Yamana, *J. Pharm. Sci.*, 1977, **66**, 1004.
9. A. Tsuji, E. Miyamoto, T. Terasaki and T. Yamana, *J. Pharm. Sci.*, 1979, **68**, 1259.
10. R. E. Notari, in 'Design of Biopharmaceutical Properties through Prodrugs and Analogs', ed. E. B. Roch, American Pharmaceutical Association, Washington, DC, 1977, p. 68.
11. J. G. Wagner, *J. Theor. Biol.*, 1968, **20**, 173.
12. G. Levy, *J. Pharm. Sci.*, 1964, **53**, 342.
13. A. Wilson and H. O. Schild, in 'Applied Pharmacology', 9th edn., Churchill, London, 1961, p. 33.
14. S. Riegelman, J. Loo and M. Rowland, *J. Pharm. Sci.*, 1968, **57**, 117.
15. D. P. Vaughan and A. Trainor, *Br. J. Clin. Pharmacol.*, 1975, **2**, 239.
16. D. P. Vaughan and A. Trainor, *J. Pharmacokinet. Biopharm.*, 1975, **3**, 203.
17. L. Endrenyi, T. Inabi and W. Kalow, *Clin. Pharmacol. Ther.*, 1976, **20**, 701.
18. D. P. Vaughan and P. G. Walton, in 'Controlled Drug Bioavailability', ed. V. F. Smolen and L. A. Ball, Wiley, New York, 1984, chap. 4, p. 103.
19. L. Shaney, L. R. Wasserman and N. R. Gevitz, *Am. J. Med. Electron.*, 1964, **3**, 249.
20. J. G. Wagner, *J. Pharmacokinet. Biopharm.*, 1976, **4**, 395.
21. R. A. Ronfeld and L. Z. Benet, *J. Pharm. Sci.*, 1977, **66**, 178.
22. D. P. Vaughan and M. J. Dennis, *J. Pharmacokinet. Biopharm.*, 1979, **7**, 511.
23. P. A. Harris and S. Riegelman, *J. Pharm. Sci.*, 1969, **58**, 71.
24. R. N. Boyes, H. J. Adams and B. R. Duce, *J. Pharmacol. Exp. Ther.*, 1970, **174**, 1.
25. D. P. Vaughan, *Eur. J. Clin. Pharmacol.*, 1977, **11**, 57.
26. D. P. Vaughan, *J. Pharm. Pharmacol.*, 1975, **27**, 458.
27. S. Riegelman and M. Rowland, *J. Pharmacokinet. Biopharm.*, 1973, **1**, 419.
28. L. Z. Benet, *J. Pharmacokinet. Biopharm.*, 1978, **6**, 559.
29. J. Z. Hearon, *Ann. N. Y. Acad. Sci.*, 1963, **108**, 36.
30. R. Bellman, in 'Introduction to Matrix Analysis', ed. R. Bellman, McGraw-Hill, New York, 1960, chap. 13.
31. O. Perron, *Math. Am.*, 1907, **64**, 248.
32. G. Frobenius, *Akad. Wiss.*, 1912, 456.
33. D. Falch, A. Teien and C. J. Bjerkelund, *Br. Med. J.*, 1973, **1**, 695.
34. A. J. Glazko, A. W. Kinkel, W. C. Alegnani and E. L. Holmes, *Clin. Pharmacol. Ther.*, 1968, **9**, 472.
35. G. W. Brice and H. F. Hammer, *J. Am. Med. Assoc.*, 1969, **208**, 1189.
36. D. P. Vaughan, *J. Pharmacokinet. Biopharm.*, 1977, **5**, 271.
37. F. H. Dost, 'Grundlagen den Pharmacokinetik', Thieme, Leipzig, 1968.
38. D. P. Vaughan, D. J. H. Mallard and M. Mitchard, *J. Pharm. Pharmacol.*, 1974, **26**, 508.
39. D. P. Vaughan, *J. Pharm. Pharmacol.*, 1976, **28**, 505.
40. J. G. Wagner and E. Nelson, *J. Pharm. Sci.*, 1963, **52**, 610.
41. J. C. K. Loo and S. Riegelman, *J. Pharm. Sci.*, 1968, **57**, 918.
42. A. Breckenridge and M. Orme, *Clin. Pharmacol. Ther.*, 1973, **14**, 955.
43. T. Suzuki and Y. Saitoh, *Chem. Pharm. Bull.*, 1973, **21**, 1458.
44. J. G. Wagner, *J. Pharmacokinet. Biopharm.*, 1975, **3**, 51.
45. D. P. Vaughan and M. Dennis, *J. Pharmacokinet. Biopharm.*, 1980, **8**, 83.
46. D. P. Vaughan, *J. Pharm. Sci.*, 1989, in press.
47. D. P. Vaughan and M. Dennis, *J. Pharm. Sci.*, 1978, **67**, 663.
48. A. Rescigno and G. Segre, 'Drug and Tracer Kinetics', Blaisdell, Waltham, MA, 1966.
49. W. Chiou, *J. Pharm. Sci.*, 1980, **69**, 57.
50. D. P. Vaughan, *J. Pharm. Sci.*, 1981, **70**, 831.
51. D. Cutler, in 'Pharmacokinetic Theory and Methodology', ed. M. Rowland and G. T. Tucker, Pergamon Press, Oxford, 1986, chap. 14.
52. D. Cutler, *J. Pharmacokinet. Biopharm.*, 1978, **6**, 243.
53. G. Forsythe, *J. Soc. Ind. Appl. Math.*, 1957, **5**, 74.
54. H. Nau, R. Zierer, H. Spielmann, D. Neubert and C. Gansau, *Life Sci.*, 1981, **29**, 2804.
55. E. Krüger-Thiemer, *J. Am. Pharm. Assoc. Sci. Ed.*, 1960, **49**, 311.
56. E. Krüger-Thiemer, *Klin. Wschr.*, 1960, **38**, 514.
57. E. Krüger-Thiemer and P. Bunger, *Arzneim.-Forsch.*, 1961, **11**, 867.
58. E. Krüger-Thiemer and B. Schlender, *Arzneim.-Forsch.*, 1963, **13**, 891.

59. E. Krüger-Thiemer, W. Diller and P. Bunger, *Antimicrob. Agents Chemother.*, 1966, **65**, 183.
60. J. G. Wagner, J. I. Northam, C. D. Alway and O. S. Carpenter, *Nature (London)*, 1965, **207**, 1301.
61. J. G. Wagner and J. I. Northam, 'Abstracts of Symposia and Contributed Papers Presented to the APLA Academy of Pharmaceutical Sciences at the 115th Annual Meeting of the American Pharmaceutical Association, Miami Beach, Florida, May 5–10, 1968', 1968, abstr. no. 11, p. 59.
62. N. P. Chau, *J. Pharmacokinet. Biopharm.*, 1976, **4**, 537.
63. G. Lam and W. L. Chiou, *J. Pharmacokinet. Biopharm.*, 1979, **7**, 227.
64. D. P. Vaughan and I. A. Hope, *J. Pharmacokinet. Biopharm.*, 1979, **7**, 207.
65. D. P. Vaughan, in 'Development of Drugs and Modern Medicines', ed. J. W. Gorrod, G. G. Gibson and M. Mitchard, Horwood, Chichester, 1986, p. 494.
66. D. Cutler, *J. Pharmacokinet. Biopharm.*, 1979, **7**, 101.
67. H. W. Smith, 'The Kidney Structure and Function in Health and Disease', Oxford University Press, New York, 1951.
68. M. K. Cassidy and J. B. Houston, *J. Pharm. Pharmacol.*, 1980, **32**, 57.
69. R. A. Moore, *Anat. Rec.*, 1931, **48**, 153.
70. L. Dettli, P. Spring and R. Habersang, *Postgrad. Med. J.*, 1970, **46**, 32.
71. L. Dettli, *Adv. Biosci.*, 1970, **5**, 39.
72. R. W. Jelliffe, *Ann. Intern. Med.*, 1968, **69**, 703.
73. B. M. Orme and R. E. Cutler, *Clin. Pharmacol. Ther.*, 1969, **10**, 543.
74. R. E. Cutler and B. M. Orme, *J. Am. Med. Assoc.*, 1969, **209**, 539.
75. A. A. Bolomey, A. J. Michie, C. Michie, E. G. Breed, G. E. Schreiner and H. D. Lauson, *J. Clin. Invest.*, 1949, **28**, 10.
76. P. J. De Schepper, A. Mullie, T. B. Tjandramaga, R. Verbeeck and R. Verberckmoes, *Br. J. Clin. Pharmacol.*, 1977, **4**, 645P.
77. K. S. Pang and M. Rowland, *J. Pharmacokinet. Biopharm.*, 1977, **5**, 625.
78. K. S. Pang and M. Rowland, *J. Pharmacokinet. Biopharm.*, 1977, **5**, 655.
79. K. S. Pang and M. Rowland, *J. Pharmacokinet. Biopharm.*, 1977, **5**, 681.
80. D. Perrier, M. Gibaldi and R. N. Boyes, *J. Pharm. Pharmacol.*, 1973, **25**, 256.
81. R. A. Branch and D. G. Shand, *Clin. Pharmacokinet.*, 1976, **1**, 264.
82. G. R. Wilkinson, *Annu. Rev. Pharmacol.*, 1975, **15**, 11.
83. G. R. Wilkinson and D. G. Shand, *Clin. Pharmacol. Ther.*, 1975, **18**, 377.
84. C. A. Goresky and C. G. Bach, *Ann. N. Y. Acad. Sci.*, 1970, **170**, 18.
85. K. Winkler, S. Keiding and N. Tygstup, in 'The Liver: Quantitative Aspects of Structure and Functions', ed. P. Paugartner and R. Presig, Karger, Basel, 1973, p. 144.
86. L. Bass, *Gastroenterology*, 1979, **76**, 1504.
87. K. S. Pang and J. R. Gillette, *J. Pharmacokinet. Biopharm.*, 1978, **6**, 355.
88. J. G. Gambertoglio, W. J. C. Amend, Jr. and L. Z. Benet, *J. Pharmacokinet. Biopharm.*, 1980, **8**, 1.
89. C. G. Regardh, L. Elk and K. J. Hoffmann, *J. Pharmacokinet. Biopharm.*, 1979, **7**, 471.
90. G. R. Wilkinson and A. H. Beckett, *J. Pharmacol. Exp. Ther.*, 1968, **162**, 139.
91. G. Levy, *Clin. Pharmacol. Ther.*, 1966, **7**, 362.
92. M. Gibaldi, *Chemotherapy (Basel)*, 1968, **13**, 1.
93. J. M. Van Rossum and J. Van Koppen, *Eur. J. Pharmacol.*, 1968, **2**, 405.
94. A. Weissler, J. R. Snyder, C. D. Schoenfeld and'S. Cohen,'*Am. J. Cardiol.*, 1966, **17**, 768.
95. J. D. Adcock and R. A. Hettig, *Arch. Intern. Med.*, 1946, **77**, 179.
96. V. E. Ziegler, J. T. Briggs, A. B. Ardekani and S. H. Rosen, *J. Clin. Pharmacol.*, 1978, **18**, 462.
97. H. Gold, *Connecticut State Medical Journal*, 1945, **9**, 193.
98. B. A. Berkowitz, S. H. Ngai, J. C. Yang, J. Hempstead and S. Spector, *Clin. Pharmacol. Ther.*, 1975, **17**, 629.
99. B. B. Brodie, *Fed. Proc., Fed. Am. Soc. Exp. Biol.*, 1952, **11**, 632.
100. P. M. Bungay, R. L. Dedrick and A. M. Guarino, *J. Pharmacokinet. Biopharm.*, 1976, **4**, 377.
101. L. B. Sheiner, D. R. Stanski, S. Vosek, R. D. Miller and J. Ham, *Clin. Pharmacol. Ther.*, 1979, **25**, 358.
102. M. Rowland, S. Riegelman and W. L. Epstein, *J. Pharm. Sci.*, 1968, **57**, 984.
103. M. Dennis, Ph.D. Thesis, Sunderland Polytechnic, 1981.
104. J. Dickerson, D. Perrier, M. Mayersohn and R. Bressler, *Eur. J. Clin. Pharmacol.*, 1978, **14**, 253.
105. T. C. Butler, C. Mahaffee and W. J. Waddel, *J. Pharmacol. Exp. Ther.*, 1954, **111**, 425.
106. B. Alexanderson, *Eur. J. Clin. Pharmacol.*, 1972, **4**, 82.
107. D. J. Greenblatt, M. Divoll, S. K. Puri, I. Ho, M. A. Zinny and R. I. Shader, *Clin. Pharmacokinet.*, 1983, **8**, 83.
108. D. F. Davies and N. W. Shock, *J. Clin. Invest.*, 1950, **29**, 496.
109. W. H. Lewis, Jr. and A. S. Alving, *Am. J. Physiol.*, 1938, **123**, 500.
110. P. R. Cheeke and L. R. Shull, 'Natural Toxicants in Feeds and Poisonous Plants', AVI, CT, 1985.
111. R. F. Keeler, *J. Range Manage.*, 1978, **31**, 355.
112. P. V. Pederson, *J. Pharmacokinet. Biopharm.*, 1977, **5**, 513.
113. P. V. Pederson, *J. Pharmacokinet. Biopharm.*, 1978, **6**, 447.

25.1

Physicochemical Principles

STEPHEN H. CURRY

University of Florida, Gainesville, FL, USA

and

KAMLESH M. THAKKER

Ciba-Geigy Corporation, Ardsley, NY, USA

25.1.1 CRYSTALLINITY AND AMORPHISM

25.1.1.1 Definitions and Examples

The physical state of a drug is a function of two attributes, namely habit and internal structure. The former describes the outer external appearance, whereas the latter refers to the internal molecular appearance or arrangement. Solute habit and internal structure have an impact on preformulation and formulation research, frequently with direct biopharmaceutical, pharmacokinetic and clinical pharmacological consequences.

Crystal habits, are divided into six distinct crystal systems, based on their geometry. They are cubic (*e.g.* sodium chloride), tetragonal (urea), hexagonal (iodoform), rhombic (iodine), monoclinic (sucrose) and triclinic (boric acid). Crystalline solids are distinguished from gases in that, unlike gases, they possess definite habits and are practically incompressible. Crystal structure consists of atomic, molecular or ionic units such as the cubic lattice of sodium ions interpenetrated by a lattice of chloride ions in the sodium chloride crystal. The electrostatic attraction of the oppositely charged sodium and chloride ions is what binds the crystal together. Weaker hydrogen bonding and van der Waals forces comprise the 'glue' that holds together molecules of organic compounds, thus predisposing such crystalline compounds to exhibit lower melting points. Covalent bonds between atoms comprise the lattice units of diamond and graphite, while naphthalene, hydrogen chloride and solid carbon dioxide crystals consist of molecular building units. Fatty acid crystals consist of dimer layers with the fatty acid chains tilted at an angle or parallel to the base plane.

In general, molecular crystals are soft and malleable with low melting points, compared with atomic and ionic crystals, which are hard and brittle with higher melting points. Metallic crystals may be hard or soft, depending on the kind of lattice defects in their crystals, and may exhibit high or low melting points. They are composed of positively charged ions in a field of freely moving electrons ('electron gas'), which makes them good electrical conductors. Liquid crystals or mesophase crystals consist of materials that have properties intermediate between the liquid and solid states. Molecules in the liquid state, unlike those in the solid crystal state, are mobile in three directional planes and rotate about three mutually perpendicular axes. In the liquid crystal state intermediate states of mobility and rotation exist, resulting in the three types of liquid crystals, namely smectic (grease or soap-like), nematic (thread-like) and cholesteric (a special case of the nematic type). The liquid crystal state results either because of solute–solvent interactions (known as lyotropic liquid crystals) or they may be formed when solids are heated (therefore called thermotropic liquid crystals). Liquid crystals are birefringent and light passing through them is divided into two different components of different velocities and refractive indices. They also undergo reproducible color changes as a function of temperature and so can be used as temperature indicators. Smectic liquid crystals are considered to be responsible for physically stabilizing emulsions and solubilized systems of water-insoluble pharmaceutical entities. Nematic mesophases are used in display systems due to their sensitivity to electrical signals. A detailed review of the literature on liquid crystals has been published by Brown.[1]

25.1.1.2 Internal Crystal Structure

Crystal habits can change, depending on the solvent used for crystallization, even while preserving the internal crystal structure, while any change in internal crystal structure will usually alter the crystal habit, *e.g.* the conversion of a free acid to its sodium salt. Halebian[2] has outlined a classification scheme for differentiating the habit and crystal chemistry for a chemical compound (Figure 1).

25.1.1.3 Amorphism

Crystalline structures consist of constituent units (atoms or molecules) in a three-dimensional lattice, while amorphous structures, as the name suggests, consist of randomly placed atoms or molecules just as seen in the liquid state. Amorphous solids are produced by supercooling of melts, by rapid precipitation or by lyophilization. They flow when subjected to the optimum temperature and pressure and are thermodynamically unstable. Amorphous solids have a propensity to revert to more stable configurations, which can be awkward when this occurs during various processing steps in the manufacture of solid dosage forms such as granulating, mixing and/or tableting or encapsulating. Amorphous forms usually possess higher thermodynamic energy than their crystalline counter-

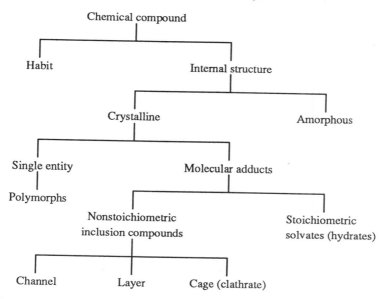

Figure 1 Outline for differentiating habit and crystal chemistry of a compound (reproduced from ref. 2 by permission of the American Pharmaceutical Association)

parts, resulting in solubilities and dissolution rates of much higher magnitudes. They also have, for the same reasons, a diffuse melting point range instead of a definite melting point. Amorphous substances are usually isotropic inasmuch as they exhibit similar properties in all directions. Crystals (other than cubic crystals, which are isotropic) are anisotropic and exhibit differences in properties such as dissolution rate, electrical conductivity and refractive index in different planes along the crystal. Beeswax and paraffin, when heated and slowly cooled, assume crystalline arrangements, while petrolatum (petroleum jelly) contains both crystalline and amorphous constituents. Amorphism can affect the therapeutic activity of a pharmaceutical entity. The amorphous form of the antibiotic novobiocin is at least 10 times more soluble than the crystalline form. Studies in dogs show that amorphous novobiocin is rapidly absorbed and is therapeutically active, while the crystalline form is not absorbed and is therapeutically inactive.[3]

Crystalline forms are further classified by Halebian on the basis of whether they contain a stoichiometric or nonstoichiometric amount of the solvent used in crystallization. Nonstoichiometric molecular adducts such as inclusion compounds or clathrates consist of a lattice or layer or cage-like structure in which solvent molecules are entrapped. If such an adduct can be reproducibly produced in terms of the habit and stoichiometry, then it would be worthwhile pursuing from a pharmaceutical product development point of view. Stoichiometric molecular adducts entrap the crystallization solvent molecules within the crystal lattice at specific sites. Crystalline hydrates are termed as anhydrous, hemihydrous, monohydrous, *etc.*, depending on the number of molar equivalents of water incorporated. Anhydrous forms generally have significantly greater dissolution rates and consequent bioavailability than the hydrous forms and it is important to preclude any such conversions during processing steps in the production cycle. The anhydrous forms of caffeine, theophylline and glutethimide dissolve more rapidly in water than the corresponding hydrous forms of these drugs.[4]

25.1.2 POLYMORPHISM

Polymorphism is the ability of a compound to exist in more than one crystalline state with different internal structures. The molecular shape may be different in the polymorphs but differences such as those seen with geometric isomers or tautomers do not constitute polymorphism. Polymorphism is differentiated from similar phenomena such as dynamic isomerism by the fact that polymorphs are identical in the liquid or vapor states even though they differ in crystal structure. Dynamic isomers on the other hand melt at different temperatures, producing melts of different composition, which upon standing revert to an equilibrium mixture of the two isomers at a given equilibrium temperature. Polymorphism can, therefore, also be defined as the ability of any element or compound to crystallize as more than one distinct crystalline species, *e.g.* carbon as cubic diamond or hexagonal graphite, with distinctly different properties. Optical and electrical properties, vapor pressure, crystal shape, hardness, density, melting point, solubility, stability, *etc.* all vary with the polymorphic form.

Nearly all long-chain compounds exhibit polymorphism. Polymorphs can be classified into two groups, namely enantiotropic, where the reversible transition from one polymorphic form to another is a function of temperature and pressure (*e.g.* sulfur), or monotropic, where one polymorph is unstable at all temperatures and pressures. The transition temperature is the temperature, at a constant pressure of 1 atm, at which two polymorphs coexist, have identical free energies, have identical solubilities in any solvent and have identical vapor pressures. Below the transition temperature the polymorph possessing the lower free energy, lower solubility and lower vapor pressure is considered to be the thermodynamically stable form. In theory, at least, it should be possible to synthesize polymorphs of a new pharmaceutical entity with attendant variation in physicochemical and biological activity. Preformulation research on new pharmaceutical entities usually includes studies on creation and detection of polymorphic forms. Parameters routinely investigated include the number of polymorphs that exist, their relative temperature stabilities, solubilities, the presence or absence of a glassy amorphous state, methods for reproducible preparation of each form, stabilization of metastable states, effect of micronization, interaction with formulation excipients and effect of processing steps during dosage form manufacture.

Halebian and McCrone[5] have outlined the following procedures to cause the crystallization of a metastable form; this is usually the initial task of the preformulation scientist when attempting to determine the existence of multiple crystalline forms.

(i) Completely melt a small amount of the compound on a slide and observe the solidification between crossed polars. If, after spontaneous freezing, a transformation is observed to occur spontaneously or if it can be induced by seeding or scratching, then it can be hypothesized that the compound probably exists in at least two polymorphic forms. Supercooling, which is essential to induce nonnucleation of the stable form, is induced by melting a small sample and holding the melt about 10 °C above the melting point for roughly 30 s and then setting it aside without any physical shock and observing it upon rapid cooling.

(ii) Heat a sample of the compound on a hot stage and observe whether a solid–solid transformation occurs while heating.

(iii) Sublime a small quantity of the compound and attempt to induce a transformation between the sublimate and the original sample by mixing the two in a drop of saturated solution of any one of them. If the two are polymorphs, the more stable will be more insoluble and will grow at the expense of the more soluble metastable form until eventually all the metastable form is converted to the stable form. If the samples are identical forms, nothing will happen. If the samples are not polymorphs, one may dissolve but the other will not grow.

(iv) A supersaturated solution of the compound in a small volume of solvent is held near the melting point of the compound. Taking care to maintain the temperature, isolate the suspended solid. Then, test the isolated material with an original sample using the procedure outlined in step (iii).

(v) Recrystallize the compound from solution by shock cooling and observe a portion of the precipitated material suspended in a drop of the mother liquor. The drop may then be seeded with the original compound to check for solution phase transformation. If the precipitate is a different polymorph, a solution phase transformation should take place.

Once the occurrence of more than one crystalline form has been determined, it is important to establish conditions under which each can be reproducibly produced. In the event that one particular polymorph is to be developed and marketed, lot to lot and intralot variations in raw material can be minimized. The recrystallization solvent used, the rate of crystallization and other factors such as the existence of metastable polymorphs and the prediction of their conversion rates within a dosage form are all factors that need investigation. For solid dosage forms such as capsules and tablets, the influence of particle size, moisture and excipients needs to be evaluated, while for suspension dosage forms, particle size, agitation, presence of seed nuclei, drug solubility in the vehicle and temperature are all possible factors influencing the rate of conversion of metastable polymorphs. The development of an analytical method which is sensitive to small amounts of the stable polymorph even in the presence of the metastable polymorph and also formulation adjuvants becomes critical. A lower limit of detection for polymorph mixtures in the range of 2–5% is considered adequate. Transition temperature determinations using van't Hoff plots are also commonly employed to distinguish between polymorphs. A free energy–temperature curve at a constant

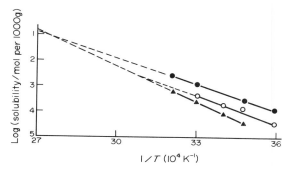

Figure 2 The van't Hoff plot of log solubility *versus* reciprocal absolute temperature for polymorphs A, B and C of chloramphenicol palmitate. Key: polymorphs A, (▲——▲); B, (●——●) and C, (○——○) (reproduced from ref. 6 by permission of the American Pharmaceutical Association)

pressure of 1 atm is constructed for such purposes. Aguair and Zelmer[6] described three polymorphic forms of chloramphenicol palmitate thermodynamically by a van't Hoff plot of solubility (indirectly determining free energy) *versus* temperature, wherein the transition temperatures are shown by the intersection of the extrapolated lines. The transition temperatures are 50 °C for forms A and C and 88 °C for forms A and B. Form A is the stable form at temperatures less than 50 °C (Figure 2). It should be noted, however, that transition temperatures obtained by extrapolation of van't Hoff plots are error-prone and direct measurements of transitions as corroborating data are advocated whenever possible.

25.1.3 QUALITATIVE AND QUANTITATIVE TECHNIQUES FOR THE CHARACTERIZATION OF CRYSTALLINE AND POLYMORPHIC FORMS OF THE SAME PHARMACEUTICAL OR CHEMICAL ENTITY

The principles behind the analytical methods used for characterization of solid forms make use of the fact that many physicochemical properties of the pharmaceutical or chemical entity vary with the internal structure of the solid drug. Quantitative analysis of these physicochemical properties such as melting point, density, hardness, vapor pressure, crystal shape and optical properties can be achieved using some of the methods listed in Table 1. It is recommended that a combination of these methods be used to ensure successful identification of all polymorphs.

25.1.3.1 Microscopy

Crystal forms exist either as isotropic or anisotropic substances. Isotropic crystals have an identical velocity of light in all directions (*i.e.* they have a single refractive index), while anisotropic crystal forms have plural velocities of light (*i.e.* they have multiple refractive indices). Amorphous substances such as supercooled glasses and cubic crystal substances like sodium chloride are examples of isotropic substances. When such substances are examined under a microscope with crossed polarizing filters, they appear dark, while anisotropic substances appear bright and birefringently multicolored. Anisotropic substances which have two principal refractive indices are called uniaxial, while those possessing three refractive indices are called biaxial. Most drugs possess

Table 1 Analytical Methods for Characterization of Solid Forms[a]

Method	Material required per sample	Method	Material required per sample
Microscopy	1 mg	X-Ray powder diffraction	500 mg
Fusion methods (hot stage microscopy)	1 mg	Scanning electron microscopy	2 mg
Differential scanning calorimetry (DSC/DTA)	2–5 mg	Thermogravimetric analysis	10 mg
IR spectroscopy	2–20 mg	Dissolution/solubility analysis	mg to g

[a] Ref. 74, p. 178.

a monoclinic, triclinic or an orthorhombic crystal form which is biaxial. Crystallographic methods are not simple and therefore require a well-trained optical crystallographer to enable proper characterization of biaxial systems. On the other hand, morphological differences between crystal forms and polymorphic transitions due to heat or solvents can usually be observed by less qualified professionals.

25.1.3.2 Hot Stage Microscopy

In this method, the polarizing microscope is fitted with a hot stage, thereby facilitating investigations of polymorphism, melting points and transition temperatures, and transition rates of metastable forms under controlled thermal and physical conditions. Hot stage microscopy has been used to facilitate the differentiation of differential scanning calorimetric endotherms.

25.1.3.3 Thermal Methods

Differential scanning calorimetry (DSC) and differential thermal analysis (DTA) have been extensively used for the identification and quantification of polymorphs. The principle governing both methods is the measurement of the temperature-mediated gain or loss of heat within a sample under uniform conditions. Examples of common exothermic (heat-emanating) processes are physical degradation and crystallization, while chemical degradation, desolvation, vaporization, solid–solid transitions, fusion, sublimation and boiling are examples of endothermic processes. Quantitative measurements of these enthalpic processes have been applied in studies on purity, polymorphism, excipient compatibility and degradation and solvation.[7–12] The area under the DSC curve can be quantified to yield the thermodynamic parameter heat of fusion (for the melting endotherm) and also the heat of transition from one polymorph to another. In general, a sharp symmetrical melting endotherm is indicative of relative purity, while broad asymmetric curves are suggestive of impurities and/or other thermal processes. It is important to heat at a uniform rate in order to influence thermal kinetics correctly and determine 'true' transition temperatures.

The atmosphere in which the sample is heated can be important in DSC and usually a continual nitrogen purge is maintained within the heating chamber. Sometimes, it is essential to conduct polymorphic transition experiments in an enclosed system in order to prevent loss of a volatile counter ion like ethanolamine or acetic acid. Conversely, venting of the atmosphere is usually needed during desolvation processes of hydrated crystal forms (Figure 3), otherwise the water vapor will cause degradation prior to the melting point of the anhydrous form. Thus, a variety of atmospheres should be tried during initial testing to determine the one best suited to quantitate the particular thermal process being studied.

Another thermal method often used for monitoring desolvation and decomposition processes is thermogravimetric analysis (TGA). TGA works on the principle of quantifying changes in sample weight as a function of temperature (for constant time) or as a function of time (isothermal). When both TGA and DSC data are obtained under identical conditions, interpretation of the underlying thermal processes is especially facilitated. Figure 3 shows how a sample containing as little as 10% of

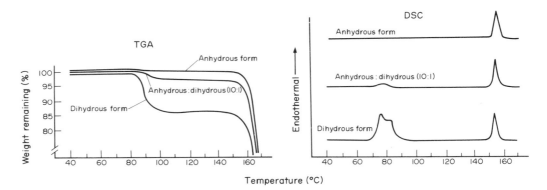

Figure 3 Thermogravimetric (TGA) and differential scanning calorimetric (DSC) analysis for an acetate salt of an organic amine that has two crystalline forms, anhydrous and dihydrous. Anhydrous/dihydrous mixture was prepared by dry blending. Heating rate was $5\,°C\,min^{-1}$ (reproduced from ref. 74, p. 179, by permission of Lea & Febiger)

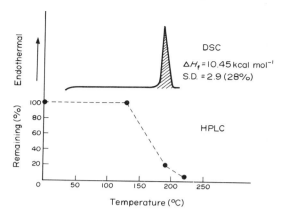

Figure 4 Differential scanning calorimetric (DSC) analysis and HPLC stability analysis of an organic amine hydrochloride salt that undergoes decomposition upon melting (reproduced from ref. 74, p. 180, by permission of Lea & Febiger)

a dihydrate was easily detected using both TGA and DSC. In the first endotherm, the dihydrate salt of an acetate salt loses 2 mol of water *via* an endothermic transition between 70 °C and 90 °C. The second endotherm at 155 °C corresponds to the melting process accompanied by weight loss due to both decomposition and vaporization of acetic acid. Degradation during thermal analysis can be detected by HPLC analysis of samples heated under similar conditions and monitoring retention of drug or appearance of decomposition products (Figure 4). Both TGA and DSC depend on thermal equilibration within the sample with sample preparation, sample homogeneity, heating rate, sample atmosphere, sample size and particle size being important variables.

25.1.3.4 Infrared Spectroscopy

Infrared spectroscopy is familiar to medicinal chemists and spectra are readily obtained routinely from modern instrumentation. IR spectra in the solid state may be obtained from mulls, alkali halide discs and by diffuse reflection.[13] Any differences observed between the spectra of different batches of a compound point to the presence of polymorphism, provided that certain other effects can be eliminated from consideration. These effects include questions of purity, solvates, crystal orientation, particle size and scattering phenomena. Purity may be investigated most readily by microanalysis, by chromatography and by spectral methods including IR spectroscopy in solution. These techniques will also indicate the occurrence of solvates. The solid-state IR spectra themselves will often show the presence of solvates, as many common solvents including water and acetone have highly characteristic spectra with strong bands. The main problem that arises is the one of recognizing polymorphism in the presence of impurities or solvation, since interstitial solvent molecules can be tenaciously held, so that the heating needed to drive off the solvent may induce a polymorphic transition. X-Ray crystallography is the surest indicator of such polymorphism if suitably large crystals of the form can be found.

Grinding is necessary to prepare mulls or discs. Sufficient grinding will remove crystal orientation and particle size effects. However the possibility of bringing about a polymorphic transition by grinding, or even on occasions by simple manipulation of an unstable polymorph, needs to be considered. Scattering is related not only to particle size and to refractive index differences between analyte and matrix but also to incorporation of air bubbles during mulling, or failure to eliminate air or crystal boundaries during the pressing of KBr discs. Good technique will avoid these problems.

Grinding is not essential for the observation of spectra by diffuse reflection, but the possibility of a dispersion effect[14] (the Christiansen filter effect) in the spectra as a result of specular reflectance at larger crystal faces must be considered. Organic compounds display weaker, less sharp bands in the overtone (near-IR) region of the spectrum[15] and so such spectra do not suffer from noticeable dispersion effects. Although the spectral patterns are much less characteristic to the eye from those in the mid-IR, multivariate statistical routines have been devised to overcome this limitation, so that near-IR reflection now offers a further possibility for the investigation of polymorphism as well as for the analysis of mixtures of polymorphs (Figure 5).[16]

A possible problem with mulls is the observation of anomalous spectra due to partial dissolution of the substance in the mulling agent. However as most substances of medicinal interest are polar,

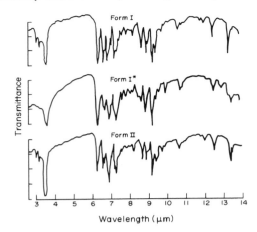

Figure 5 IR spectra of forms I, I* and II of SK&F 30 097 (reproduced from ref. 24 by permission of Mack Publishing Co.)

this is rarely a problem in pharmaceutical work. From the above discussion it can be seen that it is very desirable to obtain spectra by all available sampling techniques in the investigation of polymorphism, as well as by a range of other methods, as has always been practised by the most experienced investigator.[17,18]

Many IR spectrometers are now supplied with sample-heating devices. The change in spectrum with temperature can be very informative. Unfortunately the temperature control of some of these accessories leaves much to be desired, and so they should be calibrated with a range of substances of known melting point at varied heating and cooling rates. The ultimate goal of a thorough study should be the construction of a complete phase diagram and, to this end, cycling the temperature past the transition points is essential in distinguishing monotropic and enantiotropic transitions. This is most advantageously carried out in parallel with a similar exercise by DSC. Interpretation of the solid-state spectra can help in distinguishing dynamic and conformational[19] polymorphism from true polymorphism and can also give some indication of crystal-packing features, *e.g.* by observation of hydrogen bonding,[20] carbonyl peak shifts and band splitting. Sometimes this can supplement X-ray crystallographic investigation.[21] A further application of IR spectroscopy is through Burger's IR rule,[22] relating IR band positions to relative stabilities, which can be of help in establishing the thermodynamic relationships between the polymorphs.

25.1.3.5 X-Ray Powder Diffraction

The principle behind this technique is that the pattern of a crystal lattice in a powder sample causes the X-rays to scatter in a reproducible pattern of peak intensities at distinct angles relative to

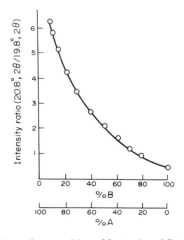

Figure 6 X-Ray intensity ratio as a function of composition of forms A and B of chloramphenicol palmitate (reproduced from ref. 44 by permission of the American Pharmaceutical Association)

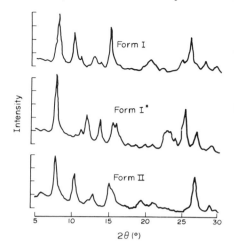

Figure 7 X-Ray diffractograms for forms I, I* and II of SK&F 30 097 (reproduced from ref. 24 by permission of Mack Publishing Co.)

the incident beam. This is due to the fact that X-rays have wavelengths of about the same magnitude as the distance between the atoms or molecules of a crystal. The X-ray diffraction pattern is photographed on a sensitive plate arranged behind the crystal form. Each diffraction pattern is characteristic for a specific crystalline lattice for a given compound.[23] An amorphous form, in contrast, does not produce a pattern. Electron density and position of atoms in complex structures such as penicillin may be determined from a mathematical study of the X-ray diffraction data. Mixtures of different crystalline forms can be analyzed using normalized intensities at specific angles unique for each crystalline form. Precise identification and description of a crystalline substance can be obtained using single-crystal X-ray analysis, while unit cell dimensions and angles help to establish conclusively the crystalline lattice system and establish specific differences between different crystalline forms of a compound. Figure 6 shows the X-ray intensity ratio as a function of composition of chloramphenicol palmitate forms A and B, while Figure 7 shows the X-ray diffractograms for forms I, I* and II of SK&F 30 097.

25.1.3.6 Dilatometry

The property of fats and fatty acids and some pharmaceutical substances to expand upon melting and contract upon passing from a metastable to a stable polymorph is the principle behind

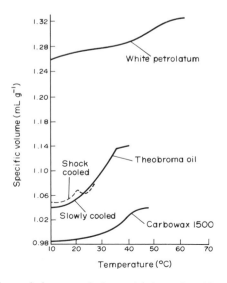

Figure 8 Dilatometric analysis of several pharmaceutical materials (reproduced from ref. 25 by permission of the American Pharmaceutical Association)

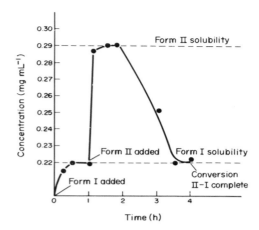

Figure 9 Powder dissolution profiles for two polymorphic forms of an organic acetate salt in acetonitrile at 25 °C (reproduced from ref. 74, p. 190, by permission of Lea & Febiger)

dilatometric analysis. The dilation of a sample confined in a dilatometer is measured as a specific volume (mL g^{-1}) and plotted against temperature to yield dilatometric curves. Dilatometric analysis, albeit extremely accurate, however, is extremely tedious and time consuming and, therefore, not widely used. Compounds of pharmaceutical interest studied by dilatometry include theobroma oil, methyl stearate and chloramphenicol.[24-27] Ravin and Higuchi dilatometrically analyzed theobroma oil under conditions of slow and rapid cooling (shock cooling by immersion in an acetone–dry ice bath after initial sample melting). The results as graphed in Figure 8 show that the shock-cooled sample of theobroma oil (cocoa butter), upon reheating, rapidly expands between 16 and 20 °C, followed by contraction between 20 and 24 °C, presumably due to the change from the metastable to the more dense polymorphic form. In contrast, white petrolatum and carbowax 1500 are not significantly affected by similar shock cooling (Figure 8). It is now known that, in general, polymorphism does not occur in substances like petrolatum fats containing complex mixtures of glycerides and fatty acids that melt over a wide temperature range. Polymorphism occurs more frequently in fats such as theobroma oil and hydrogenated corn oil, predominantly consisting of single glycerides, and also in mixtures of fats with oils of low melting point.

25.1.3.7 Dissolution and Solubility Analysis

Metastable polymorphs tend to have significantly better solubility and dissolution profiles and hence can be differentiated from thermodynamically stable forms at room temperature. Figure 9 shows the dissolution profile of an organic acetate salt in an organic solvent. When an excess of form II was added, the resulting increase in dissolution of the metastable form II, and its subsequent conversion into the stable form I, was corroborated by intrinsic dissolution rates and solubility measurements, which showed a 31% higher solubility and a 59% higher dissolution rate for form II in acetonitrile.

25.1.4 SIGNIFICANCE OF CRYSTALLINITY, AMORPHISM AND POLYMORPHISM IN THE DRUG DOSAGE FORM DEVELOPMENT SCHEME

The occurrence of polymorphism is nowadays checked in the early stage of the drug dosage form development activities. As mentioned earlier too, the accent of such preformulation research is towards understanding the physicochemical properties of the powdered drug polymorphs, while bearing in mind that the liquid forms of these polymorphs exhibit no such differences. The metastable polymorph will tend to have an increased solubility and faster dissolution than the stable polymorph. This property may become important for a drug with an inherently poor intrinsic dissolution rate profile, provided the metastable form does not revert to the stable configuration during the shelf life of the product as well as during its temporary sojourn in the GI tract.

The formulation of polymorphic forms into suspension dosage forms can be especially tricky, since ideal thermodynamic conditions exist for reverting to the stable form. Moustafa *et al.*[28]

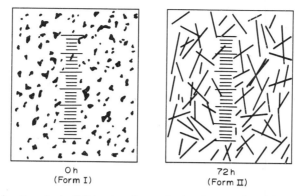

O h
(Form I)

72 h
(Form II)

Figure 10 Photomicrographs of succinylsulfathiazole polymorphs showing transformation from form I at zero time to the stable form II after 72 h (reproduced from ref. 28 by permission of the American Pharmaceutical Association)

reported six crystal polymorphs and one amorphous form of succinylsulfathiazole, of which form II was the stable configuration. In aqueous suspensions, the unstable forms gradually changed to the stable form II as shown in the photomicrographs in Figure 10; metastable form I reverts to the stable form II after 72 h. Methylprednisolone exists as two polymorphs, I and II, which exhibit significant differences in release rates both *in vitro* and *in vivo*. Polymorph II has *in vitro* and *in vivo* release rates of between 42 and 51% in excess of polymorph I.[29]

Two crystalline forms of phenytoin have been isolated, with the needle-shaped (presumably stable) form exhibiting a slower dissolution profile in water.[30] Difenoxin hydrochloride exists in two polymorphic forms with form I being more soluble than form II. The individual solubilities of the two forms were predictive of the solubilities and dissolution of physical mixtures of the two forms. Micronization improved the dissolution of tablets prepared from a lot with an approximately 2:1 form I:form II ratio and at this ratio dissolution of difenoxin hydrochloride from tablets was equivalent to that of 100 mesh pure form I.[31] Tuladhar *et al.*[32] have detailed the preparation and thermal characterization of five phenylbutazone polymorphs. Rapid heating rates produced single endothermic peaks due to melting, while slower heating rates resulted in interconversion of three of the polymorphs to the more stable form. Interconversion was also noticed upon grinding of the polymorphs. Polymorphic and particle size changes during compression and crushing and bonding mechanisms associated with such changes were also evaluated using dissolution rate testing.[32,33]

The bioavailability in dogs and 12 humans of a potential tricyclic antidepressant (SQ-10 966) was increased when the compound was administered in capsule formulations as micronized drug coated with 1% sodium lauryl sulfate or as a lyophilate with poloxamer 407 (Pluronic F-127). This observed increase *in vitro* had been predicted by *in vitro* dissolution testing in 0.1 M HCl wherein the lyophilized combination was more soluble. Characterization of the lyophilate indicated that the increase in solubility was attributable to the formation of a polymorphic form. Experiments were subsequently conducted to determine the solubility and dissolution characteristics of the two polymorphic forms A and B and of the lyophilized combination.[34] Disopyramide, too, exists in two crystal forms identified using the usual techniques. The intrinsic dissolution rates of the two forms calculated at 37 °C were not noticeably different. Capsules containing 200 mg of either disopyramide polymorph when administered to three volunteers showed no significant differences in blood levels at 1.5 and 6 h postdosing.[35]

Thakker *et al.* have shown that at least four distinct polymorphic forms of nabilone, a cannabinoid derivative, exist, depending upon the crystallization conditions and solvent. All forms appeared to be equally hydrophobic and insoluble. All the bioavailable forms tended to convert upon heating, grinding or prolonged storage to the nonbioavailable and thermodynamically stable form. An effective way of preventing this conversion is to keep nabilone dispersed in the water-soluble matrix of povidone. Administration to dogs of capsules containing a povidone dispersion at a dose of 1 mg kg^{-1} resulted in the onset of a pharmacological response 1.5 h after dosing, with the maximum intensity being seen at 3–5 h. The dispersion was also found to maintain nabilone bioavailability for at least two years at room temperature.[36]

Codeine is known to exist in three different crystal forms: I, II and III. Tablets prepared by wet granulation of either form III or I showed little or no significant difference in the dissolution rate, while tablets prepared by direct compression of form III dissolved about three times faster than that of form I.[37]

Fosinopril sodium, a new angiotensin converting enzyme inhibitor, has been reported to exist as two polymorphs, A and B. The melting points of the two forms are 197 °C (form A) and 206 °C (forms B). Using the usual techniques, further characterization was attempted. No significant differences in physical or chemical stability were observed when the pure drug in either form was stored under stress conditions. *In vitro* dissolution rates of the two polymorphs from dry filled capsules were equivalent and the two forms given to humans as dry filled capsules exhibited bioequivalence to a solution dosage form. Further, since it was found that the formation of form B polymorphs was favored when the solutions of the drug in water or alcohol were rapidly evaporated, conversion studies under normal conditions encountered in a wet granulation process were undertaken. Over the range of granulation solvent and drying conditions examined, conversion of form A and form B was ruled out by comparing the virgin powder blend and the dried granulation by X-ray powder diffraction.[38,39]

Similar crystal modification studies have recently been described by Chow and Grant[40] by crystallizing griseofulvin from various *n*-alkanoic acids. Their results suggest that the crystal structure and the properties of a sparingly soluble drug may be modified appreciably by doping with an interactive solvent through the formation of a solvate or a disrupted non-solvate. The effect of powder compaction on crystal modifications has been the subject of excellent recent reviews.[41,42] A recent report on the photostability of several crystal forms of cianidanol has also been published and shows that of five crystal forms the monohydrate (form II) showed the best photostability.[43]

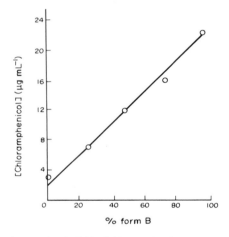

Figure 11   Correlation of 'peak' blood serum levels (2 h) of chloramphenicol *versus* percentage concentration of polymorph B (reproduced from ref. 44 by permission of the American Pharmaceutical Association)

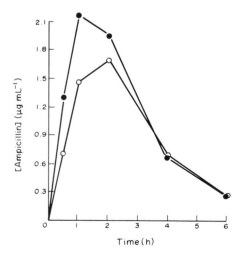

Figure 12   Mean serum concentrations of ampicillin in human subjects after oral administration of 250 mg doses of two solvate forms of the drug in suspension: ●, anhydrous; ○, trihydrous (reproduced from ref. 45 by permission of Therapeutic Research Press Inc.)

Chloramphenicol palmitate exists in four polymorphs: three crystalline forms (A, B and C) and an amorphous one. Aqueous suspensions of polymorphs A and B yielded average peak chloramphenicol concentrations in blood of 3 and 22 $\mu g\,mL^{-1}$, respectively, upon administration of 1.5 g doses to healthy volunteers (Figure 11). Similar differences were noted with respect to the extent of absorption of the two polymorphs.[44] An example of the *in vivo* importance of solvate forms is shown in Figure 12. The anhydrous and trihydrous forms of ampicillin were administered orally as a suspension to human subjects and, as seen in Figure 12, the more soluble anhydrous form ($10\,mg\,mL^{-1}$) produced higher and earlier blood serum levels than the less soluble trihydrous form.[45] Similarly, the anhydrous forms of caffeine, theophylline and glutethimide dissolve more rapidly in water than do their hydrous forms.[46] Several reports of organic solvents enhancing the dissolution of drugs have been published previously too. Thus, *n*-pentanol and ethyl acetate solvates of fludrocortisone and the *n*-pentanol solvate of succinylsulfthiazole are reported to be better dissolved.[47] Studies in man indicate that the rate and extent of absorption of griseofulvin were significantly increased after administration of the chloroform solvate compared to that observed after administration of the nonsolvated form of the drug. These findings are consistent with the observed enhanced solubility and dissolution rate of the solvate in simulated intestinal fluid.[48]

25.1.5 PARTICLE SIZE

25.1.5.1 Definitions and Examples

The science and technology of small particles is termed micromeritics. The range of particle size distributions in pharmaceutical systems covers most oral and nonenteral dispersion dosage forms such as suspensions, microemulsions, macroemulsion systems, powders and granules. The fine particles of powder and particles of suspensions and pharmaceutical emulsions can be visualized using optical microscopy, whereas some colloidal dispersions consist of particles that are too small to be seen in the ordinary microscope. Particles of coarse powders, tablet granules and granular salts fall within the sieve range (Table 2). The commonly accepted unit of particle size is the micrometer (μm), equal to 10^{-6} m. Milling or comminution is the mechanical process of reducing the particle size of solids. Milling equipment is classified as coarse, intermediate or fine, according to the final size of the milled product. Particle size is also expressed in terms of mesh or number of openings per linear inch of screen (1 inch = 25.4 mm) and a convenient classification based on mesh size is smaller than 200 mesh (fine particles), 200–20 mesh (semifine or semicoarse particles) and larger than 20 mesh (coarse particles).

25.1.5.2 Particle Size and Size Distribution

For spherical particles, the particle size is expressed in terms of its diameter. With irregular particles, however, since there is no unique diameter, the particle size is expressed in terms of an equivalent spherical diameter. The equivalent spherical diameter relates the size of the particle to the diameter of a sphere having the same surface area (d_s), volume (d_v), diameter (d_p) or Stokes' diameter (d_{st}). The Stokes' diameter describes an equivalent sphere undergoing sedimentation at the same rate as the asymmetric particle. Most powders are polydispersed with fractions of different particle size particles all mixed together in the constituent sample. The particle size distribution of such a

Table 2 Particle Dimensions in Pharmaceutical Disperse Systems[a]

Particle size (μm)	Approximate sieve size	Examples
0.5–10	—	Suspensions, fine emulsions
10–50	—	Upper limit of subsieve range, coarse emulsion particles: flocculated suspension particles
50–100	325–140	Lower limit of sieve range, fine powder range
150–1000	100–18	Coarse powder range
1000–3360	18–6	Average granule size

[a] Ref. 75, p. 492.

particulate system therefore consists of estimating both the size range and the number or weight fraction of each particle size.

25.1.5.3 Average Particle Size

Knowledge of the particle size distribution enables calculation of the average particle size for the sample. The arithmetic, geometric or harmonic mean diameter of a particle size distribution can be calculated using the following equation

$$d_{mean} = (\Sigma nd^{p+f}/\Sigma nd^{f})^{1/p}$$

where n = number of particles in a size range, d = midpoint of the equivalent particle diameters, and p = size index of an individual particle. If $p = 1$, then the index expresses particle length, if $p = 2$, it expresses surface area and if $p = 3$, it expresses volume. If $p = 0$, then the mean is expressed as the geometric mean, if p is positive, then the mean is expressed as the arithmetic mean and if p is negative, the mean is expressed as the harmonic mean. For a particular size range, the frequency with which a particle in a certain size range occurs is expressed as the frequency index nd^{f}. When the frequency index has values of 0, 1, 2, or 3, then the size frequency distribution expresses the total number, length, surface or volume of the particles, respectively. For most pharmaceutical powders the important statistical mean arithmetic diameter is given by the formula

$$d_{mean} = \Sigma nd^{3}/\Sigma nd^{2}$$

when $p = 1$, $f = 2$ and where d is the volume-surface or surface-weighted mean. The pharmaceutical importance of this parameter lies in the fact that it is inversely related to S_{w}, the specific surface area.

25.1.5.4 Particle Size Distribution

The graphical representation of the incidence frequency (by number or weight) of particles lying within a certain size range or mean particle size is called a frequency distribution curve. Such a plot offers a visual depiction of the distribution differently from the average diameter, since it is possible to have two samples with the same average diameter but different distributions. The curve also helps to determine the mode or the particle size that occurs most frequently within the sample. A typical example is shown in Figure 13.

Alternative methods of plotting include plotting the cumulative percentage over or under a particular size *versus* particle size (Figure 14). The sigmoidal curve has its mode at the greatest slope. Since a normal distribution is very rarely found in pharmaceutical powders, the frequency data are

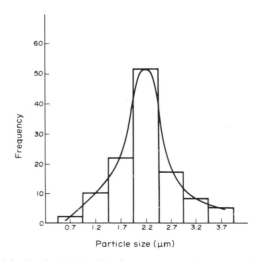

Figure 13 A histogram of a particle size distribution. The frequency curve is represented by the smooth line drawn through the histogram (reproduced from ref. 75 by permission of Lea & Febiger)

often plotted *versus* the logarithm of the particle diameter, thereby normalizing the distribution (Figure 15). When the logarithm of the particle size is plotted against the cumulative percent frequency on a probability scale, a linear relationship is observed (Figure 16) with two parameters, *i.e.* the geometric standard deviation (σ_g) and the geometric mean diameter (d_g). The geometric

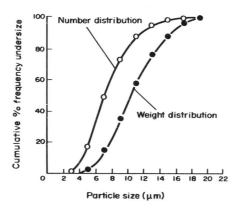

Figure 14 Typical cumulative frequency plot (reproduced from ref. 75, p. 499, by permission of Lea & Febiger)

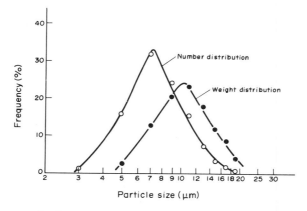

Figure 15 Typical frequency distribution plot (reproduced from ref. 75, p. 499, by permission of Lea & Febiger)

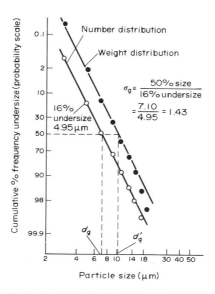

Figure 16 Typical log–probability plot (reproduced from ref. 75, p. 500, by permission of Lea & Febiger)

Table 3   Hatch–Choate Equations for Computing Statistical Diameters from Number and Weight Distributions[a]

Diameter	Number distribution	Weight distribution
Length–number mean	$\log d_{\mathrm{ln}} = \log d_{\mathrm{g}} + 1.151 \log^2 \sigma_{\mathrm{g}}$	$\log d_{\mathrm{ln}} = \log d'_{\mathrm{g}} - 5.757 \log^2 \sigma_{\mathrm{g}}$
Surface–number mean	$\log d_{\mathrm{sn}} = \log d_{\mathrm{g}} + 2.303 \log^2 \sigma_{\mathrm{g}}$	$\log d_{\mathrm{sn}} = \log d'_{\mathrm{g}} - 4.606 \log^2 \sigma_{\mathrm{g}}$
Volume–number mean	$\log d_{\mathrm{vn}} = \log d_{\mathrm{g}} + 3.454 \log^2 \sigma_{\mathrm{g}}$	$\log d_{\mathrm{vn}} = \log d'_{\mathrm{g}} - 3.454 \log^2 \sigma_{\mathrm{g}}$
Volume–surface mean	$\log d_{\mathrm{vs}} = \log d_{\mathrm{g}} + 5.757 \log^2 \sigma_{\mathrm{g}}$	$\log d_{\mathrm{vs}} = \log d'_{\mathrm{g}} - 1.151 \log^2 \sigma_{\mathrm{g}}$
Weight–moment mean	$\log d_{\mathrm{wm}} = \log d_{\mathrm{g}} + 8.059 \log^2 \sigma_{\mathrm{g}}$	$\log d_{\mathrm{wm}} = \log d'_{\mathrm{g}} + 1.151 \log^2 \sigma_{\mathrm{g}}$

[a] Ref. 75, p. 501.

standard deviation is the slope of the line and the geometric mean diameter is the logarithm of the particle size equivalent to the 50% size.

Whether particle data are collected by a weight distribution (such as sieving or sedimentation) or a number distribution (such as microscopy), it is still possible to interconvert the weight and number distributions using two methods. The first method for weight distributions is the same as outlined earlier for number distributions (Figures 14–16). It is important to distinguish size distributions on a weight and number basis since, for example, in Figure 14, 42% of the total *weight* of the particles is greater than 11 μm but this comprises only 12% of the sample by number. The second method is to use Hatch–Choate equations (Table 3), which can interconvert number and weight distributions and compute a particular average particle diameter using the relevant equation. The particle number or the number of particles per unit weight N can be calculated using the following equation (assuming spherical particles)

$$N = 6/\pi dvn^3 \rho$$

where $\pi dvn^3/6$ is the volume of a single particle and $\pi dvn^3/6\rho$ is the mass per particle (volume × density).

25.1.6   METHODS OF DETERMINING PARTICLE SIZE AND SURFACE AREA

The methods commonly used for determining the particle size of pharmaceuticals can be classified on the basis of their approximate effective particle size quantitation range. Thus, optical microscopy, sedimentation and Coulter counter (particle volume) measurements are used to determine particle sizes in the 1 to 500 μm range, while sieving determines in the 50 to 2500 μm range. Finer particles of 0.005 to 1 μm are measured using electron microscope or ultracentrifuge methods. For particle surface area quantitation, air permeability (1 to 100 μm range) or absorption (0.05 to 0.1 μm range), methods are used.

25.1.6.1   Optical Microscopy

Optical microscopy is the most direct method for particle size distribution measurement. The resolution power of the lens determines its lower limit of application. The ordinary microscope in white light can be used to measure particles from 0.4 to 150 μm. With special lenses and UV light, the lower limit could be extended to 0.1 μm. Particles in the range of 0.01 to 0.2 μm can be measured using the ultramicroscope equipped with darkfield illumination. For measurements, an undiluted or diluted emulsion or suspension is mounted on a slide or ruled cell and placed on a mechanical stage. The diameters of the particles on the slide are measured by means of a calibrated micrometer eyepiece. The hairline of the eyepiece is moved by the micrometer to two opposite ends of the particle in turn. The difference between the two micrometer readings is the diameter of the particle. All the particles are measured along an arbitrary fixed line. Eyepieces or graticules with circular or square grids are used to compare the cross-sectional area of each particle in the microscope field with one of the numbered patterns. The number of particles that best fits one of the numbered circles is recorded. The field is changed and the procedure is repeated with another numbered circle and so on until the entire size range is covered. The field can be projected on to a screen for ease of measurement or a photograph can be taken for measurement. Another method consists of splitting the image of the particle until the two images are just separated. The image-splitting device is calibrated such that the adjustments required to achieve this separation give the particle diameter.

The particulate fields are randomly selected for counting and the total number of fields selected depends on the number of particles per field. The British Standard on microscope counting recommends counting at least 625 particles. A wide particle size distribution will necessitate counting more particles, while as few as 200 particles may be sufficient if the distribution is narrow. A size–frequency distribution curve is then plotted (Figure 13) for determination of the statistical diameters of the distribution. To eliminate subjectivity and to lessen operator fatigue, photomicrographs, projections and automatic scanners could be used. The major disadvantage of the microscope method is that it is relatively slow and tedious. It is a numbered measurement and by definition only two dimensions of the particle, *i.e.* length and breadth, are estimated, not depth (thickness). The advantage of this method is that the presence of agglomerates and particles of more than one component can be detected.

25.1.6.2 Sieving

Sieving is the most popular method for measuring particle size distribution because it is a simple, rapid, inexpensive and objective method. The method utilizes a series of standard sieves calibrated by the National Bureau of Standards. The lower limit of application is usually 50 µm, although sieves produced by photoetching and electroforming techniques are available (micromesh) for extending the lower limit to 5 µm. A sieve consists of a pan with a bottom of wire cloth with square openings. The US Standard Scale uses the ratio of the width of openings in successive sieves as $\sqrt{2}$ with the size of openings being 1 mm in a wire cloth having 18 openings per linear inch, *i.e.* 18-mesh. The Tyler Standard sieves also used in the US also have the ratio of $\sqrt{2}$ but are based on the size of openings being 0.0029 inch in a wire cloth having 200 openings per linear inch, *i.e.* 200-mesh.

The USP method for testing consists of placing a definite mass of sample on the proper sieve in a mechanical shaker, shaking for a definite time period and collecting and weighing the material retained on the fine sieves. The parameters needing to be standardized include the type of motion (vibratory most efficient, followed successively by side-tap motion, bottom-tap motion, rotary motion with tap and rotary motion), and the time required to sieve a given material, which is roughly proportional to the load (thickness of powder per unit area of sieve) placed on the sieve. Alternative approaches are to assign the particles on the lower sieve the arithmetic or geometric mean size of the two screens or to assign the powder the mesh number of the screen through which it passes or on which it is retained. The data from such analysis, assuming it is log–normally distributed, can be plotted as cumulative percent by weight of powder retained on the sieves *versus* the logarithm of the arithmetic mean size of the openings of each of two successive screens (probability scale), as illustrated in Figure 16.

25.1.6.3 Sedimentation

The principle behind this method is the dependence on particle size of the rate of sedimentation as expressed by Stokes' equation

$$d_{\text{Stokes}} = \left[\frac{18\eta h}{(\rho - \rho_0)gt} \right]^{1/2}$$

where d_{Stokes} is the effective or Stokes' diameter, η is the viscosity of the dispersion fluid, h/t is the sedimentation rate or distance of fall h in time t, g is the gravitational constant, and ρ and ρ_0 are the densities of the particle and the medium, respectively. The sedimentation method may be used over a size range from 1 to 200 µm. It is applicable to free spheres falling at a constant rate independent of one another. The latter criterion is achieved if the concentration of the suspension does not exceed 2%. Stokes' law is applicable to irregularly shaped particles of various sizes, in which case the diameter obtained is a relative particle size, equivalent to that of a sphere falling at the same velocity as that of the particles under consideration. If need be, a deflocculating agent may be used to prevent agglomeration of the particles as they fall through the medium. Another assumption inherent to Stokes' law is that the flow of dispersion medium around the particle as it sediments is laminar or streamline and not turbulent. The type of flow is indicated by the dimensionless Reynolds number, *Re*, with values greater than 0.2 being considered indicative of turbulent flow. Under a given set of

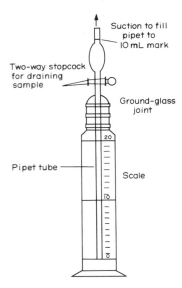

Suction to fill pipet to 10 mL mark

Two-way stopcock for draining sample

Ground-glass joint

Pipet tube

Scale

Figure 17 The Andreasen pipet apparatus for particle size measurement based on the principle of sedimentation (reproduced from ref. 76 by permission of Lea & Febiger)

density and viscosity conditions, the following equation calculates the maximum particle diameter whose sedimentation will be governed by Stokes' law (*i.e. Re* = 0.2)

$$d^3 = \frac{18Re\eta^2}{(\rho_s - \rho_0)\rho_0 g}$$

Among the different sedimentation methods available, the balance method, the hydrometer method and the pipet method, the pipet method is the most economical, accurate and easy to use. The Andreasen pipet consists of a 550 mL vessel containing a 10 mL pipet sealed into a ground-glass stopper. When the pipet is in place in the cylinder, its lower tip is 20 cm below the surface of the suspension (Figure 17). A 1 or 2% suspension of the particles in a medium containing a suitable deflocculating agent is added into the vessel up to the 550 mL mark. The stoppered vessel is shaken to distribute the particles and the apparatus with pipet is securely clamped in a constant-temperature bath. At fixed time intervals, 10 mL samples are withdrawn from a specified depth without disturbing the suspension and they are evaporated and weighed or analyzed by other appropriate means after correcting for the defloccuation agent added, if any.

The particle diameter corresponding to the different sampling periods is calculated from Stokes' law (*h* is the height of the liquid above the lower end of the pipet at the time each sample is removed). The percentage by weight of the initial suspension is calculated for particles having a size smaller than the size calculated by Stokes' equation for that time, by knowing the weight of the dried sample. The weight of each sample residue is called the weight undersize and the sum of the successive weights is known as the cumulative weight undersize. It may be expressed directly in weight units or as percent of the total weight of the final sediment (Figures 13–16). Hatch–Choate equations could be used to calculate the appropriate diameter.

25.1.6.4 Particle Volume Measurement

Particle volume measurement is achieved using the Coulter counter (Figure 18); the principle of operation is that when a particle suspended in a conducting liquid passes through a small orifice flanked on both sides by electrodes, a resultant change in the electrical resistance is registered. A known volume of a sufficiently dilute suspension is pumped through the orifice so that the particles pass through virtually one at a time. A current is produced by a constant voltage applied across the electrodes and as the particle travels through the orifice, it displaces its own volume of electrolyte, resulting in an increased resistance between the two electrodes and causing a voltage pulse (proportional to particle volume), which is amplified and fed to a pulse height analyzer calibrated in terms of particle size. The instrument electronically records all particles producing pulses that are

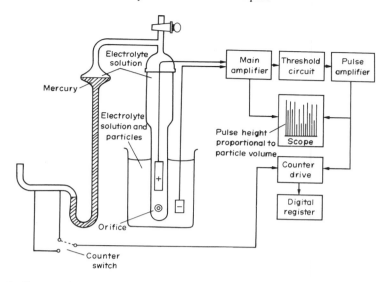

Figure 18 Schematic diagram of Coulter counter for determining particle volume (reproduced from ref. 75, p. 505, by permission of Lea & Febiger)

within the two threshold values of the analyzer. A particle size distribution is obtained by systematically varying the threshold settings and counting the number of particles in a constant particle size. The instrument can count up to 4000 particles per second and, therefore, enables data collection in a relatively short period of time. The data can, of course, be converted from a volume distribution to a weight distribution. The Coulter counter has been used to study particle growth and dissolution and the effect of antibacterial agents on microbial growth.

25.1.6.5 Laser Optic Methods

The Brinkmann Particle Size Analyzer Model 2010, which is interfaced with an IBM PC/XT and combines direct particle sizing with image analysis, has a measuring range of 0.7 to 1200 µm and measures all types of particles in all media. A He–Ne laser beam is optically shaped and focussed on to the sample and is scanned through the measuring zone by a rotating wedge prism. The time spent by the scanning beam on a particle is interpreted directly as particle size. Pulse-editing techniques validate each signal. Because individual particles are measured directly, the inconsistencies of scattering theory, Brownian motion, refractive index, viscosity variations and thermal convection are eliminated.[49] The Model 715 Granulometer also interfaces a HP 85 microcomputer and operates on the principle that the presence of particles within a coherent light beam causes diffraction, which results in the presence of light outside the geometrical limits of the beam. A parallel beam from a low power He–Ne laser lights up a cell which contains the powder suspended in a suitable liquid. The beam leaving the cell is focussed by a convergent optical system. The distribution of the light energy in the focal plane of the system is then analyzed by means of a multicell detection device. The data provided by the cells is processed in a built-in computer, which controls the display of the granulometric curve on a LED matrix and the printing on a paper tape of the average 256 measurements.[50]

25.1.7 PARTICLE SHAPE AND SURFACE AREA

A sphere has minimum surface area per unit volume, while the greater the particle asymmetry, the greater the surface area per unit volume. The specific surface area per unit volume (S_v) or per unit weight (S_w) can be given by the following two equations

$$S_v = \frac{\text{Surface area of particles}}{\text{Volume of particles}} = \frac{\alpha_s}{\alpha_{vd}}$$

$$S_w = \frac{\alpha_s}{\rho d_{vs}\alpha_v}$$

where α_s/α_v is the particle shape ratio (= 6.0 when the particle is spherical and > 6.0 when the particle is asymmetric), ρ is the true density of the particles, d is the diameter of the particle, d_{vs} is the volume–surface diameter characteristic of a specific surface. When the particles are spherical (or almost so), the surface area per unit weight is given by

$$S_w = 6/\rho d_{vs}$$

The surface area of a powder sample can be calculated from a knowledge of the particle size distribution. For direct calculation of surface area, two methods are commonly available. In the absorption method, the amount of a gas or liquid solute that is adsorbed on to the sample of powder to form a monolayer is a direct function of the surface area of the sample. The air permeability method depends on the fact that the rate at which a gas or liquid permeates a bed of powder is related to the surface area exposed to the permeant.

For more extensive reviews of particle size measurement, surface area and pore size, the reader is referred to other authors.[51-53]

25.1.8 SIGNIFICANCE OF PARTICLE SIZE OF A PHARMACEUTICAL OR CHEMICAL ENTITY IN THE PHARMACEUTICAL PRODUCT DEVELOPMENT SCHEME

The particle size of a pharmaceutical entity is a consideration in virtually all dosage form formulations meant for oral and nonoral administration. The surface area per unit weight or specific surface area is increased by size reduction, thereby aiding biodissolution and increasing the potential systemic bioavailability. Most poorly soluble drugs are marketed in microcrystalline or micronized form in order to optimize their absorption potential. For example, since the original marketing of spironolactone, the therapeutic dose has been reduced twentyfold, from 500 mg to 25 mg, by reformulation including micronization of the drug.[54] Similarly, the currently marketed formulation of micronized griseofulvin calls for a 0.5 g daily dose which is half that needed when the drug was originally marketed.[54] Reduction in the mean particle size diameter of digoxin from 20–30 μm to 3.7 μm led to an increase in the rate and extent of absorption of the drug.[55] Further, in another study, tablets containing micronized or large particle size digoxin, when administered with metoclopramide or propantheline, showed significantly decreased absorption from the large particle size tablets compared to the micronized tablets of digoxin.[56]

The absorption of medroxyprogesterone acetate from tablets containing 10 mg of micronized drug (particle size less than 10 μm, surface area 7.4 m^2 g^{-1}) compared to nonmicronized drug (surface area 1.2 m^2 g^{-1}) resulted in a twofold increase in the extent of absorption of micronized steroid.[57] Cure rates for micronized pyrvinium pamoate tablets and suspensions were similar and exceeded 90% compared to the 61% cure rate with nonmicronized drug.[58] Micronization can increase the tendency of the drug powder to aggregate and thereby decrease effective surface area. This problem can be overcome by adding a wetting agent or other formulation adjuvants. Sometimes, the decrease in particle size reduction does not translate into increased absorption. In such cases, dissolution may not be the rate-limiting step. Indeed, in the case of weak bases such as acetaminophen which dissolve readily in gastric juice, gastric emptying rather than dissolution becomes absorption rate limiting.[59] It is also possible for the advantages of micronization to be reduced or completely offset by compaction of the particles during tablet compression as was evidenced by the fact that the absorption of griseofulvin is much greater when the micronized drug is given as a suspension compared to a tablet.[60] Particle size reduction or increased surface area may also result in increased chemical instability of the drug in the GI tract. The addition of a wetting agent to a formulation of erythromycin propionate resulted in significantly lower blood levels, apparently due to increased dissolution and degradation of the antibiotic in gastric fluids.[61]

The smaller particle size of dicumarol is associated with a more rapid dissolution rate, greater inhibition of prothrombin activity and a higher plasma concentration of drug.[62] The micronized suspension dosage form of phenytoin results in higher blood levels compared to capsules.[63] The effect of particle size on the ophthalmic bioavailability of dexamethasone suspensions in rabbits showed a significant rank-order correlation between increasing drug levels and decreasing particle size.[64] An *in vitro/in vivo* correlation relating to bioavailability of four estradiol suspensions has been noted, related to particle size, with the smaller particle size being associated with greater bioavailability.[65] The dissolution rate of amorphous erythromycin estolate was much slower due to the low wettability of the amorphous form. The dissolution rate of the crystalline form decreased as the particle size increased, although it was still much higher than that of amorphous erythromycin estolate.[66] The bioavailability of unmicronized and micronized glibenclamide tablets showed a higher and faster absorption from the micronized tablet with reduced interindividual variation.[67] The control of particle size and specific surface area influences the duration of adequate serum concentration, rheology and product syringeability of penicillin procaine suspension for intramuscular injection.[68] Studies have been conducted to optimize particle size effects on bioavailability of drugs from suppositories.[69,70] The effect of particle size in inhalation aerosols on determining the position and retention of a drug in the bronchopulmonary system has also been studied.[71–73] Particle size may affect texture, taste and rheology of oral suspensions in addition to their absorption.

In general, the formulations research scientist has always to balance the particle size reduction with changes in physical properties such as flow and chemical properties such as degradation at GI pH as determinants of a physicochemically stable dosage form. On the other hand, decrease in particle size may result in increased blood levels. Thus, a delicate optimized compromise has to be the final goal of all product development efforts.

25.1.9 REFERENCES

1. G. H. Brown, *Am. Sci.*, 1972, **60**, 64.
2. J. K. Halebian, *J. Pharm. Sci.*, 1975, **64**, 1269.
3. J. D. Mullins and T. J. Macek, *J. Pharm. Sci.*, 1967, **56**, 847.
4. E. Shefter and T. Higuchi, *J. Pharm. Sci.*, 1963, **52**, 781.
5. H. Halebian and W. McCrone, *J. Pharm. Sci.*, 1969, **58**, 911.
6. A. J. Aguiar and J. E. Zelmer, *J. Pharm. Sci.*, 1969, **58**, 983.
7. E. F. Fiese and T. A. Hagen, in 'Theory and Practice of Industrial Pharmacy', ed. L. Lachman, H. Lieberman and J. L. Kanig, 3rd edn., Lea & Febiger, Philadelphia, 1986, p. 171.
8. W. P. Brennan *et al.*, in 'Purity Determinations by Thermal Methods: ASTM STP38', ed. R. L. Bline and C. K. Schoff, American Society for Testing & Materials, Philadelphia, 1984, p. 5.
9. J. K. Guillory, *J. Pharm. Sci.*, 1972, **61**, 26.
10. H. H. El-Shattawy, *Drug Dev. Ind. Pharm.*, 1981, **7**, 605.
11. H. Jacobsen and G. Reier, *J. Pharm. Sci.*, 1969, **58**, 631.
12. P. V. Allen, P. D. Rahn, A. C. Sarapu and A. J. Vanderwielen, *J. Pharm. Sci.*, 1978, **67**, 1087.
13. H. A. Willis, J. H. van der Mass and R. G. J. Miller, 'Laboratory Methods in Vibrational Spectroscopy', Wiley, Chichester, 1987.
14. A. G. Marshall, *Chemometrics Intell. Lab. Sys.*, 1988, **3**, 261.
15. C. S. Creaser and A. M. C. Davies, 'Analytical Applications of Spectroscopy', Royal Society of Chemistry, London, 1988.
16. R. Gimet and A. T. Luong, *J. Pharm. Biomed. Anal.*, 1987, **5**, 205.
17. A. Burger and A. W. Ratz, *Pharm. Ind.*, 1988, **50**, 1186.
18. M. Kuhnert-Brandstaetter and M. Riedmann, *Mikrochim. Acta*, 1987, **2**, 107.
19. J. Bernstein and A. T. Hagler, *J. Am. Chem. Soc.*, 1978, **100**, 673.
20. W. L. Duax, M. Numazawa, Y. Osawa, P. D. Strong and C. M. Weeks, *J. Org. Chem.*, 1981, **46**, 2650.
21. K. M. Harman and G. F. Avci, *J. Mol. Struct.*, 1986, **140**, 261.
22. (a) A. Burger and R. Ramsberger, *Mikrochim. Acta*, 1979, **2**, 273; (b) A. Burger, *Acta Pharm. Technol.*, 1982, **28**, 1.
23. M. Shibata, H. Kokubo, K. Morimoto, K. Morisake, T. Ishida and M. Inoue, *J. Pharm. Sci.*, 1983, **72**, 1436.
24. L. J. Ravin, in 'Remington's Pharmaceutical Sciences', 16th edn., ed. A. Osol, Mack, Easton, 1980, p. 1361.
25. L. J. Ravin and T. Higuchi, *J. Am. Pharm. Assoc. Sci. Ed.*, 1957, **46**, 732.
26. A. P. Simonelli and T. Higuchi, *J. Pharm. Sci.*, 1962, **51**, 584.
27. A. J. Aguair, *J. Pharm. Sci.*, 1969, **58**, 963.
28. M. A. Moustafa, S. A. Khalil, A. R. Ebian and M. M. Motowi, *J. Pharm. Sci.*, 1974, **63**, 1103.
29. W. E. Hamlin, E. Nelson, B. E. Ballard and J. G. Wagner, *J. Pharm. Sci.*, 1962, **51**, 432.
30. S. Chakrabarti, R. van Severen and P. Braeckman, *Pharmazie*, 1978, **33**, 338.
31. W. D. Walking, H. Almond, V. Paragamian, N. H. Batuyios, J. A. Meschino and J. B. Appino, *Int. J. Pharm.*, 1979, **4**, 39.
32. M. D. Tuladhar, J. E. Carless and M. P. Sumners, *J. Pharm. Pharmacol.*, 1983, **35**, 208.
33. M. D. Tuladhar, J. E. Carless and M. P. Sumners, *J. Pharm. Pharmacol.*, 1983, **35**, 269.
34. I. S. Gibbs, A. Heald, H. Jacobson, D. Wadke and I. Weliky, *J. Pharm. Sci.*, 1976, **65**, 1380.
35. S. R. Gunning, M. Freeman and J. A. Stead, *J. Pharm. Pharmacol.*, 1976, **28**, 758.
36. A. L. Thakker, C. A. Hirsch and J. G. Page, *J. Pharm. Pharmacol.*, 1977, **29**, 783.
37. N. A. El-Gindy and A. R. Ebian, *Sci. Pharm.*, 1978, **46**, 8.
38. R. L. Jerzewski, N. B. Jain, S. A. Varia and H. Brittain, *Pharm. Res.*, 1987, **4**, S-76.
39. R. L. Jerzewski, T. M. Wong, C. Sachs and N. B. Jain, *Pharm. Res.*, 1987, **4**, S-76.
40. K. Y. Chow and D. J. W. Grant, *Pharm. Res.*, 1987, **4**, S-76.
41. P. York and D. J. W. Grant, in 'Proceedings of the Pharm Tech '86 Conference', Aster, Oregon, 1986, p. 163.
42. H. Lenenberger, in 'Proceedings of the Pharm Tech '86 Conference', Aster, Oregon, 1986, p. 180.
43. K. Akimoto, K. Inoue and I. Sugimoto, *Chem. Pharm. Bull.*, 1985, **33**, 4050.
44. A. J. Aguair, J. Krc Jr., A. W. Kinkel and J. C. Samyn, *J. Pharm. Sci.*, 1967, **56**, 847.
45. J. W. Poole, G. Owen, J. Silveiro, J. N. Freyhof and S. B. Rosenman, *Curr. Ther. Res.*, 1968, **10**, 292.
46. E. Shefter and T. Higuchi, *J. Pharm. Sci.*, 1963, **52**, 781.
47. B. E. Ballard and J. A. Biles, *Steroids*, 1963, **4**, 273.
48. T. R. Bates, H.-L. Fung, H. Lee and A. V. Tembo, *Res. Commun. Chem. Pathol. Pharmacol.*, 1975, **11**, 233.
49. Brinkmann Instrument Co., Cantiague Road, Westbury, NY 11590.

50. Marco Scientific Inc., 1055 Sunnyvale-Saratoga Road, No. 8, Sunnyvale, CA 94087.
51. R. R. Irani and C. F. Callis, 'Particle Size Measurement, Interpretation and Application', Wiley, New York, 1963.
52. I. C. Edmundson, *Adv. Pharm. Sci.*, 1967, **2**, 95.
53. A. Martin, J. Swarbrick and A. Cammarata, 'Physical Pharmacy', 3rd edn., Lea & Febiger, Philadelphia, 1983, p. 506.
54. M. Gibaldi, 'Biopharmaceutics and Clinical Pharmacokinetics', 3rd edn., Lea & Febiger, Philadelphia, 1985, p. 55.
55. T. R. D. Shaw and J. E. Carless, *Eur. J. Clin. Pharmacol.*, 1974, **7**, 269.
56. B. F. Johnson, J. O'Grady and C. Bye, *Br. J. Clin. Pharmacol.*, 1978, **5**, 465.
57. D. L. Smith, A. L. Pulliam and A. A. Forist, *J. Pharm. Sci.*, 1966, **55**, 398.
58. R. A. Buchanan, W. B. Barrow, J. C. Heffelfinger, A. W. Kinkel, T. C. Smith and J. L. Turner, *Clin. Pharmacol. Ther.*, 1974, **16**, 716.
59. R. C. Heading, J. Nimmo, L. F. Prescott and P. Tothill, *Br. J. Pharmacol.*, 1973, **47**, 415.
60. N. Kitamori and T. Makino, *J. Pharm. Pharmacol.*, 1979, **31**, 501.
61. V. C. Stephens, J. W. Conine and H. W. Murphy, *J. Am. Pharm. Assoc. Sci. Ed.*, 1959, **48**, 62.
62. J. F. Nash, R. F. Childers, L. R. Lowary and H. A. Rose, *Drug Dev. Commun.*, 1974–75, **1**, 459.
63. L. N. Samson, L. M. Stern and J. Durham, *Med. J. Aust.*, 1975, **2**, 593.
64. R. D. Schoenwald and P. Stewart, *J. Pharm. Sci.*, 1980, **69**, 391.
65. I. Kvorning and M. S. Christensen, *Drug Dev. Ind. Pharm.*, 1981, **7**, 289.
66. J. Piccolo and A. Sakr, *Pharm. Ind.*, 1984, **46**, 1277.
67. E. C. Signoretti, A. E. Utri, E. Cingolani, U. Avico, P. Zuccaro, G. Campanari, S. Soschi and P. Fumelli, *Farmaco Ed. Prat.*, 1985, **40**, 141.
68. S. S. Ober, H. C. Vincent, D. E. Simon and K. J. Fredrick, *J. Am. Pharm. Assoc. Sci. Ed.*, 1958, **47**, 667.
69. E. L. Parrott, *J. Pharm. Sci.*, 1975, **64**, 878.
70. J. J. Rutten-Klingma, C. J. De Blaey and J. Polderman, *Int. J. Pharm.*, 1979, **3**, 187.
71. P. Paronen, M. Vidgren, A. Karkkainen and P. Karjalainen, in 'Proceedings of the 3rd European Congress of Biopharmaceutics and Pharmacokinetics', Freiburg, West Germany, ed. J. M. Aiache and J. Hirtz, University of Clermont-Ferrand, France, 1987, vol. 1, p. 357.
72. P. Koneru and M. P. Sinha, *Pharm. Res.*, 1986, **3**, 355.
73. R. W. Niven and P. R. Byron, *Pharm. Res.*, 1987, **4**, S43.
74. E. F. Fiese and T. A. Hagen, in 'Theory and Practice of Industrial Pharmacy', ed L. Lachman, H. Lieberman and J. L. Kanig, 3rd edn., Lea & Febiger, Philadelphia, 1986, pp. 178, 179, 180, 190.
75. A. Martin, J. Swarbrick and A. Cammarata, 'Physical Pharmacy', 3rd edn., Lea & Febiger, Philadelphia, 1983, pp. 492, 497, 499, 500, 501, 505.
76. E. L. Parrott, in 'Theory and Practice of Industrial Pharmacy', ed. L. Lachman, H. Lieberman and J. L. Kanig, 3rd edn., Lea & Febiger, Philadelphia, 1986, p. 28.

25.2

Formulation

ALEXANDER T. FLORENCE and GAVIN W. HALBERT
University of Strathclyde, Glasgow, UK

25.2.1 DOSAGE FORMS

The medicinal compound synthesized and produced by the chemist is in its raw state of little use to the end user, the patient; therefore, in order to be of any ultimate benefit, the raw material must be processed into a suitable dosage form. The requirement for the formulation of drugs is therefore based on such factors as patient acceptance, ease of handling and transport, stability and the need to administer accurate reproducible doses of the very potent drugs currently manufactured by the pharmaceutical industry. The drug must therefore be presented to the patient in a form that is easy to administer and contains the prescribed dose. Invariably, these restrictions mean that the drug will be formulated to provide a range of dosage forms suitable for administration by a variety of routes.[1] The route of administration controls the type of formulation possible and any specialized requirements that may be necessary (see Chapter 25.3). Parenteral products, for example, must be sterile and pyrogen free, whilst oral products do not need to meet such stringent conditions.[2]

Modern dosage forms consist of a pharmacologically active ingredient or drug and a mixture of other pharmacologically inert ingredients or excipients. The excipients perform a variety of functions, such as bulking agents, colours, antioxidants, preservatives, binders, and a diverse series of compounds may be used in any one dosage form. Even a simple tablet formulation may by necessity contain more than three or four excipients and usually the number is higher. Although the excipients may be biologically inert, they do have the possibility for modifying the patient's response to the drug by controlling the amount and the rate at which the drug is made available to the patient from the site of administration. This can be done deliberately to provide, for example, a sustained release product (see Chapter 25.4) but may also be accidentally present in other formulations in which there is an interaction either between the drug and the excipients or between the excipients alone. The process of manufacture is also important and slight variations in this can induce similar effects. The presence of this phenomenon may only be realized if the excipients are changed or a different preparation of the same drug is used with a resulting variation in the response of the patient.[3] All personnel involved in the administration of medicinal products to the public for clinical trials or for treatment should be aware that a drug substance is never administered to the patient, rather a medicine. This is an important difference, since alternative products of the same drug substance may produce totally different responses,[4] and in clinical trials a comparison of the products rather than the drug may be taking place.

The simplest formulation available would be to dispense the raw powdered drug, requiring that the patient simply measures the desired dose. This would have great advantages in the individualization of dosage regimens but few patients would have access to suitable measuring equipment. This automatically leads to the prepacking of the desired dose or the provision of the material in a form that is easily measured by the patient. A solution of the drug may fulfil these criteria, since volumetric measurement can be provided. The drug's solubility, however, will become important as will its stability in solution. The drug in solution may also provide the ideal growth medium for microorganisms so that the addition of a preservative may be required. Of perhaps greater importance to the patient are the organoleptic properties of the drug; most drugs have a bitter taste and therefore taste masking will be necessary. Even simple solution formulations by necessity contain a number of ingredients each of which is added for a specific purpose and, therefore, by their very nature are complex systems.

The science of formulation[5] is based on the utilization of detailed studies of the physicochemical properties of the drug (see Chapter 25.1). This basic information should be available before any formulation is attempted, since it may possibly rule out certain types of products. Drugs which have a limited stability in aqueous solutions, for example, cannot obviously be formulated in this fashion but may have to be formulated as granules or as a powder reconstituted before use.[6] Basic physicochemical studies are therefore very important not only to formulation but to ensure that the drug is stored correctly and that materials for initial *in vitro* and *in vivo* studies are adequately characterized. The formulator has to marry the properties of the drug with the physicochemical properties of the excipients[7] to produce a product that provides the optimal therapeutic efficacy. Each of the excipients will be very different chemically and it is not surprising therefore that the possibility of interaction between the drug and the excipients, or between excipients, exists. This possibility should be eliminated by the use of extensive preformulation studies to assess the compatibility of the drug with the excipients and also the excipients with each other.[8] Several techniques are available to measure this phenomenon and research conducted at this stage may prove beneficial in avoiding costly incompatabilities between materials that are only discovered later in the product's life. In addition to the physicochemical parameters of the dosage form, the formulator must also take account of the biopharmaceutical properties of the drug. The formulation must therefore be developed in combination with the drug's pharmacokinetic properties, such as its absorption rate from various sites of administration, and the optimal dosage form for each site and route of administration should be sought.

25.2.2 LIQUID FORMULATIONS

Liquid formulations of drugs are relatively common in the pharmaceutical industry and are administered by almost all routes from intravenous injections and oral solutions through to topical lotions.[1] This places very different requirements on the products so that they cannot simply be classed as liquids, and certain of the formulations described here are complex mixtures of immiscible liquids (emulsions) or liquids and solids (suspensions). However, there are certain general principles which govern the formulation of liquid medicines and these are applicable independent of the route of administration.

25.2.2.1 Solutions

Aqueous solutions are the most commonly used type of liquid formulation for obvious reasons, but, in some cases, oils and other solvents can be employed if the need arises. The major limitation of this type of formulation is the solubility of the drug in the chosen vehicle, which is an intrinsic property of the interaction between the two materials (see Chapter 25.1). The solubility can, however, be modified by the judicious choice of excipients or vehicles. In aqueous systems for drugs that are either weak acids, weak bases or zwitterionic the solubility may be enhanced by the simple alteration of pH to favour the ionized species, which generally has a higher solubility.[9] Folic acid (weak acid) injection is formulated at a pH of 8 to 11 by the addition of $NaHCO_3$ or $NaOH$ to give a final preparation with 5 mg mL^{-1},[10] and chlorpromazine (weak base) injection is formulated at pH of 5 to 6.5 with suitable buffering agents.[11] The type of buffering agent is controlled by the ability of the site of administration to tolerate pH values outside its normal physiological range. In eye drops, for example, buffers of a low capacity are used to prevent the induction of large changes in pH, which may give rise to eye irritation. Similar effects are important in parenteral products where changes in tissue or blood pH could have possible serious consequences. For parenteral products and products applied topically to mucous membranes the solution should also be isoosmotic with blood to prevent tissue damage due to the lysis of cells.

For nonionic drugs, other methods of enhancing the solubility are available, *e.g.* the use of cosolvents such as propylene glycol, macrogols or ethanol.[12] For nonionic poorly soluble drugs this is a useful technique and has been employed in the formulation of diazepam injection and digoxin injection, which may be formulated with either propylene glycol or alcohol[13] to obtain a suitable concentration in solution. Alternatively, surface active agents can be used to enhance solubility[14] and mixtures of cosolvents and surface active agents have been used in the formulation of the poorly soluble anticancer drug, etoposide.[15] The medicinal chemist also has a role to play in the synthesis of water soluble prodrugs,[16] which are suitable for the required routes of administration.

Unfortunately, however, drugs in solution may not always be chemically stable and this is seen especially in aqueous solutions where hydrolysis may take place. This is an intrinsic property of the drug but formulation factors may be employed to circumvent this problem. If hydrolysis is pH dependent then simple alteration of the pH can be used as a stabilization method.[17] In other cases, the vehicle can be altered to provide stability. Phenobarbitone salts are unstable in aqueous solutions but the stability can be improved to acceptable levels if glycols or alcohols are included in the solution.[18] This effectively reduces the amount of water available to hydrolyze the drug and this can be achieved by other methods such as lyophilization, which is used in the formulation of unstable injections. The product is supplied as a dry powder, which is reconstituted with water just before administration.[6] This technique is employed in the formulation of dry syrups, which contain all the ingredients for the final preparation in a dry state; this increases the shelf-life from weeks to years but the reconstituted syrup still has a relatively short shelf-life.

The processing requirements of the formulation must also be considered as they may affect the stability of the drug; the most notable example is the sterilization of products by heat. Thermolabile drugs can obviously not be processed by this method using standard sterilization cycles,[10,11] but the increasing application of validated sterilization cycles[19] may allow drugs previously sterilized by filtration to be sterilized by heat. This may have possible savings due to the high requirements for producing sterile products by aseptic means as opposed to ordinary heat sterilization.[20]

The organoleptic properties of the drug in solution are also very important in oral products, and, in most cases, some form of taste masking will be necessary. Traditionally, this has involved the incorporation of sugar (sucrose)-based syrups and flavours; however, the current trend is to noncariogenic materials, especially in paediatric products.[21] These materials not only provide a sweet-tasting vehicle but also increase the viscosity of the product to slow taste recognition processes. Taste masking can also be achieved by the use of insoluble salts but the product then will be classed as a suspension with totally different physicochemical properties.

All nonsterile liquid preparations have the possibility for sustaining the growth of microorganisms, which can cause spoilage of the product or the induction of disease in the patient. To prevent this, most nonsterile liquid preparations contain a preservative,[22] which must be carefully chosen with respect to the conditions present in the formulation (*i.e.* pH) and the types of organism that the product might encounter in use. In addition, the compatibility of the preservative and the final container in which the product will be packaged should also be evaluated as it is well known that preservatives can partition into the packaging, denuding the bulk phase of preservative, thus allowing microbial growth to take place.[23]

25.2.2.2 Suspensions

Suspension formulations consist of a finely divided solid phase, which is usually the drug dispersed in a continuous aqueous phase. These preparations are used in a variety of dosage forms, most notably for the oral route and for topical application.[1] Most pharmaceutical suspensions are coarse dispersions rather than colloidal systems, inherently unstable formulations due to the sedimentation of the solid particles in the liquid phase (Figure 1). The most important property of these preparations is that the patient is able to withdraw a reproducible dose from the container. This means that the suspension must not sediment and cake so that resuspension is difficult or impossible. Once again the physicochemical properties of the drug are important and, in this instance, density, particle size and surface properties are the most critical. The formulator has to take account of the basic physical chemistry of dispersions and combine the drug with a suitable suspending agent[5,9] to produce the desired product.

The main aim of the formulation of suspensions is to minimize any possible caking of the solid ingredients, and this is most easily performed by producing a flocculated system in the suspension. A floccule or floc is a loose cluster of particles held together in a random structure; a suspension in this state is termed flocculated. Conversely, a suspension in which the particles are not aggregated or held together is termed deflocculated. Flocculated systems generally resuspend easily but clear rapidly, making the withdrawal of an accurate dose problematical; a deflocculated suspension, however, clears only slowly but tends on prolonged storage to cake. The ideal formulation is normally a partially deflocculated suspension that clears slowly but has long-term physical stability. Controlled deflocculation may be induced in the formulation by modifications of the surface properties of the suspended particle either by altering the surface charge or by the addition of stabilizing macromolecules or polymers.

Most particles suspended in an aqueous fluid possess a surface charge due either to the ionization of surface groups or the adsorption of ions from solution. As with colloidal particles the charge provides a repulsive force due to the interaction of the electrical double layer on adjacent particles,[24] and this can be utilized as a means of stabilizing the interparticulate interactions. The surface charge (zeta potential) is easily measured[25] and by controlling the zeta potential it is possible to control the degree of flocculation in the suspension. In particles where the charge arises through the ionization of surface groups the charge is simply influenced by the pH and the surface density of the charged groups. Alternatively, ions can be added to the suspension to produce similar effects after adsorption onto the surface of the particles. This can also be accomplished by the addition of charged surface active agents, such as dioctyl sodium sulfosuccinate, which preferentially adsorb at the solid–liquid interface. These two approaches can be used in combination and the controlled flocculation of sulfamerazine by the addition of aluminum ions and dioctyl sodium sulfosuccinate has been demonstrated (Figure 2).[26]

Polymeric substances that adsorb onto the particles can also induce controlled flocculation by acting as cross-bridging materials and, if they bear charged groups, by altering the surface charge. A combination of these two effects can be quite easily achieved using charged polymers. This can produce a controlled flocculation of the particles; several types of material are available to the

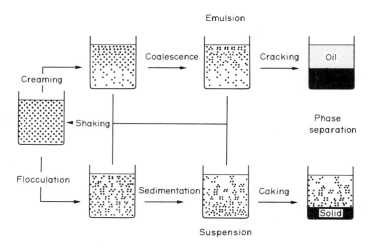

Figure 1   Stages in the breakdown of emulsion and suspension formulations

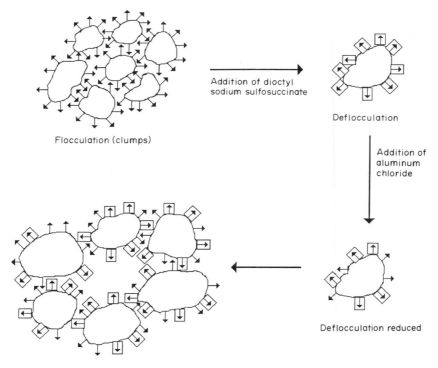

Flocculation (clumps)

Addition of dioctyl sodium sulfosuccinate

Deflocculation

Addition of aluminum chloride

Deflocculation reduced

Controlled flocculation (loose clusters)

Figure 2 Controlled flocculation of sulfamerazine suspension (adapted from refs. 9 and 26)

formulator ranging from natural macromolecules to completely synthetic polymers. Traditional agents, such as tragacanth,[7] are rarely used and semisynthetic or synthetic polymers such as povidone are more commonly employed. The addition of macromolecules to the suspension has a beneficial effect in also increasing the viscosity of the product and this further slows down clearing.

The physical stability of the preparation is one of the most important considerations, and suspension instability can arise by a number of routes. Prolonged storage may allow the formation of a sediment, which is a precursor to the formation of a caked product. Variations in temperature may induce solubility changes in the suspended material, leading to dissolution and recrystallization. It is possible therefore to get apparent increases in particle size due to the dissolution of the smaller particles in a process that is similar to Ostwald ripening in crystals.[9] This may affect the stability and performance of the preparation. The initial particle size distribution of the suspended material should be as narrow as possible to minimize this effect. The inclusion of preservatives is difficult as some preservatives will adsorb onto the particles thus reducing the concentration in solution.[27] The choice of preservative is, therefore, linked to the suspended material and formulation characteristics. If the suspended particles are lyophobic then the possibility of the adsorption of the particles onto the wall of the container exists and this would denude the bulk phase of drug. This can be eliminated by the inclusion of surface active agents, which coat the particle and the wall to prevent any interaction taking place.

Specialized suspension formulations also exist in the form of inhalation aerosols in which the drug is suspended in an organic solvent, which is usually a chlorocarbon or fluorocarbon propellant.[28] This is an entirely different system to aqueous suspensions and is usually stabilized by the addition of long chain fatty acids or Span surfactants, which are soluble in the propellant but adsorb at the solid–liquid interface. The methods of stabilization are, however, essentially the same and again basic physicochemical principles apply. The particle size distribution is also critical, since only particles of 2 μm in size will penetrate down to the level of the alveoli and be retained;[9] the drug is therefore usually micronized to produce particles with the correct size distribution.

25.2.2.3 Emulsions

Emulsions are similar to suspensions in that they contain a dispersed phase suspended in a continuous phase. Most pharmaceutical emulsions are oil in water emulsions where the continuous phase is aqueous and in this is dispersed either a vegetable or a mineral oil. Classically, this type of preparation was used for the oral administration of medicinal oils such as liquid paraffin or cod liver oil; in recent years, however, its oral use has declined, whilst it has gained favour for administration by other routes.[29] A mixture of oil and water is an inherently unstable system and the formulation of an emulsion requires an understanding of the physical chemistry of the interface between these two materials.[5,9] To disperse the oil the interfacial tension between the oil and the water must be lowered by the addition of surface active agents;[30] the type and quantity of agent used may be gauged by reference to the HLB system. The HLB (**H**ydrophile–**L**ipophile **B**alance) system allows surfactants to be classified according to the type of emulsion that they form and their utility as solubilizing agents. The surfactant in the formulation reduces the interfacial tension thus making emulsion droplet formation easier and also reducing the tendency of the droplets to coalesce by forming a protective layer on the surface of the droplet. This acts to stabilize interdroplet interactions in a fashion similar to the stabilization of suspensions (Figure 1). The physical stability of the emulsion is the most important parameter and, once produced, if properly formulated, the emulsion should remain stable for prolonged periods. Instability could arise due to dilution and a lowering of the concentration of the surfactant with subsequent loss of stabilization. Temperature variations may be able to induce this effect in the product. Ostwald ripening of the emulsion may also occur, in which the material in the smaller droplets is transferred to the larger droplets.

Intravenous emulsions are becoming a major application of this type of formulation with the administration of parenteral fat emulsions for intravenous feeding[31] being the most common use. The stability of the emulsion on admixture with these very complex systems of salts, amino acids and sugars is still governed by basic physical chemistry, and, even in these instances, it is possible to formulate stable products.[32] This type of formulation is also capable of overcoming the solubility problems posed by water insoluble drugs. The artificial blood substitute Fluasol DA is an emulsion of perfluorocarbons, materials which ordinarily would not dissolve in the blood.[33] Other insoluble drugs which cause difficulties on intravenous administration may also be formulated as emulsions with the drug dissolved in the oil phase.[34] On administration the drug slowly dissolves from the emulsion droplets, and this prevents any possible irritation due to the precipitation of the drug at the injection site.

Emulsions are also employed in aerosol preparations where the emulsion is placed in a pressurized pack and a quantity of propellant gas is included. The gas dissolves in the oil phase of the emulsion and on release from the canister the gas expands, producing a foam. This type of formulation is useful for topical application and for the medication of body cavities such as the rectum and vagina.[1]

25.2.3 SEMISOLID FORMULATIONS

Semisolid formulations are typically creams and ointments that are applied topically to the skin and mucous membranes. These formulations are useful for local medication in certain areas but can also be used to provide systemic effects (see Chapter 25.4). Although there are several types of topical semisolid formulation,[1] the basic rationale is to present the drug to the site of action in a manner that produces the desired response. The formulation should optimize the drug concentration so that all of the drug is in solution in the formulation, with the minimal possible quantity of solvent. The excipients should also be chosen to affect the permeability of the skin in a favourable manner. To further enhance penetration a large concentration of the drug is usually used to maximize the concentration gradient across the skin barrier. This will aid the penetration of the active ingredient, and the ultimate choice of formulation will again be controlled by the physicochemical properties of the drug, including its ability to penetrate the skin. The formulation should also be stable both physically and chemically and be cosmetically acceptable to the patient.[35]

25.2.3.1 Creams

The creams used in pharmaceutical products and in the cosmetics industry are complex emulsion formulations usually of the oil in water type, although water in oil creams can be used. The former type of cream is also defined as a vanishing cream, since on application the aqueous phase evaporates and the oil is absorbed into the skin. These creams do not feel greasy to apply and have the added advantage that the evaporation of the water phase further concentrates the drug at the skin surface. Unlike the simple liquid emulsion formulations, these systems are highly complex and consist of mixtures of a dispersed phase, emulsifying agent and water (Figure 3). When mixed, these constituents can be arranged in a variety of fashions and the resulting cream is not a simple homogeneous mixture.[36] Although no general approach to completely predict the stability of these formulations has been proposed, the theories that deal with the stability of emulsions are, in general, applicable to creams.[37] The type of emulsifying agents chosen are usually called emulsifying waxes for their ability to impart body or creaminess to the formulation. Cetomacrogol emulsifying wax, for example, will produce an oil in water emulsion that can be poured at concentrations below 5% (w/w of cetomacrogol), but at concentrations of 15% it forms a cream-like emulsion.[7] Emulsifying waxes are classified into two types, nonionic and ionic; the latter is further subdivided by the nature of the charged group and either cationic or anionic waxes are available.[38] To further increase the stability complex mixtures of waxes can be used,[29] which form stable interfacial films. The choice of wax should be related to the drug to be incorporated and the disperse (oil) phase, the choice, for example, of an ionic emulsifying wax may lead to incompatibility problems if it has an opposite charge to the active ingredient.

The effects of excipients on availability are probably most marked in topical formulations, and the correct formulation will require a careful choice of emulsifying wax and dispersed phase. Table 1 shows the effect of vehicle on the action of methyl nicotinate; it is easily seen that the oily cream retards action, whilst the presence of glycerol in the aqueous cream has a similar effect.[39] This can be related to the basic physicochemical properties of methyl nicotinate and the vehicle. Similar formulation studies have been conducted with the topical corticosteroids,[40] and by using a combination of *in vitro* and *in vivo* techniques the optimum formulation may be found.

The preservation of these products is difficult due to the complex nature of the formulation and the preservative needs to be chosen with great care.[22,29] Additional problems may be caused by the choice of container and most pharmaceutical products are packed in tubes, which may cause

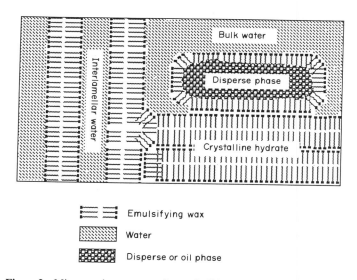

Figure 3 Microscopic structure of a semisolid cream (adapted from ref. 36)

Table 1 Effect of Topical Vehicle on the Absorption of Methyl Nicotinate[a]

| Concentration of methyl nicotinate (%) | Time to onset of Erythema (min) | | | |
	Aqueous cream BP	Oily cream BP	Aqueous cream BP + 40% glycerol	Aqueous cream BP + 60% glycerol
1.0	3.3	4.1	4.6	6.3
0.5	4.0	4.5	5.6	6.3
0.1	4.6	5.2	6.7	16.7
0.05	5.9	6.8	11.7	—
0.01	7.4	10.5	17.6	—

[a] Adapted from ref. 39.

instability problems if large quantities of the emulsifying wax adsorb onto the wall thus denuding the bulk phase and inducing instability.[37]

25.2.3.2 Ointments

The classical ointments are mixtures of fatty substances which are immiscible with water and are usually anhydrous; petrolatum (white soft paraffin) is a good example of this type of material.[7] These are often combined with emulsifying waxes to produce absorption bases and emulsifying ointments, which are capable of taking up water to form a water in oil emulsion formulation. On application to the skin these formulations feel greasy and form an occlusive layer, which will hydrate the skin and enhance drug penetration. This type of system is again unstable and the choice of excipients is based upon the physicochemical properties of the active ingredient in combination with the properties of the ointment bases. Hydrophilic ointments are also available where the base is a hydrophilic water soluble material, such as macrogols or polyethylene glycol, which can be formed with a low water content. The consistency of this type of base is easily altered by changing the relative proportions of the various molecular weight fractions present or by including other emulsifying bases. These systems are obviously stable physically as long as excessive quantities of water are not added, but the glycol can react with the active ingredient to inhibit drug release. The presence of small quantities of ethylene oxide residue from the manufacture of the excipient may also prove troublesome in some instances and the action of preservatives may be reduced by the presence, in a formulation, of large quantities of glycols.[22]

25.2.3.3 Gels

Topical gel formulations are produced by the gelation of either organic or aqueous liquids by the addition of suitable excipients which form a continuous structure in the liquid, providing solid-like properties. Again the same formulation rules apply and this type of formulation is useful for the application of preparations to mucous membranes.[1] Gelation of organic liquids may be achieved by the addition of organically substituted materials, such as the bentonites, which will gel liquid paraffin to produce a suitable ointment formulation.[41] Aqueous gels may be formed by using either water soluble polymers or materials such as colloidal silica, and, although these systems may be physically stable, chemical interactions may be a problem. The nature of the gelling agent has to be chosen carefully to avoid any interaction with the active ingredient that may limit the effectiveness of the formulation and induce physical instability.

25.2.4 SOLID FORMULATIONS

Solid formulations are the most commonly used pharmaceutical products for the obvious reasons of product stability, patient acceptability and convenience. Tablets and capsules are used to administer drugs orally to obtain a systemic effect; however, local effects may also be achieved by the administration of the solid product (pessaries) to body cavities such as the vagina.[1] The formulation and preparation of these dosage forms are dependent on the solid properties of the drug and the excipients used. Features such as the bulk density, particle size, particle shape, hygroscopicity, powder flow and compaction characteristics are of critical importance. The types of formulation and methods of production possible are controlled by these properties; suitable excipients may have to be chosen to ameliorate unsuitable properties of the drug. Hygroscopic materials, for example, cannot be incorporated into hard gelatin capsules as they would absorb water from the capsule shell making it brittle and liable to crack. The particle size of the drug is important as smaller sized particles for the same weight of drug will possess a greater surface area for drug dissolution. If the particle size of the drug is reduced then the dissolution of the drug is generally faster and this may produce higher drug blood levels in the patient.[42] Usually, in solid formulations the chemical stability of the drug is maximal since very little water is present, but solid–solid interactions can take place or the processing involved in production can induce changes in the drug that would interfere with the eventual performance of the formulation. Solid formulations lend themselves easily to alterations that can be used to control the performance of the product, either by treating the finished material or by varying the excipients used. Tablets, for example, may be coated to mask taste and improve stability, or the excipients may be changed to provide erodable matrix tablets that delay drug release (see Chapter 25.4).

25.2.4.1 Tablets

Tablets consist of hard compacted powder masses that are held together by a variety of binding agents. They have a very low water content (typically 1–3%) and are composed usually of several ingredients, each of which serves a different function in the manufacture and performance of the final product (Table 2).[43] Tablets are produced by the compaction of a powder mass between two stainless steel punches and a die at pressures of up to several tons (Figure 4). The size and shape of the die can be varied to provide an infinite number of tablet 'styles', but the majority are round with either flat or convex surfaces. Since the usual batch size of a tablet production run is in the order of millions of tablets, which are produced at the rate of up to 5000 min^{-1}, the flow and compaction properties of the powder mass are critical. Most powders will not flow fast enough for these

Table 2 Tablet Excipients[a]

Excipient	Function	Examples
Diluent	Bulking agent used to adjust the tablet weight to desired level	Lactose, dicalcium phosphate, crystalline cellulose
Binder	Adhesive agent which holds together the powder during granulation and compaction	Starch, poly(vinylpyrrolidone), cellulose derivatives
Glidant	Added to improve the flow properties of bulk powder/granule masses	Colloidal silica, starch
Lubricant	Prevents tablet adhering to punches and dies, provides lubrication for punches moving in die	Magnesium stearate, stearic acid, sodium lauryl sulfate, talc
Disintegrant	Helps tablet break up when placed in an aqueous environment	Starch, sodium starch glycollate, cross-linked poly(vinylpyrrolidone)

[a] For full list of tablet excipients see refs. 5, 7, and 45.

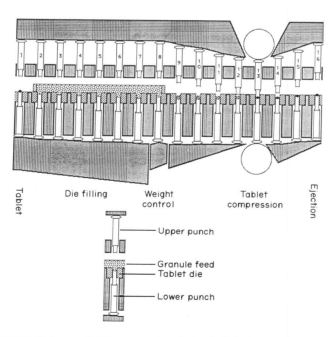

Figure 4 Diagram of a rotary tablet press. Tablet press cycle: punches 2–7, the lower punch falls to its lowest position and the space between the top of the punch and the die is filled with excess granules; the upper punches are inactive. Punch 8, the lower punch is raised to a set height and excess granulate is expelled; the tablet weight is set by the quantity of granules remaining in the die; the upper punch is inactive. Punches 9–12, the lower punch drops to prevent loss of granules; the upper punch begins to fall and enter the top of the die. Punch 13, the granules are compressed between the two punches and the compression force is set by the spacing of the rollers. If the rollers are moved closer together, the compression pressure is greater. Punches 14–1, the upper punch is withdrawn and the lower punch rises to its full extent to eject the tablet, which is then removed and the cycle restarted

processes, and to impart suitable flow and compaction properties the powder is usually granulated to a form that is similar in nature and size to instant coffee granules.[43]

The choice of formulation and method of manufacture are related to the size of dose to be administered and the compaction properties of the drug.[5,6] If the drug has poor compaction characteristics and the dose is relatively large, then granulation of the drug will probably be necessary. Classically, granulation was performed by the process of wet massing in which the drug was mixed with suitable dry diluents and a binder, usually in the form of a polymer in solution. The wet mass can then be formed into granules of the desired size by a variety of processes and dried to remove excess solvents.[44] The granulate formed is not completely dried as the adhesion properties of the binder are dependent on the residual water content of the material. This method is, however, of little use for drugs that are thermolabile or are affected by the solvents used in the granulation process. Granulation may also be performed on dry powder mixes of drug and diluents by loosely compacting the powder between rollers and then gently breaking the compact to form granules of the desired size. The final granulate formed by either method is then mixed with further excipients, such as lubricants, which are usually wax-like materials added to prevent adhesion of the tablets to the punches and to aid the movement of the punches in the dies. Glidants are used to further improve the flow of materials through hoppers, and disintegrants are added to help break up the tablet once it comes into contact with aqueous solvents. The remaining method of tablet production is by the process of direct compression, in which the drug itself either has good compaction characteristics or may be treated to provide a suitable form for compaction. In these instances the tablet may be manufactured with up to 90–95% drug content, assuming that the dose is of sufficient size to make the tablet large enough to be handled by the patient. If the dose of drug is low, then, regardless of compaction properties, it may be mixed with suitable excipients which possess good size, compaction and flow properties to provide a mixture including added lubricants and disintegrants that can be compressed directly without further treatment (Table 3).

The relative proportions of the individual ingredients in the tablet are an important factor controlling the behaviour of the tablet;[45] for example, changes in the lubricant may have marked effects on the release of the drug from the tablet (Figure 5)[46] or changing the diluent may have

Table 3 Steps in the Production of Compressed Tablets

Step	Wet granulation	Dry granulation	Direct compression
1	Mix granule ingredients	Mix granule ingredients	Mix tablet ingredients
2	Prepare solution of binder	Compress mix between rollers	Tablet compression
3	Mix binder with powder to form wet mass	Break compress to give granules	
4	Screen wet mass to provide granules	Screen to break granule aggregates	
5	Dry granules	Mix with remaining excipients	
6	Screen to break granule aggregates	Tablet compression	
7	Mix with remaining tablet excipients		
8	Tablet compression		

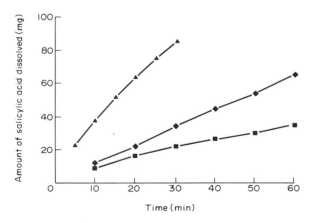

Figure 5 Effect of lubricant on the dissolution of salicylic acid from tablets. Lubricant: ▲, 3% sodium lauryl sulphate; ◆, no lubricant; ■, 3% magnesium stearate (adapted from ref. 46)

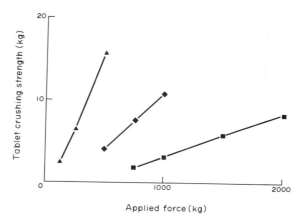

Figure 6 Crushing strength of compacts of tablet excipients. Excipient: ▲, microcrystalline cellulose (Avicel); ◆, lactose; ■, dibasic calcium phosphate (EncomPress) (adapted from ref. 45)

similar effects.[47] The formulator must be aware of the possibility of these effects and guard against their inclusion into the final product. Simple variation of the ingredients of the formulation allow for the production of different types of tablets. The use of water soluble/dispersable excipients combined with quick disintegration provide tablets that may be dispersed in water before administration. Similarly, the use of a wax matrix or polymeric substance in the tablet can slow down the release of the drug to provide controlled release formulations. In this fashion, tablet properties are easily changed to meet the requirements of the drug and the route of administration.

The physical behaviour of the tablet such as disintegration and dissolution are controlled by the ingredients and the processing applied to them (Figure 6). Several of these characteristics of tablet performance are controlled by specifications laid down in national pharmacopoeias.[10,11] The pressure with which the tablet is compacted has an obvious bearing on the ability of the tablet to disintegrate and, generally, the harder the compaction pressure the longer the disintegration time.[48] If the disintegration of the tablet is prolonged, then release of the drug from the tablet will also be delayed (see Section 25.2.6). However, a balance has to be struck between tablet hardness and disintegration since soft tablets will be friable and chip easily during transit. The choice of correct disintegrant will help obviate this dilemma and modern disintegrants are capable of breaking up all but the most poorly formulated tablets.[49]

Once manufactured, the tablet may then be coated with a variety of materials to improve its performance in specific aspects. The oldest and traditional type of coating is the sugar coat, which masks taste and improves patient acceptability but is very difficult and expensive to perform as it requires several individual coats to be built up on the tablet. The modern practice of film coating[50] is easier to apply on a large scale and the polymers used in the coat can be chosen to confer on the tablet the desired physical properties. The coat may have a purely protective function or may have a more specialized role, such as enteric coating, which is designed only to release the drug once it has passed through the stomach into the small intestine.

The choice of formulation and method of manufacture are not entirely dependent on the properties previously enumerated as the cost implications involved in the large-scale production of the formulation should also be borne in mind. In general, the lower the number of materials used and unit processes performed the cheaper the method, and, for this reason, the direct compression route is favoured if at all possible. There is great scope for variation in tablet formulation and for the introduction of problems during manufacture due to small alterations of compression pressure or raw material sources. For this reason it is essential that the original formulation is thoroughly tested and production processes rigidly controlled.[51] The final formulation will be a compromise between the flow/compaction properties, ease of manufacture, cost implications and, most importantly, the performance of the final tablet in *in vivo* situations.

25.2.4.2 Capsules

Capsules consist of a shell of gelatin into which the formulated drug is filled, and, depending on the type of gelatin and additives used, either hard or soft gelatin capsules may be formed.[52] Additives such as plasticizers, colourants and preservatives are used to improve the performance of the gelatin and provide distinguishing features. Hard gelatin capsules are supplied preformed in eight different sizes and the formulator has only to formulate the filling for the capsule.[5] As with tablets the size of capsule used is related to the dose required and the density of the fill material. The fill is usually in the form of a powder, which is mechanically packed into the capsule; the flow characteristics of the powder are therefore critical since accurate filling will be controlled by this property. To improve the flow, large quantities of glidants (see Section 25.2.4.1) are added and the main diluent is chosen to have good flow properties, which will allow the powder to pack well into the capsule. The release of drug from the capsule is related to the formulation of the filling and close attention must be paid to the ingredients used. In general, the gelatin shell will rapidly disintegrate in the stomach to expose the filling. The release of drug from the filling will be dependent on its wetting properties and on the dissolution of the drug from the solid form used. Inclusion of hydrophobic agents, such as magnesium stearate, into the capsule filling will delay release[53] as will dense packing of the fill;[54] these situations should therefore be avoided. Conversely, the inclusion of wetting agents can be used to improve poor release characteristics. Capsules may also be filled with granules or spherical pellets; the latter are preferred since they will flow easily and pack well into the capsule shell. The pellets or granules may be produced by the same method used for the production of tablet granules, and they are easily treated by coating to alter their drug release properties.[55] The size and size distribution of the pellets are crucial as they will affect the filling accuracy and monodisperse pellets are the most appropriate. This type of formulation appears to have advantages over the monolithic sustained release tablets.[56] A modern trend is the use of semisolid, or thermoliquid fills, which has the advantage of improving powder containment problems during manufacture and increasing the accuracy of the fill weight. In this type of formulation the drug is dissolved or suspended in a material which is filled into the capsule—the material chosen should not interfere with the gelatin shell. As with an ordinary powder fill the release of the drug is dependent on the fill, and variations of the materials used will provide control over the rate of drug release.[57]

In soft gelatin capsules the shell is modified by the inclusion of increased quantities of plasticizer (glycerol), which makes the gelatin soft and pliable. These capsules are not provided preformed but are formed around the fill from sheets of gelatin pressed between dies. The fill is usually in the form of a nonaqueous liquid which does not dissolve the capsule shell. Oils or low molecular weight polyethylene glycols are the most commonly used, and the drug is either dissolved or suspended in the liquid.[5] This type of formulation is suited to drugs which are themselves oily liquids or would be affected by the procedures involved in the tabletting process. The composition of the fill is again of utmost importance as it controls drug release or performance of the final product. If the drug is dissolved in a liquid fill, then the release of drug in the stomach may be equivalent to that of a solution (Figure 7),[58] and the release of drugs suspended in the fill can be enhanced by using wetting agents (polysorbate 80) in the formulation. The particle size of the solid is also important and should

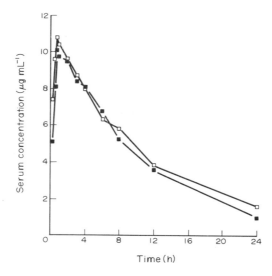

Figure 7 Absorption of theophylline from an oral solution or soft gelatin capsule formulation. Mean theophylline serum level in 14 subjects after a cross-over study comparing a soft gelatin capsule formulation with a solution. □, capsule; ■, oral solution (adapted from ref. 58)

be small enough to ensure uniform filling and to promote dissolution. The fill also must not damage the capsule shell and extremes of pH and materials which contain aldehydes should be avoided.

25.2.4.3 Moulded Products

Certain types of medicinal products, especially suppositories and pessaries, are formulated as solids which are manufactured by a process of moulding. The molten product is formed into the required shape in moulds, which usually also serve as the packing material for the product. The main excipient of these products, termed the base or vehicle, imparts the major physical properties to the product. Bases can be either hydrophilic (*i.e.* high molecular weight polyethylene glycol) or hydrophobic/fatty (*i.e.* semisynthetic glycerides) and are chosen depending on the properties of the drug to be incorporated. Water soluble drugs are incorporated into fatty bases and *vice versa* and this promotes the release of the drug from the preparation and subsequent absorption[59] in a manner similar to topical products. Again, the suitability of the excipients for the drug should be tested as it has been shown that reactions may take place between the base and the incorporated medicament.[60] In Table 4 the change in melting time during storage of aminophylline suppositories is described. It can be seen that with some bases the time increases dramatically and this would reduce the release of the medicament and any subsequent therapeutic effect. In these formulations the drug must be suspended in the molten base before the suppository is formed, and the physical stability of the suspension controls the accuracy of dosing. Particle size of the drug is also important but stability of the suspension should not be compromised. The stability of the suspension can be controlled by the addition of extra excipients but these should not affect the performance of the final product. Hydrophobic bases are chosen to melt just below body temperature in a reproducible fashion; this will allow the spread of the medicament in the rectum and so promote drug absorption. Hydrophilic bases will require a small quantity of fluid to dissolve before the drug will be released. If the base also melts at body temperatures this will further aid drug release. Product irritancy should be kept to a minimum to avoid premature evacuation of the product and the size is also an important factor in controlling this effect.

25.2.5 MARKETING, MANUFACTURING AND PRODUCTION REQUIREMENTS

Although not an exact science, the formulation of a drug into a medicinal product is a crucial stage in the development of a therapeutic substance, since alterations in the formulation, if carried out after initial clinical and stability trials or regulatory approval, can lead to costly retest procedures. The correct formulation of a drug should maximize any activity shown in initial clinical studies and allow the maximal possible benefit to be obtained from the compound. The formulation scientist may have several conflicting parameters to take account of in the final formulation. The physico-chemical properties of the drug are paramount and will control the types of excipients and formulation possible. Other limitations are also placed on the formulation by the regulatory authorities of various countries that restrict the use of certain types of excipients. The performance of the final product will also have to comply with compendial specifications,[10,11] which may be bolstered by stricter in-house limits. The marketing department of the company may require certain

Table 4 Effect of Suppository Base on the Melting Time (min at 37 °C) of Aminophylline Suppositories During Storage at 22 °C[a]

Base	Storage time (weeks)			
	0	*4*	*12*	*24*
Cocoa butter	9.2	10.8	31.2	>70
Witepsol H 15	13.1	18.5	21.4	>60
Suppocire AM	15.6	14.0	—	19.9

[a] Adapted from ref. 61.

types of product for different markets and again the formulator will have to include these requirements in the design of the product. If the product is to be used in world markets, then the stability in different environmental conditions will also have to be assessed and the formulation altered accordingly.[6] The method of packaging the final product will also have to be considered in relation to such factors as stability, patient acceptance and compliance.[62] To meet these requirements, the formulator should design quality into the product[20] by the correct choice of excipients, excipient grade and method of manufacture. The formulation should be capable of being manufactured on a large scale easily, efficiently and consistently so that minimal product variation will occur between different manufacturing plants. The above is a brief overview of the problems involved in the formulation of a drug into a simple medicinal product. The final formulation will be the best possible compromise available to the formulation scientist, which meets all of the diverse requirements that are inflicted upon the product and the formulator.

25.2.6 *IN VITRO* DISSOLUTION TESTING AND *IN VITRO–IN VIVO* CORRELATIONS

25.2.6.1 Dissolution Testing

Dissolution tests serve two purposes: (i) to show differences (or similarities) in the rate of release of active ingredients from experimental formulations *in vitro*; and (ii) to ensure batch to batch consistency in release rate during production runs of marketed formulations. In some cases a dissolution test can be used to predict differences in the bioavailability of medicines, but the dissolution test will not always be able to predict bioavailability, although at best it can. To achieve a predictive test there must have been established a linear or at least a rank-order correlation between the dissolution rate of the drug from several formulations and the extent of *in vivo* absorption from them. However, there does not need to be a linear, or even a rank-order, correlation between dissolution rate and bioavailability for the dissolution test to be a vital part of product specifications. Gross differences in the release rate may indicate problems with the formulation which are due to pharmaceutically unacceptable variation in materials or manufacturing processes. Dissolution and release tests have been devised for a variety of dosage forms. Most attention has been paid to solid oral dosage forms, which are discussed first.

25.2.6.2 Solid Oral Dosage Forms

It is important to consider the processes that occur when dosage forms interact with the gastrointestinal fluids following oral administration. Solid dose forms may or may not disintegrate, depending on their design (Figure 8). Dissolution of the drug from disintegrated dosage forms or release of the drug by other mechanisms then can occur, although frequently dissolution and disintegration can occur in parallel. The rates of disintegration (k), dissolution (k_1) and absorption (k_2) may be defined.

If k_1 is less than k_2 then dissolution will be the rate-limiting step in the absorption process. If, however, absorption is slow, then variations in dissolution rate are unlikely to influence the rate and extent of absorption, but, for many drugs, particularly those which are poorly soluble in gastric fluids, dissolution is an important rate-determining step in the absorption process, and a dissolution rate determination can be a useful guide to comparative bioavailability. A drug with a solubility of less than $10 \, \mathrm{g \, L^{-1}}$ in aqueous solution over the pH range 1–8 will, potentially, have problems with

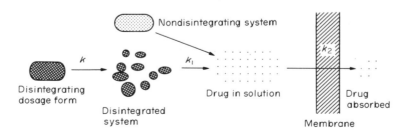

Figure 8 Schematic representation of the processes involved in the dissolution and absorption of a drug from a disintegrating solid dose form, or drug release from a nondisintegrating preparation. If $k_1 < k_2$, then dissolution or release will be the rate-limiting step in absorption

absorption. *In vitro* dissolution tests should be considered where a drug displays poor aqueous solubility, low intrinsic dissolution rate (*e.g.* when the particles are extremely hydrophobic) and where particle size has been shown or suspected to influence bioavailability, where polymorphism is a possible problem or if the tablet has a special coating which must dissolve *in vivo* to release the drug. If the drug is poorly absorbed or when there is significant first-pass metabolism, dissolution tests can confirm or deny the importance of the release of active ingredients from the dosage form in the absorption process. Where there has been a deliberate attempt to control rates of dissolution and release in specialized formulations, dissolution tests should form an intrinsic part of the development process. However, even if a drug has a high aqueous solubility, its formulations should be subjected to a dissolution test at some stage in development to check for the absence of interactions which might adversely affect release.

There is no single method to measure dissolution rate. A variety of designs of apparatus have been proposed and tested, varying from a simple beaker with stirrer to complex systems with lipid phases and lipid barrier membranes where an attempt is made to mimic the biological milieu. The extent to which this is desirable is debatable, and one maxim might be that the simplest system compatible with reality should be used. Even if the purpose of applying the dissolution test is not to predict bioavailability, the dissolution test medium should approximate to that of a physiological fluid, *i.e.* should be largely aqueous, and the system should be stirred at a rate which is not unreasonable *in vivo*. This is not to say that there are no problems with the choice of medium for drugs (*e.g.* steroids) with very low aqueous solubility. In such cases a two-phase system, where the nonaqueous phase acts as a sink equivalent to the biophase, might be adopted with advantage. Addition of surfactants to the aqueous phase can also assist dissolution, but care must be taken to ensure there is no effect of the surfactant on disintegration.

In most cases, because the dissolution test (apparatus, solvent and stirring speed) is to an extent arbitrary, the results are only valid in a comparative sense, formulation *versus* formulation or batch *versus* batch. Results are not absolute, whereas absorption data are less dependent on experimental design, provided that analytical techniques have been validated and are specific, although, of course, *in vivo* data are subject to much more variability. Analytical techniques in dissolution testing and bioavailability studies have been discussed by Munson.[63]

The principal difficulties in assessing *in vitro–in vivo* correlations rest on the questions: which *in vitro* test was used and which index of bioavailability was used? Dissolution tests have the virtue of being more controllable than *in vivo* tests and therefore have their own validity as control tests. In the last few years, attempts have been made by deconvolution techniques to assess rates of dissolution of drugs *in vivo* from pharmacokinetic data.[64] Such techniques allow a more valid critique of dissolution techniques, by direct comparison of *in vitro* and *in vivo* rates of dissolution.

Whichever technique is used it must be validated, and it must be shown to accomplish what the test is designed for: to show differences when they exist, and not to show them when they do not. If the dosage form has been designed as a modified release preparation, it may be necessary to go further and prove that the test can discriminate between formulations or batches which will produce clinically significant differences in release rates or absorption profiles. This requires more extensive *in vitro–in vivo* correlations.

25.2.6.2.1 *Background to dissolution testing*

The rate of solution of a solute from a nondisintegrating solid, in the absence of a chemical reaction between solute and solvent, is given by the Noyes–Whitney equation[9]

$$dw/dt = K(c_s - c)$$

where w is the weight of drug in solution, c is the concentration of drug in solution at time t and c_s is the saturation solubility of the solute (drug) at equilibrium. K is given by

$$K = DA/h$$

where D is the diffusion coefficient of the solute, A the surface area of the dissolving solid and h the diffusion layer thickness. Under sink conditions, where $c < 0.1c_s$, equation (1) reduces to

$$dw/dt = Kc_s$$

Intrinsic dissolution rates of an unformulated drug can be measured to a fair degree of accuracy using compressed, nondisintegrating discs of drug substance prepared in an IR or a similar press.[65]

Such work suggests that compounds with intrinsic dissolution rates greater than $1 \, mg \, cm^{-2} \, min^{-1}$ are generally not prone to dissolution-limited absorption; however, this may be the case for compounds with dissolution rates below $0.1 \, mg \, cm^{-2} \, min^{-1}$.[65] It is obvious, even with such controlled systems, that the dissolution medium assumes vital importance due to the relationship between c_s and pH in aqueous systems and c_s and the nature of the solvent in nonaqueous or mixed solvent systems. Dissolution rate increases with increasing solubility but when this is brought about by a change in bulk pH, as the interfacial pH is frequently different from bulk pH, dissolution rates do not follow the solubilities calculated on the basis of the bulk pH values. Examples include the dissolution of a sodium salt of a carboxylic acid into an acid medium, or an acid into a basic medium. The presence of buffers in formulations also influences the results as Mooney and colleagues have investigated in some detail.[66] In dissolution tests on conventional dosage forms, when disintegration usually occurs, the effect of the solvent on the disintegration characteristics of the tablet matrix or capsule formulation should not be neglected.

Philip and Daly[67] found considerable variation in the dissolution rate of bulk powdered erythromycin stearate and a good correlation between these results and dissolution of the drug from compressed tablets; they recommend this as a test to select batches of hydrophobic raw materials for tabletting.

Obviously, because of disintegration into granules or drug–excipient or simply pure drug particles, the surface area available for dissolution, A, is constantly changing during the course of the experiment. Dissolution tests on formulated products will not reveal, therefore, except with nondisintegrating systems whose size remains unchanged, the intrinsic release rates. However, if the disintegration process is consistent, as it should be, comparative dissolution rates can be readily obtained and used in correlations with absorption data.

25.2.6.2.2   *The dissolution medium*

Where possible, it makes sense to attempt dissolution testing in an aqueous medium. As the dosage form first meets the acidic environment of the stomach, a medium of pH 1–3 is a reasonable model, but where solubility problems occur, less acid and even alkaline buffers can be used. The volume of medium used can vary; frequently 200–500 mL are used but up to 18 L have been used for very insoluble drugs. Obviously, in these cases the pretence of mimicking the biological milieu founders. The justification for the dissolution medium should be made on the basis of sound knowledge of the solubility characteristics of the drug in question, in particular its pH solubility behaviour. If release is anticipated to be pH dependent from solubility studies, dissolution pH profiles can be obtained by a stepwise change in pH. Where, because of the low solubility of the drug, c_s is less than the concentration of drug in the dosage form when released into the solvent, or when sink conditions cannot be maintained during the course of the test, then a two-phase system can be adopted or a flow-through system considered. In a two-phase system an appropriate lipid phase, in which the drug has a high solubility, is chosen as the sink. The dosage form and dissolution still take place in the aqueous phase but drug escapes into the lipid reservoir.

25.2.6.2.3   *Apparatus*

Eight types of apparatus which have been suggested for the *in vitro* testing of release rate from solid dosage forms are illustrated in Figure 9.

These range from the beaker method introduced by Levy and Hayes[68] to flow-through systems as advocated by Langenbucher.[69] The official compendial methods are based on the rotating basket or paddle techniques. The Pharmacopoeias of Great Britain,[70] Japan[71] and the United States of America[72] have standardized and recommended apparatus and methodologies for the dissolution testing of solid dosage forms in an attempt to ensure the adoption of a test that can be applied widely, although individual manufacturers and laboratories may choose also to use their own in-house, and perhaps idiosyncratic, techniques. These official pharmacopoeial test systems are virtually identical. One is illustrated in Figure 10.

As many variables as possible have been controlled to minimize interexperimental and interlaboratory error such as those that arise with instrumental variables listed in Table 5.

Hanson[73] lists in detail the sources of error in the operation of tests. (For technological details of apparatus, construction and use, see ref. 73.) Here, we are concerned more with the philosophy of the dissolution test and *in vitro–in vivo* correlations.

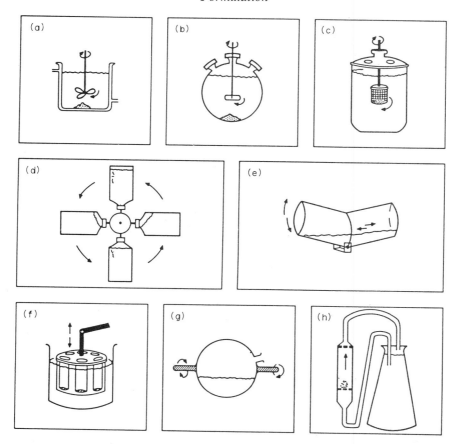

Figure 9 Eight types of apparatus that have been used for dissolution testing of solid dosage forms: (a) Levy–Hayes beaker method; (b) rotating paddle method; (c) rotating basket method; (d) rotating bottle method; (e) the Lederle oscillating system, used for the dissolution of tetracycline dosage forms; (f) modified USP/BP disintegration apparatus; (g) rotating flask, used for aspirin by Gibaldi and Weintraub; and (h) the continuous flow-through apparatus (reproduced from ref. 91 by permission of the publisher)

Table 5

Variables of the dissolution test	Dosage form variables that affect drug release
Hydrodynamic characteristics of the apparatus	Type of dosage form
types of agitation	tablet
intensity of agitation	capsule
dispersion of particles	enteric coated
Properties of the dissolution medium	sustained release
pH, buffering capacity	Release mechanism
ionic strength	diffusion controlled
presence of additives	dissolution controlled
surfactants	osmotically controlled
complexing agents	Presence of reactive substances
adsorbents	buffers
mucin, bile, enzymes	complexing agents
Operating conditions	salts
agitation intensity	Presence of other substances
volume of solvent	surfactants
exchange of solvent	lubricants
	disintegrants

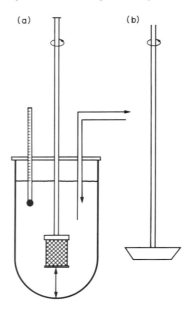

Figure 10 (a) Rotating basket and (b) rotating paddle dissolution apparatus, the dimensions of which are specified in the USP and BP. The dimensions and characteristics of the basket are specified, as are the distances between the bottom of the basket and flask. To minimize problems due to interlaboratory variation in performance of apparatus, the use of a standard tablet (*e.g.* of NaCl) has been suggested to calibrate systems

25.2.6.3   Uses of the Dissolution Test

25.2.6.3.1   *Solid dosage forms*

In vitro tests have been used in formulation development to study: (i) the effects of salt form, polymorphic form and particle size of a drug on the dissolution rate; (ii) the effect of manufacturing variables such as compression, tablet-coating thickness and excipient variation; (iii) drug release rates from dosage forms with reported problems of bioavailability;[74] and (iv) the ageing of formulations in storage trials.[75]

Finholt[76] has reviewed the influence of formulation and manufacturing variables on the dissolution rate from solid dosage forms. It is important, therefore, to assess the effect of preparation procedure on the dissolution of drugs from experimental formulations, such as capsules, where the method of filling as well as the mixing procedures adopted may well influence the rate of drug release. Magnesium stearate is well known to influence dissolution in an unpredictable way; this may or may not be of significance clinically, but should be borne in mind lest it swamps more important features of the formulation.[77] The guidelines for submission of product licence applications in the UK[78] suggest that dissolution tests should be carried out during formulation development and during stability studies on stored samples:

'A dissolution test may be required in the finished product specification where it has been shown necessary to control possible variation batch to batch. This may be related to one or more of the physicochemical properties of the active constituent, formulation or manufacturing process'.[78]

'For special formulations such as controlled release or sustained release formulations, a dissolution test is required in the finished product specification. It is particularly important to show that adequate development studies have been performed to establish the dissolution test as a suitable control test'.[78]

Results from dissolution tests are usually plotted as amount (or percentage) of drug released as a function of time. The profiles of drug release can be informative (see Figure 11) and the extent of release can be assessed. Where release is less than complete in a time which is physiologically reasonable then this might indicate an interaction between drug and excipients in the formulation. The release profile might indicate where disintegration is delayed (Figure 11).

With sustained release preparations the profile of release can be critical in determining the absorption profile and clinical performance, and it is particularly important in comparative assessments. Change in pH of the solvent during the dissolution test, or simply examining the release profile of a dosage form in acidic and alkaline media, can indicate likely abnormal behaviour of

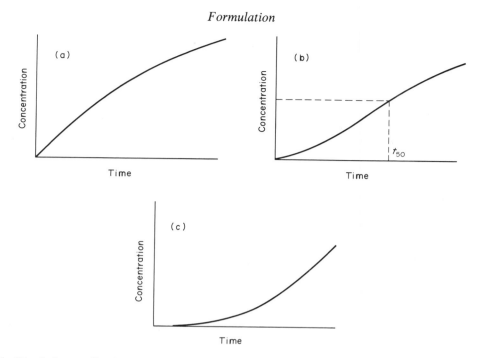

Figure 11 Dissolution profiles for (a) a nondisintegrating dosage form; (b) a disintegrating dosage form; and (c) a disintegrating dosage form, with a lag time for dissolution due to the slow disintegration of the dosage form. Dissolution behaviour is often defined by parameters such as t_{50}, the time for 50% dissolution. A dissolution profile can also be informative as the maximum percentage release can be used as individual content assay. The dissolution plot may be mathematically described by a Weibull function (see text)

either the tablet matrix or the drug itself as it passes from the stomach to the intestine. Change in pH is particularly important in the assessment of enteric coated and other pH dependent systems, such as those that might 'dose dump' as the dosage form reaches the intestine. Some theophylline products have been shown to be prone to this.[79] Figure 12 illustrates the effect of pH on some theophylline sustained release products.

'Topographical' display of dissolution characteristics of controlled release tablets at several pH values can be useful in unearthing unusual pH dependency (Figure 13).[80]

25.2.6.3.2 *Dissolution tests for systems other than tablets or capsules*

While the majority of published work has been directed towards discussion of dissolution tests for oral dosage forms, measurement of the rate of release of a drug from other dosage forms has

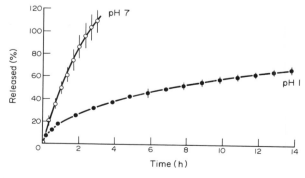

Figure 12 Dissolution profiles for a UK theophylline sustained release product (Nuelin SA) in acid and alkaline media. The faster release of theophylline in the latter is shown. Faster dissolution *in vivo* as the dosage form reaches the intestine leads to 'dose dumping', but can be avoided by appropriate design of the formulation

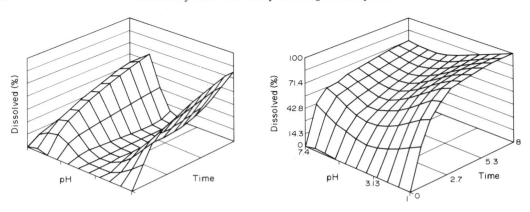

Figure 13 Topographical display of pH-dissolution–time profiles of two quinidine products readily exhibits differences in product performance *in vitro* (reproduced from ref. 114 by permission of the publisher)

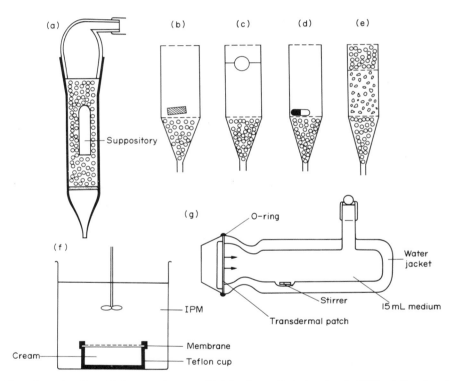

Figure 14 Flow through cells, with laminar flow provided by the packed glass beads. Without the glass beads flow is turbulent and uncontrolled. (a) A cell for suppository dissolution testing (reproduced by permission of the publisher from ref. 73); (b) tablets; (c) inlay tablet; (d) capsule; (e) a suspension formulation [from Langenbucher reproduced in Hanson (ref. 73)] showing the versatility of the system; (f) a suggested system (after ref. 115) for studying the release of a steroid from a cream vehicle, into isopropyl myristate (IPM). Release of fluocinonide showed a rank-order correlation with the vasoconstrictor index in volunteers; and (g) a diffusion cell, after Mazzo *et al.* (ref. 83) to investigate release rates from transdermal patches

considerable importance, although the choice of release test might be more problematical than for oral systems. This is particularly true of tests for dosage forms designed for administration to regions with low moisture levels, *e.g.* rectum, vagina and skin. While tests have been devised for topical preparations, including ointments and transdermal delivery patches,[81-83] no official tests have been promulgated for transdermal systems, although a supplement of the French Pharmacopoeia includes a description of one apparatus.[84] Release of active ingredients from ointments, creams and gels can be studied using apparatus such as that depicted in Figure 14.

Typical apparatus to study the release of drugs from rectal formulations[85] is also illustrated. Devices with rate-controlling membranes, such as transdermal patches, can simply be immersed in an appropriate medium, for example attached to the paddle of the USP dissolution apparatus, care being taken not to alter the barrier properties of the polymeric membrane (Figure 14).

25.2.6.4 *In Vitro–In Vivo* Correlations

In vitro tests can be used to predict the comparative absorption behaviour of a drug in a variety of formulations. They can rarely be used to assess the absorption potential of a range of drug analogues because factors other than solubility and dissolution rate, such as pK_a and lipophilicity, play their role. The absorption potential (AP) of a series of compounds has, for example, been estimated by Dressman and coworkers[86] as

$$AP = \log(PF_{non}S_oV_L/X_o)$$

where P is the octanol–water partition coefficient, F is the fraction of drug nonionized at pH 6.5, S_o the intrinsic solubility of the unionized species at 37 °C, V_L the volume of lumenal contents and X_o the dose administered. Solubility and hence dissolution rate are thus only two factors. However, given the proviso that the rate-limiting step in absorption is the rate of availability of the drug in the lumenal contents, the rates of dissolution measured *in vitro* can provide useful pointers to the rank order of bioavailability of formulations of a drug, or series of salts and esters, as in the case of erythromycin.[87] In this case, rapid dissolution *in vitro* suggests rapid breakdown of the labile molecule in the stomach, so there is an inverse relationship between AUC and rate of dissolution. Nonetheless, it is biologically significant and this case emphasizes the point that each drug must be considered in the light of known chemical, physicochemical and biological data. There are problems also because of the variety of conditions under which dissolution can be measured. It is perfectly feasible to devise a test which will not be predictive, and to devise an alternative which will be, hence some of the contradictory statements in Table 6. Optimal conditions must be achieved for correlations to be realized. A search must also frequently be made for the appropriate biological parameters.

To investigate correlations between *in vitro* performance and the *in vivo* characteristics of any dosage form, the appropriate *in vitro* and *in vivo* parameters must be selected.

There is some choice over which period of time the dissolution rate can be followed (1 h, 2 h or 12 h?), and also the *in vivo* parameter chosen—possibly the area under the plasma–time curve (AUC), but possibly also the peak plasma level of the drug, c_{max}. If the former is chosen, over what period will the AUC be calculated—0–12 h, 0–24 h or 0–48 h? Will absorption rate be used as a measure of absorption or *in vivo* dissolution rates calculated using deconvolution techniques? The variety of parameters used in correlations can be seen in Table 6.

Correlation means, according to the Joint Report of the Section of Control Laboratories and the Section of Industrial Pharmacists of FIP in their guidelines for dissolution testing of solid oral products,[88] that the entire time functions of dissolution or at least characteristic curve parameters are congruent with their counterparts *in vivo*, which is the dissolution function *in vivo*. To establish whether or not correlations exist, at least three formulations with different dissolution rates must be investigated *in vivo*, and the extent of correlation between the *in vitro* rate and some biological parameter measured. Should there be no correlation between dissolution rate, or an inverse correlation observed, the dissolution test will have little or no utility in assessing bioavailability. It will, nevertheless, retain its utility as a control test to ensure the consistency of production processes. Correlation must be judged to be poor when (i) significant differences in dissolution rate are not reflected *in vivo*; (ii) differences in *in vivo* data are not seen *in vitro*; (iii) the order of rates does not show rank ordering; and (iv) when dissolution is much faster or slower than is consistent with the *in vivo* data.

Quantitative interpretation of dissolution rate data is facilitated by the application of mathematical expressions which describe the whole dissolution curve rather than the initial approximately linear portion. As discussed above a theoretical approach to the derivation is unlikely to succeed. A general, empirical equation to describe the dissolution curve has been described by Weibull[89] and applied by Langenbucher.[69] The Weibull function is

$$F = F_\infty[1 - \exp[-(t - t_o/t_d)]]$$

where F is a dependent variable representing the fraction of the administered dose which is

Table 6 Some Published *In Vitro–In Vivo* Correlations

Drug	Comments	Ref.
Acetazolamide	Differences in rate of absorption correlated with percentage dissolution at pH 1.5, paddle method	113
Aspirin	Correlation between dissolution rate and salivary salicylate levels, using continuous flow cells and simulated gastric fluid	97
Aspirin	High buffered aspirin compared with low buffered aspirin. Levy beaker and simulated gastric juice. Correlation between dissolution and absorption, gastric pH and faecal blood loss	98
Chlorothiazide	Maximum urinary excretion rates (% dose h^{-1}) correlated with percentage dissolved in 1 min! USP rotating basket	99
Frusemide	Correlation between percentage bioavailability and t_{30}	94
Griseofulvin	18 L of pH 7.4 buffer used as dissolution medium. Serum levels at 3 h or AUC correlated with t_{30}	100
Hydrochlorothiazide	Discriminatory dissolution test	108a
Indomethacin	Correlation between percentage dissolved in 30 min (paddle method, pH 7.2, 30 r.p.m.) and C_{max}, t_{max} and AUC (0–24 h)	106
Mefenamic acid	Correlation between t_{50} and AUC and c_{max}. Paddle method, 3000 mL, pH 7.4 phosphate	101
Methenemine	Urinary excretion (% dose h^{-1}) correlated with time for 15% dissolution, in USP rotating basket method	99
Nitrofurantoin	Maximum urinary excretion rate correlated with percentage dissolved in 1 h in USP rotating basket method	99
Nitrofurantoin	Comparative dissolution rates (0–4 h) not reflected in comparative urinary excretion rates (10–24 h) of four macrocrystalline products	102
Nitrofurantoin	Percentage dissolved in 60 min correlated with maximum plasma levels or percentage excreted	107
Oxazepam	Correlation between peak levels and dissolution rate at pH 1.2	95
Oxprenolol	*In vitro* release from OROS osmotic pump similar to *in vivo* absorption rate	108a
Phenytoin	Slow dissolution rate reflected in AUC, t_{max} and c_{max}	109
Prednisolone	—	96
Prednisone	Average time to reach half-maximal plasma levels of prednisolone correlates with t_{16} and t_{50}	110
Propranolol SR	Rank-order correlation of dissolution and peak plasma levels	112
Quinidine	No meaningful correlations between absorption parameters and dissolution except for outlier product with slow dissolution and poor bioavailability. Other products nearly all 100% bioavailable	103
Quinidine	Dissolution at pH 5.4 useful for predicting bioavailability	80
Spironolactone	Rank-order correlations	93
Sulfisoxazole Tetracycline	Good particle size dependent correlation between *in vitro* and *in vivo* dissolution rates	104
Theophylline SR	USP rotating basket using simulated gastric fluid/intestinal fluid. Rank-order relationships with published bioavailability data	105
Tolbutamide	Rank-order correlation between dissolution and published bioavailability data	93
Triamterene	Dissolution characteristics can be correlated with total drug excretion	111

dissolved in time t; F_∞ is the amount dissolved at infinity and t_d and t_o the lag time for dissolution after disintegration. It can be shown mathematically that the mean dissolution time is that required for dissolution of 62.3% of the drug and is a dimensionless parameter, which has the value 0 when zero-order release kinetics are obeyed to 1 as first-order kinetics are approached, and >1 when the curve takes on a sigmoidal shape. The Weibull approach leads to the definition of four parameters to describe the often complex dissolution profile, t_o, F_∞, t_d and F_d.

Bioavailability data can be treated in a similar way if the plasma–time curve is treated by deconvolution to give absorption–time profiles (Figure 15).[90] With further refinements and assumptions, *in vivo* dissolution data can be abstracted from the absorption data. While such data are useful to gain understanding of the absorption processes, the most valuable correlations are undoubtedly with absorption rates, peak plasma levels and AUCs and with therapeutic occupancy times of sustained release formulations. There will always be difficulty in obtaining good correlations where there is extensive metabolism of the drug, or when the drug is a prodrug dependent on chemical change *in vivo* for activity, or where metabolic processes lead to a wide spread of bioavailabilities in human subjects. The argument that there is little or no correlation between plasma levels and biological activity with any drug should not be accepted as one which minimizes the relevance of the dissolution test. Frequently, the problem is the inability to measure with precision the biological endpoint (as with tranquillizers), but obviously the relevance of the dissolution test to the *in vivo*

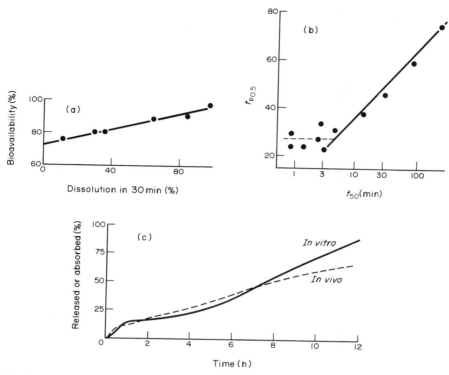

Figure 15 Typical correlations between *in vitro* release and *in vivo* performance. (a) % bioavailability *versus* % dissolution in 30 min for frusemide tablets (adapted from ref. 94); (b) correlations, for prednisone tablets, between the average time to reach half-maximal plasma levels, $t_{P0.5}$, and the time for 50% of the prednisone to dissolve (adapted from ref. 96); and (c) *in vitro* release–*in vivo* absorption correlations achieved after application of deconvolution techniques to the pharmacokinetic data on Theo-Dur. *In vitro* release determined at pH 1 (R. J. MacRae, J. Bunyan, G. W. Halbert and A. T. Florence, unpublished)

situation must always be considered in the light of the pharmacology and pharmacodynamics of the drug substance, as well as its physicochemical attributes.

Many examples exist of good *in vitro–in vivo* correlations, some of which (*e.g.* aspirin, tetracycline, salicylamide) have been discussed by Barr.[91] These serve to show the range of parameters that have been used in correlations. Examples of two-point correlations of serum levels at 10 min and percentage drug dissolved in 1.2 min and 20 min serum levels and percentage drug dissolved at 4.2 min are quoted by Wood *et al.*[92] Good correlations have also been achieved between the amount of tetracycline absorbed at time T and the amount dissolved *in vitro* at time t; this can be regarded as a continuous correlation. Rank-order correlations can readily be shown for tolbutamide, spirono-lactone and hypoglycaemic agents,[93] frusemide,[94] oxazepam[95] and prednisolone (Figure 15).[96] Table 6 illustrates the many successes and several failures to achieve correlations and indicates the wide range of parameters used to achieve such correlations. Once established, of course, these correlations mean that the often simple and highly controllable dissolution test can be used as much more than a pharmaceutical control procedure. Of course it can rarely be used instead of bioavailability studies, which are always required to establish the correlations.

25.2.7 REFERENCES

1. H. Hess (ed.), 'Pharmaceutical Dosage Forms and their Use', Huber, Berne, 1985.
2. K. E. Avis, L. Lachman and H. A. Leiberman (eds.), 'Pharmaceutical Dosage Forms: Parenteral Medications', Dekker, New York, 1984, vol. 1.
3. A. T. Florence, *Pharm. J.*, 1972, **208**, 456.
4. J. Lindenbaum, M. H. Mellow, M. O. Blackstone and V. P. Butler, *New Engl. J. Med.*, 1971, **285**, 1344.
5. M. E. Aulton (ed.), 'Pharmaceutics: the Science of Dosage Form Design', Churchill Livingstone, Edinburgh, 1988.
6. J. M. Padfield, in 'Pharmaceutical Medicine', ed. D. M. Burley and T. B. Binns, Arnold, London, 1985, p. 39.
7. The Pharmaceutical Society of Great Britain, 'Handbook of Pharmaceutical Excipients', The Pharmaceutical Press, London, 1986.
8. H. Nyqvist, *Drug Dev. Ind. Pharm.*, 1986, **12**, 953.

9. A. T. Florence and D. Attwood, 'Physicochemical Principles of Pharmacy', 2nd edn., Macmillan, London, 1988.
10. 'United States Pharmacopeia 21st Revision', United States Pharmacopeia Convention, 1984.
11. 'British Pharmacopoeia', HMSO, London, 1980.
12. S. H. Yalkowsky (ed.), 'Techniques of Solubilization of Drugs', Dekker, New York, 1981, vol. 12.
13. S. H. Yalkowsky, S. C. Valvani and B. W. Johnson, *J. Pharm. Sci.*, 1983, **72**, 1014.
14. A. T. Florence, in 'Techniques of Solubilization of Drugs', ed. S. H. Yalkowsky, Dekker, New York, 1981, p. 15.
15. N. I. Nissen, P. Dombernowsky, H. H. Hansen and V. Larsen, *Cancer Treat. Rep.*, 1976, **60**, 943.
16. V. J. Stella, T. J. Mikkelson and J. D. Pipkin, in 'Drug Delivery Systems: Characteristics and Biomedical Applications', ed. R. L. Juliano, Oxford University Press, Oxford, 1980, p. 112.
17. C. A. Janicki, R. J. Brenner and B. E. Schwartz, *J. Pharm. Sci.*, 1968, **57**, 451.
18. S. Linde, *Sven. Farm. Tidskr.*, 1961, **7**, 181.
19. D. Bell, P. S. Farrell and D. C. Small, *J. Clin. Pharm. Therap.*, 1987, **12**, 157.
20. 'Guide to Good Pharmaceutical Manufacturing Practice', HMSO, London, 1983.
21. E. Sadler and M. Brandon, *Pharm. J.*, 1987, **239**, 679.
22. J. J. Kabara, 'Cosmetic and Drug Preservation Principles and Practice', Dekker, New York, 1984.
23. D. A. Dean, *Drug Dev. Ind. Pharm.*, 1978, **4**, v.
24. J. W. Goodwin (ed.), 'Colloidal Dispersions', Special Publication, Royal Society of Chemistry, No 43, London, 1982.
25. R. H. Muller, S. S. Davis, L. Illum and E. Mak, in 'Targeting of Drugs with Synthetic Systems', ed. G. Gregoriadis, J. Senior and G. Poste, Plenum Press, New York, 1986, p. 239.
26. R. Woodford, *Pharm. Digest*, 1966, **29**, 17.
27. A. Qawas, I. Y. M. Fulayyeh, J. Lyall, J. B. Murray and G. Smith, *Pharm. Acta Helv.*, 1986, **61**, 314.
28. F. Moren, in 'Aerosols in Medicine, Principles, Diagnosis and Therapy', ed. F. Moren, M. T. Newhouse and M. B. Dolovich, Elsevier, Amsterdam, 1985, p. 261.
29. S. S. Davis, J. Hadgraft and K. J. Palin, in 'Encyclopedia of Emulsion Technology', ed. P. Becher, Dekker, New York, 1985, vol. 2, p. 159.
30. D. Attwood and A. T. Florence, 'Surfactant Systems: Their Chemistry, Pharmacy and Biology', Chapman and Hall, London, 1983.
31. A. Wretlind, in 'Parenteral Nutrition in Acute Metabolic Illness', ed. M. E. Lee, Academic Press, London, 1974, p. 77.
32. T. L. Whateley, G. Steele, J. Urwin and G. A. Smail, *J. Clin. Hosp. Pharm.*, 1984, **9**, 113.
33. R. P. Geyer, in 'Drug Design', ed. A. J. Ariens, Academic Press, London, 1976, vol. 7, p. 1.
34. R. Jeppsson, *Acta Pharm. Suec.*, 1972, **9**, 81.
35. B. W. Barry (ed.), 'Dermatological Formulations', Dekker, New York, 1983.
36. G. M. Eccleston, *Pharm. Int.*, 1986, **7**, 63.
37. G. M. Eccleston, *Cosmet. Toiletries*, 1986, **101**, 73.
38. G. M. Eccleston, in 'Materials Used in Pharmaceutical Formulation', ed. A. T. Florence, Blackwell, London, 1984, vol. 6, p. 124.
39. C. W. Barrett, J. W. Hadgraft and I. Sarkany, *J. Pharm. Pharmacol.*, 1964, **16**, suppl. 104T.
40. B. W. Barry and R. Woodford, *Br. J. Dermatol.*, 1974, **91**, 323.
41. N. F. Billups and R. W. Sager, *Am. J. Pharm.*, 1964, **136**, 183.
42. P. Finholt, in 'Dissolution Technology', ed. L. J. Leeson and J. T. Cartenson, Academy of Pharmaceutical Science, American Pharmaceutical Association, Washington, DC, 1974, p. 106.
43. H. A. Lieberman and L. Lachman (eds.), 'Pharmaceutical Dosage Forms: Tablets', Dekker, New York, 1980, vol. 1.
44. H. G. Kristensen and T. Schaefer, *Drug Dev. Ind. Pharm.*, 1987, **13**, 803.
45. G. S. Banker, G. E. Peck and G. Baley, in 'Pharmaceutical Dosage Forms: Tablets', ed. H. A. Lieberman and L. Lachman, Dekker, New York, 1980, vol. 1, p. 61.
46. G. Levy and R. H. Gumtow, *J. Pharm. Sci.*, 1963, **52**, 1139.
47. M. H. Rubinstein and M. Birch, *Drug Dev. Ind. Pharm.*, 1977, **3**, 439.
48. B. B. Seth, F. J. Bandelin and R. F. Shangraw, in 'Pharmaceutical Dosage Forms: Tablets', ed. H. A. Lieberman and L. Lachman, Dekker, New York, 1980, vol. 1, p. 109.
49. W. Lowenthal, *Pharm. Acta Helv.*, 1973, **48**, 589.
50. R. C. Rowe, in 'Materials Used in Pharmaceutical Formulation', ed. A. T. Florence, Blackwell, London, 1984, vol. 6, p. 1.
51. S. A. Hanna, in 'Pharmaceutical Dosage Forms: Tablets', ed. H. A. Lieberman and L. Lachman, Dekker, New York, 1982, vol. 3, p. 375.
52. K. Ridgway (ed.), 'Hard Capsules: Development and Technology', The Pharmaceutical Press, London, 1987.
53. D. L. Simmons, M. Frechette, R. J. Ranz, W. S. Chen and N. K. Patel, *Can. J. Pharm. Sci.*, 1972, **7**, 62.
54. J. C. Samyn and W. Y. Jung, *J. Pharm. Sci.*, 1970, **59**, 169.
55. A. D. Reynolds, *Manuf. Chem.*, 1971, **41**, 40.
56. S. S. Davis, J. G. Hardy and J. W. Fara, *Gut*, 1986, **27**, 886.
57. R. A. Lucas, W. J. Bowtle and R. Ryden, *J. Clin. Pharm. Ther.*, 1987, **12**, 27.
58. L. J. Lesko, A. T. Canada, G. Eastwood, D. Walker and D. R. Broussea, *J. Pharm. Sci.*, 1979, **68**, 1392.
59. C. J. de Blaey and J. Polderman, in 'Drug Design', ed. A. J. Ariens, Academic Press, London, 1980, vol. 9, p. 237.
60. J. F. Brower, E. C. Juenge, D. P. Page and M. L. Dow, *J. Pharm. Sci.*, 1980, **69**, 942.
61. C. J. de Blaey and J. J. Rutten-Kingma, *Pharm. Acta Helv.*, 1976, **51**, 186.
62. D. Dean, *Manuf. Chem. Aerosol News*, 1984, March 10th, 282.
63. J. W. Munson, *J. Pharm. Biomed. Anal.*, 1986, **4**, 717.
64. N. Watari and N. Kaneniwa, *Int. J. Pharm.*, 1981, **7**, 307.
65. J. H. Wood, J. E. Syarto and H. Letterman, *J. Pharm. Sci.*, 1965, **54**, 1068.
66. K. G. Mooney, M. A. Mintun, K. J. Himmelstein and V. J. Stella, *J. Pharm. Sci.*, 1981, **70**, 13, 22.
67. J. Philip and R. E. Daly, *J. Pharm. Sci.*, 1983, **72**, 979.
68. G. Levy and B. A. Hayes, *New Engl. J. Med.*, 1960, **21**, 1053.
69. F. Langenbucher, *J. Pharm. Pharmacol.*, 1972, **24**, 979.
70. 'The British Pharmacopoeia', HMSO, London, 1980, vol. I, p. A114.
71. 'The Pharmacopoeia of Japan', (English version), Yakuji Nippo, Tokyo, 10th edn., 1981, p. 729.
72. 'United States Pharmacopeia, 21st Revision', Mack, Easton, PA, 1985, p. 1243.

73. W. A. Hanson, 'Handbook of Dissolution Testing', Pharmaceutical Technology Publications, Springfield, OR, 1982.
74. L. L. Augsburger, R. F. Shangraw *et al.*, *J. Pharm. Sci.*, 1983, **72**, 876.
75. S. T. Horhota, J. Burgio, L. Lonski and C. T. Rhodes, *J. Pharm. Sci.*, 1976, **65**, 1746.
76. P. Finholt, in 'Dissolution Technology', ed. L. J. Leeson and J. Th. Carstensen, Academy of Pharmaceutical Sciences, Washington, DC, 1974.
77. Z. Chowan and L. Chi, *J. Pharm. Sci.*, 1976, **75**, 534.
78. Medicines Act 1966—Supplement to Guidance Notes on Applications for Product Licences (MAL 2), HMSO, London, 1987.
79. M. Weinberger, L. Hendeles and L. Bighley, *New Engl. J. Med.*, 1978, **229**, 852.
80. J. P. Skelly, L. A. Yamamoto, V. P. Shah, M. K. Yau and W. H. Barr, *Drug Dev. Ind. Pharm.*, 1986, **12**, 1159.
81. B. Pirotte and F. Jaminet, *J. Pharm. Belg.*, 1984, **39**, 23.
82. B. Pirotte and F. Jaminet, *J. Pharm. Belg.*, 1984, **39**, 77.
83. D. J. Mazzo, E. K. F. Fong and S. E. Biffar, *J. Pharm. Biomed. Anal.*, 1986, **4**, 601.
84. J. M. Aiche, Pro-Pharmacopoeia Technical Note No. 255, Supplement to the Pharmacopee Francaise, 10th edn., Maisonneuve, Moulin-les-Metz.
85. J. C. McElnay and A. C. Nicol, *Int. J. Pharm.*, 1984, **19**, 89; N. Senior, *Adv. Pharm. Sci.*, 1974, **4**; T. J. Roseman *et al.*, *J. Pharm. Sci.*, 1981, **70**, 646.
86. J. Dressman, G. L. Amidon and D. Fleisher, *J. Pharm. Sci.*, 1985, **74**, 588.
87. E. Nelson, *Chem. Pharm. Bull.*, 1962, **10**, 1099.
88. Guidelines for dissolution testing of solid oral products, *Pharm. Ind.*, 1981, **43**, 334.
89. W. Weibull, *J. Appl. Mech.*, 1951, **18**, 293.
90. S. Reigelman and S. A. Upton, in 'Drug Absorption', ed. L. F. Prescott and W. E. Nimmo, MTP Press, Lancaster, 1981.
91. W. H. Barr, in 'Bioavailability of Drugs', ed. B. B. Brodie and W. M. Heller, Karger, Basel, 1972.
92. J. H. Wood, *Pharm. Acta Helv.*, 1967, **42**, 129.
93. J. K. Haleblian, R. T. Koda and J. A. Biles, *J. Pharm. Sci.*, 1971, **60**, 1488.
94. M. Kingsford, N. J. Eggers, T. J. B. Maling and G. Soteros, *J. Pharm. Pharmacol.*, 1984, **36**, 536.
95. A. Pilbrant, P. O. Glenne, A. Sundwall, J. Vessmann and M. Wretlind, *Acta Pharm. Toxicol.*, 1977, **40**, 7.
96. R. L. Milsap, J. W. Ayres, J. J. MacRichan and J. G. Wagner, *Biopharm. Drug Dispos.*, 1979, **1**, 3.
97. H. Derendorf, G. Drehsen and P. Rohdewald, *Int. J. Pharm.*, 1983, **15**, 167.
98. G. Dahl, L. E. Dahlinder *et al.*, *Int. J. Pharm.*, 1982, **10**, 143.
99. M. K. T. Yau and M. C. Meyer, *J. Pharm. Sci.*, 1981, **70**, 1017.
100. N. Aoyagi, H. Ogata, N. Kaniwa *et al.*, *J. Pharm. Sci.*, 1982, **71**, 1165.
101. D. Shinkuma, T. Hamaguchi, Y. Yamanaka and N. Mizuno, *Int. J. Pharm.*, 1984, **21**, 187.
102. W. D. Mason, J. D. Conklin and F. J. Hailey, *Int. J. Pharm.*, 1987, **36**, 105.
103. I. J. McGilveray, K. K. Midha, M. Rowe, N. Beaudoin and C. Charette, *J. Pharm. Sci.*, 1981, **70**, 524.
104. N. Watari and N. Kaneniwa, *Int. J. Pharm.*, 1981, **7**, 307.
105. K. J. Simons, F. E. R. Simons, K. D. Plett and C. Scerbo, *J. Pharm. Sci.*, 1984, **73**, 939.
106. N. Aoyagi, H. Ogata, N. Kaniwa and A. Ejima, *Int. J. Clin. Pharmacol. Ther. Toxicol.*, 1985, **23**, 529.
107. J. Bron, T. B. Vree, J. E. Damsma *et al.*, *Arzneim.-Forsch.*, 1979, **29**, 1614.
108. (a) F. Langenbucher and J. Mysicka, *Br. J. Clin. Pharmacol.*, 1985, **19**, 151S; (b) K. A. Shah and T. E. Needham, *J. Pharm. Sci.*, 1979, **68**, 1486.
109. E. Cid, I. Moran, M. Monaris, Ch. Lasserre and V. Vidal, *Biopharm. Drug Dispos.*, 1981, **2**, 391.
110. T. J. Sullivan, E. Sakmar and J. G. Wagner, *J. Pharmacokin. Biopharm.*, 1976, **4**, 173.
111. V. P. Shah, M. A. Walker, V. K. Prasad *et al.*, *Biopharm. Drug Dispos.*, 1984, **5**, 11.
112. J. McAinsh, N. S. Baber, B. F. Holmes, J. Young and S. H. Ellis, *Biopharm. Drug Dispos.*, 1981, **2**, 39.
113. A. B. Straughn, R. Gollamudi and M. C. Meyer, *Biopharm. Drug Dispos.*, 1982, **3**, 75.
114. J. P. Skelly *et al.*, *Drug Dev. Ind. Pharm.*, 1986, **12**, 1177.
115. J. Ostrenga and J. Haleblian *et al.*, *J. Invest. Dermatol.*, 1971, **56**, 392.

25.3

Routes of Administration and Dosage Regimes

MAXWELL C. R. JOHNSON and SIMON J. LEWIS
Rhône-Poulenc Ltd., Dagenham, UK

25.3.1 ROUTES OF ADMINISTRATION

25.3.1.1 Introduction

For a drug to exert its therapeutic effects, it must reach the target cells in a sufficient concentration for a prolonged period of time.[1,2] This may be achieved on certain occasions by local application to a particular part of the body, *e.g.* skin, but more often the drug must be administered systemically. This requires the absorption of drug from the site of administration into the blood or lymphatic circulation.

There are many ways in which a drug can be administered and these will be discussed in the following sections. However, a knowledge of the factors involved in selecting a particular route is essential in understanding why some drugs are given orally and others by injection, for example.

There are a number of factors related to the patient which will help determine which route of administration is chosen.[3] The age of the patient will have an influence; very young children cannot easily take oral dosage forms and the parenteral or rectal routes would need to be considered.

Compliance is another factor involved. For example, a schizophrenic patient may be unwilling to take an antipsychotic drug orally each day, but can be given an oily intramuscular injection which will last for two to four weeks.[4] Conversely, injections require trained staff to administer them and are not as convenient as the oral route for an antibiotic, for example. If the patient is vomiting, the oral route will be precluded and medication would have to be given by other means,[3] so the disease state also has an effect on the choice. Lastly, some patients may be unconscious, uncooperative or uncontrollable. In such cases, the parenteral route is generally indicated.[5]

An ideal dosage form would deliver the drug to the site of action without reaching every tissue in the body. However, this is rarely possible and may preclude some routes of administration. A drug targetted in this way, for example an inhaled β-adrenoceptor agonist such as salbutamol, used in the treatment of asthma, will have less side effects associated with its use.[6] In some cases, the local administration of a drug is supplemented with a systemic dose.[3] For example, in severe cases of athletes' foot, topical and systemic antifungal agents will be administered to the patient.[3]

Perhaps the most important factors involved in administration route selection are biopharmaceutical in nature.[1] Drugs administered orally have to be stable in the acidic conditions of the stomach and must not be degraded by gastrointestinal enzymes. This is one reason why peptides cannot be administered orally. Some drugs, *e.g.* glyceryl trinitrate, are extensively metabolized by the liver after absorption. This requires such drugs to be administered by alternative routes. Some drugs are not absorbed from the gastrointestinal tract, *e.g.* sodium cromoglycate, which is administered by inhalation. Lastly, a drug may need to exert its action rapidly and the speed of onset of action is influenced by the route of administration. The route of administration also influences how long a drug will act in the body. For example, glyceryl trinitrate may need to be taken sublingually to exert its action in a few minutes. However, its effects only last for 30 min. For prevention of recurrent attacks of angina, transdermal delivery systems can be used; the onset of action is about 1 h but the duration of action is 24 h.[7]

In conclusion, the choice of route of administration depends on the patient, disease state and biopharmaceutical considerations. These must all be taken into account before selecting a particular route.

25.3.1.2 Oral Administration

The oral route is that most widely used for the administration of drugs. Under this heading is included any product taken by mouth and swallowed either alone or with a glass of water. The administration of drugs *via* the oral cavity, *e.g.* buccal or sublingual, is referred to in Section 25.3.1.7.

The reasons for the popularity of the oral route are to some extent obvious. Taking medicaments by mouth is convenient to the patient and becomes an established familiar route from an early age. The pharmaceutical characteristics of the product should ensure satisfactory appearance and taste properties and where possible provide for easy swallowing by the patient. Administration by the oral route does not require the intervention of medically trained staff, a major convenience factor from the point of view of general usage by nonhospitalized patients. Only in cases of major illness where the recipient is severely disturbed or upset is the oral route precluded.

The range of products available for oral administration is very wide and will be dealt with in some detail. Absorption processes involved in transferring the administered drug from the lumen of the gastrointestinal (GI) tract to the systemic circulation are not well understood, in spite of the popularity of oral therapy. A summary of the current level of understanding will be given, as this affects the design of oral dosage forms in terms of reproducible absorption characteristics, particularly in relation to sustained release products.

Preparations taken by mouth can be for the treatment of a variety of ailments. These can be roughly divided into drugs for the treatment of local problems in a region of the GI tract or drugs designed to be absorbed systemically, whose action is dependent on achieving the required plasma level associated with successful therapy. This latter group where systemic absorption is a prerequisite for activity constitutes the majority of drugs given orally. For example, antihypertensive compounds, cardiac drugs, antidepressants, sedatives and drugs acting on the central nervous system are all administered orally. In fact, from a development viewpoint the oral route would be the route of choice for a new product, unless there are good reasons against it. This might be because the drug molecule is unstable in the environment of the GI tract or because the activity of the compound is nonspecific and a more restricted route of administration is called for, for example in asthma therapy.

Drugs administered orally for a local activity in the GI tract include throat lozenges, gargles, cough syrups, antacids, certain types of laxatives, anthelmintics and certain antibiotics.

The efficacy of treatment of local infections of the throat and tonsils using lozenges and gargles containing antimicrobial agents is often questionable and severe infections will require antibiotics designed for systemic absorption. With cough syrups, the action of swallowing the syrup provides a local soothing effect to the throat, but any efficacy of the product will depend upon systemic absorption of the antitussive or decongestant drugs.

Antacids are preparations administered orally for the symptomatic treatment of hyperacidity associated with gastric and duodenal ulcers and reflux oesophagitis. They are also widely used for minor stomach upsets of unspecified origin. The action depends on the neutralization of the excess hydrochloric acid in the stomach by reaction with an antacid compound, such as the hydroxide or carbonate salts of aluminum, calcium or magnesium.

A further group of compounds used for local rather than systemic effects is the adsorbents. These would include kaolin, which is administered orally for nonspecific treatment of diarrhoea, and activated charcoal, which may be used for treatment of suspected poisoning.

The materials described as the bulk-forming laxatives are effective by virtue of the enormous increases in bulk volume which occurs as they take in water and swell on passage through the GI tract. The indigestible fibrous material known as bran is an example; also the high viscosity grades of methylcellulose are similarly used.

The final group of compounds with a local effect on the intestine is the chemotherapeutic agents, which are poorly absorbed and depend for their effect upon the build up of local concentrations within the gut. A good example is the anthelmintic compound mebendazole, which is practically insoluble and not absorbed to any appreciable extent from the GI tract. It is used in the treatment of tapeworm and threadworm infestations, being administered in tablet form by the oral route.

The physical form of products presented for oral administration varies widely and can range from free-flowing liquids, to semisolids, to compressed tablets. The release characteristics of oral products can also vary from those designed to dissolve immediately to products designed to release drug slowly through the GI tract, providing once daily dosing.

The major categories of liquid products for oral administration are solutions, emulsions and suspensions. One advantage of liquid products is that dose adjustment can easily be achieved by dilution, although certain restrictions as to the diluting vehicle may be necessary. Normally, the product is formulated so that the required dose is present in 5 mL or simple multiples thereof. From the recipient's point of view, the ease of swallowing is a great advantage. For children liquid products are widely used for this reason and also for geriatric patients. Also, liquid products are useful for the group of people who have difficulty in swallowing tablets or capsules. From a bioavailability viewpoint solutions will be most likely to give the highest availability by the oral route and may be used as the reference product for a comparative study involving various formulations.

Solid dosage forms for oral administration include powders, granules, tablets and capsules. The presentation of drugs as formulated powders in unit doses has declined but more popular is the unit dose satchet, which may contain either a dispersible or effervescent powder, which the patient adds to a glass of water before taking. The tablet represents the most widely used oral dosage form. Developed in the 19th century, its popularity has increased and continued to the present day. From the manufacturer's side, tablets are simple to manufacture, package and transport. For the patients, tablets provide ease of dosage and administration. An alternative solid dosage form is the capsule. These can either be the hard gelatin capsule type or the soft gelatin capsule. With hard gelatin capsules the shells, consisting of a body and cap, are purchased from specialist suppliers and then filled by the pharmaceutical manufacturer with formulated dry powders, granules or pellets. The soft gelatin capsules, which contain either a paste or oily solution of drug, are manufactured by specialist contract houses.

A significant number of products for oral administration are designed to have modified release characteristics. The most simple of these is the enteric coated tablet or granule, which has been coated with a suitable acid resistant polymer, such as cellulose acetate phthalate, to protect the product from dissolution in the acid environment of the stomach or to reduce gastric intolerance.

More sophisticated controlled release oral products are those designed to release the drug substance over either 12 or 24 h, requiring once or twice daily administration.

The development and design of such slow release products is a very complex activity and is dealt with in more detail in Chapter 25.4. In order to be acceptable the products must conform to stringent specifications regarding dissolution characteristics and the relevance of these specifications requires verification from *in vivo* performance in volunteers or patients.

Having considered the types of products which can be administered by the oral route, it is appropriate to review briefly the process of drug absorption from the GI tract.

The current level of understanding of the detailed processes involved in drug absorption is not complete by any means. It is known that absorption from the stomach is relatively insignificant for most drugs and the major site of absorption is the small intestine, starting at the duodenum, through the jejunum to the ileum. Absorption from the large intestine is less well documented and until recently had been disregarded. With the development of slow release products which manifestly provide 24 h plasma levels of the required therapeutic level, drug absorption from the colon must be involved and is therefore receiving more attention.

The pH–partition theory of drug absorption[8] provides a partial explanation for drugs which are weak acids or weak bases. The theory postulates that only the unionized species of the molecule penetrates the intestinal membrane and is thereby absorbed. The rate of absorption is dependent upon the pK_a of the drug, the pH of the environment and the relative hydrophilic/lipophilic character of the drug species, as quantified by the octanol–water partition coefficient. Deviations from the theory have been illustrated more recently [9] and possible explanations offered in terms of an aqueous boundary layer at the intestine membrane surface which also influences the absorption process.[10] The pH–partition theory has never given a satisfactory explanation of the absorption of hydrophilic molecules from the GI tract. Current thinking favours the aqueous pore or channel for absorption of hydrophilic compounds.[11] More complex models involving a combination of both theories have been put forward (see also Chapter 24.1).[12]

The effect of food on GI transit time is also of importance when considering absorption, particularly in relation to slow release products. Gastric residence time can vary, for example, from a few minutes to several hours, depending upon whether the product is administered to a fasting stomach or following a high calorie meal.[13,14] For a slow release product which is absorbed in the small intestine and colon, the consequence of being retained in the contents of the stomach for several hours could mean a long delay in the onset of activity . A much greater understanding of these factors and their relative importance in relation to product design for administration to the GI tract is being obtained from γ-scintigraphy studies.[13,14,15] The products involved, either tablets or pellets, are labelled with radionuclides of technetium and their passage through the GI tract is monitored using a γ-ray camera.

Clearly, further understanding of the mechanisms involved in drug absorption, the main sites involved and the influence of drug characteristics on the absorption process in the GI tract will lead to better design of products for administration by the oral route.

25.3.1.3 Parenteral Administration

The term parenteral is derived from the Greek words *para* (besides) and *enteron* (the gut).[16] Parenteral administration literally means delivery by routes other than the oral route. However, the description has become limited to mean administration of drugs *via* a hypodermic needle.[1] Parenteral dosage forms include sterile solutions, sterile suspensions, sterile solids for reconstitution and intravenous infusion fluids.[16]

Drugs are administered by injection for a variety of reasons.[5] These include rapid onset of action, avoidance of gastrointestinal/liver metabolism and delivery of therapeutic concentrations of drug to target areas without the risk of systemic side effects. Drugs can be administered by injection to unconscious, uncooperative or uncontrollable patients and when the oral route is precluded.

As injectable dosage forms circumvent many of the body's defences and barriers, a number of general requirements must be satisfied.[16]

Injectable preparations must be sterile, that is free from living microbiological contamination.[16] Sterility must be maintained during storage and administration. This can be achieved by steam sterilization, dry heat sterilization, aseptic filtration and various radiation techniques. Preservatives are generally included in multidose injectables to prevent contamination during removal of the dose. Freedom from pyrogens is another requirement of injectables; a pyrogen is any biological or chemical material which produces a rise in body temperature.[16]

Injectable dosage forms must be physically and chemically stable. Many drugs in solution are prone to oxidation and consequently antioxidants are included in such products.[16]

Particles present in injections can be harmful if administered as they can cause capillary blockage.[16] Injections are carefully checked for clarity during manufacture to prevent such problems.

A drug must be in solution, preferably in water, before it can be administered intravenously.[16] The dose requirements and the drug's solubility may necessitate the inclusion of water miscible cosolvents such as propylene glycol. Parenteral routes other than the intravenous route have maximum volume limitations.[5] Drugs that are insoluble in aqueous formulations can be incorporated into oils. These oily injections are generally used intramuscularly.

Ideally, the pH of an injectable product should be similar to that of biological fluids.[5] This is often not possible as drug solubility and stability may be compromised. Injections can be buffered at certain pH values but must easily adjust to blood pH on administration to avoid pain and tissue necrosis.[16]

Injections, where possible, should be isotonic with biological fluids.[5] Hypotonic solutions can cause haemolysis on injection, whereas hypertonic solutions cause blood cells to shrink (crenation).[16] Hypotonic solutions are made isotonic by the addition of tonicity-adjusting agents such as sodium chloride. Hypertonic injections must be administered slowly to permit dilution by the blood.

A wide variety of administration routes are used for parenteral products, depending on the product and its therapeutic actions.[5,16] The route itself also places constraints on the type of product which may be used.

The intravenous route is one of the most commonly used parenteral routes in hospitals today.[5] It involves the administration of injections or infusions directly into a vein. Intravenous injections should be injected slowly (over 1 min) to permit dilution of the solution in the blood.[17] The intravenous route avoids all the body's barriers to drug absorption and effective blood levels are reached almost immediately after injection. However, the duration of action is short and for prolonged actions an intravenous infusion containing the drug is used. The intravenous route is usually reserved for emergency situations or where other parenteral routes are precluded. Electrolytes and nutrients can also be supplied to the body by intravenous infusion. Only drugs in solution can be administered intravenously and particulate contamination must be avoided. Problems can ensue if a drug precipitates on dilution in blood as pulmonary embolism can occur.[17]

The intramuscular route of parenteral administration is a very popular and convenient route for the administrator and patient.[5] It consists of the injection of a solution or suspension into relaxed muscle, *e.g.* shoulder or thigh. The intramuscular route allows drugs to be absorbed over a prolonged period of time by using poorly soluble drugs in aqueous or oily vehicles. For example, one intramuscular injection of benzathine penicillin G suspension gives detectable blood levels for over one week.[17] Steroids, sex hormones and antipsychotic agents are also administered in this way. Intramuscular injection can cause pain and irritation but this can be reduced by injecting small volumes or using a local anaesthetic in conjunction with the drug formulation.[17]

Subcutaneous injections are administered into the loose connective and adipose tissue beneath the skin.[5] This route is particularly useful when the oral route is precluded but self-medication is desirable. Absorption is slower and more variable than from the intramuscular route and is determined by the subcutaneous blood flow. Subcutaneous injections are commonly used for insulin and vaccines. Absorption may be delayed deliberately by coadministration of a vasoconstrictor such as adrenaline to prolong the action of local anaesthetics.[17] Depot injections, *e.g.* insulin zinc suspension, also have prolonged actions due to slow absorption of the drug from the injection site. Absorption may be enhanced by massaging the skin around the injection site to increase local blood flow. Alternatively, hyaluronidase, an enzyme which breaks down connective tissue, may be coadministered.[17]

Most injections are administered by intravenous, intramuscular or subcutaneous injection. However, other routes are less commonly used to deliver drugs to a specific site of action in high concentrations.[17]

Intraarterial injections are used to deliver rapidly metabolized or systemically toxic drugs into the artery supplying the target organ.[19] For example, after hepatic artery infusion of the cytotoxic drug floxuridine, hepatic vein levels were found to be several times higher than after comparable intravenous dosing, yet systemic blood levels were much lower. The therapeutic index of this drug was obviously increased using this technique.[19]

Intrathecal administration is used to deliver drugs to the brain to produce a therapeutic effect with reduced systemic side effects.[19] Methotrexate, a cytotoxic agent, is administered intrathecally in the management of leukaemic involvement of the central nervous system. Epidural administration of anaesthetics is used during childbirth and morphine is given by epidural injection for chronic pain.[19]

Drugs used in the treatment of arthritic conditions can be injected into the synovial sacs of various inflamed joints.[20] This technique is known as intraarticular administration and is used for steroids, antibiotics and anaesthetics.

Eye conditions can be treated by intraocular injections when topical or systemic drug therapy has failed.[5] Such conditions include infections, inflammation and pupillary constriction.

Drugs administered by the intravenous, intramuscular or subcutaneous routes enter the general circulation *via* the lymphatic and/or venous transport systems.[5] Before entering the arterial circulation, drugs must pass through the lungs. The lungs can act as an elimination site for volatile drugs or as a reservoir if the drug partitions into lung tissue. Metabolism can also occur in the lungs.[5]

Various factors affect the distribution of drugs administered parenterally (other than intravenously) and can modulate their biological effects.[5] To enter the circulatory system, the drug must be in solution.[5,18] For suspensions, the rate of dissolution will determine the rate of absorption predominantly. The rate will be affected by particle size, the polymorphic nature of the drug, the solubility of the drug and its diffusion coefficient; viscous formulations will slow drug absorption due to the reduced diffusion coefficient.

Once a drug is in solution, its rate of removal from the injection site is largely dependent on its partition coefficient; to migrate through subcutaneous tissues and muscle it needs to be lipid soluble.[5] The lower the drug's partition coefficient, the slower it will be absorbed into the circulation.

As previously mentioned, blood flow influences the rate of drug absorption. The greater the blood flow, the more rapidly is the drug absorbed.[17]

Finally, other excipients in the formulation can affect the rate of absorption by complexing with the drug or making the injection more viscous.[5] Hypertonic injections may be absorbed more rapidly than isotonic formulations as water is drawn into the injection site by the osmotic pressure gradient.

The parenteral route of drug administration, despite its frequent usage, has a number of problems and disadvantages associated with its use. Patients do not like injections and trained medical staff are generally required to administer them.[5] Once administered parenterally, the drug has a rapid onset of action and cannot be removed from the body. This makes the treatment of accidental overdosage extremely difficult.[21] Parenteral products are manufactured under strict conditions to achieve their sterile properties and this makes them an expensive alternative to nonparenteral products.[5] Problems can also arise if air or particulates are accidentally injected into the body.

Despite these and other disadvantages, the parenteral route remains a common means of administering drugs. Indeed, one survey estimates that over 50% of patients in hospital receive an intramuscular injection.[22] For prompt onset of action, and when the oral route is precluded, the parenteral route is a vital means of drug administration which few other routes can match.

25.3.1.4 Rectal Administration

Drugs can be administered into the rectum in the form of suppositories, enemas or ointments.[23] The vast majority of drugs are administered for local actions, including the relief of pain and inflammation due to haemorrhoids, and also for the treatment of constipation. Antihaemorrhoidal preparations include local anaesthetics, analgesics, emollients and astringents.[23,24] Preparations used to relieve constipation include glycerine suppositories, which cause laxation in the rectum.[24]

Drugs are also administered rectally for systemic uses.[23] The mucous membranes of the rectum allow soluble drugs to be absorbed into the systemic circulation. Drugs which can be administered in this form include theophylline for asthmatics, prochlorperazine to relieve nausea and various analgesics for relief of early morning stiffness in arthritic patients.[23,25]

A knowledge of rectal anatomy and physiology is essential to understand the advantages and limitations of the rectal route.[26] The rectum is the terminal 15 to 19 cm of the large intestine. The large intestine has two main functions: the absorption of water and electrolytes, and the storage of faecal waste. The pH of the rectum is about 6.8 and rectal fluid has little or no buffering capacity. The rectum is usually empty as the defaecation reflex is triggered when faecal material enters into it.

Blood is delivered to the rectum by the superior rectal artery and removed *via* the superior, middle and inferior haemorrhoidal veins. Inflammation of these veins causes the development of heamorrhoids. The superior haemorrhoidal vein leads into the portal system (leading to the liver), whereas the inferior and middle haemorrhoidal veins take blood into the inferior vena cava. Obviously if a drug is absorbed and enters the superior haemorrhoidal vein, it will be subject to liver metabolism. However, if the drug is absorbed into the inferior and middle haemorrhoidal veins, it will avoid this first pass metabolism. The further away from the anus a drug is absorbed, the more likely it is to enter into the portal circulation and be metabolized in the liver. It has been estimated that 50–70% of a rectally administered drug will directly enter the systemic circulation without passing through the liver.[24] The lymphatic circulation also diverts a rectally absorbed drug from the liver.

The rectal route can be used when the oral route is unsuitable, for example because of gastrointestinal drug degradation. The rectal route is also a convenient alternative to the parenteral route when the oral route is precluded. Peptides have also been delivered by the rectal route.[27]

Various physiological factors can affect the absorption of a drug from the rectum.[23] The rectum is nonmotile when resting and there are no villi or microvilli on the rectal mucosa. However, because the mucosal membrane is well supplied with blood and lymphatic vessels, drug absorption can be high. The degree of drug absorption is influenced by the contents of the rectum. Generally absorption is optimal when the rectum is empty because of the greater contact between the drug and the absorbing surface. Evacuant enemas are sometimes used to empty the rectum before the suppository or other dosage form is inserted.

As described above, the absorbed drug can enter into either the portal or general circulation, depending on the position of the dosage form in the rectum. To avoid the portal circulation, the dosage form should not be situated too far into the rectum.

As rectal fluids have no effective buffering capacity, the form (ionized or unionized) in which a drug is administered will not change after administration. The rectal mucosa, as in the whole of the gastrointestinal tract, is preferentially permeable to the unionized form of the drug. Generally weak acids and bases are more readily absorbed than highly ionized drugs.

Other physicochemical properties of the drug can also influence its absorption from the rectum.[23,24] If a drug is to be absorbed, it must be present in solution, so a fair degree of water solubility is required. However, a drug must also possess some lipid solubility to pass through the rectal mucosa. If the drug concentration in the rectal fluid is in excess of a certain level, the rate of absorption is no longer increased by increasing the drug concentration; rate of absorption is at a maximum.

For drugs present in a suppository in the undissolved state, the drug particle size will influence the absorption rate.[23,24] The larger the drug particle size, the slower the rate of drug dissolution and the slower the absorption rate.

The base used in the rectal dosage form can have a large effect on drug absorption.[23,24] The base must be capable of melting, softening or dissolving to release the drug contents. If the drug is lipid soluble, its release from a fatty base will be slow or even negligible. For rapid drug absorption, a water soluble salt in a water soluble base should be used.

Rectal absorption of a number of types of drugs, including peptides, can be increased by the use of absorption enhancers.[27] These compounds include chelating agents such as ethylenediaminetetraacetic acid (EDTA), enamine derivatives, nonsteroidal antiinflammatory agents (NSAIDs) such as sodium salicylate, and surfactants. Their mechanisms of action are largely unknown but they can dramatically increase drug absorption from the rectum. Trials have shown that serum glucose levels can be reduced in humans using insulin suppositories containing sodium salicylate.[28] However, much work still needs to be done to ensure the safety and efficacy of such compounds.[27]

The literature is full of conflicting information concerning the effectiveness of drugs administered rectally.[24,26] In some cases, blood levels obtained were considered to be therapeutic but in other studies only subtherapeutic levels have been obtained. Consequently, one limitation of rectal administration is erratic and unpredictable absorption.

Another disadvantage is the inconvenience to patients. Rectal administration is acceptable in certain European countries, whereas in the US, suppositories only represent 1% of all medications dispensed.[24] Rectal administration of some drugs, especially if used chronically, can lead to mucosal damage. For example, indomethacin and aminophylline suppositories have been shown to cause such problems.[26]

In conclusion, the rectal route provides an alternative to parenteral administration when the oral route cannot be used. However, unless clinical efficacy and patient acceptability can be improved, it will remain an underutilized route of drug administration.

25.3.1.5 Vaginal/Uterine Administration

Medicines intended for vaginal insertion include suppositories or pessaries, tablets or inserts, creams, jellies, sprays, tampons and foams. Douches are aqueous solutions which are administered into the vagina for cleansing purposes.[23,26]

The vagina is a tubular canal between 7.5 and 15 cm in length.[26,29] It is situated behind the bladder and in front of the rectum. Blood is supplied *via* the vaginal and uterine arteries and is removed ultimately back into the vena cava. Consequently, liver first pass metabolism is avoided when drugs are administered in this way. Lymphatic circulation also drains from the vagina. The outermost vaginal cells are protected by mucus, which is produced by the cervix; this fluid is usually thick, but the viscosity is affected by hormone levels. Estrogen decreases fluid viscosity, whereas progesterone has the opposite effect. Before puberty, the vaginal epithelium is thin and the mucus is viscous and slightly alkaline in pH. At puberty, estrogen levels increase, causing thickening of the vaginal mucosa. High progesterone levels cause glycogen to be secreted and its metabolism to lactic acid maintains the vaginal pH between 3 and 5. After the menopause, the vagina returns to its prepubertal state in terms of thickness and mucus properties. The vagina is a nonsterile cavity containing a variety of microorganisms, which are, in part, responsible for the metabolism of glycogen to lactic acid.

The uterus, above the vagina, is a pear-shaped, thick-walled muscular organ. The Fallopian tubes lead into its upper end and its lower end projects into the vagina *via* the cervix.[30]

Contraception and infection are the main indications for drug administration into the vagina. Spermicidal creams, jellies and foams are used in conjunction with a diaphragm for birth control.[23,26] Various antimicrobial agents are also administered into the vagina to treat infections such as vaginal thrush. Oral systemic therapy is also used to treat vaginal infections, either alone or in conjunction with local agents.

There are a number of situations where drugs are administered into the vagina for their systemic actions. The vagina can be used as a route of administration for contraceptive steroids using impregnated vaginal rings. There are several advantages over the alternative oral route, including easy removal of the dosage form should this be required. The route also improves patient compliance as hormone levels are maintained at an effective concentration for several weeks. Progesterone and estradiol are both extensively metabolized by the liver and hence have poor oral bioavailabilities; the vaginal route bypasses this hepatic metabolism.

Another method of birth control is the use of intrauterine contraceptive devices (IUDs).[30] These can either be medicated or nonmedicated. The latter type include copper-containing IUDs and also progesterone-releasing IUDs, notably 'Progestasert', which releases drug at a controlled rate for up to one year. It is estimated that 15 million women worldwide use IUDs as the method of choice for contraception.

Termination of pregnancy by surgical methods can be avoided using prostaglandins administered into the vagina.[26,29] This is preferable to the oral route, where prostaglandins are known to cause irritation. Continuous intravaginal administration from controlled release dosage forms induces 75% of women within 15 h of prostaglandin delivery, depending on the length of the pregnancy.

Another recent development is the use of a bioadhesive dosage form to treat cervical cancer.[31] The dosage form adheres to the cervical mucosa, releasing the antineoplastic drug to the diseased site. This treatment could be developed to obviate the need for surgical removal of the uterus in the case of cervical cancer.

Vaginal absorption of peptides such as the LHRH analogue leuprolide as well as insulin has been demonstrated in rats.[32] However, the absorption of such peptides has been shown to be highly dependent on the oestrus cycle. Various organic acids, such as citric and tartaric, have been shown to act as vaginal absorption enhancers. The vagina may become a route of administration for peptides, but more work is required in this area.

The main limitations of the vaginal route are the same as those for rectal drug delivery: patient acceptability/convenience, mucosal irritation and lack of bioavailability data.[26] Absorption from the vagina can vary enormously due to changes during the menstrual cycle and also with age. Fluid characteristics, pH and mucosal thickness are all known to vary at different times of the month and various stages of life: prepuberty, child-bearing age, menopause. As yet, the full potential of the vaginal route has not yet been explored for drug administration, but this could become a useful route for certain conditions.

25.3.1.6 Nasal Administration

Medicines administered into the nose are generally intended for local actions. These products include decongestants, antibiotics and antihistamines formulated as nasal drops, sprays or ointments.[33,34]

The nose can also be used to deliver drugs into the systemic circulation. If a drug is administered orally, it may be subject to degradation in the gastrointestinal tract, poor absorption or substantial breakdown by the liver (first pass effect). However, a drug applied to the nasal mucosa is not subject to these effects and the nasal route is an attractive alternative to the parenteral route, which otherwise would have to be used.[35]

The nasal cavity is the space between the floor of the cranium and the roof of the mouth. It extends to the external nose *via* the nostrils and back to the pharynx. The walls of the nasal cavity are covered by mucous membranes, which vary in thickness and blood supply throughout the nasal cavity. The respiratory region has a large surface area due to cilia and microvilli and it is here that drug absorption occurs.[36] The vascular bed allows the rapid passage of drug molecules from the mucosa into the systemic circulation; peak drug levels are often reached within 15 min of administration.[37] The surface area of the nasal mucosa has been estimated to be 150 cm^2. The ciliated cells carry mucous towards the pharynx at a rate of 5 mm min^{-1}, which can reduce the amount of drug absorbed.[31] Experiments using nasal sprays and drops have indicated that sprays are cleared at a slower rate than drops.[38,39] In fact, drops can pass through the nasal cavity and down the pharynx without any drug being absorbed.

A wide range of drugs have been administered *via* the nasal mucosa, including β-blockers, glyceryl trinitrate and, probably of most interest, various peptides.[35] Despite this recent interest, the practice of systemic nasal administration dates back to the 1920s when pituitary extracts containing oxytocin were used to accelerate labour. Posterior pituitary gland extracts used to treat diabetes insipidus are commercially available in intranasal dosage forms (Diapid, Sandoz and DDAVP, Ferring AB).[38] Recently, a nasal spray containing buserelin has been developed for the treatment of prostate cancer (Suprefact, Hoechst).[40] Buserelin, a synthetic LHRH analogue, causes a reduction in testosterone levels, which would otherwise be achieved by castration.[41]

For some drugs, *e.g.* propranolol and progesterone, the bioavailability after nasal administration is as good as *via* the intravenous route and considerably better than *via* the oral route.[42] However, peptide bioavailability after nasal administration is often only a few per cent of the intravenous bioavailablity. Consequently, higher doses are required for similar effects.[42] However, the ease of administration often outweighs this disadvantage.

The bioavailability from the nasal route can be increased by mucoadhesion;[43,31] the drug dosage form is held in contact with the nasal mucosa, leading to a more complete absorption. Insulin bioavailability in dogs has been increased in this way.[44]

Low peptide bioavailability from the nasal route was thought to be due to the polarity of the molecule.[42] Indeed, a study with β-blockers indicated that the more hydrophobic the molecule, the better its nasal absorption would be.[45] However, clofilium tosylate, a polar quaternary ammonium compound, has a nasal bioavailability similar to that after intravenous administration in rats.[42] Poor peptide absorption is in fact probably due to enzymatic hydrolysis, which previously had been discounted.[46]

Two disadvantages with the nasal route of administration must be considered. Drugs can adversely affect the nasal mucosa; propranolol is not suitable for chronic administration due to its damaging effect on the nasal mucosa.[47] The effect of nasal conditions on drug absorption would also require evaluation for potential products. For example, rhinitis due to colds or hayfever could wash away a drug from the site of absorption. The effect of experimentally induced rhinitis on buserelin nasal absorption has been reported.[48] In this case, the rhinitis did not affect the efficacy of the product.

In conclusion, the nasal route offers a potential alternative to parenteral administration when oral administration is precluded. It will become increasingly important as more peptide drugs are developed in the future.

25.3.1.7 Buccal/Sublingual Administration

The oral cavity can be used as an absorption site for certain drugs and on certain occasions.[49] Dosage forms have been developed which can either be placed in the buccal region (between the cheek and gum) or under the tongue (sublingual).

The sublingual route is used when a rapid onset of action is required, such as with glyceryl trinitrate in angina attacks.[7] Therapeutic blood levels are reached within a few minutes of administration. For this reason, sublingual tablets tend to be small and are designed to dissolve or disintegrate rapidly, allowing speedy absorption of the drug.[50]

The buccal route is used when a drug needs to be delivered completely but not particularly rapidly. Tablets are designed to disintegrate/dissolve slowly in the buccal cavity (30–60 min) so that the drug is released slowly resulting in prolonged absorption.[50]

The lining of the buccal cavity and tongue is highly vascular and drugs applied to it can be absorbed into the sublingual or buccal capillaries and veins.[50] From here, the blood enters the jugular vein and superior vena cava and into the general circulation. Drugs absorbed by this route consequently avoid the first pass effect which is usually encountered after oral administration. In adults, the pH of saliva is about 6.4 so that acid labile drugs can be administered here without significant breakdown. Hormones such as progesterone and oxytocin can be given buccally to avoid gastric breakdown.[49] It may in the future be possible to design delivery systems for peptides to be absorbed from the oral cavity, thus avoiding the breakdown encountered after oral administration.[32] Saliva secreted by the sublingual, submandibular and parotid glands dissolves the drug, which is a prerequisite for absorption.[51] The drug must be fairly soluble as the volume of saliva is low. Another physiological factor affecting drug absorption from buccal or sublingual dosage forms is the low surface area available for absorption.

The buccal and sublingual routes have a number of advantages over other routes.[49] As previously mentioned, first pass hepatic metabolism is avoided, as is any breakdown in the GI tract. Compliance is good, especially as alternative routes would be rectal or parenteral. The onset of action can be extremely rapid as in the case of glyceryl trinitrate.

The physicochemical properties of the drug can have a large bearing on whether it is absorbed from the oral cavity (see Chapter 25.1).[50] Once a drug is in solution, it must diffuse across the mucosal membranes in the mouth. These are lipophilic in nature and, consequently, a drug needs to have a reasonable partition coefficient to be absorbed; drugs with partition coefficients of less than 20 are unlikely to be well absorbed, being too hydrophilic in nature. Similarly, drugs with partition coefficients greater than 2000 are unlikely to be absorbed due to their extremely low solubility in saliva; the concentration is too low for absorption to occur.[50]

Drugs capable of ionization will have an absorption that is influenced by their pK_a value.[50] In general, only the unionized species is absorbed and consequently the pH of the saliva has a large effect on absorption. The pH is usually between 5.6 and 7.6 but this can be altered using buffered solutions or tablets. Basic drugs, mostly unionized at pH 6 are much better absorbed than acidic drugs, which are often significantly ionized at this pH.

Two other drug properties must also be considered. Generally, the dose for a drug used buccally or sublingually should be low, *i.e.* less then 10–15 mg.[3] Also, the drug should not have an unpleasant taste as this would cause salivation and loss of drug from the oral cavity.[50]

Various drugs have been administered by the buccal or sublingual routes.[7] Apart from glyceryl trinitrate, narcotic analgesics,[52,53,54] propranolol,[47] nifedipine[55] and hormones[7] have been administered in this way. Often drugs have an enhanced bioavailability compared with the oral route. For example, the bioavailability of sublingual propranolol is over twice that obtained from the same oral dose.[47] The buccal route has been shown to give better bioavailability with morphine than the intramuscular route.[53]

Dosage forms which adhere to the oral mucosa have been developed for buccal delivery of insulin, prostaglandins and nifedipine.[31] By sticking to the buccal cavity for prolonged periods of time, the absorption of these compounds has been significantly improved.

The buccal and sublingual routes have a number of limitations associated with their use. Patients must be advised not to eat, drink or smoke while the dosage form is being used.[49] The taste of the drug must also be acceptable, otherwise compliance will be low. Lastly, the drug must be nonirritant to the mucosa, especially for chronically administered drugs.

25.3.1.8 Transdermal Administration

Drugs are generally administered topically for their local actions on the skin. They are usually formulated as ointments or creams. Drugs can also be applied in dry powders, aerosol sprays or solutions. These formulations are intended to provide prolonged local contact with minimal absorption through the skin. Drugs applied in this manner include antiseptics, antifungal agents, antibiotics, steroids and local anaesthetics.[7,56,57]

However, during the past decade, the use of the transdermal route to deliver drugs to the systemic circulation has been investigated. Previously this penetration has been envisaged as an unwanted side effect of drugs applied for their topical actions. The advantages of transdermal drug administration include the avoidance of GI drug absorption problems and the avoidance of liver metabolism (first pass effect). Transdermal administration also provides a multiday therapy with a single application, improving patient compliance, and extends the activity of drugs having short half-lives.[58] Rapid cessation of drug therapy is possible, if required.

The skin is a complex organ which offers a resistant barrier to microbial and chemical infiltration. It is also designed to prevent the loss of fluids, electrolytes, *etc.* from the underlying tissues.[56] There are two distinct layers to the skin: the outer epidermis and the inner dermis.[59] The dermis provides support for the epidermis and, because the blood supply is close to the interface between the two layers, is generally thought not to be a barrier to drug absorption. The dermis, being hydrophilic in nature, does present some resistance to the penetration of lipophilic permeants. The epidermis consists of the viable epidermis and the stratum corneum. The viable epidermis is a layer of cells which gradually differentiate into the stratum corneum, the outermost layer of skin. The latter consists of dead keratinized cells. The major source of resistance to drug permeation occurs in the stratum corneum which is lipophilic in nature and has a thickness varying from 10–15 μm in abdominal skin to 400 μm for palm skin.[60] Hair follicles and sweat ducts penetrate through the stratum corneum.

Based on this knowledge of skin anatomy, three routes of drug transport through the skin can be envisaged: transcellular, intercellular and transfollicular.[57] However, it has been estimated that the transcellular surface area accounts for 99% of the skin surface available for absorption. Consequently, investigations into improving drug penetration through the stratum corneum are generally carried out.

Various physiological factors in transdermal absorption must be considered.[57] Damaged skin is much more permeable than intact skin. The degree of hydration of the stratum corneum also affects skin permeability; generally an increase in the degree of hydration leads to an increase in drug penetration rate. Occlusive dressings lead to increased hydration and are used particularly with steroid creams to enhance permeability.

The anatomic location of the skin can have a dramatic effect on drug penetration. Using radiolabelled hydrocortisone, Feldmann and Maibach reported variations in absorption from 0.14 to 42 times that obtained from the forearm in the same subjects.[57] The age and condition of skin can also affect drug absorption. Lastly, despite consisting of dead cells, the underlying cells of the stratum corneum are capable of metabolizing drugs before they reach the systemic circulation.[57] The design of prodrugs, which are converted to the parent drug after penetration of the stratum corneum, offers increased opportunities for transdermal systemic therapy.[61]

There are a number of drug physicochemical properties which can influence a drug's permeability through the skin.[57] The most important factors are the partition coefficient and the molecular weight of the drug. For a drug to reach the systemic circulation, it must cross both the lipophilic stratum corneum and the hydrophilic viable epidermis. Obviously drugs require a degree of solubility in aqueous and oil phases to penetrate both layers of skin. Hydrophilic drugs tend to have extremely low permeability through the stratum corneum, whereas a lipophilic compound would rapidly cross but be slow in penetrating the variable epidermis. There is no simple relationship between percutaneous absorption and drug molecular weight. However, macromolecules with molecular weights in excess of 2000 Da are likely to penetrate the skin slowly, if at all. For example, peptides and proteins are not well absorbed through the skin, although such transdermal delivery has been described as the biggest challenge to date in skin permeation therapy.[62]

The nature of the transdermal dosage form can dramatically influence the absorption of a drug *via* the skin.[57] Increasing the drug concentration in the formulation generally increases drug penetration from it until the vehicle is saturated. The pH of the formulation can affect the penetration of an ionizable drug. In general, the unionized form of a drug is considerably more permeable than the ionized form. Ideally the pH should be such that the drug is in the unionized form. Surfactants have many effects on biological systems and can cause alterations in membrane permeability. Ionic surfactants are harmful to skin, whereas nonionic surfactants can enhance drug permeability without causing adverse effects to the skin.

The inclusion of penetration enhancers probably has the most dramatic effect on transdermal drug absorption. These have recently been reviewed and include any substance that increases skin permeability without causing severe irritation or damage.[63] Their mechanisms of action are largely unknown, although they may act by hydrating the stratum corneum or making it more fluid. Examples of penetration enhancers include DMSO, urea, decyl methyl sulfoxide and 1-dodecylazacycloheptan-2-one ('Azone').

The use of transdermal administration is limited to drugs which are extremely potent; the daily dose required should be less than 10–20 mg.[64] Higher doses may become possible if penetration enhancers are used. However, there are potential problems of toxicity if such excipients are used. The degree of cutaneous metabolism is also ill defined. For example, skin metabolism of glyceryl trinitrate has been estimated at 16% in the monkey. Bacteria on the skin can also reduce the amount of drug available for absorption. The use of sustained release delivery systems which maintain drug levels over prolonged periods of time is questionable; tolerance can become a problem. The development of pulsed delivery systems would be a distinct step forward. Another problem is the variability of bioavailability due to the site of application. The post-auricular site is one of high permeability and is recommended for the delivery of scopolamine.

Finally, the development of an increasing number of peptide drugs necessitates new routes of administration. As yet, the transdermal route does not look suitable unless peptide permeability can be enhanced. Consequently, transdermal delivery does not provide the answer to every delivery problem, but it is a useful route of administration for a number of drugs.

25.3.1.9 Pulmonary Administration

Medicinal products have been inhaled into the lungs for many centuries. Asthmatics once found relief from attacks by inhaling the smoke from burning mixtures of lobelia and stramonium. People suffering from common colds still find relief in inhaling vapours containing volatile oils like eucalyptus and menthol.[65]

The arrival of inhalation aerosols on to the medical scene has revolutionized the treatment of asthma.[6] Various drugs can now be delivered directly to the lower respiratory tract to treat this debilitating condition. Other indications for inhalation therapy include general anaesthesia, decongestion, breakdown of mucus secretions and systemic administration of a limited number of drugs.[66]

The lower respiratory tract consists of the trachea, which then divide into the right and left bronchi.[66] These further divide into bronchioles, which finally lead into the alveoli, where gaseous exchange with the pulmonary blood supply occurs. Mucus, produced in the walls of the trachea and bronchi, serves to moisten dry air and trap dirt and dust. This mucus film is removed toward the pharynx by cilia, hairlike structures which move back and forth clearing the respiratory tract.

One of the main advantages of inhalation drug delivery is that drugs can be delivered directly to their site of action.[66] Only small amounts are systemically absorbed generally, so that high doses can be delivered and yet side effects are minimal. For example, sympathomimetic bronchodilators like salbutamol are much more effective when inhaled than when administered orally and yet the side effects are reduced.[6] Steroid therapy is usually fraught with problems of adrenal suppression and other side effects. However, given by inhalation, steroids can exert their beneficial actions with a much-reduced risk of side effects.[66]

Rapid onset of actions is another advantage of the pulmonary route.[67] The large surface area of the alveoli (approximately 200 m^2), the thin layer of epithelial cells and the excellent blood supply ensure rapid drug absorption into the bronchial mucosa and smooth muscle. Bronchodilators exert their actions within minutes of administration, approaching the rapidity of action seen after intravenous administration.[66] Drugs administered by inhalation are not exposed to the hostile environment of the GI tract and can thus reach the systemic circulation intact.[66] Drugs with poor oral bioavailability can also be delivered *via* the lungs. Drugs absorbed into the pulmonary circulation also avoid hepatic metabolism prior to systemic distribution.

The majority of drugs delivered into the lungs are administered for local actions. Such drugs include bronchodilators, steroids and prophylactic agents for asthma. Mucolytic agents and antibiotics may also be administered to patients with severe respiratory infections.[66]

A number of drugs have been administered for their systemic actions. Glyceryl trinitrate has been administered to give rapid relief from angina attacks, but this inhalation product has not been a commercial success.[66] Peptides such as oxytocin and insulin have been delivered *via* the lungs but their absorption was irregular and variable.[65,32] Ergotamine tartrate has been delivered systemically, using an aerosol inhalation, for the treatment of migraine. It has been shown that the onset of action was between 10 and 30 min compared with 1.5 to 3 h after oral administration.[66]

The main disadvantage of the pulmonary route of drug administration is the need for dosage forms which accurately deliver the drug to its site of action. The delivery of a drug to the lungs is dependent largely on the drug's particle size distribution.[66,22] Unless particles are less than 10 µm, they impact on to the walls of the respiratory tract without reaching their target; particles tend to move in a straight line, and not with the air current, above this size. For a drug to be absorbed, its particles must impact preferably in the alveolar sacs and then dissolve. Large particles are lost higher up the respiratory tract. To reach the alveolar sacs, the particle size must be in the order of 1 µm. However, small particles are more likely to be exhaled before being impacted. The disease state of the lung can also affect drug deposition, especially when pulmonary obstruction occurs. Generally, only a small percentage of the drug actually reaches the alveolar sacs (10–20%) due to these particle size effects.[6,67]

There are a wide range of dosage forms used for pulmonary administration, depending on the nature of the drug and formulation.[68,69] Most commonly, drugs are administered to the respiratory tract as sprays from pressurized aerosol dispensers (*e.g.* Ventolin).[66,69] These contain the drug and an inert propellant, usually a low-boiling-point fluorochlorohydrocarbon. When the aerosol valve is opened, the propellant vaporizes rapidly, leaving behind a dispersion of the drug in the inhaled air.

Drugs can also be administered as micronized powders using special devices, *e.g.* Intal Spinhaler.[69,70] The drug, sodium cromoglycate, is supplied in hard gelatin capsules together with lactose. The capsule is pierced in the Spinhaler and the air drawn through it contains a dispersion of drug and lactose particles. The drug particles (1 to 10 μm) pass deep into the respiratory tract and leave the lactose particles (30 to 60 μm) in the upper airways.

Nebulizers are also used to administer drugs to the lungs, especially in the hospital environment.[69,71] A nebulizer is a device which converts an aqueous drug solution into a mist of fine particles for inhalation. This is usually achieved by bombarding the drug solution with a stream of gas, resulting in the formation of tiny droplets. Nebulizer therapy is particularly useful for patients who find it difficult to use aerosol inhalers.

Inhalation therapy has a number of disadvantages associated with its use.[72] It has been estimated that up to 75% of patients are unable to use aerosol inhalers properly.[73] A high degree of coordination of breathing and activation is required for the dose to be efficiently administered. Asthmatic patients tend to over- or under-utilize aerosols, depending on how they perceive their medical condition.[72] Another problem is the range of different devices, which often have differing instructions supplied with them. This can result in poor patient compliance, despite the apparent ease of administration. Questions of toxicity have been raised with respect to the safety of propellants used in aerosols and also the role of propellants in causing damage to the ozone layer is of serious concern.[66] The last factor to be considered is that the use of aerosols can cause a transient bronchoconstriction in some patients.[67] This is particularly undesirable in asthmatics who are suffering from pulmonary obstruction.

Despite these limitations, inhalation of drugs can be a very effective and safe means of treating various disease states.

25.3.1.10 Ocular Administration

Drugs are often applied to the eye for local actions on the surface of the eye or for actions in the interior of the eye. Such drugs include miotics for glaucoma, local anaesthetics, antiinflammatory agents and various antiseptic/antimicrobial compounds.[74]

The most common ophthalmic dosage form is the aqueous eye drop solution.[74] However, nonaqueous solutions, suspensions and ointments are also available. One of the latest developments in ocular drug administration is the use of ophthalmic inserts. These are polymeric devices containing a drug, which is released over a prolonged time, *e.g.* 7 d, at a predetermined rate.

Most ophthalmic drugs reach their receptors in the eye by permeating through the cornea to reach the anterior eye chamber. It appears that the cornea and the lacrimal system together have a great influence on the bioavailability of drugs applied to the eye.[75] The lacrimal gland secretes tears at a rate of 1.2 μL min^{-1}, which are removed *via* the lacrimal sac to the back of the nose and throat. Systemic absorption can occur from this drainage route and result in systemic toxicity; this can occur, for example, with β-blockers used in the treatment of glaucoma.[76] This problem is accentuated because a drug absorbed after such drainage is not subject to first pass liver metabolism. The capacity of the human conjunctival sac is between 25 and 30 μL, so a significant portion of eye drops are lost due to spillage out of the eye.[75] Consequently, lacrimal drainage, blinking and low capacity are responsible in part for poor ophthalmic bioavailability.

The barrier properties of the cornea are another cause of poor ophthalmic bioavailability. The human cornea has five layers consisting of the epithelium, Bowman's membrane, stroma, Descemet's membrane and endothelium.[75] The epithelium and endothelium are considerably more lipoidal in nature than the stroma. An ophthalmic drug requires lipid and water solubility to be able to penetrate the cornea. For lipophilic compounds, the stroma is the cause of poor corneal penetration, whereas for hydrophilic compounds the epithelium is the rate-determining barrier.[75]

Protein binding of drugs can reduce the effective dose of a drug applied to the eye. Similarly, extensive metabolism can occur in the cornea.[34] This high enzymatic activity can be used to deliver hydrophilic drugs as labile lipophilic prodrugs.[77] For example, the dipivalyl ester of adrenaline improves the corneal absorption of adrenaline by a factor of 10–20 and allows a reduction in dose and dosing frequency.[77] The lipophilic ester easily crosses the corneal epithelium and is cleared by esterases in the stroma to give the hydrophilic active species. This in turn rapidly reaches the aqueous humour in the anterior chamber of the eye.

Table 1 Advantages, Disadvantages and Limitations of Different Routes of Administration

Route	Advantages	Disadvantages/limitations
Oral	Easy to administer Convenient Economical Safe (overdosage can be treated generally) Self-administration Painless Controlled release Possible	Can cause nausea/vomiting Drug may be unstable in GI tract Erratic absorption for poorly soluble drugs Food can affect absorption Patient must be cooperative and conscious Drugs subject to first pass liver metabolism Slow onset of action Insufficient blood levels may be obtained
Parenteral	Useful for poorly absorbed drugs For unstable (GI) drugs Rapid onset of action Total and predictable levels Patient can be uncooperative or unconscious Exact dosage given Large doses possible Prolonged release injection possible Compliance is total No first pass effect Inject into specific area of body Peptide delivery made feasible	Requires trained staff Self-administration unusual No going back once given Disliked by patients Painful Expensive Shorter shelf-life products Injection of bacteria, particles, pyrogens or air can occur
Rectal	Little or no first pass effect Useful when oral route dosage form cannot be swallowed, *e.g.* infants Severe GI irritation is reduced Local actions possible Good alternative to oral/parenteral routes Avoids GI breakdown Can be used if patient is unconscious Peptide delivery may be possible	Unpopular in UK and USA Slow, poor and erratic absorption Rectal irritation Inconvenient for patient General paucity of relative bioavailability data
Vaginal	Avoids first pass metabolism Useful for delivey of hormones Local treatment possible Birth control devices Used for abortions instead of surgery Peptide delivery may be feasible	Inconvenient for patient Erratic and unpredictable absorption, particularly affected by menstrual/hormonal changes Irritation Lack of bioavailability data
Nasal	Avoids first pass metabolism Little or no breakdown in nose Convenient Avoids GI breakdown Suitable for peptide delivery	Irritation Absorption affected by state of nasal mucosa
Buccal/sublingual	Avoids first pass metabolism Convenient Avoids GI degradation Quick onset of action Instant cessation of treatment possible Peptide delivery may be feasible	Duration of action determined by holding dosage form in place intact Palatability of drug Limited to low dosage drugs Cannot swallow, eat, drink while taking
Transdermal	Avoids first pass metabolism Avoids GI breakdown and absorption problems Polonged action leading to good compliance Rapid cessation of treatment possible Local actions	Skin irritation Only suitable for potent drugs Tolerance may develop Erratic absorption affected by site chosen
Pulmonary	Actions localized so reduced side effects and doses Convenient Rapid onset of action No GI breakdown First pass metabolism avoided	Only small proportion reaches site of action High degree of patient coordination required Propellant toxicity Irritation to bronchi Overdosage
Ocular	Local action only Prolonged action with inserts possible	Inefficient delivery to eye Hard to administer

The transcellular mechanism is not the only way in which a drug can reach the anterior chamber of the eye.[77] Certain drugs, particularly charged ones, can diffuse between the epithelial cells; this is known as the paracellular mechanism. EDTA has been shown to facilitate this process by expanding the spaces between cells for short periods of time.

Various physicochemical properties of ophthalmic drugs can influence their corneal penetration. A knowledge of these properties enables drugs to be designed which have rapid corneal penetration.[75] The drug partition coefficient indicates the degree of lipid and water solubility that a drug possesses. For some families of drugs there is a parabolic relationship between permeability and the log partition coefficient (log P).[75] This enables the optimum partition coefficient, and hence drug analogue, to be found. Such parabolic relationships indicate that an intermediate lipophilic/hydrophilic balance must be attained to achieve rapid penetration of the cornea. This data can be helpful in the design of optimally permeable ophthalmic drugs.

Drug solubility can also play a role in corneal penetration. To be absorbed, a drug must be in solution. However, if the solubility is too great, the drug will not be absorbed because of its low lipophilicity.[75]

The ionization constant (pK_a) is another important property. The average pH of tears is about 7.4, so ideally the pK_a of a drug should be within one or two units of this so that it is mainly present in the absorbable unionized state. Most ophthalmic drugs are weak bases with pK_a values below 9 and at least 5–10% is unionized at pH 7.4. Very few weak acids are applied to the eye as they tend to be ionized at pH 7.4.[75]

A number of excipients are included in ophthalmic preparations to ensure the product is stable and remains sterile when used.[77] These additives include preservatives, buffers and polymers. Benzalkonium chloride, a common ophthalmic preservative, is also a surfactant that enhances the absorption of drugs through the cornea. The buffering capacity of tear fluids is low and the pH of an ophthalmic preparation will cause a change in tear pH.[75] This in turn can result in altered bioavailabilities as the drug may become more or less ionized. Polymers are added to ophthalmic solutions or suspensions to increase their viscosity.[75,77] This in turn prolongs drug contact with the cornea and can enhance absorption, especially for hydrophilic drugs.

Novel ophthalmic dosage forms have been developed which can improve a drug's ocular bioavailability.[77] Inserts are designed to prolong the presence of a drug in the tear fluid. The nonerodible insert Ocusert releases pilocarpine at a defined rate for a period of 7 d.[78] After this time it has to be removed and replaced. Erodible inserts are being developed which would eventually dissolve away without needing to be removed.[77] Liposomes containing ophthalmic drugs have been designed to improve the interaction between the drug and the cornea.[77] This potentially could result in better absorption. This goal has also been achieved using polymeric substances that adhere to the cornea by bioadhesion.[79] With a progesterone bioadhesive dosage form, the bioavailability in rabbits was improved by a factor of 4.2 as compared with a suspension formulation.

25.3.1.11 Summary

The advantages, disadvantages and limitations of the routes of administration discussed previously are summarized in Table 1.

25.3.2 DOSAGE REGIMES

25.3.2.1 Introduction

One of the major concerns of administering the drug to patients by any route is the safety of the drug. The immediate question in the mind of the physician is how much of the drug can be given to the patient without causing any undesirable effect and secondly how much drug is required by the patient to cure him of the illness being treated. These concepts of safety and efficacy can be described in terms of the therapeutic index and therapeutic range of the drug substance.

25.3.2.2 Therapeutic Index

A measure of the safety of a drug can be obtained from its therapeutic index. To define this index it is necessary to consider the results of animal studies. A frequently quoted measure of the toxicity of the drug is the median lethal dose (LD_{50}), the dose of drug which kills 50% of the animals in the group under test. This can be compared with the median effective dose (ED_{50}), that is the dose of drug which produces the desired therapeutic effect in half the group. The ratio of the LD_{50} to the ED_{50} is defined as the therapeutic index of the drug, *i.e.* therapeutic index = LD_{50}/ED_{50}.

The concept of a therapeutic index for drug substances is not new. Indeed, Paul Erlich in the 1890s, in developing the concepts of selectivity of action of chemotherapeutic agents, determined a therapeutic index for each of a series of arsenical compounds.[93] This involved determining efficacy of the compound in curing infected mice and assessing the toxicity to the animals.

In general the value of the therapeutic index gives a useful guide to the relative safety of the compound. If the index is unity or less, it is clearly not possible to use the compound for the effect being investigated without killing the majority of the animals. If the index is large, then the margin of safety is accordingly greater. Thus, the larger the therapeutic index, the safer is the drug in question. In practice, indices vary from slightly more than one up to several thousand. A drug may be considered as generally safe if its therapeutic index is greater than 10.[80] A further discussion of actual values for different classes of drugs is given in Section 25.3.2.7.

Clearly, the therapeutic index as described above can only be determined by testing in laboratory animals. The resulting index cannot easily be translated into something of clinical significance, relevant to the treatment of patients. The concept of therapeutic index is, however, widely referred to in the pharmacokinetic and clinical literature[80–88] and in this context can be more loosely defined as the ratio of the drug plasma concentration just producing toxic effects to the drug concentration producing the required effect. An alternative description of therapeutic index in terms of the dose would be the ratio of the maximum tolerated dose to the minimum effective dose in patients.

25.3.2.3 Therapeutic Range

For a drug to be used to produce a particular therapeutic effect in patients there has to be a recommended dose regime. This is established as a result of lengthy research programmes starting in animals and later in human volunteers and finally in patients.

It is important to realize that the therapeutic effects which are studied to monitor drug action can be quite diverse in nature. For example, a drug may produce a graded response, where the intensity of the effect being measured gradually increases with concentration, producing a characteristic dose–effect curve.[81] Eventually the concentration is approached that produces the maximal possible effect. On the other hand some pharmacological or toxic responses do not happen on a continuous basis but occur suddenly and are either present or absent. An extreme example is, of course, death. Such a response is termed all-or-none or quantal. These responses can be studied by determining the frequency with which the event occurs in relation to concentration, within a group of patients. A good example is the study of arrhythmia as a function of serum procainamide concentration.[82]

The most direct way of establishing a dose regime is to monitor plasma levels of the drug during trials in volunteers or patients and then to relate the plasma concentration to the therapeutic effect. By this means it is possible to delineate a range of plasma concentration that provides a therapeutic effect, without giving rise to toxic side effects. This is termed the therapeutic range of the drug.

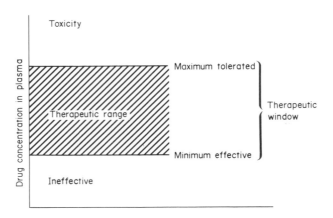

Figure 1   Therapeutic range of drug

The plasma concentration can then be related back to the dose levels corresponding to the minimal effective dose and the maximal tolerated dose, which in turn define the therapeutic dose range (Figure 1).

Thus, in general terms the therapeutic range can be described either in terms of upper or lower therapeutic doses or more exactly in terms of upper and lower therapeutic plasma levels.

25.3.2.4 Dosage Regime

From a knowledge of the minimum effective dose level and the maximum tolerated dose the concept of the average dose arises. That is the dose or dose range found to be most effective in the largest group of patients. The average dose is in effect the recommended dose quoted in data sheets or official compendia. Often an initial dose is stated, with an indication of a maximum dose to be administered within a given time frame.

It must be appreciated that the recommended dose is regarded as the dose suitable for an average adult, weighing perhaps 70 kg, with normally functioning kidneys and liver. Clearly for many groups of patients modification of the average dose level will be required and a degree of individualization is necessary.[83]

25.3.2.5 Modifications to Dosage Regime

The reasons for dose modifications are numerous. With infants and young children it is important to reduce the dose according to body weight or surface area. Children may metabolize and excrete drug at rates different to adults and also display increased sensitivity to target tissues.

Much attention is also paid to the potential problems of elderly patients with regard to dose levels. Generally lower doses are required due to impaired renal function and therefore less efficient excretion of the drug from the body. Also changes in body composition and body weight are of significance and also reduced protein binding of the drug in the plasma may occur, thereby increasing the available drug in the systemic circulation.

Disease states have a significant influence on the way in which a patient can cope with a particular dose of drug. With kidney disease the impaired renal function will give rise to the potential accumulation of toxic levels of drug within the body. Thus, the dose administered should be reduced. Similarly, in patients with liver disease the rate of biotransformation of the drug will be significantly altered and the drug may persist for longer time periods at high concentration in the circulation.

A further group of patients who may require significantly different levels are those who produce allergic or hyperreactive responses to the drug in question. This will result in the avoidance of certain classes of drug or at least extremely close monitoring of effects on further administration.

Changes to the average dose level are also necessary where tolerance to the drug develops on repeated administration, as in the case of narcotic drugs or tranquillizers.

From the above discussion it is clear that the administration of the average dose to all patients may involve considerable risks if no account is taken of age, disease state, body malfunction and inherent human variations. This becomes of greater importance in the case of drugs with narrow therapeutic range and low therapeutic index.

In order to determine the optimum dose for a particular patient for drugs with wide therapeutic ranges, it is possible to start at the average dose and make adjustments according to the changes in symptoms. This so called titration approach may be too hazardous where drugs of low therapeutic index are concerned and more sophisticated monitoring is desirable.

25.3.2.6 Drug-level Monitoring

Nowadays it is common practice and a requirement to monitor blood levels where possible. In some cases a related pharmacological or chemical effect can be measured, such as blood glucose level or clotting time. The whole science of clinical pharmacokinetics has evolved over the past 25 years with the objective of quantifying in scientific terms the events that follow drug administration.[84] (The concepts of pharmacokinetics are dealt with in depth in Part 24.) In simple pharmacological terms, the desirable therapeutic action of the drug is governed by the concentration of drug at the site of action, or more likely the drug concentration in the body fluids reaching the site of action. Since it is rarely possible to target the site of action specifically, the whole body tissues are suffused with drug, usually by oral or parenteral administration. Thus, monitoring the concentration profile of the drug in the blood circulation provides an indirect assessment of the drug concentration at the site of action, *e.g.* target tissue or organ. Clearly, the blood level concentration variation with time will be a function of the rate of drug absorption, distribution through the body and elimination. The measurement of blood levels in this way leads to the definition of a dosage in terms of a minimum blood level to produce efficacy and a maximum concentration above which toxic effects occur. The concentration region associated with therapeutic success is termed the 'therapeutic window'.[84] This is an alternative description of the 'therapeutic range' referred to in Figure 1.

25.3.2.7 Therapeutic Range and Index Relative to Drug Type

Having described the concepts behind therapeutic range and therapeutic index, it will now be useful to examine the main areas of application and in particular the classes of drugs where it is especially important to take note of these parameters.[85]

In considering a particular compound, it is important to be aware of its mode of action. In general, if the drug is being used for a life-threatening condition, it is likely that it could have disastrous effects if administered incorrectly and by implication it may well have a low therapeutic index or narrow therapeutic range. This is indeed the case with drugs used for heart conditions, such as digitalis.

From the point of view of development it may be desirable to have drugs with wide therapeutic ranges and large indices since they are inherently safer, but this does not necessarily correlate with specificity of action and efficacy. In treating certain disease states, it may be acceptable to use drugs which are potentially toxic and have very low indices if the perceived benefits are sufficiently great, such as cancer chemotherapy or the treatment of life-threatening infections. A summary of therapeutic index values for a range of drugs is given in Table 2. Comments on various major classes of drugs are given in the following sections.

25.3.2.7.1 *Cardiac drugs*

Cardiac glycosides such as digoxin and digitoxin are used for the treatment of congestive heart failure, fibrillation and other diseases of the heart. The therapeutic and toxic levels are close, as indicated by the low therapeutic index. Thus, great care has to be taken in administering the drug as cardiac arrhythmias develop if the plasma concentrations are too high. The administration is further complicated by the variable absorption and distribution kinetics of digoxin, which can lead to wide variations in response. Thus, with digoxin plasma concentration monitoring is highly desirable until a satisfactory dosage regime for the patient has been established. Digoxin is also a classic example of a drug which can be significantly influenced by the bioavailability of the product involved, thus care must be exercised when a patient changes to a new brand of product.

Another group of compounds used for the treatment of heart problems is the antiarrhythmic drugs, such as quinidine, procainamide and lidocaine. Blood level monitoring of these drugs is desirable in view of their low therapeutic indices (Table 2).

Table 2 Therapeutic Indices of Various Drugs[86,88]

Drug	Therapeutic index	Therapeutic class
Digoxin	2.2	Heart failure
Phenytoin	2	Epilepsy
Carbamazepine	2	Epilepsy
Quinidine	2.2	Cardiac arrhythmia
Procainamide	2.5	Ventricular arrhythmia
Phenobarbital	2.7	Epilepsy
Lithium carbonate	2.9	Antidepressant
Lidocaine	3.3	Ventricular arrhythmia
Theophylline	3.5	Bronchodilator
Warfarin	4	Anticoagulant
Digitoxin	4.4	Heart failure
Salicylates	2.5–5	Analgesia, antirheumatic
Propranolol	>10	Angina, hypertension
Chloropromazine	>10	Antipsychotic

A recent review article deals with a number of new inotropic drugs, which may offer advantages over the cardiac glycosides in the treatment of chronic heart failure.[86]

25.3.2.7.2 *Bronchodilators*

The major drug used for the treatment of reversible bronchoconstriction associated with asthma is theophylline. A large amount has been published on the pharmacokinetics of this compound.[87,88] An important feature of the drug is the large difference in pharmacokinetics between adults and children. Again, blood level monitoring is desirable, particularly for slow release products.[87]

25.3.2.7.3 *Drugs acting on the central nervous system*

With treatment involving antipsychotic drugs it is of advantage to have direct information such as blood level monitoring, since communication with the patient may be difficult.

The most widely used anticonvulsant for the treatment of epilepsy is phenytoin. Drug level monitoring is desirable due to the nonlinear pharmacokinetics of the drug in humans.[85]

25.3.2.7.4 *Anticancer compounds*

Antitumor drugs are often administered empirically with a fixed starting dose and subsequently increasing the dose until toxicity is observed. Anticancer compounds are usually of low therapeutic index and overdosing can lead to life-threatening toxicity. The major factor hindering routine monitoring of blood levels is often lack of rapid and specific assay methods. However, many data on a variety of compounds are being generated as summarized in a number of recent reviews.[89,90,91] The clinical pharmacokinetic aspects of many commonly used antineoplastic agents such as cisplatin, methotrexate, the nitroureas, doxirubicin and bleomycin are given in some detail[89] and undoubtedly the monitoring of plasma levels during administration will lead to more sophisticated usage of these compounds, avoiding the inherent toxicities.

25.3.2.7.5 *Antibiotics*

Many antibiotics such as the penicillins and cephalosporins have relatively large therapeutic indices and can therefore be given in high doses without risk of toxicity.

There are some important antibiotics, however, which are potentially toxic and have much smaller therapeutic indices. These are notably chloramphenicol, vancomycin and the aminoglycosides, where ototoxicity is a risk. Serum monitoring of antibiotics is reviewed in a recent article.[92]

25.3.3 REFERENCES

1. J. Wartak, 'Drug Dosage and Administration, University Park Press, Baltimore, 1983, chap. 16.
2. 'Pharmaceutical Handbook', 19th edn., Pharmaceutical Press, London, 1985, p. 294.
3. D. R. Gowley and C. T. Ueda, in 'Pharmaceutics and Pharmacy Practice', ed. G. S. Banker and R. K. Chalmers, Lippincott, Philadelphia, 1982, chap. 5.
4. 'Pharmaceutical Handbook', 19th edn., Pharmaceutical Press, London, 1985, p. 314.
5. R. J. Duma and M. J. Akers, in 'Pharmaceutical Dosage Forms: Parenteral Medication', ed. K. E. Avis, L. Lachman and H. A. Leiberman, Dekker, New York, 1984, vol. 1, chap. 2.
6. M. T. Newhouse and M. B. Dolovich, *New Engl. J. Med.*, 1986, **315**, 870.
7. H. C. Ansel, 'Introduction to Pharmaceutical Dosage Forms', 4th edn., Lea and Febiger, Philadelphia, 1985, chap. 3.
8. P. A. Shore, B. B. Brodie and C. A. Hogben, *J. Pharmacol. Exp. Ther.*, 1957, **119**, 361.
9. Y. C. Martin, *J. Med. Chem.*, 1981, **24**, 233.
10. I. Komiya, J. Y. Park, A. Kamani, N. F. H. Ho and W. I. Higuchi, *Int. J. Pharm.*, 1980, **4**, 249.
11. N. F. H. Ho, J. Y. Park, P. F. Ni and W. I. Higuchi, 'Animal Models for Oral Drug Delivery in Man: *In situ* and *In vivo* Approaches', American Pharmaceutical Association, Washington, 1983, p. 27.
12. D. C. Taylor, R. E. Pownall and W. M. Bunke, *J. Pharm. Pharmacol.*, 1985, **37**, 280.
13. S. S. Davis, J. G. Hardy, M. J. Taylor, D. R. Whalley and C. G. Wilson, *Int. J. Pharm.*, 1984, **21**, 167.
14. S. S. Davis, J. G. Hardy, M. J. Taylor, D. R. Whalley and C. G. Wilson, *Int. J. Pharm.*, 1984, **21**, 331.
15. S. S. Davis, J. G. Hardy, C. G. Wilson, L. C. Feely and K. J. Palin, *Int. J. Pharm.*, 1986, **32**, 85.

16. P. P. DeLuca and R. P. Rapp, in 'Pharmaceutics and Pharmacy Practice', ed. G. S. Banker and R. K. Chalmers, Lippincott, Philadelphia, 1982, chap. 8.
17. J. Wartak, 'Drug Dosage and Administration', University Park Press, Baltimore, 1983, chap. 18.
18. 'Pharmaceutical Handbook', 19th edn., Pharmaceutical Press, London, 1985, p. 312.
19. A. R. Gennaro (ed.), 'Remington's Pharmaceutical Sciences', 17th edn., Mack, Easton, PA, 1985, p. 764.
20. I. M. Hunneyball, *Pharm. Int.*, 1986, **7**, 118.
21. H. C. Ansel, 'Introduction to Pharmaceutical Dosage Forms', 4th edn., Lea and Febiger, Philadelphia, 1985, chap. 9.
22. S. Niazi, 'Textbook of Biopharmaceutics and Clinical Pharmacokinetics', Appleton Century Crofts, New York, 1979, chap. 4.
23. H. C. Ansel, 'Introduction to Pharmaceutical Dosage Forms', 4th edn., Lea and Febiger, Philadelphia, 1985, chap. 14.
24. L. J. Coben and H. A. Lieberman, in 'Theory and Practice of Industrial Pharmacy', ed. L. Lachman, H. Lieberman and J. Kanig, Lea and Febiger, Philadelphia, 3rd edn., 1986, chap. 19.
25. J. C. McElnay, A. J. Taggart, B. Kerr and P. Passmore, *Int. J. Pharm.*, 1986, **33**, 195.
26. W. Lowenthal and W. R. Garnett, in 'Pharmaceutics and Pharmacy Practice', ed. G. S. Banker and R. K. Chalmers, Lippincott, Philadelphia, 1982, chap. 12.
27. J. A. Fix and C. R. Gardner, *Pharm. Int.*, 1986, **7**, 272.
28. T. Nishikata, Y. Okamura, H. Inagaki, M. Sudho, A. Kamada, T. Yagi, R. Kawamori and M. Shichiri, *Int. J. Pharm.*, 1986, **34**, 157.
29. Y. W. Chien, 'Novel Drug Delivery Systems', Dekker, New York, 1982, chap. 3.
30. Y. W. Chien, 'Novel Drug Delivery Systems', Dekker, New York 1982, chap. 4.
31. T. Nagai and Y. Machida, *Pharm. Int.*, 1985, **6**, 196.
32. T. Kimura, *Pharm. Int.*, 1984, **5**, 75.
33. H. C. Ansel, 'Introduction to Pharmaceutical Dosage Forms', 4th edn., Lea and Febiger, Philadelphia, 1985, chap. 13.
34. J. R. Robinson and L. M. Goshman, in 'Pharmaceutics and Pharmacy Practice', ed. G. S. Banker and R. K. Chalmers, Lippincott, Philadelphia, 1982, chap. 10.
35. Y. W. Chien and S. F. Chang, in 'Transnasal Systemic Medications', ed. Y. W. Chien, Elsevier, Amsterdam, 1985, p. 2.
36. G. D. Parr, *Pharm. Int.*, 1983, **4**, 202.
37. J. L. Colaizzi, in 'Transnasal Systemic Medications', ed. Y. W. Chien, Elsevier, Amsterdam, 1985, p. 118.
38. K. S. E. Su, *Pharm. Int.*, 1986, **7**, 8.
39. A. S. Harris, I. M. Nilsson, Z. G. Wagner and U. Alkner, *J. Pharm. Sci.*, 1986, **75**, 1085.
40. W. Petri, R. Schmiedel and J. Sandow, in 'Transnasal Systemic Medications', ed. Y. W. Chien, Elsevier, Amsterdam, 1985, p. 161.
41. J. Sandow and W. Petri in 'Tranasal Systemic Medications', ed. Y. W. Chien, Elsevier, Amsterdam, 1985, p. 183.
42. C. H. Huang, R. Kimura, R. B. Nassar and A. Hussain, *J. Pharm. Sci.*, 1985, **74**, 608.
43. M. A. Longer and J. R. Robinson, *Pharm. Int.*, 1986, **7**, 114.
44. T. Nagai, Y. Nishimoto, N. Nambu, Y. Suzuki and K. Sekine, *J. Controlled Release*, 1984, **1**, 15.
45. G. S. M. J. E. Duchateau, J. Zuidema, W.M. Albers and F. W. H. M. Merkus, *Int. J. Pharm.*, 1986, **34**, 131.
46. A. A. Hussain, R. B. Nassar and C. H. Huang, in 'Transnasal Systemic Medications', ed. Y. W. Chien, Elsevier, Amsterdam, 1985, p. 121.
47. G. S. M. J. E. Duchateau, J. Zuidema and F. W. H. M. Merkus, *Pharm. Res.*, 1986, **3**, 108.
48. C. Larsen, M. N. Jorgensen, B. Tommerup, N. Mygind, E. E. Dagrosa, H. G. Grigoleit and V. Malerczyk, *Eur. J. Clin. Pharmacol.*, 1987, **33**, 155.
49. J. L. Colaizzi and W. H. Pitlick, in 'Pharmaceutical and Pharmacy Practice', ed. G. S. Banker and R. K. Chalmers, Lippincott, Philadelphia, 1982, chap. 7.
50. J. W. Conine and M. J. Pikal, in 'Pharmaceutical Dosage Forms: Tablets', ed. H. A. Lieberman and L. Lachman, Dekker, New York, 1980, vol. 1, chap. 6.
51. 'Pharmaceutical Handbook', 19th edn., Pharmaceutical Press, London, 1985, p. 307.
52. S. L. Wallenstein, R. F. Kaiko, A. G. Rogers and R. W. Honde, *Pharmacotherapy*, 1986, **6**, 228.
53. M. D. D. Bell, G. R. Murray, P. Mishra, T. N. Calvey, B. D. Weldon and N. E. Williams, *Lancet*, 1985, **1**, 71.
54. M. A. Hussain, B. J. Aungst and E. Shefter, *J. Pharm. Sci.*, 1986, **75**, 218.
55. G. R. Brown , D. G. Fraser, J. A. Castile, P. Gaudreault, D. R. Platt and P. A. Friedman, *Int. J. Clin. Pharmacol. Ther. Toxicol.*, 1986, **24**, 283.
56. A. W. Malick and R. E. Smith, in 'Pharmaceutics and Pharmacy Practice', ed. G. S. Banker and R. K. Chalmers, Lippincott, Philadelphia, 1982, chap. 9.
57. J. Zatz, in 'Bioavailability Control by Drug Delivery System Design', ed. V. F. Smolen and L. Ball, Wiley, New York, 1984, chap. 4.
58. H. C. Ansel, 'Introduction to Pharmaceutical Dosage Forms', 4th edn., Lea and Febiger, Philadelphia, 1985, chap. 11.
59. K. A. Walters, *Pharm. Technol.*, 1986, **10**, 30.
60. Y. W. Chien, 'Novel Drug Delivery Systems', Dekker, New York, 1982, chap. 5.
61. R. H. Guy and J. Hadgraft, *Pharm. Int.*, 1985, **6**, 112.
62. B. Barry, *Pharm. J.*, 1986, **236**, 764.
63. J. Hadgradt, *Pharm. Int.*, 1984, **5**, 252.
64. R. H. Guy, *J. Controlled Release*, 1987, **4**, 237.
65. 'Pharmaceutical Handbook', 19th edn., Pharmaceutical Press, London, 1985, p. 309.
66. R. G. Hollenbeck and T. H. Wiser, in 'Pharmaceutics and Pharmacy Practice', ed. G. S. Banker and R. K. Chalmers, Lippincott, Philadelphia, 1982, chap. 11.
67. J. Wartak, 'Drug Dosage and Administration', University Park Press, Baltimore, 1983, chap. 20.
68. W. F. Kirk, *Pharm. Int.*, 1986, **7**, 150.
69. F. Moren, in 'Aerosols in Medicine', ed. F. Moren, M. T. Newhouse and M. B. Dolovich, Elsevier, Amsterdam, 1985, chap. 10.
70. H. C. Ansel, 'Introduction to Pharmaceutical Dosage Forms', 4th edn., Lea and Febiger, Philadelphia, 1985, chap. 10.
71. H. Horsley, *Pharm. J.*, 1988, **240**, 22.
72. J. J. Sciarra and A. J. Cutie, *Aerosol Age*, March 1983, 26.

73. A. Rogers, *Pharm. J.*, 1987, **239**, 22.
74. H. C. Ansel, 'Introduction to Pharmaceutical Dosage Forms', 4th edn., Lea and Febiger, Philadelphia, 1985, chap. 12.
75. R. D. Schoenwald, in 'Bioavailability Control by Drug Delivery System Design', ed. V. F. Smolen and L. Ball, Wiley, New York, 1984, chap. 6.
76. A. G. Gilman, L. S. Goodman, T. W. Rall and F. Murad (eds.), 'The Pharmacological Basis of Therapeutics', 7th edn., Macmillan, New York, 1985, p. 9.
77. V. H. Lee, *Pharm. Int.*, 1985, **6**, 135.
78. Y. W. Chien, 'Novel Drug Delivery Systems', Dekker, New York, 1982, chap. 2.
79. H. W. Hui and J. R. Robinson, *Int. J. Pharm.*, 1985, **26**, 203.
80. J. Wartak, 'Drug Dosage and Administration', University Park Press, Baltimore, 1983, p. 78.
81. A. R. Gennaro (ed.), 'Remington's Pharmaceutical Sciences', 17th edn., Mack, Easton, PA, 1985, chap. 37.
82. M. Rowland and T. N. Tozer, 'Clinical Pharmaceutics: Concepts and Applications', Lea and Febiger, Philadelphia, 1980, p. 158.
83. J. Wartak, 'Drug Dosage and Administration', University Park Press, Baltimore, 1983, chap. 2.
84. M. Rowland and T. N. Tozer, 'Clinical Pharmaceutics: Concepts and Applications', Lea and Febiger, Philadelphia, 1980, chap. 1.
85. S. Naizi, 'Textbook of Biopharmaceutics and Clinical Pharmacokinetics', Appleton Century Crofts, New York, 1979, p. 252.
86. M. L. Rocci and H. Wilson, *Clin. Pharmacokinet.*, 1987, **13**, 91.
87. B. C. Cartstedt and W. F. Stanaszek, *U.S. Pharm.*, 1987, Feb, 100.
88. W. J. Taylor and A. L. Finn, 'Individualizing Drug Therapy: Practical Applications of Drug Monitoring', Gross Towsend Frank, New York, 1981.
89. F. M. Balis, J. S. Holcenberg and W. A. Bleyer, *Clin. Pharmacokinet.*, 1983, **8**, 202.
90. R. Canetta, M. Rozencweig and S. K. Carter, *Cancer Treat. Rev.*, 1985, **12** (Suppl. A), 125.
91. E. M. McGovern, J. Grevel and S. M. Bryson, *Clin. Pharmacokinet.*, 1986, **11**, 415.
92. M. Wenk, S. Vozeh and F. Follath, *Clin. Pharmacokinet.*, 1984, **9**, 475.
93. W. Snealer, 'Drug Development: From Laboratory to Clinic', Wiley, New York, 1986, chap. 1.

25.4

Delivery System Technology

University of Michigan, Ann Arbor, MI, USA

Syntex Research, Edinburgh, UK

and

University of California, San Francisco, CA, USA

25.4.1 DELIVERY SYSTEMS FOR THE GASTROINTESTINAL TRACT

25.4.1.1 Introduction

One of the earliest attempts to alter the release pattern of a drug in the gastrointestinal (GI) tract was the introduction of shellac as an enteric coating in 1930. Owing to its high pK_a and stability problems, it has largely been replaced by modified celluloses, such as cellulose acetate phthalate, and more recently the methacrylates.[1] Sustained release products made their debut in 1952 when Smith, Kline & French began marketing a spansule (wax-coated pellet) delivery system for dextroamphetamine sulfate. The introduction of Contac capsules in 1961 was a great commercial success, leading to a widespread awareness of the sustained release concept within the general community.[2] Subsequent research efforts have been concentrated in three main areas. Firstly, to develop a delivery system which provides a constant release rate in a reproducible manner; this effort led to the design of the elementary osmotic pump, patented in 1973, and many other devices; secondly, to find materials which can be used to manufacture matrix and coated systems having appropriate and reproducible release rates; and thirdly, to improve our understanding of the physiological factors that influence dosage form performance in the GI tract. Recent technological innovations, such as floating and bioadhesive dosage forms, have been designed to take advantage of this knowledge.

The recent market success of sustained release delivery systems such as transdermal nitroglycerin and oral theophylline products has produced a surge of interest in developing novel dosage forms for existing drugs. In addition to the scientific rationales for modifying release patterns, unique delivery systems can also be used to extend patent protection, to provide a marketing advantage and perhaps even expand existing demand for the drug, as in the case of the transdermal nitroglycerin products.[3]

This section of the chapter sets out to summarize progress since 1980 in the development of technology for altering drug release patterns within the GI tract. For the purposes of this discussion, modifying the availability of drug to the body is defined as a change in one or more of the following: lag time for onset of drug absorption, rate of absorption and extent of absorption. Due to the variety of applications for controlled drug delivery and the unique chemistry of each drug, there is no generic delivery system that can optimize the absorption profile for all orally administered drugs. Rather, the optimal formulation must be designed on an individual drug basis. As a consequence, a wide range of technologies have been developed. In this chapter, discussion is limited, by and large, to those approaches which use physical means of altering release of the drug from the dosage form rather than chemical modification of the drug itself. First, current thinking on the rationales for and design of dosage forms which can modify the rate of uptake in the GI tract are presented, and then the major categories of oral controlled release dosage forms are described. The principle of the approach is described, along with some of its advantages and limitations. The most recent developments and current research in improvements for each technological category are described, and clinical examples are given where possible. Where appropriate, and at the end of the section, current issues in controlled release dosage form design, including regulatory and testing considerations, are addressed.

25.4.1.2 Rationales for Modifying Drug Release in the Gastrointestinal Tract

It is convenient to classify the various rationales for altering *in vivo* release profiles in increasing order of general level of interest within the pharmaceutical industry.

25.4.1.2.1 *To provide a location-specific action within the gastrointestinal tract*

Oral delivery to produce an effect locally within the GI tract is similar in concept to applying drugs to the skin or eye for local action. However, because of the great absorptive capacity of the GI tract, it is something of a challenge to produce a local effect without also inducing a systemic action. Selectivity can be built into the design of the drug itself by making a drug molecule too large to permeate the gut, or by covalently attaching the drug to a polymer. Sucralfate provides an example of a compound which acts locally as a mucosal protectant in the stomach. Being a polysaccharide, it bears some structural similarity to heparin, but due to its large molecular weight it is not absorbed to a significant extent and therefore does not cause any systemic side effects.[4] The ion exchange resins used to sequester excessive bile acids in the small intestine provide another example of a nonabsorbable therapeutic agent. A second approach is to deliver the drug very specifically to the region of interest in amounts sufficient to produce the desired local action but too small to result in systemic effects. This case would apply to the selective delivery of corticosteroids to the colon in the treatment of inflammatory bowel disease. Third, a combination of formulation and inherent instability of a compound may be used to confine the activity to the desired location: pancreatic enzyme replacement therapy formulated as an enteric coated product fits into this category. The enzymes are delivered specifically to the upper small intestine where the enteric coating dissolves; then their instability in one another's presence assures that their activity is confined to this region. Glutamic acid, which is used to acidify the stomach of achlorhydric patients, achieves selectivity because it is neutralized by bicarbonate in regions distal to the stomach.

25.4.1.2.2 *To avoid an undesirable local action within the gastrointestinal tract*

Undesirable effects in the GI tract include irritation of the GI mucosa by the drug and decomposition of the drug by enzymes or the local lumenal conditions. The gastric mucosa is especially sensitive to irritants such as alcohol, bile salts, aspirin and many of the other nonsteroidal antiinflammatory drugs (NSAID), nitrofurantoin and iron salts. Another problem can arise when the drug is degraded quickly in the lumenal environment. Examples of acid sensitive drugs include methenamine, furosemide, erythromycin, and some of the tetracyclines and β-lactam antibiotics.[5] Circumventing acid degradation may entail the use of a less acid sensitive analog of the drug, as in the case of penicillins, or a dosage form which permits only a small percentage of the drug to be released in the stomach, *e.g.* low solubility salts and enteric coated dosage forms of erythromycin. Making a slowly dissolving form by increasing the drug particle size has also been employed to reduce gastric irritation, *e.g.* the macrocrystalline form of nitrofurantoin.

A somewhat greater challenge exists in attempting to deliver peptides orally, a process hampered by the presence of both lumenal and brush border peptidases. Peptide analogs resistant to enzyme attack in the gut have been developed for some therapeutic categories, *e.g.* cyclosporin, the cephalosporins, lysinopril and enalopril. It is not yet clear whether formulation approaches that can deliver peptides and their analogs to the colon rather than the upper GI tract can be used successfully to obtain absorption in therapeutically useful quantities. Peptides normally require an active transport mechanism for rapid absorption, but the presence of such carriers in the colon has not been established. The feasibility of this approach may therefore be dependent on the permeability of the colonic mucosa to passive uptake of the peptide in question.

25.4.1.2.3 *To provide a programmed delivery pattern*

The aim of changing the release rate of a drug is to match its delivery pattern to the desired profile of drug action, taking into account the distribution, metabolism and excretion kinetics. In the ideal situation, the relationships between plasma levels and pharmacodynamics are established and their interaction with the underlying temporal pattern of the disease state is known, so that the optimum pattern for drug delivery can be elucidated. In the practical situation these relationships may not be fully known, though establishing them is certainly a worthwhile aim for the delivery system formulator.

Desirable supply patterns range widely according to the disease state one is attempting to treat. A few examples include: (i) rapid delivery of the entire dose for treatment of acute pain; (ii) delayed action for prevention of late-night insomnia, or morning arthritis; (iii) steady levels for round-the-clock control of diseases such as high blood pressure, epilepsy and asthma; (iv) patterns related to meal intake for control of hyperglycemia; (v) diurnal rhythms for hormones and their analogs; and (vi) special dosing regimens for anticancer drug combinations to maximize tumor cell killing. These examples illustrate the need to be cautious about the indiscriminate application of dosage forms which attempt to provide constant plasma levels, as in many instances this would lead to less than optimal therapy.

25.4.1.2.4 *To increase the extent of absorption/bioavailability*

Many drugs are not completely absorbed when administered *via* the oral route. Some of the common reasons for poor oral bioavailability are slow drug dissolution from the dosage form due to low drug solubility and/or poor formulation, a dosage form which does not provide sufficient residence time at the absorptive site of the drug, decomposition of the drug in the GI lumen, poor permeability through the intestinal mucosa, and first pass metabolism at the gut wall and/or in the liver.

Drug dissolution can be improved by keeping the drug in regions where the drug has the greatest solubility. Examples here include poorly soluble weak bases, such as ketoconazole, which require exposure to an acidic gastric environment in order to dissolve and be adequately absorbed.[6] Another physiological factor that can be used to maximize dissolution is to administer the drug in the postprandial state when lumenal fluid volumes are higher and bile salts are present in micellar concentrations.[7,8] A potential difficulty with this approach is that the drug could adsorb to some component of the meal, countering the advantage of the fed state bile salt levels with respect to dissolution. The formation of solid drug solutions, rapidly dissolving coprecipitates or soluble complexes with compounds such as the cyclodextrins provide formulation strategies for improving the dissolution rate. Griseofulvin is an example of a poorly soluble drug for which both physiological and formulation strategies have been successfully used to improve bioavailability *via* improved dissolution.[9,10]

Drugs may exhibit site dependent absorption in the GI tract, and, once these sites are established, it behoves the formulator to design a dosage form which maximizes the residence time in this region. Compounds with regionally dependent absorption include nutrients such as riboflavin and vitamin B_{12}, and small peptides and amino acids derived from ingested proteins. Drugs which share carriers with these or other nutrients, *e.g.* cephalosporins, penicillins and catecholamine derivatives,[11,12] will also exhibit site specific absorption. Site dependent absorption can also arise if a drug has a pH of half-maximal absorption between pH 4 and pH 8,[13] since the pH in the intestine varies widely with location.

In cases where the mucosa is poorly permeable to the drug, it may be possible to modify mucosal permeability. The use of absorption enhancers has been shown to be a feasible approach for rectal drug delivery.[14] Recent research, especially by the Japanese groups (see Section 25.4.1.3.4.ii), has indicated that enhancers may also be useful for improving colonic absorption. The problems associated with the delivery of drug and enhancer together in the upper GI tract and the possibilities of mucosal irritation and the delivery of unwanted compounds (*e.g.* immunogens) have yet to be solved for the use of enhancers in the upper GI tract.

For some drugs, the limiting factor to oral availability is not dissolution, decomposition or mucosal permeability. Rather, the drug is metabolized to inactive products as it passes through the gut wall, or as it reaches the liver. These first pass effects can be minimized by delivering the drug in a high enough concentration to saturate the metabolic enzymes. The relationship between first pass effect and dosing rate has been calculated for verapamil and propranolol by Wagner.[15] An alternative approach for drugs that are metabolized primarily in the gut wall is to formulate the drug so that it is absorbed at sites where the metabolizing enzymes are present in lower quantities.

25.4.1.2.5 *To extend the time of action of the drug after administration*

There are a number of circumstances in which it is desirable to develop a sustained release (SR) dosage form of a drug. Welling and Dobrinska[16] have summarized some of these rationales. Decreasing dosing frequency, especially if one can circumvent dosing during the patient's normal sleep period, is viewed as important to improving patient compliance and subsequently therapeutic efficacy. Studies have shown that if dosing can be reduced to once or twice per day, there will be an improvement in compliance.[17] A second rationale for SR is to decrease the fluctuation in plasma levels. In many instances, steady plasma levels result in better control of the symptoms and possibly the disease state itself. The difference will be especially noticeable for short-acting drugs such as the NSAIDs, cough/cold remedies such as pseudoephedrine, and appetite suppressants. Maintaining therapeutic plasma levels between doses may be necessary for the suppression of symptoms. Drugs used in the treatment of epilepsy, to suppress gastric acid secretion, and as analgesics are candidates for SR formulation from this point of view. Absence of a rapid spike in the plasma level may also result in a reduction of side effects, which are associated with high plasma levels of drug: an example here is theophylline. Additionally, SR formulations may reduce the incidence of side effects within the GI tract because only a small fraction of the dose is in solution in the GI tract at any one time.

There are some potential disadvantages of SR dosage forms; these include the possibility of 'dose dumping'—the inadvertent rapid release of a large fraction of the dose. This would be particularly problematic for drugs with very short half-lives, in which case the SR dosage form must contain several times the single dose, and for compounds with a low therapeutic index. A further problem for drugs with short half-lives is that if the single dose is of the order of 200 mg or greater, it will be difficult to make an SR formulation which contains several doses and yet is easy to ingest. At the other extreme, it is hard to rationalize the SR formulation of compounds for which the pharmacological activity occurs over a very different timespan than that which the pharmacokinetic profile would suggest, *e.g.* MAO inhibitors and some steroids. The only advantage would be the reduction of side effects associated with rapid spikes in the plasma level. SR formulation may also do little to alter the plasma level profile of drugs with low solubility because the absorption rate is already limited by the rate of drug dissolution. The only advantage to be gained here is if the delivery rate can be changed from first-order (dissolution) to a zero-order process, as the latter is less dependent on the lumenal conditions than the dissolution process (*e.g.* the osmotic pump dosage form of nifedipine).

For drugs such as nitroglycerin, alprenolol and propranolol, which undergo saturable first pass metabolism, delivering the drug at a lower concentration may serve to increase the percentage metabolized. Another problem related to bioavailability is that of site-related decomposition. If the drug undergoes a higher rate of decomposition at more distal sites in the GI tract, *e.g.* propantheline, the SR dosage form design must avoid delivering the drug at these sites. Furthermore, if the drug is not well absorbed, SR formulation may exacerbate the absorption problem as part of the dose may be released only after the dose form has passed the locations most favorable for the drug's absorption. Even for usually well-absorbed drugs, the bioavailability can be compromised if the release time from the SR dosage form is longer than the GI transit time.

A special case is that of compounds with a low therapeutic index. One school of thought suggests that these are compounds whose therapeutic efficacy is most likely to benefit from SR formulation, since the plasma levels will be kept within a tighter range. The other school of thought takes the conservative approach that if dose dumping occurs, the ramifications for the patient will be far worse than for compounds with high therapeutic indices. The solution to this apparent conflict is to design a dosage form with a very low potential for dose dumping, *e.g.* a nonerodible matrix formulation or a multiple units system. With this type of drug it is perhaps better to err on the side of underdelivering the drug than to risk delivering it in too great a quantity.

The pharmacology of the drug candidate should also be considered. Steady levels at the site of drug action may not be the optimal situation. For example, the receptor may become fatigued and the response diminish with time,[18] and in the case of antibiotics a peak/valley drug pattern may produce more effective killing of the invading organism.

Even if none of the foregoing problems apply, there are still the questions of decreased flexibility in dose adjustment depending on the formulation design (half an elementary osmotic pump?), the difficulty in establishing meaningful *in vitro/in vivo* performance correlations and hence in developing reliable quality control procedures, and the added cost of production *vis-à-vis* a conventional dosage form.

In evaluating the potential for success of an SR dosage form, there are several important matters to consider. Firstly, can the dosage form be the rate-limiting factor to drug absorption? Secondly, what is the desirable delivery rate? Thirdly, are there regional differences in solubility, decomposition, absorption and metabolism which must be accounted for in the design? Fourthly, how long can the therapeutic effect be extended and is this a worthwhile goal for the drug under consideration? Finally, will an SR dosage form produce GI side effects or undesirable changes in the pharmacological response to the drug?

(i) Rate-controlling step

SR implies that the dosage form controls the rate of uptake to the body. However, drug release is just one of several factors which may be controlling the plasma levels achieved (Figure 1).

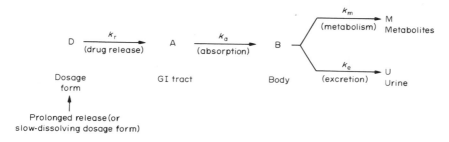

Figure 1 Schematic description of a pharmacokinetic model for drug release, absorption, metabolism and excretion[19]

Other lumenal processes that can potentially control the rate of uptake include dissolution and diffusion through lumenal fluids. In addition, the rate of decomposition may prove limiting in terms of uptake. In any of these cases, the dosage form would not really result in control of release and so one would expect a highly variable performance from the so-called controlled release dosage form. The rate of absorption is given by

$$R = k_a C_l V_l$$

where k_a is the first-order absorption rate constant, V_l is the volume of the lumenal contents and C_l is the concentration of drug in the lumen. The rate will be a maximum (R_{max}) when C_l is the drug's solubility (C_s). Note that these parameters may vary with the location in the GI tract. For the small intestine, V_l has been estimated at about 200 to 500 mL[20] and the solubility at pH 6 should provide a reasonable estimate of C_s. The absorption rate constant can be estimated from perfusion experiments, or from pharmacokinetic data following administration of an oral solution. Note that pharmacokinetically derived k_a values may be limited by the gastric-emptying rate for highly permeable drugs. In order for release from the dosage form to be the rate-limiting step to absorption, the desired release rate, k_0, will need to be substantially lower than the R_{max}, *i.e.*

$$k_0/(k_a C_s V_l) \ll 1$$

(ii) Desirable delivery rate

To calculate the desirable release rate, k_0, it is necessary to estimate the steady state plasma level, C_{ther}, which is required to provide a therapeutic effect. This information may be available from dose-ranging studies. From single dose studies, the area under the plasma/time curve (AUC) for the usual dose will be known. Then, assuming the pharmacokinetics are linear over the concentration range of interest, and that the desired dosing interval, τ, has been established, the dose can be scaled to the dose required for the SR dosage form. For a single dose, the equivalent average plasma concentration over the desired dosing interval is given by

$$C_{av} = AUC/\tau$$

The dose required to achieve the desired therapeutic average concentration over the entire dosing interval can then be calculated from

$$X(reqd) = [C_{ther} X_0(\text{single dose})]/C_{av}$$

where C_{ther} is the desired therapeutic level. The desirable release rate is then given by

$$k_0 = X(reqd)/\tau$$

An alternative method for calculation if the intravenous kinetics are known and the drug is completely absorbed after an oral dose is

$$k_0 = [C_{ther} K_{el} X_0(\text{i.v.})]/C_0$$

where X_0(i.v.) is the dose given intravenously and C_0 is the plasma level extrapolated back to the time at which the dose was given.

(iii) Are there regionally specific absorption, degradation and metabolism problems which will complicate the design options?

The rate-limiting step to absorption may change with location if the drug has pH sensitive stability or solubility. The studies necessary to evaluate these factors are normally carried out as part of the preformulation studies. Studies of bile salt solubilization may not be a part of the preformulation screen, but for poorly soluble compounds such studies may be worthwhile to determine whether dissolution is rate limiting under pre- and/or post-prandial conditions. As far as extra decomposition studies are concerned, it may be appropriate to investigate catalysis by GI enzymes, due either to a specific involvement or a general acid/general base type mechanism. For drugs that undergo first pass metabolism, the site of metabolism (gut wall or liver) and the degree of saturability should be determined. If there is gut wall metabolism, it may vary with location in the GI tract. Finally, regional absorption studies using the drug in a solution dosage form should be conducted to determine in which regions of the GI tract the drug is absorbed at a rate sufficient to maintain the desired plasma levels. The results of all these studies will serve as a guide to the feasibility of an SR dosage form and the formulation approach most likely to succeed.

(iv) Maximum extension of effect

An upper limitation to duration of action will be the period during which drug release can result in drug absorption. Usual GI transit time is of the order of 24 h. This represents a natural limit of once per day dosing. Note, though, that the range of transit times for the whole gut is quite wide, from just a few hours to several days.[21] Unless specially designed, dosage forms which release drug over a period of more than 8 h may provide highly variable amounts of drug, depending on the patient's transit time. If the drug is not well absorbed from the lower part of the GI tract, the duration of action obtainable *via* the oral route will be further limited. As a result, much effort is being directed towards the development of special gastric retention devices which can prolong the upper GI residence time (see Sections 25.4.1.3.2 and 25.4.1.3.3).

(v) Pharmacological response

Potential difficulties related to the drug's pharmacology must also be addressed. For example, will steady state levels lead to rebound effects when the drug is removed? Examples of drugs which may demonstrate rebound effects include the nasal decongestants and nitroglycerin. Undesirable side effects related to the release pattern may also occur locally within the gut. Although it would be desirable to reduce the dosing frequency of orally administered antibiotics, SR delivery systems may cause unwanted killing of colonic bacteria, which could in turn lead to diarrhea. Also related to destruction of the normal flora is the potential for overgrowth of the colon by pathogens not susceptible to the antibiotic. Finally, formulation of drugs which cause GI irritation in single-orifice osmotic pumps should be avoided in case the high concentration in the fluid exiting from the pump comes into direct contact with the GI mucosa.

25.4.1.3 Current Technologies for Modifying Oral Absorption

25.4.1.3.1 Matrix devices

These devices consist of a drug blended with some type of matrix, which retards release of the drug. They can be simple and relatively inexpensive to fabricate and many prototype dosage forms of this type have been tested for application to the controlled release of drugs (Figure 2).

The kinetics of release for such devices are dictated by the solubility and diffusivity of the drug in the matrix material, the geometry of the device and the rate at which the matrix material is eroded as it moves through the GI tract. A potential problem with matrix type dosage forms is that they do not, in the simpler embodiments, provide a constant rate of drug release.

If the drug is dispersed in the matrix at a concentration well above its solubility, release from a slab-shaped dosage form will follow half-order kinetics during approximately the first 50% of the release profile. This case was described for the analogous situation of drug release from an ointment base by Higuchi in 1961[22]

$$M_t = A(2DtC_s(m)C_0)^{1/2} \quad \text{for} \quad C_0 \gg C_s(m)$$

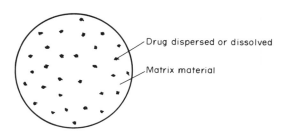

Figure 2 Schematic representation of a matrix device

where M_t is the mass of drug released at time t, A is the surface area, D is the drug diffusivity, $C_s(m)$ is the solubility of drug in the matrix and C_0 is the density of drug in the matrix at time zero. A comparison of first- and half-order kinetics with the zero-order case, as illustrated by Baker,[23] shows the degree to which the release rates vary with time between the three cases (Figure 3).

High release rates at early times for both the first- and half-order cases distinguish them from the zero-order case. Later in the release profile, the first-order rates become several fold lower than the zero-order delivery. Baker and Lonsdale[24] also showed that release from matrices in which the drug is dissolved rather than dispersed will follow approximately half-order kinetics for the early portion of the release profile, though again the release kinetics will vary according to the geometry of the matrix. For spheres and cylinders, the release begins to deviate from half-order kinetics as soon as 15–20% of drug has been released (Figure 4).

Recently, Ritger and Peppas[25,26] derived expressions giving the order of release as a function of the geometry in cases where the matrix does not change dimensions during the course of release and also in the case where release is modified by swelling of the matrix. Note that for swellable polymers, release approaches first-order kinetics for all geometries (Table 1).

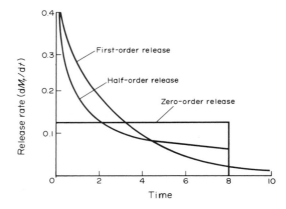

Figure 3 Zero-order, first-order and half-order release patterns from devices containing the same initial active agent content[23]

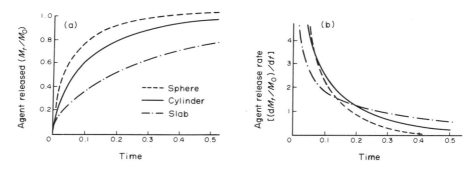

Figure 4 Theoretical fractional release and release rate *versus* time for an agent dissolved in a sphere, a cylinder and a slab $(Dl^{-2} \text{ or } Dr^{-2} = 1)$[24]

Table 1 Diffusional Exponent and Mechanism of Diffusional Release from Various Swellable Controlled Release Systems[26]

Diffusional exponent, n			Drug release mechanism
Thin film	Cylindrical sample	Spherical sample	
0.5	0.45	0.43	Fickian diffusion
$0.5 < n < 1.0$	$0.45 < n < 0.89$	$0.43 < n < 0.85$	Anomalous (nonFickian) transport
1.0	0.89	0.85	Case-II transport

Nonerodible matrix type dosage forms may fail to provide adequate control for drugs whose plasma levels must be kept within tight limits for optimal therapy. As seen in Figure 4, it is apparent that too much drug is delivered in the early part of the dosing interval, while insufficient drug is being provided at later times.

Another problem with nonerodible matrices is to ensure that the drug is delivered at a fast enough rate. The diffusivity of the drug in many insoluble polymer materials is too low to permit total release within the GI residence time. Blends of the retarding polymer with more soluble materials (an excipient, or possibly the drug itself) to create channels through which the drug can diffuse, or the use of insoluble but swellable materials which can harbor enough water to permit substantial diffusion of the drug through aqueous channels, are two approaches which have been used to obtain a release period compatible with the GI residence time.

If the matrix material is bioerodible, the rate of drug release may be dictated by the rate of matrix erosion rather than the diffusion of drug out of the matrix. The dissolution rate of the matrix material and geometry of the dosage form are important determinants of the release profile. The *in vivo* kinetics of release from bioerodible matrices may also be affected by the changing environment experienced by the dosage form as it moves through the GI tract. The mechanism by which the matrix erodes, combined with the physical properties of the drug, is therefore an important consideration in the selection of an appropriate material for bioerodible matrix type dosage forms. For example, poor drug solubility at the higher pH in the lower GI tract can be compensated for by housing the drug in a polymer which has increasing solubility with pH. Here the rate-determining step will change from diffusion in the upper GI tract to polymer erosion at more distal sites.

Table 2 lists some of the materials which have been used in matrix dosage forms. The reader is also referred to Robinson[27] for a comprehensive listing of specific drug/matrix combinations. Bioerodible matrix materials for use in the GI tract fall into three main categories. The first of these are the fatty acid esters, the erosion of which is aided by bile salt/lipase-mediated hydrolysis and which may as a result be subject to fed/fasted variation in release rate. Second are the enteric coating polymers, which may also show fed/fasted differences related to the change in GI pH profile between the fasted and postprandial states. Third are the slowly dispersable cellulosic polymers, such as the Methocels. Several groups have demonstrated that the release rate from the modified celluloses and other slowly soluble polymers is related inversely to the viscosity they develop in aqueous solution.[28-30]

Some examples of recent research involving the use of slowly soluble polymers in matrix devices include hydroxypropylmethylcellulose,[28] hydroxypropylcellulose,[30] poly(methyl methacrylate)s,[31,32] and poly(vinylpyrrolidone),[29] both alone and in combination with insoluble materials.[33,34]

Several drugs formulated as matrix products have been successfully marketed for controlled drug delivery *via* the oral route, including pseudoephedrine, potassium, iron salts and several antihistamines. A sampling of products which include some matrix release component is given in Table 3. Other examples are listed by Madan.[35]

A case example where the use of a matrix system has been successfully demonstrated *in vivo* is that of a sustained release aspirin preparation (Zorprin, Boots). The rationale behind this formulation is to provide prolonged alleviation of arthritic pain and also to avoid high local levels in the upper GI tract, where it is known that aspirin acts as a mucosal irritant. The matrix material consists of cellulose acetate phthalate and maize starch in a combined 1:20 ratio with aspirin. Wilson *et al.*[37] studied the *in vitro* and *in vivo* release rates for this formulation using a validated technetium–DPTA scintigraphic method to follow the *in vivo* kinetics, and also studied the pharmacokinetic profile of salicylate following a single dose of the formulation. They were thus able

Table 2 Examples of Materials Used to Fabricate Matrix Devices

Insoluble	Erodible	Erodible
Ethyl cellulose	Shellac	Sodium carboxymethylcellulose
Cellulose acetate	Fatty acid esters	Hydroxypropylmethylcelluloses
PVC	carnauba wax	Hydroxypropylcellulose
Silicone	beeswax and other triglycerides	Poly(methyl methacrylate)s
Polyethylene	Stearyl alcohol	Poly(hydroxybutyrate)/valerate
	Cellulose acetate phthalate and other enteric coatings	Poly(ethylene glycol)s
		Poly(vinylpyrrolidone)

Table 3 Matrix Diffusional and Dissolution Products[36]

Product	Active ingredient(s)	Manufacturer
Matrix diffusional products		
Gradumet tablets		Abbott
Desoxyn	Methamphetamine hydrochloride	
Fero-Gradumet	Iron(II) sulfate	
Fero-Grad-500	Iron(II) sulfate, sodium ascorbate	
Tral	Hexocyclium methylsulfate	
Lontab tablets		Ciba-Geigy
Forhistal	Dimethindone maleate	
Priscoline	Tolazoline hydrochloride	
PBZ	Tripelennamine	
Procan SR tablets	Procainamide hydrochloride	Parke-Davis
Choledyl SA tablets	Oxtriphylline	Parke-Davis
Matrix dissolution products		
Extentab tablets		Robins
Dimetane	Brompheniramine maleate	
Dimetapp	Brompheniramine maleate, phenylephrine hydrochloride, phenylpropanolamine hydrochloride	
Donnatal	Phenobarbital, hyoscamine sulfate, atropine sulfate, scopolamine hydrobromide	
Quinidex	Quinidine sulfate	
Timespan tablets		Hofmann-La Roche
Mestinon	Pyridostigmine bromide	
Roniacol	Nicotinyl alcohol	
Dospan tablets		Merrell Dow
Tenuate	Diethylpropion hydrochloride	
Chronotab tablets		Schering
Disophrol	Dexbrompheniramine maleate, pseudoephedrine sulfate	
Tempule capsules		Armour
Nicobid	Nicotinic acid	
Pentritol	Pentaerythritol tetranitrate	
Repetab tablets		Schering
Chlor-trimeton	Chlorpheniramine maleate	
Demazin	Chlorpheniramine maleate, phenylephrine hydrochloride	
Polaramine	Dexchlorpheniramine maleate	
Trilafon	Perphenazine	

to relate the position of the dosage form in the gut to its release rate. Although the preparation released aspirin more slowly *in vivo* than *in vitro* (90% in about 9 h compared with over 90% in 4 h *in vitro*), an approximately zero-order absorption profile was attained, as predicted from the *in vitro* results (Figures 5 and 6).

These results show that it is possible to produce a matrix type dosage form which provides a zero-order *in vivo* release pattern without using specialized geometry, but with appropriate selection of matrix materials. The discrepancy between *in vitro* and *in vivo* results, more noticeable after the first hour, was suggested to be due to differences in fluid volume and hydrodynamics in the gut *versus* the

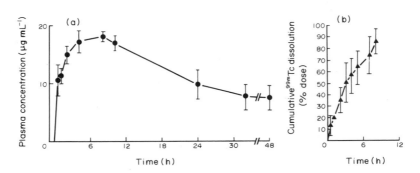

Figure 5 Comparison of the mean plasma salicylate concentration–time profile (a) with the *in vivo* release of [^{99m}Tc]DPTA, (b) from modified Zorprin tablets. Mean $\pm$ 1 S.D., $n = 5$ (from ref. 37)

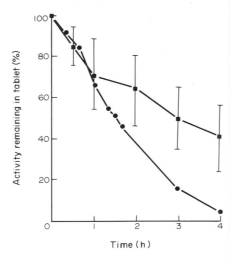

Figure 6 Comparison of the *in vivo* dissolution rate of [^{99m}Tc]DPTA (■) with the *in vitro* dissolution rate of acetylsalicylic acid and salicylic acid (●) using USP method 2. Mean ± 1 S.D., n = 5 subjects for *in vivo* determinations, data points cover the standard deviation of the mean for the *in vitro* measurements[37]

in vitro apparatus. These findings underline the need for an improvement in our understanding of the physiological conditions in the GI tract so that more predictive *in vitro* test conditions can be designed.

(i) Advantages and limitations of matrix dosage forms

One of the reasons for the popularity of matrix dosage forms is their simplicity of construction. In the case of the wax type devices the drug may be directly mixed with molten matrix material, or a spray congealing technique may be used. Simple mixture followed by compression into a suitable shape can be done using conventional tableting equipment for most of the polymer type materials. A second attraction is the wide range of materials available, and which are already approved for human use; many of them are already in use in other types of pharmaceutical products and/or the food industry. This gives the formulator flexibility in designing a dosage form with a release rate appropriate for the drug. Combinations of soluble and insoluble materials, for example, can be used to produce the desired release rate.[33,34,37] The combination of readily available matrix materials and the ability to utilize existing equipment make this type of SR dosage form relatively inexpensive to manufacture. A further advantage is that, for at least the nonerodible versions, there is very little risk of dose dumping.

Of course there are also some limitations to matrix formulations. The greatest concern is that release from the nondisintegrating types departs from zero-order kinetics, unless a special geometry or special materials are used. A further problem is that the nonerodible matrices usually release only about 85–90% of the drug, since the rate of release drops off dramatically as the matrix becomes depleted. Another factor affecting the percentage release and hence the bioavailability, is that the gastric emptying of monolithic devices tends to be quite variable. In the fasted state, a monolithic matrix is emptied from the stomach when Phase 3 of the migrating myoelectric complex (MMC) passes through. The periodicity of this contraction cycle in humans is approximately 2 h.[38] Emptying of the dosage form can therefore occur any time from immediately after ingestion up to 2–3 MMC cycles (about 6 h) later. In the fed state, the MMC is abolished, and in most cases the monolith will not be emptied until the meal has finished emptying and the fasted motility cycle is reestablished.[39] This period is determined, as a reasonable first approximation, by the calorific content of the meal, with an emptying rate of the order of 2 kcal min^{-1} (1 cal = 4.18 J) for a glucose meal.[40] There are also numerous examples in the literature to show that occasionally even quite large particles are emptied soon after a meal intake, the reasons for which are unclear at the present time.[41,42] In general, though, the emptying times for monoliths are longer when they are administered in conjunction with a meal. In the real life dosing situation, the huge variation in eating habits among the general population will result in highly variable transit times for monolithic dosage forms. A dosage form which is quickly emptied from the stomach may take as little as 4 to 5 h after ingestion to reach the colon. If the drug is not absorbed well in the colon, this rapid transit time may provide a severe limitation to the bioavailability of the drug from the matrix.

One way to get around the variability in upper GI residence time is to dose monolithic matrix systems with a standard meal, but this would result in additional difficulty in obtaining consistent compliance. A more practical approach is to use a multiple pellets matrix formulation. Non-disintegrating pellets less than about 1.5 mm in diameter empty from the stomach in a manner which is, on average, independent of fed *versus* fasted state. Better reproducibility of response from enteric coated products, where the onset of action is directly related to gastric emptying, has been demonstrated for pellets *versus* monoliths.[43] Relating these findings back to matrix dosage forms, it must be remembered that the formulation of a pellet will have to be modified from that of the prototype monolithic device to account for the greater surface area and hence the intrinsically higher release rate of the pellet relative to the monolithic dosage form.

(ii) Recent developments in matrix dosage forms

Much of the recent research in the area of matrix dosage forms has been directed at producing zero-order delivery and prolonging residence time at absorptive sites.

(a) Zero-order release pattern

One solution to the problem of matrix systems not delivering drug according to a zero-order profile is to design a dosage form geometry which offsets depletion of the matrix by providing a progressively larger surface area as a function of time. One device design which can do this is one which involves coating the outside of a hemisphere with an impermeable material, leaving a small area open on the flat surface of the hemisphere (Figure 7).

Langer and coworkers[44] have shown that such a system, with an erodible matrix core containing an evenly dispersed drug, can provide zero-order release. Recently, Kuu and Yalkowsky[45] extended this concept to a device comprising multihemispherical holes embedded in an impermeable membrane, the interior of the hemispheres being filled with a suspension of drug in a polymeric matrix. It was calculated that such a device, double-sided and with the device thickness being comparable to the distance between holes, would result in near zero-order release of drug.

de Haan and Lerk[46] described another system, comprising a housing matrix and a restraining matrix, capable of producing zero-order release. The restraining matrix contains the drug and Eudragit L or RS as well as excipients to aid granulation. The housing matrix consists of Carbopol 934 and PEG 6000, again with excipients to aid granulation. Coarse granules of the two matrices are then co-compressed into a monolithic dosage form, called the 'megaloporous' system (Figure 8). The

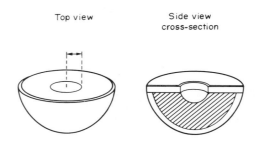

Top view   Side view
                  cross-section

Figure 7   Schematic diagram of a hemispherical matrix system[44]

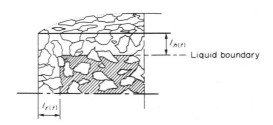

Figure 8   Schematic diagram of a partly extracted megaloporous system. Restraining matrix phase, ■; housing phase, ▨; depleted part of housing phase, □ [47]

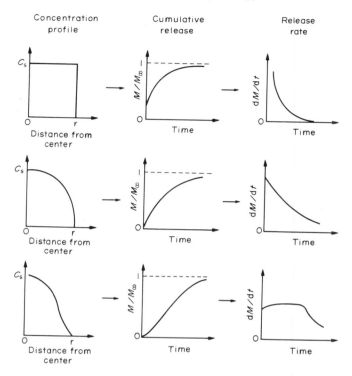

Figure 9 Theoretical profiles illustrating the characteristics of drug release from spherical matrices as a function of the initial drug concentration distribution[48]

mechanism of drug release is based on the concept of a decrease in the *rate* of surface area exposure of the restraining matrix with time, with a simultaneous increase of the *total* restraining matrix surface area contributing to the release process.[47] *In vivo* testing with theophylline as the drug indicated that approximately zero-order release could be achieved for 60–80% of the release profile, depending on the ratio of the two phases used. As with the Wilson matrix study, *in vivo* release rates were considerably slower than the *in vitro* rates, though the same form of release kinetics was observed.

Another approach to providing zero-order release from a matrix is to create a concentration profile within the matrix, so that the increased diffusion path length and decreased surface area are offset by a higher concentration of the drug in the center of the matrix compared to near the surface. This concept was developed by Lee[48] for use in hydrogels, where it is possible to use a controlled extraction process to establish the desired concentration profile, then immobilize the concentration profile using a vacuum freeze-drying step. Theoretical release profiles using different immobilized concentration profiles are shown in Figure 9. A device based on this concept has been patented by De Crosta *et al.*[49] Hard interpolymer–hydrogel shells are formed around the soft hydrogel cores by manipulating the polymerization procedure. Drug is loaded by treating the beads with a concentrated drug solution in a volatile solvent, after which the solvent is evaporated.

(b) Prolonging the GI residence time Two approaches to attaining retention of a dosage form in the stomach are to manipulate the physical properties of the device (size, density *etc.*) or to use materials which adhere to the gastric mucosa. These will be discussed in the next two subsections.

25.4.1.3.2 Gastric retention devices

Gastric retention devices can be useful in increasing the bioavailability of poorly absorbed compounds and for providing sustained release delivery patterns. The concept is to make a dosage form which, during a prolonged residence time in the stomach, gradually releases drug into the lumen. It is a particularly useful strategy for drugs whose favored site of absorption is high in the GI tract. One limitation to the application of this concept to a particular drug is for those compounds which are rapidly degraded in acid media.

Retention is theoretically achievable *via* the size of the dosage form, the density of the dosage form or bioadhesion to the gastric mucosa. If a large enough object is introduced to the stomach, it can be retained almost indefinitely, as in the case of gastric balloons used in the treatment of obesity. For pharmaceutical purposes, the dosage form must be designed so that it can be readily swallowed. This consideration has led to the development of several ingenious devices which are of acceptable dimensions prior to administration but expand to a much larger volume upon reaching the stomach. An early example of this approach was patented in 1975 by the Alza corporation[50] and consists of a drug reservoir attached to a deflated balloon which contains a substance which vaporizes and expands at body temperature (*e.g.* ether, methyl formate, halogenated hydrocarbons *etc.*), resulting in retention of the device in the stomach. The drug chamber is coated with a microporous coat to limit the rate of drug release. The intent of this approach was to provide a means of dosing once a day, once a week, or once a month (which would be useful for contraceptives). The practical problems with this device are the potential toxicity of the gas-producing substances proposed and the lack of data to suggest that the devices could be retained for as long as a month. Incomplete bioerodibility of the device also presents a problem in that spent devices could accumulate either in the stomach (resulting in bezoar formation) or in the folds of the colon.

Another approach to producing a large object in the stomach is to use a swellable polymer. The Upjohn Company[51] patented a thiolated gelatin matrix which can crosslink *in situ*, creating a swellable polymer in the stomach. An oxidizing agent such as potassium iodate is added to accelerate the crosslinking process. The functional ability of this patent, as with the Alza patent, was not verified by *in vivo* testing. Banker[52] patented a tablet dosage form containing mainly water soluble components, coated with a copolymer of methyl vinyl ether and maleic anhydride, and then crosslinked with Tween 20 or an alkylenediol. After ingestion, the water soluble materials in the tablet osmotically draw water into the dosage form, causing the coating to swell rapidly. Because of the crosslinking, the tablet coating swells to 150–1000% of its original size. An embodiment of this device was shown to increase the bioavailability of tetracycline HCl by a factor of 2.6. In a further example, the absorption of phenylpropanolamine was prolonged compared to a conventional tablet. In 1980, McNeilab[53] patented a swellable envelope device. In this device, an outer polymer envelope is used to house the expanding agent, the drug and a means of controlling the release rate. The expanding agent is either a sugar or salt capable of generating high osmotic pressure, or a swellable polymer. In the preferred embodiment, the envelope material provides the release-limiting membrane. However, data from dogs indicated that there was wide variation in the gastric emptying time, from 8 to 264 h.

In most cases, one is attempting to reduce the frequency of dosing to once or twice per day by use of a retention device. For drugs poorly absorbed from the colon, twice a day dosing may require up to a 10 h residence time in the upper GI tract. Given that the transit time through the small intestine is usually in the neighborhood of 3–5 h, the retention time in the stomach needs to be at least 5 h. This is quite feasible for dosage forms of conventional size, provided they are administered after a reasonably substantial meal. Mroz and Kelly[54] were the first to demonstrate that gastric emptying of large indigestible particles requires the passage of Phase 3 activity, which is associated with the fasted state motility cycle. This explained some earlier observations[41,55] that enteric coated dosage forms did not empty from the stomach as quickly when given after a meal rather than when given on a fasted stomach. More recent studies using monolithic dosage forms and other indigestible objects[39,56,57] have confirmed that these are retained in the stomach while it remains in the fed state.

The density as well as the size of the dosage form contributes to its retention in the stomach. Several designs utilizing floating dosage forms have been patented. Of these, the 'hydrodynamically balanced' dosage form is the only one that has been translated into a product marketed in the USA. In this patent, Sheth and Toussanian[58] described a capsule containing a drug and a hydrocolloid (high viscosity cellulose derivatives) which will hydrate and swell upon ingestion, creating air pockets. The resultant device has a relative density of less than 1 and enables the dosage form to float on the gastric contents. This process demands that the hydrocolloid remain dry during manufacture, and that voids be retained during compression. The device was demonstrated to improve the bioavailability of riboflavin and was subsequently utilized in the manufacture of a controlled release dosage form of diazepam. In a refinement of this approach, Ingani *et al.*[59] added sodium bicarbonate, calcium carbonate and citric acid in separate layers to a hydrophilic colloid formulation so that a low density in the stomach would be assured by the *in situ* production of carbon dioxide. This formulation resulted in better bioavailability of riboflavin than from a sustained release tablet, reflecting the existence of an absorption window high in the small intestine for this vitamin. The AUC was comparable to that of an immediate release tablet in the fasted state, but was not as high when the subjects were fed. However, urinary excretion remained higher than 400 μg h^{-1} for 8 h as opposed to 4–6 h with the immediate release dosage form.

Table 4 Mean Riboflavin Urinary Excretion Rates ± S.D. (µg h⁻¹) From Immediate (IRT), Sustained Release (SRT) and Floating Sustained Release (SRFT) Tablets, and Sustained Release Floating Capsules (SRFC)[59]

Time (h)	IRT	SRT	SRFT	SRFC
Fasting				
1	390 ± 118	174 ± 73	146 ± 72	120 ± 53
2	326 ± 300	239 ± 137	348 ± 118	213 ± 100
3	163 ± 156	99 ± 48	304 ± 124	145 ± 54
4	112 ± 91	96 ± 38	182 ± 44	125 ± 46
6	96 ± 65	76 ± 34	119 ± 53	103 ± 47
8	65 ± 55	54 ± 29	108 ± 40	85 ± 47
14	43 ± 22	52 ± 19	63 ± 41	82 ± 50
24	42 ± 18	37 ± 15	54 ± 20	47 ± 18
Standardized breakfast				
1	1075 ± 573	149 ± 65	90 ± 58	46 ± 36
2	2135 ± 933	159 ± 73	170 ± 79	150 ± 142
3	1115 ± 937	130 ± 77	307 ± 199	395 ± 241
4	592 ± 288	198 ± 205	375 ± 293	668 ± 390
6	289 ± 175	136 ± 81	424 ± 323	377 ± 213
8	160 ± 86	70 ± 27	457 ± 556	455 ± 553
14	97 ± 55	46 ± 10	122 ± 59	149 ± 107
24	46 ± 23	37 ± 20	76 ± 28	70 ± 29

An alternative way to produce a low density dosage form is to load the drug onto a support medium which contains a large airpocket, *e.g.* an empty capsule, or a solid foam such as polystyrene, or popcorn. The drug is contained either in the body of the foam or coated on with a retarding polymer such as ethylcellulose. These devices, patented by Watanabe *et al.*[60] of the Eisai company,

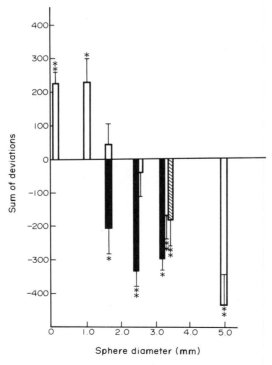

Figure 10 Mean ± S.E. values of all summed deviations (sphere emptying *versus* meal emptying) for all experiments. Unshaded bars represent spheres with a relative density of 1 and shaded bars represent spheres with a relative density of 2. The cross-hatched bar depicts emptying of 3.2 mm filter paper squares relative to the meal. Positive deviations indicate faster emptying of spheres than the meal, while negative deviations indicate that the test particles emptied slower than the meal. Asterisks indicate statistically significant differences (*$p < 0.05$, **$p < 0.01$)[62]

were shown to be retained in the stomach for at least 3 h when administered after a meal. Note that low density may also contribute to the retention of some of the expandable dosage forms described above.

As well as low density dosage forms, the use of high density has been proposed as a means of prolonging gastric residence time. Bechegaard and Pederson[61] were awarded a patent in which a multiparticulate dosage form comprising a mixture of pellets of usual and higher density were combined to provide a prolonged GI transit time. Although the clinical results used to support the patent are open to question on account of the use of ileostomy patients as subjects, later work of Meyer *et al.*[62] suggests that high density may indeed result in prolonged gastric residence in the fed state. The interaction of size with density has been studied in dogs and humans by Meyer's group.[62,63] These studies show that particles with relative densities substantially different from 1 are emptied more slowly than those with a relative density close to 1, when the same sized particles are administered (Figure 10). A dosage form consisting of a multiparticulate with a high density may therefore be a viable way of obtaining slow and more reproducible emptying from the stomach, when given in the fed state.

Note that the strategies which involve a nonexpanding dosage form of specific size and/or density rely on the patient taking the medication after eating a meal. In the situation where the patient inadvertently takes the dosage form on an empty stomach, the upper GI residence time will depend on the phase of the fasting motility cycle in which the dosage form is ingested. In the worst case, where the dosage form is swallowed immediately prior to an MMC, the upper GI residence time will be effectively that of the small intestine transit time. This could result in undermedication by as much as a half, which may be of concern especially in diseases where it is important to retain close control of the drug levels so that the symptoms are completely suppressed *e.g.* in epilepsy, angina, chronic pain, *etc.*

25.4.1.3.3 *Bioadhesives*

The rationale for incorporation of a drug into a bioadhesive matrix is to prolong its upper GI residence time. This is a particularly attractive approach in several circumstances: firstly, when a sustained delivery pattern is desirable but the drug is not well absorbed from the colon (this may be the case if the drug requires some kind of carrier uptake only available in the small intestine); secondly, if the drug is absorbed by a specific carrier mechanism restricted in its distribution in the gut, prolonging the contact time at the major absorptive site should result in more complete and reproducible absorption; and thirdly, if the drug lacks good membrane permeability anywhere in the GI tract, prolonged residence in the GI tract presents one way of improving the overall fraction absorbed.

The original basis for the use of bioadhesives in oral dosage forms was the development of denture fixatives in the early 1960s.[64,65] The Poligrip patent describes a mixture of poly(ethylene oxide)s in a petrolatum or diatomaceous earth base which can successfully adhere to the oral mucosa. The Cyr patent utilizes a gelatin dispersion in mineral oil for the same purpose. Polyethylene, pectin and carboxymethylcellulose are used to enhance the effect. Oral adhesive bandages were also developed to protect oral wounds and promote healing. A formulation consisting of gelatin, sodium carboxymethylcellulose and polyisobutylene backed by a polyethylene film was shown in clinical trials to adhere for a period of up to 24 h.[66]

The next work was in the area of delivery systems designed to adhere to the buccal mucosa. Forrest Labs[67] successfully patented a system utilizing cellulose derivatives (Dow Chemical Co.) and acrylic acid to adhere to the buccal surface, providing several hours of drug delivery. Other materials which have been used to adhere to the oral mucosa include Carbopol[68] and the polymethacrylates,[69] which were used to provide sustained release of tretinoin in the treatment of lichen ruber planus. As well as treating local conditions, buccal delivery can be used to avoid first pass metabolism, and may provide a way of improving drug absorption with the aid of absorption enhancers, since the drug and enhancer can be readily colocalized at the buccal surface. One disadvantage is that applying the formulation is less convenient than swallowing a capsule, though this may be primarily a matter of patient education.

Peppas and Buri[70] have studied the mechanism of bioadhesion to the oral mucosa, using interpenetration and fracture theories. Three possible mechanisms can be responsible for adhesion: (i) physical entanglement, which relies on the number of contact points or 'surface roughness'; (ii) the formation of intermediate strength bonds between mucosa and adherent, *e.g.* ion–dipole and hydrogen bonds; and (iii) the formation of covalent bonds. Although the formation of covalent bonds represents a permanent bond of the adhesive material to the mucosa, the turnover of the mucosa would limit the duration of dosage form residence in the gut. Therefore, it seems that all three mechanisms could be useful in designing bioadhesive dosage forms, with the latter two more likely to provide the bond strength necessary for prolonged attachment. Robinson's group has also studied polymer adhesion to the mucosa, focusing interest on the gastric and intestinal tissues rather than the oral cavity. Using a modified surface tensiometer technique,[71] it was found that the charge density, hydrophobicity of the polymer backbone and the rate and extent of hydration of the material were important to the degree of adhesion. For example, polycarbophil adheres better at pH 5–6 than at low pH, where it has a lower equilibrium volume of swelling. There is also a decrease in force required to detach the polymer at neutral pH, the mechanism for which is unknown but may be associated with a pH-related change in the mucus structure. *In vitro* results were corroborated by an increase in the GI transit time of the adhesive polymers *versus* Amberlite beads (Figure 11). In a further study, it was shown that inclusion of the poorly absorbed compound chlorothiazide into a polycarbophil matrix resulted in improved absorption in rats[72] after administration by surgically inserting the dosage form into the stomach (Figure 12). The patent arising from these studies was issued in 1986.[73] As yet, there have been no studies reported to indicate whether this bioadhesive effect can be attained in humans after a conventional administration procedure.

Another area for future study is the use of bioadhesive materials that covalently bond to the GI mucosa. With these, as well as the more loosely bound bioadhesives, there are two clinical issues that

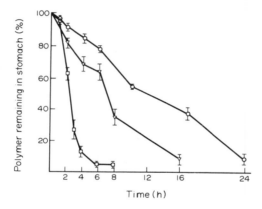

Figure 11 Rat GI transit of polymers. ○, polycarbophil (density 1.56 g mL^{-1}); ▽, poly(methacrylic acid–divinylbenzene) (density 1.36 g mL^{-1}); □, Amberlite 200 resin bead (density 1.53 g mL^{-1})[71]

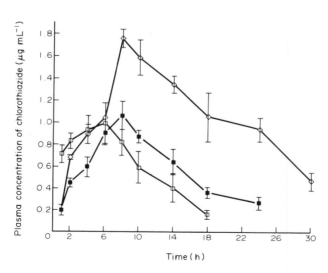

Figure 12 *In vivo* blood levels of chlorothiazide in rats. ◇, chlorothiazide in controlled release albumin beads coated with bioadhesive; □, chlorothiazide powder formulation; ■, chlorothiazide in controlled release albumin beads without bioadhesive[73]

need to be addressed. The first is whether or not reproducible adhesion can be attained when bioadhesive dosage forms are administered orally, and the second is the possibility of premature adhesion and subsequent blockage of the esophagus.

25.4.1.3.4 *Miscellaneous methods for increasing absorption in the GI tract*

As well as prolonging the upper GI residence time by the use of gastric retention devices and bioadhesive matrices, several other approaches for improving GI absorption have been studied. Two current areas of focus include the improvement of dissolution rate by incorporation into cyclodextrin complexes and the use of absorption enhancers to promote the uptake of drugs which have poor membrane permeability in the GI tract.

(i) Cyclodextrins

The cyclodextrins first came to light in the pharmaceutical literature in the 1960s through the work by Cohen and Lach on solubility enhancement[74] and by Higuchi and Connors on phase solubility techniques.[75] In 1981, the proceedings of the first international symposium on cyclodextrins was published in the Netherlands[76] and since then there has been continuing interest in the use of these compounds for enhancing drug solubility. Muller and Braun[77] provided a summary of the application of cyclodextrins as solubilizing agents. Cyclodextrins are torus-shaped oligosaccharides composed of glucose molecules, which can form inclusion complexes by taking up a guest molecule into the central cavity. A feature of these complexes is that they are stable in aqueous solution. α-, β- and γ-Cyclodextrins contain six, seven and eight glucose units respectively. The α is too small to complex with most compounds of pharmaceutical interest. The β has a rather low aqueous solubility (1.8% in water at 25 °C) because of the formation of stable intramolecular hydrogen bonds, and this limits the degree to which it can be used to improve the solubility of drugs. Recent studies have shown that inclusion of drugs such as cinnarizine[78] and ibuprofen[79] in β-cyclodextrins, and benzodiazepines[80] and digoxin[81] in γ-cyclodextrins can result in a very significant increase in the rate of dissolution. How the increase in dissolution rate translates to an increase in bioavailability depends on whether the rate and/or extent of absorption is limited primarily by dissolution when the conventional dosage form is administered. For ibuprofen, a compound which is thought to be completely absorbed from the usual dosage form, there is a modest increase in the rate, but not the extent, of absorption. In contrast, when diazepam is given as the γ-cyclodextrin complex both the rate and extent of absorption are increased.[80] The effect of cyclodextrin was even more impressive in the case of digoxin. When administered to dogs as tablets made from 100 mesh digoxin, the AUC was only one-fifth of the value as compared to that obtained when tablets containing 1:4 cyclodextrin complex were administered (Figure 13).[81] Recent developments in the use of cyclodextrins appear to be directed at developing a more soluble analog of β-cyclodextrin. Methylated cyclodextrins have been shown to effectively increase the solubility and bioavailability of a masked fluorouracil compound, carmofur, in rabbits.[82] Alternatively, β-cyclodextrin can be derivatized by hydroxyalkyl substitution, which is claimed to produce increased solubility without inducing the toxicity problems seen with the methylated analogs.[77]

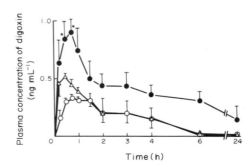

Figure 13 Plasma levels of digoxin following the oral administration of tablets containing digoxin or 1:4 digoxin/γ-cyclodextrin complex to dogs. Each point represents the mean ± S.E. of six dogs. ○, 100 µg digoxin tablet; ●, γ-cyclodextrin complex tablet containing 100 g of digoxin; △, γ-cyclodextrin complex tablet containing 50 g of digoxin; *$p < 0.01$, ● *versus* ○ (from ref. 81)

(ii) Absorption enhancers

For compounds which have poor membrane permeabilities, current thinking suggests two main ways of improving absorption using formulation techniques. The first is to increase the residence time at the site of absorption, which may be achieved by using gastric retention devices or bioadhesive dosage forms. The second approach is to identify a substance which can selectively enhance the permeability of the drug through the gut wall.

At one stage it was hoped that absorption could be improved by directing the drug to lymphatic uptake. This concept is particularly attractive for drugs which undergo hepatic first pass metabolism since the drug would enter the general circulation at the junction of the left internal jugular and left subclavian veins, thereby avoiding the liver.[83] In some early studies where the thoracic lymph was the assay site, significant drug levels were found in the lymph. The problem with this assay site is that the lymph has equilibrated with the blood, so that one measures levels resulting from distribution as well as absorption. Assay of the mesenteric lymph is a tedious process, but gives a much more accurate measure of the percentage of drug absorbed *via* the lymph. Using the latter technique, Charman and Stella were able to show that although the type of lipid vehicle used could affect the extent of uptake into the lymphatic system,[84] the lymphatic route is responsible for a significant fraction of uptake only when the compound is extremely lipophilic, *e.g.* DDT.[85]

The idea of modifying uptake *via* the choice of vehicle is not restricted to directing absorption to the lymph. Palin[86] studied the effect of oily and emulsified vehicles on the absorption of the poorly permeable cefoxitin. Fatty acid emulsions enhanced the absorption from rat intestinal loops by a factor of as much as 11 when given as a 10% emulsion, with the C_{12} acid producing the greatest effect. It was also shown that the increase was not due to lymphatic uptake. Based on the incorporation of radiolabeled fatty acid into the membrane, it was proposed that the fatty acid modified the membrane permeability directly. The medium chain glycerides have also been shown to enhance absorption of poorly permeable compounds such as ceftizoxime[87] and phenol red.[88]

Azone (1-dodecylazacycloheptanone), which has been studied extensively as an absorption enhancer for the skin, has also been tested for its ability to enhance absorption in the GI tract. Murakami[89] used a closed loop method in rats to demonstrate that Azone solubilized with HC 60, a polyoxyethylated hydrogenated castor oil, could increase the uptake of 6-carboxyfluorescein by a factor of 20 in the small intestine and 40 in the large intestine. Sodium cefazolin absorption was increased from nondetectable levels to an AUC similar to that following intramuscular injection when given with Azone in the large intestine. Further, Fukui[90] showed an additive effect when oleic acid was used in conjunction with Azone to promote the absorption of 6-carboxyfluorescein, also using a closed loop procedure. At 5 mM of both Azone and oleic acid, there was a tenfold increase in the AUC. The mechanism of the promoting effect of Azone and the fatty acids is thought to be related to the incorporation of the enhancer into the membrane, with a subsequent change in the membrane fluidity.[89]

Sodium salicylate, which has been used as an enhancer in the rectum,[14] has also been studied in the small intestine. Kajii[91] found that in epithelial cells isolated from the proximal jejunum, a combination of 30 mM sodium salicylate and 18 mM caprylate perturbed the membrane lipids and proteins. At the same time, cell viability was maintained. This combination may therefore have some potential as an absorption enhancer for the small intestine.

Surface active agents, both the naturally occurring bile salts and synthetic varieties, are also potentially useful as absorption enhancers. Studies by Miyamoto[92] indicated that, while sodium cholate and sodium taurocholate have no effect on the absorption of β-lactam antibiotics, nonionic surfactants such as poly(oxyethylene)-23-lauryl ether improves their absorption in the small intestine. Gowan[93] demonstrated a modest, 20–30%, increase in the absorption of pentobarbital when Tween 80 was coperfused in rats, noting that the effect was more pronounced at low than high concentrations of the surfactant.

In summary, it appears that the search for absorption enhancers for the small intestine is producing some promising leads. However, the ideal properties of selective action, lack of toxicity and several-fold enhancement of absorption have not yet been demonstrated for any one candidate. Azone and the fatty acids appear to increase the rate of uptake but more work needs to be done to demonstrate selectivity and the absence of toxicity. A great concern with this concept remains the ability to deliver the enhancer to the mucosa at the same time as the drug and in a concentration sufficient to enhance absorption.

(iii) Liposomes

Liposomes have been used to entrap drugs for GI delivery, the rationale being that the liposomes would protect the drug from lumenal conditions and perhaps enhance their absorption, especially if the liposomes were taken up intact. This would be particularly attractive for the delivery of peptides, which are degraded by lumenal and brush border enzymes, and, if they consist of more than three amino acid residues, usually have poor membrane permeability. Several groups have investigated the oral delivery of insulin, for example, in liposomes: conflicting results were obtained. There is a strong possibility that factors other than uptake of intact liposomes are responsible for the decrease in blood glucose levels seen in some of these studies.[94]

One problem with the liposome approach is that many formulations are degraded in the gastrointestinal environment, as shown for dimyristoyl- and dipalmitoyl-phosphatidylcholine liposomes by Rowland and Woodley.[95]

Distearoylphosphatidylcholine/cholesterol liposomes provide an example of a liposome formulation that is resistant to degradation in the GI tract. When these are administered to rats orally, they appear to be taken up by the jejunal tissue, but fail to distribute to other parts of the body.[96] Using multilamellar liposomes, prepared from L-α-distearoylphosphatidylcholine/cholesterol and egg lecithin/cholesterol, Chiang and Weiner[97] showed recently that while these formulations are reasonably resistant to simulated gastric conditions and only release 20% of entrapped glucose under pH 7/bile salt/lipase conditions, the leakage of a marker dye, carboxyfluorescein, is almost complete under simulated intestinal conditions. These results are in agreement with those of earlier studies, summarized by Chiang and Weiner. Further, they investigated the ability of these formulations to modify the uptake of both poorly- and well-absorbed compounds. Not only was there no evidence of transport of liposomes across the GI mucosa, but the absorption of glucose was prevented by entrapment into liposomes. The liposomes and their entrapped markers remained in the jejunum for the duration of the 3 h modified Doluisio experiments.[96] In summary, it appears that although some types of liposome could serve the function of retarding decomposition in the stomach, the formulations studied to date cannot provide a means of transporting low permeability compounds across the gut wall in quantitative amounts.

25.4.1.3.5 Coated dosage forms

Coated dosage forms are devices consisting of a drug reservoir surrounded by a coating material which either dissolves slowly, or at a specific location, to release the drug, or remains intact, presenting a diffusional barrier to drug release (Figure 14). The earliest versions of the coated dosage form were the type where the coating dissolves. Shellac-coated tablets were introduced as a means of providing enteric protection. The first coated dosage forms for sustained release were the Spansule dosage forms, which consisted of drug particles coated with a slowly dissolving waxy coat. By appropriate combination of particles coated with different thicknesses of the coating material, approximately zero-order release was achieved. This is the principle of Contac and several other sustained release products.

The other approach to producing a coated product with sustained release properties is to use a coating which is insoluble in the GI fluids, requiring drug to be released by diffusion through the coating material or through pores in the coating. The kinetics of release from a reservoir coated with an insoluble polymer are governed by Fick's laws of diffusion, the general expression being

$$\mathrm{d}M/\mathrm{d}t = DA(\mathrm{d}C/\mathrm{d}x)/l$$

where $\mathrm{d}M/\mathrm{d}t$ is the rate at which mass is transported across the coating, D is the diffusivity of drug in

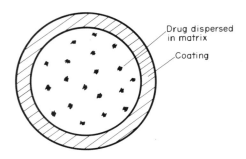

Figure 14 Schematic representation of a coated sustained release dosage form

the coating, A is the surface area of the coating, dC/dx is the concentration gradient across the coating and l is the thickness of the coating.

For a coating consisting of a homogeneous polymer material with no pores and a reservoir that is saturated with drug, the rate of transport is

$$(dM/dt)_{sat} = (DAKC_s)/l$$

where K is the partition coefficient between the reservoir and the coating and C_s is the solubility of the drug in the reservoir. The diffusivity will be a function of the rigidity of the polymer used as the coating material and will in turn depend on whether the polymer is in the rubbery or glassy state and also upon the degree of crosslinking. The molecular weight of the drug will also be an important factor in determining the diffusivity. The partition coefficient will depend on the affinity of the drug for the polymer compared to the reservoir, which will most likely consist of an aqueous dispersion once the dosage form has been ingested. The release rate also depends on the thickness of the coating, l. The geometry of the dosage form affects the form of the relationship between thickness and release rate.

If the coating contains pores, the above equation is modified to account for transport *via* the pores

$$dM/dt = (D'AK'eC_s)/l\tau$$

The diffusivity in this case is the diffusivity of the drug in the pore fluid. In most instances the pore fluid will be an aqueous solution, so the drug's diffusivity will be in the vicinity of its value in water. The partition coefficient is that between the pore fluid and the reservoir fluid, so in most cases it will be close to one. The factor 'e' accounts for the pore volume relative to the total coating volume and the tortuosity factor, τ, is used to account for the longer diffusional path length when the molecule travels through the pore as opposed to dissolving and diffusing across the polymer. A more detailed presentation of the derivations and discussion of the factors affecting the release rate from coated dosage forms has been given by Baker.[98]

For either a homogeneous or a microporous coating, release will be zero-order while the reservoir is saturated with drug. The duration of the zero-order release period is given by

$$t_{zero} = (M_0 - C_s V)/(dM/dt)$$

where M_0 is the mass of drug in the reservoir, and V is the volume of the reservoir. When the reservoir concentration drops below saturation, then the release will follow first-order kinetics, reflecting the fall in the concentration of the drug in the reservoir with time.

One of the main attractions of coated devices is that a constant release rate can be achieved over most of the release period. Several coating polymers can be used, *e.g.* ethylcellulose, cellulose acetate, polymers of methacrylic acid and its esters, and poly(vinyl chloride), to provide an insoluble coating. If necessary, the release rate can be increased by the inclusion of pore formers (water soluble excipients, usually either low molecular weight like sucrose or lactose, or water soluble polymers like hydroxymethylcellulose or PEG). For example, Lippold and Forster[99] used poly(ethylene glycol) to obtain the desired release rate for theophylline through ethylcellulose, while Kallstrand and Ekman[100] used microcrystalline sucrose to attain a release rate of 3 mg min^{-1} for potassium chloride through poly(vinyl chloride)-coated tablets.

The main potential disadvantage with coated reservoirs is that if the coating breaks then the entire dose of drug will be released, which may lead to toxicity. One way of overcoming this problem is to use a multiparticulate rather than a monolithic dosage form. Another drawback is that the coating procedure may add considerably to the unit cost of the product. Some examples of sustained release products containing coated pellets are listed in Table 5.

Table 5 Some Products Consisting of Coated Pellets[35]

Product	Manufacturer	Product	Manufacturer
Aerolate capsules	Fleming	Mol-iron chronosule capsule	Schering
Combid spansule capsules	SK&F	Nitrospan	USV
Contac capsule	Menley & James	Pavabid capsule	Marion
Ferro-sequels capsule	Lederle	Pentritol tempules	USV
Measurin tablets	Breon	Pyma Timed capsule	Fellows-Testager
Meprospan capsule	Wallace	Theobid duracap	Meyer/Glaxo

In the case of monolithic devices, the reservoir can be manufactured by a conventional tableting procedure. For multiparticulates, several manufacturing methods have been devised. Granules may be prepared by the usual procedures, but owing to their irregular shapes, the coating thickness may be uneven and require a higher coating loading than particles with even surfaces. Pellets with a narrower size distribution and which are more regular in shape may be formed in a rotor granulator or by an extrusion/spheronization process. This is a useful method if the drug loading needs to be high. For smaller doses of drug, a layer of drug can be built up on to substrate nonpareil seeds. A single step process for this method of preparation is described by Ghebre-Sellasie *et al.*[101] They used 20–25 mesh nonpareil seeds to serve as the substrate for pseudoephedrine HCl, theophylline and diphenhydramine formulations. The coating can be applied from aqueous or organic solvent solutions, in conventional coating equipment. Mehta and Jones[102] described the various methods, illustrating process-related differences with scanning electron microscopy.

A case example of the use of coated dosage forms for sustained release is that of the Theo-dur sprinkle. Gonzalez and Golub[103] describe the formulation, manufacture and pharmacokinetics of the sprinkle dosage form. The rationale for this product is to provide less frequent dosing and more consistent plasma levels of theophylline, which has a fairly narrow therapeutic plasma level range and, in a large segment of the population, a short half-life. The sprinkle pellets range in size from 600 to 800 μm and are spherical. They are prepared by coating the drug onto sugar crystals, then overcoating with the retarding polymer. The pellets are then packed in an oversize capsule, to be administered by sprinkling over the patient's meal. The modified Wagner–Nelson plot for absorption indicates that absorption is approximately zero-order for almost 90% of the release profile (Figure 15), and the multiple dose study results shown in Table 6 indicate the high level of control of the plasma level with this formulation.

As well as making pellets, coated multiparticulates can be prepared by microencapsulation. The various methods for making microcapsules have been summarized by Madan.[104] Cellulose derivatives such as ethylcellulose can be used to make sustained release microcapsules. These are usually prepared by a phase separation method. A recent example of the application of microencapsulation to sustained release formulation was described by Lin and Yang[105] for theophylline. They used varying amounts of ethylvinyl acetate copolymer to modify the release rate from ethylcellulose-encapsulated theophylline. For other polymers such as the polyamides, an alternative manufacturing procedure is by interfacial polymerization. This process has been described by Mahmoud and El-Samaligy[106] for nylon encapsulation of sulfadiazine sodium. Another recent development in microencapsulation is the ability to produce dual microcapsules.[107]

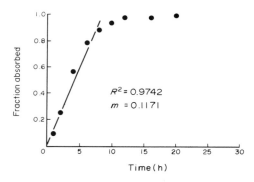

Figure 15 Wagner–Nelson plot for theophylline absorption from Theo-dur sprinkles[103]

Table 6 Serum Theophylline Concentrations Obtained after Dosing with Theo-Dur Sprinkle every 12 h for 5 Doses[103]

Time (h)	Experimental C_{ss} (μg mL^{-1})	Time (h)	Experimental C_{ss} (μg mL^{-1})
48	6.2	54	7.0
49	6.6	56	6.8
50	6.2	60	5.4
52	6.8		

(i) Enteric coated dosage forms

Enteric coated dosage forms represent a specialized type of coated dosage form, in which the coating is designed to remain intact during the gastric residence time, but dissolve at some more distal location in the gut. Much of the early literature on enteric coated dosage forms has been summarized by Chambliss.[1]

Enteric coating can be a useful approach in a number of circumstances, the most obvious of which are to prevent gastric irritation by the drug or to avoid decomposition of the drug in the stomach. Manipulation of the core materials combined with coatings which dissolve rapidly at duodenal pH may provide a means of achieving site specific delivery to the proximal small intestine. On the other hand, by appropriate choice of coating material, release can be delayed until the dosage form reaches the colon, which may improve therapeutic selectivity in treating local inflammatory disorders in the lower gut.

Four basic types of materials can be used to achieve selective release of the drug at locations distal to the stomach. These include water resistant polymers, pH sensitive polymers, materials digested by intestinal fluids and materials that slowly swell and dissolve after exposure to moisture.[1] Of these, current usage lies mainly with the pH sensitive polymer approach. Some commonly used polymers are listed in Table 7.

The dissolution pH of a particular formulation depends on the hydrophobicity of the backbone polymer, the degree of derivatization with the acidic functional group, the coating thickness and the buffer capacities in the release environment and of the core materials used in the formulation.[108-110] One must therefore take into consideration the local pH and buffers present in the gut when choosing the coating material and formulating the dosage form cores, if release at a specific site is to be achieved. Figure 16 shows average values for pH as a function of GI transit for fasted healthy humans.[111] As indicated in Table 7 there is quite a range of dissolution pH characteristics available using the existing materials. Further flexibility in formulation lies in the availability of some of these materials in forms which can be coated from aqueous solutions.

In 1930, over 3% of all prescriptions were for enteric coated products; today the percentage is much lower. Reproducibility of performance has been a problem, particularly for enteric coated

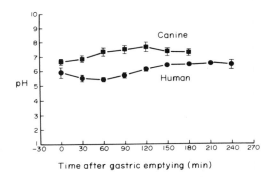

Time after gastric emptying (min)

Figure 16 Intestinal pH in fasted dogs and humans[111]

Table 7 pH Sensitive Polymers in Use as Enteric Coating Materials

Polymer	pH at which the polymer starts to dissolve rapidly
Poly(vinyl acetate/phthalate) (Colorcon)	4.7
PVAP; Coateric	
Hydroxypropylmethylcellulose phthalate (HPMCP)	
HP 50 (Shinetsu)	5.0
HP 55 (Shinetsu)	5.5
Cellulose acetate phthalate (CAP)	
(Eastman Kodak)	6.0
Methacrylic acid/methacrylic acid methyl ester polymer	
Eudragit L 100 (Rohm Pharma)	5.5
Eudragit L (Rohm Pharma)	6.0
Eudragit S (Rohm Pharma)	7.0

tablets which display a lot of variability in onset of action. Since Code and Marlett[112] first documented the fasted state motility cycle, there has been much progress in our knowledge of the factors which control transit of dosage forms through the GI tract. The variation in onset of action of enteric coated tablets given in the fasted state can be mostly attributed to variation in the motility phase in which the tablet is administered. When given on a full stomach, most enteric coated tablets are large enough to be retained in the stomach until the meal empties, further contributing to long and variable onset of action. More recently, enteric coated products have been formulated as multiparticulates so that they will exhibit more reproducible gastric emptying behavior. An illustration of the improvement using multiparticulates is the comparison of plasma salicylate profiles after administration of enteric coated multiparticulates and enteric coated tablets of aspirin,[43] as shown in Figure 17. Whereas the onset of detectable levels took anywhere from 2 to 5 h with the tablets, such levels were always attained within 2 h with the multiparticulate. Similar results were recently obtained by Duchesne *et al.*[113]

A further problem with enteric coated dosage forms is that if the polymer dissolution pH is too high, the coating may not dissolve during GI transit. Conversely, if the dissolution pH is too low, the dosage form may release the drug prematurely in the stomach. The dissolution pH window for appropriate dosage form performance may vary according to the patient population and whether the dosage form is to be given in the fasted or fed state. For example, patients with pancreatic insufficiency have a lower than normal duodenal pH, probably as a result of reduced pancreatic bicarbonate production.[114,115] The design of enteric coated dosage forms containing replacement enzymes for these patients therefore requires use of a lower pH dissolving polymer. Other patient populations in which GI pH can be significantly altered are the elderly (increased incidence of achlorhydria[116,117]) and those with GI ulcers who may be receiving H_2 receptor antagonist or antacid therapy. Another circumstance in which GI pH alters is after meal intake. The gastric contents are buffered by the meal, often to a pH in excess of pH 5 for a brief period, after which the pH gradually returns to the baseline value.[7,111] Meanwhile, the upper small intestine pH declines with time due to the influx of acidic contents from the stomach.[115] These trends are shown in Figure 18 and Table 8. Therefore, different performance criteria exist in the fed *versus* the fasted state and formulation design must take these into account.

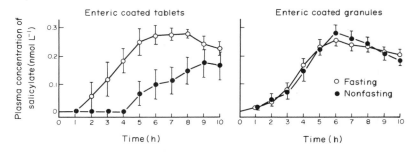

Figure 17   Mean ($\pm$S.E.M.) plasma levels of salicylic acid in eight subjects after administration of acetylsalicylic acid, 1.0 g, as conventional tablets, enteric coated tablets and enteric coated granules under fasting and nonfasting conditions[43]

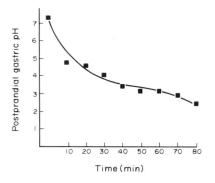

Figure 18   Postprandial gastric pH in human subjects[115]

Table 8 Time (Min) Spent at pH Levels Greater than pH 4, 5, 5.5 and 6 in each Postprandial Hour in Human Duodenum[115]

Postprandial hour	pH			
	>4	>5	>5.5	>6
1	59.8	45.0	33.8	17.9
2	56.2	31.8	23.5	6.7
3	57.7	33.6	18.4	5.1
4	55.3	28.8	16.6	2.8

Enteric coated products with a dissolution pH somewhat higher than that observed in the proximal small intestine have been studied for their ability to deliver drugs selectively to the colon. Dew *et al.*[118] showed radiologically that capsules coated with Eudragit S (a methacrylate polymer with a dissolution pH of about pH 7) released a marker compound, sulfapyridine, in the terminal ileum or proximal colon. The release correlated well with the onset of blood levels of sulfapyridine, indicating that a high dissolution pH enteric coating would be an appropriate means of delivering drugs selectively to the lower GI tract. Treatment of inflammatory diseases of the lower GI tract and circumvention of enzymatic decomposition of the drug by intestinal lumen or brush border enzymes are two potential applications of this approach.

A further application of enteric coated dosage forms is the saturation of first pass metabolism, by delivering as much of the dose as possible to the major absorptive site. For example, it has been shown that L-DOPA bioavailability can be improved by loading high concentrations in the upper intestine. Nishimura *et al.*[119] utilized an effervescent core formulation combined with a pH 5 dissolution coating to deliver L-DOPA selectively and in a rapidly dissolving manner to the proximal small intestine.

With the advent of multiparticulate enteric coated products, availability of aqueous-based coating materials and a better understanding of the pH conditions in the GI tract, enteric coated products that perform efficiently and reproducibly can be designed and produced. As a result, it is predicted that this dosage form strategy will enjoy a resurgence in interest within the pharmaceutical industry.

(ii) Coated ion-exchange resins

A very specialized example of a coated dosage form is the coated ion-exchange resin developed by the Pennwalt Corporation. Drugs that are charged can be exchanged on to an ion-exchange resin, which is then coated with a retarding polymer such as ethylcellulose to provide the appropriate release rate. Release is a three-step process comprising diffusion of the exchanging ion in through the coating, exchange with the drug, and diffusion of the drug out through the coating. The kinetics of release will depend on which of the three steps is rate limiting. The ions in the GI tract must be able to displace the drug effectively for this technique to be applicable. An advantage of this system is that high dose compounds can be incorporated since the resin can be ground to a fine particle size and a suspension dosage form prepared, *e.g.* codeine/IRP-69 resin as a cough syrup.[120] A disadvantage is that the range of application is limited to drugs ionized at GI pH. Also, it can be difficult to produce combination products if the affinities of each drug for the resin are very different. Some examples of drugs formulated using this technology are amphetamine and phenyl-*t*-butylamine (on cationic resins with no overcoat), hydrocodone and Delsym (with overcoat).

(iii) Osmotic pumps

The osmotic pump was patented by Higuchi and Theeuwes in 1973[121] and first described in the literature by Theeuwes *et al.* in 1975.[122] The general principle of osmotic pumps is that zero-order release is achieved by osmotically inducing a constant flow rate of drug-containing fluid out of the dosage form. In the simplest pharmaceutical embodiment, the elementary osmotic pump, the solid drug is surrounded by a semipermeable membrane having one delivery orifice. During pump operation, water from the environment is continuously imbibed across the semipermeable membrane by osmosis to produce the fluid/drug formulation. Osmosis may be induced by the drug or by some suitable excipient in the core formulation. Provided the membrane's structure does not allow expansion of the tablet's volume, fluid leaves the dosage form at the same rate as water is imbibed. As

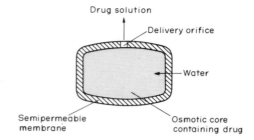

Figure 19 Schematic representation of an elementary osmotic pump in cross section[123]

the system moves through the GI tract, the fluid flows out through the orifice at a constant rate until the last of the osmotic driving agent is dissolved. Then, as the driving force starts to decline, the rate of drug release also declines (Figure 19).[123]

The rate of release from an elementary osmotic pump during the zero-order phase is given by

$$dM/dt = (Ak/h)(\pi_{in} - \pi_{out})C_s$$

where A is the surface area of the coating, h is the coating thickness, k is the permeability of the coating to water, π is the osmotic pressure and C_s is the aqueous solubility of the osmotic pressure generator. The osmotic pressure generated can be calculated from the van't Hoff equation

$$\pi = nRT/V$$

where n is the number of moles of the osmotic generator present, R is the gas constant, T is the absolute temperature and V is the volume contained in the dosage form. The osmotic pressure in the gut is about 300 mOsmol, or about 8 atm (1 atm = 101 kPa). The osmotic pressure generated inside the dosage form must be considerably higher than this to produce the required driving force. For example, a saturated solution of sucrose has an osmotic pressure of about 150 atm. Some drugs may have a high enough solubility to function as the osmotic generator without need for another osmotic agent in the formulation. If required, compounds with low molecular weight and high solubility such as fructose, mannitol, sodium chloride, or urea can be used to generate the required osmotic pressure. A limitation with very high solubility compounds is that the fraction of release by zero-order kinetics will be smaller than for less soluble compounds. The percentage of drug released according to zero-order kinetics can be calculated from

$$\% \text{ zero order} = [C_s V/(X_0 - C_s V)]100$$

where X_0 is the amount of osmotic generator initially present. Osmotic dosage forms have several important advantages over other types of sustained release preparations. The rate of delivery is governed by osmotic imbibement of water into the dosage form, leading to independence of the release rate from the pH of the environment and the degree of agitation in the environment. The rate of release measured *in vitro* will likely be predictive of the *in vivo* release rate because of this independence.[122] Provided release from the dosage form is rate controlling to absorption, the absorption profile should also mirror the *in vitro* release rate. The consistent correlations obtained between *in vitro* and *in vivo* release facilitate the design of osmotic pump dosage forms.

Indomethacin provides an example of a drug which has been formulated in an elementary osmotic pump. Potassium bicarbonate was chosen as the osmotic pressure generator because of its high osmotic pressure and buffer capacity. In addition, when the dosage form operates in the stomach, the free acid form of indomethacin formed is kept dispersed by carbon dioxide bubbles produced by the bicarbonate reacting with acid. The finely dispersed drug formulation so produced readily redissolves and remains available for absorption. Conventional tableting and coating equipment were used to make the dosage forms and the orifices were drilled using either a high speed mechanical drill or an automated laser.[124] As predicted by theory, these dosage forms exhibited release rates which were independent of stirring rate and pH *in vitro*, and which correlated well to the *in vivo* release rates (see Figure 20). Plasma levels after dosing in humans were less variable and more prolonged than with the immediate release dosage form (Figure 21).[125]

Another example of a drug formulated in an osmotic pump dosage form which has been studied in humans is oxprenolol succinate. This was formulated as an elementary osmotic pump to give an

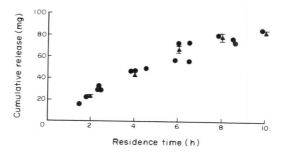

Figure 20 Cumulative amount of indomethacin released (in mg) from an elementary osmotic pump formulation *in vitro* and in the GI tract of dogs. ●, in dogs; ▲, mean, *in vitro*[124]

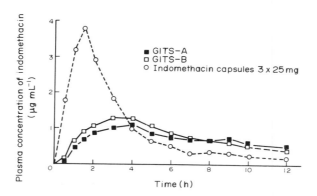

Figure 21 Mean plasma concentrations of indomethacin[125]

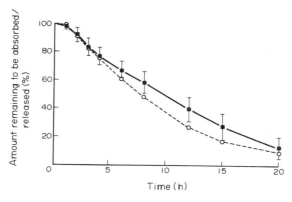

Figure 22 Mean Loo–Riegelman plot of percentage remaining to be absorbed against time for the 16/260 Oros system of oxprenolol (solid line); the vertical bars represent standard deviations of the means. The *in vitro* release rate profile, expressed in terms of the amount remaining to be released, is shown as a dashed line[126]

in vitro release rate of 16 mg h⁻¹, using a cellulosic semipermeable membrane as the coating. The osmotically pumping tablets were shown to produce an *in vivo* absorption profile which correlated closely with *in vitro* release (Figure 22)[126] and furthermore, reduction of exercise-induced tachycardia persisted for 24 h with this dosage form, in contrast to the other sustained release form of the drug, Slow Trasicor,[127] as shown in Figure 23. Other drugs currently under study or marketed in osmotic pump formulations are listed in Table 9.[128]

With the promising results obtained using osmotic pump systems, there have been many attempts to improve on the existing technology for manufacturing these devices and to extend the range of

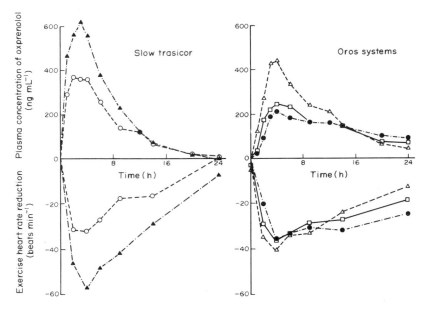

Figure 23 Mean plasma oxprenold concentration and reduction of exercise-induced heart rate in healthy volunteers after administration of Slow Transicor 160 mg ($n = 6$, ○ --- ○); Slow Transicor 2 × 160 mg ($n = 8$, ▲ –·–·–· ▲); 10/170 Oros ($n = 12$, □———□); 16/170 Oros ($n = 6$, △ --- △); and 16/260 Oros ($n = 14$, ● --- ●) systems by mouth in single doses[127]

application to less soluble drugs. For example, Bindschaedler *et al.*[129] have demonstrated that the coating material can be applied from an aqueous dispersion rather than organic solution, with retention of the required semipermeable properties. At a given level of plasticizer, membranes formed from latices were more permeable to water and swelled to a greater extent than those prepared from organic solutions. Zentner *et al.*[130] developed a membrane with controlled porosity, sponge like in appearance, which permits efflux of drug solution from the tablet. The orifices are formed *in situ* when the water soluble PEG or sorbitol component leaches out of the membrane. This has the advantage of avoiding the need for laser drilling. Another advantage is that the presence of multiple orifices should serve to minimize any irritation of the GI mucosa by the drug as it is released. A similar device using a hydrophilic polymer, PVA, as the semipermeable membrane and PEG as the water soluble component was patented in 1985 by Key Pharmaceuticals.[131]

There have also been efforts to design osmotic devices for drugs with lower water solubility. Nifedipine is an example of a drug with relatively low solubility which has been successfully formulated in a push/pull osmotic pump, illustrated in Figure 24. A lipid osmotic dosage form has also been patented for use with compounds with high lipid solubility.[132] Other recent innovations have been summarized by Eckenhoff.[128]

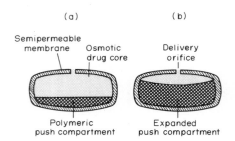

Figure 24 Cross section of the push–pull osmotic pump; (a) shows the configuration before operation, (b) shows the configuration during operation[128]

Table 9 Drugs Currently Marketed or Under Study as Osmotic Pump Formulations[128]

Type of drug system	Alza licensee	Therapeutic category	OROS system delivery
Albuterol (EOP)[a,d]	Glaxo Inc., Zebulon, NC	Antiasthmatic	Twice-daily dosage *versus* four times a day
Nifedipine (PP)[a,d]	Pfizer Inc, New York	Antianginal, antihypertensive	Once-daily dosage *versus* three times a day; more selective drug action
Bromopheniramine/pseudoephedrine[a]	Alza Corporation, Palo Alto, CA[b]	Antihistamine/decongestant (cold product)	Once-daily dosage; different, appropriate release rates for each drug; initial leading doses
Prazosin (PP)[d]	Pfizer Inc.	Antihypertensive	Once-daily dosage *versus* three times a day
Metoprolol (EOP)[d]	Ciba-Geigy Pharmaceuticals, Summit, NJ	Antihypertensive	Once-daily dosage *versus* twice a day; lower peak plasma levels with equal efficacy
Acutrim[c]	Ciba-Geigy Pharmaceuticals	Appetite suppressant	Once-daily dosage; 16 h drug delivery; no concentration spike
AcuSystem C (EOP)[c,d]	American Health Products, Ramsey, NJ	Vitamin C	Once-daily dosage

[a] Awaiting regulatory approval. [b] Developed for the OROS product limited partnership. [c] Now being marketed in the United States. [d] EOP = elementary osmotic pump; PP = push–pull system.

25.4.1.4 Current Issues in Controlled Release Dosage Forms

It has already been noted that a major objective of controlled release drug formulation is that the resultant change in the pharmacokinetics should have a positive effect on the therapeutic efficacy of the drug. This may occur due to decreased toxicity and side effects, improved absorption, more consistent levels, improved compliance *etc*. Occasionally, however, the change in release pattern can lead to decreased drug absorption, or the release rate will vary widely with the conditions under which the dosage form is administered. The ability to predict whether problems such as these will occur with a so-called controlled release dosage form is of great concern to both the formulator and the regulatory authorities.

Controlled release formulation can lead to a reduced bioavailability in several ways. The dosage form may be incapable of completely releasing the drug within its GI residence time, release may be at sites where the drug is poorly absorbed, or the first pass metabolism of the drug may be more complete because of the slower release rate.

The variation in GI transit time within subjects, between subjects and with the dosing protocol may translate into a variation in the fraction of the drug available. Total GI transit time can range from a few hours to several days, although it is normally of the order of 24 h. In subjects with rapid GI transit, a substantial proportion of the dose may still be in the dosage form at the time of its excretion, and in those subjects with average or longer transit times, most of the time is spent in the colon, which may not be a favorable site for absorption of the drug. The dosing protocol can also contribute to variability, especially for monolithic dosage forms which have a prolonged upper GI residence when administered with a meal compared to when they are administered in the fasted state.[39,42,133-135] There is also the question of release time prediction; although *in vitro/in vivo* correlations appear to be almost 1:1 for some osmotic pump formulations, matrix dosage forms tend to take longer to release drug *in vivo* than *in vitro*.

As discussed in the subsections on increasing absorption and gastric retention devices, the drug may not be equally well absorbed at all points in the GI tract. For example, a 12 h release formulation for a drug which is absorbed mainly by an active transport process, the carrier for which is found only in the small intestine, is not appropriate unless a bioadhesive or gastric retention formulation is used. On the other hand, coatings designed to release drug in the colon may not provide adequate contact with the intended site of action if the polymer pK_a is too high, in which case the dosage form may be excreted intact, or if the pK_a is too low, which would lead to drug release and inappropriate absorption/degradation in the small intestine.

If the enzyme system responsible for first pass metabolism is saturable, delivery of the drug at a slower rate may result in an increased fraction of the dose metabolized; phenytoin and propranolol are two important examples. The controlled release form of propanolol 'Inderal-SA' is only about 60% as available as the immediate release dosage form.[136] Verapamil, on the other hand, is equally well absorbed from SR and immediate release dosage forms.[15] One of the main aims of formulating a drug in a controlled release dosage form is that the absorption pattern will become more reliable. However, if the dosage form is subject to changes in release rate depending on the composition of its immediate environment, dosing at different times of day or before *versus* after meals may actually increase the variability of performance. For some drugs, bioavailability and response are dependent on the time of day at which the dose is administered,[9,137] although this may be somewhat confounded by the differences in feeding state at the different times of dosing.[136]

Of even greater concern than decreased bioavailability is the potential for sudden release of an inappropriately large fraction of the dose. This phenomenon, known as dose dumping, is usually associated with administration of the formulation with food.[136] Several groups have demonstrated differences in bioavailability when drug is administered in the fed and fasted state. This happens even for immediate release products, as shown by Chiou and Reigelman[10] for griseofulvin and Mason and Winer[138] for aspirin, but the potential for adverse effects with unanticipated rapid absorption from supposedly sustained release dosage forms makes such a situation much more serious. The effect seems to be very formulation (as opposed to drug) dependent. Whereas Hendeles *et al.*[139] demonstrated dramatic effects on the pattern of drug absorption when two theophylline sustained release dosage forms were administered with food, Jonkman *et al.*[140] found that food had no effect on a third SR formulation of theophylline. In other cases, administration of sustained release formulations has been shown to result in a delay in the onset of absorption.[141]

Food effects on bioavailability can arise from a number of sources. In addition to prolonging the upper GI residence time of monolithic dosage forms, changes in concentrations of enzymes, bile salts and in the lumenal pH, and the presence of fatty materials from the meal can all influence the dosage form performance. Not only do we have to consider partitioning of the drug into the fatty part of the meal, the potential for adsorption on to meal components, altered dissolution and decomposition of the drug (as is the case for immediate release dosage forms), but also how the release pattern from the dosage form will be affected. For example, the decreased duodenal pH in the postprandial state may result in a delay of release from enteric coated dosage forms until a more distal, higher pH location is reached. High osmotic pressure from the ingestion of sugary drinks may delay the onset of release from osmotic pump formulations. On the other hand, solubilization of a waxy coating in the presence of the fatty component of the meal or the higher bile concentrations in the small intestine may lead to faster release than seen in the fasted state.

In vitro methods or animal models which could be used to predict the *in vivo* performance of sustained release dosage forms under various dosing protocols would be of great advantage to the formulator. Recently, the biopharmaceutics research branch of the FDA published a method designed to mimic drug administration with a high fat meal.[142] Modification of current USP methods to accommodate other changes in lumenal conditions associated with meal intake should result in further improvement of *in vivo/in vitro* correlations.

A further challenge lies in the design of appropriate human studies for controlled, and particularly sustained release, dosage forms. Guidelines for testing new sustained release dosage forms for food effects were discussed at a recent workshop[143] and no doubt will be further revised as our understanding of the physiological conditions and their effects on dosage form performance become better understood.

25.4.2 TRANSDERMAL DRUG DELIVERY SYSTEMS

25.4.2.1 Introduction

Transdermal drug delivery (TDD) has been proposed as an attractive, noninvasive method to achieve drug input into the systemic circulation.[144] Yet, despite considerable effort over the past 15 years, there are, at the time of writing, only four FDA-approved drugs for transdermal administration: scopolamine, nitroglycerin, clonidine and estradiol. The overall goal of this section, therefore, is to highlight both the promise and the problems pertinent to the systemic input of therapeutic agents *via* the skin. Specifically, the objectives are as follows: (i) to review briefly the features of skin physiology pertinent to the understanding of percutaneous absorption and TDD; (ii) to identify the pharmacokinetic, pharmacodynamic and physicochemical features that should be exhibited by a drug to be considered a feasible candidate for TDD; (iii) to enumerate the advantages and drawbacks of delivering drugs transdermally for systemic effect; (iv) to review the design features and performance of the currently approved TDD systems for scopolamine, nitroglycerin, clonidine and estradiol; (v) to indicate the major problems associated with the transdermal delivery of nitroglycerin and clonidine; and (vi) to outline the potential (and pitfalls) of penetration enhancement as a means of extending the applications of TDD.

25.4.2.2 Historical Background

Topical drug application has been practised since Egyptian times. A wide variety of topical ointment and cream preparations are currently prescribed, although the vast majority are intended for local action.[145] Drug delivery from these products is poorly controlled because skin permeability varies over different regions of the body,[146,147] and the dosage (area covered and amount applied) is not easily reproduced. However, the utility of the skin as a portal for drug entry into the body has been established by such formulations.

The development of TDD devices was a logical evolution from the widespread use of medicated plasters in Japan.[148] To circumvent the dosing reproducibility problems stated above, TDD research has focused on the design of the delivery systems and the ability to modulate both the rate and duration of drug input into the systemic circulation. In an optimal device, therefore, the drug release rate should be less than the mean steady state flux of drug across the skin, thereby ensuring that the device controls the rate of drug delivery and the skin does not. This also ensures that dose dumping will not occur in patients with excessively high skin permeabilities (*e.g.* due to barrier damage). To date, a number of different systems, programmed to deliver drug for periods of between one and seven days, have been produced. The extent to which they meet the optimal criteria as specified, and the reasons for only limited success thus far, are discussed below.

25.4.2.3 Skin Structure and Function

A primary function of human skin is to provide a barrier to the ingress of xenobiotics into the body. In addressing the concept of TDD for systemic effect, therefore, it is reasonable to ask, 'Why administer compounds across a membrane which has been designed rather specifically to inhibit such transport?' The positive features of TDD which have prompted pharmaceutical scientists to undertake this challenge are presented below. It is first appropriate, however, to examine the nature and form of the resistance which the skin provides to molecular ingress into the body.

The skin is the largest organ of the human body and, on the average adult, covers an area of approximately 2 m². Skin basically consists of two layers: the epidermis and dermis, which rest on an underlying cushion of subcutaneous fat (Figure 25).

The dermis is a matrix of connective tissue embedded in mucopolysaccharide; the main functions of the dermis are to support the epidermis, and the cutaneous nervous and vascular networks, and to impart elasticity and strength to the skin.

The epidermis, the outer skin lamina, evolves from a basal layer of proliferating cells. The movement of these cells towards the skin surface takes them away from the nutrient supply provided by the upper dermal microcirculation (the epidermis being avascular). The cells change, as a result, undergoing keratinization and, ultimately, terminal differentiation. At the skin surface, the product of these biochemical changes is the stratum corneum, a thin (10–20 μm) membrane composed of 10 to 15 dead, completely cornefied, cell layers embedded in a lipid matrix. It is this, essentially

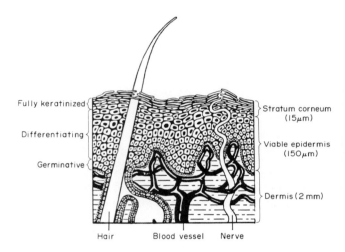

Figure 25 Schematic diagram of skin

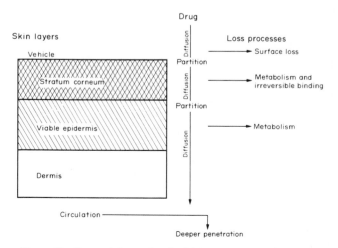

Figure 26 Sequential steps involved in percutaneous absorption

lipophilic, zone that is believed to be the major diffusional barrier to the percutaneous absorption of most topically contacting chemicals.

In the simplest sense, therefore, the skin may be represented as a bilaminate membrane; to reach the dermal vasculature (and rapid systemic distribution) a penetrant must traverse both the lipophilic environment of the stratum corneum and the aqueous environment of the underlying viable epidermis and upper dermis. The overall process of percutaneous absorption may be dissected into a series of partitioning and diffusional steps as indicated in Figure 26.

25.4.2.4 Feasibility Assessment for Transdermal Drug Delivery

The control of drug input, potentially available with transdermal delivery, is attractive. The saw-tooth profile, characteristic of conventional dosing regimens, can be damped by administration *via* the skin. A clear advantage for drugs of narrow therapeutic index is implicated.

A simple calculation can assess the feasibility of transdermal delivery for potential 'candidate' compounds. (i) Assume that a transdermal device provides zero-order delivery to the skin surface at a rate (k_0, $\mu g \, cm^{-2} \, h^{-1}$) slightly less than the maximum flux (J_m, $\mu g \, cm^{-2} \, h^{-1}$) of a model drug across the stratum corneum. Taking $J_m = 35 \, \mu g \, cm^{-2} \, h^{-1}$, then a value of $k_0 = 25 \, \mu g \, cm^{-2} \, h^{-1}$ is reasonable. (ii) If the target steady state plasma concentration of the drug is $C_{ss} \, \mu g \, ml^{-1}$, then it follows that

$$\text{Input rate} = Sk_0 = Cl \, C_{ss} \tag{1}$$

where S (in cm^2) is the area of the patch and Cl (in $mL \, h^{-1}$) is the drug clearance. Setting $k_0 < J_m$ retains drug input control within the delivery system. (iii) Equation (1) contains only S as a manipulable parameter. It follows that, if inherent skin permeability cannot be increased in some way, then the input function can be maneuvered only within the confines of $k_0 < J_m$ and that S be 'reasonable'. (iv) Table 10 is a feasibility 'screen' for a selection of drugs.[149] Given that an upper (practical and economic) limit on S is in the range of 50 cm^2, the drugs can be quickly divided into 'possibles' and 'impossibles'.

However, this simple approach leaves a number of questions unanswered: (i) is a percutaneous flux of 25 $\mu g \, cm^{-2} \, h^{-1}$ possible for all compounds? (ii) how long will be required for the attainment of C_{ss} following application of the patch? (iii) does it make sense to control the input of a drug with a half-life of $> 24 \, h$? and (iv) which of the 'feasible' candidates will elicit a local irritating effect on the skin?

Limitations also exist on the physicochemical nature of the drug to be delivered. To become systemically available, following application in a transdermal device, a drug must: (i) transport to the device–skin surface interface; (ii) partition from the delivery system into the stratum corneum; (iii) diffuse through the stratum corneum; (iv) partition from the stratum corneum into the viable epidermis; (v) diffuse through the viable tissue; and (vi) enter the cutaneous microcirculation and gain systemic access.

Table 10 Feasibility Screen for Representative Transdermal Delivery 'Candidates'

Drug	Cl ($mL \, h^{-1}$)[a]	C_{ss} ($ng \, mL^{-1}$)	S_{min} (cm^2)[b]
Acetaminophen	23	15	13 850
Aspirin	29	150	174 510
Cimetidine	49	1	1940
Clonazepam	4	0.025	4.0
Clonidine	12	0.001	0.48
Digoxin	6.8	0.002	0.54
Estradiol	67	0.0001	0.27
Indomethacin	9	0.5	181
Isosorbide dinitrate	175	0.001	7.0
Nitroglycerin	4210	0.0001	17
Propranolol	49	0.02	39
Scopolamine	43	0.0002	0.35
Theophylline	0.14	5	28

[a] Calculated for a 70 kg adult *except* in the case of theophylline, the assessment of which is for a 2.5 kg preterm infant. [b] The total body surface area of a 70 kg, 1.83 m tall adult is approximately 2 m^2. The values are calculated using equation (1) with $k_0 = 25 \, \mu g \, cm^{-2} \, h^{-1}$.

It follows that diffusion and partitioning are the key physical processes pertinent to transdermal delivery. With respect to diffusion, drug transport is determined primarily by the molecular size and the level of interaction with the medium through which diffusion is taking place. Most currently used drugs have molecular weights less than 1000 Da and the effect of size on diffusion coefficient may be adequately described by a power dependency or an exponential function.[150] With respect to partitioning, the criteria are demanding. The molecule must favor the stratum corneum over the device, and the relative affinity of the drug for the stratum corneum and viable tissue must be reasonably balanced.[150]

25.4.2.5 Advantages of Transdermal Drug Delivery

The principal advantages of TDD over traditional oral multiple dose regimens are: (i) production of a predictable, relatively nonfluctuating and sustained plasma drug concentration, in contrast to the 'peaks' and 'troughs' of multiple dosing; (ii) elimination of factors associated with inter- and intra-subject variation following drug delivery to the GI tract, such as effects of pH, motility, transit time and food intake; (iii) avoidance of GI and hepatic first pass metabolism; (iv) normalization of the rate and extent of absorption in diverse patient populations; (v) simple application and removal of the drug formulation, reduction in dosing frequency and, consequently, improved patient compliance; (vi) straightforward termination of drug input when medically indicated; (vii) enhanced convenience and safety over other methods of achieving the above benefits (*e.g.* intravenous infusion or intramuscular injection).

25.4.2.6 Limitations of Transdermal Drug Delivery

The disadvantages of drug input *via* the skin severely limit the number of agents which are suitable for TDD. These drawbacks are associated with the excellent transport barrier properties of the stratum corneum, the exquisite immunological sensitivity of the skin and the pharmacology and chemistry of the drug. (i) Percutaneous absorption is a slow process; therefore, to deliver an effective dose, from a device of reasonable surface area, requires that the drug be potent. Those agents for which the daily parenteral dose is ≤ 5 mg and the target therapeutic level in the biophase is of the order of 10 ng mL^{-1} or less, are possible TDD candidates. (ii) By its very nature, TDD leads to rather constant systemic levels over time. This pharmacokinetic pattern may be completely counterproductive from a therapeutic standpoint if tolerance develops. (iii) The drug, or a material used in constructing the device, may induce a local irritant or allergic response. (iv) In some cases, biotransformation (*i.e.* inactivation) of the applied drug can be carried out by microorganisms present on the skin surface[151] or by enzymes located within the epidermis.[152,153]

25.4.2.7 Classification of Transdermal Delivery Systems

25.4.2.7.1 Membrane-moderated

A reservoir containing the drug is enclosed on all sides, except that through which drug is released, by an impermeable laminate (Figure 27). The releasing face of the reservoir is covered by a rate-controlling polymeric membrane. Different release rates are achieved by variation of the polymer composition and the thickness of the membrane. Dose titration is facilitated by the provision of

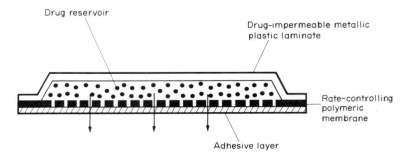

Figure 27   Membrane-moderated transdermal drug delivery system

patches of different surface areas. The devices of this type include: (i) Transderm Scop (Ciba-Geigy), the active agent is scopolamine; (ii) Transderm Nitro (Ciba-Geigy), the active agent is nitroglycerin; (iii) Catapres TTS (Boehringer-Ingelheim), the active agent is clonidine; (iv) Estraderm (Ciba-Geigy), the active agent is estradiol.

25.4.2.7.2 *Matrix diffusion-controlled*

The reservoir is manufactured by dispersing the drug in a polymer matrix which is then molded into a disc with a defined surface area and thickness. Drug release from the device into the body is controlled by diffusion through the matrix reservoir material. Devices of this type, which all contain nitroglycerin, are: (i) the Nitro-dur II (Key/Schering-Plough) system, (ii) the Deponit TTS (Pharma-Schwarz/Wyeth), (iii) the first transdermal generic; the nitroglycerin transdermal system (NTS) from Hercon Laboratories Corporation. A schematic representation of the Deponit system is shown in Figure 28.

25.4.2.7.3 *Microsealed*

The microsealed category is represented by the nitroglycerin-containing Nitro-Disc (Searle). In this device (Figure 29) the drug is dispersed throughout a silicone polymer in liquid microcompartments which act as very small drug reservoirs.

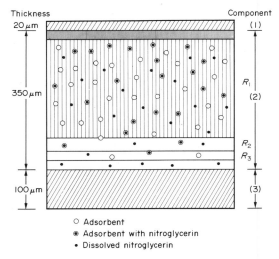

Figure 28 Example of a matrix diffusion-controlled transdermal drug delivery system

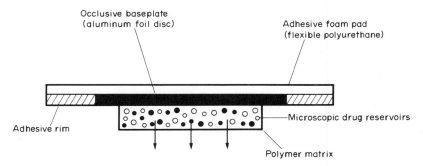

Figure 29 Microsealed transdermal drug delivery system

25.4.2.8 Transdermally Delivered Drugs—Pharmacokinetics and Pharmacodynamics

25.4.2.8.1 *Clonidine*

Clonidine is a potent antihypertensive agent used alone, or in combination with diuretics, in the treatment of mild to moderate hypertension.[154,155] Although effective at relatively low plasma concentrations (Table 10), it has to be taken two or three times daily to maintain therapeutic levels. The resultant fluctuations in plasma drug concentration are thought to be the cause of many of the adverse effects associated with oral therapy, such as drowsiness and dry mouth. Thus, the need for a sustained release product was apparent; based on a knowledge of the partition coefficient and a feasibility study similar to that in Table 10, controlled delivery *via* the transdermal route was considered possible.

The single marketed TDD system for clonidine is Catapres-TTS (Boehringer Ingelheim), which is programmed to provide steady state plasma levels for up to 7 d, representing a significant potential for improvement in patient compliance. The device is of the membrane-moderated design and consists of a reservoir of clonidine in mineral oil, polyisobutylene and colloidal silicon dioxide, surrounded on three sides by an impermeable backing layer. The underside is covered by a semipermeable polymeric membrane that serves to regulate the rate of clonidine release from the reservoir into the adhesive layer which holds the system to the skin (Figure 27); the drug within the adhesive acts as a loading dose and enables the target plasma concentration to be attained more expeditiously.[156] Different release rates of clonidine from the system have been achieved by varying the composition and/or thickness of the membrane, and the total amount released has been altered by varying the surface area of the device. Three sizes of Catapres are available; 3.5, 7.0 and 10.5 cm^2; the systems are programmed to deliver 100, 200 and 300 µg of clonidine, respectively.

The *in vivo* bioavailability of clonidine delivered transdermally was compared with oral dosing (twice a day). Mean plasma clonidine concentration *versus* time profiles following each regimen are shown in Figure 30. Transdermally, a therapeutic C_{ss} of 0.4 ng mL^{-1} was reached; however, this level was not achieved until three days after system application. This delay is in part due to the relatively long half-life (≈ 10 h) of clonidine, and to the suspected formation of a drug reservoir in the stratum corneum. The C_{ss} was maintained from day 3 through to day 7, at which point the patch was removed. Only 8 h after system removal did the plasma concentrations diminish significantly below C_{ss}. While the oral regimen resulted in a mean plasma concentration similar to that achieved following transdermal delivery, the conventional administration route was characterized by a 'peak and trough' plasma profile (Figure 30).[157] The terminal half-life following TDD is longer than that after oral dosing. Although this feature is cited as a possible benefit in preventing rebound hypertension on cessation of therapy, it may present a problem if rapid clonidine withdrawal is required.

The pharmacokinetics of transdermally delivered clonidine were extensively studied by MacGregor *et al.*,[158] who demonstrated that: (i) C_{ss} was linearly related to patch surface area; (ii) the

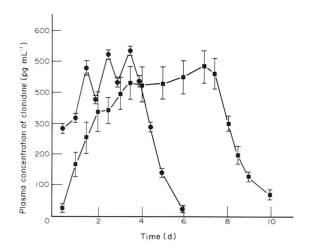

Figure 30 Clonidine plasma concentrations (mean ± S.E., *n* = 17) as a function of time; ■, during 7-d application, and for 3 d after removal of a single Catapres-TTS; ●, during, and for 2 d after, a multiple oral dosing regimen (b.i.d. for 3.5 d)

plasma concentration–time profiles following application to either the arm or the chest were not significantly different; and (iii) the C_{ss} achieved by an initial patch application was maintained by two succeeding applications. However, the plasma levels reported were associated with high variation between subjects, presumably reflecting variability in both the percutaneous absorption and the renal clearance of clonidine.

Pharmacologically, the device has been shown to control hypertension in 60 to 100% of patients studied.[159,160] In a placebo-controlled trial, the active transdermal device caused a significant reduction in blood pressure.[161] Furthermore, results from a small, unblinded study, comparing oral therapy with transdermal, suggested that the transdermal system may be substituted for oral tablets without compromising blood pressure control.[162] Another trial reports that in a small population of patients using the device for two years, the antihypertensive effect of the transdermally delivered drug did not diminish.[163] Thus, the clinical data available would suggest that the efficacy of Catapres-TTS is proven and that it has gained the approval of patients and physicians alike.[164–166] However, transdermal clonidine has been associated with cutaneous reaction in up to 35% of subjects. Clearly, this is an adverse effect specific to the route of administration. Typically, the reaction presents as a local, or occasionally generalized, contact dermatitis, which may appear anytime, from a few days to several months after application of the device. The reaction usually subsides once the patch is moved to another body site.[167,168] Occasionally, in extreme cases, the reaction can be sufficiently severe to necessitate cessation of transdermal therapy. In one study, similar reactions were observed following application of placebo patches, suggesting that a component of the device is causing the reaction, not the drug.[161] Relatedly, Hurkmans *et al.*[169] studied irritation resulting from application of TDD systems to the human back for up to 120 h. These authors concluded that sweat accumulation under the system contributed to the irritation. It is apparent that: (i) the clonidine problem is not a straightforward one; and (ii) the immunological function of skin is complex and warrants considerable further attention.

25.4.2.8.2 *Estradiol*

One therapeutic indication for the use of estradiol, an endogenous hormone, is the treatment of debilitating symptoms associated with the female menopause. The value of oral therapy has been demonstrated over 20 years. Common oral dosage forms contain micronized estradiol or conjugated equine estrogens. However, a substantial fraction of an oral estradiol dose is subject to first-pass metabolism. Consequently, estradiol is administered in relatively high doses and, as a result, the plasma concentrations of estrone and other metabolites are significantly elevated relative to premenopausal women. High estrone levels have been associated with certain drug-related side effects such as hypertension and hyperlipidemia.[170,171]

Again, the potential usefulness of the transdermal route of administration is evident and Table 10 indicates that TDD is a viable proposition for estradiol. Circumvention of first pass metabolism should permit the dose of estradiol, and hence the plasma levels of estrone, to be reduced. In a preliminary study, a 0.06% topical gel formulation of estradiol, applied every 24 h, was shown to be as efficacious as daily doses of 2 mg of the oral micronized form.[172] Subsequently, a membrane-regulated TDD device has been approved and marketed (Estraderm, Ciba-Geigy). It is similar in construction to the clonidine device described above. The drug reservoir is a 2% (1 mg per 50 mg) estradiol solution in ethanol. The ethanol is present to solubilize the drug, and it may also act as a penetration enhancer. *In vivo*, the patch is programmed to deliver estradiol at 0.21 μg cm^{-2} h^{-1} for periods of up to 4 d.[173]

A pharmacokinetic study in 14 postmenopausal women, who received, in random order, the transdermal device and two different oral formulations, demonstrated that a therapeutic C_{ss} of 20–75 pg mL^{-1} was attainable following TDD. In addition, the estradiol/estrone ratio was controlled at close to the premenopausal value of unity. The mean plasma concentration–time profiles of estradiol and estrone obtained over 3 d are shown in Figure 31. There was also a significantly linear correlation between the amount of estradiol delivered and the surface area of the device. Successive patches, applied twice weekly for three weeks, were able to maintain the desired mean C_{ss}, although peak concentrations were more than double the trough levels.[174] These results were confirmed in a later study.[175] Clinical efficacy of the estradiol TDD device over a 6 week period was demonstrated in both a large multicenter, double-blind, trial (124 patients) and in a number of smaller, unblinded studies. Topical irritation (mainly transient erythema) has been reported in up to 20% of subjects, although, generally, the patches are thought to be well tolerated.[176–178]

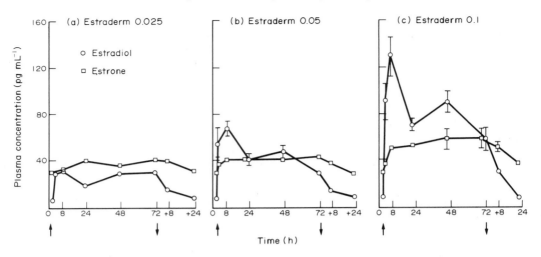

Figure 31 Estradiol and estrone plasma concentrations (mean $\pm$ S.E., $n = 14$) as a function of time following a 72-h application of Estraderm: (a) 0.025 mg 24 h^{-1}, (b) 0.05 mg 24 h^{-1}, (c) 0.1 mg 24 h^{-1}

25.4.2.8.3 *Nitroglycerin*

Nitroglycerin is a potent smooth muscle relaxant used in the treatment of angina pectoris and congestive heart failure. The drug is available as a sublingual dosage form for the rapid relief of anginal symptoms on an 'as required' basis.[179] Orally, however, the short half-life and extensive hepatic first pass metabolism of nitroglycerin limits its bioavailability to less than 1%,[180] and compromise this route, therefore, for prophylactic treatment. Thus, nitroglycerin represented an ideal candidate for TDD (Table 10) and several devices have been marketed for once daily use. The commercial success of transdermal nitroglycerin in large part accounts for the exponential growth of interest in TDD.

The original transdermal formulations of nitroglycerin were ointments from which the drug was rapidly released. Consequently, the rate of drug entry into the systemic circulation was controlled by the barrier properties of the patient's skin. Although the ointment dosage forms provide clinical effectiveness for up to 8 h,[181] there is a high inter- and intra-patient variation in drug plasma levels due to substantial differences in percutaneous absorption. Moreover, the inelegant nature of the ointment product, and the poor reproducibility in dosing (and effect), caused low patient compliance. In addition, recent studies have shown that microorganisms present on the skin surface can metabolize nitroglycerin[151] and that this effect will be more pronounced for the ointment dosage forms which place large amounts of drug in direct contact with the skin (compared to a transdermal device, in which the majority of drug, located in the reservoir, is somewhat protected).

Four of the six currently available TDD systems were developed concurrently by different manufacturers. The devices use the complete range of strategies for controlling drug delivery to the skin: one is membrane-regulated (as described previously for clonidine and estradiol), two are matrix systems, and the other is a 'microsealed' device. Briefly, the constructs of these four patches are as follows: (i) Transderm-Nitro (Ciba-Geigy), this device is a membrane-regulated system with a programmed release rate of about 220 μg cm^{-2} 24 h^{-1};[182] (ii) Nitro-dur (Schering/Key), an adhesive layer (the matrix) forms the drug reservoir;[183] (iii) Deponit (Pharma-Schwarz/Wyeth), nitroglycerin adsorbed onto lactose, is inhomogeneously dispersed in a polyisobutylene/resin base adhesive matrix, such that a drug concentration gradient exists across the layer (Figure 28). The programmed release rate of the device is 312 μg cm^{-2} 24 h^{-1};[184] and (iv) Nitrodisc (Searle), small liquid microcompartments, which act as individual rate-controlling drug reservoirs, are dispersed in a silicon polymer (Figure 29). The device is known as a 'microsealed drug delivery system' and is programmed to release drug at 624 μg cm^{-2} 24 h^{-1}.[185]

Despite differences in the *in vitro* pattern of nitroglycerin release from the different devices, the pharmacokinetic profiles observed following *in vivo* use are remarkably consistent (Figure 32). This suggests that, contrary to expectation, the skin is exerting significant control of drug input rate into the body.

Several comparative *in vivo* studies have been performed to assess the bioavailability of nitroglycerin from the various systems. In a randomized crossover experiment with eight subjects,

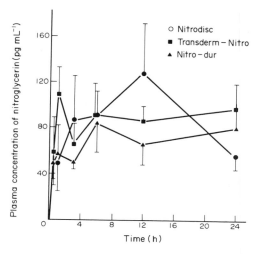

Figure 32 Nitroglycerin plasma concentrations (mean ± S.E., $n = 6$) as a function of time during 24-h applications of 10 mg 24 h^{-1} units of Transderm-Nitro, Nitrodisc and Nitro-dur

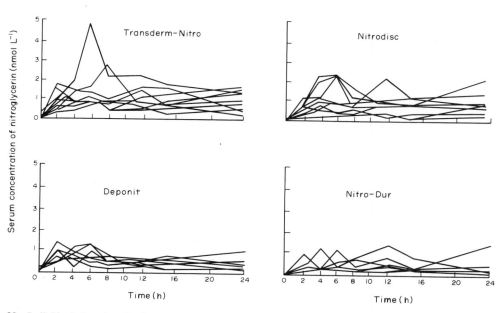

Figure 33 Individual nitroglycerin plasma concentrations as a function of time during 24-h application of four transdermal systems (10 mg 24 h^{-1})

Heidmann *et al.*[186] determined the serum levels of nitroglycerin achieved following application of the four devices, the delivery from each of which was programmed to be 10 mg in 24 h (Figure 33). The individual values of C_{ss} were subject to large variability, with a range of 10 to 200 nmol mL^{-1} observed. However, this scattered distribution fell within that found following sublingual dosing.

The devices were well received by physicians and patients and have been extensively prescribed. As with the clonidine and estradiol devices, there have been reports of contact dermatitis following application of the systems.[187] However, the irritating culprit has not been positively identified. Clinical practice presently emphasizes the importance of rotating the patch application site to minimize local skin irritation.

Unfortunately, a difficult problem has arisen with respect to chronic nitroglycerin patch usage; recent data suggest that tolerance to the effects of the drug can develop rapidly (even as early as 12 h after commencing therapy).[188,189] It has been suggested that this tolerance may be overcome by incorporation of a drug-free interval into the treatment regimen.[190,191] Preliminary studies have also demonstrated that *N*-acetylcysteine, given intravenously, can successfully reverse nitrate tolerance; an oral formulation is undergoing a clinical trial to test the effectiveness of this compound.[192] Other researchers dispute the existence of tolerance, citing small subject populations and inconsistent methods of assessing clinical efficacy.[193,194] The situation has not been helped by the fact that nitroglycerin represents a tough analytical challenge, and is metabolized in blood.[195] There is evidence that a nitroglycerin concentration gradient exists along blood vessels; ideally, therefore, plasma samples should always be taken from the same site.[196] Furthermore, the dinitrate metabolites of nitroglycerin do have activity and are present in relatively high concentrations following drug administration.[197] Details of the kinetic–dynamic relationships for these metabolites, however, are lacking.

A major, FDA-initiated, multicenter clinical study has been undertaken to evaluate the pharmacological efficacy of nitroglycerin delivered by the transdermal route.

25.4.2.8.4 *Scopolamine*

Scopolamine was the first drug to be approved and marketed as a TDD system (Scopoderm, Ciba-Geigy). The device is intended for application postauricularly to alleviate symptoms associated with motion sickness. Oral and parenteral administration of scopolamine is limited by the drug's short half-life and narrow therapeutic range. Adverse reactions, such as drowsiness, confusion and blurred vision, can occur at plasma concentrations only slightly higher than those necessary for motion sickness treatment. The aims of the TDD device, therefore, were to minimize such side effects, to maintain the plasma concentration precisely at a level within the therapeutic index, and to increase the duration of action beyond the 5 h attainable with acute oral therapy.

The scopolamine transdermal device is a membrane-regulated system with a surface area of only 2.5 cm². It is programmed to deliver 0.5 mg of drug over a 3 d period: 140 µg as an initial priming dose, superimposed upon a constant zero-order rate of 5 µg h⁻¹. This latter release rate is approximately 20% that of scopolamine's maximum percutaneous absorption flux across human skin; thus the device, rather than the skin, should control the rate of drug delivery.

Chandrasekaran[156] demonstrated that the device was able to provide an *in vivo* analog of a continuous intravenous infusion (3.7–6.0 µg h⁻¹) for 72 h; steady state following TDD was achieved after 12 h (Figure 34). The standard deviations of the *in vivo* data are relatively small compared to those observed with nitroglycerin, for example. This presumably reflects both the increased input control exercised by the scopolamine device (compared to the nitroglycerin systems) and the smaller intersubject variability in drug clearance. In contrast to the sustained effects of transdermal scopolamine, Muir and Metcalfe[198] reported that oral therapy was only able to maintain therapeutic levels for 2–4 h. The clinical efficacy of transdermal scopolamine has been tested and proven in well-controlled motion simulations[199,200] and in 'real' situations at sea.[201,202] Another use for the device, namely, to provide prolonged protection against postoperative nausea and vomiting,[203] has

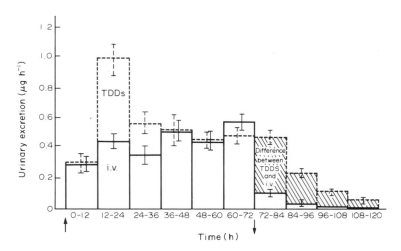

Figure 34 Urinary excretion rate profiles of scopolamine as a function of time during, and for two days after, (a) application of a single transdermal unit, and (b) slow intravenous infusions of 3.7–6.0 µg h⁻¹. Dashed line represents TDDs; solid line, i.v. administration; hatched area represents difference between TDDs and i.v.

also been discussed. However, therapeutic action was only maintained for 24 h[204] and tolerance to the antiemetic effect may develop after 72 h.[205]

Side effects reported with the use of the device are similar to those with oral therapy, and psychological problems (agitation and hallucination) have been reported in children.[206,207] Again, an approximately 15% incidence of mild irritation has been observed in users of the system.[200,202]

25.4.2.8.5 *Other drugs*

Several other drugs have been, and continue to be, considered for transdermal delivery and have undergone clinical investigation. Briefly, the main features of this work may be summarized as follows.

(i) *Azatadine*

Azatadine is an antihistamine used to relieve the symptoms associated with allergic rhinitis and urticaria. In a crossover study with 23 subjects, the therapeutic responses induced by each of two transdermal devices were similar to an oral dosage form.[208]

(ii) *β-Blockers*

Timolol has a short half-life and is subject to extensive first pass metabolism. A simple, topical formulation, applied for 30 h to the chest (25 cm^2), produced therapeutically significant plasma levels; concurrently, clinical efficacy was demonstrated by reduced systolic blood pressure.[209] Bupranolol is another rapidly eliminated β-blocker. The drug has been shown to cause a reduction in exercise tachycardia when delivered from a matrix release TDD system.[210]

(iii) *Fentanyl*

Fentanyl has been formulated as a membrane-regulated TDD device for use in postoperative pain control. The system is programmed to release 1.8 mg in 24 h. Several recent reports document that the patch can maintain analgesic levels of fentanyl comparable to constant intravenous infusion.[211,212]

(iv) *Nicotine*

Rose *et al.*[213] reported initial studies on the successful transdermal delivery of nicotine from a 30% aqueous solution under occlusion.

(v) *Testosterone*

A TDD system containing testosterone, for application to the scrotum, has been proposed for the treatment of testosterone deficiency in hypogonadal men. The plasma concentrations achieved by the device were within the normal adult male range (35–103 ng mL^{-1}) after application of 10 or 15 mg of drug.[214] Positive clinical responses to the transdermal devices, with reduced side effects, were reported.

25.4.2.9 Penetration Enhancement

To extend the therapeutic potential of transdermal drug delivery, it would be advantageous to have a reversible means of reducing the barrier to percutaneous penetration. How can this be achieved? One approach receiving considerable attention is the use of so-called penetration enhancers or accelerants. The following criteria[145] should be met by a successful penetration enhancer: (i) it should elicit no pharmacological effect; (ii) it should be specific in its action; (iii) it should act quickly, with a predictable duration, and its action should be reversible; (iv) it should be chemically and physically stable and be compatible with all components of the drug delivery system; (v) it should be odorless and colorless; and (vi) it should be nontoxic, nonallergenic and nonirritating.

These demands are stringent and the characteristics are displayed by very few chemicals. Of the approved transdermal systems, only the estradiol device contains an agent (namely ethanol) which

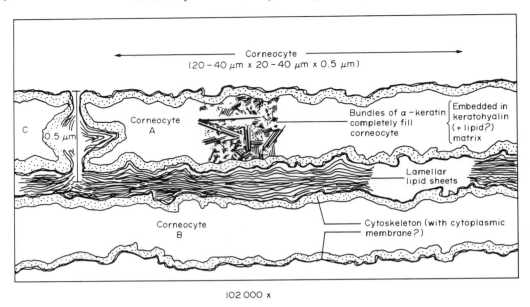

102 000 x

Figure 35 Diagrammatic cross-section of human stratum corneum emphasizing the 'brick' (corneocyte) and 'mortar' (intercellular lipid) domains

Table 11 Correlation Between Fatty Acid Structure, Stratum Corneum Intercellular Lipid Perturbation and Drug Flux.

	T_m (°C)	Frequency (cm^{-1})	Flux of salicylic acid (µg)
Stearic acid (C18:0)[a]	62.5 ± 1.0	2918.1 ± 0.4	1.21 ± 0.50
trans-Vaccenic acid (C18:1, *t*-11)[b]	61.0	2918.8	1.11
cis-Vaccenic acid (C18:1, *c*-11)[c]	57.0	2920.1	5.53

can be considered an enhancer. The mechanism of action of penetration enhancers is not fully understood.[150,215] One of the better-studied classes of accelerants is the unsaturated fatty acids. It appears that *cis*-oleic acid, for example, fluidizes the intercellular lipid domains of the stratum corneum (see Figure 35).[216] As there is increasing evidence that these lipid channels comprise the major transport pathway across the stratum corneum, fluidization of the lamellae should lead to facilitated drug passage. Transport experiments and biophysical measurements have confirmed this hypothesis (Table 11).[216,217] In Table 11, results are shown from experiments in which porcine skin was treated with each of three fatty acids. Differential scanning calorimetry (DSC) showed that only *cis*-vaccenic acid reduced the intercellular lipid phase transition temperature (T_m). IR spectroscopy showed that *cis*-vaccenic acid was again the only treatment to increase the C—H asymmetric stretching vibration associated with the alkyl chains of the intercellular lipids. Both DSC and IR results are consistent with increased fluidity of the stratum corneum lipid domains, an alteration reflected in the enhanced *in vitro* flux of the model permeant, salicylic acid.

Alternative enhancement approaches include the use of electrical current to promote ionized drug transport (iontophoresis) and the application of ultrasound. Iontophoresis has been used medically for a considerable time,[218] but the potential of the technique in transdermal delivery has not been adequately explored. Similarly with ultrasound; only very preliminary experiments have been conducted at this time.

25.4.2.10 Conclusions

1. The early promise of transdermal delivery has been realized with the successful introduction of a number of dosage forms.

2. However, the transdermal route has unique pharmacokinetic, pharmacodynamic and formulation requirements. It necessitates, furthermore, a thorough understanding of percutaneous absorption.

3. Key unknowns which remain are prediction and circumvention of local cutaneous toxicity, the significance of skin metabolism, and the techniques and wisdom of percutaneous penetration enhancement.

ACKNOWLEDGEMENTS

Supported by N.I.H. grants GM-33395 and HD-23010 to RHG, and the University of California Toxic Substances Research and Training Program. We thank Dr. J. McKie of Pfizer Central Research for Figure 35 and Andrea Mazel for manuscript preparation.

25.4.3 REFERENCES

1. W. G. Chambliss, *Pharm. Technol.*, 1983, **7**, 124.
2. *Chem. Eng. News*, 1985, 1 April, 33.
3. *M, M&M*, 1983, Sept., 96.
4. I. N. Marks, *Drugs*, 1980, **20**, 283.
5. K. A. Connors, G. L. Amidon and V. J. Stella, 'Chemical Stability of Pharmaceuticals', 2nd edn., Wiley, New York, 1986.
6. J. W. M. VanDerMeer, J. J. Keuning, H. W. Scheijgrond, J. Heykants, J. VanCutsem and J. Brugmans, *J. Antimicrob. Chemother.*, 1980, **6**, 552.
7. J.-R. Malagelada, G. F. Longstreth, T. B. Deering, W. H. J. Summerskill and V. L. W. Go, *Gastroenterology*, 1977, **73**, 989.
8. H. W. Davenport, 'Physiology of the Digestive Tract', 5th edn., Year Book Medical, Chicago, 1982, p. 211.
9. R. G. Crounse, *J. Invest. Dermatol.*, 1961, **37**, 529.
10. W. L. Chiou and S. Riegelman, *J. Pharm. Sci.*, 1971, **60**, 1376.
11. T. Okano, K. Inui, H. Maegawa, M. Takano and R. Hori, *J. Biol. Chem.*, 1986, **261**, 14130.
12. I. Osiecka, M. Cortese, P. A. Porter, R. T. Borchardt, J. A. Fix and C. R. Gardner, *J. Pharmacol. Exp. Ther.*, 1987, **242**, 443.
13. D. Winne, *J. Pharmacokin. Biopharm.*, 1977, **5**, 53.
14. T. Nishihata, J. H. Rytting and T. Higuchi, *J. Pharm. Sci.*, 1981, **70**, 71.
15. J. G. Wagner, *Clin. Pharmacol. Ther.*, 1985, **37**, 481.
16. P. G. Welling and M. R. Dobrinska, in 'Controlled Drug Delivery: Fundamentals and Applications', 2nd edn., ed. J. R. Robinson and V. H. Lee, Dekker, New York, 1987, vol. 29, p. 253.
17. U. Bergman and B. E. Wilholm, *Eur. J. Clin. Pharmacol.*, 1981, **20**, 185.
18. G. Levy, *J. Am. Pharm. Assoc.*, 1964, **NS4**, 16.
19. A. N. Martin, J. Schwarbrick and A. Cammerata, 'Physical Pharmacy', 2nd edn., Lea & Febiger, Philadelphia, 1969, p. 353.
20. R. L. Dillard, H. Eastman and J. S. Fordtran, *Gastroenterology*, 1965, **49**, 58.
21. V. A. John, P. A. Shotton, J. Moppert and W. Theobold, *Br. J. Clin. Pharmacol.*, 1985, **19** (Suppl. 2), 203S.
22. T. Higuchi, *J. Pharm. Sci.*, 1961, **50**, 874.
23. R. W. Baker, in 'Controlled Release of Biologically Active Agents', Wiley, New York, 1987, p. 3.
24. R. W. Baker and H. K. Lonsdale, in 'Controlled Release of Biologically Active Agents', ed. A. C. Tanquary and R. E. Lacey, Plenum Press, New York, 1974, p. 15.
25. P. L. Ritger and N. A. Peppas, *J. Controlled Release*, 1987, **5**, 23.
26. P. L. Ritger and N. A. Peppas, *J. Controlled Release*, 1987, **5**, 37.

27. H.-W. Hui, V. H. L. Lee and J. R. Robinson, in 'Controlled Drug Delivery: Fundamentals and Applications', 2nd edn., ed. J. R. Robinson and V. H. Lee, Dekker, New York, 1987, vol. 29, p. 373.
28. J. L. Ford, M. H. Rubenstein and J. E. Hogan, *Int. J. Pharm.*, 1985, **24**, 327.
29. N. M. Najib, M. Suleiman and A. Malakh, *Int. J. Pharm.*, 1986, **32**, 229.
30. M. Nakano, N. Ohmori, A. Ogata, K. Sugimoto, Y. Tobino, R. L. Iwaoku and K. Juni, *J. Pharm. Sci.*, 1983, **72**, 378.
31. W. Kortsako, *Pharmazie*, 1982, **37**, 272.
32. R. Martinez-Pacheco, J. L. Vila-Jato, C. Souto and T. Ramos, *Int. J. Pharm.*, 1986, **32**, 99.
33. N. Kohri, K.-I. Mori, K. Miyazaki and T. Arita, *J. Pharm. Sci.*, 1986, **75**, 57.
34. S. K. Bajeva and K. V. Ranga Rao, *Int. J. Pharm.*, 1986, **31**, 169.
35. P. Madan, *Pharm. Manuf.*, 1985, **36**, 41.
36. M. A. Longer and J. R. Robinson, in 'Remington's Pharmaceutical Sciences', 17th edn, ed. A. R. Gennaro, Mack, Easton, 1985, p. 1652.
37. C. G. Wilson, G. D. Parr, J. W. Kennerly, M. J. Taylor, S. S. Davis, J. G. Hardy and J. A. Rees, *Int. J. Pharm.*, 1984, **18**, 1.
38. G. VanTrappen, J. Jansens, J. Hellemans, N. Christofides and S. Bloom, *Gastrointest. Motil. Health Dis., Proc. Int. Symp. Gastrointest. Motil., 6th, 1977*, 1978, 1.
39. P. Mojaverian, R. K. Ferguson, P. H. Vlasses, M. L. Rocci, A. Oren, J. A. Fix, L. J. Caldwell and C. Gardner, *Gastroenterology*, 1985, **89**, 392.
40. W. Brener, T. R. Hendrix and P. R. McHugh, *Gastroenterology*, 1983, **85**, 76.
41. R. H. Blythe, G. M. Grass and D. R. MacDonnell, *Am. J. Pharm.*, 1959, **131**, 206.
42. A. Cortot and J. F. Colombel, *Int. J. Pharm.*, 1984, **22**, 321.
43. C. Bogentoft, I. Carlsson, G. Ekenved and A. Magnusson, *Eur. J. Clin. Pharmacol.*, 1978, **14**, 351.
44. W. D. Rhine, V. Sukhatme, D. S. T. Hsieh and R. Langer, in 'Controlled Release of Bioactive Materials', ed. R. Baker, Academic Press, New York, 1980, p. 177.
45. W.-Y. Kuu and S. H. Yalkowsky, *J. Pharm. Sci.*, 1985, **74**, 926.
46. P. deHaan and C. F. Lerk, *Int. J. Pharm.*, 1986, **31**, 15.
47. P. deHaan and C. F. Lerk, *Int. J. Pharm.*, 1986, **34**, 57.
48. P. I. Lee, *J. Pharm. Sci.*, 1984, **73**, 1344.
49. M. T. DeCrosta, N. B. Jain and E. M. Rudnic (E. R. Squibb & Sons), *US Pat.* 4575539 (1986) (*Chem. Abstr.*, 1986, **104**, 230490e).
50. A. S. Michaels, J. D. Bashwa and A. Zaffaroni (Alza Corporation), *US Pat.* 3901232 (1975).
51. R. H. Johnson and E. L. Rowe (The Upjohn Co.), *US Pat.* 3574820 (1971) (*Chem. Abstr.*, 1971, **75**, 40438d).
52. G. S. Banker (Purdue Research Foundation), *Br. Pat.* 1428426 (1976) (*Chem. Abstr.*, 1974, **81**, 68575u).
53. R. C. Mamajek and E. S. Moyer (McNeilab), *US Pat.* 4207890 (1980).
54. C. T. Mroz and K. A. Kelly, *Surg. Gynecol. Obstet.*, 1977, **145**, 369.
55. J. G. Wagner, W. Veldkamp and S. Long, *J. Am. Pharm. Assoc. Sci. Ed.*, 1958, **47**, 681.
56. S. S. Davis, J. G. Hardy, M. J. Taylor, D. R. Whalley and C. G. Wilson, *Int. J. Pharm.*, 1984, **21**, 331.
57. S. Sangekar, W. A. Vadino, I. Chaudry, A. Parr, R. Beihn and G. Digenis, *Int. J. Pharm.*, 1987, **35**, 187.
58. P. R. Sheth and J. L. Toussanian (Hoffman LaRoche Inc.), *US Pat.* 4167558 (1979).
59. H. M. Ingani, J. Timmermans and A. J. Moes, *Int. J. Pharm.*, 1987, **35**, 157.
60. S. Watanabe, Y. Ishino and K. Miyao (Eisai Co. Ltd.), *US Pat.* 3976764 (1976).
61. H. Bechegaard and A. M. Pederson (A/S Alfred Benzon), *US Pat.* 4193985 (1980).
62. J. H. Meyer, J. Dressman, A. Fink and G. Amidon, *Gastroenterology*, 1985, **89**, 805.
63. J. H. Meyer, J. Elashoff, V. Porter-Fink, J. Dressman and G. L. Amidon, *Gastroenterology*, 1988, **94**, 1315.
64. M. W. Rosenthal and H. A. Cohen (Block Drug Co.), *US Pat.* 2978812 (1961).
65. G. N. Cyr and H. B. Bernstein (Olin Mathieson Chemical Corporation), *US Pat.* 3029188 (1962).
66. I. W. Scopp and R. A. Heiser, *J. Biomed. Mater. Res.*, 1967, **1**, 371.
67. J. Schor and A. Nigalaye (Forrest Labs), *US Pat.* 4389393 (1983) (*Chem. Abstr.*, 1983, **99**, 76907r).
68. M. Ishida, N. Nambu and T. Nagai, *Chem. Pharm. Bull.*, 1983, **31**, 1010.
69. K.-D. Bremecker, H. Strempel and G. Klein, *J. Pharm. Sci.*, 1984, **73**, 548.
70. N. A. Peppas and P. A. Buri, in 'Advances in Drug Delivery Systems', ed. J. M. Anderson and S. W. Kim, Elsevier, New York, 1986, p. 257.
71. H. S. Ch'ng, H. Park, P. Kelly and J. R. Robinson, *J. Pharm. Sci.*, 1985, **74**, 399.
72. M. A. Longer, H. S. Ch'ng and J. R. Robinson, *J. Pharm. Sci.*, 1985, **74**, 406.
73. J. R. Robinson (BioMimetics Inc.), *US Pat.* 4615697 (1986).
74. J. Cohen and J. L. Lach, *J. Pharm. Sci.*, 1963, **52**, 132.
75. T. Higuchi and K. A. Connors, *Adv. Anal. Chem. Instrum.*, 1965, **4**, 117.
76. 'Proceedings of the First International Symposium on Cyclodextrins', ed. J. Szejtli, Reidel, Dordrecht, 1981.
77. B. W. Muller and U. Braun, *Int. J. Pharm.*, 1985, **26**, 77.
78. T. Tokumura, M. Nanba, Y. Tsushima, K. Tatsuishi, M. Kayano, Y. Machida and T. Nagai, *J. Pharm. Sci.*, 1986, **75**, 391.
79. D. D. Chow and A. H. Karara, *Int. J. Pharm.*, 1986, **28**, 95.
80. K. Uekama, S. Narisawa, F. Hirayama and M. Otagiri, *Int. J. Pharm.*, 1983, **16**, 327.
81. K. Uekama, T. Fujinaga, F. Hirayama, M. Otagiri, M. Yamasaki, H. Seo, T. Hashimoto and M. Tsursoka, *J. Pharm. Sci.*, 1983, **72**, 1338.
82. M. Kikuchi, F. Hirayama and K. Uekama, *Int. J. Pharm.*, 1987, **38**, 191.
83. W. B. Youmans, 'Human Physiology', Macmillan, New York, 1962.
84. W. N. A. Charman and V. J. Stella, *Int. J. Pharm.*, 1986, **33**, 165.
85. W. N. A. Charman and V. J. Stella, *Int. J. Pharm.*, 1986, **34**, 175.
86. K. J. Palin, A. K. Phillips and A. Ning, *Int. J. Pharm.*, 1986, **33**, 99.
87. I. Ueda, F. Shimojo and J. Kozatani, *J. Pharm. Sci.*, 1983, **72**, 454.
88. K. Higaki, I. Kishimoto, H. Komatsu, M. Hashida and H. Sezaki, *Int. J. Pharm.*, 1987, **36**, 131.
89. M. Murakami, K. Tkada and S. Muranishi, *Int. J. Pharm.*, 1986, **31**, 231.
90. H. Fukui, M. Murakami, K. Takada and S. Muranishi, *Int. J. Pharm.*, 1986, **31**, 239.

91. H. Kajii, T. Horie, M. Hayashi and S. Awazu, *Int. J. Pharm.*, 1986, **33**, 253.
92. E. Miyamoto, A. Tsuji and T. Yamana, *J. Pharm. Sci.*, 1983, **72**, 651.
93. W. G. Gowan, F. Tio, K. Lamp, C.•Gowan and S. Stavchansky, *Int. J. Pharm.*, 1986, **29**, 169.
94. C.-M. Chiang and N. D. Weiner, *Int. J. Pharm.*, 1987, **40**, 143.
95. R. N. Rowland and J. F. Woodley, *Biochim. Biophys. Acta*, 1980, **620**, 400.
96. D. L. Schwinke, M. G. Ganesan and N. D. Weiner, *Int. J. Pharm.*, 1984, **20**, 119.
97. C.-M. Chiang and N. D. Weiner, *Int. J. Pharm.*, 1987, **37**, 75.
98. R. Baker, in 'Controlled Release of Biologically Active Agents', ed. R. Baker, Wiley, New York, 1987, p. 39.
99. B. C. Lippold and H. Forster, *Pharm. Ind.*, 1982, **44**, 735.
100. G. Kallstrand and B. Ekman, *J. Pharm. Sci.*, 1983, **72**, 772.
101. I. Ghebre-Sellasie, R. H. Gordon, D. L. Middleton, R. U. Nesbitt and M. B. Fawzi, *Int. J. Pharm.*, 1986, **31**, 43.
102. A. M. Mehta and D. M. Jones, *Pharm. Technol.*, 1985, **9**, 52.
103. M. A. Gonzalez and A. L. Golub, *Drug Dev. Ind. Pharm.*, 1983, **9**, 1379.
104. P. L. Madan, *Pharm. Technol.*, 1978, **2**, 68.
105. S.-Y. Lin and J.-C. Yang, *J. Pharm. Sci.*, 1987, **76**, 219.
106. H. A. Mahmoud and M. S. El-Samaligy, *Int. J. Pharm.*, 1985, **25**, 121.
107. Battelle Development Corporation and Mitsubishi Corporation, *US Pat.* 4 532 123 (1986) (*Chem. Abstr.*, 1984, **100**, 126 886v).
108. J. Heller, R. W. Baker, R. M. Gale and J. O. Rodin, *J. Appl. Polym. Sci.*, 1978, **22**, 1991.
109. E. Shek, *Pharm. Ind.*, 1978, **40**, 981.
110. S. S. Ozturk, B. O. Palsson, B. Donohoe and J. B. Dressman, *Pharm. Res.*, 1989, in press.
111. J. B. Dressman, *Pharm. Res.*, 1986, **3**, 123.
112. C. F. Code and J. A. Marlett, *J. Physiol.*, 1975, **246**, 289.
113. J.-P. Duchesne and L. Delattre, *Int. J. Pharm.*, 1987, **34**, 259.
114. S. K. Dutta, R. M. Russell and F. L. Iber, *Dig. Dis. Sci.*, 1979, **24**, 529.
115. C. A. Youngberg, R. R. Berardi, W. F. Howatt, M. L. Hyneck, J. H. Meyer, G. L. Amidon and J. B. Dressman, *Dig. Dis. Sci.*, 1986, **32**, 472.
116. H. L. Segal, L. L. Miller and J. J. Morton, *Proc. 5th North Am. Gl. Cancer Conf.*, p. 1079.
117. P. Christiansen, *Scand. J. Gastroenterol.*, 1968, **3**, 497.
118. M. J. Dew, P. J. Hughes, M. G. Lee, B. K. Evans and J. Rhodes, *Br. J. Clin. Pharmacol.*, 1982, **14**, 405.
119. K. Nishimura, K. Sasahara, M. Arai, T. Nitanai, Y. Ikegami, T. Morioka and E. Nakajima, *J. Pharm. Sci.*, 1984, **73**, 942.
120. L. P. Amsel, O. N. Hinsvark, K. Rotenberg and J. L. Sheumaker, *Pharm. Technol.*, 1984, **8**, 29.
121. T. Higuchi and F. Theeuwes (Alza Corporation), *US Pat.* 3 845 770 (1973).
122. F. Theeuwes, *J. Pharm. Sci.*, 1975, **64**, 1987.
123. B. Eckenhoff, F. Theeuwes and J. Urquhart, *Pharm. Technol.*, 1981, **5**, 34.
124. F. Theeuwes, D. Swanson, P. Wong, P. Bonson, V. Place, K. Heimlich and K. C. Kwan, *J. Pharm. Sci.*, 1983, **72**, 253.
125. J. D. Rogers, R. B. Lee, P. R. Souder, R. K. Ferguson, R. O. Davies, F. Theeuwes and K. C. Kwan, *Int. J. Pharm.*, 1983, **16**, 191.
126. I. D. Bradbook, V. A. John, P. J. Morrison, H. J. Rogers and R. G. Spector, *Br. J. Clin. Pharmacol.*, 1985, **19**, 163s.
127. P. N. Bennett, J. Bennett, I. Bradbook, J. Francis, V. A. John, H. Rogers, P. Turner and S. J. Warrington, *Br. J. Clin. Pharmacol.*, 1985, **19**, 171s.
128. B. Eckenhoff, F. Theeuwes and J. Urquhart, *Pharm., Technol.*, 1987, **11**, 102.
129. C. Bindschaedler, R. Gurny and E. Doeller, *J. Controlled Release*, 1986, **4**, 203.
130. G. A. Zentner, G. S. Rork and K. J. Himmelstein, *J. Controlled Release*, 1985, **1**, 269.
131. A. D. Keith (Key Pharmaceuticals), *US Pat.* 4 428 926 (1985) (*Chem. Abstr.*, 1985, **101**, 43 579s).
132. T. Higuchi, G. L. Amidon and J. B. Dressman (Merck & Co.), *US Pat. Appl.* 697 105 (1985) (*Chem. Abstr.*, 1986, **104**, 197 190c).
133. S. S. Davis, J. G. Hardy, M. J. Taylor, D. R. Whalley and C. G. Wilson, *Int. J. Pharm.*, 1984, **21**, 331.
134. J. G. Hardy, D. F. Evans, I. Zaki, A. G. Clark, H. H. Tonnesen and O. N. Gamst, *Int. J. Pharm.*, 1987, **37**, 245.
135. S. Sangekar, W. A. Vadino, I. Chaudry, A. Parr, R. Beihn and G. Digenis, *Int. J. Pharm.*, 1987, **35**, 187.
136. J. P. Skelly, *Pharm. Int.*, 1986, **7**, 280.
137. P. H. Scott, E. Tabachnik, S. McLeod, J. Carreia, C. Newth and H. Levison, *J. Pediatr. (St. Louis)*, 1981, **99**, 476.
138. W. D. Mason and N. Winer, *J. Pharm. Sci.*, 1983, **72**, 819.
139. L. Hendeles, M. Weinberger, G. Melvitz, M. Hill and L. Vaugh, *Chest*, 1985, **87**, 758.
140. J. H. G. Jonkman, W. J. V. vanderBoon, L. P. Balant and J. Y. LeCotonnec, *Int. J. Pharm.*, 1985, **25**, 113.
141. M. T. Wecker, D. A. Graves, L. P. Amsel, O. N. Hinsvark and K. S. Rotenberg, *J. Pharm. Sci.*, 1987, **76**, 29.
142. P. K. Matura, V. K. Prasad, W. N. Worsley, G. K. Shiu and J. P. Skelly, *J. Pharm. Sci.*, 1986, **75**, 1205.
143. J. P. Skelly, W. H. Barr, L. Z. Benet, J. T. DoLuisio, A. H. Goldberg, G. Levy, F. D. T. Lowenthal, J. R. Robinson, V. P. Shah, R. J. Temple and A. Yacobi, *Pharm. Res.*, 1987, **4**, 75.
144. R. H. Guy and J. Hadgraft, *J. Controlled Release*, 1987, **4**, 237.
145. B. W. Barry, 'Dermatological Formulations: Percutaneous Absorption', Dekker, New York, 1983, pp. 33, 160.
146. R. J. Scheuplein and I. H. Blank, *Physiol. Rev.*, 1971, **51**, 702.
147. F. N. Marzulli, *J. Invest. Dermatol.*, 1962, **39**, 387.
148. Y. W. Chien, 'Transdermal Controlled Systemic Medications', Dekker, New York, 1987, p. 6.
149. V. M. Knepp, J. Hadgraft and R. H. Guy, *CRC Crit. Rev. Ther. Drug Carrier Sys.*, 1987, **4**, 13.
150. R. H. Guy and J. Hadgraft, *Pharm. Res.*, 1988, **5**, 753.
151. S. P. Denyer, R. H. Guy and J. Hadgraft, *Int. J. Pharm.*, 1985, **26**, 89.
152. D. A. W. Bucks, *Pharm. Res.*, 1984, **4**, 148.
153. R. J. Martin, S. P. Denyer and J. Hadgraft, *Int. J. Pharm.*, 1987, **39**, 23.
154. M. C. Houston, *South. Med. J.*, 1982, **75**, 713.
155. C. Thanaopavarn, M. S. Golub, P. Eggena, J. D. Barrett and M. P. Sambh, *Am. J. Cardiol.*, 1982, **49**, 153.
156. S. K. Chandrasekaran, *Drug Dev. Ind. Pharm.*, 1983, **67**, 1370.
157. J. E. Shaw, *Am. Heart J.*, 1984, **108**, 217.

158. T. R. MacGregor, K. M. Matzek, J. J. Keirns, R. G. A. van Wayjen, A. van den Ende and R. van Tol, *Clin. Pharmacol. Ther.*, 1985, **38**, 278.
159. D. Arndts and K. Arndts, *Eur. J. Clin. Pharmacol.*, 1984, **26**, 79.
160. M. A. Weber, *Am. Heart J.*, 1986, **112**, 906.
161. S. Popli, J. T. Daugirdas, J. A. Neubauer, B. Hockrnberry, J. E. Hano and T. S. Ing, *Arch. Int. Med.*, 1986, **146**, 2140.
162. J. F. Burris and W. J. Mroczek, *Pharmacotherapy*, 1986, **6**, 30.
163. H. Groth, P. Greminger, H. Vetter, J. Knussel and P. Baumgart, *Schweiz. Rundsch. Med./Prax.*, 1985, **74**, 10.
164. D. T. Lowenthal, S. Saris, E. Paran, N. Cristal, K. Sharif, C. Bies and T. Fagan, *Am. Heart J.*, 1986, **112**, 893.
165. J. Hollifield, *Am. Heart J.*, 1986, **112**, 900.
166. G. S. M. Kellaway, *N. Z. Med. J.*, 1986, **99**, 711.
167. H. Groth, H. Vetter, J. Knussel and W. Vetter, *Lancet*, 1984, **2**, 850.
168. H. I. Maibach, *Contact Dermatitis*, 1987, **16**, 1.
169. E. G. M. Hurkmans, H. E. Bodde, L. M. J. Van Driel, H. Van Doorne and H. E. Junginger, *Br. J. Dermatol.*, 1985, **112**, 461.
170. H. L. Judd, R. E. Cleary, W. T. Cresman, D. C. Figge, N. Kase, Z. Rosenwaks and G. E. Tagatz, *Am. J. Obstet. Gynecol.*, 1981, **58**, 267.
171. K. C. Nichols, L. Schenkel and H. Benson, *Obstet. Gynecol. Surv.*, 1984, **39** (Suppl.), 230.
172. J. Holst, S. Cajander, K. Carlstrom, M.-G. Damber and B. von Schoultz, *Br. J. Obstet. Gynaecol.*, 1983, **90**, 355.
173. W. R. Good, M. S. Powers and L. Schenkel, *J. Controlled Release*, 1985, **2**, 89.
174. M. S. Powers, L. Schenkel, P. E. Darley, W. R. Good, J. C. Balestra and V. A. Place, *Am. J. Obstet. Gynecol.*, 1985, **152**, 1099.
175. R. J. Chetkowski, D. R. Meldrum, K. A. Steingold, D. Randle, J. K. Lu, P. Eggena, J. Hershman, N. Alkjaersig, A. Fletcher and H. Judd, *New Engl. J. Med.*, 1986, **314**, 1615.
176. L. R. Laufer, J. L. DeFazio, J. K. H. Lu, D. R. Meldrum, P. Eggena, M. P . Sambhi, J. M. Hershman and H. L. Judd, *Am. J. Obstet. Gynecol.*, 1983, **146**, 533.
177. V. A. Place, M. S. Powers, P. E. Darley, L. Schenkel and W. R. Good, *Am. J. Obstet. Gynecol.*, 1985, **152**, 1092.
178. M. I. Whitehead, M. L. Padwick, M. B. Endacott and J. Pryse-Davies, *Am. J. Obstet. Gynecol.*, 1985, **152**, 1079.
179. P. W. Armstrong, J. A. Armstrong and G. S. Marks, *Circulation*, 1979, **59**, 585.
180. P. K. Noonan and L. Z. Benet, *J. Pharm. Sci.*, 1986, **75**, 241.
181. M. E. Davidov, *Angiology*, 1981, **32**, 16.
182. W. R. Good, *Drug Dev. Ind. Pharm.*, 1983, **9**, 647.
183. P. K. Noonan, M. A. Gonzalez, D. Ruggirello, J. Tomlinson, E. Babcock-Atkinson, M. Ray, A. Golub and A. Cohen, *J. Pharm. Sci.*, 1986, **75**, 688.
184. M. Wolff, G. Cordes and V. Luckow, *Pharm. Res.*, 1985, **2**, 23.
185. A. Karim, *Angiology*, 1983, **34**, 11.
186. R. Heidmann, F. Menke, H. Letzel and N. Rietbrock, *Dtsch. Med. Wochenschr.*, 1985, **110**, 1568.
187. R. G. Fischer and M. Tyler, *South. Med. J.*, 1985, **78**, 1523.
188. N. Reichek, R. E. Goldstein, D. R. Redwood and S. E. Epstein, *Circulation*, 1974, **50**, 348.
189. J. Abrams, *Ann. Intern. Med.*, 1986, **104**, 424.
190. J. O. Parker and H.-L. Fung, *Am. J. Cardiol.*, 1984, **54**, 471.
191. G. Reiniger, G. Menke, A. Boertz, F. Kraus and W. Rudolph, *Herz*, 1987, **12**, 68.
192. *Pharm. J.*, 1987, **239**, 586.
193. G. Muiesan, E. Agabiti-Rosei, L. Muiesan, G. Romanelli, P. Pollavini, C. Pasotti, G. Fiori, L. Muratori, A. Zuarini, C. Pastorini, S. Borziani, L. Bozzi and S. Marchetti, *Am. Heart J.*, 1986, **112**, 233.
194. J. R. Weber, *Am. Heart J.*, 1986, **12**, 238.
195. P. K. Noonan and L. Z. Benet, *Int. J. Pharm.*, 1982, **12**, 331.
196. S. Curry and S. M. Aburawi, *Biopharm. Drug Dispos.*, 1985, **6**, 235.
197. E. Nakashima, P. K. Noonan and L. Z. Benet, *J. Pharmacokin. Biopharm.*, 1987, **15**, 423.
198. C. Muir and R. Metcalfe, *J. Pharm. Biomed. Anal.*, 1983, **1**, 363.
199. A. Graybiel, D. B. Cramer and C. D. Wood, *Aviat. Space Environ. Med.*, 1981, **52**, 337.
200. J. I. Homick, R. L. Kohl, M. F. Reschke, J. Degionanni and N. M. Cintron-Trevino, *Aviat. Space Environ. Med.*, 1983, **54**, 994.
201. N. M. Price, L. G. Schmitt, J. McGuire, J. E. Shaw and G. Trobough, *Clin. Pharmacol. Ther.*, 1981, **29**, 414.
202. W. F. von Marion, M. C. M. Bongaerts, J. C. Christiaanse, H. G. Hofkamp and W. van Ouwerkerk, *Clin. Pharmacol. Ther.*, 1985, **38**, 301.
203. J. K. Aronson and J. W. Sear, *Anaesthesia*, 1986, **41**, 1.
204. J. Uppington, J. Dunnet and C. E. Blogg, *Anaesthesia*, 1986, **41**, 16.
205. A. Graybiel, D. B. Cramer and C. D. Wood, *Aviat. Space Environ. Med.*, 1982, **53**, 770.
206. J. G. Cairncross, *Ann. Neurol.*, 1983, **13**, 582.
207. P. A. Gibbons, S. C. Nicolson, E. K. Betts and K. R. Rosenberry, *Anesthesiology*, 1984, **57**, A330.
208. A. J. Dietz, J. D. Carlson and C. I. Beck, *Ann. Allergy*, 1986, **57**, 38.
209. P. H. Vlasses, L. G. T. Ribeiro, H. H. Rotmensch, J. V. Bondi, A. E. Loper, M. Hichens, M. C. Dunlay and R. Ferguson, *J. Cardiovasc. Pharmacol.*, 1985, **7**, 245.
210. A. Wellstein, H. Kuppers, H. F. Pitschner and D. Palm, *Eur. J. Clin. Pharmacol.*, 1986, **31**, 419.
211. R. A. Caplan, L. B. Reasy, G. L. Olsson and M. L. Nessly, *Anesthesiology*, 1986, **65**, A196.
212. W. S. Nimmo and D. J. R. Duthie, *Anesthesiology*, 1986, **65**, A559.
213. J. E. Rose, J. E. Herskovic, Y. Trilling and M. E. Jarvik, *Clin. Pharmacol. Ther.*, 1985, **38**, 450.
214. J. C. Findlay, V. A. Place and P. J. Synder, *J. Clin. Endocrinol. Metab.*, 1987, **64**, 266.
215. R. H. Guy and J. Hadgraft, *J. Controlled Release*, 1987, **5**, 43.
216. G. M. Golden, J. E. McKie and R. O. Potts, *J. Pharm. Sci.*, 1987, **76**, 25.
217. G. M. Golden, D. B. Guzek, A. H. Kennedy, J. E. McKie and R. O. Potts, *Biochemistry*, 1987, **26**, 2382.
218. P. Tyle, *Pharm. Res.*, 1986, **3**, 318.

25.5

Drug Targeting

DAAN J. A. CROMMELIN and GERT STORM

University of Utrecht, The Netherlands

25.5.1   INTRODUCTION

Our knowledge about the pathophysiology of diseases at the cellular and molecular level has grown steadily during the last decades. Scientists with different backgrounds such as cell biologists, pathologists, pharmacochemists, pharmacologists and pharmacists have been involved in the translation of these basic insights into new, improved, therapeutically active agents at the cellular level. Unfortunately, only in rare cases can the drug molecules be introduced into the body directly

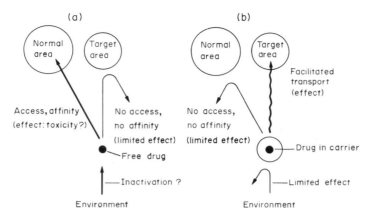

Figure 1 Pitfalls in the use of free drugs, such as their distribution over nontarget tissue followed by elimination, their metabolic inactivation and their inability to gain access to the internal cell compartments (a) could be circumvented by employing a carrier that would prevent the transported drug from reaching nontarget tissue, protect against metabolic inactivation and facilitate transport into the target cells (b)[268]

at the site of action. Usually the drug must be transported from the site of administration to the target site or sites. During this transport stage a number of unwanted effects can occur (Figure 1). The drug is distributed over the body and therefore 'diluted' and can ultimately be eliminated without reaching its target site. Metabolic degradation can add to this 'dilution' effect. Apart from the low fraction of drug reaching the target site, unwanted side effects can occur in the nontarget tissues. In addition, even if the drug reaches the target cells, then it is still an open question whether it will exert its pharmacological effect, because access to the internal cell compartments might be blocked.

For drugs with minor side effects these biopharmaceutical and pharmacokinetic inadequacies can be acceptable under normal therapeutic conditions. However, in the case of drugs with a small therapeutic index, unacceptable side effects have to be suppressed. This situation is encountered with life threatening diseases like cancer or a number of viral, microbial or parasitic infections. Drug targeting is an approach to improve the therapeutic index of drugs by manipulating the disposition of the drug in the body. To reach this goal the drug has to be protected against premature degradation, directed to the target site through the body and finally gain access into the cell interior if necessary. In this chapter progress made in the area of drug targeting will be discussed. A selection of the most promising approaches has been made, as coverage of all efforts in this area is far beyond the scope of this review.

To achieve an alteration of the natural drug disposition, and/or provide protection against deactivation, a drug carrier is used; to achieve concentration of the drug at the site of action a homing device is needed. Therefore, in general, the following functional units can be discerned in a conjugate devised for drug targeting: a drug part (to exert the therapeutic effect), a carrier part and a homing device. The homing device is the unit in the conjugate which is responsible for the selection of the target site. In special cases two, or even all three, functions can be assigned to one basic molecular structure (*cf.* Section 25.5.4.1).

25.5.2 PHYSIOLOGICAL AND ANATOMICAL CONSIDERATIONS

Figure 2 shows the pathways for site specific delivery. The success of drug targeting strongly depends on the potential of the drug delivery system to pass through the different barriers that are encountered after administration. Short cuts are possible, and by local administration high concentrations of drug can be delivered near the site of action. Examples are drugs which are administered intraocularly, intravaginally, rectally, intrabronchially, intranasally, on the skin, intraperitoneally, intraarticularly or intrathecally. Local administration of drugs, however, does not provide specificity at the cellular level. For locally delivered drugs a relatively low level of sophistication for the transport system is required to achieve a substantial improvement of the therapeutic index, as compared to systemic administration.

Unfortunately, in many cases the target tissue cannot be reached *via* local administration and delivery *via* the systemic route is necessary; then the drug delivery system faces the physiological and

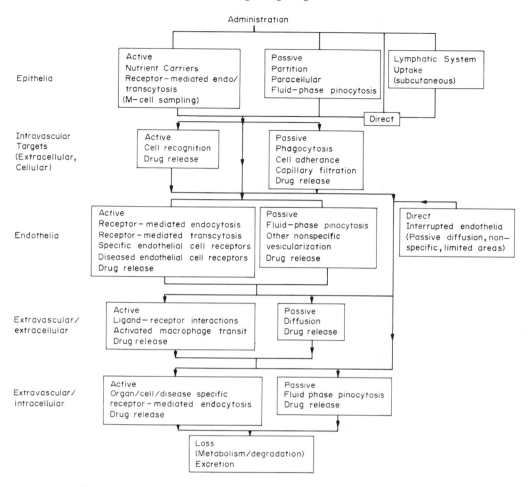

Figure 2 Anatomical and physiological pathways for site specific drug delivery[19]

anatomical barriers which have to be circumvented in order for the drug to reach the target cells. The penetration of drug laden carriers through the epithelial or endothelial barrier can be driven by an active process, or occur by passive diffusion, either through cells or through intercellular pores. A schematic representation of the different capillary wall structures is presented in Figure 3. The exact mechanism of penetration of macromolecules through the endothelial cell layer has not been fully

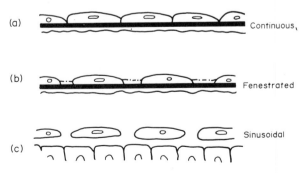

Figure 3 Schematic illustration of the structure of different classes of blood capillaries. (a) Continuous capillary. The endothelium is continuous, with tight junctions between adjacent endothelial cells. The subendothelial basement membrane is also continuous. (b) Fenestrated capillary. The endothelium exhibits a series of fenestrae which are sealed by a membranous diaphragm. The subendothelial basement membrane is continuous. (c) Discontinuous (sinusoidal) capillary. The overlying endothelium contains numerous gaps of varying size enabling materials in the circulation to gain access to the underlying parenchymal cells. The subendothelial basement is either absent (liver) or present as a fragmented interrupted structure (spleen, bone marrow) (reproduced from ref. 22 by permission of Plenum Press)

elucidated. There is experimental evidence that macromolecules ride a system of vesicles which engulf fluid and solutes on the luminal side and release their contents on the tissue side. This transport process with endosomal vesicles is called transcytosis.[1-3] The exact characteristics of the transcytosis process in terms of selectivity and capacity are still under investigation.

Basement membranes are composed of protein (*e.g.* collagen) and carbohydrate moieties. The basement membrane can act as a barrier for macromolecular transport; basement membrane resistance is related to molecular size, the charge on the molecule and specific membrane–macromolecular interactions.[1,4] The dimensions of different types of pore in the endothelial wall ('small pore' system, 'large pore' system and endothelial tight junctions) are still under investigation. However, it is clear that particulate materials with dimensions over 30 nm cannot pass through the endothelial wall except through the endothelial gaps in sinusoidal capillaries (up to about 100 nm). In the diseased state (*e.g.* in tumor tissue or inflammation tissue) the permeability of the walls in the microcirculation might be increased because of structural defects. Gaps of several hundred nanometers between endothelial cells are found under those condtions.[5] The impact of these defects in the barrier on the penetration of particulate materials is presumably limited.[6]

The penetration of drugs through the epithelial wall lining the gastrointestinal (GI) tract has been extensively studied.[7-9] Lipophilicity and molecular weight are critical parameters in the absorption process. The absorbed fraction of high molecular weight proteins such as insulin is small and variable.[10] A controversial point is the uptake of particulate materials through the wall of the GI tract. Particulate materials such as liposomes and microspheres were reported to be taken up,[11] but under normal conditions the fraction of the dose which reaches the blood is very small. The biological consequences, however, may be far reaching if the particles are immunogenic, carcinogenic, mutagenic or toxic.[12] The mechanism by which particulate materials are taken up has still not been unravelled; endocytosis and paracellular transport have been suggested.[8,12,13]

The blood–brain barrier is different from the barrier formed by systemic capillaries: (i) endothelial cells in cerebral capillaries are very tightly joined, therefore intercellular transport is severely restricted; (ii) as a rule, cerebral capillaries lack fenestrations; and (iii) there is hardly any vesicular transport within the endothelial cells.[14] It is clear that for an unhindered penetration of drug delivery systems through membrane barriers the dimensions will play a decisive role. Particulate systems have only a remote possibility of passing from one compartment to another one and this limits their potential as a carrier system.

The optimization of the therapeutic index of drugs *via* the drug targeting approach requires an insight into the pharmacokinetics of the drug and the drug–carrier–homing device. Surprisingly, only a few reports are available which address this subject *via* mathematical modelling.[15-18] Hunt *et al.*[17] defined two parameters to evaluate systems developed for drug targeting: the therapeutic availability and the drug targeting index. Their kinetic model allows one to distinguish, on the basis of the total body clearance of the drug and information on the target site anatomy and blood flow, those drugs that can benefit from site specific delivery, from those that will not.

Tissue and cell specific marker molecules are exposed on the membranes of a number of cells under normal conditions. Table 1 presents examples of cells with their receptors.[19] Membrane bound receptor molecules often consist of one or two glycoprotein structures; they have a (small) part of their structure exposed to the cytoplasm. The external ligand binding domains are usually glycosylated and contain intramolecular disulfide bridges[20] (see Section 25.5.4.2). The ligands interact with the receptors and are taken up into the cell in the form of vesicular structures called

Table 1 Reported Distribution of Some Endosomotropic Receptors (Various Species)

Cell	Receptor for
Hepatocytes	Galactose, low density lipoprotein (LDL), polymeric IgA
Macrophages	Galactose (particles), mannose–fucose, acetylated LDL, α_2-macroglobulin protease complex (AMPC)
Leucocytes	Chemotactic peptide, complement C3b, IgA
Basophils, mast cells	IgE
Cardiac, lung, diaphragm endothelia	Albumin
Fibroblasts	Transferrin, epidermal growth factor, LDL mannose 6-phosphate, transcobalamin II, AMPC, mannose
Mammary acinar	Growth factors
Enterocytes	Maternal IgG, dimeric IgA, transcobalamin-B-12/intrinsic factor
Blood/brain endothelia	Transferrin, insulin

endosomes. Then a number of possibilities exists: (i) the ligand–receptor complex dissociates and the ligand leaves the endosome for the cytoplasm, (ii) the ligand–receptor complex is transcellularly transferred, or (iii) the endosomes can fuse with lysosomes. More detailed information about the way ligand–receptor complexes are taken up and processed by cells can be found in various reviews.[1,2,20] Apart from the different receptors which are present on normal cells, specific structures can occur on diseased cells. Tumor cell associated marker molecules (*e.g.* glycoproteins or glycolipids) can be present on tumor cells in a higher density than in normal cells. Infection induced surface marker molecules might be expressed by cells infected by viruses or parasites. Antibodies against these structures can be obtained and these antibodies can act as a homing device for drug–carrier combinations. This approach has raised great hopes for the development of more effective chemotherapeutical tools for the treatment of cancer or infectious diseases. There are several problems that might limit the efficiency of the antibody–drug conjugates to reach diseased cells, *e.g.* tumor cells, without making contact with other cells.[22,23]

(i) Antigen Specificity

Only tumor cells should possess the antigen which is used as the tumor marker and a high density on the tumor cell surface is desirable. Unfortunately, the tumor cell marker molecule is usually a differentiation antigen which also occurs, although at a lower density, on certain normal cells. Oncofetal antigens like α-fetoprotein and the carcinoembryonic antigen (CEA) are a class of differentiation antigens; these molecules are normally exposed in fetal tissue and not in adult tissue.[24]

(ii) Tumor Cell Heterogeneity

A tumor contains many subpopulations of cells with different properties with respect to metastasizing capacity, sensitivity for cytostatics and, last but not least, antigens expressed. Part of the tumor heterogeneity can be ascribed to cell cycle variation, while mutation and selection can cause changes in the antigen exposition pattern. An approach to circumvent tumor heterogeneity is the use of 'cocktails' of different antibodies coupled (*via* a carrier) to different cytostatics.

(iii) Extracellular Free Antigen

Tumor cells can shed their surface antigens into the extracellular space. This means that circulating drug–carrier–antibody combinations might prematurely interact with these shedded antigens and be neutralized, thus preventing a selective interaction with tumor tissue. On the other hand, antigen shedding can be used for diagnostic purposes as high blood concentrations of antigen increase the sensitivity and therefore the reliability of the diagnostic test. Apart from receptor–ligand interactions, physical principles have been used to achieve accumulation of a drug at the target site: a magnetic field to concentrate magnetite laden particulate systems,[25] local heat treatment[6] and size (see below).

The size of particulate carrier systems is a critical parameter in determining the fate of a drug–carrier–homing device combination after intravenous injection. As mentioned above, only small sized systems can escape through the endothelial wall. If the particles are larger than around 5 μm then they will be entrapped in the first capillary system which they encounter.[27] Intraarterial injection of drug laden particles with diameters above this size upstream of an organ or a tumor causes entrapment in, or even blockade of, the capillaries in this target tissue. Then the opportunity is created for the encapsulated drug to be released close to the target cells. Intravenous injection provides a means of entrapment in and eventually blockade of capillaries in the lung. Carrier systems which are too large to escape through the endothelial wall and too small to be entrapped in the capillaries can either be disrupted in the blood circulation or become phagocytosed by mononuclear phagocytes: cells of the reticuloendothelial system (RES).

Mononuclear phagocytes originate from precursor cells in the bone marrow, which circulate as monocytes in the blood, distribute over the whole body and differentiate into macrophages. Circulating carrier systems can only contact macrophages when they pass through organs where macrophages are located in the blood compartment: the liver, spleen and bone marrow.[1,28] Macrophages play an important role in the host defense system. Based on tissue origin, different populations of macrophages can be distinguished (*e.g.* Kupffer cells in the liver, alveolar macrophages in the lung). Macrophages phagocytose pathogens, cellular debris and other materials (*e.g.* 'old' blood cells). In addition, they can cooperate with lymphocytes in the development of the immune response, *e.g.* by presenting antigenic material to lymphocytes. Lymphokines or microorganisms and their products can activate macrophages to a tumoricidal, antiviral or microbicidal state. Macrophages carry several types of receptors: a receptor for interaction with the Fc part of IgG molecules, receptors for two complement components, for mannose–fucose, for galactose,[29] for acetylated LDL,[19] and for the α_2-macroglobulin protease complex. The uptake of drug–carrier–homing device combinations depends on parameters like size, charge and surface properties (*e.g.* hydrophobicity, shape, or rigidity).[31] The rate of uptake can be increased by adhering proper sugar groups, improving the conditions for opsonization (*e.g.* a high negative charge), or coupling intact IgG or fibronectin to the surface.[1,3,28]

Several approaches have been developed to suppress the avidity of the macrophages to phagocytose carrier systems. RES function can be suppressed by predosing with colloids or macromolecules (*e.g.* dextran sulfate); the macrophages are saturated and lose (temporarily) their capacity for phagocytosis. In the clinical situation a suppression of the RES in patients, who are usually already extremely sensitive to infections, is an unrealistic option. More elegant are the attempts made to avoid uptake by the RES by manipulation of the surface characteristics. A hydrophilic surface (to reduce opsonization) with a steric repulsive barrier to minimize carrier–cell adhesion dramatically increased the circulation time of polystyrene microspheres.[30]

In many studies where targeted drug conjugates proved to have a higher therapeutic index than the 'free' drug, it has not been made clear by what underlying mechanism this positive result was obtained. Was it indeed site specific drug delivery? Or, was it site avoidance drug delivery? In the case of site avoidance drug delivery, conjugate formation changes the pharmacokinetic pattern of the involved drug and reduces active drug levels in organs where the unconjugated drug would exert its dose limiting toxic effects. In the section on liposomes (Section 25.5.6.1) an example of site avoidance drug delivery will be discussed (doxorubicin liposomes). Another question that has to be addressed is whether the homing device (*e.g.* monoclonal antibody; MAB) and the drug administered in an unconjugated form are already acting synergistically to produce a certain therapeutic effect? And, finally, is drug conjugation changing the mechanism of action of the drug involved?[1,32]

25.5.3 CARRIER SYSTEMS

Numerous carrier systems have been proposed for drug targeting purposes. The drug–carrier–homing device combinations should meet certain demands as far as pharmacological and pharmaceutical properties are concerned. The pharmacological requirements are: (i) the drug should be released in the target cells or in the target tissue and preferably stay there long enough to exert its action or induce the desired pharmacological effect over a prolonged period of time; (ii) the drug should not be released at sites other than the target tissue; and (iii) the drug should leave the target tissue in an inactive, nontoxic form. The delivery system as such and/or its breakdown products should be eliminated from the body; these entities should also be nontoxic and nonimmunogenic. The pharmaceutical requirements are: (i) it should be possible to prepare the delivery system reproducibly, on a large scale, the production process should meet the GMP (good manufacturing practice) rules and the product should be well characterized chemically and physically; (ii) the product should be sterile and pyrogen free in the case of parenteral administration; (iii) it should be easy to manipulate and administer by the patient or hospital staff; (iv) the payload of the delivery system should be high; and (v) the system should be stable on storage. Apart from the pharmacological and pharmaceutical requirements, economical considerations also play a role.

Table 2 Carrier Systems Classified on the Basis of their Physical Appearance

Macromolecular carriers	Antibodies, hormones, lectins, DNA, polymeric carbohydrates, glycoproteins
Cellular carriers	Erythrocytes, leucocytes, hepatocytes, fibroblasts, neutrophils
Particulate carriers	Liposomes, polymeric systems based on albumin, acrylates, starch, polylactic acid, carbohydrates

Today, no carrier system can, in general, satisfactorily meet all the above defined demands. In the following sections we will discuss a number of these carrier systems systematically and indicate what are their potentials and their limitations. Carrier systems can be classified on the basis of their physical appearance (Table 2).

25.5.4 BINDING STRATEGIES

The nature of the bonds which keep the drug–carrier–homing device combination together can vary widely. Homing devices are usually covalently bound either directly to the drug or to the carrier. This bond should be designed to stay intact until delivery of the combination at the target cell surface or even in the target cell. Drugs are either physically entrapped or adsorbed to the carrier or, alternatively, covalently bound. Here again, the general rule is that drug and carrier should stay together until the combination has reached the target cell or, if possible, has entered the target cell (Figure 4). After reaching the target, the drug should be released, at least partly. A certain control over this release process is desirable. This requirement of disconnection after reaching the target tissue implies that the selection of the nature of the bond should be carefully made. Advantage might be taken of special pH, temperature and enzymatic conditions in the target area. Exceptions to this rule are radioactive markers coupled to homing devices for diagnostic or therapeutic purposes; they do not have to be released to exert their action. However, for toxicity reasons adequate elimination pathways for these products should be available. Note should be taken that some conventional drugs (*e.g.* doxorubicin) have been shown to be active in a conjugated form.[33,34]

The techniques for covalent binding of drugs, macromolecules or particulate structures (see below) and homing devices together have made considerable progress during the last 10 years. The main goal was the development of efficient binding strategies under mild conditions, without reduction of the ligand–receptor binding specificity and capacity. Besides, no reduction of pharmacological activity should occur after release of the drug from the carrier. The chance for interbridging should be minimal. The functional groups regularly used for binding purposes are –NH$_2$, –CO$_2$H and vicinal –OH and –SH. A selection of binding agents and reaction products is given in Table 3. Two procedures with bifunctional coupling agents are shown in detail in Figures 5 and 6. The stability of the reaction products *in vitro* and *in vivo* is often not known. Also toxicity data on the reaction products are seldom available. It is still a point of discussion whether a spacer molecule should be introduced to keep the homing device unit and the carrier core apart. A short spacer might reduce the distance between cell surface and the drug–carrier–homing device combination and

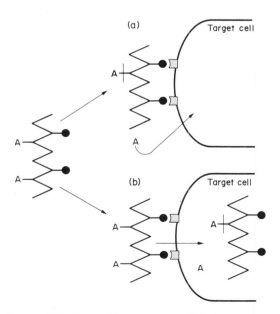

Figure 4 Release patterns of active agent (A) after reaching target tissue. (a) Release of A outside the target cell. (b) Release of A inside the target cell. Preferably the release rate is controllable

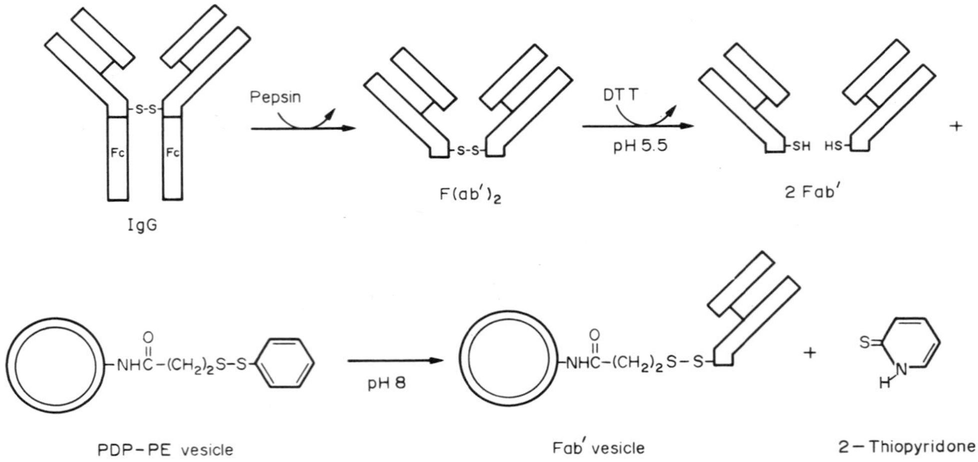

Figure 5   Covalent coupling of F(ab′) fragments to *N*-[3-(2-pyridyldithio)propionyl]-phosphatidylethanolamine (PDP-PE) vesicles. F(ab′)₂ dimers are prepared by pepsin digestion of the IgG molecules. Fab′ monomers are generated from these by reduction with dithiotreitol (DTT) at low pH. Immediately following the removal of DTT, Fab′ fragments are mixed with PDP-PE-containing vesicles and the pH is adjusted to 8.0 with NaOH. The mixture is stirred under an argon atmosphere for several hours. Coupling is the result of a disulfide exchange reaction between the thiol group on each Fab′ fragment and the pyridyldithio moiety of PDP-PE molecules present in the vesicle membranes. The chromophore, 2-thiopyridone, is released as a product of the reaction[147]

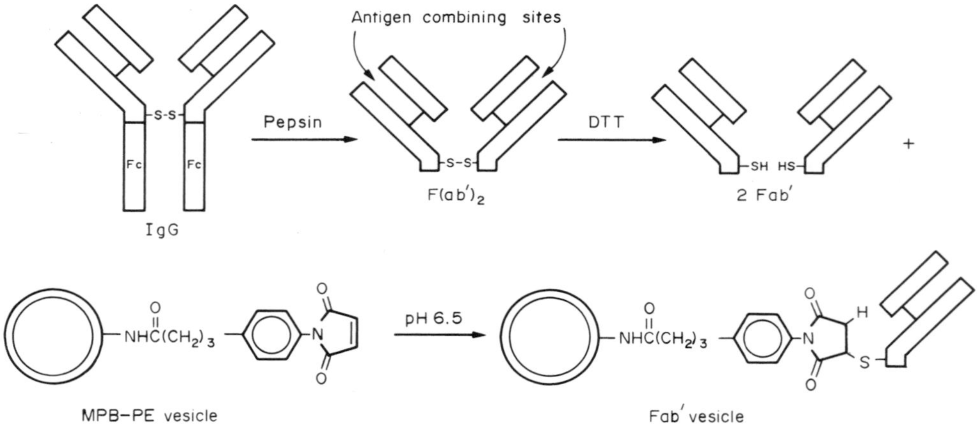

Figure 6   Covalent coupling of Fab′ fragments to *N*-[4-(*p*-maleimidophenyl)butyryl]-phosphatidylethanolamine (MPB-PE) vesicles. F(ab′)₂ dimers are prepared by pepsin digestion of IgG molecules. Fab′ monomers are generated from these by reduction with dithiothreitol (DTT) at low pH. Immediately following the removal of DTT, Fab′ fragments are mixed with MPB-PE-containing vesicles and the pH is adjusted to 6.5. Addition of the Fab′–SH to the double bond of the maleimide moiety of MPB-PE molecules present in the vesicle membranes results in a stable thioether cross-linkage[148]

therefore favor endocytic uptake. On the other hand a long spacer molecule might provide the space for optimum orientation of the homing device to interact with the receptor.[35]

25.5.5   ANTIBODIES

Antibodies can act simultaneously as carriers and as homing devices. In some cases, no extra active agent has to be attached to the antibody structure: then the antibody is also responsible for the therapeutic effect. At the end of the 1970s the small scale production of MAB on the basis of the hybridoma technique in ascites fluid from mice was introduced. Later, large scale production techniques for MAB were developed using *in vitro* cell culture techniques. MAB have several advantages over polyclonal antibodies: MAB are homogeneous, have identical affinities and antigen binding sites and can be produced on a large scale *in vitro* with minimal batch to batch variation. In

Table 3 Frequently Used Techniques for Binding Drugs or Homing Devices to Carriers[1,35,267]

Functional groups involved	Binding agent	Product
$RCO_2H + R'NH_2$	Carbodiimides	$RCONHR'^a$
$RNH_2 + R'NH_2$	Glutaraldehyde	$RN{=}CH(CH_2)_3CH{=}NR'^b$
$R(OH)_2 + R'NH_2$	Cyanogen bromide	[structure: dioxolane ring with $C{=}NR'^c$]
$RCO_2H + R'NH_2$	Mixed carbonic anhydride	$RCONHR'$
$RNH_2 + R'NH_2$	SPDP	$RNHCO(CH_2)_2SS(CH_2)_2CONHR'^d$
$RNH_2 + R'SH$	SMPB	$RNHCO(CH_2)_3$—[maleimide ring]—SR'^e

[a] Aggregate formation has been reported. [b] Changes in antibody binding specificity has been described. [c] Cyanogen bromide is very toxic and coupling reaction is highly pH dependent. [d] Bond is unstable in serum. [e] Bond is stable in serum; nonreducible sulfide bridge

contrast, polyclonal antibodies are isolated from immune serum after immunizing an animal. These polyclonal antibodies are heterogeneous in terms of affinity and antigenic determinants, and considerable batch to batch variation can occur.[36] The principle and technical details of the different production procedures are beyond the scope of this chapter; this information can be found elsewhere.[37-39]

Dependent on the subclass, MAB may not only interact with cells which expose the corresponding antigen, but they may also change the cellular function, *e.g.* kill the cell. The Fc part of the MAB then binds complement, or mediates in the binding of killer cells or macrophages *via* the MAB adhering to the target cell wall. It is also possible that MAB bind to the target cell and block essential receptor groups at the cell surface, inducing metabolic deficiencies. Finally, in animal experiments MAB were able to protect against parasitic, viral and bacterial diseases *via* passive immunization.[23,40] Information on the behavior of MAB after intravenous injection into patients has been collected in a number of clinical trials. Until now, MAB have been used in the clinic routinely only for a few therapeutic purposes. Requirements for the production and quality control of MAB intended for use in man have been formulated.[41] Upper limits for DNA levels and the degree of aggregation are defined. The product has to be free from bacterial, fungal and mycoplasmal contamination; the presence of a number of viruses for which evidence exists of a capacity to infect man or primates is not allowed. If fragments of MAB are used, then the acceptable levels of intact MAB and the enzymes used for fragmentation have to be defined. The distribution of MAB over the body after intravenous injection strongly depends on the dose: above a certain dose, saturation of tumor antigens has been described. Besides, MAB disposition is influenced by the number and strength of previous doses. MAB are commonly derived from murine cells. An antimurine IgG antibody response (AMIA) is observed frequently within 10–20 d. This response neutralizes the injected MAB and prevents binding to the surface antigens. Anaphylactic reactions are rare. Concurrent therapy may be applied to keep these side effects at an acceptable level. The immune response strongly depends on the MAB, its purity, dose regimen and the condition of the patient. The administration of high doses of pure MAB may suppress the AMIA (so called 'high zone tolerance'). The use of $F(ab')_2$ or $F(ab')$ is a way to circumvent the immune response against the Fc part. Another approach now under investigation is the design of chimeric MAB consisting of a murine variable part and a human constant part. Antiidiotypic responses still may occur.[18,23]

Usually MAB alone will fail to exert the desired therapeutic or diagnostic actions (apart from their function as homing molecules). Therefore, MAB can be conjugated with pharmacologically active agents (*e.g.* cytotoxic agents) or radioactive labels. Two types of conjugates can be discerned as shown in Figure 7. Type I systems involve coupling of the active agent directly or *via* a short spacer molecule to MAB. Examples are the immunotoxins, MAB–cytostatic and MAB–radionuclide combinations. In type II systems the MAB is bound to the carrier system. The drug can either be

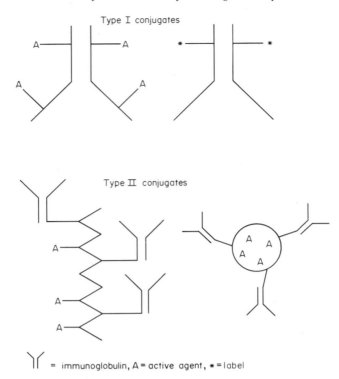

Figure 7 Types of MAB conjugates used for drug targeting purposes. Type I: direct (or *via* a short spacer molecule) coupling of active agents to MAB. Type II: MAB bound to carrier system. Drug or label is covalently bound to carrier, adheres to the carrier or is encapsulated in the carrier

bound covalently to the carrier, adhere to the carrier or be encapsulated in the carrier (liposomes, microspheres). In this type II structure there is no direct connection between active agent and MAB; a macromolecule or particulate system serves as a 'bridge'.[21,32,53]

The ultimate goal is to design highly purified MAB, which can be produced on a large scale with a high affinity for a highly specific surface antigen, and conjugated with a potent drug or radionuclide by a stable bond. After reaching the (easily accessible) target cells and, if necessary, after readily penetrating into these target cells along the desired pathway, the conjugated drug should be activated. As will be shown in the following sections, the 'state of the art' is rather different from this ideal situation.

25.5.5.1 Immunotoxins

The literature on immunotoxins (IT) is abundant and reviews have been published regularly.[23,42,44,49]

Immunotoxins are combinations of MAB or MAB fragments and toxins. The toxin is the cytotoxic part of the molecule; the MAB the homing device. Toxins are glycoproteins (containing mannose) from natural sources. Ricin has been most extensively investigated in the IT context and the discussion will therefore focus on ricin–MAB IT. Comparable results have been obtained with other toxins such as abrin and diphtheria toxin. Ricin is composed of an A chain (M 32 000) and a B chain (M 34 000) connected by a disulfide bond. The A chain kills the cells by inactivating the 60S subunit of ribosomes, thereby blocking protein synthesis. In principle one A chain molecule is able to kill a cell. The B chain has galactose recognition properties and binds the toxin to galactose-containing receptors at the cell surface. Besides, it helps the A chain to pass through the cellular membrane. This binding of the toxin is not cell specific, which precludes its use in therapy. Two approaches can be followed for the design of IT. The whole toxin can be used and coupled to MAB (or MAB fragments) (MAB–IT). It is then necessary to block the nonspecific galactose recognition properties of the B chain of the toxin. Modification of two tyrosine residues or just simple conjugation with the MAB dramatically reduces the nonspecific cell binding.[271] An alternative approach is to remove the B chain and to couple MAB (or MAB fragments) to the A chain (MAB–A–IT). The MAB (or MAB fragments) act as homing devices and help the A chain to pass through the cell membrane. The idea is that MAB–A–IT show less nonspecific toxicity than MAB–IT, while the target cell specific toxicity is preserved.

Ricin is isolated from the seeds of *Ricinus communis*. A series of chromatographic steps is necessary for purification. For the preparation of MAB–A–IT the A chain has to be isolated. Reduction with mercaptoethanol cuts the disulfide bridge; then the A chain is chromatographically separated from the B chain. The A chain fraction should be free from any detectable traces of B chain (checked, for example, by SDS–PAGE). Suggestions for safety regulations to minimize the likelihood of accidents while working with toxins are listed by Cumber *et al.*[44]

The bond between ricin or the A chain and the MAB should be stable in blood and in the extracellular compartment. Besides, the binding procedure should result in minimum levels of homopolymers, and cross-linking between the A and B chains should not occur as, once internalized, the A and B chains have to be separated to block protein synthesis effectively. The *N*-hydroxysuccinimidyl ester of chlorambucil has been used as a coupling agent. Coupling *via* disulfide or thioether bridges with bifunctional agents like SPDP or SMPB (*cf.* Table 3), respectively, has become the method of choice. The yields of the MAB–toxin conjugates in a 1:1 ratio vary between 20 and 50%.[44] Fab' can be coupled directly with the A chain *via* the sulfhydryl groups available on both molecules after reduction of F(ab')$_2$ and ricin, respectively.

Different mechanisms have been proposed for the transport of ricin IT or A chain IT to the ribosomes. In the first proposed pathway the A chain is released from the MAB (or MAB fragment) during transport through the cell membrane. Alternatively, the ricin or A chain IT are supposedly endocytosed *via* coated pits at the cell surface; this is followed by encapsulation in endosomes and finally the A chain or toxin somehow reaches the cytoplasm. It is not clear whether the A chain is released from the endosome before or after fusion of the endosome with the lysosome. The efficiency of the whole process is low. In a typical experiment it was estimated that only 1 out of 1800 IT molecules adhering to the cell actually reached the ribosomes.[45] No clear general rules for optimum therapeutic responses have emerged from the research efforts made in animal models up until now. Controversies exist on the following issues: (i) is the use of the whole ricin molecule (with masked galactose binding properties) to be preferred over the use of the A chain? (ii) are disulfide bridges more suitable than sulfide bridges? (iii) is the use of complete MAB, F(ab')$_2$ or Fab' fragments to be preferred? These issues will now be discussed.

(i) The nonspecific toxicity of A chain IT is low; however, the desired specific toxicity cannot be predicted and depends on the target cell, the MAB and the nature of the A chain involved. On the other hand, intact toxin-containing IT have a high, nonspecific toxicity which might to some extent be reduced by masking nonspecific galactose recognition sites. The LD$_{50}$ of A chain conjugates in mice is 12 mg kg^{-1} calculated on the basis of the ricin moiety, 40 µg kg^{-1} for intact ricin and 160 µg kg^{-1} for masked ricin molecules.[44]

(ii) Conflicting results have been published on the activities of conjugates with a disulfide or a sulfide bridge. Disulfide bridges are sensitive to reduction or disulfide exchange when the conjugate is still circulating in the blood: the MAB part and toxin part are then separated. On the other hand, the sulfide bond may be too stable, so the conjugate parts are neither split in the blood nor in the intracellular compartment. This latter type of conjugate has a strongly reduced capacity, or no capacity at all, to inactivate ribosomal protein synthesis when compared to similar products with a disulfide bond.[46]

(iii) Toxin–MAB conjugates have a high affinity for phagocytic cells which is related to the Fc part on the MAB and the mannose residues on the toxin, and this may cause severe side effects. To circumvent the Fc interaction F(ab')$_2$ or Fab' fragments can be used. Cross-linking of receptor molecules may enhance endocytosis and therefore the delivery of the A chain to the ribosomes; this might be an explanation why F(ab')$_2$ conjugates were more effective *in vitro* than Fab' fragments.[47] Little is known about the fate of IT after injection. Raso and Basala[46] found a biphasic blood concentration pattern after intravenous injection of murine IgG–ricin A chain conjugates with a sulfide or a disulfide bridge in rabbits (Figure 8). The first phase had a short half-life; the half-life of the second phase was around 40 h. The bridge was not a critical factor. The nonconjugated A chain was rapidly cleared (half-life ≪ 60 min). An interesting mechanistic approach to describe the kinetics of MAB distribution in solid tumors as a function of time and antibody–antigen affinity was presented by Weinstein *et al.*[18]

It is very unlikely that the IT will quantitatively accumulate in tumor tissue; side effects will therefore occur. The observed toxicity includes histological and functional changes in the GI tract, hepatic parenchyma and renal tubules. These observations parallel the toxic effects reported after

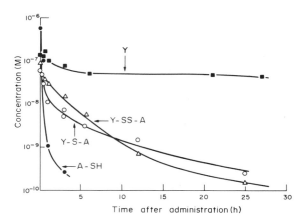

Figure 8 Serum clearance curves for antibody Y (1 mg kg^{-1} of mouse monoclonal anti-K antibody) after intravenous bolus injection into rabbits. Blood samples were taken at various times after administration from the ear vein and analyzed with a specific radioimmunoassay. Serum level of ricin A chain A-SH (■); serum level of sulfide Y-S-A (○), combination of ricin A chain and IgG; serum level of disulfide Y-SS-A (△), combination of ricin A chain and IgG[46]

administration of the A chain. Intact ricin exerts its toxic effects mainly in the vascular system and the RES.[23] The injection of murine MAB (or MAB fragments) or toxin (or toxin fragments) may trigger an immune response. The raised anti-IT antibodies may prematurely interact with the IT and neutralize them. Besides, anaphylactic reactions cannot be excluded. The approaches to reduce the immunogenicity of MAB have been discussed above. No clinical data on the immunogenicity of IT are available yet.

IT will first be tested in the area of life threatening diseases, such as cancer. The experience with IT in the clinic is limited to a few (pre)phase 1 trials. A phase 1 trial has been performed with antimelanoma antibodies coupled to ricin A. The first evaluations indicate that 10 out of 22 patients showed a clinical response.[23] Edema, weight gain and decrease of serum proteins were dose-related toxic effects.

IT are now under investigation for their potential to remove undesired cells from bone marrow *ex vivo* in autologous and allogenic bone marrow transplantations. In patients with some forms of leukemia (*e.g.* refractory B cell lymphoma and refractory acute lymphatic leukemia) tumor regression can be induced by chemotherapy and during remission bone marrow is collected. Then the patient is treated with a supralethal combination of chemo- and radio-therapy. If still present, the tumor cells in the collected bone marrow should be killed, while the pluripotent stem cells should remain unaffected. IT which meet these demands are being developed. After IT treatment the purified bone marrow is reinfused (autologous bone marrow transplantations). Several clinical trials following this procedure are in progress.[23]

IT against T lymphocytes are also being tested for their potential to prevent graft *versus* host reactions in allogenic bone marrow transplantations. Here, the reinfused bone marrow is taken from another person, and the number of T lymphocytes in the donor marrow, which may give a graft *versus* host reaction, have to be selectively reduced. Cocktails of three IT with different MAB directed against human T cell surface antigens have been used to improve the killing effectiveness. The ultimate goal, however, might be a more subtle treatment, as some subsets of donor T lymphocytes might help to kill surviving tumor cells or assist in the engraftment of donor marrow.[23,48]

In the near future clinical data will become available to evaluate the immunotoxin concept. Further reduction of the toxicity by minimizing extra-target disposition by manipulating the molecular structure is likely to be the first goal. An approach to improve the therapeutic index is the following: a ricin A chain–IgG-1 (or fragment) conjugate (A) is used in combination with a ricin B chain–IgG-2 (or fragment) conjugate (B). The IgG-2 interacts with epitopes on the IgG-1 molecule. The idea is that first A and later B is administered; A accumulates at the target cell surface and catches circulating B. Then reformation of an active unit with both cell penetration capacity (B) and high target cell toxicity (A) ('piggyback system') would be established. The systemic toxicity should be reduced because activation should preferentially occur at the target site.[49]

25.5.5.2 Conjugates containing MAB and an Effector Molecule other than a Toxin

Apart from toxins, other drugs which are being used routinely in therapy have been conjugated to MAB to increase their therapeutic index. As with IT, research has concentrated on cancer therapy. The problem with conventional cytostatics is that they are less potent on a molecular basis than toxins. Therefore, more effector molecules have to be delivered at the target site than in the case of toxins. The best candidates are therefore cytostatics with a high intrinsic cytotoxicity. Hellstrom *et al.*[50] provided the background for this conclusion with a simple calculation, based on a number of debatable assumptions, of the number of molecules that have to enter a cell to kill it. To improve the payload (number of drug molecules per MAB molecule) of the MAB another approach can be taken. The drug is not directly attached to the antibody (type I conjugate), but a carrier is used to increase the number of drug molecules delivered by one MAB (type II conjugate).

A number of questions can be raised if the drug–(carrier)–MAB approach is followed: (i) What is the effect of conjugation on the cytostatic potency of the drug? Some drugs may still be active in a conjugated form without changing their reaction mechanism (alkylating agents), others may change their mode of action (doxorubicin at the surface of particulate systems)[33,34] or lose their activity (antimetabolites) and have to be reactivated (intracellularly) by (controlled) release from the MAB or macromolecule. (ii) What is the effect of conjugation on the antibody–surface antigen interaction? The possible reduction in affinity depends both on the presence of a carrier, the nature of the binding reaction and the number of bridges between the MAB and the carrier or drug. (iii) What is the effect on the immunogenicity of the system and what is the toxicity of the carrier (whether biodegraded or eliminated intact) and bridge or spacer units? Compared to the 'state of the art' with IT, fewer answers to these questions are available and full scale clinical studies have not been undertaken yet. A listing of the studies performed up to 1981 which showed a superior response of drug–(carrier)–antibody combinations to drug or antibody alone has been presented by Ghose *et al.*[51] Hurwitz[32] and Arnon[52] connected cytostatics (*e.g.* daunorubicin, a Pt compound, doxorubicin and methotrexate) either directly, or *via* dextran, to MAB. Direct binding resulted in the coupling of up to 10 drug molecules per MAB molecule. *Via* dextran the ratio increased up to 500 drug molecules per MAB molecule. Daunomycin conjugated with hydrolyzable linkages was active *in vitro*, while systems with nonhydrolyzable linkages were inactive. Intact conjugates with daunomycin, either directly bound to a MAB, or bound *via* a carrier, were taken up by target cells. In animal studies promising results were found; the data indicated an increased therapeutic index of the conjugate compared to the 'free' drug.[32,52] Recently Rihova and Kopecec[53] synthesized MAB conjugates based on a hydrogel: poly[*N*-(2-hydroxypropyl)methacrylamide] (HPMA). Both *in vitro* and *in vivo* a specific and high activity was observed. The hydrolysis of the spacer between the carrier and the drug was essential to activate the complex.

As stated above, little is known about the pharmacokinetics, the preferred stability *in vivo* (particularly at the target cell level), the toxicity and immunogenicity of these conjugates. The question whether MAB or MAB fragments should be used has also not yet been answered.

25.5.5.3 MAB–Radionuclide Conjugates

Conjugates of MAB and radionuclides are under investigation for diagnostic (imaging) and radiotherapeutic purposes. Gamma radiation emitting nuclides are used for imaging. Candidates are ^{123}I, ^{125}I, ^{131}I, ^{111}In and ^{99m}Tc. For an evaluation of the diagnostic potential of MAB–radionuclide conjugates a low background level is essential. The elimination of intact MAB from the blood pool and their subsequent distribution takes a considerable amount of time. Therefore, clinical experiments to monitor the distribution of MAB throughout the body take between 6 and over 48 h.[23] ^{123}I and ^{99m}Tc with short half-lives of 13.3 and 6 h, respectively, are presumably less suitable markers for intact MAB; ^{125}I with a half-life of 60 d, on the other hand, causes overexposure. Therefore, in most studies with intact MAB, ^{131}I (half-life 8.1 d) or ^{111}In (half-life 67.5 h) have been monitored. Attempts have been made to accelerate the clearance of the imaging marker–MAB combinations from the blood pool by injecting an immunoliposome with MAB directed against the (first) imaging MAB some time (*e.g* 24 h) after the injection of the imaging MAB. The circulating imaging MABs are then rapidly taken up by the cells of the RES and the background from the blood compartment is strongly reduced.[54] The iodine isotopes are covalently bound to the MAB.[55] The binding process might affect the antibody–antigen binding affinity and reduce the blood elimination time considerably, and the nonspecific disposition can increase. It has been reported that iodine is released prematurely from the MAB.[23,32] As a precaution, uptake through the thyroid should be blocked before administration of the MAB–I conjugate. ^{111}In and ^{99m}Tc are bound *via* chelation; DTPA (diethylenetriaminepentaacetic acid) is used for this purpose. The resulting complex proved to be stable enough to be used *in vivo*. Their γ emissions are easily detectable with a scintillation camera.

Fab' and F(ab')$_2$ fragments have a higher elimination rate than intact MAB molecules; they are cleared through the kidneys. For MAB fragments ^{123}I and ^{99m}Tc might be interesting labels. These fragments have been reported to pass through the blood–brain barrier.[51] Full MAB, F(ab')$_2$ and Fab' were compared for their tumor imaging potential in clinical studies after intravenous injection. Nonspecific uptake in liver and spleen, poor penetration into tumor tissue, cross reactions and low antigen density on the target cells are the problems encountered. In general, the use of fragments was preferred.[23,40] From the clinical studies it can be concluded that, with the current techniques, tumors smaller than 1 cm cannot be reliably detected. An interesting field of application might be the use of MAB or (MAB fragments) for imaging of tumors in local lymph nodes after subcutaneous injection into the interstitial space. The MAB would pass through the lymph nodes draining that particular part of the body before they reach the blood circulation. In these lymph nodes they can interact with (tumor) target cells.[18] The use of labeled anticardiac myosin antibody fragments for the detection and quantification of myocardial infarcts is also under investigation.[56-58] These studies have reached the stage of clinical trials.

MAB–radionuclide combinations are also administered for therapeutic purposes. Doses with an order of magnitude of 100 mCi are injected. The radionuclide is then supposed to exert a cytotoxic effect, for example on tumor tissue. High energy emission over a short distance is now required, electrons and positrons or α-particles (helium nuclei) are potential candidates.[23] In contrast to the preferred short radiological half-life for imaging, for MAB-mediated radiotherapy isotopes with longer half-lives are, within certain limits, not unsuitable. Most animal and clinical studies done so far have applied ^{131}I, which, apart from γ radiation, also emits β particles. The β radiation has an effective range corresponding to a few cell diameters. The advantage of the use of these radionuclides for therapeutical purposes, compared, for example, to IT, is that the MAB–radionuclide conjugate does not have to penetrate the target and does not have to interact with all target cells in a tumor or metastasis to achieve the eradication of the target tissue. Nonspecific uptake of MAB ^{131}I conjugates induced considerable toxicity as only a minor fraction of the conjugate reached the target tissue. Here, as mentioned before, MAB fragments might be preferred because of their inherently lower affinity to nontarget tissue and reduced immunogenicity.[23] Tumor cell sensitivity to this form of radiotherapy varied widely.[51] The number of clinical studies on radiotherapy with MAB–radionuclide conjugates is limited. No definite conclusions can be drawn about the potential and limitations of MAB-mediated radiotherapy. It is, however, likely that MAB-mediated radiotherapy will only be applied in combination with other approaches in cancer chemotherapy. Apart from ^{131}I, other isotopes are now under investigation (^{90}Y, ^{212}Bi and ^{123}I). And, besides, a reduction of nonspecific localization of the active material by improving MAB fragment specificity and decreasing RES uptake might solve some of the problems.[23,51]

25.5.6 NEOGLYCOPROTEINS

On their surface many cells possess receptors; these receptors play a role in the selective uptake and processing of certain macromolecules. Receptor-mediated transport is strongly dependent on molecular structure. This type of transport has been reported for hormones (insulin), biological response modifiers (interleukin 2, epithelial growth factor), antibodies or colloidal structures (LDL; see Section 25.5.8) and other macromolecules with specific recognition sites (desialylated glycoproteins with terminal galactose groups, transferrin or molecules containing mannose 6-phosphate). This subject is discussed in more detail by Hopkins.[20]

The chemical and physical structure of a number of receptors has been elucidated. They consist of proteins and glycoproteins partly embedded in the membrane, with their main structure exposed to the external phase; but they are also partly located in the intracellular cytosol (*cf.* Figure 9). The part directed outwards contains the domains with disulfide bridges and the various glycosylated structures forming the binding sites.

Attempts have been made to deliver drugs specifically to hepatocytes, endothelial cells or macrophages in the liver *via* the different receptors present on the respective cells. The hepatocytes possess on their surface D-β-galactose binding sites. Macrophages recognize molecules containing D-mannose, L-fucose and *N*-acetyl-D-glucosamine. The presence of a minimal number of sugar recognition sites is necessary for triggering the endocytotic process because of receptor multivalency. Structural requirements for maximum interactions have been described both for the galactose

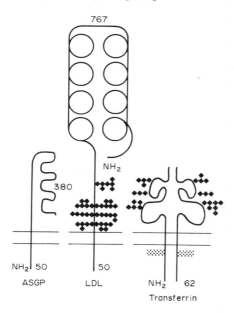

767

NH₂

380

NH₂ 50

50

NH₂ 62

ASGP

LDL

Transferrin

Figure 9 Surface receptors known to concentrate in clathrin-coated pits. The constituent polypeptides are characterized by extensive external domains, single transmembrane fragments and cytoplasmic tails of intermediate length. In each polypeptide the number of residues in the external and internal domains are indicated. ASGP, asialoglycoprotein receptors; LDL, low density lipoprotein receptor[20]

receptor on hepatocytes and the mannose receptor on macrophages.[59] Apart from hepatocytes and macrophages the liver also contains endothelial cells. These cells possess a highly active (scavenger) receptor which triggers the endocytosis of acetylated apolipoproteins of LDL[29] and other proteins with an anionic character.

The presence of the galactose receptor on hepatocytes can be utilized to target drug–macromolecule–galactose conjugates to hepatomas or hepatocytes infected by a pathogen. Fiume *et al.*[60] have investigated the possibilities of directing antiviral agents to hepatocytes infected with viruses. Initially naturally occurring desialylated glycoproteins (asialofetuin) were used as macromolecule–homing device combinations; later glycosylated albumin was used for practical reasons (*e.g.* availability and reduced immunogenicity).[61] Free drug (ara-AMP) showed more side effects in the intestine and bone marrow than the human serum albumin conjugates with comparable therapeutic effects (inhibition of DNA synthesis) in the liver after injection into mice inoculated with Ectromelia virus. The intracellular fate of drug–lactosylated albumin conjugates was not investigated. This approach might substantially improve the therapeutic index of antiviral drugs used (or yet to be used) in the treatment of hepatitis B. Extensive work on the selection of the spacer molecule to bind a drug to the carrier system has been done by Trouet *et al.*[272] The cytostatic daunorubicin (DNR) was coupled directly, or *via* mono-, di-, tri- or tetra-peptide spacer arms, to succinylated albumin. The conjugates were shown to be stable in serum and blood; only the tri- and tetra-peptide spacers were rapidly hydrolyzed in the presence of lysosomal enzymes. The spacer arm consisted of leucine (on the daunorubicin side) and alanine. The daunorubicin–leucine bond was sensitive to lysosomal enzyme hydrolysis, and alanine was introduced to maintain sufficient water solubility. The coupling procedures for the galactose residues and daunorubicin to albumin have been presented in more detail by Schneider *et al.*[62] This group performed experiments with daunorubicin–glycosylated albumin conjugates in mice implanted with human hepatoma fragments under their renal capsule. Thirty minutes after intravenous injection 70% of the conjugate was taken up by the liver; less than 1% was found in the heart (cardiotoxicity is the dose limiting side effect of the free drug), spleen and kidney. In preliminary experiments promising therapeutic effects were obtained with the conjugates on implanted tumors (Table 4) and preclinical trials were started.

A number of questions have to be answered, related to this approach. The payload of the albumin molecule is limited (*e.g.* at high drug loading levels aggregation can occur). Therefore, only potent drug molecules should be used. Conjugation of the drug may lead to partial loss of hepatocytic specificity due to uptake by another, scavenger, receptor.[43] Another important question which has not been fully answered yet concerns the receptor density on diseased cell surfaces and the endocytic

Table 4 Chemotherapeutic Activity of Daunorubicin (DNR), Free or Conjugated to Galactosylated Human Serum Albumin (DNR-gal-SA), after Intravenous Injection into Mice Implanted with Human Hepatoma Fragments under their Renal Capsule[62]

Drug[b]	Dose (mg DNR kg^{-1}) per injection	Change in tumour size (mm)[a]		
		Hep 1[c]	Hep 2[d]	Hep 3[e]
Control	—	+0.26	+0.12	+0.04
DNR	12	+0.18	−0.10	+0.03
DNR-gal-SA	19	−0.06	−0.18	−0.12

[a] Human hepatoma fragments (1 mm^3) were implanted on day 0 under the renal capsule of female BDF1 mice. The final day of the experiment was day 6 for Hep 1 and 2, and day 5 for Hep 3. [b] Drugs were given intravenously on days 2 and 4 for Hep 1 and 2, and on days 1 and 2 for Hep 3. [c] Hep 1: patient with hepatocarcinoma. [d] Hep 2: patient with hepatocarcinoma and cholangiocarcinoma. [e] Hep 3: patient with hepatocarcinoma.

capacity of these cells compared to 'normal' target cells. Little is known about the immunogenicity of albumin conjugates; the spacer structure, payload and type of drug involved (*e.g.* immunosuppressants) all play a role.

25.5.7   PRODRUGS

Prodrugs can be defined as drug molecules modified in such a way that the inactivated derivative is reactivated *in vivo*, producing the active parent compound. The prodrug concept has been successful in improving the performance of a number of active compounds. An increase in water solubility, in stability, masking of the taste, prolongation of the delivery or in the improvement of the bioavailability of a drug have been the goals to be reached in the development of a prodrug. Although the definition in principle includes the macromolecular drug–carrier combinations as discussed in previous sections, in practice the term prodrug is reserved for low molecular weight structures. Prodrugs were also developed for drug targeting purposes. Bundgaard[64] discerned prodrugs based on the idea of site directed drug delivery and site specific drug release or activation. Detailed information on prodrugs can be found in various books and reviews.[16,64–67] Stella and Himmelstein[16,68] discussed the pharmacokinetic aspects of the prodrug approach. They emphasized that for optimum performance both transport of the prodrug to the target site and also retention of the parent compound at the target site are essential.

In *site directed drug delivery* the prodrug is designed to accumulate at desired tissues in the body. This can be done *via* homing devices, as mentioned above, or, using an elegant approach, *via* a so-called 'lock in' principle, which has been tested for specific delivery of drugs to the brain (Figure 10).[67,69] Redox carriers are used. The parent drug (D) is coupled to a quaternary ammonium compound $(D–QC)^+$ and the conjugate is reduced to the dihydro form (D–DHC); D–DHC is lipophilic. *N*-Methylnicotinic acid is an example of a quaternary ammonium compound used as a redox carrier. After intravenous administration the prodrug is rapidly distributed over the different body compartments. Because of its lipophilic nature, it can pass the blood–brain barrier. Conversion to the oxidized form $(D–QC)^+$ then occurs. This conversion accelerates the excretion (rate constant k_2) from the body except from the brain; the blood–brain barrier slows down the leakage of the highly polar $(D–QC)^+$ (rate constant k_1): the prodrug is 'locked in'. In the brain the parent compound is formed by enzymatic cleavage (rate constant k_3) and the active compound will be excreted from the brain (rate constant k_4). For optimum performance of the prodrug the following relationships between the different kinetic parameters in Figure 10 should exist: $k_1 \ll k_2$, $k_3 \ll k_4$ and $k_1 \ll k_4$. Besides, the conversion to the quaternary compound should be relatively slow compared to the distribution process, but fast enough to occur before D–DHC is excreted; the cleavage of the D–DHC combination should be negligible. Because of the rapid elimination of the oxidized form from the different body compartments, the brain excluded, the parent compound D is mainly released in the brain and the rest of the body is only exposed to D that leaks through the blood–brain barrier. This approach has proven to be successful in delivering drugs to the brain in a number of animal studies.[67,69] No conclusions on the toxicity of the prodrug and the redox carrier

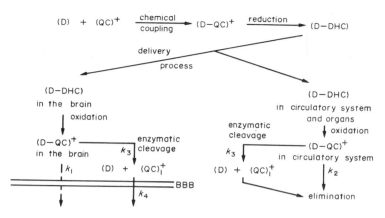

Figure 10 Schematic representation of the transport and metabolic processes occurring after administration of a dihydropyridine/pyridinium salt redox system.[69] D, drug; D–DHC, lipophilic drug–dihydro form; (D–QC)$^+$, hydrophilic drug–quaternary carrier form; (QC)$^+$, quaternary form of the carrier; BBB, blood–brain barrier

can be drawn yet. Another potential problem relates to the fact that D–DHC is distributed all over the body and might be oxidized in other compartments (*e.g.* cells) which do not readily release (D–QC)$^+$.

In *site specific drug release/activation* special properties of the target site are utilized to activate the parent compound. pH values deviating from the normal pH or the accumulation of enzymes at the site of action might provide the means for a reactivation of the parent, active, compound.[64] An example of this approach is the activation of *N*-acyl-γ-glutamyl derivatives of sulfonamides by the kidneys. For the activation of the parent sulfonamide first the prodrug has to be deacylated and subsequently the γ-glutamyl part has to be removed. Both the enzymes involved (*N*-acylamino-acid deacylase and γ-glutamyl transpeptidase) are present in the kidneys in high concentrations. A limitation of this concept is that the cleavage rate is highly dependent on the nature of the active compound.

Attempts have been made to utilize high levels of certain enzymes in tumor cells for the site specific release of cytostatics. Plasminogen activators (increasing the levels of the proteolytic enzyme plasmin), uridine phosphorylase and β-glucosidase were reported to be present in unusually high concentrations in tumor cells, and prodrugs were designed to be activated by these enzymes. However, up until now no prodrugs based on this concept have been introduced in the clinic. Certain tumor cells have a lower pH than normal cells; this acidity has been ascribed to an increased rate of glycolysis in neoplastic tissue leading to enhanced lactic acid production. The pH difference has been estimated to be as high as one unit. Prodrugs were developed (*e.g.* ketals[70] and aziridines) which are activated preferentially under acidic conditions. However, the pH difference is too small to achieve a significant site specific drug release.

Precursors of the active compound can also be considered as prodrugs. α-Methyl-DOPA and L-DOPA are the traditional representatives of this approach. A more recent successful example of the precursor concept in site specific drug activation is the antiviral drug acyclovir. This drug is converted into its monophosphate in the cell by a specific viral thymidine kinase. The monophosphate cannot leave the cell; subsequently cellular enzymes produce the active triphosphate. Activation is therefore limited to cells infected by the virus. The triphosphate blocks the viral DNA polymerase and is also incorporated into viral DNA. Acyclovir has proven to be effective in the treatment of different herpes infections.[71]

Attempts have been made to develop so-called suicide inhibitors. Prodrugs are activated by an enzyme; subsequently the activated compound is covalently bound to the same enzyme, blocking the enzyme activity. Here, in principle, both enzyme specificity and prolonged duration of the effect can be achieved. The site specificity of the drug action depends on the specificity of the localization of the catalytic process and the possibility of the prodrug reaching the target tissue. Mitomycin C and allopurinol are examples of existing drugs which are supposed to work *via* the suicide inhibition principle.[66]

The examples in the preceding paragraphs show that the prodrug concept has already been successful. A number of drugs with a certain site specificity have become available and are therapeutically highly appreciated. These examples are all based on site specific drug release/activation.

25.5.8 PARTICULATE CARRIERS

Today a variety of particulate carriers are being explored in an effort to let Paul Ehrlich's dream of the magic bullet come true: using carrier systems to deliver a drug to its required site in the body for action. Various particulate carrier systems have been investigated: liposomes, microspheres, oil/water emulsions, cellular carriers (such as erythrocytes) and carriers based on lipoprotein structures (*e.g.* low density lipoproteins). This section will consider particulate carriers and their use for the targeting of drugs to organ sites, particularly after parenteral administration. Liposomes will be discussed in more detail as most attention in the field of drug targeting using particulate systems has been directed to this colloidal carrier system.

25.5.8.1 Liposomes

Liposomes have come a long way since they were proposed as a potential drug carrier system about 15 years ago.[72] After a few years of overexposure as the ultimate answer to the problem of drug targeting, interest in liposomes seemed to fade. It became apparent that, in spite of some remarkable successes, liposomes are by no means a panacea for pharmacotherapy in general. In more recent years, however, a revival of the liposome concept in drug delivery has been observed. The purpose of this section is to describe the research activities which were the basis for this revival, and to present a balanced view on the potential and limitations of the use of liposomes as a system for drug targeting. First, a few basic aspects of the structure, preparation and *in vitro* and *in vivo* behavior of liposomes will be discussed. For detailed information about physical, pharmaceutical, physiological and pharmacological aspects of drug laden liposomes, the reader is referred to a comprehensive review published recently.[73]

Liposomes are closed vesicles consisting of one or a number of parallel bilayers surrounding an internal aqueous space or spaces (Figure 11). The backbone of the bilayer consists of phospholipids. The potential value of liposomes as drug targeting systems stems from the ability to encapsulate water soluble drugs within the aqueous interior or to incorporate hydrophobic drugs within the lipid bilayer. Liposomes can be prepared from a variety of phospholipids derived from either natural or synthetic sources. Usually phosphatidylcholine is the major lipid in the bilayer. In addition to phospholipids, large amounts of other lipids can be incorporated into the liposomal membranes. Cholesterol for instance has been extensively used to reduce the permeability to encapsulated compounds as well as to increase the stability of liposomes in plasma. To confer surface charge to

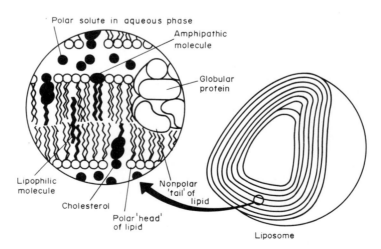

Figure 11 A simplified representation of the sites of interactions of hydrophilic, lipophilic and amphipatic molecules, cholesterol and proteins with multicompartment liposomes[265]

Table 5 Captured Volume per mg Phospholipid for
Different Liposome Preparations[274]

Liposome type[a]	Captured volume ($\mu L\ mg^{-1}$ PL)	Diameter[b] (μm; range)
REV	13.7	0.2–1
MLV	4.1	0.4–3.5
SUV	0.5	0.02–0.05

[a] Bilayer composition: phosphatidylcholine:phosphatidyl-glycerol:cholesterol 4:1:5. [b] Size range that included 90% of the vescicles; determined by negative-stain electron microscopy.

liposomes then negatively charged lipids, such as phosphatidylserine or dicetyl phosphate, or positively charged lipids, such as stearylamine, are used.

Several types of liposomes are available for application as drug delivery systems. By using different lipids and different preparation methods, a variety of liposomes can be made, differing in composition, size, surface charge and structure. Liposome types which are most commonly used will be described briefly. Multilamellar vesicles (MLV) are formed spontaneously upon hydration of dry phospholipid in excess water. These vesicles are very heterogeneous in diameter (upto several microns) but can be 'sized' by extrusion through polycarbonate filters. The size of the liposomes can also be reduced by ultrasonication, which ultimately results in a homogeneous suspension of small unilamellar vesicles (SUV) with diameters as small as 25 nm. One of the drawbacks of MLV, and more particularly of SUV, is the relatively low encapsulation efficiency for polar compounds. Reverse phase evaporation vesicles (REV) are mostly unilamellar vesicles with relatively high encapsulation efficiencies for water and polar, nonbilayer interacting drugs. Table 5 shows the captured volume [aqueous space to lipid ratio ($\mu L\ mg^{-1}$ phospholipid)] for MLV, SUV and REV with equal bilayer composition. Being composed of natural body constituents, being biodegradable and being versatile with respect to their particle characteristics, liposomes are considered to be good candidates for drug carrier systems. Toxicity and immunogenicity are not expected to be a problem.

When liposomes are injected into the circulation, the pharmacokinetics and distribution of the drug are, at least to some extent, dependent on the behavior of the liposomal carrier. Liposomes stay within the circulation and can only leave at certain sites where the capillary endothelium is fenestrated (*e.g.* in the liver, thus enabling targeting to liver parenchymal cells) or, alternatively, where the endothelium is disturbed by necrosis or inflammation (*cf.* Section 25.5.2). Liposomes are avidly taken up by phagocytic cells of the RES, including tissue macrophages (mainly by those located in liver and spleen) and circulating blood monocytes. Based on these considerations of liposome behavior *in vivo*, it is clear that the incorporation of a drug into liposomes will tend to favor drug uptake by liver and spleen, to stimulate accumulation of drug in macrophages, and to enhance drug deposition at sites of capillary damage or inflammation (either directly or *via* conveyance in inflammatory macrophages).

25.5.8.1.1 *The targeting tissue*

The *in vivo* fate of intravenously administered liposomes can be modulated to some extent by altering the physical and chemical characteristics.[73] For large liposomes ($>3\ \mu m$), there is a tendency for them to accumulate in the lungs due to a filtration effect in lung capillaries. Liposome size also plays an important role in the rate of clearance from the bloodstream; the larger the liposome particles, the more rapid their clearance. Other important factors influencing clearance rate are charge and dose. Negatively charged liposomes are removed more rapidly from the circulation than neutral or positively charged ones. An increase in liposome dose produces a relative reduction in liver uptake with a parallel increase in blood circulation time and, to some extent, in spleen and bone marrow uptake. This is indicative of a certain saturation at the hepatic level.

Two approaches to targeted drug delivery can be discerned: 'passive targeting' and 'active targeting'. The term 'passive targeting' refers to the natural localization patterns of liposomes. As already discussed, liposomes injected intravenously localize predominantly in the mononuclear phagocytes of the RES. This distribution pattern can be exploited to target liposomes to these cells, albeit in a passive manner. The preferential localization of liposomes in the RES has been considered as a major obstacle to efforts to target liposomes to other cell types. Attempts to 'block' the phagocytic uptake mechanisms of RES cells (*e.g.* by predosing with high doses of liposomes or other particles to saturate the phagocytic uptake mechanism) resulted in reduced clearance rates of circulating liposomes, but were not very successful in achieving enhanced liposome uptake by tissues outside the RES.[74]

'Active targeting' attempts to alter natural and passive localization patterns in order to direct liposomes to specific cells or tissues. The principal strategy to achieve homing to target cells involves construction of liposomes bearing ligands that will recognize determinant receptors on the surface of the target cells. Antibodies have been by far the most popular ligands adopted for this purpose to date.

An alternative approach for 'active' liposome targeting can be classified as 'physical targeting'. This strategy involves the use of liposomes with special physicochemical properties. They release the encapsulated drug when exposed to specific environmental conditions such as changes in temperature or pH. In addition, physical targeting of liposomes using external magnets in the proximity of target sites has been considered for localization of liposomes containing ferromagnetic particles.

Among the various routes of administration of liposomes the intravenous injection is the most widely applied. Therefore, in discussing therapeutic applications of liposomes in targeted drug delivery, we will confine ourselves to intravenously injected liposomes. However, at the end of this section, some applications involving injection into certain body cavities ('compartmental targeting') will be described.

At present, the passive targeting strategy is the one most extensively investigated and offers the most realistic possibilities for medical applications of liposomes. Therefore, attention will be focused mainly on this type of targeting.

25.5.8.1.2 *Passive targeting*

In recent years, fundamental questions have been addressed, such as whether liposomes are stable in blood and where liposomes go after *in vivo* administration and how this depends on parameters such as liposomal size and lipid composition.[1,75] The potentially destructive effect of serum lipoproteins on liposome integrity was recognized and it was shown that this could be overcome by selecting proper lipid compositions. Liver and spleen and, to a lesser extent, lung and bone marrow emerged as the main natural targets for intravenously and also intraperitoneally administered liposomes. Small liposomes (*e.g.* SUV) may also serve as carriers of drugs to be delivered to liver parenchymal cells by virtue of their capacity to penetrate the fenestrated liver endothelium (with gaps of up to about 0.1 µm). Based on this informaton, different pathways by which encapsulated drugs reach their ultimate site of action after intravenous injection can be proposed (Figure 12). It is possible that the drug or its metabolite(s) are released directly from liposomes still present in the blood, but also indirectly from the RES following uptake and processing by macrophages. Once taken up by macrophages *via* endocytosis, the liposomes are degraded in the lysosomal compartment. Liposome-encapsulated drugs, if resistant to the intralysosomal environment, may slowly leak out of the lysosomes into the cytosol and become available to exert their therapeutic action. In this way, internalization of liposomes by macrophages may potentiate therapeutic efficacy by enhanced delivery of drugs to the intracellular site of action. The drug may also be released by the macrophages, thereby producing a therapeutic drug level in blood and/or tissues for a prolonged period of time. In addition, passive targeting of encapsulated drugs may lead to reduced drug levels in tissues which are particularly sensitive to toxic effects of the drugs used. Thus the 'sustained release' of drugs which may occur from circulating liposomes, from liposomes adsorbed to cell membranes, as well as from macrophages which have engulfed drug-loaded liposomes, may provide an important rationale for therapeutic applications which intend to exploit the natural fate of liposome-encapsulated drugs.

(i) *Liposomes in treatment of neoplastic diseases*

(a) Liposome-encapsulated antitumor drugs. In the late 1970s, liposomes were regarded as a major new approach for improving cancer chemotherapy. This optimism was generated primarily by the idea that liposomes might be used to 'target' drugs to tumors. In the rush to demonstrate the ability of liposomes to treat neoplastic diseases more satisfactorily, a large number of antitumor drugs were trapped into liposomes and tested for their therapeutic behavior in animal tumor systems.[1,6,76] No consistent picture emerged from these studies; in some cases important successes

were claimed while in other cases disappointing results were reported. The mechanism or mechanisms underlying these results were largely obscure. Early investigators had little appreciation of the multiple barriers lying between a drug–liposome complex in the circulation and the ultimate site of action (*cf.* Section 25.5.2). At present, the use of liposomes to reduce the dose-limiting toxicity of anticancer agents without loss of antitumor activity by virtue of their ability to change the pharmacokinetics and tissue distribution (site avoidance drug delivery) is generally considered as a potentially valuable application of liposomes in cancer treatment. The most extensively studied antitumor drugs are alkylating agents,[77-79] antimitotic agents[80] and anthracyclines.[63,81-86,88,274] Perhaps the most promising example of site avoidance drug delivery is the use of doxorubicin (DXR) in liposome-encapsulated form. Liposome encapsulation of DXR reduced its notorious cardiotoxicity and maintained or even increased its antitumor activity. The use of DXR in the liposomal form is also beneficial to diminish immunosuppressive effects[81,87] and local necrotizing effects on the skin.[89] These findings prompted investigators to start clinical studies which already showed promising results as far as the acute toxicity is concerned.[83] Recently, sustained release was reported to be the primary mechanism by which liposome encapsulation of DXR induced a higher therapeutic index compared to the free drug under the selected experimental conditions. Two different pathways for sustained release were identified (as shown in Figure 12): DXR was released directly from liposomes still present in the blood, but also indirectly from the RES after uptake and processing by macrophages. The relative importance of each pathway depended on the physical characteristics of the liposome structures.[90] Peak concentrations of 'free' DXR in organs which are particularly sensitive to the toxic action of the drug, like the heart, are avoided, while the prolonged presence of relatively low 'free' DXR levels in the blood can result in sufficient exposure levels for tumor cells. These sustained release effects may be beneficial in the treatment of a variety of neoplasms (with their metastases, if any). The information collected for DXR-containing liposomes may serve to predict the potential advantage of formulating other drugs, with a small therapeutic index, into liposomes or other particulate carrier systems.

(b) Liposome-encapsulated immunomodulators. Macrophages can be activated to a tumoricidal, microbicidal and antiviral state by a wide variety of naturally occurring and synthetic agents.[91] Two major classes can be identified: lymphokines, such as macrophage activation factor (MAF), which also includes γ-interferon, and products or analogs of microbial cell wall components, such as bacterial endotoxin (lipopolysaccharide; LPS) and *N*-acetylmuramyl-L-alanyl-D-isoglutamine (muramyldipeptide; MDP). All of these agents, however, have proven to be unsuitable for *in vivo* stimulation of macrophage functions, either because of inherent toxicities or because of low potency due to rapid degradation and/or excretion. Therefore, liposomes have been suggested as drug carriers for the delivery of immunomodulating agents to macrophages in the therapy of cancer,[92] and in the therapy of viral[93,94] and fungal[95] diseases. A lipophilic analogue of MDP (*N*-acetylmuramyl-L-alanyl-D-isoglutamyl-L-alanylphosphatidylethanolamine; MTP-PE) was synthesized in order to obtain a more stable association with liposomes as MTP-PE can be inserted into the phospholipid bilayer structure.

Macrophages can protect against the proliferation and metastatic spread of tumor cells.[91] By a still unknown mechanism, activated macrophages are capable of selectively killing tumor cells; they leave normal cells unharmed. This approach may open new perspectives in the treatment of patients with metastatic cancer, which is often seriously hampered by the biological heterogeneity of the tumor cells with respect to growth rate, sensitivity to various cytotoxic drugs, *etc.* Fidler, Poste and

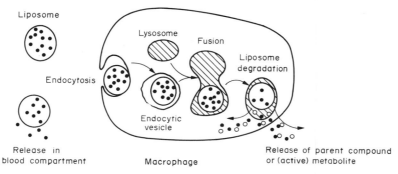

Figure 12 Proposed mode of action of liposomal drugs in the case of 'passive targeting'

coworkers have explored the use of liposome-encapsulated immunomodulators, both *in vitro* and *in vivo*.[96–98,264] The following conclusions can be drawn from their studies: (i) incorporation of lymphokines or MDP into liposomes can increase the potency of the agent in stimulating macrophage tumoricidal function *in vitro* by two to four orders of magnitude; (ii) liposomal MDP and lymphokines act synergistically when used together; (iii) liposomal MDP strongly potentiates the ability of the agent to render mouse peritoneal and alveolar macrophages tumoricidal *in vivo*; (iv) administration of liposomal MDP or lymphokine can protect mice against the metastatic spread of tumors such as B16 melanoma; and (v) the therapeutic effects seem to be largely due to actions mediated *via* macrophages, rather than *via* other cells of the immune system.[92] A number of other groups have extended these studies in various directions. The effects of liposomal MDP on the activation of various biochemical processes correlated with macrophage cytotoxic functions have been studied.[100,101] Results on the inhibition of liver metastases by intravenous administration of liposome-encapsulated immunomodulators were published by Thombre and Deodhar[102] and Daemen *et al.*[103] The therapeutic effects on liver metastases were thought to be mediated by the activation of hepatic macrophages (Kupffer cells).

Although these experimental results are encouraging, the successful application of liposome-encapsulated immunomodulators in the treatment of patients with liver metastases may be hampered by unfavourable macrophage:tumor cell ratios in many metastatic tumors. Therefore, therapeutic regimens designed to stimulate macrophage-mediated tumor cytotoxicity almost certainly will be used in combination with other treatment modalities such as surgery and chemotherapy to reduce the tumor load, while activated macrophages could eradicate surviving tumor cells (at primary as well as metastatic tumor sites).

(ii) Liposomes in treatment of infectious diseases

Infections are still a major cause of morbidity and mortality. Although effective antiinfectious drugs have been available for more than 50 years now, therapeutic failures still occur especially in immunocompromised individuals such as patients with a malignant disease or patients receiving immunosuppressive drugs. A variety of protozoal, bacterial and fungal parasites spend at least part of their lifetime within the intracellular environment, often within macrophages.[104] Obligate intracellular parasites can multiply only whilst residing within cells, whereas facultative intracellular parasites may also multiply within extracellular spaces. Viruses are considered to be obligate intracellular parasites. Although these parasites may be highly susceptible to antiinfectious drugs *in vitro*, their intracellular location *in vivo* causes therapeutic problems as they are protected from the action of these drugs because of poor penetration into affected cells and also from many immune response mechanisms. Moreover, therapy of infections in general is complicated by toxicity of the agents used and by resistance to conventional drug therapy. Application of liposomes to convey antiinfectious drugs might be an effective strategy to overcome these problems by increasing the selectivity of delivery and/or by decreasing the toxicity of applied drugs.[105]

(a) Liposome-encapsulated antiprotozoal drugs. The first reports on liposomal antiinfectious drugs were published, almost simultaneously, by three different groups using liposomes as carriers of antimonial drugs in experimental leishmaniasis infection.[106–108] The antileishmanial effect of pentavalent and trivalent antimonials and 8-aminoquinolines, as evaluated by counting the number of protozoa in hepatic or splenal smears of infected animals, was enhanced many times after encapsulation of the drugs into liposomes.[104] Similar effects were seen with other antileishmanial agents: encapsulation of antifungal agents like amphotericin B, griseofulvin and 5-fluorocytosine markedly enhanced antileishmanial efficacy.[109,110] The beneficial effects are probably related to the fact that both liposomes and protozoa are taken up by the same phagocytic cells in liver and spleen.

Liposomes have also shown considerable promise as carriers of antimalarial drugs. Human malaria is caused by *Plasmodium* species. 8-Aminoquinolines such as primaquine have the broadest spectrum of antimalarial activity. Therapeutic and prophylactic use of primaquine is hampered by its toxic side effects.[111] However, probably due to alterations in drug distribution, the acute toxicity of primaquine for uninfected mice was reduced by a factor of 3.5 after entrapment into liposomes.[112] In a murine test system infected with *P. berghei*, the therapeutic efficacy of liposomal primaquine was not different from that of the free drug. Although the therapeutic efficacy was not improved, the reduced toxicity of liposomal primaquine permitted administration of a higher dose which appeared to be 100% curative in the treatment of experimental malaria.

(b) Liposome-encapsulated antibacterial drugs. Several bacterial diseases are caused by obligate or facultative intracellular microorganisms such as *Brucella*, *Listeria*, *Salmonella* or *Mycobacterium* species.[113] Intravenous, subcutaneous, intranasal, intradermal or aerogenic inoculation of these

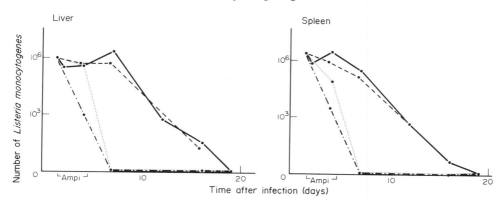

Figure 13 Effect of liposomal and free ampicillin, administered during time interval called Ampi, on the numbers of *Listeria monocytogenes* in the liver of mice: - - - -, 8 doses of 6 mg free drug; ...., 2 doses of 0.27 mg liposomal drug; –.–.–., 2 doses of 0.27 mg free drug + phospholipids; ———, untreated controls (reproduced by permission of publisher from ref. 116)

species into mice finally leads to colonization within phagocytic cells, particularly macrophages in liver and spleen.[114] Bacteria are taken up by phagocytic cells in phagocytic vacuoles which generally fuse with the lysosomal apparatus.[115] Liposomes were tested in several *in vivo* models of intracellular bacterial infections as reviewed by Bakker-Woudenberg *et al.*[266] Liposomal streptomycin administered to mice infected with *Brucella canis* was superior to the free drug in suppressing the number of surviving bacteria within the spleen.[117] Similar results were obtained with liposomal gentamicin administered to guinea pigs infected with *Brucella abortus*. Encapsulation of ampicillin resulted in sterilization of liver and spleen of mice previously infected with *Listeria monocytogenes* (Figure 13).[116] These favorable therapeutic results are presumably due to the fact that both liposomes and bacteria finally end up in phagocytic cells, mainly located in liver and spleen. Another interesting example of enhancement of antibacterial activity by liposome encapsulation concerns the use of amylopectin-modified liposomes.[118] After intravenous injection, amylopectin-modified liposomes were found to distribute with high preference to the lungs, probably due to increased uptake by circulating monocytes and alveolar macrophages. This observation was elegantly applied in the treatment of an experimental lung infection caused by *Legionella pneumophila* in guinea pigs.

(c) Liposome-encapsulated antifungal drugs. Fungal infections are a major cause of death in immunocompromised individuals such as cancer patients and renal transplant patients. For many years, the drug of choice in these cases has been the polyene antibiotic amphotericin B, a potent but extremely toxic agent. Liposomes have been used as carriers of amphotericin B in the treatment of mycotic infections such as histoplasmosis,[119] cryptococcosis[120] and candidiasis both in mice with an intact host defense system and in leukopenic mice.[99,100,121] Limited clinical trials on cancer patients who had documented systemic fungal infections were performed by Lopez-Berestein *et al.*[122] and Shirkoda *et al.*[123]

The laboratory and clinical studies indicate that use of a liposomal carrier system for amphotericin B results in a marked improvement of the therapeutic index of the drug through a reduction in toxicity with preservation of antifungal activity. Fungi are not exclusively located in macrophages;[124] therefore, the improvement of the therapeutic index after encapsulation is not solely due to passive targeting of amphotericin B to macrophages. The improved therapeutic index seems to be based primarily on a rather fundamental change in the action of amphotericin B at the cellular level. Mehta *et al.*[125] found that, while free amphotericin is extremely toxic to both fungal cells and mammalian cells *in vitro*, the liposomal drug remains toxic to fungal cells but has little effect on mammalian cells. It is well known that amphotericin B interacts with ergosterol in fungal cell membranes forming transmembrane channels, and so causing an extracellular release of ions and metabolites. On the other hand the drug also interacts with cholesterol in mammalian cell membranes, which is probably the basis of its toxicity. Incorporation of the lipophilic amphotericin B within liposomes might result in a facilitated transfer of the drug to fungal cells, while transfer to mammalian cells is hampered. This selective transfer of amphotericin B from liposomes to fungal cells may form the molecular basis of the reduced toxicity, in addition to other factors, such as altered pharmacokinetics or tissue distribution.[126]

(d) Liposome-entrapped immunomodulators in viral infections. Viral infection of nonphagocytic cells is usually accompanied by perivascular infiltration of (activated) mononuclear phagocytes,

which is supposed to be of major importance for recovery from infection. *In vitro* activated macrophages were found to lyze virus-infected cells selectively, without harming uninfected cells.[127] The selective lytic capacity of activated macrophages is possibly related to virus-induced changes in the cell surface of the infected cells.[128] Koff *et al.*[129] demonstrated that thioglycollate-stimulated peritoneal macrophages could be activated by both free and encapsulated MAF to lyze herpes simplex virus type II (HSV-II) infected murine cells *in vitro*. In a later paper they showed that free MAF and γ-interferon, and their encapsulated forms, were able to activate human monocytes to selectively lyze HSV-II-infected cells.[130] In addition, they published *in vivo* results showing that liposome-mediated delivery of MTP-PE resulted in enhanced activity against HSV-II infections in mice when compared to the free drug.[127] As measured by mean survival time and percent survival, liposomal encapsulation of MTP-PE was also reported to be beneficial in the therapy of Rift Valley fever infection in mice.[93] As this virus resides inside Kupffer cells, the possibility of direct suppression of viral replication within macrophages was indicated. Thus, liposomal encapsulation of immunomodulators may result in enhanced antiviral activity when compared to the free agent. This beneficial effect is most likely to be based on the relatively high uptake of liposomes by macrophages, leading to the activation of antiviral properties, or to direct suppression of viruses which replicate inside macrophages of the RES.[131]

25.5.8.1.3 Active targeting

Efforts have been made to assess the ability of various molecular ligands, attached to the liposomal surface, to direct liposomes to particular target cells *in vivo*.[35,42] Liposomes are certainly better suited for the assessment of their targetable properties because of the ease in modifying their surface, when compared with other particulate drug carriers. Several classes of ligands have been bound to liposome surfaces; sugars (including their lipid and protein derivatives), lectins, small haptens, antibodies and other proteins. There are already several examples of successful drug targeting with such liposomes. In the presence of lectins the association of liposomes with erythrocytes increases sharply when the liposomal membranes contain sialoglycoproteins extracted from erythrocyte membranes.[132] Liposomes with glycolipids inserted into their membranes are capable of specific recognition and binding of lectins,[133] antibodies[134] and some cells.[135] In the presence of lectins, ganglioside-containing liposomes may be bound effectively to each other[136] or to liver cells with lectin-like molecules in the membranes.[137] Animal experiments showed that the presence of desialylated fetuin in the liposomal membrane significantly increased the degree of association between bleomycin-containing liposomes and liver cells.[138] The incorporation of aminomannosyl derivatives of cholesterol into liposomes[139,140] tends to enhance uptake by macrophages, although the mechanism involved is unclear: macrophages have a receptor for mannosylated glycoproteins but it is uncertain whether this interacts with aminomannosyl compounds. This approach may enhance the spontaneous uptake of liposomes by macrophages but does not offer additional targeting possibilities.

Of particular interest to achieve effective targeting is the binding of antibodies,[138,141] especially monoclonal antibodies,[142,143] to the surface of drug-containing liposomes. Recent developments in hybridoma technology and the great versatility of antibodies in terms of target recognition have rendered antibodies the ligand of choice in most targeting work with liposomes. Therefore, the work with antibodies attached to liposomes, known as immunoliposomes,[144] will be discussed in more detail. Compared to antibody conjugates used in targeting (*cf.* Section 25.5.5), liposomes offer certain advantages. These include the ability to incorporate large quantities of a wide range of drugs, the isolation of entrapped drugs from the biological milieu (*e.g.* blood) and the transport of large numbers of drug molecules to cells, in principle by one single antibody molecule per vesicle. A firmer and more specific binding to cells may also be achieved with liposomes bearing antibodies against more than one type of antigenic determinant on the surface of the cell.

Many different techniques for the coupling of antibodies to liposomes have been described: adsorption of antibodies on the surface of preformed liposomes, association of antibodies with the liposomal membranes during their formation, covalent binding of antibodies to reactive groups on the surface of preformed liposomes and binding of antibodies modified with a hydrophobic anchor molecule to the liposomal membrane, either during or after liposome formation. Reviews of these methods have appeared elsewhere,[35,146,152] and some methods, among them those of adsorption and incorporation, are now only of historical interest and are not used any more due to their low binding efficiency. Factors which should be taken into account when considering the preparation of antibody-coated liposomes are: (i) a sufficient quantity of antibodies must be bound to the liposomal surface, (ii) the liposome–antibody binding must be stable, (iii) the homing capacity of the antibodies should remain unchanged after binding to the liposomes, and (iv) the integrity of the liposomes should be preserved during the binding process. Papahadjopoulos and coworkers reported an elegant method which seems to meet these requirements (Figures 5 and 6, and Section 25.5.4).[147–149]

Most of the early studies on antibody-mediated targeting have been done *in vitro* with liposomes bearing a hapten, usually the nitrophenylated phosphatidylethanolamines.[143] These studies demonstrated a successful induction of a specific liposome–cell association. Covalent binding of $F(ab')_2$ fragments against human erythrocytes to liposomes with exposed aldehyde groups increased liposome binding to erythrocytes 200-fold.[150] MAB against human β_2-immunoglobulin and protein A, which were bound to the liposome surface through a bifunctional reagent, led to a preparation capable of specific recognition of human cells.[143] MAB against a murine histocompatibility antigen were treated with the *N*-hydroxysuccimidic ester palmitic acid, and then incorporated into liposomes which could specifically recognize L-929 murine cells and specifically bind to them.[142] Other workers[151] have bound palmitoylated MAB against sheep erythrocytes to liposomes. This led to an increase of binding of these liposomes to erythrocytes of about 80-fold.

All these studies demonstrated that it is possible to obtain specific liposome–cell interactions *in vitro*. Literature on the *in vivo* disposition of immunoliposomes is scarce. After intravenous injection effective targeting might fail because: (i) most interesting determinants are sequestered behind endothelial or other barriers, penetration of the immunoliposomes through these barriers is required; (ii) modulation and heterogeneity in the expression of cell surface determinants is frequently encountered; (iii) the protein-coated liposomes are suspected of immunogenicity (*cf.* Section 25.5.2); and (iv) liposomes are rapidly removed from the circulation by phagocytic cells of the RES. Many of these problems are similar to those encountered with immunotoxins (*cf.* Section 25.5.5.1) or antibody-conjugated drugs (*cf.* Section 25.5.5.2). In addition, given that specific binding is obtained, it must be kept in mind that the specific association of antibody-coated liposomes with cells is not a guarantee for intracellular delivery of the liposome contents. With encapsulated radiodiagnostics intracellular delivery might not be required, but for most medical applications it is.

Considering the anatomy of the vascular system, successful targeting of circulating liposomes to extravascular sites after intravenous injection has poor prospects. Nevertheless, significant opportunities may exist for targeting liposomes to cell types within the vasculature. In active targeting of liposomes to cells other than mononuclear phagocytes within the vascular system, the preferential uptake of liposomes by the RES and circulating monocytes remains a critical issue. In fact, coupling of antibodies to liposomes may result in an even more rapid clearance by mononuclear phagocytes if antibodies are coupled in a position leaving the Fc region exposed and available for binding to Fc receptors on these phagocytic cells. Therefore, in theory, the use of coupled Fab' or $F(ab')_2$ fragments is preferred. Another problem concerning the use of antibodies to target liposomes to specific cells has been revealed in studies of Leserman and coworkers.[145,153,154] These investigators examined targeting of liposomes containing methotrexate (MTX) to mouse T and B lymphocytes using a series of MAB directed against three different cell surface antigens. MAB-mediated binding of a liposome to the cell surface did not necessarily induce effective transfer of liposomal drug. These investigators observed that the capacity of a liposome to enter lymphocytes clearly depends on the cell types (T or B lymphocyte) and the surface molecule to which the liposome binds. Binding and uptake of liposomes by the same cell type was found to differ significantly depending on the target cell antigen selected. Liposomes directed against the same target cell antigen on B and T cells also exhibited significantly different binding and internalization patterns. It was even found that liposome uptake by the same cell type was significantly different when MAB directed against different epitopes on the same target cell antigen were used as targeting ligands. Thus both the nature of the target antigen and the characteristics of the target cell population are likely to be important factors in the development of approaches to antibody-mediated targeting. An additional problem in antibody-mediated targeting may be the delivery of the liposomal contents to the cellular endosomes or lysosomes where, unless the entrapped molecule can effectively escape, it is degraded and inactivated by the lysosomal enzymes. To circumvent this problem, the use of pH sensitive immunoliposomes has been advocated.[155,156] Such liposomes, with unique lipid compositions, possess the ability to become fusogenic under mildly acidic conditions. Exposure of these liposomes to a pH between 5.0 and 6.5, which is the pH range encountered in endosomes, causes them to fuse with endosomal membranes resulting in delivery of the entrapped contents of the endocytosed liposomes to the cellular cytoplasm. The potential of pH sensitive immunoliposomes for cytoplasmic delivery has

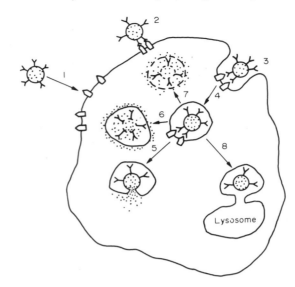

Figure 14 Schematic presentation of the interaction of immunoliposomes with the target L-929 cells. (1) Immunoliposomes bearing antibody (Y) are incubated with target cells expressing antigens (   ). (2) Binding of the immunoliposomes by the cell surface antigens. (3) Initiation of the receptor-mediated endocytosis at the coated pits. (4). Internalization of the immunoliposomes into the acidic endosomes. (5) Fusion of the pH sensitive immunoliposomes with the endosome membrane, with the liposomal contents released into the cytoplasm. (6) Leakage of the liposomal contents into the endosome with subsequent escape into the cytoplasm. (7) Rupturing of the endosomal vacuole causing release of liposome contents into the cytoplasm. (8) Fusion of the endosome with lysosomes and the delivery of the pH-insensitive immunoliposomes into the lysosomes[156]

been clearly demonstrated *in vitro*. The *in vivo* application of this concept, as well as the elucidation of the exact mode of delivery (Figure 14), requires further investigation.

As research on antibody-mediated targeting of immunoliposomes *in vivo* is in its infancy, only a few clinical studies with immunoliposomes have been published so far. The possibility to direct liposomes to specific subsets of circulating blood cells in the bloodstream and in the bone marrow has many potential clinical applications. One obvious example would be to exploit the abnormal expression of lymphocyte differentiation antigens on leukemia cells as target molecules for antibody-directed targeting of liposomes containing antitumor drugs.[22] Antibody-directed targeting of liposomes to vascular endothelial cells would also be worth investigating. Torchillin *et al.*[157] demonstrated the potential of antibody-mediated targeting for the treatment of cardiovascular diseases. They were able to direct liposomes containing the radioactive marker ^{111}In into necrotic tissue in dogs with experimental myocardium infarction by the use of monospecific anticardiac myosin antibodies, or their fragments, covalently coupled to the surface of liposomes.

25.5.8.1.4 *Physical targeting*

In physical targeting environmental characteristics are used to direct liposomes to a particular location or to cause locally selective release of their contents. It has been shown that, depending on their lipid composition, liposomes can be designed which release entrapped drugs only at increased temperatures.[158-160] This increases the potential for regional therapy because parts of the body can be made hyperthermic with various techniques. Based on observations that the pH in tumor tissue can be lower than in normal tissue, liposomes have been constructed which release their contents much more efficiently between pH 5 and 6 than at pH 7.2, with the intention that drugs will be released preferentially at the low pH at the tumor site.[161] In addition, physical targeting of liposomes using external magnets has been considered for localization of liposomes containing ferromagnetic particles within particular capillary beds.[162]

Physical targeting can solve some of the difficulties inherent in active targeting: the necessity of leaving the bloodstream to reach extravascular targets and the requirement of a relevant cell surface binding site. However, none of the described methods will enter the clinic without a lot of additional work (*e.g.* continued progress in medical physics and clinical application of hyperthermia). Besides, in cancer chemotherapy, physical targeting is not an answer to the problem of metastases. As with surgery, radiation therapy and hyperthermia itself, it is necessary to know the location of the tumor.

25.5.8.1.5 *Compartmental targeting*

In compartmental targeting, the site of administration and the site of effect are related closely. Liposomes have been administered intraperitoneally where they are subsequently drained by the lymphatic circulation as well as by the capillary circulation and then taken up into mediastinal lymph nodes. Thus, for the treatment of, for example, ovarian carcinoma, a tumor that remains confined to the peritoneal cavity for most of its natural history, as well as for the possibility of preferentially directing drugs to metastatic cells in regional lymph nodes, the intraperitoneal route has considerable therapeutic potential.[163-165] Interstitial injection of liposomes is of interest to direct liposomes to regional lymph nodes.[166,167,169,270] Intraarticular administration has been used in attempts to employ liposomes for treatment of arthritis.[178] Injection of liposomes into joints confined the liposomes within the joint and local sustained release of the liposome-associated drug was observed. Liposome-entrapped drugs are hardly distributed over the central nervous system after systemic injection, because of the low permeability of the blood–brain barrier for liposomes (*cf.* Section 25.5.2).[171] To increase the concentration of liposomes in this area, drug-containing liposomes have been injected intracerebrally.[172,173] Finally, direct administration of liposomes into bronchi and alveoli has been studied to treat respiratory distress syndrome[174] and to administer antitumor drugs.[175] Although some promising results have been reported, all projects on compartmental targeting are at a preclinical stage.

25.5.8.2 Microspheres

25.5.8.2.1 *Introduction*

Since the proposal of Kramer[176] to use heat-denatured albumin microspheres for the *in vivo* delivery of mercaptopurine, microspheres have been studied by many groups.[177] Microspheres (the smaller ones are also termed 'nanoparticles') are generally monolithic structures manufactured from biologically acceptable materials. They can differ widely in their drug release and degradation properties. They range in size from 0.01 to 100 μm. Their size and surface characteristics largely determine their fate in the body. The same rules which control the *in vivo* distribution and kinetic behaviour of liposome particles apply to these particles. Intravenously, intraarterially or intraperitoneally injected particles of 0.1 to (approximately) 2 μm in diameter will generally be recognized by the RES macrophages, which leads to accumulation in lysosomes. Particles less than 0.1 μm can reach liver parenchymal cells by passage through the fenestrations of the liver endothelium. Particles between 3 and 12 μm will distribute over the lungs, liver and spleen after intravenous injection as they become entrapped within the capillary networks of these organs.[178] Systemic administration of particles larger than 12 μm results in their entrapment in the first capillary bed encountered, which means that after intravenous injection they accumulate in the lungs. After intraarterial administration this can lead to targeting to a predetermined specific, upstream located, organ (*e.g.* liver, kidney). For tumor-bearing organs, this approach can result in selective targeting to the tumor,[179] probably due to a qualitative and quantitative difference in capillary networks of tumors compared to those of the host organ.[180] Thus, simple alteration of size and/or route of administration of microspheres allows manipulation of their localization. In addition to passive targeting, active targeting approaches are being investigated. As discussed earlier for liposomes, active targeting is based on the alteration of the surface characteristics of particles which are not entrapped in the microvasculature. Examples are the use of hydrophilic agents[181] to suppress opsonization and particle adhesion to macrophages in liver and spleen and the attachment of cell specific ligands (*e.g.* the formation of MAB–particle conjugates).[182,183] Different materials, ranging from natural materials such as albumin, gelatin and starch to synthetic polymers, have been used for the preparation of microspheres. We will confine ourselves mainly to poly(alkyl cyanoacrylate) nanoparticles and albumin microspheres, as the experimental work done with these systems adequately illustrates many general aspects of microspheres.

25.5.8.2.2 *Poly(alkyl cyanoacrylate) nanoparticles*

Poly(alkyl cyanoacrylate) nanoparticles are prepared by emulsion polymerization of alkyl cyanoacrylates following an anionic initiation mechanism.[184,185] The monomer is added to an

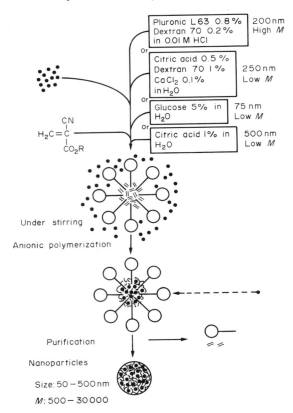

Figure 15   Schematic representation of the preparation of poly(alkyl cyanoacrylate) nanoparticles:[194] ●, drug to be linked
to nanoparticles; =, cyanoacrylic monomer; ——○, surfactant[194]

aqueous solution of a surface active agent (polymerization medium) under vigorous mechanical
stirring (Figure 15). The pH of the polymerization medium lies between 2 and 4. The alkyl
cyanoacrylate polymerizes at ambient temperature and the drug can be dissolved in the polymerization medium either before addition of the monomer or after polymerization. The pH of the medium
determines both the polymerization rate and the degree of adsorption of ionizable drug. If necessary,
the nanoparticles can be purified by ultracentrifugation of the polymeric particles and resuspending
them in an isotonic, surfactant free medium.[184] Electron microscopy of the nanoparticles shows
generally spherical ultrafine particles with a diameter of about 0.2 μm. However, smaller or larger
particles can be obtained by modifying the acidity of the polymerization medium. The inner
structure is highly porous and a large specific area is available for drug sorption.[186] Little is known
about the exact physicochemical nature of the drug binding mechanism or mechanisms.

The polymers used for the preparation of nanoparticles have been studied extensively as
hemostatic agents and tissue adhesives, and were found to be well tolerated, nontoxic and
biodegradable.[187,188] The rate of degradation *in vivo* increases with decreasing side chain length.
After intravenous injection into mice, over 75% of the poly(butyl cyanoacrylate) nanoparticle
radioactivity had left the body within 24 h, while with poly(hexyl cyanoacrylate) nanoparticles, 80%
of the radioactivity still remained in the body after 3 d.[189] The rate of degradation can be
manipulated by mixing different cyanoacrylates.[190] A major concern is the formaldehyde production observed by many authors during *in vitro* degradation of poly(alkyl cyanoacrylate)s. Therefore,
Lenearts[191] studied the contribution of the formaldehyde pathway to the degradation of
poly(isobutyl cyanoacrylate) nanoparticles *in vitro* under biological conditions. The formaldehyde-
producing degradation route occurs with a very low efficiency. The degradation mechanism was
primarily a function of enzymatic degradation reactions (hydrolysis of the ester functions of the
polymer). It is still not clear whether formaldehyde formation, indeed, is irrelevant for the *in vivo*
situation. The same group of investigators proved that drug release from nanoparticles is due to the
degradation of the polymer. An excellent correlation between the release of [³H]actinomycin D
from poly(isobutyl cyanoacrylate) nanoparticles and the degradation rate of the polymer was found.

The degradation rate depended on the length of the alkyl chain of the polymer used. Thus, degradation kinetics as well as release of the bound drug can be manipulated.

Poly(alkyl cyanoacrylate) nanoparticles have been used as carriers for antitumor agents in several studies. Brasseur *et al.*[192] observed a considerably enhanced antitumor effect against a soft tissue carcinoma (S250) with actinomycin D-loaded poly(methyl cyanoacrylate) nanoparticles, in comparison to the free drug. Nanoparticles without drug had no influence on the tumor growth. Kreuter and Hartmann[193] showed an enhanced efficacy of 5-fluorouracil against Crocker sarcoma S180 by binding to nanoparticles. The increased efficacy, however, was coupled with a higher toxicity of the drug measured by induced leukopenia, body weight loss, and premature death. Couvreur *et al.*[194] used L1210 leukemia as a tumor model. Based on the survival time of mice, doxorubicin loaded on to nanoparticles showed a higher overall effectiveness than the unbound drug. These investigators also showed that binding of doxorubicin to nanoparticles resulted in a considerable decrease in accumulation in cardiac tissue and yielded a significant decrease in acute toxicity of this drug.[195] Based on recent information on the mode of action of doxorubicin liposomes,[88,90] it may be proposed that sustained release of the drug either from circulating nanoparticles, or from macrophages which have ingested nanoparticles, is the reason for these beneficial effects. In addition to their potential use as carriers for antitumor drugs, poly(alkyl cyanoacrylate) nanoparticles were considered for the parenteral administration of insulin[196] and for the treatment of parasitic (leishmaniasis) and infectious diseases (cryptococcis, histoplasmosis).[197]

25.5.8.2.3 *Albumin microspheres*

One of the frequently utilized carrier materials for the preparation of microspheres is serum albumin from humans or animal species. There are basically two methods for the production of albumin microspheres.[178] The regularly used methods involve either thermal denaturation at elevated temperatures (95–170 °C) or chemical cross-linking in vegetable oil or paraffin emulsions. For use as a diagnostic imaging carrier, nondrug-bearing microspheres may be prepared using either a simple, one-step preparative method involving thermal degradation of a protein aerosol in a gas medium,[199] or an aerosol step followed by denaturation in an oil.[200] Although this aerosol denaturation method appears to be promising for large-scale manufacture of microspheres, the following discussion is limited to drug incorporation in albumin microspheres prepared by chemical cross-linking of the albumin in a well-defined water-in-oil emulsion, a procedure used for laboratory scale production (Figure 16). The time for preparation of the microspheres ranged between 10 min and 16 h, depending upon the degree of microsphere hardening required. The microspheres prepared according to the scheme depicted in Figure 16 are monolithic and spherical. The mean diameter is found to be highly dependent on the rate of stirring in the mix cell and the nature of the continuous oil phase used. Figure 17 illustrates the effect of mix cell stirring speed on the mean diameter of produced microspheres using pure olive oil as the continuous emulsion phase. In addition to stirring speed, the protein amount, oil and aqueous phase volumes, and amounts of cross-linker used influence the average size of the spheres. Drugs may be incorporated into albumin microspheres by simply replacing the aqueous protein solution with one which also contains the drug. After washing

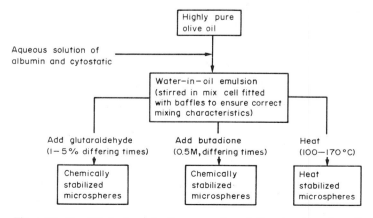

Figure 16 Simplified scheme for the preparation of albumin microspheres[27]

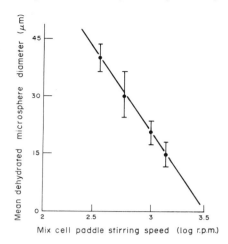

Figure 17 Effect of mix cell stirring speed on the mean diameter (as determined with a Coulter counter) of albumin microspheres[27]

and collection, the microspheres are lyophilized and stored at 4 °C. To ensure quite narrow size ranges, freeze-dried microspheres may be fractionated by first suspending them in chloroform, followed by sieving using microsieves of various sieve openings. The extent of drug incorporation depends upon the physicochemical nature of the drug (*e.g.* its solubility in the aqueous phase) and the amount of protein used. Drugs may be located within the matrix of the carrier or on the surface of the sphere. The payload of drug that can be incorporated is often higher than for liposomes of comparable size. The drug content can be as high as 87% of the total weight of formed microspheres.[201]

A large number of factors have been reported to influence the release of drugs from albumin microspheres.[178] These are connected to the nature of the drug, the nature of the microsphere and/or the environment. The release of materials *in vivo* depends on a combination of the physical and physicochemical characteristics of the drug and the microsphere and enzymatic degradation of the sphere. The challenge in the design of microspheres is to find a balance between microsphere stability (to achieve adequate product presentation and biological half-lives) and microsphere biodegradability (to permit the drug to be released and repeated dosing to occur). This balance is influenced by a lot of factors, including the extent and nature of cross-linking, the size of the spheres and the amount of drug incorporated. Release of drug from albumin is generally biphasic,[178,201] with an initial fast release phase ('burst effect') followed by a slower first-order release. The burst effect can be a problem, as between 20% and 95% of the amount of drug originally incorporated can be released within 5 min. This latter high release rate is found for highly water soluble compounds, and, although pretreatment by ultrasonication will reduce the burst effect, is definitely unacceptable. For these compounds a covalent linkage between drug and carrier is indicated, with degradation kinetics being dependent on, for example, enzymatic attack. For less water soluble compounds, and for compounds that specifically interact with serum albumin, prolonged release occurs. One of the most important parameters to control the release rate of drugs is the diameter of the albumin microspheres. The smaller the size of the microspheres, the faster the release of the drug, especially in the second phase of the release process.[178]

For therapeutic applications of albumin microspheres, basically two distinct approaches have been used: (i) the albumin microspheres were injected into the general circulation and (ii) the microspheres were administered directly into the tumor-bearing organ by means of local intra-arterial injection. Intravenously injected albumin microspheres in the size range of 0.1–1.0 μm have been considered for the delivery of immunosuppressives, for treatment of RES manifestations (*e.g.* histoplasmosis and typhoid) and for treatment of liver cancer.[19] Willmott *et al.*[202] injected doxorubicin-loaded albumin microspheres, with mean diameters between 12–15 μm, intravenously into rats, and they observed that the microspheres accumulated in the lungs by entrapment in the capillary bed, whereafter the particles released the drug slowly over a period of at least 24 h. Studies by Tomlinson *et al.*[178,198] addressed the intraarterial delivery of drug-containing albumin microspheres, in the size range of 15–25 μm, to tumor-bearing organs and for controlled release after local administration into sites such as lung and joints. Intraarterial injection of radioactively labeled microspheres in this size range leads to excellent imaging of tumors due to selective blockage of their capillaries.[203] Fujimoto *et al.*[204] reported that intraarterial treatment of rabbits bearing an implanted VX-2 carcinoma with mitomycin C-loaded albumin microspheres, with diameters of around 45 μm, increased the survival time significantly. The same mitomycin microspheres and the same mode of treatment, employing selective arterial catheterization, has been used in patients with malignant hepatic tumors.[205] In general, the hepatic tumor began to shrink a few days after start of treatment and a maximal reduction was observed after three to four weeks. Side effects were mild and temporary. These results strongly suggest that this form of chemotherapy utilizing arterial chemoembolization has great clinical potential for selected patients.

In order to circumvent the RES, Widder *et al.*[206,207] proposed the use of magnetic force for the site specific drug delivery with albumin microspheres containing magnetite (Fe_3O_4) in the treatment of localized disease processes. The magnetically responsive microspheres (approximately 1.0 μm) are infused into an artery supplying a given *in vivo* target site. A magnet of sufficient field strength to retard the microspheres solely at the capillary circulation (approaching 100% retention in some cases) is placed externally over the target area, and drug diffusion out of the microspheres may occur at or near the target site. Ideally transit of the particles from the microvasculature into the extravascular compartment will also occur (thus establishing an extravascular drug depot for sustained release in the target area). The ability to restrict drug activity to a specific body region allows administration of small doses and results in decreased side effects. Treatment of localized cancers (whether curative or palliative) or ablation of aberrant physiological functions (*e.g.* hyperthyroidism) are a few of the many fields of application in which the concentrating ability of this targeting system could be of value. Widder *et al.*[208] demonstrated selective targeting of doxorubicin-loaded magnetic albumin microspheres to Yoshida sarcoma tumors implanted in the tails of rats, and marked antitumor effects were observed. However, an obvious limitation of this approach is that it is only applicable to relatively large, easily accessible, and well vascularized tumors that can be catheterized and subjected to localized magnetic fields. Thus, it does not address the prime challenge of cancer chemotherapy, the removal of multiple occult metastases.

Albumin microspheres have also been administered *via* other routes. Royer and coworkers[209-211] administered progesterone-containing microspheres (100–200 μm in diameter) to rabbits both subcutaneously and intramuscularly. Following an initial burst of drug, sustained levels of approximately 1 ng mL^{-1} progesterone were measured in the blood of the test animals for at least 20 d after injection. Examination of the injection sites demonstrated that the microspheres administered intramuscularly were completely biodegraded by the end of two months with no adverse immunological response. Microspheres injected subcutaneously also disappeared completely but at a slower, unspecified rate.[209] Goosen *et al.*[212] entrapped insulin crystals in bovine serum albumin microspheres (containing about 20% insulin by weight). A single subcutaneous injection of these microspheres produced elevated blood insulin levels in diabetic rats for more than two months. Subsequent histological studies indicated that the microsphere implant sites were surrounded with a fibrous capsule which may have further retarded insulin release from the microspheres. Complete *in vivo* degradation of these insulin-containing microspheres took more than five months.

25.5.8.2.4 *Other types of microsphere*

Starch microspheres made of cross-linked polysaccharide derivatives have been evaluated for intraarterial drug delivery.[213,214] These microspheres are approximately 15 to 100 μm in diameter. They can be administered in combination with a drug or a radionuclide. When infused intraarterially, they form a starch gel that occludes the artery. This increases the concentration of drug in the target area because the drug is bound by the starch gel occluding the artery and is released as the starch is digested by circulating enzymes (amylase). The rate of degradation can be controlled by the degree of cross-linking. In addition to the potential use of drug-laden starch microspheres for cancer chemotherapy, there appear to be other potential uses for this technique in cancer therapy. Hypoxia resulting from the transient ischemia caused by microsphere occlusion may be effective in protecting healthy tissue from radiotherapy damage. It is also possible that the starch microspheres may be useful in hyperthermic treatment of tumors since the ability to heat selected tissues is highly dependent upon the blood flow through the tissue. A temporary reduction or arrest of blood flow by intraarterial administration of starch microspheres may reduce the cooling effect of the blood flow and thus facilitate local hyperthermia.

Starch microspheres cross-linked with polyacrylamide have been studied in order to be used in enzyme replacement therapy for the treatment of certain lysosomal storage diseases and in other forms of therapy, *e.g.* for the delivery of L-asparaginase against certain types of tumor cells.[215,216] Kato *et al.*[168,217-219] used the technique of transcatheter embolization to deliver mitomycin C-containing microspheres prepared from *ethylcellulose* (mean particle size about 225 μm) to particular organs. They reported objective tumor responses after chemoembolization of various tumor-bearing organs resulting from the slow release of drug into surrounding tumor tissue.[219] This approach is undergoing large-scale clinical evaluation in Japan. *Gelatin* microspheres (about 2 μm in size) have been employed for the intralymphatic delivery of cytostatics.[203]

Tökes and coworkers[220,221] explored the use of the anthracycline drug doxorubicin covalently coupled to *polyglutaraldehyde* microspheres (average diameter 0.45 μm). Results obtained in experiments with tumor cells in tissue culture suggest that the microsphere-bound doxorubicin does not have to enter the cell to produce its cytostatic effect. Multiple, repetitive interactions of the drug-containing microspheres with the cell surface induced cytotoxicity. Based on *in vitro* experiments with doxorubicin immobilized on agarose beads, Tritton and coworkers[222-224] also came to the conclusion that doxorubicin can be cytotoxic without even entering the cell. If this is a new mechanism for doxorubicin cytotoxicity, its severe toxicity, as well as resistance of some tumor cell types against this cytostatis, developed upon administration of the drug in its free form, might be overcome. However, successful application of this approach *in vivo* will strongly depend on the possibility for the carrier to reach target sites.

25.5.8.3 Lipoproteins

Plasma proteins have been used or suggested as potential carriers of drugs and enzymes. Indeed, they offer some attractive advantages over the particulate carrier systems that have been described. Their solubility and smaller size allow access to regions not accessible to larger insoluble particles. Another potentially favorable point is their natural origin, being plasma constituents. Virtually all plasma proteins, including antibodies, albumin, lipoproteins, fibrinogen, collagen, lectins and hormones, have been proposed at one time or another as potential carrier molecules.[1,225] In general, macromolecular carriers derived from plasma can be divided into two groups: (i) carriers that manifest specific binding to cell surface receptors, and (ii) carriers that do not. Most relevant to site specific drug delivery are the plasma proteins which belong to the first group. The most pronounced example of carriers with specific binding are antibodies, and in a separate section of this chapter (Section 25.5.5) this group of molecules has been described. Lipoproteins are another class of plasma proteins that have attracted attention for targeted drug delivery.

Lipids (cholesterol, triacylglycerols and phospholipids) are transported in the blood as components of lipid–protein complexes, the lipoproteins. The average size of the four main classes of lipoproteins is about 10 nm for high density lipoproteins (HDL), about 25 nm for low density lipoproteins (LDL), 30–90 nm for very low density lipoproteins (VLDL) and 10–100 nm for chylomicrons.[226] LDL is the major cholesterol-carrying lipoprotein in human plasma. The spherical LDL particles contain a lipid core of about 1500 cholesteryl ester molecules surrounded by a shell of free cholesterol, phospholipids and protein (called apolipoprotein).[227] Human cells express cell surface receptors for LDL.[227] Goldstein and Brown[269] have described the binding of LDL to specific LDL receptors and their subsequent internalization in a number of cell systems (including human fibroblasts and lymphocytes). Once bound to its receptor, LDL is internalized and degraded in lysosomes. The lipid core of LDL yields unesterified cholesterol, which is used for membrane synthesis, whereas the protein part of LDL is degraded to amino acids.[227] This same group[228] has also described the extraction of endogenous cholesteryl esters of LDL and reconstitution with exogenous cholesterol linoleate. It has been shown that the free and esterified cholesterol of LDL can be extracted with heptane and the particles reconstituted with exogenous hydrophobic compounds. Such reconstituted LDL binds to LDL receptors with the same affinity as native LDL and is internalized and degraded intracellularly. Taking advantage of these results, some investigators incorporated lipid soluble drugs into reconstituted LDL, attempting to direct therapeutic agents to specific cell types (extrahepatic and nonreticuloendothelial tissue) possessing specific, high affinity receptor sites for LDL (such as rapidly growing cells, *e.g.* tumor cells, which express high levels of LDL receptors). For example, leukemic cells isolated from patients with acute myelogenous leukemia are reported to have much higher LDL receptor activities than normal white blood cells and nucleated bone marrow cells.[273] Gynecological cancer cells also possess high LDL receptor activity both when assayed in monolayer culture and in membrane preparations derived from tumor-bearing nude mice.[229] An enhanced receptor-mediated uptake of LDL by tumor tissue *in vivo* was demonstrated in an animal model,[230] illustrating the potential advantages of this approach.

Vitols *et al.*[231] examined LDL as a carrier for cytotoxic drugs against the above mentioned leukemic cells isolated from patients with acute myelogenous leukemia. They incorporated the very lipophilic cytotoxic agent *N*-(trifluoroacetyl)doxorubicin 14-valerate. When white blood cells were incubated with the cytotoxic drug–LDL complex, cells with high LDL receptor activity accumulated more drug than cells with low receptor activity. The same effect was also observed when aclacinomycin A was incorporated into LDL by a mixing technique and the drug uptake studied in cultured human glioma cells.[232] However, the results indicated that in both cases a large part of the cellular drug uptake probably was due to uptake of drug leaked from the drug–LDL complex. More recently, experiments with another lipophilic derivative of doxorubicin, *N*-(*N*-retinoyl)-L-leucyldoxorubicin 14-linoleate, incorporated into LDL (100–200 drug molecules per LDL particle) showed better results.[233] When cultured normal human fibroblasts were incubated with the LDL–drug complex, there was a perfect correlation between the cellular uptake plus degradation of LDL and the cellular drug accumulation. Control experiments suggested that the drug incorporated into LDL was delivered to cells selectively by the LDL receptor pathway. Studies are now in progress to test whether the drug incorporated into the LDL complex can be targeted to malignant cells with high LDL receptor activity *in vivo*.

Of interest also is the work of Van Berkel *et al.*[29] They outlined a strategy to achieve targeting of lipoproteins to specific cell types present in one organ (in this case the liver). They employed a galactose-terminated cholesterol derivative (Tris-Gal-Chol) which was synthesized in order to utilize galactose receptors expressed on liver cells. They showed elegantly that in a complex tissue such as the liver, successful targeting of lipoproteins to the cell of choice can be achieved. Tris-Gal-Chol modified LDL was directed selectively to Kupffer cells and Tris-Gal-Chol modified HDL selectively to hepatocytes. Furthermore, acetylated LDL was introduced rapidly and selectively into endothelial cells. The authors speculated that, provided that basic knowledge of, in particular, receptor characteristics is available, similar approaches are possible for other cell types or tissues.

25.5.8.4 Cellular Carriers

25.5.8.4.1 *Erythrocytes*

Erythrocytes have been proposed as cellular drug carriers.[234–236] Since these cells are regularly used in the clinic for transfusions, much is known about the techniques for collecting and storing cells. Erythrocytes are completely biodegradable and provided that compatible cells are used in patients there is no possibility of a provoked immune response. Carefully prepared resealed erythrocyte preparations can have long circulation times (days)[237,238] and thus serve as circulating drug depots. The relatively large internal volume of red cells, as compared to other particulate systems, can allow large amounts of drug to be carried. Of particular interest is their use as carriers of therapeutic enzymes (enzymes as drugs) acting within the cellular carrier over prolonged periods of time in order to degrade potentially toxic metabolites or their metabolic precursors in the bloodstream.[239] They also have considerable potential for carrying other therapeutic agents required to act directly in the bloodstream or within the erythrocyte, for the slow and sustained release of drugs and other agents or their directed delivery to the RES.

Most encapsulation techniques used for loading materials into erythrocytes involve hypotonic lysis of the cells (so-called 'ghost' formation) followed by restoration of tonicity, which reseals the cells (Figure 18). Substances to be encapsulated are present in the hypotonic solutions. The resealed cells can be further treated by incubation at 37 °C ('annealing'). The preparation is washed prior to use to remove nonencapsulated material. Early methods were based on loading of erythrocytes by simple osmotic shock in distilled water.[240] Exposure of the cells to water resulted in rapid swelling and formation of large pores or channels in the membrane admitting the substance to be encapsulated. Upon restoration of isotonicity and stabilization by annealing at 37 °C, it was found that most of the cytoplasmic constituents were lost and the reformed cells contained only a few percent of the substance added. If the loaded cells were returned to the circulation, they were rapidly removed by the RES. Today, methods are employed which are accompanied by reduced loss of cytoplasmic constituents and provide good survival characteristics when the cells are returned into the circulation. The principle of the Preswell dilution technique is based on first swelling the erythrocytes without lysis by placing them in a slightly hypotonic solution.[235] The swollen cells are then recovered by centrifugation at low *g* and the cells taken to the point of lysis by addition of relatively small volumes of water. In the dialysis technique, packed cells are placed in dialysis tubes which are immersed in hypotonic media. The use of dialysis tubes results in the retention of cystolic contents in

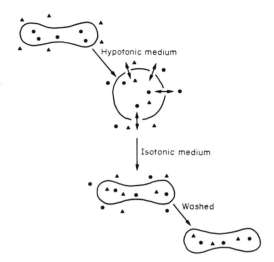

Figure 18   Scheme of an erythrocyte entrapment procedure utilizing osmotic lysis: ●, erythrocyte intracellular contents; ▲, drug[234]

the vicinity of the erythrocytes, thus reducing the loss of cytoplasmic constituents when the cells are resealed.[241] Excellent survival times *in vivo* have been found for cells loaded by dielectric stress.[238] When subjected to intense electric fields, many cells, including erythrocytes, form pores in their membranes prior to complete lysis. With this technique, the voltage as well as the pulse duration applied are important factors influencing the size of the molecule that can be loaded and the survival time *in vivo*. Polyene antibiotics such as amphotericin B increase membrane permeability to ions and metabolites. This property was used to incorporate daunomycin in erythrocytes.[242,243]

The prime field of application of erythrocyte carriers has been enzyme therapy of genetic defects[236] where enzymes are to be used as drugs to degrade circulating metabolites or where targeted delivery to the RES is required for enzyme replacement therapy of selected lysosomal storage diseases. Attempts have been made to treat Gaucher's disease with glucocerebrosidase-loaded erythrocytes.[244]. In this disease, phagocytic cells accumulate glucocerebroside which cannot be degraded because of glucocerebrosidase deficiency. The enzyme-containing erythrocytes, slightly damaged by osmotic or electric lysis, will be taken up by the RES, thus conveying the enzyme to the target site. The results of the clinical trial, however, were inconclusive. Probably, the experiments were premature as, for instance, the authors did not ascertain that an adequate amount of enzyme had been administered. Another clinical trial led to more success. Cell-entrapped desferrioxamine injected into patients with iron overload proved more efficient in chelating the metal and promoting its urinary clearance than the drug used alone.[245] As might be expected, most reported studies were performed with animals. Ihler *et al.*[246] showed that uricase-loaded erythrocytes are effective in lowering blood levels of uric acid. The rate-controlling step in the process is the rate of substrate uptake by the red blood cells. Asparaginase encapsulated in erythrocytes has potential for antileukemic therapy. In several animal species[237] this approach resulted in significantly depressed plasma levels of asparagine. Using the same system in mice, even complete removal of asparagine from plasma over a period has been reported.[247] Encapsulated corticosteroid phosphate esters were found to be superior to free steroid phosphates administered intravenously in the treatment of adjuvant-induced arthritis in the rat.[248] Kitao *et al.*[243] established an enhanced survival time for mice with the L1210 tumor if the animals received erythrocytes loaded with daunomycin. When the loaded cells were, in addition, pretreated with wheat-germ agglutinin, the survival time increased further. The authors concluded that the lectin was causing a preferential adherence of the loaded erythrocytes to the L1210 cells.

The main advantage of the use of erythrocytes over other carrier systems is that they survive much longer when administered *in vivo*. However, small changes in the erythrocyte biochemistry occurring during the preparation procedure can reduce the circulation lifetime drastically (from days to minutes).[249–251] Another problem is that erythrocytes are inherently unstable *in vitro* even under the best storage conditions, such as those used in blood banking. Thus, the normal storage time for erythrocytes used for transfusion is about three weeks. Therefore, it will be necessary to find ways to preserve resealed erythrocytes for prolonged periods of time. Additionally, although red blood cells are of interest in some examples of enzyme replacement therapy, the indications for therapy with erythrocyte carriers seem rather limited as the sites of pathogenesis in many other enzyme deficiency diseases (*e.g.* those residing in CNS cells) are not accessible to erythrocytes.

25.5.8.4.2 *Other cells*

Although erythrocytes have gained most interest, other cell types have also been considered for the delivery of drugs or diagnostic agents and are therefore worth mentioning. *Neutrophils* localize spontaneously in inflammatory lesions, and these cells, loaded with ^{111}In-oxine, have provided a method for imaging lesions in both animals and patients.[252,253] Some clinical success has also been obtained with the use of *platelets* in the treatment of patients with refractory idiopathic thrombocytopenia purpura:[254] vinblastine-containing platelets, on injection, are coated with the antiplatelet antibody present in the blood of these patients and then taken up by macrophages. This leads to macrophage dysfunction or death and avoidance of platelet destruction. Other workers have investigated islet implantation in the treatment of experimental diabetes[255,256] and cell (*e.g.* fibroblast) implantation in the treatment of enzyme deficiencies in man.[256–258] *Implanted cells* lodge in various parts of the body, from where they may secrete the missing hormone or enzyme. More recently, the possibility of using total bone marrow transplantation as a means of treating certain inborn errors of metabolism has been demonstrated.[259,260] The rationale for this treatment is that if the metabolic error is expressed in leukocytes, then the transplanted leukocytes will repopulate the recipient, displace the abnormal bone marrow and thereby provide a lasting source of leukocytes synthesizing the missing normal enzyme. As most of the examples concerning the use of other cell types as a 'drug' (enzyme or hormone) carrier are still in an early experimental stage, it is not possible to make a thorough evaluation.

25.5.9 CONCLUSIONS

By far the most attention in drug targeting using particulate systems has been directed to liposomes. Being composed of natural body constituents and being biodegradable and versatile with respect to their structure are all points in favor of liposomes. Proper selection of the bilayer components can minimize the occurrence of toxic side effects due to the lipids used. There has been considerable concern about the problem of lack of stability of drug-containing liposomes during prolonged storage *in vitro*. Therefore, more attention has to be paid in the future to the preparation of liposomes with a pharmaceutically acceptable long-term stability. New developments such as concentrating aqueous liposome dispersions or freeze drying of the dispersions show promising results, in particular for lipophilic compounds.[88,261,262]

In the case of microspheres as drug delivery systems, a wide variety of (biodegradable) materials and manufacturing methods are available. There has been little concern about the *in vitro* stability of protein or polymer microspheres; they can be freeze-dried and thus should be reasonably stable during prolonged storage. However, in some cases freeze-drying of microspheres can lead to difficulties in redispersion of the particles in aqueous media due to air adsorption or changes in the surface structure.[263] This problem can be overcome by the addition of small amounts of surfactants.

Erythrocytes have some advantages over synthetic or semisynthetic systems because the loaded cells can circulate in the blood for relatively long periods. Furthermore, they are somewhat different from other drug carriers by being naturally biodegradable cells that are nonimmunogenic if crossmatched cells or, in particular, the patient's own cells are used. However, with respect to industrial, large scale development, their main disadvantage is their inherent instability *in vitro*.

LDL have the advantage of possessing receptor specificity which may be exploited for site specific targeting. The targeting possibilities of lipoproteins are limited to lipophilic compounds.

Although liposomes, biodegradable microspheres and cells seem all relatively innocuous in terms of short-term toxicity, the issue of chronic toxicity still needs to be seriously addressed.

25.5.10 PROSPECTS

This chapter clearly indicates that during the last decade substantial progress has been made in the development of drug formulations with site specific delivery of the drug. In some cases site avoidance drug delivery has also provided beneficial therapeutic effects. A lot of essential information has become available on the anatomical, physiological and immunological barriers that may interfere with the drug targeting process. General rules concerning the behaviour of different types of carrier systems have been derived. This has allowed the development of formulations for site specific drug delivery, up to a certain degree, *via* rational design instead of *via* trial and error. Limited access to the target tissue and target cells and poor specificity for the target cell interaction were often reported to be major obstacles. The improvement of the performance of the carrier systems and homing devices with respect to targeting efficiency is the major challenge for the future. Research will be focused on the development of formulations with targeted drug delivery for life threatening diseases such as cancer and diseases caused by certain viral, microbial and parasitic pathogens.

A number of formulations for targeted drug delivery are now in the process of clinical testing (*e.g.* monoclonal antibodies, immunotoxins, liposomes). Little information is available on the long term toxicity and immunogenicity of these products. Other questions to be addressed concern typical pharmaceutical problems such as large-scale production, quality control and the long-term stability of these products.

ACKNOWLEDGEMENTS

The authors acknowledge Prof. Dr. D. K. F. Meijer, University of Groningen and Dr. E. J. E. G. Bast, University of Utrecht, for critical reading of parts of the manuscript.

25.5.11 REFERENCES

1. M. J. Poznansky and R. L. Juliano, *Pharmacol. Rev.*, 1984 **36**, 277.
2. P. Stahl and A. L. Schwartz, *J. Clin. Invest.*, 1986, **77**, 657.
3. R. L. Juliano, in 'Controlled Drug Delivery', ed. J. R. Robinson and V. H. L. Lee, Dekker, New York, 1987, p. 555.
4. A. Martinez-Hernandez, in 'Microcirculation', ed. R. Effros, Academic Press, New York, 1981, p. 125.
5. M. Bundgaard, *Annu. Rev. Physiol.*, 1980, **42**, 325.
6. G. Poste, R. Kirsch and P. Bugelski, in 'Novel Approaches to Cancer Chemotherapy', ed. P. S. Sunkara, Academic Press, New York, 1984, p. 165.
7. H. L. Duthie and K. G. Wormsley (eds.), 'Scientific Basis of Gastroenterology', Churchill Livingstone, Edinburgh, 1979.
8. J. Meier, H. Rettig and H. Hess, 'Biopharmazie', Thieme, Stuttgart, 1981.
9. L. F. Prescott and W. S. Nimmo (eds.), 'Drug Absorption', ADIS Press, New York, 1981.
10. H. M. Patel and B. E. Ryman, in 'Liposomes: from Physical Structure to Therapeutic Applications', ed. C. G. Knight, Elsevier, Amsterdam, 1981, p. 409.
11. P. P. Speiser, in 'Microspheres and Drug Therapy. Pharmaceutical, Immunological and Medical Aspects', ed. S. S. Davis, L. Illum, J. G. McVie and E. Tomlinson, Elsevier, Amsterdam, 1984, p. 339.
12. M. E. LeFevre and D. D. Joel, *Life Sci.*, 1977, **21**, 1403.
13. G. Volkheimer, *Adv. Pharmacol. Chemother.*, 1977, **14**, 163.
14. N. Bodor, in 'Methods in Enzymology 112A', ed. K. J. Widder and R. Green, Academic Press, Orlando, 1985, p. 381.
15. R. L. Juliano, in 'Liposomes: from Physical Structure to Therapeutic Applications', ed. C. G. Knight, Elsevier, Amsterdam, 1981, p. 391.
16. V. J. Stella and K. J. Himmelstein, in 'Directed Drug Delivery', ed. R. T. Borchardt, A. J. Repta and V. J. Stella, Humana Press, Clifton, 1985, p. 247.
17. C. A. Hunt, R. D. MacGregor and R. A. Siegal, *Pharm. Res.*, 1986, **3**, 333.
18. J. N. Weinstein, C. D. V. Black, J. Barbet, R. R. Eger, R. J. Parker, O. D. Holton, J. L. Mulshine, A. M. Keenan, S. M. Larson, J. A. Carrasquillo, S. M. Sieber and D. G. Covell, in 'Site-specific Drug Delivery', ed. E. Tomlinson and S. S. Davis, Wiley, Chichester, 1986, p. 81.
19. E. Tomlinson, in 'Advances in Drug Delivery Systems', ed. J. M. Anderson and S. W. Kim, Elsevier, Amsterdam, 1986, p. 385.
20. C. R. Hopkins, in 'Site-specific Drug Delivery', ed. E. Tomlinson and S. S. Davis, Wiley, Chichester, 1986, p. 27.
21. P. A. Kramer, in 'Optimization of Drug Delivery', ed. H. Bundgaard, A. Bagger Hansen and H. Kofod, Munksgaard, Copenhagen, 1982, p. 239.
22. G. Poste, in 'Receptor-mediated Targeting of Drugs', ed. G. Gregoriadis, G. Poste, J. Senior and A. Trouet, Plenum Press, New York, 1984, p. 427.
23. J. N. Lowder, in 'Current Problems in Cancer', ed. R. C. Hickey, Year Book, Chicago, 1986, vol. 10, p. 487.
24. T. P. Groen, in 'Tumor Immunology—Mechanisms, Diagnosis, Therapy', ed. W. Den Otter and E. J. Ruitenberg, Elsevier, Amsterdam, 1987, p. 13.
25. K. J. Widder and A. E. Senyei, in 'Microspheres and Drug Therapy. Pharmaceutical, Immunological and Medical Aspects', ed. S. S. Davis, L. Illum, J. G. McVie and E. Tomlinson, Elsevier, Amsterdam, 1984, p. 393.
26. R. L. Magin and J. N. Weinstein, in 'Lipsome Technology', ed. G. Gregoriadis, CRC Press, Boca Raton, 1984, vol. 3, p. 137.
27. E. Tomlinson, J. J. Burger, J. G. McVie and K. Hoefnagel, in 'Recent Advances in Drug Delivery Systems', ed. J. M. Anderson and S. W. Kim, Plenum Press, New York, 1984, p. 199.
28. J. W. B. Bradfield, in 'Microspheres and Drug Therapy. Pharmaceutical, Immunological and Medical Aspects', S. S. Davis, L. Illum, J. G. McVie and E. Tomlinson, Elsevier, Amsterdam, 1984, p. 25.

29. T. J. C. Van Berkel, J. K. Kruijt, L. Harkes, J. F. Nagelkerke, H. Spanjer and H. M. Kempen, in 'Site-specific Drug Delivery', ed. E. Tomlinson and S. S. Davis, Wiley, Chichester, 1986, p. 49.

30. S. S. Davis and L. Illum, in 'Site-specific Drug Delivery', ed. E. Tomlinson and S. S. Davis, Wiley, Chichester, 1986, p. 93.

31. S. S. Davis, S. J. Douglas, L. Illum, P. D. E. Jones, E. Mak and R. H. Muller, in 'Targeting of Drugs with Synthetic Systems', ed. G. Gregoriadis, J. Senior and G. Poste, Plenum Press, New York, 1986, p. 123.

32. E. Hurwitz, in 'Optimization of Drug Delivery', ed. H. Bundgaard, A. Bagger Hansen and H. Kofod, Munksgaard, Copenhagen, 1982, p. 251.

33. T. R. Tritton and L. B. Wingard, in 'Microspheres and Drug Therapy. Pharmaceutical, Immunological and Medical Aspects', ed. S. S Davis, L. Illum, J. G. McVie and E. Tomlinson, Elsevier, Amsterdam, 1984, p. 129.

34. Z. A. Tokes, K. L. Ross and K. E. Ross, in 'Microspheres and Drug Therapy. Pharmaceutical, Immunological and Medical Aspects', ed. S. S. Davis, L. Illum, J. G. McVie and E. Tomlinson, Elsevier, Amsterdam, 1984, p. 139.

35. P. A. H. M. Toonen and D. J. A. Crommelin, *Pharm. Weekbl., Sci. Ed.*, 1983, **5**, 269.

36. G. McKay, P. K. F. Yeung, L. Qualtiere and K. K. Midha, in 'Topics in Pharmaceutical Sciences', ed. D. D. Breimer and P. Speiser, Elsevier, Amsterdam, 1985, p. 19.

37. U. Lovborg, 'Monoclonal Antibodies: Production and Maintenance', Heinemann, London, 1982.

38. T. A. Springer, 'Hybridoma Technology in Biosciences and Medicine', Plenum Press, New York, 1985.

39. S. R. Samoilovich, C. B. Dugan and A. J. L. Macario, *J. Immunol. Methods*, 1987, **101**, 153.

40. B. L. Ferraiolo and L. Z. Benet, in 'Topics in Pharmaceutical Sciences', ed. D. D. Breimer and P. Speiser, Elsevier, Amsterdam, 1985, p. 3.

41. Committee for Proprietary Medicinal Products. Notes on Requirements for the Production and Quality Control of Monoclonal Antibodies of Murine Origin Intended for Use in Man (Draft 6, III/859/86-EN), 1987.

42. G. Gregoriadis, G. Poste, J. Senior and A. Trouet (eds.), 'Receptor-mediated Targeting of Drugs', Plenum Press, New York, 1984.

43. Van der Sluys, Ph. D. Thesis, State University Groningen, 1987.

44. A. J. Cumber, J. A. Forrester, B. M. J. Foxwell, W. C. J. Ross and P. E. Thorpe, in 'Methods in Enzymology', ed. K. J. Widder and R. Green, Academic Press, Orlando, 1985, p. 207.

45. F. K. Jansen, H. E. Blythman and D. Carriere *et al.*, *Immunol. Rev.*, 1982, **62**, 185.

46. V. Raso and M. Basala, in 'Receptor-mediated Targeting of Drugs', ed. G. Gregoriadis, G. Poste, J. Senior and A. Trouet, Plenum Press, New York, 1984, p. 119.

47. V. Raso, *Immunol. Rev.*, 1982, **62**, 93.

48. D. M. Neville, in 'Directed Drug Delivery', ed. R. T. Borchardt, A. J. Repta and V. J. Stella, Humana Press, Clifton, NJ, 1985, p. 211.

49. E. S. Vitetta, R. J. Fulton and J. W. Uhr, in 'Site-specific Drug Delivery', ed. E. Tomlinson and S. S. Davis, Wiley, Chichester, 1986, p. 69.

50. K. E. Hellstrom, I. Hellstrom and G. E. Goodman, in 'Controlled Drug Delivery', ed. J. R. Robinson and V. H. L. Lee, Dekker, New York, 1987, p. 623.

51. T. Ghose, H. Blair, P. Kulkarni, K. Vaughan, S. Norvell, and P. Belitsky, in 'Targeting of Drugs', ed. G. Gregoriadis, J. Senior and A. Trouet, Plenum Press, New York, 1982, p. 55.

52. R. Arnon, in 'Targeting of Drugs', ed. G. Gregoriadis, J. Senior and A. Trouet, Plenum Press, New York, 1982, p. 31.

53. B. Rihova and J. Kopecek, in 'Advances in Drug Delivery Systems', ed. J. M. Anderson and S. W. Kim, Elsevier, Amsterdam, 1986, p. 289.

54. B. E. Ryman, G. M. Barratt, H. M. Patel and N. S. Tuzel, in 'Optimization of Drug Delivery', ed. H. Bundgaard, A. Bagger Hansen and H. Kofod, Munksgaard, Copenhagen, 1982, p. 351.

55. J. D. Kelly, in 'Radionuclide Imaging in Drug Research', ed. C. G. Wilson, J. G. Hardy, M. Frier and S. S. Davis, Croom Helm, London, 1982, p. 39.

56. B. A. Khaw, G. A. Beller and E. Haber, *Circulation*, 1978, **57**, 743.

57. B. A. Khaw, J. A. Mattis, G. Melincoff, H. W. Strauss, H. K. Gold and E. Haber, *Hybridoma*, 1984, **3**, 11.

58. B. A. Khaw, H. K. Gold, T. Yasuda, R. C. Leinbach, M. Kanke, J. T. Fallon, M. Barlai-Kovach, H. W. Strauss, F. Sheehan and E. Haber, *Circulation*, 1986, **74**, 501.

59. T. Y. Shen, in 'Directed Drug Delivery', ed. R. T. Borchardt, A. J. Repta and V. J. Stella, Humana Press, Clifton, NJ, 1985, p. 231.

60. L. Fiume, C. Busi, A. Mattioli, P. G. Balboni, G. Barbanti-Brodano and T. Wieland, in 'Targeting of Drugs', ed. G. Gregoriadis, J. Senior and A. Trouet, Plenum Press, New York, 1982, p. 1.

61. L. Fiume, C. Busi and A. Mattioli, in 'Topics in Pharmaceutical Sciences', ed. D. D. Breimer and P. Speiser, Elsevier, Amsterdam, 1983, p. 317.

62. Y. J. Schneider, J. Abarca, E. Aboud-Pirsak, R. Barain, F. Ceulemans, D. Deprez-DeCampeneere, B. LeSur, M. Masquelier, C. Otte-Slachmuylder, D. Rolin-Van Swieten and A. Trouet, in 'Receptor-mediated Targeting of Drugs', ed. G. Gregoriadis, G. Poste, J. Senior and A. Trouet, Plenum Press, New York, 1984, p. 1.

63. Q. G. C. M. Van Hoesel, P. A. Steerenberg, D. J. A. Crommelin, A. Van Dijk, W. Van Oort, S. Klein, J. M. C. Douze, D. J. De Wildt and F. C. Hillen, *Cancer Res.*, 1984, **44**, 3698.

64. H. Bundgaard, in 'Topics in Pharmaceutical Sciences', ed. D. D. Breimer and P. Speiser, Elsevier, Amsterdam, 1983, p. 329.

65. H. Bundgaard, A. Bagger Hansen and H. Kofod (eds.), 'Optimization of Drug Delivery', Munksgaard, Copenhagen, 1982.

66. C. R. Gardner and J. Alexander, in 'Drug Targeting', ed. P. Buri and A. Gumma, Elsevier, Amsterdam, 1985, p. 145.

67. N. Bodor and T. Loftsson, in 'Controlled Drug Delivery', ed. J. R. Robinson and V. H. L. Lee, Dekker, New York, 1987, p. 337.

68. V. J. Stella and K. J. Himmelstein, in 'Optimization of Drug Delivery', ed. H. Bundgaard, A. Bagger Hansen and H. Kofod, Munksgaard, Copenhagen, 1982, p. 134.

69. N. Bodor, in 'Optimization of Drug Delivery', ed. H. Bundgaard, A. Bagger Hansen and H. Kofod, Munksgaard, Copenhagen, 1982, p. 156.

70. J. R. Dimmock and L. M. Smith, *J. Pharm. Sci.*, 1980, **69**, 575.

71. T. Visser, *Pharm. Weekbl.*, 1984, **119**, 857.

72. G. Gregoriadis and B. E. Ryman, *Eur. J. Biochem.*, 1972, **24**, 485.

73. G. Gregoriadis (ed.), 'Liposome Technology', CRC Press, Boca Raton, Florida, 1984, vols. 1, 2 and 3, p. 77.
74. G. Poste, *Biol. Cell.*, 1983, **47**, 19.
75. G. L. Scherphof, in 'Lipids and Membranes: Past, Present and Future', ed. J. A. F. Op den Kamp, B. Roelofsen and K. W. A. Wirtz, Elsevier, Amsterdam, 1986, p. 113.
76. E. Mayhew and D. Papahadjopoulos, in 'Liposomes', ed. M. J. Ostro, Dekker, New York, 1983, p. 289.
77. J. W. Babbage and M. C. Berenbaum, *Br. J. Cancer*, 1982, **45**, 830.
78. J. Khato, A. A. Del Campo and S. M. Sieber, *Pharmacology*, 1983, **26**, 23.
79. P. Large and G. Gregoriadis, *Biochem. Pharmacol.*, 1983, **32**, 1315.
80. J. Hildebrand, J. M. Ruysschaert and C. Laduron, *J. Natl. Cancer Inst.*, 1983, **70**, 1081.
81. E. A. Forssen and Z. A. Tokes, *Cancer Res.*, **43**, 546.
82. A. Gabizon. A. Dagan, D. Goren, Y. Barenholz and Z. Fuks, *Cancer Res.*, 1982, **42**, 4734.
83. A. Gabizon, D. Goren, Z. Fuks, Y. Barenholz, A. Dagan and A. Meshorer, *Cancer Res.*, 1983, **43**, 4730.
84. E. Mayhew, Y. Rustum and W. J. Vail, *Cancer Drug Del.*, 1983, **1**, 43.
85. F. Olson, E. Mayhew, D. Maslow, Y. Rustum and F. Szoka, *Eur. J. Clin. Oncol.*, 1982, **18**, 167.
86. A. Rahman, G. White, N. More and P. S. Schein, *Cancer Res.*, 1985, **45**, 796.
87. A. Rahman, A. Joher and J. R. Neefe, *Br. J. Cancer*, 1986, **54**, 401.
88. G. Storm, F. H. Roerdink, P. A. Steerenberg, W. H. De Jong and D. J. A. Crommelin, *Cancer Res.*, 1987, **47**, 3366.
89. E. A. Forssen and Z. A. Tokes, *Cancer Treat. Rep.*, 1983, **67**, 481.
90. G. Storm, Ph. D. Thesis, University of Utrecht, 1987.
91. I. J. Fidler, *Cancer Res.*, 1985, **45**, 4714.
92. I. J. Fidler and G. Poste, *Springer Semin. Immunopathol.*, 1982, **4**, 161.
93. N. Kende, A. J. Schroit, W. Rill and P. Canonico, in 'International Congress of Antimicrobial Agents and Chemotherapy, Washington,' 1983, Abstract 108.
94. W. C. Koff, S. D. Showalter, B. Hampar and I. J. Fidler, *Science (Washington, D.C.)*, 1985, **228**, 495.
95. E. B. Fraser-Smith, D. A. Eppstein, M. A. Larsen and T. R. Matthews, *Infect. Immun.*, 1983, **39**, 172.
96. G. Poste, R. Kirsh, W. E. Fogler and I. J. Fidler, *Cancer Res.*, 1979, **39**, 881.
97. I. J. Fidler, *Science (Washington D.C.)*, 1980, **208**, 1469.
98. I. J. Fidler, S. Sone, W. E. Fogler and Z. L. Barnes, *Proc. Natl. Acad. Sci. U.S.A.*, 1981, **78**, 1680.
99. G. Lopez-Berestein, K. Mehta, R. Mehta, R. L. Juliano and E. M. Hersh, *J. Immunol.*, 1983, **130**, 1500.
100. G. Lopez-Berestein, R. Mehta, R. L. Hopfer, K. Mills, L. Kasi, K. Mehta, V. Fainstein, M. Luna, E. M. Hersh and R. L. Juliano, *J. Infect. Dis.*, 1983, **147**, 939.
101. R. Mehta, G. Lopez-Berestein, R. L. Hopfer, K. Mills and R. L. Juliano, *Biochim. Biophys. Acta*, 1984, **770**, 230.
102. P. S. Thombre and S. D. Deodhar, *Cancer Immunol. Immunother.*, 1984, **16**, 145.
103. T. Daemen, A. Veninga, F. H. Roerdink and G. L. Scherphof, *Cancer Res.*, 1986, **46**, 4330.
104. C. R. Alving, *Pharmacol. Ther.*, 1983, **22**, 407.
105. F. Emmen and G. Storm, *Pharm. Weekbl., Sci. Ed.*, 1987, **9**, 162.
106. C. R. Alving, E. A. Steck, W. L. Chapman, Jr., V. B. Waits, L. D. Hendricks, G. M. Swartz, Jr. and W. L. Hanson, *Proc. Natl. Acad. Sci. U.S.A.*, 1978, **75**, 2959.
107. C. D. V. Black, G. J. Watson and R. J. Ward, *Trans. Roy. Soc. Trop. Med. Hyg.*, 1977, **71**, 550.
108. R. R. C. New, M. L. Chance, S. C. Thomas and W. Peters, *Nature (London)*, 1978, **272**, 55.
109. C. R. Alving, E. A. Steck, W. L. Chapman, Jr., V. B. Waits, L. D. Hendricks, G. M. Swartz, Jr. and W. L. Hanson, *Life Sci.*, 1980, **26**, 2231.
110. R. R. C. New, M. L. Chance and S. Heath, *J. Antimicrob. Chemother.*, 1981, **8**, 371.
111. A. R. Tarlov, G. J. Drewer, P. E. Carson and A. S. Alving, *Arch. Intern. Med.*, 1962, **109**, 209.
112. P. Pirson, R. F. Steiger, A. Trouet, J. Gillet and F. Herman, *Ann. Trop. Med. Parasitol.*, 1980, **74**, 383.
113. E. Suter, *Bacteriol. Rev.*, 1956, **20**, 94.
114. F. M. Collins, *Cornell Vet.*, 1977, **76**, 103.
115. F. M. Collins and S. G. Campbell, *Vet. Immunol. Immunopathol.*, 1982, **3**, 5.
116. I. A. J. M. Bakker-Woudenberg, A. F. Lokerse, F. H. Roerdink, D. Regts and F. J. Michel, *Infect. Dis.*, 1985, **151**, 917.
117. M. W. Fountain, S. J. Weiss, A. G. Fountain, A. Shen and R. P. Lenk, *J. Infect. Dis.*, 1985, **152**, 529.
118. J. Sunamoto, M. Goto, T. Iida, K. Hara, A. Saito and A. Tomonaga, in 'Receptor-mediated Targeting of Drugs', ed. G. Gregoriadis, G. Poste, J. Senior and A. Trouet, Plenum Press, New York, 1984, p. 359.
119. R. L. Taylor, D. M. Williams, P. C. Craven, J. R. Graybill, D. J. Drutz and W. E. Magee, *Am. Rev. Respir. Dis.*, 1982, **125**, 610.
120. J. R. Graybill, P. C. Craven, R. L. Taylor, D. M. Williams and W. E. Magee, *J. Infect. Dis.*, 1982, **145**, 748.
121. G. Lopez-Berestein, R. L. Hopfer, R. Mehta, K. Mehta, E. M. Hersh and R. L. Juliano, *J. Infect. Dis.*, 1984, **150**, 278.
122. G. Lopez-Berestein, V. Fainstein, R. L. Hopfer, K. Mehta, M. P. Sullivan, M. Keating, R. G. Rosenblum, R. Mehta, M. A. Luna, E. M. Hersh, J. Reuben, R. L. Juliano and G. P. Bodey, *J. Infect. Dis.*, 1985, **151**, 704.
123. A. Shirkoda, G. Lopez-Berestein, J. M. Holbert and M. A. Luna, *Radiology*, 1986, **159**, 349.
124. J. E. Edwards, R. I. Lehrer, E. R. Stiehm, T. J. Fischer and L. S. Young, *Ann. Intern. Med.*, 1978, **89**, 91.
125. K. Mehta, R. L. Juliano and G. Lopez-Berestein, *Immunology*, 1984, **51**, 517.
126. R. L. Juliano and G. Lopez-Berestein, *Pharm. Int.*, 1985, **6**, 164.
127. W. C. Koff and I. J. Fidler, *Antiviral Res.*, 1985, **5**, 179.
128. E. J. Shillitoe and G. Rapp, *Springer Semin. Immunopathol.*, 1979, **2**, 237.
129. W. C. Koff, S. D. Showalter, D. A. Seniff and B. Hampar, *Infect. Immun.*, 1983, **42**, 1067.
130. W. C. Koff, I. J. Fidler, S. D. Showalter, M. K. Chakrabarty, B. Hampar, L. M. Ceccorulli and E. S. Kleinerman, *Science (Washington, D.C.)*, 1984, **224**, 1007.
131. S. C. Mogensen, *Microbiol. Rev.*, 1979, **43**, 1.
132. R. L. Juliano and D. Stamp, *Nature (London)*, 1976, **261**, 235.
133. D. Boldt, S. F. Speckart, R. L. Richards and C. R. Alving, *Biochem. Biophys. Res. Commun.*, 1977, **74**, 208.
134. C. R. Alving, K. C. Joseph and R. Wistar, *Biochemistry*, 1974, **13**, 4818.
135. R. W. Bussian and J. C. Wriston, *Biochim. Biophys. Acta*, 1977, **471**, 336.
136. R. Mayet-Dana, A. C. Roche and M. Monsigny, *FEBS Lett.*, 1977, **79**, 305.

137. A. Surolia and B. K. Bachhawat, *Biochim. Biophys. Acta*, 1977, **497**, 760.
138. G. Gregoriadis and E. D. Neerunjun, *Biochem. Biophys. Res. Commun.*, 1975, **65**, 537.
139. P. S. Wu, G. W. Tim and J. D. Baldeschwieler, *Proc. Natl. Acad. Sci. U.S.A.*, 1981, **78**, 2033.
140. P. S. Wu, H. Wu, G. W. Tim, J. R. Schuh, W. R. Chroasmun, J. D. Baldeschwieler, T. Y. Schen and M. M. Ponpinom, *Proc. Natl. Acad. Sci. U.S.A.*, 1982, **79**, 5490.
141. G. Gregoriadis, E. D. Neerunjun and R. Hunt, *Life Sci.*, 1977, **21**, 357.
142. A. Huang, L. Huang and S. J. Kennel, *J. Biol. Chem.*, 1980, **255**, 3015.
143. L. D. Leserman and J. N. Weinstein, in 'Liposomes and Immunobiology', ed. B. H. Tom and H. R. Six, Elsevier, Amsterdam, 1980.
144. A. Huang, Y. S. Tsao, S. J. Kennel and L. Huang, *Biochim. Biophys. Acta*, 1982, **716**, 140.
145. L. D. Leserman, P. Machy and J. Barbet, *Nature (London)*, 1981, **293**, 226.
146. V. P. Torchillin and A. L. Klibanov, *Enzyme Microbiol. Technol.*, 1981, **3**, 297.
147. F. J. Martin, W. L. Hubbell and D. Papahadjopoulos, *Biochemistry*, 1981, **20**, 4229.
148. F. J. Martin and D. Papahadjopoulos, *J. Biol. Chem.*, 1982, **257**, 286.
149. K. S. Bragman, T. D. Heath and D. Papahadjopoulos, *Biochim. Biophys. Acta*, 1983, **730**, 187.
150. T. D. Heath, R. T. Fraley and D. Papahadjopoulos, *Science (Washington, D.C.)*, 1980, **210**, 539.
151. M. Harsch, P. Walther and H. G. Weder, *Biochem. Biophys. Res. Commun.*, 1981, **103**, 1069.
152. L. D. Leserman, in 'Liposomes in the Study of Drug Action and of Immunocompetent Cell Function', C. Nicolau and A. Paraf, Academic Press, New York, 1981, p. 109.
153. P. Machy, J. Barbet and L. D. Leserman, *Proc. Natl. Acad. Sci. U.S.A.*, 1982, **79**, 4148.
154. P. Machy, M. Pierres, J. Barbet and L. D. Leserman, *J. Immunol.*, 1982, **129**, 2098.
155. J. Connor, M. B. Yatvin and L. Huang, *Proc. Natl. Acad. Sci. U.S.A.*, 1984, **81**, 1715.
156. J. Connor, Ph. D. Thesis, University of Tennessee, 1986.
157. V. P. Torchillin, B. A. Khaw, V. N. Smirnov and E. Haber, *Biochem. Biophys. Res. Commun.*, 1979, **89**, 1114.
158. M. B. Yatvin, J. N. Weinstein, W. H. Dennis and R. Blumenthal, *Science (Washington D.C.)*, 1978, **202**, 1290.
159. J. N. Weinstein, R. L. Magin, R. L. Cysyk and D. S. Zaharko, *Cancer Res.*, 1980, **40**, 1388.
160. R. L. Magin and M. R. Niesman, *Cancer Drug Del.*, 1984, **1**, 109.
161. M. B. Yatvin, T. C. Cree and I. M. Tegmo-Larsson, in 'Lipsome Technology', ed. G. Gregoriadis, CRC Press, Boca Raton, 1984, vol. 3, p. 157.
162. R. T. Gordon, J. R. Hines and D. Gordon, *Med. Hypotheses*, 1979, **5**, 83.
163. R. J. Parker, K. D. Hartman and S. M. Sieber, *Cancer Res.*, 1981, **41**, 1311.
164. R. J. Parker, S. M. Sieber and J. N. Weinstein, *Pharmacology*, 1981, **23**, 128.
165. R. J. Parker, E. R. Priester and S. M. Sieber, *Drug Metab. Dispos.*, 1982, **10**, 40.
166. A. J. Jackson, *Drug Metab. Dispos.*, 1981, **9**, 535.
167. V. I. Kaledin, N. A. Matienko, V. P. Nikolin, Y. V. Gruntenko, V. G. Budker and T. E. Vakhrusheva, *J. Natl. Cancer Inst.*, 1982, **69**, 67.
168. T. Kato, in 'Controlled Drug Delivery', ed. S. D. Bruck, CRC Press, Boca Raton, 1982, vol. 11, p. 189.
169. J. N. Weinstein, in 'Rational Basis for Chemotherapy', ed. B. Chabner, Liss, New York, 1983, p. 441.
170. I. H. Shaw, J. T. Dingle, N. C. Phillips, D. P. Page-Thomas and C. G. Knight, *Ann. N.Y. Acad. Sci.*, 1978, **308**, 435.
171. S. I. Rapoport, 'Blood–Brain Barrier in Physiology and Medicine', Raven Press, New York, 1976.
172. D. M. Adams, G. Joyce, V. J. Richardson, B. E. Ryman and H. M. Wisniewski, *J. Neurol. Sci.*, 1977, **31**, 173.
173. H. K. Kimelberg, T. F. Tracy, R. E. Watson, D. Kung, F. L. Reiss and R. S. Bourke, *Cancer Res.*, 1978, **38**, 706.
174. H. H. Ivey, S. Roth and J. Kattwindel, *Pediatr. Res.*, 1976, **10**, 462.
175. R. L. Juliano, D. Stamp and N. McCullough, *Ann. N.Y. Acad. Sci.*, 1978, **308**, 411.
176. P. A. Kramer, *J. Pharm. Res.*, 1974, **63**, 1646.
177. S. S. Davis, L. Illum, J. G. McVie and E. Tomlinson (eds.), 'Microspheres and Drug Therapy. Pharmaceutical, Immmunological and Medical Aspects', Elsevier, Amsterdam, 1984.
178. E. Tomlinson, J. J. Burger, E. M. A. Schoonderwoerd and J. G. McVie, in 'Microspheres and Drug Therapy. Pharmaceutical, Immunological and Medical Aspects', ed. S. S. Davis, L. Illum, J. G. McVie and E. Tomlinson, Elsevier, Amsterdam, 1984, p. 75.
179. R. J. M. Blanchard, I. Grotenhuis, J. W. LaFavre and J. F. Perry, *Proc. Soc. Exp. Biol. Med.*, 1965, **118**, 465.
180. B. Lindell, K. F. Aronsen, U. Rothman and H. O. Sjoegren, *Res. Exp. Med.*, 1977, **171**, 63.
181. L. Illum and S. S. Davis, *FEBS Lett.*, 1984, **167**, 79.
182. L. Illum and S. S. Davis, *J. Pharm. Sci.*, 1983, **72**, 1086.
183. M. R. Kaplan, E. Calef, T. Bercovici and C. Gitler, *Biochim. Biophys. Acta*, 1983, **728**, 112.
184. P. Couvreur, B. Kante, M. Roland, P. Baudhuin and P. Speiser, *J. Pharm. Pharmacol.*, 1979, **31**, 331.
185. P. Couvreur, M. Roland and P. Speiser, *US Pat.* 4 329 332 (1982).
186. P. Couvreur, B. Kante and M. Roland, *J. Pharm. Belg.*, 1980, **35**, 51.
187. F. Leonard, R. Kulkarni, G. Brandes, J. Nelson and J. Cameron, *J. Appl. Polym. Sci.*, 1966, **10**, 259.
188. W. R. Vezin and A. T. Florence, *J. Biomed. Mater. Res.*, 1980, **14**, 93.
189. L. Grislain, P. Couvreur, V. Lenaerts, M. Roland, D, Deprez-Decampeneere and P. Speiser, *Int. J. Pharm.*, 1983, **15**, 335.
190. P. Couvreur, B. Kante, M. Roland and P. Speiser, *J. Pharm. Sci.*, 1979, **68**, 1521.
191. V. Lenaerts, P. Couvreur, D. Christiaens-Leyh,-E. Joiris, M. Roland, B. Rollman and P. Speiser, *Biomaterials*, 1983, **5**, 65.
192. F. Brasseur, P. Couvreur, B. Kante, L. Deckers-Passau, M. Roland, C. Deckers and P. Speiser, *Eur. J. Cancer*, 1980, **16**, 1441.
193. J. Kreuter and H. R. Hartmann, *Oncology*, 1983, **40**, 363.
194. P. Couvreur, V. Lenearts, L. Grislain, L. Vansnick and F. Brasseur, in 'Drug Targeting', ed. P. Buri and A. Gumma, Elsevier, Amsterdam, 1985, p. 35.
195. P. Couvreur, B. Kante, L. Grislain, M. Roland and P. Speiser, *J. Pharm. Sci.*, 1982, **71**, 790.
196. P. Couvreur, V. Lenaerts, B. Kante, M. Roland and P. Speiser, *Acta Pharm. Technol.*, 1980, **26**, 220.
197. P. Couvreur, V. Lenearts, D. Leyh, P. Guiot and M. Roland, in 'Microspheres and Drug Therapy. Pharmaceutical, Immunological and Medical Aspects', S. S. Davis, L. Illum, J. G. McVie and E. Tomlinson, Elsevier, Amsterdam, 1984, p. 103.

198. E. Tomlinson, in 'Site-specific Drug Delivery', ed. E. Tomlinson and S. S. Davis, Wiley, Chichester, 1986, p. 1
199. M. Przyborowski, E. Lachnik, J. Wiza and I. Licinska, *Eur. J. Nucl. Med.*, 1982, **7**, 71.
200. A. M. Millar, L. McMillan, W. J. Hannan, P. C. Emmett and R. J. Aitken, *Int. J. Appl. Radiat. Isot.*, 1982, **33**, 1423.
201 A. F. Yapel *US Pat.* 4 147 767 (1979).
202. N. Willmott, H. M. H. Kamel, J. Cummings, J. F. B. Stuart and A. T. Florence, in 'Microspheres and Drug Therapy. Pharmaceutical, Immunological and Medical Aspects', ed. S. S. Davis, L. Illum, J. G. McVie and E. Tomlinson, Elsevier, Amsterdam, 1984, p. 205.
203. E. Tomlinson, *Int. J. Pharm. Technol. Prod. Manuf.*, 1983, **4**, 49.
204. S. Fujimoto, F. Endo, Y. Kitsukawa, K. Okui, Y. Morimoto, K. Sugibayashi, A. Miyakawa and H. Suzuki, *Experientia*, 1983, **39**, 913.
205. S. Fujimoto, F. Endo, R. D. Shresta, M. Miyazaki, Y. Kitsukawa, K. Okui, K. Sugibayashi and Y. Morimoto, in 'Proceedings of the 13th International Congress of Chemotherapy, Concepts in Cancer Chemotherapy', Vienna, 1983, 258/31.
206. K. J. Widder, A. E. Senyei and B. Sears, *J. Pharm. Sci.*, 1982, **71**, 379.
207. K. J. Widder and A. E. Senyei, *Pharmacol. Ther.*, 1983, **20**, 377.
208. K. J. Widder, R. M. Morris, G. Poore, D. P. Howard and A. E. Senyei, *Proc. Natl. Acad. Sci. U.S.A.*, 1981, **78**, 579.
209. G. P. Royer (Ohio State University), *US Pat.* 4 349 530 (1982).
210. G. P. Royer and T. K. Lee, *J. Parenter. Sci. Technol.*, 1983, **37**, 34.
211. T. K. Lee, T. D. Sokolski and G. P. Royer, *Science (Washington, D.C.)*, 1981, **213**, 233.
212. M. F. A. Goosen, Y. E. Leung, S. Chou and A. M. Sun, *Biomater., Med. Devices, Artif. Organs*, 1982, **10**, 205.
213. B. Lindberg, K. Lote and H. Teder, in 'Microspheres and Drug Therapy. Pharmaceutical, Immunological and Medical Aspects', ed. S. S. Davis, L. Illum, J. G. McVie and E. Tomlinson, Elsevier, Amsterdam, 1984, p. 153.
214. R. F. Tuma, in 'Microspheres and Drug Therapy. Pharmaceutical, Immunological and Medical Aspects', ed. S. S. Davis, L. Illum, J. G. McVie and E. Tomlinson, Elsevier, Amsterdam, 1984, p. 189.
215. P. Edman and I. Sjoholm, *J. Pharm. Sci.*, 1981, **70**, 684.
216. P. Edman and I. Sjoholm, *J. Pharm. Sci.*, 1982, **71**, 576.
217. T. Kato, R. Nemoto, H. Mori, K. Iwata, S, Sato, K. Unno, A. Goto, M. Harada, M. Homma, M. Okada and T. Minowa, *J. Jpn. Soc. Cancer Ther.*, 1980, **15**, 28.
218. T. Kato, R. Nemoto, H. Mori and I. Kumagai, *Cancer*, 1980, **46**, 14.
219. T. Kato, R. Nemoto, H. Mori, M. Takahashi and Y. Tamakawa, *J. Urol.*, 1981, **125**, 19.
220. Z. A. Tokes, K. E. Rogers and A. Rembaum, *Proc. Natl. Acad. Sci. U.S.A.*, 1982, **79**, 2026.
221. K. E. Rogers, B. I. Carr, and Z. A. Tokes, *Cancer Res.*, 1983, **43**, 2741.
222. T. R. Tritton and G. Yee, *Science (Washington D.C.)*, 1982, **217**, 248.
223. T. R. Tritton, G. Yee and L. B. Wingard, *Fed. Proc., Fed. Am. Soc. Exp. Biol.*, 1982, **42**, 284.
224. J. A. Siegfried, K. A. Kennedy, A. C. Sartorelli and T. Tritton, *J. Biol. Chem.*, 1983, **258**, 339.
225. M. J. Poznansky and L. G. Cleland, in 'Drug Delivery Systems', ed. R. L. Juliano, Oxford University Press, New York, 1980, p. 253.
226. R. W. Mahley and T. L. Innerarity, *Biochim. Biophys. Acta*, 1982, **737**, 197.
227. M. S. Brown and J. L. Goldstein, *Proc. Natl. Acad. Sci. U.S.A.*, 1979, **76**, 3330.
228. M. Krieger, M. S. Brown, J. R. Faust and J. L. Goldstein, *J. Biol. Chem.*, 1978, **253**, 4093.
229. D. Gal, M. Ohasi, P. C. MacDonald, H. J. Buchsbaum and E. R. Simpson, *Am. J. Obstet. Gynecol.*, 1981, **139**, 877.
230. J. Welsh, K. C. Calmar, J. Glegg, J. M. Stewart, C. J. Parkard, H. G. Morgan and C. J. Shephard, *J. Clin. Sci.*, 1982, **63**, 44.
231. S. G. Vitols, G. Gahrton and C. O. Peterson, *Cancer Treat. Rep.*, 1984, **68**, 515.
232. M. J. Rudling, V. P. Collins and C. O. Peterson, *Cancer Res.*, 1983, **43**, 4600.
233. S. G. Vitols, M. Masquelier and C. O. Peterson, *J. Med. Chem.*, 1985, **28**, 451.
234. G. Dale, in 'Drug Delivery Systems', ed. R. L. Juliano, Oxford University Press, New York, 1980, p. 237.
235. J. D. Humphreys and G. M. Ihler, in 'Optimization of Drug Delivery', ed. H. Bundgaard, A. Bagger Hansen and H. Kofod, Munksgaard, Copenhagen, 1982, p. 270.
236. G. M. Ihler, *Pharm. Ther.*, 1983, **20**, 151.
237. S. J. Updike, R. T. Wakamija and E. N. Lightfoot, *Science (Washington D. C.)*, 1976, **193**, 681.
238. K. Kinosita and T. Y. Tsong, *Nature (London)*, 1978, **272**, 258.
239. A. R. Hubbard, U. Sprangel and R. A. Chalmers, *Biochem. J.*, 1980, **190**, 653.
240. G. M. Ihler, R. H. Glew and F. W. Schnure, *Proc. Natl. Acad. Sci. U.S.A.*, 1973, **70**, 2663.
241. J. R. DeLoach and G. M. Ihler, *Biochim. Biophys. Acta*, 1977, **496**, 136.
242. B. Deuticke, M. Kim and C. Zolinev, *Biochim. Biophys. Acta*, 1973, **318**, 345.
243. T. Kitao, K. Hattori and M. Takeshita, *Experientia*, 1978, **341**, 94.
244. E. Beutler, G. L. Dale and W. L. Kuhl, in 'Enzyme Therapy in Genetic Diseases', ed. R. J. Desnick, Liss, New York, 1980, p. 369.
245. R. Green, J. Lamon and D. Curran, *Lancet*, 1980, **2**, 327.
246. G. M. Ihler, A. Lantzy, J. Purpura and R. H. Glew, *J. Clin. Invest.*, 1975, **56**, 595.
247. H. O. Alpar and D. A. Lewis, *J. Pharm. Pharmacol.*, 1983, **35**, 48P.
248. E. Pitt, D. A. Lewis and R. E. Offord, *Biochem. Pharmacol.*, 1983, **32**, 3353.
249. J. P. Durocher, R. D. Payne and M. E. Conrad, *Blood*, 1975, **45**, 11.
250. F. Kirkpatrick, *Life Sci.*, 1976, **19**, 1.
251. C. W. M. Haest, G. Plasa, D. Kamp and B. Deuticke, *Biochim. Biophys. Acta*, 1977, **469**, 226.
252. J. G. McAfee, G. M. Gagne and G. Subramanian, *J. Nucl. Med.*, 1980, **21**, 1059.
253. J. Schlepper-Schaffer, V. Kolb-Bachofen and H. Kolb, *Biochem. J.*, 1980, **186**, 827.
254. Y. S. Ahn, J. J. Bynes, W. J. Harrington, M. L. Cayer, D. S. Smith, B. E. Brunskill and L. M. Pall, *New Engl. J. Med.*, 1978, **298**, 1101.
255. A. Andersson, *Lancet*, 1979, **1**, 581.
256. A. J. Matas, D. E. R. Sutherland and J. S. Najarian, in 'Drug Carriers in Biology and Medicine', ed. G. Gregoriadis, Academic Press, New York, 1979, p. 191.

257. M. F. Dean, R. L. Stevens and H. Muir, *J. Clin. Invest.*, 1979, **63**, 138.
258. D. A. Gibbs, E. Spellacy, A. E. Roberts and R. W. E. Watts, in 'Enzyme Therapy in Genetic Diseases', ed. R. J. Desnick, Liss, New York, 1980, vol. 9, p. 101.
259. J. R. Hobbs, *Lancet*, 1981, **2**, 735.
260. J. R. Hobbs, in 'Advanced Medicine Leeds', ed. M. S. Losowsky and R. P. Bolton, Pitman, Bath, 1983, p. 378.
261. D. J. A. Crommelin and E. M. G. Van Bommel, *Pharm. Res.*, 1984, **4**, 159.
262. E. M. G. Van Bommel and D. J. A. Crommelin, *Int. J. Pharm.*, 1984, **22**, 299.
263. J. Kreuter, *Pharm. Acta Helv.*, 1983, **58**, 242.
264. G. Poste, C. Bucana and I. J. Fidler, in 'Targeting of Drugs', ed. G. Gregoriadis, J. Senior and A. Trouet, Plenum Press, New York, 1982, p. 261.
265. J. H. Fendler, in 'Liposomes in Biological Systems', ed. G. Gregoriadis and A. C. Allison, Wiley, Chichester, 1980, p. 87.
266. I. A. J. M. Bakker-Woudenberg, F. H. Roerdink and G. L. Scherphof, in 'Liposomes as Drug Carriers: Trends and Progress', Wiley, Chichester, 1988, p. 325.
267. L. Illum and P. D. E. Jones, in 'Methods in Enzymology', ed. K. J. Widder and R. Green, Academic Press, Orlando, 1985, vol. 112a, p. 67.
268. G. Gregoriadis, *Ann. N.Y. Acad. Sci.*, 1978, **308**, 343.
269. J. L. Goldstein and M. S. Brown, *Annu. Rev. Biochem.*, 1977, **46**, 897.
270. J. Khato, E. R. Priester and S. W. Sieber, *Cancer Treat. Rep.*, 1982, **66**, 517.
271. D. McIntosh and P. Thorpe, in 'Receptor-mediated Targeting of Drugs', ed. G. Gregoriadis, G. Poste, J. Senior and A. Trouet, Plenum Press, New York, 1984, p. 105.
272. A. Trouet, R. Baurain, D. Deprez-DeCampeneere, M. Masquelier and P. Pirson, in 'Targeting of Drugs', ed. G. Gregoriadis, J. Senior and A. Trouet, Plenum Press, New York, 1982, p. 19.
273. S. G. Vitols, G. Gahrton, A. Ost and C. O. Paterson, *Blood*, 1984, **63**, 1186.
274. F. Szoka and D. Papahadjopoulos, *Proc. Natl. Acad. Sci. USA*, 1978, **75**, 4194.

Subject Index